AF801281

Textbook of Veterinary Physiology

Pradip Kumar Das • Veerasamy Sejian •
Joydip Mukherjee • Dipak Banerjee
Editors

Textbook of Veterinary Physiology

Springer

Editors
Pradip Kumar Das
Department of Veterinary Physiology
West Bengal University of Animal and Fishery Sciences
Kolkata, West Bengal, India

Veerasamy Sejian
Rajiv Gandhi Institute of Veterinary education and Research, Puducherry and Centre for Climate Resilient Animal Adaptation Studies
ICAR-National Institute of Animal Nutrition and Physiology
Bangalore, Karnataka, India

Joydip Mukherjee
Department of Veterinary Physiology
West Bengal University of Animal and Fishery Sciences
Kolkata, West Bengal, India

Dipak Banerjee
Department of Veterinary Physiology
West Bengal University of Animal and Fishery Sciences
Kolkata, West Bengal, India

ISBN 978-981-19-9409-8 ISBN 978-981-19-9410-4 (eBook)
https://doi.org/10.1007/978-981-19-9410-4

© The Editor(s) (if applicable) and The Author(s), under exclusive license to Springer Nature Singapore Pte Ltd. 2023
This work is subject to copyright. All rights are solely and exclusively licensed by the Publisher, whether the whole or part of the material is concerned, specifically the rights of translation, reprinting, reuse of illustrations, recitation, broadcasting, reproduction on microfilms or in any other physical way, and transmission or information storage and retrieval, electronic adaptation, computer software, or by similar or dissimilar methodology now known or hereafter developed.
The use of general descriptive names, registered names, trademarks, service marks, etc. in this publication does not imply, even in the absence of a specific statement, that such names are exempt from the relevant protective laws and regulations and therefore free for general use.
The publisher, the authors, and the editors are safe to assume that the advice and information in this book are believed to be true and accurate at the date of publication. Neither the publisher nor the authors or the editors give a warranty, expressed or implied, with respect to the material contained herein or for any errors or omissions that may have been made. The publisher remains neutral with regard to jurisdictional claims in published maps and institutional affiliations.

This Springer imprint is published by the registered company Springer Nature Singapore Pte Ltd.
The registered company address is: 152 Beach Road, #21-01/04 Gateway East, Singapore 189721, Singapore

This book is dedicated to all the following stalwarts who have contributed immensely towards the growth of veterinary physiology

Henry Hugh Dukes
William Harvey
James E Breazile
James G Cunningham
Elsayed Saad Eldin *Hafez*
Leslie Ernest McDonald

Foreword

In order to fully understand the mechanical, physical, and biochemical functions of an organism, one must study physiology. An understanding of what is normal provides the building blocks for the identification of what is abnormal. Veterinarians as per their training generally focus on the pathological state of an animal, rather than on the physiological state, i.e., looking at abnormal rather than normal body function. Providing specialized care for animals necessitates a thorough understanding of pathophysiology that is impossible without a solid foundation in physiology.

The *Textbook of Veterinary Physiology* is a well-conceived, new publication that bridges the gap between the theory and practice in all areas of veterinary physiology. The book would be more than suitable for the students of Veterinary Physiology since it contains all the topics related to animals' physiological systems and covers domestic, laboratory, and wild animals. While this book is geared towards veterinary students just starting out in their profession, well as seasoned veterinarians who want to brush up on their knowledge of fundamental physiology will also benefit. Students, professors, scientists, and other experts interested in animal physiology will also find this book of value.

The book's structure and content is evidence that the authors are expert physiologists. Each chapter discusses the fundamentals of veterinary physiology before moving on to clinical applications and recent breakthroughs in the discipline. The visuals are simple and straightforward, making it very apt for veterinary students to understand the basic and fundamental concepts in veterinary physiology which will enable them to prepare themselves for competitive exams. I hope that this book will serve as excellent resource material not only for the students, but also to the established teachers of physiology to provide structure to their teaching. It is expected that students, teachers, and researchers will benefit from the deep insights provided by the experienced authors in this book. I am sure the book offers a great opportunity for teachers, researchers, and students who are specialized in veterinary physiology across the globe to gain knowledge from this wonderful volume.

School of Agriculture and Food Sciences,
The University of Queensland
Gatton, QLD, Australia

John B. Gaughan

Preface

Physiology is the study of the normal functioning of the body. An understanding of the interrelationships between molecules, cells, and organs that maintain homeostasis in an individual can be achieved through physiology. Because the study of medicine is the study of abnormal functioning of the body and it is therefore very essential to understand the normal physiology to comprehend the mechanisms of diseases and their treatment. For this reason, physiology and other important sciences basic to medicine are introduced first in the veterinary curriculum.

Physiology is a relatively vast subject, and veterinary students are far too busy to study everything there is to know about it in the initial semesters. On that background, the book entitled *Textbook of Veterinary Physiology* is an attempt to provide learning resources with easy-to-understand language to the students, faculties, scientists, and related professionals in the subject concerned. This volume is an attempt to narrate physiology in 10 sections with 29 chapters. The book covers differences between various domestic animals along with laboratory animals, avian species, and wildlife at the physiological level since the scope of physiology not only encompasses the basic functional descriptions of the systems of a particular animal or breed. The various morphological descriptions of the organs, tissues, and cells involved in the physiological system have been highlighted, with a particular emphasis on species variation. Given the book's diverse users, the chapters are presented in small paragraphs with pertinent headings and sub-headings to make it easier to locate the precise information in a detailed narration.

This book is first of its kind for physiology as it comprehensively covers species diversity, thorough mechanism of action, description with clinical implications, and the most essential, the most recent breakthroughs in the field of physiology, all in one volume. The immune system, assisted reproductive technology, ovarian dynamics, environmental physiology and thermoregulation, thermoregulation in birds, and behavioral physiology are just a few of the more advanced topics covered in this book. It will fortify the readers to become a master in the subject concerned and strengthen future publications. This uniqueness will attract the diversified readers to satisfy the basic understanding of veterinary physiology under one roof rather than reading through several collections for a single subject.

The contributors of various chapters are professionals with vast experience in veterinary physiology supported by sound peer-reviewed publications. The Editorial board take this opportunity to thank all the contributors for their dedication in drafting, timely submission, and for sharing their rich knowledge and experience with others. The efforts of many others, all of those cannot be individually listed, were also very pertinent in completing this relevant and an important volume.

Kolkata, West Bengal, India — Pradip Kumar Das
Bangalore, Karnataka, India — Veerasamy Sejian
Kolkata, West Bengal, India — Joydip Mukherjee
Kolkata, West Bengal, India — Dipak Banerjee

Contents

Editors and Contributors

About the Editors

Pradip Kumar Das MSc, PhD, is a Professor in the Department of Veterinary Physiology at West Bengal University of Animal & Fishery Sciences, Kolkata, West Bengal, India. He is specialized in the area of reproduction, cardiovascular, environmental, and poultry physiology. Prof. Das has immense research contribution to repeat breeding, standardization of cardiac axis in various native breeds, and transmission of pollutants through food chain. He handled several research projects and developed a sustainable backyard poultry farming model and various poultry husbandry databases. He mentored several postgraduate and doctoral students as major guide. Prof. Das has published four books, four book chapters, and ten practical manuals for the students. He has published more than 140 research and review articles in several peer-reviewed journals. Prof. Das published 20 books for the entrepreneur and developed 7 feature films to popularize the science. He is attached to various scientific societies; he held the post of General Secretary of Society of Animal Physiologists of India. He coordinated more than 100 capacity building programs and organized about 20 national and 1 international scientific symposium. Prof. Das was an advisor to the District Planning Committee for Animal Resources Development Department, Government of West Bengal, India. He successfully served the State Agriculture Commission as the Secretary of education, research, training, and extension subcommittee. The Society of Animal Physiologists of India conferred him with the prestigious A Roy Award for his scientific and academic contribution. Presently, he is the editor of the *Indian Journal of Animal Health.*

Veerasamy Sejian MVSc, PhD, is the Dean at Rajiv Gandhi Institute of Veterinary Education and Research, Puducherry, and a Principal Scientist at ICAR-National Institute of Animal Nutrition and Physiology (NIANP), Bangalore, India. His major thrust area of research is on climate change and livestock production. Dr. Sejian established the concept of "Multiple stresses impacting small ruminant production in the changing climate scenario." His current research is focused on identifying molecular markers for different environmental stresses in small ruminants with the primary focus to develop agro-ecological zone-specific thermotolerant breed. Dr. Sejian has published more than 160 peer-reviewed research/review articles in the field of stress and climate change physiology. Apart from this, he also published 124 book chapters, 286 invited/lead papers, 173 conference papers, and 22 technical manuals. His h-index is 42 and i10 index is 116. He has organized 13 short-term training programs in the field of climate change and livestock production. For his outstanding contribution to climate change and livestock production, the Indian council of Agricultural Research (ICAR) has bestowed him with the prestigious Lal Bahadur Shastri Outstanding Young Scientist Award. Dr. Sejian is also listed in world top 2% scientist by the Stanford University, USA, for the three consecutive years 2020, 2021, and 2022. In addition, Dr. Sejian also serves as an editorial board member for 27 international journals pertaining to climate change and livestock production.

Joydip Mukherjee MVSc, PhD, is an Assistant Professor in the Department of Veterinary Physiology at West Bengal University of Animal & Fishery Sciences, Kolkata, West Bengal, India. He did an elaborate research work in the area of Lactation Physiology, particularly innate immunity of mammary gland in crossbred cattle and buffaloes. His notable work encompasses the standardization of protocol for studying the immune competence of mammary gland in terms of in vitro milk leukocyte activity and its validation during different physiological conditions. He also highlighted the molecular and endocrine control on the in vitro activity of milk leukocytes along with the modulation of in vitro immune activity through micronutrient supplementation. Presently, he is working on electrocardiography in pets and diagnostic imaging. Dr. Mukherjee has published a book, two book chapters, and four practical manuals for the students. He has published more than 50 research and review articles in peer-reviewed journals. He is a member of several scientific societies, including Indian Science Congress Association, Society of Animal Physiologists of India, and Society of Domestic Animal Biodiversity.

Dipak Banerjee MVSc, PhD, is an Assistant Professor and HOD, Department of Veterinary Physiology at West Bengal University of Animal & Fishery Sciences, Kolkata, West Bengal, India. He had received his PhD in Animal Physiology from National Dairy Research Institute, Karnal, India. After serving as Assistant Professor in Anand Agricultural University, Anand, Gujarat, India, he has joined as Assistant Professor in West Bengal University of Animal & Fishery Sciences. Presently, he is the Head of Department of Veterinary Physiology. His research interests focus on the molecular mechanism of adaptation of animals to different environmental conditions. He is the coauthor of the book *Techniques in Veterinary Physiology* (ISBN 978-81-929561-0-7). He has published more than 50 research and review articles in several peer-reviewed journals and delivered about 20 invited/lead papers and has received several honors for his research accomplishments including the Young Scientist Award from the Society of Animal Physiologist in India. Further, he is specialized in the area of stress and environment physiology and neuro-endocrine physiology.

Contributors

Dipak Banerjee Department of Veterinary Physiology, West Bengal University of Animal & Fishery Sciences, Kolkata, West Bengal, India

V. Beena Department of Veterinary Physiology, College of Veterinary and Animal Sciences, Thrissur, Kerala, India

Pradip Kumar Das Department of Veterinary Physiology, West Bengal University of Animal & Fishery Sciences, Kolkata, West Bengal, India

C. Devaraj Centre for Climate Resilient Animal Adaptation Studies, ICAR-National Institute of Animal Nutrition and Physiology, Bangalore, Karnataka, India

Prabal Ranjan Ghosh Department of Veterinary Physiology, West Bengal University of Animal & Fishery Sciences, Kolkata, West Bengal, India

Iqbal Hyder College of Veterinary Science, Sri Venkateswara Veterinary University, Garividi, Andhra Pradesh, India

Sonali Jana Department of Veterinary Physiology, West Bengal University of Animal & Fishery Sciences, Kolkata, West Bengal, India

G. Krishnan Centre for Climate Resilient Animal Adaptation Studies, ICAR-National Institute of Animal Nutrition and Physiology, Bangalore, Karnataka, India

N. Madhavan Unny Department of Veterinary Clinical Medicine, College of Veterinary and Animal Sciences, Thrissur, Kerala, India

P. Manjari College of Veterinary Science, Sri Venkateswara Veterinary University, Garividi, Andhra Pradesh, India

Ayan Mukherjee Department of Animal Biotechnology, West Bengal University of Animal & Fishery Sciences, Kolkata, West Bengal, India

Joydip Mukherjee Department of Veterinary Physiology, West Bengal University of Animal & Fishery Sciences, Kolkata, West Bengal, India

Poonooru Ravi Kanth Reddy Veterinary Dispensary, Animal Husbandry Department, Taticherla, Andhra Pradesh, India

M. R. Reshma Nair ICAR-National Institute of Animal Nutrition and Physiology, Bangalore, Karnataka, India

B. A. A. Sai Kumar Department of Veterinary Physiology, Rajiv Gandhi Institute of Veterinary Education and Research, Puducherry, India

P. Sai Mounica Division of Veterinary Microbiology, ICAR-Indian Veterinary Research Institute, Bareilly, Uttar Pradesh, India

V. Sejian Centre for Climate Resilient Animal Adaptation Studies, ICAR-National Institute of Animal Nutrition and Physiology, Bangalore, Karnataka, India

C. G. Shashank Centre for Climate Resilient Animal Adaptation Studies, ICAR-National Institute of Animal Nutrition and Physiology, Bangalore, Karnataka, India

M. V. Silpa Institute of Animal Breeding and Genetics, Justus-Liebig-UniversitätGießen, Gießen, Germany

P. Visha Department of Veterinary Physiology and Biochemistry, Veterinary College and Research Institute, Tamilnadu Veterinary and Animal Sciences University, Salem, Tamil Nadu, India

Aziz Zarina Department of Veterinary Physiology, College of Veterinary and Animal Sciences, Thrissur, Kerala, India

Abbreviations

%	Percentage
(−)	Inhibition/Negative effect
(+)	Stimulation/Positive effect
(±)	Both plus and minus
:	Ratio
=	Equals to
°C	Degree Centigrade
°F	Degree Fahrenheit
Å	Angstrom
AC	Adenylyl cyclase
Ach	Acetylcholine
AchE	Acetylcholinesterase
AI	Artificial insemination
AMP	Antimicrobial peptides
ANF	Atrial natriuretic factor
ANP	Atrial natriuretic peptide
ANS	Autonomic nervous system
AQP	Aquaporin
AR	Adrenergic receptors
ARAS	Ascending reticular activating system
AREG	Amphiregulin
ART	Assisted reproductive technologies
BBB	Blood–brain barrier
BCS	Body condition score
bGH	Bovine growth hormone
BGHI	Black globe temperature and humidity index
BGT	Black globe temperature
BNP	Brain or B-type natriuretic peptide
Ca^{2+}	Calcium ion
CaM	Calmodulin
CCK	Cholecystokinin
CKD	Chronic kidney disease
Cl^-	Chloride ion
cm	Centimeter (a unit of length equal to one hundredth (1/100) of a meter)
CNS	Central nervous system
COP	Colloid osmotic pressure
COPD	Chronic obstructive pulmonary disease
CRTZ	Chemoreceptor trigger zone
CSF	Cerebrospinal fluid
D cell	Delta cell

d	Day
DCT	Distal convoluted tubule
DHEA	Dehydroepiandrosterone
dL	Deciliter (a unit of fluid measure equal to 10^{-1} L)
DPT	Dew point temperature
e.g.	exempli gratia, the Latin phrase meaning "for example"
ECF	Extracellular fluid
ECG	Electrocardiogram
eCG	Equine chorionic gonadotropin
ECM	Extracellular matrix
EGF	Epidermal growth factor
ENS	Enteric nervous system
EOP	Endogenous opioid system
EPO	Erythropoietin
ER	Endoplasmic reticulum
ESR	Erythrocyte sedimentation rate
et al.	"et alia" meaning "and others." It is used in academic citations
etc.	"Et cetera," a Latin term, means "and other things" or "and so on"
ETT	Embryo transfer technology
FACS	Fluorescence-activated cell sorting
FFA	Free-fatty acids
FGF	Fibroblast growth factor
Fig	Figure
g	Gram (a metric unit of mass equal to 1/1000 (one thousandth) of a kilogram)
g^{-1}	Per gram
GA	Golgi apparatus
GABA	Gamma-aminobutyric acid
GC	Glucocorticoids
GFR	Glomerular filtration rate
GI	Gastrointestinal
GIT	Gastrointestinal tract
GLP-1	Glucagon-like peptide 1
GLUT	Glucose transporter
GPx	Glutathione peroxidase system
GRP	Gastrin release peptide
h	Hour
H^+	Hydrogen ion
hCG	Human chorionic gonadotropin
HCO_3^-	Bicarbonate ions
HDL	High density lipoproteins
HPA axis	Hypothalamic-Pituitary-Adrenal axis
HRC	Hormone-receptor complex
HSP	Heat shock protein
i.e.	"Id est," a Latin term means "that is" or "in other words"
ICF	Intracellular fluid
IF	Intrinsic factor
Ig	Immunoglobulin
IGF	Insulin-like growth factor
IL	Interleukin
im	intramuscular
IMM	Inner mitochondrial membrane
JG	Juxtaglomerular
K^+	Potassium ion

K_f	Filtration coefficient
kg	Kilograms (a unit of length equal to one thousand grams)
km	Kilometer (a unit of length equal to one thousand meters)
kPa	Kilopascals
L	Liter
LCT	Lower critical temperature
LDL	Low-density lipoproteins
Lp	Lipoprotein
μg	Microgram (a unit of length equaling 1×10^{-6} g)
μm	Micrometer (a unit of length equaling 1×10^{-6} m)
LPO	Lactoperoxidase
m	Meter, the fundamental unit of length in the International System of Units (SI)
mEq/L	Milliequivalents per liter
mg	Milligram (a unit of mass equal to 1/1000 (one thousandth) of a gram)
MHC	Major histocompatibility complex
min	Minute
MIT	Monoiodotyrosine
mL	Milliliter (a unit of capacity equal to 1/1000 (one-thousandth) of a liter)
mM	Millimolar
mm	Millimeter (a unit of length equal to 1/1000 (one thousandth) of a meter)
mmHg	Millimeter mercury
mmol/L	Millimole per liter
MOET	Multiple ovulation embryo transfer
mOsm	Milliosmole
MRI	Magnetic resonance imaging
mRNA	Messenger ribonucleic acid
mv	Millivolt
Na^+	Sodium-ion
NEB	Negative energy balance
ng	Nanogram (a unit of mass equal to 10^{-9} g)
nm	Nanometer (a unit of mass equal to 10^{-9} m)
NO	Nitric oxide
NPN	Non protein nitrogen
NPY	Neuropeptide Y
OMM	Outer mitochondrial membrane
OPU	Ovum pick-up
Osm	Osmole
OT	Oxytocin
P450scc	Cholesterol side-chain cleavage enzyme
P_{alv}	Alveolar pressure
P_{atm}	Atmospheric pressure
pCO_2	Partial pressure of carbon dioxide
PCT	Proximal convoluted tubule
PCV	Packed cell volume
PDGF	Platelet-derived growth factor
pg	Picogram (one trillionth of a gram, a unit of mass equal to 0.000000000001 g)
PGE2	Prostaglandin E_2
PGF2α	Prostaglandin F_2-alpha
pH	Power/potential of hydrogen (negative logarithm of the concentration of hydrogen ions or protons)
P_i	Inorganic phosphate
P_{ip}	Intra-pleural pressure

PK	Pyruvate kinase
PKA	Protein kinase A
PKC	Protein kinase C
pm	Picometer (one trillionth of a meter, a unit of length equal to 0.000000000001 m)
PNS	Peripheral nervous system
pO_2	Partial pressure of oxygen
POA	Pre-optic area
PRL	Prolactin
PTH	Parathormone
PUFA	Polyunsaturated fatty acids
PVN	Paraventricular nucleus
RAS	Reticular activating system
RAS	Renin-Angiotensin system
RBC	Red blood corpuscle
rbST	Recombinantly derived bovine somatotropin
rDNA	Recombinant deoxyribonucleic acid
RDP	Rumen degradable protein
RER	Rough endoplasmic reticulum
rGH	Recombinant growth hormone
RH	Relative humidity (RH)
RMP	Resting membrane potential
RNA	Ribonucleic acid
ROS	Reactive oxygen species
RPS	Renal portal system
rRNA	Ribosomal RNA
rT_3	Reverse-triiodothyronine
SA node	Sino-atrial node
SCC	Somatic cell counts
SCNT	Somatic cell nuclear transfer
SHBG	Sex hormone-binding globulin
SNS	Somatic nervous system
SOD	Superoxide dismutase
SR	Sarcoplasmic reticulum
StAR	Steroidogenic acute regulatory protein
STAT	Signal transducer and activator of transcription
T_3	Triiodothyronine
T_4	Tetraiodothyronine
TBG	Thyroxine-binding globulin
TBW	Total body water
Tc	Core body temperature
TCI	Thermal comfort index
Tdb	Dry bulb temperature
TEC	Total erythrocyte count
TGF	Transforming growth factor
THI	Temperature humidity index
TJ	Tight junction
TNZ	Thermoneutral zone
TPR	Total peripheral resistance
TRH	Thyrotropin-releasing hormone
tRNA	Transfer RNA
Ts	Surface temperature
TSH	Thyroid-stimulating hormone
T_{sur}	Surrounding surface temperature

UCT	Upper critical temperature
V_A/Q	Ventilation-perfusion ratio
VEGF	Vascular endothelial growth factor
VFA	Volatile fatty acid
VIP	Vasoactive intestinal peptide
VLDL	Very low-density lipoproteins
w/v	Weight by volume
WBC	White blood cell
wk	Week
×	Multiplied by
ZO	Zonula occludens
α-MSH	α-Melanocyte-stimulating hormone

Part I

Introduction

Veterinary Physiology: Past, Present, and Future Perspective

1

C. G. Shashank, Pradip Kumar Das, and V. Sejian

Abstract

Physiology is the dynamic study of normal functions in a living system. It answers how the cells, organs, organ systems work and how it is integrated at the organism level. This chapter attempts to provide the readers with updated information on animal physiology evolution from past, present, and future perspectives. Physiology is the unit of biology and zoology, covering a range of subjects, including organs, morphology, cells, and biological compounds. Studying physiology has enormous practical applications ranging from cell generation and regeneration to cell death and apoptosis. This chapter gives a glimpse of the levels of structural organisation in animals. They start with the chemical levels, where atoms combine to form molecules, followed by cell levels, where molecules combine to form organelles. Subsequent integration is seen at the tissue level; similar cells and their surrounding material form tissues. Multiple tissues together include an organ, for instance, kidney, liver, heart, and many more. Furthermore, different organs synergistically work to form an organ system. For example, kidneys and urinary bladder together forms excretory system. To maintain normal body functions, a variety of systems work together to form a structural organisation and coordinate with each other in animals. Therefore, in order to practice veterinary medicine effectively, a basic understanding of animal normal physiology is required. The knowledge of physiology can be considered as the bedrock of the medicine.

C. G. Shashank · V. Sejian (✉)
Centre for Climate Resilient Animal Adaptation Studies, ICAR-National Institute of Animal Nutrition and Physiology, Bangalore, Karnataka, India

P. K. Das
Department of Veterinary Physiology, West Bengal University of Animal & Fishery Sciences, Kolkata, West Bengal, India

© The Author(s), under exclusive license to Springer Nature Singapore Pte Ltd. 2023

P. K. Das et al. (eds.), *Textbook of Veterinary Physiology*, https://doi.org/10.1007/978-981-19-9410-4_1

Graphical Abstract

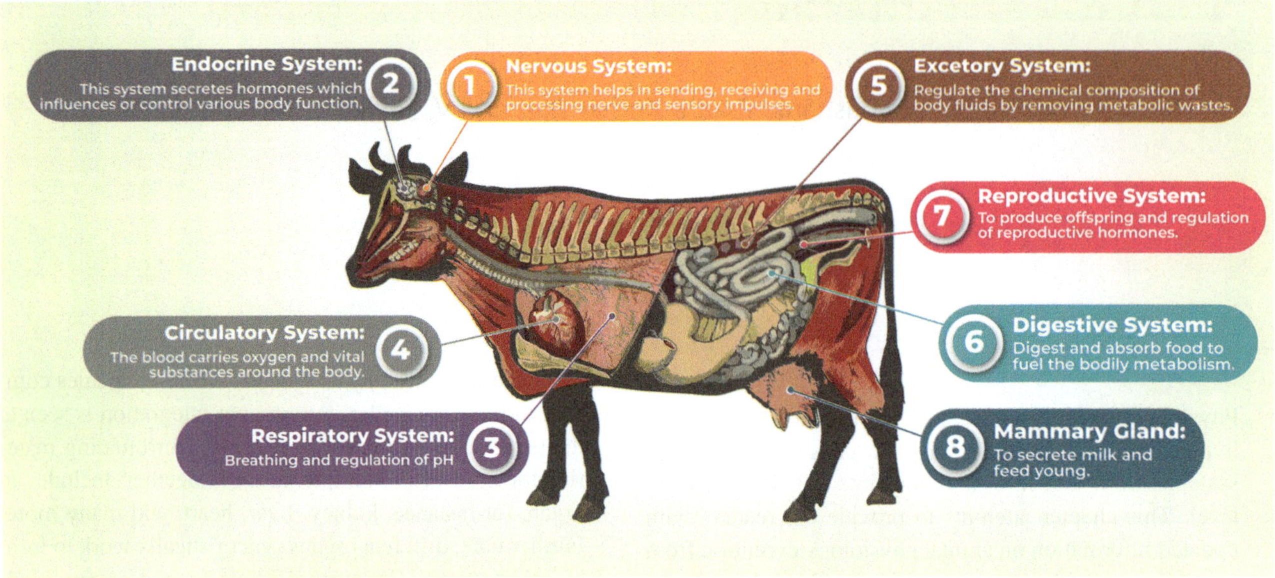

The different systems and functions in dairy cattle

Keywords

Adaptation · Cell · Circulation · Digestion · Excretion · Respiration

Learning Objectives

- To understand the fundamentals of Physiology and its evolution from past to future.
- To appreciate how different organs and organs systems work within a set point for the survivability of animals.
- To witness the advances in the field of physiology.
- To comprehend the metabolic regulation within animals in the different environments.

1.1 Introduction to Veterinary Physiology

We share the earth with millions of wonderful living things. All uniquely adapted to their environments. Yet, despite our differences, we all have one thing in common, our survival. Everything we do relies on complex internal processes that, when working well, allow us to respond to the challenges of everyday life. The study of these mechanisms collectively is called Physiology—the science of life.

A physiologist aims to understand and explain the different physical and chemical dynamics responsible for life's origin, development, and progression at different systemic levels. The study of pathology and medicine is vital to comprehend the abnormal functioning of the body; therefore, it is essential to understand the physiology if one to decipher the mechanisms of disease.

Every life form existing on the earth, from the simple virus to the biggest mammal, has its functional physiological attributes. Thus, the ocean of physiology divides into different subcategories: embryology, cellular physiology, endocrinology, immunology, nutrition physiology, reproductive physiology, lactation physiology, and many more.

Veterinary physiology deals with exploring the structure and functioning of an animal's system and the biological processes by which it interacts with its environment. Animals are exposed to several different environments; probably, the environment presents the biggest challenge. A physiologist tries to examine and explain how biological processes function, adjust, and operate under various environmental conditions, further how homeostasis is achieved such that normal processes are integrated and regulated. For example, an animal's external environment is never constant, and it keeps changing. The change in temperature, availability of feed and water, fluctuations in gas concentrations, and many more may occur daily or at regular intervals may pose a challenge to the basic functioning of animals. To acclimatise and survive in these changing environments, animals need to modify their internal environment like bodily fluids, cellular functioning, endocrine systems, and so on. This continuous maintenance of the internal state is known as homeostasis. The function of most of the organs and associated systems is to maintain homeostasis besides its regular operation. Complete details and associations between different cells, organs, and systems will be explained in further chapters.

Here is the thing, every time animals take a breath, open their eyes, or take a step; a multitude of scientific forces is at play. Thankfully advances made by physiologists and associated scientific researchers in and around the lab provide core information for veterinarians to tackle various significant health challenges animals face today.

1.2 History

Different demographic regions have their history concerning veterinary medicine and physiology. The history of physiology and medicine in a tropical country like India, dating back to the early Vedic period, is fascinating. In Rigveda (2000–4000 BC), we can find fragments of evidence indicating the presence of literature on Veterinary Science, where physicians dealt with both animals and humans comprehensively. The veterinary sciences and animal husbandry practices have been mentioned in Atharvaveda too. Furthermore, in pieces of literature like Asva-Ayurveda: about horses; Gau-Ayurveda: about cows; Hasti-Ayurveda: about elephants and Shyenka-Ayurveda: about Hawks, preservation, and breeding practices are chronicled very specifically. Salihotra, regarded as the "Father of Veterinary Sciences" in India. Palkapya (700–400 BC) dealt with elephants' anatomy, physiology, and management in detail.

Subsequently, outside Asia, the biggest chunk of physiological research was conducted using animal models to understand human anatomy and physiology better. For example, Claudius Galenus (AD 129–circa 199) performed studies to learn more about the body's mechanisms by performing dissection and vivisection on nonhuman primates, such as Barbary apes, to establish the validity of his physiological theories. Ancient Greek physiological ideas, customs, and philosophies were advanced during the Medieval period by Ibn al-Nafis (1213–1288), who discovered and characterised the heart and lungs anatomy. His revolutionary work established the crucial relationship between lungs and the aeration of the blood. The first modern anatomy textbook was written by Andrea Vesalius (1514–1564). Blood flow and cardiac contractions and relaxations were demonstrated by William Harvey (1578–1657).

In the eighteenth century, two Dutch physicians, Hermann Boerhaave and Albrecht von Haller proposed bodily functions as physical and chemical processes. Further, during the nineteenth century, the pace catches up very rapidly. Mathias Schleiden and Theodor Schwann proposed "Cell theory". Claude Bernard (1813–1878) coined the term milieu interieur (internal environment), which refers to the preservation of the internal environment in living organisms regardless of its external environment, which was refined by Walter B. Cannon (1871–1945), who coined the word homeostasis.

Per Scholander (1905–1980) was a comparative physiologist who specialised in the physiological responses of animals to extreme temperatures, such as warm-blooded animals in a cold environment. George Bartholomew (1923–2006) officially founded the section on ecological physiology, where he coupled animal behaviour, physiology, and their interface with the environment to understand animals' adaptation better. Knut Schmidt-Nielsen (1915–2007) was also an expert on environmental physiology. His primary research concentrated on the adaptation of camels to desert conditions. He pointed out the moisture recapture mechanism in exhaled air, which accounts for nearly 60% of the reduction in water loss in camels compared to other animals. Additionally, many other physiologists have made substantial contributions to this huge ocean of knowledge known as physiology.

1.3 Importance of Veterinary Physiology

Veterinary physiology is widely regarded as a foundational discipline for comprehending the distinction between animals' well-being and disease conditions. From cell differentiation and regeneration to cell death and apoptosis, the study of physiology has immense practical applications. For instance, a veterinary physician who studies heart functioning and associated diseases, it is essential to understand the blood pumping mechanism from the heart and the pressures generated with it to transport red blood cells (RBC) throughout the body. In this book, we have discussed different systems and their functionality practically so that readers get creative notions about the application part associated with them.

To demonstrate the significance of physiology, we could use the example of wild animals in captivity. Wild animals are captured and brought to captivity for numerous reasons—conservation, research, the pet trade, and many more. Sudden exposure of wild animals to confinement has resulted in various physiological stresses, like elevated glucocorticoid hormone, fluctuations in respiration rate, heart rate, core body temperature, and skin temperature. While these are adaptive, if it persists chronically, it leads to physiological problems like weight loss, improper immune system functioning, and lowered fertility rate. A study reported that beluga whales had less thyroid hormone concentration when initially exposed to captivity and continued till 10 weeks of captivity (St Aubin and Geraci 1988). In the above context, the primary work of physiologists is to comprehend and quantify the cost these wild captive animals had to spare for survival. This research may aid in a better understanding of the challenges affecting animal health and the development of ameliorative techniques to enhance their well-being while in captivity.

To illustrate, introducing a female to the cage of male brown-headed cowbirds reduced plasma glucocorticoid production and boosted testicular regrowth and many more.

To summarise, veterinary physiology is a crucial discipline for comprehending:

1. The fundamental biology of all animals
2. Animal health and disease
3. Relation between physiology and ecology, in the present and future evolutionary period
4. The relationship between humans and animals in terms of disease transmission and prevention

Know More……

First Cloned Animals Names

Cattle: Daisy; **Buffalo**: Samrupa; **Pashmina Goat**: Noori; **Dog**: Snoopy; **Camel**: Injaz; **Horse**: Prometea

1.4 Basic Living Unit of the Body: Cells

All animals are made up of a complex yet unique bunch of living units—cells (Fig. 1.1). It is the cell interconnectedness in living organisms that creates the beauty of complex systems. Each organ is a group of many different cells that are intercellularly supported through structures.

Interestingly, the surface membrane is a mere 70–100 Å thick (0.0000007–0.00001 mm). The average size of cells is around 10–100 μm, or 0.01–0.1 mm in diameter, denoting the fragile nature of the cell and membrane. This identification was made more accessible by discovering an electron microscope. Many intracellular organelles like the Golgi complex, endoplasmic reticulum, nucleus, mitochondria, vesicles, and many more organelles or structures were identified.

A typical animal cell consists of nucleus, a cell membrane, and cell cytoplasm. The nucleus is the structure that houses genetic information, generally referred to as DNA, and it controls the actions and the reactions of the cell. The cell membrane, the border of the cell, acts as the gatekeeper. It controls what enters and exits the cell and enables adjacent cells to stick to each other. The cytoplasm is where chemical reactions take place in cells.

Know More……

Cells Are Self-destructive

No, this does not imply that cells self-destruct automatically. If a cell is injured or has DNA defects, it self-destructs to avoid interfering with other cells.

There are numerous distinct cell types, each with a unique set of capabilities. For instance, cattle maintain an average of 6.3×10^{12}/L RBCs, with the major purpose of carrying oxygen from the lungs to the respective tissues. At a rough guess, cattle might have around 105–420 trillion cells. One of a kind is nerve cells which have branched endings called dendrites. This is so they can communicate with lots of other nerve cells through electrical impulses. White blood cells (WBCs) enable animals to fight infections. Their distinctive lobed nucleus can be used to identify them. WBCs have a flexible cytoplasm so that they can engulf pathogens in a process called phagocytosis. Some other WBCs are specialised in producing antibodies, and it is these antibodies fight pathogens.

Though cells have different functions, essential physical characteristics are similar. For instance, in all cells, the reaction of oxygen with carbohydrates, fat, and protein produces the energy necessary for cell function. Additionally, the chemical mechanisms by which nutrients are converted to energy are

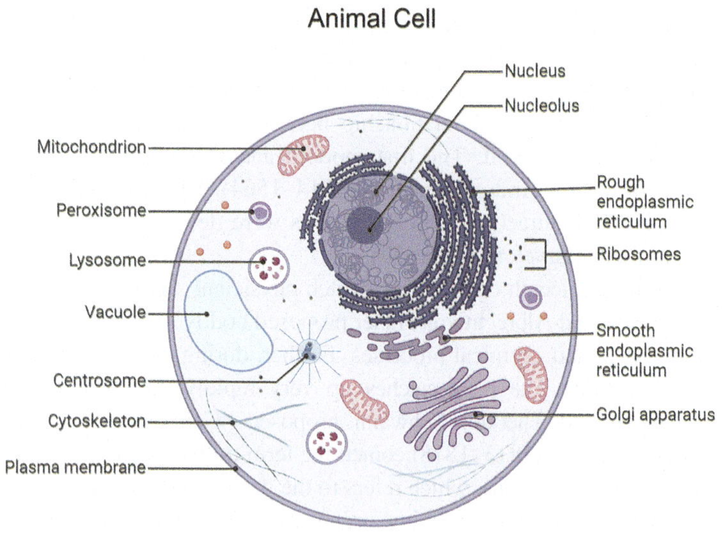

Fig. 1.1 Description of the typical mammalian cell and its components. (Courtesy: BioRender)

roughly comparable in all cells. All cells excrete the products of their chemical reactions into the adjoining fluids.

Those mentioned above are fundamental features of cells. With the improvements in technology, it is possible to generate an entire animal from a single cell using multiple techniques. For example, cattle cloning was done using the somatic cell nuclear transfer (SCNT) technique, one of the genetic engineering tools. Recently with advancements in designer nucleases, such as transcription activator-like effector nuclease (TALENs), zinc-finger nucleases (ZFNs), and clustered regularly interspaced short palindromic repeats/CRISPR-associated protein 9 (CRISPR/Cas9) made path-breaking discoveries that can create a designer offspring's, which will be beneficial to maintain elite lineage in animals. Concluding, with recent advancements in genetic engineering, one can anticipate that the agricultural and biomedical applications of genetically engineered animals that have been long envisioned will soon be recognised in the marketplace.

1.5 The Physiology of Integration

Integration is a broad term that refers to processes such as summarisation and coordination that result in incoherence and harmonious process. The term "integration" refers to combining sensory, endocrine, and central nervous system impulses to ensure the appropriate functioning of the animal. Whole animal integration consists of many systems but mainly nerve and endocrine cells, leading to smooth, coordinated movements.

1.5.1 Nervous System

The nervous system is essentially a massive, complex, body-wide communication system. To demonstrate, stimuli are relayed to the Central Nervous System (CNS) through sensory neurons; further detected by a receptor, which sends the electrical impulses along a sensory neuron to the CNS. The CNS then relays the message by motor neurons to effectors that respond, such as stimulating animals' legs to run when they are about to get hunted.

Nerve cells are known as neurons (the basic unit of the nervous system). The nervous system consists of two major systems: the central nervous system (CNS) and the peripheral nervous system (PNS). CNS comprises the brain and spinal cord, and it is rich with cell bodies (axons and dendrites) of neurons. Within CNS, there are three different types of neurons: sensory, intermediate or relay, and motor neurons. These specialised cells carry information as tiny electrical impulses and make up the nervous system. Sensory neurons carry signals from receptors to the spinal cord and then to the brain. For instance, the eyes send data to the brain about the environment. The intermediate or relay neurons carry messages from one part of the CNS to another and the motor neurons carry signals from the CNS to effectors. The PNS is a division of the nervous system that deals with all the nerves outside of the CNS. The complexes of nerves that make up the PNS are axons or bundles of axons from nerve cells or neurons. It ranges from microscopic to large size that can be easily visible to the human eye. Further divisions of PNS are the somatic nervous system (SNS), which controls voluntary movements like skeletal muscles, and the autonomic nervous system (ANS), which takes over the striated and non-striated muscles. All neurons have three main components; a cell body with a nucleus, dendron, and dendrites which are the neuron's inputs. They receive information from other neurons or the external environment and transfer it to the cell body, and other axons carry the signal away from the cell body. Nervous systems mainly consist of neurons and glial cells, connective tissue cells, and circulatory system cells.

Previously, these cells were identified through a simple microscope. Later internal structures were acknowledged through electron microscopy, and presently, different cross-section imaging technologies like functional magnetic resonance imaging (fMRI), computed tomography (CT), magnetic resonance imaging (MRI), and positron emission tomography (PET) have revolutionised neurology, where physiologists study the functioning of the nervous system. For example, common nervous neoplasia in dogs and cats are meningiomas and gliomas. These two neoplastic cells compress brain parenchyma. Demonstrating this neoplasia through MRI and CT scans is very reliable and easy compared to old techniques practised.

1.5.2 Endocrine System

Endocrine system, also known as the hormone system, comprises many tissues that regulate animals' internal environment by releasing a chemical substance called hormones into circulation to act on the target organ for the desired action. Endocrine tissues are typically ductless glands (e.g. pituitary, thyroid) that secrete hormones via capillaries that permeate the tissue (Fig. 1.2). These glands receive an abundant supply of blood. However, non-typical endocrine tissues contribute significant amounts of hormones to circulation, for example, secretion of the atrial natriuretic peptide from the heart, erythropoietin from the kidney, insulin-like growth factor from the liver leptin from fat.

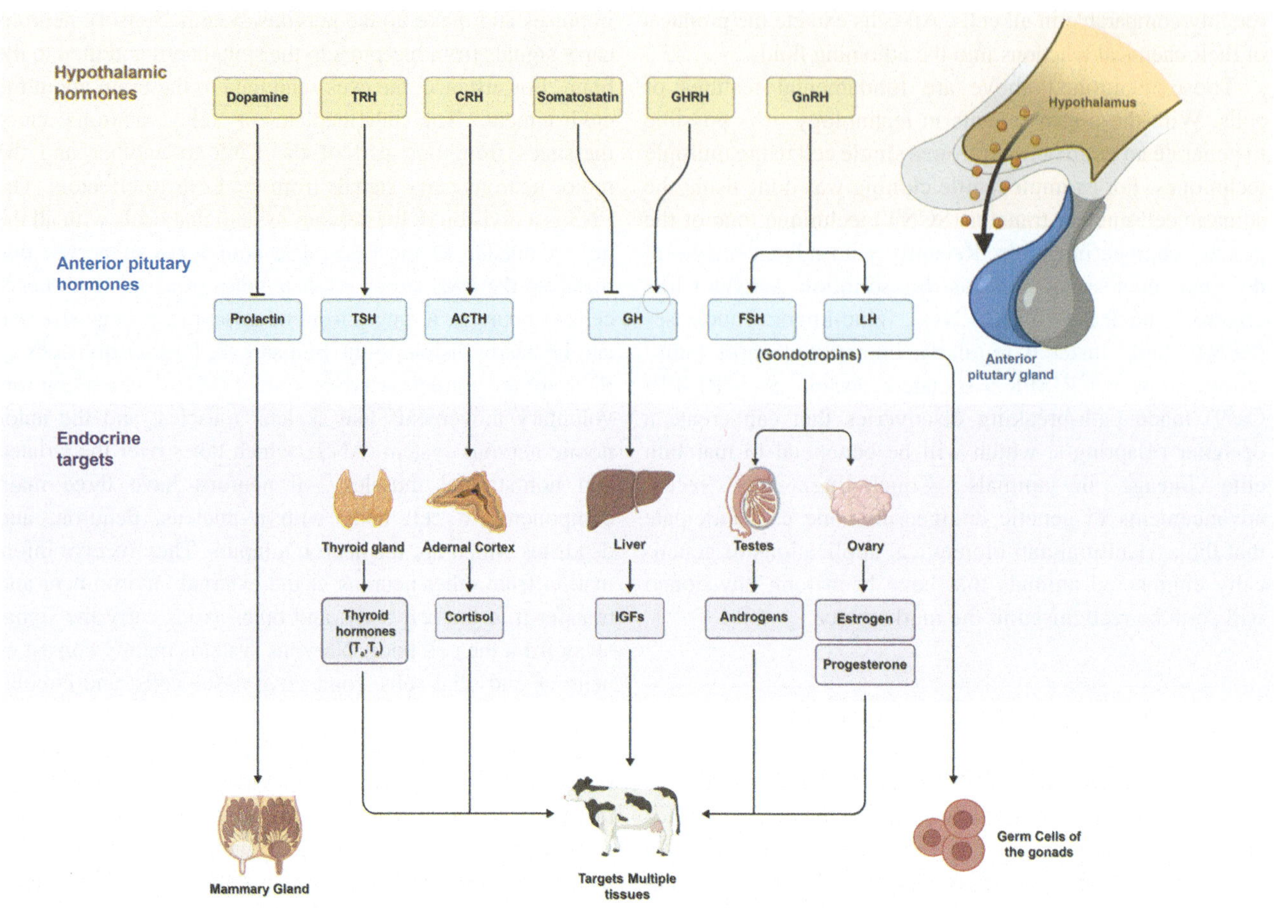

Fig. 1.2 Description of the complete endocrine system and the different constituents in cattle. (Courtesy: BioRender)

Know More......

Fun Fact

Elephants breathe out an average of 310 L of air per minute

The endocrine system controls physiological processes like growth, metabolism, reproduction, behaviour, body regulation, development, fluid, and water balance to maintain homeostasis in most animals. For instance, cows during oestrous phase, various endocrine signals cause characteristic signs like frequent micturition, bellowing, swollen vulva, and mucus discharge. These signs act as visual signs to attract ox, leading towards promoting reproduction. Nevertheless, another class of hormones named pheromones is also responsible for the mentioned example. Pheromones are chemical signals that act within species produced within the animal and then released into the environment. Pheromones also influence physiological functions like the onset of puberty and oestrous.

Several hormones are secreted by the pituitary gland located right below the hypothalamus in the brain. According to the stimuli, the hypothalamus instructs the pituitary gland to secrete related hormones. The secreted hormones make their way via the bloodstream to the target organs. They either elicit a specific reaction directly or stimulate or cause the target organ to secrete hormones. We can think of this as a post office system. Hormones or posts must reach the correct target organs or addresses for the proper response to occur. For example, when hypothalamus detects low water levels in the blood, it signals the pituitary gland to release the antidiuretic hormone (ADH) into the bloodstream. ADH travels to the kidneys, the target organ, and causes water to be absorbed, so urine becomes more concentrated and decreases output. Water consumption elevates the water content in the blood, which subsequently causes the hypothalamus to instruct the pituitary gland to secrete less ADH. Less ADH means kidneys will absorb less water, causing urine to become less concentrated and increase output.

Presently, endocrinologists play around with different hormones to get the best and fast animal stocks to fulfil the global food demand. For example, hormones like androgens and oestrogen are implanted beneath the ear skin. These implants release growth promoters over time into the bloodstream. Animals attain growth much faster compared to traditional methods, thus filling the gap. Along with the development, we need to look into welfare too. Due to urbanisation, various chemicals accumulate in environments that act as hormone agonists or antagonists, finally disrupting hormonal imbalance, leading to the endocrine disruptor hypothesis. This condition is particularly harmful to aquatic, young, and unborn animals. However, the long-term impact of this on humans is still up for debate. Developments in endocrinology depend on emerging technologies like Omics and structural biology. Looking back, one might question them; will we ever fully comprehend how hormones act at the cellular level. The quick answer is: No. Our present understanding of endocrinology at any given time will influence future hormone research.

1.6 Exchange and Transport System

Imagine if animals took a couple of hours for the oxygen that they breathe to reach their cells. Well, if the giraffe relied on diffusion alone from oxygen to get to the tip of the head, by the time oxygen reaches cells would have died. This is why large multicellular organisms have developed specialised exchange surfaces like lungs, gills, digestive systems, and specialised transport systems like circulatory systems. These systems help animals to exchange gases and nutrients with the surroundings and transport them where they are needed within the body. As we all know, animals have smaller surface areas compared to large volumes. It is slightly inefficient at exchanging substances. To cope with that, how animals have progressed exchange services and short diffusion distances are mentioned in the below sections.

1.6.1 Respiratory System

The respiration system is a significant life-supporting system with many essential functions as gas exchange, oxygen supply to cells, maintaining the balance between gases, and pH in the body (Fig. 1.3). The respiratory system provides animals with oxygen and removes metabolic by-products such as carbon dioxide. Failure of the respiratory system is related to malfunction of various organ systems, with potentially fatal implications. For example, the COVID-19 pandemic during 2020 affected not only humans but animals too. Asiatic lions were found positive for COVID-19 in Hyderabad Zoo and Etawah Safari park, India, and mink farms in the Netherlands. A virus arrested the proper functioning of the lungs by compromising the immune status of the animals, which leads to further secondary complications, finally resulting in death.

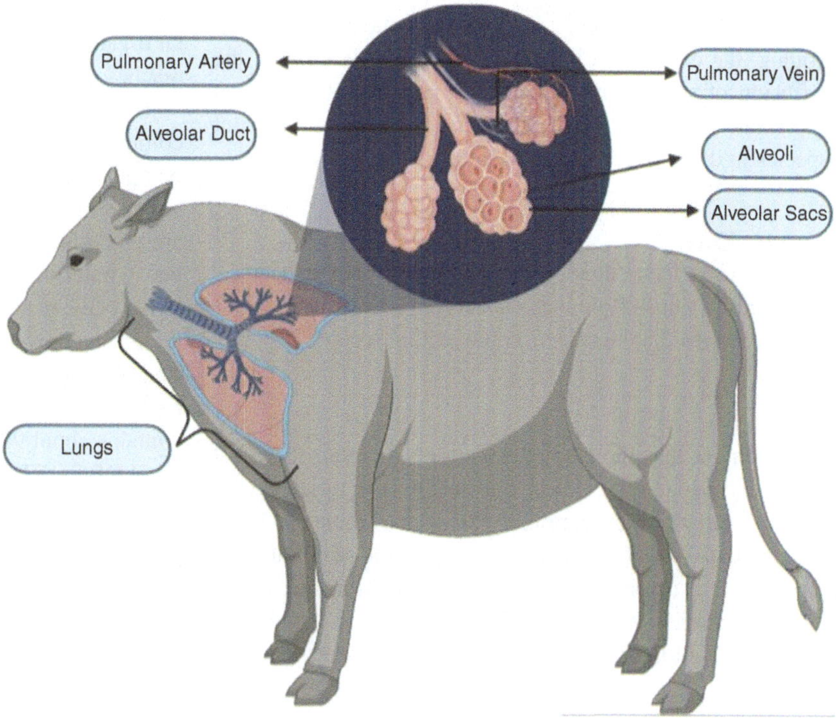

Fig. 1.3 Description of the respiratory system and its components in cattle. (Courtesy: BioRender)

Know More......

Did You Know?

Lungfish have both gills and lungs. Many researchers even believe that lungfish might be the missing link between marine animals becoming land-dwellers.

Now the question is why cells produce carbon dioxide following oxygen consumption? Animal depends on mitochondrial respiration to supply ATP to perform normal cellular activities. To summarise the whole process, animals breathe in oxygen, and lungs load up this inspired oxygen (O_2) to red blood cells (RBC), which carries O_2 to cells. Cells absorb oxygen through diffusion, where mitochondrial respiration comes into action. Mitochondria oxidise the available carbohydrates, amino acids (AA), and fatty acids to produce ATP, resulting in CO_2 production. Produced CO_2 is transported back to the lungs through RBC, then the expiration of CO_2 from the lungs. On an outer look, it looks effortless. Still, if we dive deep, we can witness the impact of different factors like haemoglobin concentration, partial pressures of various gases, lungs pressure and movement of diaphragm concentration of alveoli, and many more. Unicellular organisms and aquatic animals may utilise diffusion gradients to drive gas exchange with the environment. In comparison, gaseous O_2 requires an additional step to cross the cell membrane in terrestrial animals.

A common thought might be, what about the respiration in aquatic animals! Wouldn't it catch your attention on mentioning, the exchange of gases in fish larvae is through the diffusion of gases across their body surfaces. Readers can appreciate complete comparative mechanisms in different species in detail in respective chapters.

Know More......

Octopuses have three hearts. A blue whale's heart weighs about 400 pounds. A cheetah's heart rate can speed up to 250 beats per minute within seconds.

We know that the respiratory system is more than just lungs; it includes the nose, followed by the nasal cavity, pharynx, larynx, trachea, primary bronchus, secondary bronchus, tertiary bronchus, respiratory bronchiole, alveolar duct, alveolus, and diaphragm. But can you envision the design and fabrication of lungs outside the body? Yes, we are facing the future right in front of us. Bioengineered Lungs (BEL) is an exciting and rapidly progressing area in the veterinary biomedical field. It is an alternative for end-stage lung failures. Researchers are developing bronchial system circulation in non-immunosuppressed pigs to support BEL growth and animal survival after transplantation. Previously, researching and discovering medications for the treatment of respiratory disorders incurred very high variability and high costs. However, with the use of in silico and tissue-engineered lungs models, it is possible to understand various mechanical and biological variables that make in vivo research difficult. However, on the flip side, single-cell sequencing technologies, advancements in cellular and tissue imaging techniques, and improvements in tracking cell lineage systems have led to understanding the complex relationship between the respiratory system, cardiovascular system, and exchange of gases.

1.6.2 Circulatory System

The circulatory system works to transport oxygen and nutrients throughout the body while removing waste products such as carbon dioxide, urea, and many metabolic wastes at the same time frame (Fig. 1.4). The heart, blood vessels, and blood are the major components of the circulatory system. The circulatory system is divided into two circuits: the pulmonary and systemic. In the pulmonary circuit, deoxygenated blood is pumped from the heart to the lungs to become oxygenated. In contrast, this oxygenated blood returned to the heart is pumped to the rest of the body in the systemic circuit. Unlike other muscles in the body, the heart, a muscular organ, never tires and works very hard to ensure that blood reaches all parts from head to hoof.

Is blood always red? Wrong, blood comes in a variety of colours. It may be red in humans and other mammals, but an octopus, for example, has blue blood and oscillated eyes. Fish have completely clear blood, and in Papua New Guinea, they are green-blooded. The composition of blood is very simple yet extraordinary. Blood consists of red blood cells, white blood cells, platelets, and plasma. About 55% of the blood is the pale yellow sticky liquid found in animals called plasma. The plasma is mainly made up of water and proteins. Plasma carries nutrients, hormones, and proteins around the body. It contains about 92% water. About 45% of the blood is made up of RBCs. RBCs are tiny biconcave disc-shaped cells, also known as erythrocytes. Formation occurs in the bone marrow. It contains a special iron-containing protein called haemoglobin. Haemoglobin that makes the warm-blooded animal's blood red, the octopus, has a unique protein called haemocyanin, making their blood blue.

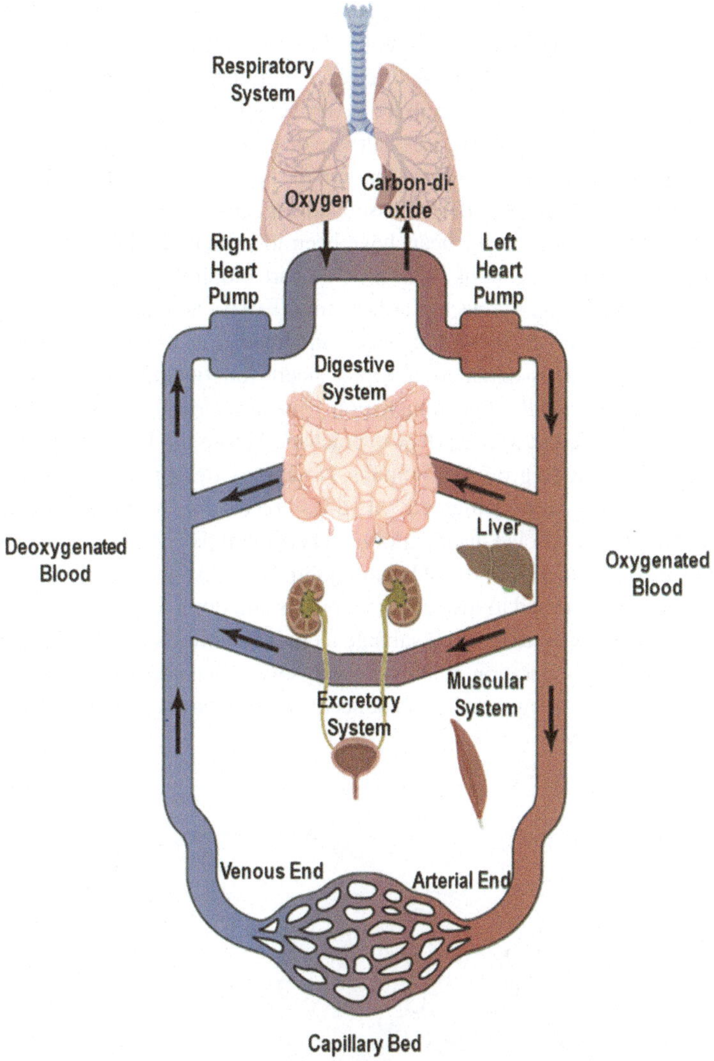

Fig. 1.4 Description of the circulatory system in animals. (Courtesy: BioRender)

> Know More......
> **Blood Groups**
> Dogs—8 groups
> Cats—3 groups
> Horse—8 major groups
> Cattle—11 groups
> Goats—5 groups
> Sheep—7 groups
> Humans—4 groups

The icefish does not have either haemoglobin or haemocyanin, which makes it blood colourless. WBCs and platelets account for less than 1% of the blood. WBCs are savage fighters. Pathogenic microorganisms are prevented from entering the animal's body by these organisms. Roughly about 70% of WBCs are phagocytes, which ingest and destroy invading pathogens in a process known as phagocytosis. Lymphocytes produce antibodies. Antibodies bind to the foreign pathogen and prevent it from spreading, whereas platelets form a minute blood fraction. Whenever an animal suffers a cut, they aid in clotting the blood, keeping the wound from becoming infected.

Transporting oxygen to all the cells in the body is the primary job of blood. The role is carried out by RBCs, which contain an essential protein called haemoglobin. So when animals breathe in, oxygen latches onto an active site in the haemoglobin with a single iron atom. We can think of it as a seat on a shuttle bus. The oxygen molecules must first find their seat and put their seat belts on before the bus can move. Once the bus moves, the oxygen molecules are released when they reach their destination, anywhere in the body. The deoxygenated blood hops on while the empty shuttle bus returns to the heart. So the deoxygenated blood has arrived at the heart. Further functioning and structure of heart, arteries, veins, capillaries, and lymphatic system are described in respective chapters.

In recent days, there have been many developments in veterinary cardiology like digital radiography, which helps diagnose congestive heart failures and monitor cardiac and noncardiac causes of cough (e.g. bronchial compression, tracheal collapse, inflammatory airway disease). Various cardiac biomarkers like cardiac troponin 1, N-terminal Pro-B-Type Natriuretic peptides released during contraction and stretch or damage to the heart have been identified. Technological advancements like cell phone heart monitor devices provide high-quality electrocardiograms (ECGs) and animal heart rate data. Moreover, a better understanding of blood transfusion is essential to interpret heart functionality. A technique called xenotransfusion means transfusions between different species. It was practised before the identification of blood groups in humans. No allergy or agglutination events were observed in dogs receiving purified polymerised porcine haemoglobin. Oxyglobin is an ultra-purified polymerised bovine haemoglobin-based oxygen-carrying solution used to treat anaemia in cats and dogs. Using a cell separator system, a cell salvage technique collects blood intra- or post-operation in animals with severe haemorrhage. The blood is filtered and reintroduced to the patients within 6 h as packed red blood cells (pRBC) suspended in saline. Prior to transfusion, anticoagulants and systemic medicines such as plasma-activated clotting factors will be purified.

1.7 Excretory System

The process of removing the metabolic waste and excess water from the body is known as excretion. Each organ, but particularly the excretory system, plays a unique role in maintaining homeostasis. Most of the products reaching the excretory system are metabolised and excreted by the body through sweat, urine, and faeces. The excretory system, also known as the urinary system, consists of kidneys, ureters, bladders, and urethra (Fig. 1.5). In addition to eliminating metabolic waste, the urinary system also regulates the pH of the blood, regulates levels of metabolites and ions such as sodium, potassium, and calcium, and regulates blood pressure and volume. Based on their evolutionary history, surroundings, and feeding and drinking habits, animals produce waste in various ways. These factors regulate water consumption in animals, and most metabolic waste must be

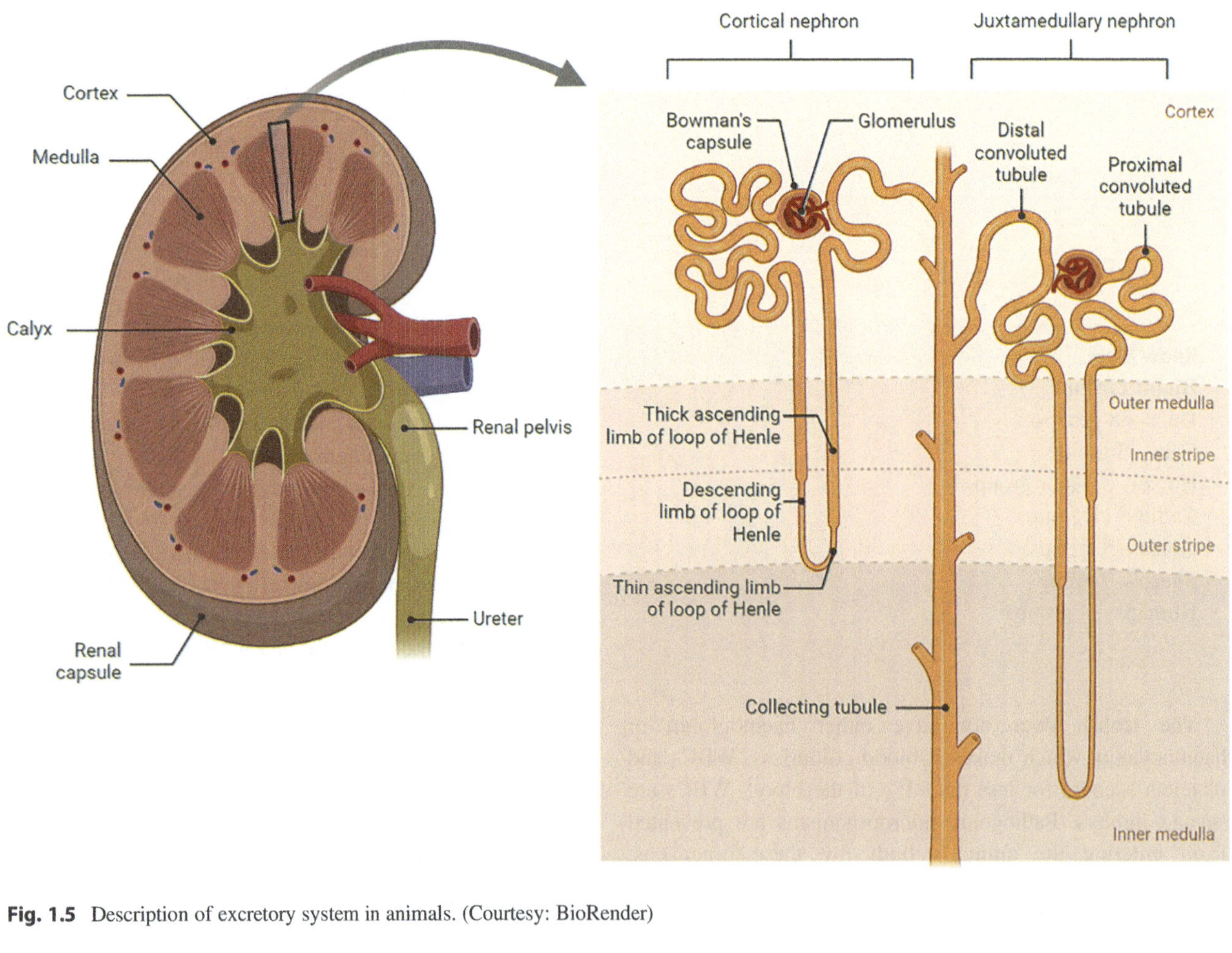

Fig. 1.5 Description of excretory system in animals. (Courtesy: BioRender)

dissolved in water before it can be expelled. The most significant by-product of food metabolism is ammonia, produced by the breakdown of proteins and is highly toxic to the body if stored. Thus, based on the amount of accessible water and the animal's ability to store it, animals transform this ammonia into either urea or uric acid.

Know More......
Fun Fact
If one kidney fails to function and is removed, the other kidney can increase in size by 50% within months to handle the entire task of filtration.

The end product of filtration in aquatic animals produces ammonia, expelled directly into the surrounding environment. In comparison, through the process of detoxification, terrestrial mammals convert ammonia-like compounds along with CO_2 into urea in their livers. Birds excrete nitrogenous wastes in the form of uric acid. Even though this process requires more energy, it allows less water to spare, and waste can be excreted as a paste. The kidneys begin by filtering out a large amount of fluid and compounds dissolved in the blood, then reabsorbing around 99% of it before excreting the remaining 1% in the form of urine. Thus, the majority of animals' excretory systems are not exclusively dedicated to excretion but are also well suited for reabsorption. Further detailed description of kidney and its working force are mentioned in the respective chapter.

There are numerous renal disorders in canines; one of the most devastating is chronic kidney disease (CKD). Previously, it was routine treatment with management practices like feeding renal diet, with mediations to prevent load on kidneys. Nevertheless, recently, with the knowledge of stem cells, researchers have bought up with adipose-derived stem cells (ADSCs) therapy. On conducting research, they could conclude that intravenous injection of ADSCs can improve the kidney recovery rate and functional capability in dogs suffering from CKD. In terms of animal welfare, stem cell therapy proved to be adequate to combat CKD.

Another field in nanoscience has the unlimited potential of overthrowing the present medication routines. Nanoscale and novel physiological engineering applications can answer the new dimensions of renal dialysis. Recently, many techniques like permeable selective membrane and nanoscale fabrication process have jumped into the market to treat dogs suffering from renal failure. Various electrokinetic methods for fluid treatment have emerged; for instance, the ion concentration polarisation (ICP) technique which is a part of an electrokinetic purification system acts as an artificial kidney. Through this, a peritoneal dialysis-based wearable artificial kidney device has been a device for end-stage renal disease dogs (ESRD). For dogs exposed to this technique, 10% of toxins were reduced by 3 h. By this, one can expect the wearable artificial kidney to advance more for quality ESRD dogs in the future.

1.8 Digestive System

The animal's digestive system is considered one of the diverse and complex systems. Organs related to the digestive system work continuously as a team to fulfil a single task: transforming the raw materials of the food into essential nutrients (e.g. carbohydrates, fats, proteins, minerals, and vitamins) and energy towards growth, maintenance, and reproduction. Generally, we can divide the digestive system into four main components. (1) Gastrointestinal tract (GIT), which transports food from mouth to rectum; (2) trio organs, i.e. liver, pancreas, and gallbladder, break down the food through enzymes or digestive juices; (3) it is a combination of enzymes, hormones, blood, and nerves, breaks down food, modulates the digestive process, and delivers the final nutrients to respective tissues and cells; (4) mesentery, a membrane that supports and holds the digestive organs to the abdominal wall.

Animals can digest food through various mechanical and chemical processes. For instance, complex food is mashed to pieces mechanically through teeth whenever food enters the mouth, whereas saliva carries out chemical digestion. Enzymes in saliva like amylase convert complex carbohydrates to simpler carbohydrates. To depict an overall pipeline of digestion, food is ingested as mentioned; it is converted into smaller pieces and mixed with digestive enzymes and transported through motor or muscular activities of GIT. In addition to protecting and lubricating the gastrointestinal tract, secretions from the salivary gland, stomach, gall bladder, and intestines also assist digestion. Digestive enzymes hydrolyse the nutrients in the stomach like carbohydrates into simpler sugars, fats into fatty acids and glycerol, and proteins into amino acids. The main job of a digestive system does not stop at the stomach. Abovementioned nutrients need to be absorbed into the bloodstream; this process occurs in the small intestine. Specialised structures called villi in the intestine increase the surface area of the intestines. More the number of villi more the absorption of nutrients. The nutrients are absorbed into the villi through diffusion and transported into the bloodstream through capillaries in the villi. Rest that is not absorbed are pushed out of the system through the rectum. The digestion process in animals varies according to species, habitat, nature of feeding, and many more. For example, ruminants (cattle, sheep, goats) get nutrients through the breakdown of cellulose of plant cells. Here animals depend on large populations of bacteria and protozoa to ferment the food and derive the nutrients to the animal's body. Whereas in simple stomach, animals such as horses and pigs, digestion of food occurs in the stomach with the help of digestive enzymes produced in the stomach. The comparative digestive process has been explained in further chapters.

1.9 Reproductive System

Evolution is the day-to-day process. According to evolutionary biologists, the three major functions of animals are growth, behaviour according to the environment, and reproduction. The biological process by which new individual organisms are produced from their parents is called reproduction. Each organism exists due to reproduction, which is a fundamental characteristic of all forms of life. The offspring represents 50% from the paternal side and 50% from the maternal side. It is because the new-born shares genetic information from both parents. This mixing comes about because of sexual reproduction, which involves the fusion of two sex cells or gametes, called fertilisation. Male and female animals (Fig. 1.6) have different reproductive systems. During puberty, the reproductive organs develop to enable the production of offspring.

The testes produce sperm, the male gametes. In most species, the testes are kept outside the body in the scrotum, except for elephants, rhinoceros, and a few marine animals, like whales and dolphins. The purpose of the external scrotum is to keep the scrotum cooler and is better to form sperm production. The sperm ducts carry sperm from the testes to the urethra, a tube running down the inside of the penis. Spermatozoa are mixed with secretion from glands to produce a liquid called semen. Semen helps carry the sperm into the female reproductive system. The female gametes eggs or ova are released from the ovaries through the process called ovulation. Immediately following ovulation, the egg travels through a tubular passageway called the fallopian tube away from the ovary and towards the uterus. During copulation, the semen is ejaculated from the penis into the female's vagina. These ejaculated spermatozoa swim up through the cervix and uterus. If a sperm manages to reach the ova in the fallopian tube, then fertilisation will occur. After ovulation, the egg only lasts 24 h, whereas the sperm can last up to 3 days (varies according to species). Once fertilisation occurs, the fertilised egg called the zygote will start to divide on reaching the uterus. This cluster of cells will settle into the lining of the endometrium of the uterus. Inside the uterus in a pregnant animal, the fertilised egg cell will continue to divide and differentiate to form different cells. Some will form structures in the embryo and others the placenta. The placenta is an external covering of the foetus where substance exchange between the mother and embryo occurs. Nutrients and oxygen will pass from the mother's blood into the embryos. Embryos receive these nutrients through the umbilical cord. Usually, the embryo is called the foetus after fertilisation at the end of week 8 (varies according to the species). The foetus continues to develop for the entire pregnancy, also called gestation. After the designated gestation period with the combination of different endocrine hormones, parturition occurs. It is the glance at the complete process of copulation to start a new life.

Various reproductive biotechnologies or assisted reproductive technologies (ART) have recently revolutionised livestock productivity to meet the increasing global population food demand. Recent biotechnologies in both males and females, particularly concerning livestock, have revolutionised the reproductive process in vitro and in vivo to improve reproduction and efficiency. Techniques like semen sexing are used to produce offspring of the desired sex. This technique is based on the flow cytometric principle of separating spermatozoa with fluorescently labelled X chromosomes from sperms and fluorescently labelled Y chromosomes. Using this technology, it can sort 15 million spermatozoa per hour into X- and Y-bearing sperms and

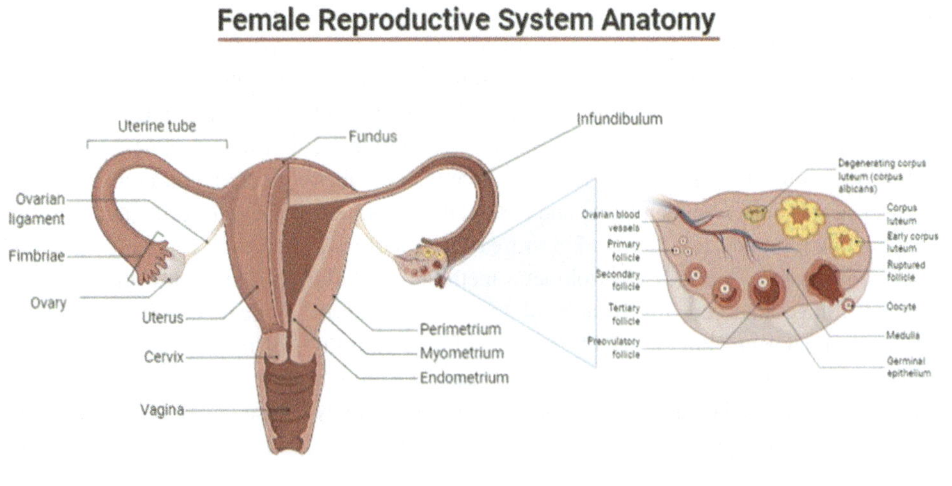

Fig. 1.6 Description of the female reproductive system in animals. (Courtesy: BioRender)

predict the gender of calves with an accuracy of between 85% and 95%. Another technique is named sperm encapsulation, which encapsulates the spermatozoa for longer term preservation in vivo. This technological innovation was intended to allow spermatozoa to stay alive longer in the body's temperature and make the release of viable spermatozoa more progressive over a longer period in domestic animals. Coming to females, Ovum pick-up (OPU) is a non-invasive approach for obtaining large quantities of high-quality oocytes from living animals without invasive procedures. In India, using this technique first buffalo calf named Saubhagya was produced. This method not only improves reproductive efficiency over time but can also be used in follicle ablation to aid in follicle turnover during the embryo transfer protocol.

Additionally, In vitro Maturation, Fertilisation, and Culture (IVMFC) involve the collection of oocytes from ovaries of slaughtered animals trailed by the production of viable embryos through in vitro maturation and fertilisation. The IVMFC approach is ideal for embryo transfer, cloning, transgenesis, and sophisticated in vitro techniques. Intracytoplasmic Sperm Injection (ICSI) is a micromanipulation procedure that involves generating healthy and desirable embryos by mechanically inserting high-quality spermatozoa into the oocyte cytoplasm. ICSI has also been performed on sexed sperm with an 80% success rate in cattle and 48–63% success rate in small ruminants using fresh and frozen-thawed sperm. There are many more techniques like embryo transfer technology (ETT), embryo cryopreservation, embryo sexing, somatic cell nuclear transfer technique, stem cell technologies, transgenesis to counter the demand of increased productivity. Although these techniques have the potential to be effective, they have been hampered by several factors, including the lack of a comprehensive database on indigenous livestock and its biodiversity (which includes traits such as production, reproduction, and disease resistance within species and breeds), which are necessary for their implementation. In the future, the use of these advanced techniques may provide additional insight into the molecular complexities of the reproductive process, including its insanity.

1.10 Lactation Physiology

The process by which mammary glands produce and secrete milk is referred to as lactation. Lactation requires synchronous physiological processes to maintain the homeorhesis of the dam and nutrient acquisition essential for milk formation. It is the most important and expensive phase in dairy animals. The lactation length is about 305 days in cattle, and it varies in different species. Mammary glands are the organs in mammals that produce milk for the sustenance of the young. Mammary glands are among the few structures in mammals that may undergo recurrent growth cycles, functional differentiation, and regression. The mammary gland is derived from the ectoderm during the embryonic stage. Mammary glands include teats, duct systems, lobes, lobules, and secretory tissue. Between puberty and parturition, the formation of ducts and milk-secreting tissue occurs. Mammary gland is modified sweat and exocrine glands, located in the inguinal region in sheep, cattle, goats, horses, and whales; thoracic region in primates and elephants; ventral surface of both thorax and abdomen in pigs, rodents, and carnivores. Delicate membranes separate the front and rear quarters. In cattle, rear quarters produce almost 60% of the milk, while the forequarters produce the remaining 40% and a lactating udder weighs around 15–32 kg. Many factors regulate milk production like age, breed, environment, hormones, and many more.

The process of synthesis and secretion of milk from the mammary alveoli is called lactogenesis. Alveoli are the grape-like clusters containing epithelial cells that absorb nutrients from the blood, transform them into milk, and discharge the milk into the alveolar cavity. Blood supply to the mammary gland is extremely important for its function. Did you know, for every litre of milk produced by a dairy cow, almost 670 L of blood flow through the udder? Further, several hormones play a major role in development of mammary gland and maintenance of milk secretion throughout lactation period. The physiological mechanisms that regulate milk production in cows are multifaceted and extensive. Several cascades and hormonal cycles in the cow's body favour the beginning and termination of milk production. Hence, the lactation stage in the life cycle of a dairy cow is extremely vital considering the production and economic conditions.

Recently, the mammary gland has been considered a bioreactor in a crude way. The mammary gland is like a factory that produces remarkable proteins. So how about genetically engineering the mammary gland after understanding the basic physiology to produce the proteins that are not produced in the mammary gland necessary for consumption. The result of this challenge is transgenesis. Using this technique, we can make the genetically engineered calf harvested for nutraceuticals or bioactive components. This concept is still under research, but the technique offers opportunities to produce milk ideally suited for a particular product, for example, milk specifically produced for cheese. This technique is challenging to harvest from human blood because of its lower concentration. Those proteins can be produced through the mammary gland of animals ranging from rabbits to cows. Anticlotting agents and drugs used to treat angioedema, emphysema, wound healing, and haemophilia have been successfully harvested. Coming to small ruminants, it is now easy to produce monoclonal and polyclonal antibodies that can be used in diagnostic products through genetically modified goats. Most people are lactose intolerant, but no worries; it can be countered by knocking out the α-lactalbumin gene, which drastically reduces the lactose concentration in the milk. Mastitis is the most significant setback in the dairy industry; a study conducted in cows found that a gene encodes lysostaphin in cow's milk to protect against *Staphylococcus aureus* mastitis.

1.11 Environment Physiology

The environment can be defined as the surroundings or conditions in which humans, animals, or plants live or operate. The environment is critical to life on the planet earth. An ecosystem refers to all the living and non-living things that exist in an ecosystem and the ecosystems' relationship to one another. It is the foundation of the biosphere, which governs the overall health of the planet. To make things simple, the environment can be classified into two parts: the biotic environment, which includes all living organisms such as animals, forests, bacteria, fungus, and so on, and the abiotic environment, which includes all non-living components such as temperature, humidity, water vapour, and air. Since industrialisation era human activities from pollution to overpopulation drive up the earth's temperature and fundamentally change the world around us. The leading cause is a phenomenon known as the greenhouse effect. Various gases surrounding the atmosphere, namely water vapour, carbon dioxide (CO_2), methane, nitrous oxide, and chlorofluorocarbons, let the sunlight enter the atmosphere but keep the radiated heat from escaping the earth's surface the glass walls of a greenhouse. The greater the concentration of greenhouse gases in the atmosphere, the greater the amount of trapped heat, strengthening the greenhouse effect and increasing the earth's temperature. Global warming has accelerated as a result of the rapid rise in greenhouse gases in the atmosphere. Climate change has repercussions on the oceans, the weather, food supplies, and human and animal health. Ice sheets such as Greenland and Antarctica are melting. Sea levels rise due to the excess water once contained in glaciers spilling out into the oceans, swamping coastal regions. Furthermore, more intense storms, flooding, heavy snowfall, and droughts incidents are getting more common. These fluctuations in the weather present difficulties; cultivating crops becomes more challenging.

As mentioned, environment plays a significant role in the productivity of an animal. In this changing environment, the animal gets exposed to different stressors, like heat stress, nutrition stress, walking stress, water stress, transportation stress, and many more. Animals have evolved mechanisms to manage short-term stressors. During the short-term exposure, the biological cost is minimal because adequate reserves of biological reserves exist to cope with the stressor and meet the impact of the stress without any disturbances on biological functions. If the animal gets challenged by multiple stressors, there will be insufficient biological reserves to satisfy the biological cost of the stress response; to counteract this, resources will be channelled from other biological functions. As shown in Fig. 1.7, when resources are sidetracked from productive functions, it leads to impairment of biological functions. For example, when multiple stresses deplete body reserves, metabolism shifts away from growth, the young animal no longer blooms, and growth is restricted. When energy is shifted from reproduction and its process, reproductive success is reduced. This metabolic maintenance behaviour of an animal's body rather than production will last until the animal restocks its resources/reserves sufficiently to re-establish normal functions.

Livestock acts as a significant contributor to global food security, particularly in fringe lands where livestock is characterised as a protein, energy, and micronutrient source. Global climate change has a considerable effect on the livestock sector, depending on different ecosystems and natural resources. Looking at the different climate change predictions, we can envision the future of the great struggle to adjust and adapt to new environmental challenges both by humans and livestock. To ensure food security, policymakers and researchers should prioritise identifying animals with superior genetic traits that are economically beneficial and identifying biomarkers for a solution to animal productivity loss due to climate change, especially when animals are exposed to multiple stressors.

As mentioned previously, homeostasis is a mechanism that maintains physiological stability through interaction between internal processes and the external environment. One of the routes to achieve is through thermoregulation.

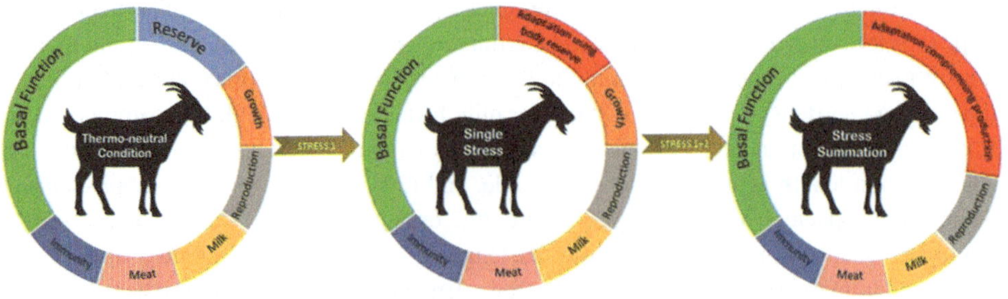

Fig. 1.7 Pictorial representation of summation effect of multiple stresses in goats. (Source: Sejian et al. 2018)

Thermo conformers and thermo regulators are the two types of organisms. The internal temperature of thermo conformers depends on the external environment, whereas thermo regulators maintain their internal body temperature within a particular limit, still being responsive to external stimuli. There are a variety of reasons that contribute to the distinction in temperature regulation among organisms. These include adaptation, mutation, and environmental stimulation. There are different adaptation strategies developed during their lifetime to maintain the desired body temperature.

On exposure to heat stress few animals loose heat through sweating (e.g. horses, humans). Rats, mice, dogs, and cats all have sweat glands on the soles of their feet. Rather than that, many mammals (e.g. dogs) pant to cool themselves. Vasodilation is another method of releasing heat from the body. Those blood vessels closest to the skin's surface expand wide and allow blood to flow through them. Blood gets cooled down as the heat radiates out of the body. In contrast to heat loss, animals employ the opposite mechanisms to retain heat when the ambient temperature falls below the core body temperature. Vasoconstriction is accomplished by subdermal capillaries, which redirect blood away from the skin and body's periphery. In extreme cold, prolonged blood rerouting away from the extremities results in numbness and cellular damage (e.g. frostbite). Animals contract minute subdermal muscles (erector pili) to erect dermal hair follicles to increase heat retention. These erect hairs form a heat-trapping insulating layer.

1.12 Omics in the Field of Physiology

Omics is a science branch that aims to characterise and quantify many biological molecules that decode into an organism's structure, function, and dynamics. Before understanding the purpose of Omics and different Omics technologies and their functional capability in physiology, let us look at the mega human genome project, the starting point for Omics. The Human Genome Project was a massive undertaking involving scientists worldwide working together to decipher the deoxyribonucleic acid (DNA) sequences that make up the human genome. The project began in 1990 and was finished in 2003, 2 years ahead of schedule. Researchers decrypted the mystery of DNA. According to them, the genome of every person on earth is 99.9% the same. It is that tiny 0.01% that makes up genes that give us our unique differences. An essential aspect of the human genome project is to identify any mutation in a gene order that could lead to disease. Scientists can use this information to extrapolate the genetic causes of diseases and develop treatments for the variants of genes or alleles associated with various inherited disorders. Genetic tests can be performed on individuals to determine whether they are carriers or sufferers of an inherited disorder. It took 13 years and billions of dollars to sequence the whole genome during the human genome project, but now researchers can do it in a few hours and for a relatively low charge.

Animal researchers have the same goals of identifying gene variants that add to good health and further increase the food production capacity and quantity of animal products. Much of the match for functional gene annotation for animal sequences comes from human orthologues. Even though advances are made in physiology and allied sciences; however, full potential has been not achieved in Omics work, for instance, Sire-based evolution using genomics for future productive offspring production.

Through a wide range of Omics techniques, it is possible to get insight into an animal's sensory system and surroundings. For instance, pufferfish, when exposed to cold stress, more than 5000 differentially expressed genes (DEG) popped up. Further, only a few differences were in proteins and metabolites. This study reflects the importance of different Omics approaches. If only transcriptomic data were be considered, it might not have reflected proteins and metabolites at a single stretch. However, when proteomics and metabolomics were combined, it gave the network analysis. It revealed that different molecular interactions were associated with immunity, metabolism of fatty acids, transport of bile salts, and lipolysis, suggesting that these processes were essential for pufferfish to withstand cold stress. This study can be the best example to weigh the multi-Omics approaches in the field of physiology.

We search for robust breeds in different species to fulfil the global food demand in this present climate change. Heat stress and nutrition stress are the major abiotic stressors for dairy cattle. Although Omics science has advanced, heat stress's direct and indirect impacts remain key obstacles in understanding the interaction between genetic polymorphism, genes, transcripts, proteins, and metabolic pathways connected to productive qualities like milk and meat output. To give a better perspective, let us imagine cattle is subjected to multiple stressors (thermal, nutritional, and walking stress). In response to the stress, the homeostatic mechanism triggers to counteract those stressors in cattle. It can be quantified using genomics, which identifies the genes, SNPs associated with stress resilience via genome-wide association studies (GWAS).

Further, epigenome studies propel us to the identification of DNA modifications that change accessibility for transcription. A transcriptomic study allows the quantification of the mRNA (gene transcripts) in different tissues in response to stress response (Fig. 1.8). It can be cross verified through proteome, examining the entire set of proteins after translation from mRNA and post-transcriptional modifications. Additionally, metabolomics gives us an idea about lipids, water-soluble and volatile molecules formed after protein

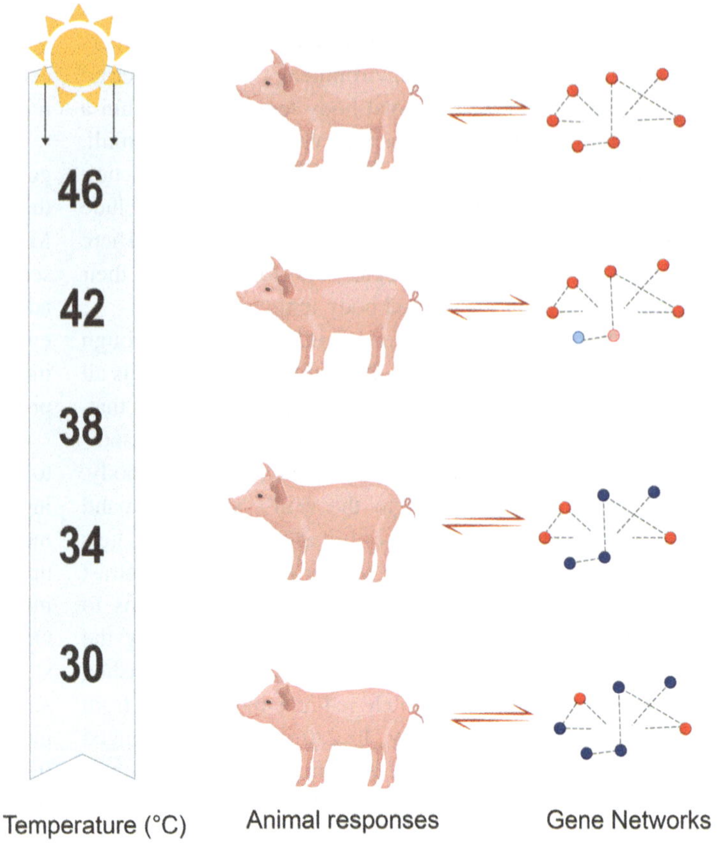

Fig. 1.8 Pig is exposed to an increase in temperature and each level transcriptomic response is recorded. Transcriptomic changes are made to improve the individual's response to heat stress. As the temperature increases, transcriptomic adjustments are made such that animal gets acclimatised to the changing environment through changes in their physiological modifications by animals. (Courtesy: BioRender)

and enzyme activity occur or formed because of these reactions.

Another area of interest for many researchers is the complex relationship between ruminants and rumen-dwelling microbiota. Based on sequencing target regions of the 16S rRNA gene, the characterisation of rumen microbiota can be done. It enables us to characterise the microorganisms and their functional contributions to the host's energy production. Nevertheless, this process will not fetch information about their functionality. However, multi-Omics science such as metagenomics, transcriptomics, metaproteomics, and metabolomics provides deeper comprehension of the ecology of rumen microbiomes, the symbiotic host–microbe relationship, and the impact of different nutritional factors manipulations on the productivity of animals. Metagenomics enables the assessment of the microbiome's diversity and potential functional capacity, whereas metatranscriptomics can shed light on the microbiome's actual function via gene expression. Metaproteomics and metabolomics, when used together with metatranscriptomics, can aid in identifying the members of an active microbial community. Additionally, they give information on differentially expressed metabolic pathways by utilising NMR or MS-based approaches to access the proteins expressed and metabolites generated. While next-generation sequencing and functional metagenomics are being used to study the rumen microbiome in tropical animals, integrating the results with other meta-Omics remains a challenge.

Learning Outcomes

- With the basic understanding of a complex yet unique bunch of living units—cells combined with recent developments, one can anticipate the future with agricultural and biomedical applications such as the production of genetically engineered animals.
- A better comprehension of the integration system includes the nervous and endocrine systems, which are primarily responsible for maintaining the animal's metabolic homeostasis. Signals from neurons are precisely targeted, but signals produced from the endocrine gland are broad-spectrum signals distributed throughout the animal's body. Coordination of both chemical and electrical systems is critical for maintaining balance within the animal.
- Exchange and transport systems, i.e. respiratory system and circulatory system, work together to supply cells and tissues with the necessary oxygen and metabolites so that they can operate optimally.

(continued)

- Kidneys act as a regulator of plasma with various functions like regulating pH, osmoregulation, and filtration of nitrogenous waste and metabolic waste products as a means of achieving homeostasis either naturally or through technological advancements.
- Deeper insights into various reproductive structures present in different types of animals, the role of hormones in gametogenesis, ovulation, and implantation and most importantly, various assisted reproductive technologies to meet global food demand.
- To appreciate the physiological mechanisms controlling the growth and development of the mammary gland—factors influencing milk secretion. To acknowledge recent developments in the field of the genetically engineered mammary gland.
- Influence of environment on various metabolic activities in animals. Causes of climate change and its impact on animal adaptation and effect of multiple stressors on animal productivity. General mechanism of thermoregulation in adapted animals.
- Basics and history of the human genome project. The advantage of multi-Omics research to increase animal productivity in this present climate change situation.

Exercises

Objective Questions

Q1. The average size of cells is ________________.

Q2. The animal nervous system is capable of a wide range of functions. The basic unit of the nervous system is ________________.

Q3. How do neurons communicate with one another?

Q4. Which is the primary glucocorticoid produced in the ruminants?

Q5. Thyroxin is responsible for ________________.

Q6. Cells absorbs oxygen through ________________.

Q7. Pigment responsible for blue blood colour in octopus is ________________.

Q8. Major excretory product in birds is ________________.

Q9. Give some examples of assisted reproductive technology.

Q10. Mammary gland is derived from the ________________ layer during embryonic stage.

Q11. Profuse sweating animal other than human is ________________.

Q12. Human genome project started in the year ________________.

Answer to Objective Questions

A1. 0.01–0.1 mm

A2. Neurons

A3. Electrically and chemically

A4. Cortisol

A5. Promoting the growth of tissues in the body

A6. Diffusion

A7. Haemocyanin

A8. Uric acid

A9. Sperm sexing, embryo transfer, and artificial insemination

A10. Ectoderm

A11. Horse

A12. 1990

Acknowledgement The figures in this chapter are created with BioRender.com and therefore the authors express their sincere gratitude for BioRender organisation for helping to create figures using their resources. The authors also thankful to the Director, National institute of Animal Nutrition and Physiology for providing necessary permission to contribute this chapter.

Further Reading

Research Articles

Sejian V, Bhatta R, Gaughan JB, Dunshea FR, Lacetera N (2018) Review: Adaptation of animals to heat stress. Animal 12(s2):s431–s444

St Aubin DJ, Geraci JR (1988) Capture and handling stress suppresses circulating levels of thyroxine T4 and triiodothyronine T3 in beluga whales Delphinapterus leucas. Physiol Zool 61:170–175

Cellular and Molecular Physiology

2

Ayan Mukherjee and Prabal Ranjan Ghosh

Abstract

A cell is the fundamental unit of life since no living organism can survive without it. Also, the cell is both a structural and a functional unit of a living system. On the basis of complexity and organization of DiseNA, cells can be classified into two types—prokaryotic cell and eukaryotic cell. Prokaryotes are unicellular organism that lack distinct nucleus and other membrane-bound organelles. The average size of prokaryotes is 2.0–2.6 μm. Eukaryotic cells have a nucleus enclosed within a nuclear envelope and different membrane-bound organelles. In a cell cytoplasm, a semisolid matrix substance is enclosed by cell wall and cell membrane. The major cell organelles are mitochondria, Golgi apparatus, endoplasmic reticulum, ribosome, lysosome, etc. In eukaryotes, the nucleus is a specialized double membrane-bound protoplasmic entity that houses all of the genetic information needed to control cellular metabolism and pass on to future generations. The cytoskeleton is the structural framework of eukaryotic cells that keeps the cell and its appendages in shape. Genetic information flows from 'DNA to mRNA to protein' within cell.

A. Mukherjee (✉)
Department of Animal Biotechnology, West Bengal University of Animal & Fishery Sciences, Kolkata, West Bengal, India

P. R. Ghosh
Department of Veterinary Physiology, West Bengal University of Animal & Fishery Sciences, Kolkata, West Bengal, India

© The Author(s), under exclusive license to Springer Nature Singapore Pte Ltd. 2023

P. K. Das et al. (eds.), *Textbook of Veterinary Physiology*, https://doi.org/10.1007/978-981-19-9410-4_2

Graphical Abstract

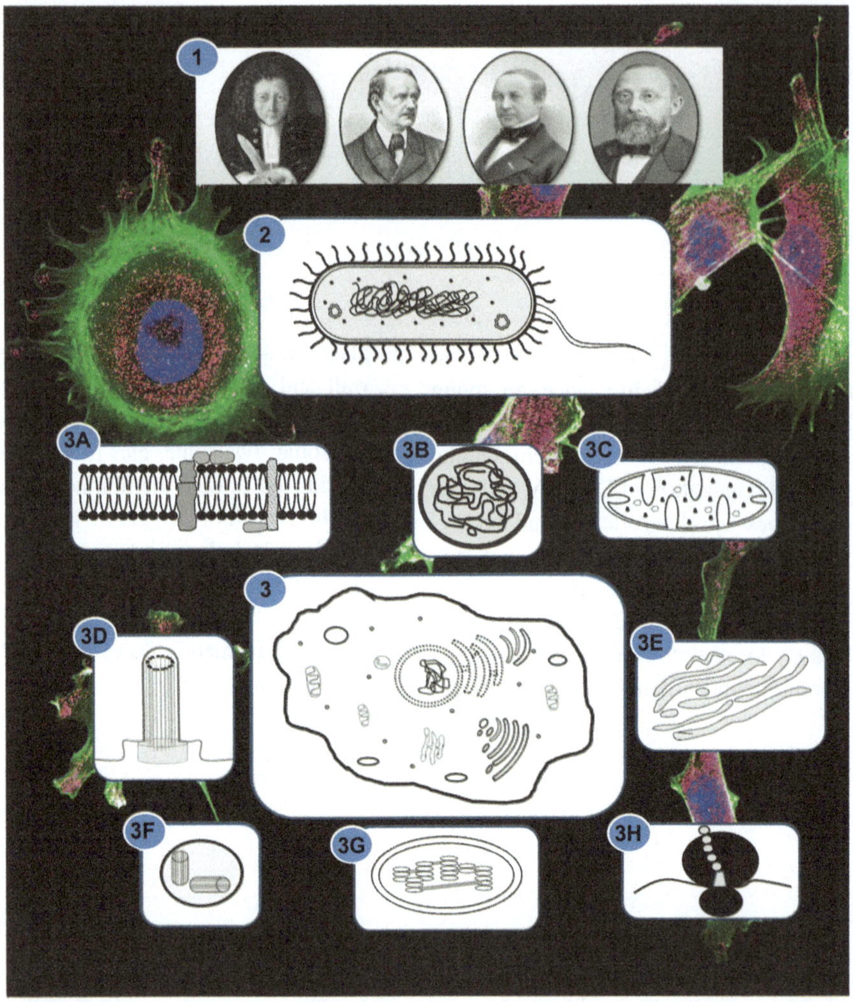

Description of the graphic: Pioneering scientists—Robert Hooke, Matthias Schleiden, Theodor Schwann, and Rudolph Virchow (from left to right) (1) discovered cell. A typical prokaryotic cell (2) and a typical eukaryotic cell with cell organelles (3) includes Biomembrane structure (3A) Nucleus (3B) Mitochondria (3C) Cytoskeletal structure (3D) Golgi Apparatus (3E) Centriole (3F) Chloroplast (3G) and Ribosome (3H)

Keywords

Prokaryotes · Eukaryotes · Cytoplasm · Nucleus · Cytoskeleton

Learning Objectives

- Development of ideas about basic structure and components of prokaryotic and eukaryotic cells.
- Structure of cell wall, cell membrane, and biomembrane.
- Cellular organelles—their structure and functions—nucleus, endoplasmic reticulum, Golgi apparatus, mitochondria, plastids, ribosomes, vacuoles, lysosomes.
- Structural organization of cytoskeleton.
- Flow of molecular-genetic information within cell.

The cell concept... is the axis around which the whole of modern science of life revolves—Paul Ehrlich in Nobel Lecture on 11 Dec 1908

2.1 Cell: The Smallest Unit of Life with Independent Existence

Life begins as a cellular unit. A cell is a fundamental unit of life since no living entity can exist without a cellular structure, and a cell is a unit of both structure and function. Numerous unicellular organisms are present in the universe, where single cell is sufficient to perform the fundamental processes of life, for example, bacteria, Protista, protozoa, and yeast. A multicellular organism like a higher animal or plant has billions of cells. For example, an average man of 70 kg weight and 1.72 m height with a body area of 1.85 m^2 has 37.2 trillion cells of about 200 types. Cells are clubbed to form tissues, tissues to organs, and organs to organ systems.

Cells are also the functional units of life. As Francois Jacob quipped (1971), 'the dream of every cell is to become two cells' a new cell is formed by the division of an existing cell. A totipotent single cell has unique capability to develop into the whole organism.

2.2 Notable Events in the Discovery of Cell

Year	Event
• 1590	Zacharias Janssen built the first microscope.
• 1661	Marcello Malpighi observed cells and named 'saccules' and 'utricles'.
• 1665	Robert Hooke improved the design of microscope. He observed honeycomb like compartments in a piece of cork of Spanish Oak. In his famous book 'Micrographia', he named these compartments as *cellulae* (Latin 'cella' means 'hollow space').
• 1671	Antonie van Leeuwenhoek first observed, described, and depicted the picture of free-living cell.
• 1831	Robert Brown discovered the nucleus in the cells of orchid root.
• 1835	Félix Dujardin discovered protoplasm in *Foraminifera* and named it 'sarcode'. Later, J.E. Purkinje termed it as 'protoplasm'.
• 1839	'Cell theory' proposed by Matthias Schleiden and Theodor Schwann.

Know More......

Cell Theory

Cell theory is one of the fundamental ideas of biology. German scientists *Theodor Schwann, Matthias Schleiden*, and *Rudolph Virchow* formulated the theory.

Tenets to the Cell Theory: Schwann and Schleiden first stated that

- The cell is the fundamental unit of structure and function in living things.
- All organisms are made up of one or more cells.

Later, Virchow stated the third tenet

- 'Omnis cellula e cellula' i.e. 'All cells arise from pre-existing cells'.

Modern Cell Theory

- Energy flow occurs within cells.
- Heredity information (DNA) is passed on from cell to cell.
- All cells have the same basic chemical composition.

2.3 Types of Cells

Based on the complexity and organization of DNA, cells can be categorized into two types—prokaryotic cells and eukaryotic cells.

2.3.1 Prokaryotic Cells

Prokaryotes are unicellular organisms that lack a distinct nucleus and other membrane-bound organelles (Fig. 2.1). The average size of prokaryotes is 2.0–2.6 μm. Based on different molecular determinants prokaryotes are divided into two domains: Bacteria (formerly Eubacteria) and Archaea (formerly Archaebacteria). The different components of prokaryotic cells are described below:

- **Nuclear material**: Naked, coiled DNA lies in the cytoplasm. This DNA is also known as gonophore or prochromosome. In some prokaryotes, extrachromosomal DNA is found as small circular plasmids.
- **Vacuoles**: Gas vacuoles are present in prokaryotes.

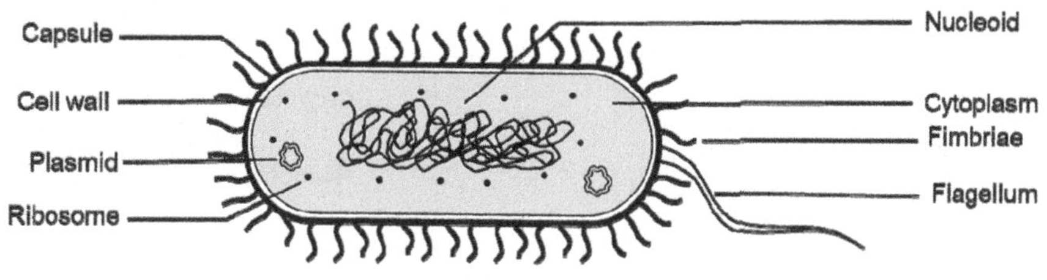

Fig. 2.1 Structure of a prokaryotic cell

- **Ribosomes**: 70S ribosomes comprising of 50S and 30S subunits are present in prokaryotes.
- **Photosynthetic thylakoids**: Thylakoids lie freely in the cytoplasm in blue-green algae and some bacteria.
- **Cell wall**: Bacteria and cyanobacteria are encased by a cell wall.
- **Capsule**: In addition to the cell wall, some bacteria have an outer protective layer. It aids in the retention of hydration, protecting cells from phagocytosis, and cell attachment.
- **Flagella and fimbriae**: In some bacteria, flagella and fimbriae are present. Flagella help in bacterial motility whereas fimbriae facilitate conjugation and attachment.

2.3.2 Eukaryotic Cells

Eukaryotic cells have a nucleus enclosed within a nuclear envelope and different membrane-bound organelles (Fig. 2.2).

Eukaryotes can reproduce asexually through mitosis and sexually through meiosis and gamete fusion. The various components have been detailed here for better comprehension of the structure and function of eukaryotic cells.

2.3.2.1 Cell Wall

The cell wall is an outer structural layer surrounding the plant cells, fungi, and a few protists providing necessary support to the cell. Its thickness ranges from 0.1 to 10 μM. Cell wall is metabolically active and can grow.

2.3.2.1.1 Chemical Composition

The principal components of the plant cell wall are cellulose, hemicellulose, pectin, and protein. In algal cell walls galactans, mannans, and calcium carbonate are also present.

2.3.2.1.2 Structure

Structurally the cell wall comprises three parts—middle lamella, primary wall, and secondary wall.

- *Middle lamella*—Middle lamella is the common layer between two adjacent cells. It is made up of calcium and magnesium pectates. This is the initial layer that is laid down during cytokinesis process.
- *Primary wall*—The extensible, thin wall (0.1–3 μm) present at the inner side of the middle lamella is primary wall. A loose network of microfibrils embedded in an amorphous gel-like matrix or ground substance makes up the primary wall. Microfibrils are generally made up of cellulose in plants. This cellulose is synthesized at cell membrane by particle rosette harbouring cellulose synthetase enzyme. The main component of fungal cell wall is repeating unit of β, 1–4 acetyl glucosamine or fungal cellulose. The wall grows by intussusception or deposition of materials inside.
- *Secondary wall*—The secondary wall is formed by gradual accumulation of materials at the inner side of primary wall on the existing structure. It is thicker (3–10 μm) than the primary wall and composed of minimum of three layers. The constituents of the secondary wall are almost similar to the primary wall. Compared to primary wall cellulose content is higher and cellulose microfibrils are longer and denser. The matrix in which cellulose microfibrils are embedded contains pectin and hemicellulose. Materials like suberin, lignin, silica, and cutin are deposited in the secondary wall.

Apart from this basic structure cell wall may possess some specialized structure like *plasmodesmata* and *pits*. *Plasmodesmata* are cytoplasmic bridges in the shape of minute pores that connect adjoining plant cells to form a symplast or continuous protoplasm (Fig. 2.3).

Plasmodesmata (diameter 40–50 nm) are encircled by plasma membrane. An endomembrane-derived structure connecting Endoplasmic Reticula (ER) of two adjacent cells (*desmotubule*) is present within plasmodesmata. Plasmodesmata regulate transport of various small molecules between two adjacent cells. *Pits* are unthickened areas present on secondary walls of plant cell. These submicroscopic pores help in rapid translocation of essential molecules between two neighbouring cells.

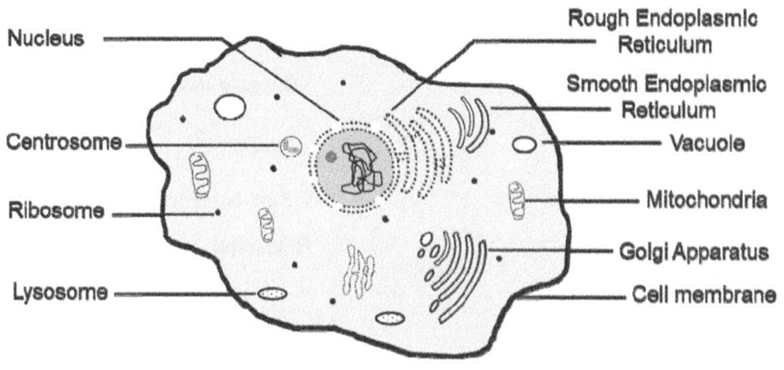

Fig. 2.2 Structure of a eukaryotic

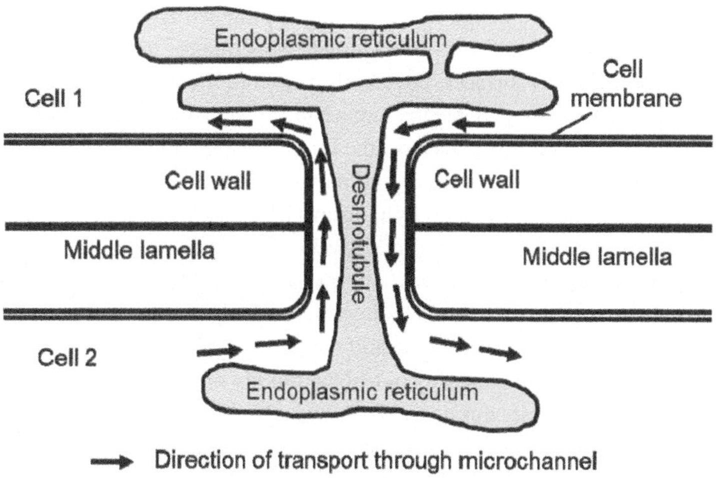

Fig. 2.3 Structure of plasmodesmata

2.3.2.1.3 Functions

(1) Cell wall provides structural firmness and shape of cell. (2) It withstands osmotic pressure and prevents rupture of cells. (3) It protects cellular protoplasm from mechanical damage and external pathogens. (4) Cell wall of sieve tubes and vessels ensure long distance transport. (5) Cell wall stores food in the form of hemicellulose in some seeds.

2.3.2.2 Cell Membrane

Cell membranes or biomembranes are dynamic, quasifluid, thin (50–100 Å) structure surrounding the cytoplasm and several cell organelles like nucleus, mitochondria, endoplasmic reticulum, Golgi bodies, lysosome, ribosome, vacuoles, etc. Biomembranes are selectively permeable for various solutes.

2.3.2.2.1 Chemical Composition

Water—20%, Carbohydrates—1–5%, Lipid—20–79%, Proteins—20–70%.

Carbohydrates—Different types of branched and unbranched oligosaccharides like hexose, fucose, hexosamine, sialic acid, etc. Proteins—globular and fibrous proteins. Lipids—phosphoglycerides or phospholipids.

2.3.2.2.2 Structure

Several models have been proposed to explain the molecular structure of cell membrane. According to the earliest lamellar models proposed by J. Danielli and H. Davson (1935), four-layered biomembrane comprises two phospholipid layers covered on either side by two layers of proteins. Later, Robertson's model proposed that outer and inner protein layers differ in distribution of protein molecules in 1959. The outer protein layer is enriched in mucoid protein whereas inner one is rich in nonmucoid proteins. Robertson's pioneering work on electron microscopic structure of erythrocyte membrane also theorized the unit membrane concept which stated that all cytoplasmic membranes are three-layered sandwich structure of protein-phospholipid-protein. Robertson called this trilaminar membrane. In the lamellar models proposed by both the groups, biomembrane has been proposed as a stable structure. However, the *fluid-Mosaic model* of Singer and Nicolson (1972), the most accepted model of biomembrane, depicted it as dynamic structure with the ability to fuse or separate, expand, or contract during diverse cellular processes (Fig. 2.4).

2.3.2.2.2.1 Characteristics of Fluid-Mosaic Model

(1) Mosaicism—Lipid molecules are present in viscous bilayer. Intrinsic and extrinsic protein molecules are present inside and outside of the lipid bilayer, respectively. *Intrinsic or integral* protein molecules penetrate the lipid bilayer at various depths and form hydrophobic interactions with lipid molecules. Several integral proteins span the complete lipid

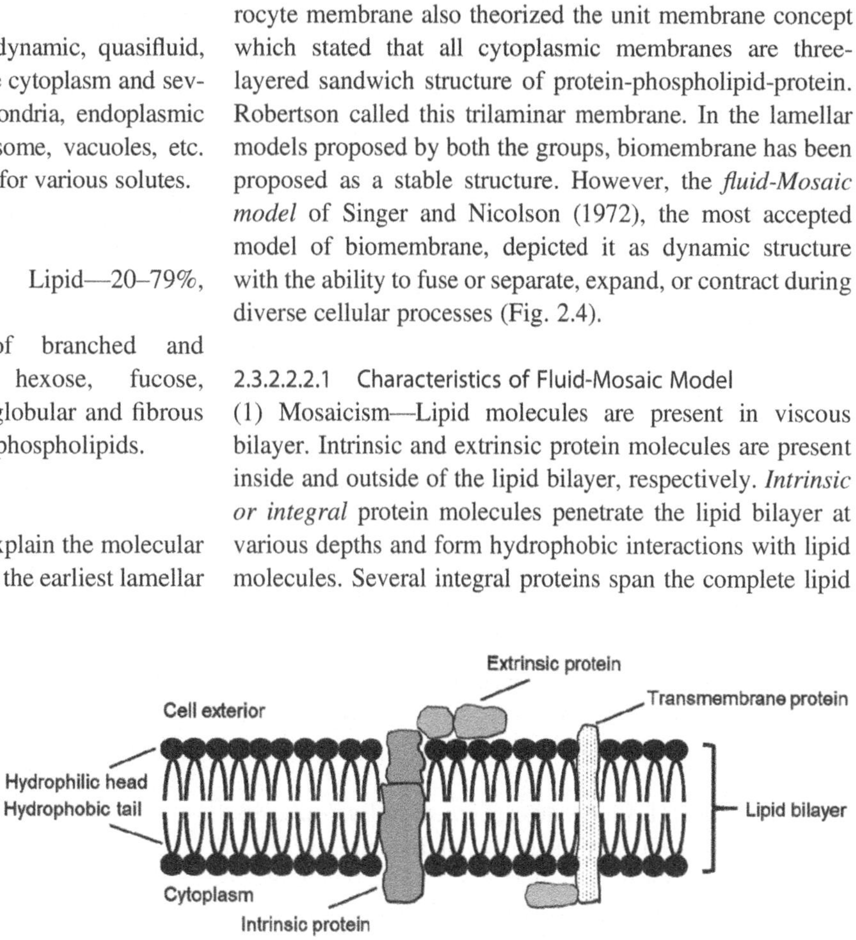

Fig. 2.4 Fluid-mosaic model of biomembrane or cell membrane

bilayer. Those are called *transmembrane* proteins. *Extrinsic* or *peripheral* proteins are found superficial to the membrane's two surfaces, covalently bound to the phospholipid head or noncovalently attached to transmembrane proteins. (2) Fluidity—Because the lipid bilayer is quasifluid in nature, membrane proteins can move laterally, giving the membrane flexibility and dynamism. Electron spin resonance, atomic force microscopy-based force spectroscopy, fluorescence, and deuterium nuclear magnetic resonance spectroscopy can all be used to determine membrane fluidity. Composition of the membrane and temperature are two most important determinants that control the fluidity of membrane. Similarly, because unsaturated fatty acids contain kinked hydrocarbon tails, lipid molecules with unsaturated fatty acids are more difficult to pack together, increasing fluidity. In animal cell membrane, cholesterol functions as a bidirectional regulator of membrane fluidity. At increased temperatures, it stabilizes the membrane and elevates its melting point, whereas at low temperatures it intercalates between the phospholipids and prevents their aggregation and stiffening, thereby increases fluidity. (3) Asymmetry—Biomembranes are asymmetric, i.e. the two sides of the biomembranes differ in the amount and types of constituent molecules like lipid, protein, and carbohydrate moieties. As for erythrocyte membrane, lecithin is present on exterior side and cephalin is rich in cytoplasmic side. Most of the membrane proteins are also abundant on inner surface. On the contrary, oligosaccharides are mostly present in outer surface and absent in inner surface of biomembrane.

2.3.2.2.3 Function

Biomembrane or cell membrane has diverse types of functions. (1) Cell membrane protects cell from mechanical stress. (2) It facilitates intracellular transport of water and different solute molecules. (3) Biomembrane structure maintains the identity and internal milieu of each cell organelle. (4) Cell membrane has receptors for different biologically significant molecules like antigens, hormones, etc. (5) Cell membrane harbours crucial enzymes like ATPase, esterase, etc. (6) In neuronal cells, cell membrane participates in neurotransmission.

2.3.2.3 Cytoplasm

Cytoplasm is a gelatinous, semi-fluid mass inside the cell containing cytosolic matrix, organelles and inclusion bodies excluding the nucleus. The components are cytoplasmic matrix, cell organelles, and cell inclusions.

2.3.2.3.1 Cytoplasmic Matrix (Cytosol or Hyaloplasm)

Cytosol is the main component of cytoplasm making 70% of the cytoplasm volume. It comprises water, cytoskeleton filaments, and dissolved molecules. Cytosol is a colloidal solution of different chemical molecules in water. Proteins are mostly present as colloidal particle, whereas lipids remain as emulsion. Outer part of cytosol is called the *cell cortex* or the *ectoplasm* whereas the concentrated inner area is called the *endoplasm*. Cytosol often exists in two states, *sol* and *gel*.

2.3.2.3.1.1 Function

(1) Cytosol functions as medium for exchange of materials between cell organelles. (2) Cellular metabolism takes place in cytosol. (3) Biosynthesis of carbohydrates, lipids, proteins, and nucleotides takes place in the cytosol. (4) Cytoplasmic streaming of matrix helps in even distribution of various essential materials inside cell.

2.3.2.4 Cell Organelles

Cell organelles are morphologically distinct subcellular structures with diverse chemical compositions and discrete functions. A cell contains diverse cell organelles like mitochondria, plastids, Golgi bodies, endoplasmic reticulum, peroxisomes, ribosomes, lysosomes, etc.

2.3.2.4.1 Mitochondria

Mitochondria are eukaryotic cell organelles that play a critical role in the generation of metabolic energy. Mitochondria are also called the *power houses* of the cell because they are the main source of energy released by oxidative phosphorylation and Kreb's cycle of aerobic respiration. Mitochondria are cylindrical in shape (length of 1.0–4.1 μm and diameter of 0.2–1.0 μm).

2.3.2.4.1.1 Chemical Composition

Proteins—60–70%, Lipids—25–35%, RNA—5–7%, DNA. Small quantity of trace minerals

2.3.2.4.1.2 Ultrastructure

A mitochondrion contains two membranes (60–75 Å thickness) and two chambers, outer and inner (Fig. 2.5).

- *Outer chamber*: The outer chamber or peri-mitochondrial space is the space (60–100 Å) between the outer and inner membranes of the mitochondrial envelope. The chamber contains different enzymes.
- *Inner chamber*: Inner chamber is filled with a semi-fluid matrix that contains ribosomes, protein particles, RNA, DNA, and enzymes of amino acid synthesis, fatty acid metabolism, Kreb's cycle.
- Mitochondria have naked, circular, or linear DNA. Mitochondrial ribosomes resemble prokaryotic ribosomes and are 55S or 70S in nature.
- *Outer membrane*: Mitochondria is surrounded by a smooth outer membrane. Several metabolites (less than 1000 Da) can make passage within mitochondria due to the presence of *porin* channels and other small pores in the outer membrane.

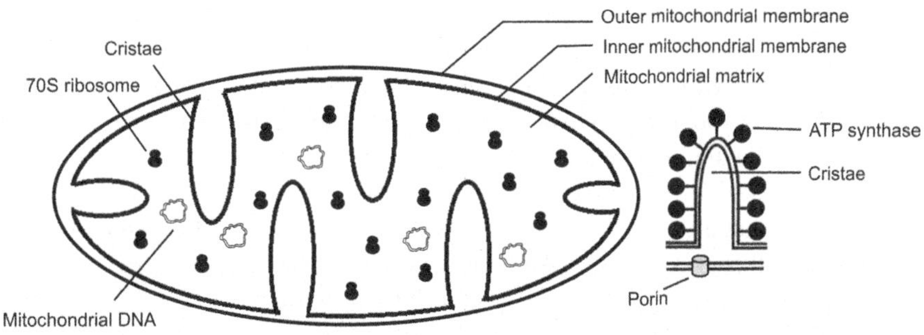

Fig. 2.5 Structure of mitochondria

- *Inner membrane*: Unlike outer membrane inner membrane is permeable to fewer metabolites. Inner membrane is rich in phospholipid molecule *cardiolipin* and many proteins. The inner membrane is folded variously to form finger-like projections called *cristae*. Presence of cristae provides adequate surface area for mitochondrial reactions to occur.

Cristae are studded with variety of proteins, including *ATP synthase* and different *cytochromes*. Inner membrane as well as cristae possess small tennis-racket like *ATP synthase dimer* (earlier known as *elementary particles* or *oxisomes*). ATP synthase in eukaryotes is F-type and consists of two major subunits, F_O and F_1. The F1 (Fraction 1) part of ATP synthase has three copies of each of subunits α and β, one each of subunits γ, δ, and ε. The γ, δ, and ε subunits constitute the central stalk or the rotor shaft (Fig. 2.6). A hexameric ring with a central cavity is formed by an alternate arrangement of 3α and 3β subunits. F_O subunit consists of a subunit c-ring and one copy each of subunits a, b, d, F6, and the oligomycin sensitivity-conferring protein (OSCP).

2.3.2.4.1.3 Functions

(1) Primary function of mitochondria is to produce energy through the process of oxidative phosphorylation. (2) It regulates the metabolic activity of the cell. (3) It plays an important role in apoptosis or programmed cell death. (4) It also stimulates cell multiplication and cell growth. (5) Mitochondria detoxify ammonia in the liver cells.

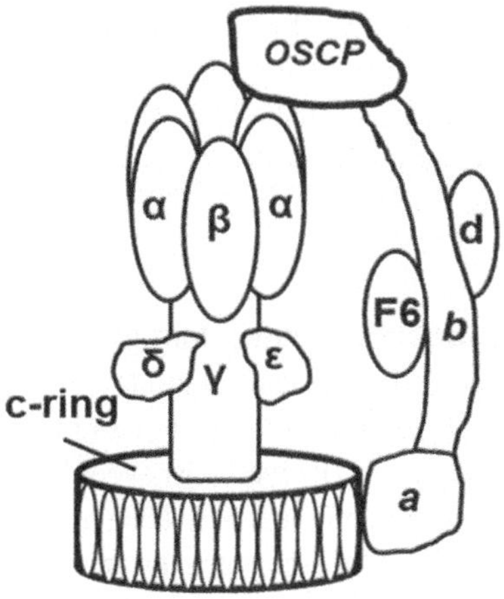

Fig. 2.6 ATP synthase subunits

2.3.2.4.2 Plastids

Plastids are a type of semi-autonomous organelle with DNA and a membrane that stores or synthesizes a wide range of biological chemicals. Plastids are present only in plant cells (except few protists like Euglena, diatoms). A. F. W. Schimper was the first to name plastids and give them a precise definition. Plastids differentiate from proplastids and are classified into three categories based on their colour: *leucoplasts*, *chromoplasts*, and *chloroplasts*.

- Leucoplasts—These are colourless plastids which generally occur near the nucleus of the cells. Based upon the content of leucoplasts, these plastids may be of different types. Elaioplasts (found in tube rose) are colourless plastids that store lipid. Amyloplasts (found in rice, wheat) contain starch within it and proteoplasts (found in maize) have protein inside the plastids.
- Chromoplasts—These are found in different flowering plants and contain coloured pigments.
- Chloroplasts—These are greenish plastids which possess photosynthetic pigments, chlorophylls.

2.3.2.4.2.1 Composition

Protein—50–60%, lipids—25–30%, chlorophyll—5–10%, carotenoids—1–2%, DNA—0.5%, RNA—2–3%, Trace minerals and vitamins.

2.3.2.4.2.2 Structure

Chloroplasts are made up of three parts: *envelope*, *matrix*, and *thylakoids*. Chloroplasts have two membranes covering them. Each membrane is 100 Å thick and divided by a gap called inter-membrane space. Matrix, often known as *stroma*, is a protein-rich semi-fluid compound. The stroma contains DNA, RNA, 70S ribosome, and many enzymes. Thylakoids are membrane-lined flattened sacs that are found in the stroma. Two to hundred thylakoids are piled to create *grana* in higher plants. In chloroplasts, there are about 40–60 grana. Membrane of thylakoids harbour photosynthetic pigments and coupling factors. Photosynthetic pigments are found in photosystems in a group. Plants have two types of photosystems: photosystem I (PS-I) and photosystem II (PS-II). PS-II is found in appressed part of grana, whereas PS-I is located within non-appressed stroma lamellae.

2.3.2.4.2.3 Function

(1) Photosynthetic reactions take place in chloroplasts. (2) Synthesis and storage of starch, lipid occurs in plastids (3) Chloroplasts contribute to photosensitivity of some algae.

Know More......

Endosymbiotic Theory

According to the endosymbiotic theory, some of the organelles in today's eukaryotic cells were previously prokaryotic bacteria. The first eukaryotic cell was most likely an amoeba-like cell. It got a nucleus when a section of the cytoplasmic membrane pinched off around the chromosomes. Some of these amoeba-like species engulfed prokaryotic cells, which gradually survived within the organism and developed a symbiotic interaction with the organism. The foreign bacteria eventually lost their cell wall and much of their DNA. Lynn Margulis proposed the theory.

Supportive evidence:

- Mitochondria and chloroplasts have their own circular DNA and own ribosome possessing 30S and 50S subunits like bacteria.
- The mitochondria and chloroplasts are the same size as prokaryotic cells and divide by binary fission. Like bacteria chloroplasts have Fts proteins and mitochondria have Fts homologue protein at their division plane.
- Primitive eukaryotic microbes, like *Giardia* and *Trichomonas*, have a nuclear membrane but no mitochondria.

2.3.2.4.3 Golgi Apparatus

The Golgi apparatus, also known as the Golgi complex or *Apparato Reticulare*, is a complex cell organelle composed of tubules connected by vesicles and vacuoles. The organelle was observed in 1867 by George and named after *Camillo Golgi*, who discovered it using the metallic impregnation method in barn owl and cat nerve cells. In cells that synthesize and secrete complex substances, the Golgi complex is abundant.

It is found in all eukaryotes except mammalian RBC, bryophyte and pteridophyte sperm, and plant sieve tubes. The Golgi complex is absent in prokaryotes. In plant cells, the Golgi complex is made up of dictyosomes, which are a collection of disconnected components. The Golgi apparatus changes shape and size depending on the cell's physiological state.

2.3.2.4.3.1 Structure

The complex is made of cisternae, tubules, vesicles, and vacuoles.

- *Cisternae*: A stack of membrane-bound cisternae or saccules of varied thicknesses makes up the Golgi complex. Cisternae lumens are 60–90 nm in diameter and contain a matrix substance. The cisternae are usually curved to give the Golgi complex a distinct polarity. Forming face, or *cis*-face, is the convex side, whereas maturing face, or *trans*-face, is the concave side. The medial Golgi stacks are the cisternae in the middle. Transition vesicles transporting secretory biomolecules from the ER enter the *cis*-Golgi, transit through the medial stacks of linked cisternae, and finally bud off as trans-face secretion.
- *Tubules*: Tubules form a complex network along the apparatus's periphery and maturing face. They are 30–50 nm in diameter.
- *Vesicles*: Vesicles are tiny sacs with a diameter of 20–80 nm. Smooth and coated types of vesicles are there in cell. The chemicals to be secreted are transported across cisternae by coated vesicles and smooth vesicles or secretory vesicles secrete the chemicals outside the cell.
- *Golgi vacuoles*: Golgi vacuoles are modified cisternae and are made from a concave surface. These vacuoles can sometimes act as lysosomes.

2.3.2.4.3.2 Function

(1) The Golgi apparatus concentrates and packages secretory macromolecules, which are then secreted outside cells via exocytosis or reverse pinocytosis. (2) Glycoproteins and glycolipids are made in the Golgi complex. (3) The Golgi complex is responsible for the formation of specialized structures such as the acrosome of mammalian sperm and the retinal pigment of the chick embryo. (4) The Golgi complex produces cell wall components such as pectic chemicals, mucopolysaccharides, sialic acid, and chondroitin sulphate.

2.3.2.4.4 Endoplasmic Reticulum (ER)

ER is an organized, interconnected labyrinth of membrane-lined channels in the form of branching tubules and flattened sacs in the cytosol. It was discovered by Porter and Thompson independently in 1945. The name ER was coined by Porter in 1953. ER is extensively branched in metabolically active cells like hepatocytes, cells of pancreas. ER is absent in oocytes, RBC, and prokaryotic cells.

2.3.2.4.4.1 Structure

Based on whether ER is studded with ribosomes or not, ER is of two types smooth and rough. Smooth Endoplasmic Reticulum (SER) does not bear ribosomes. SER is present in cells involved in glycogen synthesis and storage. SER is mainly Rough Endoplasmic Reticulum (RER) has ribosomes attached to its outer surface. Two glycoproteins ribophorin I and ribophorin II help in attachment of ribosomes to RER. There are three forms of ER—cisternae, vesicles, and tubules. SER is mostly made of vesicles and tubules, whereas cisternae is the main component of RER.

Cisternae: Cisternae are a bundled pattern of interconnecting sac-like components that remain parallel to each other. The diameter of cisternae is 40–50 nm.

Vesicles: Vesicles are oval or spherical sacs with a diameter of 25–500 nm. Microsomes are another name for vesicles.

Tubules: They are tube-like extensions that form a reticular system when attached to cisternae or vesicles. Their diameter varies between 50 and 100 nm.

2.3.2.4.4.2 Function

(1) ER provides the large surface area for different cellular reactions to occur, (2) ER supports the colloidal cytoplasmic matrix, (3) ER helps in intracellular transport, (4) ER gives rise to vacuoles, (5) ER contains a number of different metabolic enzymes, (6) SER in the form of sarcoplasmic reticulum stores Ca^{2+} required for muscle contraction, (7) SER synthesizes lipid and glycogen, (8) SER helps in metabolism of toxic chemicals and xenobiotics with the help of Cytochrome P-450.

2.3.2.4.5 Peroxisomes

Peroxisome is a small (0.5–1.0 μm in diameter) membrane-enclosed organelle that contains enzymes for several metabolic pathways. Peroxisomes were discovered by De Duve in 1965. Peroxisomes are synthesized from proteins that are translated on free ribosomes and transported to peroxisomes as mature polypeptide change. Peroxisome is considered as an 'evolutionary relic' of protoeukaryotic organelle specialized for oxidation reaction. Peroxisome is covered by single membrane and internal granular matrix contains about 50 oxidative enzymes. Peroxisomes are abundant in photosynthetic cells of plants, liver, and kidney cells. Various oxidative enzymes present in peroxisomes are urate oxidase, α-hydroxy acid oxidase, β-hydroxy acid oxidase, D-amino acid oxidase, catalase, etc.

2.3.2.4.5.1 Functions

(1) *Metabolism of fatty acids*: Long chain and branched chain fatty acids are broken down by peroxisomal enzymes. Peroxisomes generate cholesterol and dolichol in animal cells. Peroxisomes are also involved in the production of bile acids, which are produced from cholesterol, in the liver. Peroxisomes also contain enzymes that are essential for plasmalogen production. (2) *Metabolism of xenobiotics*: Various uncommon substances or xenobiotics are broken down in peroxisomes. (3) Detoxification of toxic substances like phenolic compounds, nitrites, methanol, ethanol, etc. (4) *Photorespiration* in plants: In photosynthetic plants, photorespiration or oxidative C2 cycle or the oxidative photosynthetic carbon cycle occurs in peroxisomes. Peroxisomes remain associated with mitochondria and chloroplast. Glycolate from chloroplasts is transported to peroxisomes. Glycolate is oxidized to glyoxylate by catalase in peroxisome. Subsequently, glyoxylate is converted to glycine.

2.3.2.4.6 Ribosomes

Ribosome or *Palade granule* is a complex machinery of protein or polypeptide synthesis. These are subspherical, naked ribonucleoprotein particles with a length of 200–340 Å and diameter of 170–240 Å. Ribosomes are found in cytoplasm, mitochondria, and chloroplast of all eukaryotic cells (except mammalian RBC) and prokaryotic cells. About ten million ribosomes can be found in a rapidly developing mammalian cell.

2.3.2.4.6.1 Composition

In eukaryotes, larger ribosomal subunit 60S is composed of 28S rRNA, 5S rRNA, 5.8S rRNA, and 46 proteins and smaller ribosomal subunit 40S is composed of 18S rRNA and 33 proteins. In prokaryotes, larger ribosomal subunit 50S has 23S rRNA, 5S rRNA, and 33 proteins and smaller subunit 30S has 16S rRNA and 20 proteins. The ribosomes of hepatocytes contain 5–10% lipid.

2.3.2.4.6.2 Structure

Each ribosome is made of two subunits of unequal shape. Larger subunit is dome shaped and double in size of oblate-ellipsoid smaller subunit. The ion Mg^{2+} is required for smaller and larger subunits to attach. Ribosomes can be found alone as *monosomes* or in rosettes and helical groups known as *polyribosomes* and *polysomes*. The cytosolic ribosomes are either found 'free' in the matrix or remain 'bound' to endoplasmic reticulum with the help of ribophorin protein. The size of ribosomes is expressed as 'S' or Svedberg coefficient ($S = 1 \times 10^{-13}$ s). The cytoplasmic ribosomes of eukaryotes are 80S with two subunits, 60S and 40S. In prokaryotes, 70S ribosomes have two subunits 50S and 30S. In mammalian mitochondria, ribosomes are 55S. A ribosome has four sites for specific attachment—(a) mRNA binding site, (b) *A-site* or *aminoacyl site* where mRNA codon and aminoacyl tRNA are directed during translation process, (c) *P-site* or *Peptidyl site* where peptide bond formation, elongation, and transfer of peptide chain to site A occurs, and (d) *E-site* or *Exit site* from where freed tRNA are released after translation.

2.3.2.4.6.3 Function

(1) Ribosomes are sites for translation or protein synthesis. 'Free' ribosomes synthesize structural and enzymatic proteins for use inside cells.

2.3.2.4.7 Lysosomes

Lysosomes are small membrane-bound vesicles containing hydrolytic enzymes in the form of minute crystalline or semicrystalline granules ranging in size from 5 to 8 nm. Christian De Duve discovered lysosomes in 1955. Novikoff et al. (1956) observed these organelles with electron microscopy and named them lysosome. Lysosomes have a spherical shape and a diameter of 0.2–0.8 µm, but their size can vary up to 5 µm depending on the cell type. Different enzymes like lipase, amylase, peptidase, nucleases, and acid phosphatase are present in lysosome. These enzymes are also known as acid hydrolases since they work at acidic pH of 4–5. The influx of H^+ ions into lysosomes keeps the acidic conditions there. The lysosome's covering membrane shields the remaining part of the cell from the acidic lysosomal milieu, while the membrane protects itself via glycosylation of its protein and lipid components. Lysosomes are sometimes called as suicidal bags because they contain a significant number of hydrolytic enzymes.

2.3.2.4.7.1 Function

(1) Intracellular and extracellular digestion are carried out by lysosomes. (2) Lysosomes engulf foreign antigens and harmful substances, assisting the body's natural defence system. (3) Lysosomes are involved in the transformation of numerous creatures, such as amphibians and tunicates. (4) Intracellular scavenging and autophagy occur are facilitated by lysosome.

2.3.2.4.7.2 Lysosomal Storage Diseases in Animal

Lysosomal enzymes break down a wide range of complex macromolecules. Any lysosomal enzyme deficit results in two pathogenic effects: (1) Primary accumulation, i.e. is the accumulation of partially digested insoluble metabolites within the lysosomes; or (2) secondary accumulation, i.e. the build-up of autophagic substrates within the cytoplasm due to reduced lysosomal function. As a result, the harmful proteins accumulate in cell, free radicals are generated, and cell death occurs due to damage to the plasma membrane. Dysfunctions of lysosomal catabolic pathway and subsequent accumulation of macromolecules within the cells lead lysosomal storage disease. In humans, there are about 51 lysosomal storage disorders, and many of these are also found in animals.

2.3.2.4.8 Vacuoles

Inside the cytoplasm, vacuoles are non-cytoplasmic areas separated from the latter by a particular membrane. Vacuoles originate from ER. In most cells, four types of vacuoles are found based on their content and function.

Sap vacuoles: Fluid or sap-filled vacuoles that are separated from the cytoplasm by a selectively permeable membrane (*tonoplast*). The osmotic pressure of the cell is maintained by various solutes contained in the vacuoles. Food reserves (e.g. sucrose), waste products, pigments (e.g. anthocyanins, anthoxanthins), alkaloids, and other substances may be stored in sap vacuoles.

Contractile vacuoles: These vacuoles are enclosed by a very flexible membrane in some protistan cells like Paramecium. These vacuoles get engorged after receiving water from the cytoplasm, which may or may not include waste materials. Vacuoles that have puffed up come into touch with the plasma membrane and collapse. These vacuoles are involved in osmoregulation and cellular waste excretion.

Food vacuoles: In a few protistan protozoa and phagocytes of higher animals, food vacuoles are generated by the union of lysosome and phagosome. These vacuoles play a role in digestion.

Air vacuoles: Gases are present in these submicroscopic vesicles, which provide buoyancy and mechanical strength to prokaryotic cells.

2.3.2.5 Nucleus

The nucleus is a specialized double membrane-bound protoplasmic entity that houses all of the genetic information needed to control cellular metabolism and pass on to future generations. The nucleus is the biggest cell organelle (interphase nuclei have a diameter of 5–25 m). Except for mature sieve cells of vascular plants and human RBC, all live eukaryotic cells have a nucleus. Cells typically have a single nucleus, i.e. they are uninucleate. *Paramecium caudatum* contains binucleate cells. Multinucleate cells, also known as syncytial cells, can be found in mammalian muscles, osteoclasts, helminth tegument, glass sponge siliceous spicules, and other places. Coenocytes are multinucleate cells found in plants and fungus, and are commonly seen in Rhizopus, Vaucheria, and other fungi.

2.3.2.5.1 Composition

Acid proteins, neutral proteins and enzymes—65%, Basic proteins—15%, DNA—9–12%, RNA—5%, Lipid—3%, traces of minerals like calcium, magnesium, potassium, and sodium are also present.

2.3.2.5.2 Structure

Nucleus is differentiated into five parts—nuclear envelope, nucleoplasm, nuclear matrix, chromatin, and nucleolus (Fig. 2.7).

Fig. 2.7 Structure of nucleus

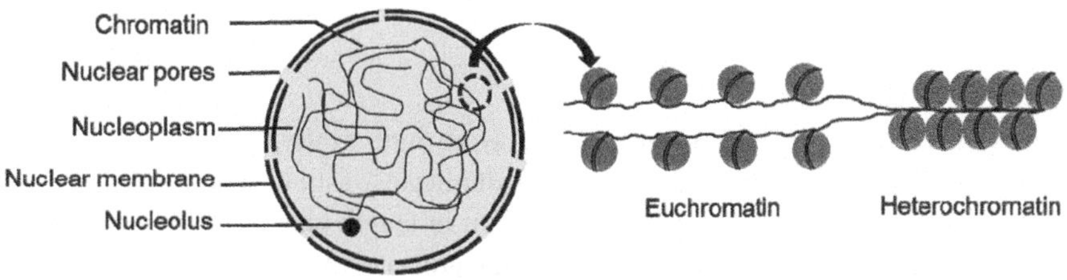

- *Nuclear envelope*: The inner and outer membranes of the nuclear envelope make up the nuclear envelope. Each membrane is 60–90 Å thick. The inner membrane is smooth, while the outer membrane is ribosome-studded. The space between inner and outer nuclear membrane is 100–500 Å thick. Nuclear envelope possesses many pores made of nucleoporin protein. The nuclear pores regulate the movement of biomolecules into and out of the nucleus.
- *Nucleoplasm*: Nucleus contains a translucent, semi-fluid, and colloidal substance. It comprises nucleosides as well as a variety of enzymes necessary for the production of DNA, RNA, nucleoproteins, and other nucleic acids.
- *Nuclear matrix*: The nuclear matrix is a network of tiny acidic protein fibrils that serve as a chromatin scaffold. Nuclear lamina, also known as fibrous lamina, is a dense fibrous layer that forms beneath the nuclear membrane. The nuclear matrix contains two types of intermediate filaments: lamin A and lamin B. The nuclear matrix and lamina give the nuclear envelope its mechanical strength.
- *Chromatin*: Chromatin is a nucleoprotein complex found in the reticular structure within nucleus. Euchromatin refers to the light-packed, transcriptionally active areas of chromatin, whereas heterochromatin refers to the condensed form of chromatin. Chromatin fibres condense to produce chromosomes during cell division.
- *Nucleolus*: The nucleolus is a membraneless, round structure within the nucleus that plays an important role in ribosome synthesis. The nucleolus organizer region (NOR) connects the nucleolus to the chromatin. In addition, the nucleolus is involved in the creation of signal recognition particles and the cellular stress response.

2.3.2.5.3 Function

(1) Nucleus is the carrier of hereditary materials called chromatin. Chromatin in the nucleus possesses genetic information necessary for all the functions of cells. (2) The nucleus harbours chromatin, which is a type of hereditary material. The genetic information needed for all of a cell's operations is stored in chromatin in the nucleus. (3) Ribosomes, the protein synthesis factory of the cells, are formed in the nucleolus part of nucleus.

2.3.2.6 Cytoskeleton

The fibrous and tubular structures that make up the cytoskeleton are the structural framework of eukaryotic cells. Cytoskeletal structures maintain the shape of the cell and its appendages and control the distribution of cell organelles, intracellular transport, and cell movement. The three types of filaments are microfilaments, intermediate filaments, and microtubules.

Microfilaments: Microfilaments are the cytoskeletal structures with the smallest diameter (approximately 7 nm). These are double-stranded polymeric fibrous actin (F-actin) molecule. F-actin is made of monomers of globular actin (G-actin). When three G-actin proteins come together to form a trimer, a microfilament begins to form. More actin binds to the barbed end after that. Autoclampin proteins, which operate as motors to assist construct the lengthy strands that make up microfilaments, aid in the self-assembly process. A microfilament is made up of two long strands of actin arranged in a spiral.

Intermediate filaments: Intermediate filaments unbranched cytoskeletal filaments that often form networks. Their average diameter falls between microfilaments and microtubules, ranging from 8 to 10 nm. Intermediate filaments are of variety of shapes and sizes: (1) Keratin filaments: They form tonofibrils and keratin of skin. (2) Neurofilaments: Filaments form lattice with bundles of microtubules in axons and dendrons of neurons. (3) Glial filaments: They are intermediate filaments of astrocytes. (4) Heterogeneous filaments: Synemin filaments, vimentin filaments, and desmin filaments are of these types. These types of intermediate filaments are found in in nucleus and centriole.

Microtubules: Microtubules are rigid hollow rods with a diameter of about 25 nm. Microtubules are composed of a single type of globular protein, called tubulin. Tubulin is a dimer made up of two 55-kDa polypeptides known as α-tubulin and β-tubulin. Microtubules are formed when tubulin dimers polymerize into microtubules, which are made up of 13 linear protofilaments arranged around a hollow core.

2.3.2.6.1 Microtubular Structures

2.3.2.6.1.1 Centrioles

Centrioles are submicroscopic structures having nine triplet fibrils and the ability to create their own duplicates, astral poles, and basal bodies. These are 0.3–0.5 μm in length and 0.15 μm in diameter. The pair of centrioles are known as diplosome and surrounding cytoplasm is called centrosphere. Centrioles are present in eukaryotic animal cells, protozoan protists (except *Amoeba*), and few eukaryotic plants like bryophytes, pteridophytes, and cycads.

Structure

A centriole possesses a cart-wheel structure in transverse section. At the periphery of the wheel, there are 9 peripheral fibrils in 9+0 arrangement. Each fibril is composed of three subfibrils C, B, and A from outside to inside. Sub-fibre A has 13 protofilaments while B and C sub-fibres have less than 13 protofilaments. The adjacent triplet fibrils are connected by C-A proteinaceous linker. The centre of centriole has a proteinaceous mass known as hub. Nine spokes around hub radiate towards the peripheral fibrils. On the outside of centrioles, dense plaques are there known as massules or pericentriolar satellites. Massules act as microtubule organizing centre or MTOC.

Function

(1) Centrioles facilitate in cell division with the formation of MTOC. (2) The distal centriole develops in axial filament or tail in spermatozoa. (3) Centrioles can develop into basal bodies from which cilia and flagella are formed.

2.3.2.6.1.2 Cilia and Flagella

Cilia and flagella are small hair-like mobile appendages of cells which can produce a current in fluid medium for locomotion or cellular transport of materials. Structurally cilia and flagella are similar and have four parts basal body, rootlets, basal plate, and shaft. The basic difference between them are cilia is smaller (5–20 μm) and more abundant while, flagella is longer (100–200 μm) but fewer (1–4 per cell).

Structure

The different parts of cilia and flagella are:

Basal body (Also known as kinetosome, blepharoplast, basal granule): It remains embedded in the periphery of the cytoplasm below the plasma membrane. The structure of basal body is similar to that of centriole.

Rootlets: The rootlets are microfilament bundles developing from the outer part of the basal body and provide support to the basal body.

Basal plate: Basal plate is dense structure above the basal body at the level of cell membrane.

Shaft: Shaft is hair-like projection differing in length. Shaft is covered by an extension of plasma membrane. Inside the shaft axoneme structure is present. Axoneme has 9 doublet fibrils at periphery and 2 singlet fibrils at the centre (9+2 arrangement) (Fig. 2.8a). Each peripheral fibril has two

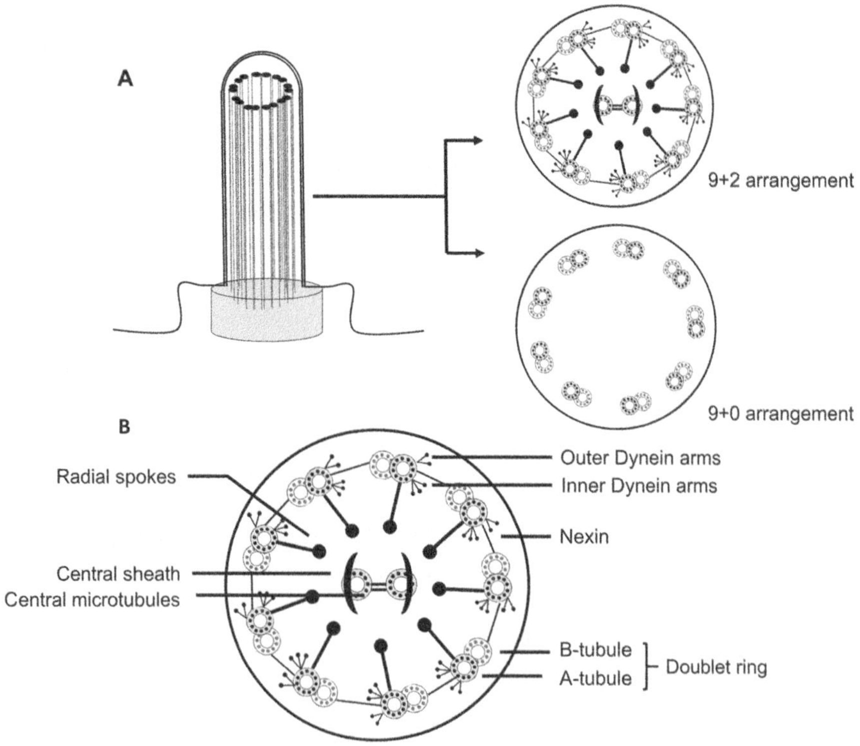

Fig. 2.8 (**a**) Structure of cilia and flagella with 9+2 or 9+0 arrangement. (**b**) Schematic diagram of cross-section of axoneme structure within cilia or flagella

microfibrils or sub-fibres A and B. Sub-fibre A has two bent arms, outer one with a hook. Arms are made of dynein protein. A-B linker made of Nexin protein makes the connection between sub-fibre-B of one fibre and inner side of A-sub-fibre of adjacent fibril (Fig. 2.8b).

Function

(1) Locomotion of flagellate and ciliates is facilitated by cilia and flagella. (2) Cilia helps in transport of gametes in reproductive tract. (3) Cilia sometimes act as sensory organs (cochlear hair cells).

Ciliopathies

Ciliopathies are diseases of human arising from the dysfunction of motile and/or non-motile cilia. Different types of ciliopathies exert phenotypes like retinal degeneration, skeletal malformation, kidney diseases, etc. Examples of ciliopathies are Alström syndrome, Polycystic kidney disease, Bardet–Biedl syndrome, Joubert syndrome, etc.

2.4 Flow of Molecular Genetic Information Within the Cell

The flow of genetic information in cell is illustrated by 'central dogma'. According to this model, information encoded in DNA is transferred from DNA to RNA and then to proteins via translation. Mechanisms for transmitting information in various forms include reverse transcription (the production of DNA from an RNA template) and replication.

DNA Replication: The basic property of any living organism is to reproduce. The only way to generate new cells is to divide existing ones. So, cell division is essential for the survival of all creatures. DNA contains the genetic information that each cell requires. When a cell splits, all of its DNA must be completely replicated in order to copy the biological information to be passed on to the daughter cell. DNA replication is the name for this process. DNA replication occurs in nucleus of eukaryotic cell.

Transcription: Transcription is the initial stage in gene expression, when information from a gene is used to build a functional product like a protein. The objective of transcription is to create an RNA copy of a DNA sequence of gene. The information needed to make a polypeptide is carried by the RNA copy, or transcript, of a protein-coding gene. Before being translated into proteins, eukaryotic transcripts must go through some processing steps. Transcription occurs in nucleus of eukaryotic cells. Transcription is regulated at various level by specific interaction of transcription factors, promoters, and other regulatory sequences of DNA which ultimately leads to gene expression control.

Reverse Transcription: Reverse transcription is the process by which double-stranded DNA is created from an RNA template with the help of reverse transcriptase enzyme. Reverse-transcribing RNA viruses (such as *retroviruses*) reverse-transcribe their RNA genomes into DNA with reverse transcriptase, becomes integrated into the host genome and replicates along with it.

RNA Processing: After transcription, newly synthesized RNA, also known as the primary transcript, is further processed before becoming functional. RNA processing differs according to the type of RNA viz. messenger RNA (mRNA), transfer RNA (tRNA), and ribosomal RNA (rRNA).

Translation: Translation is the process of deciphering messenger RNA (mRNA) and utilizing its information to produce a polypeptide, or chain of amino acids. This process is carried out in the ribosomes present in cytoplasm of the cell.

Know More......

Genetic Code

In an mRNA, groups of three nucleotides include instructions for making a polypeptide. These are called codon. Sixty-one different codons are there to code different amino acids constituting protein. Three codons (UAA, UAG, and UGA) are 'stop' codons that indicate the end of polypeptide synthesis. One codon, AUG, is considered as 'start' codon that kicks off translation process. The relationships between mRNA codons and amino acids are known as the genetic code.

Learning Outcomes

- **Cell theory**: Cell theory is one of the fundamental ideas of biology. The tenets to the cell theory are (1) the cell is the fundamental unit of structure and function in living things, (2) all organisms are made up of one or more cells, (3) 'Omnis cellula e cellula', i.e. 'All cells arise from pre-existing cells'. Theodor Schwann, Matthias Schleiden, and Rudolph Virchow are credited with the formulation of the theory.
- **Fluid-mosaic model**: This is the most accepted model of biomembrane, depicted it as dynamic

(continued)

structure with the ability to fuse or separate, expand, or contract during diverse cellular processes. The model was proposed by Singer and Nicolson in 1972.

- **Dictyosome**: In plant cells, the Golgi complex is made up of dictyosomes, which are a collection of disconnected components.
- **ATP Synthase**: Inner membrane and cristae of mitochondria possess small tennis-racket like ATP synthase dimer (earlier known as elementary particles or oxisomes). ATP synthase in eukaryotes consists of two major subunits, FO and F1. The F1 (Fraction 1) part of ATP synthase has three copies of each of subunits α and β, one each of subunits γ, δ, and ε. FO subunit consists of a subunit c-ring and one copy each of subunits a, b, d, F6 and the oligomycin sensitivity-conferring protein (OSCP).
- **Photorespiration**: In photosynthetic plants, photorespiration or oxidative C2 cycle or the oxidative photosynthetic carbon cycle occurs in peroxisomes. Peroxisomes remain associated with mitochondria and chloroplast. Glycolate from chloroplasts is transported to peroxisomes. Glycolate is oxidized to glyoxylate by catalase in peroxisome. Subsequently, glyoxylate is converted to glycine.
- **Euchromatin**: Euchromatin refers to the light-packed, transcriptionally active areas of chromatin.
- **Heterochromatin**: Heterochromatin refers to the condensed form of chromatin.
- **Nucleolar Organizer Region (NOR)**: NOR is the crucial place of chromosome for the formation of chromosome. NOR connects the nucleolus to the chromatin.
- **9+0 arrangement**: A centriole possesses a cart-wheel structure in transverse section. At the periphery of the wheel, there are nine peripheral fibrils in 9+0 arrangement.
- **9+2 arrangement**: Shaft of flagella or cilia has axoneme structure. Axoneme has 9 doublet fibrils at periphery and 2 singlet fibrils at the centre (9+2 arrangement).

Exercises

Objective Questions

Q1. Which organelle is involved in cell wall synthesis?
Q2. Which type of membrane protein spans the entire width of the membrane?
Q3. Desmosomes are concerned with ________________.
Q4. Which is the smallest cell organelle?
Q5. Which organelle is involved in glycosylation of proteins?
Q6. Amphisome is the hybrid organelle produced by fusion of __________ and ___________.
Q7. What is the marker enzyme of lysosome?
Q8. Who coined the term mitochondria?
Q9. Who built the first microscope?
Q10. Which cell organelle helps in the formation of root hair?
Q11. What is the major site of ribosomal RNA synthesis?
Q12. Who stated '*Omnis cellula e cellula*'?
Q13. Where is Photosystem I in plants found?
Q14. Which divalent ion is required for attachment of small and larger subunits of ribosome?
Q15. How many types of intermediate filaments are found in cell?
Q16. What is diplosome?
Q17. What is kinetosome?
Q18. In which plant cell centriole is found?
Q19. What are ciliopathies?
Q20. Who proposed endosymbiont theory?

Subjective Questions

Q1. What is cellular autonomy in unicellular organism?
Q2. Describe the structure of plasmodesmata with the help of a diagram.
Q3. Describe the subunit structure of ATP synthase.
Q4. Describe different types of vacuoles.
Q5. Differentiate between microtubules, intermediate filaments, and microfilaments.
Q6. Describe fluid-mosaic model of cell membrane.
Q7. Describe the structure of centriole. What are their functions?
Q8. Describe endosymbiont theory.
Q9. Name the different types of plastids. Describe the structure of chloroplastid.
Q10. Describe the ultrastructure of cilia and flagella.
Q11. What are the functions of Golgi apparatus?
Q12. What are the chemical compositions of cell membrane?
Q13. What is the difference between plasma membrane and cell wall?
Q14. Describe different components of prokaryotic cell.
Q15. What are the different functional sites of ribosomes? Write their functions.

Answer to Objective Questions

A1. Golgi Apparatus
A2. Transmembrane protein
A3. Cell adherence
A4. Ribosome
A5. Golgi complex

A6. Endosomes, autophagosomes
A7. Acid phosphatase
A8. Carl Benda
A9. Zacharias Janssen
A10. Golgi apparatus
A11. Nucleolus
A12. Rudolph Virchow
A13. Appressed part of grana
A14. Mg^{2+}
A15. Four types
A16. The pair of centrioles are known as diplosome.
A17. Basal body of cilia and flagella is also known as kinetosome.
A18. Bryophytes, pteridophytes, and cycads
A19. Ciliopathies are diseases of human arising from the dysfunction of motile and/or non-motile cilia.
A20. Lynn Margulis

Keywords for the Answer to Subjective Questions

A1. A cell of a unicellular (single celled) organism can exist independently, i.e. it does not depend upon any other cell for any function, material, or information.
A2. Plasmodesmata are cytoplasmic bridges in the shape of minute pores that connect adjoining plant cells to form a continuous protoplasm. Within plasmodesmata, an endomembrane-derived structure connecting Endoplasmic Reticula (ER) of two adjacent cells (desmotubule) is present.
A3. F_O, F_1 particles. F_1—α and β, one each of subunits γ, δ, and ε; F_O—a, b, d, F6, and the oligomycin sensitivity-conferring protein (OSCP).
A4. Sap vacuoles, contractile vacuoles, food vacuoles, air vacuoles.
A5. Diameters are different.
A6. Mosaicism, fluidity, asymmetry.
A7. Cart-wheel structure with 9+0 arrangement. (a) Centrioles facilitate in cell division with the formation of MTOC. (b) The distal centriole develops in axial filament or tail in spermatozoa. (c) Centrioles can develop into basal bodies from which cilia and flagella are formed.
A8. According to the endosymbiotic theory, some of the organelles in today's eukaryotic cells were previously prokaryotic bacteria.
A9. Leucoplast, chromoplast, chloroplast. Chloroplastid—thylakoid, stroma, grana.
A10. Basal body, rootlets, basal plate, shaft.
A11. (a) Secretion of biomolecules (b) Synthesis of glycoproteins and glycolipids (c) Formation of specialized structures (d) Synthesis of cell wall components.
A12. Water-20%, Carbohydrates-1–5%, Lipids-20–79%, Proteins-20–70%.
A13. Size, occurrence, function.
A14. Nuclear material, Vacuoles, Ribosomes, Photosynthetic thylakoids, Cell wall, Capsule, Flagella, and fimbriae.
A15. (a) mRNA binding site, (b) A-site or aminoacyl site where mRNA codon and aminoacyl tRNA are directed during translation process, (c) P-site or Peptidyl site where peptide bond formation, elongation, and transfer of peptide chain to site A occurs, and (d) E-site or Exit site.

Further Reading

Textbooks

Alberts B, Bray D, Hopkin K, Johnson AD, Lewis J, Raff M, Roberts K, Walter P (2013) Essential cell biology, 4th edn. Garland Publishing

Lodish H, Berk A, Kaiser CA, Krieger M, Bretscher A, Ploegh H, Martin KC, Yaffe M, Amon A (2021) Molecular cell biology, 9th edn. Macmillan Learning

Research or Review Articles

Brenner S, Jacob F, Meselson M (1961) An unstable intermediate carrying information from genes to ribosomes for protein synthesis. Nature 190:576–581

Danielli JF, Davson H (1935) A contribution to the theory of permeability of thin films. J Cell Comp Physiol 5(4):495–508

Koonin EV (2000) How many genes can make a cell: the minimal-gene-set concept. Annu Rev Genomics Hum Genet 1:99–116

Novikoff AB, Beaufay H, de Duve C (1956) Electron microscopy of lysosome-rich fractions from rat liver. J Biophys Biochem Cytol 2(4):179

Singer SJ, Nicolson GL (1972) The fluid mosaic model of the structure of cell membranes. Science 175(4023):720–731

Action Potential

3

Sonali Jana

Abstract

The nerve axons and muscle fibers exhibit the property of excitability by which they can transmit signals along their membranes. At rest, potential differences develop across the unit membranes owing to differences in the concentration of ions and their selective permeability towards specific ions. An action potential can be generated by a threshold stimulus that can depolarize the resting nerve and activate the voltage-gated ion channels. An action potential consists of four phases depolarization, overshoot, repolarization, and hyperpolarization. Once developed, an action potential propagates with undiminished strength until it reaches the end of the nerve fiber.

Graphical Abstract

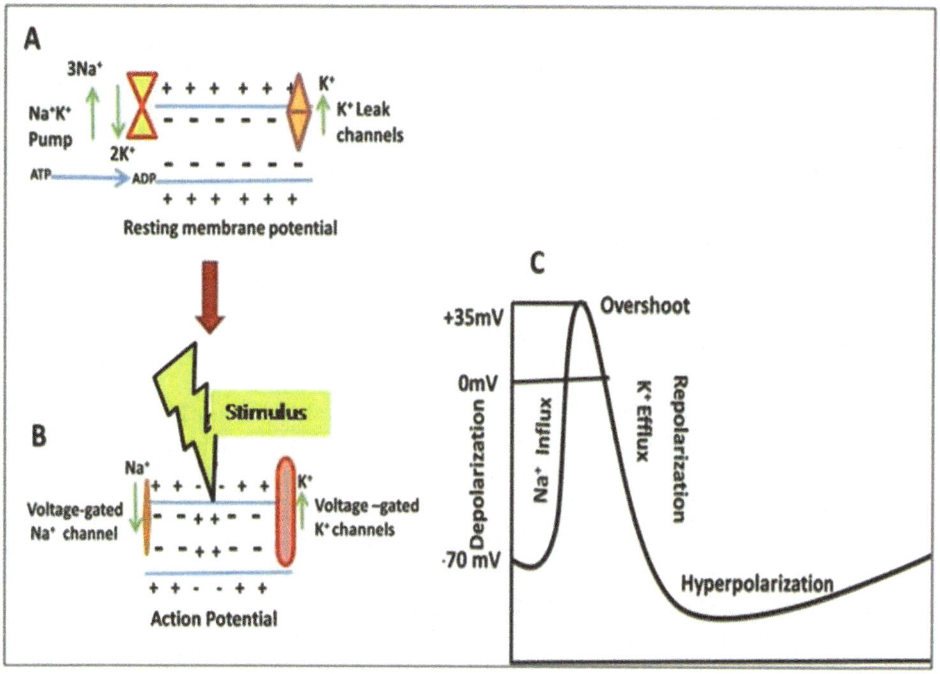

Description of the graphic: The exchange of ions and establishment of the resting membrane potential (**a**). Application of stimulus and generation of action potential (**b**). Graphical representation of action potential in nerve axonal membrane (**c**)

S. Jana (✉)
Department of Veterinary Physiology, West Bengal University of Animal & Fishery Sciences, Kolkata, West Bengal, India

© The Author(s), under exclusive license to Springer Nature Singapore Pte Ltd. 2023

P. K. Das et al. (eds.), *Textbook of Veterinary Physiology*, https://doi.org/10.1007/978-981-19-9410-4_3

Keywords

Resting membrane potential · Action potential · Polarized state · Depolarization · Repolarization

Learning Objectives

- Ionic basis of resting membrane potential (RMP) in a neuronal membrane.
- Generation of action potential and its phases.
- Characteristics of action potential.
- Basis of refractoriness in the neuronal membrane after the action potential.
- Factors that determine the propagation and conduction velocity of action potential.

Every cell of the body has electrical potentials across its cell membranes. Additionally, some cells such as the nerve axons and muscle fiber exhibit the property of excitability by which they are able to transmit signals along their membranes. The excitability or rapid changes in electrochemical impulses exhibited by certain cells of the body is governed by the principles of membrane theory or ionic hypothesis. The theory is based on the ability of the cellular membrane to respond to changes in the membrane permeability to ions. The membrane theory states that potentials develop across the unit membranes owing to differences in the concentration of ions and their selective permeability towards specific ions. The understanding of the fundamentals of nerve cell excitability are based on the finding by Hodgkin and Huxley (1952) whose experiments with the squid giant axons gave the first insights into the generation of action potential and the kinetics of ion channels. The membrane potential is an essential feature of all body cells both excitable and non-excitable. Recent studies highlight the indispensable role of membrane potentials in regulating several important functions of the body such as the biological rhythms; specifically the circadian rhythm. It denotes the events that repeat cyclically over 24 h period and is coordinated by the superior chiasmatic nucleus of the hypothalamus. The depolarization and a consequent fall in action potential firing lead to changes in the membrane potential of neurons in the superior chiasmatic nucleus. Contrary to the popular belief, here depolarization leaves the cell membrane less excitable leading to a decrease in action potential generation. The pineal glands respond to these changes by secreting melatonin which mediates the cyclic events. The membrane potential is essential for various sensory perceptions such as vision, hearing, taste, and olfaction. The neuroendocrine cells of the body located in the hypothalamus, pituitary, thyroid gland, and pancreas are excitable and secrete hormones that regulate basic body functions. The excitability of neuroendocrine cells is driven by action potential which occurs as rapid oscillation bursts induced by Ca^+ ions influx followed by a period of rest. This depolarization due to calcium ion influx leads to the secretion of hormones from the concerned cell types.

3.1 The Diffusion Potential and Membrane Potential

The diffusion of ions across the cell membrane due to their concentration gradients creates the diffusion potential. The potassium ions are of higher concentration inside the cell membrane in comparison to the outside; they diffuse outside due to this concentration gradient leaving the negatively charged ions behind. This causes electropositive charges to accumulate outside and electronegativity develops inside the membrane. The diffusion potential then develops across the membrane which prevents further efflux of potassium ions from inside despite the existence of a high concentration gradient. The value of diffusion potential is −94 mV in a nerve fiber of mammalian origin. In case the membrane is selectively permeable only to sodium ions, the diffusion of sodium ions will occur towards the inside of the membrane since the concentration of sodium ions is much more in the ECF than in the ICF. This diffusion of positive ions inside will cause electropositivity inside and electronegativity on the outer side of the cell membrane thereby diffusion potential will rise to +61 mV which will block the further influx of sodium ions. The diffusion potential at which the net diffusion of a particular ion across a membrane is prevented is known as the Nernst potential which can be determined using the Nernst Equation as given

$$E_{\text{ion}} = 2.303\frac{RT}{ZF}\log\frac{[\text{ion}]_{\text{o}}}{[\text{ion}]_{\text{i}}}$$

As this value increases, it indicates a greater tendency for the ion to diffuse which means a greater Nernst potential is required to prevent further diffusion of ions.

When the membrane is permeable to more than one ion then the diffusion potential relies on several parameters; the charge of individual ions, their concentrations on either side of the membrane, and the membrane permeability to these ions. The membrane potential on the inside of the cell membrane can then be calculated by applying the Goldman equation also known as the Goldman-Hodgkin-Katz equation as given. The ions involved are sodium, potassium, and chloride. These three ions are predominantly involved in the maintenance of the membrane potential of muscle fibers and nerve fibers.

$$V_m = -\frac{RT}{F} \ln \frac{P_{Na}[Na^+]_i + P_K[K^+]_i + P_{Cl}[Cl^-]_o}{P_{Na}[Na^+]_o + P_K[K^+]_o + P_{Cl}[Cl^-]_i}$$

The selective permeability of the membrane to particular ions determines the membrane potential. If the membrane is selectively permeable only to sodium ions and impermeable to the other ions, i.e., potassium and chloride, the membrane potential will be determined by the sodium ion concentration gradient and will be equal to the sodium ion Nernst potential. The same principle applies to other permeant ions as well.

3.1.1 Ionic Basis of Resting Membrane Potential (RMP)

Before understanding the action potential, it is very important to have an idea about the resting membrane potential. The cell membrane of most cells maintains an ionic concentration difference operated by the "*Na^+K^+ pump*" and the potassium "*leak*" channel systems. The "*Na^+K^+ pump*" actively pumps out sodium ions from the intracellular fluid to the extracellular fluid and draws in potassium ions against the concentration gradient. This pump exchanges three sodium ions for every two potassium ions and thus develops a negative potential inside the cell membrane. A large difference in ionic concentration also develops across the membrane in the resting stage by the "*Na^+K^+ pump*." The concentration of sodium and potassium ions inside the cell membrane is 14 mEq/L and 140 mEq/L while their concentration outside is 142 mEq/L and 4 mEq/L, respectively. The potassium "*leak channel*" functions opposite to the "*Na^+K^+ pump*" as this ion channel favors the movement of sodium ions into the ICF and potassium ions into the ECF by following the chemical gradient but are 100 times more permeable to potassium ions than to sodium ions. These two systems work together to maintain a steady state across membranes. This net results in a positive charge outside the cell membrane and a negative charge on the inner side. This potential difference (denoted as V_m) in the inside and outside the cell membrane at rest is known as the resting membrane potential (RMP). The magnitude of the RPM can be determined by the Nernst equation as given:

$$E_{ion} = 2.303 \frac{RT}{ZF} \log \frac{[ion]_o}{[ion]_i}$$

where $[ion]_o$ and $[ion]_i$ are the concentration of particular ions in the ECF and ICF, respectively.

In the nerve cell, the RMP is −70 mV.

The voltmeter is used to measure the membrane potential. It is a highly sophisticated instrument through which minute changes in potential difference across the membrane can be detected.

3.2 Action Potential

The Resting membrane potential is a dynamic process that functions continuously to maintain a steady ionic balance in the cell wherein the ions moving through the membranes due to their concentration gradient are again being actively pumped back against their concentration gradient. This state is also known as the "polarized state" of the cell membrane. The process occurs inherently in all excitable cells such as the nerve axon and muscle fibers. Whenever a stimulus is applied over the cell membranes at resting membrane potential, the polarity of the membrane at the particular spot of application gets reversed, i.e., the outside becomes negative and the inside becomes positive. This "depolarization state" of the cell membrane is known as action potential and it signals the onset of cellular activity.

An **action potential** is therefore a very fast and transient change in the resting membrane potential due to a stimulus. This property exhibited by neurons and muscle cells is also referred to as excitability.

3.2.1 Characteristics of Action Potential

The duration of action potential in a nerve axon is very short (about 1 ms). They follow the all-or-none law, i.e., if the stimulus is below the threshold level, the depolarization does not occur. However, once the stimulus reaches adequate strength, the action potential develops and propagates throughout the entire length of the cell membrane with undiminished strength. The size of the action potential also remains unchanged on increasing the intensity of the stimulus above the adequate level. The response is regenerative by nature since the excitation developed in a particular patch of an axon is capable of exciting the next part and so on. The impulse propagates in the form of a wave at a constant speed and amplitude along the complete length of the axon by permeability changes of the axonal membrane to sodium and potassium ions in each axonal segment. Action potential plays a vital role in transmitting information to and from the central nervous system that is mediated by the propagation of action potential through the nerve cell axons.

3.2.2 The Phases of Action Potential

The action potential is comprised of several components (Fig. 3.1). Once a stimulus reaches the threshold value it leads to the generation of an action potential. For a nerve fiber, if a stimulus of adequate strength is applied, the RMP of −70 mV can reach up to −60 mV, the so-called threshold level. The immediate response is the depolarization phase or

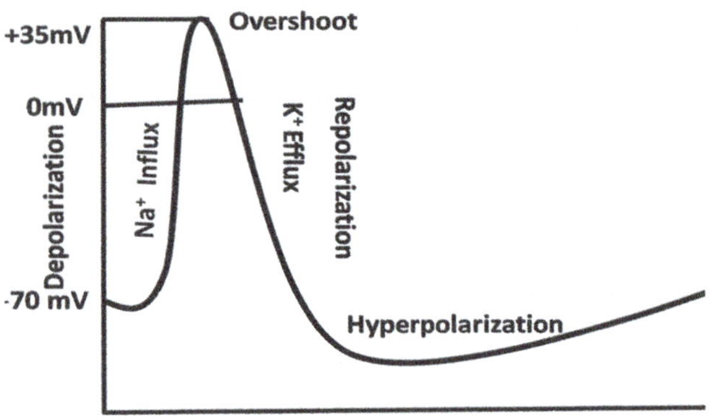

Fig. 3.1 Generation of action potential in a nerve axonal membrane. (Adapted from Guyton and Hall Textbook of Medical Physiology 12th Edition)

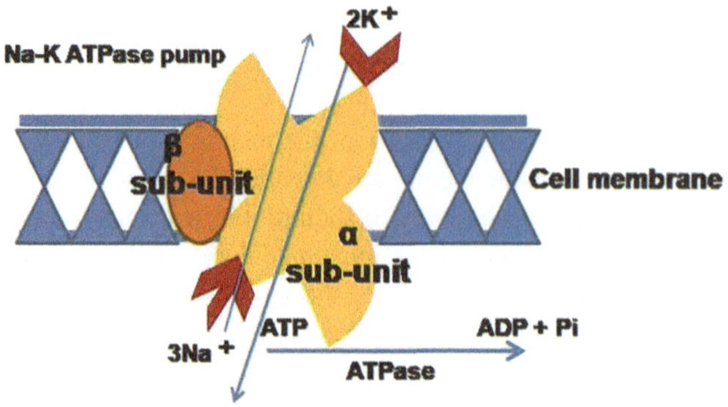

Fig. 3.2 Functioning of the Na-K ATPase pump. (Adapted from Guyton and Hall Textbook of Medical Physiology 12th Edition)

the upstroke whereby the membrane potential reaches a more positive state from a negative state. In a nerve fiber, the membrane potential of −60 mV can sharply reach 0 mV during the depolarization phase. In the next phase, the membrane potential reaches the peak amplitude beyond 0 mV. This is known as the overshoot which can be up to +30 to +40 mV. The potential then returns to the resting membrane potential of about −70 mV and is known as the repolarization phase. It may also reach a more negative phase beyond the resting membrane potential, known as hyperpolarizing after potential or the undershoot and again subsequently return to its original RMP.

3.2.3 Ionic Basis of Action Potential

The resting membrane potential of the nerve fiber is determined by the active transport of sodium and potassium ions through the "Na^+K^+ *pump*" and by the leakage of potassium ions through the K^+ *leak channels* located in the nerve membrane. The "Na^+K^+ *pump*" continuously pumps out sodium ions to the outside and potassium ions to the inside of the cell. For every three sodium ions that are pumped out, two potassium ions are taken inside leading to a net deficit of positive ions; thus, causing a negative potential inside. This phenomenon leads to the creation of a large concentration gradient for both ions. The K^+ *leak channels* are more permeable to potassium ions in comparison to sodium ions and thus potassium may leak outside. This permeability difference influences the resting membrane potential. Moreover, there are other ion channels; the voltage-gated sodium and potassium channels which specifically permit the particular ion only. The voltage-gated Na^+ ion channels which permit only Na^+ ions remain practically closed during resting conditions or at RMP.

The Na-K ATPase pump or "Na^+K^+ *pump*" is also an electrogenic pump that is responsible for maintaining a potential difference (i.e., outside positive and inside negative) across the cell membrane (Fig. 3.2). This function is essential to almost all cells of the body for regulation of cellular volume and for transmitting signals across nerve and muscle fibers. This pump also maintains the concentration of sodium and potassium ions on either side of the cell membrane. The Na-K ATPase pump is a complex structure composed of two protein subunits, the α and β subunits. The α subunit is the bigger one with a molecular weight of approx. 100,000 while the β subunit, the smaller one, has a molecular weight of 55,000. This carrier protein complex is present in the cell membrane and extends extracellularly and intracellularly. The α subunit has several important functions. The part of the α subunit that extends intracellularly has three sodium ions binding receptor sites. The α subunit also has an ATPase activity near the vicinity of receptor sites for sodium ions. While the portion that protrudes outside has two potassium ions binding sites. Whenever this carrier protein binds with three sodium ions present inside the cell and two potassium ions present outside the cell membrane, ATPase becomes activated which cleaves the ATP molecule to form ADP by releasing a high-energy phosphate bond. This energy leads to conformational alteration of the carrier protein which leads to the expulsion of three sodium ions to the outside, simultaneously drawing in two potassium ions within the cell. Thus, for each cycle of the pump, there is a net deficit of one positive ion within the cell. This creates a negative potential within the cell membrane as more positive ions are pumped out in comparison to those that enter the cell. This pump may function reversibly depending on sodium, potassium electrochemical gradients, and the relative concentrations of ADP, ATP, and phosphate.

The development of action potential is attributable to the changes in the relative permeability of the neural membrane to sodium and potassium ions. When a stimulus is applied to the nerve cell membrane, the membrane potential becomes less negative and this leads to the activation of the voltage-gated Na^+ channels which open and draw in a large amount of Na^+ ions by the electrochemical gradients earlier established by the "*Na^+K^+ pump*" and the *K^+ leak channels* at rest. As Na^+ ion enters the cell it immediately changes the relative voltage inside the cell membrane from more negative to less negative. Very few Na^+ ions channels open at the beginning marking the start of depolarization, which has a positive effect and stimulates further opening of a large number of Na^+ channels causing massive entry of positive ions into the cell. This ultimately reverses the RMP from negative to positive and a steep rise in action potential is observed. The membrane potential reaches 0 mV and then overshoots to +30 mV. Soon the voltage-gated Na^+ ion channels begin to close and the voltage-gated K^+ ion channel opens. Driven by the concentration gradient, the K^+ ions start to diffuse out of the cell into the ECF. A drop in the inflow of Na^+ ions combined with the vigorous exit of K^+ ions drives the membrane potential back toward its resting voltage, i.e., to −70 mV. The Potassium ions continue to leak out until they reach the equilibrium and this leads to hyperpolarization.

Know More......

In the cardiac muscle fibers, repolarization does not occur immediately after depolarization, instead just after the spike, the potential remains fixed for several milliseconds at a plateau stage and finally undergoes repolarization. This occurs due to the slow opening of voltage-gated calcium–sodium channels which allows the prolonged influx of calcium ions leading to the plateau formation and contraction of the cardiac muscles.

3.2.3.1 Threshold Stimulus, Chronaxie, Utilization Time, and Rheobase

Whenever a nerve fiber is stimulated by a subthreshold stimulus, no action potential is generated. However, on gradually increasing the strength of the stimulus, at a particular point, a response or action potential is produced when the stimulus is applied for an indefinite duration of time. Thus, the strength of stimulus that upon application for an indefinite period can generate a response is known as Rheobase. The minimum strength of the stimulus which can elicit an action potential is the threshold stimulus. Utilization time denotes the minimum duration of time that is just sufficient to generate excitation when Rheobasic strength of the current is applied to the excitable cell such as nerve and muscle fiber. When a stimulus having double the strength of Rheobase is applied to the nerve or muscle, the time taken for the response to happen is termed Chronaxie. While Rheobase denotes the strength of stimulus or current, Chronaxie denotes the time in seconds.

3.2.4 Refractory Period

After the occurrence of an action potential, the nerve cannot be stimulated for the generation of another action potential for a brief period of time. This nonresponding period is known as the absolute refractory period. This period overlaps with the depolarization phase and the two-third of the repolarization phase. Since all the voltage-gated sodium channels are in a transition stage from open to inactivation, it is impossible to start a new action potential at this stage. At the end of the absolute refractory period, the nerve can respond to a stimulus to generate an action potential but of much smaller size, provided the stimulus is of much larger strength, i.e., by a suprathreshold stimulus. This period of partial responsiveness is known as the relative refractory period. As all the voltage-gated sodium channels are closed now, depolarization is possible at this stage.

3.2.5 Propagation of Action Potential

The action potential developed at one particular point of the cell membrane leads to the excitation of adjacent segments which propagates the action potential along the membrane with undiminished strength until it reaches the end of the nerve fiber. Thus, new action potentials are repeatedly regenerated in a subsequent segment of the neuronal membrane. As action potential develops at a spot, the relative permeability to sodium ions increases and positive charges accumulate inside. These charges are attracted by the adjacent segment which is negative relative to the outside or at RMP. The positive charge moves to the next segment and triggers an action potential there. This is referred to as electrotonic current spread and the current is known as electrotonic current. The velocity of conduction of action potential depends on the resistance to the flow of current both internally and across the axonal membrane. It also depends on the capacitance or storage ability of the axonal membrane. The thickness of the axon determines the velocity of propagation of action potential. It is faster in myelinated nerves since salutatory conduction occurs only over the node of Ranvier and action potential jumps from one node to another. Myelination also decreases the capacitance thus conduction velocity increases. In an unmyelinated nerve, the conduction takes a much longer time since each part of the nerve membrane have to undergo depolarization for propagation. The propagation of action potential occurs only in the forward direction away from the stimulus. As the segment remains in the refractory period after the generation of action potential, backward propagation does not occur.

3.2.6 Ion Channels and Ion Channel Modulators

The membrane potential and the generation of the action potential are intimately intertwined with the functioning of ion channels. The ion channels are pore-forming proteins in the cell membrane that permit the passive flow of ions across the cell membrane. Most of the ion channels are gated, which means they can switch between open and closed states in response to particular stimuli. The voltage-gated ion channels open in response to changes in voltage or potential differences across the cell membrane. The ligand-gated ion channel opens upon binding to ligands that can be neurotransmitters or other molecules. The mechanically gated channels operate in response to changes in mechanical stimuli such as stretch and swelling. These ion channels allow the passage of cations, such as potassium, sodium, calcium, and anions such as chloride. The working of several ion channels is based on protein phosphorylation and dephosphorylation. In addition to the gated channels, the cell membrane also has leakage channels that practically remain always open permitting the selective movement of potassium ions from inside to outside the cell due to the concentration gradient. This causes the efflux of potassium ions from inside to outside the cell resulting in the development of negative potential within the cell. As soon as this reaches an equilibrium state the further efflux of potassium ions is prevented. This phenomenon is responsible for establishing the resting membrane potential of the cell. The electrochemical gradients dictate the direction of individual ion flow across the cell membrane. While sodium and calcium ion generally flows inside the cell exerting a depolarizing effect, potassium ions usually moves out of the cell causing the cell to repolarise. The chloride ions moving into the cell is responsible for causing hyperpolarization. The ion channels are composed of several protein units and subunits. Typically the pore-forming subunit is constituted by the α subunit, while the auxiliary subunits are formed by the β and γ subunits. Studies suggest that some drugs functions by binding to the α-subunits of the ion channels thereby modulating their activity. These drugs are usually small molecules that might obstruct the pore, disturb the interactions of the proteins with the several subunits, etc. Some of the drugs modulating the functions of the ion channels are enlisted here

- Amiodarone and Dofetilide are voltage-gated K^+ channel blockers that delay the repolarization of action potentials in cardiac cells and neurons and are commonly prescribed in the treatment of atrial and ventricular fibrillation.
- Nicorandil and Diazoxide are K^+ channel openers that cause hyperpolarization thereby preventing the entry of calcium ions in vascular smooth muscle cells and pancreatic β-cells. These drugs are used in treating angina and hypoglycemia, respectively.
- Some local anesthetics such as Lidocaine and Tetracaine selectively block the voltage-gated Na^+ channels thereby preventing the generation and propagation of action potential.
- Amlodipine and nifedipine are used in the treatment of hypertension and angina work by blocking the voltage-gated calcium channels which relaxes the vascular smooth muscle.

3.2.7 Channelopathies

Ion channels are gaps across cell membranes that allow the passive diffusion of ions. These can be classified into Sodium (Na^+), Potassium (K^+), Calcium (Ca^{2+}), and Chloride (Cl^-) ion channels depending upon the ions for which they are selectively permeable. Ion channels transport ions depending on the electrochemical gradient. There are various types of ion channels such as ligand-gated ion channels, voltage-gated, mechanical-gated, such as stretch-activated and also temperature-activated, i.e., heat- or cold-activated channels. They hold immense importance in regulating several important functions such as the generation of electrical signals, chemical signaling, muscle contraction, transepithelial transport of ions and molecules, hormone secretion, maintenance of cell volume, and cell proliferation. Mutations in genes that encode the proteins of ion channels such as voltage-gated potassium, sodium, calcium, and chloride channels and also autoimmune diseases are implicated in several diseases of the ion channels known as channelopathies. Table 3.1 gives an overview of the diseases due to mutations of proteins in ion channels.

Table 3.1 Genetic disorders of ion channels/channelopathies

Sl. no	Type of ion channel	Gene affected	Disease	Characteristics
1.	Voltage-gated Na^+ ion channel (skeletal muscle)	SCN4A in the α subunit of the ion channel	Hyperkalemic periodic paralysis and myotonia	In Hyperkalemic periodic paralysis, there are sudden episodes of muscle weakness or paralysis between normal muscle activities. In Myotonia there is a delay in the relaxation of muscles after contraction.
2.	Voltage-gated Na^+ ion channel (neuronal)	SCN1B and SCN1A for the β1 and α subunit of the ion channel, respectively	Epilepsy is accompanied by febrile seizures	Epilepsy is bursts of synchronized discharges causing seizures.
3.	Voltage-gated Na^+ ion channel (cardiac muscle)	SCN5A in α subunit of the ion channel leading to complete channel inactivation	Long QT syndrome	Cardiac disorder with arrhythmia and syncope leading to sudden death.
4.	Voltage-gated potassium channel (neural)	KCNA1/Kv1.1 in the α subunit of the ion channel	Episodic ataxia with myokymia syndrome	Uncontrolled muscle contractions/movements (myokymia) with brief ataxia.
5.	Voltage-gated potassium channel (renal)	KCNJ1 gene	Bartter's syndrome	Severe depletion of intravascular volume.
6.	Voltage-gated calcium channel	*CACNA1F*	Congenital stationary night blindness	
7.	Chloride channel	CFTR	Cystic fibrosis	Copious mucous secretion in the lungs and pancreas leads to constant inflammation and finally death.

Learning Outcomes

- **Resting membrane potential and action potential**: The resting membrane potential in the neuronal membrane is around −70 mV. This potential difference is created due to the activity of the "Na^+K^+ *pump*" which actively pumps out three sodium ions for every two potassium ions. A large difference in ionic concentration also develops across the membrane in the resting stage by the "Na^+K^+ *pump*." Whenever a threshold stimulus is applied to the neuronal membrane at resting membrane potential, an action potential develops that is characterized by a very fast and transient change in the resting membrane potential. The activation of voltage-gated sodium ion channels leads to a massive influx of sodium ions causing depolarization of the membrane. These ion channels inactivate spontaneously while the voltage-gated potassium channels begin to open whereby the membrane becomes much more permeable to potassium than to sodium leading to the efflux of potassium. This phrase denotes the repolarization phase of the action potential which may lead the membrane potential to a more negative value than the RMP and cause the hyperpolarization phase.
- **Refractory period**: The brief period after the occurrence of an action potential when the nerve cannot be stimulated for the generation of another action potential is the absolute refractory period. At the end of the absolute refractory period, the nerve can respond to a stimulus to generate an action potential but of much smaller size, provided the stimulus is of much larger strength, i.e., by a suprathreshold stimulus. This period of partial responsiveness is known as the relative refractory period.
- **Propagation of action potential**: The propagation of action potential occurs only in the forward direction away from the point of stimulus. The action potential developed at one particular point is repeatedly regenerated in a subsequent segment of the neuronal membrane which propagates along the membrane with undiminished strength until it reaches the end of the nerve fiber.

Exercises

Objective Questions

Q1. Which ion plays a major role in the maintenance of RMP?

Q2. What are the forces that influence ionic influx and efflux across the cell membrane?

Q3. What is the approximate value of resting membrane potential in a nerve fiber?

Q4. What causes depolarization of cell membrane?

Q5. Name the different phases of the action potential.

Q6. What causes the plateau during the action potential of cardiac muscles?

Q7. What are the two phases of the refractory period?
Q8. Why action potential cannot be generated during the absolute refractory phase?
Q9. How does the propagation of action potential occur?
Q10. Why does action potential conduction occur faster in myelinated nerve fibers?

Subjective Questions

Q1. Describe the ionic basis of resting membrane potential.
Q2. Outline the ionic basis of action potential in a nerve membrane.
Q3. Elaborate on the all or none law as applicable for action potential propagation.
Q4. Why does the action potential propagate only in the forward direction?
Q5. Explain the factors influencing the propagation of action potential in a neuronal membrane.

Answers to Objective Questions

A1. Potassium ions
A2. The concentration gradient and electrical gradient
A3. −70 mV
A4. Threshold stimulus causing activation of voltage-gated sodium channels and influx of sodium ions
A5. Depolarization, overshoot, repolarization, hyperpolarization
A6. The slow influx of calcium ions
A7. The absolute and relative refractory period
A8. Absence of resting membrane potential, inactivation of voltage-gated sodium channel
A9. Electrotonic current
A10. Less resistance to current flow due to large diameter, saltatory conduction, and less capacitance

Keywords for the Answer to Subjective Questions

A1. "Na^+K^+ *pump*," potassium "*leak*" channel.
A2. Activation of voltage-gated Na^+ channels, depolarization, activation of voltage-gated potassium pump, repolarization.
A3. Subthreshold stimulus, threshold stimulus.
A4. Refractoriness.
A5. Resistance to current flow, capacitance.

Further Reading

Choudhuri SK (2004) Membrane potential. In: Choudhuri SK (ed) Concise medical physiology, 5th edn, pp 9–15

John E (2011a) Membrane potentials and action potentials. In: Hall JE, Guyton AC (eds) Guyton and Hall textbook of medical physiology, 12th edn. Saunders, Elsevier Inc, p 53

John E (2011b) Transport of substances through cell membranes. In: Hall JE, Guyton AC (eds) Guyton and Hall textbook of medical physiology, 12th edn. Saunders, Elsevier Inc, pp 57–69

Nelson DL, Cox MM (2004) Solute transport across membranes. In: Nelson DL, Cox MM (eds) Lehninger principles of biochemistry, 4th edn, p 415

Reece WO (2005) Membrane potentials: the generation and conduction of electrical signals in neurons. In: Reece WO, Erickson HH, Goff JP, Uemura EE (eds) Dukes' physiology of domestic animals, 12th edn, pp 770–785

Reuss L (2008) Mechanisms of water transport across cell membranes and epithelia. In: Alpern R, Hebert S (eds) Seldin and Giebisch's the kidney, vol 2008, pp 147–167

Research Articles

Abdul Kadir L, Stacey M, Barrett-Jolley R (2018) Emerging roles of the membrane potential: action beyond the action potential. Front Physiol 9:1661

Cohen LB (1973) Changes in neuron structure during action potential propagation and synaptic transmission. Physiol Rev 53(2):373–418

Dworakowska B, Dołowy K (2000) Ion channels-related diseases. Actabiochim Polonica 47(3):685–703

Hille B (2011) Ionic basis of resting and action potentials. Compr Physiol:99–136

Hübner CA, Jentsch TJ (2002) Ion channel diseases. Hum Mol Genet 11(20):2435–2445

Hodgkin AL, Huxley AF (1952) A quantitative description of membrane current and its application to conduction and excitation in nerve. J Physiol 117(4):500–544

McGivern JG, Ding M (2020) Ion channels and relevant drug screening approaches. SLAS Discov 25(5):413–419

Part II

Hematology and Immune System

4 Hematology

Joydip Mukherjee, Pradip Kumar Das, and Dipak Banerjee

Abstract

Blood is a fluid connective tissue originates from mesoderm composed of corpuscles (45%) suspended in protein-rich fluid called plasma (55%). It is the main transport medium of the circulatory system responsible for transport of oxygen and nutrients to the tissue and removal of metabolic waste products away from the tissue. Among the corpuscles, erythrocytes are found in the highest proportion followed by platelets and leukocytes. Erythrocytes of mammals are the non-nucleated, non-motile, biconcave circular disk-like structure involved primarily in oxygen and carbon-dioxide transport. Leukocytes are large, nucleated cells involved in the immunity. Platelets are the smallest blood corpuscles involved in hemostasis. All the blood cells are derived from pluripotential hematopoietic stem cells mainly in the bone marrow. Hemoglobin is an iron-containing conjugated protein consists of heme and globin that is exclusively found in erythrocytes and transports oxygen. The heme portion of the hemoglobin subjected complex metabolic reactions to be eliminated from the body and iron is reutilized. Blood coagulation or hemostasis is an inherent property of the blood which enables the stoppage of bleeding from injured vessels, helps to keep the blood in a fluid state during circulation and to resolve the clot for restoring vascular integrity. The complex interaction between vascular endothelium, platelets, and various clotting factors facilitates the formation of a clot. Different animal species have different blood groups determined by polymorphic antigens that reside on the surface of erythrocytes (agglutinogens) and the antibodies (agglutinogens) present in the plasma.

J. Mukherjee (✉) · P. K. Das · D. Banerjee
Department of Veterinary Physiology, West Bengal University of Animal & Fishery Sciences, Kolkata, West Bengal, India

© The Author(s), under exclusive license to Springer Nature Singapore Pte Ltd. 2023

P. K. Das et al. (eds.), *Textbook of Veterinary Physiology*, https://doi.org/10.1007/978-981-19-9410-4_4

Graphical Abstract

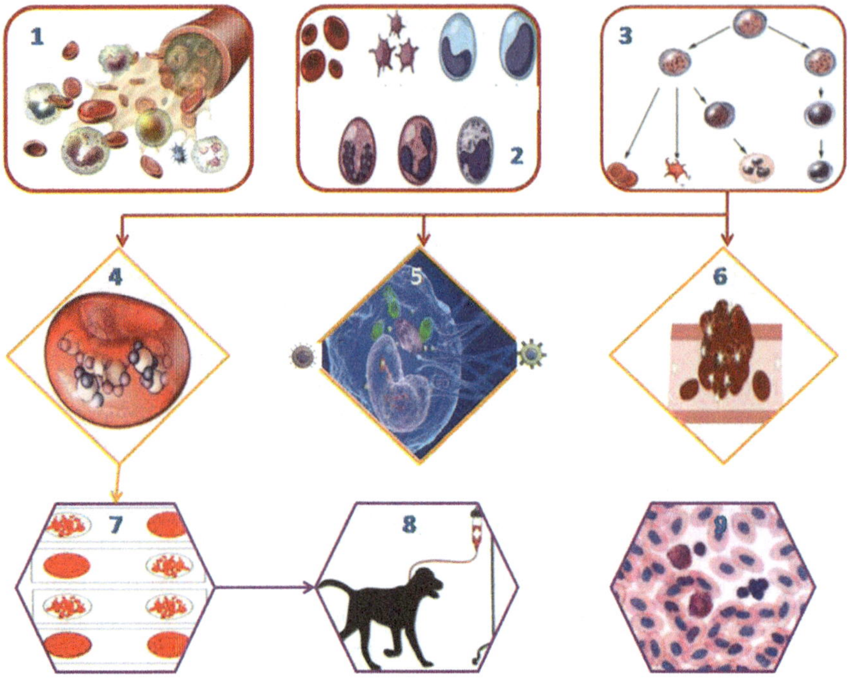

Description of the graphic: Blood is a fluid connective tissue composed of plasma and corpuscles (**1**). There are three types of blood corpuscles namely red blood corpuscles (RBC), white blood corpuscles (WBC), and platelets (**2**). All the blood corpuscles are derived from pluripotential hematopoietic stem cells through a process called hematopoiesis (**3**). Hemoglobin, an iron containing conjugated protein found exclusively in RBCs concerned with oxygen transport (**4**). WBCs provide humoral and cell-mediated immune response against pathogens (**5**). Platelets are concerned with blood coagulation with the help of vascular endothelium, platelets, and various clotting factors (**6**). Different animal species have different blood groups determined by the presence of polymorphic antigens on the surface of erythrocytes (agglutinogens) and the antibodies (agglutinogens) present in the plasma (**7**). Blood grouping is required for blood transfusion (**8**). The avian blood is differed from mammals in respect of nucleated RBCs and presence of heterophils homologous to neutrophils (**9**)

Keywords

Blood corpuscles · Plasma · Hemoglobin · Hemostasis · Blood group

Learning Objectives

- The compositions and properties of blood.
- The genesis of blood corpuscles and their functions.
- Hemoglobin and iron metabolism.
- The mechanism of blood coagulation.
- Blood grouping in animals and blood transfusion.
- Hematological disorders in animals.

4.1 Blood

Blood is a fluid connective tissue that originates from mesoderm composed of corpuscles suspended in a protein-rich fluid called plasma. It is the main transport medium of the circulatory system responsible for transport of oxygen and nutrients to the tissue and removal of metabolic waste products away from the tissue. In addition to the aforesaid functions, blood plays a pivotal role in carrying hormones and drugs to the target organs, maintenance of ionic equilibrium and pH of the body, resistance to infections and thermoregulation. All these functions performed by the blood are directed towards the maintenance of a constant internal environment called homeostasis.

4.1.1 Properties of Blood

4.1.1.1 Color

The red color of blood is imparted by the iron present in the hemoglobin within the erythrocytes. The venous blood, however, looks purplish blue due to deoxygenated hemoglobin. The plasma is yellow to colorless as it contains bilirubin, carotenoids, retinol, and tocopherol. The plasma of cattle and horses is deep yellow in color due to higher bilirubin compared to dog, sheep, and goat.

4.1.1.2 Relative Volumes of Blood Cells and Plasma/Hematocrit

Hematocrit or packed cell volume (PCV) is the ratio of the volume of red blood cells to total volume of blood. Hematocrit or packed cell volume (PCV) is directly related to erythrocyte counts and hemoglobin content and used to evaluate anemia. As PCV is the relative volume of erythrocytes to plasma, it can also be used to evaluate the degree of dehydration in animals. In dehydration, PCV is generally increased. The PCV of different animals are given in Table 4.1.

4.1.1.3 Blood Volume

Generally, blood volume is 8–10% of body weight. Indirectly, blood volume can be calculated through the following formula.

$$BV = PV/(1 - Hct)$$

where BV = blood volume, PV = plasma volume, Hct = hematocrit

Newborn of all species have a high blood volume per unit of body weight except in cattle and sheep in which age or weight have no effect on blood volume up to 2–3 months of age. Swine have the lower absolute values of blood volume due to high body fat. The blood volumes in different species have been presented in Table 4.1.

4.1.1.4 The Specific Gravity and Sedimentation of Blood

The specific gravity of plasma is lower compared to corpuscles. Among the corpuscles, erythrocytes (1.097) have higher specific gravity compared to leukocytes (1.067–1.077). Therefore, erythrocytes settle more quickly compared to leukocytes, and the rate at which erythrocytes settle per unit time is termed as the erythrocyte sedimentation rate (ESR). Erythrocytes have inherent property to adhere to each other like a stack of coins. It is called Rouleaux (singular is rouleau) formation. The unique discoid shape of vertebrate erythrocytes provides a large surface area to contact each other. The rouleaux formation is accelerated with the presence of plasma proteins, particularly fibrinogen and sialic acid. The albumin, on the other hand, hastens the rouleaux

Table 4.1 Hematological values in different animals

Species	Blood volume (mL/kg)	PCV (%)	Specific gravity	ESR	Blood pH	Alkali reserve (%)	References
Horse	76	33.4	1.053	2–12 mm after 10 min 15–38 mm after 20 min	7.20–7.55	64	Reece (2015)
Cow	55 (52–77)	40	1.052	2.4 mm after 7 h	7.38	62	
Sheep	60	32	1.051	5–6 mm after 1 h	7.44	73	
Goat	70	34	1.042				
Pig	65	41.5	1.046	0–6 mm after 30 min 1–14 mm after 60 min			
Dog	86 (79–90)	45.5	1.052	1–6 mm after 30 min 5–25 mm after 60 min	7.36		
Cat	55 (47–66)	40	1.050		7.35		
Rabbit	56 (44–70)	41.4					
Human	–	44.5	1.0621		7.39	58	
Mouse	79 (78–80)						
Tigers (*Panthera tigris tigris*)		36–45		14–26 mm after 60 min			Shrivastav and Singh (2012)
Lions (*Panthera leo*)		26.8–44.1					Maas et al. (2013)
Elephants (*Elephas maximus*)		29.4–40.7					Janyamethakul et al. (2017)

formation. ESR is increased in diabetes mellitus, renal failure, heart disease, collagen vascular diseases, and malignant tumors. Diseases with altered shape of RBC such as microcytic RBC increases ESR as microcytic RBC has lower surface to volume ratio and settles more quickly. In polycythemia (increased RBC), ESR is decreased as too many erythrocytes interfere with the compactness of rouleau. Elevation in leukocyte counts also lowers the ESR.

The specific gravity of blood and ESR in different domestic animals have been presented in Table 4.1.

4.1.1.5 Reactions of Blood

The mean pH values of blood of different species are depicted in Table 4.1.

The arterial blood is more alkaline than venous blood. The plasma is more alkaline than the corpuscles. Buffering systems, particularly hemoglobin and bicarbonate, maintain a stable pH and provide first line of defense against acidemia.

4.1.1.6 Alkali Reserve

It is the capacity of blood to combine with CO_2 and expressed as the volume of CO_2 (mL per 100 mL plasma).

The average alkali reserves of blood in different species are presented in Table 4.1.

4.1.2 Composition of Blood

Blood is composed of cellular elements (45%) suspended in plasma (44%) made of colloids and crystalloids.

4.1.2.1 Plasma

Plasma appears as a light-yellowish or straw-colored fluid composed of water (92%) and solids (8%). The solid portion of the plasma comprises of inorganic and inorganic. Plasma proteins are the most abundant plasma solutes. Besides plasma proteins, a variety of other substances are also present such as dissolved nutrients, electrolytes, respiratory gases, hormones, and vitamins. Plasma proteins and lipids are present as colloidal suspension, whereas glucose, urea, and uric acid are present as crystaloids. Plasma serves as the liquid base for whole blood to carry blood cells, nutrients, and metabolic waste products.

4.1.2.1.1 Plasma Proteins

Majorities of the plasma proteins are synthesized in the liver; however, the principal source of immunoglobulins is lymphocytes or plasma cells. The plasma proteins are mainly involved in the regulation of osmotic pressure, transport of hormones and drugs, immunity (initiation of inflammatory responses and the complement cascade system) and buffering actions. Electrophoresis is the easiest method to determine the relative proportions of plasma protein fractions as the protein fractions are separated on the basis of electrophoretic mobility determined by their molecular weight. The three major fractions of plasma proteins separated by electrophoresis are **albumin**, **globulin**, and **fibrinogen** (Table 4.2).

Table 4.2 Proportion of different plasma proteins

Name		Proportion (%)
Albumin		55.2
Globulin	α1-Globulin	5.3
	α2-Globulin	8.6
	β-Globulin	13.4
	γ-Globulin	11.0
Fibrinogen		13.4

Albumin: It is the plasma protein having the highest electrophoretic mobility with molecular weight of 69,000 Da. It is also the most abundant plasma protein with a concentration of 2.8–4.5 g/dL. Albumin is synthesized in the liver and composed of a single polypeptide chain with 610 amino acids among which lysine, valine, leucine, threonine, phenylalanine, histidine aspartic acid, glutamic acid, and arginine are predominated. Though 60% of albumin is found extravascularly, it has the ability to transport back to the circulation via. the lymphatic system. The main function of albumin is to maintain colloidal osmotic pressure. Thus, in hypoproteinemia or hypoalbuminemia water moves from vasculature and accumulates in extravascular space and leads to edema. Albumin also acts as a transport medium for different biomolecules particularly fatty acids (in the form of apoprotein), cations (Ca^{2+}, Na^{+}, and K^{+}), lipid soluble hormones, bilirubin, and drugs (phenylbutazone, warfarin, etc.). Therefore, albumin is considered as "molecular taxi."

Dehydration is probably the only cause that increases plasma albumin levels, but there were a variety of clinical conditions that caused hypo-albuminemia, either due to decreased synthesis or increased loss. Liver dysfunction, malabsorption syndrome, pregnancy, and malnutrition are the predisposing factors for decreased albumin synthesis, whereas renal impairment and burns are the two main factors for protein loss. Burn causes severe loss of albumin due to the fact that a substantial amount of albumin is stored in the skin.

Globulin: The molecular weight of globulin is about 90,000–156,000 Da, which is mainly synthesized from liver and reticulo-endothelial system. Based on the electrophoretic mobility globulins are classified into alpha, beta, and gamma globulin fractions. The functions of different fractions of globulin are summarized in Table 4.3.

Fibrinogen: It is a homo-dimeric glycoprotein having a molecular weight of 340,000 Da. It consisted of six polypeptide chains (2Aα, 2Bβ, and 2γ) linked by 29 disulfide

Table 4.3 Functions of different fractions of globulin

Name	Functions
Transferrin	Transports iron
Ceruloplasmin	Transports copper
Hemopexin	Transports heme
Haptoglobin	Transports hemoglobin
Plasminogen	Precursors of plasmin (help in clot lysis)
Prothrombin	Precursors of thrombin (help in blood coagulation)
α_2-macroglobulin	Proteases inhibitor
Immunoglobulins (γ-globulins) (IgG, IgA, IgM, IgD, IgD): Produced by plasma cells (B-lymphocytes)	Humoral immunity
Transferrin	Transports iron

bonds. The primary site for fibrinogen synthesis is the hepatocytes. The normal plasma concentration of fibrinogen is 2–5 mg/mL which can be exceed up to 7 mg/mL during acute inflammatory conditions. Fibrinogen is an integral component of blood coagulation machinery, and it is the precursor of insoluble fibrin that forms blood clot.

Acute phase proteins (APPs): These are a group of plasma proteins that originate from the liver or other extrahepatic source (epithelial cells, endothelium, and connective tissue) during acute phase reactions and used as bio-markers for early diagnosis of diseases. The predominant acute phase proteins of the blood are c-reactive proteins (CRP), plasminogen, α1 anti-trypsin, α2 macroglobulin, prothrombin, fibrinogen, ferritin, serum amyloid-A (SAA), haptoglobin (Hp), and ceruloplasmin (Cp).

4.1.2.1.2 Plasma Lipids

The plasma lipids are transported in three main forms namely free fatty acid, triglyceride, and cholesterol ester. The fatty acids are further classified in saturated (palmitic acids, stearic acids) and unsaturated fatty acids. Unsaturated fatty acids are of two types, monounsaturated fatty acids (oleic acid) and poly-unsaturated fatty acids (linoleic acid, linolenic acid, and arachidonic acid). Free fatty acids primarily originate from triglycerides of the adipocytes and transported in combination with plasma albumin. Cholesterol and triglycerides are transported in the form of lipoproteins. Lipoproteins are the complex particles in which cholesterol esters and triglycerides form a central core which is surrounded by phospholipids, cholesterol, and apolipoproteins (protein parts of lipoprotein that bids with lipids like fat and cholesterol). Dietary lipids are broken down into fatty acids and monoacylglycerol which were absorbed by enterocytes. The fatty acid-binding proteins facilitate the transport of fatty acids and monoacylglycerol to the endoplasmic reticulum

Table 4.4 Plasma cholesterol level in different species

Species	Cholesterol (mg/dL)
Dogs	135–278
Cats	71–156
Pigs	36–54
Sheep	52–76
Goats	80–130
Rabbits	10–80

Source: Kaneko et al. (1997)

of the enterocytes where they again conjugated to form triacylglycerol. The esterification of newly formed triacylglycerol together with phospholipids, cholesterol, and apolipoproteins forms the chylomicrons which transport dietary cholesterol and triglycerides to the liver and peripheral tissues.

The concentration of total cholesterol in different species has been presentation in Table 4.4.

Lipoproteins are involved to keep the lipids in solution. They appeared as complex particles characterized by hydrophobic central core (made of cholesterol esters and triglycerides) surrounded by apolipoprotein, phospholipids, and cholesterol-rich hydrophilic membrane.

Plasma lipoproteins can be classified into seven classes based on their density (Table 4.5).

LDL carries cholesterol to cells, which can lead to accumulation of plaque in the vessels and increase the occurrence of atherosclerosis. Thus, they are considered as "bad

Table 4.5 Classification of lipoproteins

Lipoproteins	Density (g/mL)	Functions
Chylomicrons	<0.930	Transportation of dietary lipids (triglycerides and cholesterol) to liver and peripheral tissues
Chylomicron remnants	0.930–1.006	Transportation of triglycerides and cholesterol to liver and peripheral tissues
Very low density lipoproteins (VLDL)	0.930–1.006	Transportation endogenous lipids from liver to cells
Low density lipoproteins (LDL)	1.019–1.063	Carry cholesterol to cells
High density lipoproteins (HDL)	1.063–1.210	Transportation of cholesterol from cells to liver (reverse cholesterol transport)
Lipoprotein (a): or Lp (a)	1.055–1.085 (composition is different from HDL in terms of apolipoproteins)	Not known

cholesterol." In contrast, HDL is involved in cholesterol excretion from body and considered as "good cholesterol."

4.1.2.2 Erythrocytes

Erythrocytes or red blood corpuscles (RBC) of mammals are the non-nucleated, non-motile, biconcave circular disk like structure that originates from bone marrow and involved primarily in transport oxygen from lung to tissue and carbon-dioxide from tissue to lungs. Erythrocytes are accounting nearly 40% of total blood volume and their number varies with species and the physiological state of animals. Erythrocytes have a unique capability to pass through microvasculature having a smaller diameter than the erythrocyte itself. However, this capability is decreased with the progression of age, and the erythrocytes become fragile. The total erythrocyte counts (TEC) in different species have been presented in Table 4.6.

Table 4.6 The total erythrocyte counts (TEC) in different species

Species	No. of RBCs ($\times 10^6/\mu L$)	References
Cat	6–8	Reece (2015)
Cattle	6–8	
Chicken	2.5–3.2	
Dog	6–8	
Horse (light)	9–12	
Horse (draft)	7–10	
Goat	13–14	
Pig	6–8	
Sheep	10–13	
Man	5–6	
Woman	4–5	
Tiger (*Panthera tigris tigris*)	4.66–9.15	Shrivastav and Singh (2012)
Lions (*Panthera leo*)	5.1–8.3	Maas et al. (2013)
Elephants (*Elephas maximus*)	1.9–3.2	Janyamethakul et al. (2017)

4.1.2.2.1 Size and Shape of Erythrocytes

In mammals, erythrocytes look like a biconcave circular disk varying in diameter and thickness depending on species, physiological stage, and nutritional status of the animals. This discoid shape of erythrocytes not only provides larger surface area to volume ratio with greater surface area for gaseous exchange but also allows minimal diffusion distance for easy exchange of blood gases. The discoid shape also helps to withstand greater osmotic swelling without affecting the integrity.

In mammals, there is marked variability in respect to erythrocyte shapes. The erythrocytes of dogs are markedly biconcave, the erythrocytes of cats and horses are slightly concave, cattle and pig erythrocytes are discoid, and erythrocytes of a camel are oval. The erythrocytes of deer and antelope are sickle shaped (also called drepanocytes). The avian erythrocytes are elliptical in shape with nuclei.

4.1.2.2.2 Structure of Erythrocytes

Erythrocytes have a diameter of 7–8 μm with a thickness of around 2 μm at the periphery and 1 μm at the center. The size of RBC in different animals is given in Table 4.7.

The erythrocytes are non-nucleated and devoid of major cell organelle. The erythrocytes are unable to synthesize proteins, the functions of proteins present in the erythrocytes during their maturation from reticulocytes and carried till their senescence. The composition of erythrocytes is given below:

- Water (61%)
- Solids (32%)
  - Hemoglobin (95%)
  - Peroxiredoxins
  - Proteins (Mostly present in cell membrane and cytoskeleton)
    Cytoskeleton proteins
    Alpha and beta spectrin
    Actin
    Ankyrin
    Membrane proteins
    Band 3 or the anion exchange protein
    Glycophorins
  - Lipids (0.4%)
    Lecithin
    Cephalin
    Sphingomyelin
    Cholesterol
    Cholesterol esters

Table 4.7 The size of RBCs in different species

	Diameter (μm)		Circumference (μm)		Surface area (μm^2)	
Species	Mean	Range	Mean	Range	Mean	Range
Bovines	5.07	4.66–5.50	18.98	17.28–20.25	21.51	18.57–26.50
Ovines	4.42	4.10–4.62	16.91	15.73–19.72	16.44	14.37–19.03
Caprines	3.39	3.09–3.60	12.94	11.60–14.65	9.50	7.61–11.28
Equines	5.66	4.71–6.07	22.46	19.01–26.99	27.70	19.38–34.32
Canines	7.02	6.30–7.71	25.67	20.63–28.31	38.67	25.55–47.05

Source: Adili et al. (2016)

Neutral fat

- Vitamins
- Carbohydrate (7%) (Glucose for energy)
- Enzymes
  - Cholinesterase
  - Phosphatases
  - Carbonic anhydrase, peptidases, and enzymes of glycolysis
- Minerals
  - Phosphorus
  - Sulfur
  - Chlorine
  - Sodium
  - Potassium

On the basis of intracellular potassium content, the RBC can be classified as high potassium (HK) and low potassium (LK) RBC. In high potassium RBCs (horses, pigs), the sodium pump is highly active which helps in the exchange of intracellular sodium for extracellular potassium. In contrast, the RBCs of sheep, goats, cattle, and dogs are having low intracellular potassium as the sodium pumps are not fully functional.

4.1.2.2.3 Erythrocyte Membrane

The semipermeable plasma membrane of erythrocytes is composed of a lipid bilayer and a mesh-like cytoskeleton which supports the plasma membrane. The composition of isolated erythrocyte membrane is as follows:

- Water (20%)
- Protein (40%)
- Lipid (35%)
- Carbohydrate (6%)

4.1.2.2.3.1 Membrane Lipids

The membrane lipids comprise of cholesterol and phospholipids. The phospholipids are arranged asymmetrically in the membrane to maintain the fluidity. The phospholipids of RBC membrane consist of a polar head group facing the exterior followed by a non-polar hydrophobic tail which aggregate in the lipid bilayer provides a hydrophobic barrier towards cytosol. The lipid composition of the erythrocyte membrane varies with species. Sheep erythrocyte membrane is devoid of phosphatidylcholine, camellias have very low phosphatidylcholine in their erythrocyte membrane, whereas rats have more phosphatidylcholine.

4.1.2.2.3.2 Membrane Proteins

The erythrocyte membrane proteins comprise the internal hydrophilic peripheral proteins, middle hydrophobic integral proteins, and external hydrophilic proteins. The main integral proteins of plasma membrane are band 3 (anion exchanger or anion channel) and glycophorins.

Band 3: It is a 93 kDa homo-dimeric protein and the most abundant proteins of erythrocyte membrane. The protein comprises three domains namely N-terminal cytoplasmic domain, hydrophobic transmembrane domain, and membrane-bound C-terminal domain. The main functions of band 3 proteins are transporting of chloride and bicarbonate ions across the membrane, membrane cross-linking (band 3 binds with band 4.2 proteins and acts as a bridge between lipid bilayer and cytoskeleton), maintaining membrane stability and associated with aging process of erythrocytes.

Glycophorins: It is a sialoglycoprotein consists of 60% carbohydrates (mainly sialic acid) with two domains. N-terminal domains are situated towards outer surface, and they act as blood group antigens (ABO and MN) whereas, C-terminal domain faces the cytoplasm and interact with the cytoskeleton. Due to high sialic acid content, glycophorins are the major contributors to the surface negativity of RBC membrane. There are several types of glycophorins characterized in human erythrocytes out of which the major four glycophorins are glycophorin A (85%), glycophorin B (10%), glycophorin C (4%), and glycophorin E (1%). Beside these, there are several other minor glycophorins such as band 4.5 protein (GLUT1), acts as a glucose transporter, Na-H exchanger (NHE1), associated with actin-binding proteins, CD44 and CD47, helps in cell-cell crosstalk. The blood group antigens are also present in the erythrocyte membrane.

The type of membrane proteins and their compositions varies with the species. In mouse, guinea pigs and cows, the membrane protein band 3 is having a higher molecular weight compared to the RBC of sheep RBCs. Band 4.1 contains two sub-bands in the erythrocytes of cow, rabbit, guinea pig, and sheep whereas there is overlapping in 4.2 and 4.1 in the erythrocytes of horse.

4.1.2.2.4 Erythrocyte Cytoskeleton

The cytoskeleton provides the structural integrity, deformability and maintains the typical biconcave shape of RBC. In the cytoskeleton, transmembrane proteins are associated with peripheral membrane proteins to form a meshwork of protein that stabilizes the membrane. The major proteins involved in the erythrocyte cytoskeleton are spectrin, ankyrin, actin, and Band 4.1R.

Spectrin: It is the principal cytoskeletal protein. It is composed of two subunits namely α-subunit (280 kDa) and β-subunit (246 kDa) oriented in opposite directions to form a tetramer ($\alpha 2\beta 2$ tetramers). It forms a complex intracellular network with actin and band 4.1R protein which then interacts with the cytoplasmic domain of glycophorin. Around 6 spectrins bind to actin and lead to a pseudohexagonal arrangement.

Ankyrin: As the name implies, it is the cytoskeletal anchoring protein. The molecular mass of ankyrin is 206 kDa. It has three domains namely N-terminal domain that binds with band 3 protein, central domain to bind with spectrin, and a C-terminal regulatory domain.

Actin: The non-muscle β-actin of cytoskeletal protein is a short helical filamentous protein of 43 kDa. It interacts with adducing that facilitates the binding of actin with spectrin. Each actin binds to six spectrin tetramer ends to pseudohexagonal lattice structure.

Band 4.1R: It is a globular protein with two domains. 30 kDa N-terminal domain of this protein binds with Band 3 and glycophorin. Another domain of 10 kDa facilitates the binding with spectrin and actin at the junctional complex.

Adducin: It is a calcium/calmodulin-binding protein exists as αβ dimer that enables capping of actin for spectrin–actin interactions at the junctional complex.

The structure of erythrocyte cytoskeleton is made of vertical and horizontal interactions (Fig. 4.1). In horizontal interactions, spectrin dimers join head-to-head and produce heterotetramer. The tails of heterotetramers bind with F-actin, protein 4.1, and actin-binding proteins like adducin, tropomyosin, and tropomodulin to produce junctional complex. This network tethered with cell membrane at two sites as vertical interactions. The vertical interaction facilitates by two complexes namely ankyrin complex and junctional complex. In ankyrin complex, the N-terminal cytoplasmic domain of band 3 protein binds with ankyrin which in turn binds with spectrin and links the cytoskeleton and the membrane. At the junction complex, the spectrin network binds with to Glycophorin C of plasma membrane. These interactions help to maintain the biconcavity of erythrocytes along with its deformability.

4.1.2.2.5 Erythrocyte Membrane Transport

The membrane of erythrocytes is impermeable to many biological molecules due to its lipid bilayer. But erythrocyte membrane proteins allow the efflux and influx of the molecules across the membrane. The functions of different membrane transporters are summarized below (Table 4.8).

4.1.2.2.6 Metabolism of RBC

The mature erythrocytes are devoid of nucleus, thus unable to synthesize nucleic acids or proteins. Absence of other cytoplasmic organelles makes the metabolic activities of erythrocytes very limited. Lack of mitochondria in mature RBC results absence of Krebs' cycle and oxidative photophosphorylation including the synthesis of lipids and heme. Fortunately, the main function of erythrocytes, i.e., the transport of oxygen and carbon-dioxide doesn't require energy. But the energy in the form of ATP is required to maintain the shape

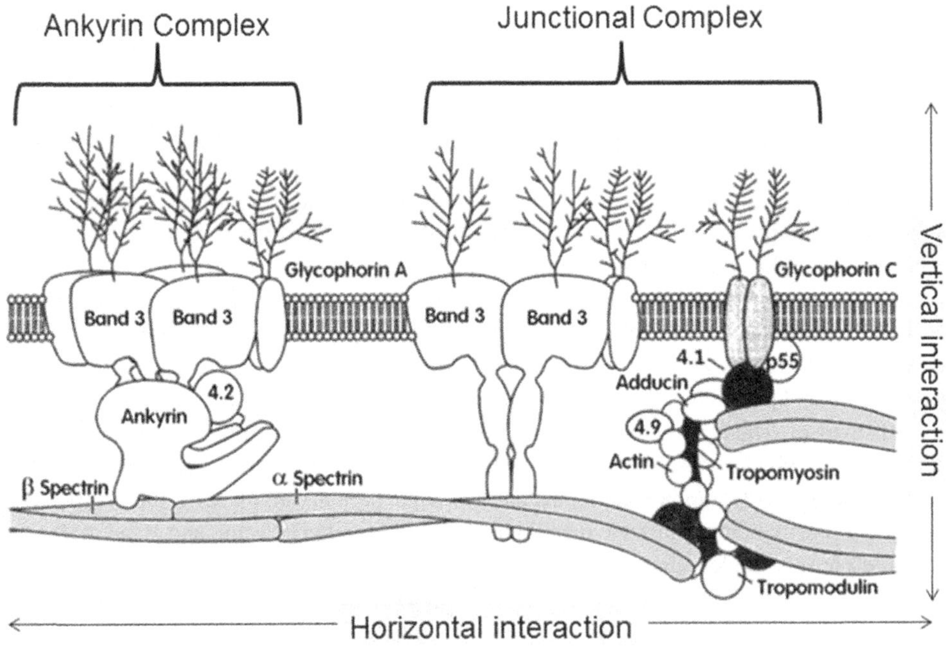

Fig. 4.1 Structure of erythrocyte cytoskeleton: The vertical and horizontal interactions erythrocyte cytoskeleton. Spectrin dimers join head to head to produce heterotetramer in horizontal interactions. The vertical interaction consists of two complexes viz. ankyrin complex and junctional complex. In ankyrin complex, N-terminal cytoplasmic domain of band 3 protein binds with ankyrin. The cytoskeleton links with the membrane through spectrin. At the junction complex, the spectrin network binds with to Glycophorin C of plasma membrane. Erythrocyte biconcavity is maintained through these interactions. (Source: Ivanov and Paarvanova 2021)

Table 4.8 Functions of different membrane transporters of erythrocytes

Membrane transporters	Functions
Aquaporins	Transports water and carbon-dioxide
Band 3	Allows the movement of anions (HCo_3^-, Cl^-), non-electrolytes
Na^+-K^+-ATPase pump	• Efflux of K^+ in exchange of Na^+ – Horse, pig: RBC is having active Na^+-K^+-ATPase pump (high potassium RBC) – Sheep, goat, cattle: Low Na^+-K^+-ATPase pump activity (low potassium RBC) – Cats, most of the dogs: Absence of Na^+-K^+-ATPase pump (Na^+ and K^+ concentration in equilibrium with plasma) – Japanese and Korean dogs: High potassium RBC with glutamate transport results high reduced glutathione (GSH) concentrations. RBCs of these dogs promote *Babesia gibsoni* replication compared to low potassium normal GSH RBCs
Ca^{2+} activated Mg^{2+}-dependent ATPase pump (activated by calmodulin)	• Efflux of calcium (high intracellular calcium causes suicidal death of RBC)
Amino acid transporters	• Efflux of amino acids during RBC maturation • Influx of amino acids for glutathione synthesis
Band 4.5 protein (GLUT1)	• Transports glucose (entry of glucose in RBC is insulin independent) – Human RBC: High glucose permeability – Cattle, Sheep, Goat: Intermediate glucose permeability – Pig: Poor glucose permeability (adult RBC lack glucose transporter)
Nucleoside transporter	• Transports adenosine and inosine – Rabbit, pig, human: More adenosine uptake – Cats, goats, and cattle: Lower adenosine uptake – Dogs: Permeable to adenosine but impermeable to inosine

and deformability, active membrane transport, protein phosphorylation, synthesis of glutathione and purin and pyramidine nucleotides. Glucose is the major metabolic fuel of RBC except in pigs which is utilized solely by anaerobic glycolysis in the Embden-Meyerhof-Parnas pathway (EMP). Additionally, two minor shunts for carbohydrate metabolism namely the pentose phosphate pathway and Luebering-Rapoport bypass provide the protection against oxidative injury and regulate the oxygen carrying capacity of RBC. In pigs, inosine produced from the liver is the major substrate for energy production. The enzymatic machineries required for the metabolic processes are gained from nucleated erythrocyte precursors. These enzymes are limited and persist till their lifetime as erythrocytes are unable to synthesize new enzymes.

4.1.2.2.6.1 Embden-Meyerhof Pathway (EMP)

It is the only source of energy production in RBCs as mature RBC lacks mitochondria. In this anaerobic glycolytic pathway, one molecule of glucose converts into two molecules of lactate with the production of two molecules of ATP. Besides ATP production, Embden-Meyerhof pathway keeps pyridine nucleotides in a reduced state.

Deficiency of pyruvate kinase (PK) interferes with the survival of erythrocytes. Energy (ATP) deficient erythrocytes can be converted to echinocytes. PK-deficient dogs and cats suffer from hemolytic anemia due to intravascular hemolysis and leads to hemoglobinuria. They also have marked iron accumulation in the liver and may die from liver impairment. Phosphofructokinase (PFK) deficiency in dogs results in decreased 2,3 DPG concentration of RBC, which results in higher intracellular pH leads to alkalemia.

4.1.2.2.6.2 Luebering–Rapoport Pathway

In this pathway, mature RBC produces 2,3-diphosphoglycerate (2,3-DPG) which helps to release oxygen from the hemoglobin to make it available for tissue utilization (Fig. 4.2). Erythrocytes of man, dogs, pigs, and horses have high concentration of 2,3DPG in contrast to cats and ruminants those are having low concentrations of

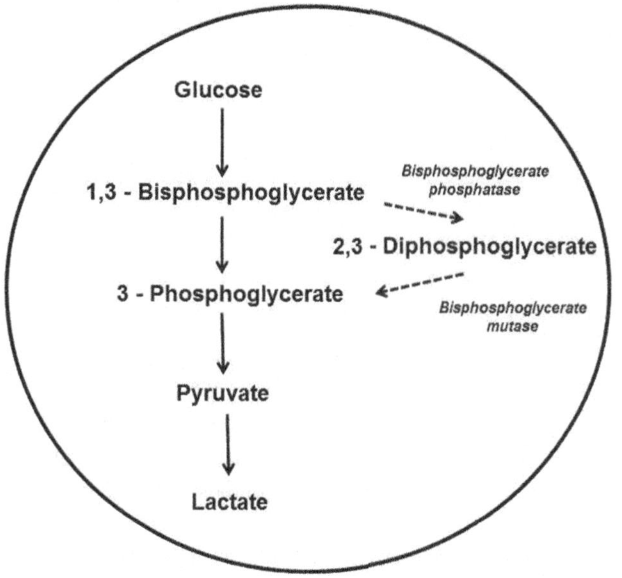

Fig. 4.2 Luebering–Rapoport pathway in the erythrocytes: In this pathway, 2,3-diphosphoglycerate (2,3-DPG) is produced by mature RBC with the help of bisphoglycerate phosphatase and bisphosphoglycerate mutase. 2,3-DPG helps to release oxygen from the hemoglobin

2,3-DPG. The formation of 2,3-DPG is stimulated by hypoxia, increased pH (metabolic alkalosis), and anemia. In all the cases, the generation of 2,3-DPG is aimed to release more oxygen.

4.1.2.2.6.3 Hexose Monophosphate Shunt/Phosphogluconate Pathway/Pentose Phosphate Shunt

In hexose monophosphate shunt, a small amount of glucose (5–13%) is utilized by oxidative metabolism not to produce energy but to generate reduced nicotinamide adenine dinucleotide phosphate (NADPH) that converts oxidized glutathione to reduced glutathione. Glutathione is the major antioxidant of RBC which protects the cell from oxidative damage.

4.1.2.2.6.4 Oxidative Injury of Erythrocytes and Its Protective Mechanisms

The molecular oxygen can be reduced to form highly reactive oxygen species (ROS) such as, hydrogen peroxide (H_2O_2), hydroxyl radical, superoxide anion, hypochlorous acid (HOCl), and nitric oxide (NO). Low concentrations of these free radicals are required for phagocytosis, signal transductions, and the biosynthesis of prostaglandins but, higher concentrations of free radicals can be detrimental to the cells in terms of DNA damage, inactivation of enzymes, oxidation of hormones, lipids peroxidation, and membrane disturbance. Glutathione peroxidase system (GPx), superoxide dismutase (SOD), and catalase are the predominant antioxidant machinery functioning in the erythrocytes (Fig. 4.3). Peroxiredoxin 2 (Prx2) is the post potent H_2O_2 neutralizer in the erythrocytes. Selenium, ascorbic acid, and vit E also help to protect erythrocytes from oxidative damage.

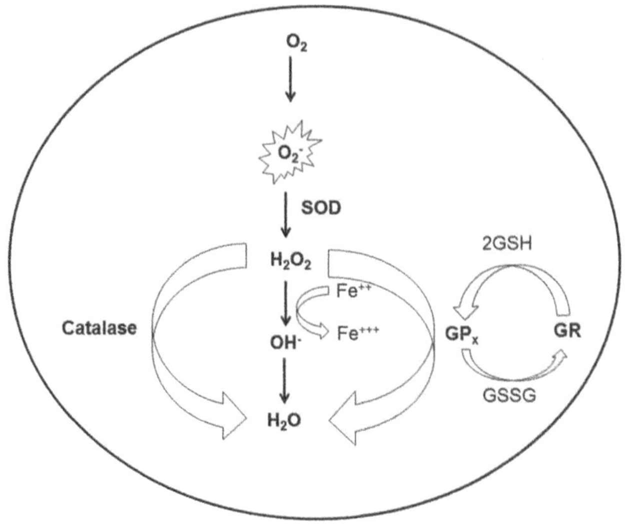

Fig. 4.3 Antioxidant machinery functioning in the erythrocytes: The reactive oxygen species are produced under oxidative stress. SOD causes dismutation of superoxide radicals to hydrogen peroxide. Catalase facilitates the breakdown of hydrogen peroxide to water and oxygen. In GPx system, two reduced glutathione (GSH) molecules accept hydrogens from hydrogen peroxide and resulting two H_2O and one glutathione disulfide (GSSG). Glutathione reductase (GR) regenerates GSH from GSSG

> **Know More......**
> Catalase deficiency disorder is called Takahara disease in human and canine catalase deficiency (CAT) in dogs. CAT is an autosomal recessive condition caused due to mutations in erythrocyte catalase gene identified in Beagle and American Fox hound breeds.

4.1.2.2.7 Destruction of Erythrocytes

The erythrocyte population in the circulating blood must remain within a limit to ensure tissue oxygenation and the delicate balance between erythropoiesis and erythrolysis needs to be maintained by body's homeostatic mechanisms. Erythrocytes in the circulation continuously suffer from oxidative injuries with a well-organized antioxidant defense system to cope up with these oxidative injuries (discussed earlier). When the RBCs are getting older, these antioxidant defense mechanisms are unable to support the erythrocytes to fight against oxidative damage and lead to membrane defects, protein denaturation, altered ion permeability, and other potential hazards against the survival of erythrocytes. The senescence signals of erythrocytes are identified by the macrophages and they phagocytose the aged RBCs by a process known as erythrophagocytosis. However, all the erythrocytes do not follow age-related destruction; some defective erythrocytes undergo premature apoptosis through a process called eryptosis. Neocytolysis is another mechanism of selective lysis of young erythrocytes during physiological polycythemia and mass destruction is required to maintain the hematocrit and blood viscosity.

4.1.2.2.7.1 Lifespan of Erythrocytes

The lifespan of erythrocytes can be measured by labeling the erythrocytes with radioactive and non-radioactive probes which should be non-toxic to the cell, non-immunogenic, should not elute before RBC destruction and can be recycled. The radioisotopes used to measure erythrocyte lifespan are ^{51}Cr and ^{14}C-cyanate, ^{15}N-glycine, ^{59}Fe and ^{55}Fe. Several other non-radioactive probes are also available which can be detectable by flow cytometry such as biotinylation of erythrocytes is now widely used which can be detected by fluorescent probes conjugated with avidin.

The erythrocyte lifespan in different species have been presented in Table 4.9. Several studies indicated that the lifespan of erythrocyte is positively correlated with the longevity of the species. The short living animals are having

Table 4.9 Lifespan of erythrocytes in different species

Species	Lifespan (days)
Human	90–140
Lab animal (rat, rabbit)	45–50
Ruminants (cattle, sheep, goats)	125–150
Horse	140–150
Dogs	100–130
Cats	70–80
Pigs	60–70

Source: Reece (2015)

higher metabolic rates compared to long lived animals and their erythrocytes are more prone to oxidative damage.

4.1.2.2.7.2 Mechanisms of Erythrocyte Clearance by the Macrophages

"Graveyards" of erythrocytes; Sites of erythrocyte destruction: In most of the domestic animals, red bone marrow is the principal site of RBC destruction whereas, in human; spleen is the main site for erythrophagocytosis. In contrast, most of the erythrocytes in birds are destroyed from liver. The preferential phagocytes for the erythrocytes are the macrophages which form phagolysosomes after engulfment. The resident macrophages of spleen are the most potent phagocytes for erythrocytes and they have well developed machineries in terms of CD163 (hemoglobin scavenger receptor) and the enzyme heme-oxygenase 1 to tackle the hemoglobin after erythrophagocytosis and recycled most of the iron in the hemoglobin to maintain iron homeostasis. With the progression of age, a variety of bio-physical, biochemical and molecular alterations occur in the erythrocytes that trigger erythrophagocytosis called engulfment signals or senescence markers as follows:

Membrane microvesiculation: It is one of the important determinants of erythophagocytosis. Microvesiculation of the aged erythrocytes lead to loss of hemoglobin and almost 50% of the erythrocyte integral membrane proteins like band 3 and glycophorin and results in decreased membrane flexibility and cellular deformability which ultimately triggers the erythrocytes removal.

Alterations in band 3: The oxidative injury to the erythrocyte causes breakdown of band 3 membrane proteins. With the breakdown of band 3, oxidative injury develops senescence neo antigens that binds with autologous IgGs together with C3 complement fraction that triggers erythrophagocytosis. This mechanism is called Band 3 mediated clearance pathway of erythrocytes.

Externalization of phosphatidylserine (PS): PS in the erythrocyte membrane has pro-phagocytic activity and can be labeled as "eat-me" signals. In young and newly formed erythrocytes, PS is situated at the inner layer of the membrane thus undetected by the macrophages for destructions. But, at the time of senescence, due to the loss of membrane architecture, PS exposes to outer membrane and can easily be detected by macrophages to initiate erythrophagocytosis termed as PS mediated clearance pathway.

Decreased expression of CD47: CD47 has anti-phagocytic activity that protects erythrocytes from phagocytosis by the macrophages. It is also labeled as "don't eat-me" signals of erythrophagocytosis. During senescence, the expression of CD47 gradually declines, making the erythrocytes more susceptible to phagocytosis by the macrophages.

Decreased expression of CD147: CD147 helps in the recruitment of erythrocytes from spleen to blood. During the time of senescence, expression of CD147 decreases which causes entrapment of erythrocytes in the spleen and facilitates their destruction by the macrophages.

Altered calcium homeostasis: Influx of calcium is highly correlated with the oxidative damage, microvesiculation, membrane deformability and apoptosis like events in erythrocytes. Thus, it is assumed that alteration in the calcium homeostasis is one of the senescence markers that trigger erythrophagocytosis.

4.1.2.2.7.3 Extravascular and Intravascular Hemolysis

Hemolysis can be broadly classified into extravascular and intravascular hemolysis based on their location. Extravascular hemolysis is a routine process of erythrocyte clearance after aging by the cells of reticulo-endothelial systems outside the blood vessel. In contrary, the lysis of erythrocytes in vivo, i.e., within the blood vascular system is termed as intravascular hemolysis. The differences between these two types of hemolytic mechanisms are summarized in Table 4.10.

4.1.2.3 Leukocytes

Leukocytes or white blood cells (WBCs) are less in circulation, when compared to red blood cells (RBC). Unlike RBSc, WBCs are colorless and thus called as leukocytes. The leukocytes can be classified into two types based on their microscopic characteristics with stains namely, granulocytes and agranulocytes. Granulocytes are characterized by their granular cytoplasm. They are of three types: eosinophils, stains with acidic stain like eosin due to the presence of basic proteins in their cytoplasm. Basophils are having acidic cytoplasmic granules and hence stain with basic dye like methylene blue. Neutrophils have both acidic and basic granules, hence, can be stained with both acidic and basic stain and look lavender color. The agranulocytes are so named as they have no granules in their cytoplasm. They are further classified monocytes and lymphocytes. The total number of leukocytes and their relative proportions in the blood has been presented in Table 4.11.

Table 4.10 Difference between extravascular and intravascular hemolysis

Parameters	Intravascular hemolysis	Extravascular hemolysis
Site of occurrence	Within the blood vascular system (in vivo)	Out site the blood vascular system (ex vivo) and within the reticulo-endothelial systems
Fate of hemoglobin	Released in the blood and binds with haptoglobin	Converted to bilirubin by the enzymatic machinery within the macrophages
Alterations in liver, spleen, and kidney	Liver and spleen remain normal but hemosiderin and iron deposition in the kidney may lead to acute kidney failure	Liver and spleen may be enlarged but kidney remains normal
Hemoglobinemia	Present	Absent
Haptoglobin and hemopexin	Decreased	Normal
Urine	Color of urine become brown due to hemoglobinuria and hemosiderinuria	Color becomes yellow due to presence of urobilinogen and urobilin

Table 4.11 Differential leukocyte counts in different species

Species	Total leukocyte counts ($\times 10^3/\mu L$)	Differential leukocyte counts (%)					References
		Neutrophils	Eosinophils	Basophils	Monocytes	Lymphocytes	
Cattle	7–10	25–30	2–5	<1	5	60–65	Reece (2015)
Goat	8–12	35–40	2–5	<1	5	50–55	
Sheep	7–10	15–30	2–5	<1	5	60–65	
Horse	8–11	50–60	2–5	<1	5–6	30–40	
Pig	15–22	30–35	2–5	<1	5–6	55–60	
Dog	9–13	65–70	2–5	<1	5	20–25	
Cat	10–15	55–60	2–5	<1	5	30–35	
Chicken	20–30	25–30	3–8	1–4	10	55–60	
Tiger (*Panthera tigris tigris*)	6.2–11.05	57–75	2–6	0–4	2–6	18–35	Shrivastav and Singh (2012)
Lions (*Panthera leo*)	7.2–25.6						Maas et al. (2013)
Elephants (*Elephas maximus*)	7.9–21.9						Janyamethakul et al. (2017)

4.1.2.3.1 Neutrophils

The neutrophils are the first line of defense against infection, facilitate the inflammatory responses and bacterial killing. They are produced from myeloid-lineage cells and contain granules with enzymes and lethal chemicals for killing pathogens. The neutrophils exist either in the circulation (circulating pools) or adhered to endothelial cells (marginal pools). In most of the animals, the neutrophil concentration in circulating and marginal pool are equal except in cats as they have threefold more neutrophils in the marginal pool compared to circulating pool.

4.1.2.3.1.1 Morphology

The mature neutrophils are motile in nature and are 10–12 μm in diameter with numerous pseudopodia in their surface. They have a segmented nucleus containing 3–5 lobes joined together by heterochromatin strands. Hence, they are called polymorphonuclear neutrophils (PMN). In females, an extra appendage in the nucleus is present occasionally, called Barr body. The immature forms of neutrophils are called "band cells or stab cells," characterized by non-lobular curved nucleus, constituting 3–5% of total WBCs in circulation. The cytoplasm of the neutrophils contains organelles like Golgi apparatus, endoplasmic reticulum, mitochondria, lysosomes, glycogen particles, and numerous granules. The granules contain an intragranular matrix surrounded by a phospholipid bilayer membrane. The neutrophils release their granular content either through exocytosis or in fusion with the lysosomes. There are four types of granules present in the mature human neutrophils.

Primary/azurophilic granules: The primary granules appear first during the promyelocyte stage. Due to their affinity for the basic dye azure A, they are also called azurophilic granules. They are the largest granules and oval or round in appearance and functionally similar to lysosomes of other cells. The primary granular contents are myeloperoxidase (MPO) catalyzes the formation of hypochlorous acid from chloride and hydrogen peroxide (H_2O_2). Other substances present in the azurophilic granules are proteases and enzymes like lysozyme, elastase, cathepsins, and acid phosphatase. The main function of primary granular contents is antimicrobial destruction.

Secondary/specific granules: These granules are unique to neutrophils, hence called specific granules. They are appeared during metamyelocyte stage. The secondary granules contain matrix metalloproteinases (MMPs) (MMP 3, 8, 9), gelatinase, collagenase, lactoferrin, alkaline phosphatase, histaminase, plasminogen activator, and β2 microglobulin. The secondary granules help in microbial destruction and neutrophil migration.

Tertiary/gelatinase granules: As the name implies, the gelatinase granules contain high concentration of gelatinase capable of tissue destruction. They usually appear during immature form. The other contents are lysozyme, acetyltransferase, acid phosphatase, and cytochrome b_{558}. The gelatinase granules help in neutrophil migration and extravasation. In most of the domestic (cattle, sheep, goats, dogs, cats, and horses) and laboratory animals (rabbits, rats, guinea pigs), neutrophils are having a third granules which are different from tertiary granules. In cattle, these third granules are larger compared to sheep and goats hence called "dense/large granules."

Secretary vesicles: They are the smallest neutrophil granules contain phosphatase, cytochrome b558, and plasma proteins. They act as the reservoir of neutrophil membrane proteins.

4.1.2.3.1.2 Functions of Neutrophils

Inflammatory Response

The first and most abundant leukocytes recruited at the site of inflammation are the neutrophils. This recruitment of leukocytes at the site of inflamed area is facilitated by the series of steps. The vasodilation and fluid exudation at the inflammatory site leads to movement of neutrophils from central to periphery of the blood vessels called *margination*. When the neutrophil reaches at the periphery, there is an interaction of neutrophil with vascular endothelial cells called *adhesion* through a group of cell adhesion molecules such as selectins and integrins binds with their specific legends (sialylated carbohydrate). The adhesion of neutrophils with the vascular endothelium and shear stress of passing erythrocytes force the neutrophils to move along the surfaces of vascular bed through a process called *rolling*. The emigration of neutrophils through the capillary into the tissue is called *diapedesis*. Diapedesis is facilitated by retraction of neutrophils with another high affinity adhesion molecules called platelet–endothelial cell adhesion molecule-1 (PECAM-1) expressed at the lateral surface of endothelial cells. The leukocytes are migrated to the injured tissues in response to some chemo-attractants (bacterial toxins, complements, and chemokines) by the process called *chemotaxis*.

Phagocytosis

Neutrophils are able to ingest pathogens and other particulate material through a process called receptor-mediated phagocytosis, and there is formation of phagosome and phagolysosome. After engulfment, the degranulation of neutrophils occurs and the particles are digested by oxygen-dependent and oxygen-independent mechanisms. The secretary vesicles release their content first followed by tertiary, secondary, and primary granules.

> **Know More.....**
> *Respiratory burst* is the process of oxygen-dependent killing of pathogens through the generation of toxic oxygen metabolites such as NO, O_2^-, and H_2O_2. It is mediated by NADPH oxidase system which in conjugation with cytochrome b_{558} transfers electrons from NADPH to generate O_2^- and followed by a spontaneous dismutase reaction which produces H_2O_2. The segregation of cytochrome b_{558} from NADPH oxidase facilitates the auto destruction of neutrophils from reactive oxygens.

Non-phagocytic Killing of Pathogens

There are two non-phagocytic strategies by which neutrophils kill the pathogens and augment host immunity.

Neutrophil extracellular traps (NETs): This is extracellular meshes composed of chromatin and neutrophil granular proteins that entraps the pathogens and immobilize them. The process is called NETosis. Beside pathogen entrapment, NET serves a variety of immune regulatory functions like promotion of inflammation and stimulation of interferon responses and potentiation of autoimmunity. NETs can capture metastatic tumors and delay diabetic wound healing. NETs sometimes occlude vasculature and lead to thrombosis and may obstruct the circulation of important organs. NETs are suicidal to neutrophils and cause their destruction.

Neutrophil-derived microparticles (NDMPs): These are spherical microvesicles of 50–1000 nm diameter containing mRNA, microRNA, cell adhesion molecules (CD11a and L-selectin), and inflammatory proteins surrounded by lipid bilayer. They are able to bind with the target cells and transfer messenger RNA (mRNA) and microRNA (miRNA) to induce cellular response. NDMPs promote pro-inflammatory effect and their numbers increase during sepsis and infection-mediated thrombosis.

Production of Pro-inflammatory Mediators

Neutrophils secret some inflammatory mediators like leukotrienes B4 (LTB4) which mediates chemotaxis of other neutrophils to the site of infection and cytokines like IL-8, IL-12, monocyte chemotactic protein (MCP)-1, and TGF-β.

Platelet Activation and Thrombosis

Neutrophils help in activation of platelets and thrombin. The interaction of platelets with the neutrophils promotes expression of tissue factor which initiate the coagulation mechanism.

4.1.2.3.2 Eosinophils

They are so named because of their affinity for anionic dyes like eosin. Eosinophils are differentiated in the and mature in bone marrow within 2–6 days.

4.1.2.3.2.1 Morphology

Eosinophils are also having polymorphonuclear nucleus but less lobulated than neutrophils. The cytoplasm of eosinophils appears pale blue in color and contains Golgi bodies, free ribosomes, mitochondria, endoplasmic reticulum, glycogen, and numerous granules. There are three types of granules in the eosinophil namely the specific granules, primary granules, and dense granules. Specific granules are the highest in number and contains cytotoxic proteins like major basic protein (MBP), eosinophil peroxidase (EPO), eosinophil cationic protein (ECP), and eosinophil-derived neurotoxin (EDN, protein X, EPX). Majority of these proteins are used to destroy parasites, protozoa, and bacteria. Eosinophil peroxidase (EPO) generates reactive oxygen to destroy pathogens. Primary granules are membrane bound and contains lysophospholipase.

4.1.2.3.2.2 Functions of Eosinophils

Parasitic defense: Eosinophils are having phagocytic capabilities. It works in conjugation with T cells. The parasitic larvae need to be opsonized first with IgG and IgE or complements; then T cell-derived perforins act over it and break the integument. Then the major basic proteins are secreted from specific granules to kill the parasitic larvae.

Allergic reactions: Eosinophils help in mast cell degranulation mediated by IgE. But the granular contents of eosinophils cause air way damage and bronchoconstriction that results in asthma.

Termination of inflammation: Eosinophils release a number of bio-active substances that help in the termination of inflammation. Histaminase released from eosinophils deactivates histamine, MBP neutralizes heparin, hypochlorous acid inactivates prostaglandins, and lysophospholipase blocks the synthesis of arachidonic acid metabolites.

Phagocytosis: Though eosinophils have lower phagocytic capability when compared to neutrophils, they are able to phagocytose mast cell granules, immune complexes, bacteria, and yeast.

Anti-cancer activity: Eosinophils are reported to have cytotoxic activity against tumor cells by promoting apoptosis.

4.1.2.3.3 Basophils and Mast Cells

Basophils are the smallest and least abundant granulocytes in the circulation constitute around 0.5% of total leukocytes under normal condition. They matured in the bone marrow within 2.5 days. Basophils are reared in the extravascular tissues and have very short half-life of 6 h in circulation. In contrast, mast cells are seen in of blood and lymphatic vessels, peripheral nerves, respiratory and gastrointestinal systems, skin and fibrous tissues.

4.1.2.3.3.1 Morphology

Like other granulocytes, basophils are polymorphonuclear cells containing granular cytoplasm. But the nucleus of mast cells appears round to oval with clumped and densely stained chromatin. The size of basophils (10–15 μm) is less compared to mast cells (10–40 μm). The cytoplasm of basophils contains Golgi apparatus, mitochondria, ribosomes, endoplasmic reticulum, and glycogen deposits. The cytoplasm of mast cells in addition contains little glycogen deposits when compared to basophils. The cytoplasmic granules of basophils are larger but fewer than mast cells.

4.1.2.3.3.2 Functions

Basophils

Allergic response: Basophils are involved in IgE-mediated immediate hypersensitivity reaction and manifest the allergic responses like asthma, allergic rhinitis, urticarial, conjunctivitis, and allergy due to insect bite.

Stimulation of T cells: Basophils stimulate T cells for Th2 response particularly during helminth infestations.

Recruitment of inflammatory cells: Basophils are reported to recruit inflammatory cells (neutrophils and eosinophils) during chronic allergy mediated by IgE.

Lipolysis: Heparin secreted from basophils activates lipoprotein lipase and promotes lipolysis.

Mast Cells

Mast cells were identified by Ehrlich in 1878 and initially named as "Mastzellen" meaning well-fed cells due to their highly granular cytoplasm. They are mesenchymal cells phenotypically similar with the basophils and derived from myeloid stem cells and residing in the skin and mucosal tissues. Mast cells play pivotal role in inflammation and allergic reactions (type I hypersensitivity). The degranulation of mast cells upon activation leads to the secretion of vasoactive substances like histamine, cytokines, chemokines, and proteases that promote vasodilation, increased vascular permeability, mucous secretion, bronchoconstriction, leukocyte recruitment, and nerve stimulation. Like basophils, they also help in T cell stimulation and promote chronic inflammatory responses by recruiting leukocytes. They also have immune protective roles against parasites.

4.1.2.3.4 Monocytes

Monocytes are the largest leukocytes and constitute around 3–8% of total leukocytes in the blood. They are differentiated to form macrophages and together make mononuclear phagocytic system (MPS).

4.1.2.3.4.1 Morphology

Monocytes possess bean shaped nucleus with blue gray cytoplasm. Sometimes large vacuoles are found in the the monocyte cytoplasm. There may be some azurophilic cytoplasmic granules in the cytoplasm of monocytes. Monocytes are differentiated in the tissues to form macrophages. The macrophages are larger in size with more intracellular organelles and higher hydrolytic enzymes making their phagocytic potential more than monocytes. Macrophages also contain large number of microvilli along their surface.

4.1.2.3.4.2 Functions

Pathogen recognition: Monocytes are involved in pathogen recognition and alert the immune system to respond during the time of infection. They have toll-like receptor on their surface which recognize PAMP and migrate to the tissues by diapedesis within 24 h of infection and differentiate to form macrophages which act as antigen-presenting cells.

Phagocytosis: Monocytes also have phagocytic property and act to remove foreign material; pathogens damaged cells from peripheral circulation.

Adaptive immunity: Monocytes interact with T and B lymphocytes to modulate adaptive immune responses.

4.1.2.3.5 Lymphocytes

Lymphocytes are the main components of the adaptive immune system. They are involved in cell mediated and humoral immune response against pathogens and retain the memory of the previous exposure. Lymphocytes have some characteristics to make them unique among other leukocytes. Unlike other leukocytes, lymphocytes are not end cells rater they are resting cells capable of producing both effector and memory cells by mitosis upon stimulation. Lymphocytes are capable of circulating from blood to tissue or vice versa. Both B and T lymphocytes have the ability to rearrange the antigen receptor gene component to express wide varieties of cell surface receptor and antibodies.

4.1.2.3.5.1 Morphology

Lymphocytes contain a round and large nucleus which covers almost all the cells leaving a very little cytoplasm. Sometimes a small clear zone can be appreciated on the side of the nucleus called perinuclear zone. Lymphocytes are having plenty of polyribosomes for synthesis of immunoglobulins and cytokines. Based on the size, lymphocytes can be categorized into small, medium, and large. The small lymphocytes are common in dogs and cats. Cows, sheep, goats, and rodents have both large and small lymphocytes. The large lymphocytes of cow are characterized by vacuolated cytoplasm and few azurophilic granules.

4.1.2.3.5.2 Lymphocyte Subsets and Their Functions

Lymphocytes can be categorized either phenotypically (expression of surface molecules) or anatomically (developmental pathways in the lymphoid tissue) and functionally. Based upon these parameters, lymphocytes can be primarily classified as T cells, B cells, and natural killer (NK) cells (Table 4.12). The most common surface markers are surface membrane immunoglobulin (SmIg), B-cell antigen receptor (BCR), and cluster of differentiation (CD).

The major T cell subsets include helper T cells (Th) (CD4 positive), cytotoxic T cells (Tc) (CD8 positive), and regulatory T cells (Treg). Th cells are further classified into Th0, Th1, Th2, Th3 (TGF-β), Th17 (IL17). The Tc cells are further divided into Tc1 and Tc2. The details of T cell subsets along with their functions are presented in Table 4.13.

Natural killer (NK) cells are identified by lipid antigen expressed by CD1d. They have two different subsets. Type I NKT cells express TCR with limited diversity and in contrast, Type II NKT cells are having more TCR sequence than Type I NKT. They function as cytotoxic cells and help in ADCC.

Table 4.12 The comparison of different surface markers of T, B, and NK cell

Surface marker	B cells	T cells	NK cells
Surface membrane immunoglobulin (SmIg)	+	–	–
T cell receptor (TCR)	–	+	–
CD3	–	+	+
CD19, CD20	+	–	–
CD16, CD56	–	–	+
Complement receptor (CR)	+	–	Part
Fc receptor	+	–	+

Table 4.13 The T lymphocyte subsets, their surface markers and functions

Cell type	Functional subsets	TCR	Surface marker	Functions
T helper cells	Th0	α β TCR	CD4+	• Precursor of other CD4+ T cell subsets
	Th1	α β TCR	CD4+	• Helps in "type 1" immune response
	Th2	α β TCR	CD4+	• Antibody production (IgE, IgG, IgA) • Helps in "type 2" immune response
	Th17	α β TCR	CD4+	• Pro-inflammatory mediators • Regulates auto immunity
Regulatory T cells	Trl (Th3)	α β TCR	CD4+	• Maintains immunological self-tolerance • Suppresses inflammation • Suppresses T cell-mediated immunity • Suppresses auto-reactive T cells that escape negative selection • Suppresses proliferation of B cells, monocytes, and other T cells • Secretes TGF-β and IL-10
–		γ δ TCR	CD4+	• First line of defense to bacterial pathogens • (May be Th 1 and 2 subsets)
Cytotoxic T cells		α β TCR	CD8+	• Antibody-dependent cell-mediated cytotoxicity (ADCC)

4.1.2.4 The Platelets/Thrombocytes

Platelets are the smallest cells of the blood (2–4 μm diameter) derived from megakaryocytes of the bone marrow and played a pivotal role in primary hemostasis. They are non-nucleated cells and possess granular cytoplasm. The circulating platelets are usually biconvex with smooth surface which allow them to flow smoothly through the vessels, but upon activation they become sticky and adhered to vessel wall. The platelet counts in different animals are represented in Table 4.14. However, only two third of the total platelets are in circulation and remaining one third resides in spleen and released in the circulation as per the need. Platelets remain in circulation for about 10 days after released from bone marrow.

4.1.2.4.1 The Structure of Platelets

The plasma membrane of platelets resembles classical biological membrane composed of phospholipid bilayer in which the polar heads direct towards plasma and cytoplasm together with the non-polar fatty acid tails towards center. These inner phospholipids play a pivotal role in platelet activation. The platelet glycocalyx situated at the membrane surface is thicker (20–30 nm) compared to other blood corpuscle. The proteoglycans present in the glycocalyx maintain a negatively charged surface that repels other platelets and blood corpuscles as well as vascular endothelium which is an important machinery to prevent hemostasis during circulation. The platelets are surrounded by microfilaments (2 nm diameter) beneath the plasma membrane and help to maintain the discoid shape of platelets. A thick meshwork of microfilaments made of action is situated between microtubules and plasma membrane that facilitate contractile property of platelet. The invaginations of the plasma membrane led to the development of open Canalicular System (OCS) or surface-connected canalicular system (SCCs). The primary functions of OCS are to uptake the particles as well as to the release the granular contents. Dense tubular system (DTS) are the cytoplasmic organelles contain enzymes like phospholipase A2, cyclooxygenase, and thromboxane synthase involved in the biosynthesis of thromboxane A2. Another important function of DTS is calcium sequestration which is facilitated by sarco/endoplasmic reticulum calcium ATPases (SERCAs).

Table 4.14 Platelet counts in different species

Species	Platelet counts ($\times 10^3/\mu L$)	References
Cattle/sheep/goats	150–450	Reece (2015)
Draft horses	150–300	
Pigs	150–350	
Dogs	170–140	Klaassen (1999)
Cats	200–500	

Platelets contain three types of storage granule namely α-granule, dense granule, and lysosomal granules.

α-granules: They are the largest and most abundant granules of the platelets (50–80/platelet). The α-granules are filled with proteins like β-thromboglobulin and thrombospondin along with some mitogenic proteins such as transforming growth factor-β (TGF-β) and platelet-derived growth factors (PDGF).

Dense granules: As the name implies, these granules look dense under microscope. Compared to α-granules, dense granules are less in numbers (2–7/platelet). Dense granules generally store ADP, serotonin, and calcium.

Lysosomal granules: The lysosomal granules contain glycosidases, proteases, and lipases responsible for digesting the cellular debris by autophagy. These granules are less in number and with large diameter (300 nm).

4.2 Hematopoiesis

Hematopoiesis is the formation of blood cells. It starts during embryonic development and continues throughout adulthood to produce and replenish the blood cellular components. Hematopoiesis either medullary (occurred in the bone marrow) or extramedullary (other than bone marrow) depending on the age of the individuals. The sites of extramedullary hematopoiesis are the yolk sac, liver, spleen, kidney, and adrenals. During the early stage of embryonic development, the hematopoiesis is mainly extramedullary, but during the late stage of gestation and soon after birth the hematopoiesis occurs in the bone marrow. In adulthood, extramedullary hematopoiesis can occur in individuals suffering from hematological disorders.

4.2.1 Pluripotential Hematopoietic Stem Cells (PHSC)

These cells are characterized by their long self-renewal capacity and pluripotency. PHSC have the capability to undergo mitotic cell divisions and a portion of these cells remain undifferentiated and maintain a constant pool of PHSC throughout life whereas another portion of PHSC is programmed for differentiation to produce specialized cells. The term pluripotency denotes the ability to differentiate into the cells of a particular lineage. The predominant source of PHSC is the bone marrow (1 in every 100,000 cells in the marrow are PHSC), but they can mobilize the circulation and a substantial number of PHSC are available in peripheral blood. Beside these, PHSC are found in umbilical cord blood, fetal hematopoietic system (yolk sac and fetal liver) and embryonic stem cells and embryonic germ cells. PHSCs migrate to the bone marrow just before birth where they permanently reside throughout life. In human, long bones are the primary site for medullary hematopoiesis up to 20 years of age. Later, the medullary hematopoiesis occurs at flat bones like sternum and vertebrae. The clinical uses of PHSC include the treatment of leukemia and lymphoma and inherited blood disorders like sickle cell anemia, autoimmune disorders and to replenish damaged cells during cancer chemotherapy.

The hematopoiesis can be broadly classified into two categories based on their time of occurrence and the nature of the precursors and progenitor cells. The primitive wave of erythropoiesis mostly seen during embryonic/fetal life in fetal hematopoietic systems characterized by generation of large, nucleated erythrocytes including macrophages and megakaryocytes. The primitive wave is transient and the erythroid progenitors thus produced are neither pluripotent nor having self-renewal property. In contrast, definitive wave is characterized by the development of all blood cell lineages from hematopoietic stem cells occurred during later stage of development. However, there is a transient definitive wave of hematopoiesis that occurs in most of the species produces erythroid-myeloid progenitors (EMPs).

4.2.2 Development of Pluripotential Hematopoietic Stem Cells

All the blood cells are derived from the mesodermal germ layer. The dorsal mesoderm gives rise to somites and notochord, whereas the ventral mesoderm is specified for blood and vasculature. The exact mechanism behind the differentiation of the mesodermal germ layer into hematopoietic stem cells is still awaiting clarification. In humans, hematopoiesis starts around day 17 of embryonic life in the yolk sac. During the course of prenatal development, the site of hematopoiesis varies considerably. The first hepatic colonization of hematopoietic site occurs around day 23 which then migrates to arterial colonization (day 27), second hepatic colonization (day 30), and finally bone marrow colonization around the 11th week of prenatal development.

4.2.3 Development of Progenitor Cells

The first primitive wave of fetal erythropoiesis occurred in the blood island of extra-embryonic yolk sac from a cluster of early endothelial cells (hemangioblast) gives rise to a large nucleated erythroid progenitor along with macrophages and rare megakaryocyte progenitors. The purpose of this primitive wave is to generate red blood cells for tissue oxygenation of a rapidly growing embryo.

A second definitive wave of hematopoiesis occurs in the fetal liver from long-term hematopoietic stem cells (LT-HSC) and generates more complex functionally competent adult-like blood cells. This wave also favors the third wave of hematopoiesis. LT-HSC are having self-renewal properties and are differentiated into short-term hematopoietic stem cells (ST-HSC), which undergoes further differentiation into common myeloid progenitors (CMP) or colony forming unit granulocyte, erythrocyte, monocyte, megakaryocyte (CFU-GEMM), and common lymphoid progenitor (CLP). CMP generates megakaryocyte, erythroid, granulocytes, and macrophages progenitors and CLP give rise to B and T lymphocytes.

The third wave of hematopoiesis occurs in the aorta-gonad-mesonephros (AGM) region of the developing embryo from autonomously generated first adult PHSCs, giving rise to the permanent adult hematopoietic system. The stages are almost similar with second wave of hematopoiesis.

4.2.4 Erythropoiesis

The genesis of erythrocytes from PHScs through series of proliferation and differentiation is called erythropoiesis. The erythropoiesis is aimed either to replenish old and damaged RBCs (basal erythropoiesis) or to cope up the hypoxia due to blood loss or anemia (stress-induced erythropoiesis).

4.2.4.1 Formation of Erythroid Progenitor Cells and Erythroid Precursor Cells

The earliest erythroid progenitor developed from CMP is burst forming unit-erythroid (BFU-E) followed by colony forming units-erythroid (CFU-Es). These erythroid progenitor cells undergo differentiation to form erythroid precursor cells termed as proerythroblast/rubriblast/pronormoblast; however, rubriblast is used as a veterinary terminology. These rubriblasts are characterized by basophilic cytoplasm along with centrally located nuclei with deeply stained chromatin granules. The rubriblasts then undergo a series of differentiation characterized by decrease in cell size with gradual increase in hemoglobin content followed by condensation of nuclear chromatin. The stages next to rubriblasts are prorubricytes which lack nucleus. From prorubricytes, the subsequent developmental stages are basophilic rubricytes/basophilic erythroblasts, polychromatophilic rubricytes/erythroblasts, and metarubricytes/orthochromatophilic erythroblasts. These stages are characterized by increasing hemoglobin content. Metarubricytes have higher hemoglobin content followed by polychromatophilic rubricytes and basophilic rubricytes.

4.2.4.2 Terminal Erythropoiesis: Enucleation and Organelle Clearance

Maturation of committed erythroid precursors to form reticulocytes is called terminal erythropoiesis. The terminal erythropoiesis involves some morphological alteration such as reduction in cell size, chromatin condensation, production of hemoglobin, and finally enucleation and elimination of cytosolic organelle such as Golgi apparatus, endoplasmic reticulum, mitochondria, and ribosomes. In early mammalian embryo, enucleation occurs at the fetal liver whereas from mid gestation and during adulthood the principal site of enucleation is the bone marrow. The erythroblastic island of the bone marrow consists of a central macrophage surrounded by erythroblasts and the interaction between macrophage and erythroblasts is essentially required for differentiation of erythroblasts.

The preparatory phages prior to enucleation include arrest in the cell cycle, chromatin condensation, and nuclear polarization. The enucleation process is facilitated by rearrangement of specific surface antigens that make erythroid cells prone to macrophage engulfment.

The removal of cell organelle is facilitated by the process of autophagy, characterized by the formation of the autophagosome, its fusion with the lysosome to form phagolysosome, and finally the degradation of organelles by hydrolytic enzymes.

4.2.4.3 Factors Affecting Erythropoiesis

4.2.4.3.1 Tissue Oxygenation: Role of Erythropoietin

Tissue oxygenation is the primary stimulus for erythropoiesis. Hypoxia-induced erythropoiesis is mediated through erythropoietin (EPO). EPO is a polypeptide (MW 34000) composed of 139 amino acid residues. During embryonic life, EPO is produced from fetal liver where it acts like paracrine–endocrine manner as liver is the site of erythropoietin synthesis as well as erythropoiesis. Later the site of EPO secretion switches from liver to kidney (mesangial cells, tuft of glomeruli, and renal tubular epithelium) and small amount from liver. During adult life, 90% of EPO is produced from the kidney. After secreted from kidney or liver, EPO reaches in the bone marrow through peripheral circulation to exert its effect. The secretion of EPO in response to hypoxia is brought about by a cellular oxygen sensor named hypoxia-inducible transcription factor-1 (HIF-1).

EPO acts through EPO receptor (EPO-R) expressed primarily on erythroid progenitor cells like BFU-E and colony forming units-erythroid (CFU-E) though their number varies on the basis of differentiation stages. The signal transduction pathway after activation of EPO-R increases calcium influx inside the cell which in turn regulates expression of proto-oncogenes, phosphorylation of transcriptional factors, or activation of calcineurin and ultimately favors the growth, survival, and differentiation of erythroid committed progenitor cells for the production of proerythroblasts/rubriblasts or by decreasing the rate of apoptosis.

4.2.4.3.2 Nutrition: Iron, Vitamin-B_{12} and Folic Acid

The nutritional status of the animals, particularly, iron, Vitamin-B_{12}, and folic acid are essentially required for erythropoiesis, and thus the deficiency of these factors leads to nutritional anemias.

Iron forms the nucleus of the iron-porphyrin heme ring and is required for hemoglobin synthesis. Iron deficiency leads to insufficient hemoglobin formation and microcytic erythrocytes (small size).

Two vitamins, namely vit-B_{12} and folic acid, are required for the maturation of RBC. Vit B_{12} and folic acid are important for the formation of thymidine triphosphate, one of the essential building blocks of DNA. The lack of these vitamins leads to formation of erythrocytes with flimsy membrane and irregular shape as well as apoptosis of erythroid cells. After entering into the circulation, the cells are rapidly destroyed due to their poor fragility.

Intrinsic factor (IF), a glycoprotein released from parietal cells, helps in absorption of vit-B_{12} by protecting it from gastric digestion. Lack of intrinsic factor causes loss of vit-B_{12} by the action of digestive enzymes and failure of its absorption.

Vitamin-C helps in iron absorption either by preventing the formation of insoluble and unabsorbable iron compounds and/or by the reduction of ferric to ferrous iron required for the iron uptake into the mucosal cells.

4.2.5 Leukopoiesis

Leukopoiesis is the development of leukocytes. PHSCs are differentiated into common myeloid progenitors (CMP) or colony forming unit granulocyte, erythrocyte, monocyte, megakaryocyte (CFU-GEMM), and common lymphoid progenitor (CLP). CMP generates megakaryocyte, erythroid, granulocytes, and macrophages progenitors and CLP give rise to B and T lymphocytes. CFU-GEMM differentiated into granulocyte-monocyte progenitor (GMP) cells which ultimately give rise to granulocytes or monocytes. CFU eosinophils (CFU-Eo) and basophils (CFU-Ba) differentiated to form eosinophils and basophils.

4.2.5.1 Granulopoiesis and Monocytopoiesis

The granulopoiesis is the development of granulocytes and monocytopoiesis is the development of monocytes. Myeloblast is the first granulocyte precursor which is differentiated into promyelocytes. Up to promyelocyte stage, the granulocyte and monocyte cell lines are similar then the cell lines develop cell lineage-specific granules and form neutrophil myelocyte, eosinophil myelocyte, and basophil myelocyte. The myelocytes then undergo differentiation to form neutrophil metamyelocyte and eosinophil metamyelocyte. Polymorphonuclear basophils are developed from basophil myelocyte. However, neutrophil and eosinophil myelocytes undergo another differentiation step called neutrophil and eosinophil meta-myelocytes. The neutrophils have one additional stage of development called "band" neutrophil metamyelocyte before the final stage, i.e., polymorphonuclear neutrophils. The granules are developed successively during the course of development. The nucleated granules are developed during early pro-myelocyte stage. Azurophil granules are developed during late promyelocyte stage and specific granules are formed during myelocyte stage. The granulocyte precursors can be divided into two pools. The proliferation pool consists of the cells that can be divided into myeloblasts, promyelocytes, and myelocytes. The metamyelocytes and band cells are unable to be divided and categorized under maturation pool. The storage pool is the subdivision of maturation pool which stores mature neutrophils. The storage pool is large in dogs but small in ruminants.

The first monocyte precursor cells developed from GMP are monoblasts which undergo differentiation to form promonocytes and monocytes. In contrast to granulocytes, monocytes don't have any storage pool, but they enter venous sinusoids and migrate to the tissue where they are differentiated into macrophages.

4.2.5.2 Development of Lymphocytes

The development of lymphocytes occurs in the central/primary lymphoid organs (bone marrow or fetal liver) and thymus for B and T lymphocytes, respectively. The lymphocytes produced in the primary lymphoid organs are then migrated to peripheral lymphoid organs/secondary lymphoid organs where they interact with the antigens. In adults, the development of T cells in the thymus is decreased, but a continual supply is maintained through division of mature T cells in the peripheral lymphoid organs. In contrast, B cells are produced continuously from the bone marrow in adults. The common lymphoid progenitor (CLP) differentiated into pro-B and pro-T cells which give rise to B and T lymphocytes, respectively.

The antigenic specificity of B and T cells is marked by the immunoglobulins and T cell receptor (TCR), respectively. The differentiation of lymphocytes therefore required genetic programming to express immunoglobulin or TCR, respectively, for B and T lymphocytes. The genes responsible for antigen receptors in B cells (immunoglobulin) and T cells (TCR) undergo specific arrangement of variable (V), diversity (D), and joining (J) regions of the gene segment through a process called V (D) J recombination. The V (D) J recombination is a process of selecting V, (D), and J chains in lymphocyte and arranging it into a single exon and generate a novel amino acid sequence for the antigen-binding regions (V regions) for immunoglobulins and TCR. The successful assembly is called a productive rearrangement and is monitored in each developmental stages and acts as a signal to progress into the next stage.

4.2.5.2.1 B Lymphocyte Development

In the bone marrow, CLP differentiated into pro-B cells without the antibody expression. The V (D) J recombination for immunoglobulins starts during this stage due to expression of recombination activating gene (Rag gene). The pro-B cells differentiated into pre-B cells once the V (D) J recombination for heavy chains is completed. In pre-B stage, the V (D) J recombination for light chains starts and once it is completed the pre-B cells are transformed to immature B cells. The immature B cells express IgM. The immature B cells then undergo negative selection to test its reactivity against self-antigens. The auto-reactive B cells either undergo apoptosis through a process called clonal deletion or there is reactivation of Rag for light chain receptor editing. The mature B cells thus ultimately produce and migrate to secondary lymphoid organs and express both IgM and IgD molecule. The time required to form mature B cells from hematopoietic stem is usually 1–2 weeks.

4.2.5.2.2 T Lymphocyte Development

The precursors of T lymphocyte are pro-T cells. Pro-T cells are differentiated into pre-T cells after V (D) J recombination of β chain of TCR. The pre-T cells express only β chain and further differentiate to form double positive T cells with both α and β chain along with CD4 and CD8 over their surface. Then they undergo positive and negative selection to form mature T cells.

4.2.6 Thrombopoiesis

The precursors of the platelets are megakaryocytes derived from bi-potent common myeloid progenitors (CMP) for both erythrocytes and platelets. CMP can be detected in the yolk sac of mouse embryo as early as day (E) 10.5. CMP differentiated into megakaryocyte colony forming cells (Meg-CFCs) with fetal liver by day (E) 11.5 which is the predominant source of megakaryocyte progenitor that produce platelets of larger size with less cytoplasmic granules. Megakaryocytes undergo nuclear endomitosis (DNA replication without cell division), disassembling of centrosomes and the translocation of microtubules to cell cortex and forms broad pseudopodia to become proplatelets in which the nucleus is extruded. A megakaryocyte can give rise to 10–20 proplatelets which protrude from megakaryocytes. The maturation of megakaryocytes to platelets required around 5 days in human and 2–3 days in rodents. Platelets may survive in blood stream for 7–10 days in human and 4–5 days in rodents.

4.3 Iron Metabolism and Hemoglobin

4.3.1 Iron Metabolism

Iron is an integral component of hemoglobin, myoglobin, and other substances such as cytochrome, cytochrome oxidase, peroxidase, and catalase. In addition to this, iron is involved at all stages of energy metabolism as it is necessary for appropriate functioning of enzymes of the electron transport chain (cytochrome oxidase, ferredoxin, myeloperoxidase, catalase, succinate dehydrogenase, and the cytochrome P-450). In the body, iron mainly exists either in form of hemoprotein (hemoglobin, myoglobin, cytochrome, peroxidase, catalase) or as non-heme iron (transferrin, ferritin, hemosiderin). In the body, hemoglobin constitutes the largest amount of iron (65%), followed by myoglobin (4%) and other heme containing irons (1%). About 0.1% of iron remains in circulation, binding with its cargo protein. The principal storage site of iron is the reticulo-endothelial (RE) system and the liver which holds around 15% of total body iron.

4.3.1.1 Absorption of Iron

The primary sites for iron absorption are duodenum and jejunum; however, in ileum, slow uptake of iron occurs. The rate of iron absorption depends upon the iron reserve and is reported to be increased during iron deficiencies. The maximum absorption of iron requires around 24 h. In carnivores and omnivores, heme iron is an important dietary source of iron which absorbs more readily than non-heme iron of vegetables and grain. Several iron transporters are expressed in enterocytes namely divalent metal transporter-1 (DMT-1) and heme carrier protein (HCP1), respectively, for non-heme and heme iron. At acidic pH in the stomach, the feed iron is released in ferric form (Fe^{3+}). But DMT1 is the most important transporter of ferrous iron. Therefore, additional machinery is required to convert Fe^{3+} to its stable form Fe^{2+} for absorption. Duodenal cytochrome B (Dcytb) or cytochrome b reductase 1, is the enzyme which catalyzes the reduction of Fe^{3+} to Fe^{2+} for its absorption. Additionally, ascorbic acid and cysteine also favor this conversion. After internalization of enterocytes, iron either conjugate with apoferritin to form ferritin for steady storage or exported into the circulation, depending on the iron pool of the cell.

4.3.1.2 Exportation of Iron from Intestine into Circulation

The iron in the enterocytes is exported to circulation by ferroportin (FPN1), situated either in the basolateral membrane of the enterocytes or in the macrophages. The expression of FPN1 is stimulated by cellular iron and suppressed by the hormone hepcidin. Hepcidin causes internalization of FPN1 and degradation by lysosomal enzymes thus decreases iron absorption in enterocytes. After iron exportation, Fe^{2+} needs to be converted to Fe^{3+} for its binding with apotransferrin, an iron carrier protein in plasma. Ferroxidases such as ceruloplasmin (Cp) helps in this conversion. The internalization of FPN1 accelerates in the absence of ferroxidase.

4.3.1.3 Iron Transportation in Blood

In circulation, iron is mainly transported by transferrin (Tf). Under normal condition, 20–40% of the total iron-binding sites in transferrin is occupied by Fe^{3+}. The saturation of iron-binding capacity of transferrin occurs as a result of iron overload with the generation of non-Tf-bound iron (NTBI). Presence of transferrin indicates iron status in animals, and thus used as diagnostic tool to measure iron deficiency or iron overload. Erythroid precursors generally use iron of transferrin whereas macrophages and hepatocytes are able to use both transferrin bound iron and NTBI. The internalization of transferrin bound iron inside erythroid precursors of macrophages is facilitated by transferrin receptor (Tfr). Tf-Tfr complex then internalize within endosomes. In acidic pH, endosomes release iron and Tf-Tfr complex localized to cell surface where Tfr dissociates from Tf-Tfr at neutral pH and get recycled.

4.3.1.4 Tissue Storage of Iron

Iron is stored mostly in the liver (50–60 mg/100 g), followed by brain (40–50 mg/100 g), spleen (15–20 mg/100 g), heart (10–20 mg/100 g), and bone marrow (4–5 mg/100 g) in the form of ferritin or hemosiderin. The lowest iron content is reported in stomach, pancreas, jejunum, colon, and urinary bladder (1–2 mg/100 g). When the renal threshold of hemoglobin exceeds, a substantial amount of iron gets accumulated in the epithelium of the convoluted tubules of nephron (20–30 mg/100 g of kidney tissue), smooth muscle and mucous membranes (3–4 mg/100 g). In healthy subjects, iron accumulates in lungs (6–7 mg/100 g) which rapidly declined during anemia (3–4 mg/100 g). However, the iron content in the tissues stated above depends on the status of the animals and the storage rapidly declined on iron deficiency within 3–4 months.

4.3.1.5 Regulation of Iron Absorption

The regulation of iron metabolism is controlled by iron reserve in the body, hypoxia, and rate of erythropoiesis. Heme iron is more efficiently absorbed (25–30%) when compared to non-heme iron (5–15%). The regulation of iron absorption can be explained by two proposed model.

Crypt programming model: This model proposes that the crypt cells of the duodenum sense body iron levels and determines the level of iron to be absorbed. Cytosolic iron regulatory proteins (IRPs) 1 and 2 are the intracellular iron censors and stimulate the expression of TfR1 or DMT1 under iron deficient state.

Hepcidin model: Hepcidin (HAMP, LEAP 1) is a cysteine-rich peptide containing 25 amino acids synthesized mainly from hepatocytes and cleared through kidney. There is an inverse relationship between hepcidin expression and iron state in animals. Hepcidin binds with ferroportin (FPN1) of enterocytes, hepatocytes, and macrophages and causes the internalization and degradation of FPN1 to block the exportation of iron.

4.3.1.6 Factors Affecting Iron Absorption

4.3.1.6.1 Factors Enhancing Iron Absorption

Ascorbic acid (Vit-C) increases the absorption of dietary non-heme iron either by preventing the formation of insoluble non-absorbable iron compounds or by the reduction of ferric to ferrous iron. Other dietary factors that enhance iron absorption are citric acid and other organic acids, alcohol, and carotene.

Citric acid chelates metal ions that interfere iron absorption. The beneficial effect of citric acid is maximum at acidic pH and the effect decreases at neutral or higher pH.

Studies have shown that consumption of alcohol at low level enhances iron absorption by increasing ferritin or TfR 1.

Carotene forms a soluble complex with iron and prevents the inhibitory effects of phytate and polyphenols on iron absorption.

Animal proteins such as meat, chicken and fish stimulate gastric HCl secretion thus promotes iron absorption.

4.3.1.6.2 Factors Inhibiting Iron Absorption

Calcium decreases luminal iron absorption affecting DMT 1 and iron exportation by inhibiting ferroportin (FPN).

Phytic acids present in grains and cereals form a complex with positively charged iron (phytic acid is negatively charged) that decreases bioavailability of iron.

Phenolic compounds present in vegetables, seed, or beverages (coffee, tea, and wine) combine with iron in lumen of intestine, making it unavailable for absorption.

4.3.2 Hemoglobin

Hemoglobin is an iron containing conjugated protein exclusively found in erythrocytes and transports the oxygen from lungs to tissues and carbon-dioxide from tissues to lungs.

4.3.2.1 Structure of Hemoglobin

Hemoglobin is one of the most exclusively studied proteins in nature. It is the first oligomeric protein (composed of two different polypeptide chains), and its complete tertiary and quaternary structures were identified by X-ray crystallography by M. F. Perutz and coworkers, awarded Nobel Prize for Chemistry in 1962. Normal adult hemoglobin molecules (HbA) have a molecular weight of 64,458 Da and are composed of four subunits, each having one polypeptide chain globin (apoprotein) and one heme group (non-protein prosthetic group).

Structure of heme: Heme molecule is composed of porphyrin molecule (Protoporphyrin-IX) with iron (Fe^{2+}) at its center. Protoporphyrin-IX formed by the fusion of four pyrrole rings, joined by the bridges of methenyl group (=CH-). The pyrole rings are named as I, II, III, IV, and the methenyl bridges are named as α, β, γ, and δ. Porphyrins are having 4 methyl, 2 propionyl, and 2 vinyl side chains. The central Fe^{2+} within the heme moiety forms six coordinated bonds. Out of which four bonds are formed with four nitrogen atoms of pyrrole ring, one with proximal histidine residue at position 87 of α-globin chain and one final bond is free for combining with oxygen molecule. The distal histidine molecule situated at position 89 close to oxygen-binding site has two important functions. Firstly, it favors the iron moiety to remain in ferrous state through steric hindrance by preventing oxidation. Secondly, it prevents the binding of carbon monoxide with Fe^{2+}.

Structure of globin chains: The globin chains of the hemoglobin are tetrameric polypeptides composed primarily of two α and two non-α (β, γ, and δ) chains. Each α and β chain are having 141 and 146 amino acids, respectively, and thus total 574 amino acids are present in hemoglobin. However, the polypeptide chains are different in different forms of hemoglobin. The β chain starts with amino acid valine and histidine at N terminal end and the C terminal residues are tyrosine (145) and histidine (146). The δ chain differs from β chain in 10 residues. The first 8 residues and C terminal 21 residues (127–146) are same in β and δ chains. The γ chains are exclusively found in fetal hemoglobin (HbF) and are differed from β chain at 39 residues. The proximal N terminal amino acids of δ chain are glycine and valine whereas C terminal amino acids are same as β and δ chains. The polypeptide chains of hemoglobin molecule are held together by hydrogen bond, hydrophobic and ionic interactions and folded in such a way that the polar residues face the external surface and the non-polar residues are internal making the hemoglobin water soluble. The heme pocket is situated internally lined with non-polar hydrophobic residues protects the ferrous iron from oxidation.

4.3.2.2 Biosynthesis of Hemoglobin

The synthesis of hemoglobin begins in the pro-erythroblast stage and continues till the reticulocyte stage. Synthesis of hemoglobin comprises two distinct steps, namely synthesis of heme and production of globin.

4.3.2.2.1 Heme Synthesis

The heme synthesis is a multistep process occurring both at mitochondria and cytosol of erythrocytes, involving at least eight enzymes out of which four works in the mitochondria and four in the cytosol.

1. Succinyl-CoA, formed in the Krebs metabolic cycle binds with glycine to form aminolevulinic acid (ALA) with the help of enzyme ALA synthase.
2. Two molecules of ALA form porphobilinogen (PBG) with the help of ALA.
3. Four molecules of PBG and produce hydroxymethylbilane by the enzyme porphobilinogen deaminase.
4. Hydroxymethylbilane produces uroporphyrinogen III with the help of enzyme uroporphyrinogen III cosynthase.
5. Conversion of uroporphyrinogen III to coproporphyrinogen III by uroporphyrinogen decarboxylase.
6. Coproporphyrinogen III is converted to protoporphyrinogen IX by coproporphyrinogen oxidase.
7. Protoporphyrinogen oxidase catalyzes the conversion of protoporphyrinogen IX to protoporphyrin IX.
8. Finally, heme molecule is produced after the incorporation of Fe in protoporphyrin IX by ferrochelatase.

Step 1 occurs in the mitochondria, steps 2–5 occur in cytosol and the final three steps (6–8) occur in mitochondria.

4.3.2.2.2 Globin Synthesis

Production of globin chain occurs in the cytosol of erythrocytes by genetic transcription and translation. Gene encoded α chains are situated in chromosome 16 and gene for the β chain are on chromosome 11.

4.3.2.2.3 Alpha Globin Locus

Two identical α globin genes namely α1 and α2 are situated in each chromosome 16. Thus, four α globin genes are present in each cell (due to paired chromosome). Zeta, a substitute for α globin gene, expressed transiently during early embryonic development in alpha globin locus.

4.3.2.2.4 Beta Globin Locus

The β globin gene cluster in chromosome 11 containing the genes are arranged sequentially from 5′ to 3′. In the first sequence, epsilon genes are situated (expressed during embryonic life), following the epsilon genes there are gamma (two copies) and delta genes (single copy). The β globin locus ends with adult beta globin gene (single copy). The protein expressed by these two beta globin genes (due to paired chromosome) matches precisely with four alpha globin genes.

4.3.2.3 Switching of Fetal to Adult Hemoglobin: Ontogeny of Hemoglobin Synthesis

During the first trimester of pregnancy, the zeta gene of the α globin gene cluster is expressed along with the ε-globin in β globin gene cluster, forming hemoglobin grower 1 (or HbE Gower-1). The primitive erythrocytes developed in the yolk

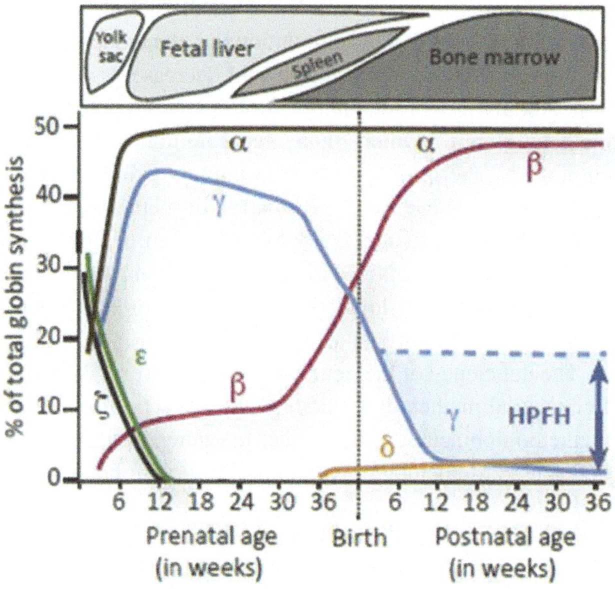

Fig. 4.4 Ontogeny of Hemoglobin Synthesis: In primitive erythroblasts, hemoglobin is synthesized in primitive erythroblasts of the embryonic yolk sac. In definitive erythropoiesis, α and β globin gene clusters express α and β globins, respectively, leading to the formation of fetal hemoglobin (HbF, α2γ2) in the fetal liver. The production of γ-globin declines at the time of birth together with the higher expression of β-globin to form adult hemoglobin (HbA, α2β2). The delta gene produces small amount of delta globin to form adult hemoglobin A2 (HbA2, α2δ2) comprising less than 5% of the total adult hemoglobin. (Source: Wienert et al. 2018)

sac which is replaced with hemoglobin grower 2 comprises alpha (2) expressed by α globin gene cluster, epsilon (2) by β globin gene cluster. During the production of first enucleated definitive erythrocytes in the fetal liver, α and β globin gene clusters express α and β globins, respectively, leading to the formation of fetal hemoglobin (HbF, α2γ2). The production of γ-globin declines at the time of birth along with the increased expression of β-globin by β-globin gene cluster and forms adult hemoglobin (HbA, α2β2). The delta gene located in β globin gene cluster on chromosome 11 between the gamma and beta genes produces small amount of delta globin which forms adult hemoglobin A2 (HbA2, α2δ2) comprising less than 5% of the total adult hemoglobin. Ontogeny of hemoglobin synthesis is depicted in Fig. 4.4.

4.3.2.4 Types of Hemoglobin

Different types of hemoglobin together with their composition and abundance are summarized in Table 4.15.

The normal range of hemoglobin in different species have been presented in Table 4.16.

4.3.2.5 Cooperative Binding of Oxygen with Hemoglobin: T and R States of Hemoglobin

The affinity of hemoglobin for oxygen is proportional to the quantity of oxygen that binds with the hemoglobin at a given time. In other words, binding of one oxygen to heme increases the affinity of hemoglobin for oxygen. Thus, affinity for first oxygen to bind with heme is 100 times more than the last oxygen. This is called cooperative binding or heme–heme interaction. This interaction is achieved due to the conformational changes in the hemoglobin structure after binding with oxygen in such a manner that favors further oxygen binding. The deoxygenated form of hemoglobin is called T-state (tense) which has less affinity for oxygen. In T state, the iron is bound to nitrogen of Histidine side chain (His E8) which pulls the iron out of the porphyrin plans but when oxygen binds with iron, the new bond pulls the iron back to heme plan. The movement of iron facilitates the movement of alpha- and beta-helix and creates a favorable condition to bind with oxygen. This state of hemoglobin is called relaxed state or R state.

4.3.2.6 Derivatives of Hemoglobin

4.3.2.6.1 Oxyhemoglobin

Combination of oxygen with hemoglobin during physical respiration forms oxyhemoglobin. One molecule of hemoglobin can bind with four molecules of oxygen. 1 g of Hb can bind with 1.34 mL of oxygen.

- Gram molecular weight of hemoglobin is 64,500 g.
- One mole of any gas at NTP occupies 22,400 mL.
- So, 4 mol of gas occupy 22,400 × 4 = 89,600 mL.
- 64,500 g hemoglobin contains 89,600 mL of oxygen.
- 1 g hemoglobin contains = 89,600/64,500 = 1.39 mL of oxygen.

Table 4.15 Different types of hemoglobin

Type	Composition	Abundance (%)	Remarks
Hemoglobin A_1	$\alpha_2\beta_2$	90	
Hemoglobin A_2	$\alpha_2\delta_2$	<5	
Fetal Hemoglobin (HbF)	$\alpha_2\gamma_2$	<2	• More oxygen affinity than adult Hb (except cat) • At birth, 41–100% of total Hb is HbF and diminished after 2–3 months • Dogs, horses, and pigs don't have structurally distinct HbF
Glycosylated hemoglobin (HbA_{1C})	$\alpha_2\beta_2$-glucose	<5%	• Marker for diabetes

Table 4.16 Normal range of hemoglobin in different species

Species	Hb (g/dL)	References
Dogs	12–18	Reece (2015)
Cats	10–15	
Cattle	8–15	
Horses	11.5–16	
Pigs	10–16	
Sheep	9–15	
Goats	8–12	
Tiger (*Panthera tigris tigris*)		Shrivastav and Singh (2012)
Lions (*Panthera leo*)	8.9–14.6	Maas et al. (2013)
Elephants (*Elephas maximus*)	9.8–15.2	Janyamethakul et al. (2017)

- 0.05 mL/g of oxygen is not capable of binding with hemoglobin as small amounts of hemoglobin exist as met-hemoglobin unable to carry oxygen.
- So, 1 g of hemoglobin carries $1.39 - 0.05 = 1.34$ mL of oxygen.

Oxygenated hemoglobin is bright red in color and deoxygenated hemoglobin is purplish red in color.

4.3.2.6.2 Myoglobin

It is a hemoprotein found in muscle tissue. Unlike hemoglobin, myoglobin has only one heme molecule; hence, it can bind with only one molecule of oxygen and a single polypeptide chain of 154 amino acids. The molecular weight of myoglobin is 17,000. Myoglobin has higher affinity for oxygen compared to hemoglobin. The main function of myoglobin is to store oxygen in muscle tissue and maintain a steady supply during hypoxia.

4.3.2.6.3 Carboxyhemoglobin

Binding of hemoglobin with carbon monoxide yields carboxyhemoglobin. The affinity of carbon monoxide with hemoglobin is 200 times greater than oxygen and carboxy hemoglobin is about 250 times more stable than oxyhemoglobin. Carboxyhemoglobin interferes with cellular respiration either by preventing the oxygen transport or by inhibiting the enzyme cytochrome oxidase. The color of carboxyhemoglobin is bright cherry red. High pressure oxygen therapy is applied as medication in carbon monoxide poisoning.

4.3.2.6.4 Methemoglobin

It is the true oxide of hemoglobin as ferrous iron is converted to ferric iron during the formation of methemoglobin. The oxidizing agents such as ferricyanide and nitrites react with hemoglobin to form methemoglobin. Like carboxyhemoglobin, methemoglobin is also unable to carry oxygen. Under normal condition, a small amount of methemoglobin is formed in the circulation but, the reducing agents such as glutathione and ascorbic acid decrease its accumulation. About 1.5% of total hemoglobin in the body remains as methemoglobin. Other than ferricyanide and nitrites; chlorates, peroxides, hydroquinone, iodine, and sulfonamides also cause the formation of methemoglobin.

The reduction of methemoglobin back to hemoglobin is mediated mainly by NADH-cytochrome b5-methemoglobin reductase. Methemoglobin is also prone to direct reduction by intracellular ascorbate and glutathione.

The deficiency of methemoglobin reductase leads to type-I congenital methemoglobinemia whereas type-II congenital methemoglobinemia occurs due to generalized reductase deficiency. Methemoglobinemia is common in horses, dogs, and cats.

Met-hemoglobinemia is common in animals exposed to nitrate (fertilizers) and chlorate (herbicides) poisoning. In rumen, nitrate is converted to nitrite and leads to the formation of methemoglobin. Methylene blue and ascorbic acid is used as medication of methemoglobinemia.

Methemoglobin itself is used as medication in cyanide poisoning. Excessive ingestion of sorghum (hydrocyanic acid) causes cyanide poisoning in animals due to formation of thiocyanate. Under these circumstances, sodium nitrite is applied to form methemoglobin which in turn combines with thiocyanate to form cyan-met hemoglobin and rapidly eliminated from the body.

4.3.2.6.5 Sulfhemoglobin

Sulfhemoglobin is formed when reduced hemoglobin combines with hydrogen sulfide. It usually occurs in the subjects working in nitrite and coal tar factories where excess amount of sulfur is handled regularly. Sulfhemoglobin at the concentration of 3–5 g/dL of blood leads to cyanosis. Sulfhemoglobinemia also occurred with the patients suffering from bacteremia with *Clostridium welchii* which rapidly oxidize the hydrogen sulfide produced during digestion in intestinal tract to sulfate. Sulfhemoglobin is not converted back to hemoglobin and persists throughout the lifespan of erythrocytes.

4.3.2.6.6 Hemin: Reactions with Acids

The action of hydrochloric acid on hemoglobin is as follows:

$$\text{Hemoglobin} \xrightarrow{\text{HCl}} \text{Globin} + \text{Heme(Ferriheme)}$$

$$2(\text{Ferriheme}) + O_2 \xrightarrow{\text{HCl}} 2\ \text{Ferriheme chloride(Hemin)}$$

This method is used in the determination of hemoglobin by Sahli's method.

4.3.2.6.7 Hematin: Reactions with Alkali

Alkali splits hemoglobin to form globin and heme. The heme rapidly oxidized to ferriheme and combines with alkali to form ferriheme hydroxide (hematin).

$$\text{Hemoglobin}\ ^{\text{NaOH}} \rightarrow \text{Globin} + \text{Heme(Ferriheme)}$$

$$2(\text{Ferriheme}) + O_2\ ^{\text{NaOH}} \rightarrow 2\ \text{Ferriheme chloride(Hematin)}$$

4.3.2.6.8 Carbaminohemoglobin

Carbaminohemoglobin is formed when carbon-dioxide reacts with free NH_2 terminal groups of the α and β chains of hemoglobin. Like oxygen, four molecules of carbon-dioxide combine with hemoglobin. About 7–10% carbon-dioxide is carried as carbaminohemoglobin by RBC and rapidly dissociates in high P_{O2} in the lung alveoli to form carbon-dioxide and carbamate with NH_2 groups.

4.3.2.7 Fate of Hemoglobin

During the intra vascular hemolysis mechanism, hemoglobin in released in the circulation and upon extra vascular hemolysis, hemoglobin is broken down by the enzymatic machinery of the macrophages. The "globin" portion of hemoglobin can be reutilized by the system but, the "heme" component needs further metabolic conversion before getting excreted from the body. Most of the iron of the heme are also reutilized by the body and believed to be the most important source of iron in iron homeostasis.

The heme undergoes several metabolic pathways for its excretion.

Trapping of free hemoglobin and heme: The free hemoglobin and its breakdown products like heme get released into the circulation and binds with plasma proteins such as haptoglobin, hemopexin, and albumin. Haptoglobin binds to free hemoglobin and hemopexin, and albumin binds with heme.

Catabolism of heme to bilirubin: Heme moiety is converted to bilirubin by two-step enzymatic breakdown in RE system. Heme oxygenase converts heme moiety to biliverdin which subsequently reduced to bilirubin by NADPH-dependent biliverdin reductase. Around 75% of the total bilirubin produced in the body comes from the senescent erythrocytes.

Transport of bilirubin to liver: Plasma albumin has strong affinity for bilirubin and acts as a vehicle to carry bilirubin from RE system to liver. The hepatocytes have organic anion transporter 2 (OATP2) which readily takes the albumin bound bilirubin. Ligandin, a cytosolic protein inside the hepocyte carries bilirubin to endoplasmic reticulum where bilirubin conjugation occurs.

Glucuronide conjugation of bilirubin: Enzyme uridine-diphosphate (UDP)–glucuronyl transferase facilitates the conjugation of bilirubin and glucuronic acid to form bilirubin glucuronide, which is the excretable form of bilirubin.

Excretion: The conjugated bilirubin produced in the hepatocytes is excreted through bile in the intestine where it is reduced to urobilinogen by the intestinal bacteria and excreted as urobilin or stercobilin, the oxidized form of urobilinogen which gives the normal color to feces. Some urobilinogen enters in the general circulation bypassing the liver and excrete in urine which subsequently oxidized to urobilin that forms the color of urine. A part of conjugated bilirubin in the intestine is hydrolyzed by intestinal beta-glucuronidase, which releases free bilirubin that enters into enterohepatic circulation and again get re-excreted through bile.

4.4 Blood Coagulation/Hemostasis

The term blood coagulation or hemostasis can be defined as a cascade of enzyme activation which enables the stoppage of bleeding from injured vessels, help to keep the blood in fluid state during circulation and to resolve the clot for restoring vascular integrity.

4.4.1 The Coagulation Machinery

The process of blood coagulation and its associated clot retraction and fibrinolytic (degradation of fibrin clot) mechanism is the result of complex interactions of some soluble clotting factors as well as platelets and vascular endothelium. The structure and the development of platelets were discussed in previous chapters. Here, we will mainly focus on the role of platelets in blood coagulation.

The Clotting Factors/Coagulation Factors The clotting factors/coagulation factors involved in hemostasis are listed in Table 4.17. Most of the clotting factors are produced in the liver except Factor-III (plasma), IV (vascular endothelium), and VIII (vascular endothelium). The post-translation modifications in terms of Vit-K dependent ϒ carboxylation of glutamic acid residues allow them to bind with divalent cations particularly calcium. Most of these coagulation factors are enzyme precursors and exist as zymogen. The nomenclatures of these clotting factors were designated by Roman numeral and the activated form of these zymogens are designated by subscript "a" to their Roman numeral recommended by the international committee for the nomenclature of blood clotting factors established in 1954. Initially

Table 4.17 Coagulation proteins/clotting factors

Clotting factor number	Name	Source	Plasma half-life (h)	Plasma concentration (mg/L)	Functions
I	Fibrinogen	Liver	90	3000	Precursor of fibrin that leads to clot formation
II	Prothrombin	Liver	65	100	Precursor of thrombin, helps in the activation of factor I, V, VII, VIII, IX, XIII, protein C, platelets
III	Tissue factor/Thromboplastin	Vascular endothelium	–	–	Acts as cofactor for F-VIIa
V	Proaccelerin/labile factor	Vascular endothelium	15	10	Acts as cofactor in prothrombinase complex
VI	Unassigned				
VII	Proconvertin/stable factor	Liver	5	0.5	Activates factor IX, X
VIII	Antihemophilic factor A	Vascular endothelium	10	0.1	Cofactor for tenase complex
IX	Antihemophilic factor B/Christmas factor	Liver	25	5	Activates factor X
X	Stuart-Prower factor	Liver	40	10	Forms prothrombinase complex with factor-V
XI	Plasma thromboplastin antecedent	Liver	45	5	Activates factor IX
XII	Hageman factor	Liver	–	–	Activates factor VII, IX, and prekallikrein
XIII	Fibrin-stabilizing factor	Liver	200	30	Joins fibrin monomers to form clot
XIV	Prekallikrein/Fletcher factor	Liver	35	–	Zymogen of serine protease
XV	High molecular weight kininogen (HMWK)/Fitzgerald factor	Liver	150	–	Initiation of coagulation and generation of vasodilator bradykinin
XVI	Von Willebrand factor (vWf)	Vascular endothelium	12	10 μg/mL	Helps in platelet adhesion
XVII	Antithrombin III	Liver	72	0.15–0.2 mg/mL	Inhibits thrombin and other coagulation proteins (IIa, Xa)
XVIII	Heparin cofactor-II	Liver	60	–	Inhibits serine protease
XIX	Protein C	Liver	0.4	2–3 ng/mL	Inactivates Va, VIIIa
XX	Protein S	Liver	48	25 μg/mL	Cofactor for protein C

factor I to IX were officially approved. Factor X, factor XI and factor XII were approved later. Factors V and VIII are also known as the labile factors because of their unstable coagulant activity in stored blood.

Know More......
It is interesting to note that factor IX (Christmas), factor X (Stuart and Prower), factor XII (Hageman), Fletcher (prekallikrein), and Fitzgerald (high-molecular-weight kininogen) were named in the memory of six patients namely Stephen Christmas, Miss Audrey Prower and Rufus Stuart, John Hageman, Fletcher family, and Allen Fitzgerald as these factors were identified in those patients were suffering from coagulopathies associated with the absence of respective factors.

4.4.1.1 Vascular Endothelium

The entire cardiovascular system including the capillaries are lined by a single layer of flattened cells known as the endothelium which separates blood from surrounding tissues. The endothelium comprised single-layered squamous cells ($1-6 \times 10^{13}$) called endothelial cells of less than 0.2 μm thick with a total surface area of 4000–7000 m^2. There are some storage granules called Weibel–Palade bodies (WPBs) situated in the inner lining of the blood vessels and heart responsible for releasing two important hemostatic factor von Willebrand factor and P-selectin. The general functions of vascular endothelium include, regulation of vascular tone, proliferation of smooth muscle, molecular exchange between blood and surrounding tissue and regulation of hemostasis. Both anticoagulant and procoagulant factors are released from vascular endothelium which provides a hemocompatible environment to keep the blood in fluid state as well as initiates coagulation mechanisms during vascular injury. The functions of anticoagulation and procoagulation factors released from vascular endothelium are listed in Table 4.18.

Table 4.18 Anticoagulant and procoagulant factors of vascular endothelium

Name of the factors	Functions
A. Anticoagulants	
Prostacyclin (PGI_2)	Vasodilation Prevents platelet aggregation with increase cyclic AMP (cAMP) production that effectively reduces the amount of thromboxane A_2
Nitric oxide	Vasodilation Inhibits platelet aggregation and adhesion to the endothelium by cyclic guanosine monophosphate (cGMP) production and subsequently mobilization of Ca^{2+} flux
Actonucleotidase (ADPase)	Degradation of ADP produced during the interaction of platelets with collagen (ADP promotes platelet aggregation)
Protein C	Degradation FVIIIa and FVa in conjugation with Protein S
Plasminogen activators	Initiates clot retraction and fibrinolytic mechanism
Tissue factor pathway inhibitors (TFPI)	Inhibits contact activation pathway or tissue factor pathway
Heparin-like substance	Inactivates thrombin
Negativity of vascular endothelium	Repels the negatively charged platelets and prevent platelet adhesion
B. Procoagulants	
Thrombin	Converts fibrinogen to fibrin with the activation of other procoagulant factors V, VIII, XI, and XIII
von Willebrand factor (vWf)	Helps in platelet adhesion
Tissue factor	Acts as cofactor for F-VIIa
Vascular cell adhesion molecule 1 (VCAM-1)	Helps in platelet adhesion
Endothelin	Vasoconstriction and stoppage of bleeding after injury to vessel
Plasminogen activator inhibitor-1 (PAI-1)	Inhibitor of tissue plasminogen

4.4.2 Mechanism of Hemostasis

The blood coagulation or hemostatic mechanism can be divided into two broad events namely primary hemostasis in which there is formation of weak platelet plug and secondary hemostasis where the reinforce of primary hemostasis is occurred with fibrin.

4.4.2.1 Primary Hemostasis

The formation of weak platelet plug in primary hemostasis is occurred through the events like vasoconstriction, platelet adhesion, platelet activation, and platelet aggregation.

Vasoconstriction: The very first response to vascular injury is the vasospasm which leads to vasoconstriction. The mediator of this vasoconstriction is endothelin-1 (ET) produced primarily from damaged endothelium. ET acts over the target tissue at the vascular smooth muscle after binding with its receptor coupled with G-protein. Binding of ET with its receptor causes the activation of phospholipase C (PLC) which converts phosphatidyl inositol bis-phosphate (PI2) into di-acyl glycerol (DAG) and inositol tri-phosphate (PI3). Both DAG and IP3 stimulate the release of calcium from sarcoplasmic reticulum which causes muscle contraction.

Platelet adhesion: In this process, the rolling platelets are attached with the subendothelial layer exposed after vascular damage. Following the endothelial damage vWf, collagen, P-selectins are exposed which act as the ligands for binding with the membrane glycoprotein (GP) receptor situated in the phospholipid bilayer of platelet membrane. The initial bond between platelet and subendothelial tissue is formed after binding of Gplb-IX receptor of platelets with vWf. The glycoprotein receptors of the platelets (GP VI) also bind with collagen. The attachment of platelet to the damaged surface facilitates platelet activation through intracellular signaling cascade.

Platelet activation: The activation of platelet is mediated through the activation of either G protein coupled receptor (GP Ib-IX-V, GP VI) or C-type lectin-like receptor 2 (CLEC-2). GP VI is the major signaling receptor for platelet activation upon binding with collagen whereas CLEC-2 activates in response to rhodocytin, snake venom. Thrombin can also stimulate platelet activation through protease-activated receptors. Activation of platelets increases platelet intracellular calcium and triggers the contraction of microfilaments and moving of microtubules inward to compress the granules and release of their contents.

Platelet aggregation: The formation of platelet aggregates in the result of platelet activation and release of granular content together with the alterations of platelet shape from discoid to multiple pseudopodal plug which increase the surface area of platelets. The release of ADP from platelet granules induces the expression of GpIIb/IIIa complex. This complex plays a dual role in platelet adhesion (binding with vWF) and aggregation (binding with fibrinogen which acts as bridge between platelet-platelet aggregations). TXA_2 produced from activated platelets intensifies platelet aggregation which ultimately develops platelet plug.

4.4.2.2 Secondary Hemostasis

Weak platelet plug generates in primary hemostasis can only resist hemorrhage temporarily and needs further strengthening by the formation of fibrin which is the ultimate aim of secondary hemostasis. The events of secondary hemostasis include activation of coagulation cascade, formation of thrombin, and conversion of fibrinogen to fibrin. The classical concept of blood coagulation described secondary hemostasis in three pathways, the extrinsic pathway, intrinsic pathway, and common pathway. The intrinsic pathway was so named because all the components required for coagulation were available in the blood itself and in contrary extrinsic mechanism required tissue factors from extravascular tissues. The detailed mechanisms will be discussed in subsequent sections. Both extrinsic and intrinsic pathways converge at the activation of factor X through different enzymatic reactions. Activated factor X then converts prothrombin to thrombin via common pathway. Thrombin initiates fibrin formation which stabilizes the clot. The entire process of secondary hemostasis requires Ca^{2+}. The pathways are diagrammatically represented in Fig. 4.5.

4.4.2.2.1 Formation of Fibrin Clot

The formation of fibrin is the end result of coagulation cascade. The fibrin is generated from the enzymatic cleavage of fibrinogen. Fibrinogen is a long flexible glycoprotein in hexametric configuration consisting of 2Aα, 2Bβ, and 2γ polypeptide chains are joined together by 29 disulfide bonds. The structure of fibrinogen revealed are three nodular structures held together by a thread. Two end nodules (termed D regions or domains) consist of Bβ and γ chains and the central nodule (termed the E region or domain) consists of two Aα alpha chains. The formation of fibrinogen to fibrin involves several steps.

Enzymatic cleavage: Fibrinogen molecule is subjected to enzymatic cleavage of N-terminal fibrinopeptides known as fibrinopeptide A and fibrinopeptide B, respectively, from both the Aα and Bβ polypeptides.

Knob-hole interaction and formation of protofibrils: The removal of fibrinopeptides exposes "knobs" on the E domain, which can interact with the "holes" on the D domains. E domain on one fibrin molecule combines with the D domains on four other fibrin molecules resulting in the formation of staggered oligomers known as protofibrils.

Formation of fibrin clot: The protofibrils continue to aggregate and lengthen to form long fiber which then wind around to form multi-stranded bundles and ultimately gives rise to 3D network of fibrin clot.

Stabilization of fibrin clot: Thrombin activates FXIII which cross-links one fibrin molecule to another forming a strong bond to strengthen fibrin clot and protects it from physicochemical damage.

4.4.3 Recent Concept of Blood Coagulation/Cell-Based Model of Coagulation

The traditional model of blood coagulation stated the intrinsic and extrinsic pathway as separate cascade for generation of FX_a. But recent experiments suggests that extrinsic and intrinsic mechanism of blood coagulation can only explain

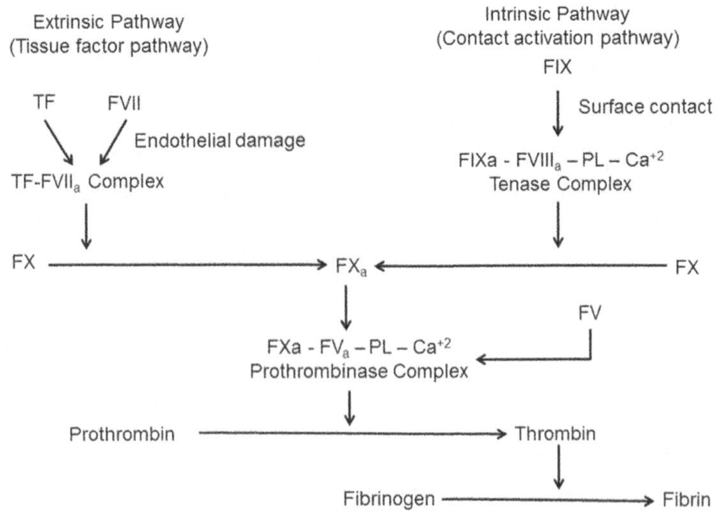

Fig. 4.5 Pathways of blood coagulation: Both intrinsic and extrinsic pathways facilitate the formation of prothombinase complex that converts prothrombin to thrombin. Thrombin converts fibrinogen to fibrin

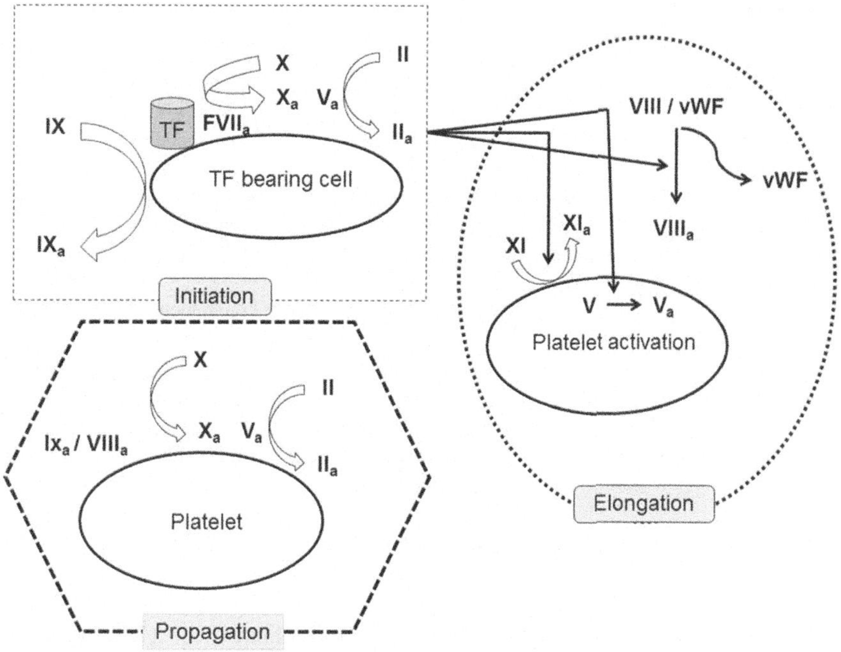

Fig. 4.6 Cell-based model of blood coagulation: In the initiation phase, small amount of thrombin (IIa) is generated on the surface of TF bearing cells. The activation of factors Va, XIa, and VIIIa are occurred. In the propagation phase, large numbers of thrombin (IIa) are produced on the platelet surface

the coagulation mechanism for in vitro laboratory evaluation, but unable to explain the coagulation mechanism in vivo. The recent investigations on blood coagulation mechanism supports that the TF released from damaged endothelium is sufficient to initiate blood coagulation and the intrinsic mechanism of coagulation does not have significant physiological role in hemostasis.

The cell-based model of coagulation describes the coagulation mechanism in the following ways (Fig. 4.6).

Initiation phase: This phase of coagulation occurs in the TF bearing cells outside the vasculature. In this phase, small amount of FIX and FX are activated by TF-$FVII_a$ complex. Factor X_a converts prothrombin to thrombin associated with its cofactor FV_a. But it plays a pivotal role to initiate coagulation; firstly, it activates platelet to release its granule contents for platelet aggregation; secondly, it activates FV and FVIII on the surface of platelet which allows dissociation of FVIIIa–vWF complex and make vWF free for platelet adhesion and aggregation at the site of injury; thirdly, it activates FXI on the surface of the platelets. It is believed that initiation phage of coagulation is always active in circulation and little amount of thrombin is continually produced independent of vascular injury.

Elongation phase: After the vascular injury, platelets leave the blood vessel and interact with the exposed collagen at the extracellular matrix of injured site to form platelet plug.

Propagation phase: The propagation of clotting mechanism allows more and more platelets to adhere at the site of injury followed by the formation of tenase and prothrombinase complex. The formation of tenase complex and prothrombinase complex allow the formation of large amount of thrombin catalyzes the formation of fibrin monomers. The monomers of fibrin then stabilize to form fibrin polymer and stable clot.

4.4.4 Clot Retraction and Fibrinolysis

Clot retraction: Clot retraction denotes squeezing of clot through the contractile force generated by platelets results the expulsion of serum. The contraction of platelets is the result of interaction between platelet contractile proteins such as actin, myosin, and thrombosthenin. The contraction of platelets further propagates through fibrin fibers to facilitate the reduction in clot volume. Clot retraction is essential to restore the blood flow at the site of injury vis-à-vis it promotes wound healing.

Fibrinolysis: The clot formation and fibrinolytic mechanism are well regulated under physiological condition to ensure the fluidity of blood during circulation. The fibrinolytic machinery consists of the activators, inhibitors, substrate and cofactors and the complex interaction amongst them facilitates fibrin degradation. The components of fibrinolytic system and their mode of actions are depicted in Table 4.19.

Table 4.19 The components of fibrinolytic system

Components			Source	Functions
Plasminogen			Liver, $t_{1/2} = 2$ days	Catalyzes fibrin to for fibrin degradation products
Plasminogen activator	Tissue-type Plasminogen Activator (tPA)		Endothelial cells, $t_{1/2} = 4$–8 min	Conversion of plasminogen to plasmin
	Urokinase Plasminogen Activator (uPA)		Fibroblast, monocytes, macrophages, epithelial cells, $t_{1/2} = 4$–8 min	Conversion of plasminogen to plasmin
Fibrinolytic inhibitors	Plasmin inhibitors	α_2-plasmin inhibitor (α_2-PI)	Liver, α-granules of platelets, $t_{1/2} = 2$–3 days	Serine protease inhibitors
		α_2-macroglobulin (a2-MG)	Endothelial cells, macrophages and α-granules of platelets	Inhibits the activity of plasmin
		Protease nexin	Liver, mesenchymal cells, vascular wall and stored in α-granules of platelets	Inhibits trypsin, thrombin, factor Xa, and plasmin
	Plasminogen activator inhibitors	Plasminogen activator inhibitors (PAI-1, PAI-2)	Endothelial cells, hepatocytes, macrophages, monocytes, adipocytes, and platelets	Inhibits tPA and uPA
		C1-esterase inhibitor	Liver	Inhibits kallikrein, FXIa, and FXIIa
Thrombin-activatable fibrinolysis inhibitor (TAFI)			Liver, platelets, $t_{1/2} = 8$ min	Attenuation of fibrinolysis

Initially plasmin acts over D-domain of the cross-linked fibrin polymer to release Aα- and Bβ-fragments from C terminal region of α and β chain. The N terminal end of β chain is further cleaved to release fibrinopeptide B (FPB). Aα- and Bβ-fragments together with FPB forms fragment X. Plasmin further cleaves three polypeptide bonds that connect D and E domain of fragment X to yield fragment D and fragment Y. The polypeptide bonds between D and E domain of fragment Y are cleaved to release individual D and E domain. The fragments thus generated by the action of plasmin upon the fibrin clot are collectively called fibrin degradation products (FDP). The principal component of FDP is D-dimer, the cross-linked product of two D fragments.

4.5 Blood Group, Blood Transfusion, and Hematological Disorders

4.5.1 Blood Grouping in Animals

The determinants of blood groups are the specific polymorphic antigens that reside on the surface of erythrocytes (agglutinogens) and the antibodies (agglutinogens) present in the plasma. There are two types of antibodies to blood group antigens.

1. ***Naturally occurring antibodies (alloantibodies)***: They are produced against the isoantigens produced by the other members of same species and destroy the isoantigens. For example: anti-A antibody in the individual with B blood group.
2. ***Acquired antibodies***: They are produced in response to the exposure of a blood group antigen mostly through blood transfusion.

Unlike humans the naturally occurring alloantibodies in animals are less immunogenic and don't induce severe hemolytic reactions. Thus, the first blood donation for an animal can be done from any animals of same species without any hemolytic reactions but thereafter, the donor blood will have to match to prevent possible complications.

4.5.1.1 Canine Blood Types

The dog erythrocyte antigen (DEA) system: The major determinant of canine blood group is dog erythrocyte antigen (DEA), and there are eight major blood groups in the dog, labeled as DEA 1–8. The different types of blood grouping systems in canines are summarized in Table 4.20.

Dalmatian (Dal) blood types: A new red cell antigen was discovered in 2007 in Dalmatian dogs. This antigen was named as *Dal* as it was first discovered in Dalmatian dogs. The prevalence of Dal antigen is about 93% dogs in the USA. Dal^+ phenotype was autosomal dominant. The highest incidence was reported in Dalmatians (85.6–100%) followed by Doberman Pinschers (78.6%). Higher incidence of Dal^-antigens was identified in Shih Tzus (57.1%).

Kai 1 and Kai 2 blood types: Two new blood groups namely Kai 1 and Kai 2 were reported in the dogs that are mostly found in North America. The word is derived from the Korean Kai meaning "dogs." These antigens were biochemically characterized through ELISA, and it was reported that both of these antigens didn't co-exist but both could be absent. Naturally occurring anti-Kai 1 or Kai 2 antibodies were absent but, mismatched transfusion in Kai 1^- and Kai 2^-dogs might develop Kai 1 and Kai 2 antibodies.

Table 4.20 DEA blood group system in dogs

Types	Factor	Phenotypes	Remarks
DEA 1.1, 1.2, 1.3 (A system)	3 (DEA 1.1, 1.2, 1.3)	4 (Aa_1, Aa_2, Aa_3, and null type)	• A particular dog exhibit only one phenotype. • The incidence of DEA 1.1 and DEA 1.2 is 45% and 20%, respectively. • Acute hemolytic transfusion reactions occur in DEA 1.1 and 1.2 and DEA 1.3 negative dogs transfused with DEA 1.1, DEA 1.2, and DEA 1.3 positive donors. • Neonatal iso-erythrolysis (hemolytic disorder in neonates) has been reported in DEA 1 sensitized DEA 1 negative bitch mated to DEA 1.1 positive male dogs.
DEA 3 (B system)	1 (DEA 3)	2 (Ba and null type)	• Incidence of DEA 3 positive dogs is 6% in the and mostly found in greyhounds (23%). • Transfusion of DEA 3 negative dogs with DEA 3 positive RBCs can induce severe acute transfusion reactions.
DEA 4 (C system)	1 (DEA 4)	2 (DEA 4 and null type)	• No naturally occurring antibody against DEA 4 has been found. • Highest population incidence (98%). • Dogs positive for DEA 4 and negative for other DEA group can act as universal donors (e.g., Greyhounds and Indian Chippiparai).
DEA 5 (D system)	1 (DEA 5)	2 (DEA 5 and null type)	• Lower incidence. • Higher incidence among Greyhounds (30%).
DEA 6 (F system)	1 (DEA 6)	2 (DEA 6 and null type)	• No reports on the naturally occurring anti-DEA 6 antibody.
DEA 7 (Tr system)	2 (Tr and O)	3 (Tr, O, and null)	• Incidence 40–54%.

4.5.1.2 Feline Blood Types

Unlike canines, cats have only one blood group system, the AB system (Table 4.21). In this blood grouping system, three blood types are available namely A, B, and AB which is based on two erythrocyte surface antigens namely antigen A and B. Antigen A is actually *N*-glycolyl-neuraminic acid which has most common incidence among cats. Antigen B is *N*-acetyl-neuraminic acid with very rare incidence. The extremely rare AB blood group contains both A and B antigens in equal amount. The naturally occurring alloantibodies are strong agglutinins and hemolysins and can cause potentially life-threatening transfusion reactions. There is no universal donor for cat because of the cat's naturally occurring alloantibodies.

4.5.1.3 Equine Blood Types

In equines and donkeys, around 30 erythrocyte antigens are available but, only seven major blood group systems are internationally recognized namely A, C, D, K, P, Q, and U. A unique erythrocyte antigen called Donkey factor not only found in mule and donkey but not found in the horse. It is responsible for neonatal iso-erythrolysis in mule pregnancies. Aa- and Qa-negative horses are the best choice as donors.

Table 4.21 AB blood group system in cats

	Group A	Group B	Group AB
Antigens in erythrocytes	A	B	A, B
Antibodies in plasma	Anti-B	Anti-A	None

4.5.1.4 Bovine Blood Types

There are 11 major blood group systems in cattle namely, A, B, C, F, J, L, M, R, S, T, and Z in which A and F are the most common. The B group itself contains around 60 different antigens. However, the J antigenic is not a true antigen as it is lipid in nature and absorbed into erythrocytes from body fluids. The newborn calves lack this antigen and gradually acquire it during first 6 months of life. In cattle, neonatal iso-erythrolysis is not a common phenomenon. Perfect cross matching before blood transfusions is very difficult in cattle but, bovines negative for J antigen should be preferred. However, first transfusions are generally of low risk in cattle.

4.5.1.5 Ovine and Caprine Blood Types

Seven blood group systems have been identified viz. A, B, C, D, M, R, and X in sheep and five major systems (A, B, C, M, and J) have been identified in goats. The B system is highly polymorphic in nature. R system of blood group is similar with J system in cattle. Heterophilic antibodies in bovine colostrum may lead to neonatal iso-erythrolysis in lambs receiving bovine colostrum.

4.5.1.6 Swine Blood Types

In pigs, eight blood group systems were identified depicted in Table 4.22.

A system has similarities with ABO system in human. A gene encodes enzyme $\alpha 1 \rightarrow 3$ *N*-acetyl-D-galactosaminyltransferase which synthesizes A antigens are lacks in pigs with O blood group.

Table 4.22 Blood grouping systems in pigs

System	Blood group
A	A, O
E	Ea, Eb, Ed, Ee, Ef, Eg
F	Fa, Fb
G	Ga
H	Ha, Hb
K	Ka, Kb
L	Lh, Lk
O	Oa

4.5.2 Blood Transfusion

Indication for Blood Transfusion

Anemia is the major indication of blood transfusion followed by hypovolemia, hypoproteinemia, and coagulopathies. Blood transfusion is also practiced in chronic inflammatory or infectious disease as well as in neoplasia. Blood transfusion generally prescribed when the PCV is less than less than 20% in dogs or 15% in cats. In Table 4.23, the need-based transfusion of different blood products has been listed.

4.5.2.1 Collection of Blood

Selection of donor: ***A donor*** should be healthy young adults that have never encountered blood transfusion. The donors must be evaluated for routine physical, hematological, and clinical chemistry and should be screened for blood parasites and other infectious diseases (*Ehrlichia* spp., *Babesia* spp., *Anaplasma* spp., and *Mycoplasma hemocanis* or *Mycoplasma haemofelis*). The permissible amount of blood to be collected in different species is summarized in Table 4.24.

Table 4.23 Blood grouping systems in pigs

Conditions	Blood products
Hemorrhage	Whole blood
Anemia	Packed cell
Burn	Plasma, whole blood
Purpura	Platelets or fresh blood
Edema	Albumin
Hemophilia	Coagulation factors, whole blood

Table 4.24 Blood collection amount in different species

Species	Amount of blood can be collected
Dogs	15 mL of blood/kg BW in every 6 weeks.
Cats	10 and 12 mL of blood/kg body weight. Healthy adult cats can donate 45–60 mL every 6 weeks.
Horse	Adult horses can safely donate approximately 6–8 L of blood. Whole blood can be collected every 15–30 days.
Cattle	Cattle can donate 8–14 mL of blood/kg of body weight.

Methods of blood collection from donors: In cats and dogs, jugular venipuncture is generally practiced for blood collection under general anesthesia, but sedation should not be performed with acepromazin as it interferes with the platelet functions. Commercially available human blood collection bags (450 mL capacity) are generally used for dogs and special pediatric collection bags (75 mL capacity) are used for cats. To minimize platelet activation during collection, blood flow should be rapid and the venipuncture should be clean. Sometimes mild vacuum pressure should be applied for the ease of collection. To accelerate hemostasis, direct pressure should be applied at the site of collection. Collected blood should be immediately refrigerated at 4 °C till transfusion or component separation. It was recommended to use desmopressin (DDAVP), a synthetic analog of arginine vasopressin to maximize the yield of vWf.

4.5.2.2 Preservation of Blood for Transfusion

The preservatives for blood are so designed to restore the normal metabolism of the corpuscles along with anticoagulant activity. The common ingredients for preserving blood along with their functions are summarized in Table 4.25.

Two anticoagulants-preservatives listed below have been preferred to preserve the blood after collection (Table 4.26).

Preservation injuries or changes in stored blood: Long-term storage or inappropriate cold chain during transportation and preservation can lead some biochemical, biomechanical, and oxidative changes as listed in Table 4.27.

Bringing the blood in the room temperature before transfusion can correct 2,3 DPG concentration. The plasma potassium level also decreased due to the activation of Na^+-K^+-ATPase pump at the room temperature. The oxidative changes are very difficult to correct.

Table 4.25 Blood preservatives and their functions

Ingredients	Functions
Dextrose	Supports ATP and 2,3-DPG generation (binds with β chain of Hb and release more O_2) by glycolytic pathways.
Adenosine	Synthesizes ATP, increases level of ATP, extends the shelf-life of red cells.
Citrate	Prevents coagulation by chelating calcium.
Sodium diphosphate	Prevents fall in pH
Inosine (hypoxanthine ribose)	Prevents decrease in 2,3 DPG

Table 4.26 Commonly used blood preservatives and their functions

Preservatives	Composition	Remarks
Citrate-phosphate-dextrose-adenine (CPDA-1)	• Dextrose: 3.19 g • Sodium citrate: 2.63 g • Citric acid: 0.38 g • Adenosine: 0.03 g • Sodium diphosphate: 0.22 g • Distilled water up to 100 mL	• It maintains higher levels of 2,3-disphosphoglycerate (2,3-DPG) and adenosine triphosphate (ATP) in collected blood. • Blood can be stored for approximately 35 days.
Acid-citrate-dextrose (ACD)	• Dextrose: 1.04 g • Sodium citrate: 1.32 g • Citric acid: 0.48 g • Distilled water up to 100 mL	• Blood can be stored for approximately 21 days.

Table 4.27 Preservation injuries of blood

Biochemical/metabolic	Biomechanical	Oxidative
• Decrease in lactate, ATP, 2,3 DPG • Increase in K^+ in plasma and Na^+ in RBC	• Altered morphology of RBC (anisocytosis) • Hemolysis	• Hb oxidation • Hb denaturation • Lipid peroxidation

4.5.2.3 Blood Transfusion in the Recipient

Cross matching technique: In cross matching system, the blood samples of both donor and recipient are subjected to separation of plasma, serum, and erythrocytes. In major cross match system, patient serum and donor RBCs are mixed whereas donor serum and patient RBCs are mixed in minor cross match system. No hemolysis after 15–30 min at room temperature indicates compatibility for transfusion. The cross matching technique is described flow diagrammatically in Fig. 4.7.

Calculations of blood volumes to be transferred to the recipient: Following formula is used to calculate the required volume of blood to be transferred to the recipient.

$$\text{Blood volume to be transfused} = k \times \text{Weight (kg)} \times \frac{(\text{Required PCV} - \text{Recipient PCV})}{\text{PCV of donated blood}}$$

where $k = 90$ for dogs and 66 for cats. Required post transfusion PCV is 20% in cats and 25–30% in dogs that are sufficient to cope up the complications related to anemia.

Route of administration: Generally, blood is administered through intravenous catheter into a cephalic or jugular vein. Transfusion of blood in the marrow is as efficient as intravenous administration, and in some cases (severe hypotension or pediatric patients) blood can also be administered into proximal femur through a syringe fitted with 18–20 gauge needle or in the trochanteric fossa with spinal needle. Intraperitoneal administration is generally avoided as the extraction rate is only 40%. A filter in the transfusion set must be used to minimize cellular aggregations and microthrombi that may lead to pulmonary edema. Under normal condition, the standard speed is 5–10 mL/kg/h in normovolemia which can be increased a maximum of 20 mL/kg/h under hypovolemia. A decreased flow rate (2 mL/kg/h) is usually recommended in patients with renal failure.

4.5.2.4 Transfusion Hazards

Monitoring of recipient health status is essentially required during transfusion period. Transfusion should be stopped immediately. Commonly occurring transfusion reactions are discussed below.

Hemolysis: Hemolytic reactions are common during mismatched transfusions. In dogs, acute hemolytic reaction with mismatched blood is rare during first transfusion due to presence of less immunogenic alloantibodies but, in cats even a small volume of mismatched blood can be life threatening due to the presence of strong naturally occurring alloantibodies. The normal lifespan of compatible transfused erythrocytes in dogs is approximately 21 days which may decrease to minutes or even up to 12 h after acute hemolytic transfusion reaction and for cat it may decrease to 1 day from normal lifespan of 30–40 days of transfused erythrocytes.

Acute hypersensitivity reactions: Allergic and anaphylactic reactions can also occur during transfusion which may lead to cardiopulmonary arrest. Allergic reactions during transfusion can be monitored by observing erythema, pruritus, urticaria, vomition, and dyspnea. In such cases, corticosteroids, antihistamines, and adrenaline are recommended.

Pyrexia: Increase in the body temperature by 1 °C or more within 3–4 h of the transfusion has been reported. It may

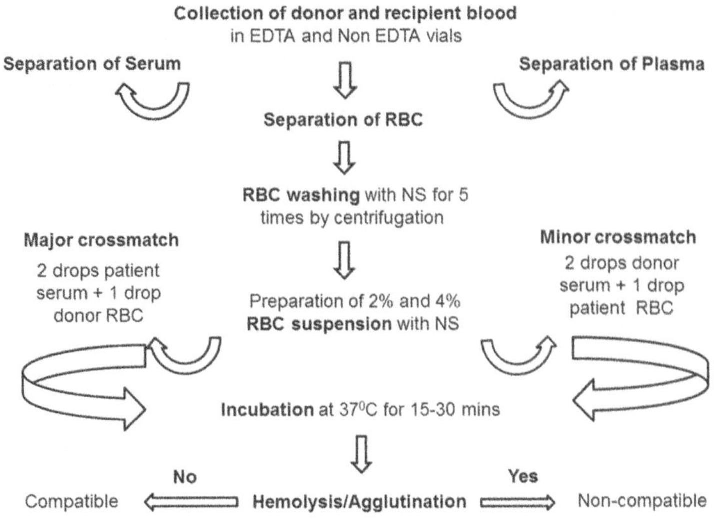

Fig. 4.7 Flow diagram of cross matching technique

be either due to bacterial contamination or reactions of antibodies with platelets and leukocytes that are undetected during typing. Non-hemolytic and non-infectious pyrexia usually requires no treatment.

Contamination: It can occur through blood bag contaminated with pathogens. Some pathogens can escape refrigeration and multiply when the blood is brought to room temperature. The signs of contamination are pyrexia, diarrhea, vomition, and abdominal pain. Severe bacterial contamination may lead to shock. In suspected cases, blood bags should be evaluated for culture and sensitivity and accordingly antibiotic therapy can be recommended. Proper screening of donors for FeLV, FIV, and Hfelis is recommended for cats and heartworm, rickettsial diseases, and blood parasites (Babesia canis) are evaluated for canines.

Hypocalcemia: Its occurrence is rare and may happen when large volume of blood with citrate anticoagulant is administered rapidly which chelates body calcium. The signs of hypocalcemia includes tremors, cardiac arrhythmia and vomition.

Circulatory overload: It is associated with rapid administration of large volume of blood in the patients with cardiac abnormalities, renal failure, and normovolemia. The clinical manifestations of circulatory overload are tachypnea, dyspnea, tachycardia, coughing, and pulmonary edema. Diuretics and oxygen therapy are generally recommended to control the complications.

4.5.3 Hematological Disorders

4.5.3.1 Anemia

It is the qualitative and quantitative decrease in RBC or hemoglobin or both with respect to age and sex of the individual. It interferes with the oxygen carrying capacity of the blood and the clinical manifestations of anemia are due to compensatory mechanism to increase oxygen saturation such as tachypnea, tachycardia, lethargy, and reduced exercise tolerance. Sometimes icterus is also associated with anemia.

4.5.3.1.1 Classification of Anemia

4.5.3.1.1.1 Based on Bone Marrow Function

On the basis of bone marrow functions, anemia can be classified as regenerative and non-regenerative anemia.

In regenerative anemia, bone marrow produces erythrocytes normally in response to decreased erythrocytes. The etiology of regenerative anemia is blood loss and destruction/lysis of erythrocytes (hemolytic anemia). Bone marrow regeneration can be evaluated through the reticulocyte response.

Hemolytic Anemia

This occurs due to intra- or extravascular hemolysis. Intravascular hemolysis is characterized by hemoglobinemia and hemoglobinuria which is absent in extravascular hemolysis. The causes of hemolytic anemia are immune mediated, drug induced (aspirin, benzocaine, propofol, levamisole, sulfonamides), or toxin (dicoumarol, naphthalene, benzene, crude oil) induced, feed borne (onions, oak, red maple), infectious (hemoprotozoa, *Clostridium* spp., *Leptospira* spp., *Mycoplasma* spp., equine infectious anemia virus, feline leukemia virus *Ehrlichia* spp.), or hereditary (pyruvate kinase deficiencies, phosphofructokinase deficiency, and porphyria).

In non-regenerative anemia, the bone marrow is affected and unable to produce sufficient erythrocytes. The main factors which cause this type of anemia are nutritional deficiencies, chronic diseases, renal failure, and the diseases affecting bone marrow.

Nutritional Deficiencies

Deficiency of iron, copper, vitamin B_{12}, vitamin B_6, riboflavin, niacin, and vitamin E leads to anemia. The deficiency of iron is the most common form of nutritional deficiency anemia affecting piglets and puppies particularly before weaning. Milk is deficient in iron thus the young animals are unable to get adequate iron during suckling. Microcytosis is common in iron deficiency anemia. In ruminants, cobalt deficiency causes normocytic, non-regenerative anemia caused by grazing cobalt deficient soil. Cobalt is required for the synthesis of cobalamin by rumen bacteria.

The deficiency of vitamin B_{12} leads to pernicious anemia.

Pernicious Anemia

It is characterized by the maturation failure of RBC due to poor absorption of vitamin B_{12} from the intestine. It occurs due poor secretion of a glycoprotein called intrinsic factor (IF) from parietal cells. It helps in the absorption of vitamin B_{12} after binding tightly with the vitamin B_{12} and protects from gastric digestion. Vitamin B_{12} then internalizes in the brush border membrane of ileum. Lack of intrinsic factor thus leads to poor absorption of vit-B_{12}. Pernicious anamia is very rare in dogs and cats and is usually manifested due to the genetic mutations in the ileal cubam receptor, often occurred in dog breeds like Beagles, Border Collies, Giant Schnauzers, and Australian Shepherds.

Anemia Associated with Chronic Disease

Chronic inflammation, tumor, liver disease, and endocrine disturbances (hyper- or hypo-adrenocorticism and hypothyroidism) lead to non-regenerative anemia in animals. Chronic inflammatory diseases lead to cytokine production which affect red blood cell survival, bone marrow's ability to regenerate and poor iron availability.

Anemia associated with chronic renal failure is common in animals. Chronic kidney disease causes less erythropoietin production which affects erythropoiesis in the bone marrow.

Anemia Associated with Bone Marrow Diseases

Diseases affecting bone marrow not only affect erythrocyte production but also the reduction in the number of all types of blood cells including leukocytes and thrombocytes. The major diseases affecting bone marrow are discussed below:

Aplastic anemia: This type of anemia is associated with decreased bone marrow response to generate erythrocytes. The underlying causes of aplastic anemia are infections (Feline leukemia virus, Feline immunodeficiency virus, *Ehrlichia*, *Mycoplasma*), drug therapy, toxins, and total body irradiation.

Myelodysplasia: It is characterized by defective formation of blood cell precursors in the bone marrow. Leukemia and thrombocytopenia are also occurred in addition to anemia in myelodysplasia.

Myelofibrosis: In this condition, normal marrow tissue is replaced by fibrous (scar) tissue. It can be occurred as a result of neoplasia, immune-mediated hemolytic or hereditary anemias, and whole-body irradiation. Immune suppression also occurred in myelofibrosis.

4.5.3.1.1.2 Based on Erythrocyte Volume and Hemoglobin Concentration

Normocytic anemia: In this condition, the size of erythrocyte is normal but the number decreases. Normocytic anemia can be classified into normocytic normochromic anemia in which MCV, MCH, and MCHC are within normal range but, the lifespan of RBC is short. Normocytic hypochromic anemia is characterized by decreased MCH and MCHC as erythrocytes are paler than normal indicating less hemoglobin. Normocytic normochromic anemia is often associated with hypothyroidism in the dogs. Blood smear of normocytic normochromic anemia shows polychromasia, anisocytosis, or nucleated red cells.

Microcytic anemia: In microcytic anemia, erythrocyte number is adequate but the size of erythrocyte is smaller than normal and decreased MCV. Iron deficiency is the most common cause of microcytic anemia. Iron deficiency leads to impaired hemoglobin synthesis and the hemoglobin-deficient erythrocyte precursors are undergoing additional mitosis to achieve normal hemoglobin level, thus generated microcytosis. In microcytic anemia, both MCV and MCH are decreased. Microcytic anemia can also occur due to copper deficiency.

Macrocytic anemia/megaloblastic anemia: Macrocytic anemia is characterized by the appearance of macrocytic erythrocytes (erythrocytes with increased size) as evident from increased MCV. Deficiency of cobalamin and folic acid is the prime cause of macrocytic anemia which interferes with DNA synthesis and leads to maturation arrest of erythrocyte precursors. In cats, feline leukemia virus (FeLV) infection and myelodysplastic syndromes lead to macrocytosis. Some poodles also have macrocytic anemia.

Sickle cell anemia: This anemia is hereditary origin and characterized by sickle shaped RBC with increased fragility. It is characterized by malformed β chain of Hb (amino acid glutamic acid at sixth position is replaced by valine) that forms thread-like structure in deoxygenated form and leads to change the shape of RBC-like sickles. In deer, the sickle shaped erythrocytes are common, but the occurrences of sickle cell anemia in other animals are rare.

Spherocytosis In this condition, the RBC loses its biconcavity and becomes spherical in shape. It is caused due to the defect in RBC cytoskeleton proteins like spectrin, ankyrin, Band 3, or Protein 4.2 or due to partial phagocytosis of cell surface. Spherocytes are prone to immune-mediated hemolysis.

4.5.3.2 Erythrocytosis/Polycythemia

An increase in the erythrocyte number together with increased hemoglobin concentration and hematocrit value is termed as erythrocytosis. Erythrocytosis and polycythemia are often synonymous but in human medicine polycythemia restricts not only to erythrocytosis but increases in leukocytes and thrombocytes also. Pathologically, erythrocytosis can be classified into two categories namely relative and absolute.

In relative erythrocytosis, increase in the hematocrit value doesn't associate with increased RBC but due to reduced plasma volume. Severe dehydration due to fluid loss is associated with diarrhea, vomition, or polyuria results in relative erythrocytosis and is most common in dogs and cats. Hemoconcentration due to catecholamine mediated splenic contraction during excitement also leads to mild erythrocytosis in horse and dogs.

Absolute erythrocytosis is characterized by true increase in erythrocyte numbers. It can be classified into primary and secondary erythrocytosis.

Primary erythrocytosis is also called polycythemia vera in which proliferation of erythroid precursors occur independent of erythropoietin. Splenomegaly, leukocytosis, and thrombocytosis are the common clinical manifestations of polycythemia vera. It is generally occurred in dogs and cats of middle age groups (6–7 years). It is interesting to note that primary erythrocytosis is common among male dogs whereas female cats mostly suffer from it. Secondary erythrocytosis results due to excess production of erythropoietin with systemic hypoxia (appropriate secondary erythrocytosis), or without (inappropriate secondary erythrocytosis) systemic hypoxia. Inappropriate secondary erythrocytosis is common in neoplasia and kidney disorders. Endocrinopathies associated secondary erythrocytosis occurs in conditions such as hyperadrenocorticism, hyperthyroidism, and acromegaly.

4.5.3.3 Bleeding Disorders

4.5.3.3.1 Platelet Defects

Disorders of platelets include thrombocytopenia (decreased number of platelets) or having thrombocytopathies (impairment of platelet functions). Both of these disorders can be congenital or acquired.

4.5.3.3.1.1 Congenital Thrombocytopenia

It is actually a hematopoietic disorder which maintains a 12 days cycle. In this condition, all types of blood cells are affected including platelets. It is very fatal as most of the affected dogs die before 6 months of age. Cavalier King Charles Spaniels suffer from benign hereditary macrothrombocytopenia which is characterized by thrombocytopenia with giant platelets.

4.5.3.3.1.2 Acquired Thrombocytopenia

The potential causes of acquired thrombocytopenia are

Rickettsial diseases: Mild to moderate degree of thrombocytopathies are seen in Ehrlichia and Anaplasma infections. It is characterized by epistaxis, gum bleeding, and gastrointestinal hemorrhage (black stools).

Immune-mediated thrombocytopenia: Disfunctions in the immune system produces antibodies against platelets or platelet precursors in the bone marrow. Repeated vaccinations of dogs with live adenovirus or paramyxovirus vaccines lead to mild thrombocytopenia.

Drug-induced thrombocytopenia: Estrogen and some antibiotics can suppress the production of platelets whereas certain drugs (acetaminophen, aspirin, penicillin) destroy the circulating platelets.

4.5.3.3.1.3 Congenital Thrombocytopathies

It can be classified into several categories

Canine thrombopathia: It is common in Basset Hounds irrespective of normal levels of platelets and von Willebrand's factor. It can be diagnosed by platelet function testing.

Glanzmann thrombasthenia: This condition is characterized by prolonged bleeding times. Blood smear reveals giant platelets with altered shape. Platelet aggregation during coagulation mechanism is inhibited in this condition. It is seen in Otterhounds and Great Pyrenees dogs.

Von Willebrand disease: Affected individuals are deficient in von Willebrand's factor. It is one of the most common inherited bleeding disorders in canines with almost all the breeds.

The disease is classified into three categories

- Type 1: Low amount von Willebrand factor with mild to moderate signs.
- Type 2: Low amount of the factor with moderate to severe signs.
- Type 3: Absence of von Willebrand factor with frequent episodes of bleeding. It is most frequent in Shetland Sheepdogs and Scottish Terriers.

4.5.3.3.1.4 Acquired Thrombocytopathies

Acquired thrombocytopathies are consequences of bone marrow tumor (multiple myeloma) and chronic kidney diseases.

Acquired thrombocytopenia also leads to platelet functional defects. It can also be drug induced.

4.6 Avian Hematology

In commercial poultry farming, serology plays a pivotal role in monitoring and disease diagnosis and the hematology assays rarely used for etiological diagnosis. In addition to this, avian hematological diagnosis has some constrains like small blood volume, the fragility of erythrocytes and hemolysis in EDTA anticoagulant, and staining variations with traditional Giemsa or Wright's stains. But hematological evaluation of avian species nevertheless can be used to evaluate health and well beings of poultry, prognosis of disease and the therapeutic responses. Since last 15 years or so, significant advances have been made in the field of avian hematology parallel to other areas like nutrition, therapeutics, wellness examinations, and surgery. Chicken is generally used as research animal model for other avian species.

4.6.1 Properties of Blood

4.6.1.1 Blood Volume

The blood volume usually ranges from 6 to 11 mL/100 kg body weight and 10% of total blood volume and can be collected safely.

4.6.1.2 The Specific Gravity and the Viscosity

The specific gravity and the viscosity of the blood of different avian species are given in Table 4.28.

The higher viscosity in male may be due to higher number of cells compared to female.

The rate of sedimentation of erythrocytes usually depends upon the cell size, specific gravity of plasma and composition of plasma by governed by the gravitational force and the functional resistance of the surrounding plasma.

4.6.1.3 Hematocrit

The hematocrit values of different avian species are given in Table 4.29.

Table 4.28 The specific gravity and the viscosity of whole blood and plasma in different avian species

	Specific gravity		Viscosity	
Species	Whole blood	Plasma	Whole blood	Plasma
Chicken (female)	1.050	1.099	3.08	1.51
Chicken (male)	1.054	1.021	3.67	1.42
Duck	1.056	1.020	4.0	1.5
Goose	1.050	1.021	4.6	1.5
Ostrich	1.063	1.022	4.5	–

Source: Sturkie (1986)

Table 4.29 The hematocrit value of different avian species

Species	Hematocrit value (%)
Chicken (sexually immature male)	29
Chicken (sexually immature female)	29
Chicken (sexually mature male)	45
Chicken (sexually mature female)	29
Turkey (male)	45.1
Turkey (male)	36.4
Ducks (mallard)	43
Quail	38
Pigeon	52

Source: Sturkie (1986)

Sex has no significant influence on the hematocrit value in birds but inter species variations are evident. An increase in hematocrit or hemo-concentration is induced by the release of epinephrine or hyperthermia.

4.6.2 Composition of Plasma

The plasma of avian species is straw yellow in color and slightly turbid in appearance. The plasma contains 80% of water along with some dissolve substances like glucose, proteins, fatty acids, vitamins, minerals, and hormones (Table 4.30). The composition varies according to sex, age, egg production, food consumption, and physiological state of birds. Unlike mammals the colloidal osmotic pressure of avian plasma depends primarily on electrolytes (95%) followed by glucose, amino acids, urea (5%), and plasma proteins (0.5%).

Plasma proteins: The females have higher plasma proteins than males due to the egg production. The plasma proteins of the avian species are classified into three categories based on electrophoretic mobility namely pre-albumin, albumin, and globulin. The pre-albumin is also called as trans-thyrectin due to its thyroid-binding ability. Albumin is involved in facilitating nutrient transport (mainly fatty

Table 4.30 The composition of plasma in avian species

Plasma constituents	Concentrations
Proteins (g/L)	39.6 ± 0.74
Albumin (g/L)	15.9 ± 0.55
Globulin (g/L)	18.8 ± 0.96
Glucose (mM/L)	15.9 ± 0.32
Sodium (mEq/L)	152.5 ± 1.13
Sodium (mEq/L)	112.6 ± 1.28
Calcium (mEq/L)	2.56 ± 0.10
Sodium (mEq/L)	3.21 ± 0.19
Uric acid (mM/L)	0.50 ± 0.04
Blood urea nitrogen (mM/L)	1.11 ± 0.09

Source: Sturkie (1986)

acids) and lipophilic hormones. The concentration of plasma albumin varies with the state of birds. Plasma albumin concentrations are reported to be decreased in laying females. Diurnal variations in the plasma albumin were also reported. Feed withdrawal and reduced photoperiod also reduce the albumin concentration. The globulins are more concentrated in birds compared to mammals; thus, birds are good antibody producers. There are three types of immunoglobulins in birds like IgA (0.3 g/L), IgM (2.7 g/L), and IgY (5.5 g/L). Immunoglobulin Y is equivalent to IgG in mammals. The globulin concentration is increased during reduced photoperiod and feed withdrawal. In layer birds, IgY are reported to be lower due to the transportation of circulating IgY into the yolk. In male chicken, α1-globulin concentration declines during growth.

Plasma lipids: The plasma of layer birds contains higher lipids (two to five-folds increase) when compared to non-laying hens.

Plasma glucose: The plasma glucose concentration is higher in birds compared to mammals. Alterations in the glucose concentration increases with hatching.

Plasma electrolytes: The calcium concentration in laying birds are more (two-folds increase) than non-laying birds. Plasma concentration of sodium and potassium are decreased during acute heat stress in chicken. In birds, urates or uric acid is one of the predominate components of plasma which are the catabolic end products of protein and non-protein.

Plasma enzymes: A number of enzymes namely alkaline phosphatase (ALP), glutamic oxaloacetic ransaminase (GOT), asparate aminotransferase (AST/SGOT), alanine aminotransferase (ALT/SGPT), cholinesterase, creatine phosphotransferase, and lactic acid dehydrogenase (LDH) are evident in the plasma of birds depending on the physiological state of the birds. In dehydration and hyperthermia, SGPT is elevated. In feed restriction, SGPT is decreased and LDH, GOT, and ALP are increased. In organophosphorus poisoning, plasma cholinesterase activity was decreased and thus it is used as bio-markers in wildlife risk assessment.

4.6.3 Erythrocytes

Unlike mammals, the circulating erythrocytes of avian species are ovoid in shape with a centrally located round nuclei and mitochondria. The morphometric characteristics of avian erythrocytes along with hemoglobin content are summarized in Table 4.31.

Table 4.31 The morphometric characteristics of avian erythrocytes along with hemoglobin content

Parameters	Value
Total erythrocyte counts (10^6/μL)	3.2
Hemoglobin (%)	10.1
Erythrocyte volume (fL)	149.4
Erythrocyte length (μm)	12.2
Erythrocyte width (μm)	7.1
Erythrocyte cross sectional area (μm^2)	68.0

Source: Sturkie (1986)

The $Na^+/K^+/2Cl^-$ cotransporters in the avian erythrocytes are involved in the transport of sodium and potassium across the erythrocyte membrane. There is higher sodium efflux in turkey erythrocytes compared to mammals. The erythrocytes potassium transport is decreased with age in avian species.

The total erythrocyte counts varied significantly between seasons with the highest in fall compared to winter and spring. Diurnal variations in erythrocyte counts are also noticed in birds. The total erythrocyte count is usually high and mid night and low around noon.

Hemoglobin is the most abundant protein inside the erythrocytes. The nuclei of avian erythrocytes also contain hemoglobin transported across the pores of nuclear membrane.

The nuclei of avian erythrocyte contain condensed chromatin associated with histones (responsible for control of transcription, DNA replication, and repair). Despite of the presence of nuclei, avian erythrocytes are unable to divide.

4.6.3.1 Erythrocyte Metabolism

The major metabolic fuel of avian erythrocytes is glucose though the entry of glucose in avian erythrocytes is low due to small number of GLUT1 (200 copies of GLUT1 per erythrocyte compared to 300,000 in humans). Chicken erythrocytes can use glycine as a substrate for energy metabolism.

The major metabolic pathway is the TCA cycle due to the presence of mitochondria. During the early embryonic life, avian erythrocytes contain high ATP concentrations along with cytidine triphosphate (CTP) and Uridine-5′-triphosphate (UTP). Another interesting feature of avian erythrocytes is the high amount of 2,3-BPG during embryonic life that increases the oxygen affinity during embryonic life.

4.6.3.2 Erythropoiesis

The site of avian erythropoiesis in the bone marrow and liver is the major site for extramedullary hematopoiesis. However, the bone marrow decreases during the development of bone air sacs. The erythroid, lymphoid, and thrombocyte precursor cells start around 2–3 days of embryonation in the para-aortic region. The hematopoiesis is a two-stage process, in stage I, PHSCs are differentiated from early mesoderm and in stage II, these cells undergo further differentiation to form committed stem cell lineages like burst-forming unit (BFU) and colony-stimulating

Table 4.32 The lifespan avian erythrocytes

Species	Lifespan (days)
Ducks	42
Chicken	35
Pigeon	48
Quail	34

Source: Sturkie (1986)

factors. Different developmental stages of avian erythrocytes are rubriblasts (or erythroblasts), prorubricytes, basophilic rubricytes, early polychromatic rubricytes, late polychromatic rubricytes, polychromatic rubricytes, and late polychromatic rubricytes. Around 1–5% of circulating erythrocytes are late polychromatic rubricytes. The cells resemble mature erythrocytes with large size with more basophilic cytoplasm and less chromatin condensation.

4.6.3.3 Lifespan of Erythrocytes

The lifespan of avian erythrocytes is less compared to mammals. This is probably due to higher body temperature, rapid metabolic rate due to consumption of more oxygen and nutrients than mammals. The lifespan of erythrocytes in different avian species have been presented in Table 4.32.

4.6.4 Hemoglobin

The hemoglobin of avian species is a tetrameric protein containing four polypeptide chains and heme unit with iron (Fe^{2+}) at the center. In chicken, six forms of hemoglobin are available summarized in Table 4.33.

Avian Hbs exist in more tense (T) conformation than the mammalian Hbs and thus have lower oxygen affinity and higher thermal stability compared to mammalian Hb. Another striking difference between mammalian and avian hemoglobin is that the avian hemoglobin binds more with inositol pentaphosphate (IP5) unlike 2,3 BPG in mammalian Hb.

> **Know More.....**
> The birds fly at high altitude (bar-headed goose and Andean goose) and emperor penguins have higher oxygen affinity due to single substitutions in α-globin chains of hemoglobin.

4.6.5 Leukocytes

The total leukocyte count in chicken is 20–30 × 10^3 cells/μL. Lymphocytes constitute largest proportion (55–60%) of blood, followed by heterophils (20–30%), monocytes (10%), eosinophils (3–8%), and basophils (1–4%). The heterophils of ostrich and pheasants is about 60–70%. Heterophils are analog of mammalian neutrophils with a diameter of 10–15 μm. The nucleus of heterophils also polymorphic with varying degree of lobulation. The cytoplasm of heterophils contains a characteristic rod-like granules.

4.6.6 Hemostasis in Avian Species

The clotting time of avian species is relatively more compared to mammals, hence avian blood clots slowly compared to mammals. The coagulation proteins are evolved from common ancestor. The overall coagulation mechanism is also same for mammals and birds.

Thrombocytes: Thrombocyte counts of the avian species usually range from 20,000 to 30,000/μL and may extend up to 50,000/μL. Avian thrombocytes are irregular in shape with pseudopodia. Intracellular organelles include mitochondria, endoplasmic reticulum, dense granules, and microtubules.

Table 4.33 Different forms of hemoglobin in avian species

Stages of development	Forms of hemoglobin	Structure	Remarks
Embryonic	Major forms	Hb P (π, α, β, and ρ globin)	The highest concentration of major Hb forms is around fourth days of embryonic development and decreases gradually and become undetectable from 15 days.
		Hb P′	
	Minor forms	Hb M (α^D globin)	Peak level is seen around 6–7 days and then decrease, but still persists up to hatching.
		Hb E (α^A globin)	
Adult	Major forms	Hb A ($\alpha^A{}_2\beta_2$ globin)	
	Minor forms	Hb D ($\alpha^D{}_2\beta_2$ globin chain)	

4.6.7 Blood Groups and Blood Transfusion

About 28 blood groups have been identified in domestic chicken under three different blood group systems (B, L, and N).

Blood transfusion is rarely practiced in avian species. The general indications for blood transfusion in birds are hypovolemic shock, acute hemorrhage, and coagulopathies in rodenticide poisoning. Blood transfusion is usually advocated when there is loss of 20% blood volume and a PCV level below 20%. The homologous transfusion is preferred in avian species. However, the first heterologous transfusion can safely be done as the birds lack preformed antibodies against blood group antigens. The mean half-life of erythrocytes in homologous transfusion is more (7 days) compared to heterologous transfusion (12 h) studied in pigeon (Sandmeier et al. 1994).

Learning Outcomes

- **Blood**: Blood is a fluid connective tissue comprising plasma and blood corpuscles. There are three types of blood corpuscles viz. red blood corpuscles (RBC), white blood corpuscles (WBC), and platelets. Blood plasma facilitates free movements of blood corpuscles. Varies inorganic and organic components constitute the plasma. Plasma protein and plasma lipids perform variety of functions. RBCs are concerned with gaseous transport. The main function of WBC is to provide immunity. Platelets are involved in blood coagulation process. Other than gaseous transport, blood facilitates nutrient transport, buffering action, thermoregulation, excretion of metabolic waste products, and transport medium of hormones and drugs.
- **Hematopoiesis**: Hematopoiesis is the formation of blood cells. It starts during embryonic development from mesodermal germ layer and continued throughout adulthood to produce and replenish the blood cellular components. The precursors of all blood cells are pluripotential hematopoietic stem cells (PHSC) characterized by their long self-renewal capacity and pluripotency. PHSCs are differentiated into common myeloid progenitors (CMP) or colony forming unit granulocyte, erythrocyte, monocyte, megakaryocyte (CFU-GEMM), and common lymphoid progenitor (CLP). CMP generates megakaryocyte, erythroid, granulocytes, and macrophages progenitors and CLP give rise to B and T lymphocytes. The platelets are derived from megakaryocytes. The generation of RBC, WBC, and platelets are termed as erythropoiesis, leucopoiesis, and thrombopoiesis, respectively.
- **Iron metabolism**: Iron is an integral component of hemoglobin, myoglobin, and other substances such as cytochrome, cytochrome oxidase, peroxidase, and catalase. In the body, iron mainly exists either in the form of hemoprotein and non-heme iron. The metabolism of iron is regulated by iron reserve in the body, hypoxia, and rate of erythropoiesis.
- **Hemoglobin**: Hemoglobin is an iron containing conjugated protein composed of heme (non-protein prosthetic group) and globin exclusively found in erythrocytes and transports the oxygen from lungs to tissues and carbon-dioxide from tissues to lungs. The synthesis of hemoglobin begins in the pro-erythroblast stage and continues till the reticulocyte stage. Hemoglobin forms different derivatives by combining with different molecules.
- **Blood coagulation**: It is cascade of enzyme activation which enables the stoppage of bleeding from injured vessels. The coagulation machinery includes clotting factors of the plasma, vascular endothelium, and platelets. The blood coagulation is divided into two broad events namely primary hemostasis in which there is the formation of weak platelet plug and secondary hemostasis where the reinforce of primary hemostasis is occurred with fibrin. The fibrinolytic mechanism facilitates dissolution of clots.
- **Blood grouping**: The determinants of blood groups are the specific polymorphic antigens that reside on the surface of erythrocytes (agglutinogens) and the antibodies (agglutinogens) present in the plasma. Different animals have different blood groups. Blood grouping and cross matching are essentially required from blood transfusion.
- **Hematological disorders**: Hematologic disorders are occurred due to pathological conditions of blood components and blood-forming organs. Hematologic diseases include genetic disorders, anemia, diseases of leukocytes, coagulation abnormalities, and transfusion hazards. Different pathogens and nutritional deficiencies are also predisposing factors for hematological disorders.

(continued)

- **Avian hematology**: The avian species have small blood volume, higher fragility of erythrocytes and staining variations. Unlike mammals, the circulating erythrocytes of avian species are ovoid in shape with a centrally located round nuclei and mitochondria. In chicken, six forms of hemoglobin are available. Lymphocytes constitute largest proportion of WBC, followed by heterophils, monocytes, eosinophils, and basophils. The clotting time of avian species is relatively more compared to mammals.

Exercises

Objective Questions

Q1. Property which allows erythrocytes to adhere with each other like the stack of coins known as ________________.

Q2. Which plasma protein helps in heme transport is?

Q3. The shape of erythrocytes of deer and antelope?

Q4. By which metabolic pathway mature RBCs produce 2,3-diphosphoglycerate (2,3-DPG)?

Q5. The third wave of hematopoiesis occurred in _________ region of the developing embryo.

Q6. What is the name of cellular oxygen sensor that induces the secretion of erythropoietin in response to hypoxia?

Q7. Heme iron is absorbed more readily than non-heme iron—(True/False).

Q8. What is the predominant source of hepcidin?

Q9. The type of hemoglobin used to monitor hyperglycemia is ______________.

Q10. Sulfhemoglobinemia occurred ______________ infection.

Q11. What is the "eat-me" signals of erythrocyte destruction?

Q12. Monocytes are the largest leukocytes—(True/False).

Q13. Eicosanoid synthesized from vascular endothelium and prevents platelet aggregation is ______________.

Q14. ________________________ species have no universal donor.

Q15. What preservative component is used in blood to prevent decrease in 2,3 DPG?

Q16. Name the clinical condition that leads to the loss of erythrocyte biconcavity.

Q17. The movement of leukocytes to the injured tissues in response to bacterial toxins is called ________________.

Q18. Which avian immunoglobulin is analogous to mammalian IgG?

Q19. What are the avian leukocytes analogous to mammalian neutrophil?

Q20. Avian Hbs exists in more tense (T) conformation than the mammalian Hb—(True/False).

Subjective Questions

Q1. "Hypo-albuminemia leads to edema"—Justify the statement.

Q2. Why chronic renal failure leads to anemia?

Q3. In what condition methemoglobin is used as medication?

Q4. Why in Vit-12 deficiencies microcytic anemia occurs?

Q5. Why physiological polycythemia is common in high altitude animals?

Q6. What are the unique features of PHSC?

Q7. What are the advantages of erythrocyte bi-concavity?

Q8. What are the non-phagocytic defense strategies of neutrophils against pathogens?

Q9. What is the fate of hemoglobin in intravascular hemolysis?

Q10. What are T lymphocyte subsets?

Q11. Write down in brief about the fibrinolytic mechanism.

Q12. What do you mean by ontogeny of Hemoglobin Synthesis.

Q13. How intact vascular endothelium prevent coagulation in vitro?

Q14. What is respiratory burst?

Q15. Write down the difference between extravascular and intravascular hemolysis.

Answer to Objective Questions

A1. Rouleaux
A2. Hemopexin
A3. Sickle shaped
A4. Luebering–Rapoport pathway
A5. Aorta-gonad-mesonephros
A6. Hypoxia-inducible transcription factor-1
A7. True
A8. Hepatocytes
A9. Glycosylated hemoglobin
A10. *Clostridium welchii*
A11. Phosphatidyl serine
A12. True
A13. Prostacyclin
A14. Feline
A15. Inosine
A16. Spherocytosis
A17. Chemotaxis
A18. Immunoglobulin Y
A19. Heterophil
A20. True

Keywords for Answer to Subjective Questions

A1. Albumin, colloidal osmotic pressure of blood
A2. Kidney, erythropoietin, erythropoiesis
A3. Cyanide poisoning, methemoglobin, cyan-met hemoglobin
A4. Vit B_{12}, DNA synthesis, Cell division
A5. High altitude, PO_2, hypoxia, erythropoietin, erythropoiesis
A6. Pluripotency, self-renewal
A7. Osmotic fragility, diffusion distance
A8. NETosis, Neutrophil-derived microparticles
A9. Hemoglobin, haptoglobin, bilirubin, liver
A10. Helper, cytotoxic, regulatory, surface marker
A11. Tissue type plasminogen activator, urokinase type plasminogen activator. Plasmin
A12. Switching of fetal to adult hemoglobin, alpha globin gene, beta globin gene
A13. Negativity, Prostacyclin (PGI_2), Nitric oxide, Plasminogen activators
A14. Oxygen-dependent phagocytic killing
A15. Site, mechanism, fate of hemoglobin

Further Reading

Books

Blackmer J, Parish S (2002) Diseases caused by allogeneic incompatibilities. In: Smith BP (ed) Large animal int. med, 3rd edn. St. Louis, MO, Mosby Elsevier Science, pp 1604–1613

Campbell TW (1995) Avian hematology and cytology, 2nd edn. Iowa State University Press, Ames

Mayer J, Donnelly TM (2013) Clinical veterinary advisor: birds and exotic pets, vol 63043. Elsevier, St. Louis, Mo. https://doi.org/10.1016/B978-1-4160-3969-3.00425-X

Reece WO (2015) The composition and functions of blood. In: Uemura EE (ed) Section II: Body fluids and homeostasis of Dukes' physiology of domestic animals, 13th edn. Wiley Blackwell, Iowa, p 122

Sturkie PD (1986) Avian physiology. Springer, New York

Wakenell PS (2010) Hematology of chicken and turkey. In: Weiss DS, Wardrop KJ (eds) Schalm's veterinary hematology, 6th edn. Blackwell Publishing, pp 958–966

Weiss DJ, Wardrop KJ (2010) Schalm's veterinary hematology, 6th edn. Wiley-Blackwell, Iowa

Research Articles

Adili N, Melizi M, Belabbas H (2016) Species determination using the red blood cells morphometry in domestic animals. Vet World 9(9): 960–963

Blais MC, Berman L, Oakley DA, Giger U (2007) Canine Dal blood type: a red cell antigen lacking in some Dalmatians. J Vet Intern Med 21(2):281–286

Euler CC, Lee JH, Kim HY, Raj K, Mizukami K, Giger U (2016) Survey of two new (Kai 1 and Kai 2) and other blood groups in dogs of North America. J Vet Intern Med 30(5):1642–1647

Ferreira CN, Sousa MO, Dusse LMS, Carvalho MG (2010) A cell-based model of coagulation and its implications. Rev Bras Hematol Hemoter 32(5):416–421

Finnegan VM, Daniel GB, Ramsey EC (1997) Evaluation of whole blood transfusions in domestic pigeons (Columba livia). J Avian Med Sur 11(1):7–14

Goulet S, Giger U, Arsenault J, Abrams-Ogg A, Euler CC, Blais MC (2017) Prevalence and mode of inheritance of the Dal blood group in dogs in North America. J Vet Intern Med 31(3):751–758

Ivanov IT, Paarvanova BK (2021) Differential dielectroscopic data on the relation of erythrocyte membrane skeleton to erythrocyte deformability and flicker. Eur Biophys J 50:69–86. https://doi.org/10.1007/s00249-020-01491-4

Janyamethakul T, Sripiboon S, Somgird C, Pongsopawijit P, Panyapornwithaya V, Klinhom S, Loythong J, Thitaram C (2017) Hematologic and biochemical reference intervals for captive Asian elephants (Elephas maximus) in Thailand. Kafkas Univ Vet Fak Derg 23(4):665–669

Kaneko JJ, Harvey JW, Bruss ML (1997) Clinical biochemistry of domestic animals, 5th ed. Academic Press, San Diego, California. ISBN 0-12-396305-2

Klaassen JK (1999) Reference values in veterinary medicine. Lab Med 30(3):194–197

Lee JH, Giger U, Kim HY (2017) Kai 1 and Kai 2: characterization of these dog erythrocyte antigens by monoclonal antibodies. PLoS One 12(6):e0179932. https://doi.org/10.1371/journal.pone.0179932

Maas M, Keet DF, Nielen M (2013) Hematologic and serum chemistry reference intervals for free-ranging lions (Panthera leo). Res Vet Sci 95:266–268

Moras M, Lefevre SD, Ostuni MA (2017) From erythroblasts to mature red blood cells: organelle clearance in mammals. Front Physiol (Abs) 8:1076. https://doi.org/10.3389/fphys.2017.01076

Oishi T (1977) Blood groups and serum protein polymorphism in pigs and their application as genetic. Markers 11(3):179–185

Sandmeier P, Stauber EH, Wardrop KJ, Washizuka A (1994) Survival of pigeon red blood cells after transfusion into selected raptors. J Am Vet Med Assoc 204(3):427–429

Sankaran VG, Orkin SH (2013) The switch from fetal to adult hemoglobin. Cold Spring Harb Perspect Med 3(1):a011643

Schepelmann K (1990) Erythropoietic bone marrow in the pigeon: development of its distribution and volume during growth and pneumatization of bones. J Morph 203:21–34

Shrivastav AB, Singh KP (2012) Tigers blood: haematological and biochemical studies. Blood Cell. https://doi.org/10.5772/50360

Siah CW, Ombiga J, Adams LA, Trinder D, Olynyk JK (2006) Normal iron metabolism and the pathophysiology of iron overload disorders. Clin Biochem Rev 27(1):5–16

Stormont C, Suzuki Y, Rhode E (1964) Serology of horse blood groups. Cornell Vet 54:439–452

Thon J, NandItaliano JE (2010) Platelet formation. Semin Hematol 47(3):220–226

Weisel JW, Litvinov RI (2017) Fibrin formation, structure and properties. Sub Cell Biochem 82:405–456

Wienert B, Martyn GE, Funnell APW, Quinlan KGR, Crossley M (2018) Wake-up sleepy gene: reactivating fetal globin for β-hemoglobinopathies. Trends Genet 34(12):927–940

Immune System

5

Joydip Mukherjee, Pradip Kumar Das, Dipak Banerjee, and Ayan Mukherjee

Abstract

Immunity is the ability of an individual to protect against invading pathogens. The machineries utilized by the immune system are intended to recognize pathogens, their destruction, and a memory to remember previous exposure. The immunity is either an inborn reflex to eliminate pathogens (innate immunity) or developed slowly after the exposure of pathogens (acquired immunity). The components of innate immunity are anatomic and physiological barriers, immune effector cells, and soluble factors like antimicrobial peptides, complement system, cytokines, and interferons. The acquired immunity is either cell mediated associated with T lymphocytes or humoral, brought about by the antibodies produced from B lymphocytes. The pattern recognition receptors (PRRs) present in the immune effector cells recognize pathogens through pathogen-associated molecular pattern (PAMP) and damage-associated molecular pattern (DAMP). The toxins released by the pathogens are processed and presented by antigen-presenting cells (APC) with the help of major histocompatibility complex (MHC) as a mark of discrimination between self and nonself. After pathogen recognition, the effector responses such as phagocytosis, complement activation, and antibody-dependent cell-mediated cytotoxicity are initiated to eliminate the pathogens. The incoordination in the different immune components leads to immune pathology such as hypersensitivity, anaphylactic shock, and autoimmune disorders.

J. Mukherjee (✉) · P. K. Das · D. Banerjee
Department of Veterinary Physiology, West Bengal University of Animal & Fishery Sciences, Kolkata, West Bengal, India

A. Mukherjee
Department of Animal Biotechnology, West Bengal University of Animal & Fishery Sciences, Kolkata, West Bengal, India

© The Author(s), under exclusive license to Springer Nature Singapore Pte Ltd. 2023

P. K. Das et al. (eds.), *Textbook of Veterinary Physiology*, https://doi.org/10.1007/978-981-19-9410-4_5

Graphical Abstract

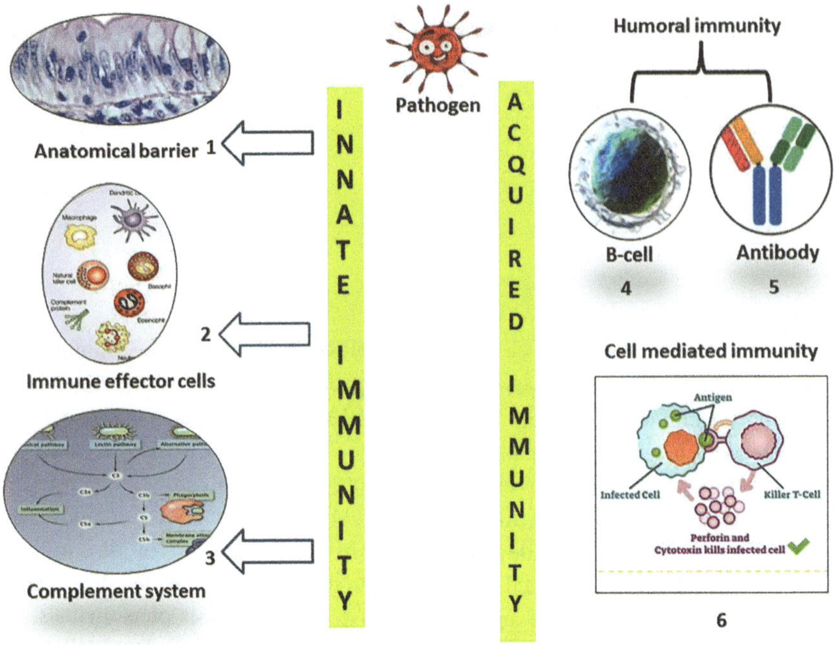

Description of the graphic: Immunity is the ability of the body to protect against pathogens. It has two important components such as innate and adaptive immunity. The innate immune system is equipped with several components like anatomical and physiological barriers (1), immune effector cells and cell surface receptors (2), and complement system (3). The adaptive immunity is of two types, humoral and cell-mediated immunity. The humoral immune response is mediated through B lymphocytes (4) capable of producing antibodies (5), which bind with antigens and make them vulnerable for destruction. The cell-mediated immune response is mediated through T lymphocytes that produce signals to activate phagocytic cells to destroy them (6)

Keywords

Antigens · Innate immunity · Adaptive immunity · Antibodies · Autoimmune disorders

Learning Objectives

- The components of immunity
- Antigens and their types
- Components of innate and adaptive immune system
- Hypersensitivity and immunological disorders

5.1 Immunity

The word immunity was derived from Latin word "immunis," meaning "exempt." Immunity is the ability of an individual to protect against invading pathogenic microorganisms and cancer through interlinking networks of cellular and biochemical mechanisms collectively known as the immune system. Two key features of the immune system are the recognition and response. Immune recognition is a highly specific process to discriminate between foreign pathogens from the body's own cells and proteins and to identify different types of foreign pathogens through their specific chemical compositions. The responses are in two ways. Firstly, the immune system neutralizes or eliminates the pathogens through a series of cellular and biochemical reactions once the recognition is completed. This is called an effector response. The immune system has the ability to remember previous exposure of foreign organisms and rapidly eliminates the pathogen after its second exposure through a rapid and strong protection called memory response.

5.1.1 Classification of Immunity

The immunity can be classified on the basis of recognition of antigens or developmental and hereditary characteristics. On the basis of immune recognition, immunity can be classified into specific and nonspecific immunity. Nonspecific immunity responds equally to all pathogens to eliminate them. In

Table 5.1 Factors determining the antigenicity

Foreignness	To elicit an immune response, an antigen must be treated as a foreign substance to the individual.
Chemical nature	Proteins have good immunogenic response followed by polysaccharides and lipopolysaccharides. The nucleic acids are poor immunogenic. The lipids usually lack immunogenicity.
Molecular size	The molecules having a molecular mass of 100,000 Da have good immunogenic potential. Substances with a molecular mass less than 5000–10,000 Da have poor immunogenicity.
Physical form	Particulate and denatured antigens are more immunogenic compared to soluble and native form, respectively.
Genetic factor	Some antigens are able to induce immunogenicity in a particular species. Bovine serum albumin is nonimmunogenic to cow, poorly immunogenic to goat, but strongly immunogenic to chicken.
Dose of the antigen	All antigens require a particular dose to elicit immune response. The optimal immune response will not be achieved above or below this dose level.
Route of administration	Intravenous administration of antigens carried first to spleen but antigens administered carried first to local lymph nodes. Therefore, subcutaneous route is better compared to intravenous route.
Age	Very young or very poor individuals have diminished immunogenic response against an antigen.
Degradability	To elicit an immune response, the antigens must be processed by antigen-presenting cells. Therefore, antigens that can be easily phagocytosed are having good immunogenic response.

contrast, specific immunity uses different strategies for different microorganisms to neutralize them. On the basis of hereditary features, immunity can be classified as innate and adaptive immunity. The innate immunity is inborn and present before the onset of infection. Adaptive immunity comes into action after the pathogenic exposure. The innate immunity thus provides the first line of defense against invading pathogen and is less specific. In contrast, adaptive immunity is highly specific. The adaptive immune responses developed slowly after the exposure of pathogens or foreign materials, but a rapid response against pathogens was achieved in innate immune system that destroys the pathogens during the first critical hours of its exposure. One of the striking features of adaptive immunity is the memory, which means the initial exposure of any pathogen results in memory response and a quick and stronger response is achieved after the second exposure of the same pathogen.

5.1.2 Antigens

The molecules that induce humoral and cell-mediated immune response are called immunogens, and their ability to induce immune response is termed as immunogenicity. Immunogens after binding with B cells generate activated B cells to produce antibody. The binding of antigens with T-cell receptors leads to activation of T cells (cytotoxic T lymphocytes). The term antigens are not synonymous with immunogens. Antigens are the molecules that bind specifically with products of immune response induced by immunogens (i.e., antibodies or cytotoxic T lymphocytes), and this property is called antigenicity. An antigen may not be immunogenic if it fails to induce an immune response. Certain molecules like haptens (discussed later) can bind with T or B cells but fail to activate them. An antigen is said to be immunogen only when it can be able to induce an immune response (activation of humoral or cell-mediated immune response). Thus, it can be said that all immunogens are antigens, but not all antigens are immunogens. Some body cells like sperm and corneal cells may cause immunoreaction if injected in the same animal.

5.1.2.1 Determinants of Antigenicity/Immunogenicity

There are several factors that determine the antigenicity (Table 5.1).

5.1.2.2 Types of Antigens

The antigens can be classified on the basis of their source and immune response.

5.1.2.2.1 Classification Based on Source

Exogenous antigens: They enter inside the body from outside and are processed by antigen-presenting cells. Some antigens are exogenous at the beginning but later become endogenous, e.g., viral antigens.

Endogenous antigens: These antigens are produced within the body. They are usually body's own cells or cell products such as corneal tissue, blood group antigens, and HLA (histocompatibility leukocyte antigens). There are several types of endogenous antigens like autoantigens (person's own self antigens, e.g., thyroglobulin, corneal tissue, DNA) and alloantigens (found in different members of the same species, e.g., red blood cell antigens A and B).

5.1.2.2.2 Classification Based on Immune Response

Complete Antigen or Immunogen: They are capable to induce immune response by their own. They are usually proteins or polysaccharide in nature having a molecular weight above 10,000 Da.

Incomplete Antigen or Hapten: They are low-molecular-weight (below 10,000 Da) substances that react with its corresponding antibody but unable to induce immune

Table 5.2 Different components of innate and adaptive immunity

Immunity	Anatomical barrier	Physiological barrier	Cellular	Humoral
Innate immunity	Skin and mucous membrane	Body temperature, pH	Dendritic cells, granulocytes, mononuclear phagocytic system (MPS), dendritic cells (DCs), innate lymphoid cells (ILCs), natural killer (NK) cells, epithelial and endothelial cells, platelets, pattern recognition receptors	Inflammatory serum proteins/acute-phase proteins (APPs), antimicrobial peptides (AMPs), complement system, cytokines, interferons
Adaptive immunity			B lymphocytes, T lymphocytes	Antibodies

response by their own. Their immunogenic property can by augmented by carrier molecules (albumin or globulin), e.g., pneumococcal capsular polysaccharide, polysaccharide C of *Streptococci*, and cardiolipin antigens.

Superantigens: These antigens are able to activate a large proportion of T cells (up to 25%) in comparison to conventional antigens that are able to induce only 1–2% T cells. The superantigens cause hyperactivation of immune system. Pyrogenic exotoxins (lead to shock) and enterotoxins (lead to food poisoning) of *Staphylococci* are the examples of superantigens.

5.1.2.3 Epitopes

Epitopes are the antigenic determinants, a small site of an antigen that can activate immune response by activating T or B cells. Based on the specificity of binding, epitopes can either be B-cell epitopes or T-cell epitopes. B-cell epitopes bind with antibody, and T-cell epitopes bind with T-cell receptor after presented with MHC molecules by antigen-presenting cells. The properties of an epitope are required to design vaccines.

5.1.2.4 Adjutants

Adjuvants are the substances that enhance the immunogenicity of an antigen by one of the following mechanisms:

- Acting as a depot and leading to sustained release of an antigen from the site of delivery
- Stimulating the production of cytokines and chemokines
- Facilitating the recruitment of immune effector cells at the site of delivery
- Helping in antigen uptake and further processing and presentation by antigen-presenting cells
- Activating inflammatory mediators

The common examples of adjuvants are Freund's complete adjuvant (containing inactivated *Mycobacterium tuberculosis* in oil, alum nonionic surfactants, and muramyl peptides). There are adjuvants licensed for human use such as MF59, AS04, and virosomes.

5.1.3 Components of Immune System

The immune system is equipped with several components like anatomical and physiological barriers, immune effector cells and cell surface receptors, inflammatory serum proteins, antimicrobial peptides, and antibodies to inhibit the entry pathogens, which resist the establishment of infection along with clearing of host and microbial debris from the site of infection. Some of the components are cellular, and some components circulate freely in the body fluids called humoral factors. Table 5.2 summarizes different components of innate and adaptive immunity.

5.2 Innate Immunity

Innate immunity is the evolutionary defensive reflex against foreign materials owned by birth. The response is nonspecific in nature. It serves as the first line of defense to present infection. Therefore, the dysfunction of the innate immune system leads to life-threatening infections or development of autoimmune disorders. Innate immunity also helps to develop adaptive immune responses. There are several components of innate immunity.

5.2.1 Anatomical Barriers

The skin and mucosal membrane restrict the entry of pathogens. The epidermis of the skin contains tightly packed epithelial cells inside and dead cells and keratin in the outside. The keratin is bacteriostatic due to the presence of esterified and nonesterified fatty acids like myristic acid, palmitoleic acid, and linoleic acid. It also contains cationic proteins that make alterations in the cell wall of pathogens making them more prone to osmotic damage. Keratinocytes in the skin also express pattern recognition receptors (PRRs) to recognize pathogens and produce cytokines and antimicrobial peptides. The sebaceous glands present in the dermis layer contain lactic acid and fatty acids and maintain the skin pH acidic that also restricts the growth of many pathogens. The tight junctions present in the epithelial surface of the skin, lung, guts, and urogenital tracts also provide physical barrier against the pathogens. The mucous layer present at the interior of the epithelial surfaces also provides protective covering against invading pathogens. Mucin and other glycoproteins secreted in the mucous layer prevent the adherence of pathogen to the epithelium and are subsequently cleared by the cilia. The antimicrobial peptides defensins present in the mucosal layer also kill the pathogens.

Know More......
Influenza virus has a unique ability to bind tightly with the mucous membrane of the respiratory tract due to the presence of a surface molecule, which enables them to escape the ciliary action of the epithelial cells.

5.2.2 Physiological Barriers

The physiological barriers of pathogens include body temperature, pH, and several other soluble factors. An increase of core body temperature to the tune of 1–4 °C was proved to be detrimental to many pathogens. Some animals have inherent capabilities to resist infections due to their high body temperature (e.g., chicken is naturally resistant to anthrax). The pyrogenic response helps to induce certain cytokines (IL-6) that help in lymphocyte trafficking. "Gastric bactericidal barrier" comprising gastric HCl has the ability to inactivate microorganisms entered during ingestion. Lysozyme of saliva and tears has the ability to cleave the cell wall of bacteria. Virus-infected cells produce interferons, a group of signaling proteins that improve the antiviral defense of neighboring cells.

5.2.3 Immune Effector Cells

These are phagocytic cells (granulocytes, monocytes/macrophages, natural killer cells, and dendritic cells), endothelial cells, epithelial cells, lymphoid cells, and platelets. The phagocytes engulf pathogens and kill them by oxygen-dependent and oxygen-independent mechanism.

Granulocytes: Neutrophils, eosinophils, basophils, and mast cells are collectively known as granulocytes due to their granular cytoplasm. These cells are involved in pathogen recognition, engulfment, and phagocytosis. They possess a variety of microbicidal enzymes like lysozyme, collagenase, and elastase.

Mononuclear Phagocyte System (MPS): MPS consists of circulating monocytes and tissue macrophages. Monocytes after maturation migrate to the tissue and differentiate into tissue macrophages, which have more intracellular organelles and increased phagocytic capabilities with higher hydrolytic enzymes compared to monocytes. There are several tissue macrophages named according to their tissue locations such as histiocytes in skin, alveolar macrophages in lungs, Kupffer cells in liver, mesangial cells in kidney, microglial cells in CNS, and osteoclasts in bone. Tissue macrophages serve a variety of functions like phagocytes and antigen-presenting cells. They also have tissue-remodeling capacity through the secretion of matrix metalloproteinases and matrix proteins like collagen and elastin. Cytotoxic factors secreted by macrophages help in tumor immunity. There are three classes of macrophages.

Type 1 activated macrophages are concerned with Th 1 immune response and destroy pathogens by nitric oxide (NO) and oxygen-dependent phagocytosis.

The alternatively activated macrophages are unable to produce NO and hence lack phagocytic properties. They produce extracellular matrix proteins and are mainly involved in tissue remodeling.

Type 2 activated macrophages are stimulated in response to IgG and secret IL-10, IL-4, TNF-α, and IL-6.

Dendritic cells (DCs): These are the antigen-presenting cells that reside in the skin and mucosal surfaces. They take the antigen by means of endocytosis, phagocytosis, pinocytosis, and macropinocytosis; carry the antigen from peripheral lymph nodes; and present it to primary lymph nodes. The antigen processing and presentation by dendritic cells are achieved through major histocompatibility complex II. Other important functions of dendritic cells include regulation of cell-mediated immune response and induction of immune tolerance at peripheral lymph nodes. Immature DCs (imDCs) and precursor dendritic cells (pre-DCs) are the two subsets of dendritic cells. imDCs are seen in bone marrow as their precursors are hematopoietic stem cells. A portion of the imDCs then migrate to the epidermis of the skin and become Langerhans cells, while other portions migrate to the dermis and other tissues and differentiate into interstitial imDCs. The mature dendritic cells are potent T-cell activators and inform T cells about the information of pathogens; thus, dendritic cells act as a bridge between innate and adaptive immune response.

Innate lymphoid cells (ILCs): These cells are involved in inflammation. They do not have antigen specificity due to lack of T-cell receptor or any other cell surface markers. Their primary role is to produce cytokines. They are subdivided into three groups. Group I cells comprise ILC1 and natural killer (NK) cells and produce type 1 cytokines. Group 2 ILCs are abundant in liver, spleen, mesenteric lymph nodes, and Peyer's patches. They produce type 2 cytokines and are associated with anthelminthic response. Group III ILCs are lymphoid tissue inducer (LTi). They are mostly present in mucosal tissue and maintain a cross talk between intestinal microbiota and gut immune system. The disruption of homeostasis between gut microbiota and gut immune system leads to severe inflammatory bowel diseases like colitis and Crohn's disease.

Natural killer (NK) cells: NK cells are responsible for cell-mediated immune response due to their cytotoxic activity. They possess a unique property called "negative recognition." The surface receptors of NK cells are inhibitory receptors. These receptors suppress the cytotoxic activity of NK cells in the presence of MHC antigens, and when the infected or malignant cells have decreased expression of MHC antigen, they are recognized by NK cells and undergo cell lysis by perforins secreted from NK cells.

Epithelial and endothelial cells: They express PRRs that recognize pathogen-associated molecular patterns (PAMPs) of pathogens. In addition, they also secrete cytokines like IL-1, IL-6, and IL-8 and antimicrobial peptides.

Platelets: Platelets are the components of blood coagulation mechanism but they also express PRR on their surface and secrete cytokines and chemokines to recruit leukocytes at the inflammatory sites. Platelets can interact with endothelium and leukocytes by P-selectin, an adhesion molecule, and initiate pro-inflammatory events.

5.2.4 Pattern Recognition Receptors

The PRRs are able to sense the pathogen-associated molecular pattern (PAMP), conserved molecular pattern of a pathogen. They are subdivided into four classes.

Toll-like receptors (TLRs) are expressed on all immune effector cells. They are able to recognize external pathogen-associated molecular patterns (PAMPs) and internal damage-associated molecular patterns (DAMPs). Till date, around ten TLRs have been identified. Some of them are expressed on cell surface (TLR-1, 2, 4, 5, and 6), and some are intracellular and localized in endoplasmic reticulum (ER), endosomes, and lysosomes (TLR-3, 7, 8, 9, and 10). The intracellular TLRs are also called nucleic acid sensors due to their ability to sense dsRNA and ssRNA of the pathogens.

C-type lectin receptors (CLRs) are mainly recognized bacterial sugar moieties but are able to identify molecules associated with dead cells. They are of two types, membrane CLRs like Dectin-1 and -2 and soluble CLRs like collectins. The ligands of CLRs are β-glucans, mannose, oligosaccharides, and other microbial carbohydrates.

The nucleotide-binding oligomerization domain (NOD) receptors (NLRs) are intracellular PRRs that recognize peptidoglycans and DAMPs and induce synthesis and secretion of cytokines.

Retinoic acid inducible gen-I (RIG)-like receptors (RLRs) are also intracellular PRRs mainly responsible for antiviral immune response. They have the ability to sense viral dsRNA.

5.2.5 Inflammatory Serum Proteins/Acute-Phase Proteins (APPs)

There are several proteins that act as the mediators or inhibitors of inflammatory process. They are also called acute-phase proteins (APPs). They are mainly synthesized in the liver and their concentrations are increased (or decreased) at the rate of 25% or more at the time of inflammation. They therefore act as a suitable biomarker of inflammation. There are two classes of APPs, viz. positive APPs and negative APPs. The concentrations of positive APPs are increased during inflammation (within 1–2 days). Based on the degree of increment, positive APPs can be categorized as major (usually present in very low concentration but may increase up to 100–1000-fold within 24–48 h and rapidly decline thereafter), moderate (increase five- to tenfold within 48–72 h and decrease at a slower rate than major APPs), or minor (increase only 50–100% above basal levels at a gradual rate). The concentrations of negative acute-phase proteins decrease by 25% upon inflammation within 24 h. Albumin and transferrin are the two main negative APPs. The species variations in terms of major and minor APPs along with their functions have been depicted in Table 5.3.

> **Know More......**
> Hp and SAA can be used as markers of early detection of subclinical mastitis in cows.

5.2.6 Antimicrobial Peptides (AMPs)

They are used by many organisms as the first line of defense against pathogens. They are multifunctional peptides with bacteriostatic, bactericidal, and cytolytic properties. They are promptly synthesized after infection and kill a wide range of pathogens. Various AMPs along with their functions have been presented in Table 5.4.

Table 5.3 Species variations in terms of major and minor APPs along with their functions

Acute-phase proteins	Functions
Positive APPs	
Haptoglobin (Hp) (Major APP in cow and minor APP in horse, pig, cat, dog, and mice)	Carries free hemoglobin after extravascular hemolysis Inhibits chemotaxis and phagocytosis
Serum amyloid A (SAA) (Major APP in cow, horse, cat, dog, and mice)	Recruits inflammatory cells at the site of inflammation via chemotaxis Stimulates the secretion of pro-inflammatory cytokines Inhibits lymphocyte proliferation Helps in lipid transport
Ceruloplasmin (Cp) (Minor APP in dog and pig)	Helps in copper transport Stimulates wound healing by collagen formation and maturation Acts as an antioxidant Inhibits endothelial attachment of neutrophils
C-reactive protein (CRP) (Minor APP in dog and pig)	Promotes bacterial attachment with complement Induces phagocytosis Stimulates cytokine release Inhibits chemotaxis
Alpha-1-acid glycoprotein (AGP) (Minor APP in cat, dog, cow, and mouse)	Anti-inflammatory agent Decreases neutrophil functions
Negative APPs	
Albumin	Regulates colloidal osmotic pressure Reduces albumin production during inflammation and increases the amino acid availability for production of positive APPs
Transferrin	Iron transport protein Decreases free iron for bacterial survival
Adiponectin	Regulates energy status of an animal Anti-inflammatory agent

Table 5.4 Antimicrobial peptides and their functions

Antimicrobial peptides	Source	Functions
Lactoferrin	Mucous membrane, biological fluids like tears, colostrum, milk, and semen	Lactoferrin binds with the lipopolysaccharide of bacterial cell wall and chelates iron (Fe^{3+}) to permeabilize membrane and cell breakdown
Lysozyme	Body secretions like tears, saliva, and milk Also produced by neutrophils and macrophages	Hydrolysis of 1,4-β-glycosidic linkages between *N*-acetylmuramic acid and *N*-acetyl-D-glucosamine of cell wall peptidoglycan
Defensins	Neutrophils, monocytes, macrophages, keratinocytes, paneth cells including mucosa of respiratory, digestive, urinary, reproductive systems	Promotes phagocytosis, chemotactic activity, cytokine production, degranulating mast cells
Histidine-rich glycoprotein (HRG)	Liver, monocytes, macrophages, and megakaryocytes	Antiangiogenic and antitumor properties, chemotaxis, cytokine production
Major basic protein (MBP)	Granules of eosinophils	Antibacterial, antihelminthic, and cytotoxic properties, induces hypersensitivity reactions
RNase 7	Skin	Broad-spectrum antimicrobial activity

5.2.7 The Complement System

The complement system is one of the major components of the innate immune response composed of several interlinked proteins that serves a wide array of functions like pathogen recognition, regulation inflammatory processes, killing of the pathogen, and removal of damaged cells. Another major function of complement system is the regulation of adaptive immune responses. Thus, complement system acts as a bridge between innate and adaptive immune responses.

5.2.7.1 Components of Complement System

The complement system consists of several proteins synthesized primarily from liver, macrophages, and neutrophilic granules. In 1963, when the complement system was first discovered, it consisted of only nine proteins labeled by the letter "C" followed by the numbers and their activated forms were designated by added symbol "a" (C1a is an activated form of C1). Till then, a variety of proteins have been identified under complement system. As per the last nomenclature recommended by the International

Complement Society (ICS), Complement Nomenclature Committee, and European Complement Network (ECN) boards, there were 50 different proteins and protein complexes (for derailed review, refer Kemper et al. 2014). The complement proteins comprise 5–10% of total plasma proteins. The sizes of complement proteins vary from 24 (D) to 460 kDa (C1q).

5.2.7.2 Mechanism of Action of Complement System

The complement system remains inactive in an uninfected animal. They can be activated either by PAMP or through antigen-bound antibodies. The activation leads to a series of reaction cascades, which ultimately produce a key protein named C3b.

5.2.7.2.1 Activation

There are three different mechanisms by which complement can be activated.

Alternative pathway: The main regulatory protein of alternate pathway is C3, which is synthesized in liver and macrophages. It has the highest abundance in serum among the complement components. The alternate pathway operates through three major steps, initiation, amplification, and regulation. In the initiation process, C3 proteins undergo autoactivation by a process called "tickover." The "tickover" of C3 facilitates the conformational changes in C3 and generates $C3(H_2O)$, which in turn binds with another factor B. The $C3(H_2O){\cdot}B$ complex undergoes cleavage by another serine protease named factor D. Factor D cleaves factor B into Ba and Bb. Bb itself acts as another serine protease that cleaves C3 into its active form C3b. Once C3b is generated, it again associates with factor B to generate more C3 and the proteolytic cycle continues. Another serum protein named properdin stabilizes these protein:protein interactions during the amplification process.

Lectin pathway: This lectin pathway activates after the recognition of oligosaccharide molecules on the surface of pathogen. There are five types of pattern recognition proteins (PRPs) that specifically bind with oligosaccharide moiety of pathogens, namely mannan-binding lectin (MBL), collectin-11 (CL-11), ficolin-1, ficolin-2, and ficolin-3. The first two PRPs bind with glucose, mannose, and *N*-acetyl-glucosamine, whereas ficolins recognize acetyl groups of bacterial membrane glycoproteins like *N*-acetyl-glycine, *N*-acetyl-cysteine, and acetyl-choline. The PRPs are complexed with MBL-associated serine proteases (MASPs), namely MASP-1, MASP-2, and MASP-3. The binding of PRP with the carbohydrate moiety results in the activation of MASPs. Activated MASP-2 splits C4 into C4a and C4b. Complement factor C2 then binds with C4b to form a complex called C4b2. The bound C2 then undergoes cleavage by MASP-2 and yields C4b2b. The protease C4b2b then splits C3 into C3a and C3b.

Classical pathway: Unlike alternative and lection pathways that involve innate immune response, the classical pathway of complement activation acts in conjunction with adaptive immune response as it is induced by antigen and antibody complexes along with other proteins like CRP, amyloid proteins, and apoptotic bodies. The major complement proteins of classical pathway are C1, C2, C4, C1 inhibitor (C1-Inh), and C4-binding protein (C4bp). The classical pathway activates when C1 binds with Fc portion of an Ag-Ab complex. C1 proteins have three subunits C1q, C1r, and C1s. C1q is like a strand, and two molecules each of C1r and C1s are located between the C1q strand. To become active, at least two C1q strands have to bind with antibody molecules. The interaction leads to conformational change in C1q, which activates C1r. C1r acts as C1s and exposes its active site to convert C1s subunits as an active enzyme. C1s then cleaves C4 into C4a and C4b. C4b acts as a receptor for C2, and C4b-bound C2 acts as a substrate for C1s and cleaves into C2a and C2b. C2a being smaller diffuses into the plasma, and larger C2b remains attached with C4b. This C4b-C2b complex cleaves C3 into C3a and C3b.

5.2.7.2.2 Amplification

The C3b thus produced by three pathways then interacts with complement factor C5 to become C3b5, which is cleaved by C3bBb into C5a and C3b5b. Factors C6 and C7 then join with C3b5b to form C5b67. Formed C5b67 then interacts with C8 to form C5b678 and further C5b6789 after combining with C9. C5b6789 is the terminal complement complex (TCC) or membrane attack complex (MAC), which forms a hole in the microbial cell membrane and induces osmotic lysis of the microbes.

5.2.7.2.3 Regulation

The regulation of alternate pathway is facilitated by factor H and factor I. Factor H blocks the binding of factor B to factor C3b, and factor I inactivates C3b to iC3b. The sialic acid blocks the alternate pathway by inducing the binding of factor H with C3b, and microorganisms lacking sialic acid are killed, but the host cells that possess a sialoglycoprotein named glycophorin A are protected. The classical pathway is regulated through C1 inactivator (C1-INH), a glycoprotein that blocks C1r and C1s. CD55 or decay accelerating factor present in all blood corpuscles and endothelial cells binds with C3 and C5 convertases and induces their decay, thus protecting the normal cells from complement attack. The

Table 5.5 Functions of different complement components

Function	Complement components
Lysis	The lytic complex (C5b6789) ruptures bacterial cell membrane
Opsonization and phagocytosis	C3b and C4b have opsonizing potential and C3b-coated microorganisms bind with CRI of phagocytes and undergo phagocytosis
Chemotaxis	Complement-derived chemotactic factors are C3a: Attracts eosinophils C5a: Chemotactic for macrophages, neutrophils, and eosinophils C567: Attracts neutrophils and eosinophils Bb: Attracts neutrophils
Activation of mast cells	C3a, C4a, and C5a activate mast cells to release histamine, and heparin causes vasodilation and increased tissue permeability
Removal of apoptotic cells	Apoptotic cells lack CD46 and CD59 complement inhibitors and bind with C1q to activate classical pathway and subsequently undergo phagocytosis
Inflammation	Complement-derived C3a and C5a stimulate the production of pro-inflammatory cytokines like TNF-α, IL-1β, and IL-6
Blood coagulation	C5a inhibits fibrinolysis and induces blood coagulation by augmenting the expression of tissue factors and plasminogen activator inhibitor
Regulation of immune functions	The adaptive immune response is increased by C3d

C4-binding protein (C4BP) in the plasma inhibits C3 convertase (C4b2a). CD35 (CR1) expressed in RBCs, phagocytic cells, T and B cells, kidney podocytes, and peripheral nerves helps to clear immune complexes and presents complement activation. There are some regulatory proteins namely protectin (CD59), clusterin, and vitronectin that interfere with TCC formation.

5.2.7.2.4 Functions of Complement System

Upon activation, complement system generates a wide array of products that ultimately leads to lysis of pathogens but the products of complement cascade have a wide range of functions detailed in Table 5.5.

5.2.8 Cytokines

Cytokines are the low-molecular-weight (smaller than 30 kDa) proteins or glycoproteins synthesized from leukocytes or other cells of the body and act as soluble mediators to regulate immunity. The cytokines exert its action after binding with its receptors in the target cells and subsequently activate the intracellular signaling cascade that ultimately alters the gene expression of target cells and causes differentiation, proliferation, and activation of the target cells. The cytokines act in autocrine, paracrine, and endocrine fashion.

5.2.8.1 Properties of Cytokines

The cytokines have properties like the following:

Pleiotropy: When a cytokine has different effects on different types of target cells, the cytokine is said to be pleiotropic. IL-4 produced from activated T_H cells causes proliferation, differentiation, and activation of B cells but only proliferates thymocytes and macrophages.

Redundancy: When two or more cytokines exert similar effect on single target cells, the effect is said to be redundant. IL-2, IL-4, and IL-5 produced from T_H cells cause proliferation of B cells.

Synergy: It is the cooperative effect of cytokines. Here, the cytokines in combinations have more pronounced effect compared to their individual effect. IL-4 and IL-5 produced from T_H cells induce B cells to produce IgE, but neither IL-4 nor IL-5 has individual effect to induce B cells for IgE production.

Antagonism: When the effect of a cytokine is inhibited by another cytokine, then the effect is called antagonism. The effect of IL-4 on B cells is inhibited by IFN-γ.

Cascade reaction: When one cytokine induces a target cell to produce one or more cytokines that in turn stimulates another target cell to produce other cytokines, it is called cascade reaction. IFN-γ produced from activated T_H cells stimulates macrophages to secrete IL-12 that in turn stimulates activated TH cells for IFN-γ, TNF, and IL-2 secretion.

5.2.8.2 Classification of Cytokines

There are six different cytokine families, namely interleukins, chemokines, interferons, tumor necrosis factors (TNF), colony-stimulating factors (CSF), and transforming growth factor-β. There are several subgroups under different families.

Interleukins: They are so named with a thought that it was synthesized by leukocytes but later, it was found that interleukins can be produced from a variety of cells. They play a pivotal role in hematopoiesis, activation, and differentiation of immune cells. They also have pro-inflammatory properties and help in leukocyte migration and adhesions. Till date, 40 interleukins have been identified and named alphabetically from IL-1 to IL-40.

Chemokines: These are chemotactic cytokines and attract the leukocytes to the site of infection. Structurally, chemokines are subdivided into four families based on the N-terminal cysteine residue.

CC chemokines: They have two adjacent cysteine residues at N-terminal region. Twenty-eight CC chemokines have been identified (named from CCL-1 to CCL-28) so far, and majority of them are chemotactic for monocytes.

CXC chemokines: They are characterized by two cysteine residues separated by an amino acid at the N-terminus. There are 17 CXC cytokines, namely CXCL-1 to CXC-17. They attract neutrophils at the site of infection.

C chemokines: They have two cysteine residues, one at N-terminal region and another at downstream. There are two C chemokines (XCL-1 and XCL-2).

CX3C chemokines: CX3C chemokines are having two cysteine residues at N-terminus separated by three amino acids. Besides their role in chemotaxis, they are also involved in cell adhesion. Till date, a single CX3C chemokine has been identified (CX3CL1).

Interferons (IFN): They emerged as antiviral proteins, but later, their roles in immunomodulation and cancer immunology have been identified. IFNs are classified into three types, type I (IFN-α and IFN-β), type II (IFN-γ), and type III (IFN-λ1, 2, and 3).

Tumor necrosis factor (TNF): TNF is produced from activated natural killer (NK) cells, macrophages, and T lymphocytes with diverse physiological functions in cell proliferation, differentiation, and carcinogenesis. It is also a pro-inflammatory cytokine. The TNF is classified into TNF-α and TNF-β. TNF-α is produced from monocytes, macrophages, and T cells and has functions as inflammatory mediators and cell adhesions. TNF-β is produced mainly from activated lymphocytes and has functions similar to TNF-α.

Colony-stimulating factor (CSF): These are responsible for differentiation of leukocytes in the bone marrow. There are four families of CSF, namely granulocyte–colony-stimulating factor (G-CSF), macrophage–colony-stimulating factor (M-CSF), granulocyte–macrophage-colony-stimulating factor (GM-CSF), and multiple-colony-stimulating factor (also called IL 3). M-CSF is responsible for differentiation of macrophage precursors. G-CSF is responsible for differentiation of granulocyte precursors. GM-CSF is produced from lymphoid and nonlymphoid cells and helps in maturation and differentiation of both granulocytes and monocytes. The role of multiple-colony-stimulating factors is the differentiation of hematopoietic stem cells into myeloid progenitor cells. There are two other CSF like erythropoietin and thrombopoietin.

Transforming growth factor-β (TGF-β): It can be produced from a variety of cells including T cells and monocytes. The main function of TGF-β is the inhibition of cellular growth and production of extracellular matrix. TGF-β also acts as a negative regulator of T-cell and macrophage activation.

5.2.8.3 Pro- and Anti-inflammatory Cytokines

The cytokines secreted in response to an infection mainly by the macrophages and that upregulate the inflammatory response are called pro-inflammatory cytokines. They are pyrogenic in nature and stimulate acute-phase reactions. IL-1 and TNF-α are predominant pro-inflammatory cytokines. The anti-inflammatory cytokines are responsible for the downregulation of inflammatory process. IL-6 is a potent anti-inflammatory cytokine that inhibits the effects of IL-1 and TNF-α. IL-4 and IL-10 are also anti-inflammatory cytokines. The overproduction of pro-inflammatory cytokines compared to anti-inflammatory cytokines leads to autoimmune diseases.

5.2.8.4 Mechanism of Action of Cytokines

The cytokines exert its effects after binding with target cell receptors. There are six classes of cytokine receptors, namely type I cytokine receptors, type II cytokine receptors, chemokine receptors, tumor necrosis factor receptor (TNFR) superfamily, TGF-beta receptors, and immunoglobulin (Ig) superfamily. The receptors are associated with tyrosine kinases called Janus kinases (JAKs) and transcription factors called signal transducer and activator of transcription (STAT). The binding of cytokine with its receptor leads to the activation of JAK by phosphorylation. Phosphorylated JAK combines with STAT, and the complex then moves to nucleus, binds with DNA regulatory site, and activates transcription. The transcription of DNA causes protein synthesis, and the response of cytokine is generated.

5.2.8.5 Functions of Cytokines

The functions of different cytokines, their sources, and target cells have been detailed in Table 5.6.

5.2.8.6 Pathogen Recognition and Inflammatory Signaling in Innate Immune System

Innate immune system, being the first line of defense, recognizes the pathogen during the initial stage of infection and subsequently eliminates the pathogens from the host body. To achieve this, the innate immune system utilizes pattern recognition receptors (PRRs) that recognize pathogen-associated molecular patterns (PAMPs), distinct structure present on the pathogen (Table 5.7).

Table 5.6 Cytokines, their sources, target cells, and functions

Name	Source	Target cells	Functions
Interleukin-1	Macrophages, monocytes, B lymphocytes, endothelium	Macrophages, neutrophils, NK cells, endothelium, and hypothalamus	• Chemotaxis of macrophages and neutrophils • Stimulation of B cells to produce antibody • Stimulation of NK cells to destroy pathogens • Stimulation of endothelium to secrete vasoactive peptides to increase vascular permeability • Stimulation of nervous system to induce including fever, anorexia, and fatigue • Stimulation of hepatic cells for production of acute-phase proteins
Interleukin-2	T_H cells	T cells, NK cells	• Stimulation of T-cell proliferation and NK cell activity
Interleukin-3	T_H cells, mast cells, and NK cells	Bone marrow, mast cells	• Stimulation of leukocyte and erythrocyte production • Stimulation of mast cell to release histamine
Interleukin-4	T_H cells, mast cells, and NK cells	B cells and macrophages	• Stimulates the differentiation of B cells into plasma cells • Stimulation of MHC expression
Interleukin-5	T_H cells, mast cells	B cells and eosinophils	• Stimulates the differentiation of B cells into plasma cells • Stimulates the proliferation and differentiation of eosinophils
Interleukin-6	Macrophages, monocytes, T_H cells, and bone marrow cells	B cells, monocytes, macrophages, hypothalamus, and hepatic cells	• Stimulates the differentiation of B cells into plasma cells • Stimulation of nervous system to induce including fever, anorexia, and fatigue • Stimulation of hepatic cells for production of acute-phase proteins
Interleukin-7	Macrophages, thymus, and bone marrow	Bone marrow stem cells and neutrophils	• Stimulates hemopoietic stem cells into progenitor B and T cells • Chemotaxis of neutrophils
Interleukin-8	Macrophages	Neutrophils	• Chemotaxis of neutrophils
Interleukin-9	T_H cells	Select T cells	• Stimulation of select T cells
Interleukin-10	T_H cells	Macrophages, APC	• Inhibits IL-1 synthesis • Downregulation of the expression of MHC
Interleukin-11	Bone marrow	PHSC, hepatocytes	• Growth and differentiation of PHSC • Produces APP
Interleukin-12	Macrophages, B cells	Cytotoxic T cells, NK cells	• Regulates T-cell response • Stimulates NK cell proliferation
Interleukin-13	T_H cells	Macrophages	• Inhibits pro-inflammatory cytokine production • Stimulates proliferation of NK cells and T cells
Interleukin-15	T cells	T cells, B cells, NK cells, and intestinal epithelial cells	• Stimulates proliferation of T cells, B cells, and gut epithelium • Stimulates cytokine production
Interleukin-16	T cells	T_H cells, eosinophils	• Induces MHC expression • Chemotaxis of eosinophils
Interleukin-17	T_H cells	Bone marrow, macrophages, splenocytes, and synovial cells	• Stimulates the release of TNF-α • Stimulates cytokine production • Stimulates the proliferation of granulocytes
Interleukin-18	Monocytes, macrophages	NK cells, monocytes, macrophages, and T_H cells	• Stimulates the release of cytokines like TNF-α, IL-1, IL-8, and IFN-γ
Tumor necrosis factor-alpha (TNF-α)	Macrophages	Tumor cells, monocytes, and macrophages	• Destroys the tumor cells • Stimulates the release of IL-1, IL-2, IL-6
Tumor necrosis factor-beta (TNF-β)	T cells	Tumor cells, neutrophils, and macrophages	• Destroys the tumor cells • Stimulates phagocytosis • Controls fatigue, pyrexia, and anorexia

(continued)

Table 5.6 (continued)

Name	Source	Target cells	Functions
Interferon-alpha (IFN-α)	WBC	Uninfected cells and hypothalamus	• Inhibits viral replication • Stimulates sickness behavior
Interferon-beta (IFN-β)	Fibroblasts	Uninfected cells	• Inhibits viral replication
Interferon-gamma (IFN-γ)	T cells and NK cells	Uninfected cells, macrophages, and B cells	• Inhibits viral replication • Increases the expression of MHC • Activates macrophages

Table 5.7 Pathogen-associated molecular pattern and pathogen recognition receptor (PRR)

Pathogens	PAMPs	PRRs
Viruses	Surface glycoproteins	TLR2 and TLR4
	Viral DNA	TLR9
	Viral ssRNA	TLR7, TLR8, and RIG-I
	Viral dsRNA	RLRs, TLR3, and NLRs
Gram (+) bacteria	Peptidoglycans	TLR2 and NLRs
	Bacterial DNA	TLR9 and NLRs
	Lipoproteins	TLR2
	Lipoteichoic acid	TLR2
Gram (−) bacteria	Bacterial DNA	TLR9 and NLRs
	Porin	TLR2
	Peptidoglycans	TLR2 and NLRs
	Lipopolysaccharide	TLR4
	Flagellin	TLR5
Fungi	Zymosan	TLR2
	β-glycans	TLR2
	Mannan	TLR2 and TLR4
Protozoa	DNA	TLR9
	GPI anchors	TLR2 and TLR4

Besides PAMPs, the PRRs of innate immune system also recognize damage-associated molecular patterns (DAMPs). DAMPs are endogenous molecules released in response to stress or tissue injury and are potent stimulators for noninfectious inflammation. Different DAMPs and their receptors are presented in Table 5.8.

Through the binding of ligands (PAMPs or DAMPs), PRRs are stimulated, facilitate downstream signal transduction, and lead to transcriptional activation of genes for pro-inflammatory cytokines, chemokines, cell adhesion molecule, and IFNs. This pro-inflammatory signaling pathway also activates the adaptive immune response.

5.2.9 Inflammation

It is a complex tissue reaction against tissue damage or pathogenic microorganisms that results in clearance of the invading pathogens through the activation of the components of the innate immune system. Rubor (redness), calor (heat), dolor (pain), and tumor (swelling) are four cardinal signs of inflammation stated by the Roman physician Celsus. These signs can be well explained by the major events of inflammation such as the following:

Vasodilation: It is mediated by nitric oxide (NO) and vasodilatory prostaglandins. Pro-inflammatory cytokines such as IL-1 and TNF-α produced from activated leukocytes stimulate inducible nitric oxide synthase (iNOS) and cyclo-oxygenase (COX-1 and -2). iNOS produces NO from L-arginine. NO in turn causes smooth muscle relaxation. Prostaglandins (PGI2, PGD2, PGE2, and PGF2α) and prostacyclins are the vasodilatory prostaglandins synthesized from arachidonic acid by the action of COX-1 and -2. Both these NO and prostaglandins cause vasodilation, and engorgement of the capillary network leads to redness (rubor) and increased tissue temperature (calor).

Increased capillary permeability: The alteration in the capillary permeability is mediated by the release of certain inflammatory mediators such as histamine, bradykinin, leukotrienes, and platelet-activating factor (PAF). Together, the increased vascular permeability and capillary hydrostatic pressure lead to leakage of protein-rich fluid (exudate) in the interstitium of the inflammatory site. Accumulation of exudates causes edema or swelling that allows the delivery of antibodies and other acute-phase proteins in inflamed site.

Leukocyte migration: See functions of neutrophils.

5.2.10 Phagocytosis

Phagocytosis is the ability to ingest or engulf other cells and particles by some specialized cells called phagocytes. The process was discovered by Élie Metchnikoff (1845–1916) during his studies on some marine organisms. He was the pioneer to develop the concept cellular immunity and was awarded the Nobel Prize in 1908 together with Paul Ehrlich for notable contribution in the field of immunology. In unicellular organisms, phagocytosis is a process of cell nutrition, but for multicellular organisms, it is a means to kill the

Table 5.8 Damage-associated molecular pattern (DAMP) and pathogen recognition receptor (PRR)

Origin	DAMPs	Receptors
Extracellular matrix	Fibronectin	TLR4
	Fibrinogen	TLR4
	Decorin	TLR2, TLR4
	Heparan sulfate	TLR4
Cytosol	Heat-shock proteins	TLR2, TLR4
	S100 proteins	TLR2, TLR4
Nuclear	Histones	TLR2, TLR4
	DNA	TLR9
	RNA	TLR3, TLR7, TLR8, RIG-I
Mitochondria	mtDNA	TLR9
	mROS (reactive oxygen species)	NLRs
Granule	Defensins	TLR4
Plasma membrane	Syndecans	TLR4
	Glypicans	TLR4

pathogen and cellular debris and thus plays an important role in innate immunity as well as tissue homeostasis. Based on these two functions, phagocytes can be classified as preferential phagocytes (neutrophils, macrophages, monocytes, dendritic cells, and osteoclasts) that act to eliminate pathogens. The nonprofessional phagocytes (epithelial cells, fibroblasts, and endothelial cells) are mainly involved in the elimination of apoptotic bodies.

A phagocyte can recognize a pathogen either directly by PRRs present on them or indirectly through some molecules that form a bridge between the phagocyte and the particle to be ingested. These are called opsonins such as antibodies (IgG) and complement components, and the process is called opsonization. The direct PRRs of phagocytes are Dectin-1, mannose receptors, CD14, and scavenger receptor A (SR-A) that recognize polysaccharides, mannans, lipopolysaccharide, and lipoteichoic acid, respectively. Fcγ receptors (FcγR) are the opsonic receptors that bind with the Fc portion of IgG molecules.

After the interaction between phagocytes and ingesting particles, a series of signaling events initiate that leads to the remodeling of membrane and cytoskeleton of the phagocytes to form pseudopods that cover the particle and a depression called phagocytic cup is formed at their point of contact. The target particle is then surrounded by the membrane, and it closes at the distal end to form phagosome. The phagosome thus formed interacts with endosomes and lysosomes and finally fuses to form phagolysosome.

The cytotoxic effects of phagocytes are achieved through oxygen-dependent and oxygen-independent mechanisms. The phagocytes produce a number of reactive oxygen and nitrogen intermediates with potent microbial activity. The production of reactive oxygen and nitrogen intermediates occurs through a metabolic process called respiratory burst, which activates peroxidase enzymes. The reactive oxygen intermediates are superoxide anion ($O_2^{\bullet -}$), hydroxyl radicals ($OH^{\bullet}$), singlet oxygen ($^1O_2$), hydrogen peroxide (H_2O_2), hypochlorous acid (HOCl), and monochloramine (NH_2Cl). The reactive nitrogen intermediates are nitric oxide (NO), nitrogen dioxide (NO_2), and nitrous acid (HNO_2). In oxygen-independent mechanism, the killing of pathogens is achieved through defensins, tumor necrosis factor, and hydrolytic enzymes.

5.3 Acquired/Adaptive Immunity

As the name implies, this type of immunity is not by birth rather acquired after previous antigenic exposures. Acquired immune responses are capable of selective elimination of pathogens. Acquired immune responses have some cardinal features like *specificity*; they have the capability to distinguish different classes of microorganisms. *Diversity* is another important feature of acquired immunity by which it can recognize a wide array of antigens and microorganisms. The acquired immune system remembers the initial exposure of an antigen and comes with higher immune resistance during its second exposure by *memory*. It also has a unique ability to *discriminate the self and nonself antigens* and to react accordingly.

5.3.1 Components of Adaptive Immunity

Adaptive immune responses are brought about by different classes of lymphocytes, namely B and T lymphocytes. The immune responses mediated through B lymphocytes are called *antibody mediated or humoral immune response* as B cells are capable of producing antibodies (or immunoglobulins) upon antigenic exposure, which bind with antigens and make them vulnerable for destructions. *The cell-mediated immune responses* are mediated through T lymphocytes that produce signals to activate phagocytic cells to destroy them.

T cells: They are so named due to their maturation in the thymus after derived from hematopoietic stem cells in bone marrow. They express antigen-binding receptors at their surface called T-cell receptor (TCR). The TCRs are able to recognize the correct antigen fragments after processed and presented through antigen-presenting cells (APC). Dendritic cells, macrophages, fibroblasts, and epithelial cells can act as APCs. These APCs express a surface protein called major histocompatibility complex (MHC) that recognizes self and nonself antigens.

Major histocompatibility complex (MHC): These are the cell surface proteins encoded by a group of genes present in chromosome 6 in humans. The main function of MHC is to discriminate between self and nonself. MHCs are of two types; MHC class I (also known as human leukocyte antigen [HLA]) is expressed in all nucleated cells and is responsible for processing and presentation of endogenous (intracellular) peptides. In contrast, MHC class II expresses only some immune effector cells like macrophages, dendritic cells, and B cells and is responsible for processing and presentation of exogenous (extracellular) peptides. The MHC-antigen complex activates TCR and stimulates the differentiation of T cells into different subsets.

B cells: B cells are matured in the bone marrow (or liver during fetal life), but in birds, the maturation of B lymphocytes takes place in the bursa of Fabricius, a lymphoid organ found near the cloaca. B lymphocytes require antigenic exposure before final maturation after which they become immunocompetent. They have unique antigen-binding receptors and are able to interact directly with the antigens without the involvement of APCs. After interaction with foreign antigens through the receptors, B cells undergo proliferation and differentiations into plasma cells that are capable of producing antibodies. Plasma cells are short lived. Therefore, a portion of B cells are differentiated into long-lived memory B cells, which are able to respond quickly on re-exposure.

5.3.2 Lymphoid Organs

The organs at which lymphocytes are produced and matured are called lymphoid organs. There are two classes of lymphoid organs.

Primary/central lymphoid organs: In primary lymphoid organs, the lymphocytes are produced and undergo maturation. The primary lymphoid organs are thymus, bone marrow, and bursa of Fabricius.

Secondary/peripheral lymphoid organs: In peripheral lymphoid organs, the lymphocytes interact with antigens. They include lymph nodes, spleen, tonsils, and gut and mucosal associated lymphoid tissues.

> **Know More.....**
> Recently, some abnormal lymph node-like structures were identified at the sites of chronic inflammation, cancers, and transplanted organs with graft rejection. These are called tertiary lymphoid organs (TLOs) or ectopic lymphoid structures (ELSs). They secrete a cytokine called lymphotoxin β (LTb), which induces the differentiation of the stromal cells into lymphoid organs. In some cases, the development of TLO at the site of tumor showed better prognosis.

5.3.2.1 Primary Lymphoid Organs

Thymus: It is a bilobed structure situated above the heart. The lobes are encapsulated and are divided into lobules by connective tissue strands called trabeculae. The lobules are divided into outer cortex and inner medulla. There are stromal cells between the cortex and medulla composed of epithelium, dendritic cells, and macrophages. The thymic epithelial cells are also called nurse cells that surround the thymocytes in the cortex.

Thymus is mainly responsible for maturation of T cells. The progenitors of T cells derived from PHSC enter into the thymus as thymocytes and become immunoreactive and antigen-competent T cells. During the course of the development of thymocytes, they express antigen-binding receptors, and the T cells capable of recognizing foreign antigens and MHC molecule will be selected and released. The selection involves two steps. In the first step, there is positive selection of T cells that recognize self MHC. The T cells that are unable to recognize self MHC molecule will undergo apoptosis. In the second step (negative selection), thymocytes with affinity receptor for self-antigen and self MHC are eliminated. So, ultimately the T cells that recognize both foreign antigen and MHC molecule are selected. This is called immune tolerance, and the thymus is called the organ of tolerance.

Bone marrow: It is a spongy tissue situated inside the bones. It is the primary hematopoietic organ. Initially, almost all the bones contain red bone marrow that creates blood cells. But, during the course of ageing, the marrow of long bones becomes fatty tissue and the hematopoiesis decreases. But the marrow of flat bones like ribs, sternum, and pelvis is active and hematopoiesis continues.

Bone marrow is the principal site for B-cell maturation. A portion of B cells enter into the secondary lymphoid organs to differentiate plasma cells upon antigenic stimulation. Other activated B cells become memory B cells or long-lived plasma cells reside in spleen and bone marrow. These are the persistent source of antibodies.

Bursa of Fabricius: It is the primary lymphoid organ of birds situated dorsal to the rectum and anterior to sacrum. It has a communication with cloaca through a short duct. The organ is composed of 12–20 longitudinal folds packed with numerous follicles separated through connective tissue layer. Each follicle contains B lymphocytes, dendritic cells, epithelium, and macrophages.

The bursa of Fabricius is the main site for antigen-committed B-cell maturation. Pre-bursal stem cells enter in the bursa around the seventh day of embryonic life and become bursal stem cells that undergo rapid differentiation in bursal microenvironment and become antigen-specific B cells and self-renewing post-bursal stem cells.

5.3.2.2 Secondary Lymphoid Organs

Lymph nodes: There are several lymph nodes strategically located in different anatomical locations to receive immunological signals from the body and provide an ideal microenvironment for immune cell communication. Each lymph node is divided into outer cortex that contains B cells. The inner medullary region contains both T and B cells. The paracortex between cortex and medulla contains T cells and dendritic cells. Both T cells and B cells enter the lymph nodes through endothelial venules and leave the node through efferent lymphatic vessels. T cells interact with dendritic cells during their movement, and their continual interactions facilitate to recognize foreign antigens entrapped in dendritic cells. Antigen-primed T cells then divide and induce immune reactions to eliminate it. Some of the dividing T cells also travel to B-cell-rich cortex and promote B-cell division and maturation to produce antibodies.

Spleen: It is one of the main secondary lymphoid organs in the left abdominal cavity beneath the diaphragm. It is the largest lymphatic organ of our body. Spleen is responsible for trapping of blood-borne antigens. Blood enters the spleen through splenic artery and leaves through splenic vein. The spleen is surrounded by a tissue capsule, which extends inward in the form of trabeculae to divide the spleen into two compartments, the red and white pulp. These red and white pulps are separated by marginal zone. The white pulp consists of mainly T lymphocytes surrounding splenic arteries and forms periarteriolar lymphoid sheath (PALS). The red pulp consists of sinusoids filled with blood and populated by macrophages. The marginal zone is populated by lymphocytes and macrophages. The red pulp consists of venous sinuses filled with blood and cords of lymphatic cells, such as lymphocytes, erythrocytes, and macrophages. The defective and old erythrocytes are destructed in the red pulp of spleen by the macrophages. When the antigens in the blood reach the spleen, they are trapped by dendritic cells situated in the marginal zone and carried to T-cell-rich PALS where the antigens are presented to TH cells with class II MHC molecules. Activated TH cells induce B-cell activation. Activated B cells and TH cells then migrate to primary follicles situated in the marginal zone. The primary follicles are differentiated into secondary follicles upon antigenic exposure. The secondary follicles are like lymph nodes containing germinal centers populated by B cells and plasma cells surrounded by lymphocytes.

Mucosal associated lymphoid tissue (MALT): They are situated in the mucous membranes of the gastrointestinal, respiratory, and urogenital systems. They are populated with plasma cells. Due to the large surface area of the mucosal lining of the different systems of our body, the populations of plasma cells in the MALTs are far more than bone marrow, spleen, and lymph nodes together. They resemble the structure of lymph nodes composed of lymphoid follicles, interfollicular region, subepithelial dome region, and follicle-associated epithelium. M cells are specialized cells responsible for transport microorganisms and soluble molecules from the intestinal lumen to the subepithelial region. There are several MALTs named on the basis of their anatomical positions such as gut-associated lymphoid tissue (GALT), bronchus-associated lymphoid tissue (BALT), nasopharynx-associated lymphoid tissue (NALT), lacrimal duct-associated lymphoid tissue (LDALT), conjunctiva-associated lymphoid tissue (CALT), larynx-associated lymphoid tissue (LALT), and salivary duct-associated lymphoid tissue (DALT) (Table 5.9). Functionally, MALTs are divided into effector sites and inductive sites. The inductive sites act as secondary lymphoid tissue where maturation of B cells occurs in response to antigen-primed T cells. GALT, BALT, DALT, and CALT are inductive sites. Effector sites are present in all mucosal tissues and contain T cells (CD4+), plasma cells specific for IgA secretions (few IgG- and IgM-secreting plasma cells are also present), and fewer numbers of B cells, dendritic cells, and macrophages.

5.3.3 Antibody

The interaction of B cells with an antigen leads to proliferation and differentiation of B cells to develop plasma cells. These plasma cells secrete antibodies specific to that particular antigen that travels in the blood to neutralize the antigens.

Table 5.9 Different MALTs and their locations

Name		Site
GALT	Peyer's patches	Mucosa and submucosa of the gastrointestinal tract with more abundance at jejunum
	Isolated lymphoid follicles	Antimesenteric border of the small intestine
	Cryptopatches	Intercryptal lamina propria of the small intestine
	Lymphoglandular complexes	Colon
NALT		Caudoventral portion of the left and right nasal passages
BALT		Bifurcation of bronchial tree between a bronchus and an artery (absent in dogs, cats, and Syrian hamsters)
Tonsils		Oro- and nasopharynx (absent in rodents)

Table 5.10 Serum levels of different immunoglobulins in different species

Species	IgG	IgM	IgA	IgE
Cattle	1700–2700	250–400	10–50	–
Sheep	1700–2000	150–250	10–50	–
Horse	1000–1500	100–200	60–350	4–106
Pig	1700–2900	100–500	50–500	–
Dog	1000–2000	70–270	20–150	2.3–4.2
Cat	400–2000	30–150	30–150	–
Chicken	300–700	120–250	30–60	–

Source: Tizard (2013)

Therefore, antibodies are antigen-binding proteins produced by plasma cells. Antibodies are found in body fluids with maximum abundance in blood serum.

5.3.3.1 Structure of Antibody

Antibodies are glycoprotein in nature. They are mostly obtained from γ-globulin fractions of plasma protein and hence called immunoglobulins. They are heterodimer consisting of two identical heavy (H) and two identical light (L) molecular weight of 50,000 and 25,000, respectively. Each L chain is linked with H chain by a disulfide bond to form a heterodimer (H-L). There are also noncovalent interactions such as hydrogen bond and hydrophobic bonds between H and L chains. The other H and L chains are joined in a similar fashion to form another H-L heterodimer. These two identical H-L heterodimers are joined by a disulfide bond and non-covalent interactions to form a "Y"-shaped heterotetramer $(H\text{-}L)^2$ by a hinge region. The tip of the "Y," the amino-terminal ends of both H and L chain containing 110–130 amino acids, varies greatly among different antibodies. This variable portion of the antibody is called V regions (V_L for L chain, V_H for H chain). The portion of V region which shows maximum variability among different antibodies is called complementarity-determining regions (CDRs) for both H and L chains. Like variable region, there is a constant (C) region at the tail of the "Y" for both L (C_L) and H (C_H) chains. Functionally, the immunoglobulin has two different regions, fragment antigen-binding (Fab fragment) and fragment-crystallizable region (Fc region) identified after digestion with the enzyme papain. The Fab region is a low-molecular-weight fraction (45,000 Da) having antigen-binding property. Another comparatively higher molecular weight fraction (50,000 Da) has no antigen-binding activity and is called Fc fragment ("fragment, crystallizable").

5.3.3.2 Immunoglobulin Classes

There are five classes of immunoglobulins identified in mammals. They are immunoglobulin G (IgG), IgM, IgA, IgE, and IgD. Different classes of immunoglobulins are identified by their amino acid sequences in the constant region of the heavy chains. Normal serum levels of different immunoglobulins in different species have been presented in Table 5.10.

Immunoglobulin G (IgG): IgG has the smallest immunoglobulin molecule with highest abundance in the blood among other immunoglobulin classes. They are having molecular weight of about 180 kDa containing two identical heavy chains (γ) and two different light chains (either κ or λ). IgG is produced from plasma cells of lymph nodes, spleen, and bone marrow. They can pass through capillary barriers and are thus found largely at the inflammatory sites when vascular permeability is increased. Based on their heavy-chain sequences, there are four IgG subclasses, namely IgG1, IgG2, IgG3, and IgG4. IgG1 has the highest concentration in the plasma, and the rest are numbered in accordance to their serum proportions, and IgG4 has the lowest concentration. Each IgG subclass also has different functional properties. IgG1, IgG3, and IgG4 are able to cross placental barriers easily and thus protect the fetus. IgG1 and IgG3 are primarily involved in opsonization due to their high affinity for Fc receptors, whereas IgG2 has the lowest affinity. The IgG3 is the most potent immunoglobulin for complement activation followed by IgG1. IgG4 is unable to activate complement system.

Immunoglobulin M (IgM): The IgM is secreted in the form of a pentamer consisting of five monomers each of 180 kDa; hence, the total molecular weight is about 900 kDa. They are primarily produced by plasma cells residing at secondary lymphoid organs. Their proportion in the serum is about 5–10% of the total immunoglobulins. Each IgM monomer consists of two light chains (κ or λ) and two identical heavy chains (μ). Five IgM monomers join together in circular fashion with the help of a small peptide chain of 15 kDa named J-chain. IgM has ten antigen-binding sites due to their pentameric structure. IgM is the first immunoglobulin produced in response to an infection, and it is also the first immunoglobulin synthesized in neonates. IgM is a potent complement activator compared to IgG and also helps in opsonization and viral neutralization, but due to their large size, they are found in very less concentration in the body fluids and rarely reach the site of inflammation.

Immunoglobulin A (IgA): They constitute around 10–15% of the total serum immunoglobulin and are mainly secreted from plasma cells located at the surfaces of the body such as skin, mammary gland, intestinal wall, and respiratory and urinary system. Thus, IgA is mostly found in the external secretions like milk, tears, saliva, and mucus. IgA is secreted as a dimer consisting of two single monomers having a molecular weight of 180 kDa, so secreted IgA has 360 kDa joined by J-chain. Each IgA monomer has two light chains and two heavy chains (α). IgA protects the mucous membrane of digestive, respiratory, and urinary systems against *Salmonella*, *Vibrio cholerae*, and *Neisseria gonorrhoeae* and polio, influenza, and reoviruses. As IgG is mostly secreted in milk, it protects the newborn from infection during pre-weaning periods.

Immunoglobulin E (IgE): It was so named as it is induced by the E antigen of ragweed pollen. It has a molecular weight of about 190 kDa composed of two heavy chains (ε chain) and two light chains. Like IgA, it is also synthesized by the plasma cells under the skin, but in serum, it has extremely low concentration. IgE is mainly involved in allergic manifestation. It binds with Fc receptors of basophils and mast cells along with antigen and activates them. The activation of basophils and mast cells causes degranulation with the release of pharmacologically active substances that mediate allergic response. IgE is also involved in anti-parasitic defense. IgE has the shortest half-life (2–3 days).

Immunoglobulin D (IgD): It has a molecular weight of about 180 kDa and is composed of two δ heavy chains and two light chains. It is predominantly a membrane-bound immunoglobulin expressed in mature B cells with a very little serum concentration (30 μg/mL). It constitutes about 0.2% of the total immunoglobin in serum. IgD induces basophils to release pro-inflammatory and antimicrobial mediators.

5.3.3.3 Immunoglobulin Variants

Isotypes: The isotypes of immunoglobulins are determined by the constant regions of the heavy chain, and the constant region determinants are called isotypic determinants. The variations in the constant region are due to variations in the genes that encode the heavy chains. Bovine IgG has three isotypes such as IgG1, IgG2, and IgG3 encoded by IGHG1, IGHG2, and IGHG3 genes, respectively. The different isotypes of immunoglobulins have different physical and functional properties. IgG2 is a potent agglutinin, whereas IgG1 does not have such a function. Subclasses of different immunoglobulins in different species have been presented in Table 5.11.

Allotypes: The allotypic immunoglobulin variants are generated due to allelic variations. In the previous section, we discussed bovine IgG isotypes like IgG1, IgG2, and IgG3 encoded by IGHG1, IGHG2, and IGHG3 genes, but there may be multiple alleles for IGHG gene which encode subtle amino acid differences and they are called allotypic determinants. IgA2 subclass has two allotypes designated as A2m(1) and A2m(2).

Idiotypes: The idiotypic immunoglobulin variants are resulted due to variations in the amino acid sequences in the variable regions of light and heavy chains.

Table 5.11 Subclasses of different immunoglobulins in different species

Species	IgG	IgA	IgM	IgE	IgD
Cattle	G1, G2, G3	A	M	E	D
Horse	G1, G2, G3, G4, G5, G6, G7	A	M	E	D
Sheep	G1, G2, G3	A1, A2	M	E	D
Pig	G1, G2a, G2b, G3, G4	A	M	E	D
Dog	G1, G2, G3, G4	A	M	E1, E2	D
Cat	G1, G2, G3, G4 (?)	A	M	E1, E2 (?)	D
Mice	G1, G2a, G2b, G3	A1, A2	M	E	D

Source: Tizard (2013)

Immunoglobulin Superfamily These are a group of proteins having immunoglobulin-like domains containing 70–110 amino acids similar to variable and constant regions of immunoglobulin molecule. They share the common ancestral primordial gene that encodes immunoglobulin domains. The proteins under the immunoglobulin superfamily are Ig-α/Ig-β heterodimer of B-cell receptor, T-cell receptor, T-cell accessory proteins (CD2, CD4, CD8, and CD28), MHC class I and II molecules, cell adhesion molecule, and platelet-derived growth factors.

5.3.4 Recognition of Antigens by T and B Lymphocytes

B lymphocytes can recognize the epitopes that present on the surface of the pathogens or soluble factors released by the pathogens without the involvement of MHCs. The immunoglobulins present in the membrane of B lymphocytes specifically bind with the antigens. These membrane-bound immunoglobulins together with disulfide-linked heterodimers called Ig-α/Ig-β are called B-cell receptor (BCR). Both Ig-α and Ig-β chains have long cytoplasmic tails consisting of 61 and 48 amino acids. The Ig-α/Ig-β heterodimer is responsible for intracellular signaling after antigen binding.

In contrast to B cells, virus-infected host cells and cancer cells are recognized by T cells in conjugation with MHC. The antigen-binding sites on the T cells are called T-cell receptors (TCRs). TCRs are analogous to membrane-bound immunoglobins in B cells and also have both V and C regions, but TCR is unable to recognize antigen directly; rather, it recognizes short peptide fragments of antigens, which bind to MHC molecules through a process called antigen processing and presentation.

5.3.5 Antigen Processing and Presentation

The recognition of antigens by the T cells requires antigen processing and presentation by the APCs. In this process, foreign antigens are degraded into smaller peptides to interact with MHC molecules. There are two different mechanisms for exogenous and foreign antigens by which APCs process and present the antigens to T cells.

The *exogenous antigens* are engulfed by the APCs through endocytosis or phagocytosis, degraded by the endocytic processing pathways, and attached with class II MHC molecules. The antigen and MHC II complex then move to the surface of APCs and are recognized by T cells displaying CD4.

The *endogenous antigens* (viral proteins or cancer antigens) are produced with the cells. They are processed with the endoplasmic reticulum and subsequently attach with class I MHC molecules within the endoplasmic reticulum. The complexes are then transported to the cell membrane and recognized by T cells displaying CD8.

5.3.6 Monoclonal and Polyclonal Antibodies

Monoclonal antibodies bind to a single epitope, and polyclonal antibodies are the heterogeneous immunoglobulin mixture against a single antigen and can bind with different epitopes of a single antigen. Monoclonal antibodies are produced from the same clone of B cells, whereas polyclonal antibodies are produced from different B-cell clones. Monoclonal antibodies have high specificity and reproducibility, but the production of monoclonal antibodies is time consuming and expensive. The polyclonal antibodies have high affinity, and cost of production is also less compared to monoclonal antibodies. Monoclonal antibodies are used in therapeutics, and polyclonal antibodies are used in research applications.

5.4 Hypersensitivity and Immunological Disorders

Hypersensitivity is the pathological consequence that leads to the fatal host responses mediated by the immune system. It occurs when a pre-sensitized immune system of the host overreacts in response to an antigen. Hyperactive immune system thus produces effector molecules that induce inflammatory responses, which is undesirable to the hosts.

P. G. H. Gell and R. R. A. Coombs classified hypersensitivity reactions into four types based on the reactions involved and mediators. They are type I or IgE-mediated hypersensitivity, type II or antibody-mediated hypersensitivity, type III or immune complex-mediated hypersensitivity, and type IV or delayed-type hypersensitivity. Among these four types, first three are categorized under humoral immune responses and the last one is under cell-mediated immune response. Types I, II, and III are also called immediate hypersensitivity due to their earlier onset.

5.4.1 Type I or IgE-Mediated Hypersensitivity

Type I hypersensitivity reactions occur in response to some antigens called allergens that bind with IgE molecules. The allergen-IgE complexes thus produced attach with the mast cells or basophils to cause degranulation and release some inflammatory mediators. It can take 15–30 min from the time of exposure to the antigen. The allergens are of different types. They may be pollens of birch tree, rag seed, or rapeseed oil; drugs such as penicillin or salicylate; foods like nuts, eggs, or seafood; and insect products like bee venom and animal hair. Type I hypersensitivity reaction is mediated by IgE, which binds with the primary cellular components of type I hypersensitivity reactions of mast cell or basophil. The other cellular components such as platelets, neutrophils, and eosinophils help to amplify the reaction.

5.4.1.1 Mechanism of Type I Hypersensitivity Reaction

Production of IgE antibody: The initial response of type I hypersensitivity is similar humoral response, except the nature of antibody produced. In exposure to normal antigens, IgG or IgM is produced, but in response to allergens, IgE is produced.

Sensitization: The IgE antibody thus produced binds with FcRI receptors present in mast cells or basophils and sensitizes them. It is also called pre-sensitization of immune system against the particular allergen.

Shocking dose of antigen: The second exposure of the allergens is called shocking dose. During the second exposure, the allergens bind with Fab region of IgE molecules attached on the surface of mast cells or basophils during the pre-sensitization stage.

Activation and degranulation of mast cell: Binding of allergens to IgE antibody causes the activation of mast cells or basophils through intracellular signaling cascade by phosphorylation, adenylation, and methylation. The activated mast cells undergo degranulation and release certain pharmacologically active substances of various functions detailed in Table 5.12.

5.4.1.2 Anaphylactic Reactions

Pharmacologically active substances released from mast cells cause vasodilation, smooth muscle contraction, mucus production, and sneezing that leads to allergic response either localized or systemic.

5.4.1.3 Examples of Type I Hypersensitivity Reactions

The manifestation of type I hypersensitivity reactions could be either localized or systemic. The generalized type I hypersensitivity leads to systemic anaphylaxis (anaphylactic shock) with a short period of time after the exposure of the allergens. Lung is the primary target organ for most of the domestic animals, but in dogs, the liver is mostly affected. The clinical signs include excitement, pruritus, salivation, vomiting, dyspnea, convulsions, or even death. The localized type I hypersensitivity includes allergic rhinitis, asthma, allergic enteritis, and atopic dermatitis.

5.4.2 Type II or Antibody-Mediated Hypersensitivity

In this reaction, the antibodies (IgG or IgM) induced against cellular antigens and this antibody-mediated immune reaction lead to cellular damage by either of the three mechanisms.

1. The antibodies bind with the target cell that expresses the antigen. The Fc portion of the antibody binds with Fc receptor of the phagocytic cells and leads to the opsonization. The opsonin then activates phagocytic cell, and the target cell is phagocytosed. In some cases, the Ag-Ab complex activates complement system and terminal complement complex (C5b6789) and lysis of target cells. This is called antibody-dependent cellular cytotoxicity (ADCC). The classical examples are autoimmune hemolytic anemia and erythroblastosis fetalis.
2. The activation of complement complex leads to the formation of C3a and C5a that are extremely chemotactic and attracts neutrophils and eosinophils towards target cells. The reactive oxygen species generated in neutrophils then destruct the cell (e.g., Goodpasture syndrome).
3. Autoantibodies cause cellular dysfunctions without inflammation or cell lysis. In Graves' disease, the

Table 5.12 Pharmacologically active substances of type I hypersensitivity reactions

Mediators	Functions
Histamine	Bronchoconstriction, mucus secretion, vasodilatation, and increased vascular permeability
Tryptase	Proteolysis
Kininogens	Vasodilatation, vascular permeability, edema
Eosinophil chemotactic factor of anaphylaxis (ECF-A)	Attracts eosinophil and neutrophils
Leukotriene B4	Attracts basophils
Leukotriene C4, D4	Similar to histamine but 100 times more potent
Prostaglandins D2	Edema and pain
Platelet-activating factor (PAF)	Platelet aggregation and heparin release

autoantibodies bind with thyrotropin receptors, which leads to overproduction of thyroid hormones.

5.4.3 Type III or Immune Complex-Mediated Hypersensitivity

In this condition, larger amount of Ag-Ab complexes causes tissue-damaging reactions. The reaction usually takes 3–10 h after exposure to the antigen in a pre-sensitized individual. The tissue damage is caused by neutrophils, macrophages, or other phagocytes. At first, the Fc region of the antibody binds with Fc receptors present in the phagocytes, and activation of the receptors leads to the production of inflammatory mediators such as prostaglandins, leukotrienes, nitric oxide, cytokines, and chemokines and promotes inflammation and migration of neutrophils at the site. When the neutrophils try to destroy the immune complexes, they deposit their granular contents in the surrounding structures like basement membrane and collagen fiber and disrupt them. In some cases, the migration of neutrophils occurs in response to chemotactic complement products (C3a and C5a). The severity of the reaction depends upon the amount of Ag-Ab complex and their site of deposition. It may cause localized tissue reactions at the site of deposition. The typical reactions are systemic lupus erythematosus and Arthus reaction in the skin. Sometimes, the immune complexes are deposited in various tissues via blood, and the reactions develop at their site of tissue deposition. The examples are lupus nephritis in kidneys, aspergillosis in lungs, polyarteritis in blood vessels, and rheumatoid arthritis in joints.

5.4.3.1 The Arthus Reaction

It is a localized type III hypersensitivity reaction due to the formation of Ag-Ab complexes after the intradermal injection of an antigen. The immune complexes form local vasculitis. It is seen very frequently after the vaccination against diphtheria and tetanus.

> **Know More......**
> Some dogs develop "blue eye," corneal edema, and opacity after infected or vaccinated with live canine adenovirus type 1. It usually takes 1–3 weeks after the exposure and resolves by its own after the elimination of the virus.

5.4.4 Type IV or Delayed-Type Hypersensitivity

It is mediated by antigen-specific T cells. The type I reaction differs from other types of hypersensitivity in terms of the time taken to respond. It usually takes 1–3 days after the antigenic exposure and hence is called delayed type. Another important unique feature of type IV hypersensitivity from other types is the involvement of T cells in contrast to antibody as in other three types. Here, in this reaction, the antigens after processing and presentation by ABP stimulate type 1 helper T (TH1) cells to secrete cytokines and chemokines like IFN-γ, TNF-α and -β, and interleukin 2 (IL-2). TNF-α and -β increase the vascular permeability, and IFN-γ causes macrophage activation to secrete their lethal contents and subsequent tissue damage. Type IV hypersensitivity reaction is important for intracellular pathogens where antibodies are unable to reach. The lethal damage to the tissue causes the destruction of pathogens together with the cells. Type IV hypersensitivity reaction is of three types.

The *contact hypersensitivity dermatitis* occurs when an exogenous antigen invades the skin and induces inflammatory reaction in dermis and epidermis. The antigen-presenting dendritic cells and Langerhans cells process the antigen and present to type 1 helper T (TH1) to induce inflammatory reaction.

Tuberculin-type hypersensitivity occurred in response to intradermal injection of purified protein derivative (PPD) named tuberculin (product of tuberculosis bacillus). It is used to detect tuberculosis in animals.

Granulomatous-type hypersensitivity occurs in more serious condition when the recruited macrophages are unable to destroy the antigens. As a result, more number of macrophages are recruited and these macrophages filled with intracellular antigen lead to granuloma.

5.4.5 Immunological Disorders in Animals

5.4.5.1 Thyroiditis

In this condition, the antibodies are generated against thyroglobulin or thyroid peroxidase. The lymphocytes invade the thyroid gland and cause epithelial cell destruction through antibody-dependent cell-mediated cytotoxicity. The clinical manifestations include dry and dull hair coat, loss of hair, scaling, hyperpigmentation, and pyoderma. Thyroiditis is common in dogs that are more susceptible, and susceptible dog breeds are Beagles, Great Danes, and Doberman.

5.4.5.2 Polyneuritis

It is an autoimmune disease against myelin proteins of nerve tissues. It occasionally occurs in horses and dogs. The characteristic symptoms are hyperesthesia and paralysis of tail, rectum, and urinary bladder. Sometimes, facial and trigeminal paralysis also occurs. In dogs, the disease is also called "coonhound paralysis" as it occurs as a result of bite or scratches from raccoons.

5.4.5.3 Uveitis/Blue Eye

It is an autoimmune disorder in response to autoantigen interphotoreceptor retinoid-binding protein and is characterized by the inflammation of uveal tract (iris and ciliary body) of the eye. It can be seen in horses and dogs. In horses, the inflammatory site is infiltrated with Th1 cells and neutrophils along with fibrin and C3 depositions, which may lead to blindness. In puppies, uveitis may develop after vaccination with canine adenovirus (CAV)-2-modified live viral vaccine and the prognosis is good. The characteristic symptoms include conjunctivitis, ulcers in cornea, vascularization, and corneal scarring.

5.4.5.4 Immune-Mediated Hemolytic Anemia

The immune-mediated hemolytic anemia occurs due to the production of autoantibodies against RBC membrane proteins like glycophorins, spectrin, and anion-exchange protein CD233. It occurs in cattle, horses, dogs, cats, and mice. The major clinical manifestations are anemia and others secondary symptoms like fever, jaundice, tachycardia, and splenomegaly.

5.4.5.5 Myasthenia Gravis

It is a skeletal muscle disease that occurs due to the formation of autoantibodies against acetyl choline receptor, which leads to blockage of the receptors and complement-mediated destruction of the receptors, which ultimately causes nerve impulse transmission. The clinical signs are abnormal fatigue, muscle weakness, and exercise intolerance. The species involved are dogs, cats, ferrets, and humans.

5.4.5.6 Systemic Lupus Erythematosus (SLE)

It is an autoimmune disorder due to the development of autoantibodies against nucleic acids, nucleoproteins, and chromatin. These autoantibodies are called antinuclear antibodies. It is a classic example of type III hypersensitivity reaction. The immune complexes are deposited in various tissues like glomeruli to form glomerulonephritis, synovial joints to form arthritis, arteriolar walls leads to fibrosis and skin leads to ulcerative lesions. It is mostly seen in dogs with symptoms like localized ulceration in the skin and mucous membrane, glomerulonephritis, lymph node enlargement, splenomegaly, nervous symptoms like lameness, and lethargy. In equines, SLE is characterized by skin diseases like dermal ulceration, alopecia, and crusting together with anemia.

5.4.5.7 Rheumatoid Arthritis

It is an immune complex-mediated hypersensitivity reaction that occurs in response to the deposition of immune complexes in the joints. It is very common in humans but is occasionally seen in dogs (mostly in toy breeds) characterized by lameness immediately after awaking in the morning in addition to anorexia, depression, and pyrexia.

5.4.5.8 Canine Leukocyte Adhesion Deficiency

It is an autosomal recessive genetic disorder due to mutation in integrins, the cell adhesion molecule. The neutrophils are unable to adhere with vascular endothelium and thus unable to migrate to the site of injury (see Sect. 5.2.9—Leukocyte migration). The affected dogs are prone to recurrent infections. Leukocyte adhesion deficiency is also seen in cows.

Learning Outcomes

- **Antigens**: Antigens are the molecules that bind specifically with the products of immune response (i.e., antibodies or cytotoxic T lymphocytes) induced by immunogens, and this property is called antigenicity. The antigenicity is determined by several factors such as foreignness, chemical structure and molecular size, doses, and route of administration. Antigens can be classified into exogenous and endogenous based on their source. Antigens can also be classified into complete antigen (immunogen), incomplete antigen (hapten), and superantigen based on immune responses.
- **Innate immunity**: Innate immunity is the evolutionary nonspecific defensive reflex against foreign materials owned by birth. It serves as the first line of defense against infection. Components of innate immunity include anatomical barriers (skin and mucosal membrane), physiological barriers (body temperature, pH, and several other soluble factors), immune effector cells (granulocytes, monocytes/macrophages, natural killer cells, dendritic cells, endothelial cells, epithelial cells, lymphoid cells, and platelets), pattern recognition receptors [Toll-like receptors, C-type lectin receptors, nucleotide-binding oligomerization domain (NOD) receptors, and retinoic acid-inducible gen-I (RIG)-like receptor], inflammatory serum proteins/acute-phase proteins (haptoglobin, serum amyloid A, ceruloplasmin), antimicrobial peptides (AMPs), complement system, and cytokines.
- **Adaptive immunity**: Adaptive or acquired immune responses are capable of selective elimination pathogens. It has some cardinal features like specificity, diversity, and memory. It also has a unique ability to discriminate the self and nonself antigens and to react accordingly. Adaptive immune responses are brought about by different classes of lymphocytes, namely B and T lymphocytes. The immune responses mediated through B lymphocytes are called *antibody-mediated or humoral immune response* as B cells are capable

(continued)

of producing antibodies (or immunoglobulins) upon antigenic exposure, which bind with antigens and make them vulnerable for destructions. *The cell-mediated immune responses* are mediated through T lymphocytes that produce signals to activate phagocytic cells to destroy them.

- **Hypersensitivity**: It is the pathological consequence that leads to the fatal host responses mediated by the immune system when a pre-sensitized immune system of the host overreacts in response to an antigen. It can be classified into type I or IgE-mediated hypersensitivity, type II or antibody-mediated hypersensitivity, type III or immune complex-mediated hypersensitivity, and type IV or delayed-type hypersensitivity. Among these four types, first three are categorized under humoral immune responses and the last one is under cell-mediated immune response. Type I, II, and III are also called immediate hypersensitivity due to their earlier onset.

Exercises

Objective Questions

Q1. Arrange the biomolecules in their ascending order of immunogenic response proteins, polysaccharides, lipopolysaccharides, and nucleic acid.

Q2. Acute-phase protein associated with copper transport is ______________.

Q3. Which biomolecule protects host cells from complement-associated lysis?

Q4. The primary lymphoid organ of birds is ___________.

Q5. Which immunoglobulin is predominant in external secretions?

Q6. Viral proteins are associated with which class of MHC?

Q7. Anaphylactic shock is an example of ________________ hypersensitivity.

Q8. Name one autoimmune disease associated with skeletal muscle.

Subjective Questions

Q1. Mention some features of acquired immunity.

Q2. Briefly describe the mechanism of action of cytokines.

Q3. How MHC helps to discriminate self and nonself?

Q4. Explain the inflammatory signs under the light of its cellular events.

Q5. "Thymus is the organ of tolerance" .... Justify the statement.

Answer to Objective Questions

A1. Proteins > polysaccharides > lipopolysaccharides > nucleic acid

A2. Ceruloplasmin

A3. Glycophorin A

A4. Bursa of Fabricius

A5. IgA

A6. Class I MHC

A7. Type I

A8. Myasthenia gravis

Keywords for Subjective Questions

A1. Specificity, diversity, memory, discrimination of self and nonself

A2. Receptor binding, intracellular signaling, response

A3. Antigen processing and presentation, exogenous antigen, MHC II, endogenous antigens, MHC I

A4. Vasodilation, redness, tissue permeability, edema

A5. T lymphocyte selection, positive, negative

Further Reading

Books

Guyton AC, Hall JE (2011) Textbook of medical physiology, 11th edn. Saunders, Philadelphia, PA

Punt J, Owen JA, Stranford SA, Jones PP, Kuby J (2019) Immunology, 8th edn. W.H. Freeman/Macmillan Learning, New York

Tizard IR (2013) Veterinary immunology, 9th edn. Elsevier Inc.

Research Articles

Cesta MF (2006) Normal structure, function, and histology of mucosa-associated lymphoid tissue. Toxicol Pathol 34:599–608

Groves E, Dart AE, Covarelli V, Caron E (2008) Molecular mechanisms of phagocytic uptake in mammalian cells. Cell Mol Life Sci 65: 1957–1976

Kemper C, Pangburn MK, Fishelson Z (2014) Complement nomenclature. Mol Immunol 61:8. https://doi.org/10.1016/j.molimm.2014.07.004

Libby P (2007) Inflammatory mechanisms: the molecular basis of inflammation and disease. Nutr Rev 65(12):S140–S146

Mogensen TH (2009) Pathogen recognition and inflammatory signaling in innate immune defenses. Clin Microbiol Rev 22(2):240–273

Pasupuleti M, Schmidtchen A, Malmsten M (2012) Antimicrobial peptides: key components of the innate immune system. Crit Rev Biotechnol 32(2):143–171

Sherwood ER, Toliver-Kinsky T (2004) Mechanisms of the inflammatory response. Best Pract Res Clin Anaesthesiol 18(3):385–405

Yalew ST (2020) Hypersensitivity reaction. Int J Vet Sci Technol 4(1):028–032

Part III

Cardiovascular, Respiration and Excretory System

6 Cardiovascular System

P. Visha and V. Sejian

Abstract

The cardiovascular system comprising the heart and the connecting blood vessels plays a vital role in maintaining the homeostasis. It integrates three basic functional components, heart as the pump that circulates the blood through a network of the blood vessels. This integrated system is able to adapt to the circulatory demands under different physiological states like rest and exercise, changes in body position, digestion, thermal stress, and emotional status. In order to meet these variable physiological demands, the entire system undertakes sophisticated regulatory mechanisms involving neuroendocrine system and renal and local circulatory mediators. Understanding the functioning of various cardiovascular system components will help the students and researchers to explore in detail the various vital factors regulating the circulatory system and identify the associated dysfunctions.

P. Visha (✉)
Department of Veterinary Physiology and Biochemistry, Veterinary College and Research Institute, Tamil Nadu Veterinary and Animal Sciences University, Salem, Tamil Nadu, India

V. Sejian
ICAR-National Institute of Animal Nutrition and Physiology, Bangalore, Karnataka, India

© The Author(s), under exclusive license to Springer Nature Singapore Pte Ltd. 2023

P. K. Das et al. (eds.), *Textbook of Veterinary Physiology*, https://doi.org/10.1007/978-981-19-9410-4_6

Graphical Abstract

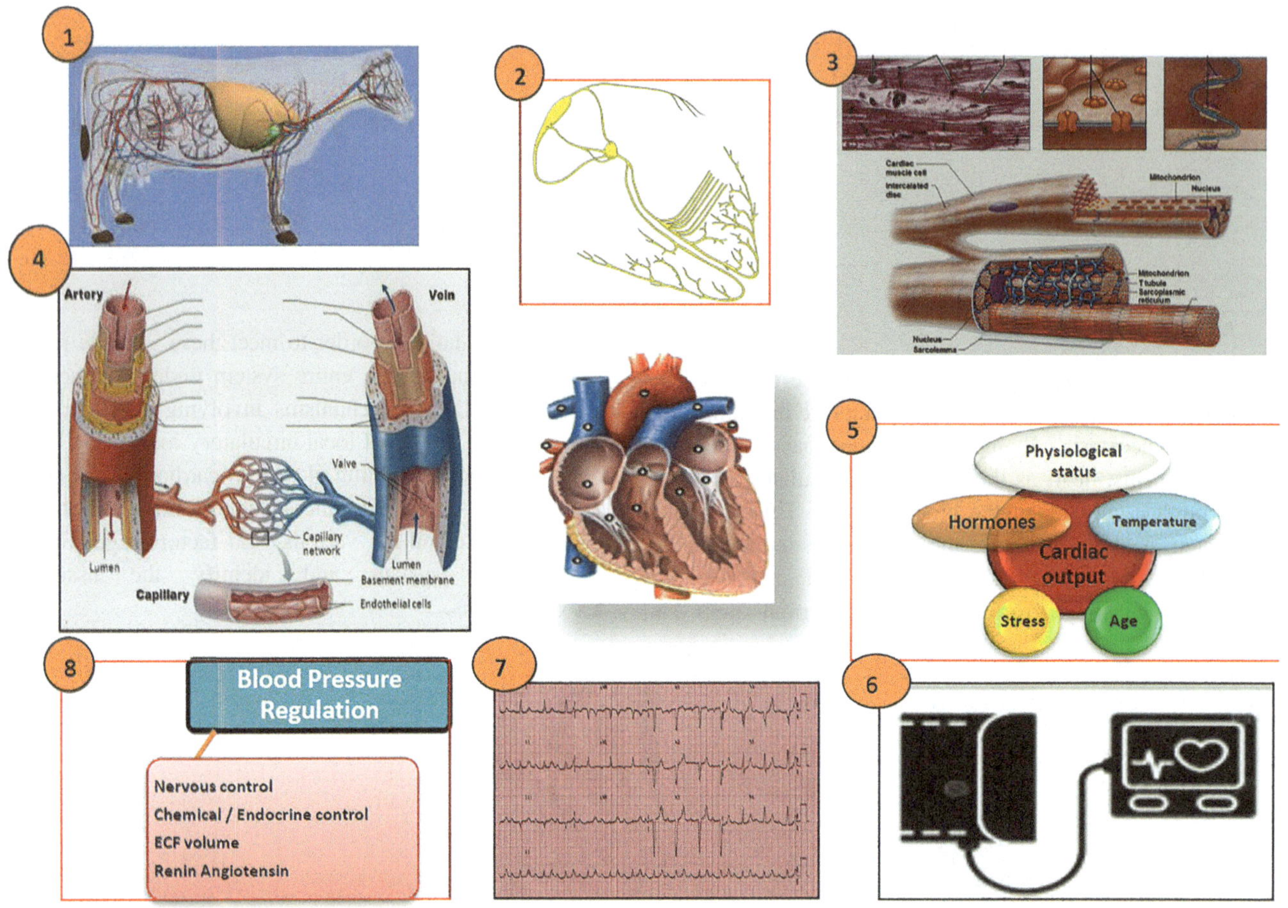

Description of the graphic: Cardiovascular system (**1**) consists of both heart and vascular systems (arterial, venous, and capillary circulations). Being the first functional organ, the development of heart begins at very early stages of life. It is made of specialized muscle cells called cardiac muscle (**2**), which are striated and involuntary muscle fibers. It has specialized excitatory and conduction systems through which electric impulses are generated and conducted throughout the heart, resulting in contraction and relaxation of the heart (**3**). It pumps oxygenated blood through blood vessels to all the body's tissues (**4**). Cardiac output and heart rate are influenced by various factors, viz. rest, exercise, stress, temperature, sex, age, emotion, and endocrine factors (**5**). The electrocardiogram (ECG) is the most important clinical tool used for diagnosing electrical dysfunctions of the heart (**6**); determination of arterial blood pressure helps in diagnosing the defects of heart and circulatory system (**7**), which is regulated by nervous system, endocrine system, regulation of ECF volume, and renin-angiotensin mechanism (**8**). The entire functional characteristics of the CV system are elaborated in this chapter

Keywords

Myocardium · Action potential · Cardiac cycle · Cardiac output · Hemodynamics

Learning Objectives

- General organization of CVS and hemodynamics of circulation
- Gross structure of heart and myocardial cells
- Electrophysiological of myocardial and nodal cells and its recording
- Integration of neuroendocrine system, renal and local control mechanisms regulating cardiac functions
- Characteristic features and functional anatomy of special circulation in vital organs
- Clinical aspects of CV system

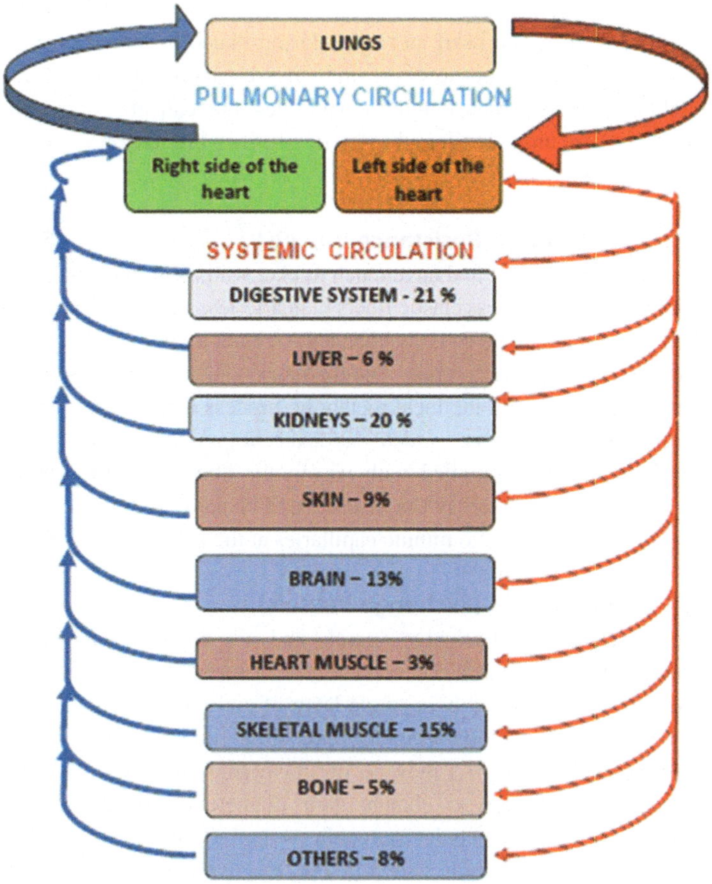

Fig. 6.1 General organization of the cardiovascular system depicting pulmonary and systemic circulation

6.1 General Organization of CVS and Hemodynamics of Circulation

The cardiovascular system consists of a central pump, the heart, and the interconnected network of blood vessels, which function efficiently to maintain homeostasis. Essential substances such as nutrients and oxygen are continuously picked up from the respective sites of availability and delivered to the cells, and end products of metabolism are continuously removed to maintain homeostasis. Furthermore, the hormones, nutrients for storage, body heat, toxic materials, and waste products are constantly taken from their production site and efficiently delivered to the site of final requirement/storage/processing/removal by the blood flowing in the blood vessels. Thus, all the cells in the body and the metabolic processes taking place within these cells constantly depend on the life-supporting blood flow that is provided during each heart contraction. This network is uniquely regulated by mechanisms to meet out the functional demands of the body. The cardiovascular system has two circulations in series: (1) the pulmonary circulation and (2) the systemic circulation. Each circulation has three major divisions: (1) the distribution system (ventricles, arteries, and arterioles), (2) the exchange system (capillaries), and (3) the collecting system (venules, veins, and atria). The general organization of the cardiovascular system is shown in Fig. 6.1.

6.1.1 Heart as a Pump

The heart is a single organ; anatomically and internally, it comprises four chambers and is divided into right and left halves. The upper chambers, the atria, receive blood returning to the heart and pump it to the lower chambers, the ventricles. The heart serves as an efficient hollow muscular pump that regulates the blood flow in the pulmonary and systemic circulation by imparting pressure to the blood to establish the pressure gradient needed for blood to flow to the organs and tissues.

6.1.1.1 Arteries (Conduit Vessels)

Arteries are thick-walled vessels containing smooth muscles, elastin, and collagen fibers. During systole, arteries expand under increased pressure to accept and temporarily store the blood ejected by the heart and then, during diastole, supply this blood to the organs downstream by passive recoil; this is facilitated by the presence of the elastin fibers that can stretch to twice their unloaded length. Arteries further branch into arterioles.

6.1.1.2 Arterioles (Resistance Vessels)

Arterioles are smaller and have much thicker walls with more smooth muscle and less elastic fibers than arteries, and hence, their diameters can be actively changed to regulate the blood flow to the peripheral organs. The diameter of the arterioles can be modulated by the local metabolic process and by the blood-borne substances such as catecholamines. The arterioles are also supplied with vasomotor fibers that are regulated by the centers in the medulla and spinal cord.

Arterioles divide into minute capillaries at the tissue level.

6.1.1.3 Capillaries (Exchange Vessels)

Capillaries are the exchange vessels and are the smallest vessels having a small blood content (5% of blood volume). These vessels are lined by a single layer of endothelial cells that separates the blood from the interstitial fluid by only approximately 1 μm. As they do not have smooth muscle, they lack the ability to change their diameters actively. Capillaries are tiny tubular network composed of only a thin layer of endothelium, which favors selective permeability of nutrients, water, oxygen, CO_2, and other metabolic waste between blood and tissues, and they are known as exchange vessels. The blood in the capillaries flows as a single layer of RBCs at the center, while WBCs are located in the peripheral side of the capillaries. The highest cross-sectional area of the circulatory system is created by capillaries. Arteriovenous shunts (A-V shunts) are the direct connections between capillaries and veins that result from blood flowing directly from the arteriole into the metarteriole. The true capillaries exchange nutrients and are connected to one another. Precapillary sphincter, a smooth muscle, is situated at the region where capillaries diverge from metarterioles. The precapillary sphincters and metarterioles are not innervated and are controlled locally by regional tissue conditions.

6.1.1.4 Venules and Veins (Capacitance Vessels)

After leaving capillaries, blood is collected in venules and veins and returned to the heart. Venous vessels have very thin walls having smooth muscles, and their diameters can actively change. Veins have very thin walls, making them very distensible. Therefore, their diameters change passively in response to small changes in transmural distending pressure. The veins continue as *vena cava*, which drains its content into the right atrium. Vena cava has the largest diameter. Veins have valves in the extremities. These valves stop blood from flowing backward, allowing for one-way blood flow to the heart. The main thoracic and abdominal veins do not have valves. Veins have about 24 times more capacitance than arteries, and about 80% of the total blood volume is located within the venous system.

6.1.2 Hemodynamics of Circulation

Blood flow in the vessels is influenced by many factors as described by the physical principles. However, as blood is not a perfect fluid but consists of both the liquid and cells and the blood vessels are not rigid tubes, the behavior of the circulation deviates, sometimes markedly, from that predicted by these principles.

6.1.2.1 Blood Velocity

It refers to the distance a blood bolus travels per unit of time (mm/time or cm/time). The average velocity of blood movement is inversely proportional to the total cross-sectional area at that point, is high in the aorta, reduces steadily in the smaller vessels, and is lowest in the capillaries.

6.1.2.2 Laminar Flow

In the straight blood vessels, the blood flow is normally laminar. Within the blood vessels, the velocity is maximum near the center and decreases near the vessel walls. When the vessels divide, the flow gets disturbed and turbulence is produced. Turbulence refers to the disruption of the laminar of streamline flow pattern. Laminar flow is silent, but turbulent flow creates sounds. The amplitude of turbulence depends on the blood velocity, diameter of the vessel, and viscosity of the blood. Laminar flow occurs at velocities up to a certain *critical velocity*. At or above this velocity, flow is turbulent. Reynold number is used to define turbulent areas:

$$\mathrm{ReN} = \frac{\rho D V}{\eta}$$

where ReN is the Reynolds number, ρ (g/mL) is the density of the fluid, D (cm) is the diameter of the vessel, V (cm/s) is the velocity of the flow, and η (poise) is the viscosity of the fluid. The blood flow is laminar when ReN is below 2000. The higher the value of ReN, the greater the probability of turbulence. In anemia, because the viscosity of the blood is lower, turbulence occurs more frequently.

6.1.2.3 Resistance and Viscosity

Resistance to the blood flow is directly proportional to vessel length and viscosity of the blood and inversely proportional to radius of the vessel. The major regulating factor for the vascular resistance is the vessel radius to the fourth power.

The radius of smaller vessels is controlled by the vascular smooth muscles. In the larger arterial vessels, the radius is controlled mainly by the amount of collagen and elastin. Increase in hematocrit causes appreciable increases in viscosity in large vessels, as compared to the smaller vessels (below 100 μm in diameter) (arterioles, capillaries, and venules). In marked anemia, peripheral resistance is decreased, because of the reduction in viscosity, whereas in severe polycythemia, the increase in resistance increases the work of the heart. The rate of blood flow through a vessel is directly proportional to the pressure gradient (as the pressure gradient increases, flow rate increases) and inversely proportional to vascular resistance (as resistance increases, flow rate decreases).

6.1.2.4 Shear Stress

Shear stress refers to the force exerted by the flowing blood on the endothelium that is parallel to the long axis of the vessel. This shear stress is proportionate to viscosity and changes in the shearing stress produce marked changes in the expression/activation of endothelial cell genes that produce substances like growth factors and integrins.

6.1.2.5 Vascular Compliance

Vascular compliance refers to the ability of a blood vessel wall to respond to the changes in pressure by expanding and contracting passively. The arterial system has less compliance as compared to the venous vessels, and hence the venous vessels hold larger volumes of blood. Vessel compliance (C) measures a vessel's capacity to distend and expand in volume when transmural pressure rises and is expressed by the change in volume (ΔV) divided by the change in pressure (ΔP):

$$C = \frac{\Delta V}{\Delta P}$$

Furthermore, the compliance of the vessel is also dependent upon the rate by which the change in volume occurs. At higher pressures and volumes, the compliance decreases (i.e., at increased pressures and volumes, vessels become "stiffer"); comparably at lower pressures (venous pressure is usually less than 15 mmHg), the venous compliance is about 10–20 times greater than arterial compliance. Hence, the veins can accommodate large changes in blood volume with only a small change in pressure. Vein compliance has increased in part as a result of venous collapse, which happens at pressures lower than 10 mmHg. Venous compliance is comparable to arterial compliance at higher pressures and volumes. This characteristic of the veins makes them suitable for use as arterial bypass grafts.

The compliance status for a blood vessel is dynamic. Even within a given artery or vein, capacitance, distensibility, and compliance are not constants in the cardiovascular system. While smooth muscle contraction improves vascular tone and decreases vascular compliance, vascular smooth muscle relaxation enhances compliance. Another example of changing compliance is the reduction in aortic compliance with age and disease as in arteriosclerosis, which in turn leads to increased aortic pulse pressure.

Characteristics associated with arteries and veins, such as vascular compliance, blood volume, and vascular resistance, affect the functioning of the heart. Vascular capacitance, compliance, and distensibility are critical determinants of the performance of the heart. During systole, the walls of large elastic arteries (e.g., aorta, common carotid, and pulmonary arteries) distend as the blood pressure rises and recoil when the blood pressure falls during diastole. Since the rate of blood entering these elastic arteries exceeds the quantity that leaves these vessels, there is a net storage of blood in the aorta and large arteries during systole, which is discharged during diastole. This feature, known as the windkessel effect, aids to decrease the load on the heart and minimizes the systolic flow and maximizes the diastolic flow in the arterioles. This effect helps in damping the blood pressure fluctuations over the cardiac cycle and assists in maintaining organ perfusion during diastole when cardiac ejection ceases. The windkessel vessels (aorta and large elastic arteries) convert the pulsatile inflow to a smooth outflow, and Otto Frank, a German physiologist, developed this concept and named this phenomenon as the windkessel effect.

6.1.2.6 Blood Pressure

This is the pressure exerted by the circulating blood against any unit area of the blood vessel. Stephen Hales, an English clergyman, demonstrated the existence of pressure in the blood vessels in 1730. Blood pressure is highest in aorta (98 mmHg), moderate in capillaries, and lowest in vena cava (3 mmHg). This pressure difference ($98 - 3 = 95$) moves the blood through systemic vessels. Perfusion pressure is the difference in pressure between the aorta and veins. The blood flows more favorably through the blood arteries due to this pressure gradient. The maximum pressure exerted in the arteries when blood is ejected into them during ventricular systole is referred to as the systolic pressure that averages 120 mmHg. It indicates the total kinetic energy imparted to the blood by the heart. Diastolic pressure is the minimum pressure within the arteries when blood is draining off into the rest of the vessels during diastole, and it averages 80 mmHg. Blood continues to exit the aorta into the tiny arteries during ventricular diastole, while the volume of blood in the major arteries declines, the arteries become less engorged, and blood exerts less pressure on the arteries.

6.1.2.7 Pulse

It is a wave of expansion and elongation followed by recoiling of the arterial walls and it is due to the forceful entry of the blood from the aorta during each heartbeat. Cardiac systole generates pressure waves that move the pliable and compliant arterial walls. It originates from aorta, spreads throughout arterial system, and disappears at the arterioles. The distension of the arterial walls due to sudden entry of blood and the subsequent increase in the pressure are marked by anacrotic limb of the pulse wave. However, the decreased pressure due to elastic recoiling of the distended arteries is referred to as catacrotic limb or declining slope in the pulse wave. Pulse pressure refers to the difference between systolic and diastolic pressure. The pulse pressure increases as blood flows from aorta to distal arteries and then becomes less and less when the blood moves toward periphery. It disappears in the arterioles and capillaries.

Mean arterial pressure (MAP) can be roughly estimated as follows:

$$\text{Mean arterial pressure} = \text{Diastolic BP} + 1/3\ (\text{Systolic BP} - \text{Diastolic BP})$$

$$\text{Mean arterial pressure} = \text{Diastolic pressure} + 1/3\ (\text{Pulse Pressure})$$

This method is useful to find out mean pressure in major arteries distal to aorta but not in aorta because the pattern of arterial pressure pulsation changes as the pulse moves away from the heart.

Arterial BP is always expressed in mmHg, whereas capillary and venous pressure can be expressed as mm H_2O.

6.1.2.8 Factors Influencing Blood Pressure

1. **Heart rate**: During systole, the arterial pressure increases. The rate at which blood enters the arterial system exceeds the rate at which it drains through arterioles and capillaries, increasing the pressure. During diastole, the BP decreases since blood passes out of arteries into capillaries. When other factors remain constant, a decrease in the heart rate causes a fall in blood pressure and an increase raises blood pressure.
2. **Peripheral resistance**: Total peripheral resistance (TPR) is the resistance to the blood flow in systemic vessels. It is caused by internal friction produced by the viscosity of blood and is mostly present in arterioles and capillaries. Blood pressure is directly proportional to the peripheral resistance. When most capillaries are open, peripheral resistance decreases and blood pressure is reduced. Resistance is inversely proportional to the fourth power of the vessel radius and varies directly with the viscosity and length of the vessel. Total peripheral resistance (TPR) can be calculated as follows:

$$\text{TPR} = \text{Aortic pressure (BP)}/\text{Cardiac output}$$

3. **Elasticity of arteries**: The elastic recoil of large arteries is responsible for the continuous flow of blood in the arterioles and capillaries. During systole when more blood enters aorta and large arteries, they expand to store the blood, i.e., potential energy is stored, and during diastole, the stored blood is released due to elastic recoil of the walls of arteries causing blood to flow into arterioles during diastole. Arterial stiffening increases systolic pressure (due to reduced expansion of artery) and decreases diastolic pressure (since less blood is stored by reduced expansion).
4. **Volume flow**: An increase and decrease in the blood volume of circulatory system cause increase and decrease of blood pressure. When hemorrhage occurs in a large artery, the blood pressure (BP) falls immediately. On moderate hemorrhage and when small arteries are cut, the fall in BP will be negligible because of the compensatory vasomotor mechanism and splenic constriction.
5. **Diameter of the blood vessels**: Arterial blood pressure varies indirectly with the diameter of blood vessel. If the diameter increases, the peripheral resistance decreases, leading to decrease in the blood pressure.

6.1.2.9 Determination of Arterial Blood Pressure

It helps in diagnosing the defects of heart and circulatory system. It can be measured by two means, the direct and indirect (clinical) methods.

The first direct measurement of mean arterial blood pressure was carried out in horse by Stephen Hales in 1730.

The usual physical examination in veterinary clinical practice does not include measurement of ABP. Pulse pressure is often indirectly assessed during physical examination by digital palpation of a peripheral artery. In veterinary practice, blood pressure cuffs are less frequently used, but the pulse is usually palpated by placing the fingertips over a major artery, such as the femoral artery. Palpation of an artery allows the clinician to sense the pulse pressure on the basis of the magnitude of pulsations that are felt in the artery. However, the character of peripheral arterial pulsations primarily reflects the pulse pressure and not the absolute level of ABP. Thus, a horse with a systolic/diastolic ABP of 100/60 mmHg may have a facial artery pulsation that is very similar to that of another horse with ABP of 180/140 mmHg. Thus, assessment of the level of ABP (i.e., diagnosis of systemic hypertension) requires measurement of ABP.

The ABP may be measured in animals by direct or indirect techniques.

Direct techniques rely on the penetration of a peripheral artery with a catheter or a needle and connection via a fluid-filled tube to a pressure transduction system. This technique generally provides an accurate and precise measure of ABP and is useful in anesthetized animals, particularly large animals. However, because anesthetics and sedatives alter ABP and physical or emotional distress can dramatically elevate blood pressure, they are not used for screening of animals for the presence of systemic hypertension and indirect techniques are usually employed in veterinary practice for this purpose. Indirect techniques rely on the placement of a cuff over an extremity (limb or tail). The cuff is inflated to occlude a peripheral artery. As pressure is automatically or manually reduced in the cuff, the restoration of arterial flow can be detected distally by a variety of methods. These devices generally utilize ultrasonic Doppler, oscillometric, or photoplethysmographic principles to detect the restoration of flow distal to the cuff. Systolic ABP may then be estimated from cuff pressure at the time of restoration of flow. Some indirect devices (e.g., ultrasonic Doppler flowmeters) are generally used only to estimate systolic arterial blood pressure (ABP), while other devices often provide an estimate of systolic, mean, and diastolic ABP.

6.2 Heart: Gross Structure and Myocardial Cells

6.2.1 Gross Structure and Size

The mammalian heart consists of the two receiving chambers—the atria—and two pumping chambers—ventricles. The size of heart is approximately 0.3–1.0% of the body weight and varies across species. The size of heart increases with body size proportionately in order to meet out with an increased metabolic demand and to deliver a large blood volume. However, within individual species, there is considerable variance, and this variability is related to the type of physical activity that the animals usually engage in. Animals that run for prolonged periods either for sports performance (e.g., thoroughbred horses and greyhounds) or for hunting (e.g., wolves) have markedly larger hearts (1.2% of body weight) than those which are physically less active (heart of sedentary pig is approximately 0.3% of body weight).

Birds possess larger hearts, higher stroke volumes, lower heart rates, and increased cardiac output than mammals of corresponding body mass, which contributes to the high aerobic energy input needed to sustain the flapping flight. The larger birds like geese, ducks, and swans tend to have proportionally smaller hearts in relation to their body mass than the smaller birds. Thus, heart mass in small birds such as the racing pigeons is about 1.1% of body mass, compared with 0.8% for the Pekin ducks. In the migratory birds, the heart becomes hypertrophic before migration, which could be due to their genetic potential to increase cardiac output and heart size either through seasonal humoral mechanisms or through natural selection over long term.

6.2.1.1 Heart Chambers

6.2.1.1.1 Atria

The thin-walled, low-pressure atria receive the venous blood, function as elastic reservoir, and act as a primer pump, thereby enhancing ventricular filling. The right atrium has an average thickness of about 2–3 mm and receives blood from three veins: the superior vena cava, inferior vena cava, and coronary sinus. A thin partition, the interatrial septum, separates the right atrium and left atrium. The presence of an oval depression (fossa ovalis), a remnant of foramen ovale, an opening in the interatrial septum of the fetal heart that normally closes soon after birth, is present in the septum. Blood passes from the right atrium into the right ventricle through the tricuspid valve (right atrioventricular valve) that has three leaflets. The left atrium is about the same thickness as the right atrium and forms most of the base of the heart. It receives blood from the lungs through four pulmonary veins. Blood passes from the left atrium into the left ventricle through the bicuspid (mitral or left atrioventricular valve), which has two cusps. The heart valves are composed of dense connective tissue covered by endocardium.

6.2.1.1.2 Ventricles

Most of the heart's weight is contributed by the ventricular myocardial mass. The right ventricle has an average thickness of about 4–5 mm. Trabeculae carneae, or elevated bundles of cardiac muscle fibers, produce a series of ridges in the inside of right ventricles. The cusps of tricuspid valve are connected to the chordae tendineae, which in turn are connected to cone-shaped trabeculae carneae called papillary muscles. Internally, the right ventricle is separated from the left ventricle by a partition called the interventricular septum. Blood passes from the right ventricle through the pulmonary valve (semilunar valve) into a large artery called the pulmonary trunk, which divides into right and left pulmonary arteries. The left ventricle, the thickest chamber of the heart (average thickness of 10–15 mm), forms the apex of the heart. Like the right ventricle, the left ventricle has trabeculae carneae and chordae tendineae that anchor the cusps of the mitral valve to papillary muscles. Blood passes from the left ventricle through the aortic valve (semilunar valve) into the aorta.

The Burmese python (*Python molurus*) shows a rapid extraordinary 40% increase in ventricular mass within 48–72 h after their biannual feeding, which could be attributed to a specific set of fatty acids that appear to promote myocardial hypertrophy, not hyperplasia, and an increased expression and activity of superoxide mutase—a cardioprotective free radical scavenger.

6.2.1.1.3 Myocardial Thickness and Functional Correlation

The thickness of the myocardial chambers varies according to their function. While the thick-walled ventricles pump blood under higher pressure over longer distances, the thin-walled atria pump blood under lower pressure into the adjacent ventricles. Despite the fact that the right and left ventricles function as two independent pumps that discharge the same amount of blood at once, the right side has a far lesser resistance and pumps blood only a short distance to the lungs at lower pressure and has lesser resistance to blood flow. The left ventricle pumps blood through great distances to all other parts of the body at higher pressure, and the resistance to blood flow is larger. Therefore, the left ventricle works much harder than the right ventricle to maintain the same rate of blood flow, and hence, the muscular wall of the left ventricle is considerably thicker than the right ventricular wall.

During systole, the right ventricular free wall moves toward the interventricular septum due to the contraction of the spiral muscles. Systole in the left ventricle also functions to assist ejection from the right ventricle by the curvature of the septum, pulling the right ventricular free wall toward the septum (called left ventricular aid).

6.2.1.2 Cardiac Valves

The four fibrous cardiac valves, namely, the atrioventricular (AV) valves that separate the atria from the ventricles and the semilunar valves positioned between the ventricles and the great arteries (pulmonary artery and aorta), aid to regulate the blood flow into cardiac chambers during the various phases of the cardiac cycle. The passive opening and closing of these valves occur in response to pressure changes produced by contraction and relaxation of the four muscular chambers, and the orientation of the valves helps to maintain a unidirectional flow of blood.

6.2.1.2.1 Atrioventricular (AV) Valves

The atrioventricular valves (tricuspid and bicuspid valves) located between the atrium and the ventricles, in open position, have the rounded ends of the cusps project into the ventricles. Hence, the relaxation of the ventricles and papillary muscles leads to loosening of chordae tendineae to open AV valves, allowing the blood to move from a higher pressure in the atria to a lower pressure in the ventricles. Similarly, when the ventricles contract, the pressure of the blood drives the cusps upward, until their edges meet and close the opening. At the same time, the papillary muscles contract, simultaneously pulling and tightening the chordae tendineae and hence preventing the valve cusps from averting (opening into the atria) in response to the high ventricular pressure. If the AV valves or chordae tendineae get damaged, blood may regurgitate into the atria whenever the ventricles contract.

6.2.1.2.2 Semilunar Valves

The aortic and pulmonary valves are known as the semilunar (SL) valves because they have three crescent moon-shaped cusps, which are attached to the arterial wall by their convex outer margin. The semilunar valves allow blood to leave the heart and enter arteries, but they stop blood from returning to the ventricles. The free borders of the cusps project into the lumen of the artery. The semilunar valves open when the ventricular pressure exceeds the pressure in the arteries, allowing ejection of blood from the ventricles into the pulmonary trunk and aorta. As the ventricles relax, blood starts to flow back toward the heart for a short period of time until the semilunar valve closes.

6.2.2 Physiology of Cardiac Muscles

Cardiac muscles contract millions of times over the lifetime of a domestic animal, which exhibits their unique endurance potential. Cardiac muscle is involuntary and striated and has sarcomeres with actin and myosin filaments and other organelles, which have their own functional and anatomical differences.

There are two specialized types of cardiac muscle cells:

1. Contractile cells, comprising 99% of the cardiac muscle cells, perform the mechanical work of pumping. These cells normally do not initiate the action potentials and are specialized for contraction and impulse conduction.
2. Autorhythmic cells are small but extremely vital for cardiac functioning. Instead of contracting, they are specialized to initiate and propagate the action potentials that cause the working cells to contract. Although the pacemaker, conduction, and working cells account for the majority (>70%) of the heart's mass, they only constitute one-third of the total number of cells in the heart. The remaining cells are fibroblasts, endocardial cells, endothelial cells, and vascular smooth muscle cells (details are discussed in Sect. 10.3.2).

6.2.2.1 Myocardial Cells

In comparison to the skeletal muscle fibers, the cardiac muscle fibers are shorter in length, branched, and less circular in transverse section. A mature cardiac muscle fiber is 85–100 μm long and has a diameter of about 15 μm. Each myocardial cell has a centrally located nucleus and is packed with contractile myofibrils consisting of sarcomeres joined end to end at their Z lines. Intercalated discs are specialized paired interdigitating membrane junctions that connect the two ends of adjacent cells in series. These intercalated discs have three types of functional specializations: fascia adherens, desmosomes, and gap junctions. Fascia adherens present on the transverse segment of the disc serves as a locus for insertion of actin myosin filaments and forms a strong connection between adjacent fibers. Desmosomes are round bodies found on the transverse segment of the disc and that mechanically hold cells together. They allow the transmission of the force of contraction and produce the mechanical syncytium. They are particularly abundant in heart, which is constantly subjected to considerable mechanical stress.

Gap junctions are found on the longitudinal segments of the discs and contain channels that allow free diffusion of ions between the cells with low electrical resistance. Gap junctions allow the entire myocardium to contract as a single, coordinated electrical syncytium. Gap junctions are elongated and numerous in Purkinje cells where conduction is rapid, whereas the gap junctions are small and sparse in SA node and AV nodal cells. Cardiac muscle thus acts as a mechanical and electrical syncytium, unlike skeletal muscle fibers, which are separate cells bound by connective tissue. Avian cardiac muscle fibers (2–7 μm in diameter) are much smaller in diameter than mammalian fibers (10–15 μm in diameter) and hence are more numerous in comparison to similarly sized hearts.

6.2.2.1.1 Sarcoplasmic Reticulum and Tubular System

The sarcoplasmic reticulum consists of a network of anatomizing thin-walled tubules that invests into the sarcomere and has the main function of providing the majority of Ca^{2+} ions required to trigger the myofilaments and to resequester Ca^{2+} from the myoplasm allowing for relaxation. The sarcoplasmic reticulum has two main regions: junctional sarcoplasmic reticulum (jSR), which confronts the surface membrane's invaginations called transverse tubules (T-tubules) directly, and extrajunctional free sarcoplasmic reticulum (fSR), which is situated near the myofibrils. jSR forms extended, flattened cisternae, which have sets of closely grouped structures ("feet") that represent the cardiac SR Ca^{2+} release channels, also known as ryanodine receptors (RyR2s), and contain the Ca^{2+}-binding protein, calsequestrin-2 (CASQ2). On the other hand, the fSR is devoid of CASQ2, and its external surface has Ca^{2+} adenosine triphosphatase (ATPase). Calsequestrin, the major Ca^{2+}-binding protein in sarcoplasmic reticulum, acts as the major Ca^{2+} storage and buffering protein and is an important regulator of Ca^{2+} release channels in both skeletal and cardiac muscle.

Phospholamban, a sarcoplasmic reticulum protein, is a key regulator of cardiac contractility. In its dephosphorylated state, it regulates SR Ca^{2+} sequestration by inhibiting the SR Ca^{2+}-ATPase (SERCA) (which transports calcium from cytosol into the sarcoplasmic reticulum). Upon phosphorylation, the inhibitory effect of phospholamban on the SERCA is removed leading to faster Ca^{2+} uptake into the sarcoplasmic reticulum.

Myocardial cells of the avian species lack transverse tubules, which are prominent in mammalian cardiac muscles. Similarly, the avian myocytes lack the M-band of the cytoskeleton structure that cross-links the myosin and titin filaments in the middle of the sarcomere.

6.2.2.1.2 Myocardial Cell Organelles

Mitochondria are more numerous in cardiac muscle and larger, and they make up about 40% of the cytoplasmic volume compared with only about 2% in skeletal muscle. In some myocardial cells, the mass of the mitochondria equals that of the myofibrils. Cardiac muscle cells stop contracting after about 30 s of oxygen deprivation, thus reflecting the reliance of cardiac muscle on aerobic metabolism. Glycogen granules are found in large numbers and are evenly dispersed in cardiac muscle. Lipid droplets, which are the predominant energy source when engaged in aerobic metabolism, are frequently found located adjacent to mitochondria.

6.2.2.1.3 Functional Syncytium

The cardiac muscle fibers do not fuse with each other and do not form a morphological syncytium, but because of their branching and bifurcations, they form a functional syncytium that allows coordinated contraction. The electrical impulse spreads from one cardiac cell to all the other cells that are joined by gap junctions in the surrounding muscle mass so that they become excited and contract as a single, functional syncytium. The atria and the ventricles each form a separate functional syncytium and contract as separate units. This division allows the atria to contract a short time ahead of ventricular contraction, which is vital for effective heart pumping. The synchronous contraction of the myocytes in the atrial and ventricular walls produces the force needed to eject the received blood. Action potentials are not conducted from the atrial syncytium into the ventricular syncytium due to the lack of gap junctions joining the atrial and ventricular contractile cells and also due to the presence of an electrically nonconductive fibrous tissue that surrounds the valves and separates the atria and the ventricles. Instead, the action potentials are conducted only by way of a specialized conductive system, the atrioventricular (AV) bundle that facilitates and coordinates transmission of electrical impulse from the atria to the ventricles, thus ensuring synchronization between atrial and ventricular pumping.

6.2.2.2 Metabolism and Energetics of Working Myocardial Cells

Under normal conditions, the metabolic system of cardiac cells relies nearly entirely on aerobic metabolism, which continuously supplies high-energy phosphate bonds for mechanical and chemical work. The cardiac myocytes have an abundance of energy-generating mitochondria, and they receive a rich blood supply, which is delivered by about one capillary for each myocardial to support their rhythmic, contractile activity. Cardiac muscle also has an abundance of myoglobin, which stores O_2 within the heart for immediate use.

The basal oxygen consumption by the myocardium is about 2 mL/100 g/min, and by the beating heart, it is about 9 mL/100 g/min at rest which increases further during exercise.

The heart can use a variety of substrates (glucose and fatty acids) to oxidatively regenerate ATP depending upon availability. Usually, the cardiac muscle cells use free fatty acids as their major fuel for metabolism. Glucose and lactate are important contributors. For each mole of O_2 consumed, there is a 53.7% higher energy production from the metabolism of glucose than from the metabolism of palmitate. The caloric value of the stored high-energy phosphate bonds produced by the oxidation of a gram of palmitate is 2.4 times greater than that produced by the oxidation of a gram of glucose, but at a greater relative expenditure of oxygen than that produced by the oxidation of glucose. Hence, when oxygen is abundant and food is scarce, utilizing fatty acids for fuel is advantageous as opposed to using glucose. Following a high-carbohydrate meal, the heart can adapt itself to utilize carbohydrates (primarily glucose) almost exclusively. The reverse occurs when food is plentiful and oxygen is scarce. The heart utilizes fatty acids (60–70%) and carbohydrates (~30%) under postabsorptive state. During exercise, lactate can be used in place of glucose. Under diabetic acidosis, amino acids and ketones are utilized as metabolic fuel. The coronary circulation is unable to provide the heart with the metabolic substrates it needs to support aerobic metabolism during ischemia and hypoxia; thus, the heart uses glycogen as a source of energy for the anaerobic synthesis of adenosine triphosphate (ATP) and the creation of lactic acid. As the cardiac muscle is highly adaptable and can shift metabolic pathways to use whichever nutrient is available, the primary danger of insufficient coronary blood flow is oxygen deficiency and not fuel shortage.

6.2.2.3 Properties of Myocardial Cells

Conductivity: Cardiac muscle cells are capable of transmitting action potentials. In spite of being separated by plasma membranes, the individual cells pass the impulse from one cell to another cell through the specialized gap junctions in the intercalated discs.

Contractility: Contractility refers to the ability of the myocardial tissue to shorten in length (contraction) after receiving a stimulus. The contractile properties of the cardiac muscle are influenced by many factors.

Following are the contractile properties of myocardial cells:

All or none principle: If at all the cardiac muscle responds to a stimulation, it contracts with maximum strength (i.e., all contractions of the cardiac muscle are maximum) under the prevailing physical and chemical environment, i.e., the heart muscle contracts maximally when stimulated with either minimum or maximum strength of stimulus. The all or none law is due to the syncytial nature of the cardiac muscle.

Staircase phenomenon: The cardiac muscle responds with increasing degree of strength for the first 3–5 stimuli when an effective stimulus is applied repeatedly with a short interval (1–2 s), after which the reaction is constant. This is referred to as staircase phenomenon or Treppe. When the interval between two stimuli of same strength is very short, the physical and chemical changes known as beneficial effects, occurring during the first response, persist and these changes facilitate the second one giving greater response. This progressive rise in tension after an increase in the heart rate was first observed by Henry Bowditch in 1871. The staircase phenomenon is caused due to an increase in sarcoplasmic reticulum Ca^{2+} content and release, which in turn occurs by three mechanisms. First, during each action potential plateau, Ca^{2+} enters the cell through L-type Ca^{2+} channels, and the more number of action potentials per minute provides a longer aggregate period of Ca^{2+} entry through these channels. Secondly, depolarization during the plateau of an action potential causes the Na-Ca exchanger (NCX1) to operate in the reverse mode, permitting Ca^{2+} to enter into the cell. At higher heart rates, these depolarizations occur at increased frequency and are accompanied by an increase in intracellular Na+, which accentuates the reversal of NCX1, both of which enhance Ca^{2+} uptake. Third, the rising intracellular Ca^{2+}, through calmodulin (CaM), activates CaM kinase II, which leads to phosphorylation of phospholamban (PLN); phosphorylation of PLN in turn stimulates sarcoendoplasmic calcium ATPase ($SERCA_2a$), thereby sequestering the Ca^{2+} in sarcoplasmic reticulum that had entered the cell as a result of the first two mechanisms.

Refractory period: When an excitable tissue is responding to a stimulus, it may not produce another response for a subsequent stimulation for a shorter period. This period of unresponsiveness of cardiac muscle is called as refractory period. The refractory period of a cardiac muscle fiber lasts longer than the contraction itself. As a result, another contraction cannot occur until relaxation is well underway. The refractory period is longer in cardiac muscle than skeletal muscles, and hence cardiac muscle cannot be thrown into continuous contractions by a series of rapid stimuli, i.e., cardiac muscle cannot be tetanized.

After initial stimulation, there is an absolute refractory phase, which is a very brief period during which there is no reaction at all. This phase is followed by a short period of partial responsiveness, during which the cardiac muscle can be stimulated by stronger stimuli, and this period is referred to as relative refractory period. The period of systole of cardiac muscle is the absolute refractory period (atria 0.15 s, ventricle 0.25 s), and relative refractory period occurs during early diastole (atria 0.03 s, ventricle 0.05 s).

Regeneration: Regeneration of cardiac muscle fibers does not occur, and if myocardial fibers die, they are replaced by fibrous noncontractile scar tissue. During the embryogenesis of skeletal muscle, once the cells have differentiated to the point where they are capable of contraction, there is no further cell division or increase in DNA content. In the adult animal, increase in myocardial mass, particularly in an adult animal, can be accomplished only through the hypertrophy of already differentiated cardiac cells, leading to the enlargement and multiplication of intracellular structures such as myofilaments and mitochondria.

Extrasystoles: Extrasystoles refer to the additional heartbeats, initiated by an irregular internal or external excitation source. Extrasystoles are induced by a hypersensibility of the autonomous nervous system or by mechanical, chemical, or pharmacological stimuli.

Electrical excitation in cardiac myocytes is linked to contraction through increase in intracellular calcium that occurs with each action potential. Most of the increase in intracellular calcium results from Ca^{2+} release from the sarcoplasmic reticulum through release channels (ryanodine receptors) (RyRs), which in turn is triggered by Ca^{2+} entry through L-type channels in the cell membrane. This process of Ca^{2+}-induced Ca^{2+} release (CICR) involves positive feedback and is potentially unstable. The cell controls this potential instability by grouping the RyRs into spatially segregated clusters. In a resting myocyte, spontaneous release of sarcoplasmic reticulum Ca^{2+} from a cluster of ryanodine receptors (RyRs) will appear as a spatially restricted (2 μm) Ca^{2+} spark that will not spread to neighboring regions. Under conditions wherein there is Ca^{2+} overload, RyRs become more sensitive to increases in intracellular calcium, wherein a spontaneous Ca^{2+} spark can trigger release from neighboring clusters of RyRs, and a regenerative Ca^{2+} wave can result. This Ca^{2+} wave, propagating roughly at 100 μm/s, raises the intracellular calcium from 100 nM to about 1 μM. This results in a delayed afterdepolarization and, if it is large, can cause an extrasystole or ectopic action potential.

Post-extrasystolic potentiation: Post-extrasystolic potentiation refers to the increase of myocardial contractility that follows a premature beat. When a premature beat occurs, most of the ryanodine receptors are refractory to activation, causing a diminished Ca^{2+} transient and thus a less forceful contraction. After the premature beat, sarcoplasmic reticulum Ca^{2+} load is increased in a number of ways. First, while less Ca^{2+} is released, Ca^{2+} loading of the sarcoplasmic reticulum continues. Next, low Ca^{2+} transient during the premature beat opposes less negative feedback to sarcolemmal Ca^{2+} influx, and this extra Ca^{2+} further increases sarcoplasmic reticulum Ca^{2+} content and all ryanodine receptor channels recover from inactivation. At the post-extrasystolic beat, all the Ca^{2+} sequestered during the previous beats is released, resulting in increased force of the post-extrasystolic beat.

6.3 Electrical Activity of Heart

6.3.1 Basic Mechanisms of Membrane Potential

All the cells in the body maintain an electrical potential difference that can be measured across their plasma membrane which is called "membrane potential." The intracellular fluid has more of K^+ ions (140 mmol/L) and less of Na^+ ions (14 mmol/L), whereas there is higher concentration of Na^+ ions (142 mmol/L) and lower concentration of K^+ ions (5 mmol/L) in the interstitial fluid. This concentration difference of ions across a selectively permeable membrane creates an electrical potential difference.

The following are the three major factors contributing to the membrane potential:

1. **Differential permeability of the membrane to diffusion of ions**: The resting membrane is 50–100 times more permeable to K^+ ions than to Na^+ ions. Therefore, positively charged K^+ ions are allowed to diffuse out of the cell through non-gated leak channels down their concentration gradient (i.e., from inside of the cell to the outside).
2. **Na^+-K^+-ATPase pump**: Cell membranes have an energy-requiring pump, which pumps three Na^+ ions out of the cell and two K^+ ions into the cell against their concentration gradient simultaneously.
3. **Trapped anions inside the cell**: Many intracellular anions are too large molecules to move outwards through the plasma membrane. Hence, they are trapped within the cell and are attracted to the inner surface of the cell membrane by the accumulated positive cations just on the outer side of the cell.

6.3.2 Cardiac Action Potential

All the living cells are excitable to stimuli. However, nerve and muscle cells are highly excitable cells than other cells.

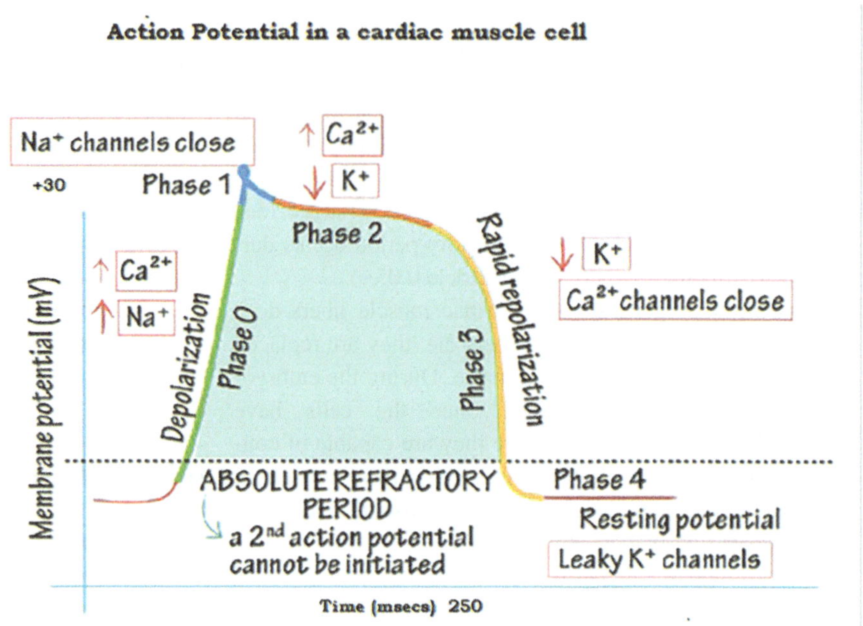

Fig. 6.2 Action potential in a cardiac muscle cell. The depolarization occurs due to the opening of the (1) fast sodium channels, (2) slow sodium-calcium channels, and (3) repolarization due to potassium channels. The action potential in a cardiac muscle is prolonged

Action potential is the rapid changes in the membrane potential from its normal negativity to positive potential that last for a period of few milliseconds, and then it returns back to its original resting potential level. The resting membrane potential in the cardiac myocytes is −90 mV (−85 to −95 mV). The cardiac muscle has slower but prolonged action potential than skeletal muscle that lasts for 150 ms in atria and 300 ms in ventricle.

Cardiac myocytes have three types of membrane ion channels that play a crucial role in causing the voltage changes of the action potential. They are (1) fast sodium channels, (2) slow sodium-calcium channels, and (3) potassium channels.

During the depolarization phase of the action potential, the membrane potential rapidly reverses to a positive value of about +20 to +30 mV (depending on the myocardial cell) as a result of activation of voltage-gated Na^+ channels and subsequent rapid entry of the Na^+ ions into the cells.

The slow channels are slow to open and remain in the open state for a few tenth of a second. The slow channels are activated at a membrane potential of −30 to −40 mV. The action potential spikes when the rapid Na channels are activated, while the action potential plateaus when the slow channels prolong the transit of Ca^{2+} ions into the cell's interior.

The inflow of Ca^{2+} ions into the cardiac muscle cells decreases K^+ efflux through voltage-gated K channels. This delays the K^+ ion permeability to outside, which in turn delays the repolarization process of the action potential in cardiac muscle. In cardiac myocytes, repolarization does not occur immediately after depolarization, but the positivity remains as a plateau near the peak of the spike potential. This plateau lasts for a few hundred milliseconds and prolongs the contraction of the cardiac muscle as shown in Fig. 6.2. The prolonged action potential makes the cardiac muscle cells to have longer contraction period than skeletal muscles.

6.3.2.1 Action Potential in Pacemaker Cells

The pacemaker cells of the SA node have an inherent property of generating their own action potentials rhythmically, independent of nerve stimulation, and they depolarize faster than any other part of the heart. The ability of pacemaker cells to self-stimulate is known as automaticity or rhythmicity. Automaticity is due to slow, spontaneous, and progressive depolarization of membrane potential until the threshold potential is reached, which initiates an action potential.

The resting membrane potential in the sinus nodal pacemaker cells (pacemaker potential) is −55 to −60 mV. The cause of this lesser negativity is that the cell membranes of the sinus fibers are naturally leaky to sodium and calcium ions. At this level of −55 mV, fast sodium channels are "inactivated," i.e., they have become blocked. This is because when the membrane potential remains less negative than about −55 mV for more than a few milliseconds, the fast sodium channels become closed. Therefore, only the slow sodium-calcium channels can open and thereby cause the action potential.

The initial decrease in resting potential is produced by natural inward leakiness of positive Na^+ ions (higher Na^+ concentration outside the nodal cells and moderately open Na^+ channels are the causes for Na^+ leaking to inside); this slowly brings the membrane potential to −40 mV at which potential the slow Na^+-Ca^{2+} channels become activated (open); Na^+-Ca^{2+} ions move inward and depolarize the membrane producing an action potential. The Na^+-Ca^{2+} channels become inactivated within about 100–150 ms after opening, and at the same time, more numbers of K^+ channels open. Therefore, influx of Na^+ and Ca^{2+} ions through the Na^+-Ca^{2+} channels stops, while large quantities of positive K^+ ions diffuse out of the cell. Both these effects reduce the intracellular potential back to its negative resting level and terminate the action potential. The K^+ channels remain open for another few milliseconds continuing the movement of positive charges out of the cell, resulting in excess negativity inside the cell; this is called hyperpolarization. The resting membrane potential goes back to about −55 to −60 mV at the end of the action potential. Within a few milliseconds after the action potential is over, K^+ channels close. The inward-leaking sodium and calcium ions once again overbalance the outward flux of potassium ions, and this causes the "resting" potential to move upward again, finally reaching the threshold level for discharge at a potential of −40 mV. Then the entire process begins again.

Stimulation of vagal fibers to heart hyperpolarizes the membrane potential since the acetylcholine increases outward K^+ conductance producing hyperpolarization of pacemaker cells. The result is a decrease in firing rate from SA node, and the heart rate is decreased. Very strong stimulation of vagus abolishes spontaneous discharge of SA node cells for some time.

Stimulation of sympathetic cardiac nerves makes the membrane potential to decrease rapidly and increases spontaneous discharge rate, and the heart rate is increased.

6.3.2.2 Specialized Excitatory and Conduction System of Heart

Impulse formation and conduction are carried out by three types of cells: nodal cells, Purkinje cells, and transitional cells.

Nodal cells are seen in sinoauricular node that is responsible for pacemaker impulse generation and in atrioventricular node, which is responsible for conduction delay.

Purkinje cells are larger cells specialized for rapid impulse conduction and are found in bundle of His, bundle branches, and Purkinje network.

Transitional cells connect Purkinje cells and contractile myocardial cells.

These specialized excitatory and conduction fibers show very feeble contractions because of the presence of very few contractile fibers.

In mammals, the sinoatrial node is a small, flattened, ellipsoid strip of specialized cardiac muscle, located at the junction of the right atrium and cranial vena cava. It has an inherent property of generating its own action potential at periodical interval. The ends of the sinus nodal fibers connect directly with surrounding atrial muscle fibers. Action potentials therefore move into these atrial muscle fibers after leaving the SA node.

SA node spreads its impulse through atrial muscular wall, interatrial bundles to atria, and through the anterior, middle, and posterior internodal pathways to AV node (atrioventricular node), which lies in the septal walls of the right atrium craniodorsal to tricuspid valve. AV node conducts action potentials to common bundle of His or AV bundle, which then runs into the ventricular septum where it divides into right and left bundle branches that run underneath the septal endocardium. The AV node and AV bundle are the only routes for the conduction of impulse from atria to ventricles. The AV node conducts the action potential very slowly. It takes 50–150 ms for an action potential to pass through the AV node. The nodal delay is contributed by the junctional fibers, which are very small fibers that connect the atrial fibers with nodal fibers. After the AV nodal delay, the impulse travels rapidly down the septum via the right and left branches of the bundle of His, and at the ventricular apex, the bundle branches finally terminate into Purkinje fibers, which are a network of conductive system in the ventricular muscle. The network of fibers in this ventricular conduction systems are specialized for rapid propagation of action potentials. These fibers coordinate and hasten the spread of ventricular excitation to ensure that the ventricles contract as a unit. The action potential is transmitted through the entire Purkinje fiber system within 30 ms. The impulse quickly spreads from the excited cells to the rest of the ventricular muscle cells by means of gap junctions. The ventricular conduction system is more highly organized. As the ventricular mass is much larger than the atrial mass, the ventricular conduction system is crucial for hastening the spread of excitation in ventricles. Rapid conduction of the action potential down the bundle of His and its swift, diffuse distribution throughout the Purkinje network lead to almost simultaneous activation of the ventricular myocardial ensuring smooth, coordinated contraction, which results in efficient pumping of the blood into the systemic and pulmonary circulations at the same time.

6.3.2.3 Propagation of Action Potential

Once initiated in the SA node, an action potential spreads throughout the rest of the heart. The right atrium begins to depolarize about 0.01 s before the left atrium. Purkinje fibers have the highest conduction velocity, while AV nodal fibers have the lowest conduction velocity. AV node delays the conduction velocity of action potentials to the ventricular musculature, which is referred to as nodal delay, and this allows the atrial contractions to occur a short time ahead of the ventricular contraction and thus facilitating the atria to discharge their blood into the ventricles before ventricular systole.

In part, the slow conduction velocity across the AV node is due to the small diameter of nodal myocytes and the complex arrangement of the myocytes, which makes the action potential follow a more tortuous path through the AV node. Additionally, the slow conduction velocity of the AV node is due to the poor expression of Na^+ channels, poor electrical coupling between the myocytes of the AV node, and lack of high-conductance connexins in the gap junction.

The self-excitation is greatest in the SA node fibers. The normal rate of discharge in SA node is 70–80/min; AV node—40–60/min; and Purkinje fibers—15–40/min. Hence, it is called as cardiac pacemaker. The SA node dominates the normal rate and rhythm of the heart, and hence the SA node normally controls the rate of the heart. In some pathological conditions of the mammalian heart, the excitatory impulses originate outside the SA node and such place is referred to as ectopic foci in which the heart rate will be less than normal. This rhythm is called as ectopic rhythm.

The velocities of conduction of impulse (m/s) of different conducting systems are atrial pathways—1; AV node—0.05; bundle of His—1; Purkinje fibers—4; and ventricular muscles—1.

6.3.2.4 Neural Control of Myocardial Rhythmicity and Impulse Conduction

The heart is supplied with both sympathetic and parasympathetic nerves. Stimulation of the sympathetic nerves releases norepinephrine at the sympathetic nerve endings, which increases the permeability of the sinus node fiber membrane to sodium and calcium ions, causing more positive resting potential and also increasing the rate of upward drift of the diastolic membrane potential toward the threshold level for self-excitation, thereby increasing the heart rate. Furthermore, increase in permeability to calcium ions increases the contractile strength of the cardiac muscle.

Stimulation of the parasympathetic nerves to the heart (vagus) causes acetylcholine to be released at the vagal endings. This hormone decreases the rate of rhythm of the sinus node and also decreases the excitability of the AV junctional fibers between the atrial musculature and the AV node by causing hyperpolarization through increasing the permeability of potassium ions into the cells of SA and AV nodes, thereby slowing the transmission of cardiac impulse into the ventricles. Weak to mild vagal stimulation slows the heart rate to one half normal. Strong stimulation of the vagi can completely stop the rhythmical excitation by the sinus node or completely block transmission of the cardiac impulse from the atria into the ventricles through the AV node, but then, ventricular escape occurs wherein the ventricular septal region of the AV bundle establishes its own rhythm and stimulates ventricular contraction at a rate of 15–40 beats per minute.

6.3.3 Electrocardiogram

The electrocardiogram (ECG) is the most widely used noninvasive clinical tool for diagnosing electrical dysfunctions of the heart, wherein two or more metal electrodes are applied to the skin surface, and the voltages recorded by the electrodes are displayed on a video screen or drawn on a paper strip.

6.3.3.1 Principles of ECG Recording

The depolarization and repolarization of the cardiac muscles generate electrical currents, which spread into the tissues around the heart and are conducted through the body fluids. A small part of this electrical activity reaches the body surface, where it can be detected using recording electrodes.

Electrocardiography provides a record of how the voltage between two points on the body surface changes with time as a result of the electrical events during the cardiac cycle. At any instant of the cardiac cycle, the electrocardiogram gives the net electrical field that is the summation of many weak electrical fields being produced by voltage changes occurring on individual cardiac cells at that instant. The ECG can be recorded by either using an active or exploring electrode connected to an indifferent electrode at zero potential (unipolar recording) or using two active electrodes (bipolar recording).

6.3.3.2 Lead Systems

ECG is recorded by placing a series of electrodes on the body surface. These electrodes, called ECG leads, are connected to the ECG machine at one end and are fixed on the left forelimb, right forelimb, and left hind limb at the other end. Electrodes on these limbs are usually envisioned as forming a triangle around the heart. Heart is in the center of an imaginary equilateral triangle (Einthoven triangle) drawn by connecting the roots of these three limbs. ECG is usually recorded in 12 leads, which are generally classified into three categories:

1. Bipolar leads
2. Unipolar leads
3. Chest leads

6.3.3.2.1 Bipolar Limb Leads (Standard Limb Leads)

Herein, two limbs are connected to the electrode and both the electrodes are active recording electrodes, i.e., one electrode is positive and the other one is negative. The right hind limb is connected to the earth.

A triangle can be formed by placing the three electrodes on right forelimb, left forelimb, and left hind limb with the heart at the center.

If electrical potential of any two of the three bipolar leads is known at a given instant, the third one can be calculated, i.e., the sum of voltage in lead I and III equals the voltage in lead II (Einthoven's law).

Standard limb leads are of three types:

1. **Limb lead I**: The voltage in the left forelimb compared with the right forelimb is called lead I. Right forelimb is connected to the negative terminal of the instrument, and the left forelimb is connected to the positive terminal.
2. **Limb lead II**: The voltage measured in the left hind limb is compared with the right forelimb. Right forelimb is connected to the negative terminal of the instrument, and the left hind limb is connected to the positive terminal.
3. **Limb lead III**: Lead III is obtained by connecting left hind limb and left forelimb. Left hind limb is connected to the negative terminal of the instrument, and left forelimb is connected to the positive terminal.

6.3.3.2.2 Unipolar Leads

In unipolar lead system, wires from two limb electrodes of the bipolar lead system are connected together and the mean electrical potential of these two leads is measured by the electrode called indifferent electrode, which is attached to the negative terminal of the ECG. Recording is taken, wherein one electrode is an active electrode and the other one is an indifferent electrode. Active electrode is positive or exploring and is placed on the third limb, and indifferent electrode serves as a negative electrode. In unipolar leads, mean electrical potential difference sensed by the indifferent and reference electrodes is measured. This is able to detect greater potential difference.

Unipolar limb leads (augmented voltage leads) are of three types:

Unipolar limb leads are also called augmented limb leads (as compared to the regular leads, the augmentation of deflections occurs). Active electrode is connected to one of the limbs, and indifferent electrode is obtained by connecting the other two limbs through a resistance.

1. **aVR lead**: Lead aVR measures the voltage from the right forelimb electrode compared with the average voltage from the other two limb electrodes. Active electrode is from right forelimb, whereas indifferent electrode is obtained by connecting left forelimb and left hind limb.
2. **aVL lead**: Active electrode is from left forelimb, and indifferent electrode is obtained by connecting right forelimb and left hindlimb.
3. **aVF lead**: Here, the voltage measured from the left forelimb (active) electrode is compared with the average voltage from the other two limb (indifferent) electrodes.

Leads I, II, and III are used routinely in veterinary electrocardiography. Recordings from the augmented unipolar limb leads (aVL, aVR, and aVF) are also often included.

6.3.3.2.3 Chest Leads

The precordial (chest) leads are used more often in human medicine than in veterinary medicine. These special additional leads are sometimes recorded by placing ECG electrodes at standardized sites on the thorax. They are helpful in the evaluation of very specific cardiac electrical dysfunctions.

6.3.3.3 ECG Waves

The ECG recording consists of waves, complexes, intervals, and segments. Deflections of normal ECG are known as waves, namely P, Q, R, S, and T.

P wave indicates the sum of all the electrical potentials produced during the depolarization of both atria and the spreading of the electrical activity from SA node throughout the atrial musculature. It starts the atrial contraction and somewhat comes before atrial systole. Due to the temporal lag between electrical and mechanical events, there is only a very brief delay between the P wave and atrial systole.

The QRS complex indicates ventricular depolarization wave initiating ventricular systole. QRS wave precedes isovolumetric contraction. It consists of three components.

Q wave is the first negative (downward) deflection of ECG; it indicates the spreading of electrical impulses from the left septal surface toward the right and from there toward the left.

R wave, a positive (upward) deflection, represents the spreading of impulses from the subendocardial termination of Purkinje system toward the epicardial surface of both ventricles via muscle fiber to muscle fiber conduction.

S wave, a negative deflection, indicates the conduction of impulses on the muscle fibers at the base of the heart and their activation.

T wave, the positive (upward) deflection, indicates the beginning of ventricular repolarization (relaxation).

ECG has no separate deflection or wave for atrial relaxation, which is due to the fusion of this deflection with QRS complex.

6.3.3.4 ECG Intervals

P-Q interval: It is the time duration between the beginning of the P wave and the beginning of the QRS wave. It indicates the time the excitation wave travels from SA node to Purkinje system and includes short AV nodal delay after the atrial contraction to permit complete ventricular filling. P-R interval indicates atrioventricular conduction time. Normally, P-R interval is 0.1 s.

Q-T interval: It is the duration from the beginning of Q wave to the end of T wave, which indicates the duration of time from initiation of ventricular depolarization to completion of repolarization or ventricular contraction.

S-T interval: Duration from the beginning of S wave to the beginning of T wave during which the ventricles remain depolarized.

From the end of P wave to the beginning of Q wave and from the end of S wave to the beginning of T wave, the ECG does not exhibit any waves; that is, the electrical potential of the cardiac musculature does not undergo any alteration (remains at isoelectric potential).

The time between successive P waves is P-P interval, and it corresponds to the time between atrial contractions.

P-R interval can be used to calculate the number of atrial contractions per minute. Similarly, R-R interval can be used to calculate the ventricular rate.

6.3.3.5 Vector Analysis and Mean Electrical Axis

6.3.3.5.1 Cardiac Vector Analysis

The standard limb leads record the potential difference between two points on the body surface, and the deflection produced in each lead at any given instant indicates the magnitude and direction of passage of cardiac potential.

Cardiac vector (cardiac axis) is the direction at which electrical potential generated in the heart travels at an instant. It indicates the magnitude and direction of the cardiac potential at any given instant. The vector for a given moment can be calculated by using any two standard limb leads. If the net potential of the QRS complex in two leads is measured, the mean vector for the ventricular depolarization can be calculated. The direction of this vector is called electrical axis of the heart.

6.3.3.5.2 Calculation of Mean Electrical Axis

An equilateral triangle is drawn, and the positions of electrodes with their polarity of each point of the standard leads are marked. In each lead, distances equal to the height of R wave minus the height of the largest negative wave in the QRS complex are measured and marked in the positive direction from the midpoint, on the side of the triangle representing that lead. All the three perpendiculars are extended, which intersect at a point within the triangle. An arrow called mean electrical axis of the heart or mean QRS vector is drawn from the center of electrical activity to the point of intersection of perpendiculars. The length of the arrow indicates the magnitude, and the arrowhead indicates the direction of the electrical activity. The normal direction of electrical axis is $-30°$ to $+110°$ (on an average of 70°). If the calculated axis falls to left of $-30°$ or right of $+110°$, left- or right-axis deviation is said to be present. Right-axis deviation indicates right ventricular hypertrophy, and left-axis deviation is due to left ventricular hypertrophy. Mean electrical axis of the dog's heart lies between $+40°$ and $+100°$ and in cats it is 0° to $+160°$.

6.3.3.6 Significance of ECG

It is a noninvasive method that aids to evaluate cardiac function; diagnose ventricular hypertrophy; evaluate conduction system blocks, myocardial infarction, drug effects, etc.; assess heart rate and rhythm; evaluate myocardial electrical conduction; and assess any myocardial abnormality.

Ventricular hypertrophy increases R amplitude (high voltages) of the QRS complex in leads II and III:

Right ventricular hypertrophy—the polarity of the R wave of QRS complex is negative instead of the normal positive in lead I.

Right or left ventricular hypertrophy—slightly prolonged QRS with high voltage.

Hypertrophy of left ventricle—abnormally high amplitude of the R wave in lead I.

Enlargement of the atria—wider P wave.

Bundle branch block—axis deviation along with prolonged QRS duration.

Cardiac tamponade, accumulation of fluid in the pericardium—low voltage of ECG waves.

Sinus tachycardia (tachycardia is increase in heart rate; sinus indicates SA node activity)—the ECG recording shows normal waves, i.e., P wave followed by QRS complex and T wave but with increased frequency of waves per minute.

Sinus bradycardia (heart rate is reduced below the normal)—the ECG recording shows normal waves, i.e., P wave followed by QRS complex and T wave but with decreased frequency of waves per minute.

First-degree AV block—the PR interval is abnormally long suggestive of long delay in the propagation of action potentials through AV node and AV bundle.

Second-degree AV block—some P waves are followed by QRS and T waves but some P waves without QRS and T waves, i.e., successive P waves occur with missed ventricular beats. Some atrial impulses fail to be conducted to ventricles.

Third-degree AV block—if QRS wave is not preceded by P wave and the frequency of P and QRS waves is different, i.e., asynchrony of P and QRS waves is observed. It occurs when conduction from atria to ventricles is completely interrupted.

6.4 Mechanical Activity of Heart

6.4.1 Cardiac Cycle and Wiggers Diagram

Cardiac cycle refers to the cardiac events that occur from the beginning of one heartbeat to the beginning of the next. Each cycle is initiated by spontaneous generated action potential in the *sinus node*. The cardiac cycle involves repeated relaxation (diastole) and contraction (systole) of the various chambers of the heart during which the heart fills with blood and pumps it out. The cycle repeats with every

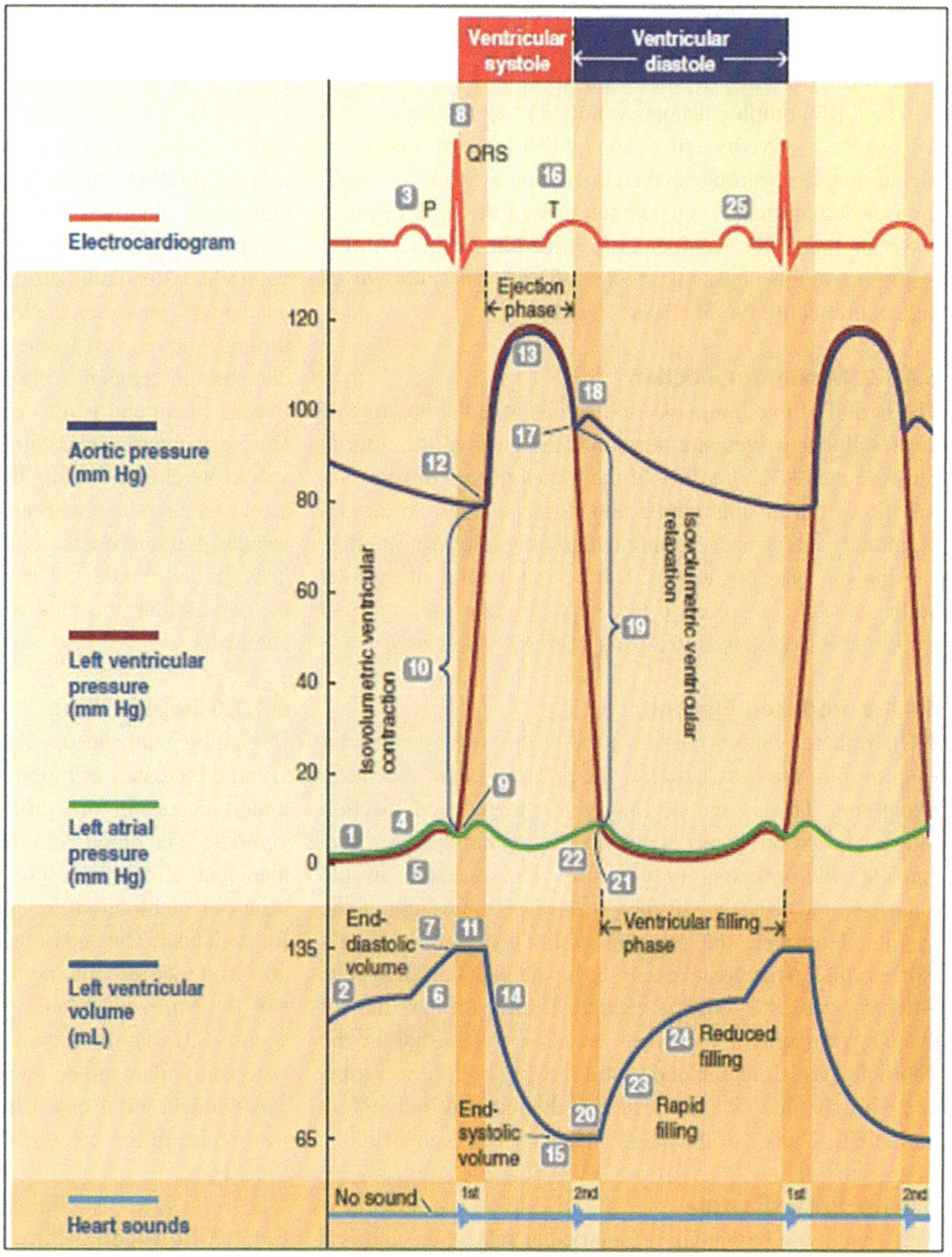

Fig. 6.3 Wiggers diagram showing the eight phases in a cardiac cycle

heartbeat and includes systole (isovolumetric contraction, ejection), diastole (isovolumetric relaxation and filling), and then back to systole. The cardiac cycle is divided into the following various stages and events (Fig. 6.3).

6.4.1.1 Isovolumetric Contraction

It is the first phase of ventricular systole in the cardiac cycle lasting for 0.05 s. Isometric contraction is characterized by increase in tension, without any change in the length of muscle fibers. Immediately after atrial systole, at the onset of ventricular contraction, the pressure in the lumen is essentially equal to that in the atria and the AV valves have floated almost into opposition. As soon as ventricular pressure exceeds atrial pressure, the AV valves close. Semilunar valves are already closed, and the ventricles contract as closed chambers, in such a way that there is no change in the volume of ventricular chambers or in the length of muscle fibers. The ventricles contract around the contained blood, which is incompressible. During isometric contraction, only the tension in the ventricular muscle increases, and as a result of this increased tension in the ventricular musculature, the pressure inside the ventricles increases significantly. This marks the end of diastole and the beginning of systole. During this phase of the cardiac cycle, the volume of the ventricles does not change and pressure increases rapidly. This phase terminates and the next begins at the moment the ventricular pressure exceeds aortic or pulmonic pressure. The semilunar valves then open, and blood is accelerated into the great arteries (aorta and pulmonary).

The QRS complex denotes ventricular depolarization and follows the P wave by a time interval (PR interval) necessary for the impulse to traverse the conduction system and reach the ventricular muscle cells. At the time of the peak of the R wave in the ECG, ventricular contraction begins. At the beginning of this phase, the closure of atrioventricular valves occurs producing the first heart sound.

6.4.1.2 Maximum Ejection

The period of maximum ejection begins with the opening of the semilunar valves and lasts until the peak of the arterial pressure curve. About 75% of the blood ejected during systole flows during this period and flows into the aorta, and pulmonary artery exceeds runoff into the peripheral arteries, causing the pressure to rise. During this period of systole, aortic pressure is exceeded by left ventricular pressure and the blood is accelerated to a peak velocity of 1–2 m/s.

6.4.1.3 Reduced Ejection

As peripheral runoff reaches equilibrium with ventricular ejection into the great arteries, the pressure curve reaches a maximum. This is the beginning of the reduced ejection phase, and blood runoff begins to exceed the ejection rate, causing the pressures to decrease. The pressure in the ventricles exceeds that in the great vessels throughout the systole. However, the pressure within the ventricle only exceeds that in the great vessels during the first half of systole when most of the blood is ejected. During the last half of systole, pressure in the great vessels exceeds that in the ventricle even though blood is still flowing out of the ventricle. This paradox occurs because of the reduced momentum and kinetic energy of the blood as it leaves the ventricle.

6.4.1.4 Protodiastole

This marks the beginning of ventricular relaxation and is a point on the ventricular pressure curve that is often difficult to identify. Due to the ejection of blood, the pressure in the ventricle continues to fall below that in the aorta and pulmonary artery. A brief retrograde flow occurs, closing the semilunar valves. This marks the end of protodiastole and the beginning of the next phase. Protodiastole is the first stage of ventricular diastole, and the duration of this period is 0.04 s. Thus, protodiastole denotes the end of systole and beginning of diastole. The semilunar valves close during this phase producing the second heart sound.

6.4.1.5 Isovolumetric Relaxation

This marks the end of ventricular systole and the beginning of the diastole. The short period of reversal of blood flow in the great vessels as the ventricles relax closes the semilunar valves and produces the incisura or dicrotic notch on the pressure wave in the great vessels. The semilunar valves keep blood from leaking back into the ventricles as the ventricular pressure drops to very low values. Since the ventricles are closed chambers, myocardial relaxation results in a steep fall in intraventricular pressure but no alteration in ventricular volume. This phase, with a rapid decrease in ventricular pressure and no blood flowing into or out of the ventricles, is isovolumetric relaxation. The aortic and pulmonary artery pressures decline during diastole as blood flows through the tissues. Isometric relaxation is characterized by decrease in tension without any change in the length of muscle fibers and is also called as isovolumetric relaxation. During isometric relaxation period, once again all the heart valves are closed. During this phase, both the ventricles relax as closed chambers without any change in volume. Intraventricular pressure decreases during this period, and the phase lasts for about 0.08 s. The T wave signifies repolarization in the ventricular muscles and relaxation of the ventricular chambers.

6.4.1.6 Rapid Filling

Beginning with the opening of the AV valves, ventricular volume increases as blood that has accumulated in the atria under increasing pressure flows quickly into the relaxed ventricle. The blood volume in each atrium is slightly greater than that of the corresponding ventricle, thus providing a reservoir of blood sufficient to fill the ventricle completely for each beat. The end of this phase is not clearly distinguishable as it merges with the next. At the transition between this and the following phase, the usually inaudible third heart sound (S3) may be recorded on a phonocardiogram. When atrioventricular valves open, there is a sudden rush of the accumulated atrial blood into the ventricles. About 70% of ventricular filling takes place during this phase, which lasts for 0.11 s.

6.4.1.7 Reduced Filling (Diastasis)

This is a period of slower filling of the cardiac chambers during which blood continues to flow into both atria and ventricles as into a common chamber. It is terminated by the onset of atrial systole. Following the sudden rush of blood, the ventricular filling becomes slow. About 20% of filling occurs in this phase, and this phase lasts for 0.19 s.

6.4.1.8 Atrial Systole

After slow filling phase, the atria contract and pump a small quantity of blood into ventricles and about 10% of ventricular filling takes place. Flow of additional amount of blood into ventricle due to atrial systole is referred to as atrial kick. In the normal heart, the initial impulse for a heartbeat arises within the SA node and quickly spreads to the two atria. The atrial wall muscles are basically arranged in a circular fashion such that the volume of blood within the atria decreases with each contraction. The ventricles are relaxed when the atria contract and blood enters the ventricle due to the pressure gradient. Atrial contraction produces only a small increase in ventricular volume and pressure. This phase ends at the onset of ventricular isovolumetric contraction, completing the cardiac cycle. In animals with atrial fibrillation, wherein functional coordinated contraction of the atria is not exhibited, a reasonable cardiac output is maintained unless the ventricular contraction rate per minute exceeds normal values. In dogs, with rapid heart rates, the atria may attribute 20–30% of ventricular filling to atrial contraction. Also, atrial contraction and relaxation are instrumental in bringing about normal closure of the AV valves. The atria depolarize, producing the P wave on the ECG, and begin to contract shortly after depolarization.

The duration of the cardiac cycle is the reciprocal of the heart rate:

$$\text{Duration (s/beat)} = \frac{60\ (\text{s/min})}{\text{Heart rate (beats/min)}}$$

For example, for a heart rate of 75 beats/min, the cardiac cycle lasts for 0.8 s.

6.4.1.8.1 Pressure Changes

Atrial pressure rises during atrial systole and continues to rise during isovolumetric ventricular contraction as the AV valves bulge into the atria (Fig. 6.3). The pressure drops quickly as the AV valves are forced down by the contracting ventricular muscle, then rises as blood flows into the atria, and finally rises as the AV valves open early in diastole. By lowering atrial capacity, the AV valves' return to their relaxed configuration also contributes to this pressure increase. Three distinct waves (a, c, and v) are produced in the record of jugular pressure as a result of the transmission of atrial pressure changes to the great veins. The "a" wave is due to atrial systole. Some blood regurgitates into the great veins when the atria contract. In addition, venous inflow stops, and the resultant rise in venous pressure contributes to this wave. During atrial contraction, the right atrial pressure normally rises by 4–6 mmHg, and the left atrial pressure increases about 7–8 mmHg; it is caused partly by slight backflow of blood into the atria at the onset of ventricular contraction but mainly by the *c wave* that occurs when the ventricles begin to contract bulging of the AV valves backward toward the atria because of increasing pressure in the ventricles.

The "c" wave is the transmitted manifestation of the rise in atrial pressure produced by the bulging of the tricuspid valve into the atria during isovolumetric ventricular contraction. The slow flow of blood from the veins into the atria during ventricular contraction causes the "v" wave, which appears at the end of ventricular contraction. Then, as ventricular contraction is over, the AV valves open, allowing this stored atrial blood to flow rapidly into the ventricles and causing the "*v*" wave to disappear.

6.4.2 Heart Sounds

Cardiac sounds are the sounds produced by mechanical activities of heart during each cardiac cycle. They are classified as either transients or murmurs. The transients are of brief duration and are named as the first, second, third, and fourth heart sounds. Murmurs refer to prolonged groups of vibrations that occur during normally silent intervals of the cardiac cycle.

The first heart sound (S1) is associated with the closure of the AV valves (the mitral and tricuspid valves). The actual closure of these valves does not produce this sound; the valve leaflets are so light and thin that their closing would be almost silent. However, there is a momentary backflow of blood from the ventricles to the atria at the beginning of ventricular systole. When this backflow of blood is brought to a sudden stop against the closing valves, brief vibrations are created in the blood and in the cardiac walls. These vibrations create the first heart sound. The other factors that contribute to the first heart sound are vibrations generated within the contracting ventricular myocardium, opening of the semilunar valves, and vibrations generated within the wall of the aorta and pulmonary artery as blood is ejected into the arteries at the onset of systole. The first heart sound is longer and has lower frequency than the second heart sound. The first heart sound is usually best heard over the cardiac apex.

In the dogs, S1 is usually more intense than S2, whereas the opposite is true for horses under basal conditions. Furthermore, a marked beat-to-beat variation in the intensity of S1 is usual in horses. Numerous extracardiac and cardiac factors influence the intensity of S1 and other sounds. The most common reasons for decreased intensity of S1 are obesity, pericardial or pleural effusions, hypervolemia, pronounced first-degree AV block, diaphragmatic hernia, peritoneal pericardial diaphragmatic hernia, and barrel conformation of the thorax. Intensity of S1 is increased in animals during periods of excitement and immediately after exercise. More vigorous closing and tensing of the AV valves related to increased sympathetic activity probably account for increased intensity of S1 under these circumstances. Other common reasons for increased intensity of S1 are deep thorax, anemia, fever, hypertension, and chronic mitral valve disease.

6.4.2.1 Splitting of First Heart Sound

Split of heart sound is a rare event in all species. Split of S1 is determined by a delayed closure of one of the AV valves. It can be occasionally heard in animals with severe right or left bundle branch block.

6.4.2.2 Second Heart Sound

The second heart sound (S2) is associated with closure of the aortic valve on the left side of the heart and the pulmonic valve on the right side of the heart. The pulmonic component of S2 normally follows the aortic component. However, they are usually heard as a single sound. S2 is a shorter higher pitched sound than S1. It is usually briefer, sharper, and higher pitched than the first heart sound. The second heart sound is caused by the echo that occurs as the valves suddenly close, stopping the brief backflow of blood into the ventricles. The aortic and pulmonic valve closure is normally simultaneous. Under certain conditions, however, the two valves close at slightly different times, and the second heart sound is heard as two distinct sounds in quick succession; this condition is called a split-second heart sound.

Decreased intensity of S2 is observed in pericardial and pleural effusion, diaphragmatic and peritoneal pericardial diaphragmatic hernia, thoracic masses, myocardial failure, and severe chronic mitral valve degeneration. The most common causes of increased intensity of S2 are fever, anemia, pulmonary hypertension, and hyperthyroidism.

6.4.2.2.1 Splitting of Second Sound

Splitting of S2 occurs when the semilunar valves close out of phase. The split of S2 is difficult to auscultate in dogs and cats due to the short interval between A2 and P2, while it is a common auscultatory finding in horses. The most common reason for splitting of S2 is delayed closure of the pulmonic valve, and for this reason, it is usually best heard over the pulmonic area of auscultation. Splitting of S2 can be physiologic or pathologic. Physiologic splitting of S2 is a respiratory-related phenomenon appearing during inspiration and disappearing during expiration. During inspiration, increased venous return to the right side of the heart occurs as a consequence of decreased intrathoracic pressure. The resultant lengthening of right ventricular ejection time causes the pulmonic valve to close later. Simultaneously, the act of inspiration hinders venous return to the left side of the heart, and this in turn abbreviates left ventricular ejection time and causes the aortic valve to close earlier. During expiration, right ventricular ejection time returns to normal and left ventricular ejection time lengthens to handle the increased amount of blood delivered to the lungs during the preceding inspiration. Therefore, S2 becomes a single sound. Respiratory-related splitting of S2 is detectable in some normal dogs, particularly if the heart rate is slow and pronounced sinus arrhythmia is present. Splitting of S2 is detectable by auscultation in most normal horses, and it tends to be fixed rather than vary with respiration. Pathologic splitting of S2 is a characteristic of certain cardiac abnormalities such as pulmonary hypertension, pulmonic stenosis, right bundle branch block, and interatrial septal defect in dogs. Pathologic split tends to be fixed differently from physiologic one.

6.4.2.3 Third Heart Sound

Third heart sound (S3) occurs early in diastole near the end of rapid ventricular filling. It is associated with sudden tensing of the chordae tendineae, deceleration of the filling wave of blood, and vibrations arising in the walls of the ventricles. Although S3 is detectable in PCGs of apparently normal dogs, it is rarely audible due to low-frequency sounds. However, the third heart sound can be easily recognized in dogs afflicted with myocardial disease such as dilated cardiomyopathy. In these patients, the presence of S3 may be the only auscultatory abnormality and is generated by increased myocardial stiffness and elevated filling pressure. An audible S3 may also be present in cats with end-stage cardiomyopathy. Occasionally, S3 is more intense than either of the two major heart sounds, S1 and S2. The third sound is readily audible in many apparently normal horses. The sound is quite intense and clicking in some horses; in others, it is soft and dull. As in dogs, S3 in horses with congestive heart failure is frequently very intense.

6.4.2.4 Fourth Heart Sound

The fourth or atrial sound is generated by vibration of cardiac structures associated with atrial contraction. Although the fourth heart sound is seldom heard in dogs, it can be recognized in normal giant breeds. Fourth heart sound is common in apparently normal horses. In many animals, S4 indicates pronounced first degree AV block. A longer AV conduction time allows for completion of the sequence of events leading to its generation. S4 is generated by transient closing and tensing of the AV valves. Pathologic S4 is present in animals with myocardial disease characterized by diastolic dysfunction such as hypertrophic cardiomyopathy, wherein there is an increased volume of blood in the atria at the onset of their contraction which induces a more energetic atrial contraction, and S4 might become audible. Other conditions associated with the presence of a detectable S4 are advanced second-degree and third-degree AV block.

6.4.2.5 Gallop Rhythm

Gallop rhythm occurs during tachycardia when S3 and S4 merge into a single heart sound. This rhythm is so called because the sequence of S1 and S2 and fusion of S3 and S4 resemble the sound of a galloping horse. As the period of diastole shortens with an increase in heart rate, atrial systole becomes superimposed upon the rapid filling phase of the ventricles. In small animals, the presence of a gallop rhythm is associated with significant myocardial failure. In cats, gallop rhythms are present in animals with hypertrophic cardiomyopathy or hyperthyroidism.

6.4.2.6 Other Systolic Sounds

Two "extra" sounds sometimes occur between S1 and S2. One of these, the ejection sound or ejection click, is an accentuation of the terminal component of S1. It often coexists with abnormalities that cause dilation of either the aorta or the pulmonary artery of the dog but is common in normal horses and in young small animals. The other extra systolic sound is the systolic click, which is often intermittent. Systolic click is a common finding in the early stages of chronic mitral valve disease in dogs. The origin of the systolic click is caused by the tensing of redundant chordate tendineae and rapid deceleration of blood against the mitral valve leaflets at maximum prolapse into the left atrium. The sound is usually midsystolic, but the timing can vary and can be closer to S1 or S2.

6.4.3 Cardiac Murmurs

A murmur is a prolonged series of auditory vibrations emanating from the heart or blood vessels that may occur at different times during the cardiac cycle. Turbulence in flowing blood is the major source of murmurs, and the turbulent blood flow may be caused due to (1) morphology alteration in heart valves (insufficiency or stenosis), (2) abnormal communication between the two sides of the heart and/or great vessels (interatrial septal defect, interventricular septal defect, or patent ductus arteriosus), (3) increased blood flow velocity through a normal valve orifice or vessel, and (4) changes to the blood viscosity, such as in severe anemia.

Murmurs are generally classified based on the following criteria: location, quality, timing, radiation, and intensity. The point of maximum intensity (PMI) of a cardiac murmur is usually located over the turbulence site and gives an indication of the origin of the murmur. The cause of murmurs can be further classified from their timing within the cardiac cycle. Murmurs occur during systole, during diastole, or during both systole and diastole. The frequency or pitch of a murmur may also aid diagnosis of the underlying cause. The direction in which a murmur radiates over the body surface also helps in localizing the site of origin. The intensity of murmurs is most commonly graded on a 1–6 scale, with a grade 1 murmur being the softest and a grade 6 the loudest. The heart murmur may be palpated as a precordial thrill in animals with a grade 5 or grade 6.

Functional (Physiological) Systolic Murmurs

Soft systolic murmurs are common in puppies, kittens, and horses having apparently normal valve orifices and great arteries. Turbulent blood flow stemming from increased blood velocity is of prime importance in the generation of these innocent murmurs. Severe anemia can also cause a functional systolic murmur by changed blood viscosity. A functional murmur is usually of low intensity (grade 1–3) and is composed of mid- or high-frequency sounds. The murmur is usually timed to early systole and ends early or in the middle of systole. The intensity of the murmur may vary with the heart rate and respiration.

Innocent Diastolic Murmurs

It is usual to find soft diastolic murmurs in apparently normal horses less than 5 years of age. These are high pitched and very brief in duration. They occur immediately after S2, and their cause is unknown.

6.4.3.1 Clinical Correlations

6.4.3.1.1 Systolic Murmurs

Systolic murmurs (between S1 and S2) occur as blood regurgitates through incompetent AV valves, or as blood is pumped through the semilunar valves or through a ventricular septal defect.

Aortic stenosis: A stenotic valve refers to a narrowed, stiff, valve that does not open completely, and hence the blood must be forced through the constricted opening at tremendous velocity, resulting in turbulence that produces an abnormal whistling sound. Subaortic stenosis is the most common form of congenital heart disease in dogs. However, aortic or subaortic stenosis is unusual in horses, and in cats the aortic stenosis is usually a consequence of hypertrophic cardiomyopathy. The principal hemodynamic consequence of aortic stenosis is an increased resistance to left ventricular outflow, with a proportional elevation of left ventricular systolic pressure if flow remains constant. Aortic stenosis usually causes a systolic murmur with the intensity of the murmur increasing until mid-ventricular systole and then decreasing during the remainder of ventricular systole. Its intensity and quality are dependent on the severity of stenosis, ranging from a soft low-intensity murmur to very loud murmurs of harsh quality.

Pulmonic stenosis: The characteristics of pulmonic stenosis are very similar to aortic stenosis, and it may be very difficult to distinguish murmurs caused by pulmonic stenosis from those caused by aortic stenosis in dogs and cats using auscultation. The murmur is systolic, and the intensity depends on the severity of stenosis, ranging from a soft low-intensity murmur to very loud murmurs of harsh quality. Right ventricular ejection time may be prolonged with more severe forms of pulmonic stenosis, and splitting of S2 may be present. Pulmonic stenosis is a common congenital lesion in dogs but is less common in cats and is extremely rare in horses.

Mitral insufficiency: A valve having edges that do not fit together properly and hence cannot close completely is referred to as an insufficient or incompetent valve. Turbulence is produced when blood flows backward through the insufficient valve and collides with blood moving in the opposite direction, creating a swishing or gurgling murmur.

Mitral insufficiency or regurgitation may be primary (i.e., caused by an abnormal mitral valve) or secondary (i.e., caused by left ventricular dilatation that leads to separation of the mitral valve leaflets). Primary mitral insufficiency is the cause of most systolic murmurs in middle-aged to old dogs and horses, and it occurs as a consequence of chronic progressive lesions of primarily the AV valves, referred to as myxomatous valve disease. It is also the most common cause of congestive heart failure in dogs and horses. Primary mitral insufficiency is uncommon in cats, but it may, like aortic stenosis, develop as a consequence of hypertrophic cardiomyopathy. The sound begins as a soft apical systolic murmur on the left side of the thorax and may be intermittent and sometimes audible only during inspiration. With further progression, the sound becomes more intense, and harsher in quality, and may radiate over to the right side of the thorax.

Tricuspid insufficiency: The characteristics of the systolic murmur of tricuspid insufficiency are very similar to mitral insufficiency. The intensity of the murmur may increase during inspiration and decrease during expiration. Generally, the intensity of tricuspid regurgitation murmurs is lower than that of mitral valve insufficiency murmurs. Conditions associated with increased right ventricular pressure, such as pulmonary arterial hypertension, can cause louder tricuspid insufficiency murmurs.

Interventricular septal defect: The systolic murmur associated with interventricular septal defect (VSD), which is the most common congenital heart defect in cats and horses wherein the murmur is generated as blood flows from the left ventricle into the right ventricle. This murmur tends to be of uniform intensity throughout its course and is usually high pitched and blowing.

Interatrial septal defect: Interatrial defect occurs as a congenital anomaly, wherein the blood flows from the left atrium to the right atrium during atrial systole (i.e., during late ventricular diastole), thereby increasing the stroke volume of the right ventricle and causing relative pulmonic stenosis. Therefore, this systolic murmur is caused by relative pulmonic stenosis rather than by blood flow through the defect itself.

Tetralogy of Fallot: Tetralogy of Fallot is a rare congenital disease in dogs, but comparatively more common in cats and horses. The murmur associated with tetralogy of Fallot is primarily caused by the pulmonic stenosis. It is a shunting defect, wherein there is usually no resistance to flow between the left and right ventricles. Hence, the blood flows to the right and left circulations proportional to systemic and pulmonary resistances. The pulmonic stenosis in tetralogy may be so severe that resistance to flow through the pulmonic valve is greater than systemic vascular resistance. Consequently, a significant amount of blood flows from the right ventricle, through the ventricular septal defect, and out the aorta. The intensity and character of the murmur depend on pulmonic stenosis characteristics.

6.4.3.1.2 Diastolic Murmurs

Murmurs heard after S2 are designated as diastolic murmurs, are extremely rare in dogs and cats, and are not uncommon in horses.

Mitral or tricuspid stenosis: Although pathology of the AV valves is virtually nonexistent in domestic animals, lesions such as interatrial septal defect, mitral insufficiency, or tricuspid insufficiency can result in an increased rate of blood flow through the appropriate AV valve during early diastole in domestic animals leading to diastolic rumble or early diastolic murmur.

Pulmonic insufficiency: Diastolic murmurs due to pulmonic insufficiency are rare in animals in spite of the fact that mild-to-moderate pulmonic insufficiency is very common in all species. The murmur of pulmonic insufficiency tends to be soft and blowing. Occasionally, dilatation of the pulmonary artery with attendant incompetence of the pulmonic valve resulting from pulmonary hypertension is the cause of a diastolic murmur in the dog. Likewise, a diastolic murmur of pulmonic insufficiency sometimes appears after surgical correction of pulmonic stenosis.

Aortic insufficiency: Like pulmonic insufficiency, mild aortic insufficiency is common in dogs, while moderate-to-severe insufficiency is comparably uncommon in dogs and cats. They are common in old horses and mostly occur as "noisy," high pitched with a blend of a wide range of vibration frequencies. It occurs either as a consequence of myxomatous lesions or, less commonly, due to bacterial endocarditis, or as a consequence of congenital aortic stenosis.

Continuous murmurs: Patent ductus arteriosus (PDA) is one of the most common congenital conditions in dogs; it occurs in cats and horses, but less commonly than in dogs. In PDA, the fundamental pathophysiologic event is shunting of blood through the patent duct. The flow direction is usually from the left (aorta) to the right (pulmonic artery) side because of the pressure gradient, but with larger ducts, which are uncommon, the flow ceases or can even be from the right to the left side (reversed PDA).

Turbulence in blood flowing through the PDA is responsible for generating the murmur. Since the abnormal blood flow is continuous during systole and diastole, the murmur is likewise continuous during systole and diastole. The intensity of the murmur increases during systole, attains a peak at the time of S2, and decreases during diastole. The murmur of PDA is frequently referred to as a "machinery" murmur. The murmur is audible over the left hemithorax region at the aortic and pulmonary valve areas.

6.4.4 Cardiac Output

Cardiac output refers to the amount of blood pumped from each ventricle. Usually, it refers to left ventricular output through aorta.

Normal cardiac output for dogs and cats is 100–200 mL/kg/min and 120 mL/kg/min, respectively. Cardiac output can be expressed in three ways:

- Stroke volume
- Minute volume
- Cardiac index

Stroke volume is the amount of blood pumped out by each ventricle during each beat. The output of the right and left ventricle per beat is approximately equal. Normal value of stroke volume is 70 mL (60 to 80 mL) when the heart rate is normal (72/minute).

Minute volume refers to the amount of blood pumped out by each ventricle in 1 min. It is the product of stroke volume and heart rate. It can be expressed as stroke volume × heart rate. The normal value of minute volume is 5 L/ventricle/min.

Cardiac index is defined as the amount of blood pumped out per ventricle/minute/square meter of the body surface area. In an adult with an average body surface area of 1.734 square meter and a normal minute volume of 5 L/min, the normal value of the cardiac index is 2.8 ± 0.3 L/square meter of body surface area/min.

Ejection fraction is the fraction of end diastolic volume that is ejected out by each ventricle. Normal ejection fraction is 60–65%.

Cardiac reserve is defined as the maximum amount of blood that can be pumped out by heart above the normal value. Cardiac reserve plays an important role in increasing the cardiac output during the conditions like exercise. Cardiac reserve is usually expressed in percentage. In a normal young healthy adult, the cardiac reserve is 300–400%. In old age, it is about 200–250%. It increases to 500–600% in athletes. In cardiac diseases, the cardiac reserve is minimum or nil.

The functionalities of the heart related to rate, contractile strength, conduction velocity, excitability, and relaxation rate can be described by the following terms:

Chronotropic effect—influence on heart rate
Inotropic effect—influence on contractile strength
Dromotropic effect—influence on conduction velocity
Bathmotropic effect—influence on excitability of heart
Lusitropy effect—rate of myocardial relaxation

6.4.4.1 Physiological Variations in Cardiac Output

Age: Cardiac output is lesser in young animals as compared to adults because of reduced blood volume. Cardiac index is more in young animals because of reduced body surface area compared to adults.

Sex: Cardiac output is less in female animals than in males because of decreased blood volume. Cardiac index is more in females than males due to reduced body surface area.

Diurnal variation: Cardiac output is low in early morning and increases in daytime. Increase in temperature above 30 °C raises cardiac output.

Stress: Stress increases cardiac output about 50–100% through the release of catecholamines, which increase the heart rate and force of contraction.

Feeding: During the first 1 h after taking feed, cardiac output increases.

Exercise: Cardiac output increases during exercise as a result of increase in heart rate and force of contraction.

High altitude: In high altitude, the cardiac output increases due to the increased secretion of adrenaline, which is stimulated by hypoxia.

Pregnancy: Cardiac output increases by 40% during later stages of pregnancy.

6.4.4.2 Methods of Cardiac Output Measurement

The following four methods are currently in use for measuring an animal's cardiac output:

1. **Transthoracic or esophageal echocardiography** is an invasive method which utilizes high-frequency sound waves (ultrasound) aimed at the heart or aorta and records the echoes reflected from the various structures. A shift in the sound-wave frequency occurs when the waves are reflected from a moving object and the magnitude of shift can be used to calculate blood flow velocity. This technique involves insertion of a flexible probe into midthoracic part of esophagus. A pulse-wave ultrasonic Doppler transducer is fixed at the tip of the probe. This transducer calculates the velocity of blood flow in descending aorta. The diameter of aorta is determined by

echocardiography. Cardiac output is calculated by using the values of velocity of blood flow and diameter of aorta. Multiplying the integration of a velocity–time curve from the ascending aorta by the cross-sectional area of the aorta and the heart rate gives an estimate of cardiac output. The advantage of this technique is that the cardiac output can be measured continuously and it can be used during cardiac surgery.

2. **Indicator dilution technique**: It is used frequently in animal research and in a few clinical situations. A known quantity of indicator is injected into a large systemic vein or, preferably, into the right atrium. This passes rapidly through the right side of the heart, then through the blood vessels of the lungs, through the left side of the heart, and, finally, into the systemic arterial system. The indicator may be a dye (e.g., indocyanine green), radioisotope, ion (e.g., lithium), or a thermal mass (e.g., cold saline). When a known quantity of the indicator is injected into an unknown volume and the diluted indicator's concentration is measured by a detector situated in the flow of blood, the cardiac output can be calculated. The concentration of the dye is recorded as the dye passes through one of the peripheral arteries, giving a curve. The cardiac output can be determined using the following formula:

$$\text{Cardiac output}\,(\text{mL}/\text{min}) = \frac{\text{Milligrams of dye injected} \times 60}{\begin{array}{l}\text{Average concentration of dye} \times \text{Duration of the curve in} \\ \text{each millilitre of blood for the seconds} \\ \text{duration of the curve}\end{array}}$$

The results obtained through this technique are accurate. The major disadvantage of this technique is that it is an invasive method involving injection of marker substance. The ultrasound velocity dilution method is an adaptation of the indicator dilution technique but requires an arteriovenous loop, which is only practical in anesthetized animals. However, it is a relatively noninvasive technique adaptable to mammals under 250 kg.

3. **Fick method**: Fick method is also the oldest, being first described by Adolph Fick in 1870, and involves the application of the law of conservation of mass. Fick method of determining cardiac output involves the measurement of animal's oxygen consumption as well as the oxygen content of arterial and venous blood.

$$\text{Cardiac output} = \frac{\text{Oxygen consumption}\,(\text{mL}/\text{min})}{\text{Oxygen content of arterial blood} - \text{Oxygen content of venous blood}}$$

The Fick principle states that the amount of a substance taken up by an organ (or by the whole body) per unit of time is equal to the arterial level of the substance minus the venous level (A-V difference) times the blood flow. This principle can be used to determine cardiac output by measuring the amount of O_2 consumed by the body in a given period and dividing this value by the A-V difference across the lungs. Because systemic arterial blood has the same O_2 content in all parts of the body, the arterial O_2 content can be measured in a sample obtained from any convenient artery. A sample of venous blood in the pulmonary artery is obtained by means of a cardiac catheter. The procedure is generally benign. Catheters are inserted through the right atrium and ventricle into the small branches of the pulmonary artery.

4. **Pressure Recording Analytical Method (PRAM)** Arterial pressure-wave contour analysis by a dedicated monitor is an old method that has recently been commercialized, but its accuracy in animals has yet to be determined. It can be used for cardiac output measurement in anesthetized animals (such as dog and swine) with a clinically stable hemodynamic status. It is a pulse contour method, wherein stroke volume and other hemodynamic parameters are estimated from the analysis of the arterial pulse waveform. The technique is based on the principle that the arterial blood pressure waveform is a result of interaction between the systolic ejection volume and the physical characteristics of the systemic vascular system (aortic impedance, vascular compliance, and peripheral arterial resistance). The monitor provides a beat-by-beat assessment of cardiac output from the arterial pressure wave. In addition to the cardiac output, heart rate, systolic, diastolic, and mean arterial pressures, pulse pressure and its variation and stroke volume and its variation are also continuously provided by the monitor. It is minimally invasive as it only requires the insertion of a regular, non-dedicated arterial catheter and does not need calibration prior to clinical use. However, due to the requirement to identify the dicrotic notch at each beat, the analytical method for pressure recording has some drawbacks. The calculations are erroneous, and the cardiac output measurements could be artifactual if the monitor is unable to accurately identify the dicrotic notch at each beat. Clinically, in the awake animal, the transthoracic ultrasound method is the most feasible, while the lithium dilution and pulmonary artery thermodilution techniques are being utilized in anesthetized patients.

6.4.4.3 Factors Influencing Cardiac Output

Cardiac output is determined by four factors: venous return, force of contraction, heart rate, and peripheral resistance.

6.4.4.3.1 Venous Return

Venous return is the amount of blood which is returned to the heart from different parts of the body. When it increases, the ventricular filling and cardiac output are increased. Thus, cardiac output is directly proportional to venous return.

6.4.4.3.2 Force of Contraction

Cardiac output is directly proportional to the contraction force. Force of contraction depends upon preload and afterload. Preload refers to the force of stretching of the cardiac muscle fibers at the end of diastole, just before contraction. It is due to the increase in ventricular pressure caused by filling of blood during diastole. Stretching of muscle fibers increases their length, which increases the contraction force and cardiac output. Afterload is the force against which ventricles must contract and eject the blood, and the force is determined by the arterial pressure. At the end of isometric contraction period, semilunar valves are opened and blood is ejected into the aorta and pulmonary artery. Hence, the pressure increases in these two vessels. Now, the ventricles have to work against this pressure for further ejection. Thus, the afterload for right ventricular pressure is determined by pressure in pulmonary artery and the afterload for left ventricle is determined by aortic pressure. Force of contraction of heart and cardiac output is inversely proportional to afterload.

6.4.4.3.3 Heart Rate

Cardiac output is directly proportional to heart rate. The resting heart rate of an animal is related to its metabolic rate and body size (Table 6.1).

- The heart rates can remarkably increase from 100 in resting state to 300 beats per minute in greyhounds, and from 30 to 150 bpm in camels, and similarly in equine cardiac outputs may rise to 250–450 L/min.
- During drought hibernation, the heart rate of the Nile crocodile (*Crocodylus niloticus*) can decrease to 2 beats per minute.

Table 6.1 Heart rate in different species

Species	Heart rate at rest (beats per minute: bpm)
Dairy cow	48–84
Ox	36–60
Goat	70–80
Sheep	70–90
Horse	28–40
Dog	70–120
Cat	120–140
Human	60–80
Rhesus monkey	160–330
Rat	250–400
Mouse	450–750

6.4.4.3.4 Peripheral Resistance

Peripheral resistance refers to the resistance offered to the flow of blood at the peripheral blood vessels against which the heart has to pump the blood. Hence, the cardiac output is inversely proportional to peripheral resistance.

Birds have evolved a high-performance efficiently functioning cardiovascular system to meet the rigorous demands of running, flying, swimming, or diving in a variety of extreme environments. Sustained increased-level activities in these environments possess severe demands on the cardiovascular system to provide adequate delivery of oxygen to working vascular beds and for efficient removal of metabolic products. Furthermore, birds being endothermic, the cardiovascular system has a major role in conserving or removing body heat.

Birds have larger hearts, bigger stroke volumes, lower heart rates, and higher cardiac output than mammals of corresponding body mass, which contribute to the high aerobic energy input needed to sustain flapping flight. Similar to the vascular changes occurring during flight, swimming is associated with changes in regional vascular perfusion with myocardium and active leg musculature blood flow increasing by 30% and 300%, respectively. These perfusion changes are quite evident in the aquatic birds that engage in surface swimming and submerged or diving swimming. The phases of the cardiovascular response to a voluntary dive include an initial tachycardia during the immediate predive followed by a diving bradycardia. After surfacing, there is usually a postdive tachycardia which aids in loading and replenishing oxygen stores and eliminating carbon dioxide. For example, during an emperor penguin's 16-min dive, its heart rate may decrease to 6 beats per minute over 5 min during the dive, with a minimum of 3 bpm. Tufted ducks' surface swimming at maximal sustainable speeds is aided with a 70% increase in cardiac output from 276 to 466 mL/min.

6.5 Regulation of Heart

6.5.1 Regulation of Cardiac Output

The changes in cardiac output required as per the physiologic conditions are brought about by altering the cardiac rate or stroke volume or both. These parameters are controlled by local intrinsic mechanisms within the myocardium and by extrinsic regulation through the autonomic nervous system, hormones, and chemicals (Fig. 6.4).

6.5.1.1 Intrinsic Regulation

Intrinsic regulation refers to the mechanisms operating within the myocardium to regulate cardiac output. Heart's inherent ability to vary stroke volume depends on the direct correlation between end-diastolic volume and stroke volume.

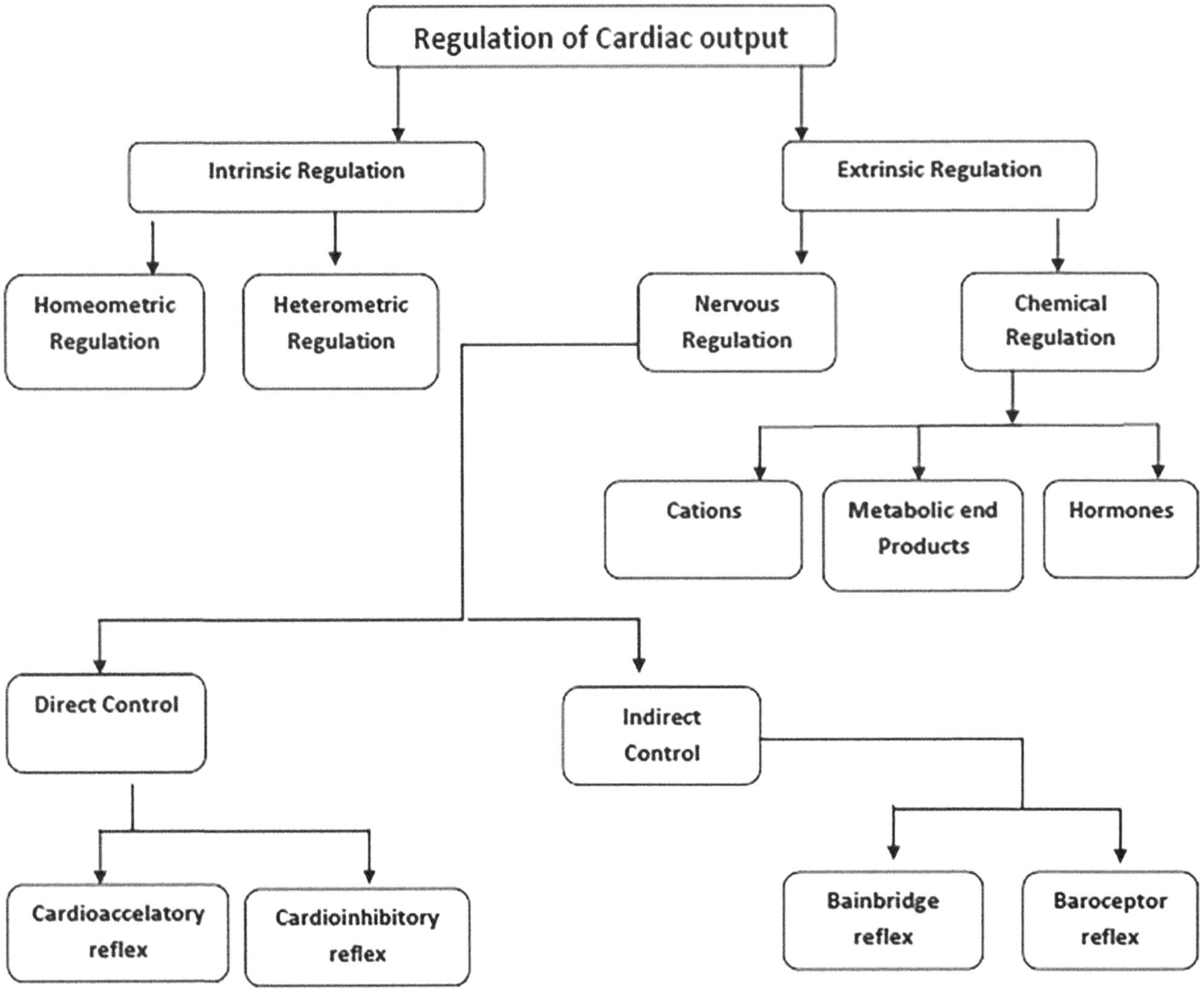

Fig. 6.4 Regulation of cardiac output. The cardiac output is controlled by intrinsic and extrinsic factors

6.5.1.1.1 Heterometric Regulation

This refers to the mechanism wherein the heart relates the stroke volume according to changes in cardiac muscle fiber length. As more blood returns to the heart, the heart pumps out more blood. The amount of blood pumped by the heart each minute is determined almost entirely by the rate of blood flow into the heart from the veins (*venous return*). This intrinsic control depends on the length–tension relationship of cardiac muscle, similar to that of skeletal muscle. This intrinsic ability of the heart to adapt to increasing volumes of inflowing blood is called the Frank-Starling mechanism of the heart, in honor of Frank and Starling, two great physiologists of a century ago. According to the Frank-Starling mechanism, the greater the heart muscle is stretched during filling, the greater is the force of contraction and the greater the quantity of blood pumped into the aorta. When an extra amount of blood flows into the ventricles, the cardiac muscle itself is stretched to a greater length causing the release of a greater amount of Ca^{2+} during systole. Further, the muscle contracts with increased force because the actin and myosin filaments are brought to a more nearly optimal degree of overlap for force generation. As a result, the ventricle automatically pumps the additional blood into the arteries as a result of its increased pumping. Additionally, stretch of the right atrial wall directly increases the heart rate by 10–20%; this, too, helps increase the amount of blood pumped each minute.

6.5.1.1.2 Homeometric Regulation

Homeometric autoregulation is referred to an intrinsic mechanism, which allows the heart muscle to adapt to changes both in heart rate (Bowditch effect) and in developed pressure (Anrep effect). The Bowditch effect is an adaptive mechanism, wherein an increase in heart rate stimulates inotropy.

Calcium is vital for cardiac muscle contraction. Calcium enters the cell during the plateau phase of the cardiac action potential occurring during each contraction. As the heart rate elevates, there is an increase in the number of action potential plateaus. In addition, elevated frequency enhances the amount of calcium entry through L-type calcium channels as well as by slowing channel inactivation. As a result of these processes, there is an increase in the amount of intracellular calcium in the myocardium. At excitation, sarcoplasmic reticulum calcium is released from the terminal cisternae as a result of the increase in intracellular calcium via the L-type membrane channels. Increased intracellular calcium from the sarcoplasmic reticulum results in augmented force of contraction. Additionally, at higher heart rates, the Na^+/K^+-ATPase pumps are unable to keep up with the sodium influx. The increased intracellular Na^+ reduces the concentration gradient of Na^+ across the sarcolemma, which reduces the inward movement of Na^+ down its concentration gradient via the sodium-calcium exchanger, which exchanges three sodium ions for each calcium ion. This, in turn, reduces the outward movement and exchange of Ca^{2+}, leading to an accumulation of intracellular calcium which causes inotropy.

The Anrep effect refers to the recovery of the ventricle from transient subendocardial ischemia induced by an abrupt increase in ventricular pressure. Following the reduction in contractility, the coronary bed's vascular autoregulation redistributes coronary flow to the ischemic regions, causing reactive hyperemia. After an abrupt increase in systolic pressure, the ventricle is in a transient state of "decompensation" owing to temporary subendocardial ischemia, which is corrected immediately by a redistribution of coronary blood to the ischemic areas. The decompensation is more severe in coronary insufficiency and left ventricular hypertension. In the healthy heart, the coronary vasodilatation is accompanied by positive inotropism.

6.5.1.2 Extrinsic Regulation

The cardiac output is regulated depending on the functional requirement of the different parts of the body by factors that are external to the heart. The extrinsic regulation of cardiac output is achieved by neural and endocrine mechanisms. Some ions also influence the cardiac output by affecting heart rate and contractility. The nerves exert their control over the cardiac output to a larger extent on the rate rather than the force of contraction either by direct or through reflex mechanisms.

6.5.1.2.1 Direct Nervous Control

Nervous regulation of the cardiac activity originates in the cardiovascular center present in the medulla oblongata. This region of the brain stem receives input from a variety of sensory receptors and also from higher brain centers, such as the limbic system and cerebral cortex. The cardiovascular center then directs appropriate output by increasing or decreasing the frequency of nerve impulses in both the sympathetic and parasympathetic branches of the autonomic nervous system (ANS) that abundantly supply the heart, thereby controlling the cardiac activity, especially pumping effectiveness of the heart.

Sympathetic nerves extend from the medulla into the spinal cord, and from the thoracic region of the spinal cord, the sympathetic cardiac accelerator nerves extend out to the SA node, AV node, and most portions of the myocardium. The sympathetic output to the heart affects both contractility and heart rate. Impulses in the cardiac accelerator nerves trigger the release of norepinephrine from the postganglionic sympathetic neurons and act on postsynaptic β1-adrenergic receptors on pacemaker cells and cardiac muscle fibers.

In SA (and AV) node fibers, norepinephrine acts on the β1-adrenoceptor and via the G protein Gs and activates the cAMP–protein kinase A pathway, which has direct action on hyperpolarization-activated cyclin nucleotide-gated channels (HCN channels) and Ca channels. The opening of HCN channels stimulates the diastolic Na^+ current into the pacemaker cells. The cAMP increases Ca^{2+} current through T-type and L-type Ca^{2+} channels. The net effect of these two changes is an increased rate of diastolic depolarization, so that these pacemakers fire impulses more rapidly; thereby, the diastole shortens and the heart rate increases.

In the contractile fibers throughout the atria and ventricles, norepinephrine enhances positive inotropic effects via protein kinase A and activates the Ca^{2+} entry through the voltage-gated slow Ca^{2+} channels, thereby increasing contractility. The net effects of these pathways are contractions that are both stronger and briefer. As a result, a greater volume of blood is ejected during systole. With a moderate increase in heart rate, stroke volume does not decline because the increased contractility offsets the decreased preload. With maximal sympathetic stimulation, however, heart rate may reach 200 beats/min in an adult human. At such a high heart rate, stroke volume is lesser than at rest due to the very short filling time. For given levels of input atrial pressure, the cardiac output can often be increased more than 100% by sympathetic stimulation.

The parasympathetic preganglionic cell bodies originate in the brain stem primarily in the dorsal motor nucleus of the vagus and nucleus ambiguous. Parasympathetic myelinated preganglionic fibers leave the central nervous system via the vagus nerve and travel to terminal ganglia on or near the epicardial surface of the heart, where they form a synapse with short postganglionic neurons in the SA node, AV node, and atrial myocardium. As in the sympathetic nervous system, ACh is released from preganglionic terminals and binds to and activates nicotinic acetylcholine receptors on postganglionic neurons. Parasympathetic nerve impulses reach the heart via the right and left vagus nerves. Parasympathetic output to the heart affects heart rate and, to a much lesser extent, contractility. ACh released by postsynaptic parasympathetic neurons binds to M2 muscarinic (i.e., G protein coupled) receptors on pacemaker cells of the SA node and on ventricular myocytes.

In pacemaker cells, ACh acts by three mechanisms: (1) ACh triggers G protein βγ subunits to directly open inward potassium channels that hyperpolarize the cell and decrease the frequency of action potentials. (2) ACh decreases the diastolic Na^+ current through HCN channels, thereby reducing the rate of diastolic depolarization. (3) ACh also decreases the Ca^{2+} current through T-type and L-type Ca^{2+} channels, thereby both reducing the rate of diastolic depolarization and making the threshold more positive. The net effect is a reduction in heart rate.

In myocardial cells, ACh exerts minor negative inotropic effect by two mechanisms: (1) Activation of the M2 receptor, via Gαi, inhibits adenylyl cyclase, reducing cAMPi and thereby counteracting the effects of adrenergic stimulation. (2) Activation of the M3 receptor, via Gαq, stimulates phospholipase C, raising Ca^{2+} and thus stimulating nitric oxide synthase. The newly formed nitric oxide (NO) stimulates guanylyl cyclase and increases cGMPi, which inhibits L-type Ca^{2+} channels and decreases Ca^{2+} influx. With maximal stimulation by the parasympathetic nerves, the heart can slow to 20 or 30 beats/min or can even momentarily stop.

A continually shifting balance exists between sympathetic and parasympathetic stimulation of the heart, and at rest, parasympathetic stimulation predominates. In humans, the resting heart rate (about 75 beats/min) is usually lower than the autorhythmic rate of the SA node (about 100 beats/min).

ACh released from vagal endings reacts with presynaptic muscarinic receptors on sympathetic nerve endings to reduce the amount of norepinephrine released from sympathetic efferent terminals. In addition to this, transmitters such as neuropeptide Y are also released, which inhibits the release of ACh from vagal nerve endings. At the level of the effector muscle cells, the two antagonistic transmitters oppose one another's effects via activation of their respective receptors and activation of second messenger systems.

The right and left vagus nerves differentially innervate the SA node. In comparison to the left vagus, stimulation of the right vagus ordinarily has a greater effect in decreasing the firing rate of the SA node (located in the right atrium) and thus decreasing heart rate (negative chronotropy). The negative inotropic effects of the vagus are primarily exerted on the atria where the vagal innervation is relatively rich. Left vagus nerve stimulation inhibits AV conduction and can produce AV block. Thus, vagal fibers have negative chronotropic (rate of contraction), inotropic (force of contraction), and dromotropic (conduction rate) actions on the heart. Vagal stimulation slows the discharge rate of the SA node, slows or blocks AV conduction, and decreases atrial and to a small extent contractility of ventricles. At rest, the vagus nerves exert a continuous or tonic restraint on the heart.

6.5.1.2.2 Endocrine Control

Exercise, excitement, and stress cause the adrenal medulla to release epinephrine and norepinephrine, which increase both heart rate and contractility, thereby enhancing the heart's pumping effectiveness. Glucagon causes positive inotropic and chronotropic effects acting via cardiac adenylyl cyclase to bring these effects. Additionally, mineralocorticoids, angiotensin II, prostaglandins, insulin, and thyroid hormones have positive inotropic effect on heart.

6.5.1.2.3 Chemical Control

The relative concentrations of three cations Ca^{2+}, K^+, and Na^+ have a large effect on cardiac function. Elevated levels of K^+/Na^+ in blood decrease heart rate and contractility. While excess K^+ prevents the generation of action potentials, excess Na^+ prevents Ca^{2+} influx during cardiac action potentials, reducing the force of contraction. A moderate increase in interstitial (and thus intracellular) Ca^{2+} level speeds heart rate and strengthens heartbeat. Oxygen, carbon dioxide, and pH have direct influence on cardiac function via actions on the carotid body and central chemoreceptors. Hypoxia and hypercapnia may depress cardiac contractility and performance through reduction in calcium sensitivity of contractile proteins. Hypercapnic effects likely occur through decreases in intracellular pH (acidosis). While changes in arterial blood gas composition may directly affect the heart, the reflex responses are likely to predominate.

6.5.2 Regulation of Arterial Blood Pressure

Under normal physiological conditions, arterial blood pressure varies; however, immediately it is brought back to normal level because of the presence of well-organized regulatory mechanisms in the body. Whenever the blood pressure undergoes change, nerves and hormones act rapidly to regulate the blood pressure; however, the body utilizes powerful mechanisms involving the kidneys to regulate the blood pressure over long periods. This long-term control of arterial pressure is closely linked with the homeostasis of body fluid volume, which is determined by the balance between the fluid intake and output. Two mechanisms regulate the arterial pressure:

1. Rapidly acting mechanism operates through nerves and hormones
2. Long-term control of arterial pressure

6.5.2.1 Short-Term Regulation of Blood Pressure

6.5.2.1.1 Neural Regulation of Blood Pressure

Nervous regulation is rapid among all the mechanisms that regulate the arterial blood pressure. Within a few minutes of a pressure change, the nervous system returns it to normal. As the nervous mechanism is rapid in action, it operates only for a short period and hence it is called short-term regulation. The nervous mechanism regulating the arterial blood pressure operates through the direct action and reflex mechanism involving vasomotor system.

6.5.2.1.1.1 Reflex Mechanism

This involves reflexes that mediate moment-to-moment adjustments in the distribution of blood flow in response to variations in regional or organ function. The cardiovascular reflexes are mediated through the sensory nerves that continuously provide the brain with information from the periphery, so that appropriate adjustments in efferent outflow and humoral secretions can be made to maintain blood pressure and meet tissue demands under a variety of physiological conditions.

1. **Baroceptor reflex**: Baroreceptors are spray-type nerve pressure-sensitive endings that lie in the walls of the arteries, carotid sinus, and aortic arch; they are stimulated when stretched, and the signals from the "carotid baroreceptors" and "aortic baroreceptors" are transmitted through very small Hering's nerves to the glossopharyngeal nerves and vagus nerve, respectively, to the tractus solitarius of the medulla of the brain stem. Additionally, secondary signals inhibit the vasoconstrictor center of the medulla and excite the vagal parasympathetic center. The net effects are vasodilation of the veins and arterioles throughout the peripheral circulatory system and decreased heart rate and strength of heart contraction. As a result, when the baroreceptors in the arteries are stimulated by high blood pressure, both peripheral resistance and cardiac output are reduced as a result of reflex action. Conversely, low pressure has opposite effects, causing the pressure to rise back toward normal by reflex action.
   The baroreceptors respond to rapidly changing pressure than to a stationary pressure. The frequency of impulses generated from baroreceptors is proportional to the blood pressure. Beat-to-beat changes in blood pressure are monitored by the brain from the impulses reaching it through aortic and sinus nerves. Increase in arterial pressure increases impulse frequency and a decrease reduces frequency.
   The brain responds to an increase in blood pressure by decreasing sympathetic activity and enhancing parasympathetic activity. Heart rate and force of contraction are decreased, and hence stroke volume is decreased. Sympathetic inhibition causes vasodilatation of arterioles, and peripheral resistance is decreased. All of these effects restore arterial blood pressure to normal. A decrease in blood pressure produces opposite effect to those considered above.
2. **Atrial volume receptor reflex**: This reflex is initiated by stretch receptors located in the walls of the left and right atria, and these receptors are called volume receptors since atrial volume decides the stretch of the atrial wall. Stretching of the atria also leads to afferent arteriolar dilation in the kidneys and signals the hypothalamus to decrease secretion of antidiuretic hormone. The kidneys' reduced afferent arteriolar resistance allows the glomerular capillary pressure to increase, which in turn leads to more fluid filtration. The reduction in antidiuretic hormone decreases the reabsorption of water from the tubules. Combination of these two effects' increase in glomerular filtration and decrease in reabsorption of the fluid increases fluid loss by the kidneys and reduces an increased blood volume back toward normal. Additionally, atrial stretch caused by increased blood volume also elicits a hormonal effect on the kidneys' release of atrial natriuretic peptide that adds still further to the excretion of fluid in the urine and return of blood volume toward normal. When blood volume is decreased (hemorrhage), the CNS receives fewer impulses from the volume receptors and CNS reflex increases the sympathetic activity to heart and blood vessels and decreases parasympathetic activity. The volume receptor reflex increases thirst sensation (acting through hypothalamus) that helps to increase blood volume, increase ADH secretion and fluid conservation through kidneys, and also activate renin-angiotensin-aldosterone system to conserve sodium.
3. **Bainbridge reflex**: This reflex operates when the venous return is increased. The stretch receptors of the atria that elicit the Bainbridge reflex transmit their afferent signals through the vagus nerves to the medulla of the brain. Then, vagal and sympathetic nerves send efferent signals back to increase the heartbeat and force of the contraction. This response thereby aids in preventing blood damming in the veins, atria, and pulmonary circulation. This reflex helps to prevent accumulation of blood in veins. Bainbridge reflex causes 40–60% increase in rate.
4. **Psychogenic responses**: Psychogenic responses originate from conscious perceptions or emotional reactions. They are eliminated by unconsciousness or general anesthesia. They involve neural pathways of the midbrain and forebrain, including the limbic system and cerebral cortex. Psychogenic responses are often triggered by sensory stimuli. Defense-alarm reaction and vasovagal syncope are two important psychogenic responses that can bring

about cardiovascular changes. The defense alarm reaction, also known as fear, fight, or flight response, is an emotional response to a threatening situation, physical injury, or trauma. It involves increased sympathetic and decreased parasympathetic activity and includes increased heart rate and stroke volume, vasoconstriction in kidneys, splanchnic organs and skin, vasodilatation in coronary vessel and skeletal muscles, and increased BP. There is enhanced secretion of ADH, angiotensin II, and adrenocorticotropic hormone (ACTH).

The defense alarm reaction ("fear, fight, or flight" response) is an emotional and behavioral response to a threatening situation, physical injury, or trauma. The response involves increased sympathetic activity and decreased parasympathetic activity. There is release of epinephrine and norepinephrine from the adrenal medulla, an increased heart rate, increased stroke volume, vasoconstriction in noncritical organs (kidneys, splanchnic organs, resting skeletal muscle), vasoconstriction in skin, vasodilation in coronary vessels and in working skeletal muscle, and increased blood pressure. The cardiovascular responses are enhanced by other circulating hormones, including ADH and angiotensin II. The resulting elevated blood pressure helps to ensure adequate blood flow for the critical organs (exercising skeletal muscles, heart, and brain). During a defense alarm reaction, the baroreceptor reflex is reset by the CNS so that it regulates blood pressure at an elevated level rather than acting to oppose the increased pressure.

Vasovagal syncope (playing dead reaction or playing opossum) is a psychogenic response that occurs in response to certain threatening or emotional situations, wherein the blood pressure decreases and involves a decrease in sympathetic activity and an increase in parasympathetic activity. These neural changes bring about a vasodilation in the noncritical organs, with a consequent decrease in total peripheral resistance. Heart rate and cardiac output also decrease, so there is a large drop in arterial blood pressure with inadequate blood flow to brain and the animal faints.

5. **Chemoreceptor reflex**: Carotid bodies located in the bifurcation of each common carotid artery and aortic bodies located in the aortic arch contain peripheral chemoreceptors, which are sensitive to oxygen lack, carbon dioxide excess, and hydrogen ion excess. The aortic bodies are supplied by the vagus nerve, and the carotid bodies are supplied by a branch of the glossopharyngeal nerve.

   The chemoreceptors excite nerve fibers that, along with the baroreceptor fibers, pass through Hering's nerves and vagus nerves into the vasomotor center of the brain stem. Each carotid or aortic body is supplied with an abundant blood flow through a small nutrient artery, so that the chemoreceptors are always in close contact with arterial blood. The chemoreceptors are activated due to decreased oxygen and an excess accumulation of carbon dioxide as a result of decreased blood flow.

   The vasomotor center is stimulated by the signals sent from the chemoreceptors, which raises the arterial pressure back to normal. However, until the arterial pressure drops below 80 mmHg, this chemoreceptor reflex is not a potent controller of arterial pressure. As a result, this response becomes crucial at lower pressures to assist in stopping a further drop in pressure.

6.5.2.1.1.2 Regulation by Autonomic Nervous System

Sympathetic and parasympathetic neurons influence the cardiovascular system through the release of the neurotransmitters norepinephrine and acetylcholine. In addition, sympathetic nerves affect the cardiovascular system by stimulating the release of epinephrine and norepinephrine from the adrenal medulla.

Epinephrine, norepinephrine, and acetylcholine exert their cardiovascular effects through activating receptor proteins, whether they are working as neurotransmitters or hormones. The receptors activated by epinephrine and norepinephrine act via two types of adrenergic receptors (α and β). The α-adrenergic receptors are subdivided into $\alpha 1$ and $\alpha 2$, which are located in the cell membranes of smooth muscle cells of the arterioles in all organs and in the smooth muscle cells of the abdominal veins. There are three subtypes of β-receptors: $\beta 1$ (located in cardiac muscle cell), $\beta 2$ (located in arterioles, particularly in the coronary circulation and in skeletal muscles), and $\beta 3$, with the first two of these being important in cardiovascular control.

Activation of the α-adrenergic receptors leads to constriction of arterioles or veins. Arteriolar vasoconstriction increases the resistance and decreases the blood flow through an organ. If one or more major body organs are vasoconstricted, the total peripheral resistance (TPR) increases, leading to an increase in arterial blood pressure. The increase in arterial pressure increases the driving force for blood flow in all organs of the systemic circulation. In effect, the sympathetic nervous system can vasoconstrict some organs and thereby direct more blood flow to other, non-vasoconstricted organs. Venoconstriction displaces venous blood toward the central circulation, which increases central venous pressure, right ventricular preload, and (by the Starling mechanism) stroke volume.

Acetylcholine activates cholinergic receptors which are of two major types: muscarinic cholinergic receptors and nicotinic cholinergic receptors. The main cardiovascular effects of acetylcholine are mediated through muscarinic cholinergic receptors located on cardiac, smooth muscle, or endothelial cells. Of the five subtypes of muscarinic receptors, the M2 and M3 receptor subtypes have the greatest cardiovascular importance.

Cardiac muscle cells of the sinoatrial and atrioventricular nodes are densely innervated by postganglionic parasympathetic neurons. Atrial cells also receive strong parasympathetic innervations, and activation of cardiac M2 receptors has effects basically opposite to those of the activation of β1-adrenergic receptors.

Parasympathetic activation powerfully slows the cardiac pacemakers, decreases cell to-cell conduction velocity, and increases refractory period. Ventricular muscle cells receive very little direct parasympathetic innervations, and hence, parasympathetic activation has only a minor, direct effect on ventricular contractility. However, parasympathetic neurons indirectly act on ventricular muscle cells and release their acetylcholine onto sympathetic neuron terminals, rather than directly onto ventricular muscle cells. This acetylcholine activates muscarinic cholinergic receptors on the sympathetic neuron terminals, which inhibits the release of norepinephrine from the terminals and thus weakens the effects of sympathetic activity on ventricular cells. Parasympathetic activation can significantly reduce cardiac output by lowering heart rate and reversing the effects of sympathetic stimulation on ventricular contractility. M3 adrenergic receptors are found in the arteries and arterioles all over the body, and acetylcholine causes these blood vessels to enlarge.

6.5.2.1.1.3 Vasomotor Mechanism

It deals with the maintenance of the diameter of the blood vessels (arteries and arterioles) to maintain blood pressure and thus regulates the blood flow to various organs or tissues according to their demand. These mechanisms regulate blood flow to different organs according to their demand, i.e., shifting of blood from one organ to another on demand, and regulate peripheral resistance which in turn influences the blood pressure.

Vasomotor system includes three components: vasomotor center, vasoconstrictor fibers, and vasodilator fibers.

6.5.2.1.1.3.1 Vasomotor Centre

Vasomotor center is bilaterally situated in the reticular formation of medulla oblongata and the lower part of the pons and includes vasoconstrictor and vasodilator centers. This center receives sensory signals through vagus and glossopharyngeal nerves. By its output signals, it regulates the activities of both the vasoconstrictor or vasodilator center. Vasoconstrictor and vasodilator centers show reciprocal inhibition. These centers are controlled by higher neurons from hypothalamus and cerebral cortex and also from sensory impulses from the peripheral organs and chemical composition of blood.

1. ***Vasoconstrictor area and vasoconstrictor fibers***
   Vasoconstrictor center is extended from the middle pons to upper spinal cord. Vasoconstrictor fibers are projected through the sympathetic nervous system. Under normal circumstances, the sympathetic vasoconstrictor nerve fibers in the vasoconstrictor region continuously transmit impulses that maintain a partial state of blood vessel contraction known as sympathetic vasoconstrictor tone.
   Small arteries, big arterioles, and veins are all innervated by sympathetic nerves (thoracolumbar output). They provide peripheral resistance and function to maintain the tonus of the arterioles, thus regulating BP. On the other hand, the capillaries and precapillary sphincters are free from sympathetic innervations. The venules have fewer adrenergic fibers than large veins, which themselves are less richly innervated than the arterioles. Sympathetic innervation to veins helps to change the volume of veins by constriction/dilatation, thus regulating BP by altering the volume of the circulatory system. The neurotransmitter of the sympathetic nerves is the norepinephrine. Sympathetic stimulation also causes the release of epinephrine from adrenal medulla.
   Both epinephrine and norepinephrine act through α- and β-receptors on blood vessels. NEP excites mainly α-receptors and causes vasoconstriction. By its action on α1-adrenergic receptors, it causes vasoconstriction in arterioles of all organs of the body. Epinephrine acts both on α- and β-adrenergic receptors. Its action on α-adrenergic receptors in cutaneous and renal arterioles causes vasoconstriction. Epinephrine causes vasodilatation in the cardiac and skeletal muscles via its effect on β-receptors. Neuropeptide Y potentiates the vasoconstrictor effect of adrenergic receptors in primates.
   All vasoconstrictor fibers are sympathetic fibers, and these fibers are in tone. Without using vasodilator fibers, the vasoconstrictor or vasodilator effect can be created by simply changing the tone of the vasoconstrictor. Vasoconstriction results from sympathetic fibers in the veins being stimulated.
2. ***Vasodilator area and vasodilator fibers***
   Vasodilator center is located medially in the floor of the ventricles of the medulla oblongata, but close to the vasoconstrictor area, and inhibits the vasoconstrictor area producing vasodilatation. Vasodilator fibers are of three types, the parasympathetic fibers, sympathetic cholinergic fibers, and antidromic fibers. These fibers do not exert tonic activity on blood vessels.
   Parasympathetic vasodilator fibers: These are the craniosacral outflow as chorda tympani (branch of facial nerve), glossopharyngeal, vagus, and pelvic nerves.
   The neurotransmitter of these fibers is acetylcholine, which acts through the cholinergic muscarinic

M3-type receptors, which are found on the endothelial cells and on the smooth muscle cells of most arterioles. M3 receptors are innervated by parasympathetic fibers in the coronary circulation and in the external genitalia and by sympathetic cholinergic fibers in skeletal muscles. The stimulation of M3 receptors on the endothelial cells both causes vasodilatation and releases nitrous oxide from endothelial cells. This nitrous oxide causes vasodilatation by altering the arterial smooth muscles. Stimulation of these fibers results in vasodilatation in coronary vessels, tongue, salivary gland, external genitalia, bladder, and rectum.

Sympathetic cholinergic vasodilator fibers: They are limited to the arterioles of active skeletal muscles and cause anticipatory increase in blood flow even before exercise to overcome fatigue (in dogs and cats). Sympathetic cholinergic fiber-innervated cholinergic muscarinic M3 receptors are stimulated by acetylcholine and cause cutaneous vasodilatation.

Antidromic fibers: These fibers originate from dorsal roots of spinal cord and show bidirectional conduction of impulses. The antidromic fibers divide at their peripheral end, one branch supplying the receptors of the skin or muscle and the other to the nearby arterioles. The receptor is sensitive to trauma, heat, cold, and frostbite. When the receptor is triggered, the impulse travels in the opposite direction (antidromic) to the affected blood artery and causes vasodilation. The response is known as the "axon reflex" because it only affects the sensory nerve and its blood vessel branch.

3. ***Sensory area***
   Sensory area is in the nucleus of tractus solitarius, which is situated in the posterolateral part of medulla and pons. This area receives sensory impulses via glossopharyngeal and vagal nerves from the periphery, particularly from the baroreceptors. Sensory area in turn controls the vasoconstrictor and vasodilator areas.

Know More……

Unique Cardiovascular Feature in Giraffe

Giraffes exhibit unique cardiovascular features. An adult full-grown giraffe may be around 5–6 m tall, and due to their long necks, their brains may be 1.6 m above their hearts. Further, standing giraffes have exceptionally high blood pressures in their legs and feet due to their tall columns of blood. Giraffes possess a remarkably well-developed left ventricle and maintain unusually high systemic aortic blood pressures. Their mean aortic pressure on standing/at rest is about 220 Hg in contrast to about 100 mmHg in most other mammals. As the giraffe lowers its head to the ground, arterial blood pressure at the level of the heart is reduced considerably, thereby maintaining a relatively constant blood flow to the brain. Further, the ability of the giraffe to regulate pressure and flow in peripheral vessels other than those to the head is also particularly very crucial for renal functioning.

6.5.2.1.1.3.2 Vasomotor Reflexes

These are of two types:

1. ***Pressor reflex***: When the blood pressure falls to less than normal, the impulse frequency passing through the buffer nerves decreases, which stimulates cardio accelerator and vasoconstrictor centers of the medulla, the stimulation of which increases BP to normal level.
2. ***Depressor reflex***: This reflex produces a fall in blood pressure. A rise in blood pressure above normal causes increased sensory impulse frequency through buffer nerves to the medullary vasomotor centers, where it causes inhibition of vasoconstrictor and sympathetic center but stimulates vagal center. These effects result in vasodilatation and decreased heart rate to establish normal blood pressure.

6.5.2.1.2 Endocrine Regulation

The epinephrine and norepinephrine released from sympathetic system and adrenal medulla activate adrenergic receptors on the cardiac muscle cells and on the endothelial or smooth muscle cells of blood vessels, act through α-receptors, and cause arteriolar vasoconstriction. Resistance to blood flow is increased, and total peripheral resistance increases. The hormones also cause blood to be pushed to the central circulation, an increase in ventricular preload, and venoconstriction (α-receptor activation). The hormones also cause the heart's receptors to contract more quickly and forcefully, increasing the volume of the stroke. The increase in preload, stroke volume, and total peripheral resistance raises blood pressure. There are several different hormones that have an impact on blood pressure.

The following hormones cause an increase in blood pressure:

Adrenaline: Adrenaline is secreted by the adrenal medulla and is also released by sympathetic postganglionic nerve endings. Through its effects on the heart and blood vessels, adrenaline controls blood pressure. It increases systolic pressure by increasing the heart rate and cardiac output. By lowering the total peripheral resistance, it lowers diastolic pressure. Adrenaline causes constriction of blood vessels through α-receptors. It also causes dilatation of blood vessels through β2 receptors in some areas of the body like skeletal muscle, liver, and heart. So, the total peripheral resistance is reduced leading to decrease in diastolic pressure.

Noradrenaline: Noradrenaline is secreted by the adrenal medulla. It is also released by sympathetic postganglionic nerve endings. Because of its overall vasoconstrictor impact, noradrenaline causes a rise in diastolic pressure. It has stronger effects on blood vessels than on the heart. It causes constriction of all blood vessels throughout the body via α-receptors. Noradrenaline increases the total peripheral resistance, increases diastolic pressure, and causes slight increase in the systolic pressure by increasing the force of contraction of heart.

Thyroxine: Thyroxine secreted from thyroid gland increases systolic pressure but decreases diastolic pressure. It increases the systolic pressure by increasing cardiac output due to increase in the blood volume and force of contraction of the heart. Thyroxine has indirect action on diastolic pressure. Large quantities of metabolites are produced during increased metabolic activity induced by thyroxine. These metabolites cause vasodilatation, leading to decrease in peripheral resistance. It causes decrease in diastolic pressure.

Aldosterone: Aldosterone, secreted from adrenal cortex, causes retention of sodium and water and thereby increases the ECF fluid volume and blood volume, leading to increase in blood pressure.

Vasopressin: Vasopressin produced by the supraoptic nucleus of hypothalamus has vasoconstrictor and antidiuretic effect. However, the physiological level of ADH in plasma is too low to produce vasoconstriction and its physiological role is related to long-term regulation of BP brought about by water reabsorption from the kidneys. In hemorrhage, large amounts of ADH are released, which brings about vasoconstriction.

Angiotensins: Angiotensins II, III, and IV obtained from angiotensinogen cause constriction of systemic arterioles and elevate blood pressure.

Serotonin (5-hydroxytryptamine): Serotonin is present in highest concentration in blood platelets and in the gastrointestinal tract, where it is found in the enterochromaffin cells and the myenteric plexus. It is also found within the brain and spinal cord. It increases the blood pressure by vasoconstriction.

Endothelin (ET): Endothelin (ET), a 21-amino acid polypeptide produced primarily by the vascular endothelium, is a potent vasoconstrictor that has three isoforms of which ET1 is most prominent that causes vasoconstriction and vascularization, induces the release of norepinephrine and serotonin during the regulation of vascular tone, and participates in the redistribution of blood flow during exercise. ET1 functions in a paracrine and autocrine fashion in pulmonary and systemic arteries and veins. Endothelins are produced by stretching of blood vessels. These peptides act by activating phospholipase, which in turn activates prostacyclin and thromboxane A2. These two substances cause the constriction of blood vessels and increase the blood pressure.

Thromboxane A2 (TXA2): Thromboxane A2 released from platelets causes vasoconstriction, especially when blood vessels become traumatized or ruptured.

Vasoactive intestinal polypeptide: Vasoactive intestinal polypeptide (VIP) is secreted in the stomach and small intestine. A small amount of this hormone is also secreted in large intestine. VIP is a vasodilator and causes dilatation of peripheral blood vessels and decrease in blood pressure.

Atrial natriuretic factor (ANF): Atrial natriuretic factor is a peptide hormone having 24–28 amino acids. Stretching of atria due to increased blood volume stimulates its release. It acts on kidney tubules and favors increased GFR and decreased Na^+ reabsorption (natriuresis), diuresis, and vasodilatation. By its inhibitory action, it modulates the activity of renin, aldosterone, and ADH. Neuropeptide Y has direct vasoconstrictor property and regulates the release of atrial natriuretic factor and angiotensin II. It potentiates the effects of norepinephrine as well as vasoconstrictor actions of serotonin and K^+, whereas it inhibits renin release. It provides moment-by-moment regulation of blood pressure and blood flow.

Histamine: Histamine is secreted in nerve endings of hypothalamus, limbic cortex, and other parts of cerebral cortex. Histamine is also released from mast cells and basophils during allergic conditions, inflammation, or damage. Histamine causes vasodilatation and decreases blood pressure.

Prostaglandins: Prostaglandins are local hormones, synthesized by vascular endothelium from arachidonic acid. The prostaglandin PGE2 and prostacyclin (PGI2) are potent vasodilators.

Nitric oxide (endothelium-derived relaxing factor) (EDRF): Nitric oxide is released by the vascular endothelium and brain. It is synthesized from arginine and brain nitric oxide synthase. Nitric oxide synthesis is stimulated by acetylcholine, bradykinin, VIP, substance P, and platelet breakdown products. As nitric oxide is a vasodilator, deficiency of this leads to constant vasoconstriction and hypertension.

Adenosine: Adenosine released during tissue anoxia stimulates adenosine A2 receptors, activates cAMP mechanism, and results in profound vasodilatation. ADP and ATP cause release of nitrous oxide from endothelial cells and act on P2 receptors. The nitrous oxide as a vasodilator substance causes vasodilation.

Bradykinin: Bradykinin is a vasoactive peptide causing functional hyperemia of the salivary glands and the pancreas, and its activity includes stimulation of nitric oxide synthase. Bradykinin is produced in blood during conditions like inflammation. During such conditions, the enzyme in the blood called kallikrein is activated. It acts on α2-globulin to form kallidin, which is converted into bradykinin. Bradykinin is a vasodilator substance and causes reduction in blood pressure.

6.5.2.1.3 Local Control of Blood Flow (Control of Capillaries)

1. ***Myogenic theory of autoregulation (pressure autoregulation)***: Blood supply to organs is maintained almost constant even when arterial pressure changes. When blood pressure rises, the blood vessels dilate due to increased pressure and stretch the vascular smooth muscles surrounding the vessel, which contract because of stretching, and when pressure decreases, these muscles relax. In addition, when perfusion pressure (arteriovenous pressure difference) is increased above normal, additional blood flows through the organ due to increased pressure, which accelerates removal of metabolic products and increases oxygen delivery to the tissues. Hence, the concentration of vasodilator metabolic products decreases. These two mechanisms help to maintain normal blood flow even when arterial pressure is increased or decreased. Pressure regulation is important to maintain blood supply to brain, heart, and kidneys.
2. ***Metabolic theory of autoregulation***: Capillaries regulate the local blood supply to tissues according to the need of the tissues. Metabolic control of blood flow is the most important local control mechanism as it matches the blood flow in a tissue to the metabolic rate of the tissue, and it is stimulated by chemical changes within the tissue. An increase in tissue blood flow in response to increased metabolic rate is called active hyperemia. When the metabolic rate of a tissue increases, its consumption of oxygen increases, and there is an increased rate of production of metabolic products, including carbon dioxide, adenosine, and lactic acid. Also, some potassium ions (K^+) escape from rapidly metabolizing cells, and these ions accumulate in the interstitial fluid. Therefore, as the metabolism of a tissue increases, the interstitial concentration of oxygen decreases, and the interstitial concentrations of metabolic products and K^+ increase. All these changes relax the arteriolar smooth muscle; thereby, the arterioles and precapillary sphincters dilate, vascular resistance decreases, total capillary surface area for diffusional exchange increases, and more blood flows through the tissue, and hence more O_2 is delivered and accumulated metabolic products are removed. Reactive hyperemia refers to a temporary increase in blood flow to above normal in a tissue after a period of restricted blood flow. Autoregulation of blood flow is also a metabolic control phenomenon.

6.5.2.2 Long-Term Regulation of Blood Pressure

Kidneys play a vital role in the long-term regulation of arterial blood pressure. Slow blood pressure changes cause the nervous system to become accustomed to the new pressure and lose sensitivity to the changes. It cannot regulate the pressure anymore. In such conditions, the renal mechanism operates efficiently to regulate the blood pressure. Hence, it is called long-term regulation. Kidneys regulate arterial blood pressure by two ways:

1. By regulation of ECF volume
2. Through renin-angiotensin mechanism

6.5.2.2.1 Regulation of Extracellular Fluid Volume

When the blood pressure increases, kidneys excrete large quantities of water and salt, particularly sodium, by means of pressure diuresis and pressure natriuresis. Even a slight increase in blood pressure doubles the water excretion. Because of diuresis and natriuresis, there is a decrease in ECF volume and blood volume, which in turn brings the arterial blood pressure back to normal level. The reabsorption of water from renal tubules increases as blood pressure drops. This in turn raises cardiac output, blood volume, and ECF volume, restoring blood pressure.

6.5.2.2.2 Renin-Angiotensin Mechanism

In response to decreases in blood pressure, cells in the juxtaglomerular apparatus of the kidney produce an enzyme, renin. Renin acts on angiotensinogen, an α2-globulin produced by the liver and released into the circulation, and this results in the production of angiotensin I, a decapeptide. Angiotensin I is further hydrolyzed to angiotensin II, an octapeptide, by angiotensin-converting enzyme. Angiotensin II stimulates the zona glomerulosa to produce mineralocorticoid aldosterone released from adrenal cortex and conserves body sodium and water. The vasoconstriction and sodium conservation elevate blood pressure to normal. Angiotensin II also increases peripheral resistance of the blood vascular system by causing vasoconstriction of smooth muscle of the blood.

6.6 Circulation

6.6.1 Capillary Circulation/Microcirculation

The capillary bed is the main portion of the microcirculation disposed between arterioles and venules, where the exchange or transport of nutrients and waste occurs.

6.6.1.1 Functional Anatomy and Types

On the basis of the completeness of their endothelial walls, capillaries are classified into three types as continuous, discontinuous, and fenestration capillaries.

Continuous capillaries with complete endothelial walls and basement membranes are found in adipose tissue; smooth, skeletal, and cardiac muscle; placenta; lungs; and central nervous system. They have pinocytic vesicles (60–70 nm in diameter) along their luminal and basal borders, form tight junctions with adjacent cells, and have pores or intercellular clefts between cells that allow the passage of water-soluble ions and molecules across the capillary wall. Many of these capillaries also possess gap junctions that allow for cell-to-cell communication along the capillary wall.

Discontinuous capillaries or sinusoids have gaps between the endothelial cells and incomplete or absent basement membranes. They are found in the liver, spleen, and bone marrow and allow passage of whole cells, macromolecules, and particles across the capillary wall. Although capillaries are generally considered passive, contractile properties of endothelial cells have been observed in liver sinusoids.

Fenestrated capillaries contain small openings of 0.1 μm or less in diameter that are closed (except for glomerular capillaries) by a thin diaphragm. These holes or fenestrae allow the rapid diffusion of solutes and water across the capillary wall. Fenestrated capillaries are found in endocrine and exocrine glands, gallbladder, synovial membrane, ciliary body, and choroid plexus and in countercurrent flow systems as found in the renal medulla.

6.6.1.2 Exchange Across Capillary Wall

Velocity of blood flow is least (about 0.05 cm/s) in capillaries. The capillaries have the most important function of exchange of substances between blood and tissues. Oxygen, nutrients, and other essential substances enter the tissues from capillary blood; carbon dioxide, other end products of metabolism, and other unwanted substances are removed from the tissues by capillary blood. Exchange of materials across the capillary endothelium occurs by the processes of diffusion, filtration, and pinocytosis.

Diffusion is the main process for exchange of gases, water, glucose, sodium, urea, and many other substances. These substances diffuse through the intercellular clefts in the endothelial wall of the capillaries because of concentration gradient. Filtration occurs through the slit pores present in capillary endothelium, which depends upon the net filtration pressure. Larger molecules are transported across the capillary endothelium in the form of vesicles by the process of pinocytosis.

6.6.1.3 Capillary Pressure and Regulatory Factors

Blood pressure decreases from the arterial to the venous end of the capillary, while capillary diameter usually increases. In systemic capillaries, pressures are 35 and 15 mmHg, respectively. The mean hydrostatic blood pressure for systemic capillaries is about 25 mmHg.

Capillary blood flow is controlled by both nervous and chemical factors. Capillaries are mainly supplied by the sympathetic vasoconstrictor fibers. Many chemical factors such as excess of carbon dioxide, lack of oxygen, increased hydrogen ion concentration, histamine, and metabolites like lactic acid cause dilatation of capillaries. Serotonin causes constriction of capillaries.

6.6.2 Lymphatic Circulation

6.6.2.1 Organization of Lymphatic System

Lymphatic system arises from tissue spaces as a meshwork of delicate vessels called lymph capillaries. Lymph capillaries start from tissue spaces as enlarged blind-ended terminals called capillary bulbs, which are surrounded by muscle fibers. These bulbs contain valves, which allow flow of lymph in only one direction. These muscle fibers cause tonic and phasic contractions of bulbs so that lymph is pushed through the vessels. The lymphatic pressure pulses are generated by contractions of the lymphatics themselves that occur in response to pacemaker activity located within the smooth muscle layer of the wall. The lymphatics are present as a series of contracting chambers or lymphangions, demarcated by the lymph unidirectional valves formed by endothelial cells. Lymph capillaries unite to form large lymphatic vessels. Lymphatic vessels become larger and larger because of the joining of many tributaries along their course. Lymph capillaries are lined by endothelial cells, which lie overlapping on one another and are more porous. This allows the fluid to move into the lymph capillaries and not in the opposite direction. At sites of open intercellular junctions, there are anchoring fibers attached to the outer membranes of endothelial cells that allow the formation of passages and the movement of fluid into the initial lymphatic. Initial lymphatics are larger than capillary tubes. As the lymphatic network is followed centrally toward the collecting vessels, specialized smooth muscle cells appear.

The fluid flow through the lymphatic vessels that roughly parallels the venous circulation is one-way because of the presence of valves. Moreover, because it is not pressurized, the lymph is dependent on passive movement and some impact of changes in the diaphragm. These vessels with branches coming from throughout the body progressively coalesce and empty into the general circulation in the cranial vena cava. The beginnings of this circulatory pathway are essentially thin, blind-ended pouches that collect interstitial fluid that drains to localized tissue lymph nodes and the regional nodes and large lymph vessels that parallel the trunk of the body. Lymph drainage from the forelimbs, neck, and head also drains into the venous blood via the thoracic and right lymphatic ducts.

The lymphatic fluid, containing the plasma proteins, flows to the thorax, where the fluid reenters the bloodstream at the subclavian veins. The role of lymphatic flow in counteracting the accumulation of excessive interstitial fluid is especially important in the lungs. Lung capillaries are more permeable to plasma proteins than are most capillaries in the systemic circulation.

6.6.2.2 Formation of Lymph

The blood vascular system transports compounds such as nutrients and metabolites to and from the blood–tissue exchange system at the capillary level. The interstitium filled with gel-like matrix and the lymphatics constitute an extravascular flow system on which the blood capillary–tissue exchanges depend. The steady state of the interstitium depends on the passage of materials in and out of the blood capillaries and the passage of materials into the lymph system and then back to the bloodstream. The excess of capillary filtration over reabsorption is normally balanced by lymph flow. Lymph is formed from interstitial fluid, due to the permeability of lymph capillaries. As the blood passes via tissue capillaries, 9/10th of fluid passes into the venous end of capillaries from the arterial end and the remaining 1/10th of the fluid passes into lymph capillaries.

Large molecules, such as plasma proteins, cannot be reabsorbed into the capillaries against their concentration gradients that enter the lymphatic system. The blind beginnings of the lymph vessels are adapted for the intake of large molecules, and concentration and pressure gradients favor this route. Water-insoluble fats are absorbed from the intestine into the lymphatics. Lymphocytes enter the circulation principally through the lymphatics, which play a role in the immune mechanism also.

6.6.2.3 Regulation of Lymphatic Circulation

Lymph progresses through the channels by contractions of the lymph vessels and by a massaging action of muscles that overlie lymph vessels. Forward movement of lymph lowers the pressure in the part of the vessel evacuated and, because there is no backflow of lymph, entry of lymph from the backward parts is favored. There is no central pump, such as the heart, to facilitate lymph circulation, and disturbances in lymph flow can cause accumulation of interstitial fluid in low-lying body parts. The main determinants of lymph flow are filling pressure (preload) and outflow resistance (afterload). When afterload, preload, or both are increased, the lymph vessel is stretched; it responds by increasing the rate and strength of contraction.

Hormones, vasoactive substances, and nerves affect the rate and strength of contraction in lymph vessels. Norepinephrine, epinephrine, and α-adrenergic sympathetic nerves stimulate motor activity and local lymph flow. Stimulation of β-adrenergic receptors results in the opposite effect. Similarly, a decrease in contractile activity is observed in response to stimulation with acetylcholine, which causes nitric oxide synthesis by endothelial cells. Certain endotoxins can paralyze lymphatics. Ordinarily, gravity has little effect on lymph flow because the column of fluid in the lymphatic is not continuous and a hydrostatic gradient is not present as it is in veins.

6.6.2.4 Functions of Lymph

Lymph is a clear and colorless fluid composed of 96% water and 4% solids. Lymph returns the proteins from tissue spaces into blood and is responsible for redistribution of fluid in the body. Bacteria, toxins, and other foreign bodies are removed from tissues via lymph, and it plays an important role in immunity by transport of lymphocytes. Lymph flow is responsible for the maintenance of structural and functional integrity of tissue. Obstruction to lymph flow affects various tissues, particularly myocardium, nephrons, and hepatic cells. Lymphatic flow serves as an important route for intestinal fat absorption.

6.6.2.5 Edema

Edema is an abnormal accumulation of interstitial fluid accompanied by swelling. Factors that result in edema are an increase in capillary pressure, increased capillary permeability, a decrease in the concentration of plasma proteins, and obstruction of the lymphatic vessels. An increase in capillary pressure due to increased venous pressure (because of obstruction of the veins) or due to excessive vasodilation of precapillary resistance vessels favors filtration and forces more fluid out through the capillary wall into the interstitium. Heart failure often leads to increased capillary pressures and edema due to an increase in venous pressure. An increase in capillary permeability may occur with severe burns, resulting in an increase in the permeability of the capillaries that allows protein to leak into the interstitium pulling water out and causing edema. A decrease in the plasma protein concentration due to renal or hepatic damage also upsets the balance of forces across the capillary wall, resulting in increased filtration of fluid out of the capillary tubes. Commonly occurring edemas are either pulmonary or peripheral edema.

Peripheral edema: Peripheral edema in horses is caused by a reduced venous massage to aid in the return of venous blood from pendant blood capillaries and inability of the lymphatic system to remove this excessive interstitial fluid. With exercise, muscle massage decreases venous pressure and aids in lymph return. With exercise, stocking edema soon disappears. The lymphatics ordinarily handle minor increases in tissue flow, preventing the formation of edema. The edema that occurs in immunologically mediated tissue reactions (e.g., urticaria) and in various renal diseases is apparently caused by damage to the capillary basement membrane.

Pulmonary edema: In the lungs, where edema is particularly dangerous, the rapid removal of capillary filtrate by the lymph system plays an important role in preventing fluid accumulation in the alveoli when capillary hydrostatic pressure increases or when plasma protein concentration decreases. One of the most common causes of pulmonary edema is left-sided heart failure or mitral valvular disease. This results in a large increase in pulmonary venous pressure and pulmonary capillary pressure and flooding of the interstitial spaces and alveoli with fluid. Another common cause of pulmonary edema is damage to the capillary membrane by infections such as pneumonia. This results in the rapid leakage of both plasma protein and fluid out of the capillaries into the interstitial spaces and the alveoli.

6.6.3 Venous Circulation

The venous system completes the circulatory circuit. Blood leaving the capillary beds enters the venous system and is transported back to the heart. At the microcirculatory level, capillaries drain into venules, which progressively converge to form small veins that exit the organ. In contrast to the arterioles, venules have little tone and resistance. Extensive communication takes place via chemical signals between venules and nearby arterioles. This venoarteriolar signaling is vital for matching capillary inflow and outflow within an organ. Veins have a large radius, so they offer little resistance to flow. Furthermore, because the total cross-sectional area of the venous system gradually decreases as smaller veins converge into progressively fewer but larger vessels, blood flow speeds up as blood approaches the heart. In addition to serving as low-resistance pathways to the returning blood from the tissues to the heart, systemic veins also serve as a blood reservoir. Because of their storage capacity, veins are often called capacitance vessels. Veins have thinner walls with less smooth muscle than arteries do. Also, in contrast to arteries, veins have little elasticity because venous connective tissue contains considerably more collagen fibers than elastin fibers and also venous smooth muscle has little inherent myogenic tone. Because of these features, veins are highly distensible, or stretchable, and have little elastic recoil. They easily distend to accommodate additional volumes of blood with only a small increase in venous pressure. Veins containing an extra volume of blood simply stretch to accommodate the additional blood without tending to recoil. In this way, veins serve as a blood reservoir. Under resting conditions, the veins contain more than 60% of the total blood volume. When the body is at rest and many of the capillary beds are closed, the capacity of the venous reservoir is increased as extra blood bypasses the closed capillaries and enters the veins. When this extra volume of blood stretches the veins, the blood moves forward through the veins more slowly because the total cross-sectional area of the veins has been increased as a result of the stretching. Therefore, the blood spends more time in the veins. As a result of this slower transit time through the veins, the veins essentially store the extra volume of blood because it is not moving forward as quickly to the heart to be pumped out again.

When the stored blood is needed, such as during exercise, extrinsic factors reduce the venous reservoir capacity and drive the extra blood from the veins to the heart so that it can be pumped to the tissues. Increased venous return leads to an increased stroke volume, in accordance with the Frank-Starling law of the heart. In contrast, if too much blood pools in the veins instead of being returned to the heart, cardiac output gets abnormally diminished.

The term venous return refers to the volume of blood per minute entering each atrium from the veins. By the time the blood enters the venous system, blood pressure averages only 17 mmHg. However, because atrial pressure is near 0 mmHg, a small but adequate driving pressure still exists to promote the flow of blood through the large-radius, low-resistance veins.

In addition to the driving pressure imparted by cardiac contraction, other factors also enhance venous return: sympathetically induced venous vasoconstriction, skeletal muscle pump, venous valves, respiratory pump, and cardiac suction. Additionally, low pressure in the veins, unidirectional blood flow by valves, higher intra-abdominal pressure, and tonicity of the skeletal muscles aid in the venous return.

Know More......

Venous Return Aided by Extrinsic Pumping Involving Skeletal and Connective Elements

Some mammals with hooves (Perissodactyls such as rhinoceros and horse) have a cartilaginous plate, the "frog" or cuneus ungulae, at the base of the hoof that is flexed upward with each footstep. This flexing pushes on an overlying elastic cushioning tissue that is compressed and moves outward, expanding nearby cartilage and hoof walls. It also compresses nearby veins, sending blood into the leg above. When the foot is lifted, elastic rebound of these tissues draws blood into the local veins. This pumping action is particularly important in horses, with their long, thin legs, because they do not have enough skeletal muscle mass below the knee to act as extrinsic muscle pumps.

6.6.3.1 Sympathetically Induced Venous Vasoconstriction

Veins are not very muscular and have little inherent tone, but venous smooth muscles are abundantly supplied with sympathetic nerve fibers. Hence, sympathetic stimulation produces venous vasoconstriction, which modestly elevates venous pressure; this, in turn, increases the pressure gradient to drive more of the stored blood from the veins into the right atrium, thus enhancing venous return. Since veins often have such a broad radius, the mild vasoconstriction caused by sympathetic activation barely affects flow resistance. Veins still have a significant radius and are low-resistance conduits even when they are restricted.

Arteriolar vasoconstriction immediately reduces flow through these vessels because of their increased resistance, whereas venous vasoconstriction immediately increases flow through these vessels and causes increased venous return. The increased venous return initiated by sympathetic stimulation leads to increased cardiac output because of the increase in end-diastolic volume. Sympathetic stimulation of the heart also increases cardiac output by increasing the heart rate and increasing the heart's contractility.

6.6.3.2 Skeletal Muscle Pump

Many large veins in the extremities are located between the skeletal muscles, so muscle contraction compresses the veins. This external venous compression reduces venous capacity and increases venous pressure, in effect squeezing blood in the veins forward toward the heart. This pumping action, known as the skeletal muscle pump, is another way in which extra blood stored in the veins is returned to the heart during exercise. The skeletal muscle pump also counters the effect of gravity on the venous system.

6.6.3.3 Venous Valves

Venous vasoconstriction and external venous compression both drive blood toward the heart, which is aided by the venous valves. Large veins are equipped with one-way valves spaced at 2–4 cm intervals; these valves let blood move forward toward the heart but keep it from moving back toward the tissues. These venous valves also play a role in counteracting the gravitational effects of upright posture by helping to minimize the backflow of blood that tends to occur when an animal stands up and by temporarily supporting portions of the column of blood when the skeletal muscles are relaxed.

6.6.3.4 Respiratory Pump

The pressure within the chest cavity averages 5 mmHg less than atmospheric pressure due to the respiratory activity. Blood is exposed to this subatmospheric pressure when it passes through the chest cavity on its way to the heart from the lower parts of the body. An externally applied pressure gradient exists between the lower veins (at atmospheric pressure) and the chest veins because the venous system in the limbs and abdomen is susceptible to normal atmospheric pressure (at less than atmospheric pressure). This pressure difference pushes blood from the lower veins to the chest veins, promoting increased venous return. This mechanism of facilitating venous return through respiratory activity is called the respiratory pump. Increased respiratory activity, skeletal muscle pump, and venous vasoconstriction, all enhance venous return during exercise.

Effect of Cardiac Suction on Venous Return During ventricular contraction, the AV valves are drawn downward, enlarging the atrial cavities. As a result, atrial pressure transiently drops below 0 mmHg, thus increasing the vein-to-atria pressure gradient so that venous return is enhanced. In addition, rapid expansion of the ventricular chambers during ventricular relaxation creates a transient negative pressure in the ventricles so that blood is "sucked in" from the atria and veins, that is, the negative ventricular pressure increases the vein-to-atria-to-ventricle pressure gradient, further enhancing venous return. Thus, the heart functions as a suction pump to facilitate cardiac filling.

6.6.3.5 Venous Pressure and Flow

The pressure in the venules is 12–18 mmHg. In the larger veins, it decreases gradually to roughly 5.5 mmHg in the major veins outside the thorax. The pressure in the great veins at their entrance into the right atrium (central venous pressure) averages 4.6 mmHg but fluctuates with respiration and heart action. Peripheral venous pressure, like arterial pressure, is affected by gravity. Thus, on a proportional basis, gravity has a greater effect on venous than on arterial pressures. When blood flows from the venules to the large veins, its average velocity increases, as the total cross-sectional area of the vessels decreases. In the great veins, the velocity of blood is about one-fourth that in the aorta, averaging about 10 cm/s.

6.6.4 Special Circulations

6.6.4.1 Cerebral Circulation

The brain receives 15% of the cardiac output. Brain tissues need adequate blood supply continuously, and stoppage of blood flow to the brain for 5 s leads to unconsciousness and for 5 min leads to irreparable damage to the brain cells in contrast to the other tissues, which can be deprived of a blood supply for extended periods and recover to normal function when blood supply resumes. The tolerance of an adult brain to hypoxia is much lower than the tolerance of a newborn brain.

Anatomic considerations: The main blood supply to the brain comes from the paired internal carotid artery in horses and cats and from maxillary artery in the ruminants. The internal carotid artery in the horse and dog and the cerebral carotid artery in the other domestic mammals penetrate the dura mater and form a vascular ring located ventral to the hypothalamus called as cerebral arterial circle or circulus arteriosus cerebri or circle of Willis.

In dogs, where the circle of Willis is complete, blood flow to the brain is maintained even in the face of an obstruction at any single point in the vascular circle. In ruminants and cats, the internal carotid arteries connect with other vessels in the head before entering the arterial circle. These vascular connections form a complex network of vessels, the rete mirabile. The rete mirabile facilitates maintenance of a constant brain temperature in the face of either high or low temperature stress.

Blood–brain barrier: One of the most characteristic features of the brain vasculature is the blood–brain barrier, which prevents the solutes in the lumen of capillaries from having direct access to the brain extracellular fluid as these capillaries have very limited permeability. The blood–brain barrier consists mainly of tight junctions that seal together the endothelial cells of brain capillaries, along with a thick basement membrane around the capillaries. The astrocytes' processes press up on the capillaries and produce substances that keep the tight junctions' permeability properties constant. Lipid-soluble substances, such as oxygen, carbon dioxide, alcohol, and most anesthetic agents, can readily diffuse across the cerebral capillaries. Glucose crosses more slowly via facilitated diffusion. Polar and water-soluble compounds cross the blood–brain barrier slowly, and the ability of proteins to cross the barrier is extremely limited. However, trauma, tumors, certain toxins, and inflammation can cause a breakdown of the blood–brain barrier. In specialized areas of the brain, i.e., the circumventricular organs including median eminence, pineal gland, area postrema, and posterior pituitary, the capillaries are fenestrated and have permeability characteristics similar to those of capillaries in the intestinal circulation.

6.6.4.1.1 Blood Flow Dynamics and Regulatory Mechanisms

Local control mechanisms: As the brain is extremely intolerant to ischemia; cerebral blood flow is regulated primarily by local mechanisms, particularly in response to changing levels of local metabolism as well as systemic alterations in blood gases. Neural activity and increased metabolism result in reductions in ATP accompanied by adenosine release as well as decreases in tissue PO_2, increased PCO_2, and decreased pH contributing to cerebral vasodilation.

Autonomic vascular regulatory mechanisms play only a secondary and comparatively minor role. Coupling of flow to metabolism in the brain involves a unique interaction among neural activity, cerebral vasculature, and action of astrocytes. Astrocytes contact neurons, and their endfeet form a discontinuous sheath around capillaries. Neurotransmitters such as glutamate activate astrocytes, ultimately producing vasodilation due to direct relaxation of vascular smooth muscle as well as release of vasodilators such as NO and metabolites of arachidonic acid.

Hypoxia-stimulated cerebral vasodilation: Cerebral blood flow increases in response to hypoxia, due to increased concentrations or release of a wide variety of vasodilators, including adenosine, potassium and hydrogen ions, prostaglandins, excitatory amino acids, and NO. Additionally, hypoxia also has direct effects on cerebrovascular myocytes, including modest reductions in ATP, activation of the KATP, and other potassium channels which cause cerebral vasodilation. The relative importance of the many mechanisms leading to cerebral vasodilation depends significantly on the size of the vessel.

Hypercapnia-stimulated cerebral vasodilation: Carbon dioxide is one of the most potent dilators of cerebral vessels, and cerebral resistance vessels are extremely sensitive to even minor elevations in arterial pCO_2. For example, breathing 7% CO_2 is capable of doubling cerebral flow. NO plays a role in hypercapnia-induced cerebral vasodilation, although it may be permissive and modulatory rather than primary in nature. CO_2-induced changes in cerebral flow appear to be related primarily to alterations of pH in extracellular fluid within the brain.

Neural control factors: Cerebral blood vessels are innervated primarily by the sympathetic nervous system, which mediates vasoconstriction due to release of norepinephrine and also neuropeptide Y (NPY). Sympathetic control of cerebral blood vessels is relatively weak due to low density of adrenergic receptor in cerebral blood vessels. Parasympathetic nervous system has a minor effect on blood flow. The distal cerebral vessels are innervated by sensory nerves containing substance P and calcitonin gene-related peptide (CGRP), which are potent vasodilators. Local perturbations result in their reflex release from perivascular nerves, through activating ATP-sensitive potassium channels, thereby producing hyperpolarization and relaxation of cerebral arteries.

Brain ischemia elicits the Cushing reflex, which involves a large increase in sympathetic nerve activity, peripheral vasoconstriction, and arterial blood pressure, presumably in an attempt to increase cerebral flow.

Autoregulation: Autoregulation is a significant feature of the cerebral circulation. The range of perfusion pressures over which autoregulation occurs is not fixed, and the upper limit of autoregulation is promoted by increased activity of cervical sympathetic nerves. Cerebral flow autoregulation during arterial hypotension is achieved by vasodilation, and the autoregulation ability is diminished by hypercapnia. Calcitonin gene-related peptide, adenosine, endothelium-dependent hyperpolarizing factor, vasoactive intestinal peptide (VIP), cyclic AMP, prostacyclin, and norepinephrine-stimulated β-adrenoceptors also vasodilate cerebral vessels, via activation of KATP channels. Despite the importance of metabolic factors, none of these can completely account for autoregulation of cerebral blood flow. Thus, myogenic mechanisms contribute importantly to cerebral autoregulation. Resistance vessels in the cerebral circulation respond robustly to increases in pressure with vasoconstriction and to decreases in pressure with relaxation.

6.6.4.2 Coronary Circulation

6.6.4.2.1 Anatomical Considerations

Cardiac muscles are supplied by two coronary arteries, viz. right and left coronary arteries, which arise from the aorta at the level of the sinus of Valsalva (aortic sinus), thus comprising the first branches of aorta. As these arteries encircle the heart in the pattern of a crown, the coronary arteries derive their name from the Latin word corona meaning crown.

The main coronary arteries lie on the heart surface, and smaller arteries penetrate from the surface into the cardiac muscle mass. The myocardium receives its nutritive blood supply almost entirely through these arteries, and the inner 1/10 mm of the endocardial surface receives nutrition directly from the blood inside the cardiac chambers, which contributes only to a minuscule amount. The coronary circulation outlay is as shown in Fig. 6.5.

6.6.4.2.1.1 Branches of Coronary Arteries

Coronary arteries divide and subdivide into smaller branches, which run all along the heart surface. Smaller branches are called epicardiac arteries and give rise to further smaller branches known as final arteries or intramural vessels. Final arteries run at the right of the wall of the heart.

The right coronary artery supplies the right atrium and ventricle, while the left coronary artery supplies primarily the left atrium and ventricle and interventricular septum, although there can be overlap. The majority of cardiac veins drains into the right atrium through the coronary sinus.

6.6.4.2.1.2 Venous Drainage

Venous drainage from the heart muscle is by four types of vessels, coronary sinus, anterior coronary veins, arteriosinusoidal vessels, and Thebesian veins.

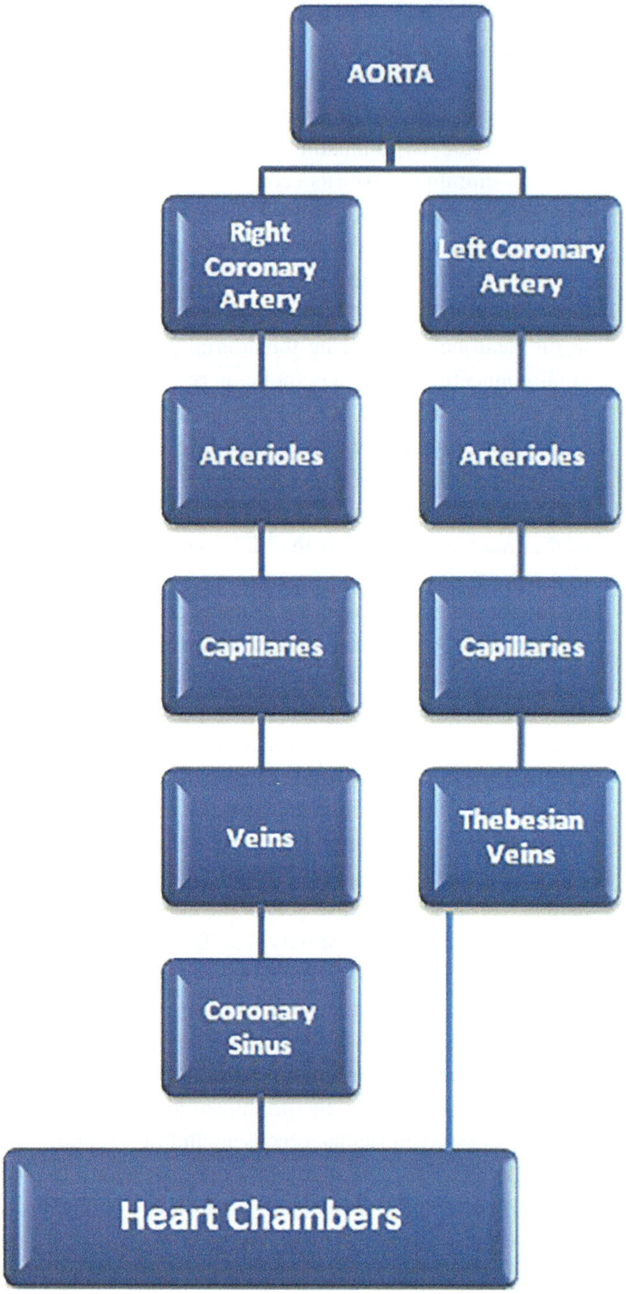

Fig. 6.5 Coronary circulation showing the blood flow in the coronary vessels and back to the heart chambers

Most of the venous blood is returned to the heart through the coronary sinus and anterior cardiac veins, which drain into the right atrium.

Coronary sinus is the larger vein draining 75% of the total coronary flow. It drains blood from the left side of the heart and opens into right atrium near the tricuspid valve.

Anterior coronary veins drain blood from the right side of the heart and open directly into the right atrium. In addition, there are other vessels that empty directly into the heart chambers.

Arteriosinusoidal vessels are sinusoidal capillary-like vessels that connect arterioles to the chambers.

Thebesian veins connect capillaries to the chambers and drain deoxygenated blood from myocardium, directly into the concerned chamber of the heart.

A few arterioluminal vessels which are small arteries drain directly into the chambers. A few anastomoses occur between the coronary arterioles and extracardiac arterioles, especially around the mouths of the great veins. Anastomoses between coronary arterioles in humans only pass particles less than 40 m in diameter, but these channels enlarge and increase in number in patients with coronary artery disease.

6.6.4.2.2 Species Variation

Cattle—Coronary sinus blood consists of a mixture of venous blood from the heart and from the azygous vein, which carries venous blood from noncardiac tissues and empties into the great cardiac vein.

Dog—More than 80% of coronary sinus blood arises entirely from the left ventricle, with the remaining 20% originating from other parts of the heart. Consequently, the collection and analysis of coronary sinus blood have been a valuable tool for the study of left ventricular metabolism. However, caution must be used in extending this technique to other species because of anatomic differences.

Venous coronary blood from the right ventricle of the dog heart predominantly drains into the right atrium through the right cardiac vein. Only a very small amount of venous blood drains from myocardial tissue directly into the cardiac chambers, primarily into the right atrium and ventricle, through the very small Thebesian vein.

6.6.4.2.3 Collateral Coronary Arteries

The coronary arteries lack collateral arteries in most species, i.e., vessels connecting major arteries without an intervening capillary bed are absent. However, there are species, within-species, and breed variation.

Coronary collateral arteries in the dog heart are typically thin-walled vessels of very small diameter and therefore conduct only a very low volume of collateral coronary blood flow. However, collateral coronary arteries are absent or so small in size or number that collateral blood flow is inadequate to prevent infarction following sudden coronary occlusion.

The purebred Beagle dog tends to have an extensive collateral coronary circulation. Similarly, the guinea pig has abundant collateral arteries, so that tissue perfusion is maintained following abrupt occlusion of a coronary artery, preventing myocardial infarction (an area of necrosis resulting from inadequate coronary arterial flow) or tissue damage. Non-canine species vary widely in their ability to develop coronary collateral vessels and in the location and nature of these collateral vessels. In humans and pig, collateral vessels arise as microvascular connections between capillary beds of an occluded and a nonoccluded artery and are located within the myocardium or in the subendocardium. The development of collateral vessels in porcine and human hearts occurs through a process of angiogenesis or the sprouting of new capillaries from preexisting capillaries. Capillaries develop in response to factors such as vascular endothelial growth factor (VEGF), which is stimulated by hypoxia, in part via hypoxia-mediated adenosine release.

For example, most of the mammals can develop collateral coronary arteries if a major coronary artery is occluded slowly and gradually, as occurs in atherosclerosis even though they lack significant preexisting coronary collateral circulation. Initially, this newly developed collateral circulation has little reserve capacity or ability to increase flow in response to an increase in myocardial oxygen demand. However, collateral development and expansion continue well beyond the time of occlusion, so collateral flow capacity 6 months after occlusion may approach that of the coronary artery prior to occlusion.

Collateral coronary arteries in the dog heart develop through a process of enlargement and expansion of preexisting epicardial arterioles (arteriogenesis). Arteriogenesis is stimulated by shear stress due to increased flow velocity proximal to the occlusion and involves participation of adhesion molecules and growth factors. The process of collateral development by arteriogenesis is not significantly influenced by the presence or absence of physical exercise.

6.6.4.2.4 Control of Coronary Blood Flow

Normal blood flow through coronary circulation is about 200 mL/min. It is about 65–70 mL/min/100 g of cardiac muscle or about 225 mL/min, which is about 4–5% of the total cardiac output.

6.6.4.2.4.1 Basal Tone

The heart exhibits a high level of oxygen extraction; yet, coronary blood flow in most mammals is relatively high at rest (around 100 mL/min per 100 g of tissue). While α-adrenergic stimulation can produce coronary vasoconstriction, the importance of neural mechanisms in establishing or maintaining basal coronary tone is minimal as compared to tissues such as skin and gastrointestinal vascular beds.

6.6.4.2.5 Regulatory Mechanisms

6.6.4.2.5.1 Autoregulation

Heart also has the capacity to regulate its own blood flow by autoregulation like any other organ. Coronary blood flow is not affected when mean arterial pressure varies between 60 and 150 mmHg. Coronary blood flow is regulated mainly by local vascular response to the needs of cardiac muscle.

That is, whenever the vigor of cardiac contraction is increased, the rate of coronary blood flow also increases. Conversely, decreased heart activity is accompanied by decreased coronary flow. The factors regulating coronary blood flow are:

1. Oxygen demand
2. Metabolic factors
3. Coronary perfusion pressure
4. Nervous factors

1. ***Oxygen demand***: Blood flow in the coronary arteries usually is regulated almost exactly in proportion to the oxygen need by the cardiac musculature. Normally, about 70% of the oxygen in the coronary arterial blood is removed as the blood flows through the heart muscle. Because not much oxygen is left, very little additional oxygen can be supplied to the heart musculature unless the coronary blood flow increases. Fortunately, the coronary blood flow does increase almost in direct proportion to any additional metabolic consumption of oxygen by the heart. Muscle cells release vasodilator molecules when the oxygen content in the heart drops, and these compounds widen the arterioles. Amount of blood passing through coronary circulation is directly proportional to the consumption of oxygen by cardiac muscle. Even in resting condition, a large amount of oxygen, i.e., 70–80%, is consumed from the blood by heart muscle than by any other tissues. In conditions associated with increased cardiac activity, the need for oxygen increases enormously. Thus, the need for oxygen, i.e., hypoxia, immediately causes coronary vasodilatation and increases the blood flow to heart. Asphyxia and hypoxia increase coronary blood flow 200–300% in denervated as well as intact hearts, and the feature common to these three stimuli is hypoxia of the myocardial fibers. Oxygen is the most important factor maintaining blood flow through the coronary blood vessels.
2. ***Metabolic factors***: Coronary vasodilatation during hypoxic conditions occurs because of some metabolic products, which increase the coronary blood flow by vasodilatation.

   Reactive Hyperemia

   Reactive hyperemia is the increase in blood flow due to the vasodilator effects of metabolites.

   Metabolic products increase the coronary blood flow.

   Adenosine is a potent vasodilator, and it increases the blood flow to cardiac muscle. In the presence of very low concentrations of oxygen in the muscle cells, a large proportion of the cell's ATP degrades to adenosine monophosphate; then small portions of this are further degraded and release adenosine into the tissue fluids of the heart muscle, with resultant increase in local coronary blood flow. After the adenosine causes vasodilation, much of it is reabsorbed into the cardiac cells to be reused.

   Other substances which increase the coronary blood flow by vasodilatation are potassium, hydrogen, carbon dioxide, lactate, and prostaglandins.
3. ***Coronary perfusion pressure***: Perfusion pressure is the balance between mean arterial pressure and venous pressure. Thus, coronary perfusion pressure is the balance between mean arterial pressure in aorta and right atrial pressure. Since right arterial pressure is low, the mean arterial pressure becomes the major factor that maintains the coronary blood flow.
4. ***Nervous control of coronary blood flow***: The coronary circulation is relatively independent of central neural regulation; however; coronary blood vessels are innervated by both sympathetic and parasympathetic nervous system. Sympathetic stimulation, by releasing norepinephrine and epinephrine, increases both heart rate and heart contractility and increases the rate of metabolism of the heart. In turn, increased metabolism of the heart sets off local blood flow regulatory mechanisms for dilating the coronary vessels, and the blood flow increases approximately in proportion to the metabolic needs of the heart muscle. In contrast, vagal stimulation, with its release of acetylcholine, slows the heart and has a slight depressive effect on heart contractility. These effects in turn decrease cardiac oxygen consumption and, therefore, indirectly constrict the coronary arteries.

When the systemic blood pressure falls, the overall effect of the reflex increase in noradrenergic discharge causes increased coronary blood flow secondary to the metabolic changes in the myocardium at a time when the cutaneous, renal, and splanchnic vessels are constricted. In this way, the circulation of the heart, like that of the brain, is preserved when flow to other organs is compromised. Thus, sympathetic nervous system stimulation generally produces a mild α-receptor-mediated vasoconstriction in the coronary circulation. Parasympathetic effects on the coronary vasculature are minor, and vagal activation primarily has an effect to decrease heart rate.

An unusual feature of the coronary circulation is the effect of mechanical compression of the blood vessels during systole in the cardiac cycle. This compression causes a brief period of occlusion and reduction of blood flow. When the period of occlusion (i.e., systole) is over, reactive hyperemia occurs to increase blood flow and oxygen delivery and to repay the oxygen debt that was incurred during the compression.

6.6.4.3 Splanchnic Circulation

6.6.4.3.1 Anatomic Considerations

The splanchnic circulation includes the vascular beds of the gastrointestinal tract, spleen, pancreas, and liver. The splanchnic circulation contains approximately 15% of the total blood volume, with the majority contained in the liver. Blood flow to the splanchnic bed has two primary functions. It is important both in supplying oxygen and nutrients to the tissues and in supporting the absorption of substances from the gastrointestinal tract. This circulation is unique in that it contains two capillary beds in series; venous flow from the capillary beds of the gastrointestinal tract, spleen, and pancreas combines to provide a major source of blood flow, through the portal vein, to the liver capillaries.

6.6.4.3.2 Intestinal Circulation

The celiac artery is the primary blood supply to the stomach, pancreas, and spleen. The superior and inferior mesenteric arteries supply the large and small intestines, as well as parts of the stomach and pancreas. The superior mesenteric artery is the largest of all the splanchnic branches from the aorta, carrying >10% of the cardiac output. Extensive interconnections between the arterial branches provide multiple collateral pathways through which blood can reach each portion of the intestines. Small arteries course through the various muscle layers and reach the submucosa as arterioles, and then some enter into the intestinal villi. The incoming arteriole courses up the center of the villus, branching into many capillaries along the way to the tip of the villus. Capillaries converge into venules and carry blood back to the base of the villus. Capillaries also interconnect the arteriole and the venule all along the villus. These microvessels of villi are highly permeable to solutes of low molecular weight, thereby facilitating the absorption of nutrients, and this countercurrent exchange system enables permeable solutes to move from the arteriole to the venule without having to traverse the entire length of the villus, particularly when blood flow to the villi is low.

Because the capillaries in the villi are fenestrated and have a large surface area, they are well suited for absorbing nutrients from the intestinal lumen. The venous blood carries away the majority of water-soluble nutrients absorbed from the gut, eventually delivering them to the portal vein.

Intestinal blood flow is directly linked to metabolic activity. Flow is low in unfed animals but can increase as much as eightfold, following food ingestion (postprandial hyperemia). Throughout the gastrointestinal tract, blood flow in each layer of the gut wall closely correlates with the local metabolic activity occurring during digestion and absorption. Intestinal blood flow at rest, in the fasting state, is typically 30 mL/min for each 100 g of tissue. However, flow can reach 250 mL/min for each 100 g during peak hyperemia after a meal. These activities consume O_2 and produce vasodilator metabolites (e.g., adenosine and CO_2) that increase blood flow locally. Digestive activity increases metabolism, increasing concentrations of local vasodilator mediators such as adenosine and CO_2 and, consequently, augmenting blood flow. Further, the absorption of nutrients generates hyperosmolality in both the blood and lymphatic vessels of the villus, which stimulates an increase in blood flow. Also, during digestion, the gastrointestinal tract releases several hormones, such as cholecystokinin and neurotensin, that promote intestinal blood flow. The intestinal epithelium also releases various kinins (e.g., bradykinin and kallidin), which are powerful vasodilators. In the case of ruminants, various products of rumen fermentation, such as CO_2 and fatty acids, also act as vasodilators of the rumen circulation. Mechanical stimulation of the small intestine of the cat increases blood flow through the release of serotonin from enterochromaffin cells. Serotonin, in turn, stimulates the release of the vasodilator VIP from nerve endings. Additionally, under specific circumstances, especially during inflammation or noxious stimulation of the gut lumen, a wide variety of other chemical mediators such as substance P and CGRP released from neurons cause vasodilation.

6.6.4.3.3 Neural Control Factors

Reciprocal changes in the autonomic nervous system also contribute to postprandial hyperemia. Feeding and digestion are associated with increased parasympathetic nervous system activity, leading to increased secretory activity, motility, and metabolism in gastrointestinal tissues and increased release of acetylcholine, which indirectly produces vasodilation. Concomitantly, there is diminished sympathetic nerve activity, thereby reducing α-adrenoceptor-mediated vasoconstriction. Splanchnic blood flow is autoregulated despite fluctuations in arterial blood pressure by mediators such as adenosine, K^+, and increased osmolality and via myogenic response.

6.6.4.3.4 Hepatic Circulation

The total blood supply and blood volume contained in the liver constitute approximately 25% of cardiac output, and the liver accounts for 20% of total body oxygen consumption. The hepatic circulation is characterized by a dual blood supply from the hepatic artery and the portal vein, which supplies approximately 75% of its flow. Another unique aspect of the hepatic circulation is that the capillary network in the liver is composed of sinusoids, which converge to form hepatic venules. The sinusoids are highly fenestrated and hence very permeable, allowing rapid exchange between the blood and hepatocytes. However, they are also very sensitive to changes in central venous pressure that alter capillary pressure and thus produce large alterations in fluid exchange through the sinusoids. For example, the increase in central venous pressure that occurs in right-sided heart failure produces substantial movement of fluid across the sinusoids into the peritoneal cavity, resulting in ascites.

6.6.4.3.4.1 Autoregulation

Hepatic artery blood flow is autoregulated mainly by adenosine in an effort to maintain total hepatic perfusion at a constant rate. The liver is unique in that it responds to increased metabolic demand through increased oxygen extraction rather than increased blood flow.

6.6.4.3.5 Functional Significance of Splanchnic Circulation

The splanchnic circulation serves as an important blood reservoir and site of vascular resistance, thus contributing to cardiovascular homeostasis. The splanchnic vascular bed of the dog normally contains more than 20% of the blood volume, particularly on the venous side of the circulation. Sympathetic α-adrenoceptor stimulation significantly reduces venous capacitance, without a change in hepatic arterial blood flow, and may mobilize up to half of this volume in response to severe hypoxia, heavy exercise, or hemorrhage. The spleen, particularly in species such as the dog, horse, sheep, cat, and guinea pig, is an important component of this response and transfers red cell-rich blood to the central circulation. Splenic contraction associated with heavy exercise in the dog, horse, and sheep increases the hemoglobin concentration of circulating blood by 20–50%, with a corresponding increase in oxygen-carrying capacity. Sympathetic vasoconstriction of the splanchnic bed also contributes significantly to circulatory adjustments to exercise and hemorrhage by redistributing the flow to other vascular beds such as brain and cardiac and skeletal muscle.

6.6.4.4 Skeletal Muscle Circulation

Skeletal muscles receive 15% of the cardiac output at rest. Skeletal muscles are well supplied with blood vessels, and an artery and one or two veins supply a skeletal muscle. Capillaries are plentiful in muscular tissue. Blood vessels within skeletal muscle receive both sympathetic adrenergic and sympathetic cholinergic innervation. The cholinergic system, acting directly via muscarinic receptors, relaxes vascular smooth muscle cells and causes rapid vasodilation. This vasodilation in skeletal muscle occurs in the fight-or-flight response and during the anticipatory response in exercise caused by extensive activation of the sympathetic division. Blood flow to skeletal muscle varies with muscle fiber type. For example, slow-twitch high oxidative muscle exhibits greater flow and capillary density compared with fast-twitch glycolytic muscle in conscious mammals. Blood flow is low at rest in most muscles, although it is relatively high in postural muscles of animals maintaining posture (standing). An important aspect of the skeletal muscle circulation is that the volume of blood flow to skeletal muscle is intimately linked to the level of muscle metabolic activity. Muscle blood flow can increase by 100-fold during strenuous exercise.

Autoregulation is a characteristic feature of the skeletal muscle circulation at rest and during exercise that appears to depend on several mechanisms, with the significance of each mechanism varying from rest to exercise and with the level of exercise. For example, the myogenic response appears to play an important role in autoregulation in the skeletal muscle circulation at rest, while metabolic factors play a greater role during exercise.

6.6.4.5 Pulmonary Circulation

The pulmonary circulation is a relatively short, low-resistance, low-pressure system, which conducts blood to and from a single but very dense capillary bed enveloping the pulmonary alveoli. It consists of the right ventricle, pulmonary arteries, pulmonary capillaries, pulmonary veins, and left atrium. Because the pulmonary vessels are very distensible, they serve not only as a channel but also as a reservoir between the right and left ventricles. The pulmonary vasculature is unique in that it accommodates a blood flow that is almost equal to that of all the other organs in the body. The pulmonary capillaries are large, and there are multiple anastomoses, so that each alveolus sits in a capillary basket.

6.6.4.5.1 Anatomical Consideration

The special considerations in the pulmonary circulation are (1) the position within the negative but rhythmically variable pressure of the thorax; (2) the position between the right ventricle and the left atrium; and (3) the relatively great distensibility and collapsibility of the vessels. There are species differences in the morphology of the pulmonary vessels that are extremely important in the response to the inspired hypoxia found at altitude.

The mean pressure in the pulmonary artery is approximately one-sixth that of the systemic arteries. This low-pressure system has two major advantages: (1) it minimizes the work of the right heart and (2) allows a very thin blood–gas barrier suitable for high rates of gas exchange. The pressures vary with age as well as with species. The short lengths and large radii of the pulmonary arteries yield a small value for resistance, approximately one-fifth to one-tenth of the resistance to flow observed in the systemic arteries.

Approximately 9% of total blood volume is contained in the pulmonary vessels. It is distributed about equally among the arteries, capillaries, and veins. Increases in pulmonary blood volume of 25–50% may occur when the pulmonary circulation serves a reservoir function to accommodate increases in total blood volume or when extensive systemic arterial and venous constriction causes a shift of blood volume from the systemic to the pulmonary circuit.

Capillary pressures within the lung are low compared with the systemic circulation. Capillary hydrostatic pressure is influenced by gravity, being lower at the apex and higher at the base (mean value of 9 mmHg). Because of the prevailing negative intrathoracic pressures, generally the lung interstitial fluid pressure is subatmospheric (about −12 mmHg).

In the lung capillaries, the net outward forces slightly exceed the inward force, causing a very small excess of fluid to filter out of the capillaries. This is returned to the circulation through the lymphatics. The exceptionally rich lymphatic network prevents pulmonary edema until capillary hydrostatic pressure exceeds 25 mmHg. The negative interstitial fluid pressure favors passage of fluid across the alveolar membranes into the interstitial fluid spaces, thus preventing accumulation of fluid in the alveoli. The lymphatic vessels, which are contractile, serve as skimming pumps for maintaining the extravascular fluid volume. The alveolar epithelial membrane keeps fluid from entering the alveolar gas spaces. The alveolar epithelium is less permeable than the capillary endothelium, and therefore fluid does not leak into the alveoli unless the epithelium is damaged.

6.6.4.5.2 Regulatory Mechanisms

6.6.4.5.2.1 Autoregulation

Autoregulation is well developed in the pulmonary vascular system and helps match blood flow to ventilation. The major vasomotor activity of resistance vessels depends on local autoregulatory mechanisms rather than central neural reflex systems. The action of alveolar hypoxia is limited to very short segments of arteries/arterioles less than 200 μm in diameter that are immediately adjacent to the alveolus. Carbon dioxide acts on longer segments. It is the effect that PCO_2 has on the local hydrogen ion concentration rather than the PCO_2 per se, which stimulates vasoconstriction. The site of action is important in that it renders autoregulation effective down to the level of the alveoli.

> Know More……
>
> **Left–Right Difference in Ventricle Pumping Force**
>
> The pulmonary circulation is a low-pressure, low-resistance system, whereas the systemic circulation is a high-pressure, high-resistance system. Therefore, even though the right and left sides of the heart pump the same amount of blood, the left side performs more work because it pumps an equal volume of blood at a higher pressure into a higher resistance system.

6.6.4.5.2.2 Vasoactive Substances

A wide variety of vasoactive substances are synthesized, stored, or activated by cells of the lung. Histamine, widely distributed in mast cells, is a powerful but rapidly inactivated pulmonary vasoconstrictor, although it has a vasodilator action on the neonatal bovine circulation. Serotonin (5-hydroxytryptamine) is found in mast cells and blood platelets and is also a potent vasoconstrictor. Angiotensin II, formed from angiotensin I by a lung-converting enzyme, is also a vasoconstrictor. Bradykinin, a vasodilator, is both generated and destroyed in the lungs. The prostaglandin vasodilators PGE1 and PGE2 are synthesized and stored in the lungs, but the vasoconstrictor PGF2α is most abundant in lung parenchyma. Nitric oxide (NO) is a vascular smooth muscle-relaxing factor produced by the action of nitric oxide synthase within vascular endothelial cells.

> Know More……
>
> **Story of Nitric Oxide Leading to the Nobel Prize**
>
> Ferid Murad, an American physician and pharmacologist, studied how nitroglycerin activated an enzyme that formed cyclic guanosine monophosphate (cGMP), which in turn caused blood vessels to expand, and in 1976, he was able to show that nitroglycerin produced this effect by emitting nitric oxide (NO). The discovery represented a new principle for transferring signals between cells; a gas as a signal-transferring molecule had never been observed before. Robert Furchgott, Louis Ignarro, and Ferid Murad jointly received the Nobel Prize in 1998 for their discovery that nitric oxide signals blood vessels to dilate.

6.6.4.5.2.3 Vasomotor Nerves

The small pulmonary vessels have muscular coats and dual nerve supply, sympathetic and parasympathetic; however, the predominant supply is via adrenergic sympathetic vasoconstriction. Stimulation of the sympathetic pulmonary nerves increases the pulmonary vascular resistance and hence pulmonary blood pressure. Stimulation of baroreceptors results in increased pulmonary blood flow and decreased pulmonary arterial pressure. The capacious pulmonary vessels constitute one of the body's blood reservoirs, and the vasoconstrictor fibers function more in the reflex mobilization of blood (e.g., in hemorrhage) than in pressor or depressor pulmonary responses. Small doses of epinephrine produce either minimal vasoconstriction or vasodilatation. Larger doses produce definite vasoconstriction.

6.6.4.5.2.4 High Altitude

Unlike most other blood vessels, the vessels of the lungs constrict in response to hypoxia. At high altitudes, airway or ventilatory hypoxia occurs in the absence of any hypercapnic acidosis or CO_2 retention. Ventilatory hypoxia elicits pulmonary arterial vasoconstriction and consequently an elevated pulmonary arterial pressure. Alveolar hypoxia, but not pulmonary arterial hypoxemia, causes this response. The hypertension is reversible when the hypoxia is relieved. The magnitude of the pressor response varies among species, being most pronounced in those (such as calves and pigs) having the most vascular smooth muscle. A local hypoxia (e.g., as occurs when a bronchiole is partially or wholly occluded, decreasing or preventing ventilation to the dependent lung region) is helpful in regulating the distribution of blood flow by causing vasoconstriction, which would divert blood from the anoxic region to vessels in better aerated parts of the lung.

6.6.4.5.3 Clinical Correlations

Emphysema (heaves) in horses: Chronic obstructive lung diseases, such as chronic obstructive pulmonary disease (COPD), chronic emphysema, and bronchitis, are characterized by increased hindrance to airflow out of the lungs. The primary lung disease results in destruction of vessels and thus a reduction in the overall radius of the pulmonary vascular system. Increased vascular resistance, pulmonary hypertension, and an apparent decrement in the density of capillaries occur in emphysematous horses. In emphysema, airway hypoxia and hypercapnia result from ventilation–perfusion mismatch that favors pulmonary vasoconstriction.

Heartworm disease in dogs: Pulmonary hypertension is characteristic of heartworm disease in dogs. Right-heart failure is common in dogs infested with *Dirofilaria immitis*. Mature worms may be found in the venae cavae, lumens of the right atrium and ventricle, and orifices of the tricuspid and pulmonary semilunar valves. They obstruct flow into and out of the right side of the heart and hinder normal function of cardiac valves. Pulmonary vascular resistance is increased due to obstruction, narrowing, or closure of pulmonary vascular pathways. In large vessels, mature worms occlude their lumens. In smaller vessels, mechanical obstruction is caused by thrombi containing fragments of disintegrating parasites, by emboli, and by fibroplasia involving the walls of arteries.

Exercise-induced pulmonary hemorrhage: Exercise-induced pulmonary hemorrhage (EIPH) occurs in racehorses during sprint racing and is characterized by pulmonary hypertension, edema in the gas-exchange region of the lung, rupture of the pulmonary capillaries, intra-alveolar hemorrhage, and presence of blood in the airways. The enormous cardiac output demanded by the racehorse associated with maximal recruitment and distension of the pulmonary capillaries also contributes to hypertension. During maximal exercise, the pulmonary arterial pressures may exceed 120 mmHg in thoroughbred, and it is observed that the exercise-induced pulmonary hemorrhage occurs above a mean pulmonary artery pressure of about 90 mmHg.

6.6.4.6 Cutaneous Circulation

The major functions of the skin include thermoregulation, storage of blood, protection, cutaneous sensations, excretion and absorption, and synthesis of vitamin D. The dermis has an extensive network of blood vessels that carry 8–10% of the total blood flow in resting state, thus acting as a blood reservoir. In skin, capillaries reach only as superficially as the dermis; the epidermis does not have a blood supply. The venules that are part of a plexus of vessels near the dermal-epidermal border contain an appreciable volume of blood. Total blood flow through the cutaneous circulation is composed of both nutritional flow perfusing the capillary beds of the skin and flow directed through arteriovenous anastomoses (AVAs), which shunt flow between arteries and veins. The skin is normally overperfused in relation to its nutritional requirements. Thus, local metabolic control of skin blood flow is of little functional importance. Local vasodilator metabolites, sympathetic neuronal regulation mediated by $\alpha 1$ adrenoceptors, and sensory cues (e.g., temperature, touch, pain), all work together to regulate local nutrient flow through the precapillary sphincters and capillaries.

The skin is classified into apical and nonapical skin based on the blood flow. The apical skin at the extremities of the body has a very high surface-to-volume ratio that favors heat loss. Circulation to these apical regions has an unusual feature of arteriovenous anastomosis (AV) called glomus bodies. Glomus bodies of the skin are tiny nodules found in many parts of the body, including the ears, the pads of the fingers and toes, and the nail beds.

The AV anastomoses, which are involved in heat exchange, are in parallel with the capillaries of the skin, which are involved in nutrient exchange. The anastomotic vessels are under neural control, rather than the control of local metabolites. In these apical regions, blood flow is under the control of sympathetic fibers that release norepinephrine and thereby constrict the arterioles, anastomotic vessels, and venules. Therefore, the increase in sympathetic tone that occurs in response to decreases in core temperature elicits vasoconstriction in the AV anastomoses mediated by both $\alpha 1$ and $\alpha 2$ adrenoceptors, a fall in blood flow, and a reduction in heat loss. On the other hand, when the core temperature rises, the withdrawal of sympathetic tone leads to passive vasodilation; there is no active vasodilation. Thus, sympathetic tone to the vasculature of apical skin is substantial at rest under cool environments to minimize heat loss.

6.6.4.6.1 Nonapical Skin

The nonapical skin almost completely lacks AV anastomoses, and there are two types of sympathetic neurons innervating the vessels of the skin which release norepinephrine and acetylcholine. Vasoconstriction occurs in response to the release of norepinephrine. Vasodilation in nonapical skin occurs in response to sympathetic neurons that release acetylcholine. The acetylcholine stimulates eccrine sweat glands to release kallikrein, a protease that converts kininogens to kinins, and these kinins act in a paracrine fashion on nearby blood vessels to relax the vascular smooth muscle cells. Cholinergic sympathetic neurons may cause vasodilation by means of a second pathway involving the co-release of vasodilatory neurotransmitters (e.g., calcitonin gene-related peptide, vasoactive intestinal peptide) that act directly on vascular smooth muscle cells (VSMCs), independently of sweat gland activity. In addition to sympathetic nerve fibers, mammalian skin contains small-diameter unmyelinated and thinly myelinated nerve fibers with nociceptive receptors that can contribute to vasodilation in response to pain or injurious stimuli. Activation of these receptors markedly increases CGRP release, which contributes to local vasodilation. In addition to neurally mediated effects, the cutaneous circulation dilates or constricts in response to local heating or cooling, respectively.

6.6.4.7 Placental and Fetal Circulation

6.6.4.7.1 Circulation in Fetus

The placenta has a low vascular resistance and receives 45% of the cardiac output through the umbilical arteries in the umbilical cord to the placenta, which serves as the "fetal lung." The umbilical veins drain the placenta toward the liver. The fetal circulation is capable of considerable regulation, particularly as the fetus matures. Fetal hypoxia can stimulate vasoconstriction in the skeletal tissues, gut, kidneys, and vasodilation in the heart and brain. If the fetus is hypoxic, there is severe constriction in the fetal pulmonary circulation to divert more blood through the ductus arteriosus to the systemic tissues.

The fetal circulation has three shunts; two of these shunts, the foramen ovale and the ductus arteriosus, cause the fetal right and left ventricles to operate as parallel pumps rather than pumps in series as in the adult.

The third shunt in the fetal circulation is the ductus venosus, a low-resistance channel that allows a significant fraction of relatively oxygenated blood in the umbilical vein to bypass the fetal liver and directly enter the caudal vena cava.

In species such as sheep, a low-resistance pathway, the foramen ovale, connects the right and left atria, and a structure known as the crista dividens directs the better oxygenated blood from the posterior vena cava through the foramen ovale to the left atrium. The poorly oxygenated blood returning to the right atrium in the cranial vena cava is directed into the right atrium and right ventricle. Most of the right ventricular output does not go through the lungs, however, because fetal lungs have a high vascular resistance. Another low-resistance channel, the ductus arteriosus, connects the pulmonary artery with the aorta and allows blood to bypass the lungs. The better oxygenated blood enters the left ventricle, from which it reaches the brachycephalic vessels and the front of the animal. The less well-oxygenated blood from the ductus arteriosus enters the aorta downstream from the brachycephalic vessels.

Relatively oxygenated blood from the ductus venosus joins blood from the lower extremities and hepatic veins in the caudal vena cava and continues to the heart. Pressure in the fetal right atrium is normally higher than pressure in the left atrium, allowing blood to flow through an open flap in the foramen ovale from the right to the left atrium. Anatomically, the foramen ovale lies in the pathway of blood from the caudal vena cava carrying the relatively well-oxygenated blood from the ductus venosus. The tendency for this relatively oxygenated blood from the caudal vena cava to preferentially stream toward the foramen ovale is further accentuated by the crista dividens of the interatrial septum. Consequently, the majority of blood from the caudal vena cava is directed through the foramen ovale into the left atrium and subsequently into the left ventricle. As a result, the PO_2 and oxygen saturation of blood in the fetal left ventricle is relatively high. In contrast, oxygen saturation of blood in the cranial vena cava is much lower due to high oxygen consumption in the developing brain. The anatomic location of the entry of the cranial vena cava into the right atrium leads to preferential streaming of the majority of the cranial vena cava blood into the right ventricle. Consequently, oxygen saturation of blood in the right ventricle is lower than that in the left ventricle.

The ductus arteriosus forms a vascular conduit between the pulmonary artery and aorta. In the fetus, the ductus arteriosus allows blood to flow from the high-pressure pulmonary artery to the lower pressure aorta. Physiologically, the ductus arteriosus provides a pathway for blood in the fetal pulmonary artery to bypass the high-resistance vascular bed of the fetal lung and instead to flow into the aorta distal to the origin of the coronary arteries and the brachiocephalic trunk. In fact, only 10–12% of the blood flows through the lungs in the fetus.

The roles of the foramen ovale and ductus arteriosus in the fetus are closely interrelated. During ventricular systole, the relatively well-oxygenated blood in the left ventricle is ejected into the aortic root. The first vessels arising from the aorta include the coronary arteries and the brachiocephalic trunk, and they primarily receive this flow, so that the developing heart and brain benefit from receiving comparatively well-oxygenated blood. This relationship would seem teleologically to be advantageous to these organs, but it is not essential for fetal survival or continued development, since fetuses lacking this preferential direction of flow due to congenital malformations continue to develop to term. The foramen ovale is also important for normal left ventricular development. By increasing the volume provided to the left ventricle, flow through the foramen ovale is important in promoting the normal growth and development of the fetal left ventricle.

6.7 Clinical Aspects of CV System

6.7.1 Circulatory Adjustments During Exercise

Physiological adaptation associated with exercise physical conditioning is a mechanism by which exercise capacity can be optimized. Progressive stress produces remarkable adaptations, enabling the individual to cope with increased physical demand and to attain maximum performance, e.g., the greyhound is capable of attaining speeds near 1000 m/min over 400 m with peak speeds reported over 1300 m/min. In contrast, the Siberian husky has tremendous endurance capacity: it is capable of speeds of 12–15 mph while racing over 1700 km in 12–14 days. The thoroughbred horse can attain speeds up to 55 mph over 400 m.

During strenuous exercise, the metabolic needs of working muscle increase dramatically. The pumping ability of the heart to eject sufficient blood to meet the needs of the exercising horse and provide effective redistribution of the blood to working skeletal muscle is essential. During exercise, increases in heart rate are brought about by increases in sympathetic activity. This sympathetic activation also increases cardiac contractility, so the ventricles empty more completely with each beat. In addition, sympathetic activation shortens the duration of systole, which helps to preserve diastolic filling time. In summary, under sympathetic action, the heart not only contracts more frequently (increased rate) and more forcefully (increased contractility), but also contracts and relaxes more quickly (helping to preserve diastolic filling time).

Cardiac output: Exercise demands an increased cardiac output to meet the oxygen requirements to fuel working muscle energetics. Because maximal heart rate is attained during severe exercise, stroke volume may limit the increase in cardiac output during exercise. During submaximal exercise, cardiac output increases close to linearly with workload, and this is principally due to increased heart rate. However, during maximal work in horses, cardiac output can increase 5–16 times than at rest and the elite greyhound may be better still. A high cardiac output is also aided in the elite athlete by a high ratio of heart weight to body weight (g/kg). Heart weight to body weight may range between 0.9% and 2%, for the thoroughbred and greyhound, compared with only 0.4–0.8% for humans.

Heart rate: The substantial increase in cardiac output is primarily due to the very high heart rates. In the trained thoroughbred, average resting heart rate is around 35 bpm. During maximal exercise, heart rate can increase to 240–250 bpm in the racing thoroughbred. In the dog, resting heart rates can be less than 100 bpm, particularly in the racing greyhound, rising to 300 bpm or more during maximal exercise. Heart rate rises rapidly at the onset of exercise, reaching a maximum in 30–45 s, and then often drops before reaching a plateau during steady-state work. In addition, heart rate during submaximal exercise is affected by apprehension and anxiety. The psychogenic component of the heart rate response to exercise is proportionately larger at lower, relative workloads.

Stroke volume: Stroke volume is increased or unchanged during exercise in the dog and increased during submaximal exercise in the horse. Maintenance of stroke volume during exercise occurs by several physiological mechanisms. Increased sympathetic nervous activity during exercise results in both tachycardia and reduced end-systolic ventricular volumes by increasing myocardial contractility, thereby making the ventricular emptying more effective. Venous return during exercise is supplemented by mobilization of the splenic reserve of blood volume, increased negativity of intrathoracic pressure, and muscle movement. Increased stretch of myocardial fibers within physiologic limits leads to an increase in developed pressure and stroke volume. The increases in left ventricular end-diastolic pressure and contractility during exercise, along with the increased venous return, assist in the maintenance of stoke volume, despite the decreased filling time associated with shortened diastole at increased heart rates. During severe exercise, increases in end-diastolic left ventricular diameter and pressure have been observed, with a reduction in end-systolic diameter. Therefore, stroke volume generally increases during exercise to aid in the augmentation of cardiac output and oxygen delivery to the body.

Myocardial contractility: During strenuous exercise in the dog and pony, marked augmentation of myocardial contractility is observed, along with pronounced increases in both ventricular preload (increased end-diastolic volume) and afterload (mean arterial pressure). The net result is an increase in myocardial oxygen consumption, which is met by increases in both coronary blood flow and increased oxygen extraction.

Blood flow: The main changes in distribution of blood flow during exercise are (1) enhanced pulmonary blood flow from opening of previously closed pulmonary capillaries, (2) coronary vasodilation resulting in increased coronary flow to provide oxygen for myocardial contraction, (3) vasodilation in working skeletal muscles that elevates capillary red blood cell flux, (4) vasoconstriction in the nonworking muscles and the splanchnic vasculature, and (5) increased blood flow to the skin. These cardiovascular adaptations elevate the oxygen supply to tissues with increased oxygen requirements during exercise and body thermoregulation. Blood flow to the skin is dependent on body temperature as well as environmental temperature and humidity.

Blood pressure: During submaximal exercise, systemic arterial blood pressure is maintained relatively constant by arterial baroreceptors. During strenuous exercise, cardiac output increases up to 16-fold in the horse and significant increases in mean systemic arterial pressure occur.

6.7.2 Cardiac and Circulatory Changes at Parturition

During a normal birth, the newborn emerges from the birth canal at about the time the placenta is detaching from the uterine wall. Placental gas exchange probably continues well into third-stage labor. Parturition is associated with marked and rapid changes from the fetal hemodynamic pattern. Expansion of the lung and the associated increase in alveolar and arterial PO_2 after birth contribute to a marked reduction in pulmonary vascular resistance and pulmonary arterial blood pressure. Pulmonary arterial pressure continues to decline over the next 1–2 weeks. In contrast, systemic arterial pressure rises after birth. The umbilical vessels are highly sensitive to trauma, catecholamines, angiotensin, bradykinin, and changes in PO_2. They constrict strongly at birth, reducing the risk of hemorrhage in newborn animals. An immediate increase in systemic arterial pressure at birth is due partly to elimination of the low-resistance placental vascular bed and partly to an increase in cardiac output. A continued rise in systemic arterial blood pressure over a period of weeks after birth is due largely to a gradual increase in peripheral vascular resistance.

The newborn heart, especially the left ventricle, must cope with significant hemodynamic challenges including a rapid rise in systemic arterial blood pressure and a marked increase in pulmonary venous return to the left atrium. In spite of the structural and biochemical immaturity of the newborn heart, the left ventricle supports a two- to threefold increase in cardiac output, increased stroke volume, and high heart rate. This response is supported by an increase in the left ventricular contractile state.

The newborn left ventricle demonstrates a high level of contractility but operates more nearly at maximal contractile capacity than does the adult heart. Hence, the contractile reserve of the newborn ventricle is much less than that in the adult heart.

This limits the ability of the newborn heart to respond to further increases in diastolic volume or arterial blood pressure. Myocardial mass increases rapidly during the neonatal period due to hypertrophy.

At birth, the thickness of the right ventricular wall may be equal to that of the left ventricular wall, reflecting the high right ventricular pressure in the fetus. The left ventricle gradually increases in thickness after birth, related to general body growth and to increasing arterial pressure, cardiac output, and left ventricular workload. Consequently, the normal relationship of the adult heart, where left ventricular muscle mass is approximately double the right ventricular muscle mass, is gradually established over a period of weeks after delivery.

After birth, the fall in pulmonary vascular resistance due to inflation and oxygenation of the lung leads to a dramatic increase in pulmonary arterial blood flow and, consequently, to a marked increase in pulmonary venous flow returning to the left atrium. The increase in left atrial volume, combined with increased peripheral vascular resistance and blood pressure, causes left atrial pressure to increase above right atrial pressure. At about the same time, the umbilical vessels rupture because the animal struggles to stand or the umbilical cord is torn by the mother. Umbilical blood flow is arrested by local vasoconstriction in the umbilical vessels. The loss of the low-resistance placental circulation increases systemic vascular resistance, which results in an increased pressure in the aorta, left ventricle, and left atrium. As a result of these changes, aortic pressure exceeds pulmonary arterial pressure, and left atrial pressure exceeds right atrial pressure. Therefore, blood flow through the ductus arteriosus and foramen ovale reverses. Flow reversal in the foramen ovale causes a flap valve to close and occlude the foramen. Over succeeding days to weeks, this valve becomes adherent to the wall of the atrium, thus permanently closing the foramen physiologically or as functional closure. True and permanent anatomic closure of the foramen ovale is due to fibrosis, which requires a period of weeks after birth. In a minority of instances, true closure fails to occur, resulting in a patent foramen ovale. It is important to note that a patent foramen ovale may not be functionally open during life, provided that left atrial pressure continues to exceed right atrial pressure. Reversal of flow in the ductus arteriosus exposes the ductus wall to well-oxygenated blood. This causes constriction of smooth muscle in the wall of the ductus, thus arresting blood flow. Ductus closure involves a decrease in the concentration of vasodilator prostaglandins. When the ductus has constricted and flow has been arrested, the ductus is gradually converted into a fibrous band of scar tissue. The ductus arteriosus and ductus venosus normally close shortly after parturition, although the precise timing of closure varies among species. In species, such as the pig and horse, the ductus venosus disappears early in gestation. The primary mechanisms of constriction of the ductus arteriosus involve changes in blood oxygen, cytochrome P450 system, and endothelin 1 and a decrease in prostaglandins from the placenta. Additionally, generation of reactive oxygen species inhibits voltage-gated K^+

channels, producing depolarization. This increases calcium influx across the vascular smooth muscle cell membrane via voltage-dependent calcium channels, producing smooth muscle cell constriction and physiological closure of the ductus arteriosus. The placenta normally releases prostaglandin E2 (PGE2), which contributes to ductus relaxation in the fetus. An abrupt decrease in PGE2 immediately after birth, accompanied by decreased responsiveness to PGE2, is critical for ductus closure. Physiological closure of the ductus arteriosus is followed by anatomic closure from scarring and sclerosis over a period of weeks and results in formation of the ligamentum arteriosum.

6.7.2.1 Patent Ductus Arteriosus

Abnormal, persistent patency of the ductus arteriosus after birth is termed patent ductus arteriosus and is one of the most frequently noted forms of congenital cardiovascular disease in the dog. If the ductus arteriosus remains patent after delivery, changing pressures in the pulmonary artery and aorta change the direction of blood flow across the ductus arteriosus from a right-to-left pattern (pulmonary artery to aorta) in the fetus to a left-to-right pattern (aorta to pulmonary artery) in the neonate. The ductus venosus closes before or at birth depending on the species.

6.7.2.2 Canine Puerperal Tetany (Eclampsia)

Puerperal hypocalcemia is an acute, life-threatening condition most often seen in small-breed bitches with large litters and usually occurring at peak lactation, 2–3 weeks after whelping. Hypocalcemia most likely results due to inadequate dietary calcium intake and from loss of calcium into the milk. This imbalance in calcium metabolism occurs because of the decreased calcium mobilization from bone into the serum pool together with the efflux of calcium leaving through the mammary glands. In bitches, the excitation–contraction coupling is maintained at the neuromuscular junction. Low concentration of calcium in the extracellular fluid has an excitatory effect on nerve and muscle cells, because it lowers the threshold potential, hence requiring a stimulus of lesser magnitude to depolarize. Tetany results due to spontaneous repetitive firing of motor nerve fibers. Tachycardia, prolongation of the QT interval, and ventricular premature contractions may be seen on the ECG.

6.7.3 Heart Failure

Heart failure is cardiogenic circulatory failure with sustained inability of the heart to produce a stroke volume to adequately meet the tissue metabolic demands. Heart failure refers to any condition which limits the ability of the heart to deliver a normal cardiac output due to depressed cardiac contractility. Depressed cardiac contractility can result from coronary artery disease, cardiac hypoxia, myocarditis, toxins, drugs, or electrolyte imbalances. If the decrease in contractility affects both sides of the heart, the condition is called bilateral heart failure, and if failure is restricted primarily to either the left ventricle or the right ventricle, it is called left-sided heart failure or right-sided heart failure.

It is useful from a pathophysiological perspective to categorize heart failure on the basis of where the primary defect occurs in the cardiac cycle. Systolic dysfunction or systolic failure refers to heart failure in which diastolic filling of the ventricle is normal but cardiac output (usually stroke volume) is still decreased. Diastolic dysfunction or diastolic failure refers to heart failure due to abnormal cardiac filling with normal ventricular contractility (normal systolic function).

Cardiac failure may actually be classified according to whether stroke volume is reduced (low-output cardiac failure) or increased (high-output cardiac failure). Some clinical patients with heart failure are affected with conditions characterized by excessive need for tissue perfusion. Heart failure may be present in these patients because a normal or even high cardiac output cannot meet tissue needs. Feline hyperthyroidism, chronic anemia, congenital left-to-right shunts (patent ductus arteriosus), and arteriovenous fistulas are examples of conditions observed in veterinary patients that may result in high-output heart failure.

6.7.3.1 Causes of Systolic Dysfunction

A frequent cardiac disease producing systolic failure in veterinary medicine is dilated cardiomyopathy, which may be heritable (e.g., Doberman Pinscher cardiomyopathy) or acquired (e.g., dietary taurine deficiency in cats). Other acquired causes of myocardial failure, such as myocarditis or myocardial infarction, are occasionally observed. Valvular insufficiencies can reduce stroke volume by allowing retrograde flow of blood, abnormal shunting of blood (e.g., arteriovenous fistula or patent ductus arteriosus), or tissue perfusion needs may rise excessively (e.g., feline hyperthyroidism producing high-output cardiac failure).

6.7.3.2 Causes of Diastolic Dysfunction

Some conditions reduce cardiac output by interfering with ventricular filling, producing diastolic dysfunction. These may be lesions (e.g., vena caval thrombosis or atrioventricular valvular stenosis) that reduce inflow during diastole, or excessive hypertrophy of ventricular muscle mass in hypertrophic cardiomyopathy can reduce chamber compliance and limit preload. Similarly, pericardial diseases (e.g., constrictive pericarditis or pericardial hemorrhage) may limit cardiac expansion and reduce cardiac filling.

6.7.3.3 Compensatory Responses to Heart Failure

In response to the cardiac failure, many compensatory mechanisms are mediated through the activation of arterial baroreceptors and renin-angiotensin-aldosterone mechanism.

6.7.4 Shock

Shock refers to a state of peripheral circulatory failure characterized by inadequate peripheral tissues perfusion resulting in cardiovascular collapse, leading to organ and tissue dysfunction. It is generally characterized by systemic hypotension, inadequate tissue perfusion, oliguria, cellular hypoxia, and generalized dysfunction of cells, tissues, and organs. Shock and its consequences are often reversible with appropriate therapy in its early stages. Eventually, however, sustained tissue hypoperfusion leads to irreversible cell injury progressing to cell death and organ dysfunction. In the absence of effective interventional therapy, shock generally progresses through defined stages, frequently resulting in fatality. Shock usually results from inadequate cardiac output, decreased tissue perfusion, and reduced venous return. Cardiac abnormalities that decrease the pumping ability of the heart include cardiac arrhythmias, myocardial infarction, and severe heart valve dysfunction.

6.7.4.1 Classification of Shock

Circulatory shock is classified into three main types: cardiogenic, hypovolemic, and septic shock.

Cardiogenic shock: It is the end-stage result of progressive heart failure, with the primary causative mechanism being a failure of cardiac output that cannot be compensated by other factors. Cardiogenic shock is caused by a severe decline in cardiac output. There are a variety of feedback mechanisms that serve to maintain arterial blood pressure within normal limits, despite a decline in cardiac function. However, if the disease process causing heart failure progresses in severity, these compensatory mechanisms may fail to sustain arterial blood pressure. The consequent systemic hypotension can lead to tissue hypoperfusion and cellular hypoxia, initiating the early stages of cardiogenic shock. In the absence of effective intervention, death may result.

Hypovolemic shock: Low cardiac output due to a reduction in circulating blood volume (e.g., severe dehydration or hemorrhage) leads to hypovolemic shock. Hypovolemic shock results due to a precipitous fall in cardiac output caused by the decrease in blood volume (30% or more of total blood volume) upon hemorrhage, fluid loss, or fluid sequestration. All these factors lead to decrease preload and stroke volume. However, circulatory collapse from decreased effective circulating blood volume may also be caused by peripheral vasodilation with venous pooling of blood. Mechanisms causing this latter form of hypovolemic shock include neurogenic shock, which may occur secondary to central nervous system injury, anaphylactic shock caused by a systemic allergic response associated with IgE-triggered histamine release, and anesthetic shock caused by anesthetic overdose.

The proximate cause of the cardiovascular collapse in hypovolemic shock is the inadequacy of blood volume to sustain venous return and cardiac preload. With a marked decrement in preload, stroke volume declines quickly and tachycardia is unable to restore cardiac output. In the early stage of hypovolemic shock, peripheral vasoconstriction sustains perfusion of vital organs (i.e., brain, heart, and kidney) at the expense of nonvital tissues (e.g., skin and abdominal viscera). Later, cell and organ dysfunction ensue and, untreated, hypovolemic shock is generally terminal.

Septic shock: This refers to a bacterial infection that is widely disseminated through the blood from one tissue to another and causing extensive damage by the blood-borne microbes or microbial toxins. It may be caused due to peritonitis, generalized infection, or generalized gangrenous infection. Circulatory shock caused by endotoxic Gram-negative bacilli is relatively common in all veterinary species. Circulatory shock produced by endotoxemia is referred to as endotoxic shock. Endotoxic shock is the most common form of septic shock. It is initiated by a complex interaction between endotoxin and monocytes that results in a cascade of events. The earliest stage of endotoxic shock is generally a hyperdynamic in which increased cardiac output predominates. Generally, peripheral vasodilation occurs secondary to a variety of inflammatory mediators produced by activated monocytes, the condition being referred to as systemic inflammatory response syndrome (SIRS). Although cardiac output may be normal or elevated initially, extreme peripheral vasodilation and cardiac depressant factors eventually lead to hypotension and cardiovascular collapse as the syndrome progresses, ultimately leading to death (in the absence of effective intervention).

Anaphylactic shock and histamine shock: Anaphylaxis refers to the allergic condition in which the cardiac output and arterial pressure often decrease drastically. It results primarily from an antigen-antibody reaction that takes place immediately after an antigen to which the individual is sensitive enters the circulation. The principal effects include the release histamine or a histamine-like substance from the basophils in blood and mast cells in the pericapillary tissues. The histamine causes (1) increased capillary permeability, leading to rapid loss of fluid and protein into the tissue spaces; (2) arteriolar dilation, resulting in greatly reduced arterial pressure; and (3) venous dilation leading to an increase in vascular capacity, thus causing a marked decrease in venous return. The net effect is increased reduction in venous return, and sometimes it is so serious that the animal dies within minutes.

Shock may also be classified on the basis of the level of cardiac output as low- or high-output shock. Low-output shock is generally either cardiogenic or hypovolemic in origin. High-output shock is generally associated with septicemia or endotoxemia (septic shock). Once circulatory shock reaches a critical level of severity, despite its initiating cause, the shock itself promotes and aggravates further shock. That is, the inadequate blood flow causes the body tissues including the heart and circulatory system to undergo deterioration, resulting in greater decrease in cardiac output, and a vicious circle ensues, with progressively increasing circulatory shock, poor adequate tissue perfusion, increased shock, and so forth until death.

The sympathetic reflex compensations have an important role in maintaining the arterial pressure. After hemorrhage, decrease in arterial pressure as well as decreases in pressures in the pulmonary arteries and thoracic veins cause powerful sympathetic reflexes. These reflexes activate the sympathetic vasoconstrictor system, resulting in enhanced heart activity and arteriolar constriction throughout the systemic circulation, thereby increasing the total peripheral resistance. The veins and venous reservoirs constrict, thus helping to maintain adequate venous return despite diminished blood volume.

6.7.4.2 Stages of Shock

Shock has been divided into three stages based on the response to therapy:

Stage 1: Compensated circulatory shock

In nonprogressive stage (compensated stage), without the external therapy, the normal circulatory compensatory mechanisms eventually cause full recovery. In this stage, tissue perfusion is inadequate. With cardiogenic and hypovolemic shock, hypotension is also present. In compensated septic shock, cardiac hyperfunction is generally present and arterial blood pressure is normal (or possibly elevated). In this stage, neurohumoral responses maintain adequate tissue perfusion to vital organs, preventing their hypoxia. Adapting control mechanisms include the arterial baroreceptor system, renin-angiotensin-aldosterone system, and antidiuretic hormone. The result is tachycardia, heightened sympathetic stimulation of the ventricular myocardium, peripheral vasoconstriction with increased total peripheral resistance, reduced venous capacitance, and oliguria with renal retention of salt and water. Untreated, an animal in this stage of shock often advances to the second, progressive stage.

If shock is not severe enough to cause its own progression, the animal eventually recovers. Therefore, this lesser degree shock is called nonprogressive shock. Further on, it is also called compensated shock, as the sympathetic reflexes and other factors provide sufficient compensation to stop the circulation from getting worse.

Stage 2: Progressive circulatory shock

This refers to the progressive stage, wherein without therapy, the shock becomes steadily worse. With failure of adequate treatment of an animal with compensated shock or with further insult to the circulatory system, progression to stage 2 often occurs. Compensatory mechanisms now fail to sustain arterial blood pressure, and tissue perfusion falls precipitously. Without intervention, cellular hypoxia and organ dysfunction will predominate. At this stage, appropriate therapy (e.g., intravenous fluid or transfusion therapy in hypovolemic shock; therapy for heart failure in cardiogenic shock; or intravenous fluid and antimicrobial therapy in septic shock) may restore cardiovascular function. Untreated, this stage advances to terminal cardiovascular collapse.

Stage 3: Irreversible circulatory shock

During this stage, due to extensive progression of the shock, it becomes life threatening. Without therapeutic intervention, shock tends to progress inexorably toward this irreversible stage and death. In stage 3, widespread cellular injury due to hypoxia causes a failure of vascular smooth muscle, endothelial cells, and ventricular myocardium. This leads to a loss of vascular tone, extravasation of fluid into the intestinal lumen in some species (e.g., dogs), and stasis of blood in vascular beds. Blood stasis causes intravascular activation of the clotting cascade producing a syndrome known as disseminated intravascular coagulation (DIC). Intestinal ischemia disrupts the mucosal barrier, leading to entry of bacteria or bacterial by-products (e.g., endotoxin) into the circulation, ultimately superimposing endotoxic shock on all forms of shock in this terminal, irreversible stage. Despite efforts at therapeutic intervention at this stage, death of the patient is the result.

6.7.5 Hypertension

Systemic hypertension refers to persistently elevated systemic arterial blood pressure. In dogs and cats, secondary hypertension commonly occurs wherein the cause is associated with another disease. Examples of conditions associated with the development of hypertension in dogs and cats include chronic kidney disease, hyperthyroidism, diabetes mellitus, hyperadrenocorticism, pheochromocytoma, and hyperaldosteronism and medications such as corticosteroids cyclosporine, phenylpropanolamine, and erythropoietin. Most cases of systemic hypertension in dogs and cats are associated with chronic kidney disease, wherein abnormal patterns of intrarenal blood flow or renal artery stenosis may lead to reduced pressure within renal afferent arterioles, causing increased renin release and activation of the RAAS and finally increased arterial blood pressure. Episodic release of adrenaline and noradrenaline from adrenal medulla in Pheochromocytomas leads to peripheral vasoconstriction, tachycardia, and hypertension. In dogs suffering with Cushing syndrome or hyperadrenocorticism, overproduction of glucocorticoids occurs, leading to increased blood volume and overproduction of renin, contributing to the development of systemic hypertension. In cats, the most common adrenocortical disorder is primary hyperaldosteronism, wherein excess production of aldosterone leads to sodium retention and potassium depletion resulting in expansion of blood volume, increased stroke volume, and thus arterial blood pressure. Hyperthyroidism due to benign tumor is relatively common in geriatric cats, wherein the thyroid hormone enhances cardiac function and sensitivity of the myocardium to catecholamines leading to tachycardia and increased stroke volume, resulting in systemic hypertension. Animals with diabetes mellitus may develop systemic hypertension as a result of blood volume expansion associated with hyperglycemia and increased production of renin.

6.7.6 Hemorrhage

Hemorrhage refers to the excess blood loss due to rupture of blood vessels. Hemorrhage is classified into four categories, based on the cause: capillary hemorrhage, internal hemorrhage, accidental hemorrhage, and postpartum hemorrhage.

6.7.6.1 Integrated Response to Hemorrhage

Compensatory responses to progressive blood loss involve activation of multiple reflex systems involving both neural and humoral components, which work in concert to restore cardiac output and perfusion pressure. The integrated response to hemorrhage is shown in Fig. 6.6. An initial blood loss up to 10% of total blood volume results in little or no decrease in arterial blood pressure. Reduced blood volume after mild hemorrhage decreases venous return, ventricular filling, and cardiac output. The decrease in blood volume and venous return reduces stretch and thus "unloads" volume receptors on the low-pressure side of the circulation (cardiopulmonary receptors).

During hemorrhage, the arterial blood pressure falls and baroreceptors stop discharging impulses, which increases the vasomotor tone and finally leads to vasoconstriction. Heart rate, cardiac contractility, and total peripheral resistance increase, and venoconstriction promotes venous return to the heart. The arteriolar constriction in response to hemorrhage is most pronounced in the reservoir organs such as cutaneous, skeletal muscle, and splanchnic beds, favoring maintenance of blood flow into systemic circulation, especially in the cerebral and coronary circulations. Arteriolar constriction decreases the capillary pressure, which facilitates the tissue fluid to enter the capillaries, thereby helping to compensate the blood loss. As hemorrhage progresses, increased sympathetic activity, especially to the kidney, results in vasoconstriction, decreased glomerular filtration rate, and decreased urine volume. Decreased stretch of the atrium due to reduced blood volume results in diminished secretion of ANP, while decreased renal perfusion pressure promotes secretion of renin and stimulation of the renin-angiotensin-aldosterone system. In addition, both cardiopulmonary and arterial baroreflex mechanisms promote increased circulating levels of angiotensin II, vasopressin, and aldosterone. These humoral changes promote sodium and water reabsorption in the kidney. As levels of ADH and angiotensin II continue to rise, direct vasoconstrictor effects of these peptides contribute to increased total peripheral resistance. With more severe hemorrhage, peripheral chemoreceptors sense hypoxia due to inadequate blood flow to the carotid body and contribute to further increases in sympathetic outflow. In addition, increased ventilation due to chemoreflex activation assists in promoting venous return. If cerebral ischemia occurs, elevated $PaCO_2$ and decreased blood pH activate chemosensitive neurons in the brain, which results in a massive activation of the sympathoadrenal systems. The compensatory mechanisms restore arterial pressure and cardiac output following mild-to-moderate hemorrhage. The longer term processes are critical for complete restoration of blood volume.

Pronounced renal and splanchnic vasoconstrictions during severe hemorrhage help maintain adequate perfusion of the heart and brain. However, if prolonged, vasoconstriction in these circulations can result in irreversible damage. A patient may survive the initial blood loss, but die several days later due to acute renal failure due to falling of arterial blood pressure and damage to renal tubules. Prolonged intestinal ischemia may result in liver damage, an increase in intestinal blood loss, and the release of potent vasodilatory endotoxins

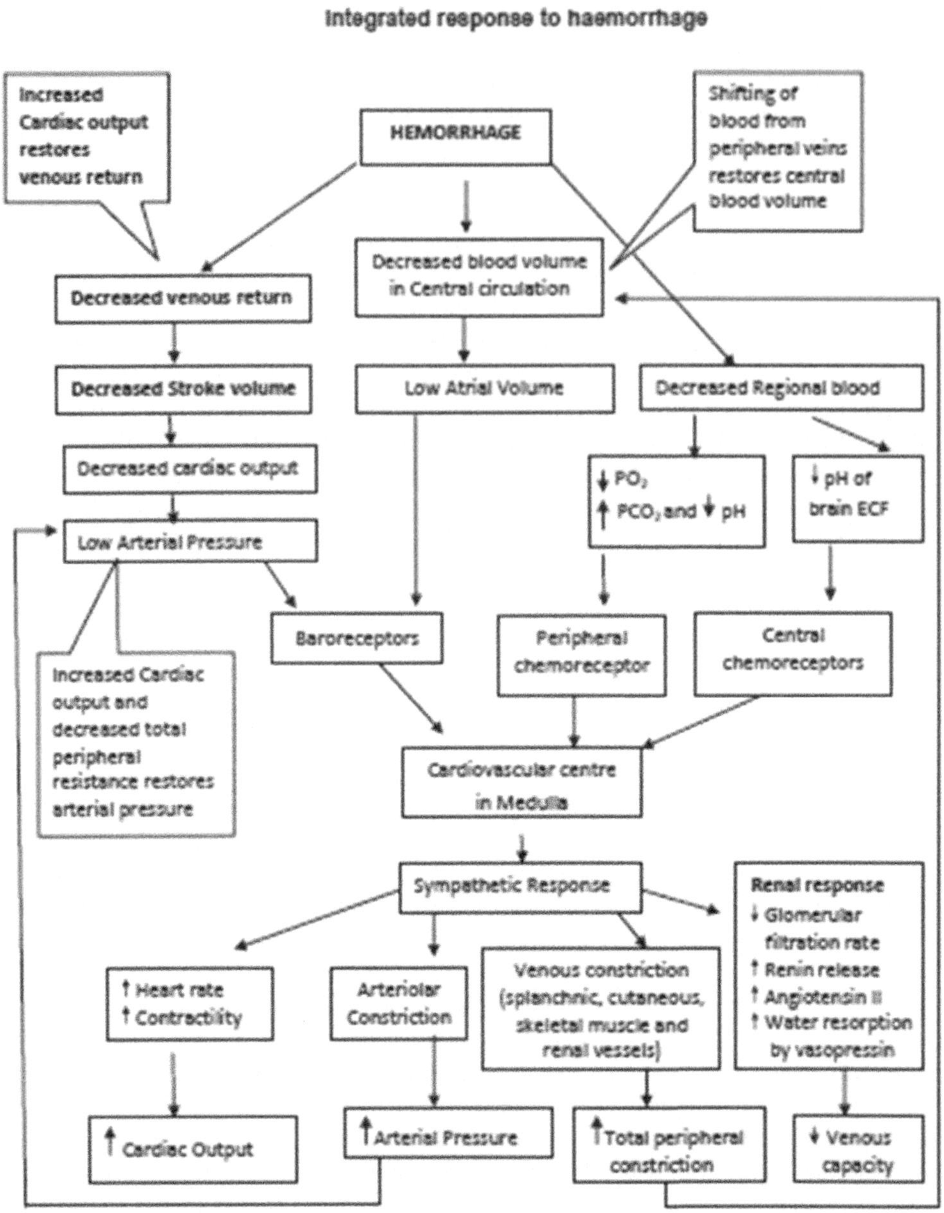

Fig. 6.6 Integrated response to hemorrhage. The central and peripheral chemoreceptors and baroreceptors signals to the cardiovascular center in the medulla, which in turn initiates the sequential events leading to the compensatory mechanisms for hemorrhage

into the general circulation. Decompensation during hemorrhagic shock is the irreversible process which aggravates hypotension and leads to circulatory failure and death.

Learning Outcomes

- **General organization of CVS and hemodynamics of circulation**
  The circulatory system is the transport system of the body. The three basic components of the circulatory system are the heart (the pump), the blood vessels (the passageways), and the blood (the transport medium). The heart functions as a dual pump that provides the driving pressure for blood to flow through the systemic circulation (between the heart and peripheral organs/tissues) and pulmonary circulation (between the heart and lungs).
- **Electrical activity of heart**
  The self-excitable heart initiates its rhythmic contractions. Autorhythmic cells are 1% of the cardiac muscle cells; they do not contract but are specialized to initiate and conduct action potentials.

(continued)

The other 99% of cardiac cells are contractile cells that contract in response to the spread of an action potential initiated by autorhythmic cells.
The cardiac impulse originates at the SA node, the pacemaker of the heart, and spreads throughout the right and left atria and ventricles facilitated by specialized conduction pathways.

- **Regulation of heart and mean arterial pressure**
  Cardiac output and total peripheral resistance determine the heart activity and the blood pressure. Regulation of cardiac output, in turn, depends on the heart rate and stroke volume regulation, whereas total peripheral resistance is influenced primarily by the degree of arteriolar vasoconstriction. Short-term regulation of blood pressure is carried out mainly by the baroreceptor reflex. The baroreceptors present in the carotid sinus and aortic arch continuously monitor MAP. Long-term control of blood pressure involves renal maintenance of proper plasma volume and renin-angiotensin-aldosterone system.
- **Regional circulation**
  Coronary blood vessels are the dedicated blood vessels that supply the heart muscles. During systole, the contracting heart muscles compress the coronary vessels and hence most coronary blood flow occurs during diastole. The pulmonary circulation is a high-flow, low-resistance pathway for the blood to flow between the lungs and heart to provide oxygenation of the venous blood. Fetal circulation differs from the adult in that the fetus receives oxygenated blood and nutrients from the placenta, and fetal lungs and liver are bypassed.

Exercises

Objective Questions

Q1. Which type of channel is most closely with the movement of cations during the depolarization phase of the atrioventricular (AV) nodal action potential?
Q2. What is the resting membrane potential of the SA node?
Q3. Which phase of cardiac cycle is associated with the slow rate of blood flow from the atria to the ventricles?
Q4. What is the cause of the "c" wave of atrial pressure?
Q5. Which component of the circulatory system has the largest distribution of distribution of blood volume?
Q6. What causes elevation of blood pressure during a mass sympathetic discharge in dog?
Q7. What is meant by lusitropic reserve of cardiac muscle?
Q8. Which channels are activated by cholinergic M2 receptors in the heart?
Q9. Name the heart sound that occurs during the period of isovolumetric contraction.
Q10. Identify the property of SA node which makes it to act as a pacemaker.
Q11. What is the amount of oxygen extracted by the cardiac muscles from the blood?
Q12. Which of the vessels does not have sympathetic control?
Q13. What causes positive bathmotropic effect on heart?
Q14. On what factor does the force of contraction within physiological limit in the heart directly depend on?
Q15. What is the influence of increased vagal tone on heart?
Q16. Name the physiologists who proposed that an increase in pressure inside most small arteries unexpectedly results in a vasoconstriction.
Q17. An increase in the heart rate during inspiration and a decrease during expiration are known as?
Q18. Which organ receives the maximum amount of cardiac output under resting condition?
Q19. Which volume is directly related to preload in heart function?
Q20. Name the reflex that causes increase in heart rate due to increase in venous return or blood volume.

Subjective Questions

Q1. Discuss the functional anatomy of the myocardial and pacemaker cells and the special properties of the cardiac muscle cells.
Q2. Write briefly on the autorhythmicity and conduction system of the heart.
Q3. Explain the various phases of cardiac cycle and associated events.
Q4. Discuss the different control mechanisms of cardiac output regulation.
Q5. Explain the short-term regulation of blood pressure.
Q6. Describe in detail about the long-term regulation of blood pressure.
Q7. Write a note on the role of capillary circulation and exchange dynamics.
Q8. Describe in detail the various methods to estimate the cardiac output.
Q9. Discuss in detail about the lymphatic circulation.
Q10. Write briefly on the cerebral circulation.
Q11. Write note on the coronary circulation.
Q12. Discuss in brief on the fetal circulation.
Q13. Discuss in detail about the special features of pulmonary circulation.
Q14. Describe the etiology of shock and the various stages involved.

Q15. Describe the various circulatory changes occurring during exercise.

Answer to Objective Questions

A1. Slow voltage-gated calcium channels
A2. −55 to −60 mV
A3. Diastasis
A4. Bulging of AV valves into the atrium
A5. Veins
A6. Contraction of the capacitance vessels
A7. Ability to relax
A8. K^+ channels
A9. First heart sound
A10. Low resting membrane potential and leakiness to Na^+ ions
A11. 15 mL O_2/100 mL
A12. Cerebral vessels
A13. Stimulation of sympathetic nerves
A14. Initial length of the cardiac muscle
A15. Bradycardia
A16. Sir William Bayliss
A17. Respiratory sinus arrhythmia
A18. Liver
A19. Venous return
A20. Bainbridge reflex

Keywords for Answer to Subjective Questions

A1. Contractile myofibrils, syncytium, desmosomes, gap junctions, conductivity, contractivity, all or none principle staircase phenomenon, refractory period, extrasystoles
A2. Leaky sodium channels, low resting membrane potential, nodal cells, Purkinje cells, transitional cells, sinoatrial node, atrioventricular node, Purkinje fibers, Bundle of his
A3. Isometric contraction, maximum ejection, reduced ejection, protodiastole, isovolumetric relaxation, rapid filling, reduced filling, atrial systole, pressure changes, volume changes, sound changes, electrical changes
A4. Intrinsic regulation, heterometric regulation, homeometric autoregulation, extrinsic regulation, nervous control, reflex control, chemical regulation, humoral control
A5. Nervous regulation, reflex mechanism, baroceptor reflex, atrial volume receptor reflex, Bainbridge reflex, psychogenic responses, local control myogenic theory, metabolic theory
A6. Regulation by extracellular fluid volume renin-angiotensin mechanism
A7. Continuous capillaries, discontinuous capillaries, fenestrated capillaries, diffusion
A8. Transthoracic or esophageal echocardiography, indicator dilution technique, Fick method, pressure recording analytical method (PRAM)
A9. Lymph capillaries, lymphangions, thoracic and right lymphatic ducts
A10. Circulus arteriosus cerebri, blood–brain barrier, hypoxia-stimulated cerebral vasodilation
A11. Right and left coronary arteries, coronary sinus, anterior coronary veins, arteriosinusoidal vessels, collateral coronary arteries, and Thebesian veins
A12. Ductus arteriosus, foramen ovale and ductus arteriosus, ductus venosus
A13. Low-resistance, low-pressure system, relatively great distensibility and collapsibility of the pulmonary vessels
A14. Circulatory failure, cardiogenic, hypovolemic and septic shock, anaphylactic shock and histamine shock, compensated circulatory shock, progressive circulatory shock, irreversible circulatory shock
A15. Sympathetic activation, increased end-diastolic volume, increased pulmonary blood flow, increased coronary flow, increased blood flow to the skin, coronary vasodilation

Further Reading

Textbooks

Berne RM, Levy MN (2000) Cardiovascular physiology, 8th edn. Mosby Year Book, St Louis, MO, pp 5–52
Boron WF, Boulpaep EL (eds) (2008) Medical physiology, 2nd edn. Saunders Elsevier, Philadelphia
Boulpaep EL (2009) Organization of the cardiovascular system. In: Boron WF, Boulpaep EL (eds) Medical physiology, 2nd edn. Saunders Elsevier, Philadelphia
Braunwald E, Ross J Jr, Sonnenblick EJ (1968) Mechanisms of contraction of the normal and failing heart. Little, Brown, Boston
Cohen PF (1985) Cardiac pumping action and its regulation. In: Cohen PF, Brown EJ Jr, Vlay SC (eds) Clinical cardiovascular physiology. W.B. Saunders, Philadelphia
Dzialowski EM, Crossley DA (2015) Chapter 11 - The cardiovascular system. In: Scanes CG (ed) Sturkie's avian physiology, 6th edn. Academic, pp 193–283
Erickson HH, Detweiler DK (2004) Control mechanisms of the circulatory system. In: Reece WO (ed) Dukes' physiology of domestic animals, 12th edn. Cornell University Press, Ithaca, NY, pp 275–302
Ganong WF, Barrett KE, Barman SM, Boitano S, Brooks HL (2015) Review of medical physiology. McGraw-Hill Medical, New York
Grant AO (2009) Cardiac ion channels. Circ Arrhyth Electrophysiol 2: 185–194
Hall JE (2015) Guyton and Hall textbook of medical physiology, 13th edn. W B Saunders
Katz AM (2011) Physiology of the heart, 5th edn. Wolters Kluwer Lippincott Williams & Wilkins, Philadelphia
King AS (1999) The cardiorespiratory system. Blackwell Science, Oxford

Mohrman DE (2010) Cardiovascular physiology, 7th edn. McGraw-Hill, New York
Monfredi O, Maltsev VA, Lakatta EG (2013) Modern concepts concerning the origin of the heartbeat. Physiology 28:74–92
Pappano AJ, Wier WG (2013) Cardiovascular physiology, 10th edn. Mosby, Philadelphia
Reece WO, Erickson HH, Goff JP, Uemura EE (2015) Dukes' physiology of domestic animals, 13th edn. John Wiley Blackwell

Research Articles

Animal athletes: a performance review (2016) http://veterinaryrecord.bmj.com
Marengo FD, Márquez MT, Bonazzola P, Ponce-Hornos JE (1999) The heart extrasystole: an energetic approach. Am J Physiol 276(45): H309–H316

Physiology of Respiration

7

Sonali Jana, P. Manjari, and Iqbal Hyder

Abstract

Respiration is one of the body's vital functions that facilitate gas exchange, thereby ensuring oxygen supply and removing carbon dioxide from the body. Different domestic animals have evolved different features while undertaking various facets of respiration; for example, in birds, the mechanics of respiration is dominated by air sacs instead of the lungs of mammals. The diffusion of gases at the alveoli level is followed by their transport via blood. The transport of blood gases is dependent upon the type of gas. The dominant system for oxygen is via binding to haemoglobin; in the case of carbon dioxide, it is transported through plasma that predominates through a considerable extent of the gas that is also transported by haemoglobin. Various integrated physiological processes mediate the regulation of respiration. In the CNS, the centre controlling respiration is present in the pons and medulla. The sensory inputs for respiratory centre are relayed through central and peripheral chemoreceptors and mechanoreceptors. The respiratory system can develop various anomalies, which may be congenital or acquired, that can severely hamper the respiration process. This chapter focuses on all the basic physiological features of mammalian respiration, including the features in foetal and neonatal stages, with updated advances in avian species.

S. Jana
Department of Veterinary Physiology, West Bengal University of Animal & Fishery Sciences, Kolkata, West Bengal, India

P. Manjari
Department of Veterinary Physiology, College of Veterinary Science (Sri Venkateswara Veterinary University), Proddatur, India

I. Hyder (✉)
Department of Physiology & Biochemistry, College of Veterinary Science (Sri Venkateswara Veterinary University), Garividi, Andhra Pradesh, India

© The Author(s), under exclusive license to Springer Nature Singapore Pte Ltd. 2023

P. K. Das et al. (eds.), *Textbook of Veterinary Physiology*, https://doi.org/10.1007/978-981-19-9410-4_7

Graphical Abstract

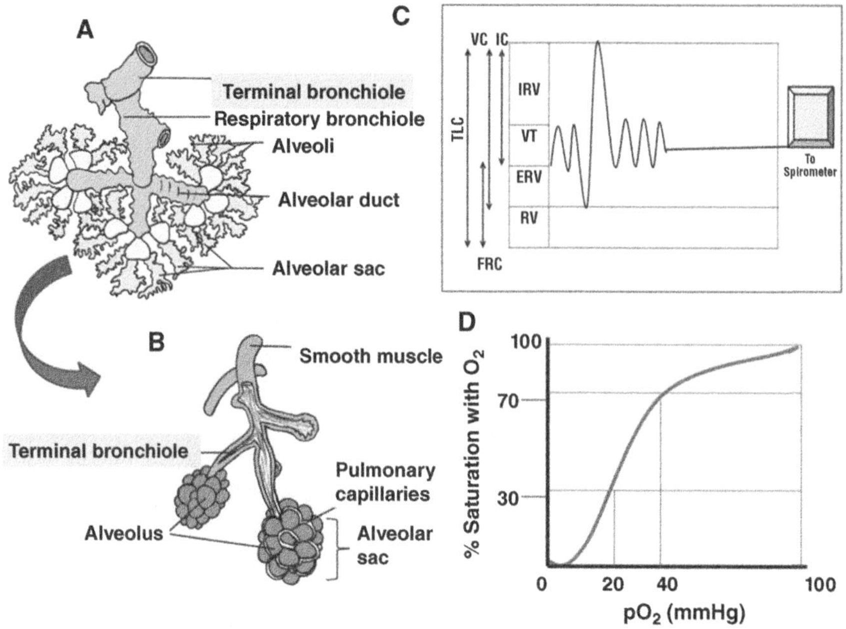

Description of the graphic: (**a**) The process of respiration occurs in the lungs, where the functions of gaseous exchanges occur in the zone of respiration; (**b**) respiratory zone: terminal bronchioles, alveolar sac and alveoli; (**c**) the lung function is evaluated by measuring the pulmonary volumes and capacities through *spirometry*; and (**d**) tissue supply of oxygen is determined by oxygen-haemoglobin dissociation curve

Keywords

Respiration · Ventilation · Lung volumes · Pulmonary circulation · Reflexes

Learning Objectives

- Respiratory system—anatomical structure and functional components
- Transport of respiratory gases (oxygen and carbon dioxide) through blood during respiration
- The control or regulation of respiration by the respiratory centres, reflexes and chemical changes in blood
- Avian respiratory system—structural design, its salient features and control mechanism
- Fundamentals of respiratory dysfunctions

7.1 Overview of Respiratory System

The respiratory system serves one of the vital functions of the animal body by delivering oxygen from the surroundings to each cell of the body and removing carbon dioxide to be released to the surroundings. Animals have specialised systems that help them do this successfully and efficiently. The process of respiration involves many independent but well-coordinated and regulated events.

7.1.1 Components of Respiratory Apparatus

Mammals' respiratory systems are divided into two main functional sections—(1) the respiratory component, which is responsible for exchanging oxygen for carbon dioxide in the lungs, and (2) the conducting and conditioning part, which includes respiratory airways. The respiratory tract is made up of two parts: (1) the upper respiratory tract, which starts with the nasal cavity, and (2) the lower respiratory tract, which is represented by the terminal part's consecutive bronchiole branching.

The lungs are present in chest or thoracic cavity, more or less symmetrically about the heart. Each lung is surrounded by a double layer of pleura, internal and external forming a pleural cavity. Around and below the root of the lung, the pleural layers are continuous. It is not a real cavity because there is no physical space between the two pleural walls. Instead, a tiny layer of fluid that is sandwiched between the two membranes lubricates them when breathing. The lung is held to the surface of the pleural membrane by surface tension. The bronchus, pulmonary artery, pulmonary vein, bronchial arteries and veins, pulmonary nerves, lymphatic vessels, and bronchial lymph nodes make up the root of the lung. Each lung is further divided into a number of separate lobes, depending on the species. In cattle, the right lung has four lobes, while the left lung has two.

7.1.1.1 Airways to Lung

Air is taken into the respiratory tract through external nares (nostrils) and passed forward through nasal cavities. The quantum of air entering the nasal cavity is determined by the degree of flexibility, termed "pliability" of nares, which differs between species. It is highly pliable in equines but more rigid in porcine species, which implies that the pliable nostrils are an evolutionary feature of horses.

The epithelium of nasal cavities is pseudostratified-ciliated-columnar epithelium containing aggregates of lymphoreticular tissue, lymphocytes and tubuloalveolar glands. The nasal cavities are endowed with complex structure for carrying out its functions. One such structure is conchae, which are spiral bone laminas coated with mucosa present inside the nasal cavity. The mouse has two nasal conchae—dorsal and ventral conchae. In some species (e.g. small laboratory animals), a distinct nose-associated lymphoid tissue (NALT) is present in the nasal cavity. Bidirectional air movement in nasal cavities keeps the surface cool that allows further cooling of nasal venous blood temperature as compared to the general body temperature. The internal carotid artery passes through the cavernous sinuses, where the cool venous blood decreases blood temperature of the internal carotid artery in the countercurrent transfer mechanism. This arrangement keeps the temperature of brain 1–3 ° C lesser as compared to general body temperature.

Air passes the nasopharyngeal tube to reach the pharynx, which is the intersection point with the digestive tract. The mucosa of nasopharynx has pseudostratified-ciliated-columnar epithelium and goblet cells for secretion of mucosa. Waldeyer's tonsillar or lymphatic ring, a ringed arrangement of lymphoid organs, is also present. This organ is vital for inducing immunity at the mucosal layer. The larynx links the nasopharynx with the lower respiratory tract and forms the landmark of its beginning. The larynx also contains epiglottis and vocal cords (*plica vocalis*). The mucosa of larynx (except epiglottis and vocal cords) is lined by a pseudostratified-columnar epithelium and isolated goblet cells; in the submucosa of the epiglottis, *plica aryepiglottica* and *vestibulum laryngis*, lymphatic nodules are present.

The trachea is comprised of 32–36 "C"-shaped rings of hyaline cartilage with strong fibroelastic membranes bridging the gaps between the rings. The trachea is placed ventrally in the neck and traverses the *apertura thoracis cranialis* to enter into the thoracic cavity. As trachea reaches the level of heart, it ramifies many times into several smaller airways (bronchi and bronchioles). The trachea is lined by pseudostratified-ciliated-columnar epithelium and isolated goblet cells as compared to that present in extrapulmonary bronchi. The mucosa of trachea, also called *lamina propria mucosae*, contains small islets of lymphoreticular tissue and combined tubuloalveolar glands.

Know More......

Vitamin A plays a vital role in developing the lungs in the foetal stage. Studies have shown that specific retinoic acid (RA) receptor proteins are organised in a particular way during the formation of branches and growth of airways in the foetus. This retinoic acid can regulate other genes affecting lung development, such as homeobox genes, certain growth factors and matrix molecules. A deficiency of retinol can result in similar pathology as bronchopulmonary dysplasia in premature babies.

The large bronchial tubes subdivide each lung like a tree branch. Functionally, the airway system in lungs can be categorised into a conducting zone (trachea, bronchi, bronchioles and terminal bronchioles), a transition zone (respiratory bronchioles) and a gas-exchange region (alveolar ducts, sacs and walls). This structural arrangement enhances respiratory efficiency by influencing the distribution of air and blood. The bronchial tree ends are called the terminal bronchioles, and the parts lying after the terminal bronchioles are involved in gaseous exchange with the blood and are also called the respiratory zone (Fig. 7.1a). These parts are the respiratory bronchioles, alveolar ducts, sacs and pulmonary alveoli (Fig. 7.1b). As the branching increases in the bronchial tree, the diameter decreases, but the cross-sectional area increases exponentially. This aids in drastically reducing airflow resistance along the airway path following Hagen-Poiseuille's law. Also, in accordance with the aerodynamic and hydrodynamic principles of Murray's law, as the airways decrease in diameter by 2–1/3 ($\approx$0.79), there is lesser energy loss in fluid transport through branched conduits.

The respiratory bronchioles are an extension of terminal bronchioles and have the same diameter as terminal bronchioles, i.e. about 0.24 mm. About 5 or 6 alveolar ducts (alveolar duct diameter is about 0.19 mm) are attached to the respiratory bronchioles. About 3–6 alveolar sacs are attached to each alveolar duct. The pulmonary alveoli are hemispherical out-pouching of the alveolar sacs. Their diameter ranges from 0.075 to 0.125 mm, and their total number is estimated to be about 750 million. The inner lining of pulmonary alveoli is made up of a single layer of epithelial cells.

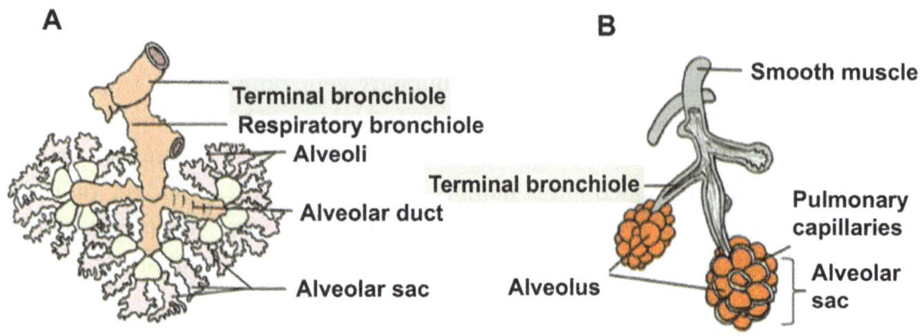

Fig. 7.1 "Respiratory zone" demonstrating its various parts where gaseous exchange occurs (**a**). The zone beyond terminal bronchiole comprises the site of gaseous exchanges (**b**)

The alveoli walls also contain elastic fibre and a massive capillary network that aids in gaseous exchange. The entire set of dichotomous branching is called *respiratory tree.*

7.2 Functional Components of Respiration

Respiration is to replenish tissues and each body cell with oxygen and carry away carbon dioxide. Respiration serves several important functions. These are (1) the swapping (exchange) of gases in the alveoli, also called as pulmonary ventilation that comprises inspiration and expiration; (2) transfer of gases, i.e. oxygen by diffusion from the alveoli to blood and carbon dioxide from blood to alveoli; (3) transport of oxygen and carbon dioxide, respectively, to and from the tissues via blood; and (4) controlling the overall process of ventilation.

7.2.1 Pulmonary Ventilation

The process of breathing air in and out of the lungs is known as pulmonary ventilation. The process comprises two distinct phases: (1) inspiration (or inhalation), which enables air to enter the lung, and (2) expiration (exhalation), which enables air to move out of the lungs. One cycle of inspiration followed by expiration is termed as respiratory cycle.

Two major groups of muscle take part in regular inspiration: (1) diaphragm and (2) external intercostals. Apart from these, other muscles are also engaged depending on the requirements during a deeper breath. The diaphragmatic contraction generates a large space within the thorax as it is pushed down the abdomen, creating more room for the lung. Similarly, contraction of external intercostals increases the size/volume of the thoracic cavity by causing outward and upward rib movement. Since the lung is surrounded by pleural fluid, the force of expansion in thoracic cavity expands the lungs too. Expansion of lungs causes intra-alveolar pressure to fall below atmospheric pressure; the resultant pressure gradient draws air into the pulmonary system from the ambient conditions. As the inspiratory muscles relax after completion of inspiration process, the elastic recoil tendency of the lung tissues contributes towards evacuation of the lungs albeit not completely. The expiration is a passive process in mammals during which both the lung and thoracic volume decrease causing an increase in inter-pulmonary pressure. The rise in inter-pulmonary pressure beyond P_{atm} generates a gradient of pressure that propels air out of lungs.

As described above, two distinct kinds of movements are associated with breathing. When the visible abdominal movements predominate in breathing, it is called abdominal breathing, and if the breathing is predominated by rib movements, it is called costal breathing. In dogs and cats, both diaphragm and respiratory muscles coordinate respiratory movements and the respiration is termed as costo-abdominal respiration. In cases involving loss of diaphragmatic function (e.g. rupture or other causes), bulging of the abdomen alone does not support inspiration. Under such situations, the abdominal circumference usually decreases during inspiration and the resulting respiration pattern is called as pendulous respiration.

In severe diseases involving the lungs, its airways or the heart, animals or humans will show shortness of breath; this is known as dyspnoea or difficulty breathing. In contrast, normal breathing is termed eupnoea in a healthy state.

7.2.2 Pressures Driving Ventilation

The mechanical process of ventilation is dependent upon three factors: (1) the atmospheric pressure (P_{atm}), (2) the alveolar pressure (P_{alv}) and (3) the intrapleural pressure (P_{ip}).

7.2.2.1 Atmospheric Pressure

According to Dalton's law, in a mixture that contains two or more gases that do not react with each other, the combined resultant pressure is same as the sum of the individual partial pressures of each gas present in that mixture. Hence, the barometric pressure (BP) and fractional concentration in the gaseous mixture play a key role in determining the partial pressure of oxygen (PO_2), which implies that the altitude of a place in comparison to mean sea level is critical in determining the barometric pressure. With rise in altitude, there is a corresponding decrease in the number of gas molecules per unit volume, so the air density is lesser than that at sea level. Human life depends on oxygen, this gas being acquired from the atmosphere where the partial pressure of oxygen ($P_{Atm}O_2$) within the troposphere depends on BP according to the Dalton's law:

$$P_{Atm}O_2 = 0.21 \cdot 760\,\text{mmHg} = 159\,\text{mmHg}$$

7.2.2.2 Alveolar Pressure

Alveolar pressure (P_{alv}) is the pressure exerted by air inside the lung alveoli. If we assume the value of atmospheric pressure as zero, then the alveolar pressure (cm H_2O) can be depicted as positive or negative with respect to P_{atm}. During the process of inspiration, the respiratory muscles cause the P_{alv} to drop below P_{atm} (−1 cm of water). This causes air to move inside lungs along pressure gradient, thus storing potential energy in the elastic structures. At the end of inspiration, the respiratory muscles relax leading to elastic recoil of the respiratory system. This causes the P_{alv} to be positive (+1 cm of water) relative to atmospheric pressure, and hence, expiration occurs.

7.2.2.3 Pleural Pressure

Pleural pressure (also called intrathoracic pressure), or P_{pl}, is the pressure surrounding the lung within the pleural space. Similar to intra-alveolar pressure, intrapleural pressure also varies during the different phases of breathing. During rest or quiet breathing, the pleural pressure assumes a negative value. The intrapleural pressure always assumes a negative value with respect to the atmospheric pressure and the intra-alveolar pressure (approx. −4 mmHg), with minor fluctuations during both phases of respiration, i.e. inspiration and expiration.

The negativity of the intrapleural pressure arises out of opposing forces within the thoracic cavity. The centripetal forces that pull the lungs inwards, tending to collapse the alveoli, are elastic forces of the lung tissues plus the alveolar fluid surface tension. The thoracic wall and fluid contribute to the forces acting in the opposite direction within the pleural space. The pleural layer lining the inner thoracic wall, i.e. the parietal pleura, is strongly adhered to this wall which counters the inward-acting forces of the lungs. Therefore, the intrapleural pressure always remains lesser or negative (−4 mmHg), whereas the outward forces remain greater than the centripetal pull though only to a smaller extent. The difference between the intra-alveolar and intrapleural pressure is called transpulmonary pressure, and this pressure determines lung size.

Pneumothorax refers to air or gas in the pleural cavities, resulting in atelectasis because negative intrapleural pressure cannot be maintained.

7.2.3 Ventilatory Asynchronism

In horses, intrapleural pressure values vary in upper and lower thoracic cavity. In a standing horse, the pressure in the uppermost part of thorax is more sub-atmospheric than in the lowermost parts. Consequently, dorsal part of equine lung is highly distended and less compliant than ventral part. Air preferentially moves to more compliant regions, resulting in a vertical gradient of air movement in a standing horse. However, when viewed in terms of factors affecting the distribution of ventilation, this phenomenon of relative lung distention of different regions in horses is just one among many factors. The air distribution within the lung is also influenced by local lung compliance and airway resistance. This altered distribution of air in lungs of horses is called *ventilator asynchronism*.

7.2.4 Respiration Rate

The respiration rate, also called the respiratory frequency, is defined as the number of breaths per minute and is often used as a measure of health status of any animal or individual. The respiration rate varies widely among different species of animals. In addition, it also depends on several parameters like age, health conditions, ambient temperature, body size, pregnancy, excitement and exercise. As the ambient temperature rises, the respiration rate increases in animals to aid in thermoregulation. An overview of the respiration rate of several animals under varied conditions is presented in Table 7.1.

7.3 Pulmonary Compliance

The compliance of any system is defined as the change in volume per unit change in the pressure of the system. It is synonymous to ease with which an elastic structure can be stretched and hence is a measure of the elasticity of a system. Pulmonary compliance (C) measures the total compliance of both the lungs. It is the increase in the volume of the lungs for each unit increase in the transpulmonary pressure (assuming that the system is in equilibrium). Lung compliance can be calculated as follows:

$$\text{Lung compliance} = \text{Changes in the lung volume}/\text{Change in transpulmonary pressure}$$

The total compliance of both lungs in a healthy adult human is about 200 mL/cm H_2O. The reciprocal of compliance is elastance.

Table 7.1 Respiration rate of different species of animals under varied conditions

Animal	Conditions	Respiration rate (range)
Dairy cow	At rest (standing position)	26–35
Dairy cow	Sterna recumbency	24–50
Sheep	Ruminating (standing position at 18 °C)	20–34
Sheep	Ruminating (standing position at 10 °C)	16–22
Pig	Lying down (23–27 kg bd. Wt)	32–58
Dog	Sleeping	18–25
Dog	Standing	20–34

Adapted from Duke's physiology of domestic animals, 12th edition

7.3.1 Compliance Diagram

On plotting the volume changes occurring in lungs with change in transpulmonary pressure changes, two pressure-volume curves are obtained; one is the inspiratory compliance curve, and the other is the expiratory compliance curve. The whole image is called the compliance diagram (Fig. 7.2), and the characteristic curve obtained is due to lung elasticity and elastic forces contributed by the surfactants lining the alveoli. The lung volume varies even at a given transpulmonary pressure depending on the inspiration or expiration phases. From the curves, it is evident that lung volume is higher when the lung deflates during expiration at the same pressure. The inspiration curve thus lags behind the expiration curve, so these curves are also termed as hysteresis curve. Therefore, the compliance is high at low lung volume, both during inspiration and expiration, making the curve steeper. However, at high lung volume, the compliance falls, making the curve flatter.

Lung compliance is measured by two different methods: static and dynamic compliance.

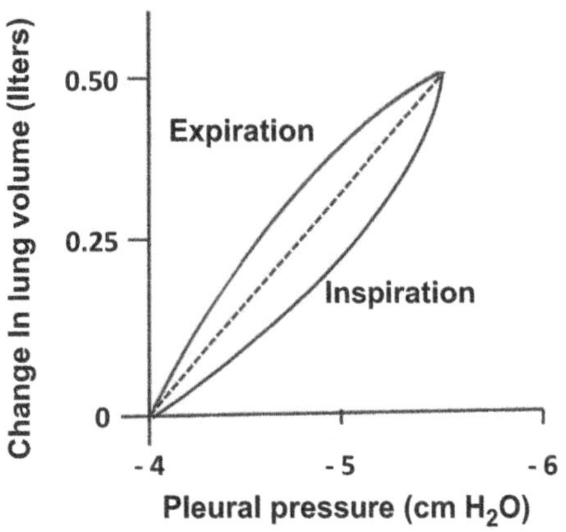

Fig. 7.2 Compliance of a healthy lung. [Even at the same transpulmonary pressure, the lung volume differs during inflating and deflating. The inspiration curve always lags behind the expiration curve, so these curves are also known as the hysteresis curve.]. (Adapted from Guyton and Hall Textbook of medical physiology, 12th edition)

1. **Static compliance**: Pulmonary compliance measured at a fixed specific volume of lungs under relaxed conditions of no airflow is called static compliance. Static compliance measures the elastic resistance of lungs that occurs when transpulmonary pressure equals the recoil pressure of the lungs.
2. **Dynamic compliance**: The compliance that is assessed during flow or when breathing. As frequency rises, dynamic lung compliance falls, implying that some airways and subtending alveoli are becoming more constricted.

Low lung compliance indicates that it needs to work more for inflating the lungs for inspiration. Bovine lungs have relatively low lung compliance due to more lobulation and greater interstitial tissues.

7.3.2 Factors Determining Lung Compliance

Two essential factors determine lung compliance. One is the presence of elastic forces, which is determined by *elastin* and *collagen* fibres present in lung parenchyma in an interwoven manner. In deflated lungs, these fibres are in a relaxed state being contracted together. With the expansion of the lungs, the fibres attain a stretched position from a relaxed state.

The second factor that determines lung compliance is the presence of pulmonary surfactant. According to Laplace law, i.e. pressure = 2 × T (surface tension)/*R* (radius), the P_{alv} within the smaller alveoli would be higher than the pressure within the large alveoli. The pressure can collapse the tiny alveoli, which, however, does not occur in a true sense, as the surfactants lower the surface tension preventing alveolar collapse.

A surfactant is a substance produced by type II alveolar epithelial comprising primarily of phospholipids, proteins and calcium ions. Smaller alveoli have a small surface area but have higher concentration of the surfactants, which eventually lowers the surface tension. So, by modulating surface tension, surfactants indirectly affect lung compliance.

The surfactant is rich in phospholipids and surfactant-associated proteins, viz. SP-A, SP-B, SP-C and SP-D. Among the proteins, SP-A and SP-D are hydrophilic and SP-B and SP-C

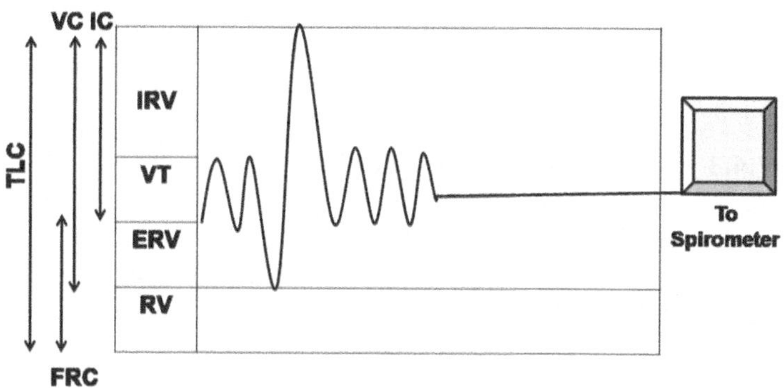

Fig. 7.3 Pulmonary volumes and capacities as recorded by spirometry. *IRV* inspiratory reserve volume, *VT* tidal volume, *ERV* expiratory reserve volume, *RV* residual volume, *TLC* total lung capacity, *FRC* functional residual capacity, *VC* vital capacity, *IC* inspiratory capacity. (Adapted from Duke's Physiology of Domestic Animals. 12th edition)

are hydrophobic. The surface tension-lowering function is attributed to dipalmitoylphosphatidylcholine, a unique phospholipid, and two hydrophobic SP-B and SP-C proteins. The hydrophilic SP-A and SP-D proteins maintain the pulmonary immune defence by clearing invading pathogens and modifying the immune responses. Besides, the surfactant can aid in preventing plasma exudation, help the smooth muscles lining the airways to relax and prevent the respiratory surfaces from adhering. The lung compliance increases in the presence of surfactant and stabilises alveoli, thus reducing the respiratory workload. Surfactant constitutes about 10% of the alveolar surface. In humans, it begins to appear in the lung at the 26th week of gestation, and deep breathing and cortisol stimulate its production. The lack of surfactant in neonatal lungs results in a condition called respiratory distress syndrome or hyaline membrane disease.

7.4 Pulmonary Volumes and Capacities

Pulmonary volumes are important parameters that reflect overall lung health or functioning when complete pulmonary function tests (PFTs) are conducted. The volume of air in the lung at any time point of the respiratory cycle is known as the pulmonary volume. Pulmonary capacities are obtained when at least two pulmonary volumes are summed up.

Pulmonary ventilation can be studied by a simple non-invasive method called "*spirometry*". It involves the simple process of recording the volumes of air inhaled and exhaled by the individual. The instrument comprises a drum that is placed upside down over a water chamber. The drum consists of either air or oxygen as a breathing gas, and the gas chamber has a tube with mouthpiece for blowing of the air. The drum moves up and down in response to a subject's inhalations and exhalations, and this movement is accurately captured on a moving piece of paper.

7.4.1 Pulmonary Volumes

Volume of air in the lungs varies with the events happening during the stages of pulmonary ventilation. For the purpose of better understanding of lung volume, the volumes of air in the lungs have been subdivided into the following volumes and capacities, as depicted in Fig. 7.3.

1. **Tidal volume**: The volume of air inhaled or exhaled in each normal breath; it is about 500 mL in adult humans (male).
2. **Inspiratory reserve volume (IRV)**: The extra volume of air that can be inspired during forceful inspiration beyond the tidal volume. In adult humans (male), it is about 3000 mL in volume.
3. **Expiratory reserve volume (ERV)**: It is the extra maximal volume of air that can be forcibly expired after the end of a tidal expiration; this is about 1100 mL in adult humans (male).
4. **Residual volume**: The volume of air that remains in the lungs after maximum forceful expiration: it is about 1200 mL.

7.4.2 Pulmonary Capacities

The two or more pulmonary volumes can be combined to obtain the pulmonary capacities as described below:

1. **Inspiratory capacity**: The total tidal volume along with the IRV is called inspiratory capacity. It is the maximum amount of air an individual can breathe after a resting state and is about 3500 mL.
2. **Functional residual capacity**: It is the total of the expiratory reserve volume plus the residual volume, an amount of air remaining in the lungs at the end of normal expiration (2300 mL).
3. **Vital capacity**: The maximum amount of air that can be expelled out of the lungs after a maximum deep inhalation and includes inspiratory reserve volume plus the tidal volume and expiratory reserve volume (4600 mL).
4. **Total lung capacity**: The maximum volume of air the lungs can accommodate with the greatest inspiratory

effort, which is about 5800 mL and is equal to the sum of vital capacity and the residual volume.

7.4.3 Minute Respiratory Volume (MRV)

It is the total volume of air that can be inhaled or exhaled from the respiratory passage in 1 min, which is arrived at by multiplying the tidal volume with the respiratory rate per minute. The normal tidal volume and respiratory rate in humans are 500 mL and 12 breaths/min, respectively; hence, the MRV amounts to about 6 L/min.

7.4.4 Alveolar Ventilation

About 30% of the total tidal volume of air remains in the anatomical dead space and does not participate in gaseous exchange at the alveolar surface. During each cycle of respiration, 70% of air is involved in gaseous exchanges at the alveoli, alveolar sacs, alveolar ducts and respiratory bronchioles. Under normal conditions, the tidal volume is 500 mL, of which only 350 mL takes part in gaseous exchange. This implies that under the normal respiratory rate of 12 breaths/min, the alveolar ventilation amounts to 4.2 L/min.

7.4.5 Dead Space

The parts of respiratory tree like nose, pharynx and trachea which come in contact with air but are not involved in gaseous exchange are called as anatomical dead space, and the air present in those parts is called as dead-space air. In some instances, poor blood perfusion in some alveoli may render them not useful or only partly beneficial for gaseous exchanges. Such alveoli form the alveolar dead space. Thus, the summation of all dead spaces, i.e. anatomic dead space and alveolar dead space, is the total dead space where gaseous exchange never occurs, also known as the physiologic dead space.

7.5 Pulmonary Circulation

It is the system of circulation connecting the heart and lungs, which comprises the widely interconnected blood vascular system and lymphatics. They perform certain unique functions, like exchanging gases in the lungs apart from acting as a reservoir for blood storage. The pulmonary circulation has the following three major components.

1. ***Arterial circuit***: Arterial circuit begins with the pulmonary artery arising from the right ventricle and then divides into the right and left branches, followed by further divisions into many branches to form an extensive network of small arteries and arterioles, which finally ramifies into capillaries. The pulmonary arteries are much wider (having a large diameter) and thinner, which confers them with enhanced distensibility and compliance (approximately 7 mL/mmHg) features. It is advantageous for accommodating a larger blood volume in a limited space.
2. ***Venous circuit***: It begins with the finest venules that merge into smaller veins and eventually join the main pulmonary veins, which drain blood into the left atrium. Compared to the systemic veins, the pulmonary veins are thinner with enhanced distensibility, like the arteries, making them more compliant that enables them to accommodate more blood.
3. ***Lymphatics***: The lymphatics originate near the terminal bronchioles and help maintain an excess fluid accumulation near alveoli that may otherwise affect the function of gas exchange and finally drain into the mediastinal lymphatics that ultimately terminate near the right lymphatic duct.

The pleural space between the two pleural membranes is also supplied by a similar lymphatic system that aids in pleural fluid drainage and provides a frictionless viscous environment for lung movement during respiration. The lymphatic system also contributes to a highly negative (approx. −4 to −7 mmHg) pleural pressure that prevents alveolar collapse.

Bronchial Vessels Small bronchial arteries branch away from the systemic circulation and supply oxygenated blood to the lungs. Although this constitutes only a tiny fraction (1–2%) of cardiac output, it plays an essential role in supplying oxygenated blood to various lungs, including the connective tissues and bronchi. It later drains into the left atrium via the pulmonary veins. It contributes to the slightly increased (1–2%) inflow and output of the left atrium and left ventricle, respectively, in comparison to the output of the right ventricle.

7.5.1 Pressures in the Pulmonary Artery

The average systolic, diastolic, and mean pulmonary arterial pressure is about 25, 8, and 15 mmHg, respectively, in humans, while the mean pulmonary capillary pressure is 7 mmHg. A mean pressure of 2 mmHg exists in the left atrium and the major pulmonary veins. Pulmonary arterial pressure is only about one-sixth of systemic arterial pressure.

The capillary and venous pressures do not vary much in the two circulations. Thus, there is only a minute fall of pressure along the pulmonary arterioles and, therefore, a reduced potential for active regulation of the distribution of the pulmonary blood flow. It is also responsible for arterial pressure wave's minimal dampening and pronounced pulsatile nature of the capillary blood flow in lungs.

7.5.2 Blood Volume of the Lungs

At any given point of time in physiological conditions, only about 9% of total blood volume is present in the lungs, which can increase by four- to sevenfold during heavy exercise. The lungs have the capacity to increase the flow by (1) threefold increasing the number of open capillaries, (2) twofold distension of capillaries to increase the flow rate of individual capillary and (3) increasing the pulmonary arterial pressure.

Constriction of blood vessels takes place when the oxygen concentration in the alveolar air falls about 70% below normal (73 mmHg PO_2), a response strikingly different from systemic vessels, which dilate in low oxygen concentrations. The systemic vessels act oppositely by dilating in low concentrations of oxygen. This feature of pulmonary vascular resistance in response to low oxygen is practically important since this promotes the distribution of blood to areas where there is better alveolar oxygen pressure. Table 7.2 demonstrates the partial pressure of various gases, atmospheric, humidified, alveolar and expired air.

7.6 Respiratory Unit

The *respiratory unit or the respiratory lobule* is the basic structural unit where gaseous exchange takes place. It comprises a respiratory bronchiole along with its associated alveolar ducts, atria and alveoli. The thickness of the alveolar walls is extremely less and is endowed with fine interconnecting capillaries between the alveolar walls. Such extensive capillary plexus distribution ensures the flow of blood precisely as a thin sheet in the alveolar wall. It also allows the alveolar gases to remain in very close proximity to the capillary blood.

It is pertinent to note that alveoli are not the only sites of gas exchange but the exchange of gases occurs throughout the membranes of the respiratory unit. All the membranes of the respiratory unit that take part in gaseous exchange are known as the *respiratory* or *pulmonary membrane* (Fig. 7.4).

7.6.1 Respiratory Membrane

From inside to outside, the sequence of different layers of respiratory membrane (thickness of 0.2–0.6 μm) is (1) layer of fluid containing surfactant lining the alveolus; (2) a thin layer of alveolar epithelium; (3) basement membrane; (4) an interstitial space between the alveolar epithelium and capillary membrane; (5) the basement membrane of capillary, which at some places may form a continuous membrane by fusing with the alveolar epithelial basement membrane; and (6) the membrane endothelium of the capillaries.

Table 7.2 Partial pressures of respiratory gases as they enter and leave the lungs (at sea level)

	In atmosphere (mmHg)	In humidified air (mmHg)	In alveoli (mmHg)	Exhaled air (mmHg)
N_2	597.0 (78.62%)	563.4 (74.09%)	569.0 (74.9%)	566.0 (74.5%)
O_2	159.0 (20.84%)	149.3 (19.67%)	104.0 (13.6%)	120.0 (15.7%)
CO_2	0.3 (0.04%)	0.3 (0.04%)	40.0 (5.3%)	27.0 (3.6%)
H_2O	3.7 (0.50%)	47.0 (6.20%)	47.0 (6.2%)	47.0 (6.2%)
Total	760.0 (100.0%)	760.0 (100.0%)	760.0 (100.0%)	760.0 (100.0%)

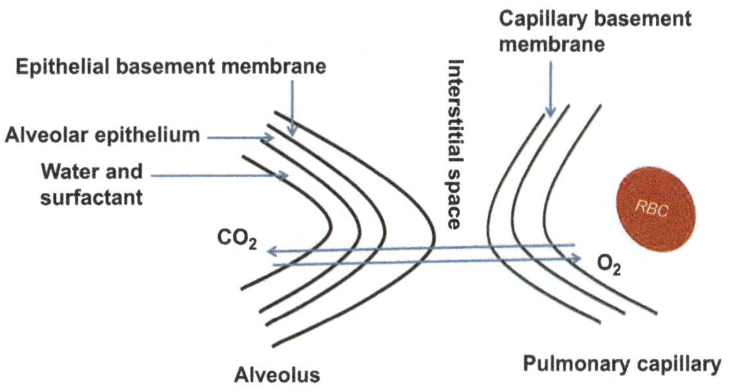

Fig. 7.4 Schematic diagram of the respiratory membrane showing the gas exchange between the alveolus and the RBC. The cross section of the ultrastructure of the pulmonary membrane consists of various layers that the gases must diffuse through before reaching the pulmonary capillary

7.6.2 Diffusing Capacity of the Membrane

The respiratory membrane's diffusing capacity is a measure of the pulmonary membrane's capacity to permit gaseous exchange between the alveoli and the pulmonary blood. This is defined as the volume of a gas (either O_2 or CO_2) that can diffuse through the membrane at a pressure difference of 1 mmHg in 1 min.

The factors that determine the diffusion capacity are (1) the distance of diffusion determined by membrane thickness; (2) the surface area available for diffusion; (3) the diffusion coefficient of the gas, which depends upon the solubility of a gas and its molecular weight; and (4) the partial pressure difference of the gas between the two sides of the membrane.

7.6.2.1 Diffusing Capacity (Dl) for Oxygen

It is also called the "*transfer factor*", and it is the measure of rate of gas transfer from alveolar space to alveolar capillary blood. The diffusing capacity depends upon the number of functioning alveolar-capillary units, i.e. the surface area available for gas exchange and the volume of blood or haemoglobin available in the pulmonary capillaries. In a human being, the diffusing capacity for oxygen during rest is approximately 21 mL/min/mmHg, increasing up to three times during heavy exercise, and for carbon dioxide, it is 400–450 mL/min/mmHg and 1200–1300 mL/min/mmHg during rest and heavy exercise, respectively.

7.6.3 Ventilation-Perfusion Ratio

The ventilation-perfusion ratio expressed as (V_A/Q) is a quantitative measurement or ratio of all the air entering the alveoli to the total blood flowing to both the lungs per minute. Under normal conditions, each alveolus perfuses well, and the blood flow (Q) is normal in each alveolus. The alveolar ventilation, V_A, is also normal. The normal ventilation-perfusion ratio will ideally be 1. The collapsed alveoli will remain unventilated albeit with the intact blood vessels, allowing normal perfusion to occur. Under such conditions, this ratio will be zero, whereas in the lung, the blood supply gets blocked because of emboli, and the ventilation will occur but without perfusion. This ratio has a value of infinity. Different diseases can indicate a high or low V_A/Q value.

When the alveolar ventilation and perfusion are normal, an optimum exchange of respiratory gases occurs via the pulmonary membrane. If (V_A/Q) is subnormal, there is inefficient oxygenation of the deoxygenated blood in the alveolar capillaries due to poor ventilation. Under such conditions, some fraction of blood fails to become oxygenated, which is termed as shunted blood. The physiologic shunt measures the total quantity of shunted blood formed in 1 min. A higher value of physiologic shunt is indicative of the quantum of blood that fails to be oxygenated.

Normoventilation refers to normal ventilation, whereby the P_aCO_2 is maintained at 40 mmHg. Hyperventilation occurs when the alveolar ventilation increases, which lowers the P_aCO_2 below 40 mmHg. In hypoventilation, the alveolar ventilation decreases, elevating the P_aCO_2 above 40 mmHg. Hypoventilation is associated with respiratory acidosis, which disturbs the body's acid-base equilibrium and blood pH. Hyperventilation leads to respiratory alkalosis, wherein excess deep breathing exhales out more CO_2 leading to a rise in pH due to a fall in H^+ ions as given in the equation. The reverse happens during hypoventilation, where excess carbon dioxide is retained by the body, elevating the H^+ ion concentration, and reduces blood pH causing respiratory acidosis:

$$H_2O + CO_2 \rightleftharpoons H_2CO_3 \rightleftharpoons H^+ + HCO_3^-$$

The reversible reaction influences the rise and fall in H^+ ions with the increase and decrease of carbon dioxide.

7.6.4 Common Anomalies of the Respiratory System

Atelectasis: A congenital or an acquired defect whereby the alveoli fail to open with air entry at birth or the alveoli have collapsed after inflating. Congenital atelectasis is seen in newborns when the lungs fail to inflate after inhaling a few breaths resulting from airway obstructions with clogged amniotic fluid and meconium. In premature newborns with inadequate quality and quantity of pulmonary surfactant, it is termed infant respiratory distress syndrome. This defect is also observed in piglets and foals, where newborns show a characteristic sign of gasping commonly termed "barkers". Acquired atelectasis most commonly occurs owing to obstructions caused by abscesses and tumours in the pleural cavity, the other reasons being pneumothorax, hydrothorax and bloat.

***Acute respiratory distress syndrome* (ARDS)**: It occurs in adult humans and animals characterised by several pathological manifestations such as intravascular accumulation of neutrophils, pulmonary hypertension, diffuse damage of the alveoli and acute lung injury accompanied by oedema with the formation of hyaline membranes. Several factors contribute to the pathogenesis of ARDS, resulting in diffuse alveolar damage from systemic diseases or from lung injury that generates a cytokine storm triggered by TNF-α, IL-1, IL-6 and IL-8, causing the release of cytotoxic enzymes and free radicals from neutrophils, which damages the lungs.

Pneumonia: Pneumonia occurs in all species and is characterised by acute inflammation of the lungs from several causes. The capillaries get filled with blood, and serous fluid occupies the alveoli. Eventually, the RBC, leukocytes and fibrin mix with this fluid. In the final stage, the debris is liquefied and removed along with repair and regeneration of the alveolar epithelium.

7.7 Transport of Oxygen

Oxygen is vital for life-sustaining aerobic respiration. Generally, there are two types of oxygen transfer in the body: convection and diffusion. Convection refers to the active process where oxygen is moved in circulation through mass transport. Diffusion refers to the passive movement of oxygen along a concentration gradient, for example from the microvasculature to tissues.

7.7.1 Convective Transport of Oxygen

7.7.1.1 Oxygen Uptake into the Bloodstream

Oxygenation of deoxygenated venous blood takes place in the pulmonary capillaries, as oxygen diffusion takes place through the alveolar-capillary membrane depending on the concentration gradient.

7.7.1.2 Haemoglobin and Its Role in Oxygen Transport

Oxygen is transported in the blood in two forms: firstly bound to haemoglobin (Hb) and secondly as a gas dissolved in plasma.

After diffusion through the alveolar membrane, oxygen binds to haemoglobin to form oxyhaemoglobin in the pulmonary capillaries. One molecule of haemoglobin can carry up to four molecules of oxygen and can be released in a reversible manner when needed. The binding of an oxygen molecule to haeme changes the shape of the globin chain, thereby changing the quaternary structure of haemoglobin. It makes it easier for subsequent oxygen molecules to bind to haemoglobin molecules with greater affinity, a phenomenon called cooperativity. This phenomenon can be described using a sigmoid-shaped oxyhaemoglobin dissociation curve (ODC). The curve has some characteristic features with specific physiological significance.

Major events during oxygen dissociation are the following: (1) The curve is S-shaped with a plateau showing nearly 98.3% saturation. (2) At a pO_2 of 70 mmHg, Hb is fully saturated even though the alveolar pO_2 is 100 mmHg. It is advantageous in situations where Hb can bind to enough oxygen even though the oxygen level can fall to low levels, such as at high altitudes and in some disease conditions. (3) Between the arterial pO_2 of 100 mmHg and venous pO_2 (at rest) of 40 mmHg, only a small amount of oxygen is unloaded by Hb. (4) Between the two extremes of normal venous pO_2 at rest (40 mmHg) and under conditions of strenuous exercise, the working muscles and tissues can get a lot of oxygen. Since this portion of the curve has a steep slope, a small reduction in pO_2 causes a release of large amounts of O_2; that is, with an increase in the demand for O_2, a lot of oxygen is given to the tissues.

Haemoglobin exists in two forms: The two forms are termed taut (T) and relaxed (R). Taut (T) has a low affinity for oxygen, and relaxed (R) has a high affinity for oxygen. In the tissues, where the environment is rich in carbon dioxide and low in pH, the taut form of haemoglobin commonly occurs, favouring oxygen delivery to the tissues being dissociated from the haemoglobin. In the reverse conditions, especially in the alveoli where the carbon dioxide is low and the pH is higher with the high partial pressure of oxygen, the relaxed form of haemoglobin is found to be bound strongly with oxygen. This phenomenon is known as the *Bohr effect*.

The oxygen capacity of haemoglobin is expressed by the maximum volume of oxygen that 1 g of haemoglobin can combine with, also known as Hüfner's constant. The theoretical maximum oxygen carrying capacity of haemoglobin is 1.39 mL O_2/g Hb. In fact, the oxygen capacity of haemoglobin is lower than the calculated value, and according to Nunn's Applied Respiratory Physiology, 1.306 (or 1.31) mL/g is the accepted value for clinical purpose. This is partly due to altered forms of haemoglobin, such as methaemoglobin and carboxyhaemoglobin, which reduce the capacity of haemoglobin to carry oxygen.

Haemoglobin oxygen saturation is the percentage of the number of occupied oxygen-binding sites out of the maximum number of available oxygen-binding sites. P50 is the partial pressure of oxygen at which haemoglobin is 50% saturated. It is a marker of haemoglobin oxygen affinity and is used to compare changes in curve position. The position of ODC changes in the face of various chemical and physiological factors (Table 7.3) and different haemoglobin present in different species.

7.7.1.2.1 Role of 2,3-DPG

An organic phosphate, 2,3-diphosphoglycerate (2,3-DPG) is produced during glycolysis and found in red blood cells that supports the release of oxygen from haemoglobin. High concentrations of oxyhaemoglobin in the erythrocytes suppress the production of 2,3-DPG by inhibiting the enzyme that forms 2,3-DPG. However, when oxyhaemoglobin levels are low, 2,3-DPG synthesis increases. It occurs with chronic hypoxia due to high altitudes and anaemia, and 2,3-DPG decreases the affinity of haemoglobin for oxygen. Under the anaemic condition, elevated production of 2,3-DPG occurs, which shifts the ODC to the right, thereby increasing oxygen delivery to the tissues and minimising the hypoxic effects. On the contrary, in banked donor blood, 2,3-DPG is lost by

Table 7.3 The left and right shifts of the oxygen dissociation curve under different physiological conditions

Left shift of haemoglobin (↓P_{50})	Right shift of haemoglobin (↑P_{50})
↑pH; ↓P_aCO_2; ↓2,3-diphosphoglycerate; ↓temperature	↓pH; ↑P_aCO_2; ↑2,3-diphosphoglycerate; ↑temperature
Increased affinity of haemoglobin to oxygen, increased binding of oxygen	Decreased affinity of haemoglobin to oxygen, increased release of oxygen in tissues
Carbon monoxide poisoning, foetal haemoglobin, methaemoglobin	Adult haemoglobin

↓ indicting the decrease, ↑ indicating the increase

metabolism as a result of reduced transfusion capacity of oxygen delivery to the tissues.

Blood oxygen content is the amount of oxygen carried in every 100 mL of blood.

This can be estimated by the sum total of O_2 transported as bound Hb and the O_2 dissolved in solution form = (1.34 × Hb × SpO_2 × 0.01) + (0.023 × P_aO_2):

[SO_2 = saturation of Hb with oxygen in percentage; Hb = gram haemoglobin concentration (Hb in grams/ 100 mL blood). pO_2 = partial pressure of oxygen (0.0225 mL of O_2 dissolved per 100 mL plasma per kPa, or 0.003 mL per mmHg)].

When applied for a healthy adult male, the oxygen content of arterial blood is as given:

The arterial oxygen saturation (SpO_2) = 98.3%, Hb = 15 g/100 mL, and arterial partial pressure of oxygen (P_aO_2) = 13.3 kPa.

The oxygen content of arterial blood (CaO_2) is CaO_2 = 19.758 + 0.3 = 20.058 mL/100 mL.

Similarly, the oxygen content of mixed venous blood can be calculated.

The normal values of mixed venous oxygen saturation (SvO_2) = 75% and partial venous pressure of oxygen (PvO_2) = 6 kPa, so CvO_2 = 15.2 + 0.1 = 15.2 mL/100 mL.

7.7.1.3 Foetal and Neonatal Oxygen Transport

The respiratory system plays a vital role in transporting oxygen to the tissues that, in turn, is utilised in aerobic metabolism for the supply of energy during the foetal and neonatal stages. The oxygen delivery to the foetal tissues depends mainly on the oxygen consumption by the tissues, the pressure gradient and the affinity between haemoglobin and oxygen. Oxygen is transported by blood either in the dissolved state or bound to haemoglobin. Foetal haemoglobin plays an essential role in transporting and delivering oxygen during the foetal and neonatal stages. In humans, the foetal haemoglobin, denoted by HbF, comprises two alpha and gamma chains ($\alpha 2\gamma 2$). Oxygen has greater affinity for this haemoglobin than adult haemoglobin (HbA), favouring oxygen binding with haemoglobin across the placenta. After birth, the oxygen demand greatly increases in the newborn, which the foetal haemoglobin cannot meet since enough oxygen cannot be diffused from HbF. Therefore, in the postnatal stage, the HbF gradually decreases and eventually replaces the adult form HbA. The HbF is reduced to 2% from about 75% during the first year of life. During this transition period, there is an increase in 2,3-DPG concentration, which helps decrease the oxygen affinity and fulfils the demand for increased oxygen supply to the tissues.

7.7.1.4 Non-classical Role of RBC in Oxygen Delivery to Tissues

During circulation, the red blood cells can sense the oxygen status of tissues through their degree of deoxygenation and use this sense to stimulate the release of vasodilatory compounds such as nitric oxide (NO) or ATP, which stimulate blood flow to hypoxic tissues. There are three mechanisms by which this happens: (a) release ATP that stimulates endothelial cells for the release of NO; (b) upon deoxygenation, it triggers *S*-nitroso-Hb^- to release nitric oxide; and (c) deoxyhaemoglobin reduces nitrite (NO_2^-) to vasoactive NO by means of nitrite reductase activity.

7.8 Transport of Carbon Dioxide

The purpose of breathing is to supply oxygen to the tissues and get rid of the carbon dioxide (CO_2) produced during metabolism, which can sometimes be at the rate of 200 mL/ min, thus stabilising the biochemical environment necessary to maintain the vital metabolic process.

Carbon dioxide occurs in three different forms in the body, i.e. (a) dissolved, (b) as bicarbonate and (c) as carbamate.

7.8.1 Transport as Dissolved Carbon Dioxide

Only 5% of total arterial content is present in dissolved CO_2, and the contribution of dissolved CO_2 to the total arteriovenous CO_2 concentration difference is only 10%. During heavy exercise,. the CO_2 in the dissolved state can rise up to sevenfold that may contribute to one-third of the total CO_2 exchange.

Though carbon dioxide is 20 times more soluble than oxygen because of its high solubility and diffusing capacity, carbon dioxide partial pressure of alveolar and pulmonary end-capillary blood is virtually the same. At the same time, arterial blood will contain about 2.5 mL of dissolved carbon dioxide per 100 mL and venous blood 3 mL per 100 mL.

7.8.2 Transport as Bicarbonate

The majority of CO_2 is transported as HCO_3. The *Henderson-Hasselbalch* equation gives the ratio of HCO_3 over dissolved CO_2:

$$pH = pK'_a + \log \frac{[HCO_3^-]}{SCO_2 + PCO_2}$$

In humans, the plasma value of pK_a is 6.10 at 37 °C, which varies with changes in temperature and ionic strength.

Carbon dioxide and water can easily enter the red blood cell by diffusion. They form carbonic acid, a reversible reaction favoured by carbonic anhydrase, dissociating into hydrogen and bicarbonate ions. The ratio of H_2CO_3 to HCO_3^- is 1: 20 at a physiological pH of 7.4. However, during exercise, the lactic acid produced in addition to CO_2 will reduce the blood pH, resulting in a drop in the ratio of H_2CO_3 to HCO_3^- from 1:20 to 1:13:

$$CO_2 + H_2O \underset{}{\overset{\text{Carbonic anhydrase}}{\rightleftarrows}} H_2CO_3 \rightleftarrows H^+ + HCO_3^-$$

The bicarbonate ions formed in RBC diffuse out to the plasma, and chloride ions move in. This phenomenon is known as the chloride shift (or Hamburger effect). This is facilitated by an ion-exchange transporter protein in the cell membrane called capnophorin or Band 3 for Cl–HCO_3 that allows chloride shift at a 1:1 ratio. As the process of carbonic acid production and further dissociation continues, there is a danger of building up hydrogen ions in the RBC, which can further prevent this process. But this does not happen since hydrogen ions bind to reduced haemoglobin formed by unloading oxygen at the tissue level. This reduced haemoglobin is less acidic than oxygenated haemoglobin and can accommodate hydrogen ions. Therefore, the degree of oxygenation determines the CO_2-binding capacity of haemoglobin, a phenomenon known as the Christiansen-Douglas-Haldane effect (CDH effect) or simply the *Haldane effect*, named after the three physiologists who first demonstrated this phenomenon in 1914.

The chloride shift has a major impact on the volume of the red blood cells. The chloride shift increases the intercellular osmolarity of the cells which along with the hydrogen ion buffering increases the volume of the red blood cells by drawing in water. As a result, the mean corpuscular volume (MCV) rises. This process gets reversed as the blood passes through the lungs.

At the level of the lung, as O_2 enters RBCs and binds to Hb, it promotes the release of Hb-bound CO_2 and H^+ that deoxy-Hb buffered. The H^+ binds to HCO_3^- to form H_2CO_3, which dissociates to form CO_2 and water. The CO_2 is expired out by the lungs, thereby decreasing the CO_2 and H^+ concentrations, which increases the affinity of Hb for oxygen in the RBCs passing through pulmonary capillaries. Because CO_2 and H^+ effects are interrelated and additive, the combined change in Hb oxygen affinity has been called the classical *Bohr effect*. In contrast, the change in Hb oxygen affinity produced only by H^+ is called the Bohr effect:

$$O_2 + HbH + HCO_3^- \leftrightarrow HbO_2 + H_2O + CO_2$$

7.8.3 Transport as Carbamate

Carbon dioxide tends to bind rapidly to terminal uncharged amino groups ($R\text{-}NH_2$), resulting in the formation of carbamino compounds. The combination of CO_2 with NH_2 groups is called *carbamate*. When carbamate is formed from haemoglobin, it is called *carbaminohaemoglobin*. The carbamino form comprises a very small amount of the carbon dioxide transported by the blood, about one-third of the difference in the amount of carbon dioxide between the veins and arteries.

The amount of CO_2 bound as carbamate to haemoglobin or plasma proteins depends on several factors such as the saturation of haemoglobin with O_2, the 2,3-diphosphoglycerate (2,3-DPG) concentration of RBC and the concentration of H^+ ions in red blood cells and plasma. Deoxygenation of haemoglobin facilitates the binding of the increased amount of CO_2 to haemoglobin, while an increase in H^+ ions or acidification decreases the carbamate formation by haemoglobin.

In general, carbaminohaemoglobin has a lower affinity for oxygen. In sheep and goats, a type of haemoglobin HbC exists prominently in newborns and anaemic and hypoxemic adults, which binds twice as much CO_2 as HbA in goats. This produces a marked reduction in the binding affinity of oxygen to HbC in goats and sheep than that in normal adult Hbs.

The carbamino formation also produces H^+, which further lowers Hb oxygen affinity.

7.8.4 Carbon Dioxide Dissociation Curve

The carbon dioxide dissociation curve depicts variations in the total CO_2 carried by the blood with the changes in partial pressures of carbon dioxide. The curve shifts to the right when the Hb is oxygenated, which means that the blood begins to release carbon dioxide as the Hb becomes oxygenated. It is known as the Haldane effect.

7.9 Control and Regulation of Respiration

Respiration is well regulated both at central and peripheral levels in the body. The centre for respiration, or the respiratory centre, is situated in the medulla oblongata and pons that controls breathing.

7.9.1 Medullary Control

The *dorsal respiratory group* (DRG) and the *ventral respiratory group* (VRG) comprise the medullary control centres of respiration.

Dorsal respiratory group: The major component of the dorsal respiratory group is the nucleus tractus solitarius (NTS). It contains inspiratory or I neurons which form the centre of inspiration. The NTS receives afferent inputs from some regions of VRG neurons, inputs that modulate respiration from peripheral arterial chemoreceptors, upper airway and lungs. The various respiratory inputs are integrated into the NTS. When this centre is stimulated, the inspiratory or I neurons fire and send impulses to the neurons supplying the inspiratory muscles. Contraction of these muscles occurs, which begins the process of inspiration.

Ventral respiratory group: These respiratory neurons are situated in the ventral part of the medulla and are involved in both inspiration and expiration. This area primarily consists of expiratory neurons in addition to some inspiratory neurons. When the ventral respiratory group of neurons is stimulated, it leads to expiration. The VRG encompasses several adjacent compartments that include the pre-Bötzinger complex (preBötC), the Bötzinger complex (BötC) and retrotrapezoid/parafacial nucleus (RTN/pFG). Whereas the pre-Bötzinger complex (preBötC) is a neural network responsible for inspiration during respiratory activity and a central pattern generator with respect to respiration (respiratory rhythm generator), the retrotrapezoid nucleus (RTN) is located on the ventral surface of the brain and contains chemosensitive cells that can detect pH alterations.

7.9.2 Pontine Respiratory Centres

The *pneumotaxic* and the *apneustic* centres help set up the breathing rate by influencing the depth of breathing. The pneumotaxic centre inhibits the inspiratory area and thus helps to stop inspiration. The apneustic centre stimulates the inspiratory area, prolonging inspiration (apneusis or deep sigh). But the pneumotaxic centre is dominant over the apneustic centre, so when the pneumotaxic centre is active, the apneustic centre is overridden by it.

The pneumotaxic centre: The pneumotaxic centre is situated in the upper part of the pons and is comprised of the parabrachial and Kölliker-Fuse nuclei. The pneumotaxic centre is responsible for controlling the rate and breathing pattern. This centre limits the inspiration, providing an inspiratory off-switch (IOS) by restricting the firing of action potentials in the phrenic nerve, thereby decreasing the tidal volume. Lesions in this centre are associated with a reduced respiratory rate with an increase in the depth of respiration. It has connections with the DRG of the medulla; hence, when needed, it can send signals to increase the breathing rate.

The apneustic centre: Apneusis, or apneustic breathing, is an abnormal breathing pattern that is characterised by a full inspiration followed by a prolonged pause. This centre in the lower pons promotes prolonged inspiration by continuously stimulating the neurons in the medulla oblongata. The medulla (dorsal group) receives signals from the apneustic centre to prolong the inspiratory off-switch (IOS) signal generated by the pneumotaxic centre. These positive impulses to the inspiratory neurons, in turn, control the intensity of breathing. Impulses from the pulmonary stretch receptors and pneumotaxic centre inhibit the apneustic centre. Thus, the centre, in turn, can also inhibit the pneumotaxic centre.

7.9.3 Other Higher Centres Controlling Respiration

The spontaneous rhythmicity-generated impulse in the medullary respiratory centre can be completely overwhelmed (at least temporarily) by influences from the higher brain centres (hypothalamus, limbic system, cortex).

1. **Hypothalamus** is involved in the respiratory modifications during pleasure, anger and fever.
2. **Limbic system** is involved in the emotional responses.
3. **Cortex controls** maximum voluntary ventilation and respiratory modifications during the speech, singing, playing wind instruments and emotional states.

7.9.4 Reflexes and Chemical Control of Respiration

Breathing is modulated by several reflexes emanating from different sites in the body. Some of them are physical reflexes, and others are chemical reflexes.

7.9.4.1 Physical Reflexes

7.9.4.1.1 Herring-Breuer Reflex

In 1868, two physiologists, Herring and Breuer, reported that enlargement of the lungs of anaesthetised animals decreases the frequency of inspiratory efforts, whereas collapse has the opposite effect, thus concluding that lung receptors modulate the pattern of breathing. This phenomenon is termed as Herring-Breuer reflex. If it takes place during inspiration, it is called inflation reflex (HBIR), and if it takes place to terminate further expiration, it is called deflation reflex (HBDR). Slow-adapting receptors (SARs) in the lungs and pleural tissues are stimulated by the pulmonary stretch during inspiration-associated inflation to terminate further inspiration and overinflation (inspiratory switch off). Similarly, pulmonary stretch extends the period between subsequent breaths, and pulmonary deflation quickens the next inspiration. The vagus nerve conveys the afferent impulses to pump cells present in and around the ventrolateral nucleus solitary tract. These cells project to inspiratory neurons in the lateral respiratory column and halt the respiration by releasing inhibitory neurotransmitters. In other words, it is often considered an inhibitory sensory feedback loop that shapes the respiratory motor pattern.

This reflex is strongest in neonates, especially at birth, but decreases during the first year of life. It occurs due to an excessively compliant chest wall in newborns that can collapse at volumes below functional residual capacity during expiration, thereby activating HBDR. The HBR becomes less significant due to postnatal maturation of the cardio-respiratory control circuits in mammals. Nevertheless, during exercise and vocalisation, this reflex regulates breathing patterns according to the changes in behaviour and emotions.

7.9.4.1.2 J Reflex

In 1970, an Indian physiologist, Autar Singh Paintal, first described the J reflex. These receptors are located within alveolar septa and are juxtaposed to the pulmonary capillaries and are hence called juxta-pulmonary capillary receptors or J receptors. Due to their widespread presence in most tissues, they are also known as pulmonary C fibre receptors and are non-myelinated afferent fibres. They are considered irritant receptors, responding to noxious stimuli like chemical irritants or dust. They also react to inflammation or accumulation of fluid within the pulmonary interstitium and trigger tachypnoea.

7.9.4.2 Chemical Control of Respiration

The chemical changes controlling respiration are mediated by chemoreceptors, which are sensors that can detect alterations in oxygen, carbon dioxide and pH. They are located either in the brain (central chemoreceptors) or other peripheral areas of the body (peripheral chemoreceptors).

7.9.4.2.1 Central Chemoreceptors

They occur in various parts of the brain, viz. cerebellum, midbrain, hypothalamus and brainstem. Central chemoreception has two major physiological functions: (1) maintenance of a constant, normal arterial PCO_2 and (2) maintenance of a constant pH through the exchange of CO_2 by ventilation to correct the acid-base disturbances. The central chemoreceptors also control a more comprehensive range of physiological processes. The central and peripheral chemoreceptors are CO_2/H^+ sensitive that can sense the level of CO_2 in arterial blood and alveolar air and provide instant feedback to the brainstem respiratory control system. This PCO_2 value is determined by the ratio of metabolic CO_2 production at tissues and the amount of alveolar ventilation. A decrease in alveolar ventilation with a constant rate of CO_2 production results in an increase in arterial and alveolar PCO_2 and vice versa. An increase in PCO_2 would stimulate chemoreceptors, increasing alveolar ventilation and decreasing PCO_2, correcting the initial increase. CO_2/H^+-sensitive chemoreceptors provide excitatory or inhibitory afferent input to the respiratory control system, a classic feedback control loop depending on the metabolic status relative to the alveolar ventilation.

Central chemoreceptors monitor brain interstitial fluid (ISF) pH, which is determined by tissue PCO_2 and bicarbonate. Three factors determine tissue PCO_2: the arterial PCO_2, the rate of CO_2 production by medullary tissue and the cerebral or medullary blood flow (CBF). In this view, central chemoreceptors may detect arterial PCO_2 and serve as a chemical feedback loop in the control of breathing and changes in tissue pH that result from acid-base disorders that arise either in the periphery or centrally.

Increased PCO_2 vasodilates cerebral vessels, and CBF increases; decreased PCO_2 vasoconstricts cerebral vessels, and CBF decreases. Increased arterial PCO_2 increases ISF H^+, stimulates central chemoreceptors and vasodilates cerebral vessels. The resultant increase in CBF decreases tissue PCO_2 widening the arterial tissue PCO_2 difference and minimising the initial stimulus intensity at the chemoreceptors. Conversely, decreased arterial PCO_2 decreases ISF H^+, inhibits central chemoreceptors and vasoconstricts cerebral vessels. The resultant reduction in CBF increases tissue PCO_2 diminishing the arterial tissue PCO_2 difference and minimising the degree of central chemoreceptor inhibition. Thus, the responses of CBF to changes in PCO_2 serve both to maintain ISF pH relatively constant and to modulate the central chemoreceptor response to a level appropriate for the ISF pH stimulus intensity.

7.9.4.2.2 Peripheral Chemoreceptors

The peripheral chemoreceptor, especially the carotid body, was first anatomically described in 1762 by Albrecht von Haller. But only in 1900 was its physiological function described by Kohn.

The carotid bodies are located bilaterally in the neck, at the anterior end of the left and right common carotid arteries. Similarly, there are other chemoreceptor structures called aortic bodies in the aortic arch region, which are structurally similar to the carotid bodies. The glossopharyngeal nerve and the vagus nerve carry the chemoreceptor input to the brainstem, making synapses with neurons in the DRG.

The carotid body is comprised of type I and type II cells. Type I are the most abundant cells and are referred to as "glomus", "chief", "epithelioid" or simply "chemoreceptor" cells. Type II cells are termed "sustentacular or sheath" cells or "pericyte". The organisation of these cell types within the carotid body is non-uniform, with type II cells forming 20% of the total population being closely connected with type I cells which form small groups with 3–5 cells. The carotid body primarily responds to hypoxia and responds to many other respiratory and nonrespiratory stimuli, including CO_2, pH, glucose, proinflammatory cytokines, circulating hormones, K^+, osmolarity and temperature. These observations raise the question of whether the carotid body is a "polymodal" sensory receptor or whether these stimuli modify some elements of the O_2 transduction pathway. Future studies addressing the carotid body response to these "other" stimuli may prove to be as relevant for human health as the response to hypoxia.

In response to falls in arterial PO_2, the glomus cells depolarise due to a decrease in K^+ efflux and ensuing activation of the voltage-gated Ca^{2+} channels. Carotid body produces a characteristically non-linear, graded chemo afferent discharge in the carotid sinus nerve. The in vivo increase in discharge frequency gradually reduces below P_aO_2 of 20–30 mmHg and may even fall sometimes, often with the failure to maintain adequate systemic blood pressure.

The carotid body helps to maintain normal level of ventilation in sleep. During sleep, the apnoea that occurs within seconds of transient hyperventilation has been attributed to hypocapnia sensed in the carotid body.

Central and peripheral chemoreceptors are interdependent for functional activity rather than being separate entities. Under its ability to maintain neuronal excitability, the tonic drive influences the ventilation and regular respiratory rhythm in adults. Drive can arise from many sources, e.g. the activity of the reticular formation, excitatory input from afferents involved in respiratory control and inputs related to CO_2 sensed by central (and peripheral) chemoreceptors.

Based on their location in the body, two types of chemoreceptors have been recognised: central and peripheral. The differences between the central and peripheral chemoreceptors are outlined in Table 7.4.

7.9.5 Regulation of Respiration During Exercise and at High Altitudes

The brain simultaneously sends motor impulses to the muscles during exercise as well as excitatory impulses to activate the respiratory centre in the brainstem.

When one individual begins to exercise, the first response is the increase in ventilation before any changes in blood chemicals occur. It is presumed that neurogenic signals transmitted directly into the brainstem respiratory centre produce most of the increase in respiration, which also reaches the muscles and results in muscle contraction. However, the nervous and respiratory control signals can be too strong or weak in some instances. In such cases, the chemical factors play a significant role in maintaining equilibrium. The adjustment of respiration brings the body fluids' carbon dioxide, oxygen and hydrogen ion concentrations to normal levels.

When a person ascends to a high altitude slowly, several compensatory mechanisms develop in the body. At higher

Table 7.4 Differences between central and peripheral chemoreceptors

Central chemoreceptors	Peripheral chemoreceptors
• Located in the central nervous system in or around the medulla oblongata.	• Present as aortic bodies in the aortic arch wall and as carotid bodies in the wall of the carotid sinus. The carotid bodies are supplied by the sensory fibres of the glossopharyngeal nerve, and aortic bodies are supplied by the sensory fibres of the vagus nerve. Carotid bodies are more important than the aortic bodies as respiratory regulatory organs.
• More sensitive towards increasing H^+ ion concentration or pCO_2 (*hypercapnia*). CO_2 is more lipid soluble and can diffuse into the cerebrospinal fluid (CSF) from the capillaries in the central nervous system (it can cross the blood-brain barrier readily). In the CSF, it forms H_2CO_3 by the action of carbonic anhydrase, which dissociates to form H^+ and HCO_3^- ions. The H^+ ions stimulate the chemoreceptors, which stimulate the inspiratory area to cause an increase in the rate and depth of breathing so that the increased pCO_2 can be brought down to the normal levels.	• Are sensitive to the levels of O_2, CO_2 and H^+ ions. Whenever there is an increase in pCO_2 or H^+ ions or reduction in pO_2 (only drastic reduction in case of pO_2 because a slight reduction in pO_2 around higher values of pO_2 would not affect as the Hb is 90% saturated even at a pO_2 of 60 mmHg), it causes these chemoreceptors to be stimulated, which in turn stimulates the inspiratory area to increase the rate and depth of breathing (hyperventilation) so that normal O_2, CO_2 and H^+ ion levels can be restored.
• Not stimulated by H^+ ions generated by other sources, e.g. lactic acid, because H^+ ions themselves cannot cross the blood-brain barrier so readily.	• Respond to pO_2 in the plasma and not oxygen bound to Hb, so there is no change in the respiratory rate in response to anaemia.

altitudes, the partial pressure of oxygen in the arterial blood decreases due to hypoxia. It causes the stimulation of peripheral chemoreceptors, which leads to hyperventilation. Hyperventilation, in turn, increases the partial pressure of oxygen in the arterial blood. The other acclimatisation responses are polycythaemia leading to a rise in RBC count due to erythropoietin production by the kidneys under a hypoxic state. The 2,3-DPG concentration of the RBC increases at high altitude, which shifts the oxygen dissociation curve to the right, leading to oxygen supply to the tissues by dissociation from haemoglobin. However, these compensatory mechanisms do not develop in persons who ascend to high altitudes at high speed, such as travelling via aircraft from sea to mountain leading to unconsciousness from high-altitude sickness.

7.9.6 Panting

The animal responds to an increase in core body temperature through regulatory mechanisms involving the respiratory centre, whereby the metabolic needs are balanced by an increase in respiratory frequency, which increases the dead-space ventilation, thereby increasing the evaporative heat loss accompanied by a fall in tidal volume. Panting is an important mechanism of controlling body temperature, particularly in smaller animals, whereas for larger mammalian species, sweating regulates body temperature. During panting, the alveolar ventilation remains unaffected, maintaining a constant partial carbon dioxide pressure in the alveolar air. Panting may occur in three patterns: (1) where the air is inhaled and exhaled through the nose; (2) air is inhaled through the nose and exhaled through both nose and mouth; and (3) where the air is inhaled and exhaled through both nose and mouth. It is to be mentioned here that maximum cooling can be achieved when the air is directed through the nose and leaves through the mouth and the minimum occurs when the air is inhaled and exhaled through the nose. When the ambient temperature is above 30 °C, (2) and (3) types of panting are observed in dogs. While air moves through the nasal cavity and mouth, the evaporative heat loss occurs mostly over the nasal mucosa and the tongue. The nasal mucosa receives a continuous supply of water from the nasal and orbital glandular secretions, which provides for this evaporative cooling by increasing the secretions with an increase in ambient temperature.

7.10 The Avian Respiration

Birds are an elite taxon of animals that have evolved the ability for powered flight. The respiratory system of birds is markedly different from other vertebrates in having relatively small lungs and the air sacs that are important for respiration.

7.10.1 Basic Design of the Respiratory System in Birds

The avian lung is non-lobulated and deeply marked by the vertebrae. Unlike mammals, there is no dichotomously branched bronchial system, and the airways do not terminate blindly. Avian lungs have a broad dorsal and a thin ventral aspect. The avian lung is rigid and wedge shaped that alters its volume by only 1.4% between the respiratory cycles. Birds have a larynx, which does not produce sounds. Instead, an organ termed the syrinx functions as the voice box. Lung volume in a bird is approximately 26% smaller, compared to animals of equivalent body mass, respiratory surface area (RSA) about 15% greater, harmonic thickness of the blood-gas barrier (τht) around 62% thinner and PCBV approximately 22% greater.

The tracheal rings are complete in birds rather than incomplete as in mammals. In comparison to mammals, the trachea in the bird is 2.7 times longer and 1.29 times wider, resulting in similar air resistance in the trachea as in mammals. Still, the dead-space volume in the trachea is about 4.5 times greater than in mammals. A lower respiratory frequency accompanied by a larger tidal volume compensates for the large dead space in the trachea.

The lungs are the gas-exchange structures, but they do not contract and expand during respiratory cycles and are relatively small and fixed to the ribs. Their ventilation depends on bellows-like extensions from the lungs and the air sacs, which expand and contract during respiratory cycles. The primary bronchus of birds extends from tracheal bifurcation to the ostium of the abdominal air sac and has only two dusters of secondary bronchi. From the cranial end of the bird's intrapulmonary primary bronchus (IPPB) arise four ventrobronchi, while its caudal segment gives rise to 7–14 variably sized dorsobronchi and a variable number (<6) of laterobronchi. The lining epithelium of the primary bronchi and the initial segments of the secondary bronchi are similar to those observed in larger mammalian airways, i.e. pseudostratified ciliated epithelium with mucous-secreting goblet cells.

At the endpoint of the mammalian bronchial system, there are many small-sized, blind-end alveoli serving as gas-exchange and tidal expansion sites. In sharp contrast, birds have a nearly constant volume and flow-through lung where the location of gas exchange is the parallel tertiary bronchi, i.e. parabronchi (few hundred to <2000 depending on taxa), connected between the ventrobronchi and the dorsobronchi or laterobronchi. The parabronchi form a complete airway loop from the caudal primary bronchus to the primary cranial bronchus (via secondary bronchi), permitting a unidirectional fresh gas flow.

7.10.1.1 Gas-Exchange Tissues

The parabronchi are densely packed in a hexagonal array simulating the honeycomb. The parabronchi are classified into two categories. The majority types of parabronchi are paleopulmonic parabronchi, connected between ventro- and dorsobronchi, which allows unidirectional flow of air. The second type is neopulmonic parabronchi, which varies from 0% to 20% of lung volume depending on the species. They conduct air in an oscillatory pattern and are variously connected between any two of the following: primary bronchus, ventrobronchi, dorsobronchi, laterobronchi and air sacs. The lumen of parabronchi is lined by non-stratified, non-secretory and non-ciliated epithelium.

The air capillaries arranged as serial parallel units communicate with the parabronchial lumen. Surfactant is also present in the parabronchial lumen and air capillaries. The pulmonary blood flows radially into the mantle of exchange tissue from the arteries located along its periphery towards the parabronchial lumen, along which are located the pulmonary veins that drain the oxygenated blood back towards the periphery.

7.10.1.2 Air Sacs

The bronchial system in avian species communicates with a system of poorly vascularised air sacs that are the sites of tidal volume expansion that aids to circulate air through the avian lung-air sac respiratory system. The wall of air sacs is formed by simple, non-stratified epithelium with a few clusters of ciliated and secretory cells over a diffuse elastin network. The air sacs have little function in gaseous exchange. Several air sacs communicate with bones and soft tissues, i.e. large pectoral or flight muscles. These diverticula and the small cervical air sacs lying along the neck do not contribute to ventilation.

Most birds have nine air sacs, four paired and one unpaired. These are the unpaired interclavicular sacs, one pair of cervical sacs (absent in some species), one pair of anterior thoracic sacs, one pair of posterior thoracic sacs and one pair of abdominal sacs.

Functionally, the nine air sacs are broadly classified as anterior and posterior sacs. The interclavicular, cervicals and anterior thoracics comprise the anterior sacs, while the posterior thoracics and abdominals constitute the posterior sacs.

These thin-walled air sacs have scanty vascular supply and, being mucoserous, have little function in direct gaseous exchange. They are primarily involved in allowing a unidirectional airflow, thereby ensuring a supply of oxygen-rich fresh air in the lungs.

So, in birds, more oxygen is available, reaching the blood system. In contrast, mammals have a bidirectional airflow in the lungs, and consequently, the air coming into the lungs is mostly mixed-up air which eventually has less oxygen.

Airway divisions serve the lungs and air sacs from the trachea known as primary, secondary and tertiary bronchi. The tertiary bronchi are also known as parabronchi. The parabronchi give rise to out-pocketing (atria), extensions from the atria (infundibuli) and extensions from the infundibula known as air capillaries.

The blood capillaries make intimate contact with the air capillaries and provide for the gas exchange in the lung's mantle.

7.10.2 Mechanics of Respiration in Birds

Birds lack a diaphragm, so pressure changes in the air sacs move the air in and out of the respiratory system. Air enters the respiratory system when chest muscles push the sternum outward, creating a negative pressure in the air sacs. Unlike mammals, the process of expiration in birds is an active process that requires muscular contraction, which can raise the pressure and force the air out of the air sacs.

7.10.2.1 The Respiratory Cycle of a Bird

The inspiration begins with the air travelling through the nares or nostrils surrounded by a fleshy tissue in some birds called the *cere*. The air then moves into the nasal cavity and passes down the larynx and the trachea. It passes to the syrinx, finally being divided into two as the trachea divides. Instead of moving directly to the lung, it moves into the caudal (posterior) air sacs, only a small amount being directed from here to the lungs. During the first expiration, the air travels back into the lungs from the posterior air sacs, being moved through the ventrobronchi and dorsobronchi. The bronchi ramify into smaller diameter air capillaries as blood capillaries flow through the air capillaries to exchange gases. During the second inspiration, the air moves to the *cranial* air sacs. The air moves out of the cranial air sacs during the second expiration and then passes over the syrinx into the trachea. Moving through the larynx reaches the nasal cavity and exits through the nostrils.

7.10.3 Exchange of Gases

In the avian lung, the exchange of gases occurs by simple diffusion. Oxygen diffuses into the blood from the "air capillaries", and carbon dioxide diffuses into the "air capillaries" from the blood. There is a cross-current flow of air in avian species. As the air passes through the parabronchi, blood moving through capillaries travels at right angles. Cross-current exchange is a very efficient system, which enables the pressure gradients of oxygen and carbon dioxide to be maintained along the entire length of the parabronchus-capillary "connection". Like in mammalian lungs, in avian lungs, too, O_2 in inspired air diffuses passively into the pulmonary capillary blood, and CO_2 diffuses in the opposite direction. However, there are certain special features in the avian parabronchial lung, different from the mammalian bronchoalveolar lung. The oxygen uptake and loss of carbon dioxide are not affected by the direction of gas flow through parabronchi. Air containing oxygen flows into the parabronchial lumen, from which it diffuses radially into the air capillary network, followed by pulmonary capillaries. The CO_2 moves in the opposite direction. Blood flow in pulmonary capillaries occurs in the opposite direction to that of O_2 diffusion in the air capillaries. The systemic arterial blood is a mixture of blood drawn from all individual air-blood capillary units that can contribute to a greater arterial PO_2 as compared to end-expired PO_2. This is unique to avian species only and never observed in the mammalian lung.

Another countercurrent-like system also exists in the avian respiratory system along with the cross-current system. It is comprised of the inward and outward directional flow of blood in the blood capillaries and air in the air capillaries. This system is not an efficient system of gas exchange because of complex, tortuous arrangement and short contact points between blood and air, which prevents sufficient gaseous exchanges. The blood-gas barrier (BGB) in birds is formed by the fusion of type I epithelial and endothelial cells over the basement membrane, having an almost uniform thickness.

7.10.4 Control of Ventilation in Birds

Like mammals, the central control area of ventilation in birds is located in the pons and medulla oblongata, which are higher brain control centres. The frequency of respiration and the time duration of inspiration and expiration are controlled by feedback received from several receptors located within the lungs and peripheral chemoreceptors, the receptors located within and in the vicinity of the air sacs, viz. the mechanoreceptors and thermoreceptors (in the hypothalamus and spinal cord).

Unlike mammals, birds have CO_2 receptors in their lungs (intrapulmonary receptors) that detect carbon dioxide levels in lung air. With a fall in the PCO_2 in the lung, the receptors in the lung get stimulated and increase their rate of discharge. As the rate of discharge increases, ventilation decreases and hence fine-tuning occurs.

7.10.5 Special Adaptations

There are two types of haemoglobin in adult birds, HbA and HbD, which vary in their affinity for oxygen. It is advantageous for birds that have to cope with large variations in the partial pressure of oxygen as they move from one habitat to another. HbA is more common and has a lesser affinity for oxygen, which can readily deliver oxygen to the tissues. Avian Hb also shows more cooperativity with oxygen than mammalian Hb. The Hills coefficient indicates the degree of cooperativity, which is 2.8 for mammals and 4 for birds.

In mammals, hyperventilation causes a decrease in P_aCO_2, which causes vasoconstriction and a decrease in cerebral blood flow, leading to cerebral ischaemia. Mammals can tolerate P_aCO_2 of 20 mmHg, but birds can maintain cerebral blood flow even at 8–10 mmHg. It is important for the survival of birds, which hyperventilate during flight.

Surfactant SP-B is especially important for maintaining airflow through the "tubes" of the respiratory system in birds, which can be ascribed to the phospholipids and proteins present in it. Present only in the mesobronchi, the surfactant SP-A contributes towards innate defence mechanism and regulation of inflammatory responses and has a prime role in the mesobronchi because airflow is slower and small particles could accumulate there.

7.11 Nonrespiratory Functions of the Lung

Apart from its well-studied classical function, the nonrespiratory function of the lungs is also multidimensional, as outlined in Table 7.5.

Learning Outcomes

- Respiration is vital to the survival of any animal species. The respiratory system is comprised of a conducting zone and respiratory zone, the latter taking part in gaseous exchange in the lungs. In birds, the air sacs play a prominent role in respiration. The respiratory system serves basic functions of providing pulmonary ventilation that involves the exchange of air between the atmosphere and alveoli of the lungs. Next, it facilitates the diffusion of gases such as carbon dioxide and oxygen between the alveolar membrane and blood, followed by the transport of oxygen and carbon dioxide dissolved in plasma and combination with Hb to and from the tissues. The respiration is regulated by the respiratory centres located in the medulla oblongata and pons. The medullary respiratory centre is divided into the dorsal group of respiratory neurons in the nucleus of tractus solitarius, which controls inspiration. The other part is the ventral group of respiratory neurons that causes expiration when stimulated. The respiratory centre sends impulses after receiving sensory inputs from peripheral chemoreceptors and mechanoreceptors. The respiratory system also performs nonrespiratory functions such as providing immune protection, neuroendocrine and filtering functions.

Table 7.5 Nonrespiratory functions of the lung

Functions	Significance	Description
Circulatory reservoir	*Recruitment and distension* Helps cope with the alterations in the cardiac output (during extraneous exercise)	A transition from the stage of incomplete perfusion of the pulmonary vascular bed during the resting phase to complete perfusion occurs during the high cardiac output phase. It is accomplished by recruitment of under perfused pulmonary vasculature followed by distension (smooth muscle relaxation) to accommodate increased blood flow. The pressure in the pulmonary circulation is six times less than systemic. This phenomenon helps in altering the blood volume by 500–1000 mL.
	Change instance and circulatory redistribution of blood flow	The blood volume in pulmonary vessels decreases by 50% and increases by 100% during forced expiration and inspiration, respectively. The blood flow in the pulmonary vessels is equivalent to right ventricular output. Around 70–100 mL of this flow is in pulmonary capillaries in gas exchange.
Immunity	Pulmonary alveolar macrophages (PAMs) are majorly responsible for lung immune function	• Engulfs the particles in the alveolar region and directs them to blood or lymph for removal: • Antigen presentation. • T-cell activation. • Neutrophil activation triggers the release of trypsin and elastase with high antibacterial activity and damages the self-mucociliary lining. The binding of these proteases with alpha-anti-trypsin may result in its inactivation (advocated in the treatment of pulmonary emphysema).
	Other immune mediators	• Epithelial cells in the air passage secrete lactoferrin, nitric oxide, defensins, mucin, lysozyme, etc., which are reported to have function against a diverse range of microbes. • Reactive oxygen species (ROS), cytokines like tumour necrosis factor-alpha (TNR-α), interleukins (IL-1β) and platelet-activating factors are potential mediators of inflammation. IgA in bronchial secretions resists infection.
Natural purifier against inspired substances	Pulmonary epithelium forms the first line of defence against inspired chemicals. The pseudostratified epithelium of air passages is carefully encased under a protective covering called a "mucous blanket". It is made up of mucopolysaccharide gel (secreted by goblet and mucous cells) and forms the first line of defence in air passages.	• Mucociliary escalator is the unique phenomenon in lungs characterised by cilia beating with 10–15 Hz frequency that moves the overlying mucosa @ 1 mm/min in the pharynx and 20 mm/min in the trachea. • Particulate matter of 5 μm deposits on larger airways, while that in 2–5 μm deposits on smaller airways. The cilia with these entrapped particles shift the mucus from peripheral to central airways from where the mucus is either swallowed or expectorated. Particles <2 μm reach alveoli by Brownian movement. • Defects in the mucociliary escalator may lead to pulmonary infections, chronic sinusitis, bronchiectasis, etc. • Deviations in either mucus production or ciliary moment may impair this phenomenon. Anaesthetic drugs (local or general), thermal stress, dehydration, atropine, etc. depress ciliary motility.
Biological filter	Protective functions	• Lungs have been demonstrated to have resistance against certain blood-borne substances by filtration but to a limited extent. However, emboli based on gas and fat can pass through the capillary filter. These microemboli can result in the activation of neutrophils, which further causes alveolar oedema in pulmonary capillaries.

(continued)

Table 7.5 (continued)

Functions	Significance	Description
		• The lung controls coagulation or its degradation based on biological necessity. The rich source of fibrinolysin activator can lyse the clots in pulmonary capillaries, while the heparin and thromboplastin can hasten the clotting process.
Neuroendocrine	Pulmonary neuroendocrine system	• Lungs have a well-developed *pulmonary neuroendocrine system* that encompasses the neuroendocrine cells and neuroepithelial bodies in the airway mucosa and promotes cell growth, differentiation and organ morphogenesis. • Pulmonary capillaries efficiently remove serotonin, epinephrine, bradykinin, endothelins, atrial natriuretic peptide, prostaglandins (E, E2, F2α), leukotrienes, ATP, ADP and AMP. Angiotensin I from blood gets converted to angiotensin II with the ACE present in the capillaries.
Pharmacokinetic action	Pulmonary endothelial cells play an important role in drug metabolism in the lungs.	• Upon administration of intravenous drugs, pulmonary extraction allows uptake of drugs from systemic circulation when excess and binds to them in pulmonary endothelial cells (pulmonary buffering). Medicines will be released slowly back into the system as unchanged, thereby responsible for the regular maintenance of drug concentrations. • The administration of beta-blockers and antidepressants disrupts this function and causes a dangerous rise in drug concentrations.
	Prodrug action	• Administration of steroid beclomethasone-dipropionate (an inactive form) becomes 17-beclomethasone monopropionate (active form) by esterases in the lungs to disperse its biological actions, which is none other than a prodrug feature of lungs.

Exercises

Objective Questions

Q1. Specify the gas-exchange site of the respiratory system.
Q2. Which type of cell produces surfactant in the lung alveoli?
Q3. What happens to the lung compliance when surface tension increases in the alveoli?
Q4. Where are the maximum inspiratory neurons found?
Q5. What is the numerical value of the partial pressure of carbon dioxide in the alveoli?
Q6. Where is the anatomic dead space located?
Q7. Where does the cross-current flow of air occur in birds?
Q8. Name two types of avian haemoglobin.
Q9. What is the Hills coefficient for oxygen cooperativity in avian species?
Q10. Where is surfactant SP-A found in birds?

Subjective Questions

Q1. Describe the role of surfactants in the lungs.
Q2. Describe the physiology of oxygen transport by haemoglobin.
Q3. What is the Haldane effect?
Q4. Describe the process of control of respiration in brief.
Q5. Write out the salient features of respiration in birds.

Answers to the Object Questions

A1. The respiratory zone
A2. Type II
A3. Decreases
A4. Dorsal group of respiratory neurons
A5. 46 mm of Hg
A6. Conducting zone of lungs
A7. Parabronchi and blood in capillaries
A8. HbA and HbD
A9. Hills coefficient—4
A10. Mesobronchi

Keywords for the Answer to Subjective Questions

A1. Surface tension, immune protection

A2. Oxyhaemoglobin dissociation curve
A3. Oxygenation of haemoglobin
A4. Medulla, pons, peripheral chemoreceptors
A5. Air sacs, parabronchi

Further Reading

Books

John E (2011) Pulmonary ventilation. In: Hall JE (ed) Guyton and Hall textbook of medical physiology, 12th edn. Saunders, Elsevier Inc, p 467

Reece WO (2005) Renal and respiratory function and acid-base balance: the generation and conduction of electrical signals in neurons. In: Dukes' physiology of domestic animals. 12th ed. p 120

Research Papers

Davis RP, Mychaliska GB (2013) Neonatal pulmonary physiology. Semin Pediatr Surg 22(4):179–184. https://doi.org/10.1053/j.sempedsurg.2013.10.005

Joseph D, Puttuswamy RK, Krovvidi H (2013) Non-respiratory functions of lung. Continuing Educ Anaesth Crit Care Pain J 13: 98–102

Lopez A, Martinson SA (2017) Respiratory system, mediastinum, and pleurae, Pathologic basis of veterinary disease. Elsevier, pp 471–560.e1

Prohl A, Ostermann C, Lohr M, Reinhold P (2014) The bovine lung in biomedical research: visually guided bronchoscopy, intrabronchial inoculation and in vivo sampling techniques. J Vis Exp 89:51557. https://doi.org/10.3791/51557

Zachman RD (1995) Role of vitamin A in lung development. J Nutr 125(6 Suppl):1634S–1638S. https://doi.org/10.1093/jn/125.suppl_6.1634S

Fluid and Electrolyte Balance

8

N. Madhavan Unny, Aziz Zarina, and V. Beena

Abstract

Many of the body's cellular operations use water as a medium. When water intake equals water loss, the body's water equilibrium is maintained. Water disperses throughout the body in distinct compartments. Several factors influence the distribution of water in different fluid compartments. Water and electrolyte intake and outflow are more strictly regulated to maintain total body water and total body osmolarity. Water intake is controlled by water consumption, whereas water loss is controlled by urine output. Electrolytes serve a critical function in controlling bodily water content, distribution, and osmolarity. Calcium, sodium, chloride, and bicarbonates are the major electrolytes in extracellular fluid; potassium, magnesium, phosphates, proteinates, and sulphates are the major electrolytes in intracellular fluid. Sustaining acid-base balance is another essential part of maintaining homeostasis. The chemical buffer system, respiratory system, and renal system are the three primary systems that manage the acid-base balance in the body. Clinical disorders impacting hydration, acid-base balance, and electrolyte status can have serious, even life-threatening repercussions; therefore, it is critical to recognise and treat them.

N. Madhavan Unny
Department of Veterinary Clinical Medicine, College of Veterinary and Animal Sciences, Thrissur, Kerala, India

A. Zarina · V. Beena (✉)
Department of Veterinary Physiology, College of Veterinary and Animal Sciences, Thrissur, Kerala, India
e-mail: beenav@kvasu.ac.in

© The Author(s), under exclusive license to Springer Nature Singapore Pte Ltd. 2023
P. K. Das et al. (eds.), *Textbook of Veterinary Physiology*, https://doi.org/10.1007/978-981-19-9410-4_8

Graphical Abstract

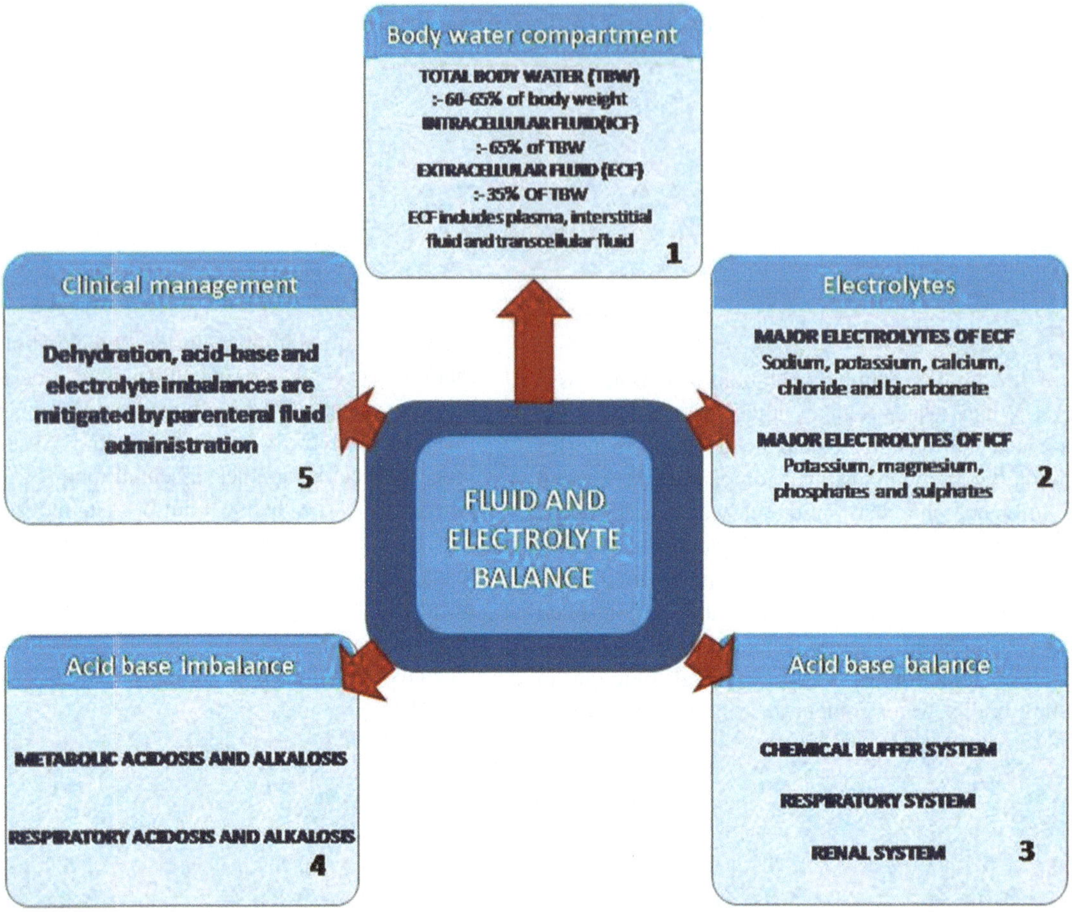

Description of the graphic: Total body water (TBW) constitutes 60–65% of body weight (**1**). About 65% of TBW is ICF and 35% is ECF. Electrolytes are differentially distributed with respect to the concentration in both ECF and ICF, sodium being the major extracellular cation and potassium being the major intracellular cation (**2**). Along with the chemical buffer system, respiratory and renal systems are also involved in the acid-base regulation of the body (**3**). Respiratory or metabolic alterations can cause acidosis or alkalosis (**4**). Clinical management by parenteral fluid administration is decided by the level of dehydration, pH, and electrolyte imbalance (**5**)

Keywords

Acid-base balance · Electrolytes · Dehydration and fluid therapy · Total body water · Transcellular fluid

Learning Objectives

- To know about the different body water compartments and regulation of water balance
- The physiological importance of electrolytes and various causes of electrolyte disturbances
- Different types of transcellular fluids and their physiological importance
- Various systems involved in maintaining acid-base balance
- Causes for acid-base balance disturbances
- Clinical management of dehydration, acid-base imbalance, and electrolyte disturbances

8.1 Water, Electrolytes, and Body Fluids

8.1.1 Introduction

Osmoregulation is **osmotic balance** maintained across membranes within the body's fluids. Osmotic balance is contributed by the electrolytes and non-electrolytes.

Electrolytes* are the** solutes that dissociate into ions during water dissolution, and ***non-electrolytes do not dissociate into ions during water dissolution. Electrolytes in living systems include zinc, sodium, potassium, magnesium, chloride, copper, bicarbonate, manganese, calcium, iron, phosphate, molybdenum, and chromium. Sodium, bicarbonate, phosphate, potassium, calcium, and chloride are the major electrolytes involved in various body functions. These electrolytes perform a variety of functions in the animal body such as transmission of electrical impulses in neurons and muscles, release of hormones, stabilisation of protein structure, acid-base regulation, and also osmoregulation of body fluids. The hydrostatic pressure and osmotic pressure control the movement of water along the cell membrane. Among these two physical factors, osmosis can only be directly controlled by the movement of electrolytes. Maintenance of an electrical and chemical balance within and outside the cell is accountable for an electro-chemical gradient existing between extracellular and intracellular fluids. This gradient is actually made use of in the movement of electrolytes and also in the movement of water across the cellular compartments.

Osmoconformers are organisms that maintain their internal salinity same as their environment (e.g., most marine invertebrates). *Osmoregulators* tightly maintain body osmolarity even after fluctuations in salt levels in the environment and are more common in the animal kingdom.

Osmosis is a physical process, by which diffusion of water occurs across a semipermeable membrane, from an area of high water potential (low solute concentration) to an area of low water potential (high solute concentration) without input of energy. The force per unit area that halts osmosis is called the *osmotic pressure* of the solution. Colloid osmotic pressure, also known as plasma oncotic pressure, is the effective osmotic pressure exerted by plasma proteins on fluid flow between the two compartments. A colloid is a term that refers to the big molecular weight (MW >30,000) particles that are present in a solution. Plasma proteins are the most common colloids found in plasma. The osmotic pressure of a solution is a colligative feature. It is determined by the number of particles dissolved in a unit volume of solvent, not by the particle's valence, weight, or shape. Because most biological membranes are semi-permeable, osmosis is vital in living systems. Large-molecule solutes, such as polysaccharides, are impermeable, but water and tiny, uncharged solutes are permeable. The number of molecules of dissolved particles is represented by the osmole. Each osmole (Osm) has 6.023×10^{23} molecules since one osmole is defined as 1 g molecular weight of any non-dissociable material. (According to Avogadro's law, 1 mol of any substance contains the same number of particles 6.023×10^{23}, independent of its molecular weight.) *Osmolarity* is the measure of solute concentration in 1 L solution (osmol/L). The osmoles of solute per kilogramme of solvent (osmol/kg) are a measure of osmolality. However, unlike osmolarity, osmolality is the preferred physiological term because it is unaffected by the temperature of the solution. Weight is a temperature-independent variable, whereas volume is dependent on temperature. Electrolyte concentrations are commonly represented in terms of milliequivalents per litre (mEq/L), which is equal to the ion concentration (in millimoles) multiplied by the number of electrical charges on the ion. Since electrolytes create ions in aqueous solutions, the milliequivalent unit takes into account the ions present in the solution as well as the charge on the ions. One milliequivalent is equal to 1 millimole for ions with a charge of one, and one milliequivalent is equal to 0.5 millimoles for ions with a charge of two (such as calcium). The milliosmole (mOsm), which is the number of milliequivalents of solute per kilogramme of solvent, is another unit for expressing electrolyte concentration. The osmotic pressure of body fluids is normally kept between 280 and 300 mOsm.

When a solution has a higher concentration of solute than a solution with a lower concentration of solute, it is referred to as hyperosmotic. If two solutions are having the same solute concentration, they are said to be isosmotic.

Tonicity is determined by the number of solutes that cannot penetrate the membrane and the effective osmotic pressure gradient. To put it another way, tonicity refers to the relative concentration of dissolved solutes in a solution that determines the direction and extent of diffusion. It describes what a solution would do to a cell's volume at equilibrium if the cell was placed in the solution. The solutes that are unable to penetrate the cell dictate the tonicity of an infused solution. As a result, the terms isosmotic, hyperosmotic, and hyposmotic are not interchangeable with tonic, hypertonic, and hypotonic. Isosmotic refers to the osmolalities of various physiological fluids having the same value, such as plasma versus cerebrospinal fluid. The terms *isotonic*, *hypotonic*, and *hypertonic* are used to describe the osmolalities of solutions used in clinical practice to restore bodily fluid losses. When compared to plasma, a hypertonic solution has a higher effective osmotic pressure. A hypotonic solution has lower effective osmotic pressure compared to plasma. A solution having effective osmotic pressure identical to that of plasma is called isotonic solution. Plasma has an osmolality of 0.3 Osm or 300 mOsm. 0.15 M solution of NaCl is equal to 0.3 osmol solution because NaCI ionises into two ions, Na^+ and Cl^-; half the molar NaCI (0.15) solution contributes to 0.3 osmole. Isotonic saline solution for clinical applications is prepared with sodium chloride concentration of 9 g/L of water or 0.9% (w/v), i.e. 0.167 M NaCI.

If water or hypotonic saline is added to extracellular fluid (ECF) or plasma, the osmotic concentration of this compartment is lowered and water moves into cells—this may interfere with normal metabolism of cells or may even lead to death of cells—this over-hydration is known as *water intoxication*. When isotonic saline solution is given to ECF, it spreads evenly throughout the ECF but has no effect on ICF. When a hypertonic saline solution is added to ECF/plasma, the osmotic concentration of ECF rises, and water moves from within the cells (ICF) to ECF; however, the cell membrane is less permeable to sodium ions, so sodium ions accumulate within the cells, raising intracellular osmotic pressure; this is known as *cellular dehydration*.

8.1.2 Body Water Compartments

Water is the most important component of life and the body's main component. The body's water is divided into compartments separated by a selectively permeable membrane. The chemical composition of these compartments varies. The compartments are as follows: Intracellular fluid (ICF) is the water contained within the cells and accounts for approximately (two-thirds of body water) 65% of total water; extracellular fluid (ECF) is the water outside the cells and accounts for approximately (one-third of body water) 35% of total water. Intravascular fluid or plasma (one-fifth of ECF), interstitial fluid (four-fifths of ECF), and transcellular fluid make up extracellular fluid. The fluid that surrounds the cell is known as interstitial fluid. The fluid contained in body cavities such as cerebrospinal, peritoneal, synovial, ophthalmic fluids, and fluids in the gastrointestinal system (which is the major component in ruminants) is known as transcellular fluid ("third space," which is generally neglected in calculations). Urine and bile are also considered as transcellular fluids.

8.1.2.1 Total Body Water (TBW)

The sum of water in intracellular and extracellular body compartments is referred to as total body water. It is influenced by factors, like species, age, size, gender, and nutritional status. Water makes up 60–65% of the body weight of an adult animal with little fat. Water makes almost 70% of a lean cow's body weight, but it makes up just approximately 40% of a fatty cow's. Water content is highest in newborn and lowest in aged adult animals. Males have higher water content than females. About 60% of body weight is considered water in adult males and 55% in adult females. Water makes up around 75% of the weight of lean muscle tissue. Water makes up 95% of blood, 14% of body fat, and 22% of bone. Skin also contains much water.

Out of 60% TBW, intracellular fluid accounts for 40% of total body weight, whereas extracellular fluid accounts for 20%. The amount of water in each of these compartments is controlled by the body. To keep the amount of water in each compartment generally constant, water is transported across most of the cell membranes as needed.

8.1.2.2 Water Balance

Homeostasis (coined by W. B. Cannon) is the existence and preservation of stability within the internal environment. The term internal environment or internal milieu was put forth by Claude Bernard. When daily water intake equals daily water loss, the body's water equilibrium is maintained. In most cases, the body's water content varies very little.

Ingestion and metabolic water (end product of cellular metabolism) provide water to the body. One gram of carbohydrate, fat, and protein on oxidation supplies 0.6 mL, 1.1 mL, and 0.4 mL of water, respectively. Metabolic water provides 5–10% of total body water requirements. The importance of metabolic water in many desert rats cannot be overstated. It may provide 100% of their water consumption, allowing them to subsist on dry food and no water. One of the examples is the kangaroo rat.

Water is lost from the body through various routes like urine, skin, expired air, faeces, and milk (lactating animals). The body's natural regulatory mechanism regulates water losses through several routes in order to maintain equilibrium (Table 8.1).

8.1.2.2.1 Regulation of Water Intake

Water intake is intermittent, whereas water loss is continuous. A gradual dehydration problem is constantly present in an animal's life. The osmolarity in all fluid compartments is almost equal. It ranges from 301.5 to 303 mOsm/kg of water in human. Water will be shifted from the cell into the ECF during gradual dehydration since the ECF is the immediate source of water loss. Water ingestion is used to replenish this water deficiency on a regular basis. Water consumption is influenced by habit or a daily routine of eating and drinking without being thirsty.

Any significant loss of bodily fluid results in a sensation of thirst as well as a behavioural desire to drink. When there is a water shortage, both thirst and the desire to drink grow. There are a number of regulating mechanisms in place to ensure that the amount of water consumed equals and corrects a bodily water deficit.

Table 8.1 Water balance in a cow

	Intake (L)				Output (L)				
Animal status	Drinking	In feed	Metabolic water	Total	Urine	Faeces	Evaporation	Milk	Total
Non-lactating	26	1	2	29	7	12	10	0	29
Lactating	51	2	3	56	11	19	14	12	56

Dehydration causes a decrease in blood volume and blood pressure, as well as an increase in blood osmolarity. The thirst centre of hypothalamus reacts to a variety of dehydration signals, including angiotensin II (subjected to release in reaction to low blood pressure), vasopressin (released in response to increased blood osmolarity), and osmoreceptor cues (neurons in hypothalamus monitoring extracellular fluid osmolarity).

When the thirst centre receives the signals, it inhibits salivation by activating the sympathetic nerve supply to the salivary gland. Salivation is also reduced in dehydration due to decreased capillary filtration produced by low blood pressure and high blood osmolarity. Reduced salivation causes the animal to feel thirsty, which leads to water consumption. Drinking water cools and moistens the mouth while also causing stomach and intestinal distension. These will suppress thirst for a brief period of time.

Water consumption rehydrates the blood, lowering osmolarity. It reduces osmoreceptor response and enhances capillary filtration. As a result, saliva becomes more watery and profuse. Long-term thirst inhibition is caused by water absorption from the small intestine and a decrease in blood osmolarity.

8.1.2.2.2 Regulation of Water Output

The only means by which water output can be significantly controlled is through regulation of urine volume. Antidiuretic hormone (ADH) is involved in controlling water output. In dehydration, elevated blood osmolarity stimulates the osmoreceptors in the hypothalamus. As a result of the stimulation of osmoreceptors, the posterior pituitary is stimulated to release ADH. In the kidney, ADH interacts with V_2 receptors present on the basal side of epithelial cells of late distal convoluted tubule, collecting tubule, and collecting duct. Binding of ADH with V_2 receptor will cause formation of second messenger cAMP. cAMP will in turn increase the number of water channels (aquaporin 2) in the epithelial cytoplasm. Aquaporin 2 is inserted on the luminal surface of the epithelium and thus increases water permeability. Water is reabsorbed till osmotic equilibrium is reached. Thus, kidney will increase water reabsorption resulting in reduced urine output. ADH helps in elevating blood volume and lowering blood osmolarity in dehydration.

On the other hand, if blood osmolarity is very low or blood volume and blood pressure are very high, ADH release is blocked. The renal tubule's ability to reabsorb water decreases, resulting in an increase in urine volume. This will lower the total blood water level and return it to normal.

Adjustments in sodium reabsorption are also linked to urine volume regulation. Sodium reabsorption and excretion are accompanied by a proportionate amount of water. With regard to electrolyte balance, maintaining water balance by regulating sodium excretion is better understood.

8.1.2.3 Measurement of Body Water

The difference in the weight of fresh carcasses and dried carcasses was used to compute the total body water at earlier times. The total body water was afterwards estimated using the dilution approach. The volume of bodily water can be determined by injecting a known-concentration indicator chemical into the bloodstream, allowing it to diffuse equally throughout the plasma, and then evaluating the amount of dilution (Table 8.2). The total body water (TBW) can be calculated from the formula given below:

Table 8.2 Different substances used to determine body fluid volume

Body fluid	Substance used to measure the body fluid	Characteristics of substance used
Total body water	Radioactive water (tritium), heavy water (deuterium), antipyrine, urea, thiourea, sulphanilamide	Must penetrate the cell membrane and uniformly disperse in ECF and ICF
ECF	Radioactive chloride, radioactive iothalamate, thiosulfate, inulin, sucrose, thiocyanate	Must disperse in plasma and interstitial fluid but do not penetrate the cell membrane
Plasma volume	Serum albumin labelled with radioactive iodine, Evans blue dye (T-1824)	Does not penetrate capillary membrane but remains in vascular system
Blood volume	RBC labelled with radioactive material like chromium or phosphorus	
ICF	ICF = Total body fluid − ECF	

ECF extracellular fluid, *ICF* intracellular fluid, *RBC* red blood cells

$$\text{TBW} = \frac{\text{Volume of indicator solution injected} \times \text{Concentration of indicator solution injected}}{\text{Concentration of indicator solution after equilibration}}$$

An indicator solution that equally spreads in plasma and interstitial fluid but does not infiltrate the cell membrane can be used to assess the volume of extracellular fluid. Because the intracellular fluid volume cannot be directly determined, it must be calculated by subtracting the extracellular fluid volume from total body water. A material that does not readily permeate capillary membranes but persists in the circulatory system after injection is used to quantify plasma volume. Similarly to intracellular fluid volume, interstitial fluid volume cannot be measured directly. It is computed by subtracting the volume of plasma from the volume of extracellular fluid.

The haematocrit (the fraction of total blood volume made up of cells) can also be used to compute the blood volume using the equation below:

$$\text{Total blood volume} = \frac{\text{Plasma volume}}{1 - \text{Haematocrit}}$$

Another approach to estimate blood volume is to insert radioactively labelled red blood cells into the circulation. The radioactivity of a mixed blood sample can be measured after extensive mixing of these cells in circulation, and the total blood volume can be determined using the dilution principle. Radioactive chromium (^{51}Cr), which binds firmly to red blood cells, is a common material used to identify red blood cells.

Newer non-invasive techniques using body composition have been developed to measure total body water. These techniques include bioelectrical impedance analysis, air displacement plethysmography, dual-energy X-ray absorptiometry, and nuclear magnetic resonance. Bioelectric impedance analysis is economical and simple to use. These newly developed techniques measure total body water using empirical equations. These empirical equations are developed by comparing the measurements obtained by new methods with measurements made using reference methods.

8.1.3 Electrolytes

Electrolytes play a crucial role in animal health. Chemically reactive, they have a role in metabolism. They are responsible for determining the electrical potential across the cell membrane. They have a significant impact on the osmolarity of body fluids and the content and distribution of body water. Because ICF and ECF have the same osmolarity, cells do not bulge or shrink, but in terms of electrolyte concentration (Table 8.3), electrolytes are more numerous than non-electrolytes; they regulate water osmosis between bodily compartments. Sodium and calcium are the predominant cations in extracellular fluid, whereas chloride and bicarbonates are the major anions. Blood plasma will contain other proteins. Potassium and magnesium are the most abundant cations in intracellular fluid, while phosphates, proteinates, and sulphates are the most available anions. The erythrocytes of cats, dogs, cattle, and sheep have higher sodium ions than potassium. There is a little osmotic activity difference between plasma and interstitial fluid. The concentration of positively charged ions in plasma is slightly higher (approximately 2% higher) than the interstitial fluid (Table 8.4).

8.1.3.1 Electrolyte Transportation

The electrolytes transfer across a cell membrane in two ways: passive and active transportation. Transportation follows a concentration gradient called passive transportation that requires no energy input. Simple diffusion, facilitated diffusion, filtration, and osmosis are examples of passive transportation. In simple diffusion, electrolytes flow across the cell membrane according to their concentration gradient, from areas of high concentration to areas of low. In facilitated diffusion, electrolytes migrate into or out of the cells down to their concentration gradient through protein channels present in the cell membrane. Molecules move across the cell membrane due to a pressure gradient in the filtration process. Hydrostatic pressure is generally applicable here. The movement of a solvent through a selectively permeable or semi-

Table 8.3 Osmotically active substances in human body fluid (mOsmol/kg H_2O)

Substances	Plasma	Interstitial fluid	Intracellular fluid
Sodium	146	142	14
Potassium	4.2	4.0	140
Calcium	2.5	2.4	0
Magnesium	1.5	1.4	31
Chloride	105	108	4
Bicarbonate	27	28.3	10
Phosphate	2	2	11
Sulphate	0.5	0.5	1
Glucose	5.6	5.6	–
Proteins	1.2	0.2	4
Urea	4	4	4
Other organic substances	3.4	3.4	83.2
Total osmolality	302.9	301.8	302.2
Osmotic pressure at 37 °C (mmHg)	5453	5430	5430

Table 8.4 Plasma concentrations of electrolytes in dogs and cats

Substance	Dog	Cat
Sodium (mEq/L)	140.3–153.9	145.8–158.7
Potassium (mEq/L)	3.8–5.6	3.8–5.3
Ionised calcium (mEq/L)	5.4	5.1
Total calcium (mg/dL)	8.7–11.8	7.9–10.9
Total magnesium (mg/dL)	1.7–2.7	1.9–2.8
Chloride (mEq/L)	102.1–117.4	107.5–129.6
Bicarbonate (mEq/L)	21	20
Phosphate (mg/dL)	2.9–6.2	4.0–7.3
Proteins (g/dL)	7	7
Lactate (mg/dL)	15	15

permeable (cell) membrane, from higher to lower, is called osmosis.

Active transportation necessitates the use of energy to transport molecules into a cell. Primary active transportation and secondary active transportation are two types of active transportation. The transport protein comprises an ATPase, which hydrolyses ATP to create the energy required for transport in primary active transportation. It may also be called as an ion pump. There is no direct coupling of ATP in secondary active transportation; instead, the potential difference established by pumping ions out of the cell via primary active transport is used. Multiple electrolytes are moved across the membrane via secondary active transportation, which combines the uphill movement of one electrolyte (s) with the downhill movement of the other(s). It is called symport or co-transport when electrolytes flow in the same direction and antiport or counter-transport when they move in the opposite direction. ABC transporters, P-type ATPases, and solute carrier family are the three main membrane transporters that transport electrolytes. ABC transporters are key active transporters that transfer a wide spectrum of ions. P-type ATPase enzymes are used in primary active transportation to move cations. Ca^{2+}-ATPases and Na^+,K^+-ATPases are examples of this family. Transporters in the solute carrier family use secondary active transport and facilitative diffusion to move solutes. The Na^+/H^+ exchanger is an example of the solute carrier family.

8.1.3.2 Sodium

Sodium is the primary cation in extracellular fluids. Nearly 45% of sodium in the body is present in the extracellular fluid, 45% in the bones, and the rest in the intracellular fluid. Sodium is a key solute in determining total body water volume and water distribution among fluid compartments. Sodium is responsible for 90–95% of ECF osmolarity, and it accounts for half of the osmotic pressure differential that exists between the inside of cells and their surroundings. The kidneys are primarily responsible for salt excretion. Sodium is freely filtered through the kidneys' glomerular capillaries, and the majority of it is reabsorbed in the proximal convoluted tubule leaving only a small amount in urine.

In ECF, sodium content is relatively constant. In animals, dietary deficiency of sodium is rare, but adequate sodium excretion by the kidney is of primary concern. Four hormones are mainly involved in the regulation of sodium; they are ADH, atrial natriuretic factor, aldosterone, and angiotensin II.

The antidiuretic hormone regulates water absorption and excretion independently of sodium excretion. Osmolarity of ECF increases when sodium concentration rises above the normal level in ECF. The osmoreceptors in the hypothalamus detect it, causing the release of ADH from the posterior pituitary. ADH increases water reabsorption from renal tubules and stimulates the thirst centre. In contrast, the release of ADH is inhibited by a drop in sodium content below normal in ECF, causing diuresis with the excretion of more water followed by the rising of sodium content in the ECF.

The hormone aldosterone (also known as the salt retention hormone) regulates the rate of sodium excretion. The adrenal cortex is directly stimulated to release aldosterone when sodium levels are reduced or potassium levels are elevated. The renin-angiotensin system promotes aldosterone secretion when blood pressure is reduced. The major cells that line the renal tubules, particularly at the later part of the convoluted tubule and collecting duct, are affected by aldosterone. The basement membrane allows aldosterone from the blood to enter the main cell. It induces gene transcription by binding to its receptors in the cytoplasm. The proteins formed cause three effects: (1) attach new sodium-potassium ATPase pumps on the basal surface of principal cells, (2) attach ENaC (epithelial sodium channels) on the luminal surface of principal cells, and (3) activate the existing sodium-potassium ATPase pump on the basal surface and sodium and potassium channels on the luminal surface. Sodium from the tubular fluid is reabsorbed into the main cell and transported to the blood via sodium-potassium ATPase pumps due to these three processes. At the same time, potassium from the blood enters into the principal cell via sodium-potassium ATPase pumps on the basal surface and is excreted into the tubular fluid of the renal tubule via potassium channels on the luminal surface. Sodium, from the blood, enters the principal cell via sodium-potassium ATPase pumps on the basal surface. As a result, aldosterone causes an increase in salt absorption and a decrease in potassium excretion. Along with sodium and water, chloride is also reabsorbed. The blood volume, blood pressure, and sodium and potassium concentration are restored to normal.

An increase of sodium with the rise in blood volume results in the release of the atrial natriuretic factor (ANF) due to distension of the heart's atria. ANF increases salt and water excretion through the kidneys while inhibiting ADH and renin secretion. As a result, the kidneys excrete more salt and water. It aids in restoring normal blood volume and sodium levels.

Hyponatraemia is a condition in which sodium concentration in the blood is lower than normal. It is mainly caused by an excess of water in the body, which dilutes the sodium. Reduced sodium intake combined with constant excretion through urine, severe sweating, vomiting, diarrhoea, and diseases that cause diuresis, such as diabetes and acidosis, causes extreme hyponatraemia. Relative hyponatraemia can occur as a result of excessive water retention in oedema or congestive heart failure.

Hypernatraemia is a condition in which blood sodium levels are abnormally high. It can be caused by the loss of water from the blood, which causes the haemo-concentration of all blood elements. It can also be caused by hormonal abnormalities involving ADH and aldosterone.

8.1.3.3 Potassium

Potassium is the most abundant as the intracellular cation, and it aids in generating action potentials and establishing the resting membrane potential after depolarisation in neurons and muscle fibres. Potassium, unlike sodium, has minimal effect on osmotic pressure. The sodium-potassium pumps in cell membranes, which maintain appropriate potassium concentration gradients between the ICF and ECF, are responsible for the low potassium levels in the blood and cerebrospinal fluid. The renal tubules, particularly at the distal convoluted tubule and collecting ducts, discharge potassium both actively and passively. Under the effect of aldosterone, potassium participates in the exchange of sodium (discussed earlier).

Aldosterone aids in the control of potassium levels in the ECF. The adrenal cortex will release aldosterone in response to a slight increase in potassium content. The tenfold increase in potassium concentration results in a threefold increase in aldosterone. Aldosterone raises potassium excretion in the urine and returns the ECF potassium concentration to normal.

Hypokalaemia is a condition in which the potassium level in the blood is abnormally low. Hypokalaemia can develop due to loss of potassium in the body or a relative reduction in potassium in the blood due to potassium redistribution. Reduced intake, which is commonly associated with famine, vomiting, diarrhoea, and alkalosis, can cause potassium loss. The redistribution of potassium causes a relative decrease in the blood in some insulin-dependent diabetic individuals. When insulin is given and glucose is taken up by cells, potassium travels through the cell membrane with the glucose, lowering the amount of potassium in the blood and interstitial fluid, leading to hyperpolarisation of neuron cell membranes and reduced sensitivity to stimuli.

Hyperkalaemia, or an abnormally high potassium level in the blood, can harm skeletal muscles, neurological system, and heart. Increased potassium intake in the diet can cause hyperkalaemia. Increased potassium concentration in the ECF can cause partial depolarisation (excitation) of the plasma membrane of skeletal muscle fibres, neurons, and cardiac cells and an inability to repolarise the cells. In such circumstances, the heart will not relax after a contraction, thus seizing and ceasing to pump blood, resulting in death within minutes. An individual with hyperkalaemia may have mental disorientation, numbness, and weaker respiratory muscles due to the effects on the neurological system.

8.1.3.4 Chloride

The most common extracellular anion is chloride. Chloride contributes significantly to the osmotic pressure difference between the ICF and the ECF, and it is essential for optimal hydration. Chloride maintains the electrical neutrality of the ECF by balancing cations in the fluid. Chloride ion secretion and reabsorption in the kidneys follow the same pathways as sodium ions.

Hypochloraemia, lower-than-normal blood chloride levels, occurs mainly due to faulty renal tubular absorption. Hypochloraemic dogs and cats have chloride concentrations of less than 100 mEq/L and 110 mEq/L, respectively. Hypochloraemia can also be caused by vomiting, diarrhoea, or metabolic acidosis. Dehydration, excessive ingestion of food salt (NaCl) or swallowing seawater, renal failure, renal tubular acidosis, diabetes mellitus, congestive heart failure, chronic lung disease, and other factors can cause *hyperchloraemia*, or higher-than-normal blood chloride levels. Pseudo-hyperchloraemia often occurs during serum examinations in the laboratory. It causes excessive loss of water and leads to chloride loss, lipemic serum, and pigments like bilirubin and haemoglobin in the serum. Injection of potassium bromide can also cause pseudo-hyperchloraemia.

8.1.3.5 Bicarbonate

The second most abundant anion in the blood is bicarbonate. Its primary role is to regulate the body's acid-base balance by acting as a component of buffer systems. In body fluids, a small amount of CO_2 may dissolve. It leads to the production of approximately 90% of CO_2 into bicarbonate ions ($HCO3^-$) following the reaction as:

$$CO_2 + H_2O \leftrightarrow H_2CO_3 \leftrightarrow HCO_3^- + H^+$$

The bidirectional arrows indicate that depending on the concentrations of the reactants and products, the reactions can proceed either way. In tissues with a high metabolic rate, considerable volumes of carbon dioxide are generated. Carbon dioxide converts into bicarbonate in the cytoplasm of red blood cells by the action of an enzyme known as carbonic anhydrase, and then it enters into the bloodstream. The CO_2 is regenerated from bicarbonate in the lungs, causing a reverse reaction and expelling as metabolic waste.

8.1.3.6 Calcium

Calcium, the most abundant mineral in bones and teeth (calcium reservoirs), is responsible for its hardness. Muscle contraction, enzyme function, and blood coagulation require calcium ions (Ca^{2+}). Calcium also aids in the stabilisation of cell membranes and is necessary for the release of neurotransmitters and hormones from endocrine glands.

Nearly 30% of the total calcium in the bone is made up of amorphous salts that can easily be exchanged with ECF. The amount equates to around 5–10 g in total. The amorphous calcium crystals have a wide surface area that can easily absorb extra calcium when hypercalcaemia occurs. The amorphous salts are easily carried into the bloodstream if hypocalcaemia occurs. In about 70 min, any alterations in calcium concentration in the blood are restored to normal levels by this buffering mechanism. Parathyroid hormone, calcitonin, and vitamin D play a role in calcium homeostasis. These hormones regulate eucalcaemia by their effects on bone deposition and bone resorption, urinary excretion, and intestinal calcium absorption. Bone is a long-term regulator of eucalcaemia. When bone is saturated with or depleted with calcium salts, the intestine and kidney regulate eucalcaemia.

Hypocalcaemia, abnormally low blood calcium levels, is seen in hypoparathyroidism that can occur during the dysfunction of the thyroid gland as four nodules of the parathyroid gland are lodged within it. Renal illnesses, insufficient dietary calcium, vitamin D deficiency, low magnesium levels, pancreatitis, hypoparathyroidism, and certain drugs, including anticonvulsants and corticosteroids, cause hypocalcaemia. Primary hyperparathyroidism is characterised by *hypercalcaemia*, or unusually high calcium blood levels. Hypercalcaemia is a side effect of several cancers. Magnesium levels are closely linked to calcium levels; hence, it is common to fix and treat magnesium levels before treating calcium levels.

8.1.3.7 Phosphate

Dihydrogen phosphate ($H_2PO_4^-$), monohydrogen phosphate (HPO_4^{2-}), and phosphate (PO_4^{3-}) are the three ionic forms of phosphate found in the body. HPO_4^{2-} is the most prevalent kind. Calcium-phosphate salts, bone, and teeth bind up 85% of the body's phosphate. Phospholipids, such as those that make up the cell membrane, ATP, nucleotides, and buffers, all include phosphate. They play a crucial role in maintaining acid-base equilibrium by functioning as buffers.

Hypophosphataemia, or abnormally low phosphate blood levels, can occur due to excessive antacid usage or malnutrition. The kidneys generally conserve phosphate when faced with phosphate depletion, although this conservation is substantially hampered by hunger. *Hyperphosphataemia*, or unusually high phosphate levels in the blood, occurs when renal function is impaired or acute lymphocytic leukaemia is present. Phosphate is a major component of the ICF; hence, any considerable cell death might result in phosphate being dumped into the ECF.

8.1.4 Transcellular Fluid

The transcellular fluid is found in epithelial cell-lined bodily cavities. It includes the cerebral fluid, synovial fluid, peritoneal fluid, pleural fluid, pericardial fluid, aqueous humour, and vitreous humour of the eye, bile, and fluid from the digestive, urinary, and respiratory tracts.

8.1.4.1 Cerebrospinal Fluid

Cerebrospinal fluid (CSF) is a unique fluid found in and around the brain and spinal cord. It presents in the brain's ventricles, the spinal cord's central canal, and the subarachnoid region. The choroid plexus of the brain's lateral and third ventricles produces the majority of the cerebrospinal fluid (two-thirds). Ependymal cells that line the ventricles and spinal canal create cerebrospinal fluid. The pia mater, which covers the central nervous system, produces a minor amount. A layer of pia mater and choroid epithelial cells covers the choroid plexus, which resembles a cauliflower-like proliferation of blood vessels (modified ependymal cells). Microvilli cover the apical surface of choroid epithelial cells. Many fenestrae in the capillary endothelium's wall allow many tiny molecules to flow through. Tight connections connect adjacent choroid epithelial cells preventing water-soluble compounds from passing into the cerebrospinal fluid. The blood-cerebrospinal barrier (BCB) or blood-brain barrier (BBB) is made up of several tight junctions. Blood pressure and cerebrospinal fluid pressure both have little effect on cerebrospinal fluid secretion, which is an active process. The sodium is actively transported into the ventricles by epithelial cells. Chloride and bicarbonate are diffused into the ventricles to preserve electrical neutrality. As a result, the concentration of sodium chloride in the ventricles rises, causing osmosis to sip water into the ventricles. By facilitating diffusion, carrier proteins will aid in moving essential chemicals from the blood into the cerebrospinal fluid.

Table 8.5 Biochemical constituent of cerebrospinal fluid of cow

Constituent	Cow
Total proteins (mg/dL)	23.4–66.3
Albumin (mg/dL)	8.21–28.71
Creatine kinase (U/L)	2–48
Lactate dehydrogenase (U/L)	2–25
Magnesium (mg/dL)	1.8–2.11
Potassium (mEq/L)	2.7–3.2
Sodium (mEq/L)	13.2–14.2
Glucose (mg/dL)	37–51

Cerebrospinal fluid is a colourless, transparent liquid. The cerebrospinal fluid has plasma's specific gravity, pH, and osmolarity. It contains a tiny amount of protein, the same amount of plasma sodium, 15% more chloride than plasma, 40% less potassium, and 30% less glucose than plasma (Table 8.5). In comparison to plasma, cerebrospinal fluid contains less urea. Except for a few lymphocytes, the cerebrospinal fluid lacks biological components.

The cerebrospinal fluid is generated in the lateral ventricles and enters into the third ventricle through the foramen of Monro. A small volume of CSF can infuse into the third ventricle. The cerebrospinal fluid will subsequently pass through the aqueduct of Sylvius and into the fourth ventricle. A minute volume of CSF can also enter into the fourth ventricle. The fourth ventricle's cerebrospinal fluid will enter the spinal cord's central canal. A portion of the cerebrospinal fluid from the fourth ventricle will reach the subarachnoid space through the foramen of Luschka and the foramen of Magendie.

The cerebrospinal fluid is replaced by new cerebrospinal fluid four to five times a day. The rate of formation of the CSF varies in different species (Table 8.6). Arachnoid villi absorb the cerebrospinal fluid. Microscopic extensions of the arachnoid membrane into the dorsal sagittal sinus are known as arachnoid villi. Arachnoid granulation is a macroscopic structure formed by the aggregation of these villi. The arachnoid villi operate as a valve, allowing cerebrospinal fluid to flow quickly into the sagittal sinus while preventing backflow. The cerebrospinal fluid pressure is 1.5 mmHg higher than that of the plasma.

Cerebrospinal fluid helps cushion the central nervous system against shock, thus protecting the brain against a blow to the head. Cerebrospinal fluid significantly lowers the brain weight by providing a buoyancy effect. It helps to maintain the consistent extracellular environment of cells of the nervous system. It is an effective waste control system that can remove potentially harmful cellular metabolites. It transports and distributes some peptide hormones and various substances of the brain into general circulation. It serves partially as nutritive media for the brain and spinal cord.

Table 8.6 Rate of cerebrospinal fluid formation in various species

Species	Cat	Dog	Sheep	Goat	Cow	Human
Rate (μL/min)	20–22	47–66	118	164	290	350–370

8.1.4.2 Synovial Fluid

Synovial fluid is a thick, viscous liquid found in the cavities of joints, tendon sheath, and bursae. A thin layer of synovial fluid surrounds the articular cartilage and penetrates its interior regions. The synovial fluid within the auricular cartilage acts as a reserve. The reserve synovial fluid is forced out of the cartilage during joint movements to keep a fluid layer on the cartilage surface.

Synovial fluid is generated by ultrafiltration from blood plasma. The pH of synovial fluid is usually between 7.2 and 7.4. It contains proteins acquired from plasma through filtration and synthesised by synovial cavity cells. It contains small amounts of albumin, globulin, mucin, proteinase, collagenases, prostaglandins, and hyaluronic acid, but no fibrinogen. Thus, synovial fluid does not clot. Small molecules like electrolytes and glucose have similar concentrations in the synovial fluid to plasma. Large molecules are found in lower concentrations in synovial fluid than in plasma. The synovial fluid contains hyaluronic acid produced by fibroblast-like cells (type B cells) in the synovial membrane. Hyaluronan is a lubricant that enhances the viscosity and flexibility of articular cartilage. Lubricin, a glycoprotein, is secreted by chondrocytes on the surface of the articular cartilage in the synovial joint. Lubricin is involved in lubrication and helps regulate synovial cell growth.

The synovial fluid contains only a few phagocytic cells (mainly mononuclear cells). These cells remove germs and debris from the joints caused by wear and tear. Less than 10% of these cells are neutrophils, and the remaining cells present are lymphocytes, monocytes, and macrophages. The usual synovial fluid volume in dogs and cats is 0.24 mL.

Synovial fluid acts as a lubricant in the joints, reducing friction. Because of its rheopectic characteristics, it functions as a thick absorbent. It provides oxygen and nutrition to the synovial tissues, nourishing them. It gets rid of metabolic waste. In the joint, it also serves as a molecular sieve.

8.1.4.3 Peritoneal Fluid

Peritoneal fluid is present between the peritoneal layers that line the abdominal cavity. It separates the peritoneum into two layers having an odourless, non-turbid, and clear or pale yellow colour. The pH ranges between 7.5 and 8.0. Peritoneal fluid has a specific gravity of less than 1.016. It is a blood ultrafiltrate, and water is the most important component. Simple diffusion allows electrolytes and tiny compounds to enter the peritoneal cavity. Electrolyte concentrations in the peritoneal fluid are similar to those in plasma. The total protein content is less than 2.5 g/dL, and the cell count is less than 3000–5000/μL. The cells present are leukocytes and desquamated mesothelial cells. The packed cell volume of the fluid in horses is less than 1%. The volume of peritoneal fluid increases during pregnancy. Lymphatics drain peritoneal fluid from the peritoneal cavity. The drainage is proportionate to the rate of its production. The purpose of peritoneal fluid is to lubricate abdominal organs and reduce friction between them during digestion and movement.

8.2 Fluid Balance

8.2.1 Acid-Base Balance

Sustaining acid-base balance is one of the most crucial parts of maintaining homeostasis. The negative logarithm of hydrogen ions (H^+) determines the pH of a solution. In mammals, the pH varies from 7.0 to 7.8. A mild alteration in pH threatens various physiological functions, impairs cellular functions, and affects the structure and functions of macromolecules. Acids are constantly generated in the body. The formation of bases balances the acids produced. Hence, the acid-base balance is maintained. An acid is any substance that donates a proton (releases H^+ in a solution). Among acids, some are strong acids, and some are weak. Strong acids freely ionise by giving up most of their hydrogen ions; thus, they reduce pH substantially. Weak acids have a minor effect on pH because they ionise only slightly, maintaining most of the H^+ in chemically bound form. Any substance that can take proton is referred to as a base (accept an H^+ ion). Bases can be strong or weak bases. A strong base affects the pH markedly by raising the pH as it has a stronger tendency to bind with hydrogen ions. Because it binds with fewer hydrogen ions, a weak base has little effect on pH.

Arterial blood has a pH of 7.36–7.44 (7.4), while interstitial fluid and venous blood have 7.35. The pH of the intracellular fluid lies between 6.0 and 7.4. (7.0). The body can maintain pH balance when the pH ranges from 7.0 to 7.8. The body tries to bring back the pH to a normal physiological level through various compensatory mechanisms whenever there is an alteration in pH beyond the normal range. Three basic systems regulate the concentration of hydrogen ions in body fluid:

1. Chemical buffer system: Mixes with acid or base right away to prevent excessive hydrogen ion concentration shifts.
2. Respiratory system: Controls carbon dioxide removal from extracellular fluid.
3. Kidney: Excretes acid or alkali, bringing the hydrogen ion concentration in the extracellular fluid back to normal.

Within a fraction of a second, the chemical buffer system (first line of defence) reacts to reduce the change in hydrogen ion concentration. The respiratory system (second line of defence) will intervene within minutes to keep hydrogen ion concentrations from fluctuating too much. Both the first and second lines of defence do neither remove nor add hydrogen ions to the body; instead, they bind them until a balance can be restored. Kidneys, the body's third line of defence, react slowly and remove excess acid or base. Kidneys are the most potent acid-base balance mechanism, and they are responsible for the final correction of acid-base balance.

8.2.1.1 Chemical Buffer System

A buffer prevents pH fluctuations by converting a strong acid or base to a weak one. A chemical buffer system is made up of a weak acid and the conjugate base of that acid. Bicarbonate phosphate, protein, and haemoglobin are all essential chemical buffer systems in the body.

8.2.1.1.1 Bicarbonate Buffer System

It is a mixture of carbonic acid (a weak acid) and bicarbonate ions in a protonated state (an unprotonated substance—a weak base). In extracellular fluid, it is the most significant buffer system. Carbonic acid is formed when carbon dioxide is hydrated in the presence of carbonic anhydrase, which dissociates into HCO_3^- and H^+:

$$CO_2 + H_2O \leftrightarrow H_2CO_3 \leftrightarrow HCO_3^- + H^+$$

Carbonic acid behaves as a weak acid as the reaction progresses to the right, releasing H^+ and reducing pH. When the reaction moves to the left, $HCO3^-$ functions as a weak base, binds H^+, and raises the pH. When the pH drops, the reaction shifts to the left to raise the pH and restores it to normal. When the pH rises, the reaction will move to the right to lower the pH and return it to normal.

The p*K* of the bicarbonate system (6.1) and the pH of the extracellular fluid are very different (7.4). As a result, the bicarbonate system is less effective than other chemical buffers. Compared to other chemical buffer systems, it plays a critical role in maintaining the pH of bodily fluids. The kidney and respiratory systems regulate this buffer system's bicarbonate and carbon dioxide components separately. These two regulating systems operate continuously and simultaneously, resulting in a more productive and efficient bicarbonate buffer system.

8.2.1.1.2 Phosphate Buffer System

It is the combination of hydrogen phosphate, HPO_4^{2-} (weak base—unprotonated substance), and dihydrogen phosphate, $H_2PO_4^-$ (weak acid—protonated substance). It works the same as the bicarbonate system:

$$H_2PO_4^- \leftrightarrow HPO_4^{2-} + H^+$$

When the reaction goes to the right, H^+ is liberated and pH is lowered, and when the reaction goes to the left, H^+ is bound and pH is raised. The optimal p*K* for this system is 6.8, which is close to the pH of body fluids. Hence, phosphate buffer has a more substantial buffering effect than an equal $HCO3^-$ buffer. But phosphates are less in extracellular fluid than bicarbonates, and they are important in renal tubules and intracellular fluid.

8.2.1.1.3 Protein Buffer System

Proteins are more concentrated in intracellular fluid than bicarbonates and phosphates. Intracellular proteins are responsible for 60–70% of chemical buffering within cells. The capacity to buffer is attributed to the side groups of their amino acid residues. Some have a carboxyl (-COOH) side group that releases H^+ as pH rises, lowering pH:

$$-COOH \leftrightarrow -COO^- + H^+$$

Others have amino (-NH_2) side groups, which bind H^+ when pH drops too low, thus raising pH towards normal. The most important buffering amino acid is histidine:

$$-NH_2 + H^+ \leftrightarrow NH_3^+$$

The protein buffer system is more powerful in the plasma since their p*K* value is very near to 7.2 due to their high concentration in plasma.

8.2.1.1.4 Haemoglobin Buffer System

It is the most effective protein buffer and the second most essential blood buffer. Because haemoglobin has a higher concentration than plasma proteins, it has a sixfold better buffering capacity. Haemoglobin in the form of reduced haemoglobin (Hb^-) and their weak acid (HHb) form buffer. It works the same as the bicarbonate system:

$$H^+ + Hb^- \leftrightarrow HHb$$

The chemical buffer system as a whole works together. Whenever the concentration of H^+ in extracellular fluid changes, the equilibrium of the buffer system changes as well, and the isohydric principle describes this phenomenon.

8.2.1.2 Respiratory Regulation

The bicarbonate buffer system equation demonstrates that adding carbon dioxide to bodily fluid boosts H^+ concentration and reduces pH, while removing carbon dioxide has the opposite effect. It is the foundation for the respiratory system's high buffering capacity. This technology neutralises 2–3 times the amount of acid that a chemical buffer system can balance:

$$CO_2 + H_2O \leftrightarrow H_2CO_3 \leftrightarrow HCO_3^- + H^+$$

Pneumatic ventilation is stimulated by an increase in partial pressure of carbon dioxide (PCO_2) and a decrease in pH. As a result, excess carbon dioxide is ejected. The reaction shifts to the left. As a result, the concentration of H^+ is reduced, and free H^+ becomes a member of the water molecule. When H^+ levels decline, pH rises and inhibits pulmonary ventilation. Carbon dioxide builds up as a result of this process. As the reaction progresses to the right, the pH decreases. Chemoreceptors are responsible for this impact. Based on their location, chemoreceptors are divided into central chemoreceptors and peripheral chemoreceptors.

Central chemoreceptors are chemosensitive areas on the ventral surface of the medulla. They stimulate the respiratory centres, generating a rise in tidal volume and breathing rate. These receptors are sensitive to variations in the concentration of H^+ ions in the brain's interstitial fluid and cerebrospinal fluid (CSF). Because H^+ ions have a difficult time crossing the blood-brain barrier (BBB), changes in H^+ ion concentration in the blood have a much smaller effect on stimulating the chemosensitive area. Carbon dioxide passes through BBB easily.

Carbon dioxide may easily cross the BBB; hence, increased blood PCO_2 raises the PCO_2 of interstitial fluid and cerebrospinal fluid. CO_2 is quickly hydrated in the interstitial fluid and CSF, forming carbonic acid. Carbonic acid breaks down into H^+ and HCO_3^-. Chemoreceptors detect a rise in hydrogen ion concentration, which causes the respiratory centres to be stimulated. Increased alveolar ventilation is caused by an increase in breathing rate and depth.

Aortic and carotid bodies are peripheral receptors. The aortic arch contains a cluster of chemoreceptors known as aortic bodies. The vagus nerve provides a signal to the dorsal respiratory group in the medulla oblongata. Carotid bodies are oval nodules found in the left and right common carotid arteries' walls. The glossopharyngeal nerve provides an impulse to the dorsal respiratory group. The aortic and carotid bodies will be stimulated by an increase in PCO_2 and hydrogen ions in the blood. They then activate the brain's respiratory regions, resulting in enhanced breathing.

8.2.1.3 Renal Regulation

The kidneys neutralise more acid and base than the respiratory system or chemical buffers. The kidneys remove hydrogen ions from the body, and the other buffer systems can lower their concentration by attaching it to other molecules. Three primary mechanisms regulate the extracellular fluid hydrogen ion concentration: (1) hydrogen ion secretion,

(2) reabsorption of filtered bicarbonate ions, and (3) bicarbonate ion generation. The rate of H^+ secretion by renal tubules is primarily determined by the intracellular pH of renal tubular cells. An increase in pH lowers H^+ secretion by the kidney, and lowering of pH increases hydrogen ion secretion. Intracellular pH of cells of renal tubule changes as blood pH or PCO_2 changes. Therefore, acidaemia and hypercapnia increase hydrogen ion secretion. In contrast, alkalaemia and hypocapnia reduce hydrogen ion secretion.

The proximal convoluted tubule secretes 85% of hydrogen ions. In contrast, the intercalated cells of the second half of the distal convoluted tubule, collecting tubule, and collecting duct secrete the remaining 15%.

Secondary active transport secretes hydrogen ions in the proximal convoluted tubule (sodium-hydrogen counter-transport). Carbon dioxide is either diffused into tubular cells or produced by cell metabolism. In the presence of carbonic anhydrase, it will mix with water to generate carbonic acid, which will then dissociate into HCO_3^- and H^+. Sodium-hydrogen counter-transport transports hydrogen ions into the tubular lumen. Carbon dioxide and water will result from the reaction of H^+ with filtered bicarbonate. Carbon dioxide penetrates tubular cells and becomes carbonic acid when it reacts with water.

Primary active transport secretes hydrogen ions in intercalated cells (second half of distal convoluted tubule, collecting tubule, and collecting duct). Carbonic acid is formed when dissolved carbon dioxide in the cell reacts with water to create bicarbonate ions (which are reabsorbed in the blood) and hydrogen ions, released into tubules by the hydrogen ATPase process.

As a result, whenever hydrogen ions are secreted into the renal tubules, the same amount of filtered HCO_3^- is reabsorbed. Only a small percentage of surplus H^+ in ionic form can be eliminated in the urine when more hydrogen ions are produced than HCO_3^- filtered into the tubular fluid. Urine can only be acidified to roughly a pH of 4.5, and it means that most of the hydrogen ions expelled must be bound by bases rather than being free in solution.

When hydrogen ions are titrated with bicarbonate ions in the tubular fluid, a bicarbonate ion is reabsorbed for each hydrogen ion released. In the tubular fluid, excess hydrogen ions will mix with buffers other than bicarbonates, such as ammonia buffer and phosphate buffer, resulting in new bicarbonate, which enters the blood. When extracellular fluid includes surplus hydrogen ions, kidneys reabsorb filtered bicarbonate from the tubular fluid and generate new bicarbonate ions. Although urea and citrate buffer systems exist, they are of minor consequence.

8.2.1.3.1 Phosphate Buffer

As the tubular fluid is acidified with hydrogen ion secretion, hydrogen phosphate (HPO_4^{2-}) takes up and binds hydrogen ions to form dihydrogen phosphate, $H_2PO_4^{2-}$, predominantly. Part of the cation (Na) that electrically balances $H_2PO_4^{2-}$ in the glomerular filtrate is exchanged with a secreted hydrogen ion and thus is returned to the blood.

8.2.1.3.2 Ammonia Buffer

Tubular epithelial cells produce ammonia, which diffuses into the tubules. In the renal tubular fluid, ammonia interacts with hydrogen ions to generate ammonium ions and then combines with chloride ions to form ammonium chloride. Each time an ammonia molecule combines to generate ammonium, the concentration of ammonia in tubular fluid decreases, which causes more ammonia to diffuse from epithelial cells. Chloride ions make up the majority of anions in the tubular fluid. The tubular fluid will fall below 4.5 if all hydrogen ions are carried with chloride ions. Still, ammonia mixes with hydrogen ions and chloride ions to generate ammonium chloride, a weak acid.

Glutamine produced in the liver is transferred to the proximal convoluted tubule, thick ascending limb of the loop of Henle, and distal convoluted tubule epithelial cells. A single glutamine molecule is digested inside the cell to produce two ammonium ions and two bicarbonate ions. The sodium-ammonium counter-transport mechanism secretes ammonium ions into the tubular lumen. This method reabsorbs two bicarbonate ions into the bloodstream for every glutamine molecule digested. As the level of acidity rises, the amount of glutamine digested by collecting duct tubular cells increases.

Urine pH is used to determine the amount of hydrogen ions present in the urine. It reflects the acid-base state of an animal. The ability of the kidneys to regulate hydrogen ion and bicarbonate concentrations in the blood determines the pH of urine. The pH of a dog's or cat's urine is between 6.0 and 7.5. Dairy cows have an average urine pH of 8.10, with a range of 7.27–8.71, and the mean urine pH of beef cows is 7.73, with a range of 7.42–8.12. The pH of an animal's urine fluctuates based on its diet. Urine produced by high-protein diets, such as those consumed by carnivores, is neutral to acidic. The urine of herbivores is more alkaline than that of carnivores. Forages having high K-salt concentrations cause a high dietary cation–anion difference resulting in alkaline urine. Further, with the buffering that happens in reaction to gastric acids, any animal can produce alkaline urine shortly after eating.

8.2.2 Acid-Base Balance Disturbances

The rate of the conjugate base to their weak acids determines the pH of the ECF. Buffer base refers to the overall amount of buffer base in whole blood, including bicarbonate, haemoglobin, and other minor bases (BB). These bases are called metabolic components, and they play a role in setting blood pH. Acid-base disruption occurs when the ECF gains or loses strong acid or base (Cl^- or HCO_3^-). At a pH of 7.4, the ratio of bicarbonate to carbonic acid in the extracellular fluid is 20:1. When carbonic acid levels rise, the ratio changes, resulting in a lower pH. Acidosis occurs when the pH goes below 7.5 due to a lack of bicarbonate or an increase in carbon dioxide partial pressure in the blood. In contrast, alkalosis happens when the pH rises above 7.4 due to an excess of bicarbonate or a decrease in carbon dioxide partial pressure in the blood.

Hydrogen ions diffuse into the cells to maintain electrical neutrality in acidosis, while potassium flows out of the cell. Intracellular proteins buffer hydrogen ions that enter the cell. As a result of the exchange between hydrogen and potassium, the cell loses a net amount of cation. Hyperpolarisation occurs when a cation is lost from a cell. The formation of the action potential in muscle cells and neurons is hindered. Acidosis reduces the activity of both the central nervous system and the muscles. Severe acidosis can result in unconsciousness and death. In alkalosis, hydrogen ions diffuse out of the cell, and potassium enters the cell. The membrane potential becomes more positive with the net gain of cations in the cell. As a result, neural tissue hyperexcitability and muscular overstimulation occur, resulting in tetany, convulsions, or respiratory paralysis.

Respiratory disturbances are acid-base imbalances caused by changes in the partial pressure of carbon dioxide in the blood. Acid-base imbalances due to alterations in bicarbonate levels are called metabolic disturbances. Metabolic acidosis, metabolic alkalosis, pulmonary acidosis, and respiratory alkalosis are acid-base abnormalities.

8.2.2.1 Metabolic Acidosis

Metabolic acidosis is defined as a gain of strong acid or a loss of base from the ECF. In metabolic acidosis, acidaemia will be present. It happens in ketosis, diabetes mellitus, and renal acidosis, where bicarbonate is lost in the urine due to tubular reabsorption failure. It also occurs in diarrhoea, where bicarbonate is lost. Due to a decrease in bicarbonate ions, the pH drops.

As a result, all blood buffer bases drop. In most cases, the partial pressure of carbon dioxide in the plasma does not vary. A drop in pH causes increased alveolar ventilation and reduced carbon dioxide partial pressure. Reduced carbon dioxide partial pressure will restore the natural ratio of conjugate base to weak acid. However, the total bases will be lower than usual, necessitating renal correction, i.e. H^+ ion excretion and plasma HCO_3^- restoration.

8.2.2.2 Metabolic Alkalosis

ECF results in the acquisition of base (OH^- or HCO_3^-) or the loss of strong acid. The symptoms of metabolic alkalosis are chronic vomiting (loss of stomach acid), potassium deficit (due to excessive renal excretion of hydrogen ions), and oxidation of organic acids. The parenteral introduction of bicarbonate solutions also causes metabolic alkalosis.

There is an increase in HCO_3^- in ECF, increasing the base content in all of these situations. The body's reaction is the polar opposite of that seen in metabolic acidosis. Alkalaemia causes a rise in pH, reducing lung ventilation and raising carbon dioxide partial pressure. Respiratory compensation brings the pH back to normal. Kidneys correct the condition by decreasing the secretion of H^+ ions and increasing the excretion of HCO_3^-.

8.2.2.3 Respiratory Acidosis

When the rate of CO_2 clearance by the lungs falls below the rate of CO_2 creation in the body, respiratory acidosis develops. It raises the partial pressure of carbon dioxide in the blood (hypercapnia). The inability of the lungs to exhale CO_2 at a regular pace is the primary cause of respiratory acidosis. It can occur by a lack of ability to enlarge the thorax due to a defect in the chest wall or respiratory muscles or any obstruction in the respiratory system that limits normal gas movement in the lungs.

A rise in PCO_2 causes an increase in H_2CO_3, and buffer reaction prevents the fall of pH caused by the increase in H_2CO_3. Renal compensation then follows. With a surge in plasma HCO_3^-, low pH enhances H^+ secretion into the urine.

8.2.2.4 Respiratory Alkalosis

In alveolar hyperventilation, the rate of removal of CO_2 exceeds the rate of creation in the body developing respiratory alkalosis. Low plasma PCO_2 (hypocapnia) and alkalaemia will be present. Increased alveolar ventilation is induced by aberrant activation of respiratory centres in the brain, either directly (as in ammonia poisoning) or indirectly (through peripheral chemoreceptors) through lower partial pressure of oxygen. Even when the partial pressure of carbon dioxide falls, there will be no change in the plasma concentration of bicarbonates at first. Non-bicarbonate buffers cause an immediate reaction. Thus, HCO_3 falls, and haemoglobin proteinate ions increase. Alkalaemia depresses H^+ ion secretion by renal tubules and increases the outflow of filtered HCO_3^- within a few hours, causing renal compensation. These result in further lowering of plasma HCO_3^-, and the ratio of HCO_3^- to H_2CO_3 moves back to normal.

8.2.3 Dehydration and Clinical Management

Clinical conditions affecting the hydration, acid-base, and electrolyte status are common in veterinary practice. As these conditions may result in harmful, often life-threatening consequences, recognition and management are vital.

8.2.3.1 Dehydration and Its Management

In small animal practice, dehydration is frequently linked to gastro-enteric diseases such as vomiting and diarrhoea that change electrolyte and acid-base status. Neonatal calf diarrhoea is a severe illness that causes severe dehydration in newborns. Dehydration status is assessed by physical examination and laboratory tests. Skin elasticity (skin turgor) is a valuable guide for evaluating dehydration. It can be carried out in the forehead in dogs and the neck region in cattle. With dehydration of about 5–6%, the loss of skin elasticity is mild, whereas in 10% dehydration, the skin often remains 'tented'. With higher percentages of dehydration, the animal becomes moribund. If the dehydration level is less than 5%, it cannot be reliably assessed by clinical findings. The relatively higher percent of body water in neonates and the variation in skin elasticity in older animals make the skin turgor test a less reliable tool in these age groups. Obesity can also affect skin tenting. Tacky mucous membrane on examination is suggestive of early stages of dehydration. With dry mucous membranes, the dehydration will be more than 6%. Eyeballs sunken in orbit are also noticed as dehydration increases. Rapid and weak pulses, coldness of extremities, animal appearing depressed, and prolonged capillary time indicate severe dehydration; shock may manifest. These clinical findings are noticed with more than 12% dehydration and have a grave prognosis. Eyeball recession (mm) and skin tent duration (seconds) are good indicators of calves' percentage dehydration and fluid replacement requirement. Dehydration is measured in the lab using packed cell volume (PCV) and total solids. Dehydration causes a rise in PCV and total solids. An increase in urine specific gravity can also detect dehydration. The aetiology of the existing disease should also be considered when interpreting laboratory findings, as disorders such as anaemia and hyperproteinaemia can cause variances. Fluid therapy for the management of dehydration has a quantitative aspect that is based on the correction of existing deficiencies, ongoing losses, and maintenance requirements.

The existing deficit is calculated as:

$$\text{Deficit (hydration) in litres} = \%\text{of dehydration (in decimals)} \times \text{body weight in kg}$$

Based on the formula used, a 300 kg cow with 8% dehydration would require about 24 L to correct the existing deficit. For ongoing losses, general thumb rules dictate the volume of fluid that needs to be replaced and may vary based on age and species. To meet daily needs, a volume of 50 mL/kg is required in dogs. Owing to the higher percent of extracellular fluid in young animals, their maintenance requirements are greater. A rate of 5 mL/kg/h or 120 mL/kg/day is required in calves, almost double the adult maintenance needs. The guidelines of the American Animal Hospital Association–American Association of Feline Practitioners suggest 2–6 mL/kg/h as maintenance rate in dogs; the formula suggested for 24 h is 132 × body weight in $kg^{0.75}$. Correction of ongoing losses depends on the type of loss (e.g. vomiting) and the number of episodes. The primary aim is to correct ongoing losses in 2–3 h. In 24 h, the patient's hydration status should be restored based on the total volume needed. Careful monitoring of the patient is essential during fluid therapy for signs of fluid overload. In such a case, reassess the status and adjust the rate of fluid administration. Tachypnoea crackles on auscultation, and watery nasal discharge suggests fluid overload. The volume requirement and rate of administration would be considerably different in animals with diseases affecting the organs like kidney and heart and in shock states.

Common routes of administration include intravenous, subcutaneous, and oral routes. Severe dehydration warrants intravenous fluid therapy. In patients with minimal dehydration, subcutaneous fluid administration can be considered. Isotonic fluids (normal saline and Ringer's lactate) are utilised for subcutaneous delivery. If vomiting is not present, the oral route may be used. Oral rehydration treatments are useful in preventing dehydration and electrolyte loss in calves. It is best to keep the amount of fluid given to calves to roughly 1–1.5 L at a time. Because of the practical challenges in intravenous fluid delivery in terms of the volume that needs to be provided, oral rehydration salts are now frequently suggested in adult cattle to overcome dehydration. However, if the animal is recumbent, the volume that can be administered orally will be reduced. An oral rehydration mix for cattle is sodium chloride 7 g, potassium chloride 1.25 g, and calcium chloride 0.5 g added to 1 L of water. This preparation is not an alkalinising solution, like other oral rehydration therapy preparations in calves. Calves with diarrhoea develop metabolic diseases in many instances due to hypovolaemia or specific diseases. Oral rehydration formulas for calves primarily have sodium and potassium, glucose, and chloride. Sodium bicarbonate, magnesium, acetate, and propionate are also included in some preparations for calves. Acetate and propionate act as metabolisable bases, which are converted to bicarbonate in the liver. These bases are considered superior to direct administration of sodium bicarbonate.

Moreover, they can act as an energy source and support sodium and water transport out of the small intestine. Fluid is administered intraosseously in paediatric patients and small dogs and cats when access to the intravenous route is difficult. The intraperitoneal route is also considered in such patients, provided that conditions like ascites and peritonitis are absent.

8.2.3.2 Types of Parenteral Fluids

The fluids utilised in clinical practice are divided into crystalloids and colloids. In veterinary medicine, crystalloid fluids are often used to treat dehydration. Fluids are classed as isotonic, hypertonic, or hypotonic based on their osmolality. Fluids having an osmolality similar to that of extracellular fluids (about 270–310 mOsmol/L) can be regarded as isotonic for all practical purposes. Normal saline (0.9% NaCl) and lactated Ringer's solution are two common examples. These fluids 'seep' into other body compartments and are redistributed within extracellular compartments. Only less than one-third of the total volume of fluids administered intravenously will be present in circulation after 1 h of administration. When administered, hypertonic fluids are useful to draw large quantities of fluid into circulation and are preferred in conditions like gastric dilatation and volvulus in dogs. Hypertonic fluids should not be used in cases of dehydration. Dextrose 50% is a hypertonic crystalloid used to manage ketosis in bovines. Sodium bicarbonate as a 5% solution is employed to treat carbohydrate engorgement of ruminants. Hypertonic saline (3% or 7% NaCl) is used in veterinary practice to manage intracranial pressure in head trauma conditions. Hypertonic saline is also used in hypovolaemic shock management, as the volume required for resuscitation is relatively less than isotonic fluids. Hypertonic saline should not be used in dehydrated patients. Preparations like 0.45% NaCl and dextrose 5% are hypotonic fluids. Crystalloid fluids can further be classified, based on usage, as replacement fluids and maintenance fluids. Replacement fluids (e.g. normal saline) have higher sodium concentration and lower potassium levels than maintenance fluids and are indicated in cases of ongoing fluid and electrolyte losses, as in vomiting. A combination of half-strength dextrose (2.5%) and NaCl (0.45%) is also isotonic and is used as a maintenance fluid along with potassium chloride supplementation. It can be used after the ongoing electrolyte imbalances, and dehydration is corrected. The pH of the fluids may also vary. Normal saline has a pH of 5.5 and that of lactated Ringer's is 6.5.

> **Know More......**
>
> **Dehydration Management in Birds**
>
> Panting during periods of increased ambient temperature can lead to respiratory alkalosis in birds as excessive carbon dioxide losses occur. Dietary electrolyte (Na^+ + K^+-Cl^-) balance and electrolyte [(K^+ + Cl^-)/Na^+] ratio in the feed need to be monitored to alleviate the physiological and metabolic changes of heat stress. These electrolytes are considered important in managing acid-base balance and osmotic pressure of body fluids. In heatstroke, cooling the bird is an emergency measure that owners can try before veterinary aid is available. It involves the use of tap water or tepid water, and cold water should not be used for the purpose. The birds can also be misleading with water making sure that the water has good contact with skin. Moistening the feet and beak is also required.

Colloids can be broadly classified into natural colloids and synthetic colloids. Blood and blood products (albumin) are examples of natural colloids. Hydroxyethyl starch and dextran are synthetic colloids. Due to their higher molecular weight, these intravenous preparations remain in circulation for more extended periods ('crystalloids seep fast'). A veterinary product of hydroxyethyl starch available in India has a molecular weight of 130 kDa. Indications for the use of synthetic colloids are in the management of acute hypovolaemia and maintenance of plasma oncotic pressure. The volume required for fluid resuscitation when plasma expanders are used would be considerably less than that of crystalloids. A clinician needs to be aware of the potential signs of hypersensitivity and organ injury when natural and synthetic colloids are used (Table 8.7).

8.2.3.3 Acid-Base Imbalances and Electrolyte Abnormalities

Acid-base imbalances and electrolyte abnormalities often have life-threatening effects on animals. The disorders vary from carbohydrate engorgement in ruminants to hypokalaemia associated with diabetic ketoacidosis in dogs. The use of blood gas and electrolyte analysers would be beneficial in detecting these variations and monitoring treatment.

Carbohydrate engorgement of cattle (lactic acidosis) results in metabolic acidosis and dehydration, requiring intravenous sodium bicarbonate therapy to manage the acidosis and intravenous fluids to correct the dehydration. Base deficit measurement is the ideal method for deciding on the bicarbonate quantity to be administered. In a severe case of metabolic acidosis, sodium bicarbonate required (in mmol) is base deficit (from blood analysis) × 0.5 (or 0.3) × body weight in kg. Half the calculated dosage needs to be administered for 3–4 h, and the patient values need to be reassessed before administering sodium bicarbonate further. In dogs with chronic renal disease, bicarbonate medication may be required to keep the bicarbonate level between 18 and 24 mmol/L. Thumb guidelines are also employed in ruminant practice to determine the amount of sodium bicarbonate to deliver in cases of lactic acidosis, depending on the clinical severity, because direct access to the laboratory may not be possible in many farms. In urea toxicosis of ruminants, dilute

Table 8.7 Types of dehydration and fluids administered

Type of dehydration	Fluids preferred[a]
Isotonic dehydration (normal serum sodium levels)	Isotonic fluids like normal saline and Ringer's lactate
Hypertonic dehydration (elevated serum sodium)	Fluids with 'free water' (dextrose 5%)
Hypotonic dehydration (Low serum sodium—Not commonly encountered in clinical practice)	Normal saline

[a] Fluid administration also depends on the severity of the condition, primary aetiology, metabolic status, and electrolyte imbalances. Oral rehydration can be tried in less severe cases, especially when vomiting is absent

acetic acid (vinegar) is administered orally to manage ruminal alkalosis. Hypercapnia can result from airway obstruction and pulmonary disease; respiratory acidosis manifests. In dogs, tracheal collapse, brachycephalic syndrome, and chronic bronchial diseases can result in respiratory acidosis. Two or more separate acid-base abnormalities characterise mixed acid-base disorders. Interpretation of the results is essential in deciding the treatment options in such cases.

Diseases, conditions like vomiting, and drugs can contribute to electrolyte imbalances. Renal failure and hypoadrenocorticism can result in hyperkalaemia in dogs. Hypochloraemia and hyponatraemia were reported in hypoadrenocorticism. Hypokalaemia and hyponatraemia can be associated with diabetic ketoacidosis. In ruminants, hypochloraemic, hypokalaemic alkalosis occurs in left abomasal displacement. Administration of furosemide, a loop diuretic, can cause hyponatraemia. Hypokalaemia, hypocalcaemia, and hypomagnesaemia can also result in this drug's administration. Hyperphosphataemia that arises in many cases of chronic kidney disease may warrant dietary phosphate restriction and the use of phosphate binders. Loss or excess of electrolyte management is challenging in many clinical settings. Intravenous potassium chloride administration is carried out after dilution in normal saline and needs to be monitored carefully due to the cardio-toxic effects of potassium. Fluid and electrolyte therapy will have to be tailored based on the primary disease and the body system involved.

Learning Outcomes

- The water in the body is divided into intracellular and extracellular fluid compartments (plasma, lymph, and interstitial and transcellular fluids). The plasma and interstitial fluid in vertebrates are similar in composition, but the ECF and ICF in all animals are significantly different, with NaCl prevailing in the ECF and potassium and organic molecules dominating in the ICF.
- ECF volume and osmolarity are both regulated in mammals to maintain fluid balance. Controlling ECF osmolarity prevents hyper- or hypotonicity from causing variations in ICF volume. The baroreceptor reflex and plasma–interstitial fluid shifts regulate ECF volume in the short term, which is critical in the long-term regulation of blood pressure. Water and salt balances are used to regulate osmolarity and volume.
- In acid-base balance, the management of free hydrogen ions in physiological fluids is critical to survival. Free hydrogen ions are liberated by acids, whereas bases accept free hydrogen ions. The hydrogen ion concentration is expressed using the pH scale. Hydrogen ion fluctuations affect neuron, enzyme, and potassium ion activity. From metabolic activities, hydrogen ions are constantly added to bodily fluids.
- The major ECF buffer is the bicarbonate buffer system. Intracellularly, the peptide and protein buffer system, which includes haemoglobin in erythrocytes, is crucial. Buffers are only a temporary solution because they do not remove excess hydrogen ions; thus, the second and third lines of defence are required. The second line of defence is the respiratory system, which regulates hydrogen ions by adjusting ventilation. Carbon dioxide is removed when breathing occurs more deeply, but it is retained when breathing happens less deeply. Excretory systems control both bicarbonate and hydrogen ions in the ECF and constitute the third line of defence that aid in acid-base homeostasis. Cells in kidneys can release hydrogen ions and reabsorb bicarbonate, while other cells can do the opposite. Some cells release ammonia trap hydrogen ions as ammonium in acidosis.
- Hypoventilation causes respiratory acidosis, caused by an increase in carbon dioxide. Hyperventilation causes respiratory alkalosis, caused by a reduction in carbon dioxide. A decrease in plasma bicarbonate is linked to metabolic acidosis induced by acute diarrhoea, diabetes, intense exertion, or uraemia. Hyperventilation causes respiratory alkalosis caused

(continued)

by a reduction in carbon dioxide. Vomiting can cause metabolic alkalosis, marked by an increase in bicarbonate.

- Body fluid, electrolyte, and acid-base homeostasis must be maintained and regulated to the sustained body's functions. The body has several compensating processes to keep fluid, electrolytes, and acid-base balance; if its compensating systems fail to maintain homeostasis, it can have profound, even life-threatening implications. Fluid treatment can help with these problems.

Exercises

Objective Questions

Q1. Which is the most effective buffer system in the intracellular fluid?
Q2. Which is the major cation of the extracellular fluid?
Q3. Which is the first line of defence in acid-base regulation?
Q4. What does the isohydric principle state?
Q5. How many bicarbonate ions are returned to the blood for each glutamine metabolised within tubular epithelial cells?
Q6. What are the only means by which water output can be significantly controlled?
Q7. Which hormone is released from the heart when there is an increase in sodium level and blood volume?
Q8. What amount of fluid is required to correct 8% dehydration in a 300 kg cow?
Q9. Which fluid compartment has the major part of the body's water?
Q10. Which electrolyte is the chief determinant of cellular volume and intracellular osmolarity?
Q11. Where does the reabsorption of cerebrospinal fluid occur explicitly?
Q12. Which are the major electrolytes involved in the total osmolarity of the interstitial fluid and plasma?
Q13. Which condition is indicated when there is increased excretion of ammonium chloride in urine?
Q14. What amount of water forms when one gram of fat is oxidised?
Q15. Which type of acid-base imbalance is observed in the carbohydrate engorgement of cattle?

Subjective Questions

Q1. Explain in detail body water compartments.
Q2. Which are the methods for measuring body water?
Q3. What is cerebrospinal fluid? Explain the formation, absorption, and functions of cerebrospinal fluid.
Q4. How is acid-base regulated in the body?
Q5. Which are the acid-base imbalances?
Q6. What is dehydration, and how is it managed?
Q7. Which are the common electrolyte abnormalities?
Q8. Which are the important electrolytes in the body fluids?
Q9. How is water intake and water output regulated?
Q10. Which are the different types of parenteral fluids?

Answer to Objective Questions

A1. Protein
A2. Sodium
A3. Chemical buffering
A4. The buffers of the blood and body fluids do not act independent of each other, but rather react in unison
A5. Two
A6. Through regulation of urine volume
A7. Atrial natriuretic factor
A8. 24 L
A9. Intracellular fluid
A10. Potassium
A11. Arachnoid villi
A12. Sodium and chloride ions
A13. Acidosis
A14. 1.1 mL of water
A15. Metabolic acidosis

Keywords for Answer to Subjective Questions

A1. Intracellular fluid, extracellular fluid, transcellular fluid
A2. Indicator dilution technique, haematocrit
A3. Choroid plexus, arachnoid villi, buoyancy
A4. Chemical buffers, respiratory system, kidney
A5. Metabolic acidosis, metabolic alkalosis, respiratory alkalosis, respiratory acidosis
A6. Diarrhoea, skin turgor test, Ringer's lactate
A7. Hyperchloraemia, hyperkalaemia, hypernatraemia, hypocalcaemia, hypochloraemia, hypokalaemia
A8. Sodium, potassium, phosphate, chloride, hydrogen, bicarbonate, calcium
A9. Antidiuretic hormone, thirst, urine, atrial natriuretic factor
A10. Crystalloids, colloids, isotonic, hypertonic, isotonic

Further Reading

Textbooks

Hall JE (2011) Acid–base regulation. In: Hall JE (ed) Guyton and Hall textbook of medical physiology, 12th edn. Saunders Elsevier, Philadelphia, pp 379–396

Recee WO (ed) (2015) Duke's physiology of domestic animals. Wiley

Reece WO (2009) The urinary system. In: Reece WO (ed) Functional anatomy and physiology of domestic animals, 4th edn. Wiley-Blackwell, pp 344–347

Rudloff E (2015) Assessment of hydration. In: Silverstein DC, Hopper K (eds) Small animal critical care medicine, 2nd edn. Elsevier, St. Louis, MO, pp 307–310

Research Articles

Adrogue HJ, Madias NE (2012) The challenge of hyponatraemia. J Am Soc Nephrol 23:1140

Bhave G, Neilson EG (2011a) Body fluid dynamics: back to the future. J Am Soc Nephrol 22:2166

Bhave G, Neilson EG (2011b) Volume depletion versus dehydration: how understanding the difference can guide therapy. Am J Kidney Dis 58:302–309

Brooks A, Thomovsky E, Johnson P (2016) Natural and synthetic colloids in veterinary medicine. Top Comp Anim Med 31:54–60

Damkier HH, Brown PD, Praetorius J (2013) Cerebrospinal fluid secretion by the choroid plexus. Physiol Rev 93:1847

Davis H, Jensen T, Johnson A, Knowles P, Meyer R, Rucinsky R, Shafford H (2013) AAHA/AAFP fluid therapy guidelines for dogs and cats. J Am Anim Hosp Assoc 49:149–159

DiBartola SP (2012) Introduction to acid-base disorders. In: DiBartola SP (ed) Fluid, electrolytes, and acid-base disorders in small animal practice, 4th edn. Saunders, St. Louis, MO, pp 231–252

Hallowell G, Remnant J (2016) Fluid therapy in calves. In Pract 38:439–449

Hamilton PK, Morgan NA, Connolly GM, Maxwell AP (2017) Understanding acid-base disorders. Ulster Med J 86(3):161–166

Hamm LL, Nakhoul N, Hering-Smith KS (2015) Acid-base homeostasis. Clin J Am Soc Nephrol 10(12):2232–2242

Hew-Butler T, Holexa BT, Fogard K, Stuempfle KJ, Hoffman MD (2015) Comparison of body composition techniques before and after a 161-km ultramarathon using DXA, BIS and BIA. Int J Sports Med 36:169–174

Koeppe BM (2009) The kidney and acid–base regulation. Adv Physiol Educ 33:275–228

Marshall WJ, Bangert SK, Lapsley ML (2012) Hydrogen ion homeostasis and blood gases. In: Marshall WJ, Bangert SK, Lapsley ML (eds) Clinical chemistry, 7th edn. Mosby, Edinburgh, pp 41–62

Monnig AA (2013) Practical acid-base in veterinary patients. Vet Clin N Am Small Anim Pract 43:1273–1286

Perotta JH, Lisboa JAN, Pereira PFV, Ollhoff RD, Vieira N, Faliban KKMC, Filho IRB (2018) Fractional excretion of electrolytes and paradoxical aciduria in dairy cows with left displaced abomasum. Pesq Vet Bras 38:840–846

Popkin BM, D'Anci KE, Rosenberg IH (2010) Water, hydration, and health. Nutr Rev 68:439–458

Roussel AJ (2014) Fluid therapy in mature cattle. Vet Clin N Am Food Anim Pract 30:429–439

Sam R, Feizi I (2012) Understanding hypernatraemia. Am J Nephrol 36: 97

Taylor JD, Rodenburg M, Snider TA (2016) Comparison of a commercially available oral nutritional supplement and intravenous fluid therapy for dehydration in dairy calves. J Dairy Sci 100:4839–4846

Thomovsky E (2017) Fluid and electrolyte therapy in diabetic ketoacidosis. Vet Clin N Am Small Anim Pract 47:491–503

Excretory Physiology

9

V. Beena

Abstract

In all organisms, the excretory system maintains the internal aqueous and ionic environment for homeostasis. The organs of the renal system, constituted by paired kidneys, ureters, urinary bladder and urethra, help in the excretion of waste products and hydrogen ions, harmful drugs and toxins. Nephrons, the kidney's functional units, are concerned with the glomerular filtration, tubular reabsorption and tubular secretion to form urine. Glomerular filtration is a passive, non-selective process where fluids and electrolytes are filtered through the three layers of the glomerular membrane into Bowman's space under the influence of physical forces. Kidneys maintain relatively constant renal blood flow and glomerular filtration rate (GFR) within a wide range of mean systemic arterial pressure by autoregulatory mechanisms. Tubular reabsorption is a highly selective process by which water and solutes are reabsorbed to the peritubular capillaries. The concentration of urine occurs by the countercurrent multiplier and countercurrent exchange systems existing in the hypertonic renal medullary interstitium. Urine formed in the kidneys is conveyed to the urinary bladder through ureters for temporary storage and removed from it periodically through the urethra by a process called micturition. The avian excretory system is structurally and functionally modified to avoid excess water loss and eliminate the major nitrogenous waste product, the uric acid.

V. Beena (✉)
Department of Veterinary Physiology, College of Veterinary and Animal Sciences, Thrissur, Kerala, India
e-mail: beenav@kvasu.ac.in

© The Author(s), under exclusive license to Springer Nature Singapore Pte Ltd. 2023

P. K. Das et al. (eds.), *Textbook of Veterinary Physiology*, https://doi.org/10.1007/978-981-19-9410-4_9

Graphical Abstract

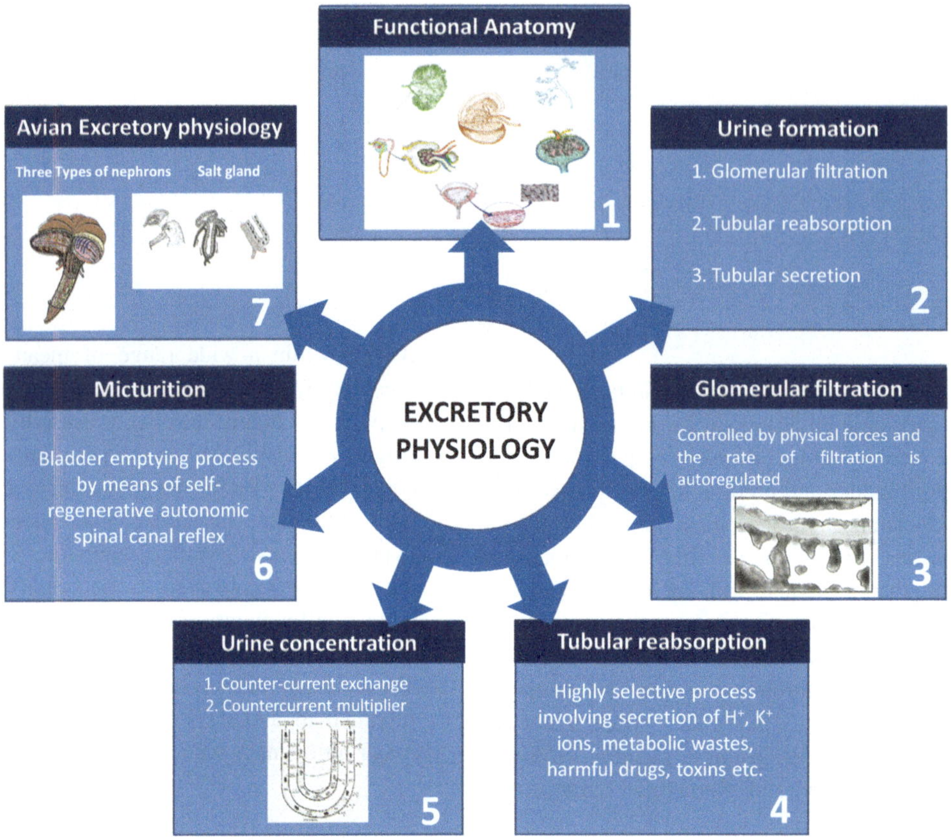

Description of the graphic: The renal system is structurally organised (**1**) to eliminate end products of metabolism and other foreign waste materials. The process of urine formation (**2**) involves glomerular filtration, tubular reabsorption and tubular secretion. Glomerular filtration (**3**) is controlled by physical forces acting across the glomerular capillary wall and autoregulated by the kidney. Tubular reabsorption (**4**) is a highly selective process that occurs passively and actively. The tubular filtrate is concentrated by countercurrent exchange and countercurrent multiplier mechanisms (**5**). The urine formed is voided out by a process known as micturition (**6**). The avian renal system (**7**) is also structurally and functionally modified to eliminate the major nitrogenous end product, uric acid

Keywords

Avian excretion · Micturition · Nephron · Urine formation

Learning Objectives

- Importance of excretory system and physiological functions of the kidney
- General and functional ultrastructural morphology of the renal system
- The process of urine formation includes glomerular filtration, tubular reabsorption and secretion
- Urine concentration mechanisms following countercurrent multiplier and countercurrent exchange mechanisms
- Renal function tests
- Urinary bladder and its functions
- Peculiarities of avian excretion

9.1 Introduction

Maintaining an internal aqueous environment with consistent water and solute composition is vital for homeostasis. Loss of water and electrolytes during digestion, metabolism, thermoregulation and elimination of waste products always creates imbalances in fluid and electrolyte composition in the internal environment. In animals, the urinary or renal excretory system counteracts such imbalances by effectively eliminating metabolic wastes and selective retention of water and solutes. Thus, the renal system is considered a major excretory system in the body. The other systems involved in excretion are the respiratory system, digestive system and integumentary system and have a minor role in animals.

Protein and nucleic acid metabolism always produces nitrogenous waste products in the body. Ammonia (NH_3), a potentially toxic base that can bind with protons to become an ammonium ion (NH_4^+), constitutes the major end product of protein metabolism. If present in excess, these ions can also interfere with Na^+/K^+ ATPase substituting for K^+, causing morphological changes in neurons, interrupted ion conduction and disrupted neurotransmitter metabolism. Thus, NH_3 and NH_4^+ must either be highly diluted and rapidly excreted or be converted into a less toxic form. The most common of these forms are urea and uric acid.

Most aquatic animals are ammonotelic, and they rely on ammonia excretion through gills in the most diluted form as water is freely available for them. Most terrestrial mammals used metabolic energy to convert ammonia to urea using ATP in the liver and excreted through kidneys, known as *ureotelic*. Urea is 10–100 times less toxic than ammonia. Thus, it can be accumulated in too much higher concentrations and has the benefit of removing two nitrogens per molecule. It takes about ten times less water to excrete a given amount of nitrogen as urea than ammonia. In reptiles, birds and insects, ammonia is converted to uric acid (*uricotelic*) before excretion. It requires more ATP for the production of uric acid but is less toxic as it is highly insoluble and has the added benefit of removing four nitrogens per molecule. It is converted as urates into the hindgut and excreted in a semisolid form. It takes about 50–100 times less water to excrete a given amount of nitrogen as uric acid than ammonia. Compared to urea, which is essentially infinitely soluble in solution, uric acid is highly insoluble (precipitates at concentrations greater than about 0.4 mM). Urea and uric acid are excreted through renal system, which is constituted by kidneys and other organs with functional and structural modifications in mammals and birds.

9.2 Physiological Functions of the Kidney

Excretion of waste products of metabolism and unwanted foreign materials: The major metabolic waste products include urea, creatinine, bilirubin, uric acid and metabolites of protein, nucleic acid, haemoglobin and various hormones. Kidneys also aid in eliminating pesticides, drugs and food additives that enter by different means to the body.

Maintenance of water and electrolyte balances: The kidneys can alter either the rates of absorption or excretion or the rates of both water and ions like sodium, potassium, chloride, calcium, hydrogen, magnesium and phosphate to regulate the fluid and electrolyte balance in the body.

Regulation of arterial pressure: Arterial pressure is maintained on a long-term basis by adjusting the kidneys' excretion of sodium and water. Angiotensin II, a vasoactive peptide, has a vital role in the short-term maintenance of blood pressure produced in the body with the help of renal proteolytic enzyme renin.

Acid-base regulation: Kidneys can regulate excess hydrogen ions and acids, like sulphuric and phosphoric acids, to maintain the pH of body fluids.

Erythropoietin production: In hypoxic conditions, kidneys can produce hormone-like erythropoietin (EPO), which stimulates erythrocyte production.

Calcium homeostasis by calcitriol formation: The kidneys can produce the metabolically active derivative of vitamin D-1,25, dihydroxy cholecalciferol (calcitriol or vitamin D_3). It is essential in maintaining the normal calcium deposition in bones and reabsorption of calcium from the gastrointestinal tract.

Glucose synthesis: When the blood glucose level lowers, kidneys can synthesise glucose by gluconeogenesis.

9.3 Functional Morphology of Kidney

The mammalian urinary system has a pair of bean-shaped kidneys located on the dorsal side of the lower abdominal cavity on either side of the vertebral column (Fig. 9.1). Both kidneys together constitute 0.4% of the body weight. Each kidney is externally covered by a thick connective tissue renal capsule and internally divided into two parts, the outer cortex and an inner *medulla*. The *hilum* is an indentation seen on the medial aspect of the kidney through which pass renal blood vessels (renal artery and vein), lymphatics and nerves. The medulla has triangular structures called *pyramids* or *papillae* extending into *calyces* (singular calyx) channels. The minor calyces drain urine from the tubules of each papilla and discharge it to the major calyces. The calyces open into a wide conical central cavity called the *pelvis*. Some mammalian species like humans, dogs, small ruminants and rabbits have uni-pyramidal kidneys, whereas the kidneys of large ruminants and pigs are multi-pyramidal in nature.

There are two tubular structures called *ureters* carrying urine from the pelvis to the urinary bladder for temporary storage. Urine is removed from the *urinary bladder* periodically through the urethra.

The nephrons are the basic functional units of the kidney. Several nephrons of a kidney bind together by connective tissue. The number of nephrons in each kidney varies in different species of mammals, for example, about 4 million in cattle, 1.25 million in pigs, 1 million in humans, 0.5 million in dogs and 0.25 million in cats. Each nephron has a *renal corpuscle* and a *renal tubule*. The renal corpuscle comprises a tuft of capillaries called *glomerulus* surrounded by a cup-shaped *Bowman's capsule* (Fig. 9.2). The Bowman's capsule encloses a space between the two walls

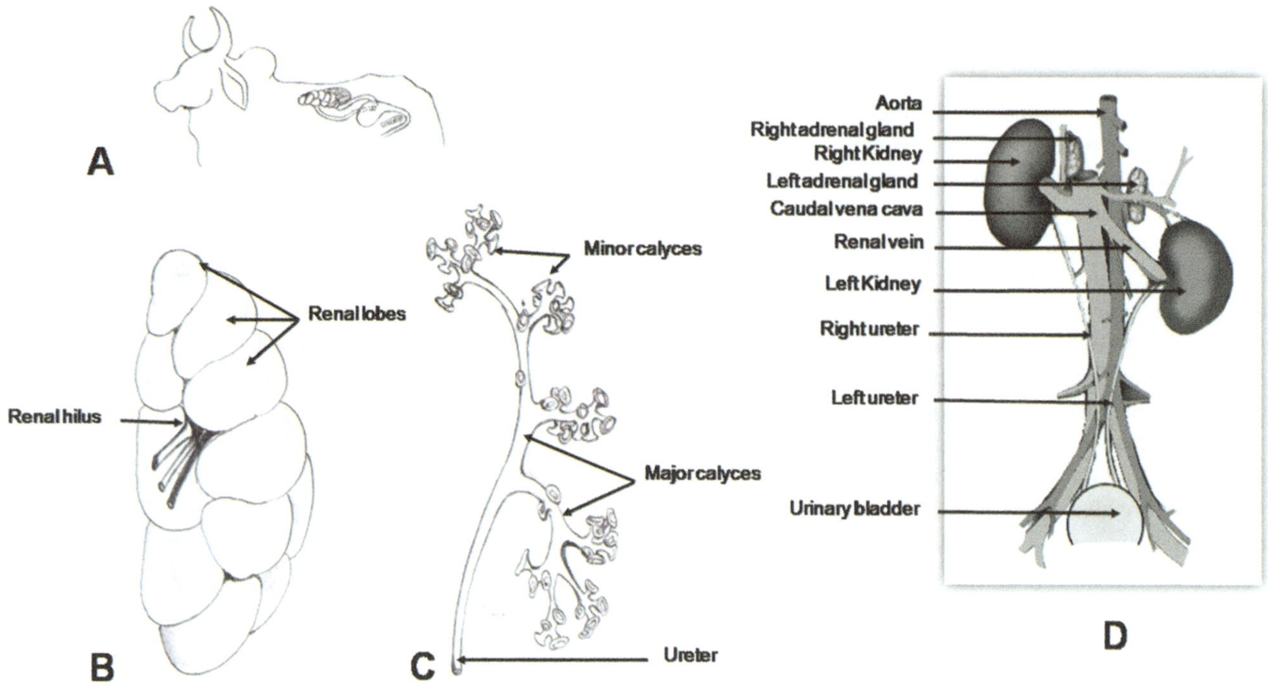

Fig. 9.1 General organisation of the kidneys and the urinary system. (**a**) Bovine kidneys in situ, (**b**) bovine kidney showing pseudo-lobulation, (**c**) corrosion cast depicting major and minor calyces, (**d**) schematic diagram of the excretory system of dog

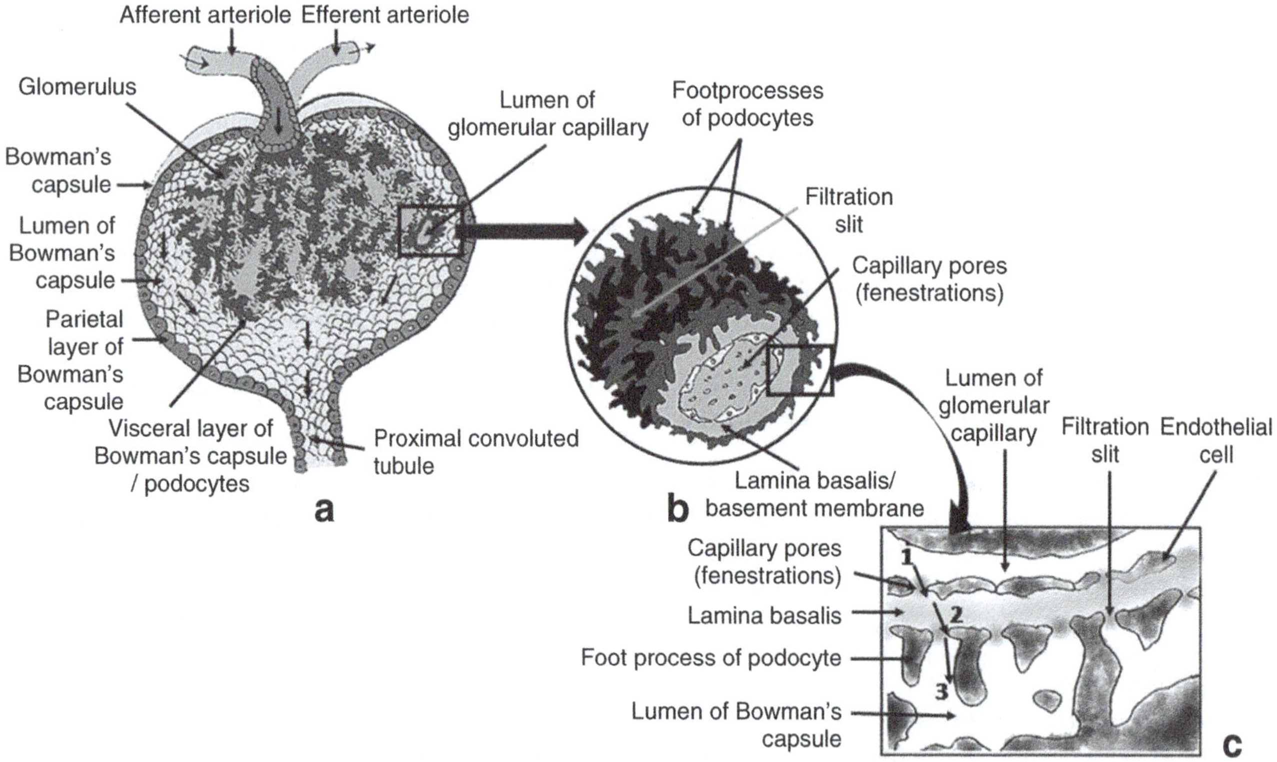

Fig. 9.2 Schematic diagram showing the major features of the renal corpuscle. (**a**) Renal corpuscle, (**b**) structure of glomerular capillary, (**c**) enlarged structure of glomerular capillary wall showing the intercellular pores (fenestrations) between the endothelial cells lining the glomerular capillary, acellular basement membrane and filtration slits situated between the foot processes of the podocytes constituting the visceral layer of Bowman's capsule

(visceral and parietal) of it called Bowman's space, which continues as the renal tubule.

The renal tubule has three structurally and functionally different divisions; the *proximal convoluted tubule (PCT)* located in the cortex, the *loop of Henle* (with its thin descending limb, thin ascending limb and thick ascending limb) located in the medulla and the *distal convoluted tubule (DCT)* present in the cortex. In mammals, up to 8 DCTs coalesce to form a *collecting duct* (CT), which runs back to the medulla (Fig. 9.3). The portion of the CT that lies in the cortex is the cortical collecting duct, which dips into the medulla called the medullary collecting duct. The collecting ducts plunge deep into the medulla to evacuate their contents to the renal pelvis.

In mammals, there are two types of nephrons: *cortical nephrons* and *juxtamedullary nephrons* (Fig. 9.4). The glomerulus of cortical nephrons lies in the peripheral cortex, and that of the juxtamedullary nephrons lies in the cortex's inner layer, adjacent to the medulla. The arrangement of glomeruli in the cortex gives the cortex a granular appearance.

The position of the loops of Henle varies in two types of nephrons; the terminal hairpin loop of Henle dips slightly into the medulla, whereas the longer loops of juxtamedullary nephrons pass deep into the medulla.

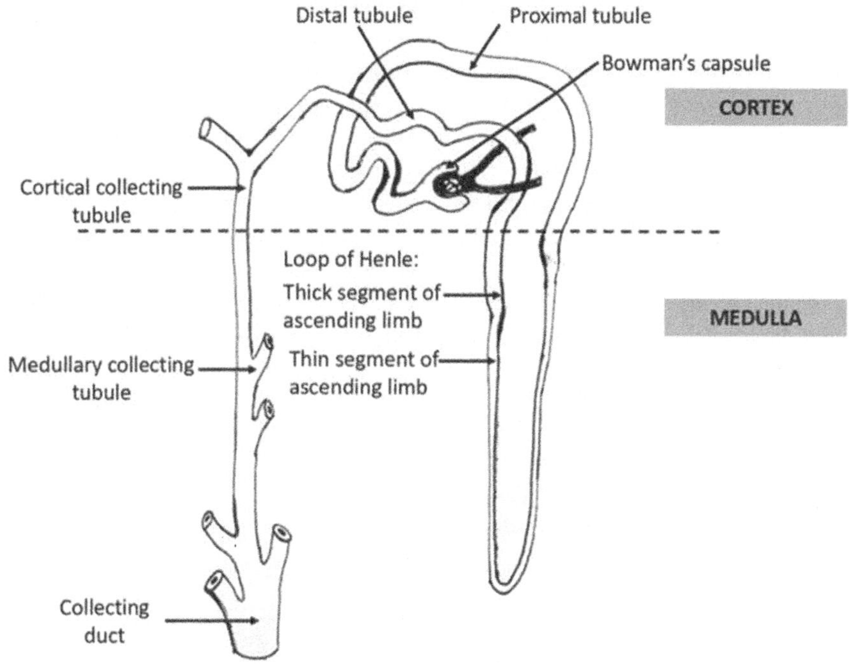

Fig. 9.3 Structure of a nephron. Nephron, the functional unit of kidney, is structurally modified as PCT, the loop of Henle, DCT, collecting tubule and finally collecting duct. The various parts of nephron are differentially placed in the renal cortex and medulla

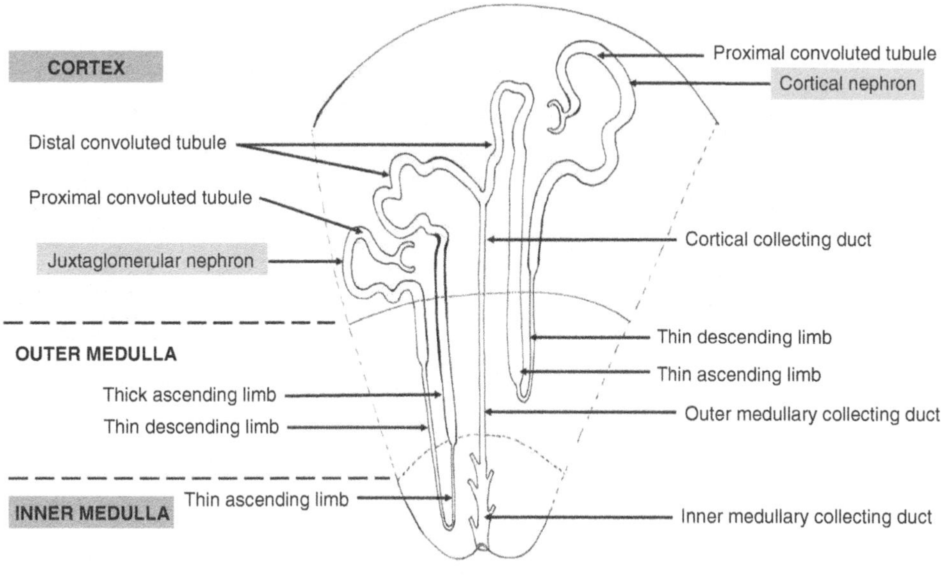

Fig. 9.4 Relative positioning of cortical and juxtamedullary nephrons. The glomeruli of both the nephrons are placed in the cortex, but the loop of Henle of juxtamedullary nephrons dips deep into the medulla

9.3.1 Renal Blood Supply

In a human being with a 70 kg body weight, 1100 mL of blood flows through both kidneys per minute, constituting 22% of cardiac output. The arterial blood reaches the kidney through the renal artery through the hilum. It then branches progressively to form the interlobar arteries, arcuate arteries, interlobular arteries (radial arteries) and short *afferent arterioles*, one of which supplies blood to the capillary tuft of each nephron. The capillaries of the glomerulus reunite at the exit point as efferent arteriole. The *efferent arteriole* then subdivides into the second capillaries called *peritubular capillaries* surrounding the proximal and distal convoluted tubules in the renal cortex. These capillaries ultimately rejoin to form the vessels of the venous system, which run parallel to the arteriolar vessels and progressively form the *interlobular vein*, *arcuate vein*, *interlobar vein* and *renal vein*, which leaves the kidney at the hilus. The peritubular capillaries of juxtamedullary nephrons form hairpin loops in close association with the loops of Henle. These vascular hairpins are known as *vasa recta*. When passing through the medulla, the collecting ducts of both cortical and juxtamedullary nephrons always run parallel in proximity to the ascending and descending loops of Henle of juxtamedullary nephrons and vasa recta. The parallel arrangement gives the striated appearance to the medulla.

9.3.2 Juxtaglomerular Apparatus

The thick ascending limb of the loop of Henle enters the cortex from the medulla, passes between the afferent and efferent arterioles and continues as a distal tubule (Fig. 9.5). The epithelial cells of the tubules that come in contact with the afferent and efferent arterioles facing the angle between the blood vessels are collectively known as *macula densa*. The presence of epithelial cells in the densa marks the beginning of the distal tubule. The specialised granular smooth muscle cells of afferent arterioles that contact macula densa are called *juxtaglomerular* (JG) cells. From these cells, proteolytic enzyme *renin* is produced. Mesangial cells and the matrix secreted by the *mesangial cells* occupy the space between the afferent arteriole, efferent arteriole, macula densa and glomerular capillaries. Those cells located between the arterioles and macula densa are *lacis* cells or extraglomerular mesangial cells. Mesangial cells exhibit contractile property to aid blood flow through glomerular capillaries. Besides, they also secrete prostaglandins, maintain basement membrane and possess phagocytic properties. The macula densa, juxtaglomerular cells and lacis cells are together known as the *juxtaglomerular (JG) apparatus*. This apparatus has functional importance in renal haemodynamics and glomerular filtration rate (GFR).

9.3.3 Nerve Supply

Sympathetic fibres of the autonomous nervous system enter the kidneys through the hilus along with the renal artery and vein and innervate blood vessels of the kidney, nephron segments and JG cells. Motor activity of these nerves produces alterations in renal haemodynamics and composition of the tubular fluid. The kidney contains afferent sensory nerve fibres sensing stretch, located mainly in the renal pelvic wall. The renal pelvic pressure increases due to the accumulation of urine, and it causes stretching of the pelvis wall that stimulates the mechanoreceptors present in between the smooth muscles. This sensation results in the activation of sensory afferent nerves of the ipsilateral kidney, which in turn causes decreased contralateral motor renal nerve activity. The

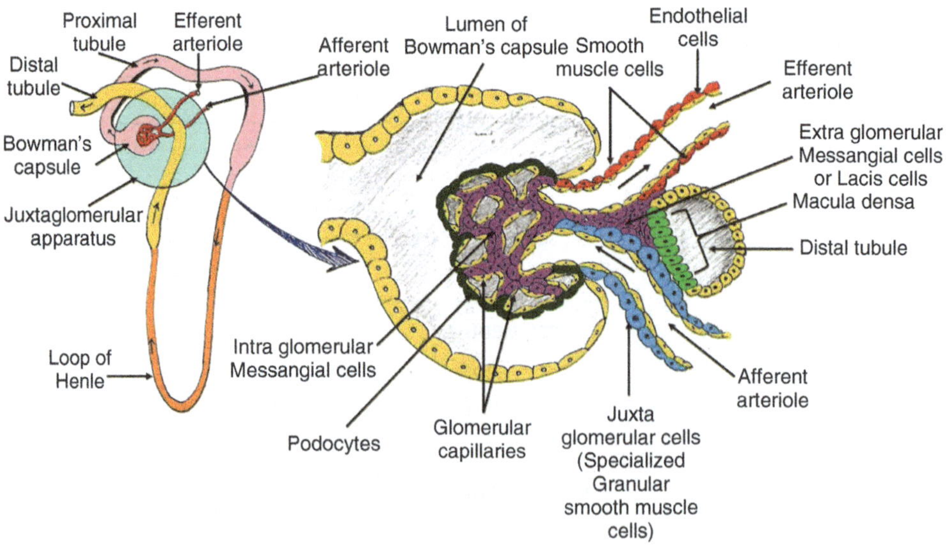

Fig. 9.5 Juxtaglomerular apparatus. It comprises specialised granular afferent arteriolar JG cells, macula densa of DCT and extraglomerular mesangial cells or lacis cells. JG apparatus is concerned with controlling the dynamics of renal circulation and tubular reabsorption of sodium ions

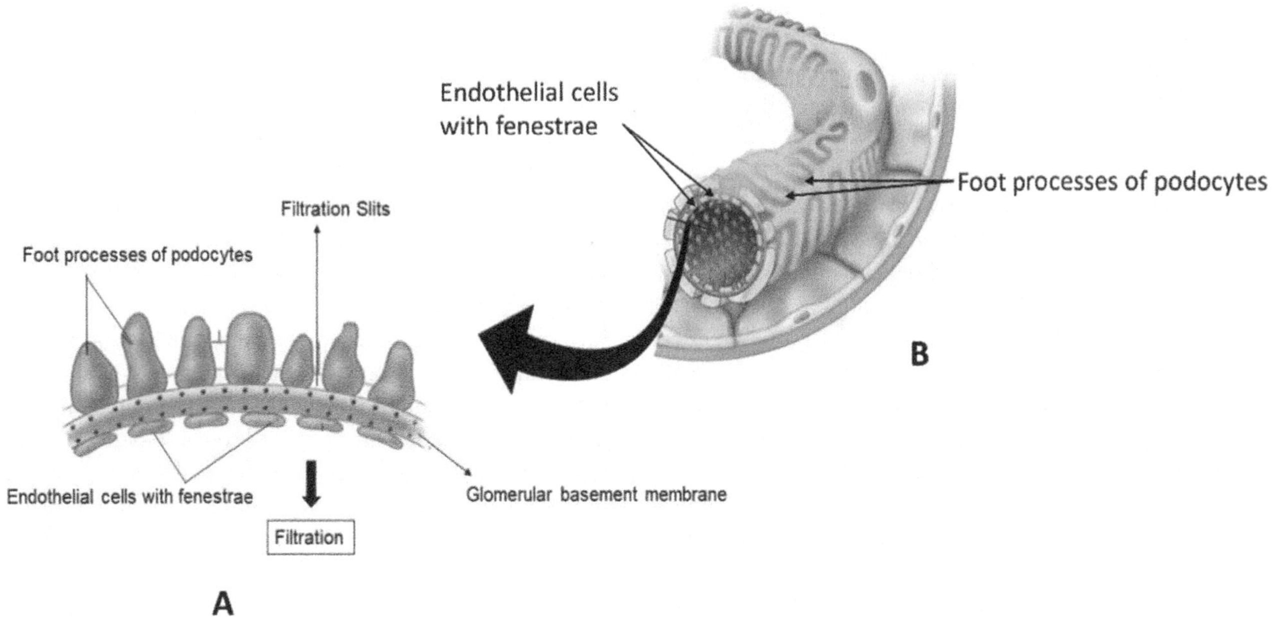

Fig. 9.6 Glomerular membrane. (**a**) Filtration surface of the glomerular membrane with podocytes, glomerular basement membrane and endothelial cells. (**b**) The structure of a glomerular capillary, which is placed in close apposition with the visceral layer of Bowman's capsule

diminished efferent renal sympathetic nerve activity in the contralateral kidney causes excess sodium ions and water excretion as a compensatory mechanism. The *renorenal reflex* coordinates the functions of the two kidneys and thus facilitates the physiological regulation of sodium and water balance to maintain homeostasis.

9.3.4 Ultrastructure of Glomerulus

The glomerular capillary membrane has three layers, the innermost highly *fenestrated endothelial layer*, next to that a *basement membrane* and an outermost epithelial layer made of *podocytes* (Fig. 9.6). Podocytes are flattened epithelial cells with foot processes, which interdigitate with adjacent foot processes, leaving *filtration slits* in between and encircling the complete glomerular tuft. The basement membrane is composed of collagen and glycoprotein, and the collagen provides structural strength to the membrane. The glycoproteins repel the plasma proteins, if present, with their negative charge. Less than 1% of albumin is completely excluded from the filtrate usually.

9.4 Urine Formation

The physiological events involved in urine formation can be categorised into three stages: glomerular filtration, tubular reabsorption and tubular secretion. The urine is formed and excreted due to the combining effect of the three stages.

$$\text{Urine excreted} = (\text{Quantity of glomerular filtrate} - \text{substances absorbed by the tubule}) + \text{secretions of the tubule.}$$

9.4.1 Glomerular Filtration

Glomerular filtration is a passive, non-selective process where fluids and electrolytes are filtered through the three layers of the glomerular membrane into Bowman's space under the influence of specific physical forces. It is called non-selective and passive because filtration is indiscriminate without expending energy. The fluid collected in the Bowman's space is called *glomerular filtrate*. The amount of glomerular filtrate formed per minute is known as the *glomerular filtration rate* (GFR). The composition of this filtrate is similar to blood except for the presence of blood cells and plasma proteins. The plasma flow rate through both the kidneys per minute is called *renal plasma flow* (RPF). Usually, about 20% of plasma that enters the glomeruli is filtered, producing 170 L of glomerular filtrate with an average GFR of 125 mL/min or 180 L/24 h by both kidneys in humans. The fraction of renal plasma flow that is filtered is called the filtration fraction:

$$\text{Filtration fraction} = \text{GFR}/\text{renal plasma flow}$$

The amount of any substance present in plasma reaching the kidneys per minute is the *plasma load* for that substance.

The plasma load that filters into the capsular space is known as the *tubular load* of the substance.

The physical forces involved in glomerular filtration are (1) glomerular *capillary hydrostatic pressure* that favours filtration, (2) *colloid osmotic pressure* (COP) of plasma proteins that oppose filtration, (3) *hydrostatic pressure* of fluid in the Bowman's capsule which opposes the filtration and (4) *Bowman's capsule osmotic pressure* that favours the filtration. The glomerulus' hydrostatic capillary pressure is higher than the pressure in other capillaries because the amount of blood that enters through the wider afferent arteriole is subjected to the increased resistance offered by the comparatively narrow efferent arteriole. The difference in diameter of the arterioles makes more or less the same capillary blood pressure all along the capillary tuft of the glomerulus. The hydrostatic pressure of the glomerular capillary was estimated to be about 60 mmHg, which was higher than the capillary pressure elsewhere. It forms the major driving force of fluid from the glomerulus to the Bowman's space. The COP of plasma proteins (32 mm of Hg) of glomerular capillary blood and hydrostatic pressure of capsular fluid (18 mm of Hg) together exert an opposing force of filtration of magnitude 50 mm of Hg. The COP of capsular fluid is negligible because very little amount of protein is present in the capsular space and the fluid remains in the capsular space for a very short duration due to the forward propulsion by the hydrostatic pressure. So, the *net pressure of filtration* becomes 10 mm of Hg (60 − 50 mm of Hg).

Besides the net filtration pressure, GFR also depends on the glomerular capillary surface area and the permeability or the hydraulic conductivity of the capillary membrane. The product of these two factors is known as the filtration coefficient (K_f). The value of K_f cannot be estimated directly but can be calculated as

$$K_f = \text{GFR}/\text{Net filtration pressure}$$

So, from the values mentioned above, K_f can be calculated as 125/10 = 12.5 mL/min/mm of Hg of filtration pressure.

Chronic uncontrolled hypertension and diabetes mellitus affect the glomerulus's permeability characteristics, reducing the K_f value and thereby reducing GFR.

9.4.2 Physiologic Control of GFR and Renal Blood Flow

Among the physical forces controlling filtration, plasma colloid osmotic pressure and hydrostatic pressure of capsular fluid are usually not regulated for controlling glomerular filtration. Instead, by extrinsic sympathetic nerves, humoral factors and autoregulatory mechanisms, the hydrostatic pressure of glomerular capillaries can be controlled to optimise GFR.

In acute conditions, like severe haemorrhage and ischaemia, sympathetic stimulation causes vasoconstriction of renal arterioles resulting in reduced GFR. Apart from that, it can also contract the mesangial cells, reducing the surface area of the glomerulus participating in filtration resulting in the reduction of K_f value and GFR. Under normal resting conditions, the role played by these nerves in the regulation of GFR is meagre.

Humoral factors, like epinephrine, norepinephrine and endothelin, cause vasoconstriction and reduction in GFR, whereas prostaglandins (PGE_2 and PGI_2) and bradykinin cause vasodilatation, an increased renal blood flow and increase in GFR. Angiotensin II is a peptide showing preferential vasoconstrictor property with the efferent arteriole of the kidney. This peptide is produced from angiotensinogen, a plasma protein. Angiotensinogen is converted to angiotensin I on proteolytic cleavage with renin of the JG apparatus. Angiotensin I in the lungs is converted to angiotensin II by the angiotensin-converting enzyme.

Autoregulation: Autoregulation is a process involving various intrinsic mechanisms by which the kidneys maintain relatively constant renal blood flow and GFR within a wide range of mean systemic arterial pressure. Two mechanisms are mainly involved, *myogenic response* and *tubuloglomerular feedback*.

Myogenic response: When the blood enters the afferent arteriole, the stretch receptors of smooth muscles on the walls of the blood vessels experience increased or decreased tension depending on the hydrostatic pressure exerted by blood on the walls. The increased pressure within the vessel causes the vessel wall to stretch which inherently contracts the vessel wall to restrict the blood flow. At the same time, the inherent relaxation of an unstretched afferent arteriole increases blood flow when the pressure decreases.

Tubuloglomerular feedback: Tubuloglomerular feedback is another mechanism of autoregulation to maintain optimum filtration pressure when the kidney experiences altered perfusion pressure. When GFR increases due to elevated glomerular hydrostatic pressure, macula densa cells sense an increase in the concentration of Na^+ and Cl^- in the tubular fluid. This sensitisation of macula densa cells, by some unknown means, constricts the afferent arteriole lowering hydrostatic pressure and also contracts the mesangium lowering the filtration surface area. Both these effects reduce GFR. Macula densa cells can sense the reduced concentration of Na^+ and Cl^- in the tubular fluid when GFR decreases due to lower hydrostatic pressure. Then renin is released from the JG cells, ultimately causing the formation of angiotensin II. Angiotensin II selectively constricts the efferent tubule offering more resistance to blood flow, and thus GFR is increased.

Know More......

Transmembrane Proteins in Excretion

Polycystin 1 and polycystin 2 are the two transmembrane proteins coded by *PKD1* and *PKD2* genes expressed in different locations of renal epithelial cells, including the primary cilia present on the apical membrane of these cells. These cilia protrude from the cells to the lumen and act as tubular fluid flow sensors. They transduce the alterations in the tubular flow into a cellular response regulating fluid and electrolyte transport. Whenever increased tubular fluid flow occurs, there will be bending of cilia, which activates PKD1/PKD2-dependent calcium ion influx into the cell and results in potassium ion secretion. It is also reported that functional loss of polycystins results in renal cyst formation.

9.4.3 Tubular Reabsorption

The water and solutes of the tubular fluids when transported to the peritubular capillaries throughout the nephron, including the collecting duct, are called tubular reabsorption. The substances reabsorbed across the tubular epithelial layer reach the renal interstitial, are then absorbed into the peritubular capillaries and finally get into the systemic circulation. Unlike glomerular filtration, tubular reabsorption is a highly selective process. The transportation occurs by two routes, the *transcellular* route across the epithelial cells and the *paracellular* route across the junctional spaces. Tubular reabsorption can either be active or passive. Water is normally reabsorbed by a passive diffusion process in a concentration gradient through both transcellular and paracellular routes.

Along with water, soluble solutes like potassium, magnesium and chloride ions and organic solvents are also taken to the interstitium through a process known as solvent drag. About 65% of filtered sodium, chloride, bicarbonate, magnesium and potassium and almost all filtered amino acids and glucose are absorbed through PCT. The epithelial cells of PCT are provided with numerous mitochondria for meeting the increased metabolic demand associated with transport mechanisms. The microvilli's extensive surface area provided on the luminal surface also favours bulk reabsorption.

Capillary dynamics in the peritubular capillaries favours reabsorption by bulk flow. In the peritubular capillaries, hydrostatic pressure and COP are 17 and 30 mm of Hg, respectively, whereas in the interstitial fluid, the values are 6 and 10 mm of Hg, respectively. Hence, the reabsorption pressure ($30 + 6 = 36$ mmHg) exceeds filtration pressure ($17 + 10 = 27$ mmHg) by 9 mmHg ($36 - 27 = 9$ mmHg), favouring reabsorption to peritubular capillaries. The tubular fluid remains isosmotic with plasma as both solutes and water are reabsorbed.

The amount of a substance filtered through the glomerular filtrate and presented to the tubule per minute is known as the *tubular load* of that particular substance. The maximum rate at which the substance is reabsorbed from the tubular lumen to the peritubular fluid is the *tubular transport maximum* (Tm). The renal threshold is the plasma concentration of a substance at which it first appears in the urine. The property of tubules to increase reabsorption by the increased tubular load due to increased GFR is known as *glomerulotubular balance*.

Na^+-K^+ ATPase present on the basolateral side of the tubular epithelial cells hydrolyses ATP. The released energy is used to transport sodium ions from the tubular cells to the interstitium. Potassium ions are taken in return into the interior of the cells from the interstitial space. This is known as *primary active transport*. The increased transport of sodium ions out of the cell creates an intracellular potential of −70 mV. This negative potential and decreased intracellular sodium concentration favour sodium diffusion into the cell from the tubular fluid in a concentration gradient. The energy released during primary active transport is used to transport another substance known as *secondary active transport*. For example, glucose or amino acid is transported along with sodium. Hence, secondary active transport is known as co-transport. If a secondary secretion occurs along with primary active sodium transport, that is known as *counter-transport*. The inward influx of sodium ions accompanied by the outflow of hydrogen ions is an example of counter-transport.

Sodium reabsorption mainly (65%) occurs at the proximal PCT as co-transport and counter-transport. But chloride-driven Na^+ transport takes place from the distal portions of the PCT. In the first two modes of transportation, sodium-coupled carrier molecules are involved, whereas in the chloride-driven Na^+ transport, both Cl^- and Na^+ ions are transported through the leaky tight junctions. About 25% of Na^+ present in the tubular fluid is absorbed in the thick ascending limb of both cortical and medullary segments of the loop of Henle. Sodium is transported by co-transport using Na^+-K^+-$2Cl^-$ carriers present on the luminal surface of the loop of Henle. Nearly 5% tubular Na^+ is absorbed from the proximal segment of the distal tubule along with Cl^- co-transport. The second half of the distal tubule has principal cells and intercalated cells.

The principal cells reabsorb sodium and water from the lumen and secrete potassium ions into the lumen. The intercalated cells absorb potassium ions and secrete hydrogen ions. The principal cells are the site of action of the adrenal cortical hormone, aldosterone. The proteolytic enzyme renin is released from JG cells, and it affects the production of angiotensin II, which stimulates the adrenal cortex to release aldosterone. Aldosterone favours sodium reabsorption by increasing the number of Na^+-K^+ ATPase on the basolateral membrane of tubular epithelial cells. Aldosterone also increases potassium secretion by principal cells. The remaining 5% of sodium ions in the tubular fluid are absorbed under the control of aldosterone, depending on the body's requirement. Atrial natriuretic peptide (ANP) inhibits the renal absorption of sodium by exerting its influence on the principal cells.

The excess positive charge generated due to absorption of sodium ions is neutralised up to 75% by chloride ion transport. Chloride transport occurs through tight junctions, and chloride can also be passively absorbed in a concentration gradient at the time of solvent drag. From the thick limb of the loop of Henle and the proximal segment of the distal tubule, chloride ions are absorbed by way of secondary active transport or co-transport with sodium.

Glucose and amino acids are reabsorbed by co-transport with sodium ions. They are released from the carrier molecules and transported to the peritubular space by facilitated diffusion inside the cells. In human beings, if GFR is 125 mL and plasma concentration of glucose is 1 mg/mL (100 mg/dL), tubular load of glucose will be $125 \times 1 = 125$ mg/min. The transport maximum or tubular maximum (Tm) value for glucose is estimated to be an average of 375 mg/min for an adult human being. It is the maximum milligrams per minute at which a substance is transported from the tubular lumen to the interstitial fluid. Beyond the Tm value, the same increment in the serum level of a substance will be excreted through urine. When the blood glucose level increases to 2 mg/mL from 1 mg/mL, the tubular load becomes 250 mg/min. In this stage, a trace amount of glucose may appear in urine because some individual nephrons may have lower Tm values, and some other nephrons may not absorb to their maximum capacity. At a tubular load of 375 mg, both kidneys' nephrons absorb glucose at their maximum capacity. The increased urinary concentration of any substance will reflect the increased plasma level of that substance. The Tm value for amino acids is 1.5 mM/min (Table 9.1).

Table 9.1 Renal threshold for glucose in different domestic species

Species	Renal threshold (mg/dL)
Dog	180–200
Cat	280–290
Horses	160–180
Cattle	100–140

Lower thresholds may occur in diabetes in cats, but stress causes hyperglycaemia and glycosuria.

Proteins of molecular weight less than 69,000 will be entirely absorbed from the PCT by active *pinocytosis*. Inside the cells, they will be degraded by cellular lysozyme to amino acids and these amino acids are transported through the basolateral membrane to the peritubular space. The peptides are hydrolysed at the luminal brush border, and the amino acids formed are transported by co-transport.

When water is reabsorbed osmotically, that will facilitate urea transport to the peritubular space in a concentration gradient. In the inner medullary collecting ducts, facilitated diffusion occurs through urea transporters. Half of the amount filtered will be reabsorbed, and the remaining half is excreted through urine. Since the tubular membrane is impermeable to creatinine, the amount filtered will be excreted as a whole through urine.

Water is reabsorbed extensively from the PCT. Although the descending thin segment of the loop of Henle is permeable to water, it is highly impermeable to solutes. The ascending segments (including both thin and thick) are highly impermeable to water. In the presence of vasopressin or antidiuretic hormone (ADH), the late distal tubule and the collecting ducts are made permeable to water (Fig. 9.7). Whenever the extracellular fluid volume decreases or osmolarity increases, ADH is released from the posterior pituitary. The water reabsorption by ADH is mediated through an intracellular protein called aquaporin-2 (AQP-2). When ADH binds to the plasma membrane receptors of late distal tubules, collecting tubules and collecting ducts, there will be increased formation of cAMP, activation of protein kinases and translocation of intracellular AQP-2 proteins from the interior to the plasma membrane. Fusion of AQP-2 to the luminal plasma membrane results in the opening of water channels in these regions, resulting in water entry to the interior of the cell. Water exits the cell through a different water channel (either AQP-3 or AQP-4) permanently positioned at the basolateral border and then enters the blood, in this way being reabsorbed.

About 50% of the plasma calcium is ionised, and the remainder binds to the plasma proteins or exists in combination with anions such as phosphate. So, the glomerulus can filter only about 50% of the plasma calcium. Usually, about 99% of the filtered calcium is reabsorbed by the tubules, and only about 1% of the filtered calcium is excreted. About 65% of the filtered calcium is reabsorbed in the proximal tubule, about 25–30% is reabsorbed in the loop of Henle and 4–9% is reabsorbed in the distal and collecting tubules. Parathyroid hormone increases calcium reabsorption, especially from the distal tubules, and magnesium reabsorption from the loop of Henle and decreases phosphate reabsorption by PCT.

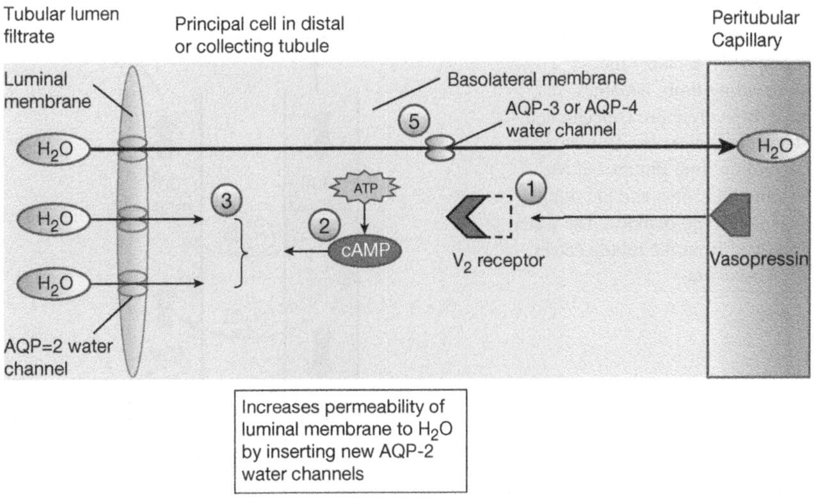

Fig. 9.7 ADH-mediated water reabsorption from renal tubules. Dehydration causes the release of vasopressin (ADH) from the posterior pituitary, which attaches to the basolateral plasma membrane of epithelial cells of the late distal tubule, collecting tubule and collecting duct. This attachment causes the translocation of intracellular water channels AQP-2 to the luminal membrane increasing water permeability

Know More……

Aquaporins

Aquaporins (AQPs) constitute a family of proteins located in the plasma membrane and mediate water transport. A total of 13 proteins (AQP1–12A, B) are included in the AQP family in humans. Though the majority of AQPs facilitate water reabsorption, AQPs (like AQP3, AQP7 and AQP9) also play a significant role in glycerol transportation, thus being referred to as aquaglyceroporins. Because of AQPs' fundamental roles in essential water homeostasis, the distribution of AQPs was initially regarded as ubiquitous from prokaryotes to eukaryotes.

Renal excretion of magnesium is increased markedly during increased magnesium ion levels but decreases to almost nil during times of its depletion in the blood. Regulation of magnesium excretion is achieved significantly by changing tubular reabsorption. The proximal tubule usually reabsorbs only about 25% of the filtered magnesium. The loop of Henle is the primary site of reabsorption, where about 65% of the filtered load of magnesium is reabsorbed. Only a minimal amount (less than 5%) of the filtered magnesium is reabsorbed in the distal and collecting tubules. Increased extracellular fluid magnesium concentration, extracellular volume expansion and increased extracellular fluid calcium concentration may increase magnesium excretion.

9.4.4 Tubular Secretion

Hydrogen ions are secreted to the PCT in return to sodium ions, produced by the dissociation of H_2CO_3 formed by the hydration of CO_2 inside the cell. The intercalated cells of the late distal tubule and cortical and medullary collecting ducts secrete hydrogen ions by an active hydrogen-ATPase mechanism. The principal cells of distal and cortical collecting tubule segments of nephrons secrete potassium under the influence of aldosterone whenever dietary potassium intake is high. Organic acids and bases, like bile salts, urates and catecholamines, are secreted from the proximal tubule. In addition, some waste products of metabolism, harmful drugs and toxins are also secreted to the tubular fluid of PCT.

9.5 Urine Concentration

The kidneys can dilute or concentrate urine without altering the amounts of solutes reabsorbed or excreted. Kidneys can eliminate excess water in the body even by diluting urine to the lowest limit of 50 mOsm/L. At the same time, if the body faces acute dehydration, kidneys can conserve water by increasing the osmolarity of urine excreted to about 1200–1400 mOsm/L in humans. In these two conditions, extracellular fluid osmolarity is maintained at the normal level of 300 mOsm/L. A human being of 70 kg body weight must excrete a minimum 600 mOsm of solutes every day. At the maximum osmotic concentration of 1200 mOsm/L, the obligatory minimum quantity of urine that must be produced is 600 mOsm/1200 mOsm/L = 0.5 L. A desert animal, an Australian hopping mouse, can concentrate urine up to 10,000 mOsm/L, whereas an aquatic animal beaver can concentrate only up to 500 mOsm/L and has the minimum urine-concentrating capacity.

The capacity of the kidney to concentrate urine is attributed to the special arrangement of juxtamedullary nephrons with their loops of Henle dipping deep into the hyperosmotic medullary interstitium. Two mechanisms are involved in the process of urine concentration. One creates hypertonicity of the medulla by the countercurrent multiplier system, and the other maintains it by the countercurrent exchange system.

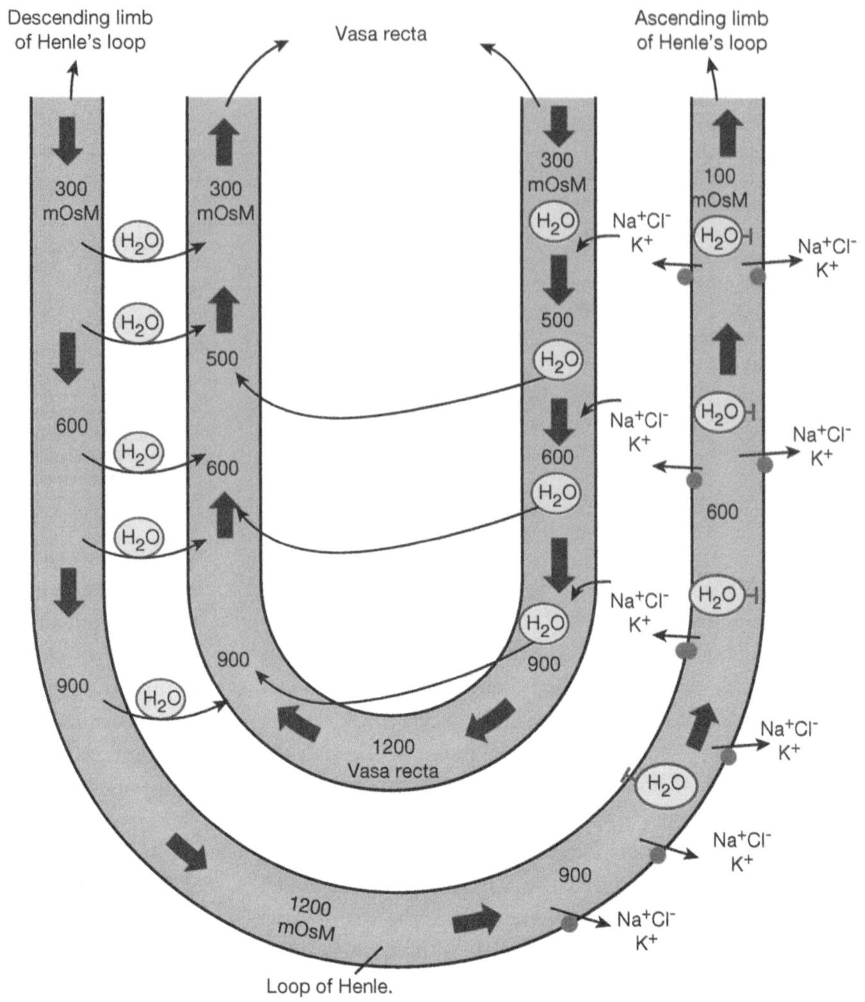

Fig. 9.8 Countercurrent multiplier system. The tubular fluid, when it enters the descending limb, becomes progressively more concentrated due to loss of water. But the ascending limb pumps out Na^+, K^+ and Cl^- ions, and the filtrate becomes hypo-osmotic. The water removed from the tubule enters the vasa recta

9.5.1 Countercurrent Multiplier System

The descending loop of Henle carries the tubular fluid downward from the cortex to the medulla. The tubule's U-shaped arrangement enables the fluid to flow in opposite directions (countercurrent) in the two tubules (Fig. 9.8). In the thick ascending limb of the loop of Henle, sodium is actively transported from the tubular fluid to the peritubular space. Although chloride, potassium and other ions are co-transported with sodium, the thick ascending limb is impermeable to water. Passive reabsorption of sodium and chloride ions also takes place from the thin ascending limb of the loop of Henle. The descending limb is permeable to the water simultaneously and impermeable to solutes. Since the medullary interstitium gets more and more concentrated with the sodium chloride diffused from the ascending limb, there is increased osmotic outflow of water from the descending limb as it dips deep into the medulla. More sodium chloride will be added to the descending limb from the proximal tubule as a continuous process. The newly arrived sodium chloride gets added to the already existing sodium chloride in the interstitium, thus multiplying the osmotic concentration of the medullary interstitium. Hence, the physiological processes involved in the multiplication of medullary interstitial hypertonicity with specialised transport mechanisms of solutes and water when the countercurrent flow of tubular fluid occurs through the loop of Henle are known as *countercurrent multiplier system*.

The fluid leaving the loop of Henle has an osmolarity of 100 mOsm/L, which is about one-third of plasma osmolarity. Suppose further reabsorption of water is not taking place under the influence of ADH. In that case, the osmolarity can even become 50 mOsm/L since additional reabsorption of solutes occurs from the distal tubule and collecting ducts. The plasma concentration of ADH influences water reabsorption from the late distal tubules, collecting tubules and ducts. ADH-dependent water reabsorption is more pronounced in the cortical collecting tubules than in the collecting ducts. The limited absorption of water in the medullary collecting duct also helps to maintain hypertonic medullary interstitium.

9.5.2 Countercurrent Exchange System

Countercurrent exchange system is a U-shaped blood vessel (vasa recta) that parallels Henle's loop. Blood flow through the vasa recta constitutes 5% of total renal blood flow, meeting the metabolic demands of the interstitium, at the same time preserving the hypertonicity of the renal medulla. The colloid osmotic pressure of plasma proteins and hydrostatic pressure of blood flowing through these capillaries are favoured by reabsorption. As the descending limb dips deep into the medulla, it loses water and solutes, increasing its tonicity. When the blood in the ascending limb flows in the opposite direction, it gains water and loses solutes, gradually decreasing tonicity.

9.6 Obligatory Urine Volume

The maximal concentrating ability of the kidney decides to a minimum how much urine volume must be excreted each day to eliminate the body of waste products of metabolism and ions that are ingested. A normal 70 kg human being excretes about 600 mOsm solute in 1 day. Maximal urine concentration is 1200 mOsm/L, characterised by the *minimal volume* of urine called the *obligatory urine volume*. It can be calculated as 600 mOsm/day/1200 mOsm/L = 0.5 L/day. This minimal loss of volume in the urine contributes to dehydration and water loss from the skin, respiratory tract and gastrointestinal tract, especially when the animal is deprived of water.

9.7 Urea Recycling

The thick ascending limb of the loop of Henle, distal tubule and cortical collecting ducts (top and middle portions of collecting ducts) are impermeable to urea. When ADH concentration of plasma is very high, water reabsorption from the distal parts of the nephron takes place at a faster rate, resulting in urine concentration. The innermost portion of the collecting duct contains many membrane transporters for urea called UT-As, upregulated by vasopressin. The concentration difference favours urea diffusion out of the duct into the interstitial fluid, creating a high urea concentration in the inner medullary interstitial fluid. When the kidney is forming maximum concentrated urine, the urea contributes to about 50% of the osmolarity of the medullary interstitium (Fig. 9.9). A portion of the urea reabsorbed from the inner medullary ducts diffuses into the thin loop of Henle and is re-circulated. This re-circulation helps the concentration of urea to be excreted so that excretion is possible with minimum water loss. A high-protein diet always increases the capacity to concentrate urine because of the comparatively more urea formation as a nitrogenous waste. Malnutrition always reduces the ability to concentrate urine.

> **Know More......**
>
> **Shipwreck Victims**
>
> The maximum osmotic concentration of the renal medulla is 1200 mOsm/L. The seawater has an osmolarity between 1000 and 1200 mOsm/L with a sodium chloride concentration of about 3.0–3.5%. If an individual drinks 1 L of seawater with an osmotic concentration of 1200 mOsmol/L, that would provide a total sodium intake of 1200 mOsm/L. So, for every litre of seawater drunk, a volume of 2 L of urine would be voided out to get rid of 1200 mOsm of solutes from the body in addition to other obligatory solutes such as urea. The shipwreck victims would have a net fluid loss of 1 L for every litre of seawater drunk, explaining the rapid dehydration in such individuals.

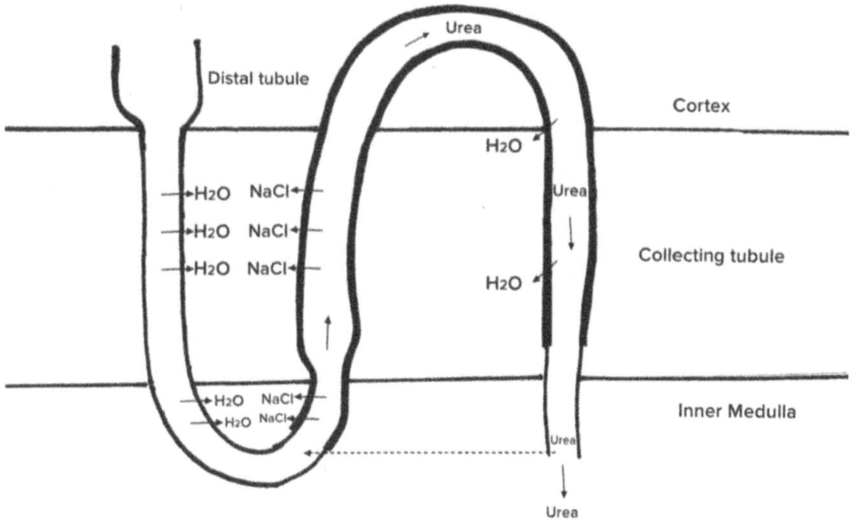

Fig. 9.9 Urea recycling. The urea is transported from the collecting duct to the medullary interstitium, and from there, 50% is re-circulated through the thin limb of the loop of Henle

9.8 Renal Clearance

Renal clearance of a substance is plasma volume that completely clears a particular substance per minute.

$$\begin{aligned}&\text{Clearance rate of a substance}\\&= [\text{Urine concentration of that substance(amount/mL)}\\&\times \text{Urine flow rate(mL/ min)]/}\\&\text{Plasma concentration of that substance(amount/mL).}\end{aligned}$$

If a substance is freely filtered but not reabsorbed or secreted, the plasma clearance rate will equal the glomerular filtration rate (GFR). But no endogenous substance is completely cleared from plasma through the kidneys. For example, the glucose is freely filtered, complete reabsorption takes place usually and glucose clearance will be zero. Since 50% of filtered urea is reabsorbed, the plasma clearance of urea is 50% of GFR. The H^+ is freely filtered as well as secreted but not reabsorbed. So, the plasma clearance rate of H^+ is always greater than GFR.

An exogenous biologically inert substance, *inulin* obtained from tubers of a plant, is neither reabsorbed nor secreted but freely filtered. So, inulin clearance rate will be a measure of GFR. Determination of plasma inulin clearance requires the continuous intravenous infusion of inulin to maintain a constant plasma concentration. Because of this reason, renal clearance of an endogenous substance, *creatinine* (end product of muscle metabolism), is usually used clinically to give a rough estimate of GFR. Creatinine is freely filtered, not reabsorbed but slightly secreted. So, creatinine clearance is not an absolute reflection of GFR but gives a close approximation.

A substance must be freely filtered, and the remaining should be secreted for it to be completely cleared from total plasma reaching kidneys because GFR constitutes only 20% of renal plasma flow. Such a substance can only be used for the measurement of renal plasma flow. But no known substance is completely cleared from plasma. However, *para-aminohippuric acid* (PAH) is the only substance that is cleared up to 90% from plasma. So, renal PAH clearance can be used to approximate renal plasma flow measurement.

9.8.1 Assessment of Renal Function

The renal function tests may be classified into (1) tests which measure glomerular filtration rate and (2) tests which study the tubular function. In clinical biochemistry, certain physical and chemical constituents are checked when reporting on a urine sample (Table 9.2). The physical constituents have volume, appearance, odour, colour and specific gravity. The chemical characteristics usually checked are pH, proteins, blood, reducing sugars (glycosuria), ketone bodies, bile pigments and nonprotein nitrogen compounds like urea, creatinine and uric acid. Clearance tests are usually performed to

Table 9.2 Characteristics of urine in different species

Parameters	Cattle	Sheep	Goat	Horse	Dog	Cat	Rabbit	Humans
Urine volume	16–50 mL/kg	10–40 mL/kg	10–40 mL/kg	8–30 mL/kg	14–50 mL/kg	18–25 mL/kg	20–350 mL/kg	1–2 L/day
Colour	Pale yellow–dark brown yellow	Pale yellow–dark brown yellow	Pale yellow–dark brown yellow	Ochre	Pale yellow–dark brown yellow	Yellow–strong dark yellow	Pale yellow–red brown	Colourless–umber
Transparency	Clear	Clear	Clear	Turbid	Clear	Clear	Clear	Clear
Odour	Aromatic	Indifferent aromatic	Indifferent aromatic	Aromatic	Garlicy	Sharp	n.s.	Coffee, saffron, onion
Specific gravity	1.020–1.040	1.020–1.040	1.020–1.040	1.020–1.040	1.001–1.065	1.001–1.080	1.003–1.036	1.003–1.030
pH value	7.0–8.4	7.5–8.5	7.5–8.5	7.6–9.0	5.5–7.0	5.0–7.0	8.2	4.6–8 {7}
Protein	Negative	Negative	Negative	Negative	Negative	Negative	Negative	0–20 mg/dL
Glucose	Negative	Negative	Negative	Negative	Negative	Negative	Negative	Negative
Ketones	Negative	Negative	Negative	Negative	Negative	Negative	Negative	Negative
Bilirubin	Negative	Negative	Negative	Negative	Negative–weak positive	Negative	Negative	Negative
Urobilinogen	Negative–weak positive	Negative–weak positive	Negative–weak positive	Negative–weak positive	Negative–weak positive	Negative–weak positive	Negative–weak positive	0.2–1 mg/dL
Blood	Negative	Negative	Negative	Negative	Negative	Negative	Negative	Negative
Leucocytes	Negative	Negative	Negative	Negative	Negative	Negative	Negative	0–2/high-power field (Hpf)

assess the GFR. The measurement of specific gravity or osmolality indicates tubular function. The tubular efficiency for urine concentration can be evaluated using *water deprivation* and *ADH response tests*.

9.8.2 Diuresis

It is the increased urine formation by the kidneys resulting in excessive urination. The reasons for diuresis are increased water intake, action of diuretic drugs and certain diseases such as diabetes mellitus and diabetes insipidus. In diabetes mellitus, the extra glucose present in the tubular fluid build-up increases osmotic pressure and causes osmotic diuresis. In diabetes insipidus, impaired water reabsorption occurs due to lack of sufficient ADH and excessive water loss through urine. The increased water loss may lead to dehydration and polydipsia.

9.9 Micturition

The process of bladder emptying is known as *micturition* or urination. The wall of the mammalian urinary bladder is made up of smooth muscle layers of the detrusor muscle. The smooth muscle fibres extend in all directions and fuse so that low-resistant electrical pathways exist from one muscle to another. The epithelial lining of the bladder is an impermeable transitional epithelium that can accommodate the changes in the bladder size. The highly folded bladder wall can flatten out to increase bladder storage capacity. In animals, the small triangular area on the dorsal aspect of the urinary bladder (posterior aspect in human beings) lying immediately above the bladder neck is *trigone*. The ureters enter the bladder at an oblique angle in the trigone to form a functional valve to prevent the backflow of urine at the time of bladder filling (Fig. 9.10).

The urethra is a tubular structure conveying urine from the neck of the bladder to the exterior. It has two sphincters, an internal urethral sphincter and an external urethral sphincter. Since the internal sphincter is a continuation of the smooth muscle fibres of the detrusor muscle, it is involuntary in action. Its peculiar anatomical arrangement keeps it closed until the pressure inside the bladder exceeds a critical threshold. The external sphincter is lying beyond the bladder and is composed of skeletal muscle, which is under cortical control by a voluntary motor neurone.

Micturition reflex is an autonomic spinal cord reflex (involving the spinal cord's sacral segment) initiated by the stretch of the receptors in the bladder wall during filling. Afferent impulses are sent through the sensory fibres of pelvic nerves and efferent impulses through the parasympathetic fibres of the same nerve. Thus, contraction of the detrusor muscle begins, and no special mechanism is needed to open the internal sphincter; changes in the shape of the bladder during contraction mechanically pull the internal sphincter open. When urine reaches the neck of the bladder and the external sphincter, afferent impulses are sent to another reflex centre located in the pons. Efferent impulses from this centre (through the pudendal nerve) prevent contraction of the bladder and relaxation of the external sphincter allowing the urine to flow through the urethra. Once the micturition reflex begins, it is self-regenerative, and as the bladder continues to fill, the generation of reflex becomes more frequent and powerful. The external sphincter relaxes when the pressure exceeds the critical threshold, and urine gets voided out. Voluntary inhibition of micturition is possible by the tonic contraction of the external sphincter until a convenient time comes with the involvement of cortical centres of the brain.

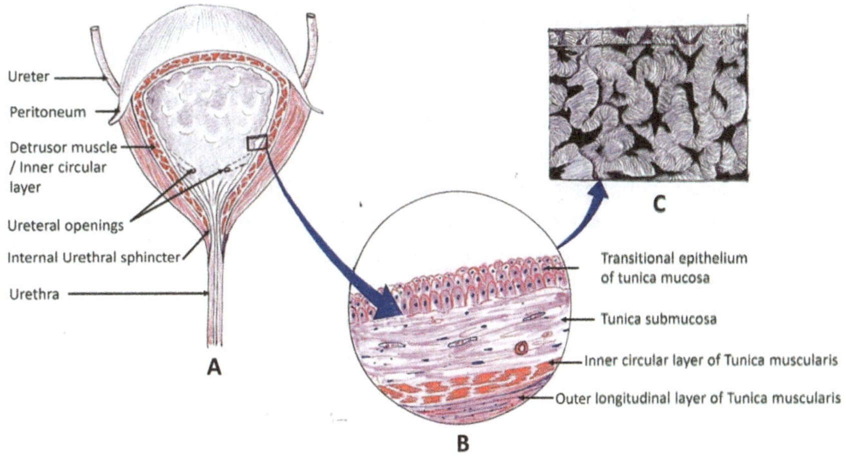

Fig. 9.10 Structure of urinary bladder. (**a**) Gross morphology of urinary bladder. (**b**) Cross section of the bladder wall (haematoxylin and eosin staining). (**c**) Scanning electron microscopic view of the inner lining of urinary bladder showing corrugated appearance in the non-distended state

9.10 Descriptive Terms

The normal condition of storing urine in the bladder during filling is known as *urinary continence*. *Urinary incontinence* is the frequent dribbling of urine due to improper functioning of the external sphincter. *Polyuria* is increased urine output, and *oliguria* is decreased urine output. The condition of no urine output is known as *anuria*. *Dysuria* is difficult or painful urination, and *stranguria* is painful, drop-by-drop and slow discharge of urine.

Uraemia is a clinical state in which the blood urea nitrogen level, an indicator of nitrogen waste products, is elevated. It results due to certain factors, like (1) renal failure; (2) increased production of urea in the liver due to a high-protein diet, drugs and increased breakdown of protein; (3) decreased elimination of urea due to reduced blood flow to the kidney, and obstruction of urinary tract; (4) dehydration; and (5) chronic infection of the kidney.

Uraemia is a severe condition that can even become fatal because very high nitrogen in the blood is toxic to the body. Symptoms of uraemia include mental confusion, loss of consciousness, decreased urine production, dry mouth, debility, paleness of skin or pallor, bleeding problems, increased heart rate (tachycardia), oedema and increased thirst. Treatment includes dialysis or a kidney transplant.

9.11 Avian Renal Physiology

The avian urinary system consists of a pair of kidneys and ureters that transport urine to the urodeum of the cloaca. The urinary bladder is absent in birds. The kidney lies in a cavity formed by the ventral surface of the synsacrum. The mass of the two kidneys is proportional to (body mass)$^{0.9}$ or 0.8% or 1.8% of the body weight. The external appearance of the kidney is elongated and tri-lobed with anterior, middle and posterior divisions. Within each division, the kidney is divided into numerous subunits.

In avian species, a dual afferent blood supply is present in the kidney via a 'high-pressure' (160/120 mm of Hg) renal artery and a 'low-pressure' (25 mm of Hg) supply via a renal portal system (RPS). It is estimated that 1/2 to 2/3rd of blood supplied to avian kidneys is through the renal portal system. The renal artery supplies glomerular areas of the kidney; peritubular areas are partly by efferent glomerular arterioles and also by the venous return from the legs communicating with the RPS. The magnitude of the renal portal supply reaching the peritubular regions appears to be controlled by a smooth muscle valve called the *renal portal valve*. When the valves close by contraction, the peritubular areas of kidney are perfused with blood. When the valve is opened, the blood is directly shunted to posterior vena cava.

The functional unit of the kidney is the nephron. Avian kidney has two kinds of nephrons. One kind is reptilian type with no loops of Henle, located in the cortex, and another is mammalian type with long- or short-length loops located in the medulla (Fig. 9.11). Only a small percentage of nephrons (15–25%) are mammalian-type nephrons in birds.

Avian kidneys usually alternate between the uses of mammalian- and reptilian-type nephrons. When birds concentrate urine, they opt for mammalian-type nephrons and completely shut down about 80% of the reptilian-type nephrons. Thus, the GFR is reduced to 40% of the normal value. During diuresis, 75% of the filtrate comes from the reptilian-type nephrons.

The role of the kidney in the bird is filtration, absorption and secretion, as in the case of other vertebrates. They filter water and water-soluble substances from the blood, including waste products of metabolism and ions, and are voided out through urine. Kidneys also have an important role in conserving body water and reabsorption of needed substances, viz. glucose.

Blood enters nephrons via afferent arterioles, as in the case of mammals. In the glomerulus, the blood under high pressure gets filtered through the walls of the capillaries and the capsule walls. The filtrate entering the proximal tubules is plasma without protein since protein molecules are generally not filtered due to their large size. In the proximal convoluted tubules, the vital substances in the filtrate, such as vitamins and glucose, are reabsorbed into the blood. The kidney tubules can reabsorb almost 98% of the glucose that filters into the tubule even in a carbohydrate-rich dietary state.

Birds can conserve body water by producing urine of more osmotic concentration than plasma, as in the case of mammals. But the urine concentration capacity is limited in birds compared to mammals. On water deprivation, mammals can concentrate urine 5–10 times more than plasma; some mammals can do it about 20–25 times. But birds in water deprivation can only produce 1.4–2.8 times more concentrated urine than plasma. This '*concentrating capacity*' is a feature of the medullary cones.

Solutes, like sodium chloride, are actively transported out of the ascending limb of the loop of Henle, and they become concentrated in the medulla (medullary cones). Unlike in mammals, only sodium chloride is responsible for maintaining medullary hypertonicity in birds. When the filtrate passes through the osmotic gradient in the medulla, water gradually leaves the tubules by osmosis, and the filtrate becomes concentrated. Because only the looped mammalian nephrons contribute to the intramedullary osmotic gradient, the presence of reptilian (loopless) nephrons limits the ability of the kidneys to produce hyperosmotic urine. Thus, the birds have limited urine concentration ability than mammals.

Usually, more water accompanies the solutes that travel from the kidneys through the ureters to the cloaca due to the

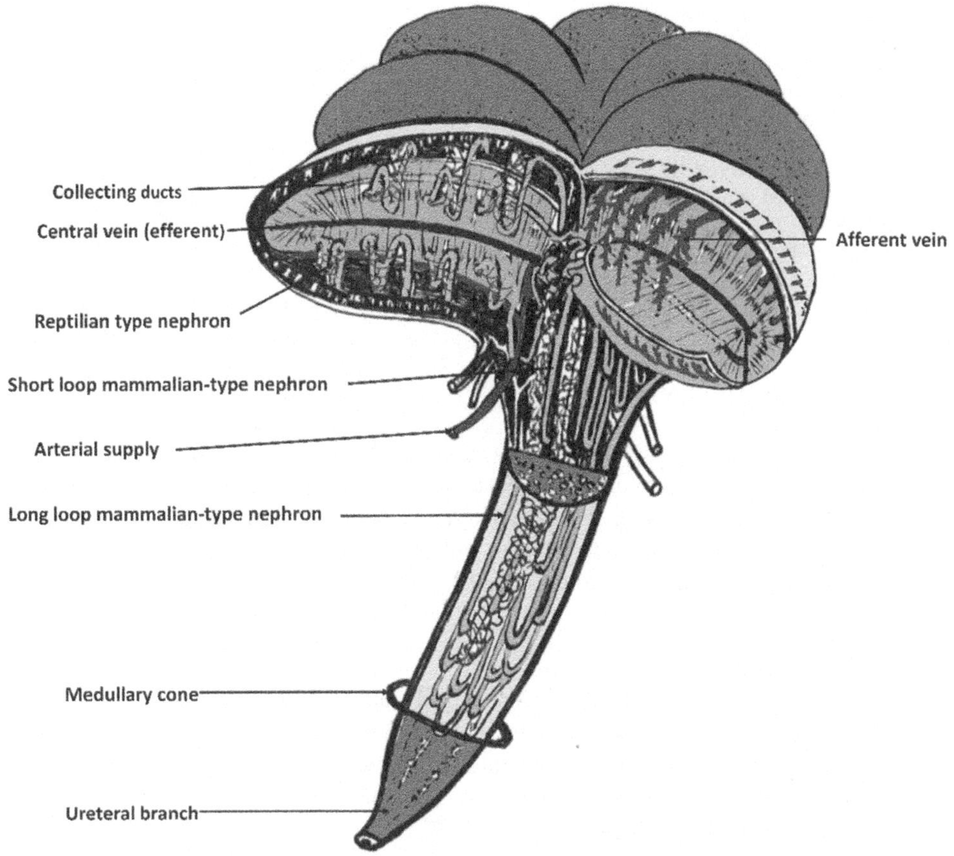

Fig. 9.11 Avian kidney. It has both reptilian- and mammalian-type nephrons, and mammalian-type nephrons can be long-loop or short-loop nephrons. The renal medulla has different medullary cones

reduced capacity of avian kidneys to concentrate urine (compared to mammals). A mechanism exists in water-deprived birds for reducing the amount of water leaving the kidneys. When dehydration occurs, the pituitary gland releases a hormone called *arginine vasotocin* (AVT) into the blood. AVT causes a decrease in the glomerular filtration rate in the avian kidneys, reducing the amount of water moving from the blood into the kidney tubules. Besides, AVT aids in the opening of protein water channels called aquaporins and thus increases the permeability of the walls of collecting ducts to water. Due to the increased permeability of the collecting ducts, more water leaves by osmosis out of the collecting ducts to the hypertonic medullary cones. From there, water is reabsorbed by kidney capillaries. Studies suggest that the effectiveness of AVT in reducing urine production or water reabsorption varies among species, but, in general, AVT is considered to be less effective in conserving water than the mammalian equivalent antidiuretic hormone (ADH). Therefore, water-deprived birds tend to lose more water from the kidneys than similar sized water-deprived mammals.

Uric acid is the end product of nitrogen metabolism in terrestrial reptiles and birds. In these species, the embryo's development takes place in eggs that have shells impermeable to water. So, the embryo is provided only with a limited supply of water. Hence, these organisms deposit excretory products as insoluble substances that do not require water to minimise water usage. Although uric acid is freely filtered at the glomerulus, secretion in the tubules accounts for 90% of the uric acid excreted in the urine of birds. Precipitation of uric acid occurs when the quantity of uric acid present in the tubules exceeds its solubility. Uric acid sediment moves through the tubules and appears as a white coagulum in the urine. Since uric acid is not present in the solution, it does not contribute to the osmotic pressure of the urine and thus avoids the obligatory loss of water.

Salt glands of birds are supposed to be evolved from the nasal glands of reptiles. They lie immediately under the skin in supraorbital depressions of the frontal bone in the skull of Charadriiform birds. Still, in other birds, they may be located above the palate or within the orbit of the eye (Fig. 9.12). The marine birds (and some desert and Falconiform birds) secrete excess sodium chloride via the salt glands using less water than is consumed, thus saving water. So, birds are not physiologically affected by the high salt load.

Salt glands have a countercurrent blood flow system to remove and concentrate salt ions from the blood. It is paired and crescent-shaped glands. Each gland contains several longitudinal lobes approximately 1 mm in diameter, and each lobe contains a central duct from which radiated

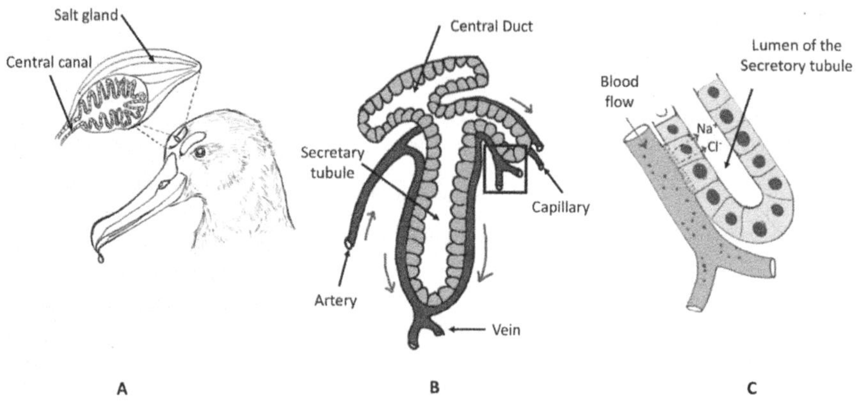

Fig. 9.12 Salt gland function. (**a**) Salt gland located in the supraorbital depression of frontal bone. (**b**) Structure of the longitudinal lobe of a salt gland. (**c**) The pattern of salt excretion from the capillaries, wherein Na^+ and Cl^- are secreted into the tubular lumen

thousands of tubules are enmeshed in blood capillaries. These tiny capillaries carry blood along the tubules of the gland, which have walls just one cell thick and form a simple barrier between the salty fluid within the tubules and the bloodstream. The salt excretion occurs in this gland.

Learning Outcomes

- Metabolic waste products are eliminated, and water and electrolytes are retained in the body at optimum concentrations by effective glomerular filtration, tubular reabsorption and tubular secretion occurring inside the kidney's nephrons.
- The kidney can auto-regulate its functions by inherent mechanisms, like myogenic response and tubulo glomerular fluid. The countercurrent multiplier and countercurrent exchange systems are responsible for creating and maintaining graded hypertonicity in the medullary interstitium, which is essential for concentrating urine. Urea recycling also helps to maintain medullary hypertonicity.
- The urine formed is conveyed to the urinary bladder through ureters for temporary storage until it is voided out by the process of micturition.
- Structural and functional modifications are present in the avian renal system to eliminate the major metabolic end product of birds, the uric acid. Salt glands provide a means of eliminating excess salt from the body, especially in marine species of birds.

Exercise

Objective Questions

Q1. Inulin clearance study is used to measure _______.

Q2. The plasma concentration of a substance at which it first appears in urine is known as ______.

Q3. Which type of collecting duct cells is involved in acid secretion?

Q4. Which condition stimulates the kidney to release erythropoietin?

Q5. Which is the major force favouring filtration across the glomerular capillary wall?

Q6. Which water channel is responsible for ADH-induced water reabsorption from the collecting duct?

Q7. Which tubular part of the nephron is completely impermeable to water?

Q8. Which hormone increases renal reabsorption of calcium?

Q9. What is the major end product of nitrogen metabolism in birds?

Q10. The property of tubules to increase reabsorption following the increased tubular load due to increased GFR is known as ______.

Q11. Which mechanism helps glucose reabsorption from PCT?

Q12. What are the target cells of aldosterone in renal tubules?

Q13. Which organic compounds are responsible for 50% renal medullary interstitial hypertonicity?

Q14. Where are the urea transporters located?

Q15. The bulk of glomerular filtrate is reabsorbed in which part of the nephron?

Q16. PAH clearance is used to study in ______.

Q17. What amount of mammalian-type nephrons are present in the birds?

Q18. The mode of Na^+ reabsorption from Henle's loop is ______.

Q19. Which proteolytic enzyme affects the release of aldosterone after secreting from JG cells?

Q20. Which specialised tubular epithelial cells are involved in monitoring sodium ion concentration in the tubular fluid?

Subjective Questions

Q1. What is tubuloglomerular feedback?

Q2. Describe the dynamics of glomerular filtration.

Q3. Explain micturition.

Q4. How does ADH-mediated water reabsorption occur in the kidney?

Q5. How does sodium reabsorption occur from the renal tubules?
Q6. Describe the urine concentration mechanisms.
Q7. What is urea recycling?
Q8. Describe renal clearance.
Q9. Write the speciality of avian excretion.
Q10. Write the dynamics of tubular reabsorption.

Answers to Objective Questions

A1. GFR
A2. Renal threshold
A3. Intercalated cells
A4. Hypoxia
A5. Hydrostatic pressure of capillary blood
A6. Aquaporin-2
A7. Thin portion of the loop of Henle
A8. PTH
A9. Uric acid
A10. Glomerulotubular balance
A11. Co-transport/symport/secondary active transport
A12. Principal cells
A13. Urea
A14. Collecting duct
A15. Proximal tubule
A16. Renal plasma flow
A17. 15–25%
A18. Co-transport with K^+ and Cl^-
A19. Renin
A20. Macula densa

Keywords for Answer to Subjective Questions

A1. Filtration pressure, macula densa, afferent arteriole, mesangium
A2. Hydrostatic pressure, colloid osmotic pressure, net filtration pressure
A3. Micturition reflex, reflex centres of pons, relaxation of external sphincter
A4. Dehydration, ADH, aquaporin-2
A5. Co-transport, counter-transport, chloride-driven sodium transport, transport from DCT
A6. Countercurrent multiplier, countercurrent exchange
A7. Inner medullary collecting duct, urea transporters, interstitial hypertonicity
A8. GFR, inulin clearance, creatinine clearance, renal plasma flow, PAH clearance
A9. Uric acid, reptilian and mammalian nephrons, AVT, renal portal system
A10. Hydrostatic pressure, colloid osmotic pressure, capillary blood, interstitial fluid

Further Reading

Hall JE (2011) Urine concentration and dilution: regulation of extracellular fluid osmolarity and sodium concentration. In: Hall JE (ed) Guyton and Hall textbook of medical physiology, 12th edn. Elsevier Saunders, Philadelphia, pp 345–360

Magouliotis DE, Tasiopoulou VS, Svokos AA, Svokos KA (2020) Aquaporins in health and disease. Adv Clin Chem 98:149–171

Parrah JD, Moulvi BA, Gazi MA, Makhdoomi DM, Athar H, Din MU, Dar S, Mir AQ (2013) Importance of urinalysis in veterinary practice – a review. Vet World 6(9):640–646. https://doi.org/10.14202/vetworld.2013.640-646

Reece WO (2009) The urinary system. In: Reece WO (ed) Functional anatomy and physiology of domestic animals, 4th edn. Wiley-Blackwell, pp 344–347

Reece WO (ed) (2015) Duke's physiology of domestic animals. Wiley

Part IV

Neuro-Muscular System and Special Senses

Muscular System

10

Dipak Banerjee, Pradip Kumar Das, and Joydip Mukherjee

Abstract

Muscle is a soft tissue, having a special ability for contraction. The contraction of the muscle fibers creates force and that causes movement of body as well as visceral organs. Three types of muscles, namely skeletal, cardiac, and smooth muscles, are present in the body. The skeletal muscle is voluntary, while the cardiac muscle and smooth muscle are involuntary. Muscle fibers have some special properties, i.e., excitability, contractility, extensibility, and elasticity. The skeletal muscle fibers have a long cylindrical structure with many nuclei located in the periphery. The active contractile unit of muscle is known as sarcomere. Each myofibril contains several types of protein cells, called myofilaments. During contraction, action potential propagates through the sarcolemma and travels down the T-tubules causing the sarcoplasmic reticulum to release Ca^{2+} ions to the sarcoplasm. The myosin head then attaches to the binding site of the G-actin molecule, and the formation of crossbridges occurs. The following power stroke occurs, which forces and leads to the shortening of muscle fiber. Muscle relaxation occurs when the release of the neurotransmitter stops at the neuromuscular junction. In smooth muscle, cells are small and have one central nucleus. No neuromuscular junctions exist; instead, varicosities transmit the nerve impulse to cells. Contraction and relaxation are slower than the skeletal muscle, and less energy is required for contraction. Cardiac muscle cells are small and branched and have a single nucleus. An intercalated disc is present at the junction between two cells. Gap junction located at the intercalated disc spreads action potential from one cell to another.

D. Banerjee (✉) · P. K. Das · J. Mukherjee
Department of Veterinary Physiology, West Bengal University of Animal & Fishery Sciences, Kolkata, West Bengal, India

© The Author(s), under exclusive license to Springer Nature Singapore Pte Ltd. 2023

P. K. Das et al. (eds.), *Textbook of Veterinary Physiology*, https://doi.org/10.1007/978-981-19-9410-4_10

Graphical Abstract

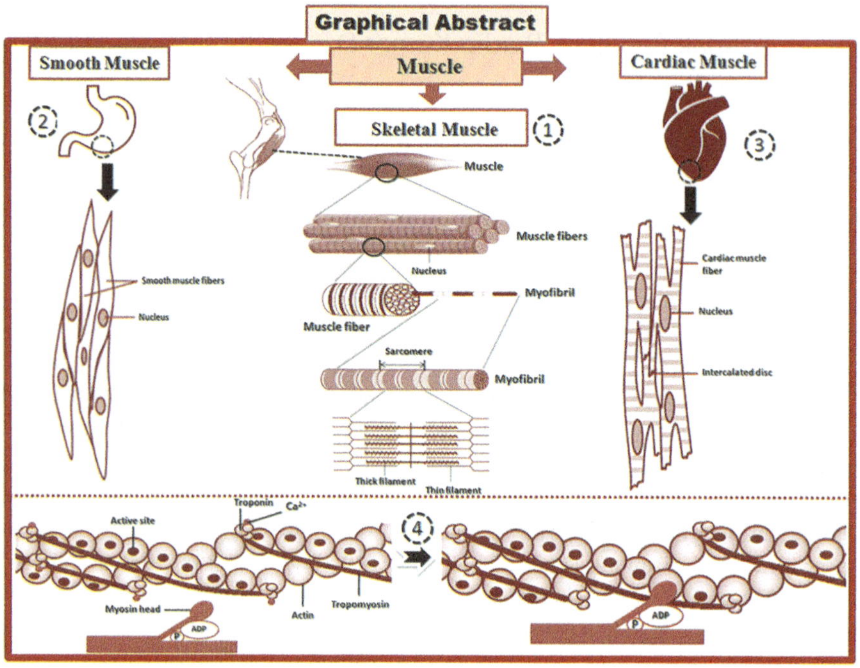

Description of the graphic: (1) Skeletal muscle fibers have a long cylindrical structure with many nuclei located in the periphery. The active contractile unit of muscle is known as sarcomere. Each myofibril contains several types of protein cells, called myofilaments. (2) In smooth muscle, cells are small and have one central nucleus. Two types of smooth muscles are single-unit smooth muscle and multiunit smooth muscle. (3) Cardiac muscle cells are small and branched and have a single nucleus. An intercalated disc is present at the junction between two cells. Gap junction located at the intercalated disc spreads action potential from one cell to another. (4) During contraction of skeletal muscle, action potential propagates through the sarcolemma and travels down the T-tubules causing the sarcoplasmic reticulum to release Ca^{2+} ions to the sarcoplasm. The myosin head then attaches to the binding site of the G-actin molecule, and the formation of crossbridges occurs

Keywords

Skeletal muscle · Cardiac muscle · Smooth muscle · Muscle contraction · Muscle fibers · Crossbridge

Learning Objectives

- Functions, properties, and types of muscle tissues
- Microscopic structure of a skeletal muscle
- Major phases of skeletal muscle contraction and relaxation and their neural control
- Sources of energy for skeletal muscle contraction, mechanism of muscle fatigue, and rigor mortis
- Structure and contraction mechanisms of smooth and cardiac muscle
- Different muscular disorders of domestic animals

10.1 Basic Characteristics of Muscle

Muscle is a soft contractile tissue, originated from the embryonic mesodermal layer. Muscle consists of muscle cells or muscle fibers. Contraction of muscle fibers generates force and that causes motion (i.e., locomotion or movement of visceral organs). The word "muscle" derived from the Latin word "musculus," which means "little mouse." It may be due to the shape of muscles like mouse, or contracting muscles look like mouse moving under the skin. Muscle fibers contain contractile filaments myosin (also known as thick filament) and actin (also known as thin filament). These protein filaments slide over one another and produce contraction. According to structure, situation, and function, muscles are generally classified into skeletal muscle (attached to bones), smooth muscle (present in the visceral organs), and cardiac muscle (found in heart). On the basis of action, muscles are classified into voluntary and involuntary muscle. Muscles which can be controlled by animal's own will are voluntary muscle like skeletal muscle, whereas muscles which are not under voluntary control are called involuntary muscle like smooth and cardiac muscle. Depending on the presence of striation, muscles can be classified into striated and nonstriated muscles. Skeletal muscle and cardiac muscle are striated muscles, whereas smooth muscle is a nonstriated muscle. The energy for muscle contraction is provided by ATP molecules, which are generated mainly through oxidation of fats and carbohydrates, but anaerobic reactions also occur.

10.1.1 Functions of Muscle

1. ***Movement of body or locomotion***: The major function of muscular tissue is locomotion or movement of body. The movement is of two types, i.e., gross movement and fine movement. Gross movement includes large, coordinated movements like walking, running, and swimming. Fine movement includes smaller movements, which occur in the limbs. The movements are mainly under voluntary control, but some movements are reflexive.
2. ***Maintenance of posture of the animal***: Skeletal muscles maintain the body in the right position when an animal is in sitting or standing condition. Posture of an animal depends on strong and flexible muscles, whereas stiff, weak, or rigid muscles result in abnormality in posture. This abnormality in posture for a long time may cause pain in joint and muscles.
3. ***Joint stability***: Tendons help in joint stability. Tendons of knee and shoulder joint are very important for stabilization, and muscles of the abdomen, back, and pelvic region are also involved in stabilization of the body and help in different activities.
4. ***Respiration***: Breathing process in animals depends on the contraction of respiratory muscles, mainly diaphragm and intercostal muscles. During inspiration, contraction of diaphragm results in movement of air into the lungs down the pressure gradient, whereas during expiration, relaxation of the diaphragm leads to increase in pressure which pushes air out of the lungs. Forced breathing or deep breathing requires help from other muscles, like muscles of abdomen, back, and neck.
5. ***Circulation***: Cardiac muscle helps in contraction and relaxation of heart, which results in circulation of blood throughout the body. Smooth muscles are present in blood vessels and regulate the constriction and relaxation of blood vessels, blood flow, as well as blood pressure.
6. ***Digestion***: In the gastrointestinal tract (GI tract), contraction of the smooth muscle helps in the movement of the food particle by a wavelike motion called peristalsis. This process also helps in the mixing of food particles with stomach acid and enzymes. Smooth muscles also help to pass the undigested food out of the body as feces.
7. ***Urination***: In the urinary system, smooth as well as skeletal muscles are there and the muscles along with nerves work together to hold and release urine from the urinary bladder. Some abnormal cases of urinary system like poor bladder control or retention of urine are caused by damage of the nerves that carry signals to the muscles.
8. ***Parturition***: During parturition, contraction of uterine smooth muscles helps in expulsion of fetus as well as fetal membrane. Oxytocin hormone initiates the contraction process.
9. ***Vision***: Muscles adjacent to the eye control the movements of eyeball. These muscles around eyes also help the eyes to maintain a stable image, scan the surrounding area, and track moving objects.
10. ***Protection of organ***: Muscles provide protection to different parts of the body, bones, as well as visceral organs. Muscle absorbs shock and protects a visceral organ or bone.
11. ***Thermoregulation***: Muscular system has a significant role in thermoregulation. During heat stress, relaxation of smooth muscles of blood vessels results in increased peripheral circulation and quick loss of heat from the surface of the body. However, during cold stress, contraction of smooth muscles of blood vessels reduces peripheral blood circulation and helps in the reduction of heat loss and maintaining body temperature. During cold stress, shivering thermogenesis also helps in body temperature maintenance.
12. ***Communication***: Muscles help in communication between animals as well as between birds by making sounds and different types of activities.

Except these functions discussed above, muscles are also involved in different physiological functions. These functions are discussed in details later in this chapter (in the discussion portion of individual muscles).

10.1.2 Properties of Muscle Tissue

Muscle fibers have some special properties, which help them to carry out their functions and differentiate them from other types of cells in the body.

The properties are excitability, contractility, extensibility, and elasticity:

1. ***Excitability***: The ability of muscle cells to respond to a stimulus is known as excitability. The stimuli are

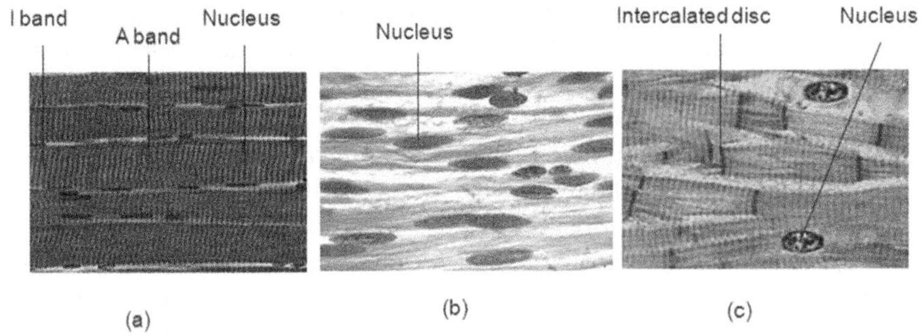

Fig. 10.1 Histological structure of three types of muscles. (**a**) Skeletal muscle. Muscle fibers showing striations with multiple peripherally located nucleuses. (**b**) Smooth muscle. No striations are present. Each cell contains single centrally located nucleus. (**c**) Cardiac muscle. Muscle fibers are branched having intercalated disc and one nucleus per cell

neurochemical, mechanical, and chemical in nature. When the muscle fibers are properly stimulated, then the muscle will respond to the stimulus.

2. ***Contractility***: Contractility is the ability of a muscle to contract and generate pulling force when properly stimulated.
3. ***Extensibility***: Muscle cells can lengthen in response to stretch, which is called extensibility. This property is more evident in smooth muscle compared to skeletal muscle.
4. ***Elasticity***: It is the ability of muscle fiber to recoil to its original resting length once stretched.

10.1.3 Types of Muscles

Three types of muscles are there in the body, i.e., skeletal muscle, smooth muscle, and cardiac muscle.

1. ***Skeletal muscle***: Skeletal muscles are mainly attached with bones (via tendons), maintain posture of the animal, and control the movement or locomotion. Some skeletal muscles are directly attached with other muscles or skin like in the face where different muscles control facial expression.
   Skeletal muscle is under voluntary control and innervated by somatic motor neurons but can maintain posture or balance even in subconscious state. Muscle fibers are striated, elongated, and tubular in shape with multiple nuclei located peripherally (Fig. 10.1).
2. ***Smooth muscle***: Smooth muscle is present in the walls of visceral organs, like organs of digestive system, respiratory system, blood vessels, glands, uterus, eye, and skin. Smooth muscles regulate the functions and movement of such systems, such as movement of food through GI tract via peristalsis or expulsion of fetus during parturition, propel urine, dilate/constrict pupils, regulate blood flow, etc. In some locations, they are autorhythmic like in GI tract. Smooth muscle is controlled involuntarily by endocrine and autonomic nervous systems. No striation is present in the muscle, and cells are uninucleated (Fig. 10.1).
3. ***Cardiac muscle***: Cardiac muscle is located only in the heart, controls the cardiac contractions, and pumps blood all over the body. Like skeletal muscle, cardiac muscle is also striated. The contraction is slow and rhythmic and involuntary. Cells are branched, and intercalated discs are present. Lengths of cardiac myocytes are lesser than skeletal muscle fiber and contain 1–2 centrally located nuclei (Fig. 10.1 and Table 10.1).

10.2 Skeletal Muscle

Skeletal muscles are attached to the bones, are voluntary in nature, and have striations, so they are also called striated muscle. About 40% of the body weight is comprised of skeletal muscle. The major function of skeletal muscle is the movement or locomotion.

Constant little contractions of the skeletal muscle are essential to hold the body upright in any position, even at rest. Skeletal muscles also maintain skeletal stability and protect the skeletal structure from any damage. They act as an external barrier to the body and protect the bones as well as visceral organs from external shock or trauma. They also support the weight of the organs. They also help in the generation of heat by shivering thermogenesis (Fig. 10.2).

10.2.1 Skeletal Muscle: Gross and Microscopic Structure

During embryonic development, embryonic stem cells produce immature muscle cells known as myoblasts (blast = "precursor") (Fig. 10.3). Later, several myoblasts fuse together to produce one long muscle cell or muscle fiber and so each muscle fiber contains multiple nuclei.

Table 10.1 Comparison between skeletal, cardiac, and smooth muscles

	Skeletal muscle	Smooth muscle	Cardiac muscle
Location	Attached to bones	Present in the walls of visceral organs, blood vessels, eye, glands, uterus, skin	Present in the heart
Functions	Responsible for different types of movement like body movement and locomotion, posture, communication, facial expression, and breathing	Propel urine, mix food in digestive tract, dilate/ constrict pupils, and regulate blood flow	Helps in cardiac contractions and pumps blood all over the body. Involuntary and controlled by autonomic nervous systems and hormones
Appearances	Fibers are striated and tubular with peripherally located multiple nucleus	Nonstriated, smooth appearance, and mononucleated cells	Striated mononucleated cells
Control	Voluntary in nature and controlled by somatic motor neurons	Controlled involuntarily by endocrine and autonomic nervous systems	Involuntary, controlled by autonomic nervous system
Contraction	Both contraction and relaxation are very fast	Slow contraction and relaxation, can maintain for extended period	Moderate contraction and relaxation
Fatigue	Easily fatigue	Do not fatigue	Do not fatigue

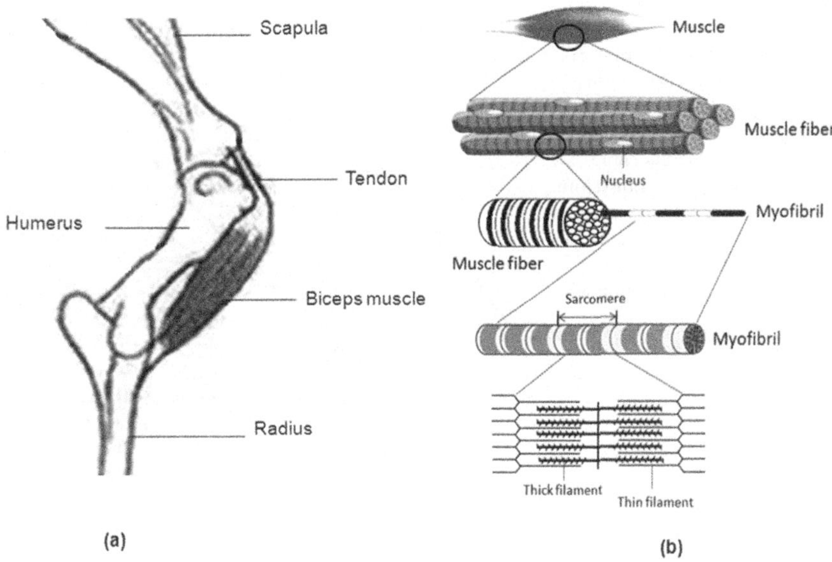

Fig. 10.2 (**a**) Biceps muscle of horse. (**b**) Structure of skeletal muscle. Muscle fibers are long cylindrical with multiple nucleuses. Myofibrils are bundles of rodlike contractile elements made up of myofilaments—thick and thin filaments. The muscles are striated due to the regular arrangement of thick and thin filaments

Each muscle fiber is enclosed by a fine layer of loose (areolar) connective tissue called endomysium ("endo"—inside) (Fig. 10.4). Several muscle fibers form a bundle known as fascicles. In fascicles, blood vessels and nerves are present. Fascicles are covered by a connective tissue known as perimysium ("peri"—around). The fascicles are bundled together and form a muscle. The entire muscle is enclosed by a layer of dense fibrous connective tissue called epimysium ("epi"—outside, and "mysium"—muscle).

Different layers of connective tissue extend away from the ends of the muscle fibers themselves and form the tendons or aponeurosis, which connects muscles to bone. Each muscle fiber is covered by its plasma membrane, known as sarcolemma.

The cytoplasm or sarcoplasm contains a huge amount of glycogen (polysaccharide of glucose), which is utilized for energy. Myoglobin, which is a red color pigment, is also present in sarcoplasm. The major portion of the sarcoplasm (almost 80% of the intracellular space) is occupied by myofibrils, which are rodlike cylindrical contractile proteins. Myofibrils run longitudinally. Around 100–1000 myofibrils are present in each muscle fiber. Myofibrils are compactly

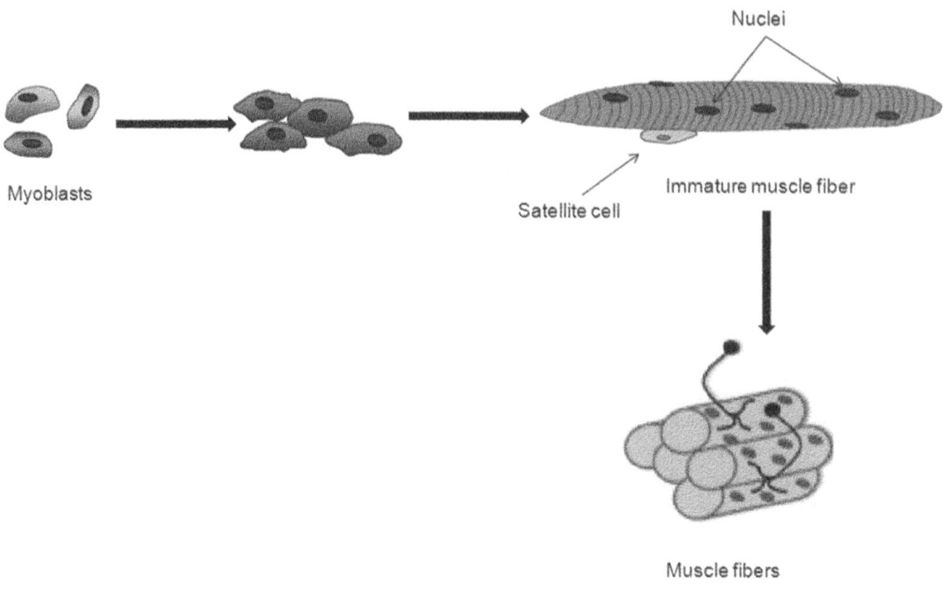

Fig. 10.3 Development of skeletal muscle fiber. During embryonic stage, a number of uninucleated myoblasts fuse to each other and form multinucleated skeletal muscle cell or muscle fiber. Cells which do not fuse remain as satellite cells and function as muscle stem cells

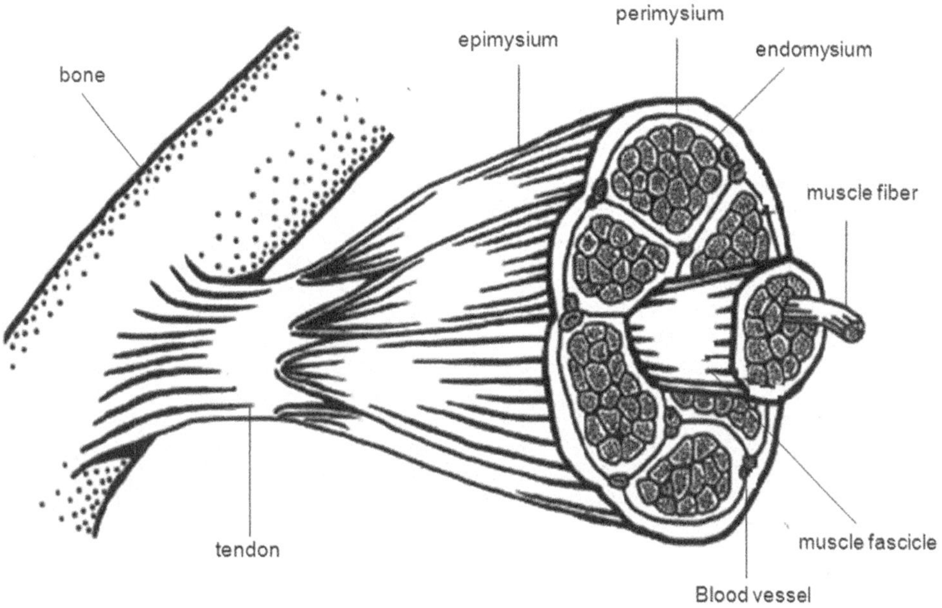

Fig. 10.4 Cross section of skeletal muscle. Individual fiber is surrounded by **endomysium**. A number of muscle fibers form a **fascicle** which is surrounded by **perimysium**. Several fascicles form a muscle, and the entire muscle is covered by **epimysium**. All the connective tissue layers of muscle unite at the end and form a **tendon**, which helps in the attachment of muscle with **bone**

packed, and the cell organelles remain sandwiched between them. The nuclei get pressed to the periphery of the cell just under the sarcolemma.

10.2.1.1 Description of a Single Muscle Fiber of Skeletal Muscle

An individual muscle cell or muscle fiber has the following parts:

1. ***Sarcolemma***: Sarcolemma is the plasma membrane of individual muscle fiber. Sarcolemma is enclosed by the basement membrane and endomysium (Fig. 10.4). It is an excitable membrane. Sarcolemma has many properties similar with nerve cell membrane. Regular invaginations are seen in the sarcolemma, which form tubes that remain around the myofibrils, known as transverse tubules or T-tubules. During muscle contraction, action potential spreads through the sarcolemma, and through T-tubules, it reaches the sarcoplasmic reticulum.
2. ***Sarcoplasm***: The cytoplasm in the muscle fiber is known as sarcoplasm. Like that of other cells, different cell organelles are present in the sarcoplasm. Large numbers of glycosomes are present in the sarcoplasm. Small fat droplets and large amount of myoglobin are also present. Myofibrils fill the sarcoplasm. Mitochondria are present in between the myofibrils, near the Z line and A bands.

3. ***Sarcoplasmic reticulum***: Sarcoplasmic reticulum (SR) is the smooth endoplasmic reticulum (SER) present in the muscle fiber (Fig. 10.5). They form a network of tubes surrounding the myofibrils and remain closely associated with myofibrils. SR stores the Ca^{2+}. During muscle contraction, this stored Ca^{2+} ions are released out from the SR to the sarcoplasm and reabsorbed during relaxation. The membrane of SR has "pumps" (active transport) for the transport of Ca^{2+}. Along with the "pumps," the SR membrane has special types of openings, or "gates," for transport of Ca^{2+} ions. During relaxation of muscle, these gates remain closed and Ca^{2+} ions are unable to pass through the membrane, and the Ca^{2+} concentration is very high in SR and very low in the sarcoplasm. During muscle contraction, when an impulse reaches the sarcolemma, it propagates through the T-tubule to SR membrane. The action potential initiates the opening of Ca^{2+} "gates" and Ca^{2+} comes out of the SR into the sarcoplasm.
4. ***Transverse tubules***: Transverse tubules or T-tubules are narrow tubelike structures formed due to invaginations of the sarcolemma (Fig. 10.5). These tubules extend into the interior of the muscle fiber and encircle each myofibril but never open inside the muscle fiber and form sarcotubular system. They carry action potentials deep into the muscle fiber. The T-tubules are present at the junction between the A and I bands. SR remains parallel to the myofibrils. SR of skeletal muscles forms right-angle enlargements at the junctions of A and I bands near the T-tubules. These enlargements of the SR are called terminal cisternae ("end sacs"). One T-tubule along with two terminal cisternae is known as the triad. Triad plays a very important function in skeletal muscle contraction. The membrane of T-tubule has a number of voltage-dependent proteins, known as dihydropyridine (DHP) channels or L-type calcium channels. But these channels do not allow calcium to move through them. They are physically associated with the calcium-release channels on the terminal cisternae called ryanodine receptor channels (RyR). When action potential comes to sarcolemma and sarcolemma becomes depolarized, the DHP channel identifies the depolarization and causes opening of the RyR channels, resulting in the release of calcium from the terminal cisternae of the SR.

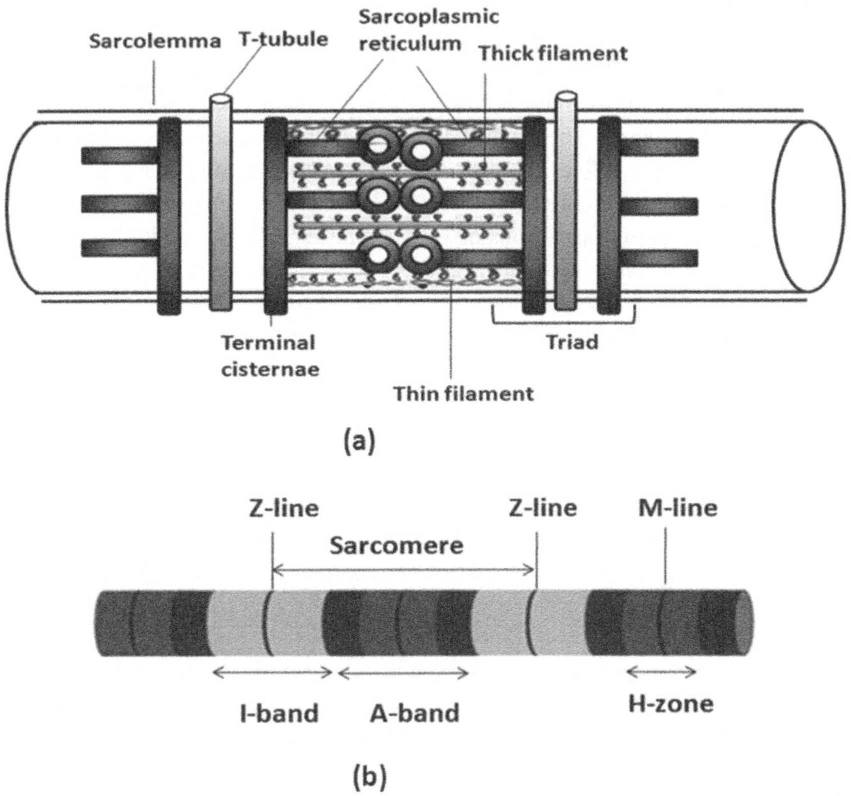

Fig. 10.5 Skeletal muscle fiber. (**a**) Muscle fiber containing sarcoplasmic reticulum and T-tubules. One T-tubule along with two terminal cisternae forms the triad. (**b**) Sarcomere of muscle fiber (the sarcomere is the unit of skeletal muscle fiber, which is the length between two consecutive Z-lines). Z-disc or Z-line: It is a line that separates two adjacent sarcomeres; **A band**: they are dark band known as anisotropic bands or A bands (includes overlapping myosin and actin filaments). **I bands**: the area where only actin filaments are present; the light bands are known as isotopic bands or I bands. **H zone**: a lighter zone in the middle of each dark band or **A band** (the area where only myosin filaments are present). **M-line**: a darker line at the middle of each H zone (central line of the sarcomere where myosin filaments are anchored)

5. ***Myofibril***: Myofibrils are bundles of rodlike contractile elements made up of myofilaments. Almost 80% of the muscle volume is occupied by myofibrils. They are composed of different types of proteins, which form the myofilaments. Contractile elements are present in the myofilaments which help in contraction. The functional contractile unit of skeletal muscle is known as sarcomere ("sarc" means muscle, "mere" means part). Sarcomere is the region of a myofibril between two consecutive Z-lines (Fig. 10.5). The striated appearance in skeletal muscle is produced due to regular, organized arrangement of myofilaments. So light and dark striations are present in each cell. The dark areas in muscle fiber are known as anisotropic bands or A bands, and the light areas are called isotopic bands or I bands. Each myofibril contains several varieties of protein molecules, called myofilaments. The larger or thick myofilaments are made up of the protein, myosin, and the smaller thin myofilaments are chiefly made up of the protein, actin.

   Z-disc or Z-lines are fine dense lines that appear in the middle of each I band. Z-line separates two adjacent sarcomeres from each other. In the middle of each dark band or A band, a lighter zone is present known as H zone (H for "helle"—"bright"). Each H zone has a darker line known as M-line (M for "middle"), which runs right down the middle of the A band. Sarcomere of muscle fiber: Sarcomere is the unit of skeletal muscle fiber, which is the length between two consecutive Z-lines. Z-disc or Z-line: It is a line that separates two adjacent sarcomeres; A bands are dark band known as anisotropic bands or A bands (include overlapping myosin and actin filaments). I bands are the area where only actin filaments are present; the light bands are known as isotopic bands or I bands. H zone is a lighter zone in the middle of each dark band or A band (the area where only myosin filaments are present). M-line is a darker line at the middle of each H zone (central line of the sarcomere where myosin filaments are anchored).
6. ***Myofilaments***: Myofilaments are fine stringlike contractile filaments of myofibrils; they consist of thick filament (myosin) and thin filament (actin). Myosin and actin are contractile proteins, which interact with each other to generate force, resulting in shortening of muscle fiber. Two major regulatory proteins troponin and tropomyosin bind to actin and regulate the attachment of myosin head with actin during muscle contraction.

Thick filaments: Thick myofilaments are mainly made up of the protein, myosin (myosin II). Each thick myofilament is approximately 15 nm in diameter composed of about 300 myosin molecules. Each myosin is made up of six protein subunits, two heavy chains and four light chains. The shape of heavy chains is similar to a golf club, with a long shaft-like structure connected to globular myosin head (Fig. 10.6). The heavy chains of myosin are twisted over one another forming a double-helix structure.

The link between the head and the shaft of the myosin molecules remains as a hinge, and so it is known as hinge region. This hinge region is able to bend and generate power stroke during muscle contraction. The centers of the thick filaments are comprised of the shaft portions of the heavy chains. Each head of myosin has two light chains. Each myosin head has a binding site for actin and an ATPase, which hydrolyzes ATP during muscle contraction.

Thin filaments: The thin filaments contain three different proteins, i.e., actin, tropomyosin, and troponin (Fig. 10.7). Each actin molecule has an active site for attachment with myosin head during muscle contraction. Other two proteins, tropomyosin and troponin, are

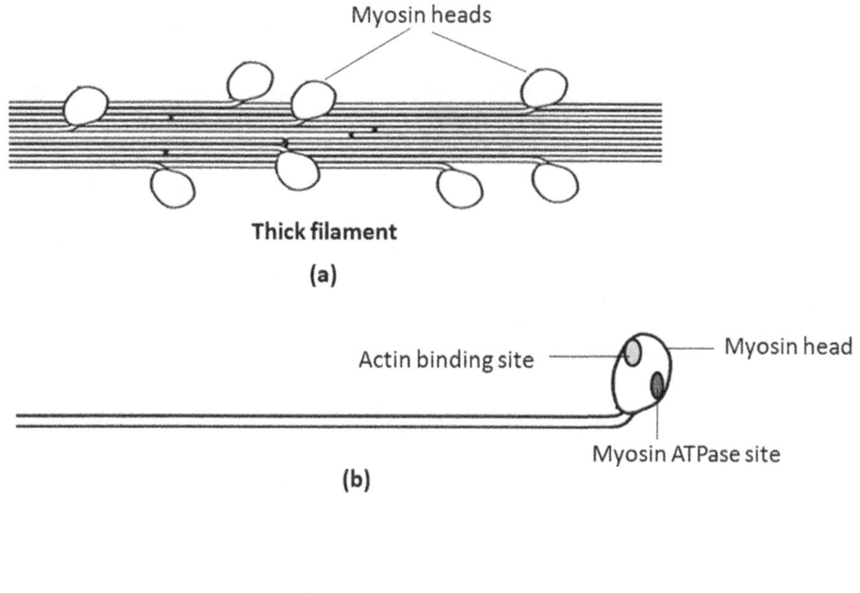

Fig. 10.6 Thick filament and myosin heavy chain. (**a**) Half of the myosin molecules have their heads remain towards one end of the thick filament, and the other half remain in the opposite direction. The heads of the myosin bind to the active sites on the actin during muscle contraction. (**b**) Myosin heavy chain, with a long shaft-like structure connected to globular myosin head. Myosin head has a binding site for actin and an ATPase, which hydrolyzes ATP during muscle contraction

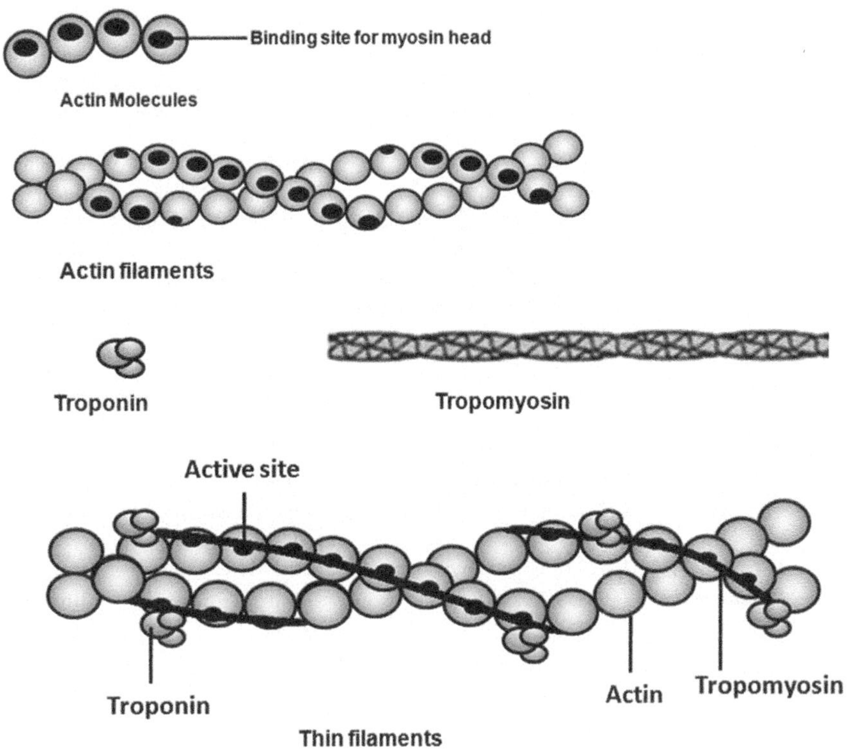

Fig. 10.7 Thin filaments: actin, troponin, and tropomyosin. Strings of globular actin (G-actin) twisted over one another like a double-helical structure. Tropomyosin made up of coiled-coil dimer and the two strands run diametrically opposite to each other along the actin filaments and cover the active sites on actin. Troponin is a complex of three different globular protein subunits, i.e., troponin C, troponin T, and troponin I

regulatory subunits which are bound to actin. Actin is a globular protein called globular actin or G-actin (free monomeric units). Each F-actin (filamentous actin which is the polymer form) is formed by two strings of globular actin (G-actin) twisted over one another like a double-helical structure, which looks like twisting two strands of pearls with each other where individual G-actin molecule is like a pearl necklace. Tropomyosin is another component of thin filament. It is a long threadlike polypeptide that remains parallel to each F-actin strand and covers the active sites of each G-actin molecule when the muscle remains in relaxed state, whereas during contraction, tropomyosin is replaced from its position and the active site on the actin is exposed to which myosin head binds. Tropomyosin has a structural similarity with that of the myosin tail, being a coiled unit of two protein chains. Tropomyosin made up of coiled-coil dimer and the two strands run diametrically opposite to each other along the actin filaments. Another component of thin filament troponin is a complex of three different globular protein subunits, i.e., troponin C, troponin T, and troponin I.

Troponin C has a receptor for Ca^{2+} ions and binds to calcium ions (released from the sarcoplasmic reticulum) on activation of the muscle contraction. Another subunit troponin T is the tropomyosin-binding subunit of troponin, which binds with tropomyosin and keeps it in this position on F actin strands. Troponin I binds to actin, holds the troponin-tropomyosin complex in proper position, and inhibits binding of myosin head with actin. It inhibits the interaction between myosin and actin.

During skeletal muscle contraction, Ca^{2+} binds to troponin C, which results in a conformational change in the entire complex, and tropomyosin is released from its position and myosin-binding sites on the G-actin subunits become exposed for attachment with myosin.

Titin or connectin is another important structural protein that functions like a big rubber band in muscles. It is a long elastic protein that runs within the thick filament and extends from the Z-disc to the M. Titin is the third most abundant protein in the muscle after myosin and actin. It is the largest known protein in the body and has around 30,000 amino acids. Titin acts as a molecular spring in the skeletal muscle and prevents overstretching as well as damage of muscle. Titin helps to return the muscle to its normal length when the muscle is stretched.

Dystrophin is another muscle protein, which connects the cytoskeleton of a muscle fiber to the extracellular matrix through the cell membrane. Dystrophin is located between the sarcolemma and the outermost myofilaments. Mutation of the gene coding for dystrophin is one of the major causes of a class of muscle diseases collectively known as muscular dystrophy (MD).

As a complex structure, sarcomere contains a number of proteins; few of them are listed in Table 10.2.

Table 10.2 Different proteins present in the sarcoplasm of vertebrate striated muscles and their properties

Sl no.	Sarcoplasmic protein	Molecular weight and subunits	Location	Functions	Related diseases
1	Actin	42 kDa, globular monomer	Thin filament (~360 molecules), helical polymer	Filament formation, myosin ATPase activation, filament sliding. Binds myosin, tropomyosin, troponin, nebulin, α-actinin	FHC, NM
2	α-Actinin	190 kDa (homodimer 2 × 95 kDa); CH, spectrin-like, EF hand domains	Z filaments linking actin and titin filaments	Integrates Z-line. Binds actin, titin, CapZ, myopalladin, myozenin, myotilin, ZASP/Cypher, synemin	
3	Cap Z (β-actinin)	68 kDa, heterodimer (36 and 32 kDa subunits) 1 per filament	Caps barbed end of thin filament, in Z-line	Length stabilization. Binds actin, α-actinin	
4	Desmin/vimentin	~55 kDa, α-helical core, nonhelical ends	Surrounds and runs between Z-lines	Sarcomere strengthening and connection with each other and cell membrane	Desmin myopathy
5	FATZ (calsarcin-2, myozenin)	32 kDa	Z-line	Binds α-actinin, γ-filamin, telethonin	
6	γ-Filamin	CH domain, Ig repeats	Z-line	Binds myozenin, myotilin	
7	MM creatine kinase	86 kDa, dimer (2 × 43 kDa)	Line M4 and M4′ of M-line	Buffers [ATP], bridges thick filaments	
8	M protein	165 kDa, Ig and Fn domains	Line M1 of M-line	Bridges thick filaments. Binds myosin	
9	Myomesin	185 kDa, Ig and Fn domains	M-line	Binds myosin and titin	
10	Myopalladin	145 kDa	Z-line	Anchors nebulin in Z-line. Binds α-actinin, nebulin, and CARP	
11	Myopodin	80–95 kDa	Z-line	Bundles actin filaments. It is a zyxin-binding protein, has capabilities to regulate cell growth and motility	
12	Myosin	~520 kDa, hexamer, 2 heavy chains (223 kDa), 4 light chains (~20 kDa)	Thick filament (~300 molecules), helical polymer	Filament formation, ATPase, filament sliding, modulation of contraction. Binds actin, titin, MyBPs, M protein, myomesin	FHC
13	MyBP-C (-X) and MyBP-H	140 kDa (C, X), 86 kDa (H) modular (Ig and Fn domains)	Stripes 3–11 (C, X), 3 (H), 43 nm apart in each half of A band	Myofibrillogenesis, filament stabilization, modulation of contraction. Binds myosin, titin	FHC
14	Myotilin	57 kDa	Z-line	Binds α-actinin, γ-filamin	MD
15	Nebulin (nebulette)	800 kDa (nebulette 109 kDa), single chain. Modular (35-amino acid actin-binding modules)	Extends from Z-line (C terminus) to filament tip (N-terminus)	Thin-filament length determination and stabilization. Binds actin, tropomyosin, tropomodulin, myopalladin	NM
16	Nestin	220–240 kDa, IF protein	Z-line periphery, with desmin	Similar to synemin but mainly in developing muscle	
17	Paranemin	180 kDa, IF protein	Z-line periphery, with desmin	Similar to synemin	
18	Plectin	High molecular weight, α-helical coiled coil	IFAP, at and between Z-lines	Connects Z-line IFs to actin filaments, cell membrane, and organelles. Binds actin, IFs	MD
19	Skelemin	~200 kDa, modular structure, splice variant of myomesin	Periphery of M line	Connects myofibrils at M-line. Binds myosin, IFs, and integrins	
20	Synemin	230 kDa	Z-line periphery; co-polymer with desmin	Links between Z-lines and cell membrane. Binds α-actinin, vinculin	
21	Syncoilin	64 kDa, IF protein	Z-line and sarcolemma	Links IFs to sarcolemma via dystrophin complex.	
22	Telethonin (T-cap)	19 kDa	Z-line, at N-terminus of titin	Binds titin, myozenin, cell membrane K channel	MD
23	Titin (connectin)	~3 MDa (single polypeptide). Modular (Ig and Fn domains, PEVK segment)	Extends from Z-line (N-terminus) to M-line (C-terminus)	Developmental sarcomeric template, muscle elasticity. Binds myosin, MyBP-C, α-actinin, myomesin, telethonin	FHC
24	Tropomodulin (Tmod)	40 kDa, monomer, 1 or 2 per filament	Caps pointed end of thin filament	Thin-filament length stabilization. Binds actin, nebulin, tropomyosin	
25	Tropomyosin	65 kDa, coiled-coil dimer of 2 α-helices (32 kDa each)	Thin filament, ~50 molecules 38 nm repeat	Filament stabilization and regulation. Binds actin, troponin, nebulin, tropomodulin	FHC, NM

(continued)

Table 10.2 (continued)

Sl no.	Sarcoplasmic protein	Molecular weight and subunits	Location	Functions	Related diseases
26	Troponin	80 kDa, complex of TnC (18 kDa), TnI (20–24 kDa), TnT (31–36 kDa)	Thin filament, one per tropomyosin, 38 nm repeat	Regulation of contraction. Binds actin, tropomyosin	FHC
27	ZASP/Cypher	~32 kDa, PDZ-motif protein	Z-line	Binds actinin	Myopathy

Source: Craig and Padr6n (2004)

KEY: *CH* calponin homology, *MD* muscular dystrophy, *NM* nemaline myopathy, *FHC* familial hypertrophic cardiomyopathy

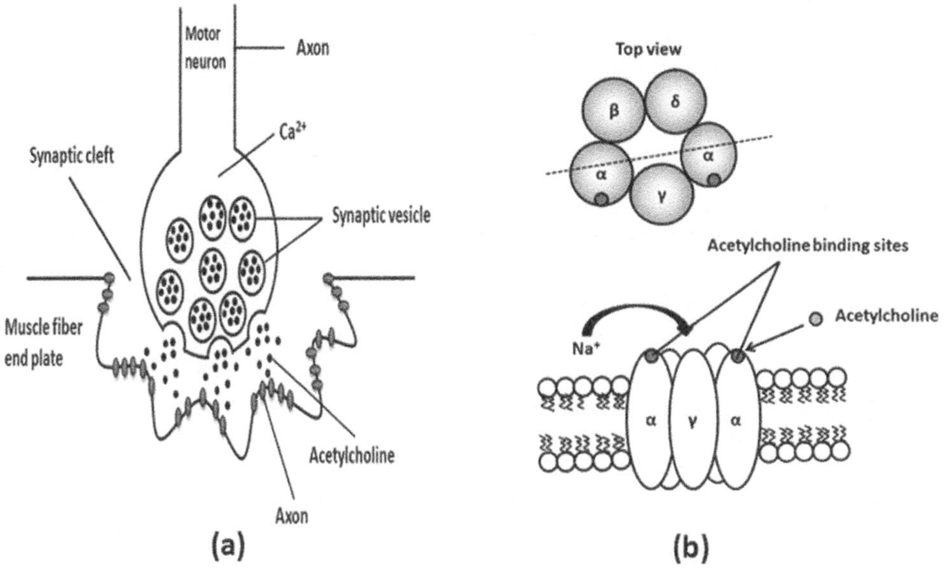

Fig. 10.8 Neuromuscular junction and acetylcholine receptor channel. (**a**) Neuromuscular junction and (**b**) acetylcholine receptor channel: The nicotinic acetylcholine receptor is a ligand-gated ion channel, composed of five subunits arranged symmetrically around a central conducting pore. Upon binding acetylcholine, the channel opens and allows diffusion of sodium (Na^+) and potassium (K^+) ions through the conducting pore

10.2.2 Neuromuscular Junction

Nerve impulse or action potential travels through a motor neuron to a skeletal muscle fiber to trigger the contraction of that muscle. The site attachment between the nerve ending and the skeletal muscle is known as neuromuscular junction. It is like that of the synapse between two neurons. This junction is a chemical synapse formed by an axon terminal of the neuron and motor end plate of a skeletal muscle fiber (Fig. 10.8). The motor neuron can have a number of terminal branches; each of these nerve endings attaches with a separate muscle fiber.

10.2.3 Motor Unit

In the muscle, motor neuron forms many branches and each branch innervates a single muscle fiber. The neuron along with the muscle fiber (innervated by that motor neuron) is known as the motor unit. The size of the motor unit depends on the function of that muscle it innervates. Muscles of limbs and postural muscles are attached with largest motor units, in which one axon supplies many muscle fibers, whereas the smallest motor units, in which one axon may supply only a few muscle fibers, are seen in association with eye movements.

During contraction of a muscle, a number of motor units frequently work together and act like a group and these motor units within a muscle are called motor pool. These muscle fibers within a motor unit are of same type and contract together when activated. The number of motor units also controls the force of contraction.

The terminal branch of the axon does not actually make contact with the muscle fiber but is separated from it by a gap of approximately 50 nm wide called synaptic cleft.

The membrane of nerve terminal releases neurotransmitter (acetylcholine), which has a receptor on postsynaptic membrane. The nerve ending has membrane-bound vesicles containing neurotransmitter. These vesicles are synthesized from the cell body of the neuron and come to nerve ending as an empty bag of proteins.

In the nerve endings, the membrane has choline transporter, which transports choline from outside of the neuron to inside (Fig. 10.9). Then in the nerve endings, mitochondria synthesize acetyl-CoA, and this acetyl-CoA attaches with choline with the help of the enzyme choline acetyl transferase

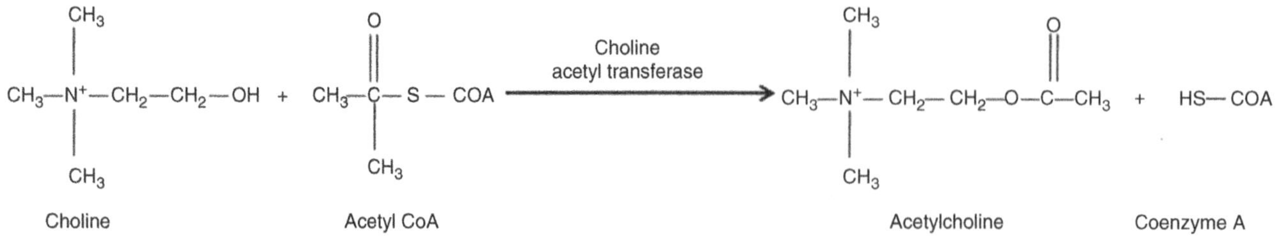

Fig. 10.9 Molecular pathway of acetylcholine synthesis

and forms acetylcholine. The enzyme choline acetyl transferase is also synthesized from the cell body of the neuron.

Then acetylcholine enters into the synaptic vesicles through the transporter on the membrane of the vesicle and is stored in the synaptic vesicle.

10.2.4 Transmission of Action Potential Through Neuromuscular Junction

1. Nerve impulses or action potential travels from the brain or spinal cord to initiate the muscle contraction of skeletal muscle. An action potential propagates through the motor neuron and reaches the axon terminal.
2. During this propagation of action potential through the membranes of nerve endings, the voltage-gated sodium channel opens. This results in the influx of Na^{2+} inside the nerve endings.
3. As Na^{2+} enters inside the nerve ending, the charges of the membrane change from the resting membrane potential to depolarization, and it activates the depolarization-sensitive calcium channel. Now voltage-gated calcium channels open and Ca^{2+} diffuses into the terminal.
4. The axon terminal contains membrane-bound synaptic vesicles, which are filled with the neurotransmitter acetylcholine (ACh).
   Calcium entry causes the synaptic vesicles to release acetylcholine neurotransmitter through exocytosis process. The membrane of synaptic vesicles contains a calcium-sensitive protein called synaptobrevin, which is an intrinsic membrane protein of small synaptic vesicles. Synaptobrevin is a specific secretory organelle of neurons, which accumulates neurotransmitters and participates in their calcium-dependent release by exocytosis.
   Another calcium-sensitive protein is also present on the membrane of nerve endings called syntaxin (syntaxins are a family of membrane-integrated Q-SNARE proteins participating in exocytosis). As soon as the calcium ions bind with the calcium-sensitive protein, i.e., syntaxin on presynaptic membrane, they transform into active configuration and then calcium-sensitive protein on synaptic vesicle, i.e., synaptobrevin attached with it, and a fusion of membrane of synaptic vesicle with the membrane of axon terminal occurs. Then the membrane is dissolved at the site of attachment and results in the release of acetylcholine in the synaptic cleft through exocytosis.
5. Acetylcholine diffuses across the synaptic cleft and binds to the acetylcholine receptors on the motor end plate, which is a ligand-gated cation channel (Fig. 10.9).
   Acetylcholine moves from nerve membrane to motor end plate through the synaptic cleft. On the motor end plate, there are receptors (also called nicotinic cholinergic receptor) for acetylcholine attachment (the channel composed of pentameric proteins 2 α, β, γ, and δ). There are two attachment sites for acetylcholine on a single channel.
6. ACh binding causes ligand-gated cation channels to open. These ion channels are permeable to both Na^+ and K^+.
7. Na^+ enters the muscle fiber, and K^+ exits the muscle fiber. More Na^+ moves inside, and K^+ goes outside. The greater influx of Na^+ relative to outward flux of K^+causes the membrane potential to less negative and the local potential is generated which is also called motor end plate potential.
8. Then the entry of more number of Na^+ ions changes the end plate potential to threshold potential.
   Then this threshold potential causes opening of voltage-gated Na channel nearer to motor end plate. Then more number of Na^+ ions enter inside the cell through the voltage-gated Na^+ channel, and as a result, an action potential generation occurs which propagates along the sarcolemma.
9. The action potential travels across the sarcolemma and is propagated down the T-tubules. The T-tubules are filled with extracellular fluid, high in sodium (Na^+) and low in potassium (K^+) ions.
10. Before the discussion of spreading of action potential and starting of muscle contraction, let us describe how the stimulation of the ACh receptors is terminated.

For relaxation of muscle, ACh should be removed from the synaptic cleft. It is initiated when ACh is cleaved (split) by an enzyme called acetylcholinesterase, which remains in the synaptic cleft. Acetylcholinesterase splits ACh into acetate (acetyl) and choline.

The acetate diffuses out of the synaptic cleft and choline, which is an essential nutrient in the vitamin B group (B4), and is taken up by the axon terminal, where it is recycled to make more acetylcholine.

10.2.5 Muscle Contraction

A series of molecular events occur during muscle contraction known as the crossbridge cycle. The following steps occur during contraction (Fig. 10.10).

10.2.5.1 Crossbridge Formation

The action potential spreads through the sarcolemma, and it reaches the T-tubules. The action potential triggers SR. The resultant change in potential causes the voltage-gated channels in the T-tubule to respond. In the T-tubule, these channels are known as dihydropyridine channels (DHP) or L-type Ca^{2+} channels. They are mechanically linked to ryanodine receptor channels (RyR), which are calcium channels located in the membrane of sarcoplasmic reticulum. When the membrane potential changes, then the DHP channel opens and Ca^{2+} ions come out of the sarcoplasmic reticulum and diffuse into the sarcoplasm. Then Ca^{2+} ions bind to the troponin C. The binding of Ca^{2+} ions causes conformational change in troponin. Then tropomyosin which covers the active site of the actin moves from its position, which results in the exposure of active site of G-actin molecule. Now myosin head attaches with the binding site on G-actin molecule and the formation of crossbridges occurs.

10.2.5.2 Power Stroke Generation

When muscle remains in relaxed state, the myosin head is "cocked." ADP and phosphate (Pi) remain attached with myosin head. As the myosin head is attached with actin, the Pi detaches from the myosin head and energy is released. This energy results in bending of myosin head. The bending or power stroke forcefully pulls the actin past the myosin. ADP is also released from the myosin during the power stroke. Myosin heads pull the thin filament towards the middle and sarcomere shortens.

10.2.5.3 Crossbridge Detachment

For considerable shortening of muscle fiber, the myosin heads must be detached from the actin and reattached with the next actin molecule. When another ATP attaches with myosin head, the attachment of myosin head with actin becomes weaker and myosin head detaches.

10.2.5.4 Reactivation of Myosin Heads

The ATPase present in the myosin heads hydrolyzes the ATP into ADP and Pi, which causes the head to "recock" (the recovery stroke), preparing it for the next power stroke.

The hydrolysis of ATP releases energy which re-energizes the myosin head for the next power stroke. During muscle contraction, each myosin molecule undergoes the entire

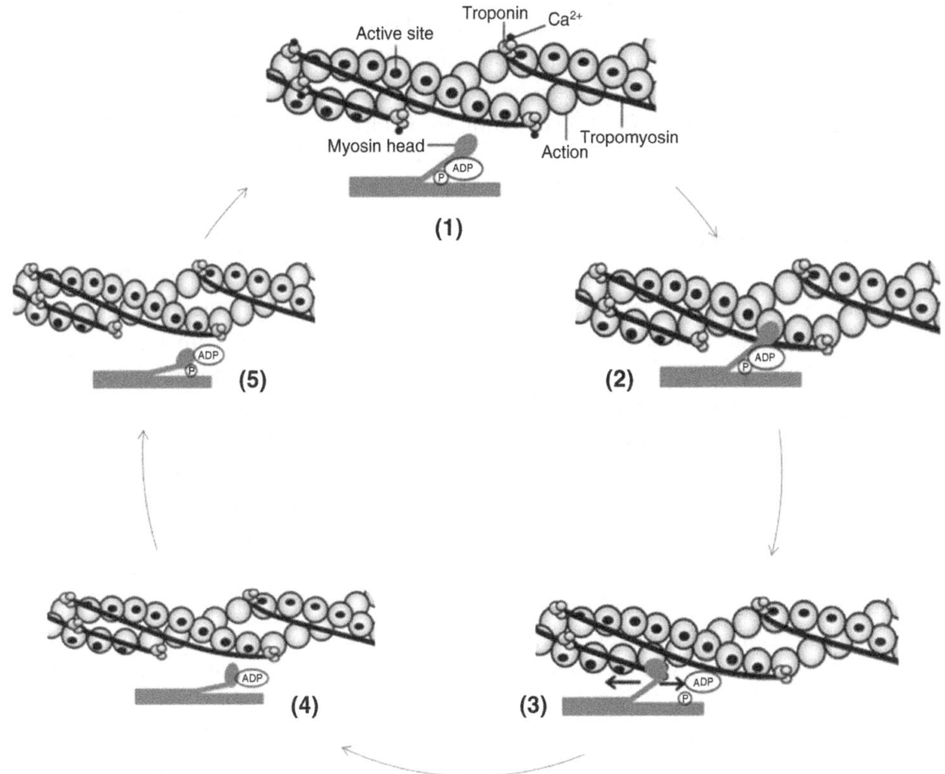

Fig. 10.10 Pathway of skeletal muscle contraction. (1) The active site on the actin is exposed as Ca ions bind to troponin. (2) The myosin head forms a crossbridge with actin. (3) During the power stroke, the myosin head bends, and ADP and phosphate are released. (4) A new molecule of ATP attaches to the myosin head, causing the crossbridge to detach. (5) ATP hydrolyzes to ADP and phosphate, which returns the myosin to the "cocked" position

crossbridge cycle numerous times, so the process is known as crossbridge cycling. The crossbridge cycle will repeat as long as the active site in actin is exposed. As the cycle repeats, the sliding of thin filaments over the thick filaments occurs and Z-lines come closer, resulting in shortening of sarcomere. This shortening causes the whole muscle to contract. As long as Ca^{2+} is present and the active sites are exposed, the process will continue.

10.2.6 Muscle Relaxation

During relaxation of muscle, the release of neurotransmitter (acetylcholine) stops. The remaining acetylcholine is broken down into acetate and choline by acetylcholinesterase. This stops the release of Ca^{2+} from the sarcoplasmic reticulum (SR). Then Ca^{2+} ions diffuse away from troponin C and are pumped back into sarcoplasmic reticulum (SR) by the ATP-dependent Ca^{2+} pump in SR membrane.

Tropomyosin returns back to its original position and covers the active site on the individual G-actin molecules. This prevents crossbridges from reforming. A new ATP binds to the myosin head. Binding of actin and myosin stops, and relaxation of muscle fiber takes place.

> **Know More.....**
> The sliding filament model of muscle contraction explains the fact that when skeletal muscle fibers contract, the individual proteins (actin and myosin) do not shorten. Rather, they slide over each other. ATP is necessary for detachment of myosin heads from actin. Also it is interesting that when a sarcomere contracts, both the H zone and the light I band shrink in width, while the dark A band does not appear to narrow.

10.2.7 Muscle Tone

Muscle always maintains a tension or resistance to stretch (Fig. 10.11). This tension or resistance to stretch is called muscle tone. Skeletal muscles are seldom completely relaxed, or flaccid, even at that time of rest when a muscle does not produce any movement. During rest, little contraction is present in the muscle fibers, which are essential for maintaining the posture, balance of the body, generating the reflexes, and controlling functions of different organs. Muscle tone is also seen in cardiac and smooth muscles. A complex interaction of nervous system and muscles is required for the activation of a few motor units at a time.

That is why muscles not at all fatigue completely because some motor units can recover from fatigue when others are active. The absence of the skeletal muscle tone results in the absence of low-level contractions that lead to loss of resistance to passive stretching muscle. This type of muscle tone is called hypotonia. Hypertonia occurs due to any damage of central nervous system (CNS), such as the cerebellum, or due to loss of innervations to the skeletal muscle. Hypotonic muscles show a flaccid appearance and exhibit functional impairments, such as weak reflexes.

Excessive muscle tone is called hypertonia, which results in hyperreflexia (excessive reflex responses). Hypertonia often occurs due to the damage of upper motor neurons in the CNS. Hypertonia is seen in muscle rigidity or spasticity. This type of condition is seen in neurological disorders in the body like Parkinson's disease.

10.2.8 Types of Muscle Contraction

Force (tension) and length (shortening) are two important variables for description of skeletal muscle contraction. The force which is exerted by a muscle on an object during contraction is known as muscle tension, whereas the force that is exerted by an object to a muscle is known as load (weight of the object). On the basis of the force of contraction and change of muscle length, muscle contractions are of two types.

10.2.8.1 Isometric Contraction

When the tension of muscle increases but the length of the muscle remains the same, then the contraction is known as an isometric contraction (iso = same, metric = length). In this type of contraction, muscle provides force but no movement occurs at the joint and muscle length remains unchanged. Isometric contractions of muscles are very important for maintaining posture or stabilizing a joint. Examples of activities where muscles use isometric contraction include pushing an object that was initially stationary or holding a weight in a certain place above the ground.

10.2.8.2 Isotonic Contraction

When the muscle length changes but the muscle tension remains unchanged, then the contraction is known as an isotonic contraction (tonic = tension). Isotonic contraction is seen during walking, running, and different types of activities.

Based on the pattern of muscle length changes, the isotonic contraction is classified into concentric and eccentric contractions.

If the entire muscle shortens during contraction, then it is called concentric contraction. For example, during lifting a weight, the concentric contraction of the biceps muscle causes the arm to bend at the elbow and lifting the weight towards the shoulder.

If the total length of a muscle increases when tension is produced, then the contraction is called eccentric contraction.

Fig. 10.11 Multiple motor unit recruitment and stimulus intensity. Stimulating the whole nerve with higher and higher voltage produces stronger contractions. More motor units are being recruited called multiple motor unit summation

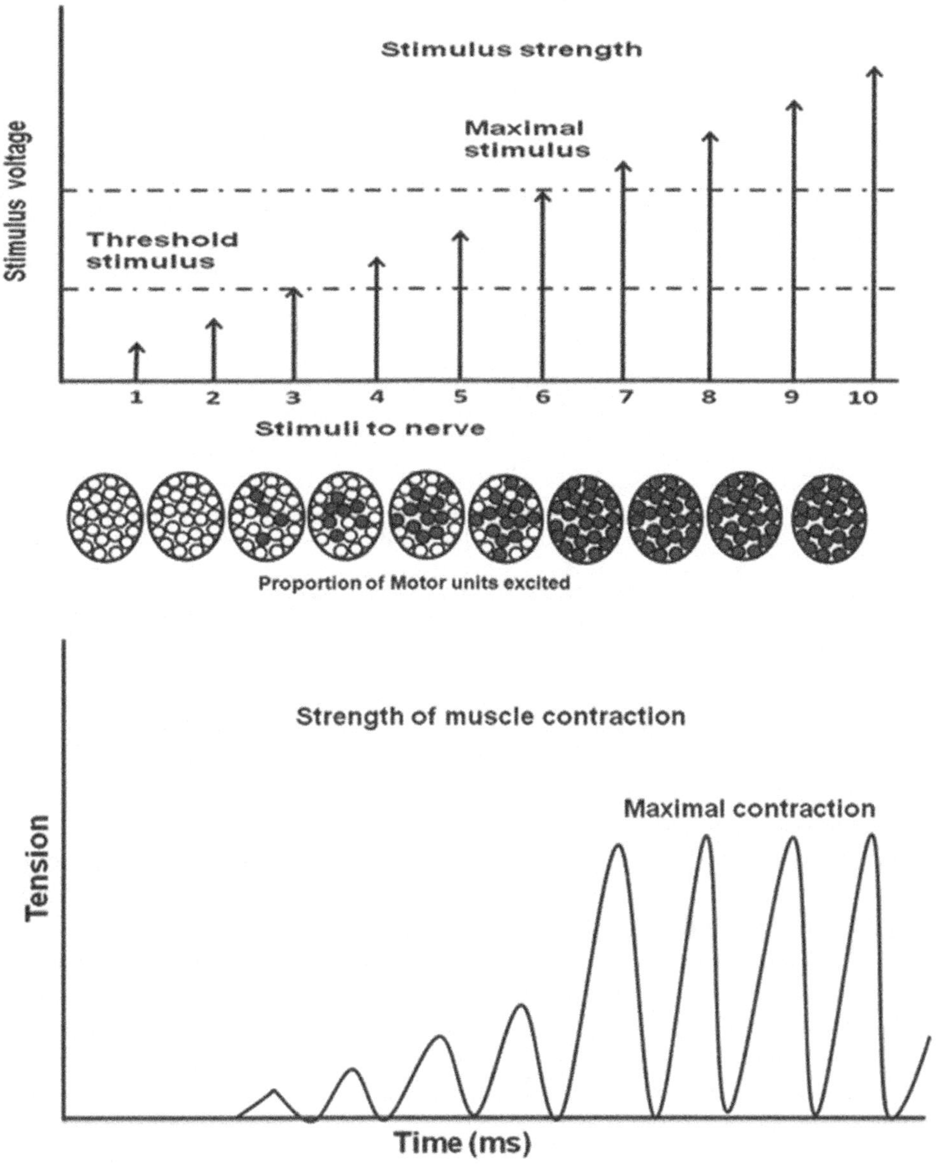

For example, the lowering phase of a biceps curl shows an eccentric contraction. In eccentric contraction, muscles are able to generate greater forces than in isometric or concentric contractions.

10.2.9 Muscle Twitch

Muscle contraction due to a single action potential is known as twitch contraction (Fig. 10.12). When a single action potential travels through the motor neuron and reaches the muscle fiber of that unit, it initiates the contraction of that single muscle fiber. This isolated contraction is known as muscle twitch. The duration of muscle twitch depends on the type of the muscle, and it can last for a few milliseconds to 100 ms.

Myogram is the graphical representation of the phenomena of muscular contractions (Fig. 10.13). A single muscle twitch has three phases, i.e., latent period or lag phase, contraction phase, and relaxation phase. The first phase is called latent period or lag phase (1–2 ms), which is the short time between the application of stimulus and starting of muscle contraction. During this period, propagation of action potential in the sarcolemma occurs and Ca^{2+} ions are released from the SR for binding with the troponin C. Then tropomyosin moves from its position, myosin head is attached with the actin, crossbridges are formed, and as a result shortening of the muscle fiber occurs. The last phase of twitch is the relaxation phase.

During relaxation phase, muscle contraction stops, Ca^{2+} ions are pumped back to the SR by calcium pump, and muscle returns back to its original resting length. The

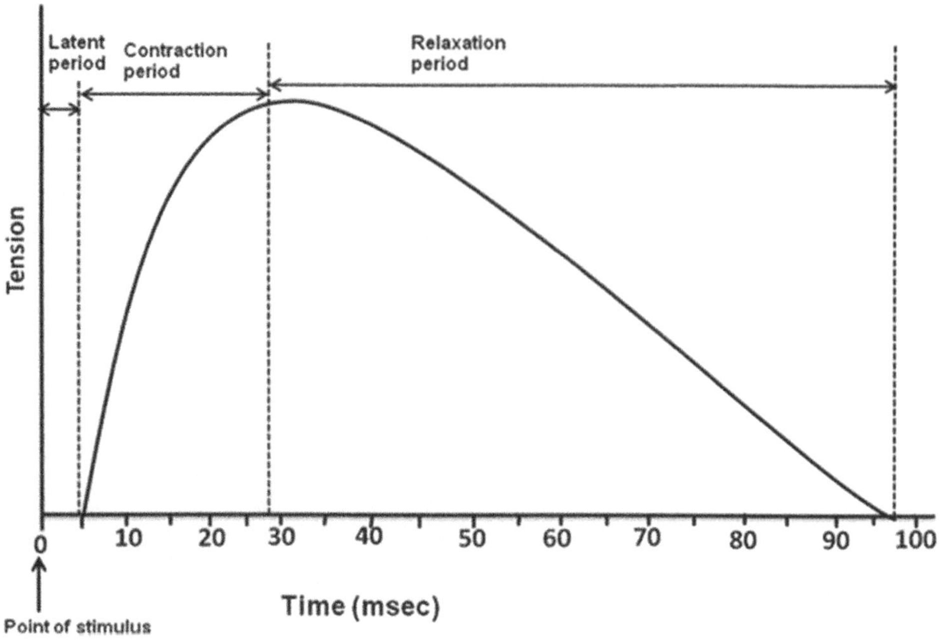

Fig. 10.12 A myogram of a muscle twitch. A single muscle twitch has three phases, i.e., a latent period between the point of stimulus and the starting of contraction, a contraction phase when tension increases, and a relaxation phase when tension decreases

duration of twitch varies between different types of muscle and ranges from 10 to 100 ms.

The **refractory period** is the time immediately after application of a stimulus. If a stimulus is applied during the contraction stage of the muscle, then muscle will not respond to this second stimulus.

10.2.9.1 Factors Influencing Force of Muscle Contraction

Based on the type of work, muscles can generate different levels of force during contraction. Some actions need much more force generation, whereas some work requires less force like lifting a heavy load requires more force compared to lifting a light object.

10.2.10 Multiple Motor Unit Summation or Recruitment

Different ranges of motor units are prudent in a skeletal muscle, and nervous system has a wide range of control over the muscle (Fig. 10.13). Small motor units are innervated by smaller motor neurons with lower threshold. These motor units generate relatively small degree of contractile strength (tension).

Larger motor units are also present with bigger motor neurons having higher threshold. These neurons activate larger muscle fibers and are used when more strength is required. So, increased activation of motor units results in increase in muscle contraction, which is known as recruitment. Motor unit summation is the recruitment of extra motor units within a muscle to develop additional force. The summation of motor units occurs until sufficient force is developed by recruitment of more numbers of motor units within that muscle to move a load. The maximum contraction is generated when all the motor units within a muscle are activated.

The muscle contraction becomes progressively stronger due to recruitment of more number of muscle fibers. In some skeletal muscles, the largest motor units can generate a contractile force of 50 times greater than the smallest motor units in that muscle. The greater the load an animal is carrying, the more number of motor units are activated.

However, at the time of generation of the maximum force, animals are only able to use about 1/3 of total motor units at one time. All muscle fibers do not fire at the same time, which helps in the generation of maximum force and prevents the muscles from fatigue. When muscle fibers begin to fatigue, they are replaced by other fibers, resulting in maintenance of the force. However, under extreme conditions, animals are able to recruit even more motor units at a time to perform a heavy work.

10.2.10.1 Wave Summation

In a muscle fiber, the tension depends on the rate of firing action potential by a motor neuron to that muscle. If the muscle is stimulated before the end of previous twitch, the second twitch will be stronger, and this phenomenon is called wave summation (Fig. 10.14). Wave summation occurs when a given set of muscle fibers is stimulated repeatedly without complete relaxation. The second stimulus causes the release of more number of Ca^{2+} ions from the SR. These Ca^{2+} ions are utilized for the activation of additional sarcomeres while

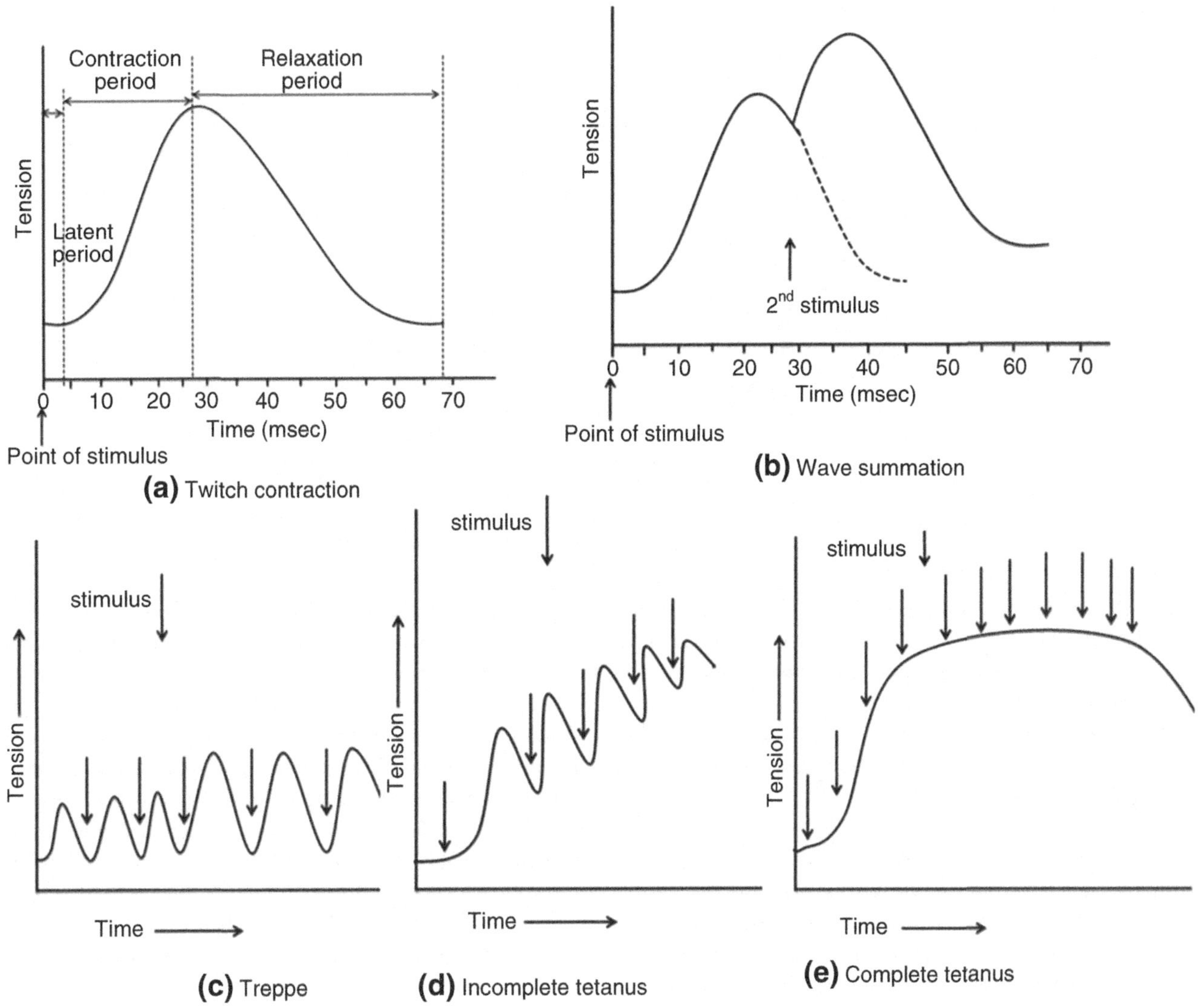

Fig. 10.13 Multiple motor unit summation or recruitment. (**a**) Twitch contraction, (**b**) wave summation, (**c**) treppe, (**d**) incomplete tetanus, (**e**) complete tetanus

the muscle is still contracting from the first stimulus. So, summation results in greater contraction of the motor unit.

10.2.10.2 Treppe

When a muscle is stimulated with repeated stimuli and the stimuli are given just after the completion of the previous contraction, then the tension of the muscle increases in a graded manner till a maximum height is reached, which looks like a staircase (Fig. 10.14). This phenomenon is known as treppe or staircase effect. The frequency of stimuli should be just below the tetanizing frequency.

In this condition, due to a steady stream of signals from the motor neuron, the concentration of Ca^{2+} ions in the sarcoplasm becomes very high.

10.2.10.3 Incomplete Tetanus

If a muscle is given repeated stimuli during contraction phase, then the contraction mechanism will start repeatedly before any relaxation has occurred. Increasing the frequency of motor neuron signaling increases summation, and tension in the motor unit keeps on rising until it reaches a peak. The tension at this time is several times more than the tension of a

Fig. 10.14 A schematic depicting the sarcomere length-tension relationship

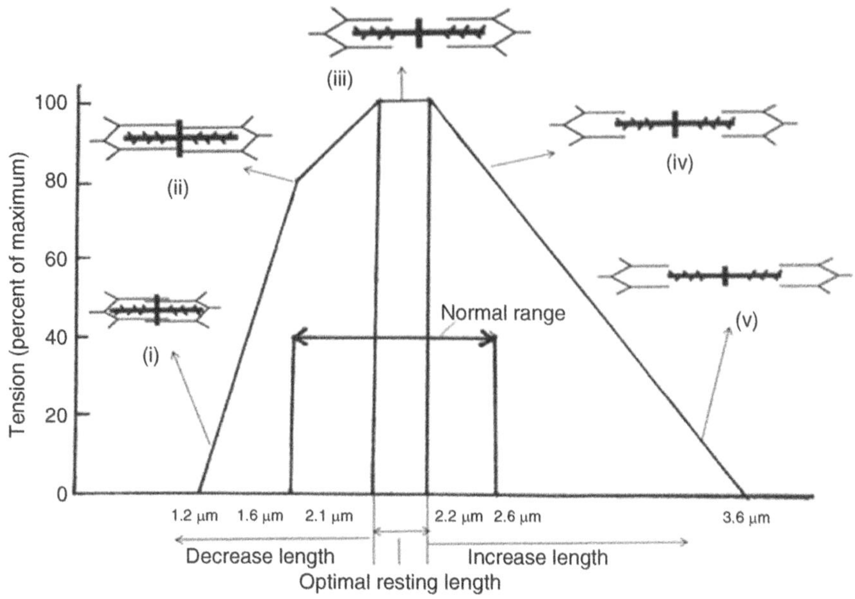

single muscle twitch. This state of muscle is called incomplete or unfused tetanus. During incomplete or unfused tetanus, the muscle goes through quick cycles of contraction with a short relaxation phase for each contraction.

Incomplete tetanus occurs due to repeated stimulus, when there are phases of incomplete relaxation between the summated stimuli (Fig. 10.14).

10.2.10.4 Complete Tetanus

If the frequency of the stimuli is very high and the muscle will not get time for relaxation, then the phenomenon is called complete tetanus (Fig. 10.14). In complete tetanus, the relaxation phases are absent and the contractions become continuous. In complete tetanus, the individual responses fuse and form one continuous contraction. Tetany is the sustained contraction resulting from high-frequency stimulation.

During tetanus, the concentration of Ca^{2+} ion remains very high in the sarcoplasm and that allows nearly all of the sarcomeres to form crossbridges and shorten, and the contraction continues uninterrupted (until the muscle becomes fatigue and is not able to produce tension).

10.2.11 Length-Tension and Force-Velocity Relationship

10.2.11.1 Sarcomere Length-Tension Relationship

A direct relationship is there between the initial length of muscle fibers and the tension in the muscle or force of contraction.

The initial length of the sarcomere influences the force of the contraction, which a muscle can generate (Fig. 10.14). If the sarcomere length is optimum, the isometric tension is maximum due to the position of thin and thick filaments forming maximum number of crossbridges in sarcomere. If the initial sarcomere length is very short, then the thick filaments will already be pushing up against the Z-disc. In this situation, there is no chance of further shortening of the sarcomere as the latter is already short and muscle will not be able to generate much force.

Similarly, if the muscle is stretched very high, then the myosin heads can no longer be able to contact the actin and less force will be generated.

So, maximum force is produced if the muscle is stretched to the point which allows every myosin head to contact with the actin and when the sarcomere has the maximum distance to shorten, i.e., the thick filaments are at the very ends of the thin filaments. This is applicable only in isometric contraction. During dynamic contraction, length-tension relationship must be combined with force-velocity relationship to determine the effect that both length and velocity have on muscle tension.

10.2.11.1.1 Application of Length-Tension Relationship

When applying to muscle joint system, sarcomere length is not same throughout. So, at a particular joint position, there are sarcomeres at many different lengths corresponding to different points of length-tension relationship. During movement, torque produced at joint is not only due to muscle force but also due to function of moment arm (MA) of muscle. So,

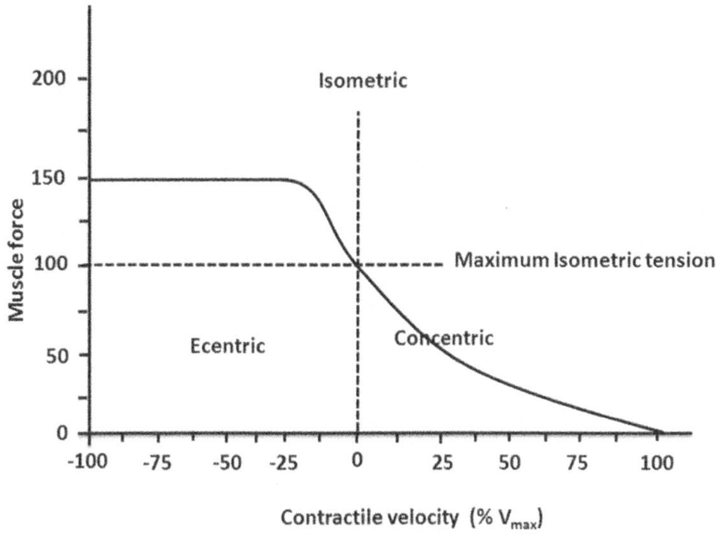

Fig. 10.15 A schematic depicting muscle force-velocity curve

at particular joint position, muscle length may be short but has long MA, maintaining higher torque.

10.2.11.2 Force-Velocity Relationship

The speed of shortening of myofilaments also affects the tension development. Speed of shortening depends on the type and length of muscle fiber. Force-velocity relationship describes the relation between the velocity of muscle contraction and the force produced (concentric and eccentric muscle contraction) (Fig. 10.15). The force which is generated during muscle contraction is the function of velocity of contraction.

For example, in concentric contraction, if speed decreases, the tension increases.

In isometric contraction, the shortening speed is 0, but the tension reduced is more than concentric contraction. In eccentric contraction, as the lengthening speed increases, the tension increases and then plateaus.

10.2.12 Skeletal Muscle Energetics and Metabolism

Muscle contractions require plenty of energy. The major portion of this energy is utilized for the crossbridge cycles, and some portion is also utilized for propelling the Ca^{2+} ions back into the SR from the sarcoplasm during relaxation of the muscle and propelling Na^+ and K^+ ions through the sarcolemma.

ATP is the instant source of energy ($ATP \rightarrow ADP + Pi + energy$) for muscle contraction. Continuous supply of ATP is required for muscle contraction. For muscle contractions, there are four different ways through which muscles get the ATP.

1. ***Cytosolic stored ATP***: Very little amount of ATP remains inside the muscle fiber as stored ATP. This cytosolic stored ATP can instantly provide energy for contraction and does not require oxygen. Very little amount of ATP is stored in muscle fibers, which can provide energy for muscle contraction for a few seconds. So, it is not enough for long-term contraction. This cytosolic ATP provides energy for contraction of eye muscles which contract constantly and quickly but for a very little period.
2. ***Creatine phosphate***: Muscles cannot obtain ATP from the blood or other tissues. They can produce it as per need. ADP (2 molecule), inorganic phosphate (Pi), and energy from other chemical sources are required to generate a single molecule of ATP by rephosphorylation of ADP.
   When the cytosolic stores of ATP are utilized, muscle fiber initiates another rapid energy source, i.e., creatine phosphate (CP). Creatine phosphate is a high-energy compound, which can rapidly transfer its phosphate to an ADP molecule for synthesis of one molecule of ATP (Fig. 10.16). The process is called phosphagen system, and it does not require oxygen. Creatine kinase or creatine phosphokinase (CPK) enzyme catalyzes the reaction. Creatine kinase enzyme is present on the M-line of the muscle fiber.
   But the energy available from the stored creatine phosphate is also limited, which is sufficient for another 5–8 s. This source is also termed as the immediate energy source and is a very important source of energy for activities like jumping, hitting, and throwing.

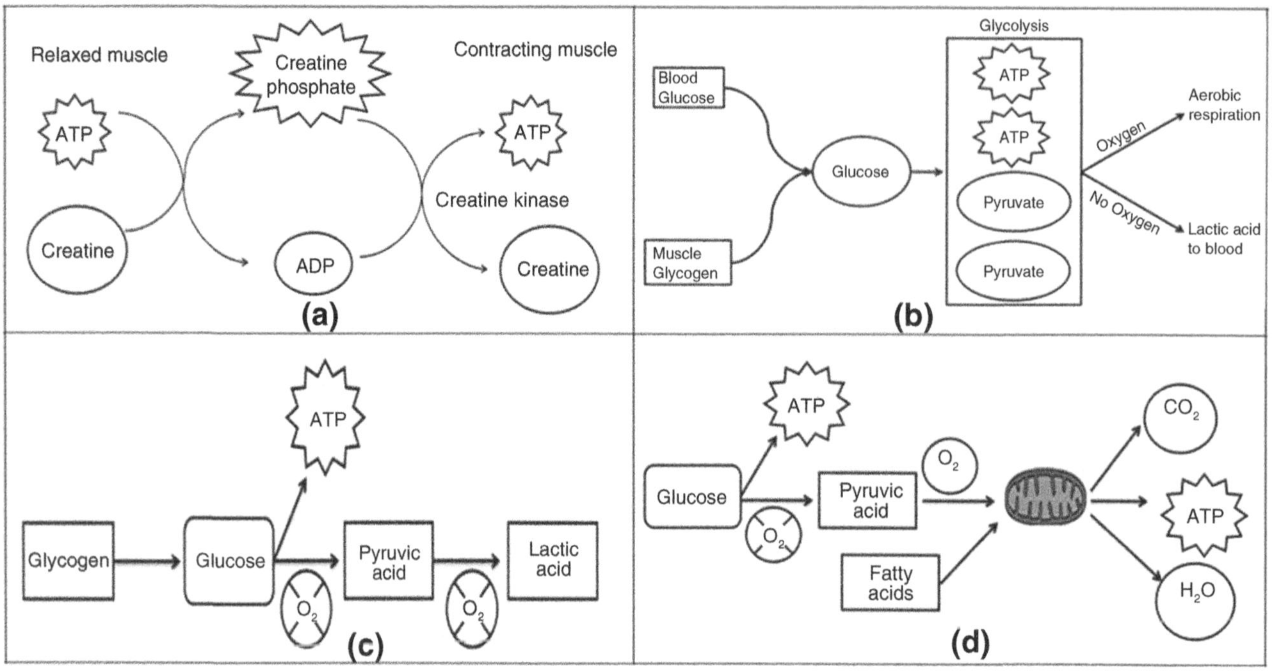

Fig. 10.16 Sources of energy for muscle contraction. (**a**) Molecular pathway of creatine phosphate synthesis. At the time of muscle contraction when muscles require ATP, the reaction is represented by the above equation that runs from left to right. When the muscle is at rest and excess ATP is available, the reaction is represented by the above equation that proceeds from right to left. (**b**) Glycolysis pathway. (**c**) Anaerobic mechanism (glycolysis and lactic acid formation). (**d**) Aerobic or oxidative respiration

Know More......

Creatine phosphate or phosphagen system is rapidly replenished during recovery. It requires about 30 s to replenish about 70% of the phosphagens and 3–5 min to replenish 100%. During intermittent work (short periods of activity followed by rest periods), much of the phosphagen can be replenished during the recovery period and thus be used over and over again.

3. ***Glycolysis***: Glycolysis is the metabolic pathway which breaks down glucose into pyruvate and a hydrogen ion (H^+). Glucose molecules reach the muscle tissues through blood. Glucose is also produced in muscle cells through breakdown of stored glycogen (Fig. 10.16). The glycolysis in skeletal muscle generates ATP molecules required for contraction. Glycolysis which occurs in the absence of oxygen is called anaerobic glycolysis, which is the main source of ATP during anaerobic activity. Glycolysis process is very fast and can supply energy for intensive muscular activity, but it can supply energy for about a minute before the muscles begin to fatigue.
   Glycolysis occurs in the cell cytosol and can produce molecules 2 ATP by converting a glucose molecule into pyruvate. In anaerobic glycolysis, the pyruvic acid is converted to lactic acid. The accumulation of lactic acid reduces the pH and makes it more acidic, which produces the stinging feeling in muscles during exercise. This slows down further anaerobic respiration and brings fatigue. Now when the activity of the muscle slows down, oxygen becomes available and lactate is converted back into pyruvate.
4. ***Aerobic or oxidative respiration***: The different mechanisms discussed above are able to supply ATP for about a minute.
   But the different activities like walking and running which continue for a long duration require a constant supply of ATP. So, in these activities where constant ATP is required, the cells utilize aerobic or oxidative respiration occurring in the mitochondria (Fig. 10.16). This aerobic respiration can supply adequate ATP to the muscle cells for hours, but this process of metabolism is slower than anaerobic mechanisms and is not fast enough for intense activity.
   It is an important source of energy for muscle contractions of athlete animal and for endurance exercise required of migrating animals, where repetitive skeletal muscle contractions continue for hours or days. Though glucose can be used as an energy source for aerobic respiration, the primary fuel for muscle contractions during prolonged endurance exercise is fatty acids rather

than glucose. These fatty acids are broken down to acetyl-CoA and enter the citric acid cycle and produce ATP.

Muscle contractions are 50–70% efficient in regard to completion of a work, and the nonwork portion is dissipated as heat. This heat source is very important for the maintenance of body temperature. During cold stress, shivering results in the production of heat in the body.

10.2.13 Muscle Fatigue

If a muscle is used exhaustively, then the performance of the muscle decreases progressively, which mostly recovers after a period of rest. This phenomenon is known as muscle fatigue. This process is temporary and reversible state of muscle. When a muscle is contracted for a long time, then muscle fatigue occurs. If fatigue starts in a muscle, then the force of the muscle decreases and the response of that muscle to stimuli reduces and as a result the activity levels decrease. In fatigue condition, the muscle is unable to contract optimally. Muscle fatigue starts at the time of heavy muscular activity or exercise. As the muscle starts fatigue, the speed and force of contraction reduce, relaxation time prolongs, and a period of rest is required to restore normal function. The factors which influence the muscle fatigue are the following:

1. ***Ionic imbalance within the muscle***: Muscle contraction requires Ca^{2+} ions to interact with troponin for exposing the actin-binding site to myosin head. If deficiency of Ca^{2+} occurs in the body, then muscles do not get the required Ca^{2+} ions for contraction.
2. ***Nervous fatigue and loss of desire***: The contraction of a muscle is controlled by nerves. In central fatigue or "psychological fatigue," brain feels tired. This leads to fatigue.
3. ***Metabolic fatigue***: Depletion of ATP or glycogen results in fatigue due to unavailability of energy. Metabolites like Mg^{2+} ions induce fatigue by inhibiting the release of Ca^{2+} ions or reducing sensitivity of troponin to Ca^{2+} ions.
4. ***Exercise and aging***: With aging of animal, the levels of ATP, CTP, and myoglobin decrease and it reduces the muscle's ability to function. Training and exercise increase the metabolic capacity of a muscle, which delays the onset of muscle fatigue.
5. ***Lactic acid accumulation***: Lactic acid is the by-product of anaerobic respiration, which strongly contributes to muscle fatigue.

10.2.13.1 Effects of Muscle Fatigue

Muscle fatigue causes different types of sign and symptoms in the body like muscle pain, burning, fast breathing, vomiting, and stomach pain.

10.2.13.2 Types of Muscle Fatigue

There are two types of fatigue, i.e., central fatigue and peripheral fatigue:

Central fatigue—It occurs due to the decrease in the capacity to voluntarily activate a muscle during a maximal effort. It occurs due to a decrease in motor unit recruitment levels or a reduction in motor unit firing rates or both.

Peripheral fatigue—It occurs due to the decrease in the capacity of a muscle to produce force even if it is receiving signals from the nervous system.

10.2.14 Types of Skeletal Muscle Fiber

Skeletal muscle fibers can broadly be classified into type I or slow-twitch fibers and type II or fast-twitch fibers on the basis of their contraction speed and fatigue resistance. The relative proportions of these types of muscle fibers are basically determined by genetic factors and influenced by physiological, hormonal, and nutritional factors.

1. ***Slow-twitch muscle fibers***: The slow-twitch muscle fibers are also known as type I muscle fibers. They have myoglobin content and are red in color. Their contraction speed is less than the other types of muscle fibers. The muscle fibers are smaller and produce tension slowly. Their capacity to produce force or power is less. But the main advantage is that these types of muscle fibers are slow to fatigue. They have high myosin ATPase activities, capillary density, and mitochondrial density and have low power. Slow-twitch fibers depend on oxygen for energy, and they can continue the activity for long duration. These types of muscle fibers are generally associated with endurance activities, and highly active animals or athlete animals have higher proportion of these types of muscle fibers in the body.
2. ***Fast-twitch muscle fibers***: Fast-twitch muscle fibers are also known as type II muscle fibers. The speed of contraction is faster than the type I muscle fibers. The duration of contraction is short. Type II muscle fibers are more powerful than type I muscle fibers and are associated with activities such as lifting a heavy weight, which requires more power. This type of fibers gives major strength to the

animal, but they become fatigue very quickly. These fast-twitch fibers can be further classified into fast-twitch oxidative-glycolytic fibers (type IIa) and fast-twitch glycolytic fibers (type IIb) in rodents and pigs. In large mammals including humans and ruminants, type IIx replaces type IIb as the dominant fast-twitch fibers.

***Type IIa muscle fibers*—Type IIa muscle fibers** depend on oxidative glycolysis for energy and produce lactic acid. But the duration of contraction is very short. Animals that are associated with powerful activities have higher proportions of type IIa fibers in their muscles.

Fast-twitch glycolytic (type IIx) fibers—Type IIx muscle fibers are faster and more powerful than type IIa muscle fibers. They also fatigue very quickly. They are associated with the activities of very short duration and which require more power.

In cheetahs and domestic cats, several muscles have high proportion of type IIx fibers and low proportion of type I fibers. In beagle dogs, the percentage of type IIa fibers is very high (Table 10.3).

10.2.15 Rigor Mortis

The word rigor mortis came from two Latin wards, i.e., "rigor" means "stiffness" and "mortis" means "of death." Rigor mortis or postmortem rigidity of muscles is an important sign of death of an animal. It starts a few hours after death. After death of the animal, the muscles become rigid, it is difficult to move or manipulate the body, and the state is irreversible. In the muscle, continuous actin-myosin interaction is seen in rigor mortis. After death of the animal, cellular respiration stops and results in stop of synthesis of adenosine triphosphate (ATP). So, ATP is not available for relaxation of muscle and muscle remains in contraction state.

10.2.15.1 Mechanism of Rigor Mortis

1. Absence of ATP → no reuptake of Ca^{2+} into the SR as Ca^{2+} uptake also requires ATP-dependent Ca^{2+} pump → Ca^{2+} level of sarcoplasm ↑ → continued binding of Ca^{2+} to troponin C → abnormal, rigid, and uninterrupted contraction.
2. No ATP → no relaxation, a new molecule of ATP must attach to the myosin head for detachment of actin-myosin interaction → thus, when no ATP is present, then myosin heads cannot detach themselves from actin.

In humans, rigor mortis starts after about 3–4 h after death. It reaches maximum stiffness after about 12 h and gradually dissipates until approximately 48–60 h (3 days) after death. The onset of rigor mortis depends on the ambient temperature. The warm conditions can speed up the process of rigor mortis. Rigor mortis ends when contractile proteins of the muscle like other body tissues undergo autolysis caused by enzymes released by lysosomes.

10.2.15.2 Factors Affecting Rigor Mortis

Ambient temperature: Cold temperature inhibits rigor mortis and the onset of rigor becomes slower, whereas hot temperature accelerates the process and faster onset and faster progression of rigor mortis occur.

Activity before death: Anaerobic exercise before death accelerates the rigor mortis process because lack of oxygen to muscle buildup of lactic acid and higher body temperature accelerate rigor. Sleep before death slows the process as fully oxygenated muscles exhibit rigor more slowly.

Body mass: In obese animals, the rigor mortis process is slow because fats store oxygen. In thin animals, the process is fast as the body loses oxygen quickly.

Table 10.3 Comparison between three types of muscle fibers

Characteristics	Type I	Type IIA	Type IIX
Myosin ATPase activity	Slow	Fast	Fast
Fiber length	Small	Medium	Large
Duration of contraction	Long	Short	Short
Fatigue	Slow	Quick	Very quick
Energy utilization	Aerobic/oxidative	Both	Anerobic/glycolytic
Capillary density	High	Medium	Low
Availability mitochondria	High numbers	Medium numbers	Low numbers
Color of the fiber	Red (contain myoglobin)	Red (contain myoglobin)	White (no myoglobin)
Force production	Low	High	Very high

10.3 Smooth Muscle and Cardiac Muscle

10.3.1 Smooth Muscle

Smooth muscle is involuntary and nonstriated. Smooth muscle is present mainly in the visceral organs, so this type of muscle is also known as visceral muscle. They are found in GI tract, urinary tract, blood vessels, airways, different glands, inside eye, etc.

Smooth muscle helps in many vital functions in the body like passage of food bolus in the digestive tract through peristalsis, elimination of execratory products through urinary system, and regulation of blood flow through the blood vessels. Smooth muscle is under the control of autonomic nervous system and endocrine system. Muscle cells are small and spindle shaped with one centrally located nucleus. No neuromuscular junction is present, and instead of that, varicosities help in the transmission of nerve impulse to the cells. The contraction and relaxation are slower than the skeletal muscle. In some organs, smooth muscles have pacemaker cells. Less energy is required for their contraction, and they become fatigue slowly.

10.3.1.1 Cellular Structure

Smooth muscle fibers are small and spindle shaped with one centrally located nucleus (Fig. 10.17). The average diameter of fibers is 2–10 mm. The elasticity is more in smooth muscle than striated muscle. Elasticity is very important for visceral organs like urinary bladder. In smooth muscle, myofibrils are absent and thick and thin filaments are not arranged in sarcomere pattern; that is why they are nonstriated.

Smooth muscle fibers contain three types of filaments, i.e., thick myosin filaments, thin actin filaments, and intermediate filaments. Thick myosin filaments are longer in smooth muscle than skeletal muscle. Thin filaments actin and tropomyosin are present, but troponin is absent. During contraction, calcium ions attach with calmodulin instead of troponin. The intermediate filaments do not directly participate in contraction, and they only form part of cytoskeletal framework that supports cell shape. Dense bodies are button shaped and present throughout the cell. Dense bodies contain the same protein found in Z-lines. Actin filaments are attached with these dense bodies.

Sarcoplasmic reticulum stores Ca^{2+} ions, which are essential for contraction. Cells are usually arranged in sheets within muscle and organized into two layers (longitudinal and circular) of closely apposed fibers, which have essentially the same contractile mechanisms as skeletal muscle.

10.3.1.1.1 Structural Differences: Smooth Muscle with Striated Muscle

Smooth muscle is nonstriated, and myofibrils and sarcomeres are absent. Thick and thin filaments are not arranged like skeletal muscle. Thick filaments are scattered throughout sarcoplasm, whereas thin filaments are attached to dense bodies. T-tubule is absent. Loose network of sarcoplasmic reticulum is present in the cytoplasm. Troponin is not present in smooth muscle, and instead of troponin C calcium ions attach with calmodulin during contraction. In single-unit smooth muscle, muscle cells are attached with one another with gap junctions. Gap junction is an electrical junction, which helps in the transmission of nerve impulse from one cell to another. In smooth muscle, connective tissues never unite to form tendon. Thick and thin filaments are arranged in slightly diagonal chains, which are attached to the plasma membrane or dense bodies. During contraction, when action potential reaches the cells, thick and thin filaments slide past each other. In smooth muscle, neuromuscular junction is absent. Instead of neuromuscular junction, varicosities help in the transmission of nerve impulse to the smooth muscle cells.

10.3.1.2 Types of Smooth Muscle

Smooth muscle can be broadly classified into single-unit smooth muscle and multiunit smooth muscle.

1. ***Single-unit smooth muscle***: In single-unit smooth muscle, muscle cells are connected to each other through gap junctions (Fig. 10.18).
   Through these gap junctions, action potential transmits from one cell to another. The cells are stimulated in a

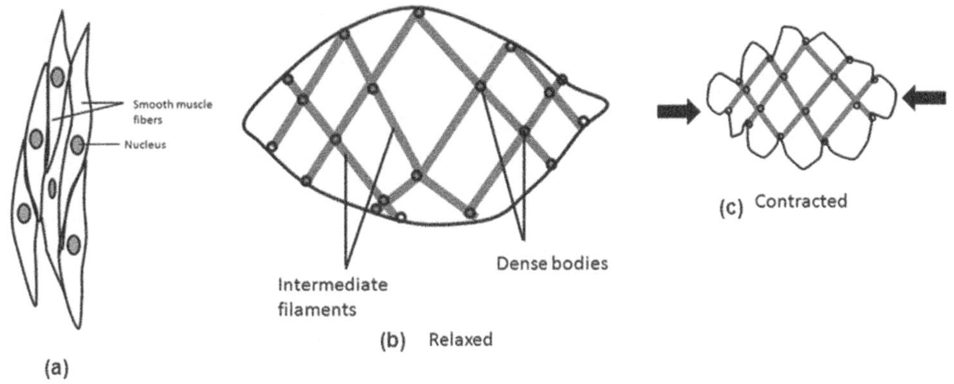

Fig. 10.17 Smooth muscle tissue. (**a**) The cells are spindle shaped with a centrally located nucleus. Anatomy of a relaxed (**b**) and contracted (**c**) smooth muscle cell

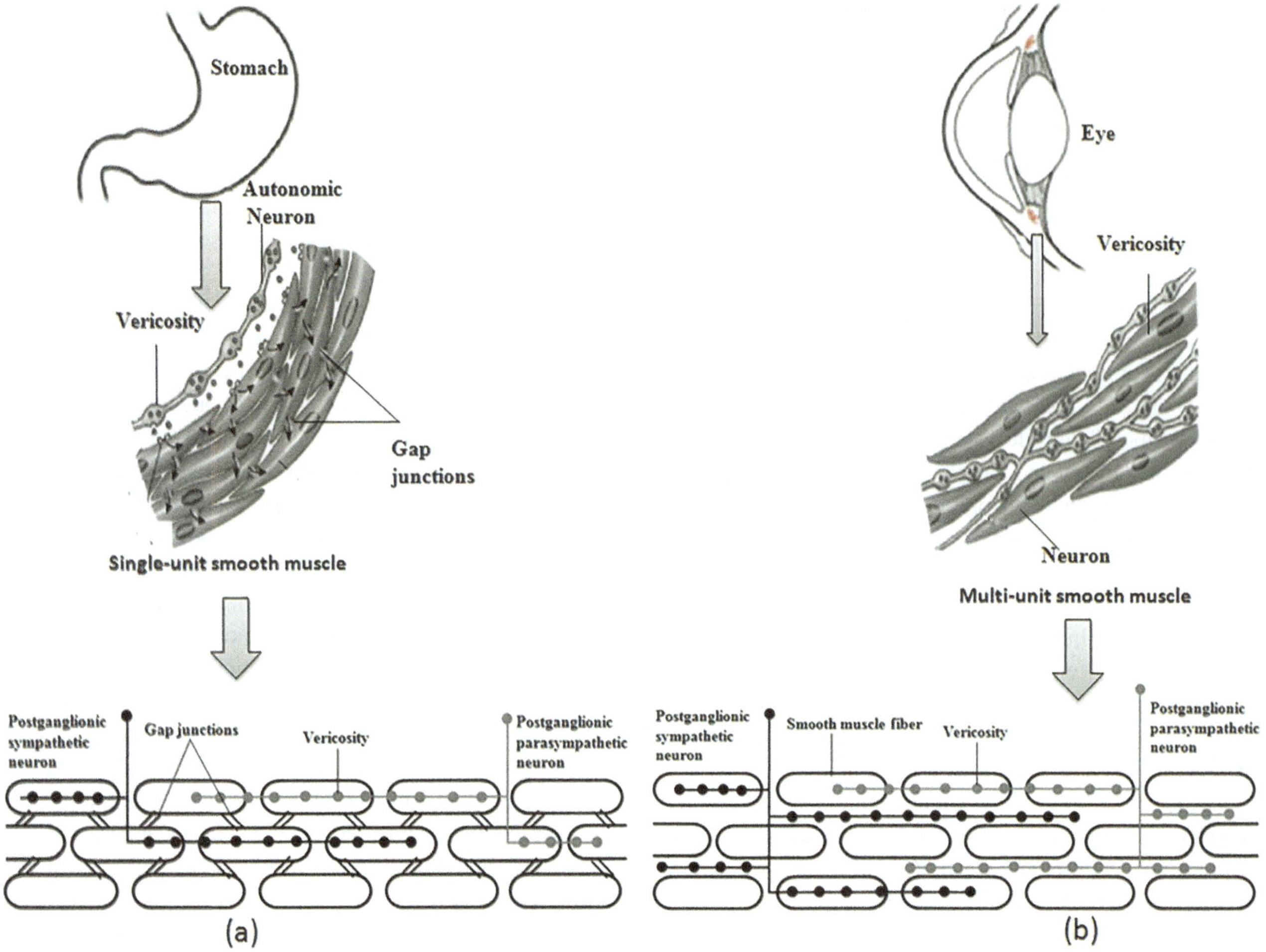

Fig. 10.18 Single-unit smooth muscle and multiunit smooth muscle. (**a**) Innervations of a single-unit smooth muscle through varicosities. (**b**) Innervations of multiunit smooth muscle through varicosities

synchronous pattern from only one synaptic input, and action potential spreads to all the cells and contraction of all the cells occurs at a time like a functional syncytium. Single-unit smooth muscle is myogenic with pacemaker potentials.

2. ***Multiunit smooth muscle***: In multiunit smooth muscle, each cell receives its own synaptic input through single varicosity (Fig. 10.18). This allows for the multiunit smooth muscle to have a much finer control. No gap junction is present. So, cells are not electrically connected and hence selective activation of muscle fibers occurs. Multiunit smooth muscle is neurogenic.

10.3.1.3 Function of Smooth Muscle

Smooth muscle is involved in the movement of different visceral organs and glands; thus, smooth muscle serves a variety of functions in the body.

The basic functions of smooth muscle are the following:

1. Smooth muscle in gastrointestinal tract helps in the movement of the food bolus through peristalsis.
2. The smooth muscles in blood vessels regulate the blood flow and blood pressure through vascular resistance.
3. In urinary system, smooth muscle regulates the urine flow and smooth muscle in urinary bladder regulates the micturition.
4. Smooth muscle of reproductive tract helps in gamete transport, and contraction of uterus helps in parturition.
5. The contraction of smooth muscle of air passages of respiratory tract regulates the diameter bronchiole and passage of air.
6. The contraction of integument causes piloerection and helps in shivering thermogenesis during cold stress.
7. The smooth muscles in eye regulate the dilation and constriction of the pupil and regulate the entry of light

through pupil. Smooth muscles in eye also change the shape lens as required.

10.3.1.3.1 Innervations of Smooth Muscle

Smooth muscles are innervated by postganglionic autonomic neurons. In smooth muscle, neurotransmitter remains in varicosities. When an action potential reaches the varicosity through the axon the neurotransmitter releases from the varicosities and attached with the receptors on the plasma membrane of muscle fibers. In single-unit smooth muscle, the innervation is restricted to a few fibers in the muscle and action potential is transmitted from one cell to another through gap junctions.

10.3.1.3.2 Stimuli Initiate Smooth Muscle Contraction

Different stimuli which influence the smooth muscle contraction are the following:

1. The spontaneous electrical activity in the plasma membrane of the smooth muscle fiber
2. Release of neurotransmitter by autonomic neurons
3. Different hormones
4. Some local changes in the chemical composition like paracrine agents, acidity, oxygen, osmolarity, and ion concentrations of extracellular fluid surrounding the muscle fibers
5. Stretch

10.3.1.4 Mechanism of Smooth Muscle Contraction

When an action potential reaches the sarcolemma through the neurotransmitter released from the varicosities, it causes the depolarization of membrane of smooth muscle cell (Fig. 10.19).

The depolarization of membrane or activation of neurotransmitter results in entry of Ca^{2+} ions through the L-type voltage-gated calcium channel located in the plasma membrane. This increase in Ca^{2+} ions stimulates the release of Ca^{2+} ions from sarcoplasmic reticulum by the way of ryanodine receptors and IP3.

This process is known as Ca-induced Ca release. Then the Ca^{2+} ions bind with calmodulin, which results in the activation of calmodulin. Now the activated calmodulin activates the enzyme myosin light-chain kinase (MLCK). MLCK phosphorylates the light chains in myosin heads and increases myosin ATPase activity. Then myosin binds with actin.

Now crossbridge cycling occurs, which leads to muscle tone. The ATPase activity is less in smooth muscle than in skeletal muscle. That is why the speed of contraction is slow in smooth muscle.

10.3.1.5 Mechanism of Smooth Muscle Relaxation

Smooth muscle contraction ends with the dephosphorylation of myosin light chains.

Unlike skeletal muscle, the depolarization in smooth muscle occurs during its activation. That is why simply reducing calcium ion concentration will not produce the relaxation of smooth muscle. Here, myosin light-chain phosphate is responsible for dephosphorylation of myosin light chain, which ultimately leads to relaxation of smooth muscle.

In smooth muscle, action potentials are slower and they can last for a long time. This may be due to slow opening of calcium channels. Repolarization of smooth muscle is also slow as potassium channels are also slow to react. Some smooth muscle cells act as pacemaker cells and generate action potential. These types of cells are seen in the intestines. It has been seen that in some smooth muscles, they contract without any action potential.

In multiunit smooth muscle, action potentials generally do not occur. Like in smooth muscle, iris depolarization occurs by norepinephrine and ACh, which is known as junctional potential. These neurotransmitters cause the contraction of smooth muscle. The junctional potential results in an influx of calcium through L-type channels into the cell.

Sometimes, the neurotransmitters activate a G-protein, which activates phospholipase C generating IP3. IP3 then initiates the release of calcium from the sarcoplasmic reticulum. Smooth muscle contraction is required to last for a long time. If the contraction occurs like skeletal muscle, then the energy demand will be high for this type of sustained contraction and muscle will become fatigue as intracellular ATP is depleted.

But this phenomenon does not occur because of a special mechanism known as latch state, which allows the smooth muscle to maintain high tension at low energy consumption. The smooth muscle tone remains high even if there is decrease in myosin light-chain kinase.

10.3.2 Cardiac Muscle

Cardiac muscle is only present in the heart. They are striated like skeletal muscle but involuntarily. Cardiac muscle is mainly controlled by autonomic nervous system and endocrine glands. The pacemaker cells of heart generate action potential, and the heart beats rhythmically. Heart supplies blood through the body, and it is possible because of well-organized contraction of cardiac muscle cells.

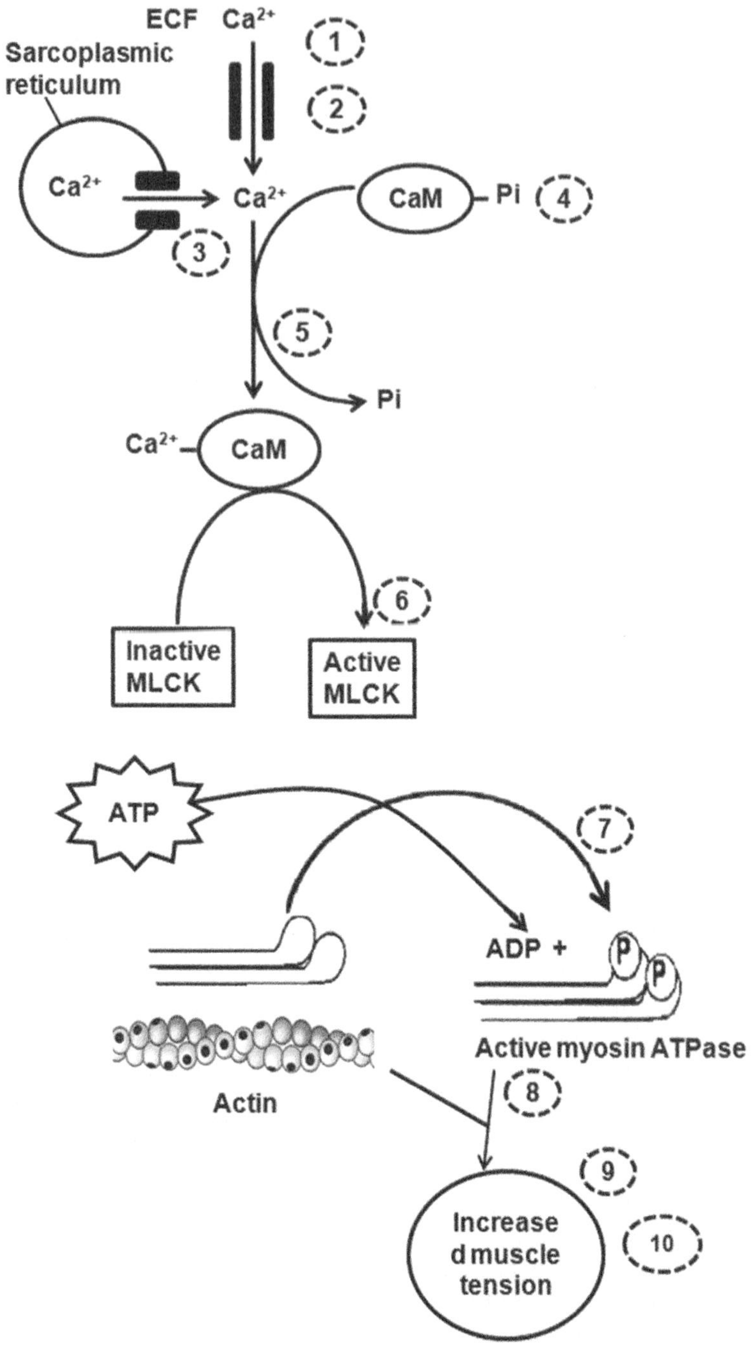

Fig. 10.19 Mechanism of smooth muscle contraction. (**1**) Depolarization of cell membrane or activation of hormone/ neurotransmitter, (**2**) opening of L-type voltage-gated calcium channels, (**3**) calcium-induced calcium release from sarcoplasmic reticulum, (**4**) increased intracellular calcium level, (**5**) calcium binds with calmodulin, (**6**) activation of myosin light-chain kinase (MLCK), (**7**) phosphorylation of myosin light chain, (**8**) increase in myosin ATPase activity, (**9**) myosin-P binds actin, (**10**) crossbridge cycling leads to muscle tone

10.3.2.1 Basic Organization of Cardiac Muscle Cells

Like skeletal muscle, cardiac muscle is also striated, and the organizations of sarcomeres are also similar with actin and myosin filaments (Fig. 10.20). But some differences are also there in the structures and arrangement of the muscle fibers, which allow their coordinated function. The cells of cardiac muscle are smaller than the cells of skeletal muscle and exhibit branching. The cells are interconnected to each other by intercalated discs and form functional syncytia. Intercalated disc has two types of membrane junctions, i.e., desmosomes and gap junctions. Desmosomes are mechanical junctions between the cardiac muscle cells. Gap junctions are electrical junctions which connect one cell to another and allow the propagation of action potential between cells. Action potential is generated from the cardiac cells, and the electrical impulse spreads from one cell to another through the gap junction. So, all the cardiac cells become excited at a time and contract as a single syncytium. The thick and thin filaments are arranged like that of skeletal muscle, which gives a striated appearance. Repeated dark and light bands, i.e., A bands and I bands, are seen when viewed under

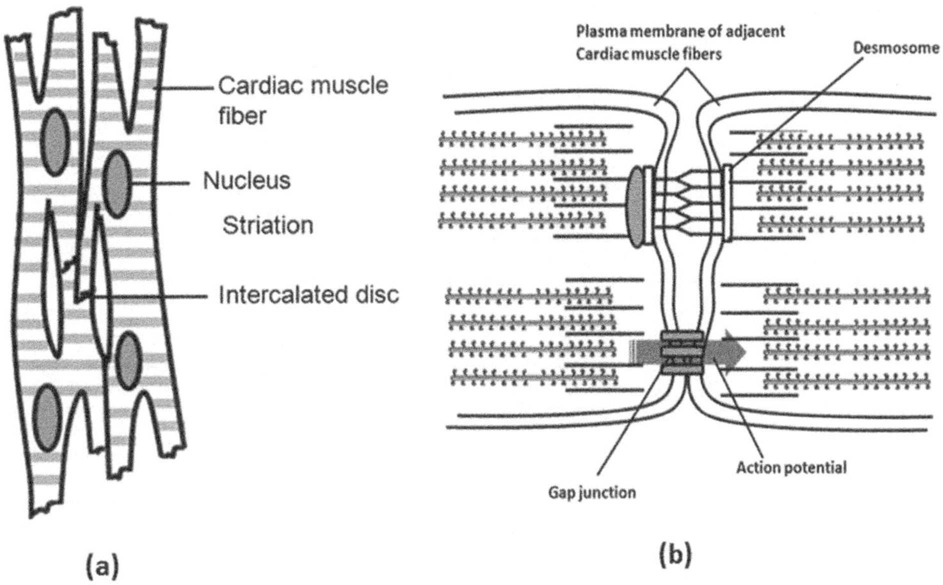

Fig. 10.20 Cardiac muscle tissue. (**a**) Cardiac muscle cells are branched and are interconnected to each other by intercalated discs. (**b**) Intercalated disc

electron microscope. The Z-lines are present at the lateral border of the sarcomere. Thin filaments are composed of actin, troponin, and tropomyosin. Thick filaments are made up of myosins, which extend from the center of the sarcomere towards the Z-lines. The amount of connective tissue is more in cardiac muscle than in the skeletal muscle. This high amount of connective tissue prevents not only muscle rupture, but also overstretching of the heart.

10.3.2.2 Mechanism of Contraction

Like that of skeletal muscle, cardiac muscle contraction is thin filament regulated, with an elevation in intracellular Ca^{2+} required to promote actin-myosin interaction (Fig. 10.21). Action potential transmits through the plasma membrane of cardiac contractile cells. Then it travels down to T-tubule. The action potential causes opening of plasma membrane L-type Ca^{2+} channel in the T-tubules. Ca^{2+} enters cytosol from T-tubules. Increase of cytosolic Ca^{2+} concentration leads to release of large amount of Ca^{2+} from SR through ryanodine release channels. This process is called Ca^{2+}-induced Ca^{2+} release. Increase in cytosolic Ca^{2+} causes binding of Ca^{2+} to troponin C.

This binding of Ca^{2+} with troponin C results in a conformational change in the troponin-tropomyosin complex, tropomyosin is removed from its position, and the active site on actin becomes exposed. Now myosin head binds with actin and crossbridge formation occurs like skeletal muscle. Thin filaments slide inward between thick filaments.

Cardiac muscle can regulate the rise in intracellular Ca^{2+} ions, and by this process, force of contraction is regulated. In heart, all the muscle cells are activated during contraction, so recruiting more muscle cells is possible.

Moreover, tetany of cardiac muscle cells would prevent any pumping action and thus be fatal. Consequently, the heart relies on different means of increasing the force of contraction, including varying the amplitude of the intracellular transient Ca^{2+}.

10.3.2.3 Mechanism of Relaxation

In cardiac muscle, relaxation starts due to re-accumulation of Ca^{2+} by the SR through the action of the SR Ca^{2+} pump (SERCA). It plays an important role in the relaxation process and decreases the cytosolic Ca^{2+}, but the process is more complex in cardiac muscle.

The refractory period in cardiac muscle is long, and the plateau phase is also prolonged. Because of these reasons, cardiac muscle never tetanizes.

10.3.2.4 Cardiac Muscle Metabolism

Like skeletal muscle, myosin uses energy in the form of ATP during contraction. So, the ATP pool is continuously replenished. The source of this ATP is aerobic metabolism, including the oxidation of fats and carbohydrates. During ischemic condition, the creatine phosphate pool, which converts ADP to ATP, may decrease and like skeletal muscle the creatine phosphate pool is small.

10.3.2.5 Cardiac Muscle Hypertrophy

Regular exercise such as regular running causes hypertrophy of individual cardiac muscle cells, which ultimately results in increased heart size.

This is an example of "physiological hypertrophy," and it is beneficial for the animal. In contrast to this, pathological hypertrophy is also seen. If heart is remaining in constant

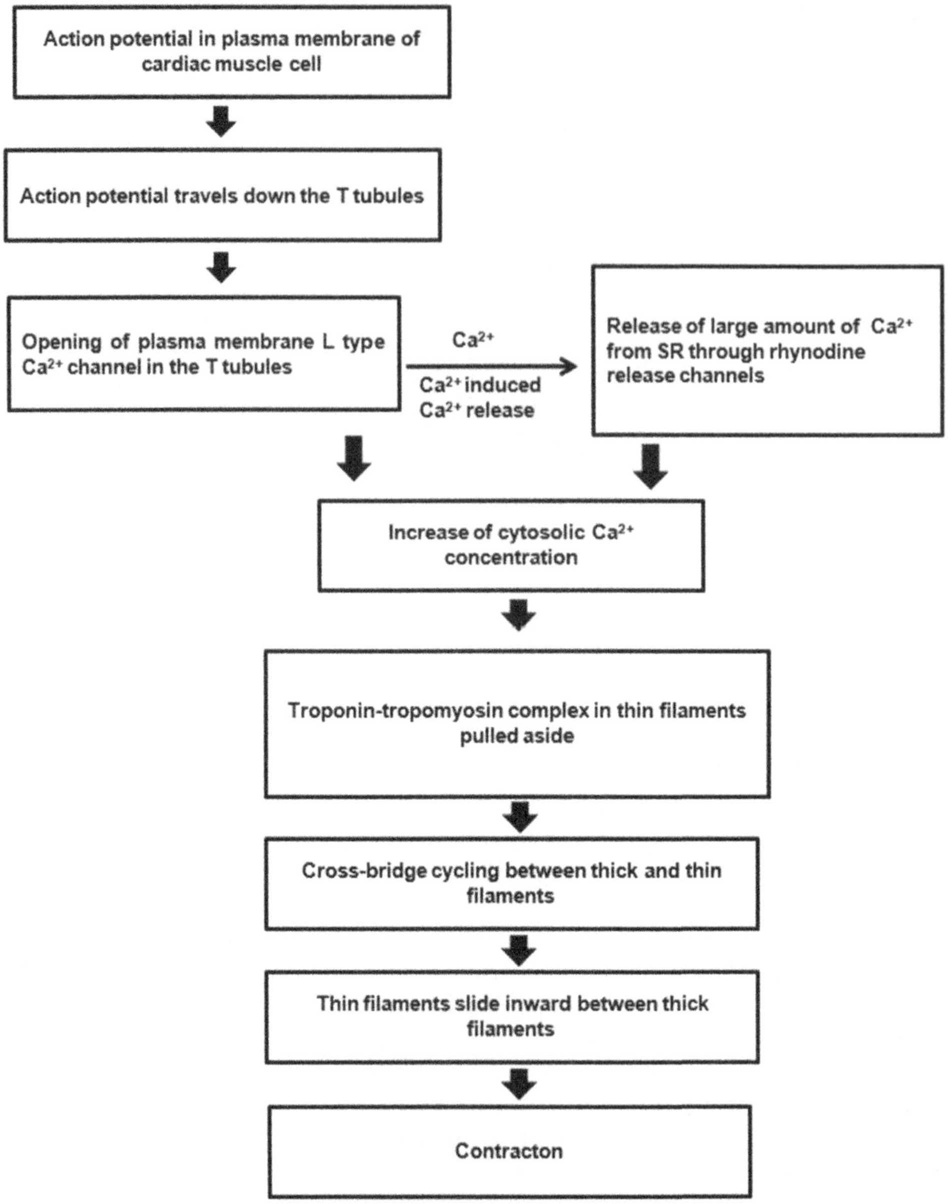

Fig. 10.21 Mechanism of cardiac muscle contraction

chronic pressure overload, it may undergo either concentric left ventricular hypertrophy or dilated left ventricular hypertrophy, with impaired functional consequences.

10.3.3 Muscular Disorders of Domestic Animals

Different diseases influence the typical structure and functions of muscle. Various infections, toxins, or congenital origin causes primary muscular dysfunctions, leading to complete paralysis, paresis, or ataxia. But the major cause of the muscular disorder is dysfunction of the nervous system, e.g., rhinopneumonitis, tetanus, protozoal myelitis, and canine distemper. Some disorders that affect the neuromuscular junction, like hypocalcemia, hypermagnesemia, and myasthenia gravis, can lead to muscular weakness, fatigue, and paralysis. Some antibiotics, toxins (e.g., venoms, botulinum toxins, tetanus toxins), and some muscle-relaxing drugs also affect the neuromuscular junction. The disorder in muscle membrane and muscle fiber is known as myopathies. The disorders in muscle membrane may occur due to hereditary problems (e.g., congenital in goats, myotonia) or acquired (e.g., hypothyroidism, vitamin E and selenium deficiency, hypokalemia). In muscle fiber, various diseases happen, like polymyositis, muscular dystrophy, white muscle disease, and eosinophilic and myositis myopathy.

Muscle trauma is prevalent in the horse and may be from external or extreme activities leading to muscle rupture. In horses, fibrotic myopathy in the rear limb is a mechanical lameness caused due to the trauma and subsequent fibrosis or ossification of the muscle. Different laboratory tests, viz. determining serum enzyme levels, histopathological examination, and electromyographic studies, are used to diagnose muscular diseases.

Learning Outcomes

- The muscle is a contractile tissue consisting of muscle cells or muscle fibers. Contraction of muscle fibers generates force, and that causes motion. Three types of muscles are there in the body, i.e., skeletal muscle, smooth muscle, and cardiac muscle. Skeletal muscles are mainly attached to bones, smooth muscle is present in the walls of visceral organs, and cardiac muscle is located in the heart.
- The skeletal muscle fibers have a long cylindrical structure with many nuclei located in the periphery. The active contractile unit of muscle is known as the sarcomere. Each myofibril contains several types of protein cells called myofilaments.
- During contraction, action potential propagates through the sarcolemma and travels down the T-tubules causing the sarcoplasmic reticulum to release Ca^{2+} ions. The myosin head then attaches to the binding site of the G-actin molecule, and the formation of crossbridges occurs. Muscle relaxation occurs when the release of the neurotransmitter stops at the neuromuscular junction.
- Smooth muscle fibers are tiny and spindle shaped with one centrally located nucleus. Smooth muscle fibers contain three types of filaments, i.e., thick myosin filaments, thin actin filaments, and intermediate filaments. During contraction, calcium ions attach with calmodulin instead of troponin. The intermediate filaments do not directly participate in contraction, and they only form part of the cytoskeletal framework that supports cell shape.
- Cardiac muscles are striated like skeletal muscle but involuntarily, mainly controlled by the autonomic nervous system and endocrine glands. The cardiac muscle cells are smaller than the cells of skeletal muscle and exhibit branching. The cells are interconnected by intercalated discs and form functional syncytia. Like skeletal muscle, cardiac muscle contraction is thin filament regulated, with an elevation in intracellular Ca^{2+} required to promote actin-myosin interaction.

Exercises

Objective

Q1. What is the layer of connective tissue that separates the muscle tissue into small sections?
Q2. What is the name of loose connective tissue that surrounds individual muscle fibers?
Q3. Where are the crossbridges involved in muscle contraction located?
Q4. During smooth muscle contraction, Ca^{2+} is attached to which protein?
Q5. What is the zone's name in the central portion of A band of skeletal muscle where thin filaments are absent?
Q6. Into what does the neuron release its neurotransmitter at the neuromuscular junction?
Q7. What type of muscle is found in the eyes' irises and the blood vessels?
Q8. What is the function of varicosities?
Q9. What is a T-tubule?
Q10. What is titin?
Q11. What are the principal proteins of muscle contraction?
Q12. What is sarcomere?

Subjective Questions

Q1. What is rigor mortis?
Q2. What is muscle atrophy?
Q3. What is muscle hypertrophy?
Q4. What is muscle fatigue?
Q5. Differentiate between the single-unit and multiunit smooth muscle.
Q6. Differentiate between isotonic and isometric contraction.
Q7. Describe the energy sources for skeletal muscle contraction.
Q8. Describe the events of skeletal muscle contraction (flow diagrammatically).
Q9. Describe different types of skeletal muscle fibers.
Q10. Explain—Cardiac muscle cannot be tetanized in vivo.

Answer to Objective Questions

A1. Perimysium
A2. Endomysium
A3. On the myosin myofilaments
A4. Calmodulin
A5. H zone
A6. Synaptic cleft
A7. Multiunit smooth muscle
A8. Varicosities innervate the smooth muscle
A9. The T-tubules are an invagination of the muscle cell's sarcolemma
A10. Titin is a molecular spring attached to thin filaments
A11. Actin and myosin
A12. The sarcomere is the portion between two successive Z-lines

Keywords for the Answer to Subjective Questions

A1. The word rigor mortis came from two Latin words, i.e., "rigor" means "stiffness" and "mortis" means "of death." Rigor mortis or postmortem rigidity of muscles is an important sign of animal death.

A2. Muscle atrophy is the decrease in the size of the muscle due to a reduction in muscle mass.

A3. Muscle hypertrophy is the increase in the size of the muscle due to increase in muscle mass.

A4. If a muscle is used exhaustively, then the muscle's performance decreases progressively and mostly recovers after a period of rest. This phenomenon is known as muscle fatigue.

A5. In a single unit of smooth muscle, the muscle cells are connected through gap junctions. Through these gap junctions, action potential transmits from one cell to another.

Each cell receives its synaptic input through single varicosity in multiunit smooth muscle. No gap junction is present. So, cells are not electrically connected, and selective activation of muscle fibers occurs.

A6. Isotonic contraction: When the muscle length changes but the muscle tension remains unchanged, the contraction is known as an isotonic contraction (tonic = tension). Isotonic contraction is seen during walking, running, and different types of activities.

Isometric contraction: When the muscle's tension increases but the muscle's length remains the same, then the contraction is known as an isometric contraction (iso = same, metric = length). In this type of contraction, muscle provides the force, but no movement occurs at the joint and muscle length remains unchanged.

A7. Sources for skeletal muscle contraction: (1) cytosolic stored ATP, (2) creatine phosphate, (3) glycolysis, and (4) aerobic or oxidative respiration.

A8. Events of skeletal muscle contraction: crossbridge formation; power stroke generation; crossbridge detachment; reactivation of myosin heads.

A9. Types of skeletal muscle fiber: (1) slow-twitch muscle fibers or type I muscle fibers; (2) fast-twitch muscle fibers or type II muscle fibers: type IIa muscle fibers, fast-twitch glycolytic (type IIX) fibers.

A10. Cardiac muscle cannot be tetanized due to the long refractory period of its action potential.

Further Reading

Allen DG, Lamb GD, Westerblad H (2008) Skeletal muscle fatigue: cellular mechanisms. Physiol Rev 88:287–332

Basmajian JV (1978) Muscle alive: their functions revealed by electromyography. Williams & Wilkins

Bradley K (2012) Cunningham's textbook of veterinary physiology, 5th edn. Elsevier

Cooper R (1980) EEG technology. Butterworths, London

Craig R, Padrón R (2004) Molecular structure of the sarcomere. Chapter 7. In: Engel AG, Franzini-Armstrong C (eds) Myology, 3rd edn. McGraw-Hill, Inc., New York, pp 129–166

Frandson RD, Lee WW, Dee FA (2009) Anatomy and physiology of farm animals, 7th edn. Wiley–Blackwell

Goto M, Kawai M, Nakata M, Itamoto K, Miyata H, Ikebe Y, Tajima T, Wada N (2013) Distribution of muscle fibers in skeletal muscles of the cheetah (Acinonyx jubatus). Mamm Biol 78(2):127–133

Gunning P, O'Neill G, Hardeman E (2008) Tropomyosin-based regulation of the actin cytoskeleton in time and space. Physiol Rev 88:1–35

Klemm WR (1969) Animal electroencephalography. Academic, New York

Luna VM, Daikoku E, Ono F (2015) "Slow" skeletal muscles across vertebrate species. Cell Biosci 5:62

Reece WO (2015) Dukes' physiology of domestic animals, 13th edn. Wiley Blackwell

Schiaffino S, Reggiani C (2011) Fiber types in mammalian skeletal muscles. Physiol Rev 91:1447–1531

Scott W, Stevens J, Binder-Macleod SA (2001) Human skeletal muscle fiber type classifications. Phys Ther 81:1810–1816

Smith RF (1978) Fundamentals of neurophysiology. Springer

Swenson MJ, Reece WO (2005) Duke's physiology of domestic animals. Panima

Talbot J, Maves L (2016) Skeletal muscle fiber type: using insights from muscle developmental biology to dissect targets for susceptibility and resistance to muscle disease. Wiley Interdiscip Rev Dev Biol 5: 518–534

Wan JJ, Qin Z, Wang PY, Sun Y, Liu X (2017) Muscle fatigue: general understanding and treatment. Exp Mol Med 49:e384

Nervous System 11

Dipak Banerjee, Pradip Kumar Das, and Joydip Mukherjee

Abstract

Nervous system is the major communication system in animals. Neurons are the structural and functional unit of the nervous system. There are numerous specialized contact areas known as *synapses*, which mediate signals from one neuron to others. Neurotransmitters are the chemical messengers liberated at the nerve endings and help to transfer the nerve impulses in the presynaptic neuron to adjacent postsynaptic neurons or muscle or glands. The nervous system has two main subdivisions, the central nervous system (CNS) and peripheral nervous system (PNS), which act together in a synchronous pattern with each other. The central nervous system is the major processing center in the body, which is composed of the brain and spinal cord. The sensory receptors are present throughout the body, which continuously monitor the external as well as the internal environment and send the information to the CNS via PNS. The information is then analyzed in the CNS, which sends signals to the target organ through PNS. Then the particular organ takes necessary action according to the need. The two main subdivisions of PNS are somatic nervous system and autonomic nervous system. The somatic nervous system is associated with the voluntary movement of skeletal muscles, whereas the autonomic nervous system regulates the involuntary functions of organs and tissues.

D. Banerjee (✉) · P. K. Das · J. Mukherjee
Department of Veterinary Physiology, West Bengal University of Animal & Fishery Sciences, Kolkata, West Bengal, India

© The Author(s), under exclusive license to Springer Nature Singapore Pte Ltd. 2023

P. K. Das et al. (eds.), *Textbook of Veterinary Physiology*, https://doi.org/10.1007/978-981-19-9410-4_11

Graphical Abstract

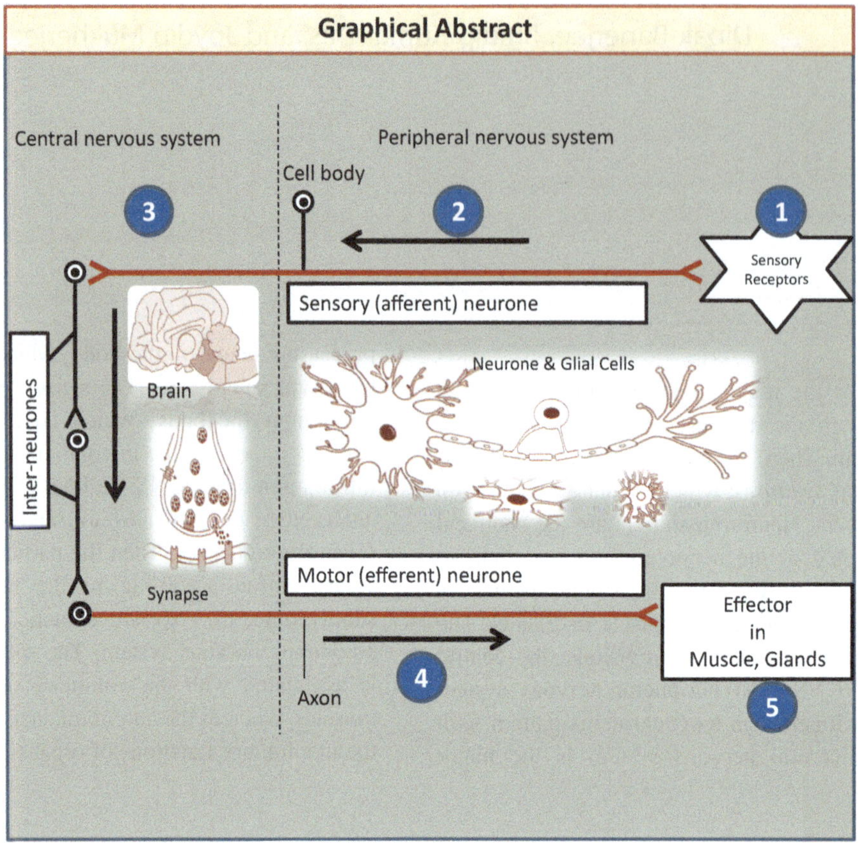

Description of the graphic: Receptors are sensory structures to detect changes in the internal or external environment (**1**). The sensory division of the PNS brings information to the CNS from receptors in peripheral tissue and organs (**2**). Information processing includes the integration and distribution of information in the CNS (**3**). The motor division of the PNS carries motor commands from the CNS to peripheral tissue and systems (**4**). Effectors are target organs whose activities change in response to neural commands (**5**)

Keywords

Neuron · Nerves · Central nervous system · Autonomic nervous system · Brain

Learning Objectives

- Structural as well as functional divisions of the nervous system
- Structure, classification, and functions of two major types of cells of nervous system—neuron and glial cells
- Neurotransmitter and synaptic transmission of action potential
- Anatomical structure and functions of the two main divisions of the nervous system—central nervous system and peripheral nervous system
- Somatic and autonomic nervous systems

11.1 Functional Morphology of Nerve Tissue

Animals respond with their internal and external environment by two types of communication systems, the chemical communication and neural communication. The nervous system can recognize the environmental changes, which can influence the body and work in tandem with the endocrine system to respond to the events. It is a special system which coordinates to collect the information from the external as well as the internal environment and sends to the respective center(s) to generate responses through the motor system. It receives the information through receptors. It coordinates voluntary as well as involuntary functions in the body by receiving and sending information through secretion of various glands, endocrine system, and musculoskeletal system. The change in behavior of any living beings in accordance with the change in the internal as well as external environment is very important for maintenance of body homeostasis as well as for existence in the world. This is possible due to the intracellular communication system present in the body. The neural communication works speedily and is involved in the coordination of different specialized functions in the body, which is better than chemical communication system.

The nervous system is the highly complex and specialized part of the body, which harmonizes the animal's behavior. The system is involved in thinking, making decision, creation, and invention. The input system is fast and can selectively receive information from the body as well as from the external environment through different receptors and afferent paths. The information then reaches the brain and is stored in the memory according to the body's need. The nervous system is morphologically and functionally divided into two components, central nervous system (CNS) and peripheral nervous system (PNS). The central nervous system consists of the brain and spinal cord. The peripheral nervous system comprises sensory and motor nerves that run throughout the body. Neurons are responsible for sending, receiving, and interpreting information from all parts of the body.

11.1.1 Nerve Tissue

The nervous system is made up of a large number of cells (over 100 billion). The cells are mainly of two types, the *neurons* and neuroglia or *glial cells*. Neurons (Greek neuron, nerve) are specialized types of excitable cells and carry electrical impulses. Thus, they are also called conducting cells. Neuroglia (Greek glia, glue) or glial cells are nonconductive and supporting cells of nervous system. Neurons are also called structural and functional units of the nervous system. These neurons are composed of simple elements but are interconnected in a complex way. There are numerous specialized contact areas known as *synapses*, which mediate signals from one neuron to others. Synapses play a vital role in the formation of complex neuronal networks designed for information processing. Neuron transmits information between cells. Neurons with a particular function are found in a particular location in the nervous system. The cell division of neurons generally stops within a few months after birth. So, nerve damage involves cell bodies, resulting in neuronal death. It causes permanent change in the structure and functions of the affected areas. But, unlike neurons, glial cells can continue to divide. This property of glial cells is crucial for their structural and functional support of neurons. Neurons as well as glial cells need a chemically stable environment. The endothelial cells of the CNS and the choroid plexus of brain help to maintain such an environment by regulating molecules secreted into the interstitial fluid and cerebrospinal fluid (CSF). The neurons have a variable number of cytoplasmic processes attached to them. These processes are of two types, axons and dendrites.

11.1.1.1 Structure of Neuron

A nerve cell with all its processes is called a neuron. Neurons are the basic units of the nervous system as the functions of the nervous system are carried out by the neurons. Neurons are specialized types of conducting cells. Neurons are the longest living cells in the body, and the life span of neurons is nearly 100 years. The rate of metabolism in neurons is very high. It is almost 20% of the total body's utilized energy. So, the requirement of glucose and oxygen is very high in the neurons. With the lack of supply of nutrient and oxygen in brain, the body will shut down immediately. The phenomenon like faint may occur. Most of the neurons are amitotic, so they are unable to divide. Exceptions are seen in olfactory neurons and regions of hippocampus in brain. The recovery rate from severe brain or spinal cord injuries is very less. A neuron has two main parts, a cell body or soma and nerve processes or extensions of neuron (Fig. 11.1). Nerve processes are of two types, dendrites and axon. Nerve processes help in receiving and sending information. Axons carry information away from the cell body, but dendrites carry information toward the cell body.

A neuron has similar functional characteristics like other cells of the body along with some specialty. It is enclosed by unit membrane, which also encloses the dendrites and the axon processes. The membrane contains receptors, ion channels, and pumps necessary for the activities of the neurons.

Cell Body The part of a neuron which surrounds the nucleus is known as cell body or soma. Cell body has the major role in protein synthesis. Cell body contains the cellular components and cell organelles like other cells of the body. The nucleus is large with prominent nucleoli and has Barr bodies. The nucleus is present at the center of the cell body. The chromatin material is comparatively more active, and there is continuous transcription. The cell body contains several cytoplasmic organelles such as the mitochondria, Golgi apparatus, endoplasmic reticulum, secretory granules, ribosomes, and polysomes. Centrioles are absent in cell body. Hence, neurons are unable to divide. The cell body or soma contains rough endoplasmic reticulum, free ribosomes, and Golgi apparatus, all having usual functions as in other somatic cells. There are sufficient mitochondria to run the tricarboxylic acid (TCA) cycle and for generation of required energy. The cell body produces proteins, which are required for the construction of other parts of the neuron.

Nerve Processes Nerve processes are "fingerlike" cytoplasmic extensions from the cell body. They are also known as nerve fibers. They are able to conduct and transmit signals.

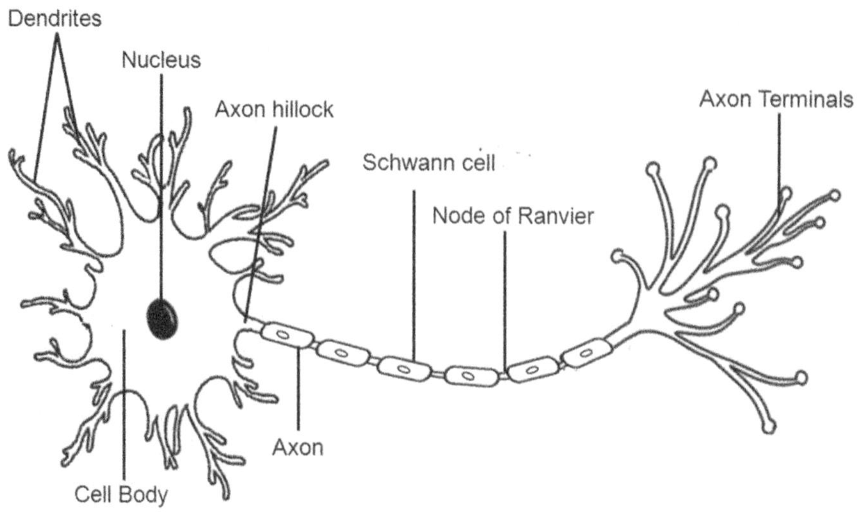

Fig. 11.1 Structure of a typical neuron

There are two types of nerve processes, the dendrites and axon.

Dendrites: Dendrites are short, branched processes, which extend from the cell body and normally carry nerve impulses toward the cell body. Hence, they are also referred to as afferent processes. These processes are usually many in number in a single neuron, but may be single or absent altogether. Dendrites are generally shorter than the axon and more branched. They form many synapses with nearby neurons for receiving nerve signal. The branching facilitates to increase the surface area of the cell for attachment with a large number of other neurons. They receive information through many receptors present in their membranes, which bind to chemicals, known as neurotransmitters. The number of dendrites on a neuron varies. The dendrites form contacts with other neurons. There are plenty of small projections; the dendritic spines are present on their surface. It makes a complex dendritic connection, which is important for integrative functions of CNS. One neuron can attach with more than 1000 neurons through synapse. Dendrites provide input to neuron not in the form of action potential, but as local electronic potentials.

Axon: A neuron has a single axon which extends from the cell body. The axon originates from a specialized cone-shaped area of neuronal cell body known as axon hillock. The summation of the excitatory as well as inhibitory activity occurs in the axon hillock. Generally, the neuronal action potential is formed at the axon hillock. Axon is generally longer than dendrites, and the axon can extend for more than a meter. They carry nerve impulse away from the cell body and that is why they are called an efferent process. Axons convey signals to various other neurons, muscles, glands, or other cells. At the termination, the axons as well as the axon collaterals form many short branches known as telodendria. Ends of the telodendria are slightly enlarged known as synaptic bulbs. Many axons are wrapped in a segmented, white, fatty insulating coat called the myelin sheath. The myelin sheath is produced by the glial cells called oligodendrocytes and Schwann cells. Myelination increases the speed of nerve impulse; that is why the speed of nerve impulse in myelinated fibers is more than the nonmyelinated fibers. The white matter in the CNS is due to the presence of myelinated fibers, whereas unmyelinated fibers make the grey matter. In the myelinated fibers, the unmyelinated regions between the myelin segments are known as the nodes of Ranvier. In the CNS, myelin is produced by oligodendrocytes, whereas in PNS, the myelin is produced by Schwann cells. In the peripheral nervous system, the cytoplasm, nucleus, and outer cell membranes of the Schwann cell form a tight covering around the myelin and around the axon itself at the nodes of Ranvier. This covering is the neurilemma. It plays an essential role in the regeneration of nerve fibers. In the CNS, myelin sheath is produced by oligodendrocytes and neurilemma is absent, and that is why fibers within the CNS do not regenerate. At the end of each axon, it terminates into multiple endings known as axon terminals, which take part in synapse. The axon terminal is able to convert the electrical signal into a chemical signal in a process called synaptic transmission.

11.1.1.2 Classification of Neurons

Neurons are classified into three types on the basis of their structure as well as on the basis of the functions.

11.1.1.2.1 Structural Classification

In structural classification, the neurons are classified on the basis of the number of processes of a neuron. Structurally, neurons are of three types—unipolar, bipolar, and multipolar neurons.

Unipolar neurons: Unipolar neurons have one process, which arise from the cell body and then branch into two parts, which extend in opposite directions. One part of the process extends peripherally and is associated with sensory reception, which is known as the peripheral process. Another part of the process extends toward the CNS, which is known as the central process. This type of neurons is mainly found in the afferent division of the PNS.

Bipolar neurons: Bipolar neurons have two processes, one axon and other being dendrites. The processes extend in opposite directions from the cell body. These types of neurons are found in the retina of the eye and the olfactory system.

Multipolar neurons: Multipolar neurons have multiple processes, one of which is axon and the rest are dendrites. Multipolar neurons are the major neuron type found in the CNS as well as the efferent division of the PNS. In humans, more than 99% of the neurons are multipolar.

11.1.1.2.2 Functional Classification

Neurons are also classified on the functional basis of the direction of the signal, in relation to the CNS. On this basis, there are three different types of neurons: sensory neurons, motor neurons, and interneurons (Fig. 11.2).

Sensory neurons: Sensory neurons or afferent neurons carry input to the CNS for processing. They carry information from sensory receptors present in the skin or in the visceral organs to the CNS. They are mainly unipolar and have very long axons. The information from different sense organs, muscles, and other organs reaches the brain through the sensory neurons. The nerves associated with vision, hearing, taste, and smell are cranial nerves. They do not use the spinal cord. The nerves associated with touch (pressure, temperature, and pain) move through the spinal cord to reach the brain. So, sensory neurons are associated with the incoming of messages from the external environment as well as from the internal organs.

Motor neurons: Motor neurons or efferent neurons transmit the output of CNS to the periphery (to the muscles or glands). Motor neurons are mainly multipolar, and they carry signals from the CNS to the effectors present in muscles and tendons all over the body. They may be somatic or autonomic. Somatic neurons are again subdivided into upper motor and lower motor neurons. The autonomic neurons are divided into preganglionic and postganglionic neurons. So, the motor neurons are associated with the outgoing messages from the brain or spinal cord to act upon the outer environment as well as to different organs. Axons of motor neurons are generally long.

Interneurons: Interneurons remain in between two neurons and relay the information in between them with necessary modification. Interneurons may be inhibitory or excitatory. Interneurons are small and have short axon. Most of the interneurons are located within the CNS. They are mainly present in the brain, spinal cord, and eye. The numbers of interneurons are higher than sensory or motor neurons.

11.1.1.3 Glial Cells or Neuroglia

Glial cells or neuroglia are supporting cells in the nervous system. They are not able to conduct nerve impulses. Their main functions are nourishment and protection of neurons. The number of glial cells is more than neurons, and they occupy half of the volume of the brain. Unlike neurons, glial

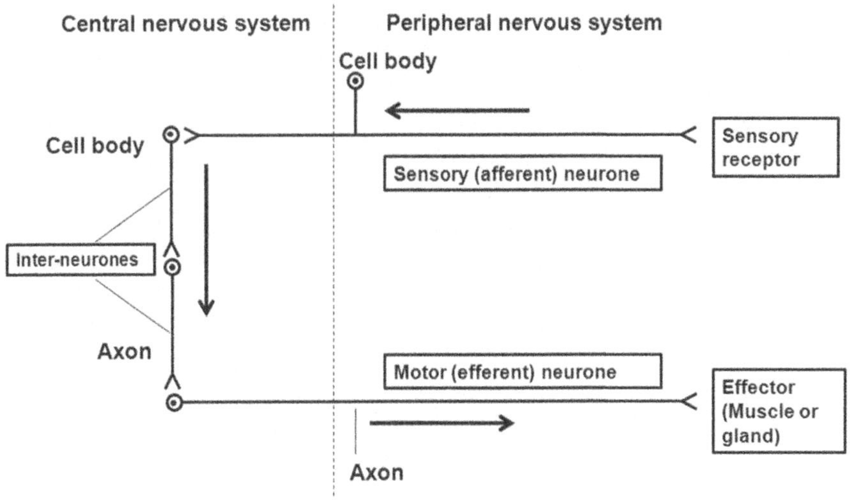

Fig. 11.2 Types of neuron. The figure depicts the **afferent**, **efferent**, and **interneurons**

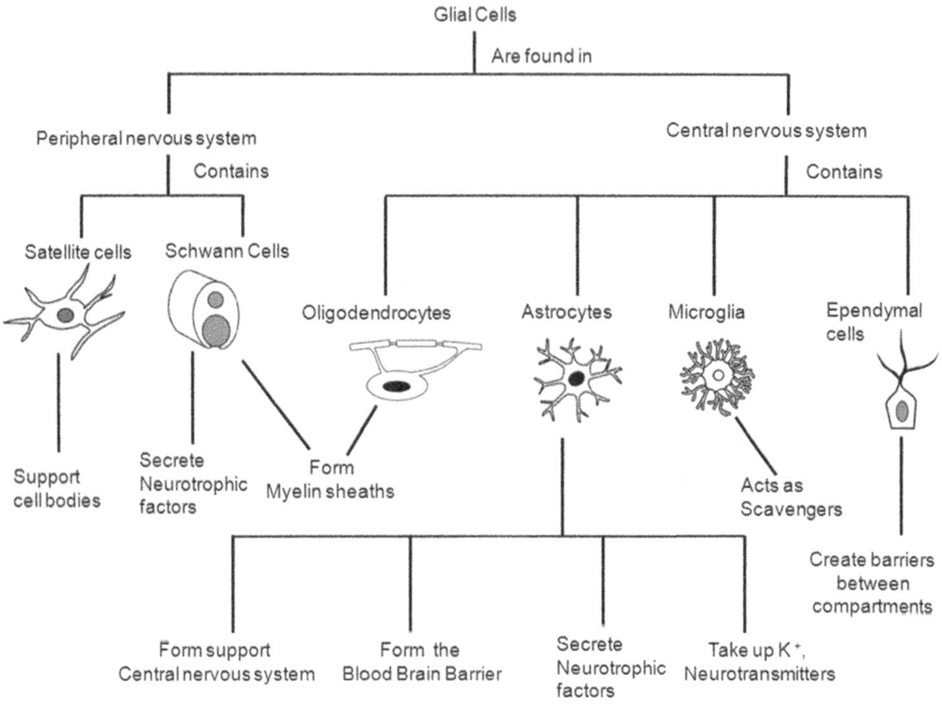

Fig. 11.3 Different types of glial cells, their locations, and major functions

cells are capable of mitosis. Hence, glial cells can be replaced when damaged. Four types of glial cells are present in the CNS. These are astrocyte, oligodendrocyte, ependymal cells, and microglia (Fig. 11.3). In PNS, two types of glial cells are present, Schwann cell and satellite cell (Fig. 11.3).

11.1.1.3.1 Glial Cells of CNS

1. ***Astrocyte***: Astrocytes are star-shaped cells with multiple radiating cytoplasmic processes remaining in the CNS. Large number of processes wrap around the blood vessels and neurons. They have all the organelles of the somatic cell. Their position facilitates to control and adjust the extracellular environment around the neurons. Astrocytes are of two types, fibrous and protoplasmic. The former type has more fibrils, and the latter has more cytoplasm. The fibrous astrocytes are present in the white matter, while the protoplasmic astrocytes are present in the grey matter. Astrocytes attach their processes (foot process) to the capillaries and also to the neurons and synapses. They have a role in providing nutrition to the neurons and also contribute to blood-brain barrier. Astrocytes remove K^+ and some neurotransmitters (GABA), which are liberated due to the actions of the neurons. Thus, astrocytes keep the vicinity of the neurons suitable for normal activity.
2. ***Oligodendrocytes***: Oligodendrocytes are comparatively smaller and have less process than astrocytes. They form the myelin sheath in the nerve fibers in CNS; but, unlike the Schwann cells, they form the myelin sheath of many fibers at a time. It helps in the propagation of electrical impulses through the axon without being spread to other axons. They are also present in the grey matter around the cell bodies of the neurons. Oligodendrocytes wrap several times around a section of an axon. The intermittent gaps in the myelin sheaths of axons where the portion of axon is exposed are known as node of Ranvier. Myelination of nerve fibers increases the speed of nerve impulses through the axon. The propagation of action potentials through myelinated axons from one node of Ranvier to the next node increases the velocity of conduction of action potentials. It is called salutatory (Latin *saltare*, to hop or leap) conduction. Myelinations also bring about the clustering of voltage-gated Na^+ channels at the nodes. Oligodendrocytes also help in the regulation of pH of the CNS.
3. ***Microglia***: Microglia are small cells with long thin tortuous processes, which look like spines. These are believed to be derived from blood. These are phagocytic in function and are motile. Microglial cells are quickly activated in response to injury and infection or disease in CNS. These cells can proliferate and change shape. Microglial cells also play an important role in presenting the antigens to lymphocytes in response to any infection. Although these cells are an essential component of the CNS, it is believed that their activity is also toxic to neurons, causing long-term damage. As a result, medical intervention in response to brain injury often involves factors that inhibit microglial activity.
4. ***Ependymal cells***: The lining cells of the ventricles of the brain and the central canal of spinal cord are called ependyma or ependymal cells. These cells are ciliated

columnar type and are situated between brain extracellular fluid and cerebrospinal fluid (CSF). Thus, they form a blood-CSF barrier. They also line the outer surface of choroid capillaries, which are fenestrated, and the blood-CSF barrier is formed mainly by ependyma cells.

11.1.1.3.2 Glial Cells of the PNS

1. ***Schwann cell***: Schwann cells are the myelinating cells of the PNS. Unlike oligodendrocytes, a Schwann cell provides myelin sheath for a single segment of an axon; but the appearance and function of the myelin sheath in the PNS are just same as that of CNS.
2. ***Satellite cell***: Satellite cells in the nervous system help in regulation of the external chemical environment around the neurons of the PNS. Satellite cells have functions very similar to the functions of astrocytes of the CNS. In addition, they are very sensitive to injury and inflammation.

11.1.2 Synapse

The communication of neurons between each other and with other cells of the body, like muscle and glandular cells, occurs very fast, at specialized junctions called *synapses* (Greek, "junction" or "to bind tightly"). Synapse is also known as neuronal junction. The synaptic junction between a neuron and a muscle cell is called neuromuscular junction.

Structurally, two types of synapses are present in the body, the *chemical synapse* and *electrical synapse*. Both types of synapses help in the transmission of nerve impulse, but the mechanisms of transmission are different.

In electrical synapses, adjoining cell membranes are attached to each other and gap junctions appear as direct points of contact between the cytoplasm of adjacent neurons. Gap junctions permit the movement of ions from one cell to another. This type of synapses is found in cardiac muscle and single-unit smooth muscle.

In chemical synapses, cell membranes of the neurons adjoin very close to each other but remain distinct, leaving a space. In this type of synapses, neural communication occurs using chemical messengers, which are known as neurotransmitters. Neurotransmitter helps in the transmission of nerve impulses from one cell to another. Here, synapse is the junction between two nerve cells. Most of the synapses in the nervous system are chemical synapse. Discussion on the chemical synapse is emphasized in the chapter. The synapse is a site of attachment between a presynaptic element of one neuron and a postsynaptic membrane of another neuron (or an effector organ). The presynaptic axon enlargement releases neurotransmitter molecules, which diffuse across a synaptic cleft and bind to receptor present in the postsynaptic membrane.

11.1.2.1 Synaptic Anatomy

Generally, synapses are comprised of three major elements. These are the axon terminal or presynaptic nerve terminal. It transmits the information to the next part of the synapse, the synaptic cleft. The third element is the dendrite or postsynaptic element, which receives the information.

1. ***Presynaptic nerve terminal***: Presynaptic nerve terminal contains synaptic vesicles rich in neurotransmitter, which is released during the time of synaptic transmission. The vesicles fuse with the presynaptic membrane during synaptic transmission.
2. ***Synaptic cleft***: It is the narrow gap between presynaptic and postsynaptic membranes into which neurotransmitter molecules are released.
3. ***Postsynaptic element***: It receives the neurotransmitter. It may be a dendrite, a cell body, or a target cell receiving the synaptic input. The neurotransmitter molecules bind with the receptor protein molecules, which are embedded in the postsynaptic plasma membrane.

11.1.2.2 Classification of Synapse

The synapse is classified into four types according to the part of the neurons that are involved in the synapse. These are axodendritic, axosomatic, axoaxonic, and dendrodendritic.

1. ***Axodendritic***: Synapse having the axon of one neuron attaching with the dendrite of another neuron is called axodendritic synapse.
2. ***Axosomatic***: The axon of one neuron attaching with cell body or soma of another neuron forms axosomatic synapse.
3. ***Axoaxonic***: Two axons attaching with each other form the axoaxonic synapse.
4. ***Dendrodendritic***: Dendrites of two nerve cells attaching with each other form dendrodendritic synapse.

11.1.2.3 Synaptic Transmission

When an action potential travels through the axon and reaches the axon terminal, the adjacent presynaptic membrane becomes passively depolarized (toward zero transmembrane potential). Then voltage-gated Ca^{2+} channels open. The Ca^{2+} enters the presynaptic neuron and signals to neurotransmitter vesicles (Fig. 11.4).

Neurotransmitter molecules are released in proportion to the amount of Ca^{2+} influx. The Ca^{2+} influx is in turn proportional to the amount of presynaptic membrane depolarization. The elevated Ca^{2+} causes mobilization of synaptic vesicles contacting neurotransmitter and docking of the vesicles with the plasma membrane. A number of synaptic vesicles fuse with presynaptic membrane, which results in the release of neurotransmitter molecules in the synaptic cleft via exocytosis. The neurotransmitter molecules then diffuse across the

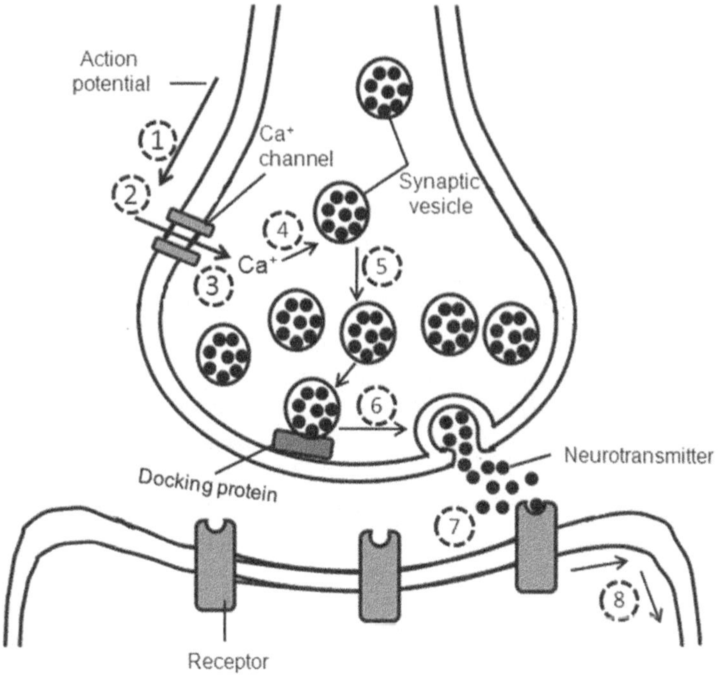

Fig. 11.4 Synaptic transmission. (1) Axon potential arrives at the axon terminal, (2) voltage-gated Ca^{2+} channels open, (3) Ca^{2+} enters the presynaptic neuron, (4) Ca^{2+} signals to neurotransmitter vesicles, (5) vesicles move to the membrane and dock, (6) neurotransmitters released via exocytosis, (7) neurotransmitters bind to receptors, (8) signal initiated in postsynaptic cell

synaptic cleft and ultimately bind with receptor proteins on postsynaptic membrane. Then the signal is initiated in postsynaptic cell. After the work is over, the neurotransmitter molecules are eliminated from synaptic clefts via pinocytotic uptake by presynaptic or glial processes and/or via enzymatic degradation at the postsynaptic membrane. The neurotransmitter molecules are recycled and subsequently presynaptic plasma membrane repolarizes (due to K^+ channel conductance).

When neurotransmitter binds with the postsynaptic membrane, a proportional ion flux across the postsynaptic membrane occurs. The excitability effect depends on the nature of the ion flux. The ion flux depends on the nature of the ion channels in the particular postsynaptic membrane. In the resting state, postsynaptic plasma membrane remains polarized and at that time voltage-activated K^+ channels dominate conductance.

When neurotransmitter molecules bind to ligand-gated receptors, which opens ion channels directly or by means of second messengers activation of [Na^+ and K^+] channels that leads to depolarization toward zero potential and activation of Cl^- or K^+ channels. This results in hyperpolarization of postsynaptic membrane. A postsynaptic potential (PSP) occurs from the altered membrane conductance.

On the basis of the type of ion, the effect on the postsynaptic cell may be depolarizing (excitatory) or hyperpolarizing (inhibitory) and thus the excitatory response is known as *excitatory postsynaptic potential* (EPSP) and an inhibitory response is known as *inhibitory postsynaptic potential* (IPSP). From their names, it is clear that EPSP results in an excitatory response, or depolarization of membrane, and an IPSP elicit in an inhibitory response, or hyperpolarization of membrane.

EPSP and IPSP The cell body of neuron forms multiple synapses on it and on its dendrites. Within these synapses, some change the membrane potential of cell body nearer to threshold potential, whereas other synapses change membrane potential of the cell body moving further from threshold potential (hyperpolarization). The synapses which move membrane potential closer to threshold potential are known as excitatory postsynaptic potential, and the synapses which move the potential further from threshold are known as inhibitory postsynaptic potential. As a result, the net effect of all the EPSPs and IPSPs occurs at the axon hillock. If the potential reaches the threshold, then an action potential will generate and that will continue down the axon.

The aim of an EPSP is to initiate the change in the membrane to generate an action potential. On the other hand, the IPSP prevents the generation of an action potential. EPSP or IPSP lasts for a few milliseconds, and then the membrane returns to the original resting membrane potentials. Sometimes, a single EPSP is not sufficient to generate an action potential and then many EPSPs from different synapses combine at the cell body and result in much larger voltage change, which can exceed threshold potential and cause generation of action potential. This phenomenon is known as *spatial summation*. The combined phenomenon of occurrence of multiple EPSPs from the same synapse in rapid succession is known as *temporal summation*.

11.1.2.3.1 Summation

The generation of EPSP or IPSP depends on the type of neuron and their receptors. Receptors can be divided into two broad categories, the chemically gated ion channels and second messenger systems. After activation, chemically gated ion channels allow certain ions to move across the membrane. The type of ion will determine the generation of EPSP or IPSP. Activation of second messenger system results in a cascade of molecular interactions in the postsynaptic cell. The type of cascade which occurs will result in the response being either excitatory or inhibitory.

Excitatory Synapses: The neurotransmitters mostly used in excitatory synapses in the brain are glutamate or aspartate. They bind to nonselective cationic channels which allow for Na^+ and K^+ to pass. A number of EPSPs from these types of synapses cause the depolarization of postsynaptic neuron. When it reaches the threshold, the action potential is generated.

Inhibitory Synapses: Inhibitory neurotransmitters are essential for controlling the excitability of neurons. This is regulated by a balance between excitation and inhibition. The major inhibitory neurotransmitters are GABA and glycine. They bind to their receptors resulting in an increased conductance of Cl^-. The negatively charged Cl^- that usually moves into the cell results in inhibition of depolarization and keeps the membrane to move away from threshold.

Modulatory Synapses: The activity of many synapses is influenced by neuromodulators. They are able to respond more powerfully to other inputs. For example, norepinephrine acts as a neuromodulator. Norepinephrine has little effect on synaptic transmission, but when a cell is exposed to norepinephrine first, it will react more powerfully to glutamate.

11.1.3 Neurotransmitter

Neurotransmitters are the chemical transmitter or chemical messenger substances liberated at the nerve endings and help to transfer nerve impulses in the presynaptic neuron to an adjacent cell (neighboring postsynaptic neurons or muscle or gland cells). They are endogenous chemical messengers that help in communication within the nervous system as well as between the nervous system and the rest of the body. Synapses relay information between the neurons and ultimately regulate a wide range of bodily functions.

Different types of neurotransmitters with different functions and mechanisms of action are present in the nervous system. Their levels and function are very important for maintaining the homeostasis, and any alteration in their levels or functions can lead to diseases. Chemicals secreted by neurons enter in blood to act as hormones, which are often called neurohormones. ADH and GnRH are neurohormones.

11.1.3.1 Characteristics of Neurotransmitter

A chemical to be liberated as a neurotransmitter should have the following criteria:

1. The chemical must be synthesized in the neuron concerned.
2. It should be stored in the presynaptic terminal.
3. It should be released at the synapse in amounts sufficient to exert a defined action.
4. It should have its specific receptors on postsynaptic membrane.
5. It should be removed quickly by the specific mechanism as soon its action is over.

The chemicals which are secreted from the nerve endings on the target organ or into the ECF are called neurosecretion. Chemicals liberated at the neuromuscular junction are also called neurotransmitters. Another term neuromodulator is used to name the chemicals which are used to modify the activities of postsynaptic neuron.

11.1.3.2 Mechanism of Action of Neurotransmitter

Neurotransmitters transmit signals through the synapse at various locations, such as from one neuron to another neuron, at the neuromuscular junction (NMJ) like a target muscle cell or a gland.

Generally, neurotransmitter releases at a low (basal) level without any stimulation. The level increases in response to threshold action potentials. The binding of neurotransmitters to the postsynaptic membrane results in either excitation or inhibition depending on the neurotransmitter and the binding receptor. Details of mechanism of action have been explained in Sect. 11.1.2.3 and in Fig. 11.4 of this chapter.

11.1.3.3 Classification of Neurotransmitter

There are hundreds of neurotransmitters, but they can be grouped into certain classes depending on their structure and function.

On the basis of structure, neurotransmitters can be classed as follows:

1. ***Monoamines***: The dopamine, noradrenaline, adrenaline, histamine, and serotonin belong to monoamine group.
2. ***Amino acids***: The glutamate, gamma-aminobutyric acid (GABA), glycine, aspartate, and D-serine are in amino acid group.
3. ***Peptides***: The opioids, endorphins, somatostatin, oxytocin, and vasopressin are differentiated as peptide group.

4. ***Other***: The acetylcholine (ACh), adenosine, and nitric oxide are considered in the other group of neurotransmitter.

Neurotransmitters and neuromodulators can also be classified into two main categories, as *small-molecule transmitters* and *large-molecule transmitters*. Small-molecule transmitters include monoamines, catecholamines, and amino acids. Large-molecule transmitters include a large number of peptides called *neuropeptides* including substance P, enkephalin, vasopressin, and a host of others.

Some other substances are also released in the synaptic cleft, which acts as either a transmitter or a modulator during synaptic transmission. The purine derivatives, like adenosine and adenosine triphosphate (ATP), and a gaseous molecule, like nitric oxide (NO), are examples of such substances.

Neurotransmitters can be classified on the basis of their function:

1. ***Excitatory neurotransmitters***: This type of neurotransmitter increases the electrical excitability on the postsynaptic side through modulation of the transmembrane ion flow to facilitate transmission of an action potential.
2. ***Inhibitory neurotransmitters***: This type of neurotransmitter decreases electrical excitability on the postsynaptic side to prevent the propagation of an action potential.
3. ***Neuromodulators***: The neurotransmitters which can alter the strength of transmission between neurons by altering the amount of production and release of it are considered as neuromodulators.

11.1.3.3.1 Neurotransmitters in the Central Nervous System

A number of neurotransmitters act in the central nervous system (CNS), like ACh, amines, serotonin, dopamine, norepinephrine, epinephrine, glutamate, aspartate, glycine, γ-aminobutyric acid (GABA), peptides, and nitric oxide. Acetylcholine is synthesized from choline and acetyl coenzyme A (acetyl-CoA) in the axon terminal. Neurons that release ACh are called *cholinergic neurons*. The amine neurotransmitters (like dopamine, norepinephrine, epinephrine, serotonin, histamine, tyrosine) are derived from amino acids. Dopamine, norepinephrine, and epinephrine are synthesized from tyrosine. Neurons that release norepinephrine or epinephrine are called *adrenergic neurons*. Serotonin (or 5-hydroxytryptamine or 5-HT) is derived from the amino acid tryptophan and histamine from histidine. Glutamate and aspartate are the excitatory neurotransmitters of the CNS.

The primary inhibitory neurotransmitters in the CNS are GABA and glycine. Peptides that act as neurotransmitters include substance P and opioid peptides such as enkephalins and endorphins. Substance P is involved in pain pathways, and enkephalins and endorphins mediate analgesia.

An unusual neurotransmitter, nitric oxide (NO), diffuses freely into the target neuron to bind to intracellular proteins. Nitric oxide is synthesized from oxygen and the amino acid arginine.

11.1.3.3.2 Neurotransmitters in the Peripheral Nervous System (PNS)

The neurotransmitters present in the peripheral nervous system (PNS) are acetylcholine (ACh), norepinephrine, and epinephrine.

11.1.3.4 Fate of Neurotransmitter

The neurotransmitters are detached from their receptors and are removed very fast from the synaptic cleft immediately after the completion of its action.

There are two processes for removal of the neurotransmitters after their action. These are the following:

(1) Enzymatic inactivation in the synaptic cleft and (2) diffusion away from the synaptic cleft.

1. ***Enzymatic inactivation***: In this process, the neurotransmitter is inactivated by a specific enzyme in the synaptic cleft. The uptake constituents occur by the presynaptic terminal used for resynthesis of neurotransmitter.
   Example: ACh quickly detaches after the action is over and is broken down into choline and acetate by the enzyme acetylcholinesterase (AChE), which is present on the postsynaptic membrane. Then choline is actively transported back into the presynaptic terminal for resynthesis.
2. ***Diffusion***: In this process, neurotransmitters enter the circulation or are transported back into the neuron or into astrocytes.
   Example: Glutamate is transported back into the presynaptic terminal or astrocytes after the completion of its work at synapse. Glutamate is repackaged into synaptic vesicles in the presynaptic terminal. In the astrocytes, glutamine synthetase acts on glutamate to convert it into glutamine, which is then transported back to the presynaptic terminal by glutamine transporters. Then it is repackaged into synaptic vesicles and used as neurotransmitter again.

Any abnormality in the release of neurotransmitters and their activity results in various neurological disorders and diseases. In schizophrenia, dysfunction of dopamine, glutamate, and GABA has been reported, whereas reductions in levels as well as activities of norepinephrine and serotonin have been reported in persons with depression. In Parkinson's disease, decreased levels of dopamine have been reported due to the loss of so-called dopaminergic neurons.

11.2 Organization of the Nervous System

The nervous system has two main subdivisions, the central nervous system (CNS) and peripheral nervous system (PNS). The central nervous system is the major processing center in the animal body, which is composed of the two major parts, the brain and spinal cord. Both the parts are protected by bones. The brain remains inside the skull, and the spinal cord is covered by the vertebral column. Structurally, the brain is divided into three main components, the forebrain, midbrain, and hindbrain.

The nervous tissue, except the brain and spinal cord, is known as the peripheral nervous system (PNS). PNS consists of the nerves, ganglia, and sensory receptors.

number of neurons in the enteric nervous system is more than the entire spinal cord.

Know More......

Nucleus: Nucleus is the collection of neuron cell bodies in the CNS.

Tract: Nerve tract is the collection of axons in the CNS.

Ganglia: Ganglia is the collection of neuron cell bodies in the PNS.

Nerve: Nerve is the collection of axons of neurons in the PNS.

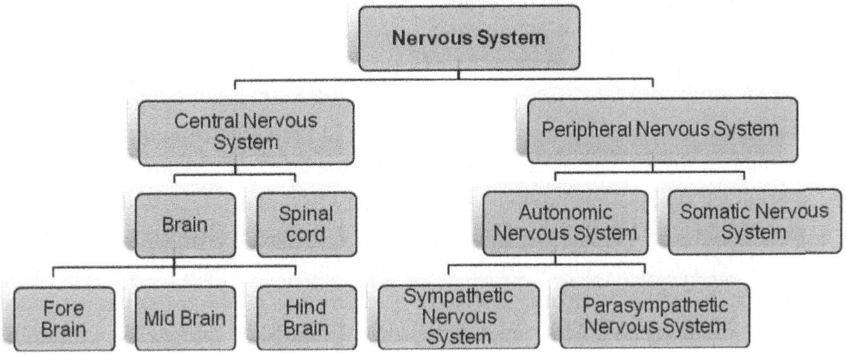

The CNS and PNS act together in a synchronous pattern with each other. The sensory receptors are present throughout the body, which continuously monitor the external as well as the internal environment and send the information to the CNS via PNS. The information is then analyzed in the CNS which sends signals to the target organ through PNS. Then the particular organ takes necessary action according to the need. Some of the functions are completely restricted within the CNS, viz. dreaming, thinking, and storage of information.

The two main subdivisions of PNS are somatic nervous system and autonomic nervous system. The somatic nervous system is associated with the voluntary movement of skeletal muscles, whereas the autonomic nervous system regulates the involuntary functions of organs and tissues. The autonomic nervous system has elements in both the central and peripheral nervous system, and the major subdivisions are sympathetic nervous system and parasympathetic nervous system. Sometimes, enteric nervous system is considered as another subdivision of the PNS. It is a semi-independent system, which controls the activities of gastrointestinal tract. The

11.2.1 Central Nervous System

Central nervous system is composed of brain and spinal cord. Brain is the main part of CNS, covered and protected by the skull. Morphologically, the brain is divided into three parts, forebrain or prosencephalon, midbrain or mesencephalon, and hindbrain or rhombencephalon. The major parts of forebrain are cerebrum, thalamus, and hypothalamus (part of the limbic system). The main parts of midbrain are the tectum and tegmentum, and the hindbrain consists of the cerebellum, pons, and medulla oblongata. The midbrain, pons, and medulla together are considered as the brain stem.

11.2.1.1 Meninges

The intact CNS is enclosed by the connective tissue covering known as *meninges*. Meninges has three layers, viz. dura mater, arachnoid mater, and pia mater (from outside to inside). Dura mater is composed of an outer endosteal layer and an inner meningeal layer, and the dural sinus is present in between the layers. Arachnoid mater covers the brain just above the pia mater. Pia mater is attached to the brain by astrocytes and wraps brain tightly. The space between the arachnoid mater and dura mater is referred to as subdural space, whereas the space between arachnoid mater and pia mater is known as subarachnoid space. Both the spaces are filled with cerebrospinal fluid (CSF). Subdural space and subarachnoid space are frequent sites of intracranial hemorrhage. In the spinal cord, the dura covering is single layered. The space underneath the dura is known as subdural space, and the space external to it is called epidural space.

11.2.1.2 Ventricles of Brain

There are four ventricles present in the brain (Fig. 11.5). Two lateral ventricles are present in each cerebral hemisphere, appeared with a cavity filled with cerebrospinal fluid (CSF). The third ventricle is present in between the right and left thalamus and connected with the lateral ventricles by the foramen of Monro. The fourth ventricle is on the back of the brain stem. It is connected with the third ventricle through the **cerebral aqueduct and continued** below with the central canal of spinal cord. The ventricles are lined by **ependymal cells, which form the** choroid plexus and secrete CSF.

11.2.1.3 Choroid Plexuses

The choroid plexus is formed by capillaries as a complex network. It is lined by specialized types of ependymal cells. These cells produce cerebrospinal fluid (CSF). Choroid plexus acts as a barrier and separates the blood from the CSF and is thus known as the blood-CSF barrier. Choroid plexus also secretes different growth factors, which maintain the stem cell pool in the sub-ventricular zone. It is involved in brain development and gives protection against pathogenic microorganisms and toxic materials.

11.2.1.4 Cerebrospinal Fluid

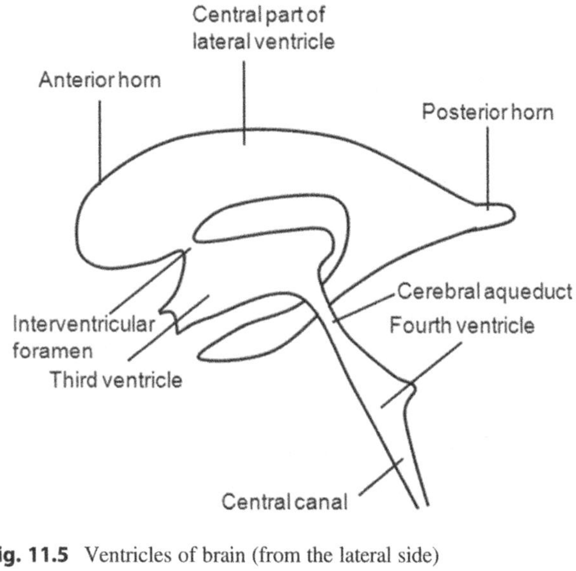

Fig. 11.5 Ventricles of brain (from the lateral side)

Cerebrospinal fluid (CSF) is a colorless fluid found in the ventricles of the brain, central canal of the spinal cord, and subarachnoid space surrounding the outer surface of the brain and spinal cord. The CSF contains very less quantity of protein and almost no blood cells due to selective tight junction barrier.

11.2.1.4.1 Formation, Absorption, and Composition of CSF

CSF is secreted by the ependymal cells located in the choroid plexus. The plexuses consist of tufts of capillaries covered by a layer of ependymal cells. These ependymal cells unlike the cells lining the rest of the ventricle form a selective tight junction barrier to the secretions of the leaky capillaries and to other surrounding fluids. Membrane transporters and selective channels regulate the passage of ions and molecules across the ependymal cell barrier, effectively controlling the composition of the CSF being synthesized in the ventricle. In CSF, 99% is water and rest of the 1% is made up of glucose, proteins, neurotransmitters, and ions. The water is secreted by the choroid plexus into the ventricles of the brain due to generation of an ion gradient on apical and basal surfaces of choroid epithelial cells. In the choroid epithelial cells, water is dissociated into hydrogen (H^+) and hydroxyl (OH^-) ions. Then OH^- ions attach with intracellular CO_2, which is produced by metabolism of cells, and synthesize bicarbonate ions (HCO_3^-). Then, H^+ ions are exchanged for extracellular sodium ions (Na^+) at the basal surface of the cells from the blood, and Na^+ is pumped out through the apical surface into the ventricles. The positive charge in the ventricles increases due to the entry of a large number of Na^+ ions. For neutralizing the excess positive charges, chloride (Cl^-) and HCO_3^- ions enter into the ventricles. Water diffuses into the ventricles to balance the osmotic pressure. Along with water and ions, micronutrients such as vitamin B6 (pyridoxine), folates, and vitamin C enter the brain through the CSF. The rate of formation as well as the flow and absorption of CSF are high. Thus, it is replaced several times daily.

CSF is absorbed and returns to the venous system mainly into dural venous sinuses, which are present intracranially between the endosteal layer and meningeal layers of the dura mater. Majority of the CSF is absorbed from the subarachnoid space into the dural sinus through arachnoid villi. The absorption occurs due to the difference of pressure between the arachnoid mater and venous sinuses. CSF is also drained into lymphatic vessels. Reabsorption of CSF occurs through sheaths of cranial and spinal nerve and through ependymal cells.

11.2.1.4.2 Functions of CSF

The major function of CSF is to act as a cushion and provide supports to different structures of CNS. CSF protects the brain and spinal cord from any physical injury and in any significant variation in the local environment. The specific gravity of the brain and CSF is similar, and thus the brain floats in the CSF. So, the force of a blow to the head is buffered by the CSF instead of being transferred directly to the brain tissue. CSF also helps in the transport of different materials in the nervous system. It supports to maintain a consistent extracellular microenvironment for the neurons and glia of the CNS. CSF also acts as an efficient waste control system by removing harmful cellular metabolites and helps in the transportation of several polypeptide hormones and growth factors.

11.2.2 Cerebrum

Cerebrum is the topmost and largest part of brain. It is composed of two cerebral hemispheres separated incompletely by the median longitudinal fissure, and the hemispheres are joined with each other by the corpus callosum. Cerebrum controls the higher mental functions, which include the conscious thoughts and experience. Cerebrum processes the somatic sensory and motor information. The surface layer of cerebrum is of grey matter (cerebral cortex). Cerebrum is attached with the rest of the brain through the cerebral peduncles. Cerebral peduncles are interconnected in between them and also the subarachnoid cisternae through the foramina of Luschka and Magendie.

The surfaces of hemisphere are highly convoluted having numerous elevations and depressions known as gyri and sulci, respectively. These convoluted appearances increase the surface area for accommodation of large number of cells. The deep grooves are called fissures. The superficial layer of cerebral hemisphere is rich in cell bodies of neurons, known as cerebral cortex. The cerebral hemispheres regulate the major functions of the body. It is also the site for integration of different somatic functions. Cerebral cortex is grey in appearance due to the presence of large numbers of cell bodies of the neurons, so the cortex is known as grey matter. Beneath the cerebral cortex, the nerve fibers are grouped together and form the white matter.

> Know More......
> **Corpus callosum**: Corpus callosum is a broad band of nerve fibers that join the left and right hemispheres. It is the largest white matter structure in the brain and allows the two hemispheres to communicate.

11.2.2.1 Lobes of the Brain

Each of the cerebral hemispheres has been divided into **four lobes, viz. frontal lobe, parietal lobe, temporal lobe, and occipital lobe**. Each lobe is associated with different functions. Frontal lobe is located at the front part or anterior part of cerebral hemispheres of the brain and extends up to the central sulcus. This lobe is associated with reasoning, planning, making decisions, and controlling the behavior, parts of speech, movement, emotions, etc. Parietal lobe is located just behind the frontal lobe and is associated with the sensory information like touch, spatial awareness, and navigation. This lobe controls the movement, orientation, recognition, and perception of stimuli. Temporal lobe is located on either side of the brain and just above the ears in human. This lobe is related to perception and recognition of auditory stimuli, memory, and speech. Occipital lobe is present at the back portion of brain and is associated with visual processing. Basal ganglia are the masses of grey matter, which are stacked lateral to the hypothalamus inside each hemisphere below the lateral ventricles but lateral to the third ventricle.

11.2.3 Diencephalon

Diencephalon is a derivative of prosencephalon and located just below the cerebral hemispheres. It is mainly composed of the thalamus, epithalamus, hypothalamus, and third ventricle. Diencephalon connects the cerebrum with the rest of the brain.

11.2.3.1 Thalamus

The thalami are two egg-shaped structures present on either side of the midline. It acts as an important relay center for the nerve fibers to connect the cerebral hemisphere with the brain stem, cerebellum, and spinal cord. All the sensory information passes through the thalamus and reaches the cerebral cortex for processing; the only exception is the smell sensation, which does not pass through the thalamus. The thalami also involve to process the information.

11.2.3.2 Epithalamus

The epithalamus is located just dorsal to the thalamus and forms the roof of the third ventricle. It contains the pineal gland, an endocrine gland that secretes melatonin hormone. Melatonin plays an important role in the regulation of circadian rhythms and sleep-wake cycle. It also regulates the breeding seasons in seasonal breeding animals. Epithalamus also includes the choroid plexus of the third ventricle and involves in the formation of the CSF.

11.2.3.3 Hypothalamus

The hypothalamus is present just ventral to the thalamus and surrounds the ventral part of the third ventricle. It terminates in a sharp angle where the pituitary gland is attached. The hypothalamus forms the floor of the diencephalon. Hypothalamus is an important center of autonomic nervous system. Hence, hypothalamus is termed as the "*captain of autonomic nervous system*." Hypothalamus plays an important role in the regulation of body temperature, water balance, metabolism, and emotions. Thus, it is an essential part of the limbic system. Hypothalamus also controls the endocrine system and plays an indispensable role in the regulation of homeostasis. The hypothalamic hormones regulate the secretion of pituitary gland. Hypothalamus is discussed in details in Sect. 11.2.5.

11.2.3.4 Mammillary Bodies

The mammillary bodies are small round-shaped paired structures present just inferior to hypothalamus. It is a part of diencephalon and connected to hippocampus, thalamus, and tegmental nuclei of the midbrain. Mammillary bodies act as a relay center for olfaction and are associated with memory.

11.2.4 Brain Stem

The brain stem, like true stem, holds the cerebrum at its top and cerebellum posteriorly. Brain stem processes information between spinal cord and cerebrum or cerebellum. It controls autonomic behavior necessary for survival. It is divided into midbrain, pons, and medulla oblongata, from top to downwards. It is associated with 10 of the 12 pairs of cranial nerves. Outer surface of the brain stem, unlike the cerebrum and cerebellum, contains white matter which opens up posteriorly for the fourth ventricle. There are many grey matter masses in its substance serving various important functions. Some of these nuclei connect with the central nervous system.

11.2.4.1 Mesencephalon or Midbrain

Mesencephalon or midbrain is a small component of brain, which remains in between the thalamus and pons. Midbrain has two major parts, i.e., the tectum and tegmentum. Tectum from the roof and tegmentum form the floor of midbrain. Four round-shaped protrusions or colliculi (singular = colliculus) are present on the dorsal side of the midbrain known as corpora quadrigemina. They consist of right and left rostral colliculi and right and left caudal colliculi. The rostral colliculi are associated with orienting the eyes to a sound or touch stimulus, and the caudal colliculi are responsible for auditory reflex. Two cerebral peduncles, also called crura cerebri, are present in the midbrain. They consist of bundles of nerve fibers, which connect the spinal cord and brain stem to the cerebral hemispheres. These peduncles mainly comprise descending motor fiber tracts.

11.2.4.2 Cerebellum

The cerebellum (Latin for "little brain") is located caudal to the cerebral cortex and dorsal to the brain stem and serves important functions in the regulation of balance and equilibrium. It is connected by three pairs of peduncles with the brain stem. Cerebellum is the second largest part (almost 10%) of the brain; but, due to the highly convoluted structure, it contains more than half of all the brain's neurons. The outer layer of cerebellum is called cerebellar cortex, which consists of grey matter and highly regular arrangement. Two large pairs of white matter stacks called cerebellar peduncles mainly carry axons into the cerebellum, and a third pair of cerebellar peduncles carries axons out from the cerebellum. A number of cerebellar nuclei are present within the cerebellar white matters, which are the principal origin of the axons leaving the cerebellum. Cerebellum controls the timing and coordination of movement by adjusting and modulating the output of the motor cortices, corticospinal tract, descending brain stem motor pathways, and spinal cord. The intended movement is continuously compared with the actual movement by the cerebellum, and then suitable adjustment is done. The vestibulocerebellum adjusts the coordination of vestibular reflexes. This part of the cerebellum was the first to appear in vertebrate evolution; hence, it is sometimes called the archicerebellum. The spinocerebellum extends rostrocaudally through the medial portion of the cerebellum, which helps to coordinate muscle tone as well as limb movement. It recognizes and predicts subconsciously (as does all processing that occurs outside the cerebral cortex). In some of the motor learning processes which require balance, the cerebellum activity is high during learning, and when they become automatic, cerebellum is no longer involved. Any disease in cerebellum results in abnormalities of movement, which further illuminates cerebellar function.

11.2.4.3 Pons

Pons (Latin for "bridge") is a part of brain stem located in between the midbrain and medulla oblongata. The white matter on the anterior surface forms the bridge, whereas the grey matter which remains beneath the white matter is the extension of the tegmentum from the midbrain. The grey matter contains neurons which receive descending inputs from the prosencephalon and send it to the cerebellum. Pons is associated with the somatic and visceral motor control. Four cranial nerve nuclei are present in the pons, viz. V, VI, VII, and VIII. Pons also contains the nucleus for reticular formation. Several nuclei and tracts to cerebellum and other parts of CNS pass through the pons. The nerves associated with pons are involved in the regulation of hearing, maintenance of equilibrium, taste, and facial sensations of touch and pain.

11.2.4.4 Medulla Oblongata

It is the most inferior part that connects brain to spinal cord. It plays an important function in the transmission of signals between the spinal cord and the higher parts of the brain and controls some autonomic functions like heartbeat and respiration. The medullary pyramids are two longitudinal ridges formed by corticospinal tracts. Medulla oblongata regulates autonomic functions. It regulates arousal, heart rate, blood pressure, pace for respiration, and digestion. Cranial nerves, IX, X, XI, and XII, come off or enter from the medulla oblongata. The nerve fibers related to the reticular formation pass through the medulla oblongata.

11.2.4.4.1 Medullary Nuclei

Some regulating centers are present in medulla. These are the following:

1. Cardiovascular control center: It adjusts force and rate of heart contraction.
2. Respiratory center: It controls the rate and depth of breathing.
3. Additional center: It regulates vomiting, hiccupping, swallowing, coughing, and sneezing.

> Know More……
>
> **Basal ganglia**: This is involved in the control of voluntary motor movements, procedural learning, and decisions about which motor activities to carry out. Diseases that affect this area include Parkinson's disease and Huntington's disease.
>
> **Broca's area**: This small area on the left side of the brain (sometimes on the right in left-handed individuals) is important in language processing. When damaged, an individual finds it difficult to speak but can still understand speech. Stuttering is sometimes associated (Trusted Source) with an underactive Broca's area.

11.2.5 Limbic System

Limbic system is referred to the entire neuronal components, which control the olfaction, emotional behavior, motivational drives, and memory. The limbic system is composed of several parts of the brain present on both sides of the thalamus, just under the cerebrum. The structures are hypothalamus, hippocampus, amygdala, and other parts of the brain in nearby areas. The major function of limbic system is control of emotions and formation of memories. It is also involved in the regulation of homeostasis, olfaction, and many other psychologic functions.

11.2.5.1 Hypothalamus

Hypothalamus is the major part of the limbic system and regulates most of the vegetative and endocrine functions and emotional behaviors (anger and aggressive). It is involved in the homeostasis. The thermoregulatory, appetite, satiety, and thirst centers are located in the hypothalamus. Hypothalamus regulates different functions of the body in two ways, either through the autonomic nervous system or via secretions of different hormones from the pituitary gland. Hypothalamus controls different vital functions like blood pressure, heart rate, breathing, digestion, sweating, and arousal in response to emotional situations through autonomic nervous system. It also regulates all the sympathetic and parasympathetic functions. Hypothalamus controls the secretion of pituitary hormones by discharging the releasing or inhibitory hormones. Hypothalamus also regulates many important functions, like response to pain, levels of pleasure, and sexual satisfaction.

11.2.5.2 Hippocampus

The hippocampus is present in the temporal lobe and medial to the inferior horn of the lateral ventricle of each cerebral hemisphere. The name hippocampus came from the Greek word for seahorse because of its structure. Hippocampus is associated with different important functions; but it is best known for its role in memory. The hippocampal formation, a prominent C-shaped structure located in the temporal lobe, consists of hippocampus and the adjacent cortex called the parahippocampal gyrus and a strip of grey matter in between the two structures known as the dentate gyrus.

The hippocampal gyrus areas are called the entorhinal cortex and subiculum, which are both involved in the flow of information through the hippocampus. The hippocampus is anatomically subdivided into four regions, i.e., CA1 through CA4 (the CA stands for cornu ammonis). The hippocampus receives information from the rest of the cerebral cortex primarily via the perforant pathway, which originates in the entorhinal cortex and projects to the dentate gyrus. Fibers then leave the dentate gyrus, which is part of the hippocampal formation, and project to neurons in the CA3 region of the hippocampus. Neurons in CA3 then send axons to neurons in the CA1 region, which projects to neurons in the subiculum. The subiculum can be considered the main output region of the hippocampal formation. Fibers from the subiculum project back upon neurons in the entorhinal cortex. Fibers from the entorhinal cortex travel out to a variety of areas in the cerebrum. Output fibers also leave the subiculum and hippocampus and enter the fornix, a fiber bundle that connects the hippocampus with a variety of subcortical areas like the thalamus and hypothalamus.

The sensory information causes activation of hippocampus and distributes signals to the anterior thalamus, hypothalamus, and other parts of the limbic system, especially through the *fornix*, a major communicating pathway. So, the hippocampus acts as an additional channel which transmits incoming sensory signals that can initiate behavioral reactions related to different purposes. Hippocampus is associated with different behavioral patterns like other parts of the limbic system.

11.2.5.3 Amygdala

The amygdales, two almond-shaped structures, are the collection of neurons present deep in each temporal lobe on either side of the thalamus and adjacent to the hippocampus. It is connected with the hypothalamus as well as other parts of the limbic system. It receives information from different parts of the limbic system and from the neocortex of the temporal, parietal, and occipital lobes specially from the auditory and visual association areas. Amygdala is primarily associated with emotion, memory, and fight-or-flight response. Because of these multiple connections, the amygdala has been called the "window" through which the limbic system sees the place of the individual in the world. When amygdala is stimulated, animals respond with aggression.

11.2.5.4 Other Related Parts of Limbic System

The limbic system is intimately connected with some nearby structures, like cingulated gyrus, ventral tegmental area, basal ganglia, and prefrontal cortex. Cingulate gyrus is a part of the cerebrum and is present just above the corpus collosum. It makes a pathway between the thalamus and hippocampus. Cingulate gyrus is associated with emotional events and associating memories to smells and to pain.

The ventral tegmental area is a part of the brain stem present just below the thalamus. It is a part of the limbic system. Ventral tegmental area has dopamine pathways, which are associated with pleasure. The basal ganglia are also a part of the limbic system and remain over and to the sides of the limbic system. Basal ganglia are firmly connected with the cerebral cortex above them. Basal ganglia consist of caudate nucleus, putamen, globus pallidus, and substantia nigra. These parts are responsible for repetitive behaviors, reward experiences, and focusing attention.

Another part of limbic system is the prefrontal cortex. It is a part of the frontal lobe of cerebrum located in front of the motor area. Prefrontal cortex is associated with thinking about the future, making plans, and taking action. It is also involved in the dopamine pathways as the ventral tegmental area and plays a part in pleasure and addiction.

11.2.6 Reticular Formation

The *reticular formation* is the phylogenetically primitive network of small neurons and occupies the midventral portion of the medulla and midbrain, extending throughout the brain stem and into the spinal cord. The reticular formation is a very complex structure, which contains a number of nuclei, which are interconnected, and nerve tracts. The nerve fibers form a netlike appearance in the central core of the brain stem. But the part is not anatomically well defined as the nuclei are present in different parts of the brain. It is also known as the reticular activating system (RAS). The neurons form a complex network, and they are extended from the upper part of the midbrain to the lower part of the medulla oblongata. The ascending pathway of reticular formation toward the cortex is known as ascending reticular activating system (ARAS), whereas the descending pathway toward the spinal cord is known as reticulospinal tracts.

Various neural clusters and fibers of it have discrete functions. For example, it contains the cell bodies and fibers of many of the serotonergic, noradrenergic, adrenergic, and cholinergic systems. It also contains many of the areas concerned with regulation of heart rate, blood pressure, and respiration. Some of the descending fibers in it inhibit transmission in sensory and motor pathways in the spinal cord; various reticular areas and the pathways from them are concerned with spasticity and adjustment of stretch reflexes. The RAS is a complex polysynaptic pathway arising from the brain stem reticular formation with projections to the intralaminar and reticular nuclei of the thalamus, which, in turn, project diffusely and nonspecifically to wide regions of the cortex. The system is therefore *nonspecific*, whereas the classic sensory pathways are *specific* in that the fibers in them are activated by only one type of sensory stimulation. The axons of the reticulospinal tract are associated with spinal reflex activity and can modulate the sensory input by controlling the gain at synapses in the spinal cord. The reticulospinal tract also carries axons that modulate autonomic activity in the spinal cord.

Reticular formation is the major regulator of the state of consciousness and arousal. Several neurons of reticular formation are serotonergic; that is, they use serotonin as neurotransmitter. Fibers from these nonspecific thalamic nuclei extend to most of the cerebral cortex and thus control the activity of large numbers of neurons. Reticular formation is mainly associated with the regulation of arousal and consciousness of animals. A variety of sensory stimuli like auditory, visual, somatosensory, and visceral sensory stimuli excite the neurons of the reticular formation. Different regions of the cerebral cortex are associated with the arousal. As different features of the external environment (viz. color, shape, location, sound of various external stimuli) are represented in different areas of the cortex, it has been suggested that "binding" of neural activity in these different areas is involved in consciousness. The reticular system is also associated with the regulation of respiration, heart rate, and blood pressure and modulation of nonspecific sensory information. Reticular system is also associated with the adjustment of reflexes and postural control.

11.2.7 Blood–Brain Barrier

The CNS has some special kind of barrier, which isolates neuronal tissue of CNS from the general circulation known as the blood–brain barrier (BBB) (Fig. 11.6). The existence of the BBB has been observed using the dye like trypan blue. The trypan blue can stain all tissues in the body except tissues of brain and spinal cord when infused intravenously. But if the stain is injected directly in the ventricles, then brain tissue takes the stain.

11.2.7.1 Structure of BBB

The barrier is formed by a network of tight junctions between endothelial cells of CNS capillaries and by feet of astrocyte processes. Entry of the substances is generally restricted through BBB due to special morphological features of the capillary endothelium, the structural basis of the blood–brain barrier. The endothelial cells interact with the surrounding layer of astrocytic "end feet" to form the special barrier (Fig. 11.6).

The blood capillaries of the CNS have some unique characteristics, like the following: (1) the capillaries have continuous tight junctions, which seal neighboring endothelial cells; (2) absence of fenestrations; and (3) presence of very few numbers of small pinocytotic vesicles. The capillaries of the brain have a more number of mitochondria, which helps in the operation of the transporters.

11.2.7.2 Restricted Movement Through BBB

The body water-soluble compounds easily pass through open clefts present between capillary endothelial cells. However, most of the compounds that pass through intercellular clefts of various tissues are blocked by tight junctions, and entry of brain capillary blood solutes is very much selective. Generally, the molecules which are relatively small, uncharged, lipid soluble, and unbound to plasma proteins (e.g., O_2, CO_2, N_2O, ethanol, nicotine) can easily pass through the capillary endothelium of the BBB and glucose, and some amino acids are able to pass by specific carrier-mediated transport mechanisms.

Some smaller neutral amino acids such as glycine, alanine, serine, cysteine, proline, and γ-aminobutyric acid (GABA) are synthesized in the CNS, and these are transported mainly from the brain to the blood circulation. The transport of these amino acids requires an energy-dependent and Na^+-dependent symport carrier located at the abluminal side of the endothelial cell membrane. The lopsided transport of the neurotransmitters across the blood–brain barrier ensures that neurotransmitters will not accumulate in the brain, preventing the potential neurotoxic glutamate effect and unwanted inhibition of neurons by glycine and GABA. Few degradative enzymes like monoamine oxidase are expressed by brain capillary endothelium, which gives an additional restriction on substances to pass the BBB.

11.2.7.3 Non-barrier Regions in the Brain

There are some specialized areas in the brain, like choroid plexus, hypophysis, median eminence, pineal gland, and area postrema, where blood capillaries are fenestrated, with lack of interaction between astrocytes and endothelial cells. These

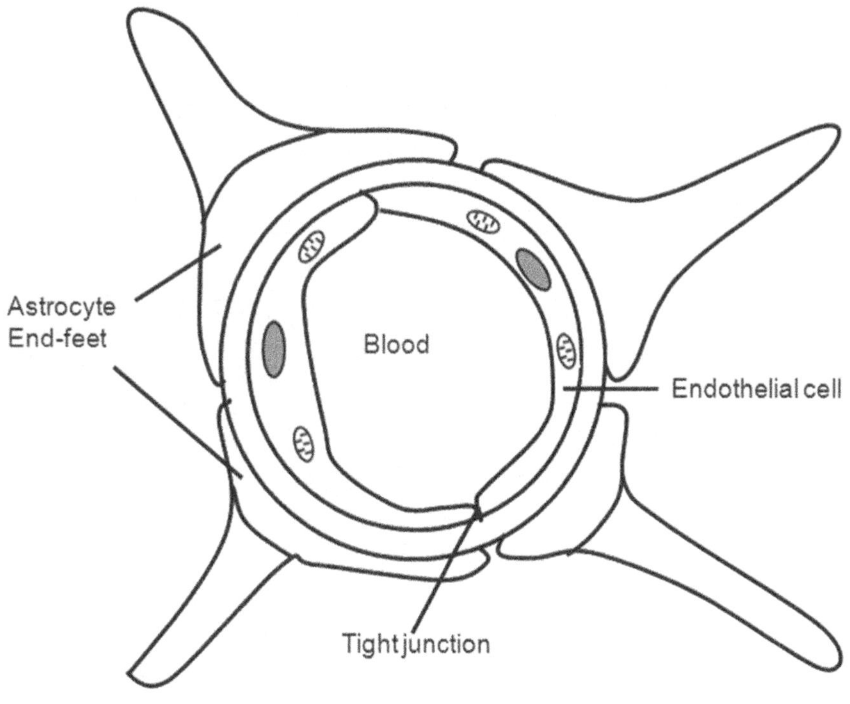

Fig. 11.6 Blood–brain barrier. In brain, the cells of the capillary walls are joined by tight junctions, which restrict the passage of different materials. End feet astrocytes also help in the formation of tight junction and limit the entry of different materials in the brain

areas are considered as non-barrier regions. It facilitates the related organs to maintain their normal functions, like release of hormones into the circulation and monitoring circulating molecules. The interaction between astrocytes and endothelial cells is disturbed in some pathological conditions, viz. neoplasia, hypertension, dementia, epilepsy, infection, multiple sclerosis, and trauma.

11.2.8 Spinal Cord

The spinal cord is the caudal continuation of the medulla oblongata protected in the vertebral canal. Spinal cord contains the central canal in the middle around which there is a mass of grey matter, and in the periphery, there is the white matter. The spinal cord has 31 functional segments, each of which is connected with a pair of spinal nerves. In the spinal cord, the dorsal and ventral roots unite and come together as a nerve at the point where the axons exit and enter the vertebral canal (Fig. 11.7). Sensory neuronal cell bodies are present as a group called dorsal root ganglia lateral to the spinal cord. The neurons within these ganglia are pseudounipolar. They give rise to processes that enter the dorsal horn of the spinal cord.

Other fibers which unite with motor fibers form the ventral horn neurons to become the spinal nerve extending into the periphery. The processes which extend from the spinal nerve to the spinal cord form the dorsal root.

The ventral root of the spinal nerve consists of motor fibers that arise from the nerve cells primarily in the ventral horn of the spinal cord. The dorsal and ventral roots unite to form the spinal nerve close to the intervertebral foramen between adjacent vertebrae. The dorsal root ganglion remains very close to the joining of dorsal and ventral roots. The spinal cord is extended throughout the full length of the back and carries information between the brain and the other parts of the body. Throughout the length, the spinal cord is connected with different nerves of peripheral nervous system. The sensory information from different tissues runs through the spinal cord to reach the brain. The motor commands from the brain run in the course of the spinal cord to the muscle. The spinal cord is also associated with reflexive responses: for example, the sudden involuntary movement of the arm takes place if fingers come in contact with a flame.

11.2.9 Peripheral Nervous System

The peripheral nervous system (PNS) is composed of the nerves and ganglia present outside the CNS. This includes the 12 pairs of cranial nerves and 31 pairs of spinal nerves along with different ganglia and plexuses. The function of PNS is to convey sensory information to the brain and spinal cord and to produce movement of muscle as well as secretion from glands via its motor nerves. Function of PNS is to connect different parts of the body with CNS. The nerves in the PNS are the processes of the neurons whose cell bodies are situated mostly in the CNS, and some are situated in the dorsal root ganglia of spinal nerves, ganglia of cranial nerves, autonomic ganglia, etc.

The enteric nervous system (ENS) is another division of peripheral nervous system (PNS), which controls the activity of gastrointestinal tract. ENS is the largest section of the autonomic nervous system. The ENS is able to control the gastrointestinal function independently of central nervous system (CNS) input.

11.2.9.1 Ganglia

A ganglion is the collection of neuron cell bodies in the PNS. On the basis of the primary functions, ganglia are of two types, sensory ganglia and autonomic ganglia. The dorsal

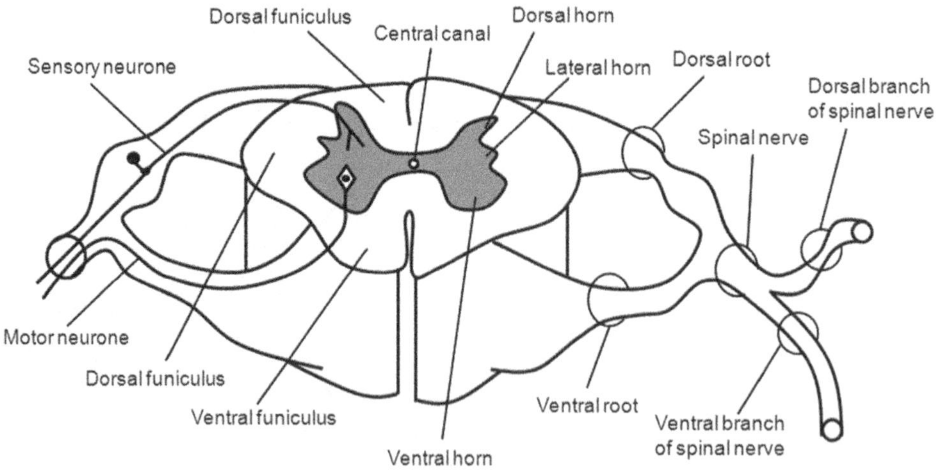

Fig. 11.7 Cross-sectional view of the spinal cord and spinal nerves

(posterior) root ganglion is the most common sensory ganglion. In ganglia, cell bodies are present and the axons of the neurons act as sensory fiber endings in the periphery, like in the skin. It extends into the CNS through the dorsal nerve root. In dorsal root ganglion, the neurons are unipolar and small round nuclei of satellite cells are also seen.

11.2.9.2 Nerves

A number of axons form a bundle in the PNS called nerve. In CNS, collection of axons is called a tract. Nerves have connective tissues in their structure. Blood vessels supply nourishment to the nerve tissues. Epineurium is the fibrous connective tissue layer surrounding the outer surface of a nerve. Within the nerve, a number of axons form a bundle called the fascicles. Each fascicle is surrounded by fibrous connective tissue layer called perineurium. Individual axons are surrounded by a loose connective tissue known as the endoneurium. These layers of connective tissue surrounding a nerve are similar to the connective tissue coverings of muscles. Nerves are of two types depending upon the regions to which they are connected, i.e., cranial nerves that are connected to the brain and spinal nerves which are connected to the spinal cord.

11.2.9.2.1 Cranial Nerves

The nerves associated with the brain are called cranial nerves. These nerves come out from the part of CNS present within the cranium. These nerves are mainly responsible for the sensory and motor functions of the head and neck (one of these nerves targets organs in the thoracic and abdominal cavities as part of the parasympathetic nervous system). Twelve pairs of cranial nerves are present in the nervous system. The first and second are connected to the forebrain, and rest of the nerves are connected to the brain stem.

The nerves are classified as sensory nerves, motor nerves, or a combination of both (known as mixed nerve). Within the 12 pairs of cranial nerves, 3 pairs of the nerves (cranial nerves I, II, and VIII) are exclusively composed of sensory fibers; 5 pairs of nerves (III, IV, VI, XI, XII) are strictly motor; and the remaining 4 pairs (V, VII, IX, X) are mixed nerves (Table 11.1).

11.2.9.2.2 Spinal Nerves

The nerves which arise from the spinal cord are called spinal nerves. Their arrangement is much more regular than that of the cranial nerves. Each nerve has both sensory and motor fibers, which separate into two nerve roots. Generally, a pair of spinal nerves (one right and another left) come out caudal to the vertebra of the same number and name (exceptions are cervical and caudal nerves). For example, the first pair of thoracic nerves originate from the intervertebral foramina between the last thoracic and first lumbar vertebrae, and the first pair of lumber nerves originate from the foramina between the first and second lumbar vertebrae.

The sensory axons of the spinal nerve enter into the spinal cord as the dorsal root, and the motor fibers of the spinal nerve, both somatic and autonomic, come out as the ventral root. The dorsal root ganglion is an enlargement of the spinal nerve. There are 31 pairs of spinal nerves attached to the spinal cord by two routes. The nerves are named according to the name of the structure from where it originates, viz. cervical, thoracic, lumbar, sacral, and coccygeal nerve. There are

Table 11.1 Cranial nerves with their origin and functions

Number	Name	Type	Arises from	Major function
I	Olfactory	Sensory	Olfactory bulb	Sense of smell
II	Optic	Sensory	Diencephalon	Vision, papillary light reflexes
III	Oculomotor	Motor	Midbrain	Parasympathetic innervation to the iris sphincter and ciliary muscles for constriction of pupil and accommodation reaction of lens, respectively
IV	Trochlear	Motor	Midbrain	Dorsal oblique muscle, rotates the dorsal portion of eye medioventrally
V	Trigeminal	Mixed	Pons	Motor: muscles of mastication Sensory: Face rostral to ear
VI	Abducens	Motor	Medulla	Lateral rectus and retractor bulbi muscles for lateral movement of eye, retraction of eye if it exits orbital fissure
VII	Facial	Mixed	Medulla	Muscles of facial expression and taste (rostral two-thirds of tongue) for cutaneous sensation of tongue, and cutaneous sensation of inner surface of pinna
VIII	Auditory (vestibulocochlear)	Sensory	Medulla	Equilibrium and hearing
IX	Glossopharyngeal	Mixed	Medulla	Sensory and motor to pharynx and palate, parasympathetic to zygomatic and parotid salivary glands
X	Vagus	Mixed	Medulla	Sensory and motor to pharynx and larynx, thoracic and abdominal viscera
XI	Accessory	Motor	Medulla	Trapezius and parts of sternocephalicus and brachiocephalicus muscle
XII	Hypoglossal	Motor	Medulla	Movement of tongue

8 pairs of cervical nerves (C1–C8), 12 pairs of thoracic nerves (T1–T12), 5 pairs of lumbar nerves (L1–L5), 5 pairs of sacral nerves (S1–S5), and 1 pair of coccygeal nerves.

11.2.9.2.3 Nerve Plexuses

Spinal nerves originate from the spinal cord, come out through the vertebral column, and enervate the periphery. The nerve fibers of different spinal nerves come together to form a bundle and again come out and give rise to systemic nerves. These nerve bundles are called nerve plexus. The formation of nerve plexus is seen at four places in the body. The other spinal nerves directly correspond to nerves at their respective levels. So, nerve plexus is the network of nerve fibers without any associated cell bodies.

Out of the total four nerve plexuses, two are found at the cervical region, one at the lumbar region, and one at the sacral region. The axons of cervical nerve C1–C4 form the cervical plexus and again branches are distributed in the head and neck region. Spinal nerves C3, C4, and C5 together form the phrenic nerve, which is connected with diaphragm. The C5, C6, C7, C8, and T1 come together and form the brachial plexus. Brachial plexus innervates the arms. Three nerves originating from the brachial plexus are the medial nerve, radial nerve, and ulnar nerve. The spinal nerves L1 through L4 form the lumbar plexus and give rise to the nerves enervating the pelvic region and thigh region of legs. Femoral nerve is a major nerve that arises from the lumbar plexus, which gives rise to saphenous nerve that extends through the lower part of legs. The sacral plexus is formed by the lumbar nerves L4 and L5 and the sacral nerves S1–S4. The most important systemic nerve which arises from this plexus is the sciatic nerve. It is a combination of the tibial nerve and the fibular nerve. The sciatic nerve runs across the hip joint. The sacral plexus supplies nerves to the posterior leg.

The systemic nerves that arise from the different nerve plexuses have fibers for the function of both the sensory and motor activities. The sensory fibers extend from cutaneous or other peripheral sensory surfaces and send information to the CNS. The sensory neurons in the dorsal root ganglia enter the spinal cord through the dorsal nerve root. The motor neurons of the anterior horn of the spinal cord, which emerge in the ventral nerve root, send action potentials to cause skeletal muscles to contract in their target regions. The spinal nerves of the thoracic region, T2 through T11, do not form any plexuses, but they give rise to the *intercostal nerves* present in the intercostal space.

11.2.10 Myelin Sheath

Several nerve fibers are covered by a lipid envelope called myelin sheath, which acts as an insulator and helps in the transport of the action potential faster (saltatory conduction). The myelin sheath is deposited by Schwann cells in PNS and by oligodendrocytes in CNS. The process by which myelin sheath is formed or deposited on nerve fibers is called myelinogenesis, which occurs as follows: the nerve fiber to be myelinated is first invaginated by the cell membrane of the Schwann cell and thus a double-layered mesaxon is formed. Then, the Schwann cell rotates around the nerve fiber and several layers of this mesaxon (actually two layers of the cell membrane).

11.2.11 Sensory Receptors

Receptors which are related to the nervous system are called sensory receptors. Sensory receptors are biological transducers, which can convert various forms of energy into action potential in the sensory nerves to which they are connected. Sensory receptors are present in everybody's tissue except the nervous system itself. These receptors receive different types of information from inside and outside the body to maintain homeostasis. The sensory receptors correctly sense other types of changes, which occur inside and outside the body.

11.2.11.1 Classification of Sensory Receptors

Based on structure, sensory receptors are of three types:

1. *Free nerve endings*: Most sensory receptors are axon terminals of primary sensory neurons. Those receptors which do not have any modification are called free nerve endings. Free nerve endings are nonmyelinated and widely distributed in the body. They form many branches in the tissue. They are receptors for pain, touch, temperature, etc. Some sensory receptors have expanded nerve endings; that is, their nerve endings are thickened to form a specialized structure to detect the sensory stimuli. For example, in Merkel's corpuscle, a mechanoreceptor detects pressure.
2. *Encapsulated nerve endings*: In some sensory receptors, sensory terminals are covered by a connective tissue capsule, classified as encapsulated nerve endings. They are mechanoreceptors and are not myelinated. They are mainly seen in the inner dermis, fasciae, mesenteries, skeletal muscles, and some viscera and comprise Pacinian corpuscles, Meissner's corpuscles, and Ruffini's corpuscles.
3. *Specialized receptors*: These specialized receptors have distinct structural components for interpreting particular types of stimuli, for example, retinal photoreceptor cells. Some special mechanoreceptors, Golgi tendon organs, are present in skeletal muscle tendons and muscle spindles. They are responsible for the awareness of kinesthesia (i.e., joint position direction and velocity of joint movements). They are called proprioceptors.

The receptors can also be classified into three types based on the origin of stimuli:

1. *Exteroceptor*: For stimuli outside the body. For example, somatosensory receptors on the skin.
2. *Interoceptor*: For stimuli inside the body, i.e., from the internal tissue and organs. For example, receptors sense the blood pressure in the aorta or carotid sinus.
3. *Proprioceptors*: Proprioceptors respond to the stimuli originating in the muscles and tendons.

According to the nature of the stimulus, sensory receptors are of the following types:

(1) Mechanoreceptors: Physical stimuli, such as pressure and vibration, and sensation of sound and body position (balance), are interpreted through a mechanoreceptor. (2) Nociceptors: Receptors for pain. The potential damage to the tissues by noxious stimulation gives rise to the sensation of pain. The pain receptors in the skin and other tissues are free nerve endings. Pain sensation may be of three types, i.e., pricking pain, burning pain, and aching pain. (3) Chemoreceptors: Receptors for chemical changes. Important chemoreceptors are (a) receptors for taste buds; (b) receptors of the olfactory epithelium for detection of smell; and (c) chemoreceptors in the carotid body and aortic arch.

(4) Thermoreceptors: Thermoreceptors are stimulated by temperature changes. When responding to decreased temperature, some of these thermoreceptors are called cold receptors, and others that respond to increased temperature are called warm receptors. (5) Osmoreceptors: Receptors for osmotic changes. They respond to solute concentrations of body fluids. (6) Photoreceptors: The photoreceptors are found within rod and cone cells of the retina. When light falls on the photoreceptors, the photopigments are transformed leading to changes in the membrane potential of photoreceptors.

11.2.12 Reflex Action

The sudden rapid involuntary effector response to a sensory stimulus is a reflex. Reflexes are the simplest example of the general function of the nervous system: a collection of sensory input, integration, and motor output. Marshall Hall, an English physician, first observed this type of action in 1833. Neural reflex involves sensory fibers to CNS and motor fibers to effectors. Reflexes help maintain homeostasis of different autonomic functions like heart rate, breathing rate, BP, and digestion. Reflexes also perform other important automatic actions like swallowing, sneezing, coughing, and vomiting. Reflexes also help in maintaining the balance and posture of the body.

11.2.12.1 Properties of Reflexes

Reflexes have the following properties:

1. Reflexes are spontaneous reactions.
2. Reflexes are automatic.
3. Reflexes are a short-lived response.
4. Reflexes are a mechanical action.
5. Spinal cord is predominately involved in this (though the brain is also involved, e.g., cranial reflex).

11.2.12.2 Reflex Arc

The path through which the reflex action takes place is known as a reflex arc (Fig. 11.8). The reflex arc is essential for maintaining the posture and locomotion of the animal and is very useful for the clinical diagnosis of nervous disorders. Reflexes are vital for both survival and different critical behaviors. A reflex is composed of five primary components, and abnormality in any one of these components alters reflex response.

Fig. 11.8 Basic components of the reflex arc

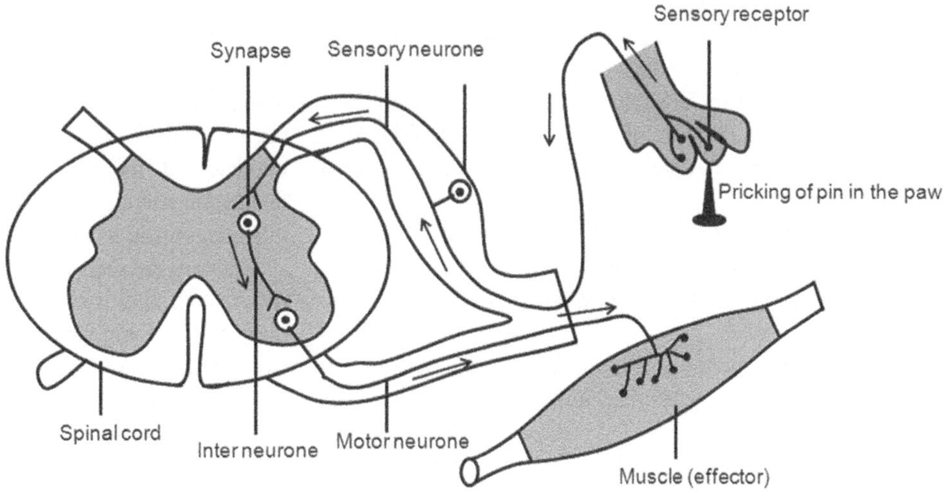

11.2.12.3 Basic Components of the Reflex Arc

1. **Receptors**: Reflex arcs begin with a sensory receptor that receives the stimulus and generates impulses. These receptors send the signal to the next component of the reflex arc, i.e., sensory neuron. Different sensory receptors are present in the body. Still, all have a common function, i.e., transduction of different types of stimuli (like light, heat, cold, pressure, taste) into a cellular response that directly or indirectly produces action potentials along the sensory neuron.
2. **Sensory neuron**: Sensory neuron or afferent neuron carries the nerve impulse in the form of action potential to the interneurons of the brain or spinal cord. Sensory neurons enter the spinal cord through the dorsal roots or enter the brain through cranial nerves.
3. **Interneuron**: Interneurons act as a processing center, process the information, and generate responses. In this part of the reflex arc, synapse formation is seen. The majority of the reflexes are polysynaptic. However, a few reflex arcs originating from muscle spindles are monosynaptic.
4. **Motor neuron**: Motor neurons or efferent neurons transmit the brain or spinal cord response to the effector organs. These neurons leave the spinal cord through the ventral roots and leave the brain through the cranial nerves.
5. **Effectors**: The last part of a reflex arc is the effector. The effector shows the effect of the reflex action. It may be an organ, muscle, or gland. In "knee jerk," the effector is the quadriceps muscle of the leg.

11.2.12.4 Types of Reflex Action

Reflexes can be classified in several ways as follows:

1. Based on the control center reflex, actions are of two types: cranial reflexes and spinal reflexes.
   (a) **Cranial reflexes**: It is under the control of cranial nerves and takes place in the facial or head area. Cranial reflexes are slow in response, and hence there is no emergency. The brain generally regulates cranial reflex. Constriction of the pupil in response to light is an example of cranial reflex. The release of saliva after the sight or smell of food is another example.
   (b) **Spinal reflexes**: It involves only the spinal nerves, and the response in spinal reflex is quick. Examples are the stretch reflex, knee-jerk, or patellar reflex.
2. Based on the previous experiences, reflex actions are classified into two types, i.e., unconditional reflexes and conditional reflexes:
   (a) **Unconditional reflex**: These are innate responses that are the same among the same species members. It is inborn, and they do not require any previous experience. This type of response helps the particular animal to adapt to a stable living condition. Examples are sneezing, coughing, hiccupping, and yawning.
   (b) **Conditional reflexes**: It is also called an acquired reflex. This type of reflex develops after birth through conditioning or learning: for example, secretion of saliva after seeing known food.
   (c) **According to the number of synapses in the reflex path**

According to the number of synapses in the reflex path, the reflex action is classified into monosynaptic reflex and polysynaptic reflex:

1. **Simple monosynaptic**: There is only one synapse in the reflex path; that is, it involves only sensory and motor neurons, for example, knee-jerk reflex.
2. **Complex polysynaptic reflex**: There is more than one synapse in the reflex path; that is, it involves sensory interneuron and motor neuron, for example, cycling and swimming.

11.2.12.4.1 Clinical Classification

Clinical classification is essential and is used in clinical practice to examine the nervous system to diagnose any abnormality in the nervous system.

(1) Superficial reflexes: planter reflex, abdominal reflex, etc. (2) Deep reflexes.

Generally, the majority of the reflexes are polysynaptic. Even if a sensory neuron participates in a monosynaptic reflex arc, it will often give off branches in the CNS that participate in polysynaptic reflex circuits. Even the simplest mammalian reflex responses often involve the excitation of a given muscle or muscles and the inhibition of another (usually antagonistic) muscle or muscles. For example, in the knee-jerk reflex, some sensory neurons make excitatory monosynaptic connections with motor neurons that activate the quadriceps muscle. In addition to that, other terminal branches of that same sensory neuron participate in a disynaptic circuit that inhibits motor neurons innervating the antagonistic hamstring muscle.

Reflexes are seen in most parts of the nervous system. So, reflexes are vital for clinical examination of animals for any diseases involving the nervous system, for example, pupillary light reflex, muscle stretch (knee jerk) reflex, and flexor reflex. Any abnormality of any component of the reflex arc results in an altered reflex action.

11.3 Autonomic Nervous System

Autonomic nervous system (ANS) is a part of the peripheral nervous system, which is associated with the involuntary functions of the body. Autonomic nervous system controls the visceral organs and endocrine glands, and thus it is associated with the maintenance of homeostasis. Autonomic nervous system is a motor system; hence, it is also called *visceral efferent motor* system. The ANS regulates the activity of heart, gastrointestinal (GI) tract, respiratory system, and salivary glands. ANS is also associated with perspiration, dilation of pupil, micturition, and sexual arousal. All the organs are innervated by the ANS, except in skeletal muscle. Skeletal muscle is supplied with the somatomotor nervous system.

11.3.1 Divisions of Autonomic Nervous System

There are two major subdivisions of autonomic nervous system: the sympathetic nervous system and parasympathetic nervous system (Fig. 11.9). The functions of these two divisions are opposite to each other. Generally, these two systems work together, typically in antagonistic manner (seems complementary to each other), which helps in the maintenance of homeostasis, as happened in regulation of heart rate and in respiratory cycles. The enteric nervous system is considered as the third subdivision of the autonomic nervous system. The enteric nervous system is composed of interconnected sensory, motor, and interneurons located in the GI tract, which control the activity of the gut. However, the neurons in the GI tract can also be influenced by the CNS through input from the sympathetic and parasympathetic subdivisions. Most of the visceral organs receive innervation from both sympathetic and parasympathetic divisions, but some organs, like sweat glands (both apocrine and merocrine type), adrenal medulla, blood vessels, pancreatic islets, pineal gland, pupillary dilator muscles, vascular smooth muscle of skeletal muscle, skin, and arrector pili, are innervated only by the sympathetic division. However, some organs like tissues of sphincter muscle of the iris, ciliary muscle, and nasopharyngeal glands receive only parasympathetic innervations. Autonomic nervous system has slow and long-lasting effect. Some visceral organs like heart and GI tract have intrinsic neuronal system, which helps in rhythmic movements. The major regulatory center for ANS is the hypothalamus; that is why it is called the *captain of autonomic nervous system*. Generally, autonomous functions are involuntary, but few actions can work alongside some degree of conscious control. The hypothalamic control is related with the reticular formation, which is located in the brain stem. The efferent tracts of the reticular formation reach the sympathetic and parasympathetic nuclei located in the brain stem and spinal cord. Hypothalamus releases the releasing and inhibitory hormones, which control the release of pituitary hormones. So, neuronal as well as hormonal influences are the basis for hypothalamic regulation of different major functions of the body, like heart rate, respiration rate, blood pressure, body temperature, conjugate eye movement, locomotion, swallowing, vomiting, micturition, defecation, water balance, food intake, circadian rhythms, and emotion. The cerebral cortex also has some influence on the ANS; for example, the sight of food triggers secretion of saliva and anticipation of a walk increases the heart rate of a dog.

11.3.2 Organization of the Autonomic Nervous System

In dogs, the sympathetic fibers arise from the thoracic (T1–T13) and lumbar (L1–L3) spinal cord. Hence, sympathetic division is called thoracolumbar division, whereas

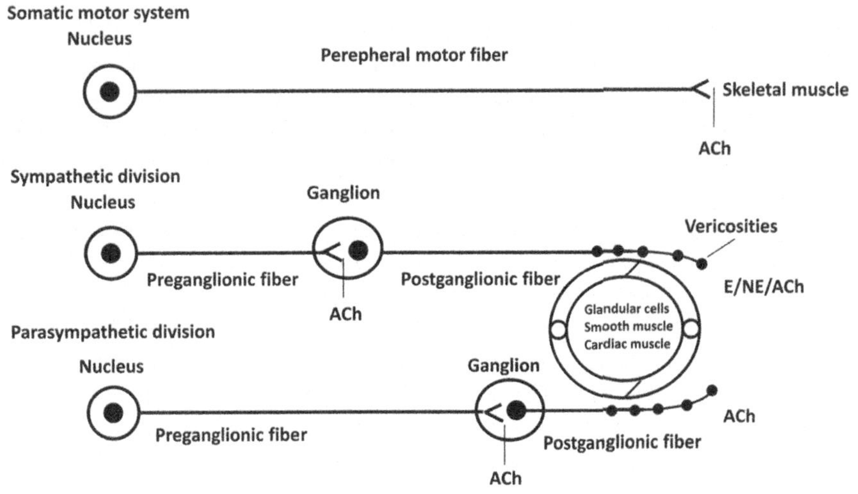

Fig. 11.9 Organization of the autonomic nervous system

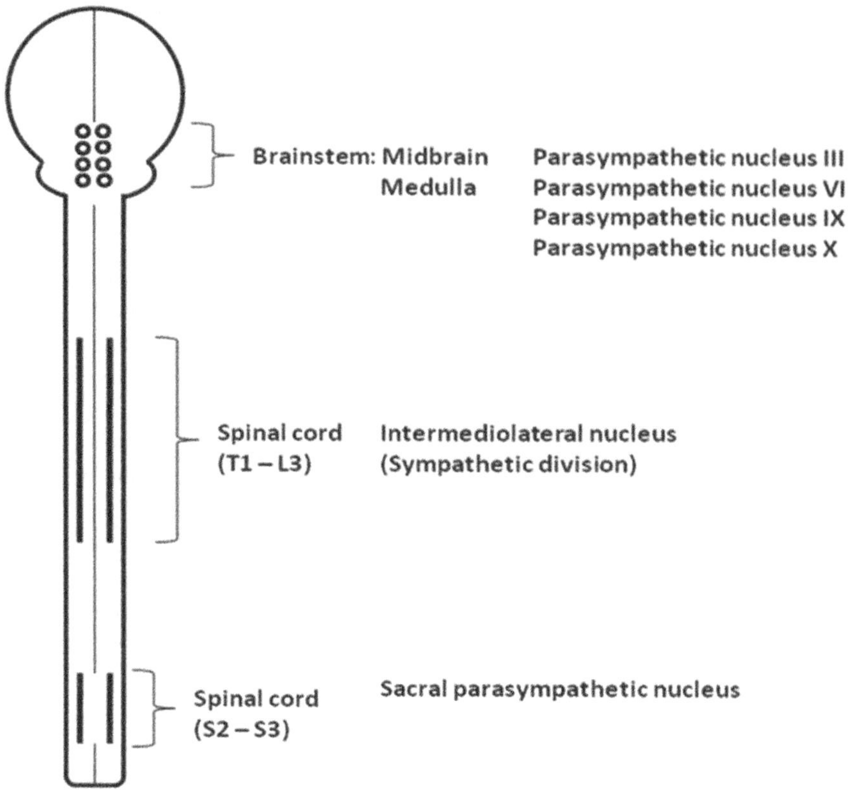

Fig. 11.10 Origin of sympathetic and parasympathetic fibers in dogs. Sympathetic fibers arise from the **thoracic (T1–T13)** and **lumbar (L1–L3)** spinal cord. The parasympathetic fibers originate from the brain stem and **sacral cord segments (S2–S3)**

parasympathetic fibers originate from the brain stem and sacral cord segments (S2–S3) (Fig. 11.10). Thus, the parasympathetic division is called craniosacral division. In the somatic nervous system, the cell body of the neurons remains in the central nervous system (CNS) and the axon fiber extends to the skeletal muscle. Neuromuscular junction is present at the junction between the nerve fiber and the skeletal muscle.

In the autonomic nervous system, both the divisions require a two-neuron chain between the nucleus, the CNS, and the peripheral target organ (Fig. 11.9). The synapse between the two neurons occurs outside the CNS in the ganglia. The presynaptic axon is called preganglionic fiber, and the postganglionic axon is called a postganglionic fiber. The cell body of preganglionic neuron remains in the CNS, whereas the cell body of the postganglionic neuron remains in the ganglion. The postganglionic fibers transmit impulse to the target tissue and organs. Chemical synapses are present both between the preganglionic and postganglionic neurons and between the postganglionic neuron and the cells of its target organ. Generally, the preganglionic fibers are myelinated, whereas the postganglionic fibers are nonmyelinated.

Role of sympathetic and parasympathetic nerves is summarized in Table 11.2. The apocrine sweat glands are innervated adrenergic sympathetic postganglionic fibers and secrete a viscous fluid that may contain pheromones. The merocrine (or eccrine) sweat glands are supplied by sympathetic cholinergic postganglionic fibers. Merocrine glands are located in the skin of footpads. The adrenal medulla is innervated by sympathetic preganglionic fibers that synapse with chromaffin cells, the vestigial postganglionic neurons of the adrenal medulla. Most of the chromaffin cells release epinephrine and some norepinephrine. Blood vessels are only supplied with the sympathetic division; they are not innervated by the parasympathetic division. So, constriction and dilation of the blood vessels occur due to decreasing or increasing sympathetic stimulation, respectively. The response of different tissues and organs to sympathetic and parasympathetic stimulation varies differently. During any adverse situation, the sympathetic stimulation results in vasoconstriction, which causes increased blood pressure, increased heart rate, increased airflow through the lungs, and epinephrine release from the adrenal gland, whereas vasodilation occurs in the heart, lungs, and skeletal muscles to supply the needed oxygen. The parasympathetic effects are suppressed during the fight-or-flight response.

11.3.2.1 Sympathetic Nervous System

The sympathetic system helps the body to cope up with the adverse conditions and strengthens the body's defense against those conditions by expenditure of energy. The sympathetic system does it either through fight with that adverse

Table 11.2 Influence of sympathetic and parasympathetic nervous system on different tissues and organs

Organ	Tissue	Sympathetic nervous system	Type of receptor	Parasympathetic nervous system
Skin	Apocrine sweat glands	Increase in secretion	β_2	–
	Merocrine sweat glands	Increase in secretion	M_3	–
	Arrector pili	Erection	α_1	–
Eye	Iris: dilator muscle	Dilation of pupil	α_1	–
	Sphincter muscle	–	–	Constriction of pupil
	Ciliary muscle	–	–	Contraction
Lung	Bronchiolar muscle	Relaxation	β_2	Contraction
Heart	SA node	Increase in heart rate	β_1	Decrease in heart rate
	AV node and Purkinje fibers	Increase in conduction velocity	β_1	Decrease in conduction velocity
	Atria, ventricle	Increase in contractility	β_1	Decrease in contractility
Arteries	Skin and mucosa	Constriction	α	–
	Salivary glands	Constriction	α	–
	Cerebral	Slight constriction	α	–
	Skeletal muscle	Dilation	β_2	–
	Coronary	Dilation	β_2	Slight dilation
	Pulmonary	Dilation	β_2	–
	Abdominal viscera	Constriction	α	–
Veins		Constriction, dilation	α, β_2	–
Gastrointestinal system	Stomach, intestinal tract	Decrease in motility	α, β_2	Increase in motility
	Sphincters	Contraction	α_1	Relaxation
	Gastric gland	Decrease in secretion	α_2	Increase in secretion
	Gallbladder	Relaxation	β_2	Contraction
	Liver	Glycogenolysis, gluconeogenesis	α_1, β_2	Glycogen synthesis
	Pancreatic acini	Decrease in secretion	α	Increase in secretion
	Pancreatic Islets	Decrease in secretion	α_2	–
Adrenal medulla		Secretion of E and NE	N	–
Kidney		Renin secretion	β_2	–
Urinary bladder	Detrusor muscle	Relaxation	β_3	Contraction
	Trigone and sphincter	Contraction	α_1	Relaxation
Reproductive organs	Penis	Ejaculation	α_1	Erection
	Uterus (pregnant)	Contraction	α_1	Variable
	Uterus (nonpregnant)	Relaxation	β_2	Variable
Glands	Lacrimal	Slight secretion	α	Increase in secretion
	Salivary	Slight viscous secretion	α	Increase in watery secretion
	Nasopharyngeal	–	–	Secretion
	Pineal	Melatonin synthesis	β	–

situation or through flight from that situation. That is why the sympathetic nervous system is also known as the "fight-or-flight" system. In the sympathetic nervous system, preganglionic sympathetic neuron cell bodies are present in a small lateral horn of the spinal cord grey matter between dorsal and ventral horns. The axons leave through the ventral root, enter the spinal nerve, and then leave it just outside the intervertebral foramen to join a longitudinal chain of autonomic ganglia.

In the head and neck region, the sympathetic innervation is mediated by the cranial cervical ganglia. Preganglionic fibers from spinal cord segments T1–T5 (some fibers may even come from T6 and T7) join the vagosympathetic trunk to reach the cranial cervical ganglion.

Postganglionic fibers arise from the cranial cervical ganglion innervate salivary glands, nasal glands, smooth muscles (blood vessels, periorbital area, eyelids, pupillary dilator), carotid body, carotid sinus, and thyroid gland (Table 11.2). Postganglionic fibers may also join the cranial laryngeal branches and pharyngeal branch of the vagus nerve.

The sympathetic trunk ganglia caudal to T4 innervate the rest of the body wall and extremities. Postganglionic fibers join the spinal nerves by way of the rami communicantes to innervate blood vessels, sweat glands, and arrector pili. These structures do not receive parasympathetic innervation, so they are an exception to the dual innervation. Increased activity of the sympathetic system results in the contraction of arteriolar smooth muscles. This leads to an increase in peripheral vascular resistance and subsequent increase in blood pressure. In contrast, decreased activity of the sympathetic system decreases vascular resistance due to relaxation of the arteriolar smooth muscle. Thus, decrease in sympathetic activity lowers blood pressure.

The organs of thoracic cavity are mainly innervated by the cervicothoracic and middle cervical ganglia. The ansa subclavia may also contribute some fibers. Preganglionic fibers originate from spinal cord segments T1–T4. They reach postganglionic neurons in the cervicothoracic ganglion by way of the rami communicantes. Preganglionic fibers also ascend to the ansa subclavia and middle cervical ganglion, where they synapse with postganglionic neurons. These fibers innervate the vasculature and smooth muscle of the respiratory airways and the lung. The sympathetic stellate cardiac nerve mediates relaxation of the smooth muscle of the respiratory airways and blood vessels, whereas the vagus nerve contracts the smooth muscle. Stimulation of the sympathetic fibers results in an increase in heart rate by increasing the pacemaker activity of the sinoatrial (SA) node cells, impulse conduction at the atrioventricular (AV) node, and contractile force of atrial and ventricular muscle fibers.

The abdominal and pelvic viscera are innervated by spinal cord segments T5–L3. To reach the celiac ganglion in the abdominal cavity, preganglionic fibers from the thoracic spinal segments caudal to T5 descend in the sympathetic trunk and emerge as the greater thoracic splanchnic nerve at the level of the T13 sympathetic trunk ganglion. From the celiac ganglion, postganglionic fibers accompany the arteries to the stomach, duodenum, pancreas, liver, gallbladder, spleen, and adrenal glands. Motility of the gastrointestinal tract is enhanced by parasympathetic fibers of the vagus nerve. However, how the sympathetic division controls the gastrointestinal tract is not clear. It has been speculated that adrenergic fibers synapse on inhibitory α-adrenergic receptors on parasympathetic postganglionic cells of the myenteric plexus and inhibitory β-adrenergic receptors on smooth muscle fibers. Thus, peristaltic action can be decreased by the sympathetic system. The adrenal medulla receives sympathetic preganglionic fibers from cord segments T4 (or T5) to L1 (or L2). The adrenal medulla is composed of chromaffin cells, which are vestigial postganglionic neurons. Chromaffin cells secrete catecholamines (mostly epinephrine and some norepinephrine) into the bloodstream in response to signals from sympathetic preganglionic neurons. Thus, the sympathetic system regulates functions of endocrine cells in the adrenal gland. Preganglionic fibers from spinal segments L1–L3 reach the abdominal and pelvic ganglia via the sympathetic trunk. They either leave the sympathetic trunk ganglia at the level they enter or descend in the sympathetic trunk before exiting. Each sympathetic trunk ganglion of the lumbar segments gives rise to a lumbar splanchnic nerve. It is named after the level from which it arises. The first five lumbar splanchnic nerves supply one or more of the following collateral ganglia: celiac, cranial mesenteric, renal, and gonadal ganglia.

11.3.2.2 Parasympathetic Nervous System

Parasympathetic nervous system is associated with the conservation and restoration of energy. Hence, the parasympathetic nervous system is known as the *"rest-and-digest"* system. The parasympathetic nervous system helps in the conservation of energy sources of the body. Sympathetic nervous system decreases the heart and blood pressure. It also decreases the force of contraction and thus it reduces the energy expenditure in the body. It also increases the digestive function and increases the blood flow to the GI tract. It increases the GI motility and increases the secretion of digestive enzymes. The parasympathetic division also regulates the micturition process by contracting the urinary bladder.

The preganglionic fibers of parasympathetic division arise from the cranial nerves and sacral segments of the spinal cord; that is why the parasympathetic division is also known as the craniosacral division of the ANS. The fibers of the cranial portion are distributed via four cranial nerves: the oculomotor (III), facial (VII), glossopharyngeal (IX), and vagus (X) nerves. Within these four nerves, the first three nerves supply parasympathetic fibers to smooth muscle and glands of the head, whereas the vagus nerve supplies parasympathetic fibers to the visceral organs of the thorax and neck. It also supplies nearly all the abdominal viscera.

Parasympathetic fibers, which arise from the oculomotor nerve, supply the iris and ciliary body of the eye and control the contraction of the pupillary sphincter and the ciliary muscle. The parasympathetic stimulation results in the constriction of pupil, and the lens becomes more convex, allowing greater refraction of light for near vision. The parasympathetic fibers of the facial nerve supply the submandibular and sublingual salivary glands. It also innervates the lacrimal glands.

The parasympathetic fibers of the glossopharyngeal nerve supply the parotid and zygomatic salivary glands. The parasympathetic innervation causes increase in watery secretions from these glands. The parasympathetic fibers of the vagus nerve innervate most of the thoracic and abdominal viscera.

The distal part of the digestive tract (including the transverse colon and the area caudal to it) and the pelvic viscera are innervated by parasympathetic fibers from the sacral portion of the parasympathetic nervous system originating from cord segments S2 and S3 in dogs and S1–S3 in cats. These pelvic fibers intermix with sympathetic nerves to form the pelvic plexus. The preganglionic fibers run through the ventral root of the S2 and S3 spinal nerves; then, they together form the pelvic nerve which is present on the lateral wall of the distal rectum. Then the pelvic nerve forms a plexus. The plexus also receives the sympathetic fibers of the hypogastric nerve. The preganglionic fibers of the sacral segment either terminate in the pelvic ganglia of the pelvic plexus or pass through the plexus to terminate in the terminal ganglia in the wall of the pelvic viscera. The pelvic nerve is essential for erection, ejaculation, urination, and defecation.

11.3.3 Neurotransmitters and Their Receptors in Autonomic Nervous System

Acetylcholine (ACh) is the preganglionic neurotransmitter for both sympathetic and parasympathetic systems. The ACh is also a parasympathetic postganglionic neurotransmitter. Hence, the parasympathetic division is also known as the cholinergic division. In sympathetic division, norepinephrine acts as the postganglionic neurotransmitter, and the sympathetic division is called the adrenergic division. Exceptions are found in the merocrine (or eccrine) sweat glands, which are innervated by cholinergic sympathetic postganglionic fibers. The terminal portions of the postsynaptic axons form a number of beadlike swellings known as varicosities. They are also known as boutons en passage. Varicosities contain neurotransmitters and remain close to the surface of the effector cells.

11.3.3.1 Cholinergic Receptors

Based on the selective response to nicotine or muscarine, the cholinergic receptors are of two types: the nicotinic receptor and muscarinic receptor. Nicotinic acetylcholine receptor (nAChR) is mainly found in the neuromuscular junction as well as in all autonomic ganglia. The binding of ACh to the nAChR results in opening of the ion channels, which allows the influx of both Na^+ and K^+ according to the electrochemical gradient. Then the neurons are depolarized, as the driving force for Na^+ to enter the cell far exceeds that of K^+ to leave the cell. The activation of nicotinic receptors results in the generation of excitatory postsynaptic potentials (EPSPs) of the postsynaptic neuron.

Muscarinic acetylcholine receptors (mAChRs) are present in effector tissues innervated by parasympathetic postganglionic fibers. They are also present in the merocrine sweat glands innervated by cholinergic sympathetic fibers. Different types of muscarinic receptors are present, like M1, M2, and M3. All of these receptors are coupled to G proteins linked to second messenger systems. The binding of ACh to mAChR results in generation of either excitatory or inhibitory postsynaptic potentials (EPSPs or IPSPs). The postsynaptic response depends on the receptor subtype activated by ACh and subsequent opening of ligand-gated channels for specific ions. So, the action of ACh at a synapse depends on the muscarinic receptor subtypes present in the tissue. For example, binding of ACh to M2 receptors in the heart results in decreasing heart rate and contraction force, whereas binding of ACh to M3 receptors in bronchioles and urinary bladder causes contraction of bronchiolar and bladder smooth muscles. Atropine is an anticholinergic drug, which blocks parasympathetic effects. Atropine is used to dilate the pupil or to suppress salivation and respiratory secretions.

11.3.3.2 Adrenergic Receptors

There are two types of adrenergic receptors, the *alpha* (α) and *beta* (β). The *α-adrenergic receptors* are mainly excitatory, which cause vasoconstriction of blood vessels, raising blood pressure, constricting sphincters of the gastrointestinal tract, contracting urethral smooth muscle, and dilating the pupils. The *β-adrenergic receptors* are of different types, like β1, β2, and β3. The *β1-adrenergic receptors* in the heart (cardiac muscle, pacemakers) increase heart rate and contraction force. Beta-blockers that act on the β1 receptors of the heart reduce heart rate and prevent arrhythmias. The *β2-adrenergic receptor* is present in smooth muscle of the gastrointestinal tract. These β2 receptors relax gastrointestinal smooth muscle. The β2-adrenergic receptor is also present in the vascular smooth muscle of the heart and skeletal muscle, as well as bronchiolar smooth muscle. The *β3-adrenergic receptors* are present in the urinary bladder and relax detrusor muscle in response to norepinephrine released from the sympathetic hypogastric nerve.

11.3.4 Enteric Nervous System

The enteric nervous system (ENS) is considered as the third division of the ANS. Enteric nervous system is the network of motor and sensory neurons located within the walls of the gastrointestinal tract and its accessory glands (viz. pancreas, liver). It is influenced by both the parasympathetic and sympathetic divisions of the ANS, but the system is functional without input from outside the viscera. It is comprised of two well-organized neural plexuses. The submucosal plexus is present just deep to the inner lining of the gut. The another is myenteric plexus, found within the muscular layer, involved in the control of digestive tract motility. It senses the environment of the lumen and regulates gastrointestinal blood flow and epithelial cell function.

The enteric nervous system contains large number of neurons almost equal to the number of neurons present in the entire spinal cord. Enteric nervous system contains all the elements of the nervous system; hence, it is also known as the "mini brain." It contains the sensory neurons, interneurons, and motor neurons. The sympathetic as well as parasympathetic nerves connect the CNS to the ENS or directly innervate the GI tract. Though the ENS can act autonomously, normal function of GI tract often requires communication between the central nervous system and the enteric nervous system.

Learning Outcomes

- Nervous system is a special system which coordinates to collect the information from the external as well as the internal environment and sends to the respective center(s) to generate responses through the motor system. It is morphologically and functionally divided into two components, central nervous system (CNS) and peripheral nervous system (PNS). The central nervous system consists of the brain and spinal cord. The peripheral nervous system comprises sensory and motor nerves that run throughout the body.
- The nervous system is made up of a large number of cells (over 100 billion). The cells are mainly of two types, the neurons and neuroglia or glial cells. Neurons are the basic unit of the nervous system as the functions of the nervous system are carried out by neurons. Neurons are specialized types of conducting cells.
- The communication of neurons between each other and with other cells of the body, like muscle and glandular cells, occurs very fast, at specialized junctions called *synapse*. Neurotransmitters are the chemical transmitters or chemical messenger substances liberated at the nerve endings and help to transfer the nerve impulses in the presynaptic neuron to an adjacent cell (neighboring postsynaptic neurons or muscle or gland cells).
- Autonomic nervous system (ANS) is a part of the peripheral nervous system, which is associated with the involuntary functions of the body. Autonomic nervous system controls the visceral organs and endocrine glands, and thus it is associated with the maintenance of homeostasis. There are two major subdivisions of autonomic nervous system: the sympathetic nervous system and parasympathetic nervous system.
- The sympathetic system helps the body to cope up with the adverse conditions and strengthens the body's defense against those conditions by expenditure of energy. The sympathetic system does it either through fight with that adverse situation or through flight from that situation. Parasympathetic nervous system is associated with the conservation and restoration of energy. Hence, the parasympathetic nervous system is known as the "rest-and-digest" system.
- The enteric nervous system (ENS) is considered as the third division of the ANS. Enteric nervous system is the network of motor and sensory neurons located within the walls of the gastrointestinal tract and its accessory glands (viz. pancreas, liver).

Exercises

Objective Questions

Q1. Which was the first neurotransmitter discovered?
Q2. Unipolar neurons are mainly found in which part of the nervous system?
Q3. Which structure produces cerebrospinal fluid in the ventricles of brain?
Q4. Vomiting center is located in which part of the brain?
Q5. What is the major function of mammillary bodies?
Q6. Thermoregulatory center is located in which part of the brain?
Q7. Almost all of the cerebral cortex has direct two-way communication with which subcortical structure?
Q8. What is ganglion?
Q9. The summation of the excitatory as well as inhibitory activity occurs in which part of the neuron?
Q10. Free nerve endings are sensory receptors for which sensations?

Subjective Questions

Q1. Describe the different phases of synaptic transmission.
Q2. Write the basic characteristics of neurotransmitter.
Q3. Write the functions of different types of glial cells.
Q4. Write the fate of neurotransmitter.
Q5. Describe the different layers of meninges.
Q6. Write the formation and absorption of CSF.
Q7. Classify sensory receptors on the basis of structure.
Q8. Write the functions of cerebellum.
Q9. Describe the components of reflex arc.
Q10. Write the difference between sympathetic and parasympathetic nervous system.

Answer to Objective Questions

A1. Acetylcholine
A2. The afferent division of the PNS
A3. Choroid plexus
A4. Medulla oblongata
A5. Act as a relay center for olfaction and are associated with memory
A6. Hypothalamus
A7. Thalamus
A8. A ganglion is the collection of neuron cell bodies in the PNS
A9. Axon hillock
A10. Temperature and pain

Keywords for Answer to Subjective Questions

A1. (1) Axon potential arrives at the axon terminal, (2) voltage-gated Ca^{2+} channels open, (3) Ca^{2+} enters the presynaptic neuron, (4) Ca^{2+} signals to neurotransmitter vesicles, (5) vesicles move to the membrane and dock, (6) neurotransmitters released via exocytosis, (7) neurotransmitters bind to receptors, (8) signal initiated in postsynaptic cell

A2. (1) Must be synthesized in the neuron, (2) should be stored in the presynaptic terminal, (3) should be released at the synapse in amounts sufficient to exert a defined action, (4) should have its specific receptors on postsynaptic membrane, (5) should be removed quickly by the specific mechanism as soon its action is over.

A3. Astrocytes: form support in CNS, form the blood–brain barrier, secrete neurotrophic factors, take up K^+, neurotransmitters. Oligodendrocytes: form myelin sheaths in CNS. Microglia: phagocytic in function. Ependymal cells: create barriers between compartments. Schwann cells: form myelin sheaths in PNS. Satellite cells: support cell bodies.

A4. There are two processes for removal of neurotransmitters after their action. These are (1) enzymatic inactivation in the synaptic cleft and (2) diffusion away from the synaptic cleft.

A5. Meninges have three layers, viz. dura mater, arachnoid mater, and pia mater (from outside to inside).

A6. CSF is secreted by the ependymal cells located in the choroid plexus. CSF is absorbed and returns to the venous system mainly into dural venous sinuses, which are present intracranially between the endosteal layer and meningeal layers of the dura mater.

A7. (1) Free nerve endings, (2) encapsulated nerve endings, (3) specialized receptors.

A8. Postural adjustments in order to maintain balance.

A9. (1) Receptors, (2) sensory neuron, (3) interneuron, (4) motor neuron, (5) effectors.

A10. The sympathetic system helps the body to cope up with the adverse conditions and strengthens the body's defense against those conditions by expenditure of energy. The sympathetic system does it either through fight with that adverse situation or through flight from that situation. That is why the sympathetic nervous system is also known as the "fight-or-flight" system. Parasympathetic nervous system is associated with the conservation and restoration of energy. Hence, the parasympathetic nervous system is known as the "rest-and-digest" system.

Further Reading

Bradley K (2012) Cunningham's textbook of veterinary physiology, 5th edn. Elsevier

Frandson RD, Lee WW, Dee FA (2009) Anatomy and physiology of farm animals, 7th edn. Wiley–Blackwell

Guyton AC, Hall JE (2005) Textbook of medical physiology, 11th edn. W.B. Saunders, Philadelphia, pp 764–767

Orešković D, Klarica M (2014) A new look at cerebrospinal fluid movement. Fluids Barriers CNS 11:16

Reece WO (2015) Dukes' physiology of domestic animals, 13th edn. Wiley Blackwell

Sakka L, Coll G, Chazal J (2011) Anatomy and physiology of cerebrospinal fluid. Eur Ann Otorhinolaryngol Head Neck Dis 128(6):309–316

Swenson MJ, Reece WO (2005) Duke's physiology of domestic animals. Panima

Wright BL, Lai JT, Sinclair AJ (2012) Cerebrospinal fluid and lumbar puncture: a practical review. J Neurol 259(8):1530–1545

Special Senses 12

P. Visha, Pradip Kumar Das, Joydip Mukherjee, and Dipak Banerjee

Abstract

The senses of sight, smell, hearing and touch help the animals to perceive the environment and modulate their general behaviour, learning ability and social relationship. The eyes are the sensory organs that receive visual information from the environment and transmit them to the visual sensory area of the brain for interpretation. The eyes, equipped with an adjustable pupil and a lens, capture the illumination patterns in the environment as an 'optical picture' on a layer of light-sensitive photoreceptor cells in the retina. The auditory system perceives the frequency of sounds as pitch and their amplitude as loudness. The external and the middle ears conduct the sound to the auditory receptors (organ of Corti) in the cochlea of the inner ear. The auditory receptors are the hair cells embedded in the basilar membrane and apical surface and convert sound wave signals into nerve impulses. Olfaction (smell) is an animal's primary special sense perceived by the main and the accessory olfactory systems with receptors in the vomeronasal or Jacobson's organ located near the external nares. The odorant molecules combine with odorant receptors to mediate the signal transduction mechanism leading to a series of electrical events to facilitate the sense of smell. Gustation or taste is a chemical sensation perceived through chemoreceptors in the taste buds. The skin and its derivative or accessory structures, called the integumentary system that provides a physical barrier and antimicrobial protection, involve thermo- and immuno-regulation, excretion, secretion, pigmentation, sensation and locomotion of the animal.

P. Visha
Department of Veterinary Physiology and Biochemistry, Veterinary College and Research Institute, Tamil Nadu Veterinary and Animal Sciences University, Salem, Tamil Nadu, India

P. K. Das (✉) · J. Mukherjee · D. Banerjee
Department of Veterinary Physiology, West Bengal University of Animal & Fishery Sciences, Kolkata, West Bengal, India

© The Author(s), under exclusive license to Springer Nature Singapore Pte Ltd. 2023

P. K. Das et al. (eds.), *Textbook of Veterinary Physiology*, https://doi.org/10.1007/978-981-19-9410-4_12

Graphical Abstract

Description of the Graphical Abstract: The **special senses** are specialised structures in the body which provide information about the body and its environment and help the animals to modulate their general behaviour, learning ability and social relationship. It includes **vision** (for which eyes are the sensory organs, which receive visual information and transmit them to the visual sensory area of the brain for interpretation), **hearing** (the sense of sound perception occurs by the ears; it also helps in the maintenance of balance), **taste** (that is perceived through chemoreceptors in the taste buds of the tongue), **smell** (perceived by the main olfactory system and the accessory olfactory system with receptors in the nasal passages) and **touch** (a perception resulting from the activation of neural receptors, generally in the skin including hair follicles, but also in the tongue, throat and mucosa).

Keywords

Vision · Hearing and equilibrium · Smell · Taste · Skin

Learning Objectives

- Visual system, its components and photochemistry of vision
- Auditory system and mechanism of hearing and equilibrium
- Olfactory system and mechanism of olfaction
- Gustatory system and mechanism of taste sensation
- Skin (integumentary system) and its function

12.1 Vision

The eyes' photoreceptor cells are responsive to the electromagnetic frequencies with the wavelengths ranging from nearly 380 to 750 nm, the visible light to the humans, domestic animals and birds. The ultraviolet (UV) rays and the rays of short wavelengths than the visible light are generally absorbed by the ozone, and the cloud absorbs rays of higher wavelengths. The frequencies of the visible light are 4×10^{14} to 8×10^{14} hertz (Hz). The generated electrical potential within the sensitised photoreceptor cells sends the stimulation to the brain's visual centre and recognises the image developed by the combinations of various frequencies of light. About 96% of the nine million animal species, including all the vertebrates, have eyes with species-specific morphological characteristics and a diversified nervous system. Hence, different animals and birds have a unique world of vision. This chapter considers the basic similarities in morphology and mechanism of vision of the domestic animals, birds and wild animals.

12.1.1 Visual System

The eyes are the sensory organs that receive visual information from the environment and transmit them to the visual sensory area of the brain for interpretation. The eyes, equipped with an adjustable pupil and a lens, capture the illumination patterns in the environment as an 'optical picture' on a layer of light-sensitive photoreceptor cells in the retina. The retina facilitates feature analysis of the image. It transmits visual signals through the steps of visual processing to the various structures of the brain, where it is finally perceived.

12.1.2 Structure of the Eye

The eye is a ball-shaped sensory organ enveloped by three layers: the outer fibrous *tunic*, the *sclera* modified anteriorly into a stratified squamous epithelial layer and the transparent *cornea* (Fig. 12.1). When preserved at a low temperature, the cornea absorbs water and expels water at a high temperature. Hence, corneal grafting is done immediately after expiration without any preservation process. The fibrous tunic provides mechanical support and protection to the eye. Inner to the sclera lies the vascular tunic consisting of the *iris*, *ciliary body* and *choroid*, which are vascularised and highly pigmented. The choroid provides nutrition to the ocular tissue. The melanocytes present within absorb light and prevent the light that escapes past the *retina* from being reflected into the retina, where it would burr the sharpness of the image. In nocturnal animals, these pigmented layers contain a reflecting pigment called *tapetum lucidum* (Latin; tapetum: carpet, lucidum: bright). It allows the retina to make optimal use of available light, but *visual acuity* reduces. The reflection of light from the tapetum causes the 'night shine' from the nocturnal animal's eye. The innermost layer is a neuroepithelial tunic. The retina consists of the photoreceptor cells (rod and cone cells) and other visual processing cells, such as *bipolar, ganglion, horizontal, amacrine and pigment epithelial cells*. These specialised photoreceptors absorb light and transduce the light energy into neural signals. The processing of these neural signals that begins in the retina continues as it passes along the pathways to the brain's occipital lobe, where the visual cortex processes these signals.

The eye's interior comprises two fluid-filled anterior and posterior cavities, separated by an elliptical *lens*. These structures are transparent to permit light to pass through the eye from the cornea to the retina. The lens, composed of jelly-like crystalline unique proteinous substance, is the primary structure responsible for vision accommodation. Disturbances in the oxygenation of the lens-forming substances cause more metabolites, including carbon dioxide, leading to opaque in the lens, called *cataracts*. The lens suspends from the suspensory ligament. The convexity of the lens can be altered by the ciliary muscles of the ciliary body. The retina facilitates focusing of images of objects located at varying distances from the eye, and ciliary muscle contraction increases the convexity and focuses near objects. The ability of the eye lens to adjust its focal length by changing curvature for focusing the image to the retina for both near and distant vision is called *accommodation*.

The space between the cornea and the lens is called the anterior chamber and posterior chamber, which lies between the iris and the suspensory ligament. These chambers fill with a clear, water-like fluid called *aqueous humour*. The ciliary processes of the ciliary body in the posterior chamber produce aqueous humour, and it flows into the anterior chamber through the *pupil*. It is absorbed into the venous system at the cornea and the iris angle. Disturbances in drainage of the fluid or contraction of extraocular muscle, particularly retractor bulbi, increase intraocular pressure, called *glaucoma*. A diaphragm of varying sizes separates the anterior and posterior chambers called the iris. The iris is a pigmented structure containing a dilator and constrictor smooth muscles that alter the pupil's diameter, which is the hole in the iris through which light passes to the retina. Behind the iris is the lens suspended in the eye by the suspensory ligaments. The suspensory ligaments attach to the lens and the ciliary body, a muscular structure located at the base of the iris. The intraocular fluid drains out through the trabecular network and canal of Schlemm.

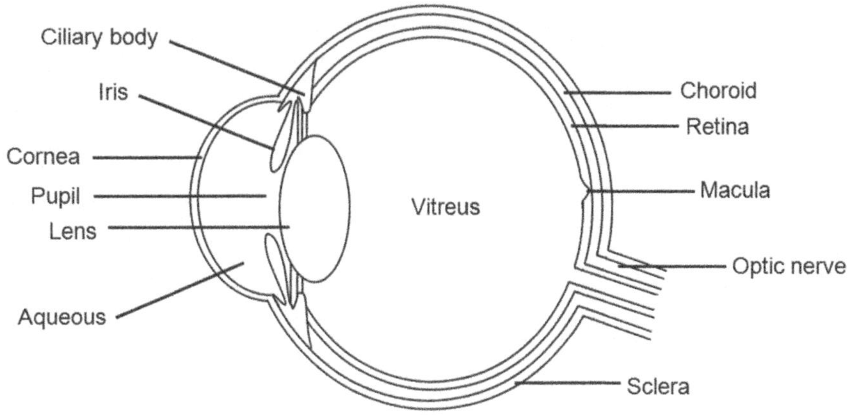

Fig. 12.1 Internal structure of the eye

Know More...

Pupil's World

The shape of a pupil and eye structure depends on the activities of animals. The round pupil appears in large and wild cats and lions, and also in humans, to find the prey animals from a large field, while domestic cats have an elliptical and vertical pupil to track close to the ground. Elliptical and horizontal pupils present in prey animals, viz. goats, sheep, horses and deer, have wider and narrower horizontal pupil for a broad field of vision. Usually, large-sized daylight predators have round pupils and forward-facing eyes. In contrast, small cat-like animals that hunt during the day and night comprise vertical slit pupil facilitating depth perception and night vision.

Behind the lens is a chamber filled with a hydrogel called *vitreous humour* that contains hyaluronic acid and collagen fibres to supply the retina's nutrition. The eye equipped with a pupil adjusts aperture diameter and possesses a lens that focuses light on the retina, where photoreceptor cells receive images. However, the retina is not just converting the image into nerve impulses; it also facilitates feature analysis of the captured image. Feature analysis and visual information processing progressively occur as visual signals are passed to the thalamus, rostral colliculus of the midbrain and visual cortex.

Lacrimal glands near the lateral canthus of the eye produce the tear in response to parasympathetic nerve stimulation. Tears flow over the cornea and are drained into the nose by the nasolacrimal duct. The *nictitating membrane* or the third eyelid is found in the medial canthus, and it aids in protecting the eye, and its glands also produce tears. The tear is also produced by the harderian glands, one of the lacrimal glands found in many birds and mammals except in carnivores.

The photoreceptor cells are named due to their shape and have four main functional components: the outer segment, inner segment, nucleus and synaptic body. The light-sensitive outer segment contains photochemical rhodopsin in the rod cells and iodopsin in the cone cells, stimulated by low and high light intensity, respectively. These photopigments are incorporated in as many as 700–2000 discrete disc membranes in humans, which are invaginations of cell membranes in each rod or cone cell. The retina contains approximately 80–110 million rod cells and four to five million cones in humans. The rod and cone cells differ in their distribution, with most cone cells present in the area centralis where the sharpest images are obtained. The inner segment of the photoreceptor contains mitochondria that play an important role in providing energy to the photoreceptor cells. The synaptic body, the terminal portion of the photoreceptor cell, connects with the bipolar cells that form the following linking structure in the vision chain. The bipolar cells are of two types: OFF and ON, which express different glutamate receptors and respond in opposite ways to the glutamate released by the photoreceptors. The OFF and ON bipolar cells synapse on OFF-centre and ON-centre ganglion cells, respectively. Ganglion cells are the only cells that fire action potentials and send visual information out of the retina. The ganglion cells fire in all lighting conditions, but the relative firing rate encodes information about light. A move from dark to light will cause OFF-centre ganglion cells to decrease their firing rate and ON-centre ganglion cells to increase their firing rate. In humans, the photoreceptors (nearly 100 million) transmit signals to bipolar cells (36 million), which send signals to the ganglion cells (one to two million cells per eye), indicating that as information flows centrally, the number of cells carrying the information decreases. The axons of ganglion cells unite to form the optic nerve and, along with blood vessels, leave the retina at the point known as the optical disc or blind spot, as no images are detected in this area.

12.1.3 Photochemistry of Vision

The photoreceptor cells contain photosensitive pigments that decompose on exposure to light and, in the process, cause the excitation of the nerve fibre leading from the eye. Photopigments are made up of opsins, a protein (in rods, the opsin is scotopsin, and in cones, the opsin is iodopsin) and an aldehyde of vitamin A, the retinal (retinene). Cones are colour-sensitive photoreceptors associated with day vision, and there are three kinds of cone opsins—blue cone (sensitive at 430 nm), green cone (535 nm) and red cone (575 nm). Rhodopsin and iodopsin are embedded in the disc membrane of the outer segment. Rod and cone cells contain a G protein called transducin, which gets stimulated by the light-activated photopigments, and transduction of visual signals occurs via opsins. The plasma membrane of the photoreceptor's outer segment contains cyclic guanosine monophosphate (cGMP)-gated Na^+ channels that open in the dark and close in response to light.

Additionally, K^+ channels present in the inner segment membrane of the rod and cone cells allow the leaking out of K^+, thus helping in maintaining proper K^+ levels. Na^+/K^+ ATPases located in the inner segments of the photoreceptor cells maintain intracellular concentrations of Na^+ and K^+. Generated electrical impulses in the photoreceptor and nerve cells can be measured by placing the electrode on the cornea and the skin, called *electroretinography*.

12.1.3.1 Photoreceptor Activity in the Dark

In the dark, the 11-cis-retinal fits into a binding site within the interior of the opsin portion of rhodopsin. With the concentration of cGMP being high, an inward Na^+ leak depolarises the photoreceptors. This passive spread of depolarisation from the outer segment (where the Na^+ channels are located) to the synaptic terminal (where the photoreceptor's neurotransmitter is stored) keeps the synaptic terminal's voltage-gated Ca^{2+} channels open. Ca^{2+} entry triggers the release by exocytosis of the neurotransmitter glutamate from the synaptic terminal while in the dark.

Glutamate release from the photoreceptor terminal in the dark has opposite effects on the two types of bipolar cells (ON and OFF bipolar cells) because they have different types of receptors that lead to different channel responses on binding with glutamate. The ON bipolar cells express inhibitory metabotropic glutamate receptors, and the OFF bipolar cells express excitatory ionotropic glutamate receptors.

In the dark, glutamate released by the photoreceptor activates the ionotropic receptors, leading to sodium flow into the cell and depolarising the membrane potential in the OFF bipolar cells, whereas in the ON bipolar cells, glutamate released by the photoreceptor binds to the metabotropic receptors. The G proteins close cation channels in the membrane, stopping the influx of sodium and calcium and hyperpolarising the membrane potential. Thus, the glutamate hyperpolarises (inhibits) ON-centre bipolar cells and depolarises (excites) OFF-centre bipolar cells.

12.1.3.2 Photoreceptor Activity in the Light

When light strikes the photoreceptor, rhodopsin's photopigment begins to decompose, leading to an instantaneous change of cis-retinal to all-trans-retinal, which starts to pull away from the scotopsin. The partially split combination of all-trans-retinal and scotopsin is bathorhodopsin, which decays to lumirhodopsin, which decays to metarhodopsin I, followed by metarhodopsin II (activated rhodopsin). The activated rhodopsin acts as an enzyme to activate many molecules of transducin. The activated transducin then activates the intracellular enzyme phosphodiesterase, which degrades cGMP. The hydrolysed cGMP thus moves away from the cyclic guanosine monophosphate (cGMP)-gated Na^+ channels resulting in the closure of many Na^+ channels. This channel closure stops the depolarising Na+ leak, thereby causing hyperpolarisation that passively spreads from the outer segment to the synaptic terminal of the photoreceptor. Here, the potential change leads to the closure of the voltage-gated Ca^{2+} channels and a subsequent reduction in neurotransmitter glutamate release from the synaptic terminal. This reduction/absence of glutamate causes the ionotropic receptors to close, preventing sodium influx and hyperpolarising the membrane potential in the OFF bipolar cells. In the ON bipolar cells, the absence of glutamate results in the ion channels being open, allowing cation influx and depolarising the membrane potential. Thus, in the light exposure, the reduction of glutamate depolarises (stimulates) ON-centre bipolar cells and hyperpolarises (inhibits) the OFF-centre bipolar cells. The bipolar cells pass on the information about patterns of illumination to the subsequent neurons in the processing chain, the ganglion cells, by changing their rate of neurotransmitter release following their state of polarisation-increased neurotransmitter release on depolarisation and decreased neurotransmitter release on hyperpolarisation. Thus, the hyperpolarising potential and subsequent decrease in neurotransmitter release are graded according to the light intensity. The brighter the light, the greater the hyperpolarising response and the greater the reduction in glutamate release. Thus, photoreceptor, horizontal, bipolar and amacrine cells have graded membrane potential but do not produce action potentials. Ganglion cells, however, produce action potentials that travel along their axons down the optic nerve.

12.1.3.3 Reformation of Rhodopsin

The short-lived active form of the photopigment quickly dissociates into opsin and retinal. The all-trans-retinal is reconverted to 11-cis-retinal by retinal isomerase with energy expenditure. The 11-cis-retinal normally recombines with scotopsin to reform rhodopsin. In the dark, enzyme *rhodopsin kinase*, rejoins opsin with 11-cis retinal, restore the photopigment to its initial inactive conformation, and the entire cascade turns back to the normal state with open sodium channels.

> Know More...
>
> **Cat's Vision**
>
> The domestic cat has tapetum lucidum to amplify the available light into 130 times more than the human fundus, adapted minimum light detection threshold up to seven times lower than that in humans and large quantities of light-sensitive rod photoreceptors. The large-sized cornea allows more light to enter, and the pupil can dilate about 6 mm more than the human. A distantly located lens produces a smaller but brighter image. All such adaptation facilitates the domestic cat as an efficient adapted domestic animal in nocturnal vision. But it has visual acuity with a range of 20/100 to 20/200; that is, it can see any object at 20 ft distance that a human can see at 100 or 200 ft.

12.1.3.4 Dark and Light Adaptation

In the dark, rhodopsin accumulates in the rods, and in about 20–40 min in humans, the rods become maximally sensitive to light, called *dark adaptation*. Similarly, when exposed to light, within 5 min, the concentration of rhodopsin decreases in rods, and they become insensitive to light, and the vision is caused by cone stimulation known as *light adaptation*.

12.1.3.5 Processing of Visual Information in the Retina

The processing of visual information in the retina involves the formation of three images. The effect of light on the photoreceptors changes the first image in the bipolar cells to a second image and, in the ganglion cells, converts to a third image. The horizontal cells alter the signal during the second image, which is further modified by the amacrine cells in the formation of the third image. A little change occurs in the impulse pattern in the lateral geniculate bodies when the third image reaches the occipital cortex.

12.1.4 Types of Vision

Primates, birds, reptiles, amphibians and fish perceive colour to a greater extent than domestic animals.

Monocular vision or periscopic vision: In monocular vision, animals with laterally placed eyes view the objects with one eye at a time independently of the other eye. It occurs due to the wider visual angle between the optic axis and median eyeline, e.g. amphibians and reptiles.

Binocular vision: Primates, carnivores and birds have the power of converging the eyes, thus viewing the same object simultaneously with both eyes. It occurs due to a parallel optic axis and median line, which provide an overlap of the field of vision (Fig. 12.2).

Stereo vision: Primates, cats and other felines have a three-dimensional view because of the small angle between the optic axis and median line of the eye, with each eye viewing the same object at a different angle. The left and right eyes show dissimilarity in viewing an object. These two retinal images are fused in the brain's visual centre, giving information about the object's height, width and depth—the '3'-dimensional picture.

Animals have a wider peripheral vision than humans because the visual fields of each eye do not completely overlap. In the dog, about 50% overlap of the visual fields occurs, which perceives the middle half of the field of vision. This area of visual overlap provides a binocular vision for the judgment of distances. The field of vision outside the binocular zone is the monocular zone. Binocular vision varies significantly in different animals, reflecting the position of the eyes in front of the head. Both eyes move as a unit to maintain clear binocular vision. In dogs, the binocular field is 60°, and they can see about 240° around their nose, which is 65° and about 360° in horses and 140° and 200° in cats (Fig. 12.2). The binocular field in humans is about 120°.

In equines, the binocular overlap of the visual field is located down the nose of the horse and is limited to between approximately 65° and 80°. The lateral position of the eye provides the horse with a wide panoramic vision, about 340°–360°, that facilitates maximum detection of predators at the expense of the advantages of binocular overlap. Two blind spots have been identified within the visual field, one in front of the forehead and the other directly behind the horse.

> Know More...
>
> **The Best Eyesight in the Animal Kingdom**
>
> The eagle has the best eyesight in the animal kingdom, with a visual acuity of 20/5, which means that the eagle can see an object at 5 ft distance while the human see at 20 ft. The four-time stronger visual acuity occurs in eagles due to two foveae having one million cones per millimetre, whereas humans have one fovea with 0.2 million. The eagle has a long-range focal length lens with the compatible structured fovea, which improves long-range vision.

12.1.5 Diversification of Vision

Humans can distinguish nearly one million colours as they have a different permutation of blue, green and red cones in the eyes. Dogs have two types of cone cells sensitive to the wavelength of yellow and blue. Hence, they are acquainted with the images of combinations of these two colours. Cats have three types of cone cells like humans but contain fewer combinations of these cells, so they cannot distinguish the colour combination as humans can do. The red colour appears as dark, and the green colour senses white or grey to the dog and cat.

Domestic animals usually cannot interpret the sharp image due to fewer cones and wide pupils, which make a blur vision resulting in reduced visual acuity than humans. The visual acuity in dogs is 0.2–0.4 times that of humans (20/50–20/100), 0.6 in horses (20/33) and 0.2 in cats (20/100), considering the human as 20/20.

Rod cells infer light. Hence, animals bearing more rod cells can see in dim light. Owls have the best night vision due to their larger eye and possession of almost one million rod cells in a square millimetre, about five times greater density than humans. The rod photoreceptors can finely detect the motion and shape of the object. Hence, nocturnal animals perceive movable objects nicely, particularly in dim light. A

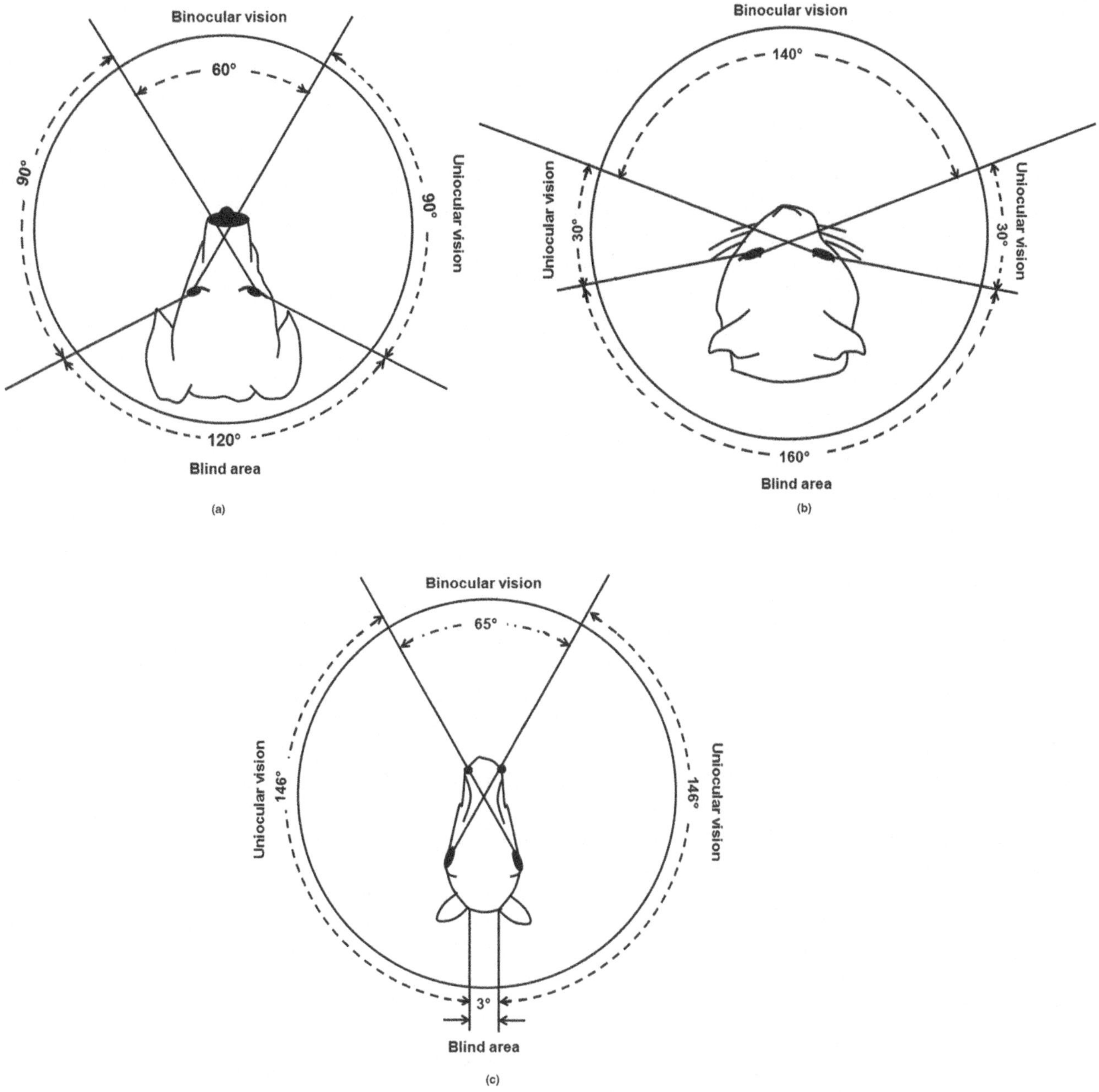

Fig. 12.2 Visual field of domestic animals. (**a**) The dog's visual field shows 60^0 binocular and 90° monocular fields with a blind area of 120°. (**b**) Visual field of the cat has 140° binocular and small 30° monocular fields and wider 160° posterior blind area. (**c**) The horse's visual field presented 65° binocular and comparatively larger, about 146° panoramic monocular fields with a little 3° blind areas

dog can recognise a movable object at nearly 800 m and a stationary entity at about 500 m.

12.2 Auditory System

The auditory system forms the basis for communication for the animals, and it is designed to detect, analyse sound in the surrounding environment and react accordingly. For the location of the sound, the auditory system requires both the ears to detect the difference in the time of arrival and sound intensity approaching the two ears, whereas hearing requires at least one ear. Further, animals' sense of hearing is also enhanced by their ability to move their ears around to scan the environment for different sounds and locate where the sound is coming from.

Sounds are pressure waves in the air with given frequencies and amplitudes. The auditory system perceives the frequency of sounds as pitch and their amplitude as loudness.

12.2.1 Structure of Auditory System

Hearing involves the external ear, middle ear and inner ear where the sensory receptor (i.e. the organ of Corti) is located. The sound waves directed by the external ear enter into the ear canal and induce vibration of the tympanic membrane that separates the external ear from the middle ear.

The external ear is composed of pinna (auricle) and external auditory canal. The pinna is a cartilaginous flap that is used to guide sound waves through ear canal. Pinna is absent in birds, reptiles and amphibians. The external auditory canal is composed of both bone and cartilages. The inner end of external auditory canal is shut by tympanic membrane. The air wax secreted by the sebaceous gland is deposited in the external auditory canal.

The external and the middle ears conduct sound to the auditory receptors (organ of Corti) in the cochlea of the inner ear. The eardrum separates the external and the middle ear. The middle ear consists of air-filled tympanic cavity containing a chain of three auditory ossicles (malleus, incus and stapes) and the Eustachian tube (auditory tube). As the auditory tube connects the middle ear to the nasopharynx, pressure in the middle ear gets equalised with external atmospheric pressure and also facilitates clearing of fluids from the middle ear. In horse, there is a ventral diverticulum located in the auditory tube called as guttural pouch.

The vibrations of the tympanic membrane caused by the sound waves are transmitted through the middle ear by the three ossicles. The malleus (hammer) is attached to the tympanic membrane, while the footplate of the stapes contacts the oval (vestibular) window membrane, in the cochlea of the inner ear, thereby providing a mechanical linkage between the tympanic membrane and the oval window. The unique design and the relative size difference between the bones and the tympanic membrane magnify the vibrating pressure of the tympanic membrane to the stapes which is essential, as the sound waves travel from air to the fluid perilymph in the inner ear. This arrangement also decreases the amplitude of sound waves transferred to the perilymph, thereby providing protection to the sensitive sensory cells of the organ of Corti.

Two small skeletal muscles are located in the middle ear, which alter the transmission of vibrations between the eardrum and the oval window. In birds, only one bone—columella—is present in the middle ear, which connects the eardrum and the oval window, and hence the transmission of sound is less efficient.

The inner ear or labyrinth consists of an acoustic part, the cochlea and a non-acoustic part, the vestibular organ. The inner ear is made of a bony labyrinth; within it is located the membranous labyrinth. The cochlea is formed by the coiling of three fluid-filled tubes known as scala vestibuli (vestibular duct), scala media (cochlear duct) and scala tympani (tympanic duct). The scala vestibuli and scala tympani are connected by a narrow channel, the helicotrema. The receptor cells of the auditory system are located within the cochlea. The scala media is separated from the scala vestibuli by the Reissner's membrane or vestibular membrane, while the scala media is separated from the scala tympani by the basilar membrane. Along the floor of the scala media, on the basilar membrane lies the hair cell receptor system, the organ of Corti. The scala tympani and scala vestibuli are filled with Na^+ ion-rich fluid called the perilymph. The scala vestibuli at the basal end faces the oval window, whereas the scala tympani faces the round window. Thus, pressure on the oval window due to the movement of the stapes is equalised by fluctuation of the connective tissue sheath covering the round window. The scala media is filled with endolymph, which has a concentration of K^+ ions.

The cochlea of birds is short, almost straight, and hair cells are not arranged in rows. The hair cells, tectorial membrane and basilar membranes with cochlear nerve terminals form the organ of Corti.

The auditory receptors are the hair cells embedded in the basilar membrane and their apical surface contains hair-like cilia—stereocilia that project into the endolymph-filled scala media. The apical surface of the hair cells has 50–100 stereocilia, which are connected by filamentous material called the tip link. The tip link is attached to a K^+ channel and thus, when the bending of the stereocilia pulls the tip links, the K^+ channels are opened. The tectorial membrane, the specialised gelatinous flap of the basilar membrane composed primarily of glycoprotein, overhangs the cilia of hair cells. Due to this structural arrangement, vibrations of the basilar membrane cause bending of the hair cell stereocilia, which is translated into voltage changes of the hair cell membrane. The terminals of the cochlear nerve fibres synapse with the basal ends of each hair cell. The auditory impulses are transmitted through vestibulocochlear nerve to higher brain centres.

12.2.2 Mechanism of Hearing

Sound waves from the external environment cause vibrations of the tympanic membrane. These vibrations are transmitted through the middle ear by the auditory bones and are presented to the oval window. This sets up travelling waves in the perilymph of scala vestibuli, which in turn cause vibrations in the basilar membrane. Since the stereocilia of the hair cells are embedded in the tectorial membrane, the up-and-down movement of basilar membrane causes the stereocilia of hair cells to bend. This bending pulls the tip links, which in turn sets up an action potential in hair cells by inducing opening of K^+ ion channels at the tip of the stereocilia, opening of voltage-gated Ca^{2+} channels at the base of the cells and subsequent release of neurotransmitter into the synaptic cleft between sensory hair cells and cochlear nerve terminals. As the basilar membrane has graded stiffness and width being narrow near the base, and wider near the apex of the cochlea, the location of maximum displacement of the basilar membrane is related to the frequency of the tone. Low-frequency tones tend to distort the entire basilar membrane with maximum displacement of the membrane occurring near the apex of the cochlear duct, intermediate tones distort the basilar membrane from the base to an intermediate region and high-frequency tones selectively distort the basilar membrane close to the base of the cochlea. Hair cells at different locations along the basilar membrane respond to different frequencies of sound waves. Stimulation of nerve fibres from different regions of the basilar membrane provides the central nervous system about the frequency of the sound. Humans can detect sounds in the range of 20–20,000 Hz. In dogs, the upper limit is about that of humans. Frequencies of about 98,000 Hz produce potential changes in the cochlea of bats.

12.2.3 Auditory Pathway

The vestibulocochlear nerve terminates in the cochlear nucleus of the medulla, and then the impulses are transmitted through superior olivary nuclei, inferior colliculus and median geniculate body to the temporal lobe of the primary auditory cortex. Each cochlea is mapped bilaterally in the auditory cortex, wherein the decoding and feature extraction of complex auditory information occur. The primary auditory cortex is surrounded by association cortical areas that integrate various sensory stimuli and are thus critical in understanding the surrounding environment.

On exposure to loud sounds, the stereocilia may be forced to move excessively leading to their destruction and resulting in hearing loss. This type of hearing loss is prevented by reflex contraction of the tensor tympani and stapedius muscles in response to loud sound, thereby protecting the sensory cells. This middle-ear reflex (acoustic stapedius reflex) protects the sensory cells by reflexively contracting the tensor tympani and stapedius muscles in response to loud sound. The movement of the tympanic membrane and stapes is limited by the contraction of these muscles, thereby reducing the force and amplitude of sound applied to the organ of Corti. The motor nuclei of the trigeminal and facial nerves are involved in mediating this reflex. As it takes time for these muscles to contract fully, the protection offered by this reflex against loud sound is only partial. The reflex mediated by bilateral loud sound to one ear also triggers the reflex in the opposite ear. This is done by sending signals from the cochlear nuclei to the contralateral dorsal nucleus of the trapezoid body. Hearing is a function of the cerebral cortex, whereas the auditory reflex, such as turning the head in response to sound, is mediated by the brainstem.

12.2.4 Equilibrium

The vestibular apparatus of the ear helps to maintain the posture and equilibrium of the body. It is mainly controlled by a gravity sensor (vestibular apparatus) in response to visual and proprioceptive information. The body movements cause the stimulation of hair cells, and afferent nerve impulse travels from vestibular nuclei to cerebellum and thalamus for the adjustment of posture and positioning of the body, respectively. The third nerve nuclei help to fix the eyes during head movement. Animal species like squirrel, hawk and mountain goat and birds have well-developed vestibular system for their survival.

12.3 Olfactory System

Olfaction (smell) is an animal's primary special sense, and their sense of smell is far more sensitive than that of humans. Olfaction is essential for the localisation of food, reflex stimulation, secretion of digestive enzymes, detection of danger, finding of direction, seeking of prey or avoiding of predators, avoiding of poisonous and spoiled feeds, social recognition of kin and sexual attraction to a mate. For example, mice sniff for potential mates while avoiding those that are infected with parasites, and hermit crabs find buried snail shells by smelling calcium leaching from them. The sense of taste and smell is important to discriminate between desirable and undesirable feed, and by acting together, they contribute to the palatability of foods by detecting flavours that influence appetite.

12.3.1 Components of Olfactory System

Olfaction in animals is mediated by two important sensory systems—the main olfactory system with receptors in the dorso-caudal part of the nasal cavity and the accessory olfactory system with receptors in the vomeronasal or Jacobson's organ located near the external layers. The olfactory mucosa, present in the ceiling of the nasal cavity, occupies a relatively larger area in dogs (100 cm^2) in comparison with that in humans (about 5 cm^2). The mucosa has three cell types: basal cells, supporting cells and olfactory receptor cells. The basal cells being the precursor cells constantly replace the olfactory receptor cells. The supporting cells secrete mucus that forms a protective coat over the nasal passages. The olfactory receptors are bipolar cells with a single dendrite at one end that terminates in the olfactory mucosal surface as an expanded olfactory knob. Approximately, 10–20 cilia arising from each olfactory knob spread across the olfactory mucosal surface. The cilia have sensory receptors that act as transducers for the olfactory stimulus. The membrane of cilia is covered with many G protein-coupled olfactory receptors.

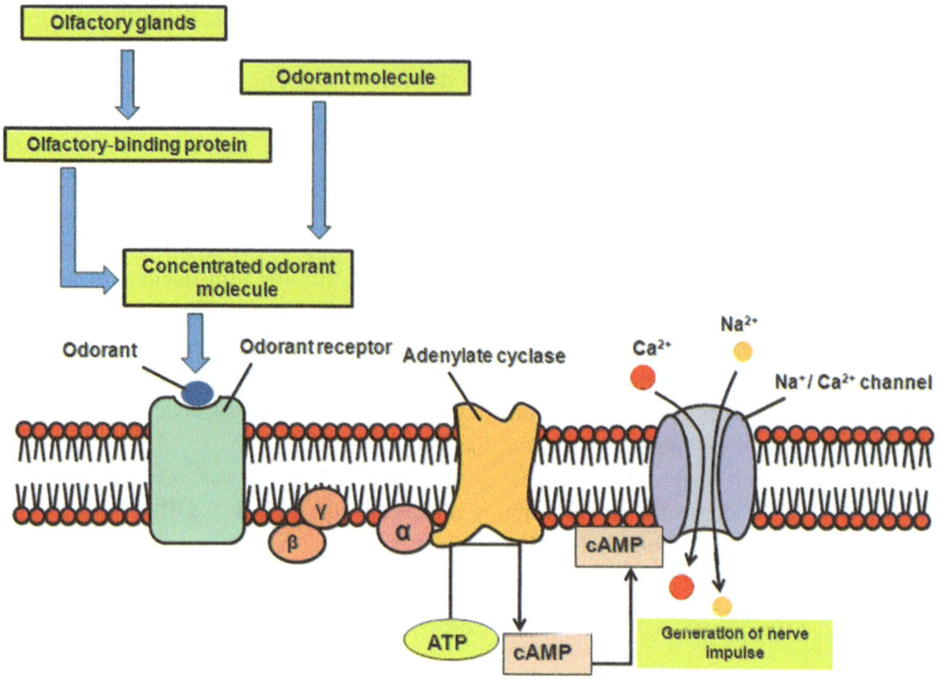

Fig. 12.3 Mechanism of taste. [The odorant molecules combine with olfactory binding proteins and concentrate to bind with odorant receptors (ciliary G protein-coupled receptor). Binding odorant molecule to the receptors activates G protein (G_{olf})-mediated signal transduction mechanism that leads to a series of electrical events]

12.3.2 Mechanism of Olfaction

The volatile and water-soluble odorant molecules entering the nasal cavity are concentrated by the olfactory binding proteins secreted by the olfactory glands and are bound to the ciliary sensory G protein-coupled receptors (Fig. 12.3). Binding of an odorant molecule to the ciliary G protein-coupled receptor activates G protein (G_{olf}) that unites with guanosine triphosphate (GTP). The GTP–G_{olf} complex activates phospholipase C, generating inositol triphosphate (IP3) that opens Ca^{2+} channels and opens Na^{+} and Ca^{2+} channels through adenylyl cyclase-generated cAMP. Thus, the GTP–G_{olf} complex leads to a series of electrical events (i.e. increase in intracellular Ca^{2+}) that result in the generation of excitatory postsynaptic graded potentials (EPSPs) in the cilia, which ultimately results in the propagation of action potential along the axons of the olfactory cells to the olfactory bulb that is dispersed to wide areas of the cortex. The second set of neurons from the olfactory bulb divide into lateral and medial and project to the limbic system including hippocampus, frontal and temporal cortices, thalamus, amygdala, hypothalamus and reticular formation, thereby regulating feeding, sexual and emotional behaviours. Thus, olfactory signals, unlike other sensory systems, do not directly project to the thalamus. The projection of olfactory signals suggests that emotional reaction to olfaction is carried out by the entorhinal cortex, hippocampus, septal nuclei and amygdala of the limbic system.

Nonmyelinated axons arising from the olfactory receptor synapse with the mitral and tufted cells in the olfactory bulb in the olfactory lobe of the cerebral cortex. The specific affinity of olfactory receptor for each odorant and the differential electrical discharge rates of the mitral and tufted cells generate a unique activation pattern and subsequently a unique sense of odour. A dog, for example, has more than 220 million olfactory receptors in its nose, while humans have only five million.

A unique feature of olfactory transmission is its rapid adaptation to stimulus; that is, the initial discharge of axons in response to stimulation is followed by quick decline to a steady-state discharge of lower amplitude. Although the olfactory system is sensitive and highly discriminating, it is also quickly adaptive. Sensitivity to a new odour diminishes rapidly after a short period of exposure to it, in spite of the continued presence of odour source; this could be attributed to the presence of the odorant-clearing enzymes in the nasal mucosa. This mechanism might serve the dual purpose of clearing the olfactory mucosa of old odorants and transforming potentially harmful chemicals into harmless molecules. Such detoxification helps to avoid the entry of toxicant through the open passageway between the olfactory mucosa and the brain.

12.3.3 Vomeronasal Organ (VNO)

In addition to the olfactory mucosa, the vertebrate nose has accessory olfactory system with receptors in the Jacobson's or vomeronasal organ (VNO) that mediates sex odours. The VNO plays a significant role in governing reproductive and social behaviours, such as identifying and attracting a mate. VNO is the major site for pheromone detection. The VNO is open at one end and forms a blind sac at the other end. The location of the opening is variable; in rodents, the opening is into the nasal cavity and in cows it opens into the oral cavity. The VNO is also involved in the perception of large, non-volatile molecules that could not reach the main olfactory system. The receptors are similar to those of the main olfactory system, and they project to the olfactory bulb. The axons of the receptor from the olfactory nerve terminate in the olfactory bulb. The second set of neurons from the olfactory bulb divide into lateral and medial and project to limbic system including hippocampus, frontal and temporal cortices, thalamus, hypothalamus and reticular formation, thus regulating feeding, sexual and emotional behaviours. Hence, odour signals are not just for olfaction. Processing of odour signals by the limbic system is the basis for forming olfactory memories, and olfaction can evoke strong emotional reactions. In animals, the mother recognises its newborn young ones by odour. The olfactory system helps in the regulation of reproduction—animals are attracted by pheromones produced by opposite sex; individual recognition, mother-young interaction and establishment of dominance are some social behaviours affected by olfactory signals. Ungulates and rhesus monkeys detect the female in heat by the odour of vaginal secretions. Male ungulates, such as stallions, put their lips into the urine of an oestrus mare and curl their upper lip in the flehman position, which partially blocks the nostril opening. Specialised pumping mechanism aided with deep breathing serves to suck molecules (pheromones) into the organ and to carry the urine into the VNO, where the stallion determines sexual receptivity of the mare by the concentration of pheromone. Sex pheromones are also produced in the urine of female elephants about to ovulate, which initiates the flehman response in reproductively active males. Males first identify the urine through the sense of smell and then detect the pheromone by placing urine from the tip of their trunk to the opening of ducts leading to the VNO. *Androstenone*, a steroid in the saliva of male pigs and human male sweat, is one of the few mammalian pheromones that have been characterised. Release of this potent pheromone in pigs causes sows to assume the lordosis (mating) position. Androstenone elicits this behaviour only in oestrus sows, suggesting that the response to a *releaser* pheromone is specific and contextual dependent. Animals differ in their ability to detect odours. Animals like dogs are very sensitive in the detection of odour and are called as macrosmatic; those that can detect odour but with less sensitivity like birds are termed microsmatic; those that lack olfactory apparatus like dolphins and whales are termed anosmic.

12.4 Gustatory System

The sense of taste plays a significant role in animals including influencing their behaviour as compared to humans. Gustation or taste is a kind of chemical sensation that is perceived through taste receptors in the taste buds. All five basic taste sensations, namely sweet, bitter, salty, sour and umami, can be perceived by humans. Ruminants like cattle, sheep and goats also have the lingual receptors for all five basic tastes; however, they are more tolerant to bitter taste than other mammals as they have fewer genes responsible for coding the bitter receptors. Cats lack genes Taslr2 and Taslr3 that encode sweet taste and hence they are unable to taste sweets.

12.4.1 Taste Buds

In higher vertebrates, the chemoreceptors for taste sensation are located in taste buds which are present on the tongue, roof of the mouth, soft palate and pharynx, with the greatest percentage on the upper surface of the tongue. There are about 15,000 taste buds in pigs and rabbits, 550 in lizards and only 24 in chickens. The taste buds are located on the apical ends of the papillae. There are primarily three different types of papillae that are located on various areas of the tongue, fungiform papillae (distributed throughout the dorsal surface of the rostral two-thirds of the tongue, especially along the lateral margins and the tip), vallate papillae (present on the caudal portion of the dorsal tongue) and foliate papillae (present on the dorsolateral part of the caudal part of the tongue). Each taste bud has about 40–50 elongated spindle-shaped receptor cells and sustentacular or supporting cells. The taste cells are arranged around the taste pore and several microvilli, or taste hairs present on their apical region extend into the taste pore. Individual taste receptor cells are constantly replaced once in 5–10 days by new receptor cells derived from the basal cells. The basal surface of the receptor cell is innervated by the terminals of gustatory nerves.

12.4.2 Mechanism of Taste Sensation

The plasma membrane of the microvilli contains receptors that bind selectively with dissolved chemical molecules in ingested liquids or solids and evoke the sensation of taste. Binding of a taste-provoking chemical, a tastant, with a receptor cell ultimately alters the cell's ionic channels to produce a depolarising receptor potential. This receptor potential, in turn, initiates action potentials within terminal endings of afferent nerve fibres with which the receptor cell synapses. The mechanisms that generate membrane depolarisation depend on the taste molecules that bind to their specific receptors.

12.4.3 Stimulants for Taste Sensation

Mammals have the ability to discriminate among thousands of different taste sensations; however, most of the taste sensations are considered to varying combinations of the primary five taste categories, which include salty (sodium), sour (or acidic), sweet, bitter and umami (meaning delicious in Japanese and having taste of monosodium glutamate and aspartate, commonly present in proteinaceous food) taste. The tip of the tongue is sensitive to sweet stimuli, salty and sour tastes are sensed by the sides of the tongue, whereas the back of the tongue is sensitive to bitter substances.

Each taste receptor cell responds in varying degrees to all the primary tastes but is generally preferentially responsive to one of the taste modalities. These five tastes and their distinct transduction mechanisms appear to be separately localised in different taste receptor cells. Two of these mechanisms are ionotropic (for salty and sour), and the remaining three are metabotropic, mediated by GPCRs.

Salty taste is stimulated by chemical salts, especially Na^+, and helps in obtaining sufficient dietary NaCl, which is critical for osmotic balance and electrical signals. Salty taste is particularly prominent in herbivores, whose plant diet is low in sodium. Transduction is direct, due to entry of positively charged Na^+ through specialised Na^+ channels that are either simple leak channels or gated channels in the receptor cell membrane, and subsequently receptor potential is generated. Even though the dogs have sense of all the common tastes, they do not have highly sensitive salt receptors.

Sour taste is caused by acids, which contain a free hydrogen ion, H^+. The citric acid content of lemons, for example, accounts for their distinctly sour taste. Strong acid taste might indicate spoiled food or unripe fruit. Sour tastants cause the depolarisation of the receptor cell by H^+ entry or when H^+ blocks K^+ channels in the receptor cell membrane.

Sweet taste is evoked by the interaction of sugar molecules with sweet receptor-binding sites. Signalling begins with the binding of sweet taste-evoking chemical to a G protein-coupled receptor that leads to either the cAMP second messenger pathway or the IP3 pathway. The second messenger pathway then results in either phosphorylation and blockage of K^+ channels in the receptor cell membrane, leading to a depolarising receptor potential, or IP3-induced release of Ca^{2+} from the endoplasmic reticulum. The taste-selective cation channels (TrpM5) are calcium sensitive, and IP3 plays a key role in activating TrpM5. Subsequent Na+ influx mediated by TrpM5 leads to the generation of a depolarising receptor potential in receptor cells. Intracellular elevation of Ca^{2+} combined with membrane depolarisation results in ATP release via gap junction channels in the plasma membrane. The released neurotransmitter ATP acts on sensory nerve endings, inducing generator potentials, which may be followed by action potentials if generator potentials reach the threshold potential. Cats cannot taste sweets because their sweet receptor gene has presumably become a pseudogene.

Bitter taste helps to detect toxic and noxious feed. For example, alkaloids and many plant derivatives such as caffeine, nicotine, strychnine, morphine as well as many poisonous plant compounds all taste bitter. Bitter taste is elicited by a chemically diverse group of tastants that is elicited by the presence of multiple receptors and multiple signalling pathways mostly involving G protein-coupled receptors in each of the taste sensory cells for detecting a wide array of potentially harmful chemicals. The first G protein in taste gustducin involved in one of the bitter signalling pathways has been found to be structurally very similar to the visual G protein, *transducin*.

Umami taste is elicited by glutamate, which binds to G protein-coupled receptor and activates a second messenger system, subsequently leading to the generation of action potential.

Terminal afferent endings of several cranial nerves synapse with taste buds in various regions of the mouth. Signals in these sensory inputs are conveyed uncrossed via the brainstem and thalamus to the cortical gustatory area in the parietal lobe. The fibres involved in taste sensation also project to the hypothalamus and limbic system to associate affective dimensions as to whether the taste is pleasant or unpleasant and also to process the behavioural aspects with taste.

Sensory receptors of these systems respond to chemical molecules mixed in the air or saliva, and the two systems complement each other for better interpretation of what animals eat and smell. Most importantly, taste and smell determine flavours, the sensory impressions of food or other substances.

12.5 The Skin (Integumentary System)

The outer surface covering of the animal is the skin. The skin and its derivative or accessory structures are called the integumentary system. It is the largest organ of the animal's body, about 16% of the body weight in humans. It continues with the mucous membranes lining the body. It gives a physical barrier, and antimicrobial protection involves thermo- and immuno-regulation, excretion, secretion, pigmentation, sensation and locomotion of the animal. Vitamin D is produced in the skin. The skin has different appendages, claws, nails, hoofs, horns, antlers and feathers. These modified skins are used for protection or defence, locomotion, hunting or holding of the feed and other physiological activities in animals

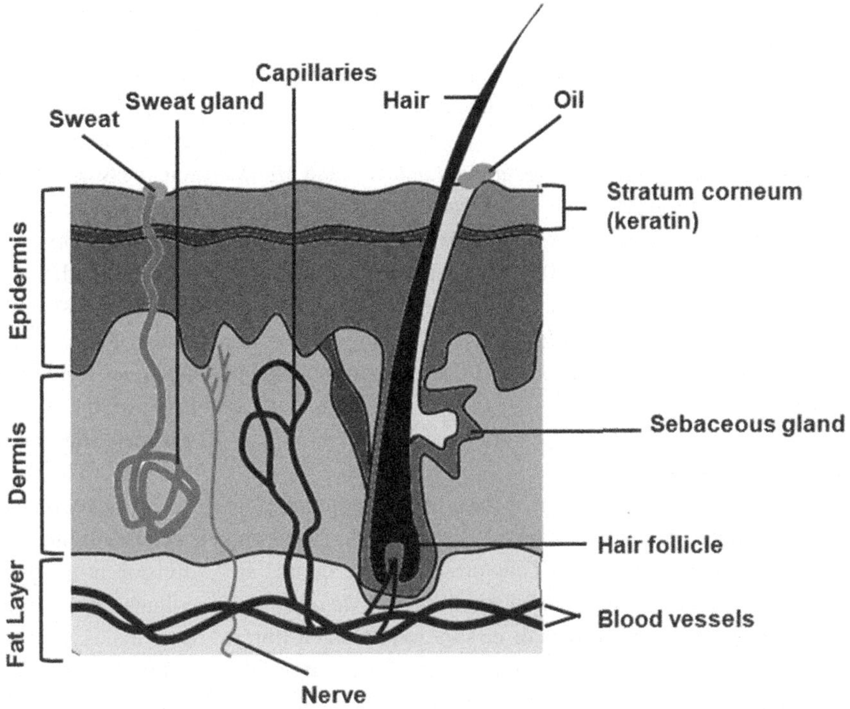

Fig. 12.4 Morphology of the skin. [The cross section of the skin showing the epidermis, dermis and hypodermis with various glands]

and birds. Skin's morphological features of all animals and birds are not the same.

12.5.1 Morphology of the Skin

The skin consists of three layers (exterior to inner side): epidermis, dermis and hypodermis (Fig. 12.4). Dermis appears from the word 'Derma', which means skin, 'epi' denotes over and 'hypo' signifies below.

12.5.1.1 Epidermis

The outermost epidermis layer contains 4–5 cell strata (Fig. 12.5). The innermost single-cell layer, *stratum germinativum* or *basale*, is present over the basement membrane of the dermis and includes precursors of *keratinocytes*. The major cells of the epidermis are keratinocytes, which are generated and increase persistently, forming the second layer, *stratum spinosum*. The spiny-like multiple (8–10 layers in humans) cells of stratum spinosum are transitory and transformed to the third layer, *stratum granulosum*. Keratin filaments (fibrils) and matrix with keratohyalin granules are generated in the keratinocytes within the grainy-appearing multi-cell stratum granulosum. The layer secretes lipids to moisten the skin surface. The keratinocytes become non-nucleated, dense and dead when they reach the fourth layer, *stratum lucidum*. The stratum lucidum is intensely available in the footpads of ungulates, dogs and cats, whereas it is thin in abdominal areas. The lucidum is densely packed with a translucence protein, the eleiden, providing a lucid

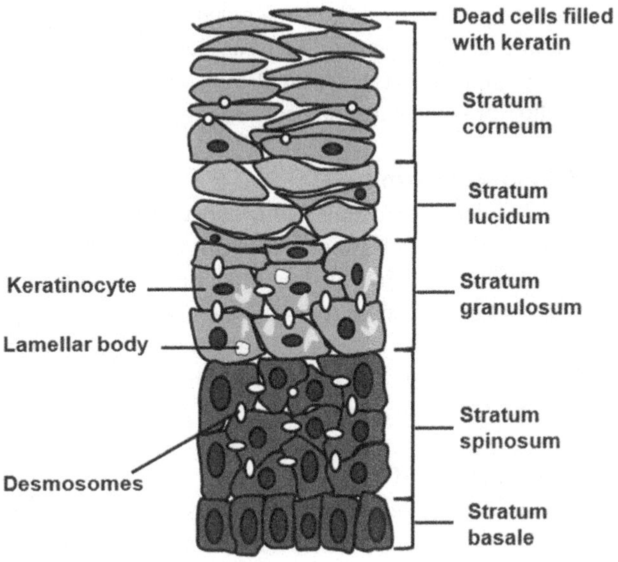

Fig. 12.5 Morphology of the skin. [Various layers of the epidermis are depicted]

appearance in the region. The dead keratinocytes reach the superficial layer, the fifth layer, named *stratum corneum*, which contains about usually 15–30 layers of cells in humans. It is horny due to the completion of *keratinisation* or *cornification* of the keratinocytes. The dead cells of this layer appear as a protein mass, the keratin. The intracellular insoluble fibrous protein-rich water-resistant keratins in the stratum corneum provide strength, rigidity, elasticity, pliability and physical protection. The stratum corneum protects the body against invasion of microorganisms as the cells of the layer are periodically (within about 4–6 weeks) shed and replaced by cells from the immediately lower layers. Keratin is the principal component of hair, nails, hoofs and other appendages.

The epidermis layer also contains *melanocytes* and *Merkel cells* at stratum basale and *Langerhans cells* at stratum spinosum. Melanocytes are responsible for skin colour, and ultraviolet (UV) radiation (100–400 nm) damages the living cells by producing melanin. Merkel cells act as sensory receptors to sense touch and pressure. The Langerhans cells are the *antigen-presenting cells* (APCs) with immunogenic functions.

12.5.1.2 Dermis

The middle layer of the skin is the dermis, composed of collagen fibres, ground substances, hair follicles, sweat and sebaceous glands. The major cells present in the dermis are *fibroblasts*, *mast cells*, *histiocytes* and *dendrocytes*. Fibroblasts are the precursors of elastin and collagen fibres, giving the skin strength. Mast cells are involved in the inflammatory process releasing histamine and heparin. Histiocytes have phagocytic activity, and dendrocytes are the antigen-processing cells. The ground substances contain hyaluronic acids, glycosaminoglycans and proteoglycans, providing strength, elasticity and pliability and water-holding capability for maintaining homeostasis.

12.5.1.3 Hypodermis

The lower-most hypodermis is the subcutaneous fat tissue layer. Differentiation between the dermis and hypodermis is very difficult; hence, hypodermis is not firmly considered as part of the skin. It contains adipose tissues, areolar connective tissue, blood and lymph vessels and nerves. The profound adipose or fat tissues involve insulation, storing various substances and energy reserve.

12.5.1.4 Glands of the Skin

The skin contains several exocrine glands. The *sebaceous gland* secretes *sebum* to moisten the skin, and the *sweat gland* produces *sweat* involved in thermoregulation. Sweat is odourless, but has hormones; the odour may appear in certain conditions, viz. kind of feed, infection, medications and diabetes-like diseases. The sweaty smell usually develops the products of reactions of secreted sweat with the surface bacteria. The *preen gland*, mostly available near the base of the bird's tail, secretes oil to keep the feathers in sound condition. The wax-producing gland is present in the ears. Mammary glands present only in mammals are considered modified sweat glands.

12.5.1.5 Thickness of the Skin

The thickness of skin varies depending on the species, breed, sex, age, strata position and location on the body, ranging from 0.5 to 5 mm (average 1.03 mm) in the dog, 0.4–2 mm in the cat and 1.5–5 cm in rhinoceroses. The average skin thickness of Devon is 8.2 mm, 5.8 mm in Zebu and 5.5 mm in Jersey breeds of cattle. The thickest skin is present at the back and dorsal neck regions, thinner at the abdomen and thinnest at the inguinal and axillary areas. The abdominal skin thickness in the cat is 15 μm, whereas it is nearly 2 mm in the dog's footpad. Generally, thickness is inversely proportional to the density of the hair. In humans, the thinnest layer of the epidermis (0.1 mm) is present in the eyelid, whereas the palms and soles have the thickest (about 1.5 mm). The dermis is thickest on the back, about 30–40 times thicker than the immediate over the epidermis. The skin becomes thinner in advanced age due to less mitotic cell division in the germinativum layer. Androgens influence the collagen formation in the skin; hence, females' skin is softer than males. The thickness of various stratum also varies in different appendages. In *full-thickness grafting*, the entire epidermis and dermis with their appendages transfer, whereas only the epidermis and parts of the dermis replace in *split-thickness grafting*.

12.5.2 Major Functions of the Skin

12.5.2.1 Protection Mechanism of the Skin

Physical protection governs by the hair of the skin. The water-repellent outermost keratinised layer protects the body fluid, and regular sloughing of this layer regulates the debris and invasion of microorganisms. The skin performs antimicrobial and antifungal activities through immunoglobulins, interferons, various antimicrobial peptides (AMPs), lipids and organic acids present in sebum (sebaceous gland product) and sweat. Phagocytosis of the histiocytes and immunogenic dendrocytes is also involved in the antibacterial activity of the skin. The Langerhans cells trap antigens, like microorganisms and foreign proteins, and present them to T-helper lymphocytes.

12.5.2.2 Sensation

Skin can sense the pain, temperature (heat or cold) and amplification of itching by the Merkel cells. The receptor-like Merkel cell senses the external stimuli attaching to the sensory nerves to interpret the stimulus in the brain. The Merkel cell acts as thermo-receptors and nociceptors related to irritation and itching. Merkel cells are densely found in the digits, lips, oral cavity and outer root sheath of the hair follicle. *Pacinian corpuscles* act as mechanoreceptors to detect pressure and vibration, *Vater Pacini corpuscles* are mostly present in the genital organs involved in pressure detection and *Meissner's corpuscles* are touch sensitive. Specific mediators, like histamine, interleukin (IL-2, IL-31), nerve growth factor and endothelins, regulate the sensation process.

12.5.2.3 Pigmentation

The pigments like melanin, produced by *melanocytes*, *carotene* and *haemoglobin*, are responsible for skin pigmentation. Melanin exists in eumelanin (black and brown) and pheomelanin (red) forms. The abundant eumelanin forms of melanin pigment are stored in the melanosomes of keratinocytes. The melanin formation (melanogenesis) is stimulated by sunshine, specific genes and various hormones, viz. melanocyte-stimulating hormone (MSH) and adrenocorticotropic hormone (ACTH), and inhibited by carbon monoxide, hydrogen sulphide and various organic sulphur compounds. The lack of melanin occurs because of *albinism*, the consequence of a genetic disorder. The inability of melanocytes in certain regions to form melanin causes local colourless patches, the *vitiligo*. Excess melanin will interfere with *vitamin D* production and interfere in calcium absorption. Liver disease (jaundice) causes accumulation of yellow pigment bilirubin over the skin. Yellowish coloured skin may also occur due to excess ingestion of a yellow-orange pigment-containing carotene-rich feed. Carotene is deposited in the stratum corneum of the skin. Reddening in the skin may occur due to allergy and inflammation leading to the accumulation of more blood or haemoglobin in the dermis. Pale-coloured skin appears in anaemia and low blood pressure conditions. Prolonged lack of oxygen develops dark red deoxyhaemoglobin in the blood leading to cyanosis, a blue-coloured skin.

12.5.2.4 Thermoregulation

Blood vessels in the dermis form a complex network, and superficial and deep plexus are involved in thermoregulation by vasodilation during hot and vasoconstriction in a cold environment. The secretion and evaporation of sweat from the skin surface facilitate homeostatic body core temperature during hot and dry conditions. The coiled tube-like sweat glands are profusely available in ruminants and horses; poorly developed in dogs, cats and pigs; and absent in birds. The gland opens directly on the skin surface. The epithelial cells initially secrete an isotonic sweat in response to a thermal stimulus; later, sodium is reabsorbed, resulting in the hypotonic solution, which is evaporated.

It contains little salt and waste products, like urea; hence, it involves excretion. But the total excretory amount is negligible compared with the excretion process of the kidney. Thus, it is not considered a part of the excretory system.

12.5.2.5 Synthesis of Vitamin D and Biochemical Reactions

The *cholecalciferol* (vitamin D3), a form of vitamin D, is synthesised from steroid cholesterol (7-dehydrocholesterol) in the skin in the presence of sunlight. The cholecalciferol converts to calcidiol in the liver and further transforms into an active form of vitamin D, the calcitriol, in the kidney. The formation of cholecalciferol reduces during advanced age, leading to an increase in the risk of osteoporosis. The skin can exert various biochemical reactions over some steroid hormones, like oestrogens, progestogens and glucocorticoids, and vitamin A through its specific receptors.

12.5.3 Role of Hair and Sebaceous Gland

12.5.3.1 Types of Animal Hair

Animal hairs are generally of three types: *guard hair*, *fur* or *wool* and *tactile hair*. Guard hairs, and outer coat, protect from physical injury, trauma and UV radiation. It also has a role in thermoregulation. The fur or wool hairs are the inner coat and mainly involve insulation. The tactile hair or whiskers present on the face of some animals (feline) provide sensory perception. Other hairs are also present in some animals, like mane hair in horses and tail hair.

12.5.3.2 Structure of Hair

Hair has two structures: the follicle remains within the skin, and the shaft or scalp exists outside the body. The follicle extends into the dermis. The papilla resides at the follicle's base, contains blood capillaries and engages in nourishing the cells. The living cells surround the papilla called the bulb. Compound hair follicles with clusters of primary hairs encircling smaller secondary hairs are found in dogs and cats. Two sheaths cover the follicle. The inner sheath continues to the hair shaft and terminates below the opening of the sebaceous gland, and the outer sheath (cuticle) goes on with the gland. The central core is called the medulla. The outer sheath attaches to the *arrector pili* muscle through a connective tissue at the upper dermis below the sebaceous gland. Contraction of pili muscle causes the erection of hair and secretion of the sebaceous gland, the sebum.

12.5.3.3 Distinguishable Features in Animal Hair

The cuticle of the hair configures with the combination of its three structures: coronal (crown-like), spinous (petal-like) and imbricate (flattened). Coronal appearance is most common in small rodents and bats, with triangular shaped spinous or petal-like scales present at the upper part of the hair of mink and fur hairs of seals and cats. The coronal and spinous shaped scales are hardly found in human hairs. The hairs of the humans and some animals have an imbricate or flattened-scale type. The medulla of the human hair is unstructured, whereas very regular and well defined in the animal. Microscopically, the animal hair can be distinguished into three broad classes. The scale pattern is the identifying trait having a regular diameter with a wave or crimp pattern that exists throughout the hair in the deer family and antelopes. A wide variation in diameter (20–150 μm) throughout a single hair with a banded appearance is common in the hair of commercial fur animals. Diversified pigmentation, root and medullary structure with a unique character predominate in particular domestic animals.

12.5.3.4 Hair Cycle

Hair contains three vital components: keratin, melanin and a trace amount of metallic elements. It grows and sheds in four stages: *anagen* (growth), *catagen* (regression), *telogen* (resting or relative quiescence) and *exogen* (shedding). A new hair shaft forms in every cycle, and activation of hair-specific epithelial stem cells at the bulb controls the cycle or generation of the new hair shaft. Various factors, like genetics, hormones, neurotrophins, photoperiod, nutrition, cytokines and some intrinsic factors, regulate the cyclic activity of the hair. Alopecia or abnormal hair loss may occur due to autoimmune reaction to anagen follicles or destruction of hair follicles after inflammation and other causes.

12.5.3.5 Sebaceous Gland

Sebum is a lipid-rich compound secreted from the sebaceous gland. Major lipid constituent of sebum is species specific and provides a unique scent to a particular species. For example, the sebum of goats contains caproic acid, and sheep contain lanolin. The glands are densely present in the facial region. Usually, sebum has a role in softening the hair and skin surface and bacteriostatic property. The modified sebaceous glands can produce a specific odour called a scent gland. The scent acts as pheromones, like territory marking and communication, mainly during the breeding season. Testosterone and progesterone influence the growth of the gland.

Learning Outcomes

- **The sense of vision:** The eyes are the sensory organs that receive visual information from the environment and transmit them to the visual sensory area of the brain for interpretation. The eyes, equipped with an adjustable pupil and a lens, capture the patterns of illumination in the environment as an 'optical picture' on a layer of light-sensitive photoreceptor cells in the retina. The retina facilitates feature analysis of the image and transmits the visual signals through the steps of visual processing to the various structures of the brain where it is finally perceived.
- **The sense of hearing:** The auditory system perceives the frequency of sounds as pitch and their amplitude as loudness. The external and the middle ear conduct the sound to the auditory receptors (organ of Corti) in the cochlea of the inner ear. The auditory receptors are the hair cells embedded in the basilar membrane and their apical surface and convert sound wave signals into nerve impulses. The vestibular apparatus of the ear helps to maintain the posture and equilibrium of the body. It is mainly controlled by a gravity sensor (vestibular apparatus) in response to visual and proprioceptive information.
- **The sense of smell:** Olfaction (smell) is an animal's primary special sense perceived by the main olfactory system and the accessory olfactory system with receptors in the vomeronasal or Jacobson's organ located near the external nares. The odorant molecules combine with odorant receptors to mediate signal transduction mechanism that leads to a series of electrical events to facilitate the sense of smell.
- **The sense of taste:** Gustation or taste is a kind of chemical sensation that is perceived through chemoreceptors in the taste buds. Binding of a taste-provoking chemical, a tastant, with a receptor cell ultimately alters the cell's ionic channels to produce a depolarising receptor potential for taste sensation.
- **The skin and integumentary system:** The skin and its derivative or accessory structures are called integumentary system that provides physical barrier and antimicrobial protection and involves in thermo- and immuno-regulation, excretion, secretion, pigmentation, sensation and locomotion of the animal.

Exercise

Short Questions

1. The opaque lens condition is termed ______.
2. The '3'-dimensional picture occurs in which types of vision?
3. Which layer of the skin does not contain blood vessels?
4. Why Langerhans cells are called antigen-presenting cells?

5. The scala vestibuli and scala tympani are connected by a narrow channel called ________.
6. __________________ acts as a gravity sensor during the equilibrium.
7. Animals that are very sensitive in detection of odour are called as______________.
8. Umami taste is elicited by ________________.
9. The antigen-presenting cells (APCs) of the skin are __________________.
10. __________________ act as mechanoreceptors to detect pressure and vibration.

Subjective Questions

1. How does a panoramic vision occur in horses?
2. How can a cat's hair be distinguished from a human's hair in the forensic investigation?
3. Write the role of vomeronasal organ in sexual behaviour.
4. Briefly describe the mechanism of olfaction.
5. How skin helps in thermoregulation?

Answers to the Short Questions

1. Cataract
2. Stereo vision
3. Epidermis
4. They trapped antigens and presented them to the T-helper lymphocytes
5. Helicotrema
6. Vestibular apparatus
7. Microsmatic
8. Glutamate
9. Langerhans cells
10. Pacinian corpuscles

Keywords for the Answer to Subjective Questions

1. Field of binocular vision, monocular vision, position of blind spots
2. Structure of cuticle and scale appearance, the structure of medulla and pigmentation, the structure of the follicle
3. Vomeronasal organ, perception of pheromone
4. Odorant molecules, ciliary G protein-coupled receptor, signal transduction, change in membrane potential
5. Vasodilation during hot and vasoconstriction of blood vessels, secretion and evaporation of sweat

Further Readings

Choi EH, Daruwalla A, Suh S, Leinonen H, Palczewski K (2021) Retinoids in the visual cycle: role of the retinal G protein-coupled receptor. J Lipid Res 62:100040. https://doi.org/10.1194/jlr.TR120000850

Heo S, Hwang T, Lee HC (2018) Ultrasonographic evaluation of skin thickness in small breed dogs with hyperadrenocorticism. J Vet Sci 19(6):840–845. https://doi.org/10.4142/jvs.2018.19.6.840

Kolarsick PAJ, Kolarsick MA, Goodwin C (2011) Anatomy and physiology of the skin. J Dermatol Nurses Assoc 3(4):203–213. https://doi.org/10.1097/JDN.0b013e3182274a98

McKnight G, Shah J, Hargest R (2022) Physiology of the skin. Surgery (Oxford) 40(1):8–12. https://doi.org/10.1016/j.mpsur.2021.11.005

Willough CE, Ponzin D, Ferrari S, Lobo A, Landau K, Omidi Y (2010) Anatomy and physiology of the human eye: effects of mucopolysaccharidoses disease on structure and function—a review. Clin Exp Ophthalmol 38:2–11. https://doi.org/10.1111/j.1442-9071.2010.02363.x

Part V

Digestive System

Physiology of Digestion

13

Iqbal Hyder, Poonooru Ravi Kanth Reddy, and Joydip Mukherjee

Abstract

Digestion is a process by which foods are broken down chemically and mechanically into smaller units that can be then absorbed. The organs of digestive system facilitate this process via movement of nutrients, water, and electrolytes from the external environment into the body's internal environment. The broad functions of digestive tract include secretory and motility functions that ultimately aid in digestion and absorption. Apart from the enzymes secreted from the gut itself, there is significant contribution from other organs like liver and pancreas in the process of digestion. Both secretory and motility functions of gastrointestinal (GI) tract are tightly regulated by intrinsic control mechanism via enteric nervous system apart from direct control of vagus nerve. In addition to the neural control, the GI tract is also controlled by hormones secreted by GI tract itself that predominantly act in autocrine and paracrine manner. The avian digestive system is modified to accommodate flight. This chapter focuses on all of the above discussed aspects with additional excerpts on recent advances.

I. Hyder
College of Veterinary Science, Sri Venkateswara Veterinary University, Garividi, Andhra Pradesh, India

P. R. K. Reddy
Veterinary Dispensary, Animal Husbandry Department, Taticherla, Andhra Pradesh, India

J. Mukherjee (✉)
Department of Veterinary Physiology, West Bengal University of Animal & Fishery Sciences, Kolkata, West Bengal, India

© The Author(s), under exclusive license to Springer Nature Singapore Pte Ltd. 2023
P. K. Das et al. (eds.), *Textbook of Veterinary Physiology*, https://doi.org/10.1007/978-981-19-9410-4_13

Graphical Abstract

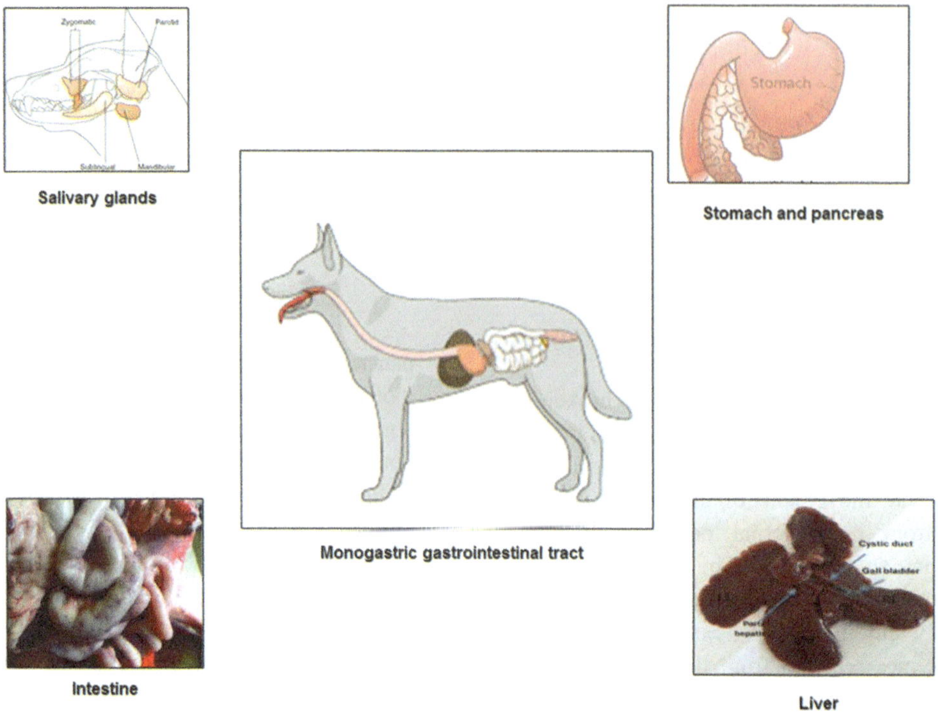

Description of the graphic: Monogastric digestive tract is a long hollow tube extending from the mouth to anus together with some accessory organs in between. The digestion process starts at the mouth cavity by mechanical force of mastication/chewing. The saliva lubricates feed particles for swallowing. The enzymes present in the saliva partially digest the feed. The feed enters into the stomach through oesophagus. Stomach secrets gastric HCl and enzymes to digest proteins. The final stage of digestion is occurred at the small intestine with brush boarder enzymes. The absorption of nutrients occurs at the small intestine. The liver produces bile that helps in the digestion and absorption of lipids.

Keywords

Monogastric digestion · Secretions of GI tract · Absorption · Avian digestion

Learning Objectives

- general overview of digestive system in monogastric digestion
- gastrointestinal motility and its control
- secretory functions of digestive system and its control
- digestion and absorption of nutrients
- physiology of avian digestive system

13.1 Monogastric Digestion

13.1.1 Overview of Monogastric Digestion

The digestion is a complex process of feed intake, conversion of the complex feed into their simplest form by mechanical and biochemical processes, absorption of the nutrients and their assimilation together with the removal of undigested feed materials. The process of digestion starts at the oral cavity where mastication reduces particle size of the ingested feed and incorporates saliva into ingesta for swallowing. The stomach facilitates grinding and mixing of the food along with digestion of proteins with the help of acid and enzymes. Once the chime passes into the small intestine, it is mixed with pancreatic enzymes and membrane bound enzymes in the enterocytes to convert the complex feed materials into their simplest form for absorption. The gastrointestinal system is the portal through which nutritive substances, vitamins, minerals, and fluids enter the body.

13.1.1.1 Functional Anatomy of Gastrointestinal Tract in Different Domestic Animals

The digestive tract of different species shows numerous structural and functional modifications from their primitive forms to accommodate their wide range of diets and habitats (Table 13.1). The elementary canal or the gastrointestinal [GI] tract is a hollow tube comprises mouth, pharynx, oesophagus, stomach, small intestine, large intestine, and rectum along with the accessory organs like salivary glands, pancreas, and liver. Animals are classified according to their diet in natural conditions as herbivores, omnivores, and carnivores.

The wall of the entire GI tract is made up of four basic layers, from the lumen towards the outside viz. mucosa, submucosa, muscularis, and serosa.

13.1.1.1.1 Mucosa

There are three layers of mucosa:

1. Innermost layer of epithelium made up of non-keratinized stratified squamous cells lines along the lumen to provide protection against wear and tear particularly in the mouth, oesophagus, and the anal canal. In the stomach and intestine, the stratified squamous cells gradually turn into columnar cells to carry out the specific functions of secretion and absorption.
2. In the middle, a layer of areolar connective tissue called lamina propria contains many blood vessels and lymphatic vessels to supply the mucosa and carry the absorbed nutrients of digestion. The lamina propria also contains many lymphatic nodules called mucosa-associated lymphoid tissue (MALT) that facilitates protection against microbes of food origin. MALTs are prevalent in tonsils, small intestine, appendix, and large intestine.
3. The smooth muscle layer arranged in outer longitudinal and an inner circular layer responsible for the local movements of the mucosal layer is called muscularis mucosa. This layer makes the mucosal membrane of the stomach and small intestine into folds to increase the surface area for digestion and absorption.

13.1.1.1.2 Submucosa

This connective tissue layer contains blood vessels and nerves in the form of a plexus apart from lymphatic tissue and glands. The submucosal plexus is highly developed in intestines whereas in stomach and oesophagus such ganglionated plexus is poor to sparsely developed. In larger mammals, the intestinal submucosal plexus is of two types. The inner Meissner's plexus is situated at serous end of muscularis mucosa and outer Schabadasch's or Henle's plexus is adjacent to circular muscle layer at the luminal side.

13.1.1.1.3 Muscularis

The muscularis layer consists of muscle fibres arranged in two layers viz. an outer longitudinal and an inner circular. In the mouth, pharynx and upper one third of the oesophagus it has voluntary skeletal muscle fibres that regulate swallowing.

Table 13.1 Modifications in GI tract according to the type of animal

Characteristics	Herbivore	Carnivore	Omnivore
Facial muscles	Well developed	Reduced to allow wide mouth gape	Reduced
Jaw motion	No shear; more side to side and front-to-back	Shearing, minimal side to side	Shearing, minimal side to side
Major muscle	Masseter and Pterygoid	Temporalis	Temporalis
Mouth opening	Small	Large	Large
Teeth (Incisors)	Broad, flattened, and spade shaped	Sharp and pointed	Sharp and pointed
Teeth (canines)	Short or long (for defence) or none	Long, sharp, and curved	Long, sharp, and curved
Teeth (molars)	Flattened with cusps	Sharp, jagged, and blade shaped	Sharp blades and/or flattened
Stomach type	Simple or multiple chambers	Simple	Simple
Acidity of the stomach	Less acidic (pH 3–4)	Highly acidic (pH 1)	Highly acidic (pH 1–2)
Length of small intestine in Comparison to body length	12–27 times	4–6 times	10–14 times
Caecum	Very well developed	Reduced	Moderately developed
Colon	Long, complex, sacculated	Simple, short, and smooth	Simple, short, and smooth
Length of body: Length of GIT	Horse—1:12 Cattle—1:20 Sheep—1:27	Dog—1:6 Cat—1:4	Pig—1:16

The rest of the portions are lined by smooth muscle fibres regulated by the autonomic nervous system. The muscularis layer is responsible for the movement of the GI tract to facilitate mixing and propulsion of food.

13.1.1.1.4 Serosa

This is the outermost layer consisting of areolar connective tissue and simple squamous epithelium (mesothelium). It is also called adventitia in the oesophagus where it is made up of only the areolar connective tissue without mesothelium.

13.1.1.2 Mechanical Factors Involved in Digestion

The mechanical factors are principally required for physically breaking down feed particles into their smaller forms for effective chemical digestion. It involves prehension, followed by mastication and deglutition (Swallowing of food).

13.1.1.2.1 Prehension

Prehension is the grasping and conveying of food into the oral cavity. The act of prehension varies between species. Cattle use protrusible tongue and incisors of lower jaw for prehension, horses use upper lip, tongue, and incisor teeth to collect food. In sheep and goat, the mobile upper lips are involved in prehension. Pigs use lower lips for prehension while the dogs and cats grasp their prey with forelimbs and carry into the mouth by the movements of head and jaw. In cats, papillae of the tongue (dorsal lingual spicules) help in pushing the feed into the oral cavity.

13.1.1.2.2 Drinking

The drinking is facilitated by suction of fluids by creating negative pressure in horse, cattle, sheep, and goat. Dog and cat use their ladle shaped tongue for drinking. It is vigorously extended and retracted to carry the liquid into the mouth. The negative pressure inside the mouth cavity created by backward tongue movement forces the milk to enter inside the mouth during suckling.

13.1.1.2.3 Mastication (Chewing)

Mastication is the act of chewing by the movement of jaw, tongue, and cheeks that facilitates grinding, moistening, and lubricating the food after mixing with the saliva. Mastication increases the surface area of the feed particles for better enzymatic digestion. Mastication involves rhythmic movements of mandible accompanied by extension of tongue called linguo-mandibular reflex. The presence of feed in the oral cavity stimulates tongue and oral receptors. The sensory inputs via trigeminal, facial, and glossopharyngeal nerves are carried to brainstem and efferent inputs via trigeminal nerve reach to masticatory muscles. In herbivores, mastication is facilitated by lateral movement of the lower jaw. Chisel shaped molar teeth and sharp-edged lower teeth help in grinding of feed particles. Incisor teeth are used for cutting the food. In ruminants, the modified dental pads together with the lower jaw help to cut the feed materials due to the absence of upper incisors. Masseter and pterygoids muscles are very prominent in herbivores. Masseter muscle helps to close the jaw and facilitates forward movement. Pterygoids muscles help to grind the feed by side-to-side jaw movement. Temporal muscles are prominent in carnivores that help to close the jaw and allow the teeth to sink into pray. The lateral and forward jaw movement are restricted in canines due to smaller masseter and pterygoid muscles.

13.1.1.2.4 Deglutition

It is a highly complex reflex that delivers ingesta or fluids from mouth to the stomach through pharynx and oesophagus. It starts as a voluntary act then becomes an involuntary reflex during its execution. Deglutition is facilitated by coordinated motor activities involving the muscles of tongue, pharynx, and oesophagus. Deglutition centre is situated at the medulla. The complex mechanism of deglutition is controlled by lower motor neuron, vagus, hypoglossal, glossopharyngeal nerves, and motor parts of trigeminal nerve. In the voluntary phase of deglutition, the ingested feed materials are converted into bolus by the tongue and pushed back into the pharynx. The pharyngeal pressure receptors (sensory nerve endings) detect the presence of bolus and stimulate deglutition centre to initiate swallowing reflex (involuntary phase). At the beginning of the involuntary phase of deglutition, the breathing is completely stop followed by the elevation of soft palate to close the pharyngeal opening of the nasopharynx. That restricts the entry of feed into the internal opening of the nostrils. To close the oral opening of pharynx, tongue is pressed against hard palate. The glottis is pulled under the epiglottis to ensure the blocking of laryngeal opening followed by the constriction of arytenoids cartilage that prevents the feed to enter into the respiratory passage. After the closure of all pharyngeal openings, the muscular contractions along the wall of the pharynx, push the bolus towards oesophageal opening. The feed enters into the oesophagus after the relaxation of upper oesophageal sphincter.

13.1.1.3 Gastrointestinal Motility

The motility of GI tract results from coordinated contractions of smooth muscle to propel, retain, and mix the ingesta. The motility of the GI tract also facilitates the movement of ingesta around the absorptive surface for efficient absorption. The GI motility can be of three types, propulsive motility to propel the ingesta in forward direction. The propulsion of feed materials is achieved through wave-like muscle contractions called peristalsis. It is occurred through contraction and relaxation of circular and longitudinal smooth muscles of gastrointestinal tract. Peristalsis is of two types.

Primary peristalsis is induced by swallowing and secondary peristalsis is induced by oesophageal distension. The peristalsis is achieved by alternating relaxation and contraction of distal and proximal muscles of GI tract. Retentive motility ensures the retention of feed at a particular segment of GI tract. A combination of both propulsive and retentive motility is also occurred. The time taken by the ingesta to travel from one portion of the GI tract to another is called transit time. Propulsive motility decreases transit time and retentive motility decreases it.

13.1.1.3.1 Motility of the Oesophagus

The oesophagus is a muscular tube extends from the pharynx to the stomach. The upper oesophageal sphincter is formed in part by the cricopharyngeal muscle and lower oesophageal sphincter is surrounded by the crural diaphragm. Upon deglutition, the relaxation of the upper oesophageal sphincter allows the passage of the food bolus into the oesophagus. The upper portion of the oesophagus consists of striated muscles and their activities are regulated by central controlling mechanisms like swallowing reflex. The lower part of the oesophagus is made of smooth muscles that exhibit peristaltic movements under central and intrinsic controlling mechanisms. Normally, the lower oesophageal sphincter is tightly closed under the influence of gastrin and vagal parasympathetic stimulation to prevent stomach contents and acid from entering the oesophagus. In most species, opening of the lower oesophageal sphincter is mediated by vasoactive intestinal polypeptide (VIP) accompanied by peristaltic waves that propels the bolus into the stomach. The peristalsis in oesophagus is of two types, primary and secondary. Primary peristalsis refers to bolus-induced oesophageal contractions upon swallowing whereas secondary peristalsis refers to distension-induced contractions independent of swallowing.

13.1.1.3.2 Gastric Motility

The motility of the proximal stomach is characterized by continuous weak contraction that allows gentle propelling of feed into the distal stomach. The proximal stomach also has adaptive relaxation property as it stores feed. Adaptive relaxation facilitates to accommodate large volume of feed without increasing the intraluminal pressure. Vagal stimulation suppresses the muscular contraction in the proximal stomach to facilitate adaptive relaxation process. It is of great importance in carnivores like wolves and lions to ingest large volume of meat from prey available only once every few days. In contrast, horse has a relatively small stomach with a very limited capacity for distension.

The distal stomach facilitates the grinding of feed by intense slow wave activity with frequent muscular contractions. The propulsive motility begins at the junction of proximal stomach and moves towards pylorus where finely ground and liquefied materials pass through the duodenum. Sometimes the ingesta are propelled back to the proximal stomach for proper grinding.

Reflexes Associated with Gastric Emptying

Gastro-gastric Reflex: The distension of the gastric reservoir initiates excitatory reflexes to stimulate antral contractions. In contrast, the inhibitory reflexes are induced by antral distension for relaxation of the stomach.

Duodenal Control: Gastric emptying is inhibited by nutrients entering the duodenum designated as "Duodenal control". The feedback-inhibition of gastric emptying is elicited by various stimuli. Hydrochloric acid, osmolality of the chyme, and an increased amount of nutrients entering the small intestine reduce the rate of gastric emptying. The afferent vagal fibres act as the receptors for glucose, osmolality, hydrochloric acid, amino acids, and long-chain fatty acids. Gastrointestinal hormones are also involved in the feedback regulation. One of the most important hormones is cholecystokinin (CCK) that mainly causes relaxation of the reservoir and delays emptying. Peptide YY and the glucagon-like peptide (GLP-1) also inhibit gastric emptying. Secretin and gastric inhibitory polypeptide (GIP) also reduce gastric motility whereas gastrin increases it.

13.1.1.3.3 Motility of Small Intestine

The peristaltic waves at the proximal part of intestine are rapid and far spreading which gradually become shorter and slower towards the distal gut to achieve different transit rates along the intestine. Under physiological conditions, small intestine exhibits five different contractile patterns.

Peristaltic waves: These are circular constrictions propagating aborally associated with an aboral relaxation or inhibition of the muscle, respectively, to facilitate an aboral transport of chime. In dogs, the propagation velocities of the peristaltic waves are 7–12 cm/s in the duodenum, 4–7 cm/s in the jejunum, and 0.7–0.8 cm/s in the ileum.

Stationary contractions (segmenting contractions): It occurs as segmental contractions at single sites. By means of segmenting contractions, the chyme is pushed orally and aborally at a localized area for the mixing of the luminal contents.

Clusters of contractions: These are complex contractile patterns characterized by several short repetitive contractions pushing the chyme a few centimetres aborally followed by a partial backflow during the period of relaxation. These types of contractions are required for mixing of chyme and frequently seen after a fat meal.

Migrating motor complex (MMC): It is a cyclic motor pattern of the GI tract exhibited during the inter-digestive state. It is appeared as clusters of contractions divided into four phases that propagate over a longer intestinal segment.

Phase I is called quiescent phase as no contractions occur during this phase. In phase II, random contractions occur. Phase III is characterized by a rapid contraction with the highest amplitude and duration that occur suddenly. The amplitude and duration of contractions are decreased in phase IV. MMC is present in rats, sheep, rabbits, pigs, dogs, and cows. MMC is generally occured during inter-digestive periods, but ad libitum feeding has no effect of MMC in sheep, pigs, and rabbits. But MMC is disrupted when the animals ingest feed once or twice a day. Four phases of MMC are not recognizable in rats and mice; hence, in this species MMC can be described as phase-I-like and phase-III-like contractions that occur in every 12–15 min. In dogs, large particles such as bones, stones of peaches, or insoluble tablets are forced into the intestine by the onset of the inter-digestive motility.

Giant contractions: These are characterized by large amplitude and a long duration. These types of contractions are seen in the ileum of dogs, pigs, and horses during inter-digestive period. In pigs, giant contractions are also observed during digestive period. The giant contractions completely occlude the intestinal lumen and propagate slowly to aboral direction pushing the luminal contents distally and cleaning the intestine; hence, they are also called "stripping wave". Aboral giant contractions of the small intestine are the typical contractile pattern in diarrhoea.

13.1.1.3.4 Motility of Large Intestine

The large intestine serves as fermentation chambers wherein microbial digestion takes place. The faeces are also produced in the large intestine after the absorption of water. These two functions require proper mixing and transport of digesta. Different parts of large intestine show different contractile patterns.

1. peristaltic and antiperistaltic waves
2. aborally migrating segmenting contractions
3. haustral movements and
4. aborally propagating giant contractions

Peristaltic and antiperistaltic waves: These are characteristic motor patterns of the caecum and proximal colon. Waves with shallow circular constrictions followed by low retropulsion is causes by an intensive mixing of chyme.

Aborally migrating segmenting contractions: These are unique contractile patterns of the large intestine frequently seen in the species producing faecal boli, also in carnivores. In dogs and horses, they are called "colonic motor complex" (CMC). The segmenting contractions separate the digesta into boli. In contrast to the segmenting contractions of the small intestine, the segmenting contractions of large intestine represent long-lasting circular constrictions that occur simultaneously at adjacent sites with slow distal movement.

Haustral movements: Haustra are the small-segmented pouches of large intestine. Movements of the haustra are characterized either by alternating contractions and relaxation resulting in mixing of digesta or by an oral or aboral rolling movement causing transport of liquids in a definite direction. Haustral movements are frequently associated with the migrating segmenting contractions.

Aborally propagating giant contractions: These are characterized by large amplitude, long duration, and slower propagation velocity in comparison to peristaltic waves. They facilitate pronounced aboral transport of digesta.

13.1.1.3.5 Defecation

It is the act of expelling faeces from the digestive tract through the anus. Defecation requires a complex and synchronized interactions between gastrointestinal system, nervous system, and musculoskeletal system. The anal opening is guarded by internal anal sphincter and external sphincter made of involuntary circular smooth muscle and voluntary striated muscle, respectively. The faecal contents are channelized into the rectum by peristalsis of colon. The filling of rectum stimulates mechanoreceptors of the wall of the rectum to initiate defecation reflexes. The reflexes are of two types.

Intrinsic reflex: It is mediated by enteric nervous system (myenteric plexus) after the distension of rectal wall. It causes peristaltic waves in descending colon, sigmoid, and rectum followed by relaxation of internal anal sphincter to allow a small amount of faeces to pass through to the anal canal. It is called the recto-anal inhibitory reflex frequently used for anal sampling.

Defecation reflex (Parasympathetic): In the reflex, the signals for rectal filling first transmitted into the spinal cord and then back to the descending colon, sigmoid, rectum, and anus through parasympathetic nerve fibres in the pelvic nerves. These parasympathetic signals result strong peristaltic waves followed by relaxation of the internal anal sphincter to clear the bowl. Strong contractions generate a pressure gradient between the rectum and anal canal for defecation. After the defecation, external anal sphincter regains its normal tone and maintains continence.

13.1.1.3.6 Emesis

Vomiting or emesis is the forceful oral expulsion of gastrointestinal contents by the contractions of the gut and the thoraco-abdominal musculature. The urge to vomit is called nausea. Vomiting is the act of defence intended to remove toxins, drugs, and pathogens entered into the body through enteral or parenteral route. Vomiting centre is situated at the medulla oblongata includes the reticular formation and the nucleus tractus solitarius. The vomiting centre receives inputs from four principal areas namely gastrointestinal tract, vestibular region, chemoreceptor trigger zone (CRTZ), and cerebral cortex. Out of these four regions, the CRTZ is closest to the vomiting centre as it lies between the medulla and the floor of the fourth ventricle. CRTZ is devoid of blood–brain barrier thus the irritants can easily pass through it. The peripheral stimuli such as toxic substances and pathogens and pathology of GI tract induce the release local emetic neurotransmitters like serotonin to stimulate vomiting centre. The motion sickness and opioid analgesics act via vestibular region to stimulate vomiting centre by releasing histamine and acetylcholine. Pain and anxiety induce stimulate vomiting centre through thalamus and cerebral cortex. CRTZ has receptors for neurokinin, mu/kappa opioids, and dopamines. The mechanism of vomiting includes relaxation of muscle of stomach and lower oesophageal sphincter followed by closing of pylorus. The intra-thoracic pressure is decreased due to expansion of chest cavity and closure of glottis. Finally, the upper oesophageal sphincter closes and the gastrointestinal contents are expelled by antiperistalsis activity. Carnivores and most omnivore mammals are emetic species. But rodents are non-emetic species that lack a vomiting reflex.

Know More...
The rodents have long abdominal oesophagus, and they lack neurological component for vomiting reflex. Hence, they are unable to vomit.

13.1.1.4 Control of GI Functions

The gastrointestinal (GI) tract function is very well regulated at various levels. These controls can be classified as myogenic, neurogenic, and endocrine controls.

13.1.1.4.1 Myogenic Control

The contractions of GI smooth muscles are derived from the electrical activity across the membranes of smooth muscle cells. The resting membrane potential of smooth muscle cells is between −50 and −60 mV. In contrast to nerves and other types of muscle cells, the membrane potential of smooth muscle cells fluctuates spontaneously. The electrical activity of GI smooth muscles is initiated from interstitial cells of Cajal (ICC) that surround the circular and longitudinal smooth muscles. ICC resembles Purkinje cells of heart with rhythmic oscillating properties hence called "pacemakers of the guts". The characteristics features of ICCs include small cell bodies with elongated processes, numerous mitochondria, abundant intermediate filaments, few ribosomes, and endoplasmic reticulum. The gap junctions facilitate the communications between ICCs and other smooth muscles to propagate the electrical signals. There are morphologically distinct ICC in different locations of GI tract. Majorities of ICCs are abundant in the myenteric plexus (Auerbach's plexus) called ICC of the myenteric plexus (ICC-MY or ICC-MP) or ICC of Auerbach's plexus (ICC-AP). IC-SM (submucosal ICCs) are located at the submucosal surface of the colon; IC-DMP are cells along the intestinal deep muscular plexus; IC-IM are the intramuscular cells in oesophagus, stomach, and colon. ICCs generate a spontaneous rhythmic membrane potential between 65 and 45 mV called basic electrical rhythm (BER) by the activation of L-type voltage-dependent calcium channels that allow the entry of calcium into smooth muscle cells. The BER itself does not cause muscle contraction.

The electrical activity across the membranes of GI smooth muscles shows two patterns namely slow waves and spike potentials. The slow waves are generated as a result of partial depolarizations when the membrane potential fluctuates between 5 and 15 mV. Slow waves sweep along the digestive tube for long distances. The frequency of slow waves varies along the different sections of the digestive tract, it is 10–20 times per minute in the small intestine and 3–8 times per minute in the stomach and large intestine in dogs. Slow wave activity is the intrinsic characteristics of smooth muscle independent to nervous stimuli.

Slow waves are generated as partial depolarization; they are unable to elicit contractions. Rather, they coordinate or synchronize muscle contractions in the gut to initiate a second type of depolarization event called "spike potentials". Spike potentials are true action potentials generated when the slow waves pass over the area of GI smooth muscle sensitized with neurotransmitters of the enteric nervous system. The spikes contain both depolarizing and repolarizing components result due to calcium influx and potassium efflux, respectively.

13.1.1.4.2 Neurogenic Control

The GI system is innervated by the enteric nervous system and the autonomic nervous system.

The enteric nervous system (ENS): It is the intrinsic nervous system of the gut together with pancreas and gall bladder, composed of mesh-like system of 500 million neurons. It is also called "Gut brain" as it is independent of the central nervous system. ENS receives information form the mechanical and chemical receptors stimulated by sight smell and presence of feed. ENS is composed of an outer myenteric or Auerbach's plexus situated between longitudinal and circular muscle layer. Another inner submucosal plexus known as Meissner's plexus situated at the submucosa of the intestinal wall. Both the excitatory and inhibitory motor neurons of myenteric plexus innervate the intestinal smooth muscles and secretomotor neurons project to the mucosa. The secretomotor activities are initiated through various neurotransmitters. The neurotransmitters for relaxation of GI smooth muscles are nitric oxide, pituitary adenylate cyclase-activating peptide, vasoactive intestinal peptide (VIP), and purine. Whereas tachykinins and acetylcholine are responsible for intestinal contraction.

The submucosal ganglia are composed of different types of neurons like intrinsic primary afferent neurons (IPANs), interneurons, secretomotor, and vasodilator neurons.

With these neural components, the enteric nervous system (ENS) exhibits neural reflexes to control and coordinate motility, secretion, and blood flow.

The autonomic nervous system (ANS): This system includes sympathetic and parasympathetic systems. The sympathetic nervous system is inhibitory to GI muscles and glands. However, the sympathetic system regulates blood flow in the GI system. The parasympathetic nervous system has both excitatory and inhibitory control over the gastric functions. The parasympathetic system comprises vagus (oesophagus, stomach, pancreas, upper large intestine) and pelvic nerves (lower portion of large intestine, rectum, and anus). The vagus nerves synapse with myenteric motor neurons and control them by nitric oxide (inhibitory actions) and acetylcholine/neurokinins (excitatory actions).

Hypothalamic Control of GI Functions

The hypothalamus plays a pivotal role in regulating appetite and energy expenditure by sensing the metabolic signals from leptin, amylin, cholecystokinin (CCK), glucagon-like peptide-1 (GLP-1), peptide YY (PYY), and ghrelin. Distinct hypothalamic areas play a critical role in controlling GIT functions, especially with respect to feeding and satiety. Lateral hypothalamus (LHA) acts as "feeding centre" to promote food intake and weight gain whereas ventromedial hypothalamic nucleus (VMH) acts as "satiety centre" that favours weight loss. The sub-populations of arcuate (ARC) neurons express a variety of neuropeptides including orexigenic (appetite stimulant) Agouti-related protein (Agrp), Neuropeptide Y, and the anorexigenic (appetite suppressor) pro-opiomelanocortin (POMC). These neurons are highly responsive to metabolic status and regulate energy intake by modulating melanocortin-4 receptor (MC4R). In response to these metabolic signals, hypothalamus coordinates with multiple brain regions through neuronal circuits. The area postrema (AP), on the caudal brainstem, regulates satiation in response to metabolic signals.

13.1.1.4.3 Endocrine Control

The GI hormones can be classified into endocrines, paracrine, and neurocrine based on their mode of delivery. Hormone gastrin, secretin, cholecystokinin (CCK), motilin and gastric inhibitory polypeptide (GIP), or glucose-dependent insulinotropic peptide (GIP) are secreted from the enteroendocrine cells into the blood stream hence categorized under endocrines. Somatostatin and histamine are secreted as paracrine fashion to their local target tissue. Some hormones act through both endocrine and paracrine mechanisms like pancreatic polypeptide, glucagon-like peptide-1 (GLP-1), and peptide YY. Neurocrine hormones like enkephalins, vasoactive intestinal peptide (VIP), and gastrin release peptide (GRP) are secreted from postganglionic non-cholinergic neurons of the enteric nervous system. The roles of different hormones in GI functions are summarized in Table 13.2.

13.1.2 Secretory Functions of GI Tract

The primary functions of the GI tract are the digestion and absorption of feed. Different parts of GI system secrete a wide range of chemical substances to assist digestive and regulatory processes of GI function. Salivary glands, stomach, pancreas, gall bladder, and intestine are the predominant organs that contribute GI secretions. There are several anatomically distinct glands in the epithelial surface of GI tract. Single cell mucous glands like goblet cells contribute to mucous secretion in response to irritation. Small intestine is equipped with specialized secretory cells at the epithelial invaginations called Crypts of Lieberkühn. The glandular part of stomach contains deep tubular cells (oxyntic gland) that secret acid and pepsinogen. The salivary glands and the pancreas are complex acinar glands situated outside of the elementary canal, but their acinar secretions are poured into the GI tract. The glands of GI system are stimulated by direct contact of food. The tactile, chemical, and wall distension activates ENS that stimulates the glands for secretion. Parasympathetic stimulation increases the secretions of glands of upper GI tract. The GI secretions are also influenced by endocrine factors.

13.1.2.1 Salivary Secretions

Saliva is the collective secretions of three major salivary glands (Table 13.3) and numerous minor salivary glands situated at the mucous membrane of oral cavity. Shape, size, and number of salivary glands are varied among animal species. The functional capabilities of salivary glands are also varied. In majority of species, it lubricates the feed bolus, aids digestion, and protects the oral cavity. But, in some arthropods, saliva is used to prepare threads for the cocoons.

Other than three main salivary glands, sheep, and cattle have paired inferior molar salivary glands. There are some minor salivary glands in animals like buccal, palatine (palate), labial (lips and cheeks), and pharyngeal (pharynx). Dorsal buccal gland of dogs is also called zygomatic salivary gland. It is a sero-mucous gland situated at the zygomatic arch and ducts open at caudal parotid papilla in the oral cavity.

Salivary glands can also be classified on the basis of their secretory contents. Parotid, inferior molar, palatine, and buccal glands secret more bicarbonate hence called alkaligenic glands. In contrast, submaxillary, sublingual, and pharyngeal are called mucogenic glands as they secret more mucin.

Table 13.2 Roles of different hormones in GI functions

Name	Source	Functions
Gastrin	G cells in the stomach and duodenum	Stimulates parietal cells in the stomach for acid secretion Growth of intestinal mucosa Inhibition of secretin and GIP
Cholecystokinin (CCK)	I cells in the duodenum and jejunum	Contraction of the gallbladder Inhibition of gastric emptying Stimulation of pancreatic enzymes and bicarbonate secretion
Secretin	S cells in the duodenum	Inhibition of gastrin and acid secretion Stimulation of biliary secretion Stimulation of pancreatic bicarbonate secretion
Glucose-dependent insulinotropic polypeptide (GIP) (Previously known as gastric inhibitory polypeptide	K cells in the duodenum and jejunum	Stimulation of insulin secretion Induction of satiety Stimulation of lipoprotein lipase
Glucagon-like peptide-1 (GLP-1)	Intestinal L-cells	Inhibition of gastric emptying Induction of satiety
Somatostatin	D cells of stomach, duodenum, and pancreatic islets	Inhibition of pancreatic and gastric exocrine functions Inhibition of the motility of stomach and the gut
Histamine	Intestinal enterocytes	Stimulation of gastric acid secretion
Peptide YY	L cells at the distal GI tract	Reduction of feed intake
Enkephalin	Postganglionic non-cholinergic neurons of the enteric nervous system	Inhibition of intestinal fluid and electrolyte secretion
Vasoactive intestinal peptide (VIP)	Postganglionic non-cholinergic neurons of the enteric nervous system	Relaxation of GI smooth muscles Stimulation of pancreatic and biliary secretions Inhibition of gastric acid secretion
Gastrin release peptide (GRP)	Postganglionic non-cholinergic neurons of the enteric nervous system	Stimulation of gastrin release
Motilin	Entero-endocrine cells (Mo cells) in the upper small intestine	Increases gastrointestinal motility

Table 13.3 Types of major salivary glands

Name	Anatomical Position	Type	Name of duct	Site of secretion	Nerve supply
Parotid	Under the ear and vertical ramous of mandible	Serous	Duct of Stensen	Either side in the vestibule of mouth cavity	Glossopharyngeal and Trigeminal nerve
Submaxillary or Mandibular	Intramandibular space	Mixed (Sero-mucous)	Ducts of Wharton	Either side of frenulum of tongue	Lingual and facial nerve
Sublingual	Base of tongue	Mucous	Duct of Bartholin Duct of Rivinus	Either side of frenulum of tongue	Lingual and facial nerve

13.1.2.1.1 Functional Anatomy of Salivary Gland

The glands are lobular and each lobule comprises acini and duct. The individual salivary secretory unit is called salivon. Each salivary acinus is lined by glandular epithelial cells surrounding a central lumen. The secretions from these glandular cells are poured into these lumens. The acini are of three types based on their nature of secretions viz. serous, mucous, and mixed (sero-mucous). The serous types of cells appear dark in haematoxylin and eosin (HE) stain due to plenty of zymogen granules. In contrast, mucous-secreting cells look empty under HE stain. Serous demilunes is the characteristic feature of mixed type of glands where serous part remains compressed at the periphery of a mucous acinus. Serous glands produce thin, watery, and enzyme-rich secretions. Mucous cells secret thick and viscoid mucous. There are specialized modified smooth muscle cells called myoepithelial cells surrounding the acinus. These cells have long cytoplasmic processes spread like a basket, hence called basket cells. The contraction of myoepithelial cells increases the ductal pressure that leads to salivary secretions.

There are four generations of salivary ducts viz. intercalated, striated, excretory, and main collecting duct. The striated duct is active and involved in reabsorption of electrolytes.

13.1.2.1.2 Composition and Rate of Salivary Secretion

The composition of saliva varies greatly among species and types of glands (Table 13.4). In non-ruminants, the secretions of submaxillary and parotid glands are hypotonic during basal or unstimulated secretions. The concentration of sodium chloride and bicarbonate increases with the secretory flow rates and at maximum flow rate it is isotonic in nature. But the ruminant saliva is isotonic at any flow rate though reciprocal changes in phosphate and bicarbonate concentration is noted at increased flow rates. The secretions of the parotid glands are continuous. The flow rate of parotid glands is about 2 mL/min at rest and 30–50 mL/min during rumination. Total salivary gland flow rate is 60–160 L/day in cow and 6.0–16 L/day in sheep. Free flow of submaxillary and sublingual glands is seen during chewing of normal meat in dogs. Parotid secretion occurs only during feed intake in horse. Ruminants have numerous minor salivary glands viz. buccal, pharyngeal, palatine, inferior molar, and labial. The composition of saliva is depicted in Fig. 13.1.

Table 13.4 Ionic composition of saliva (mmol/L) in different species at maximum flow rate

	Parotid gland			Submandibular gland		
Species	Na^+	K^+	HCO_3^-	Na+	K+	HCO_3^-
Sheep	160–175	9–10	113–140	20	7	23
Dog	80–110	6–14	50	70–100	12–15	10–30
Cat	–	–	–	40–51	9–10	26
Rabbit	110–140	10	12–30	50–100	10–40	25

Hornbuckle et al. (2008)

13.1.2.1.3 Control of Salivary Secretion

The salivary glands receive both efferent innervations of sympathetic and parasympathetic nervous system which mainly act synergistically on the salivary glands.

Sympathetic supply: The pre-ganglionic fibres originate from thoracic nerves and terminate at superior cervical ganglion after passing cervical sympathetic chain. The postganglionic adrenergic nerves originate from superior cervical ganglion and supply the blood vessels of salivary glands. The sympathetic stimulation causes myoepithelial cell contraction by norepinephrine. But, at the later phase of secretion, the sympathetic stimulation leads to thick mucin-rich salivary secretion mediated by vasoconstriction.

Parasympathetic supply: Parasympathetic stimulation comes from superior and inferior salivary nucleus located in the pons and medulla. Superior salivary nucleus supplies submandibular and sublingual salivary glands and inferior salivary nucleus supplies parotid glands. The sensory stimulation is carried by trigeminal and glossopharyngeal nerves. The afferent impulses are carried to the salivary glands by facial and glossopharyngeal nerves to cause vasodilation and

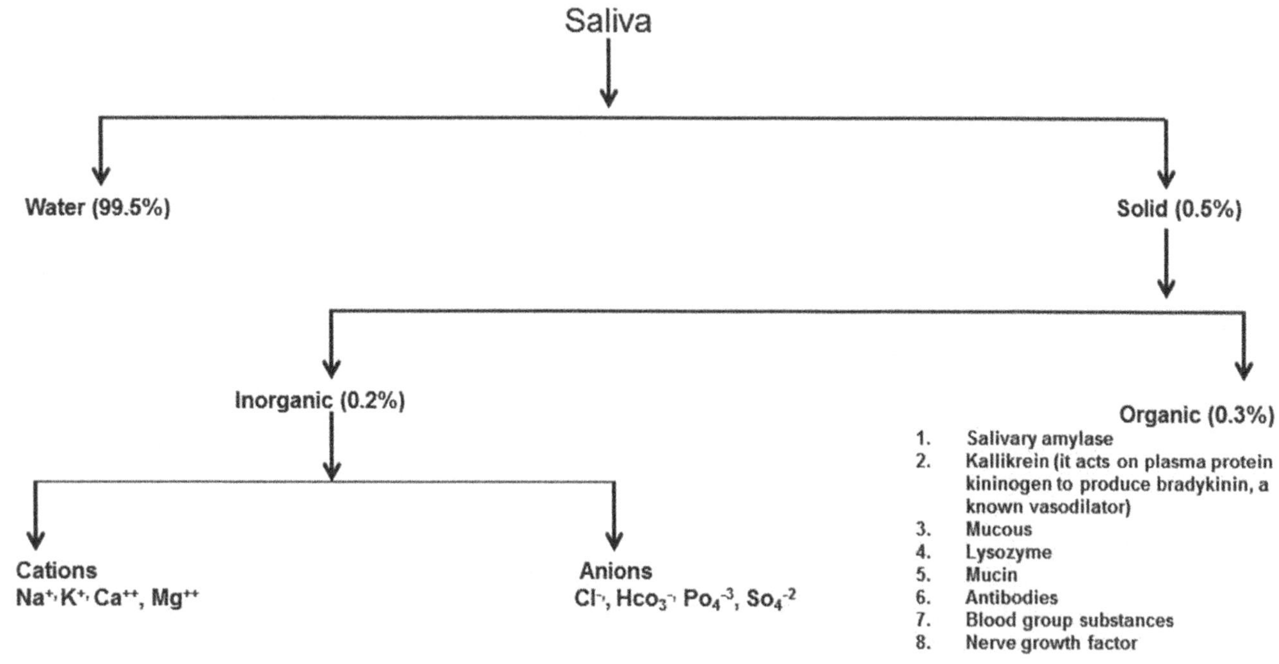

Fig. 13.1 Composition of saliva

copious salivary secretion rich in water and HCO_3, but low content of protein.

The main stimulus for salivary secretion is the feed intake that activates a number of taste receptors, olfactory receptors, mechanoreceptors, and nociceptors. All forms of taste sensations (sour, salt, sweet, and bitter) are able to stimulate salivary secretion, but sour has more pronounced effect. The movement of teeth during mastication stimulates mechanoreceptors present in periodontal ligaments. Olfactory receptors at the roof of the nasal cavity stimulate salivary secretion. The sensation is perceived through sniffing that increases airflow to the receptor area. Olfactory salivary reflex is common for submandibular salivary gland but absent in parotid glands. Spicy food activates nociceptors for salivary secretion. Dryness of the oral cavity also stimulates salivary secretion. Growth hormone, sex steroids, and thyroid hormones affect salivary gland metabolism and secretory property. Decreased level of sex steroids leads to hyposalivation in postmenopausal women. In sheep, aldosterone induces sodium uptake (without water) by salivary ductal cells and leads to decreased sodium concentration in the saliva. But aldosterone has little effect on ductal sodium reabsorption in human. Gastrin induces salivary secretion during gastric phase of salivary secretion whereas cholecystokinin and melatonin stimulate salivary secretion during intestinal phase. Ageing results atrophy of salivary glands and decreased saliva secretion. The loss of teeth, olfactory, and noci receptors during ageing also cause hyposalivation. Secretory pattern of saliva also follows circadian and circannual rhythm. Salivary secretion is lower in the morning and gradually increases till afternoon to reach highest around late afternoon. Flow of saliva is higher during winter than summer. Circadian pattern of salivary secretion is regulated by peripheral clock mechanism (clock gene) as in other organs. Secretion of salivary substances also follow circadian pattern. Salivary IgA concentrations show circadian rhythmicity and reaches peak during sleep.

Phase of Salivary Secretion

Secretion of saliva occurs in two phase basal and reflex action. Secretion of saliva without any stimulation is called basal salivary secretion. The reflex action is of two types. In unconditioned reflex, salivary secretion occurs due to the presence of food in mouth. Oesophago-salivary reflex and gastro-salivary reflex are seen when the food is present at oesophagus and stomach. The salivary secretion through conditioned reflex is brought about by thinking of food, sight, or smell. This reflex can be elicited even by non-physiological stimuli like ringing of a bell if properly conditioned.

13.1.2.1.4 Functions of Saliva

Taste: Saliva helps in dissolution of feed particles and helps in the perception of taste. The hypotonic nature of saliva helps in dissolution of nutrient particles. Further, saliva contains a protein named gustin required for the growth and maturation of these buds.

Protection and Lubrication: Saliva protects the oral tissue against irritating agents by forming a seromucosal layer. Mucins of salivary secretions facilitate lubrication and maintenance of salivary viscosity. They also prevent the adhesion of pathogens to the oral mucosa and prevent colonization. In addition, they protect against proteolytic attacks by microorganisms. The lubricating action of saliva also helps in mastication and deglutition.

Dilution and Cleaning: Saliva helps in mechanical cleaning of oral cavity and clears the residues such as nonadherent bacteria and cellular and food debris. Saliva limits the sugars utilization by biofilm microorganisms after eliminating excess carbohydrates.

Buffering Action: The bicarbonate and phosphate present in the saliva act as buffering agents. The bicarbonate is the major buffering agent of the saliva and phosphate contributes a little. Buffering action of saliva helps to maintain a stable oral pH and protect the mouth from pathogenic microorganisms. Further, saliva neutralizes the acids produced by acid forming microorganisms, thus, preventing demineralization of tooth enamel.

Integrity of Tooth Enamel: Saliva plays a key role in maintaining the physical-chemical integrity of tooth enamel through various ways. Firstly, salivary glycoproteins help to form acquired dental pellicle (a thin protein film over the surface enamel and dentin). Secondly, it neutralizes acid by dilution and buffering action and protects the enamel from erosion. Saliva contains calcium, phosphate, and fluoride necessary for remineralization of enamel.

Digestion: In omnivores such as rats and pigs, saliva contains a starch-digesting enzyme α-amylase (ptyalin). This enzyme is usually absent from saliva of carnivores such as cats and dogs. Lingual lipase present in young animals helps to digest lipids during milk diet and disappears as the animals mature.

Absorption of Vit-B12: Salivary glands produce a glycoprotein called "Haptocorrin" or "Cobalophilin" that binds with Vit-B12 and protects it from acid digestion in stomach. However, in duodenum, the Vit-B12 once again becomes free to bind, this time to another molecule called "Intrinsic Factor" forming a B12–IF that is absorbed in the ileum.

Tissue Repair: Epidermal growth factor produced by the submandibular glands has a role in wound contraction.

Antibacterial Properties: Secretory immunoglobulin A (IgA) of saliva neutralizes viruses, bacterial, and enzyme toxins. Among the non-immunologic salivary protein components, there are enzymes (lactoferrin, lysozyme, and peroxidase), proline-rich proteins, histatins, mucins, statherins, and cystatins. Lysozyme hydrolyses the cellular wall of some bacteria. The histatins, a family of histidine-rich peptides, have antimicrobial activity against some strains of

Streptococcus spp. They neutralize the lipopolysaccharides of the external membranes of Gram-negative bacteria and are potent inhibitors of *Candida albicans* growth and development.

Thermoregulation: Some animals such as rats spread their saliva on their body so that it evaporates and provides a cooling effect on the body. The parotid glands of dogs are capable of secreting at ten times of the rate of parotid glands in human during panting.

Special Functions in Ruminants: In ruminants, saliva provides a proper media for the bacterial growth and activity in the rumen. Further, bicarbonates and other contents of alkaline saliva (pH 8.1) neutralize the volatile fatty acids produced during microbial fermentation and maintain a stable rumen pH. Apart from buffering action, saliva acts as anti-foaming agent due to its mucin content. Urea is nonprotein source supplies nitrogen for the bacterial growth and microbial protein synthesis. Phosphates are utilized for nucleoprotein and phospholipid synthesis.

13.1.2.1.5 Pathophysiology of Salivary Gland

Ptyalism: Hyper secretion of salivary glands lead to a condition called ptyalism characterized by drooling of saliva. Ptyalism is secondary to swallowing disorders in animals. The causes of ptyalism are toxins, drugs, poison such as organophosphorus compounds, glossitis, stomatitis, convulsive disorders, nervousness, motion sickness, linear foreign body ingestion, and oral tumour. The conformational defects like pendulous lips may also result ptyalism. In rabies, ptyalism is very characteristics hence care should be taken to examine patient with ptyalism.

Sialadenitis: The inflammation of the salivary gland is called sialadenitis. It rarely occurs in dogs and cats.

Xerostomia (dry mouth): Xerostomia or dry mouth is a clinical condition develops due to hyposalivation. It is uncommon in dogs and cats but can occur in animals under frequent radiation exposure. Xerostomia causes discomfort during eating and oral infections. Administration of drugs like atropine, severe dehydration, fever, and anaesthesia may also cause hypo salivation. Immune-mediated keratoconjunctivitis sicca in canines can also lead to xerostomia.

13.1.2.2 Gastric Secretion

The stomach lies between the oesophagus and duodenum at the left side of the abdominal cavity. Stomach stores food which is then mixed with acid, mucus, and pepsin; and released at a controlled and steady rate into the duodenum. Gastric juice contains hydrochloric acid (HCl), lipase, and pepsin that help in the digestion of proteins and fats. Gastric acid also helps to inactivate microorganisms thus acts as first line of defence against infection.

13.1.2.2.1 Functional Anatomy of Stomach

The stomach can be divided into four distinct functional compartments based on the distribution of gastric mucosa however, not all species have all four compartments.

Oesophageal stomach: The portion of stomach that lies just below the oesophagus is called oesophageal stomach. It is lined by stratified squamous epithelium. This portion of stomach is non-glandular in nature as no mucus, acid, or proteolytic enzymes are produced from this area. The horse has a rather large oesophageal stomach compartment, but it is very smaller than the dog, pig, and cow.

Cardia stomach: It is situated just below the oesophageal stomach. Here, the gastric mucosa changes from stratified squamous to simple columnar epithelium. It is considered a glandular stomach that produces thick mucus and buffer. The mucous and buffer protect the epithelium from corrosive actions of gastric acid the proteolytic enzymes. The portion of cardia is very large in pigs, very small in dogs. Cardia is almost absent in horse and cow.

Fundic stomach: This portion is the largest compartment, and all the animals have fundic stomach. It is glandular in nature and mucosal lining of this area has very deep invaginations lined by a variety of cells that produce acid, proteolytic enzymes, hormones, and mucus.

Pyloric stomach: It is the terminal portion of the stomach joined with duodenum and guarded by pyloric sphincter. Pyloric stomach is glandular with moderately deep glands lined by epithelial cells. These glands produce only mucus and buffer without acid or proteolytic enzymes. The G cells present at the pyloric region produce the hormone gastrin in response to gastric distension or in increased stomach pH. All mammals have a pyloric stomach.

13.1.2.2.2 Gastric Mucosa

The fundic stomach contains gastric pits lined with mucus-secreting cells at the luminal surface. The mucous protects the gastric mucosa from acid and proteolytic enzymes by forming a gel. The mucous gel also entraps bicarbonate ions for neutralization of gastric HCl. Each gastric pit has deep gastric gland that extends to reach the submucosal layer. There are about 35 millions of gastric glands in human stomach. Each gastric gland has three parts namely neck, body, and base. The junction of glands and gastric pits is called Isthmus. The cells have rapid regenerating property to mature within 2–3 days. About 0.5 million cells are desquamated per hour and replaced by new cells. There are different types of cells in the gastric pits.

Chief cells (peptic cells or zymogenic cells): Chief cells are predominant at the base of glands throughout the fundus of the stomach. They contain plenty of rough endoplasmic reticulum and dense zymogen granules. They secret pepsinogen, a proteolytic enzyme precursor into the lumen of the gastric gland. The zymogen granules move towards the apical surface to fuse with plasma membrane and release pepsinogen. Pepsinogen is converted to its active from pepsin by the hydrochloric acid. Chief cells also produce rennin, a proteolytic enzyme required to curdle milk. Renin is important in neonates to digest milk proteins.

Parietal cells (Oxyntic cells): These cells are pyramidal in shape with plenty of mitochondria, lysosomes, and tubulovesicles. The canaliculi are projected from the apical surface and lined by actin filaments and microvilli. The cells are tubulovesicular in nature under resting (non-secretary) condition. Upon excitation, the microcanalicular system extends up to the interior of the cells. These cells produce gastric HCl that facilitates hydrolytic breakdown of proteins and also kills many of the bacteria ingested through food. In most species, parietal cells also produce a protein known as intrinsic factor that helps in absorption of Vit-B12. Intrinsic factor tightly binds with Vit-B12 and facilitates absorption of the intrinsic factor–vitamin B12 complex through endocytosis in ileum.

Enterochromaffin (Enteroendocrine cells): These are small polygonal cells found predominantly in the small intestine and appendix, but also scattered in the colon, rectum, and stomach. In gastric mucosa, these cells are found among the parietal and chief cells. They are the predominant neuroendocrine cells of GI tract and produce serotonin which controls gastric acid and proteolytic enzyme secretion in endocrine and paracrine fashion.

Mast cells: They are found in the lamina propria of GI tract and comprise 2–5% of mononuclear cells. Mast cells are activated by substance P and release inflammatory mediators like serotonin, histamine, proteases, prostaglandin D, and other pro-inflammatory cytokines.

Delta cells (δ-cells or D cells): These are somatostatin-producing cells of the stomach, intestine, and pancreatic islets. D cells have close connection with gastrin-producing G cells and somatostatin inhibits gastrin release.

Gastrin cells (G cells): These are flask-shaped cells with microvilli at the apical surface found in the pyloric mucosa. G cells secret gastrin in response to peptides and amino acids. The neurotransmitters help in gastrin release are gastrin-releasing peptide (GRP) and bombesin.

13.1.2.2.3 Composition of Gastric Juice

Gastric secretion has two components. Basal secretion is the continuous secretion produced from epithelial cells and other mucus-producing cells. The electrolyte composition of basal secretion is similar to plasma ultrafiltrate which is neutral or slightly alkaline in nature. The basal secretion also has mucous that protects the gastric epithelium. Upon stimulation, cells of gastric glands secret HCl and pepsinogen. The flow rate of gastric secretion under resting or basal condition is about 5 mL/h which can go maximum up to 80 mL/h under stimulated secretion. The composition of gastric juice is depicted in Fig. 13.2.

13.1.2.2.4 Gastric Acid Secretion

HCl is actively secreted by the parietal/oxyntic cells of the fundic glands (Fig. 13.3). H^+ and Cl^- are secreted separately. The CO_2 diffuses from plasma into parietal cells and combines with cellular water to form carbonic acid

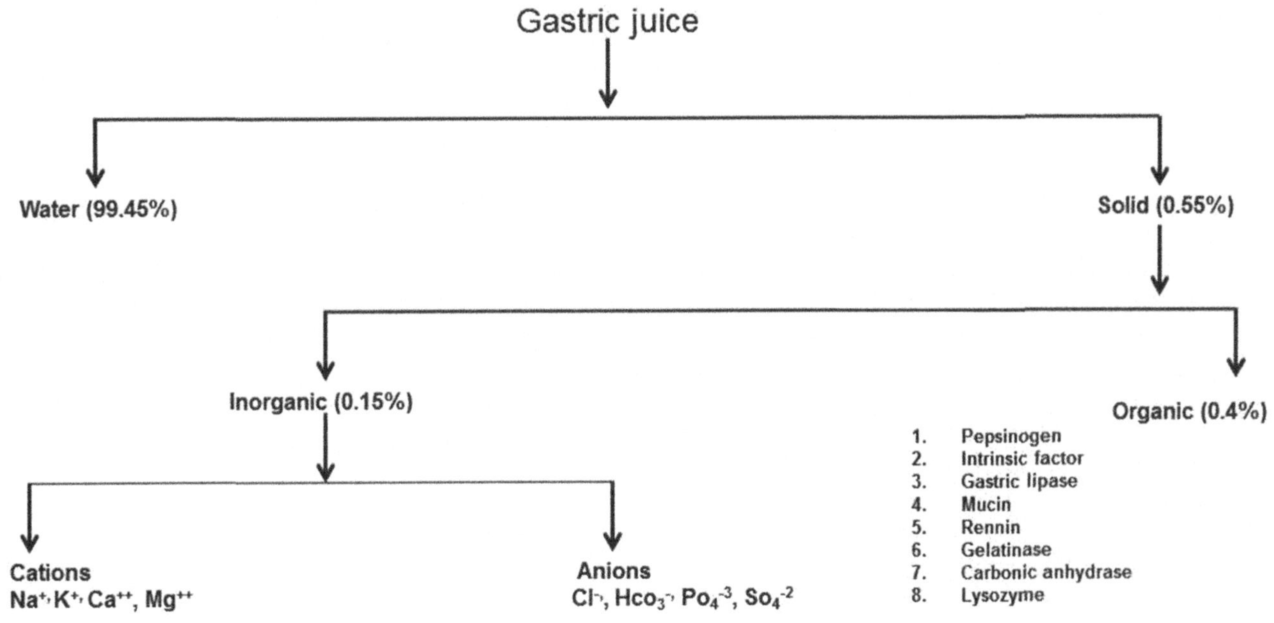

Fig. 13.2 The composition of gastric juice

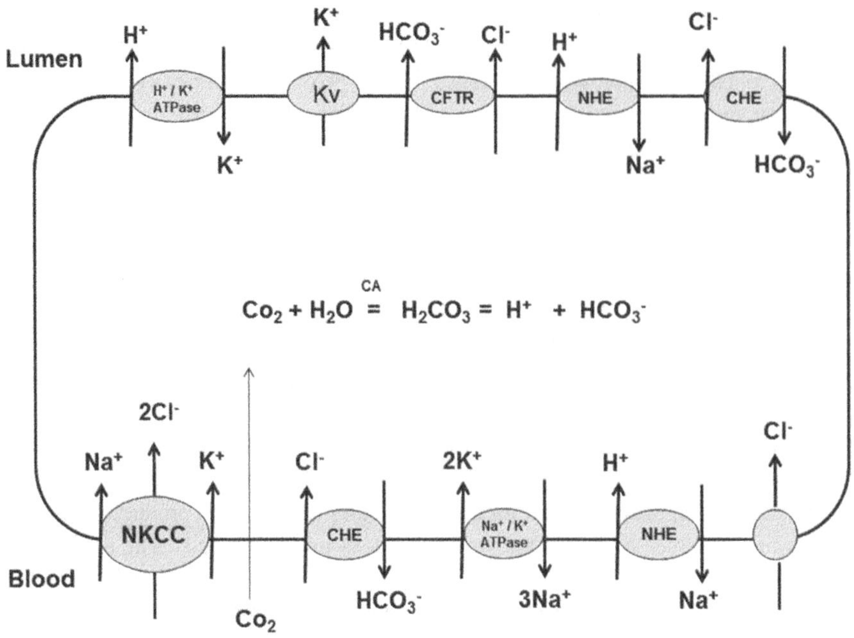

Fig. 13.3 Mechanism of gastric acid secretion. [The CO_2 diffuses from plasma into parietal cells and combines with cellular water to form carbonic acid (H_2CO_3) under the influence of carbonic anhydrase (CA). Carbonic acid then dissociates into HCO_3^- and H^+. The H^+ is extruded from oxyntic cells by K^+-stimulated ATPase that acts as an H^+/K^+ exchange pump. The transport of Cl^- into the lumen of stomach is mediated by Cl^-/HCO_3^- exchange (CHE)-transporter and cystic fibrosis transmembrane conductance regulator (CFTR). Recycling of K^+ is mediated by voltage-gated K^+ channel (Kv) and Na^+-K^+-Cl^- cotransporter (NKCC)

Table 13.5 Ion transporter systems and their roles in gastric HCl secretion

	Transporter/exchange system	Location	Function
Extrusion of H^+	ATP-dependent H^+/K^+ exchange	Apical membrane	Extrudes H^+ in exchange for K^+
	Na^+/H^+ exchange (NHE)	Apical membrane	Transport Na^+ against concentration gradient in exchange for H^+
Extrusion of Cl^-	Cl^-/HCO_3^- exchange (CHE) transporter	Basolateral and apical membrane	Extrudes Cl^- in exchange for HCO_3-
	Cystic fibrosis transmembrane conductance regulator (CFTR)	Apical membrane	Transport Cl^- against its electrochemical gradient into the gastric lumen
K^+ recycling	Voltage-gated K^+ channel	Apical membrane	Sustains the activity of H^+/K^+ exchange ATPase
	Na^+-K^+-Cl^- cotransporter (NKCC)	Basolateral membrane	Helps in the inward transport of one N^+, one K^+, and two Cl^-

(H_2CO_3) under the influence of carbonic anhydrase. Carbonic acid then dissociates into HCO_3^- and H^+. The H^+ is extruded from oxyntic cells by K^+-stimulated ATPase acts as an H^+/K^+ exchange pump. The transport of Cl^- into the lumen of stomach is mediated by Cl^-/HCO_3^- exchange (CHE)-transporter and cystic fibrosis transmembrane conductance regulator (CFTR). The key features of HCl secretion are extrusion of H^+ and Cl^- together with K^+ recycling mediated through a numbers of ion transporter systems (Table 13.5).

Alkaline tide: It is a state of metabolic alkalosis develops after heavy meal due to the diffusion of HCO_3^- into the venous blood. During every hydrogen ion secretion, one bicarbonate ion enters into the blood in exchange for chloride by Cl^-/HCO_3^- exchange (CHE) transporter.

13.1.2.2.5 Pepsinogen Secretion

Pepsinogen exists as zymogen granules in association with divalent cations like Ca^{+2} within the cells and secrets by exocytosis. The granules are fused together and also with the plasma membrane. Fusion results in a small pore formation through which Ca^{+2} is released and leads to vesicular swelling. A large pore is formed and pepsinogen is released through this pore.

13.1.2.2.6 Gastric Mucosal Barrier

The concentration of gastric HCl (150 mM) is three to four million times greater than plasma. Therefore, some protective mechanisms are required to prevent the back diffusion of the HCl that may damage the surrounding tissue. Gastric

Table 13.6 Components of gastric mucosal barrier

Pre-epithelial protection (first line of defence)	Mucus gel	Protects the gastric mucosa by forming a visco elastic gel
	Bicarbonate ion	Bicarbonate ions are entrapped within the mucous gel and neutralize hydrogen ions. It creates a pH gradient in the mucus gel and maintains the pH of epithelial surface near neutral
Epithelial protection	Luminal cell hydrophobicity	Amphoteric phospholipids in the luminal cell membrane increases the hydrophobicity and prevent the water-soluble agents to reach at the epithelium
	Sulphydryl compounds (reduced glutathione)	Neutralizes reactive-free radicals like superoxide, hydrogen peroxide, and hydroxyl radicals
	Rapid cell turnover	The proliferation rate of epithelium is very high and steady state. Human gastric epithelial cells can divide once in every 36 and matured within 48 and 96 h. This allows rapid renewal of damaged epithelial surface
	Restitution	It is the process of migration of new cells from the gastric pits to replace the damaged cells within a very short period of time
Sub-epithelial protection	Mucosal blood flow	The disposal of hydrogen ions and other deleterious agents are achieved by mucosal blood flow

mucosal barrier protects the stomach mucosa against gastric acid and other noxious agents. There are three levels of protective mechanism, pre-epithelial, epithelial, and sub-epithelial (Table 13.6).

Role of prostaglandin in gastric mucosal defence: Prostaglandin stimulates the secretion of mucus and bicarbonate to promote pre-epithelial protection against gastric injury. Endogenous prostaglandins increase mucosal blood flow and help in the disposal of hydrogen ions and harmful agents. They also increase the hydrophobicity of epical cell membranes and prevent the exfoliation of mucosal cells. Therefore, decreased prostaglandin secretion may result in gastric damage. Hence, antacids are prescribed along with non-steroidal anti-inflammatory (NSAID) drugs that primarily act by inhibiting prostaglandin synthesis.

13.1.2.2.7 Control of Gastric Secretion

Gastric functions are controlled by an integrated mechanism involving neural, endocrine, and paracrine pathways. The neural control is brought about by the enteric nervous system (ENS) with cholinergic and vagal inputs. Hormones are released into blood and control the secretion by classical endocrine pathways. Paracrine factors like histamine and somatostatin diffuse into the target cells to control their functions.

Neural control: The neural control of gastric functions can be divided into cephalic, gastric, and intestinal phases.

Cephalic Phase: It is stimulated by three reflexes, unconditioned, conditioned, and vagal. Cephalic phase accounts for 1/3 to 1/2 of total gastric acid secretion through cholinergic and vagal mechanisms. The unconditioned reflexes are brought about by sight, smell, taste, and swallowing of food. The conditioned reflex results from thought of food. The cephalic phase is entirely. The response is mediated by the vagus nerve which involves three mechanisms that stimulate gastric acid secretion.

1. The vagal efferent fibres synapse with postganglionic cholinergic neurons that innervate the parietal cell. Acetylcholine (ACh) released from postganglionic cholinergic neurons increases gastric acid secretion via M3 muscarinic receptors.
2. Vagal efferent fibres synapse with enteric neurons which secret gastrin-releasing peptide (GRP) or bombesin. GRP stimulates G cell to secret gastrin which reaches the oxyntic cells via circulation to stimulate acid secretion.
3. Somatostatin is the main inhibitor of HCl secretion. Vagal efferent fibres synapse with inhibitory neurons innervating the somatostatin cell and somatostatin release in inhibited.

The cephalic phase of gastric secretion is absent in ruminants.

Gastric Phase: It begins with the presence of food in the stomach and involves both vagal and local neurone reflexes (due to gastric distension). The vagal efferent pathway is mediated by postganglionic cholinergic neurons release acetyl choline and bombesin to stimulate G cells for gastrin release. Gastrin in turn stimulates acid secretion. Acetylcholine (ACh) also acts directly over the parietal cell by muscarinic M3 receptors to release gastric acid. The distension of stomach stimulates stretch receptors that cause the release of gastrin from G cells and histamine from enterochromaffin cells (ECLs). Both gastrin and histamine stimulate parietal cells to release HCl.

Intestinal phase: It starts after the food leaves the stomach and enters into the duodenum. It is mediated by duodenal cholecystokinin (CCK) and gastrin. CCK is a full agonist of gastrin and stimulates H^+ secretion in cats but partial agonist and competitive inhibitor in dogs. The intestinal phase also contains a cholinergic component to stimulate gastric secretion. However, most of intestinal responses are inhibitory to gastric secretion.

Endocrine and Paracrine Control: In addition to neural reflexes, some endocrine and paracrine factors involved in regulation of gastric acid secretion (Table 13.7).

Inhibitors of Gastric Acid Secretion

Several drugs are used to inhibit gastric acid secretion as therapeutic interventions in acid reflux disorders. They are classified on the basis of their mode of actions (Table 13.8).

Table 13.7 Endocrine and paracrine factors involved in regulation of gastric secretion

Endocrine and paracrine factors	Source	Mechanism of action	Functions
Gastrin	G cells of gastric mucosa, duodenum, jejunum, ileum, and pancreas in response to proteins, peptides, and amino acids	Gastrin acts via CCK-2 receptor (G-protein-coupled receptor) in parietal and enterochromaffin-like cells (ECL cells) to release histamine from ECL cells	Gastrin stimulates histamine release that causes acid secretion from stomach (role of histamine on acid secretion is discussed later) Gastrin promotes growth of gastric oxyntic cells by increasing the expression of fibroblast growth factor, epidermal growth factor receptors, and mitogen-activated protein kinase
Histamine	Enterochromaffin-like cells (ECL cells)	Histamine acts via H2 receptors to generate cAMP. It causes translocation and activation of H^+/K^+-ATPase (proton pump). Inhibits somatostatin release via H3 receptors	Efflux of H^+ into the lumen of gastric lumen Inhibition of somatostatin indirectly stimulates gastric acid secretion
Somatostatin	D cells of gastric mucosa. The primary stimuli for somatostatin release are gastric acid and gastrin. Gastric HCl activates calcitonin gene-related peptide (CGRP) neurons to release somatostatin	It acts through somatostatin receptor subtype 2 (sst2) with multiple signalling molecules	Somatostatin inhibits acid secretion directly acting on the parietal cells. Indirect actions of somatostatin include inhibition of histamine and gastrin secretion
Acetylcholine	Postganglionic neurons in Meissner's plexus	Acts over ECL cells through M1 receptors Activation of proton pump by M3 muscarinic receptors on parietal with increased intracellular calcium Inhibition of somatostatin via of M2 and M4 receptors on D cells	Stimulates histamine release Increases acid secretion Inhibits somatostatin release
Prostaglandins (PGE_2)	Endothelial cells and macrophages	Acts through surface receptors in parietal and gastric mucosal cells	Decreases histamine release and reduces acid secretion
Transforming growth factor-alpha (TGF-alpha)	Epithelial cells		Inhibits gastric acid and mucous secretion
Peptide YY	Ileum and colon	Inhibits gastrin stimulated histamine release	Inhibits gastric acid secretion
Cholecystokinin (CCK)	I cells in the duodenum and jejunum	Stimulates acid secretion via CCKb receptors on parietal cells Increases serotonin secretion via CCKa receptors on mucosal D cells	It performs dual role. It stimulates acid secretion through CCKa receptors and inhibits acid secretion via CCKb. But inhibition is the predominate effect
Secretin	S cells in the duodenum	Inhibits gastrin release	Decreases acid secretion
Neurotensin	Ileum and nerve terminals in the myenteric plexus in response to fat diet		Inhibits acid secretion
Glucagon-like peptide 1 (GLP-1)	L cells of duodenum		Stimulates somatostatin release hence decreases acid secretion
Oxyntomodulin	L cells of duodenum		Stimulates gastric acid secretion
Ghrelin	Entero-endocrine cells of gastric mucosa	Acts via growth hormone receptors in the oxyntic cells	Stimulates acid secretion
Orexin	Hypothalamus and gastric mucosa	Acts through orexin-1 receptors (OX1R) at anterior hypothalamus and ventromedial nucleus	Stimulates gastric acid secretion
Adrenomedullin	ECLs of the gastric mucosa	Stimulates gastric somatostatin release and decreases serotonin level	Inhibits gastric acid secretion

Table 13.8 Drugs used to inhibit gastric acid secretion

Class		Examples
Inhibitors of H^+/K^+-ATPase (Proton pump)		Verapamil, omeprazole, vanadate
Inhibitors of carbonic anhydrase		Acetazolamide
Inhibitors of cell activation	Calcium channel blockers	Verapamil, lanthanum
	Prostaglandin E2	
Receptor antagonists	H_2 receptor antagonists	Ranitidine, cimetidine
	Gastrin receptor antagonists	Proglumide, benzotript
	Anticholinergic drugs	Atropine
Calmodulin inhibitor		Trifluoperazine

13.1.2.2.8 Functions of Gastric Juice

Functions of HCl: Gastric HCl converts pepsinogen to its active form pepsin and facilitates the digestion of protein. It also promotes optimum environment (pH) for the action of pepsin. It can also slightly hydrolyse sucrose. It helps to destroy pathogens and acts as physiological barrier against infection.

Gastric mucin: Gastric mucin forms a gel (95% water, 5% mucin, and electrolytes) over the gastric mucosa to protect it from the corrosive action of gastric acid and other harmful agents. It also entraps bicarbonate to neutralize HCl and maintains a pH gradient to protect gastric mucosa.

Functions of gastric enzymes: Pepsinogen is a proteolytic enzyme secreted from chief cells and mucus neck cells. It is converted to its active form pepsin by HCl. Secretion of pepsinogen is enhanced by ACh, CCK, and gastrin. The optimal pH for the action of pepsin is between 1.6 and 2.5. Pepsin cleaves the proteolytic bonds involving amino acids such as tyrosine, phenylalanine, and leucine.

Gastric lipase is a lipid digesting enzyme secreted by chief cells. It is capable of hydrolysing 20% of triglycerides in the feed. In dogs, gastric lipase is secreted under the influence of histamine, prostaglandin E2, pentagastrin, and secretin. The activity of gastric lipase is independent of gastric pH.

Rennin is a proteolytic enzyme secreted from gastric mucosa. It is usually seen in newborn calves under milk diet. Rennin converts casein into paracasein which in turn combines with calcium to form an insoluble coagulum.

Function of Intrinsic Factor (IF): It is secreted from parietal cells and helps in Vit-B12 absorption. IF forms a complex with Vit-B12. The complex then binds with cubilin receptor in the ileal mucosa and absorbed by the enterocytes through endocytosis. In the enterocytes, Vit-B12 is released from IF.

13.1.2.2.9 Pathophysiology of Gastric Acid Secretion

Achlorhydria: It is a clinical condition in which the stomach is unable to produce hydrochloric acid. It is caused due to a variety of reasons such as pernicious anaemia, *Helicobacter pylori* infection, gastric bypass, hypothyroidism, radiation exposure of gastric mucosa, and gastric cancer.

Hypochlorhydria (HCH): It is characterized by reduced secretion of gastric acid. The predominant causes of HCH are chronic atrophic gastritis, *Helicobacter pylori* infection, or autoimmune disorders.

Hyperchlorhydria (sour stomach/acid stomach): It is a clinical condition in which the gastric HCl production is more than normal. It usually occurs due to higher gastrin production.

13.1.2.3 Pancreatic Exocrine Secretion

Pancreas is a lobulated gland comprises two distinct components exocrine and endocrine. Both exocrine and endocrine components are distinct structurally and functionally. The pancreas is appeared as discrete organ containing a right (proximal to the duodenum) and left limb in species like dogs and cats. Pancreas in large animals like cattle and horse is appeared diffused within the mesentery close to duodenum. The exocrine part of the pancreas constitutes more than 90% of the total pancreatic mass. The exocrine pancreas is an acinus gland structurally similar with salivary glands. The pancreatic enzymes are stored in the acinar cells in the form of zymogen granules and released upon activation. Pancreas also secrets HCO_3^- ions that neutralizes acid chime entering into the duodenum.

13.1.2.3.1 Structure of Exocrine Pancreas

The exocrine pancreas is structurally similar with salivary gland. The secretory units are called acini along with ductules for drainage. Each pancreatic acinus is composed of pyramidal glandular cells. There are centro-acinar cells situated at the junction between acini and duct. Acinar cells synthesize and store digestive enzymes at the apical region of acinar cells. The acinar cells contain nucleus and plenty of rough endoplasmic reticulum. There are numerous microvilli at the apical surface of the cells. The acinar cells are connected through tight junctions and act as a barrier for the large molecules but allow paracellular transport of water and ions. The secretions are poured into the lumen of the acinus and drained through duct system. The duct system comprises ductules and interlobular (intercalated) ducts. The ductules carry acinar secretion into the intercalated ducts. The intercalated ducts drain into main pancreatic duct called duct of Wirsung. The accessory pancreatic duct is called duct of Santorini. Both the main and accessory pancreatic

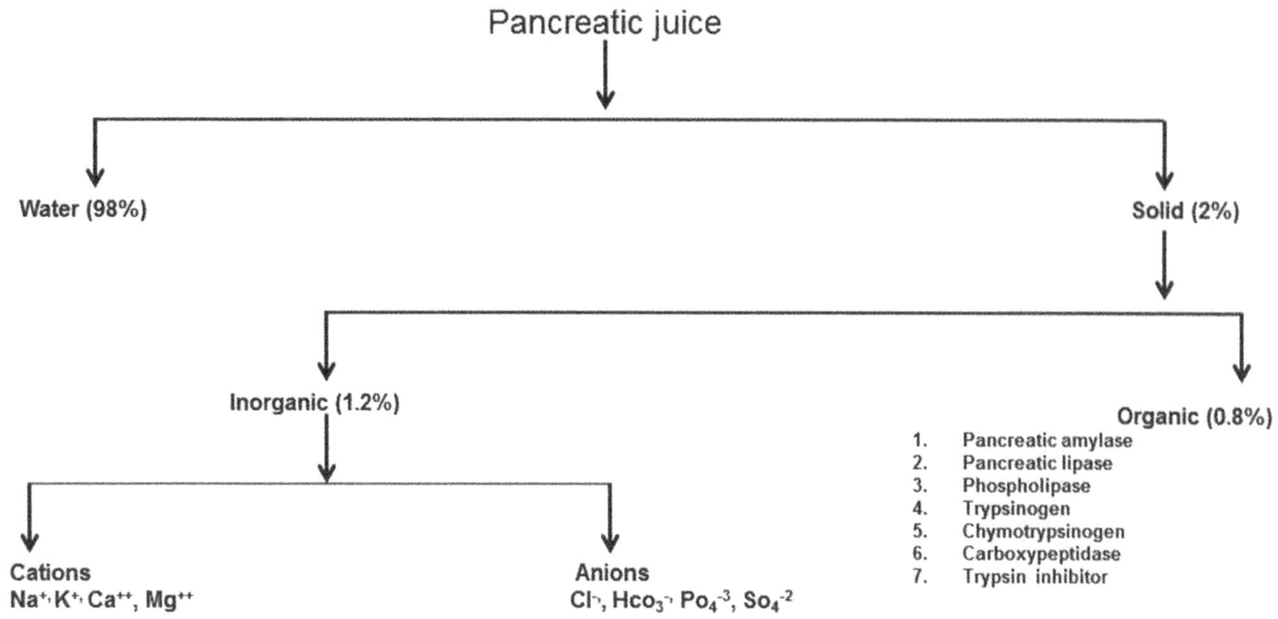

Fig. 13.4 Composition of pancreatic juice

Table 13.9 Transporters involved in pancreatic bicarbonate secretion

Transporter	Location	Function
Electrogenic Cl^-/HCO_3^- exchanger	Luminal surface of pancreatic duct cells	Extrudes HCO_3^- in exchange for Cl^-
Na^+/H^+ exchanger	Basolateral surface of pancreatic duct cells	Extrudes H^+ in exchange for Na^+
Na^+-HCO_3^- cotransporter	Basolateral surface of pancreatic duct cells	Entry of HCO_3^- at basolateral side
Na^+, K^+-ATPase	Basolateral surface of pancreatic duct cells	Maintains inward Na^+ and outward K^+ gradient
K^+ channel	Basolateral surface of pancreatic duct cells	Maintains membrane potential
Cystic fibrosis transmembrane conductance regulator (CFTR)	Luminal surface of pancreatic duct cells	Transport Cl^- into the lumen against its electrochemical gradient
Water channels (Aquaporin) AQP1, AQP5	Both basolateral and luminal surface	Water transport

ducts drain separately into the duodenum along with common bile duct guarded by sphincter of Oddi.

13.1.2.3.2 Composition of Pancreatic Juice

The pancreatic juice is alkaline in nature due to high HCO_3^- content (113 mEq/L). About 1.500 L pancreatic juice is secreted per day in human, 3–5 L/100 kg/day in cow, 0.5–1 L/100 kg/day in sheep, and 2–3 mL/min dog. The composition of pancreatic juice is depicted in Fig. 13.4.

13.1.2.3.3 Mechanism of Pancreatic Bicarbonate Secretion

Carbonic anhydrase catalyses the reaction between cellular water and diffused carbon-di-oxide to form carbonic acid. The dissociation of carbonic acid yields HCO_3^-. It extrudes from the cells into the lumen in exchange for Cl^- by electrogenic Cl-/HCO_3– exchanger. Continuous supply of Cl^- in the lumen is maintained by secretin-regulated Cl^- channel. The H^+ is removed from the cell by Na^+/H^+ exchanger at the basolateral membrane. The Na^+ gradient is achieved through Na^+, K^+-ATPase system. There are several transporters involved in the pancreatic secretory process (Table 13.9 and Fig. 13.5).

13.1.2.3.4 Mechanism of Pancreatic Enzyme Secretion

The enzymes are stored as zymogen granules in the acinar cells and secreted via exocytosis. All the enzymes are secreted as inactive form to avoid tissue digestion. Proteasomes are involved in the pancreatic enzyme secretion.

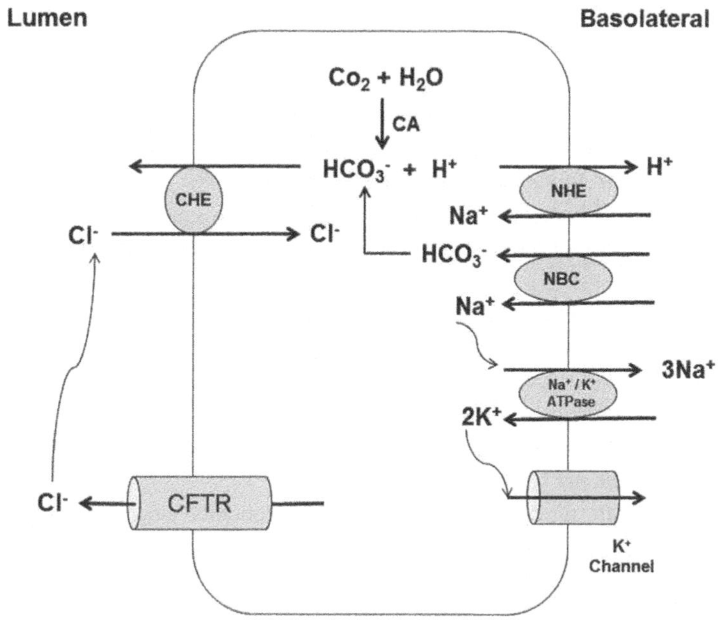

Fig. 13.5 Mechanism of pancreatic bicarbonate secretion. [Carbonic anhydrase (CA) catalyses the reaction between cellular water and diffused carbon-di-oxide to form carbonic acid. The dissociation of carbonic acid yields HCO_3^-, which extrudes from the cells into the lumen in exchange for Cl^- by electrogenic Cl^-/HCO_3^- exchanger (CHE). Continuous supply of Cl^- in the lumen is maintained by CFTR. The H^+ is removed from the cell by Na^+/H^+ exchanger (NHE) at the basolateral membrane. The Na^+ gradient is achieved through Na^+, K^+-ATPase system. The membrane potential is maintained by K^+ channel]

In mouse pancreas, anti-factor 4 of the 26S proteasome regulates the secretion of digestive enzymes. In cows, high leucine concentration decreases pancreatic amylase secretion by inhibiting proteasomes. The protein synthesis in the acinar cells is induced by mammalian target of rapamycin complex (mTOR), phosphatidylinositol-3 kinase (PI3K)-RAC alpha serine/threonine-protein kinase (Akt), and the general amino acid control repressor 2 (GCN2) signalling mechanism. The initiation of translation is occurred through the activation of AktmTOR and ribosomal protein S6 kinase beta-1 (S6K1) induced by leucine and phenylalanine. The exocytosis of zymogen granules is mediated by calcium ions. CCK and acetylcholine stimulate the acinar cells to release zymogen granules through two second messengers namely inositol-tri-phosphate (IP3) and nicotinic acid adenine dinucleotide phosphate (NAADP). Both of these two compounds induce the release of calcium from the sarcoplasmic reticulum via inositol 1,4,5-trisphosphate and ryanodine receptors, respectively.

13.1.2.3.5 Control of Pancreatic Secretion

Pancreatic secretion occurs in three phases viz. cephalic phase, gastric phase, and intestinal phase. The cephalic phase of pancreatic secretion is under the control of ANS. The gastric and intestinal phase is controlled by hormones and enteropancreatic reflex.

Cephalic phase: It is induced by sight, taste, and smell of food. Acetylcholine is released from vagal nerve endings and acts through muscarinic receptors to increase intracellular calcium and granular exocytosis (see Sect. 13.1.2.3.4 Mechanism of pancreatic enzyme secretion). Cephalic phase constitutes 20% of the total pancreatic enzyme secretion. But little of the secretion can reach to the intestine as little amount of water and electrolytes are secreted along with the enzymes.

Gastric phase: Gastric phase is initiated after the presence of food in stomach. Gastric distension stimulates stretch receptors and to initiate vago-vagal reflex or gastro-pancreatic reflex. Gastric phase is accounting for another 5–10% of pancreatic enzymes secretion but like cephalic phase; little can be reached to duodenum due to unavailability of water.

Intestinal phase: The copious secretion of pancreatic enzymes along with water and electrolytes occurs during the intestinal phase. The enzyme secretion is mediated through CCK under the influence of proteoses and peptones and constitute around 70–80% of total enzyme secretion. CCK has two receptor subtypes CCK1 and CCK2. In calf, both the receptors are expressed in the pancreas, but CCK 2 is the main receptor in adulthood. CCK increases intracellular calcium to release zymogen granules of the acinar cells (see Sect. 13.1.2.3.4 Mechanism of pancreatic enzyme secretion).

The secretion of water and bicarbonate is mediated by secretin. It acts through secretin receptor (SR) located basolaterally in the acinar cells. Secretin activates cAMP-dependent anion Cl/HCO_3^- exchanger and CFTR at the apical membrane of pancreatic acinar cells.

There are several factors affecting exocrine pancreas secretion in animals. In ruminants, age is the primary factors to affect pancreatic exocrine functions. The amount of pancreatic juice secretion increases with age. In calf, the amount of pancreatic juice is 150 mL/day on fourth day after birth compared to 1000 mL/day at 3 months of age. The flow rate is also increased with age. It is about 7.9 mL/kg bwt in 3-week-old calves compared to 14.2 mL/kg bwt at 3 months of age. Types of feed also affect the pancreatic enzyme secretion. The nature of dietary carbohydrates affects the secretion of pancreatic amylase. In sheep, corn feeding stimulates more amylase production compared to hay. In goats, the amylase secretion is increased with dietary starch content. The starch entering the intestine after bypassing the rumen has tremendous influence over pancreatic amylase secretion. Leucine, isoleucine, and phenylalanine stimulate the secretion of amylase, trypsin, chymotrypsin, and lipase.

13.1.2.3.6 Functions of Pancreatic Juice

Neutralization of acid chyme: The pH of the pancreatic juice is alkaline due to high HCO_3^- concentration. The pH of pancreatic juice in dogs ranges from 7.4 to 8.3. Pancreatic juice neutralizes the acid chime entering into the duodenum to raise the pH 6.0 to 7.0.

Role in digestion: Pancreatic juice contains enzymes responsible for digestion of carbohydrates, proteins, and lipids.

Pancreatic α-amylase causes hydrolysis of α-1,4-glucosidic bonds present in starch and glycogen. Pancreatic amylase is activated by Cl^-. The optimum pH for α-amylase action is 6.7–7.2. Newborn calves and pigs have lower amylase than mature animals.

Pancreatic juice contains proteolytic enzymes for protein hydrolysis. There are two classes of proteolytic enzymes. Endopeptidases cleave peptide bonds along the peptide chains whereas exopeptidases act at the amino terminal or carboxyterminal ends of polypeptide chains. The proteolytic enzymes of pancreatic juice and their functions are depicted in Table 13.10.

Pancreatic lipase hydrolyses dietary triglycerides into glycerol, monoglycerides, and fatty acids. But lipase requires emulsified fat as dietary substrate. Bile salts help in the emulsification of fats. In addition to lipase, pancreatic juice also contains phospholipase A that converts lecithin to lysolecithin. The detergent action of lysolecithin favours emulsification of fats.

13.1.2.4 Hepatobiliary Secretion

The hepatobiliary system comprises liver, gall bladder, and bile ducts. Liver is the largest gland in the body that performs wide ranges of physiological functions including metabolism of macro- and micronutrients, blood volume regulation, lipid and cholesterol homeostasis, immunity, endocrine control of growth signalling pathways, and detoxification of xenobiotic compounds, drugs, and hormones. The hepatocytes of the liver produce bile that transports through bile ducts into the gall bladder. The bile is concentrated and stored at the gall bladder. Bile helps in emulsification of dietary lipids for enzymatic actions.

13.1.2.4.1 Structure of Hepatobiliary System

Hepatobiliary system comprises two components. The intrahepatic components lie within the liver, whereas extrahepatic components are situated outside of the liver. Hepatic cords comprising hepatocytes are arranged radially around the central vein. In between the hepatic cords, there are endothelial lined spaces called sinusoids. The portal triads viz. portal vein, hepatic artery, and bile ducts are situated at the periphery of hepatic lobules. Portal arteries and veins carries allow the blood to flow centrally and bile duct facilitates to drain bile peripherally into ductules and finally to bile duct in portal triads. The liver sinusoids contain Kupffer cells (Littoral cells) having phagocytic activity. They are the components of mononuclear phagocytic system or reticuloendothelial system. The space between sinusoidal endothelium and hepatocytes is called space of Disse. The ions and nutrients have to cross this place before entering into hepatocytes. The stellate cells are also situated at the space of Disse and help to form fibrous scar tissue to prevent the spreading on infections. Thus, it protects the hepatocytes from toxins. The large pores along the sinusoidal epithelium allow unrestricted passage of albumin from sinusoid into the extravascular fluid.

Bile is synthesized in the hepatocytes and secreted into the bile canaliculi. Bile canaliculi exist as a groove of plasma membrane between two hepatocytes. Bile canalicular membrane represents about 13% of the total hepatocyte plasma membrane. It has numerous microvilli to increase the surface area. The bile canaliculi empty into the canals of Hering which gives rise to the biliary tress comprising intra- and interlobular ducts. The intralobular ducts join to form interlobular bile ducts which ultimately give rise to right and left hepatic ducts. The right and left hepatic ducts give rise to common hepatic duct. The cystic duct arising from the gall bladder communicates with common hepatic duct to form common bile duct that opens into the duodenum. The opening of duodenum and common bile duct is guarded by the

Table 13.10 Proteolytic enzymes of pancreas

Class	Enzymes	Functions
Endopeptidase	Trypsin	Cleaves peptide bonds on carboxyl side of basic amino acids (arginine or lysine)
	Chymotrypsin	Cleaves peptide bonds on carboxyl side of aromatic amino acids
	Elastase	Cleaves the peptide bonds of aliphatic amino acids at carboxyl side
Exopeptidase	Carboxypeptidase A	Cleaves carboxyl terminal amino acids that have aromatic or branched aliphatic side chains
	Carboxypeptidase B	Cleaves carboxyl terminal amino acids that have basic side chains

sphincter of Oddi. Common bile duct unites with the main pancreatic duct before entering the duodenum. Sphincter of Boyden is situated in the common bile duct just before the joining of pancreatic duct. Sphincter of Oddi is closed during inter-digestive period but, when the gastric chyme enters into the duodenum, the sphincter is relaxed under the influence of cholecystokinin (CCK).

There are a population of epithelial cells that forms a three-dimensional network of bile ducts called cholangiocytes. The hepatic bile is modified at the biliary tract through the secretion and reabsorption by the cholangiocytes.

The sphincter of Oddi is less developed in ruminants and pigs and bile flow is continuous into the duodenum. Similar mechanism occurs in horse due to lack of gall bladder. In dogs and cats, continuous secretion of bile is unnecessary as they feed only once or twice a day so the bile is stored in gall bladder.

Gall Bladder

It is a pear-shaped organ that helps to store bile until the body needs it for digestion. It is the component of extrahepatic biliary system. Gall bladder is connected with liver and duodenum by biliary tract. The wall of the gallbladder is made of several layers. The innermost mucosal layer is composed of columnar epithelium with microvilli. The epithelial lining is characterized by recesses called Aschoff's recesses, which form pouches inside the lining. Under epithelium there is a layer of connective tissue followed by a muscular wall that contracts in response to cholecystokinin, a peptide hormone by the duodenum. The main function of the gall bladder is to store and concentrate bile. The hormone cholecystokinin (CCK) mediates the contraction of gall bladder. CCK is released from I cells of duodenum and jejunum under the influence of fatty acids and amino acids. Beside gall bladder contraction, CCK also relaxes the sphincter of Oddi to release bile into the duodenum. Agents increase the bile flow by contracting the gall bladder is called chalogogues such as CCK.

13.1.2.4.2 Bile

Bile is a complex lipid-rich hepatic secretion. It is synthesized in the hepatocytes and modifications occur at the bile duct epithelium through secretory and absorptive mechanism. Gall bladder acts as the temporary storage site of the bile and bile is released into the duodenum after the contraction of gall bladder.

Formation of Bile

Bile is produced in the hepatocytes and subsequently modified in bile ductules by cholangiocytes. The secretion requires active transport systems in the biliary tree.

Synthesis of bile acids: Bile acids are synthesized in the pericentral hepatocytes from its cholesterol precursor. Bile acids derived from hepatocyte are called primary bile. The secondary bile acids are formed by gut microbes through dehydroxylation, dehydrogenation, oxidation, epimerization, and esterification. Formation of bile acids is a complex process that involves 17 different reactions catalysed by 16 enzymes. The hydrophobic cholesterol is converted to an amphipathic hydrophilic compound through hydroxylation of sterol ring and side chain oxidation. There are two pathways of bile acid synthesis. In the classical pathway, modification of the steroid nucleus occurs before the side chain. Whereas in the alternative pathway side chain modifications occur prior to changes in the steroid nucleus. The rate limiting enzyme of classical pathway is cholesterol 7α-hydroxylase (CYP7A1). In human and rodents, the classical pathway is predominant pathway of bile synthesis accounts for 90% and 70% bile formation, respectively. The alternative pathway is mostly seen in human neonates due to lack of CYP7A1 expression. After the synthesis, bile acids are conjugated with glycine and taurine. It is a two-step process. Firstly, bile acids are converted to bile acid-COA by bile acid-CoA synthase followed by amidation with taurine or glycine with the help of enzyme bile acid-CoA: amino acid N-acyltransferase (BAAT). The amino acid conjugation makes bile acids more resistant to hydrolysis by pancreatic carboxypeptidases so, taurine and glycine-conjugated bile acids escape the cleavage in the intestinal lumen.

Sinusoidal uptake of bile acids: Conjugated bile acids are taken up by the hepatocyte through an active process. The process involves both Na^+-dependent and Na^+-independent mechanisms. There are different ion channels involved in this process. Sodium taurocholate cotransporting polypeptide (Ntcp) facilitates entry of conjugated bile salts after utilizing transmembrane sodium gradient maintained by Na+, K+-ATPase at the plasma membrane. The Na^+-independent mechanism is brought about by organic anion transporting polypeptide 1 and 2 (oatp1 and oatp 2) also helps in the uptake of conjugated bile salts along with other organic anions.

Intracellular transport: The intracellular transport of bile salts to the canalicular pole of the hepatocyte is mediated by intracellular binding proteins. Two cytosolic bile acid binding proteins namely 3-hydroxysteroid dehydrogenase and dihydrodiol dehydrogenase are identified in rat and human liver cells.

Transport of bile salts into canaliculi: The secretion of bile acids from the hepatocytes into the canaliculus requires an active transport system due to high osmotic and chemical gradient between hepatocytes (5 μM) and canalicular space (1000 μM). There is a linear relationship between bile acid secretion rate and bile flow. When the flow of bile depends upon the osmotic force of bile acids, it is called bile acid-dependent flow which accounts for 30–60% of spontaneous basal bile flow. When the secretion of bile occurs via the osmotic force generated by other than bile acids, it is called bile acid-independent flow accounts 30–60% of basal bile flow. In bile acid-dependent flow, bile salts enter into the canaliculus by two transporter system namely bile salt export pump (BSEP) and multi-drug resistance protein 2 (MRP2). BSEP facilitates apical excretion of taurine and glycine amidated bile salts and MRP2 mediates the transport of sulphated and glucuronidated bile salts along with non-bile-salt organic anions (bilirubin glucuronides). After entering into the bile canaliculus, bile salts create an osmotic gradient to allow influx of water and electrolytes. In bile acid-independent flow, the osmotic gradient is created by HCO_3^- and reduced glutathione. HCO_3^- enters into the canalicular space by anion exchanger 2 (AE2) and MRP2 mediates the transport of reduced glutathione. The osmotic gradient created by HCO_3^- and reduced glutathione allows the influx of sodium and water.

Modifications to bile in the biliary tree: Biliary epithelial cells modify bile by adding fluid and electrolytes into the canalicular bile. The secretion of epithelium is stimulated by secretin and accounts 30% of basal bile flow. Somatostatin decreases epithelial secretion. Both secretin and somatostatin act through their receptors at basolateral surface of cholangiocytes. Secretin stimulates intracellular cAMP production which in turn activates low conductance chloride channel like cystic fibrosis transmembrane conductance regulator (CFTR). The efflux of chloride depolarizes cholangiocyte and allows the entry of bicarbonate via electrogenic sodium-bicarbonate cotransporter. Increased intracellular bicarbonate stimulates apical chloride/bicarbonate exchanger which facilitates the secretion of bicarbonate in exchange with chlorides. The water movement across the canalicular membrane is mediated by water channel aquaporin at the apical and basolateral membrane. Bile ductules have reabsorptive function and water reabsorption occurs at the distal to the canaliculi under the influence of prostaglandin.

Regulation of Bile Acid Synthesis

The synthesis of bile acids is mainly regulated by bile acids through farnesoid X-receptor (FXR) pathway. In this mechanism, bile acids activate FXR in the enterocytes followed by synthesis of fibroblast growth factor (FGF15/19). EGF15/19 then transport to the hepatocytes by portal circulation and binds with fibroblast growth factor receptor-4 (FGFR4) and Klotho protein complex to inhibit the rate limiting enzyme of bile acid secretion (CYP7A1). There are several proteins involved in the regulation of bile acid synthesis (Table 13.11).

Composition of Bile

Bile comprises nearly 95% water with dissolved solids like bile salts, lipids, pigments, enzymes, vitamins, heavy metals along with exogenous drugs, xenobiotics, and environmental toxins (Fig. 13.6). There are two forms of bile, hepatic bile and gall bladder bile. The gall bladder bile is more concentrated (Table 13.12). The solutes that are pumped actively in the bile canaliculus are called primary solutes. Apart from bile salts, glutathione, conjugated bilirubin, bicarbonate, glucuronide, or sulphate conjugates of xenobiotics are called primary solutes. Water, electrolytes, glucose, low-molecular weight solutes are called secondary solutes as they flow passively into the canaliculi by the osmotic force generated by primary solutes. The predominant compound of bile is bile acids (67%) followed by phospholipids (22%), proteins (4.5%), cholesterol (4%), and bilirubin (0.3%). Flow rate and electrolyte concentrations of hepatic bile varies with species (Table 13.13).

Bile acids: The primary bile acids synthesized in the hepatocytes includes cholic acid (CA) and chenodeoxycholic acid (CDCA). These two are the primary bile acids in most of the animal species. In pigs, hyocholic acid (HCA) is the primary bile acid derived from chenodeoxycholic acid after hydroxylation at 6α position. Secondary bile acids are produced by the intestinal bacteria by removing the α-hydroxyl group at the position 7 of CA or CDCA. The secondary bile acid includes deoxycholic acid (DCA), lithocholic acid (LCA), and ursodeoxycholic acid (UDCA) or ursodiol.

Bile phospholipids: Lecithin is the sole phospholipid in the bile. Lecithin is secreted as vesicles into bile. The translocation of vesicles is mediated by a transmembrane protein called MDR3 P-glycoprotein.

Bile cholesterol: The main route of cholesterol excretion is the through bile. The bile salts and lecithin solubilize hydrophobic cholesterol in the form of micelles. Under abnormal conditions, the cholesterol may precipitate in the

Table 13.11 Regulatory proteins of bile acid synthesis

Protein	Tissue	Functions
Farnesoid X-receptor (FXR)	Intestine, liver, kidney	Inhibits the rate limiting enzyme of bile acid secretion (CYP7A1)
Hepatocyte nuclear factor 4α (HNF4α)	Intestine, liver	Increases hepatic bile acid synthesis by stimulating the expression of CYP7A1
Small heterodimer partner (SHP)	Liver, intestine	Negative feedback regulation of bile synthesis
Pregnane X-receptor (PXR)	Liver, intestine	Detoxification of bile acids
Vitamin D receptor (VDR)	Intestine	Detoxification of bile acids
Fibroblast growth factor (FGF15/19)	Intestine	Inhibits the rate limiting enzyme of bile acid secretion (CYP7A1)

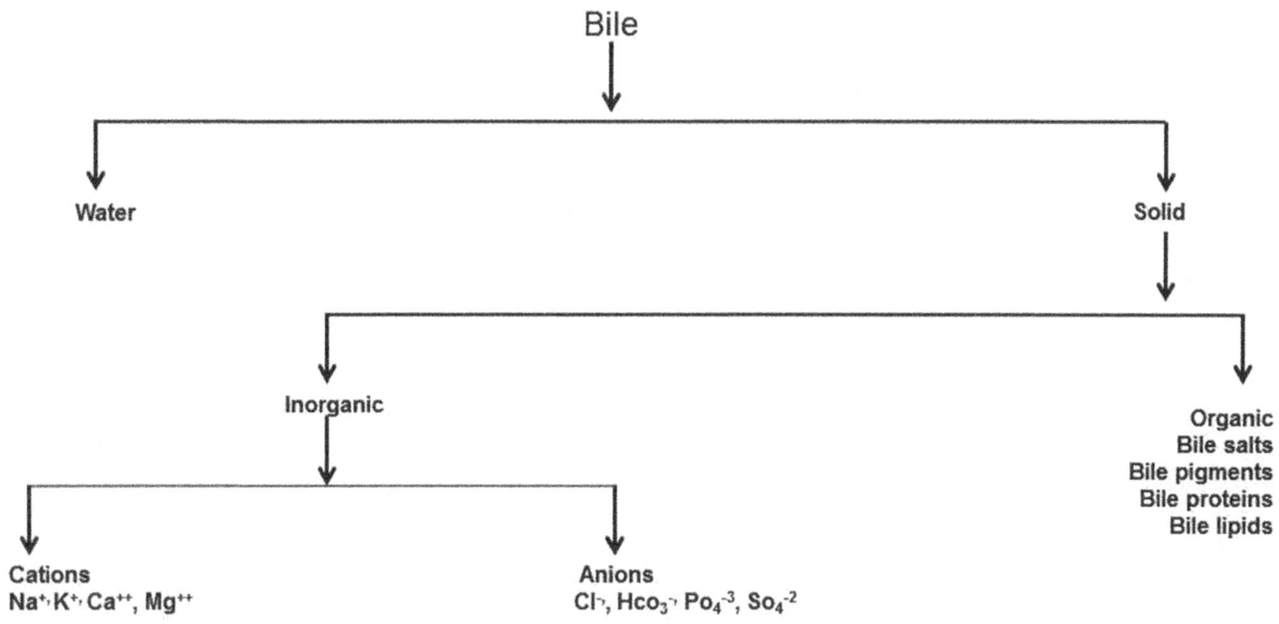

Fig. 13.6 Composition of bile

Table 13.12 The composition of hepatic and gall bladder bile

Constituents	Hepatic bile	Gall bladder bile
Water	97.5 gm%	92 gm%
Bile salts	1.1 gm%	6 gm%
Bilirubin	0.04 gm%	0.3 gm%
Cholesterol	0.1 gm%	0.3–0.9 gm%
Fatty acids	0.12 gm%	0.3–1.2 gm %
Lecithin	0.04 gm%	0.3 gm %
Na^{+}	145 mEq/L	130 mEq/L
K^{+}	5 mEq/L	12 mEq/L
Ca^{+}	5 mEq/L	23 mEq/L
CI	100 mEq/L	25 mEq/L
HCO_3	28 mEq/L	10 mEq/L
pH	7.1–7.3	6.9–7.7

Table 13.13 Flow rate and electrolyte concentrations of hepatic bile varies in different species

		Concentration (mmol/L)						
Species	Flow (μL/min/kg)	Na^{+}	K^{+}	Ca^{++}	Mg^{++}	Cl^{-}	HCO_3^{-}	Bile acids
Man	15–15.4	132–165	4.2–5.6	0.6–2.4	0.7–1.5	96–126	77–55	3–45
Dog	10	141–230	4.5–11.9	1.5–6.9	1.1–2.7	31–107	14–61	16–187
Sheep	9.4	159.6	5.3	–	–	95	21.2	42.5
Rabbit	90	148–156	3.6–6.7	1.3–3.3	0.15–0.35	77–99	40–63	6–24
Rat	30–150	157–166	5.8–6.4	–	–	94–98	22–26	8–25
Guinea pig	115.9	175	6.3	–	–	69	49–65	–

Erlinger (1992)

gallbladder, resulting in the formation of cholesterol gallstones.

Bile proteins: Plasma proteins are transported into bile by two routes viz. trans-hepatocyte or paracellular pathways. The bile proteins can be classified into several groups like transport proteins (albumin, ceruloplasmin, transferrin, haptoglobin, apolipoprotein), enzymes (lysosomal enzymes, alkaline phosphatase, amylase), hormones (insulin, CCK, epidermal growth factor), and immunoglobulins (IgG, IgA, and IgM).

Bile pigments: Bilirubin and biliverdin are the two main bile pigments. Bilirubin is derived from the metabolism of haemoglobin. Bilirubin is carried to the hepatocytes albumin and stored in the hepatocytes by binding to the Y protein. In hepatocytes, bilirubin is conjugated with glucuronic acid to form bilirubin diglucuronide which is excreted into the canaliculi. The green colour of bile is chiefly due to conjugated bilirubin. Intestinal bacteria convert conjugated bilirubin into urobilinogen and stercobilinogen to excrete through faeces. A portion of intestinal urobilinogen is reabsorbed into the blood and removed by the kidney. The urobilinogen gives urine a distinct yellow colour.

Functions of Bile

Digestion and absorption of dietary lipids: Bile promotes the digestion and absorption of dietary lipids such as cholesterol and long-chain fatty acids. Bile increases the solubility of lipids for better enzymatic actions. The micelle also helps in the absorption of lipid digestion end products by increasing their diffusion across the intestinal epithelium. Bile also helps in the absorption of fat-soluble vitamins (A, D, E, K).

Bile also facilitates intestinal absorption of dietary proteins by causing denaturation and hydrolysis of proteins by pancreatic proteolytic enzymes.

Cholesterol homeostasis: Bile acid increases absorption of biliary and dietary cholesterol in the intestine together with elimination of cholesterol from the body. The hepatic secretion of cholesterol is enhanced by bile acids which enable the removal of cholesterol from liver to intestine.

Gut antimicrobial defence: The bacteriostatic actions of bile acid-fatty acid mixed micelles prevent bacterial infection in the GI tract. Bile causes membrane-damaging effects over the microbes. Further, bile induces some antimicrobial peptides such as cathelicidin to regulate intestinal inflammation.

Prevention of calcium gallstone formation: Bile prevents the formation of calcium gallstone and calcium oxalate kidney stones after binding calcium.

Signalling molecules: Bile activates farnesoid X receptor alpha (FXR), vitamin D receptors (VDR), G-protein-coupled receptors, pregnane X receptor (PXR), and epidermal growth factor receptor (EGFR) to regulate gut motility, hepatic functions, and energy homeostasis.

13.1.2.4.3 Control of Bile Secretion

Agents that increase the secretion of bile are called choleretics. The synthesis and secretion of bile are not under direct control of ANS. However, the hepatic duct system, sphincter of Oddi, and gall bladder are innervated by ANS. Catecholamines increase bile flow through beta receptors. CCK is the only known hormone that contracts the gallbladder along with relaxation of the sphincter of Oddi. The parasympathetic nervous system regulates gallbladder muscle tone. Vagotomy (disruption of the parasympathetic system) decreases bile flow by decreasing cAMP production. Blocking of the sympathetic nervous system also results in decreased bile flow. The secretion of bile can be inhibited by α2-adrenergic receptor agonist whereas α-1 adrenergic receptor agonists increase bile secretion. Secretin has stimulatory role in bicarbonate secretion by activating cAMP synthesis followed by phosphorylation of protein kinase A (PKA). Protein phosphorylation results in activation of Cl^-/HCO_3^- exchanger and opening of CFTR. Several endocrine factors are involved in the regulation of bile secretion (Table 13.14).

13.1.2.4.4 Enterohepatic Circulation of Bile Acids

The bile salts secreted in the duodenum are absorbed in the ileum and return back to the liver again by portal veins for reutilization. This effective recycling of bile salts is called enterohepatic circulation. The anatomical components of enterohepatic circulation comprise liver, gallbladder, biliary tract, small intestine, and portal venous circulation. The absorption of bile acids in the intestine is facilitated by both active (ileum) and passive (down the intestinal length) transport mechanism. Around 95% of intestinal bile acids are absorbed by the enterocytes of the ileum through apical sodium-dependent bile acid transporter (ASBT). From the enterocytes, bile acids are refluxed into the sinusoids by organic solute transporter and heterodimer (OST/OST) transports. From the sinusoids, bile acids are taken up by the hepatocytes by Na^+-dependent taurocholate cotransport peptide (Ntcp).

13.1.2.5 Secretory Functions of the Intestine

Intestinal secretion is necessary to dilute and solubilize nutrients for better digestive and absorptive functions. In the intestine, continuous water and electrolyte recirculation

Table 13.14 Endocrine factors regulating bile secretion

Name	Effect on bile secretion	Mechanism
Somatostatin	Inhibitory	Inhibits cAMP through somatostatin receptor 2 (SSTR2)
Gastrin	Inhibitory	Decreases the expression of secretin receptor (SR) It also interacts with CCK-B receptors and activates protein kinase C (PKCα)
Endothelin-1	Inhibitory	Inhibits cAMP production via endothelin (ETA) receptors
Insulin	Inhibitory	Decreases cAMP level by stimulating PKCα and inhibiting PKA
D2 dopamine agonists	Inhibitory	Decreases cAMP level by stimulating PKC γ and inhibiting PKA by D2 receptors
Acetylcholine	Stimulatory	Activation of the Cl^-/HCO_3^- exchanger

Table 13.15 The length of intestine in different species

	Species						
Average length (m)	Horse	Cattle	Sheep/goat	Pig	Dog	Cat	Rabbit
Small intestine	22.44	46.00	26.20	18.29	4.14	1.72	3.56
Caecum	4.39	0.88	0.36	0.23	0.08	0.35 (large intestine)	0.61
Colon	3.08	10.18	6.17	4.99	0.60		1.65
Total	29.91	57.06	32.73	23.51	4.82	2.07	5.82

Stevens (1977)

occur from the lumen of the intestine to the enterocytes. The osmotic gradient generated by Na^+-H^+ exchange and solutes entering into the absorptive cells facilitate water influx by diffusion. The fluid movement allows the intestinal content to become isotonic for efficient absorption.

13.1.2.5.1 Anatomical Considerations

The intestine of domestic animals has variability among species (Table 13.15). The small intestine comprises duodenum, jejunum, and ileum. Duodenum is the first portion of small intestine extends from pyloric sphincter to the ligament of Treitz, a fibrous band that demarcates duodenum and jejunum. The portion of small intestine between Treitz and the ileocecal sphincter are jejunum (first one third segments) and ileum (remainder segments). Like other components of GI tract, intestine also have four distinct layers namely mucosa, submucosa, muscularis, and serosa. The mucosa provides physical defence against microorganism and digestive enzymes. The small intestine has a single mucosal layer, whereas colon consists of an inner more viscous and outer less viscous mucosal layer. The mucosal layer is renewed through by the secretion of mucous from the goblet cells. The functional activity of the intestinal lumen depends upon the intestinal epithelial cells (IECs). There are different intercellular junctions made of gap junctions, desmosomes, and adherent junctions comprising different proteins like occludins, claudins, and zonula occludens (ZO-1, ZO-2). They maintain the integrity of intestinal epithelial layer and facilitate the paracellular transport of ions and macromolecules. The epithelial cell lining is rapidly regenerated by $Lgr5^+$ stem cells. These stem cells are differentiated into varies epithelial cell subsets via progenitor cells or transit amplifying cells. Different classes of epithelial cells possess different functions in relation to digestion, neuroendocrine, and immunity. The cell types include enterocytes, enteroendocrine cells, M cells, Paneth cells, and tuft cells.

Surface area of the small intestine is characterized by large folds called plica circularis that increases the surface area. There are finger-like epithelial projections called as villi to increase the further increase surface area by about 10 to 14-fold. The villi are covered with microvilli to form brush border which further increases the surface area. Crypts of Lieberkühn are the glandular structure at the vase of villi. Glycocalyx, a jelly-like layer of glycoproteins covers the microvilli which contains digestive enzymes that project into glycocalyx. The part of enterocytes towards the lumen is called apex and the part opposite to lumen is called basolateral membrane. Nutrients are absorbed into enterocytes through apex and exit through basolateral membrane before entering into blood.

The large intestine is divided into caecum, colon, rectum, and anus from proximal to distal end. The colonic enterocytes are morphologically identical with intestinal enterocytes. Unlike small intestinal mucosa, the villi are absent in large intestine but contains deep tubular pits extends up to muscularis layer. The colonic mucosal cells include absorptive colonocytes, goblet cells, M cells, and Paneth cells.

Goblet cells are situated between enterocytes. They secret mucins that provide physical protective barrier against intestinal pathogens. The mucous entraps secretory IgA that further helps to destroy intestinal pathogens.

Microfold (M) cells specialized epithelial cells found in the intestinal lymphoid tissue (Peyer's patches). M cells help in antigen transport across the epithelial cells. M cells engulf luminal pathogens and their antigens by phagocytosis and presented to dendritic cells at lamina propria.

Paneth cells are situated at the base of the intestinal crypt responsible for the secretion of antimicrobial peptides (AMPs) like alpha defensins, secretory phospholipase A2 (sPLA2), lysozyme, cathelicidins, C-type lectin regenerating islet-derived protein IIIγ (RegIIIγ) and angiogenin4. These AMPs help to protect the epithelial barrier.

13.1.2.5.1.1 Blood and Lymph Supply to the Intestine

Numerous small arteries enter the base of the villus to form a capillary network under its epithelium. Veins arise at the tip of the villus from a capillary network that runs downward. The venules and veins, empty into the portal vein which enters into the liver and venous blood is mixed with that of hepatic arterial blood. The hepatic vein conveys the blood from the liver to the posterior vena cava. Monosaccharides, amino acids, free glycerol, short-chain fatty acids, water and inorganic salts and are absorbed through this route.

The lymph capillaries originate as lacteals near the villus tip and enter into a lymph plexus inner side of the muscular coat. The branches of these plexus enter into the submucosa to form a loose plexus of large lymphatics and pass into mesentery. The lymph capillaries drain their content into large lymph vessels, which empty into the mesenteric vessels. The mesenteric connect with mesenteric lymph nodes. The contents of the mesenteric vessels empty into the cisterna chyli which is continued as thoracic duct and finally empties into the venous system anterior to heart. Glycerides, long-chain fatty acids, cholesterol, and the immunoglobulins are absorbed by the lymphatic system. The rate of lymph flow increases after a meal.

13.1.2.5.2 Intestinal Secretion

Water and Electrolytes

During the digestive process, large quantities of water are secreted into the small intestinal lumen. But all the water is also reabsorbed simultaneously in the small intestine. The movement of water is facilitated by osmotic gradients by two distinct processes.

Osmotic gradient due to digested feed materials: The feed that enters into the intestinal lumen are not so hypertonic. But upon digestion their osmolality increases rapidly. The starch into the intestinal lumen has limited osmolality but, when it is digested into maltose, its osmolality increases tremendously. So, as the digestion continues, the osmolality of intestinal lumen gradually increases and pulls the water. But the digested end products such as glucose, maltose, and amino acids are absorbed, the osmolality decreases and leads to water absorption.

Osmotic gradient due to ion channels: The bicarbonate enters into the duodenal epithelium by paracellular mechanism via leaky tight junctions. Apart from this, duodenal epithelium contains different ion channels to regulation electrolyte movement across the duodenal epithelium. The movement of HCO_3^- is mediated via sodium-bicarbonate cotransporters (NBC), anion exchanger (AE), Na^+/H^+ exchange (NHE) transporter, Na^+-K^+-Cl^- (NKCC) channels, Na^+-K^+-ATPase pump, and Na^+/HCO_3^- cotransporter (NBC). The movement of Cl^- ion is facilitated by cystic fibrosis transmembrane conductance regulator (CFTR). The secretagogues (agents which increase the intestinal secretion) activate adenylyl cyclase and generation of cyclic AMP. Increased cAMP activates CFTR to increase the secretion of chloride ions into the lumen. The accumulation of Cl^- into the lumen generates an electrical gradient that pulls sodium into the lumen. As a result, NaCl is secreted and osmotic gradient is created to draw water.

Know More...
Cholera toxins cause abnormal activation of (CFTR) in crypt cells that results in massive secretion of water to manifest severe diarrhoea.

Neuroendocrine Factors

Intestinal glands secret a variety of neuroendocrine factors that regulate GI tract functions (Table 13.2).

13.1.2.5.3 Regulation of Intestinal Secretion

The factors that stimulate intestinal secretion are wall distension, luminal acidity, glucose, and bile salts. Stretch reflexes act via parasympathetic and enteric nervous system. The Ach released from cholinergic nerve endings increases bicarbonate secretion by increasing intracellular Ca^{++}. The sympathetic neurotransmitters like noradrenaline (NA) and neuropeptide Y (NPY) inhibit secretion. The enteric nervous system releases somatostatin which inhibits duodenal secretion by decreasing cAMP via somatostatin receptor (SSTR1).

One of the most important secretory agents of the intestine is serotonin (5-hydroxytryptamine, 5-HT). The receptors of serotonin are distributed from duodenum to the colon. It inhibits sodium Na^+/H^+ exchanger and prevents influx of Na^+ into the enterocyte.

Both vasoactive intestinal polypeptide (VIP) and prostaglandin E2 (PGE2) increase HCO_3^- secretion via G-protein-coupled receptor by increasing cAMP production. The secretion of mucin is associated with HCO_3^- secretion through CFTR.

Substance P (SP) alters water exchanges in the intestinal epithelium together with increased blood flow and intestinal motility.

Kinins act on the Na^+-K^+-ATPase pump and transmembrane conductance regulator (CFRT) to regulate duodenal bicarbonate secretion.

Some dietary bioactive peptides also regulate intestinal secretion. They can escape the hydrolysis and return to GI tract through circulation. One of such substance is carnosine (β alanyl-L-histidine) exerts vasodilatory function by cGMP production.

13.1.3 Digestion and Absorption of Nutrients

The digestion of nutrients in monogastric species is predominantly enzymatic with a minor microbial digestion in the large intestine. The characteristics features of enzymatic digestion is the hydrolysis of glycosidic bonds (carbohydrates), peptide bonds (proteins), ester bonds (lipids), and phosphodiester (nucleic acids) by the insertion of water molecule. The enzymatic digestion occurs in two phases namely luminal and membranous phase. The luminal phase of digestion is occurred in the lumen of GI tract and facilitates incomplete hydrolysis of nutrients leads to the production of short-chain polymers of original macromolecule by salivary, gastric, and pancreas glands. The membranous phase of digestion is catalysed by the enzymes situated at the apical surface of enterocytes. These enzymes help in

Table 13.16 Carbohydrate splitting enzymes of brush border and their functions

Enzyme	Substrate	Product
Maltase	Maltose, maltotriose, α-dextrins	Glucose
Lactase	Lactose	Glucose and galactose
Sucrase	Sucrose; also maltotriose and maltose	Fructose and glucose
α-Dextrinase	α-Dextrins, maltose maltotriose	Glucose
Trehalase	Trehalose	Glucose

the final breakdown of the substrates derived from luminal phase of digestion followed by absorption of end products of nutrients across the intestinal epithelium.

13.1.3.1 Digestion and Absorption of Carbohydrates

13.1.3.1.1 Dietary Substrate

The principal dietary carbohydrates are polysaccharides, disaccharides, and monosaccharides. The predominant dietary polysaccharides are starch, glycogen, and cellulose. Glycogen and starch are composed of glucose chains connected by α-1,4-glucosidic linkage and branching chains are joined by α-1,6-glucosidic linkage. Sucrose and lactose are main dietary disaccharides composed of glucose and fructose or glucose and galactose, respectively.

Fibrous like hemicellulose and cellulose consists of β-1,4 glucose unit. Ruminants can effectively utilize cellulose by hydrolysing them with ruminant microbes. The monogastric herbivores are unable to utilize cellulose as efficient nutrient source except horse. The large intestine of equines is well developed and microbial digestion of cellulose takes place in their large intestine.

13.1.3.1.2 Digestion of Carbohydrates in the Mouth and Stomach

Saliva contains salivary α-amylase or ptyalin (where present) which mixes with the feed during chewing. Salivary amylase hydrolyses starch into maltose and other small glucose polymers containing 3–9 glucose residues. However, only 5% of starch is digested in the mouth as the feed remain in the oral cavity for a very short period of time.

The action of salivary amylase that continues till the feed enter into the gastric body and fundus. The gastric acidity decreases the activity of salivary amylase as the optimum pH for α-amylase is 6.8 in comparison to gastric pH of 4.0. But, before complete blockade of salivary amylase, 30–40% of the starches will have been hydrolysed to form maltose.

13.1.3.1.3 Digestion of Carbohydrates by Pancreatic Amylase

Pancreatic α-amylase is identical with the salivary α-amylase of saliva, but its potency is several times more than salivary amylase. It causes the hydrolysis of almost all the starches within 15–30 min. The enzyme has specificity for α-1, 4 linkage and results in the combination of disaccharides and trisaccharides and α-limit dextrin.

13.1.3.1.4 Digestion of Carbohydrates by Intestinal Enzymes

The enterocytes of the small intestine contain brush boarder enzymes such as maltase, lactase, sucrase, and dextrinase specific for dietary disaccharides (e.g. maltose, lactose, sucrose) and limit dextrins (Table 13.16). They convert disaccharides and limit dextrins into hexose (glucose and galactose) or pentose (fructose) molecules ready for the absorption. Sucrase and isomaltase exist as an enzyme complex responsible for the hydrolysis of the products of amylase digestion. The enzyme lactase has two forms. The primary lactase involved in the digestion is associated with the brush border of the enterocytes with strong lactase activity. Another nonspecific β-galactosidase associated with cell lysosome hydrolyses lactose slowly at an optimal pH of 3. Maltase, sucrase, and isomaltase are rarely found in the intestine in newborn calves and pigs and their activity increases with the age. In contrast, lactase activity is highest during neonatal period and gradually decreases with age.

13.1.3.1.5 Absorption of Carbohydrate

The end products of the carbohydrate digestion (monosaccharides) are absorbed through the enterocytes and transported to the portal circulation. The absorption of glucose is occurred through specific transporters called sodium-linked glucose transporter (SGLT-1) at the brush border of the enterocytes of the duodenum and jejunum. Galactose is also absorbed through SGLT-1. SGLT uses sodium gradient for transport of glucose. The sodium gradient is created by Na^+, K^+-ATPase pump. This pump maintains a low intracellular sodium concentration by expelling 3 Na^+ in exchange for 2 K^+ entering inside the cell. The binding of two sodium ions with SGLT-1 facilitates conformational changes to allow glucose binding. Glucose together with two sodium ions then enter into the enterocytes. SGLT-1 then undergoes another conformation change to release glucose followed by sodium. The sodium ion then expelled through Na^+, K^+-ATPase pump. The glucose then leaves the enterocytes via glucose transporter (GLUT-2). A small amount of glucose is utilized by the enterocytes. Phlorizin blocks SGLT1 and inhibits sodium-dependent glucose transport into the enterocytes.

The fructose is absorbed by facilitated diffusion utilizing glucose transporter (GLUT-5) on the apical membrane independent of concentration gradient. GLUT-2 and GLUT-5 present at the basolateral membrane transfer fructose to portal circulation.

The rate of absorption of glucose and galactose are highest among the monosaccharides. The rate of absorption of fructose is almost half of the glucose. The rate of absorption of mannose is lowest (one-tenth of glucose).

13.1.3.1.6 Regulation of Carbohydrate Digestion

The levels of disaccharidases particularly sucrase-isomaltase (SI) and maltase-glucoamylase are increased in response to high carbohydrate diet. There is a transcriptional regulatory mechanism involving different transcriptional proteins. Three are several transcriptional regulatory proteins involved in the regulation of SI protein transcription. Hepatocyte nuclear factor (HNF-1), caudal-related homeodomain proteins (Cdx) and GATA type zinc finger transcription factors are involved in the upregulation of SI protein transcription by binding at the DNA regulatory regions at the 5′ flanking region of SI gene located at chromosome 3. The down regulation of SI protein transcription is mediated by the presence of glucose.

Epidermal growth factor (EGF) increases glucose transport by promoting insertion SGLT into the luminal membrane.

Peptide YY (PYY) increases glucose absorption by increasing energetic efficiency of glucose transport.

Somatotropin increases glucose absorption from the intestinal epithelium.

The genetic factors also influence the glucose absorption. Studies in broilers reported that birds genetically selected for higher growth had less intestinal absorptive epithelium.

13.1.3.2 Digestion and Absorption of Proteins

13.1.3.2.1 Dietary Substrate

The protein substrates available for digestion are of two types, exogenous and endogenous proteins. The exogenous or dietary proteins are long chains of amino acids bound together by peptide linkages. The characteristics of each protein are determined by the types of amino acids and their sequence of arrangements. The majorities of dietary proteins are easily digestible by the proteolytic enzymes (Table 13.17). However, the proteins conjugated with

Table 13.17 Principal proteolytic enzymes and their functions

Source	Enzymes (proenzymes)	Activator	Substrate	Product
Stomach	Pepsin (pepsinogen)	HCl	Proteins and peptides	Cleaves peptide bonds adjacent to aromatic amino acids
Exocrine Pancreas	Trypsin (trypsinogen)	Enterokinase	Proteins and peptides	Cleaves peptide bonds on carboxyl side of basic amino acids (arginine or lysine)
	Chymotrypsins (chymotrypsinogens)	Trypsin	Proteins and polypeptides	Cleaves peptide bonds on carboxyl side of aromatic amino acids
	Elastase (proelastase)	Trypsin	Elastin, some other proteins	Cleaves peptide bonds on carboxyl side of aliphatic amino acids
	Carboxypeptidase A (procarboxypeptidase A)	Trypsin	Proteins and polypeptides	Cleaves carboxyl terminal amino acids that have aromatic or branched aliphatic side chains
	Carboxypeptidase B (procarboxypeptidase B)	Trypsin	Proteins and polypeptides	Cleaves carboxyl terminal amino acids that have basic side chains
	Ribonuclease		RNA	Nucleotides
	Deoxyribonuclease		DNA	Nucleotides
Intestinal mucosa	Enterokinase		Trypsinogen	Trypsin
	Aminopeptidases		Polypeptides	Cleaves amino terminal amino acid from peptide
	Dipeptidases		Dipeptides	Amino acids
	Carboxypeptidases		Polypeptides	Cleaves amino terminal amino acid from peptide
	Endopeptidases		Polypeptides	Cleaves between residues in midportion of peptide
	Nuclease and related enzymes		Nucleic acids	Pentoses and purine and pyrimidine bases
Cytoplasm of mucosal cells			Di-, tri-, and tetrapeptides	Amino acids

carbohydrates may restrict effective proteolysis. The dietary proteins can also be damaged by heating during processing. Endogenous proteins include digestive gland secretions, desquamated cells, and small amounts of plasma proteins.

13.1.3.2.2 Digestion of Proteins in the Stomach

The digestion of dietary proteins begins at the stomach as saliva lacks proteolytic enzymes. The predominant proteolytic enzyme of the stomach is pepsin. It is secreted as inactive pepsinogen and activates under the influence of gastric HCl. In human gastric mucosa, two different forms of pepsinogen are available. Pepsinogen I is secreted from acid-secreting regions, whereas pepsinogen II is also found in acid secreting as well as pyloric region. Maximal acid secretion correlates with pepsinogen I levels. Pepsinogen hydrolyses peptide bonds adjacent to aromatic amino acids (phenylalanine or tyrosine) and yields polypeptides of different sizes. The optimum activity of pepsinogen is achieved at the pH range of 2.0–3.0. The gastric HCl provides favourable environment form pepsinogen activity. The enzyme is inactive at a pH above 5.0.

13.1.3.2.3 Digestion of Proteins in Intestine

Most protein digestion occurs at the duodenum and jejunum, under the influence of proteolytic enzymes from pancreatic secretion and peptidases from brush border.

Digestion of Proteins by Pancreatic Enzymes: In the small intestine, the partial protein breakdown products of gastric digestion are hydrolysed by major proteolytic enzymes of pancreas like trypsin, chymotrypsin, carboxypolypeptidase, and proelastase. Trypsinogen is activated by enterokinase and trypsin in turn activates chymotrypsin, elastase, and carboxypeptidase. The endopeptidases (like trypsin, chymotrypsin) yield peptides with C-terminal amino acids which are subsequently hydrolysed by exopeptidases (carboxypolypeptidase). The end products of trypsin hydrolysis contain C-terminal amino acids having basic side chains, which become the substrates for carboxypeptidase B. The chymotrypsin produces peptides with aromatic or branched aliphatic side chains which are further hydrolysed by carboxypeptidase A. Elastase hydrolyses elastin fibres. The proteins digested by the pancreatic juices are mostly dipeptides and tripeptides.

Digestion of Peptides by Peptidases of Mucosal Epithelial Cells: The last stage of protein digestion is mediated by the peptidases of mucosal epithelial cells. The epithelial cells contain aminopeptidases at cytosol and brush border. The brush border proteolytic enzymes show more tripeptidase activity (50%) compared to dipeptidase activity (less than 10%) compared to cytosolic enzymes. The tetrapeptidases are present in the brush border. The cytosolic peptidases are also capable of hydrolysing proline-rich peptides. But the leucine-rich peptides are hydrolysed by brush border aminopeptidase. Initially, long-chain peptides are hydrolysed to di- and tripeptides which then enter into the enterocytes for further hydrolysis by intracellular peptidases to form free amino acids.

13.1.3.2.4 Absorption of Amino Acids

Amino acids are absorbed active transport processes against concentration and electrochemical gradients and transported to the portal circulation. There are separate transport mechanisms for different groups of amino acids. The mechanisms are classified into Na-dependent and sodium independent. In sodium-dependent mechanism, Na^+ is required for absorption of amino acids. There is a different sodium-dependent transport mechanism for different amino acids (Table 13.18).

The sodium-independent amino acid transport mechanism involves ϑ-glutamyl cycle with the involvement of glutathione. The extracellular amino acids form a non-covalent binding with the plasma membrane and interact with ϑ-glutamyl moiety to form ϑ-glutamyl-amino acid complex to enter the cell. Inside the cytosol, the ϑ-glutamyl-amino acid complex is split by ϑ-glutamyl cyclotransferase to release amino acids.

13.1.3.2.5 Absorption of Immunoglobulins in Neonates

The maternal immunoglobulins are absorbed from the small intestine in animals like cattle, horse, sheep, goat, dog, and cat through colostrum maternal immunity transfer. Protein enters in the enterocytes by pinocytosis and transports to the lymphatics. The ability to absorb immunoglobulins diminishes soon after birth. In piglets, the ability decreases within 1–2 days. However, in rodents the absorption continues up to 3 weeks.

Absorption of antigens, bacterial and viral proteins is mediated through large microfold cells or M cells in conjugation with lymphoid tissue (Peyer's patches). M cells transfer the antigens to the lymphoid cells to active them. The activated lymphoblasts are entered into the circulation and later migrated to the intestinal epithelium to secret igA in response to the same antigenic exposure.

13.1.3.3 Digestion and Absorption of Lipids

13.1.3.3.1 Dietary Substrate

The dietary lipids include triglycerides (TG), phospholipids (PL), and sterols. Triglycerides are the predominant dietary lipid component chiefly found in the lipids of animal origin. TG provides 90–95% of the total energy derived from dietary lipids. The main phospholipid is in the diet is phosphatidylcholine (PC) derived from diet and the bile as well. Bile is the main source of PC in the intestinal lumen. The plant and animal origin dietary cholesterol include β-sitosterol and cholesterol, respectively. Other form of dietary lipids includes fat-soluble vitamins.

Table 13.18 Amino acid transport system in the intestine

Amino acid transport system	Transported amino acids	Mechanism	Rate of transport
Neutral (monoamino, monocarboxylic)	**Aromatic** (tyrosine, phenylalanine tryptophan) **Aliphatic** (glycine, serine, alanine, leucine, isoleucine, valine, threonine, histidine, methionine, cysteine, asparagine, glutamine)	Na-dependent active transport	Very rapid
Dibasic (diamino)	Arginine, lysine, cystine, ornithine	Na-dependent (partial) active transport	10% of neutral transporters (rapid)
Dicarboxylic (acidic)	Aspartic acid, glutamic acid	Carrier-dependent active transport. Partially Na dependent	Rapid
Amino acids and glycine	Proline, hydroxyproline, and glycine	Na-dependent active transport	Slow

13.1.3.3.2 Digestion of Lipids

Oral Cavity and Stomach The lipid digestion starts at oral cavity by the enzyme lingual lipases secreted from Ebner's glands. Before enzymatic hydrolysis, lipid droplets are broken down into smaller particles by mechanical force exerted by chewing. It increases the surface area for better enzymatic action. Lingual lipase cleaves TG into fatty acids and glycerol.

Hydrophobic lipids are clustered together in large droplets hence required special machinery called emulsification to facilitate digestion in aqueous medium. The stomach performs emulsification of lipids by the mechanical force generated by contraction. The grinding action is performed by the antrum of the stomach followed by the retropulsion of the content back to corpus. The process continues for several times for generating fine lipid droplets. The emulsification is further reinforced in the duodenum by the bile salts. The detergent-like action of bile salts reduces the surface tension and prevents the aggregation of lipids particles to form droplets. Gastric lipase causes hydrolysis of TG into glycerol and fatty acids.

The abundance of gastric and lingual lipase shows species specificity. In rats, lingual lipase predominates gastric lipase, but reverse is true for primates and human. In stomach, the activity of lipase is restricted in the fundus region. The optimum pH for both lingual and gastric lipase is 4 (hence called acid lipase) but the enzymes may remain active at pH 6–6.5. Both lingual and gastric lipase is unable to hydrolyse PL and cholesterol esters.

In neonates, both gastric and lingual lipase play important role in digesting milk for two reasons. Firstly, milk fat contains considerable proportion of medium chain TG compared to long-chain TG and the acid lipases are specific for medium-chain TG. Secondly, the pancreatic lipase is not developed during the neonatal period.

Small Intestine

In the jejunum, TGs are digested by pancreatic lipase. Phospholipids are hydrolysed by pancreatic phospholipase A2 (pPLA2) to yield free fatty acids and lysophosphatidylcholine. Most of the cholesterol in the diet are free cholesterol and can be absorbed through micelles. Only the esterified cholesterol (10–15%) requires enzyme hydrolysis by cholesterol esterase to release free cholesterol.

Pancreas also secrets a colipase which protects lipase from inactivation and facilitates the interaction of lipase with its substrate.

13.1.3.3.3 Absorption of Lipids

The enterocyte brush border membrane is separated from the lumen by an unstirred aqueous layer. The fatty acids and monoacylglycerol generated by the enzymatic hydrolysis of lipids are unable to reach in the intestinal brush border membrane due to their poor solubility in the aqueous membrane. Micelle formation is required to bring the lipid molecules close to the microvilli.

Micelle Formation: Micelles are the aggregation of phospholipids and cholesterol together with bile salts in such a manner that the hydrophobic ends of the lipid molecules are inside and hydrophilic ends are outside the aggregate to keep the lipids in aqueous solution. The micelles are further combined with monoacylglycerol, free fatty acids, and fat-soluble vitamins to form mixed micelles. Mixed micelles then move to intestinal mucosal cells and release contents near the brush border. The bile salts are recycled for emulsification and micelle formation.

Lipid Uptake by the Enterocytes: The uptake of lipids by the enterocytes is mediated through different transport proteins. Niemann-Pick C1 like 1 (NPC1L1), a glycosylated protein, is involved in the cholesterol uptake. It is located in the enterocyte brush border membrane. Ezetimibe blocks

Table 13.19 Transporters for water-soluble vitamins in the intestine

Vitamin	Transporter	Regulatory mechanism
Thiamine (vitamin B1)	Thiamine transporters (THTR-1 and THTR-2)	Transcriptionally regulated mechanism and intracellular Ca^{++}/calmodulin regulatory pathway
Riboflavin	RF transporters (RFVT-1 and RFVT-3)	Transcriptional regulated mechanism
Niacin and nicotinic acid	Monocarboxylate transporters (MCTs). pH-dependent	Ca^{++}/calmodulin regulatory pathway
Biotin (vitamin B7) and pantothenic acid	Na-dependent multivitamin transporter (SMVT) and accessory protein PDZ domain-containing protein 11 (PDZD11)	Intracellular protein kinase C (PKC)-mediated pathway
Folic acid (Vitamin B9)	Reduced folate carrier (RFC) and the proton-coupled folate transporter (PCFT)	Intracellular cAMP-mediated pathway and protein tyrosine kinase-mediated pathway

NPC1L1 and used in hypercholesterolaemia. Fatty acid (FA) transport protein (FATP) and cluster-determinant 36 (CD36) are involved in the transport of fatty acids into the enterocytes. Inside the cytosol of enterocytes, FA-binding protein (FABP) facilitates the transport of fatty acids.

Esterification and Chylomicron Formation: Glycerol and fatty acids (short- and medium chain) are directly absorbed into the blood stream. But, long-chain fatty acids, monoglycerides, cholesterol and fat-soluble vitamins are esterified within the prior to their delivery into the blood stream. In the enterocytes, the cholesterol and monoacylglycerols (MAG) undergo esterification by Acyl-CoA: cholesterol acyltransferase (ACAT) and diacylglycerol acyltransferase (DGAT) situated at the membrane of endoplasmic reticulum.

The esterified products are conjugated with apolipoproteins to form chylomicron. The chylomicrons are the large molecules composed of cholesterol and triglycerides at the core covered by a phospholipid membrane interspersed with apolipoprotein. The outer hydrophilic membrane allows the chylomicron molecules to travel in the aqueous medium.

The chylomicrons are transported to the Golgi vesicle in cis-Golgi in prechylomicron transport vesicles (PCTVs). These PCTVs are released into lymph vessels to reach into the blood stream. Hepatic fatty acid binding proteins (FABP) are required for this process.

The basolateral membrane of the enterocytes also contains ATP-binding cassette (ABC) transporter A1 to facilitate cholesterol efflux.

13.1.3.4 Absorption Vitamins

Water-Soluble Vitamins

The water-soluble vitamins can be obtained by two sources. The water-soluble vitamins of dietary origin (ascorbic acid) are absorbed in the small intestine. The vitamins synthesized by intestinal microbes (Vitamin B7) are absorbed at large intestine. There are some dual origin vitamins (niacin, thiamine, riboflavin, folic acid, biotin, and pantothenic acid) absorbed at both small and large intestine. Niacin is synthesized in the body from tryptophan except cat. Most of the water-soluble vitamins like thiamin, niacin, riboflavin, pyridoxine, biotin, ascorbic acid, and pantothenic are absorbed at the upper small by specific carriers (Table 13.19). In contrast, Vit-B12 and folate absorption are Na^{+}-independent and major site of absorption is ileum.

Vit-B12 requires two different glycoproteins namely haptocorrin and intrinsic factor (IF) secreted in saliva and gastric juice, respectively. Vit-B12 first binds with haptocorrin which protects it from stomach environment. In the duodenum, haptocorrin is cleaved by pancreatic protease to release free Vit-B12 which then binds with IF. The complex of vitamin B12–IF binds with cubilin receptors at the ileum and taken up by the enterocytes through endocytosis.

Fat-Soluble Vitamins

Absorption of vit E: Vitamin E remains in two forms viz. tocopherols and tocotrienols. Both these forms vary in their bioavailability and antioxidant property. Like the absorption of other lipid molecules, vit E absorption is also required micelle formation. The entry of vit E into the enterocytes is mediated by scavenger receptor class B type I (SR-BI). The vit E is carried to the lymphatic system via chylomicron. ABC transporters (ABCA1) are also involved in the absorption of vit E.

Absorption of vit A: Dietary vit A consists of retinyl esters (animal origin) or carotenoids (fruits and vegetables). The retinyl esters require luminal hydrolysis by lipase to release retinol. The carotenoids can be absorbed intact or cleaved to retinol by β-carotene-15,15′-dioxygenase in the liver or intestine. The free retinol is transported to enterocytes by the enterocytes and binds with cellular retinol-binding protein (CRBP). The esterification of retinol is catalysed by lecithin: retinol acyltransferase (LRAT). Retinyl esters are then secreted as chylomicrons.

Absorption of vit K: Two forms of vit K are available namely K1 (phylloquinone) and K2 (menaquinones). The dietary sources of vit K1 are green leafy vegetables and vegetable oils. Vitamin K2 is found in fermented products. Gut microbes are also able to produce menaquinones. NPC1L1, SR-BI, and CD36 are involved in the uptake of vit K by the enterocytes.

Absorption of vit D: Vitamin D exists in two physiological forms namely vit D3 (cholecalciferol) and vit D2 (ergocalciferol). The major dietary form of vit D is cholecalciferol. The absorption of vit D follows similar mechanism as other lipophilic compounds. The carrier mediated transport of vit D into the enterocytes are mediated by SR-BI, CD 36, and NPC1L1.

13.1.3.5 Absorption of Minerals

The dietary minerals exist as complex structures with major nutrients like proteins, carbohydrates, and fats. The digestion of minerals includes the hydrolysis to release minerals from the modular units. The absorption of minerals occurs via transcellular (uptake of minerals by the enterocytes, intracellular transport, and efflux through basolateral membrane into the circulation) and paracellular (applicable only for metals in aqueous phase). The solubility of some minerals is pH independent (Solubility does not depend on pH, for example, Na^+, K^+, Mg^{+2}, Ca^{+2}), and they are soluble throughout the gastrointestinal pH range (1–8) while some minerals (Cu^{+2}, Fe^{+2}, Mn^{+2}, Zn^{+2}) requires acidic medium for their solubility and remain insoluble hydroxy-polymers at neutral or alkaline pH.

Calcium: Calcium is absorbed from the intestinal lumen by two distinct mechanisms. In the duodenum, calcium is absorbed through active transcellular mechanism under low calcium intake. The calcium uptake by the enterocytes is mediated by voltage-gated transient receptor potential (TRP) channels. The transport of calcium across the cells is facilitated by a carrier protein called calbindin. The calcium is pumped out from the enterocytes by calcium-ATPase. The transport of calcium is enhanced by vit D which stimulates calbindin synthesis. The hormone-induced stimulation of calcium absorption in the gut is called transcaltachia. In the ileum, jejunum, and colon the calcium absorption occurs through passive paracellular mechanism at moderate to high calcium level. The ionized calcium diffuses through tight junctions at the basolateral membrane into the blood. The paracellular route of calcium absorption is predominating in ruminant (50% of total dietary calcium). Less soluble calcium complexes such as calcium carbonate ($CaCO_3$) are mixed with HCl in the abomasum and their solubility increases. Thus, the concentration of ionized calcium is more at the duodenum. Further the absorption of water duodenum and jejunum increases calcium concentration to facilitate paracellular absorption of calcium. The chelating agents decrease calcium absorption.

Phosphate: The absorption of phosphate occurs at all sections of small and large intestine. In ruminants, phosphate absorption occurs in the rumen also. There are two distinct mechanisms of phosphate absorption. Under low phosphate concentration, the absorption is mediated by active transcellular transport system. The uptake of calcium by intestinal and ruminal epithelium occurs via type II Na^+-coupled phosphate cotransporter proteins (NaPi-II) situated at the apical membrane. NaPi-II utilizes Na^+ ions to transport an HPO_4^{-2} anion. The phosphate efflux occurs at the basolateral membrane by facilitated diffusion with the help of phosphate channel. The Na^+ ions enter inside the cells are pumped out by Na^+/K^+ ATPase pump at the basolateral membrane. Vit D helps in the absorption of phosphate by stimulating the production of NaPi-II. Passive paracellular transport of phosphate occurs when soluble HPO_4^{-2} or $H_2PO_4^-$ are more in the intestinal lumen.

Copper: The first step of copper absorption is the reduction of dietary copper to cuprous form by apical metalloreductases called 6-transmembrane epithelial antigen of prostate (STEAP) family of proteins. The cuprous form is taken up by the gut epithelium through *Saccharomyces cerevisiae* copper transport protein (Ctr1p). The intracellular copper trafficking is mediated by Cu chaperone proteins (Atox1). The efflux of copper from the cells into the blood stream is facilitated by copper transporting ATPases (ATP7A or ATP7B).

Zinc: The intestinal absorption of zinc occurs at duodenum and ileum. The zinc is imported to the epithelium from the intestinal lumen through Zrt-, Irt-like protein (ZIP4) at the apical site. The zinc is exported from enterocytes into the portal blood by Zinc transporter 1 (ZnT-1). Animal proteins stimulate zinc absorption. Phytates chelate zinc and inhibit its absorption.

The absorption iron is discussed in haematology section.

13.1.3.6 Absorption of Water, Sodium, Potassium, and Chloride

Bi-directional movement of water across the mucosa of the small and large intestines occurs in response to osmotic gradients. About 98% water is reabsorbed in the intestine to reduce the faecal water loss (only 2%). The osmotic gradient is created by absorption of solutes and that promote the water uptake in the intestine.

The uptake of sodium is mediated by secondary active transport together with glucose, amino acids, and other compounds. Both Na^+ and Cl^- enter into the enterocytes by Na^+-K^+-$2Cl^-$ cotransporters. The sodium is exported from the enterocytes by via sodium pumps. The secretion of Cl^- is facilitated by chloride channels.

The movement of K+ across the gastrointestinal mucosa is due to diffusion. K+ moves actively down its electrochemical gradient by H^+-K^+ ATPase in the luminal membrane. The K^+ is secreted at the basolateral membrane by K^+ channels. Loss of ileal or colonic fluids in chronic diarrhoea leads to severe hypokalaemia.

13.1.4 Avian Digestion

On the basis of feed habits, birds can be classified into granivorous or seed eating birds (e.g. domestic pigeons, budgerigar, zebra finch, canary), omnivores but tending to be granivorous (e.g. sulphur-crested cockatoo, rosella, African grey parrot, macaw, chicken), omnivorous, but tending to be frugivorous or fruit eating birds (e.g. mynah, toucan, barbets), omnivores, but tending to be nectarivorous and frugivorous (e.g. lorikeets, lories), and carnivorous (e.g. owls, falcons, eagles, vultures). The avian digestive system has some striking differences in comparison to mammals. Most of these modifications are intended to reduce the weight to fly. The teeth are absent in birds. The jaw muscles are lighter in weight. In most of the species, the soft palate is absent and a cleft at the hard palate communicates with the nasal cavity. The muscular stomach or gizzard is located at the body's centre of gravity for balancing during flight. The intestine is shorter in length compared to mammals. Some additional features of avian digestive system compared to mammals are paired caecum and shorter colon which links the ileum and cloaca. The avian species is unique for its common passage for excretory and digestive waste products.

13.1.4.1 Anatomical Consideration

The avian digestive system begins at the mouth and ends at the cloaca and has several intervening organs in between.

Beak/Mouth

Birds obtain their feed by the beak and received in the mouth. There is huge variability in the shape and size of beak in different avian species according to food habits. In general, the beak covered with keratin. The birds don't have teeth to chew its feed. Instead, the tongue is used to push feed to the back of the mouth so that it can be swallowed. The palate forms the roof of the mouth cavity containing two clefts. The long median cleft or choana connects with the nasal cavity followed by a short infundibular cleft opens at the auditory tubes. The numbers of salivary glands are less in birds and secret only mucous type saliva which helps in lubricating the feed bolus. Oral sacs are seen in certain species of birds. They help to carry food and act as sexual display apparatus. The laryngeal mound is an elevation behind the tongue. It has papillae that help in deglutition. It connects with glottis with a narrow slit like opening.

Oesophagus and Crop

Oesophagus is a flexible tube that connects pharynx with the stomach. The flexibility of the oesophagus is due to longitudinal folds. The wall of the oesophagus consists of mucosal, submucosal, muscularis, and serosal layers. In the muscular layers, circular smooth muscles are predominant. The epithelium is keratinized stratified squamous in nature. Unlike mammals the upper and lower oesophageal sphincters are absent in birds. The oesophagus in birds can be divided into cervical and thoracic parts. Oesophageal sacs, bilateral diverticulum of cervical oesophagus is a characteristic feature of cervical oesophagus in certain species of male birds. It is used during breeding season for sexual display. The thoracic oesophagus is characterized by numerous mucous glands that provide lubrication for food for swallowing. The innervation of cervical oesophagus is parasympathetic whereas vagus and the coeliac plexus innervate thoracic oesophagus. The crop is a pouch-like structure originates as a dilatation of the cervical oesophagus with food storage function. The crops are well developed in omnivores and herbivores/granivorous birds compared to carnivorous birds. The crop is absent in ostriches, gulls, owls, geese, and penguins. In chicken, the crop has single pouch whereas pigeon's crop has two pouches. The crop is lined by keratinized stratified squamous epithelium. The feed is stored in the crop till the ventriculus becomes empty to receive it and it is controlled by a sphincter. The empty crop also sends hunger signals to the brain so that more feed is consumed. Apart from the storage functions, the crop secrets "crop milk" to feed newly hatched chicks in doves and pigeons. Crop also provides favourable environment for probiotic microbes.

> **Know More**
> In hoatzin (*Opisthocomus hoazin*), an obligate folivorous (eating leaves) crop is the largest part of its digestive tract.

Stomach

The stomach in avian species consists of two parts viz. proventriculus (pars glandularis) and gizzard or ventriculus (pars muscularis). The proventriculus resembles mammalian stomach, but ventriculus is unique to bird that facilitates mechanical grinding of feed. The size of proventriculus depends upon the feed habits of the birds. It is larger in aquatic carnivorous birds compared to granivorous species. The birds that lack crop (ostrich) have large proventriculus to store feed. The proventriculus contains oxyntic cells that secrete HCl, pepsin, and mucous.

Ventriculus or gizzard is the muscular stomach composed of thick and thin smooth muscles. The gizzard is covered by a sandpaper-like membrane called koilin membrane. The sandpaper-like appearance of the koilin membrane is due to solidification of mucous. Koilin membrane protects the gizzard from acid and enzymes secreted in proventriculus. The sandpaper-like appearance of koilin membrane also facilitates the mechanical grinding of feed. The bile pigments are refluxed from the duodenums that give characteristics greenish brown colour of the koilin membrane.

Small stone particles (grit) are ingested by some birds (mostly granivores) along with feed. These stones are used to grind feed in the gizzard.

Proventriculus is innervated by vagus and perivascular nerves from mesenteric and coeliac plexus. The muscles and blood vessels are innervated by cholinergic and noradrenergic fibres, respectively.

Small Intestine

The small intestine is divided into duodenum, jejunum, and ileum. The length of intestine is shorter in carnivores, frugivorous compared to herbivorous and granivorous. The division between duodenum and jejunum is demarcated by Meckel's diverticulum or vitelline diverticulum, a remnants of vestigial yolk sac. The intestinal mucosa is appeared progressively thinner from the duodenum to the ileum. The intestinal villi become shorter at the jejunum. The villi are covered by enterocytes with microvilli.

Large Intestine

The large intestine is composed of the caecum and the rectum (colon). The caecum has wide variability in their structures between different species of birds. The caecum is single in the birds of Ardeidae family (herons), paired in herbivores, granivores, and owls and double paired in secretary birds (*Sagittarius serpentarius*). The caecum is rudimentary in birds of Columbiformes and Piciformes order. The birds under Psittaciformes, Apodiformes, and Piciformes order do not have caecum. The right and left caecum (in birds with paired ceca) originate at the junction of the small and large intestines called ileocecal junction. The caecum is divided into three distinct portions namely basis ceci near ileocecal junction, corpus ceci at the medial region, and apex ceci at distal position. The caecum performs several functions. It helps in the microbial digestion of cellulose and reabsorption of water. The efflux of urine from colon into the caecum is mediated by antiperistatical movement. The urine acts as the source of nitrogen for cellulolytic microorganisms.

The rectum (colon) is joined with the cloaca which is the common passage for digestive, reproductive, and urinary systems. Unlike mammals, the colon in avian species contains numerous villi and goblet cells. The cloaca is divided into three compartments viz. into coprodeum, urodeum, and proctodeum. The cranial most coprodeum compartment stores faecal materials. Coprodeum is separated from urodeum by the coprourodeal fold. The middle compartment or urodeum stores urine. The ductus deferens open into the urodeum. There is antiperistaltic movement of urine from urodeum into the coprodeum and large intestine to facilitate reabsorption of water. The urodeum and proctodeum is separated by uroproctodeal fold. The caudal most portion is the proctodeum where cloacal bursa opens. The proctodeum opens outside through vent.

The cloaca bursa, a fabricius structure projecting dorsally from proctodeum. It is the lymphoid organ of birds and involved in B lymphocytes preprocessing.

Liver and Pancreas

The liver is usually bi-lobed. The right and left liver lobes are joined cranially at the midline. The right liver lobe is larger than left lobe. In domestic fowl and turkey, the left lobe has dorsal and ventral portions. Gallbladders are present in most of the species except ostrich, pigeon, and parrots. The bile drains into the duodenum by two ducts, hepatocystic and cysticoenteric duct.

The pancreas is situated between duodenal loop and divided into dorsal, ventral, and splenic parts. Pancreas is shorter in carnivores and granivores compared to piscivores (eating fish). Pancreas secrets amylase, lipase, proteolytic enzymes, and bicarbonate.

13.1.4.2 Gastrointestinal Motility

The prehensile organ of birds is beak. The processing of feed after ingestion is facilitated together by beak and tongue. Saliva lubricates the feed for swallowing. Rapid posterior movements of the tongue propel the feed bolus into the pharynx. Primary peristalsis within the oesophagus moves the feed into stomach. Oesophago ingluvial fissure controls the movement of feed inside the crop. During fasting, it is closed and restricts the entry of feed into the crop. But when the proventriculus is filled, it relaxes and allows the feed to enter inside the crop for temporary storage. The contraction of crop wall allows the feed to enter into the stomach. Motility of stomach is characterized by slow waves of circular smooth muscle (most birds lack birds lack longitudinal smooth muscles) generated from interstitial cells of Cajal. The peristaltic waves of the stomach move aborally into the gizzard and small intestine. Egestion, the oral expulsion of undigested materials, is seen in carnivorous birds. It is different from regurgitation or vomiting. The gastric emptying is controlled by enterogastric reflexes. The migrating myoelectric complex (MMC) is characteristics feature of small intestine similar to mammals. The filling of ceca is mediated by rectal antiperistaltic and ileal peristaltic waves. Continuous antiperistaltic movements of the help to carry urine from the urodeum into the ceca.

13.1.4.3 Secretary Functions of Digestive System

The primary role of the saliva is the lubrication of feed. The species that eat dry food have large numbers of salivary glands. Different types of salivary glands in birds are maxillary (roof of the mouth), palatine (either side of nasal opening at roof of the mouth), spheno-pteryoid glands (roof of pharynx on each side of eustachian tube), anterior and posterior submandible glands (inter-mandibular space), lingual glands (tongue), and crico-arytenoid glands (glottis). The salivary glands secret mostly mucous in majority of the species but, in poultry, saliva also secret amylase. The flow rate of saliva is 7–30 mL per day in chicken.

Crop milk that contains 50–60% protein, 32–45% fat (cholesterol, phospholipids, triglycerides, free fatty acids), and 1–3% carbohydrate. The ash content of crop milk is 4.4–4.8% which comprises mostly calcium, phosphorus, sodium, and potassium.

Proventriculus contains two types of glands. The simple mucosal glands are responsible for the mucous secretion and the compound submucosal glands resemble chief and the parietal cells of mammalian gastric mucosa secrets HCl and pepsinogen in addition to mucous. The pH range of gastric juice is about 0.5–2.5. The pH is higher in herbivores than carnivores. The flow rate of gastric juice is higher in birds compared to dogs, human, and monkeys. It ranges from 8 to 10 mL/kg body weight/h in chicken. The pepsin secretion rate (2400–2500 IU/kg body weight/h) is higher in avian species compared to mammals. The main stimulus for gastric secretion is histamine.

The pancreas juice contains digestive enzymes and bicarbonate ions. The predominant pancreatic enzymes are pancreatic amylase (28–30%) followed by chymotrypsin (20%) and trypsin (10%). Other pancreatic enzymes include procarboxypeptidases (A and B), proelastase, lipase, and trypsin inhibitor (prevent the autolysis of epithelium). The buffering action of pancreatic juice helps to neutralize the acidic chime to bring the pH around 6–8. The flow rate of pancreatic juice is more in birds compared to dogs, rats, and sheep.

Biliary secretion rate in chicken is about 24.2 μL per minute. Predominant bile salts in the chicken and turkey are chenodeoxycholyltaurine and cholyltaurine. Whereas, in ducks, chenodeoxycholyltaurine and phocaecholyltaurine are predominant. Some species of birds contain amylase in their bile. The avian species also exhibit enterohepatic circulation of bile.

13.1.4.4 Digestion and Absorption of Nutrients

Carbohydrate

The digestion of starch takes place in the upper jejunum by pancreatic α-amylase into maltose, maltotriose, and α-limit dextrins. In the unstirred water layer, the maltose, maltotriose, and α-limit dextrins are hydrolysed by maltase, isomaltase, and α-limit dextrinase into hexose (glucose and galactose) or pentose (fructose). The brush border enzymes (Table 13.16) split disaccharides into monosaccharides that are ready for absorption. The majority of monosaccharides are absorbed through active transport mechanism by utilizing the sodium gradient.

Protein

Protein digestion begins at the ventriculus (gizzard) pepsin secreted from proventriculus. The second stage of protein digestion occurs at intestinal lumen by trypsin, chymotrypsin, and elastase secreted from pancreas. These pancreatic enzymes hydrolyse large protein molecules into small oligopeptides and dipeptides. Final stage of protein digestion is facilitated by pancreatic (carboxypeptidases A and B) and brush (aminopeptidases and dipeptidases) enzymes. The absorption of amino acids in birds is similar to mammals. Different amino acid transporters are used to transport amino acids.

Lipids

The digestion of lipids begins with the emulsification at the gizzard as the upper GI tract doesn't secret lipase. The process of emulsification is accelerated in the duodenum when the diet is mixed with bile. The enzymatic hydrolysis of dietary lipids is initiated with the activation of lipase by pancreatic colipase. The hydrolysis of phospholipids is mediated by pancreatic phospholipase A. The products of lipid hydrolysis undergo micelle formation for solubilization. The mechanism of incorporation of micelle into the mucosal cells is not known. However, some theories suggest disruption of micelle before their entry to the mucosal cells. Jejunum is the principal site for lipid absorption in chicken and turkey. But the linoleic acid, stearic acid, and palmitic acid are absorbed in the ileum. The uptake of lipids by the brush border membrane is passive and the rate of absorption depends upon the saturation. The transport of lipids in the cytosol is mediated by fatty acid binding proteins. These proteins have high affinity for unsaturated fatty acids than saturated fatty acids. The medium- or short-chain fatty acids are unable to bind with these proteins. The fatty acids undergo esterification within the enterocytes and incorporated into portomicrons to secret into the hepatic portal blood supply in contrast to mammals where the lipids are entered into the lymphatic circulation.

Utilization of Yolk

Yolk contains 50% lipids that are major source of energy in the newly hatched fowl. The digestion of yolk lipids is mediated by lipase secreted from internal surface of the yolk sac. The yolk lipids are absorbed by three routes viz. yolk sac membrane, yolk stalk epithelium, and the intestinal mucosa. During early embryogenesis, the endodermal cells the yolk sac membrane absorbed and packaged the yolk lipids to release it into the blood. This process is accelerated during last week of incubation and may continue after hatching. The yolk assimilation through yolk stalk is started few hours after hatching and may continue up to 5 days in turkeys. Yolk lipids reach into the intestine through yolk stalk are hydrolysed and absorbed. The yolk stalk is occluded by the lymphocyte aggregation 4 days after hatching and the yolk stalk converts to lymphopoietic tissue after 14 days and acts as a site for extra medullary haematopoiesis. The remnant of yolk sac is called Meckel's diverticulum.

Learning Outcomes

- **Overview of monogastric digestion:** The digestion is a complex process of feed intake, conversion of the complex feed into their simplest form by mechanical and biochemical processes, absorption of the nutrients and their assimilation together with the removal of undigested feed materials. The process of digestion starts at the oral cavity where mastication reduces particle size of the ingested feed and incorporates saliva into ingesta for swallowing. The stomach facilitates grinding and mixing of the food along with digestion of proteins with the help of acid and enzymes. Once the chime passes into the small intestine, it is mixed with pancreatic enzymes and membrane bound enzymes in the enterocytes to convert the complex feed materials into their simplest form for absorption. The gastrointestinal system is the portal through which nutritive substances, vitamins, minerals, and fluids enter the body. The functions of digestive system are under neural and endocrine control.
- **Secretory functions of GI tract:** Different parts of GI system secrete a wide range of chemical substances to assist digestive and regulatory processes of GI function. Salivary glands, stomach, pancreas, gall bladder, and intestine are the predominant organs that contribute GI secretions. The salivary glands and the pancreas are complex acinar glands situated outside of the elementary canal, but their acinar secretions are poured into the GI tract. The glandular part of stomach contains deep tubular cells (oxyntic gland) that secret acid and pepsinogen. Small intestine is equipped with specialized secretory cells at the epithelial invaginations called Crypts of Lieberkühn. Single cell mucous glands like goblet cells contribute to mucous secretion in response to irritation. The glands of GI system are stimulated by direct contact of food. The tactile, chemical, and wall distension activates ENS that stimulates the glands for secretion. Parasympathetic stimulation increases the secretions of glands of upper GI tract. The GI secretions are also influenced by endocrine factors. The hepatobiliary system secrets bile that helps in digestion and absorption of lipids.
- **Digestion and absorption of nutrients:** The digestion of nutrients in monogastric species is predominantly enzymatic with a minor microbial digestion in the large intestine. The characteristics features enzymatic digestion is the hydrolysis of glycosidic bonds (carbohydrates), peptide bonds (proteins), ester bonds (lipids), and phosphodiester (nucleic acids) by the insertion of water molecule. The enzymatic digestion occurs in two phases namely luminal and membranous phase. The luminal phase of digestion is occurred in the lumen of GI tract and facilitates incomplete hydrolysis of nutrients leads to the production of short-chain polymers of original macromolecule by salivary, gastric, and pancreas glands. The membranous phase of digestion is catalysed by the enzymes situated at the apical surface of enterocytes. These enzymes help in the final breakdown of the substrates derived from luminal phase of digestion followed by absorption of end products of nutrients across the intestinal epithelium.
- **Avian digestion:** The avian digestive system has some striking differences in comparison to mammals. Most of these modifications are intended to reduce the weight to fly. The teeth are absent in birds. The jaw muscles are lighter in weight. In most of the species, the soft palate is absent and a cleft at the hard palate communicates with the nasal cavity. The muscular stomach or gizzard is located at the body's centre of gravity for balancing during flight. The intestine is shorter in length compared to mammals. Some additional features of avian digestive system compared to mammals are paired caecum and shorter colon which links the ileum and cloaca. The avian species is unique for its common passage for excretory and digestive waste products.

Exercises

Objective Questions

1. Deglutition centre is situated at ______________.
2. MMC is generally occurs during __________________ period.
3. Pacemakers of the guts is ________________________.
4. Somatostatin inhibits the motility of stomach and the gut. (True/False).

5. Zygomatic salivary gland is present in ______________________ species.
6. The inflammation of the salivary gland is called ________________.
7. The cephalic phase of gastric secretion is absent in ______________.
8. The rate limiting enzyme of classical pathway of bile acid synthesis is ___________.
9. Arrange these monosaccharides on the ascending order of their rate of absorption. Glucose, fructose, mannose. ____________________.
10. Predominant bile salts in the chicken and turkey are ________________ and ____________________.
11. Agents increase the bile flow by contracting the gall bladder are called _____________.
12. The conversion of trypsinogen to trypsin is accelerated by enzyme ____________________.

Subjective Questions

1. Why motion sickness leads to nausea?
2. How saliva helps in vitamin B12 absorption?
3. Antacids are prescribed along with non-steroidal anti-inflammatory (NSAID) drugs. Justify the statement.
4. Achlorhydria is associated with pernicious anaemia. Justify the statement.
5. Why metabolic acidosis occurs after heavy meal?
6. Gall bladder bile is more concentrated than hepatic bile. Justify the statement.
7. Why emulsification is required prior to lipid digestion.

Answers to Objective Questions

1. Medulla
2. Inter-digestive
3. Interstitial cells of Cajal (ICC)
4. True
5. Canine
6. Sialadenitis
7. Ruminants
8. Cholesterol 7α-hydroxylase (CYP7A1)
9. Glucose > Fructose > Mannose
10. Chenodeoxycholyltaurine and cholyltaurine
11. Chalogogue
12. Enterokinase

Keywords for Answer to Subjective Questions

1. Motion sickness, vestibular apparatus, vomiting centre
2. Saliva, Haptocorrin
3. Prostaglandin, mucosal blood flow, gastric mucosal barrier
4. Parietal cell, hydrochloric acid, intrinsic factor
5. HCl secretion, alkaline tide
6. Gall bladder, reabsorption of water
7. Lipid droplet, hydrophobic, surface area

Further Reading

Books

Hornbuckle WE, Simpson KW, Tennant BC (2008) Gastrointestinal function. In: Kaneko J, Harvey J, Bruss M (eds) Clinical biochemistry of domestic animals, 6th edn. Elsevier Publishing, pp 413–457

Reece WO (2015) Digestion, absorption, and metabolism. In: Section VII: Dukes' physiology of domestic animals, 13th edn. Wiley-Blackwell, Ames, IA

Stevens CE (1977) Comparative physiology of the digestive system. In: Swenson MJ (ed) Dukes physiology of domestic animals, 9th edn. Comstock, Ithaca, New York and London, pp 216–232

Sturkie PD (1986) Avian physiology. Springer Verlag, New York

Research Articles

Dehority BA (2002) Gastrointestinal tracts of herbivores, particularly the ruminant: anatomy, physiology and microbial digestion of plants. J Appl Anim Res 21(2):145–160

Erlinger S (1992) Bile secretion. Br Med Bull 48(4):860–876

Guo L, Yao J, Cao Y (2021) Regulation of pancreatic exocrine in ruminants and the related mechanism: the signal transduction and more. Animal Nutr 7:1145–1151

Martinsen TC, Fossmark R, Waldum HL (2019) The phylogeny and biological function of gastric juice—microbiological consequences of removing gastric acid. Int J Mol Sci 20:6031

Wapnira RA, Teichberga S (2002) Regulation mechanisms of intestinal secretion: implications in nutrient absorption. J Nutr Biochem 13: 190–199

Ruminant Digestion

14

Poonooru Ravi Kanth Reddy and Iqbal Hyder

Abstract

This chapter reviews the ruminant digestion with a special emphasis on the mechanical factors, gastro-intestinal tract structure, and nutrient digestibility. Ruminants possess large compartmental gastro-intestinal tract viz. rumen, reticulum, omasum, abomasum, and intestine, which favors handling large amounts of fibrous plant materials. Among the four-rumen compartments, abomasum occupies large space in newborn ruminants; however, the growth rate of the rumen and reticulum will be faster compared to abomasum as the age advances. In adult ruminants, the rumen harbors vast range of microbes enabling microbial fermentation of ingesta before exposing to gastric juices of abomasum. Ruminant digestion involves mechanical processing of feed stuff. Among various mechanical factors, rumination aids in complete digestion of feed stuff and include regurgitation, remastication, reinsalivation, and redeglutition. The rumen microbiota, consisting of bacteria, protozoa, fungi, and archea degrade the ingested fiber-based diets and aids in nutrient fermentation. The fermentation of complex carbohydrates produces short-chain fatty acids (acetate, propionate, and butyrate), isoacids (valeric, isovaleric, isobutyric, and 2-methylbutyric acids), and gases such as CO_2, CH_4, and H_2. About 70% of the ruminant animal's energy supply will be met by the produced volatile fatty acids. High fiber diet induces the production of acetate while the starch and sugars yield propionate as end product. Milk fat synthesis requires acetate and hence, low fiber diets lead to milk fat depression. Similarly, propionate contributes to most of the energy required for weight gain and lactose production. Rumen pH is an important factor to be considered; low pH suppresses the growth of certain bacteria sensitive to pH-causing rumen dysfunction and subacute rumen acidosis. The protein metabolism in ruminants depends upon the ability of rumen microbes utilizing ammonia. More than 80% of the rumen bacteria utilizes ammonia as nitrogen source for growth and yields microbial protein. For every 1 kg organic matter digested, the microbial yield ranges from 90 to 230 g, which is sufficient for growth and production to certain extent. Fat digestion in ruminants is unique in that the ruminal bacteria split the fatty acids and sugars from glycerol backbone through lipolysis. The metabolism of lipids by rumen microbes involves a four-stepped process viz. hydrolysis of esterified fatty acids, biohydrogenation of unsaturated fatty acids, lipid biosynthesis in the rumen, and metabolism of phytal to phytanic acid. Further, incomplete biohydrogenation generally produces conjugated linoleic acids (CLA), which are proven to benefit human health.

P. R. K. Reddy
Division of Animal Sciences, College of Agriculture, Food, and Natural Resources, University of Missouri, Columbia MO, United States of America

I. Hyder (✉)
Department of Physiology & Biochemistry, College of Veterinary Science (Sri Venkateswara Veterinary University), Garividi, Andhra Pradesh, India

© The Author(s), under exclusive license to Springer Nature Singapore Pte Ltd. 2023

P. K. Das et al. (eds.), *Textbook of Veterinary Physiology*, https://doi.org/10.1007/978-981-19-9410-4_14

Graphical Abstract

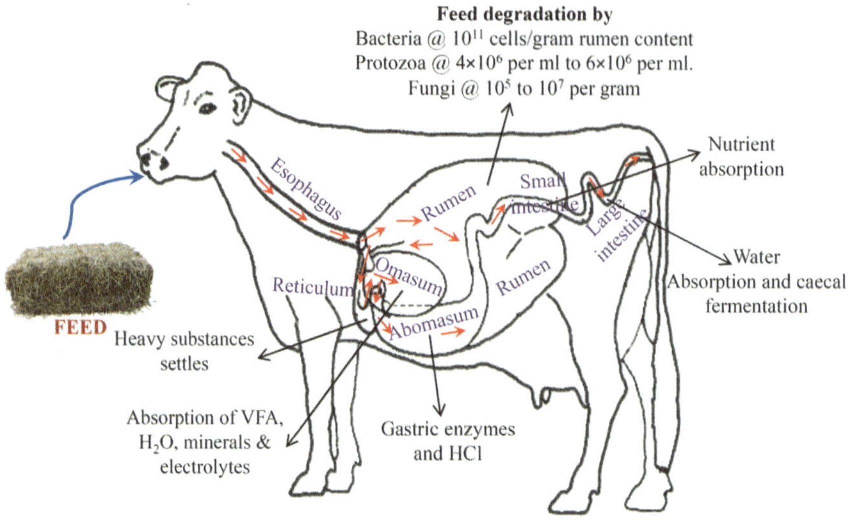

Description of the graphic: The digestion in ruminants is fermentative, i.e., the nutrients are subjected to fermentation in a specialized compartment of stomach is called rumen. The specialized environments in the rumen favors the growth of protozoa, bacteria, and fungi required for fermentative digestion. The motility of the rumen facilitates continuous mixing of the ruminal content and eructation of gases. The partially degraded feed undergoes regurgitation and the cud reaches ventral rumen, followed by reticulum, omasum, and abomasum. The carbohydrates are hydrolyzed and converted to volatile fatty acids and utilized by the body after absorption. The dietary proteins are converted to microbial crude proteins in the rumen and digested in the abomasum. Abomasum acts as true stomach and favors enzymatic digestion. Further digestion takes place in small intestine, where absorption of nutrients occurs through villi. Ultimately, the undigested feed will be excreted as feces.

Keywords

Ruminant · Digestive system · Rumen fermentation · Subacute rumen acidosis

Learning Objectives

- The structure of rumen and its environment
- Mechanical factors involved in the ruminant digestion
- Significance of ruminal microbes in modulating nutrient digestibility
- Fermentative digestion of nutrients and utilization of fermentation end products

14.1 Overview of Ruminants' Stomach

Ruminants are the even-toed ungulate herbivorous mammals capable of regurgitating food from their stomach for remasticating. They possess large compartmental gastrointestinal tract (rumen, reticulum, omasum, abomasum, and intestine), which favors handling of large amounts of fibrous plant materials. The rumen harbors vast range of microbes enabling microbial fermentation of ingesta before exposing to gastric juices in abomasum. The rumen microbiota, consisting of bacteria, protozoa, fungi, and archea degrade the ingested fiber-based diets. The mechanical activity of rumen, reticulum, and omasum supports this degradation. Esophagus opens into the rumen through cardia. The rumen is composed of cranial sac, ventral sac, and ventral blind sac, which are freely communicated with one another. The rumen wall is covered with many finger-like projections known as papillae with 5 mm in length and 3 mm wide in cattle.

In ruminants, abomasum is analogous to monogastric stomach and hence called as true stomach. Similar to the monogastric's stomach, abomasum secretes gastric enzymes and HCl. Among the four-rumen compartments, abomasum occupies large space in newborn ruminants. However, the growth rate of the rumen and reticulum will be faster compared to abomasum as the age advances. After completing the growth, rumen and reticulum, omasum, and abomasum occupies 69%, 8%, and 23% of the stomach portion. The pictorial representation of bovine stomach development from birth to maturity is presented in Fig. 14.1.

The omasum component is not well developed in small ruminants and is completely absent in the animals belonging to the suborder tylopoda (Camel and Llama). Esophageal groove or reticular groove, a gutter like invagination, extends from the cardia to reticulo-omasal orifice. The stimulation of sensory receptors in the pharynx and mouth causes closure of reticular groove to bypass milk directly from esophagus into

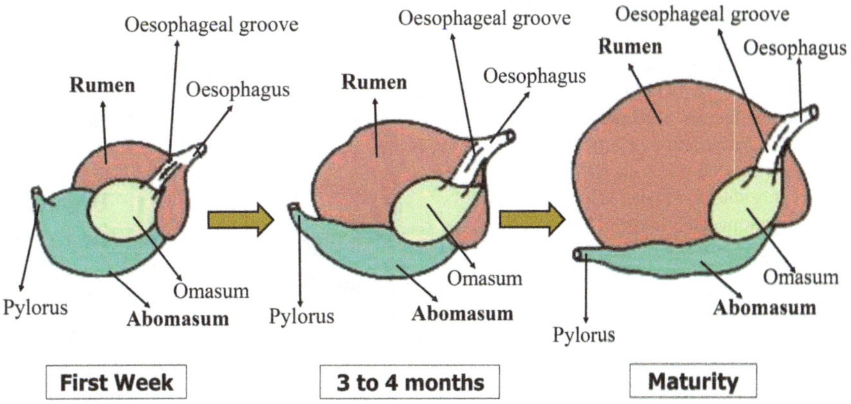

Fig. 14.1 Bovine stomach development from birth to maturity. [The size of rumen increases as the age advances and reaches maximum size at maturity]

reticulo-omasal orifice avoiding rumen and reticulum. Closure of esophageal groove is mediated by psychological responses, behavioral patterns, and chemicals such as sodium chloride, sodium bi-carbonate, copper sulfate, and sugar solutions. Among these chemicals, copper sulfate is less effective in calves and older ruminants and more effective in sheep.

14.2 Features of Digestion in Ruminants

Ruminants' gastro-intestinal tract holds numerous colonies of microorganisms. The type of microbes depends upon the diet and modifies accordingly as the age advances. A complex interaction exists between the host animal and variety of microbes. The gastro-intestinal tract of ruminants is unable to digest cellulose due to the lack of the degrading enzymes, hence completely relies on metabolic activities of gut microbes in utilizing the complex carbohydrate-based feed such as roughage. The fibrous materials retain in the gut for longer period to support the slow fermenting property of microbes. Among the fibrous particles, larger ones are retained at the reticulo-omasal orifice for mechanical digestion. Microbial fermentation produces volatile fatty acids, mainly acetate, propionate, and butyrate, which are of high value to the host ruminant system. Nearly 70% of the energy supply will be met by the produced volatile fatty acids. The ruminal microbes can also use non-protein nitrogen compounds such as ammonia to synthesize amino acids. The host proteolytic enzymes later digest the synthesized microbial protein. The gases produced by fermentation viz. CO_2 and CH_4 are expelled by eructation. In ruminant digestive system, saliva plays an indispensable role for buffering action against VFAs and monitoring the rumen pH. Therefore, physical effective NDF, a fraction of fiber that stimulates chewing activity and saliva production, is an important parameter to be considered while feeding the animals. Fermentation also allows detoxification of toxic substances before reaching small intestine.

14.3 Mechanical Factors Involved in Ruminants' Digestion

The mechanical factors involved in ruminant's digestion include mastication, deglutition, rumination, and eructation (Fig. 14.2). Rumination is a procedure of retrieving the food from upper part of rumen to the mouth for mastication. Rumination aids in complete digestion of feed stuff and include regurgitation, remastication, reinsalivation, and redeglutition.

14.3.1 Regurgitation

Heavy substances such as grains, rocks, or nails settle into the reticulum after ingestion, whereas lighter substances (roughage) enter the rumen. The saliva and fermentative gases accompany the lighter substances. Based on the specific gravity, the ruminal substances partitions into three zones viz. gas (upper), lighter roughage pieces (middle), and grain and fluid-saturated roughage (Bottom) (Fig. 14.3). Freshly eaten forages whose particle size is too great to be suspended in the rumen fluid for extensive maceration are not regurgitated immediately. The fermentation led constantly proliferating microbes reduce the feedstuff into micro-sized pieces. The continuous ruminal contractions push lighter roughage pieces into middle layer and denser substances into cranial sac of rumen and reticulum. The elevated soft palate closes glottis and the inspiratory effort with tongue drops intra-esphageal and intrathoracic pressure. The negative intrathoracic pressure opens cardia and caudal esophageal sphincters and forces the cud from middle layer of rumen into esophagus. The retrograde peristaltic wave originated from the terminal part of the thoracic esophagus carries cud to oral cavity. The lighter roughage pieces return to the mouth in cud form causing remastication. The regurgitated cud consists of small particulate matter highly mixed with liquid, which sinks to bottom layer within the rumen.

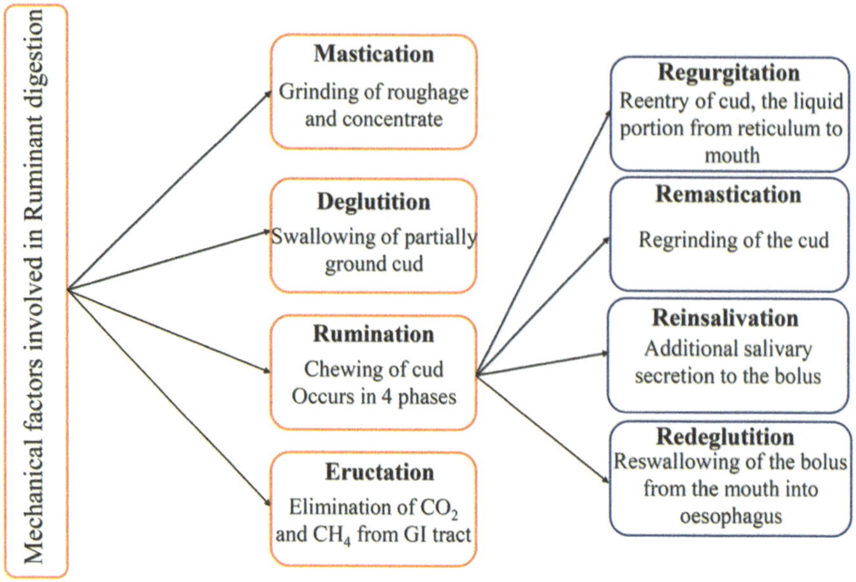

Fig. 14.2 Mechanical factors involved in ruminant digestion. [Mastication, deglutition, rumination, and eructation are the mechanical factors of ruminant digestion. Rumination in turn could be divided into four steps viz. regurgitation, remastication, reinsalivation, redeglutition]

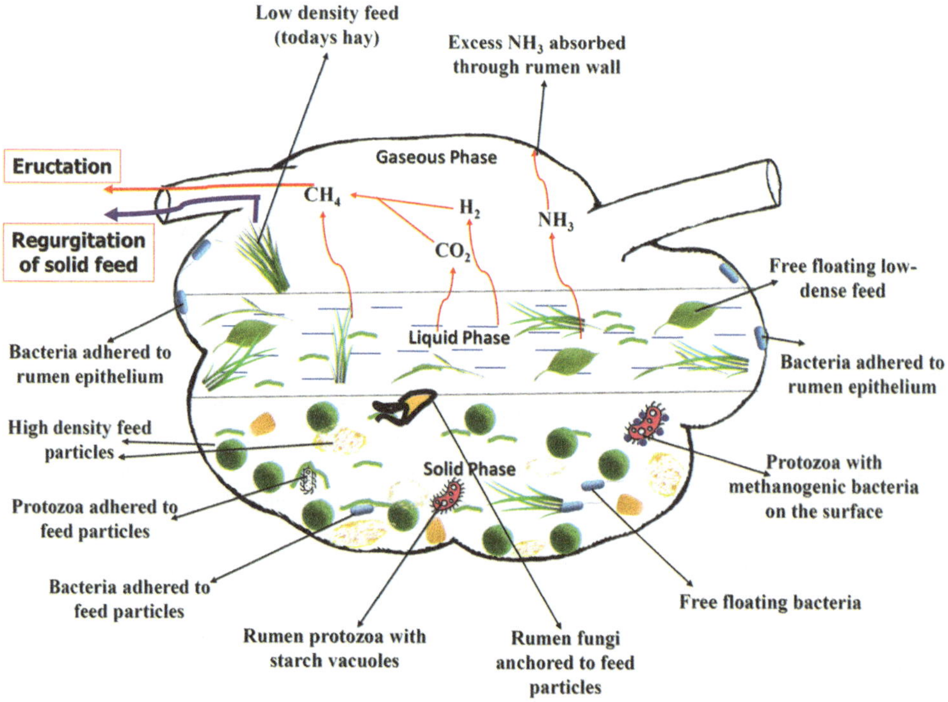

Fig. 14.3 Rumen ecology with the three partitions. [Rumen consists of complex ecosystem with diverse microorganisms such as bacteria, protozoa, and fungi. The gaseous phase of rumen contains CH_4 and NH_3, liquid phase comprises of free-floating low-dense feed, and the solid phase consists of high dense feed particles]

14.3.2 Remastication

After regurgitation, the rigid portion of cud is masticated for about 30–70 s. Entire cycle is repeated with an interval of 2–4 s between the remastication of two boluses. While chewing, the liquid portion is swallowed spontaneously. Remastication reduces the particle size and provides more surface area for the attack of microorganisms. Optimum remastication time is essential to lessen the risk of acidosis and improve the fiber degradation within the rumen. The chewing time is related to physically effective NDF portion of the diet.

14.3.3 Reinsalivation

As the cud is remasticated, parotid glands secrete saliva, facilitating reswallowing of the chewed cud. The remastication of the solid cud is accompanied by reinsalivation and redeglutition. Saliva is a significant buffering agent for rumen and hence reinsalivation plays a pivot role in maintaining optimum rumen pH of 6–7.

14.3.4 Redeglutition

Redeglutition is an act of reswallowing the cud. The reswallowed cud directly reaches rumen for increased microbial action on the complex carbohydrates including cellulose and hemi-cellulose.

Several intrinsic factors such as sex, age, and body size, and extrinsic factors including diet, time, and season affect the time of rumination. On an average, the rumination time for cattle is 10 h per day on complete hay-based diet. Grinding the roughage may decrease rumination time to 3 h per day.

14.3.5 Eructation

Eructation is the expulsion of fermentation gases like carbon dioxide and methane accumulated in the rumen.

14.4 Rumen Fermentation

The distinctive feature of ruminant digestive system is the fermentative digestion of feed materials through microbes, which occurs in rumen and reticulum. Besides, the fermentative digestion of feed can also be seen in pseudo-ruminants such as llamas, camels, and hippopotamus. The major microbes in rumen include ciliate protozoa, non-spore forming anaerobic bacteria, and anaerobic fungi followed by few facultative anaerobic bacteria. About 3.6% of the strained rumen liquor is composed of microbes with equal weights of bacteria and ciliate protozoa. The amount of rumen fungi is insignificant, but their activity is of huge importance. Both the bacteria and protozoa grow on the substrates of structural and non-structural carbohydrates, which are hydrolyzed by microbial enzymes. The gases generated by fermentation (carbon dioxide, methane, and traces of hydrogen) maintains anaerobic environment. The little amount of oxygen released into rumen is utilized by facultative anaerobes to maintain anaerobic condition.

14.4.1 Rumen as Microbial Habitat

Rumen provides congenial environment for the growth and multiplication of microbes. The rumen maintains a constant temperature of 40 °C.The HCO_3 andHPO_4 buffers of saliva provide a constant pH of 6–7. The saliva secretion also provides aqueous environment, thereby supplying substrates for continuous microbial activity. The primary contractions of rumino-reticulum aids in proper mixing of ruminal contents and the secondary contractions cause eructation (Fig. 14.4).

Rumen microbes use the host ruminants' feed stuff constituting cellulose, hemi-cellulose, pectin, soluble sugars, and starch to synthesize their energy for growth. Consequently, the fermentation produces acetic, butyric, propionic, and lactic acids along with gases such as H_2, CO_2, and methane. The fermentative end products act as inhibitors of fermentation and are removed continuously from the rumen. Although the calves are devoid of rumen microbes, they later attain the microbial population because of the dams' rumination ability. Rumination aids in regurgitating feed and rumen contents back into the mouth thus salivating and contaminating the feed consumed by the young calves. The rumen microbes could also be passed directly to calves during grooming.

Abomasum is analogous to monogastric stomach and causes the hydrolysis of protein of both dietary and microbial

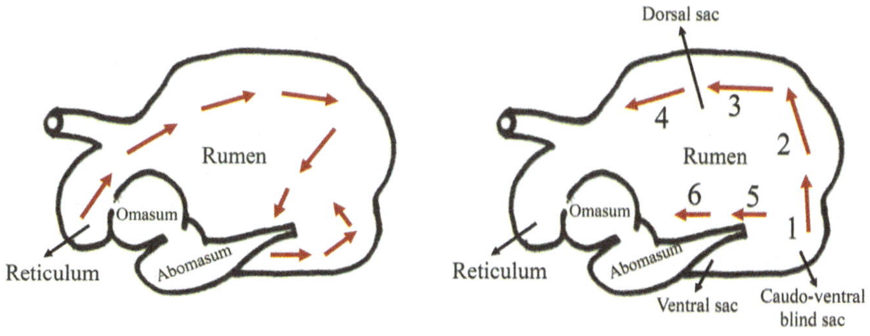

Primary contraction **Secondary contraction**

Fig. 14.4 Primary and secondary contractions. [The primary contraction occurs every 60 s and includes two contractions of the reticulum, which reaches the rumen. It leads to the ingesta flow from reticulum to cranial ruminal sac and later to ventral sac. The secondary contraction causes eructation. It leads to the ingesta flow from the caudo-ventral blind sac to the dorsal blind sac followed by dorsal sac (causes eructation) and ventral sac in sequence] origin, which is later absorbed in small intestine. The HCl and gastrin secretion are stimulated by a rise in abomasal pH and short-chain fatty acid levels. Gastric secretion occurs both from fundic and pyloric glands with the former as a major secretory source. The secretion from fundus region contains pepsin and HCl with pH close to 1.0, while that from pyloric glands is slightly alkaline with slight peptic activity.

14.4.1.1 Rumen Bacteria

Among the diverse microorganisms of rumen, bacteria are the predominant microbes contributing to nitrogen and carbohydrate metabolism through fermentation (Table 14.1). The rumen content comprises as high as10^{11} cells per gram of rumen content with more than 200 species. Although the total volume of small bacteria is same as ciliate protozoa, the metabolic activity of bacteria is far greater than protozoa, presumably because of the greater surface area. The rumen bacteria metabolize ingested feed material into volatile fatty acids, vitamins, and microbial biomass, which are later utilized by the host tissue. Based on the environmental existence, bacteria inhabiting the rumen have been classified into five groups. They include free-living bacteria associated with rumen liquid phase, bacteria loosely associated with feed particles, bacteria firmly adhered to feed particles, bacteria associated with rumen epithelium, bacteria attached to the

Table 14.1 The types, examples, substrates, and fermentative end products of rumen bacterial species

Type	Example	Substrate	Fermentative end product
Cellulolytic species	*Fibrobacter succinogenes*	Cellulose	Acetate, Formate, and Succinate
	Butyrivibrio fibrisolvens	Cellulose	Acetate, Butyrate, Formate, Lactate, H_2, CO_2
	Ruminococcus albus	Cellulose	Acetate, Formate, H_2, CO_2
	Clostridium lochheadii	Cellulose	Acetate, Formate, Butyrate, H_2, CO_2
Hemicellulolytic species	*Butyrivibrio fibrisolvens*	Xylans	Acetate, Butyrate, Formate, Lactate, H_2, CO_2
	Ruminococcus sp.	Xylans	Acetate, Formate, H_2, CO_2
	Bacteroides ruminicola	Xylans	Acetate, Formate, Succinate, CO_2
Pectinolytic Species	*Butyrivibrio fibrisolvens*	Pectin	Acetate, Butyrate, Formate, Lactate, H_2, CO_2
	Bacteroides ruminicola	Pectin	Acetate, Formate, Succinate, CO_2
	Succinivibrio dextrinosolvens	Pectin	Acetate, Succinate
Amylolytic species	*Bacteroides amylophilus*	Maltose	Formate, acetate, Succinate
	Selenomonas ruminantium	Oligosaccharides	Formate, acetate, Succinate
	Succinomonas amylolytica	Oligosaccharides	Acetate, Propionate, Succinate
	Streptococcus bovis	Starch substrates	Lactate at pH less than 5.5 Acetate, Formate, Ethanol at pH more than 6.0
Ureolytic species	*Succinivibrio dextrisolvens*	Urea with sugar or starch source	Acetate, Succinate
	Selenomonas sp.	Urea with sugar or starch source	Formate, acetate, Succinate
	Butyrivibrio sp.	Urea with sugar or starch source	Acetate, Butyrate, Formate, Lactate, H_2, CO_2
	Bacteroides ruminicola	Urea with sugar or starch source	Acetate, Formate, Succinate, CO_2
Lipolytic species	*Anaerovibrio lipolytica*	Triacylglycerols	Free fatty acids, Glycerol
	Micrococcus sp.	Triacylglycerols	Free fatty acids, Glycerol
Lactate utilizing sps.	*Selenomonas lactilytica* *Selenomonas ruminantium*	Lactic acid	Acetate, Succinate
	Megasphaera elsdenii	Lactic acid	Acetate, Propionate, Butyrate
Methane-producing species	*Methanobrevibacter ruminantium*	Cellulose, hemi-cellulose	H_2, CO_2, Formate, and the ultimate end product CH_4
	Methanobacterium formicicum	Cellulose or hemi-cellulose	H_2, CO_2, Formate, and the ultimate end product CH_4
	Methanomicrobium mobile	Cellulose or hemi-cellulose	H_2, CO_2, Formate, and the ultimate end product CH_4
Sugar-utilizing species	*Treponema bryantii*	Sugar	Acetate, Propionate
	Lactobacillus sp.	Sugar	Lactic acid
Proteolytic species	*Bacteroides amylophilus*	Proteins	Acetate, Propionate, Butyrate, CO_2, NH_3
	B. ruminicola	Proteins	Acetate, Propionate, Butyrate, CO_2, NH_3
	Butyrivibrio fibrisolvens	Proteins	Acetate, Propionate, Butyrate, CO_2, NH_3
	Streptococcus bovis	Proteins	Acetate, Propionate, Butyrate, CO_2, NH_3

surface of protozoa or fungal sporangia. Depending on the utilized substrates and end products, rumen bacteria are categorized into cellulolytic, hemi-cellulolytic, pectinolytic, amylolytic, ureolytic, methane producing, sugar utilizing, acid utilizing, proteolytic, ammonia producing, and lipid utilizing species.

The rumen bacteria and their activities are known to be influenced by several factors, revealing the possibility of their manipulation. These factors include, but not limited to, feeding regimen, diet changes, antibiotic usage, animal's age and health, season, stress level, geographic location, photoperiod, and environment.

14.4.1.2 Rumen Protozoa

Ciliates are the most abundant protozoa representing two physiologically and morphologically different groups viz. entodinomorphs and holotrichs, whereas flagellates occupy the niche to a very limited extent (Fig. 14.5).

The anaerobic rumen ciliates aid in digestion of plant material and ranges from 4×10^6 per mL to 6×10^6 per mL. On the basis of their substrates, the rumen protozoa were classified as starch degraders, soluble sugar utilizers, and lignocellulose hydrolyzers. The large quantities of reserve starch stored in protozoan vacuoles could be used on exhaustion of exogenous energy supply. Larger protozoa prefer structural polymers while smaller protozoa ingest sugars and storage polymers. Generally, holotrichs use soluble sugars and entodinomorphs utilize starch and other plant materials. The protozoal count is affected by ruminal pH, composition of diet, digestibility of diet, frequency of feeding, and season. Protozoa contribute 19–28% of cellulase activity of the total rumen fibrolytic activity. Protozoa are also a good source of lipids and roughly 27% of total lipids are thought to be contributed by holotrichs. The ruminal protozoa help in stabilizing the ruminal fermentation by ingesting feed particles and storing reserve polysaccharides. However, protozoa reduce the bacterial biomass by ingesting the ruminal bacteria. Because of the decreased bacterial protein availability, protozoa decrease the protein to energy ratio and increase the protein requirement by the host. Besides, the protozoa reduce the rate of bacterial colonization and feed degradation.

14.4.1.3 Rumen Fungi

The ruminal anaerobic fungal inhabitants range from10^5 to 10^7 per gram and include *Neocallimastix frontalis, Sphaeromonas communis*, and *Piromonas communis*. Rumen fungi degrade unlignified components of plant cell walls by producing cellulases, hemi-cellulases, and particularly xylanases. High roughage diet increases the fungal proportion, consequently increasing the adhesion and degradation of plant cell wall. The uniqueness in fungi lies in their ability to penetrate the cuticle.

14.4.2 Fermentation of Carbohydrates

The carbohydrate fermentation aids rumen microbial population to attain energy for growth and multiplication. About 75% of the plant tissue dry matter comprises carbohydrates. Microbial fermentation breaks carbohydrates into simple sugars. The end products of carbohydrate fermentation include volatile fatty acids (acetate, propionate, and butyrate) and gases (carbon dioxide and methane).

The speed of fermentation depends on the structure and solubility of carbohydrates. Glucose is a simple sugar with a molecular formula $C_6H_{12}O_6$. Starch contains amylose and amylopectin as polymer chains. Cellulose is beta 1,4 glucose linkage polysaccharide, and hemi-cellulose is composed of beta-linked xylose units and few hexoses. Pectin is beta-linked galacturonan (polysaccharide based on galactose with uronic acid). Lignin, a phenolic compound, is resistant even to microbial enzymatic digestion. Majority of the lignin is indigestible. However, rumen fungi are able to degrade the lignin to a certain extent. Based on the fermentation speed, the Cornell Net Carbohydrate and Protein System (CNCPS) classified soluble sugars as rapidly fermented, starch as less rapidly fermented, and cellulose and hemi-cellulose as slowly fermented carbohydrates. The carbohydrates in roughages are structural (cellulose, hemi-cellulose, lignin, and pectin)

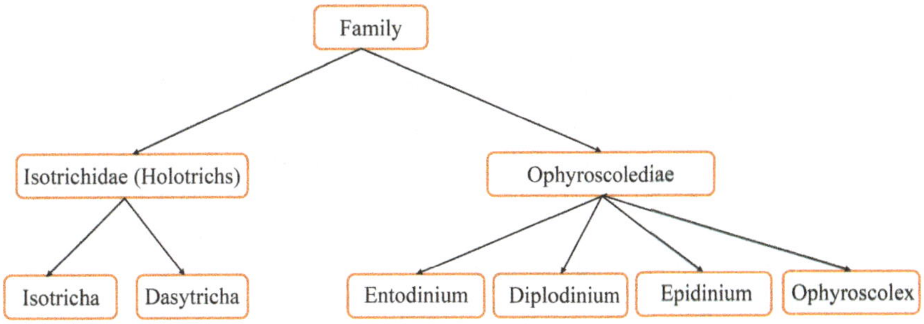

Fig. 14.5 Classification of rumen protozoa. [The Isotricha and Dasytricha genera belongs to Isotrichidae family while the genera Entodinium, Diplodinium, Epidinium, and Ophyroscolex belongs to Ophyroscolediae family]

while those in concentrates are non-structural (sugars and starch).

The extent of carbohydrate fermentation and the end products depends on the type of diet, maturity status, ruminal pH, anti-nutritional factors, and type of microbes. Matured forages are less digestible due to the higher proportion of lignin. Similarly, young grasses are more digestible due to the lower lignin quantity and higher fructosans fraction. Feeding roughage-rich diets leads to the production of acetate at higher proportion and concentrate-rich diet produces higher amount of propionate as end product.

Degradation of carbohydrates involves four steps.

14.4.2.1 Adherence

The bacteria adherence process plays a crucial role in fiber digestion. In the first phase, bacteria transport to fibrous substrate. Later the initial nonspecific adhesion of bacteria to substrates is followed by the specific adhesion of bacteria with digestible tissue. Finally, the attached bacteria proliferate to form colonies on specific sites of the plant tissue. Among various bacteria, coccoids prefer to attach plant cell wall. Attachment helps the bacteria to retain for a longer time and facilitates sustained action. Further, the adherence renders the produced enzymes to come into contact with the substrate and ensures that resulting degradation products are preferentially available. The adherence will be maximum at 40 °C, decreases at a pH below 5.0, and is facilitated at the pH of 5.5–7.8. Certain rumen fluid factors such as phenyl propanoic acid and phenylacetic acid stabilizes the bacterial adherence. The lignin and soluble cellulose derivatives like carboxy methyl cellulose are found to inhibit bacterial attachment.

14.4.2.2 Disaggregation

The fibrous feeds soak in the rumen fluid breaking them into small pieces. Disaggregation increases the degradable ability by rumen microbes. For instance, the starch granules are easily attacked on grounding.

14.4.2.2.1 Extracellular Degradation

The rumen liquor is the best source of bacterial and protozoal enzymes. The enzyme activities in rumen fluid are diverse. They include, cellulases, xylanases, β-glucanases, pectinases, amylases, proteases, phytases, and toxin-degrading enzymes such as tannases. Many of these microbial enzymes act on the soaked and disaggregated feed substances within the rumen, degrading them into short-chain oligosaccharides and sugars. Most of the crystalline cellulose is degraded through extracellular fungal cellulases.

14.4.2.2.2 Intracellular Degradation

The bacteria engulf simple sugars produced through the degradation of oligosaccharide and disaccharides. The intracellular enzymes of microbes metabolize mono- and disaccharides through phosphoroclastic cleavage forming pyruvate, phosphoenol pyruvate, volatile fatty acids, CO_2, and methane (CH_4). The bacterial enzymes degrade starch to maltose and glucose. Maltose is fermented to glucose, which gets converted to pyruvic acid through a metabolic pathway known as glycolysis. The anaerobic glycolysis yields two ATP molecules, contributing the energy source for rumen bacterial maintenance and growth. The degradation of amorphous form of cellulose occurs in anaerobic cellulolytic bacteria by producing enzymes such as endo-β-glucanohydrolase, glucosidase, and endo-xylanase. The hemi-cellulases are highly degradable compared to cellulose and require bacterial cellulases. The β-glucosidase hydrolyzes cellobiose and cellodextrins, producing hexose; however, the enzymatic degradation of hemi-cellulose yields pentoses. Pectin, a polymer of galacturonic acid, will be finally converted to short-chain fatty acids.

14.4.2.3 Formation of Volatile Fatty Acids

The pyruvate, an intermediate compound of carbohydrate fermentation, yields volatile fatty acids, CO_2 and CH_4. The metabolic pathways of pyruvate degradation are presented in Fig. 14.6. The yielded short-chain fatty acids act as major energy sources in ruminants.

14.4.2.3.1 Acetic acid formation

The acetic acid formation occurs in two pathways:

1. *Oxidative decarboxylation of pyruvic acid*
   The pyruvic acid formed during glycolysis enters into mitochondrial matrix and gets converted to acetyl CoAby removal ofCO_2 and H_2, in the presence of thiamine pyrophosphate (TPP) and lipomide. The reaction is catalyzed by pyruvic dehydrogenase. The Acetyl-CoA yields acetic acid by removal of thioester bond and coenzyme A.
2. *Phophoroclastic split*
   The phosphoroclastic reaction of pyruvate cause the formation of acetic acid and formic acid from two molecules of pyruvic acid yield. Being the simplest carboxylic acid, the formic acid (H_2CO_2) is dehydrogenated to CO_2 and H_2. A portion of the generated H_2 is utilized for the production of succinate, propionate, butyrate, and lactate and biohydrogenation of unsaturated fatty acids. Remaining portion will be utilized by methanogenic bacteria for methane production.

14.4.2.3.2 Propionic Acid Formation

The propionic acid formation occurs in two pathways:

1. *By carbon dioxide fixation:* The pyruvic acid combines with CO_2 forming oxalo-acetic acid, which is further

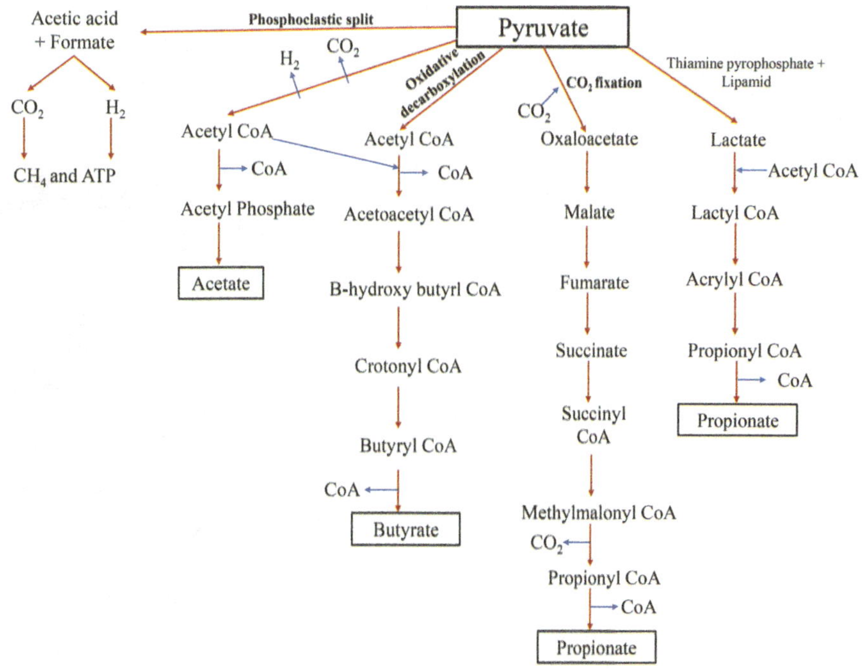

Fig. 14.6 The metabolic pathways of pyruvate degradation. [The phosphoclastic split of pyruvate yields acetic acid consequently producing CH_4. The propionates act as H sink and competes with CH_4, thereby indirectly regulating the CH_4 production whereas the acetate and butyrate formation releases hydrogen]

reduced to malic acid. The resultant malic acid is converted to fumaric acid on removal of one water molecule. The hydrogenation of fumaric acid produces succinic acid followed by its decarboxylation yielding propionic acid.

2. *By acrylate pathway*: The pyruvic acid produces lactic acid on hydrogenation and the resultant lactic acid is converted to acrylic acid on removing water. The hydrogenation of acrylic acid yields propionic acid.

14.4.2.3.3 Butyric Acid Formation

The different pathways of butyric acid formation are two molecules of acetyl-CoA condense to yield acetoacetyl-CoA and $2H_2$ by 3-ketoacyl-CoA thiolase. The acetyl-CoA is converted to beta hydroxybutyrl CoA by reduction. The resultant beta hydroxybutyrl CoA is converted to crotonyl CoA on removal of one H_2O molecule. Reduction of crotonyl CoA leads to formation of butyrl CoA along with one molecule of ATP. The butyrl CoA yields butyrate.

14.4.2.4 End Products of Carbohydrate Fermentation

The end products of carbohydrate fermentation include short-chain fatty acids (acetate, propionate, and butyrate), isoacids (valeric, isovaleric, isobutyric, and 2-methylbutyric acids), and gases such as CO_2, CH_4, and H_2. The CO_2 accounts for 40% of the total rumen gas, CH_4 accounts nearly 30–40% and hydrogen about 5%. The extra hydrogen should be removed from the rumen to maintain pH and rumen ecosystem. Methane acts as hydrogen sink and is considered as net energy loss as most of the CH_4 is lost as eructation. On an average, 4.5 g CH_4 is produced for every 100 g carbohydrate digested.

The total volatile fatty acid content of ruminants ranges from 60 to 120 mEq/L. The individual concentrations of VFA depends upon substrate composition, rumen ecosystem, and health status. The volatile fatty acids proportion changes according to the diet fed. The ratio of acetate, propionate, and butyrate ranges from 70:20:10 for high forage diets to 60:30:10 for high grain diets. The rumen liquor of ruminants fed with normal mixed diet contains 60–65% acetate, 15–20% propionate, and 10–15% butyrate.

14.4.2.5 Absorption of Volatile Fatty Acids

The VFA are directly absorbed from the rumen, reticulum, omasum, and large intestine. Undissociated acids absorb directly by simple diffusion in pH conditions of less than or equal to 6.7. The rate of absorption of VFA increases with the decreased pH, thereby regulating the rumen pH. The absorption of VFA results in accumulation of CO_2 and HCO_3 ion concentration. Lactacidemia is observed because of the lactate absorption on feeding high starch diets. Increased lactic acid formation in the rumen. Increased lactic acid formation in the rumen reduces rumen pH, leading to lactic acidosis and increased *Streptococcus bovis*. Low pH suppresses the growth of other types of bacteria sensitive to pH-causing rumen dysfunction and dehydration. The spiral flow chart of the rumen acidosis sequel is provided in Fig. 14.7.

14.4.2.6 Utilization of VFA in Ruminants

The fermentation of fiber yields acetate as main end product. Low energy and high fiber diets such as roughage leads to

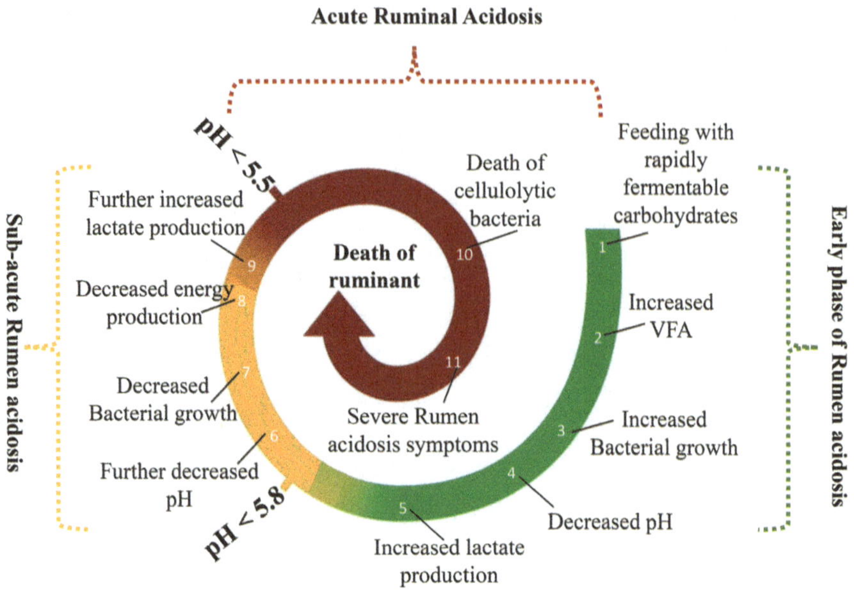

Fig. 14.7 Spiral flow chart of the rumen acidosis sequel. [Feeding rapidly fermentable carbohydrates lead to severe rumen acidosis and death of the animal]

increased ratio of acetate to propionate. Milk fat synthesis require acetate and hence low fiber diets lead to milk fat depression. Starch and sugars yield propionate as end product. The propionate converts to succinate and enters Krebs cycle producing glucose through gluconeogenesis. Propionate contributes to most of the energy required for weight gain and lactose production. Rapidly fermentable carbohydrates such as cereal grains lead to increased propionate proportion. Feeding inadequate amount of grain-based concentrate may decrease the lactose and overall milk production. As the propionate is glucogenic, the acetate and butyrate are ketogenic producing ketone bodies such as acetone, acetoacetic acid, and beta hydroxy butyric acid, ultimately contributing the energy needs of ruminant animals. The ketone bodies are used by skeletal muscles and other body tissues as a source of energy for fatty acid synthesis. Butyrate acts as energy source for rumen epithelium. It stimulates epithelial cell proliferation, consequently improving feed utilization. The concentration of butyrate significantly increases with increased concentrate feeding.

14.4.3 Protein Digestion in Ruminants

The rumen microbes utilize nitrogen and prepare their own sequence of amino acids for their growth and multiplication. The protein metabolism in ruminants depends upon the ability of rumen microbes utilizing ammonia. In ruminant nutrition, proteins can be divided into rumen degradable protein and non-degradable protein. The non-protein nitrogen substances are entirely degradable proteins. Of the protein consumed, depending upon the source, 20–100% will be degraded to ammonia. The rumen degradable protein fraction is hydrolyzed by extracellular proteolytic activities yielding short-chain peptides. Energy is a limiting factor determining the fate of absorbed peptides and amino acids. In the case of energy availability, the amino acids will be transaminated and used for microbial protein synthesis. In the event of energy deficit, the amino acids will be deaminated with the resulting carbon skeleton fermented into volatile fatty acids. Deamination causes the release of ammonia and carbon skeleton; the latter enters into various steps of VFA pathways, consequently producing acetic, propionic, and butyric acids. The rumen undegradable protein escapes ruminal microbial degradation reaching small intestine for enzymatic digestion.

14.4.3.1 Nitrogen Metabolism in Rumen

The protein requirement of ruminants met by the microbial protein. On total nitrogen basis, rumen bacteria contain about 65% protein. For every 1 kg organic matter digested, the microbial yield ranges from 90 to 230 g, which is sufficient for growth and production to certain extent. The peptides are generally absorbed by microbial cells. The efficiency of nitrogen incorporation into bacterial protein is higher for peptides. Whereas the individual amino acids will be subjected to rapid deamination producing NH_3 for bacterial growth along with CO_2 and volatile fatty acids. The pathways of digestion and metabolism of nitrogenous compounds in ruminants is presented in Fig. 14.8.

The bacteria synthesize protein by utilizing certain portion of true protein and entire non-protein nitrogen compounds such as urea. The urea includes urea from diet, saliva, and rumen epithelium. The protein degradation depends on dietary (structure, solubility, number of disulfide bonds and cross linkages between amino acid) and ruminal (type of bacteria, species, ammonia concentration, and pH)

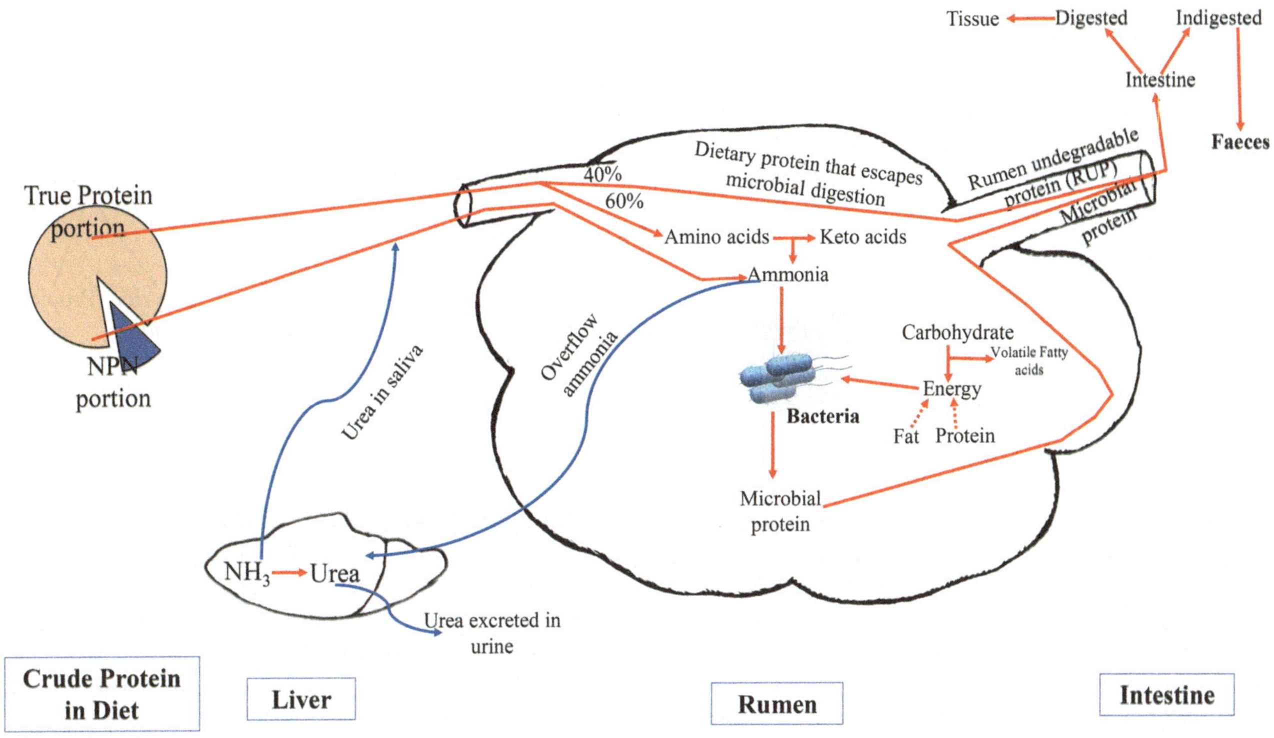

Fig. 14.8 The pathways of digestion and metabolism of nitrogenous compounds in ruminants. [The proteolytic bacteria break down the amino acids into ketoacids and ammonia, which in turn is used to prepare microbial protein]

conditions. The bacteria producing highest proteolytic enzyme concentration include *Butyrivibrio* sps., *Bacteroides* sps., *Selenomonas* sps., *Succinivibrio dextrisolvens*, and *Megasphaera elsdenii*.

More than 80% of the rumen bacteria utilizes ammonia as nitrogen source for growth. The concentration of ammonia nitrogen in rumen liquor varies with the diet from as low as 2 mg/dL in low-protein diets and as high as 100 mg/dL in high-protein diets. Urea in diets is converted by ureolytic bacteria to ammonia. Although the ruminants are able to utilize NPN compounds such as urea, the urea poisoning is not an uncommon phenomenon. Urea poisoning is mainly because of consuming higher quantities of urea, consequently increasing rumen pH and ammonia absorption rate into blood stream.

14.4.3.2 Metabolism of Amino Acids

Certain reactions occur for synthesis of non-essential amino acids, interconversion of amino acids, energy production, and ammonia excretion. These reactions include transamination, deamination, and decarboxylation.

14.4.3.2.1 Transamination

Transamination refers to a process whereby amino groups are removed from amino acids and transferred to acceptor ketoacid without the intermediate formation of ammonia. The most common transaminases are alanine transaminase and aspartate transaminase.

14.4.3.2.2 Deamination

Deamination refers to a process of removal of an amino group from an amino acid. The reaction is catalyzed by deaminases. They are of either oxidative or non-oxidative type.

Oxidative deamination: Oxidative deamination is a form of deamination involving oxidation in the conversion of amino acid to ketoacid and amino group to ammonia.

Non-oxidative deamination: Non-oxidative deamination refers to the deamination process involving non-oxidative steps and is catalyzed by amino acid dehydratase.

14.4.3.2.3 Decarboxylation of Amino Acids

Decarboxylation refers to reactions involving the removal of a carboxyl group from amino acids releasing biogenic amines and CO_2. The decarboxylases may be either specific or nonspecific.

14.4.4 Lipid Metabolism in Rumen

The uniqueness of lipid metabolism in ruminant calves is the presence of pregastric esterases in saliva, providing the

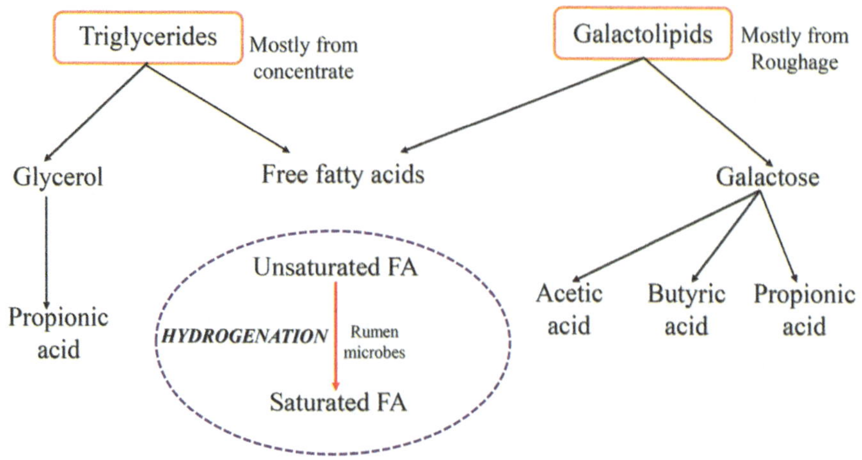

Fig. 14.9 Ruminal metabolism of lipids. [Lipolytic bacteria of rumen cause the breakdown of triglycerides into glycerol and free fatty acids and the galactolipids into galactose and free fatty acids]

ability to start digestion of milk fat from mouth. Dietary lipids include structural lipids of forages and storage lipids of oil seeds. Majority of the lipids in forages are phospholipids, whereas the oil seeds mainly comprise lipids as free fatty acids. A typical ruminant diet contains unsaturated fatty acids at higher proportion. They may be either from the galactolipids of forages or triglycerides of cereal grains and oil seed cakes. The rumen microbes hydrolyze the galactolipids and triglycerides to free fatty acids and glycerol. Glycerol is fermented to propionic acid. The ruminal metabolism of lipids is shown in Fig. 14.9.

The metabolism of lipids by rumen microbes involves a four-stepped process.

14.4.4.1 Hydrolysis of Esterified Fatty Acids

The triglycerides are hydrolyzed to fatty acids through hydrolysis. The lipids are subjected to hydrolysis by microbial lipases viz. cell bound esterases and lipases produced by rumen bacteria. Feeding concentrates at higher levels leads to production of higher concentration of unesterified fatty acids. Less than 10% of polyunsaturated fatty acids escapes the ruminal hydrogenation.

14.4.4.2 Biohydrogenation of Unsaturated Fatty Acids

The unsaturated fatty acids are biohydrogenated to saturated fatty acids. The linolenic acid of grasses is rapidly converted in rumen producing stearic acid, cis-transmonoenoic acid, and cis-trans dienoic acid as end products. Incomplete biohydrogenation generally produces conjugated linoleic acids (CLA), which are proven to benefit human health. Although the biohydrogenation ability is found in both bacteria and protozoa, the extent varies with higher ability in ruminal bacteria such as *Ruminococcus albus* and *Butyrivibrio fibrisolvens*. The biohydrogenation procedure is continuously monitored by the presence of metabolic hydrogen as end products of carbohydrate fermentation.

14.4.4.3 Lipid Biosynthesis in the Rumen

The ruminal fauna, especially bacteria synthesize odd chain fatty acids from propionate and branch chain fatty acids from valine, leucine, and isoleucine. The presence of odd chain and branch chain fatty acids in milk and higher stearic and oleic acids of ruminant fat depots are related to the biohydrogenation and rumen synthesis of fatty acids.

14.4.4.4 Metabolism of Phytal to Phytanic Acid

Phytal is an isoprenoid alcohol present in the chlorophyll of leaves. On consuming forages, the ruminant bacteria hydrogenate phytal to dihydrophytal, consequently producing phytanic acid on oxidization. The resultant phytanic acid is incorporated into rumen organisms and is reported to activate the transcription factors.

14.4.5 Lipid Digestion in Small Intestine

The short-chain fatty acids are mostly absorbed from rumen wall. The lipids leaving the rumen include 85–90% free fatty acids and 10–15% phospholipids. The neutral pH conditions render most of the free fatty acids assaults of calcium, sodium, and potassium. Reaching the acidic abomasal pH conditions dissociates the free fatty acids from the minerals. The free fatty acids adsorb on the degraded feed particles and pass to duodenum through pylorus.

In non-ruminants, monoacylglycerols play an important role in the formation of micelles. However, in ruminants, lysophosphatidyl choline acts as emulsifying agent. Micelle of saturated fatty acids forms under the influence of bile salts and lysolecithin. The pancreatic phospholipase hydrolyzes lecithin into a fatty acid and highly polar lysolecithin. The higher percent of lipid absorption occur in lower part of the jejunum. The bile salts are absorbed in ileum and reaches back to liver to contribute to bile. After entering into mucosal cells, resynthesis of triglycerides occurs via the glycerophosphate pathway. The triglycerides combine with the proteins inside the Golgi body to form chylomicrons. The chylomicrons and very low-density lipoproteins (VLDL) are carried to adipose tissue by capillaries.

Learning Outcomes

Ruminants possess large compartmental gastro-intestinal tract viz. rumen, reticulum, omasum, abomasum, and intestine, which favors handling large amounts of fibrous plant materials. In adult ruminants, the rumen harbors vast range of microbes enabling microbial fermentation of ingesta before exposing to gastric juices of abomasum. The fermentation of complex carbohydrates produces short-chain fatty acids (acetate, propionate, and butyrate), and gases such as CO_2, CH_4, and H_2. The protein metabolism in ruminants depends upon the ability of rumen microbes utilizing ammonia to produce microbial proteins. Ruminal bacteria split the fatty acids and sugars from glycerol backbone through lipolysis. The metabolism of lipids by rumen microbes involves a four-stepped process viz. hydrolysis of esterified fatty acids, biohydrogenation of unsaturated fatty acids, lipid biosynthesis in the rumen, and metabolism of phytal to phytanic acid.

Exercises

Objective Questions

1. The juice which plays an important role in the digestion of fats is ___________.
2. The feedstuff that is regurgitated and remasticated in mouth of ruminants is ______.
3. Important parameter that stimulates chewing activity and saliva production is ________.
4. An example for lipolytic bacteria is ___________.
5. Rumen Holotrichs use _________ and entodinomorphs utilize _________for survivability.
6. ___________ are able to penetrate the cuticle and degrade plant cell wall.
7. The roughage fraction composed of beta-linked galacturonan structure is _______.
8. Feeding roughage and concentrate-rich diets leads to the production of_________ and___________ as fermentation end products, respectively.
9. The first step of bacterial degradation of carbohydrate is ______________.
10. ________________ reaction of pyruvate causes the formation of acetic acid and formic acid from two molecules of pyruvic acid.
11. _____________ is an example for hydrogen sink in rumen.
12. The ratio of acetate, propionate, and butyrate ranges from ______ for high forage diets.
13. Feeding rapidly degradable starch substances at huge level may leads to __________.
14. __________ is the desired carbohydrate fermentation end product for milk fat synthesis.
15. __________ is the desired carbohydrate fermentation end product for weight gain and lactose production.
16. __________ acts as energy source for rumen epithelium.
17. On total nitrogen basis, rumen bacteria contain about ________ protein.
18. During lipid metabolism, glycerol is fermented to _____________ volatile fatty acid.
19. _________ is an isoprenoid alcohol present in the chlorophyll of leaves.
20. The short-chain fatty acids are mostly absorbed from ____________.

Subjective Questions

1. Explain in detail about the mechanical factors involved in ruminant digestion.
2. Write about the microbial habitat of rumen and classify the bacteria according to the substrate.
3. Elucidate the metabolic pathways of pyruvate degradation.
4. Explain clearly the pathways of digestion and metabolism of nitrogenous compounds in ruminants.
5. Describe the role of rumen biohydrogenation procedure in lipid metabolism.

Answers to Objective Questions

1. Bile juice, pancreatic juice
2. Lighter roughage pieces
3. Physically effective NDF
4. Micrococcus sps.
5. Soluble sugars and starch
6. Fungi
7. Pectin
8. Acetate and propionate
9. Adherence
10. Carbon dioxide fixation
11. Propionate, sulfate, and nitrate
12. 70:20:10
13. Subacute rumen acidosis
14. Acetate
15. Propionate
16. Butyrate
17. 65%
18. Propionic acid
19. Phytal
20. Rumen

Keywords for Answer to Subjective Questions

1. Mastication, deglutition, rumination, eructation, regurgitation, remastication, reinsalivation, redeglutition
2. Rumen fermentation, nitrogen metabolism, carbohydrate metabolism, cellulolytic bacteria, proteolytic bacteria, lipolytic bacteria
3. Phosphoclastic split, oxidative decarboxylation, acetyl Co-A, lactyl Co-A, acetate, propionate, butyrate
4. Non protein nitrogen, urea, microbial protein, rumen degradable protein, rumen undegradable protein
5. Triglycerides, galactolipids, glycerol, galactose, unsaturated fatty acid

Further Reading

Books

Duncan AJ, Poppi DP (2008) Nutritional ecology of grazing and browsing ruminants. In: Gordon IJ, Prins HHT (eds) The ecology of browsing and grazing. Ecological studies, vol 195. Springer, Berlin, Heidelberg

Hyder I, Reddy PRK, Raju J (2017) Alteration in rumen functions and diet digestibility during heat stress in sheep. In: Sejian V, Bhatta R, Gaughan J, Malik P, Naqvi S, Lal R (eds) Sheep production adapting to climate change. Springer, Singapore

McDonald P, Edwards RA, Greenhalgh JFD, Morgan CA, Sinclair LA, Wilkinson RG (2011) Animal nutrition, 7th edn. Pearson Education Limited, London, UK

Reddy DV (2021) Principles of animal nutrition and feed technology, 3rd edn. Oxford and IBH Publishing Co. Pvt. Ltd., New Delhi, India

Research Articles

Baldwin RL, Vi McLeod KR, Klotz JL, Heitmann RN (2004) Rumen development, intestinal growth and hepatic metabolism in the pre-and postweaning ruminant. J Dairy Sci 87:E55–E65

Fiorentini G, Carvalho IP, Messana JD, Canesin RC, Castagnino PS, Lage JF, Arcuri PB, Berchielli TT (2015) Effect of lipid sources with different fatty acid profiles on intake, nutrient digestion and ruminal fermentation of feedlot Nellore steers. Asian Australas J Anim Sci 28(11):1583–1591

Reddy PRK, Kumar DS, Rao ER, Seshiah CV, Sateesh K, Rao KA, Reddy YPK, Hyder I (2019a) Environmental sustainability assessment of tropical dairy buffalo farming vis-a-vis sustainable feed replacement strategy. Sci Rep 9:16745

Reddy PRK, Kumar DS, Rao ER, Seshiah CV, Sateesh K, Rao KA, Reddy YPK, Hyder I (2019b) Assessment of eco-sustainability vis-à-vis zoo-technical attributes of soybean meal (SBM) replacement with varying levels of coated urea in Nellore sheep (*Ovisaries*). PLoS One 14(8):e0220252

Shen H, Lu Z, Xu Z (2017) Associations among dietary non-fiber carbohydrate, ruminal microbiota and epithelium G-protein-coupled receptor, and histone deacetylase regulations in goats. Microbiome 5:123

Part VI

Endocrine System

General Endocrinology and Hormones of Hypothalamus and Pituitary

15

Sai Kumar B. A. A and Sai Mounica P

Abstract

The endocrine system comprised of ductless glands that secrete hormones. Along with the nervous system, it regulates various homeostatic mechanisms in an animal's body and serves as a communication channel between various organ systems. Based on their chemical nature, hormones can be classified as polypeptides, proteins, amines, or derivatives of fatty acids. They bind to specific receptors and elicit biological effects in the target organs. Cranial endocrine glands such as the hypothalamus, pituitary, and pineal gland secrete neurohormones that are synthesised by distinct neuroendocrine cells. Hypothalamic releasing and inhibitory hormones carried by the hypophyseal portal system serve to regulate different endocrine activities of the pituitary gland. The pituitary gland in turn controls various endocrine glands, forming a functional hierarchy in the endocrine system. The hypothalamus and pituitary hormones are responsible for maintaining normal growth, metabolism, reproduction, and animal behavioural patterns. The pattern of their secretion will be regulated chiefly by negative feedback mechanisms based on the changes in the concentration of a specific target hormone or a metabolite in circulation. Apart from the hypothalamus and the pituitary gland, melatonin secreted from the pineal gland regulates circadian rhythm, sleep, seasonal breeding and ensures optimum fertility in farm animals.

S. K. B. A. A (✉)
Department of Veterinary Physiology, Rajiv Gandhi Institute of Veterinary Education and Research, Kurumbapet, Puducherry, India

Sai Mounica P
Division of Veterinary Microbiology, ICAR-Indian Veterinary Research Institute, Bareilly, Uttar Pradesh, India

© The Author(s), under exclusive license to Springer Nature Singapore Pte Ltd. 2023

P. K. Das et al. (eds.), *Textbook of Veterinary Physiology*, https://doi.org/10.1007/978-981-19-9410-4_15

Graphical Abstract

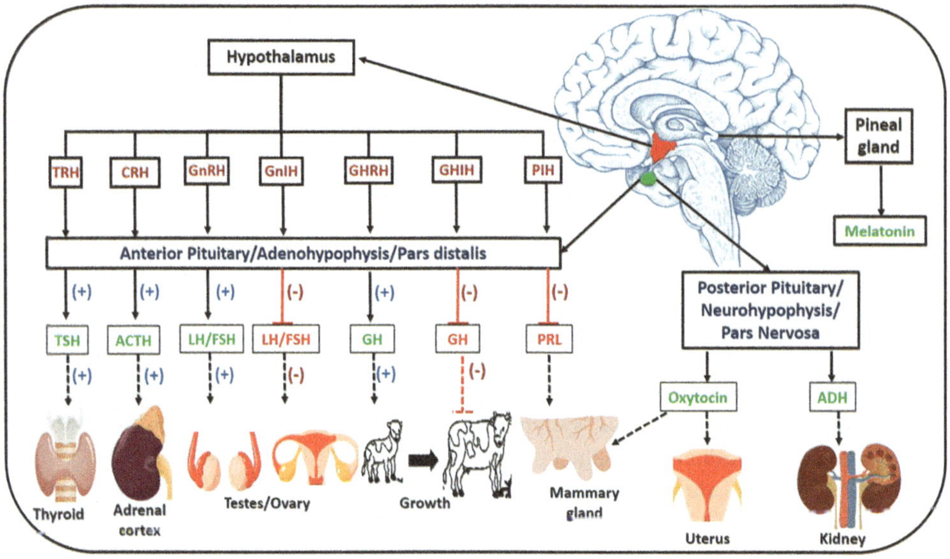

Description of the graphic: Functional hierarchy of the endocrine system. The releasing and inhibitory hormones secreted from the hypothalamus will act on the anterior pituitary lobe to affect its hormone secretion. The pituitary hormones further affect the secretory activities of different endocrine glands situated in the body. *TRH* thyrotropin-releasing hormone; *CRH* corticotropin-releasing hormone; *GnRH* gonadotropin-releasing hormone; *GHRH* growth hormone-releasing hormone; *GnIH* gonadotropin-inhibiting hormone; *GHIH* growth hormone inhibitory hormone; *GH* growth hormone; *LH* luteinising hormone; *FSH* follicle-stimulating hormone; *ACTH* adreno-corticotrophin hormone; *PRL* prolactin; *ADH* anti-diuretic hormone; (+) stimulate; (−) inhibit

Keywords

Hormone · Hypothalamus · Hypophyseal portal system · Pituitary gland · Pineal gland

Learning Objectives

- General properties of hormones
- Classification of different types of hormones
- General mechanism of action of peptide and steroid hormones
- Hypothalamus and its hormones
- Pituitary gland, its cell types and hormones
- Pineal gland

15.1 General Endocrinology

15.1.1 Introduction

The communication between different cells across various organ systems is a pre-requisite way to maintain a wide range of physiological mechanisms, which helps in governing any multi-cellular organism to function as a singular entity. Numerous chemical messengers that help in achieving the coordination between cells to bring about a particular homeostatic or homeorhetic response are identified (Fig. 15.1). Broadly, the nervous and endocrine systems function as the major relay channels to achieve cellular communication. Even though distinct, both the systems are intricately interwoven in the hypothalamus. The former system operates as an acute channel, whereas the latter works slowly to bring a sustained response. The endocrine system is an amalgamation of several ductless endocrine glands, attributed with the manufacturing of a specialised class of chemical messengers known as hormones. The hormones are secreted into blood as a result of an appropriate stimulus, carried and act on a specific target organ or an organ system, thereby bringing a well-defined biological response.

15.1.2 Brief History

Claude Bernard has used the word “internal secretion” to denote the secretion of glucose from liver. However, it is widely extended to denote any bio-molecule that is released into blood. The French physiologist Charles Brown-Sequard demonstrated that the ablation of adrenal gland is fatal and also claimed that injecting testicular extracts has a rejuvenating effect in men. Another exciting discovery by a British neurosurgeon named Victor Horsley was the onset of

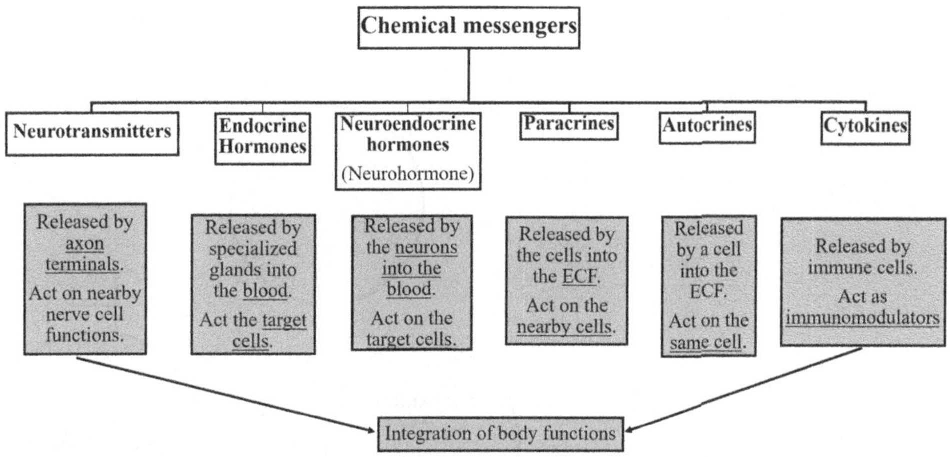

Fig. 15.1 Types of chemical messengers. Chemical messengers released from various cells help in the communication and coordination between different tissues culminating in the execution of different physiological functions

myxoedematous signs upon removing the thyroid gland in monkeys. Ernest Starling (1866–1927) coined the term "hormone", derived from the Greek word "*ormao*", which means to "excite" or "stir-up". Ernest Starling in collaboration with William Bayliss had isolated the first hormone "Secretin" (1902), and their discovery has revolutionised the branch of endocrinology.

15.1.3 Glands and Classification

Glands are defined as a group of cells that are structurally and functionally organised to work in unison to synthesise a product, viz. either an enzyme, sweat, saliva, milk, or hormone, and secrete into the duct or bloodstream. The presence or absence of a duct is often used as a distinguishing feature in classifying different glands into three major classes, namely:

1. **Exocrine glands:** Consists of secretory acini, which empty their products into a specialised duct.
   E.g. Liver (Bile), salivary gland (Saliva), and mammary gland (Milk)
2. **Endocrine glands:** Glands that lack ducts and the synthesised products are often secreted directly into the bloodstream.
   E.g. Hypothalamus (Somatostatin), pituitary gland (GH), thyroid gland (T_3/T_4)
3. **Mixed glands:** Glands that perform both exocrine and endocrine functions.
   E.g. Pancreas (exocrine: enzymes, endocrine: insulin), testes (exocrine: spermatozoa, endocrine: testosterone)

15.1.4 Hormones and Their General Properties

Hormones are defined as the chemical substances secreted by specialised endocrine glands or cells in minute quantities, and conveyed to distant or nearby target cells via the bloodstream or interstitial fluid respectively (components of ECF). The definition has been evolved to incorporate all the chemical messengers that are qualified as hormones irrespective of their source and site of action.

The salient properties of hormones (Fig. 15.2) are enlisted below:

1. **Bind to specific receptors:** Every hormone binds to a specific high-affinity receptor that ensues the formation of a hormone-receptor complex (HRC) before eliciting a target response. Hence, the mere presence of the receptors to a particular hormone dictates the type and various target tissues that it can act up on to get desired effects. Biochemically, all the hormone receptors are protein in nature. The receptor concentration on each target cell is never static and nonetheless, it is negatively related to the hormone concentration. The phenomenon wherein the receptor number increases due to the lower circulating levels of a hormone is termed upregulation whilst the contrary is known as downregulation. Put together, they act as amplifiers, transducers, and selectors of hormone signalling to produce a target response that is either stimulatory or inhibitory in nature depending on the cell type.
2. **Slow onset of action:** The onset of biological effects of various hormones is slow ranging from a few hours to even days, generally attributed to their action on the genetic machinery to initiate transcription of several effector genes.
3. **Absence of enzymatic activity:** They lack inherent enzymatic activity and cannot catalyse any intracellular enzymatic reaction directly.
4. **Signal transduction:** A phenomenon characterised by the activation of a cascade of intracellular reactions by several hormones upon binding to the target cell receptors. Hormones are widely recognised as "first messengers"

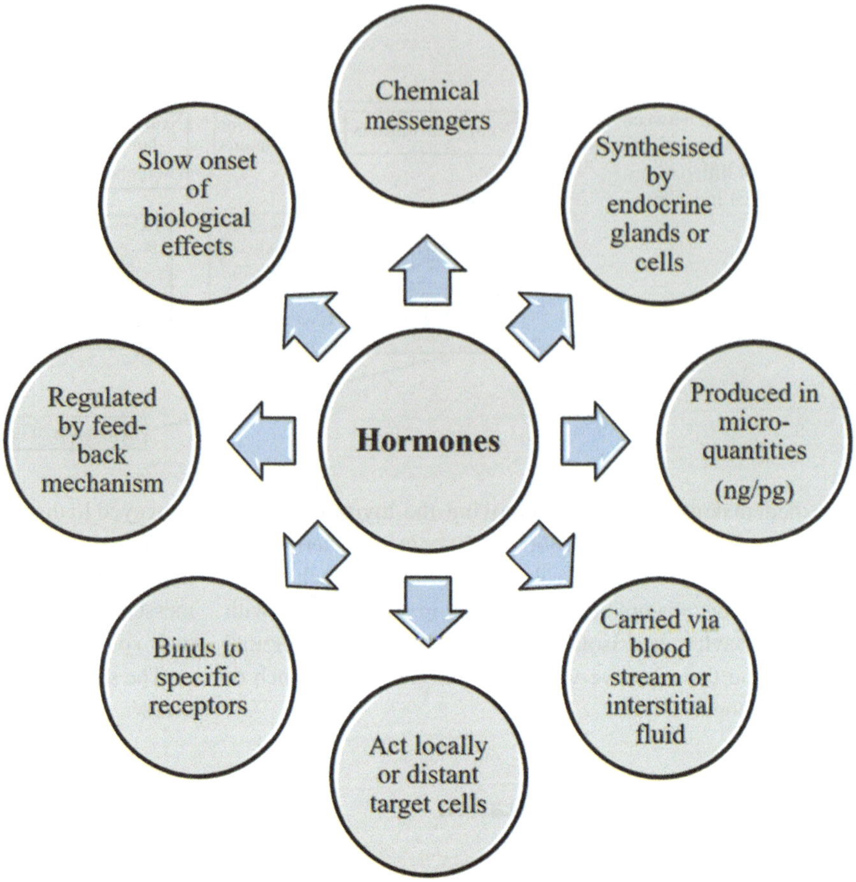

Fig. 15.2 General properties of hormones. [Synthesised in specialised endocrine cells and carried in the bloodstream, hormones bind to specific receptors present upon reaching the target cells to elicit specific biological effects]

and often lead to the synthesis of intracellular "secondary messengers" in the target cells.

5. **Feedback mechanism:** The rate of secretion and concentration of every hormone are regulated within a narrow range mostly by the means negative feedback mechanism.
6. **Metabolic clearance rate:** After eliciting desired biological effects, they are rendered inactive by the target tissues and/or eliminated from the circulation by excretory actions of the liver and kidney. The rate of elimination is inversely correlated to the affinity by which they bind with the carrier plasma proteins.

15.1.5 Classification of Hormones

The source of secretion, chemical nature, physiological action, mechanism of action, and degree of solubility are the few criteria used in classifying different hormones.

1. **Based on their source of secretion:** Classified as pituitary and non-pituitary hormones based on their site of production.
   (a) **Pituitary hormones:** Synthesised and secreted from different lobes of the pituitary gland.
   E.g. Growth hormone (GH), prolactin (PRL), luteinising hormone (LH), follicle-stimulating hormone (FSH), adrenocorticotropic hormone (ACTH), melanocyte stimulating hormone (MSH), anti-diuretic hormone (ADH), & oxytocin.
   (b) **Non-pituitary hormones:** Include hormones produced by other endocrine glands except from the pituitary gland.
   E.g. Somatocrinin (GHRH) from the hypothalamus, insulin from the pancreas, aldosterone from the adrenal cortex, thyroid hormones (T_3/T_4) from the thyroid gland, etc.
2. **Based on their chemical nature:** Depending on the bio-chemical structure, they are categorised as protein, polypeptide, amine, steroid, and fatty acid derivatives in nature.

Table 15.1 List of amine hormones and their site of synthesis

S. No.	Hormone	Source
1.	Dopamine (Prolactin-inhibiting hormone)	Tyrosine
2.	T_3/T_4	Tyrosine
3.	Epinephrine/nor-epinephrine	Tyrosine
4.	Melatonin	Tryptophan

(a) **Protein hormones:** Those hormones which are composed in excess of 50 amino acids are termed as protein hormones. Further, the presence of glycosylated amino acid residues is used to sub-categorise the protein hormones into simple protein and glycoprotein hormones.
   i. **Simple protein hormones:** Comprised of amino acids joined together by peptide bonds. **E.g.** GH, Insulin, PRL, etc.
   ii. **Glycoprotein hormones:** A carbohydrate moiety is conjugated to the protein. **E.g.** LH, FSH, TSH, and chorionic gonadotropins (eCG/hCG)

(b) **Polypeptide hormones:** Composed of less than 50 amino acid residues. **E.g.** Calcitonin, TRH, glucagon, etc.

(c) **Amine hormones:** Hormones derived from decarboxylated amino acids (Table 15.1).

(d) **Steroid hormones:** Comprised of hormones derived from cholesterol. Majorly produced from gonads (e.g. oestrogen, testosterone, etc.) and adrenal cortex (e.g. aldosterone, cortisol, etc.)

(e) **Fatty acid derivatives:** Synthesised from fatty acids. Includes hormones of prostaglandin family (e.g. $PGF_2\alpha$, PGE_2) derived from arachidonic acid.

Know More...

- The smallest peptide hormone is TRH (thyrotropin-releasing hormone) composed of three amino acids (Glutamate-Histidine-Proline).
- The Swedish physiologist Ulf Von Euler discovered prostaglandins in human semen (1935) and thought to be secreted from prostate gland.

3. **Based on their physiological action:** The integration of endocrine and nervous systems is crucial in controlling most basic physiological events such as the growth, reproduction, intermediary metabolism, stress response (fight or flight response, emotional/physical stress, environmental stress), feeding responses, water and electrolyte balance (Fig. 15.3). Hence, the appropriate functioning of the endocrine system is vital for animal survival, optimum reproduction, production, and adaptation to various environmental conditions.
4. **Based on their degree of solubility:**
   (a) **Water-soluble/hydrophilic hormones:** The polypeptide/glycoprotein hormones are soluble in water

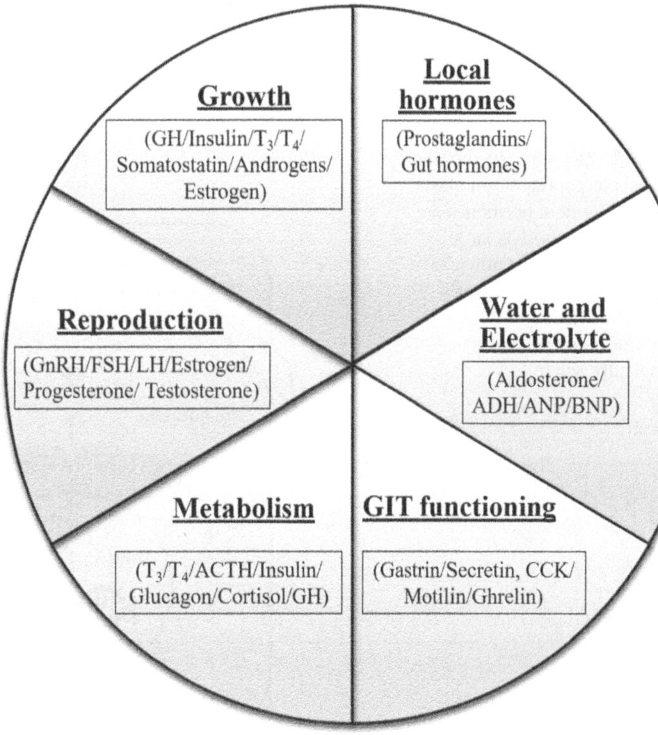

Fig. 15.3 Physiological actions of hormones. [Hormones are implicated in regulating a variety of physiological functions such growth, metabolism, appetite, reproduction, fluid and electrolyte balance. [*GnRH* gonadotropin-releasing hormone; *GH* growth hormone; *LH* luteinising hormone; *FSH* follicle-stimulating hormone; T_3 triiodothyronine; T_4 tetraiodothyronine; *ADH* anti-diuretic hormone; *ANP* atrial natriuretic peptide; *BNP* B-type natriuretic peptide; *CCK* cholecystokinin]

and hence does not require carrier proteins for their transport in blood. In addition, they cannot freely pass through the lipid bi-layer membrane and require membrane-bound receptors or transporter molecules. **E.g.** LH, catecholamines, GH, etc.

(b) **Water-insoluble/lipophilic hormones:** Hormones derived from cholesterol, fatty acids, and thyroid gland are insoluble in water and require carrier proteins for their transport in the systemic circulation. These hormones possess intracellular (cytosolic/nuclear) receptors as they can freely cross the cell membrane. **E.g.** T_3/T_4, testosterone, aldosterone, etc.

15.1.6 Synthesis of Hormones

The altered levels of a circulating hormone or a metabolite (including ions) along with the sensory stimuli are vital factors in regulating the synthesis of different hormones. Additionally, the releasing and inhibiting hormones from the hypothalamus exert potent effects on the synthesis of various hormones. An appropriate stimulus increases the transcription and subsequent translation of the gene encoding a target hormone. However, the general mechanism for the biosynthesis and post-translational modifications varies between different peptide/protein and steroid hormones.

15.1.6.1 Peptide or Protein Hormones

During the translation, the cellular organelle rough endoplasmic reticulum (RER) produces large precursor proteins known as "pre-prohormones" and is regarded as the initial site of synthesis for various peptide and protein hormones. The signal peptide present in a pre-prohormone is cleaved by the signal peptidases of RER to produce prohormones. Followed by the site-specific endopeptidases (also called as prohormone convertases) in the golgi apparatus (GA) finally transform prohormones into mature hormones, which are then stored as cytoplasmic vesicles. Furthermore, the enzymatic cleavage mediated by carboxypeptidase and aminopeptidase along with simultaneous post-translational modifications (Table 15.2) occurring in the secretory vesicles is responsible for conferring the biological activity to a specific hormone. The increased intracellular concentration of calcium (Ca^{+2}) and cAMP produced due to a particular stimulus ensues the fusion of storage vesicles with the cell membrane to release a hormone into the extracellular fluid and the whole process is commonly known as exocytosis (Fig. 15.4).

Table 15.2 Types of post-translational modifications seen in peptide hormones

S. No.	Type of post-translational modification	Hormone
1.	Glycosylation	TSH, FSH,LH
2.	Acetylation	β-Endorphin, α-MSH
3.	Sulfation	CCK, insulin
4.	Amidation	Vasopressin, oxytocin

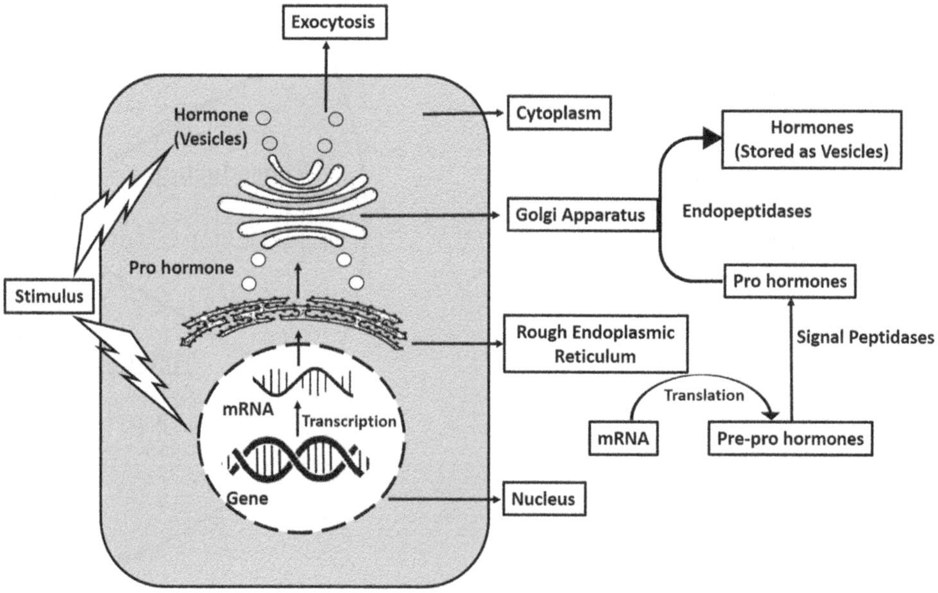

Fig. 15.4 Steps involved in the synthesis of peptide hormones. Peptide or protein hormones are initially synthesised as large precursor molecules known as pre-pro hormones, undergoes proteolytic cleavage to yield mature hormones that are biologically active

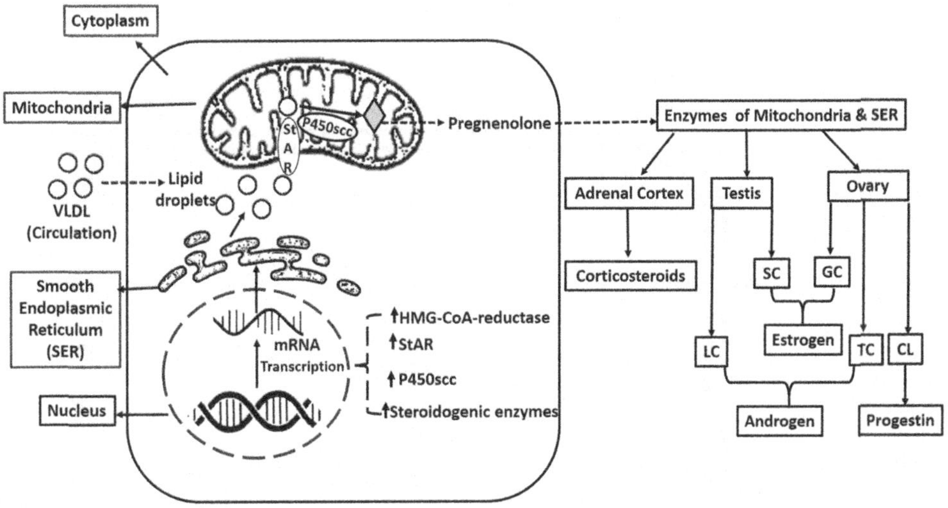

Fig. 15.5 Biosynthesis of steroid hormones. Derived from cholesterol, steroid hormones are manufactured in the mitochondria and smooth endoplasmic reticulum. Conversion of cholesterol to pregnenolone by P450scc is the common step occurring in the steroid biosynthesis; subsequently tissue specific hydroxylases convert pregnenolone in to different types of steroids. [*StAR* steroidogenic acute regulatory protein; *P450scc* cholesterol side chain cleavage enzyme; *SC* Sertoli cell; *CL* corpus luteum; *GC* granulosa cell; *TC* theca cell; *LC* Leydig cell]

Know More...

- The specific post-translational modifications seen in different peptide hormones confer or potentiate their biological activity. In addition, they play a role in determining the duration of action and half-life of peptide hormones.
- More than 50% of the mammalian hormones undergoes amidation in order to exhibit their biological activity.

15.1.6.2 Steroid Hormones

All the steroid hormones are derived from cholesterol that is obtained from the circulation or synthesised de novo from the condensation of acetyl-CoA in the smooth endoplasmic reticulum (SER) with the help of a rate-limiting enzyme HMG-CoA-reductase. Cells that have the ability to secrete steroid hormones have abundant SER and in turn lipid droplets in the cytosol. Most often, tropic peptide hormones stimulate the steroid hormone-producing cells by increasing the uptake, de novo synthesis of cholesterol, and the activation of downstream enzymatic machinery. An appropriate stimulus triggers the steroidogenic acute regulatory protein (StAR) and elicits an acute steroidogenic response marked by a rapid mobilisation of cholesterol into the inner mitochondrial membrane (IMM) from the outer mitochondrial membrane (OMM). The steroidogenesis is initiated by the transformation of cholesterol into pregnenolone by a cytochrome enzyme commonly known as the cholesterol side chain cleavage enzyme (P450scc or CYP11A1). The former step is considered as the rate-limiting step common for all classes of steroid hormones. The subsequent conversion of pregnenolone into different steroid hormones depends entirely on the mitochondrial enzymes and SER possessed by the particular cell (Fig. 15.5).

15.1.7 General Mechanism of Action

Hormones act on the target cells/tissues by binding to its specific receptor to form a HRC. The cellular localisation of receptors varies according to different hormone types. The receptors for various peptide hormones are present on the cell membrane, whereas the receptors for steroid hormones localise in both cytosol and nucleus. Thyroid hormones are unique that they bind to nuclear receptors in order to elicit their biological effects. The HRC initiates the downstream cellular signalling pathways that either directly or indirectly culminate in modulating the cellular metabolic and transcriptional activities.

15.1.7.1 Mechanism of Action of Peptide Hormones

The hydrophilic nature of peptide and protein hormones necessitates the presence of membrane-bound receptors. They act on the target cells by binding to any of the two different types of receptors, i.e. G-protein coupled receptors (GPCRs) and receptor tyrosine kinases (RTK).

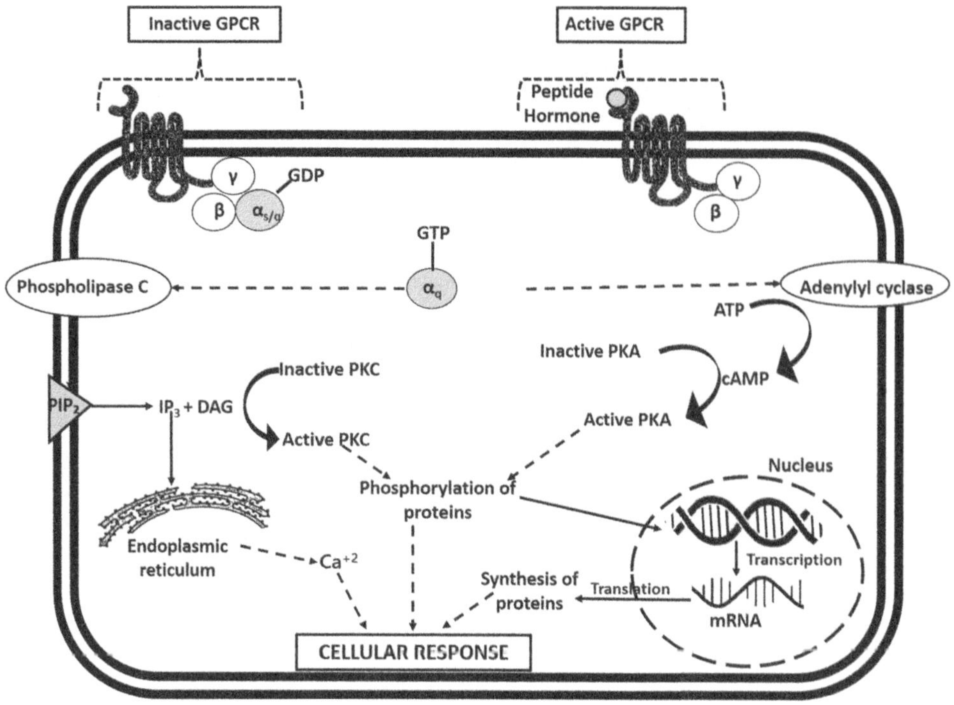

Fig. 15.6 Mechanism of activation of adenylate cyclase and phospholipase C systems by peptide hormones through GPCRs. [Peptide hormones bind to membrane-bound receptors such as GPCRs to exert biological effects on the target cells. G_s and G_q are stimulatory GPCRs, upon their activation generate the production of secondary messengers such as cAMP and Ca^{+2} that initiate the downstream signalling pathways. [*GPCRs* G-protein coupled receptors; *ATP* adenosine triphosphate; *cAMP* cyclic adenosine monophosphate; *PKA* protein kinase A; *PKC* protein kinase C; *PIP_2* phosphatidylinositol 4, 5-bisphosphate; *IP_3* Inositol triphosphate; *DAG* di-acyl glycerol]

15.1.7.1.1 Mechanism of Hormone-Responsive GPCR Signalling

The polypeptide/protein hormones initiate the signal transduction predominantly by binding to the membrane-bound GPCRs. They are proteins consisting of three distinct regions, i.e. extracellular (N-terminus), transmembrane, and intracellular domains (C-terminus). The extracellular domain acts as a receptive site for binding with the hormone (first messenger). The transmembrane segment is mainly composed of a protein with seven membrane-spanning α-helices, responsible for linking and anchoring the other two domains. The intracellular domain, which extends into the cytoplasm, is coupled to a heterotrimeric G-protein with three sub-units, α, β, and γ. In addition, inactive GPCRs have a GDP molecule attached to α sub-unit of the G-protein. The formation of HRC leads to the activation of GPCRs characterised by phosphorylation of GDP molecule to GTP and consequent dissociation of GTP bound α sub-unit ($G_α$-GTP). The $G_α$-GTP then either activates or inhibits the membrane-bound enzymes such as adenylate cyclase (AC), guanylate cyclase (GC), and phospholipase C (PLC) or ion channels. The G-proteins that activate a specific membrane-bound enzyme/ion channel are known as stimulatory G-proteins (G_s, Gq) whereas the opposite is true with inhibitory G-protein (G_i). G_s stimulates the production of the secondary messenger cAMP, wherein Gq activation leads to the production of Di-acyl glycerol (DAG) and inositol triphosphate (IP_3). A surge in the production of secondary messengers leads to the activation of serine/threonine kinases such as protein kinase A (PKA) and protein kinase C (PKC). They modulate the activity of several enzymes that affect cellular metabolism, transcription, and reproduction, thereby producing a well-defined target effect (Fig. 15.6).

15.1.7.1.2 Receptor Tyrosine Kinases (RTK) in Peptide Hormone Signalling

Receptor tyrosine kinases (RTK) are transmembrane receptors that includes receptors with an inherent tyrosine kinase domain (also known as tyrosine kinase receptors) and receptors associated with proteins possessing tyrosine kinase activity (also called as tyrosine kinase-associated receptors).

15.1.7.1.2.1 Tyrosine Kinase Receptors

The tyrosine kinase receptors dimerise upon binding to a hormone with the subsequent activation of the intracellular kinase domain by transphosphorylation. Subsequent to the activation of tyrosine kinase domain, the phosphorylation of tyrosine moieties residing in the receptor's intracellular domain takes place. This ensues the binding of proteins or enzymes that contain SRC homology domain (SH2) resulting in the activation downstream signalling pathways such as Ras-MAPK and Ras/PI3K/AKT to modulate the cellular metabolism, proliferation, and differentiation. Insulin is a classic example for a peptide hormone that acts via tyrosine kinase receptors.

15.1.7.1.2.2 Tyrosine Kinase-Associated Receptors

Tyrosine kinase-associated receptors are receptors coupled with protein tyrosine kinases (PTKs) to complement the lack of an intrinsic protein kinase domain. Peptide hormones such

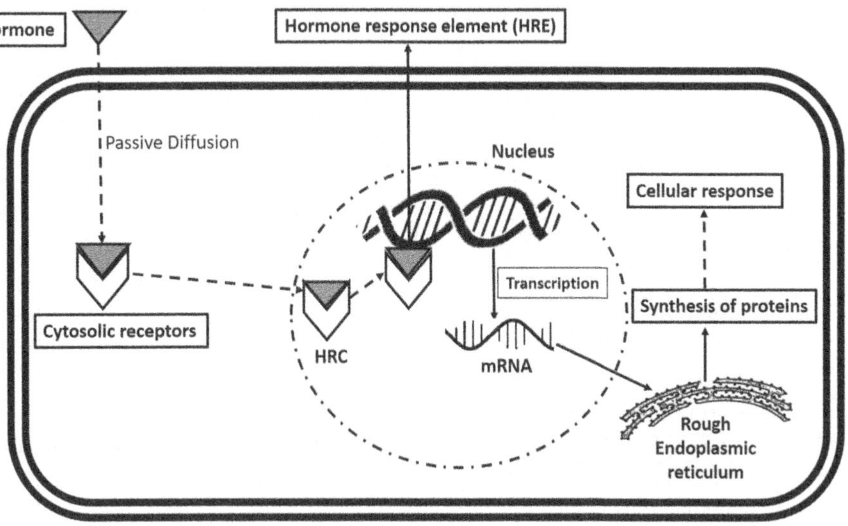

Fig. 15.7 Mechanism of action of steroid hormones [Steroid hormones bind to specific intracellular receptors to form HRC, which subsequently migrates to nucleus and binds to specific regions in the genome known as hormone response element (HRE) to initiate the transcription and translation process in the target cell. [*HRC* hormone-receptor complex]

as GH, PRL, leptin, etc. exert their biological action by binding to these receptors with a majority of PTKs belonging to the Janus kinase (JAK) family and initiate classical JAK-STAT signalling pathway. The dimerization of receptors when bound with a hormone leads to a conformational change in its intracellular domain with the subsequent activation of an associated JAK. The kinase domain of an activated JAK phosphorylates the tyrosine residues present in the receptor's intracellular region. This enables the binding, phosphorylation and dissociation of various signal transducer and activator of transcription (STAT) proteins. The phosphorylated STATs migrate in to nucleus and effect the transcription process by binding to specific gene promoter regions. In addition, an active JAK phosphorylates various cellular kinases such as SRC family kinase (SFK) that activate other signalling pathways to alter the cellular metabolism and functions.

15.1.7.2 Mechanism of Action of Steroid Hormones

The receptors for a majority of steroid hormones reside in the cytoplasm, and few hormones such as oestrogen (E2) have receptors localised in the nucleus. These intracellular receptors are characterised by having three major domains, namely: ligand-binding domain (LBD) at C-terminus, DNA binding domain (DBD), and amino terminal domain (NBD). The LBD serves as binding site for steroid hormones and contains a region known as activation function 2 (AF2). The DBD enables them to bind to specific regions of genome known as hormone response element (HRE) or steroid response element (SRE), thence directly modulating the transcriptional rate of specific genes. The NTD consists of activation function 1 (AF1) region that determines optimum transcriptional activity. The cytosolic steroid receptors are bound with heat shock proteins (HSP) in their inactive state. The removal of HSP along with the phosphorylation of the receptor happens when bound to a steroid hormone, subsequent migration and binding of HRC to SRE/HRE regions in the DNA contribute the classic steroid hormone-signalling pathway. However, instead of directly binding to DNA, AF2 can bind with DNA bound transcription factors such as AP1 or SP1 there by indirectly regulating the cellular transcription (Fig. 15.7).

Know More...

- Although the steroid receptors initially believed to be intracellular, recent studies on the rapid non-genomic effects of steroid hormones suggest their presence on the cell membrane affecting various cellular signalling pathways. Membrane-bound receptors for oestrogen, androgen, glucocorticoid, and mineralocorticoid play a role in activating kinases, Ca+2 influx, and G-protein activation.
- The cytosolic steroid receptors can also be stimulated in the absence of hormones on phosphorylation due to the activation of intracellular kinases, a phenomenon that is termed as "ligand independent activation".

15.2 Hypothalamus as an Endocrine Gland

15.2.1 Introduction

The higher centres for regulating body temperature, appetite, sexual behaviour, sleep-wake cycle, and emotional states reside in the hypothalamus. In addition, it secretes various hormones with a wide range of afferent neural pathways, thereby playing a pivotal role in integrating nervous and endocrine systems. These neurohormones further regulate the endocrine activities of the pituitary gland.

Table 15.3 Different hypothalamic nuclei and their secretory hormones

S. No	Hypothalamic nuclei	Hormone(s)
1.	Para-ventricular nucleus (PVN)	CRH,GHIH,TRH,ADH
2.	Pre-optic area (POA)	GnRH
3.	Dorsomedial nucleus (DMN)	GnIH
4.	Arcuate nucleus (ACN)	GHRH, DA (PIH)
5.	Supra-optic nucleus (SON)	OT

15.2.2 Mechanism of Secretion

The hypothalamic hormones are synthesised by specialised neuroendocrine cells that are organised as distinct nuclei (Table 15.3). Magnocellular (MC) and parvocellular (PC) neurons are the two different categories of neuronal cells in hypothalamus that confer its neuroendocrine activity. The nerve terminals of MC neurons end in a specialised area known as median eminence. The parvocellular neurons are characterised by long axons that traverse through pituitary stalk reaching the posterior pituitary lobe, they are responsible for secreting anti-diuretic hormone (ADH) and oxytocin (OT). Hence, ADH and OT are commonly categorised as posterior pituitary hormones.

Know More...

- Although oxytocin is mainly synthesised in SON, it is also produced in little quantities from PVN. In the same way, SON acts as a minor source of ADH.
- Popa and fielding were the first to describe about "Hypophyseal portal system".

15.2.3 Hypothalamic-Hypophyseal Portal (HHP) System

The median eminence is the site of origin for "hypothalamic-hypophyseal portal system". The portal system consists of a hypophyseal artery derived primary capillary bed, which embeds median eminence, and a secondary capillary bed formed from the long portal veins in the anterior pituitary lobe. This acts as a functional connecting circuit between the hypothalamus and anterior lobe of pituitary (Fig. 15.8).

15.2.4 Hypothalamic Hormones

The hypothalamic hormones released by MC nerve terminals located in the median eminence are carried by the HHP system to act on their target cells in the pituitary gland. With the pituitary gland as their only target organ, they act on different neuronal cell types to either stimulate or inhibit its endocrine activity. Therefore, the hypothalamic hormones are broadly categorised as releasing and inhibiting hormones based on their effect on pituitary gland.

15.2.4.1 Mechanism of Action

They act by binding to their respective GPCRs, result in either stimulation or inhibition of adenylyl cyclase (AC) or phospholipase C (PLC) systems. This results in affecting the levels secondary messengers, initiating various downstream signalling pathways to either produce or inhibit synthesis and secretion of pituitary hormones (Table 15.4).

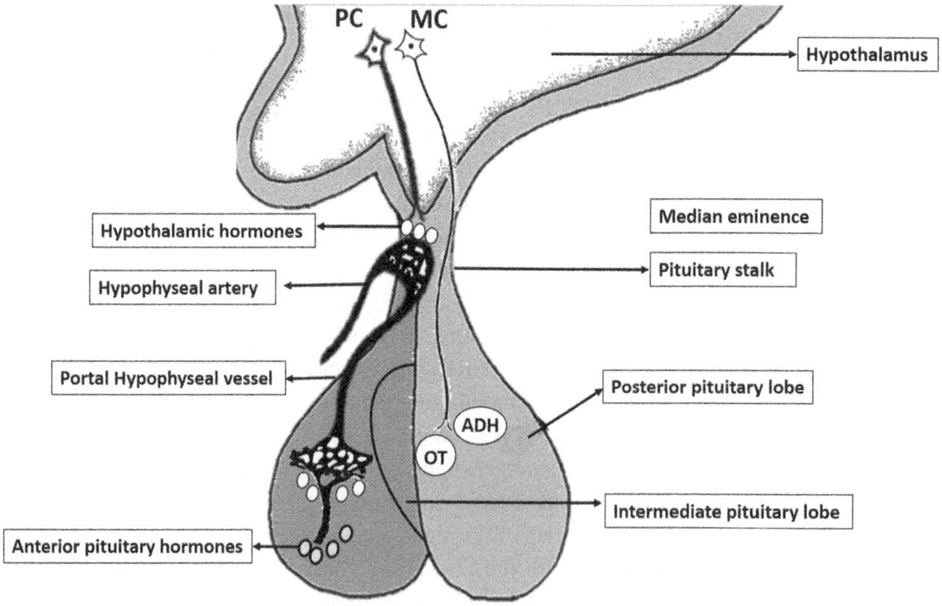

Fig. 15.8 Hypothalamic-hypophyseal portal system. [Hypothalamic hormones released in median eminence enter the hypophyseal portal system and stimulate the neuroendocrine cells in the anterior pituitary to release tropic hormones into the systemic circulation. [*MC* magnocellular neuron; *PC* parvocellular neurons; *OT* oxytocin; *ADH* anti-diuretic hormone]

Table 15.4 List of hypothalamic hormones, their respective chemical structure, mechanism of action and function [↑ = increase, ↓ = decrease]

S. No	Hormone	Chemical structure	Target cell	Mechanism of action	Target effect
1.	Thyrotropin-releasing hormone (TRH)	Peptide (3 a.a)	Thyrotropes	Stimulate phospholipase C system (↑DAG & IP_3)	Stimulate TSH secretion
2.	Corticotropin-releasing hormone (CRH)	Peptide (41 a.a)	Corticotropes	Stimulate adenylyl cyclase system (↑cAMP)	Stimulate ACTH secretion
3.	Gonadotropin-releasing hormone (GnRH)	Peptide (10 a.a)	Gonadotropes	Stimulate phospholipase C system (↑DAG & IP_3)	Stimulate gonadotropins (LH &FSH) secretion
4.	Growth hormone-releasing hormone (GHRH/Somatocrinin)	Peptide (44 a.a)	Somatotropes	Stimulate phospholipase C system (↑DAG & IP_3)	Stimulate GH secretion
5.	Gonadotropin-inhibiting hormone (GnIH)	Peptide (12 a.a)	Gonadotropes	Inhibit adenylyl cyclase system (↓cAMP)	Inhibit Gonadotropin secretion
6.	Growth hormone inhibitory hormone (GHIH/Somatostatin)	Peptide (14 a.a)	Somatotropes	Inhibit adenylyl cyclase system (↓cAMP)	Inhibit GH secretion
7.	Prolactin inhibiting hormone (PIH/Dopamine)	Tyrosine derivative	Lactotropes	Inhibit adenylyl cyclase system (↓cAMP)	Inhibit PRL secretion

15.2.4.2 Regulation of Secretion

Pituitary tropic hormones and their target hormones regulate the secretion of hypothalamic hormones. From systemic circulation, they can enter the brain via fenestrated capillaries found in circum-ventricular organs or choroid plexus into cerebrospinal fluid (CSF). These factors then act directly or indirectly on the neuroendocrine cells of hypothalamus and modulate their secretions.

Know More...

- The hypophyseal portal system prevents the entry of hypothalamic hormones into the systemic circulation. It helps in rapid delivery of hormones to the pituitary gland without being diluted. Hence, their concentration cannot be measured in the general circulation.
- Kisspeptin, a neuropeptide that acts on GnRH producing neurons plays major role in the onset of puberty and modulating hypothalamo–pituitary–gonadal (HPG) axis. Currently, it is being used to advance puberty and to develop novel oestrus synchronisation protocols in animals.
- GnIH was first identified in Japanese quail.

15.3 Pituitary Gland

15.3.1 Introduction

Widely regarded as a "master gland", the pituitary gland originates from the ectoderm and functions to secrete hormones that are regulated by hypothalamic stimuli. It comprises three distinct anatomical and functional lobes known as anterior pituitary (adenohypophysis or pars distalis), intermediate (pars intermedia), and posterior pituitary (neurohypophysis or pars nervosa). The pituitary hormones influence secretory activities of different target organs and are hence known as tropic hormones. With close interdependency, the hypothalamus and pituitary act in concert to regulate homeostatic mechanisms that drive growth, metabolism, and reproduction.

15.3.2 Anterior Pituitary

The anterior lobe comprises two-thirds of the pituitary gland and consists of neuroendocrine cells responsible for the synthesis of major tropic hormones.

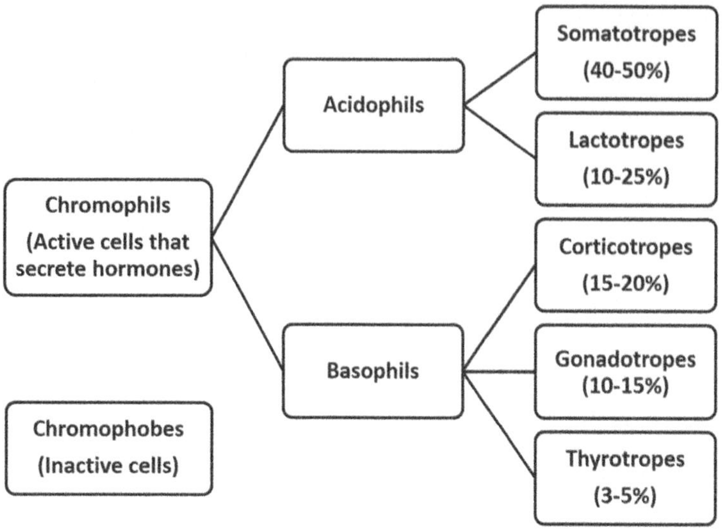

Fig. 15.9 Classification of cell types present in the anterior pituitary gland. [Chromophils are the active neuroendocrine cells present in the anterior pituitary. Further, they are categorised into acidophils and basophils based on their affinity for a particular type of stain]

Table 15.5 Different cell types in anterior pituitary, their distribution, hormones, and target organs

S. No.	Cell type	Population (%)	Hormone produced	Target organ
1.	Somatotropes	40–50	Growth hormone (GH/Somatotropin)	Diffused action in body
2.	Lactotropes	10–25	Prolactin (PRL)	Mammals: Mammary gland Birds: Crop milk, plumage maternal behaviour
3.	Corticotropes	15–20	Adrenocorticotropic hormone (ACTH)	Adrenal cortex
4.	Gonadotropes	10–15	Follicle-stimulating hormone (FSH) Luteinising hormone (LH)	Gonads
5.	Thyrotropes	3–5	Thyroid stimulating hormone (TSH)	Thyroid gland

15.3.2.1 Cellular Types and Their Hormones

Based on the ability to take up general histological stains, cells in adenohypophysis are classified broadly into two categories, namely the chromophils and chromophobes (Fig. 15.9). Depending on their ability to take up acidic or basic stains, the chromophils are further categorised as acidophils or basophils. They are active neuroendocrine cells characterised by the presence of stainable cytoplasmic secretory granules and are responsible for the secretion of tropic hormones (Table 15.5). In contrast, chromophobes comprised of inactive reserve cells, undifferentiated stem cells, and degenerated chromophils.

15.3.2.2 Growth Hormone (GH)

Also referred to as Somatotropin, derived from the Greek words "Soma" meaning body and "tropikos" refer to turn or change. It is implicated in both pre-natal and post-natal animal growth that varies according to each physiological states such as pre-pubertal phase, pubertal phase, post-pubertal, and senescence. Regulated by genetics, it plays a prime role in attributing the phenotypic characteristics to different species in the animal kingdom.

15.3.2.2.1 Chemical Structure and Mechanism of Action

GH is a single-chain protein hormone synthesised and secreted by the somatotropes as a prohormone with 217 amino acids, while the mature hormone consists of 191 amino acids with two intramolecular di-sulphide bridges (Molecular weight: 22 KDa). They bind to specific membrane-bound growth hormone receptors (GHRs) which belong to the family of tyrosine kinase-associated receptors. The activation of GHRs initiates signal transduction mechanism primarily by JAK-STAT pathway with a subsequent effect on cellular genetic machinery. Furthermore, it also activates other signalling pathways such as Ras/MAPK and PI3K. Together, they regulate cell cycle, proliferation, gene expression, growth, and differentiation in various organs.

15.3.2.2.2 Biological Effects

Mainly considered as a hormone that regulates the body metabolism to suit different physiological states. However, it affect several organs including liver, bones, skeletal muscle, adipose tissue, and gonads (Fig. 15.10).

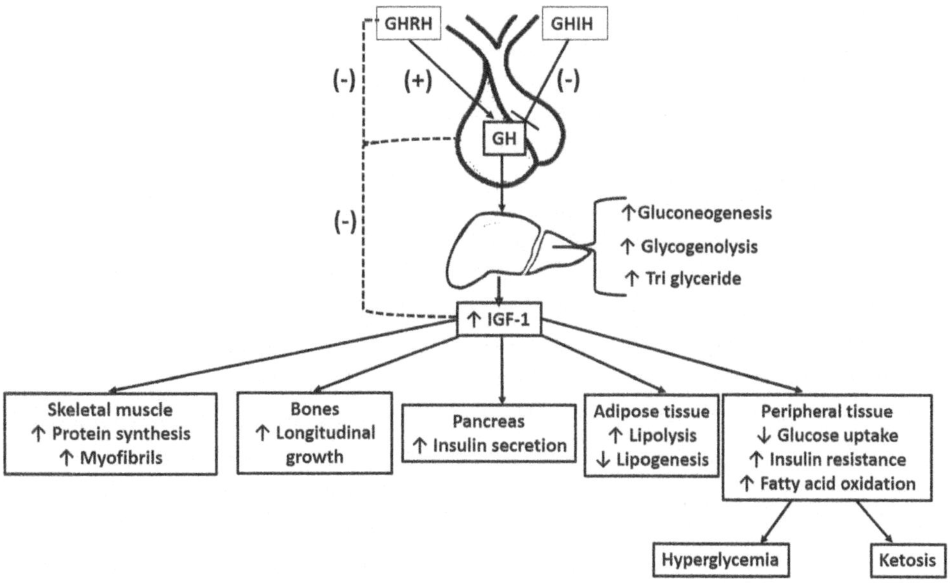

Fig. 15.10 Biological effects of growth hormone [Somatotropin released from the adenohypophysis results in the growth of an animal by stimulating hyperglycaemia, protein anabolism, growth of bones, and lipolysis. Majority of the biological effects of GH are mediated by the insulin-like growth factor produced by the liver, which exerts a negative feedback inhibition on the secretion of both GHRH and GH. [*GHRH* growth hormone-releasing hormone; *GHIH* growth hormone inhibiting hormone; *GH* growth hormone; *IGF-1* insulin-like growth factor 1; (+) stimulate; (−) inhibit; (↑) increase; (↓) decrease]

15.3.2.2.2.1 Effects of GH on Liver

Stimulating the hepatocytes to produce somatomedins (via JAK-STAT pathway) remains the most crucial target effect of GH. Somatomedins or insulin-like growth factors (IGFs) are regarded as the vital extra-hepatic mediators of GH actions. Somatomedin-A (IGF-2) and Somatomedin-C (IGF-1) are the two types of somatomedins implicated in pre-natal and post-natal growth, respectively. They mediate the effects of metabolic and functional changes that are observed in target organs. Moreover, GH stimulates hepatic glucose production by increasing the rate of gluconeogenesis and glycogenolysis. The production of transcription factors that belong to sterol regulatory element binding proteins (SREBs) stimulate the synthesis of lipid, sterols and their oxidation. Along with their increased synthesis, it also leads to the increased secretion of tri-glycerides. Furthermore, the regenerative capacity of liver is believed to be dependent on GH.

15.3.2.2.2.2 Effects of GH on Carbohydrate Metabolism

Growth hormone stimulates the production of glucose from liver with a concurrent decrease in its utilisation by skeletal muscle and adipose tissues. The decreased peripheral utilisation of glucose is attributed to the attenuation of insulin effects such as increased glucose uptake, utilisation and decreased gluconeogenesis, producing GH-induced insulin resistance. This leads to increased circulatory levels of glucose and commonly known as diabetogenic actions of GH. In addition, GH act on β-cells of pancreas to stimulate the synthesis and secretion of insulin to offset the insulin resistance.

15.3.2.2.2.3 Effects of GH on Protein Metabolism

Growth hormone stimulates the rate of transcription, translation with a concomitant rise in the uptake of amino acids. Moreover, reduced cellular dependence on gluconeogenesis for meeting energy demands is also prominent with GH stimulation. Overall, the above listed cellular processes favour the protein anabolism and supress the amino acid catabolism resulting in the accumulation of proteins in various cells especially the skeletal muscles.

15.3.2.2.2.4 Effects of GH on Fat Metabolism

The GH stimulation of hepatocytes and adipocytes leads to increased circulatory levels of free-fatty acids (FFA), TG, and cholesterol. This helps the target cells to utilise fats to meet their energy requirements while sparing the carbohydrates and amino acids.

15.3.2.2.2.5 Effects of GH on Skeletal Muscle

Growth hormones stimulate the protein accumulation, subsequently used for synthesising myofibrils and collagen leading to hypertrophy of skeletal muscles. The GH-IGF1 axis dependent skeletal muscle growth and development plays a crucial role in post-natal growth of animals.

15.3.2.2.2.6 Effects of GH on Bones

The IGF-1 produced from liver or locally produced is a chief regulator of the longitudinal growth of bones in young animals. The GH-IGF1 axis stimulates the rate of chondrogenesis in growth plate leading to the formation of new cartilaginous tissue. This increased deposition of cartilaginous tissue between the shaft and epiphysis results in longitudinal bone growth. The osteoblasts that are stimulated by GH-IGF1 axis further help in the ossification of newly laid cartilage. The rate of GH stimulated chondrogenesis and consequent longitudinal bone growth is maximum at pubertal and peri-pubertal periods. Whereas, GH stimulation during the post-pubertal phase results in thickening of bones due to lateral ossification.

15.3.2.2.2.7 Effects of GH on Cardiovascular System

The GH-IGF1 dependent of amino acids uptake, specific gene transcription, deposition of proteins, and collagen results in the hypertrophy of cardiomyocytes. Along with hypertrophy, inhibition of apoptosis in cardiomyocytes contributes to an increase in cardiac mass. Furthermore, GH stimulates the expansion of vascular system by increased expression of angiogenic factors that result in the endothelial cell proliferation and tube formation. Hence, both increased cardiac mass and expansion of vascular system help in meeting the circulatory requirements for rapid growth and development of an animal.

15.3.2.2.2.8 Effects of GH on Gonads

The presence of a blood–testis barrier limits the access of GH to testicular cells. The positive regulation of GH on testicular growth, development, steroidogenesis, and gametogenesis is thought to be dependent on IGF-1,although the exact underlying mechanisms are largely unknown. In addition, it plays an important role in the maintenance, development of ovarian follicles and promotes the initial steps of steroidogenesis in corpus luteum.

15.3.2.2.2.9 Effects of GH on Adipose Tissue

Growth hormone causes lipolysis by activating various classes of lipases like hormone-sensitive lipase (HSL), adipose triglyceride lipase (ATGL), and monoacylglycerol lipase (MGL). The increased lipolysis is necessary to shift the cellular metabolism to lipids. The rise in oxidation of free-fatty acids (FFA) leads to an increased production of ketone bodies, which is termed as ketogenic effect of GH. GH-induced insulin resistance leads to a decrease in the insulin dependent glucose uptake by adipocytes. Therefore, the increased mobilisation of lipids with a concurrent inhibition of lipogenesis in the adipose tissue results in the depletion of fat reserves.

15.3.2.2.2.10 Regulation of Secretion

The hypothalamic hormones GHRH and GHIH regulate the pulsatile secretion of GH. In addition, the increased circulatory levels of IGF-1 strongly inhibit the GH secretion. Along with them exercise, starvation, deep sleep, and stress act as potential stimulators for GH secretion.

Know More...

- **Gigantism:** A pathological condition due to the excess secretion of GH in pre-pubertal life, characterised by abnormal longitudinal growth of long bones.
- **Dwarfism:** The abnormal stunting of longitudinal growth due to the deficiency of GH.
- **Acromegaly:** The pathological condition characterised by excess thickening of bones (bone deposition) due to hypersecretion of GH in post-pubertal life.
- **Somatopause:** The gradual decrease in the secretion of GH and IGFs due to ageing is referred to as somatopause.

15.3.2.3 Prolactin (PRL)

The lactation promoting effect of administering bovine pituitary gland extract in rabbits has led to an inception of the term "Prolactin". Primarily known as a hormone for lactation, it regulates diverse functions in animals including their behavioural patterns.

15.3.2.3.1 Chemical Structure and Mechanism of Action

Prolactin is a single-chain polypeptide hormone with 199 amino acids and three intramolecular disulphide bonds, synthesised from a prohormone consisting of 229 amino acids after the removal of signal sequence (1–30 amino acids). Transmembrane prolactin receptors (PRLR) belong to the family of tyrosine kinase-associated receptors and initiate signal transduction up on binding to PRL. The classical JAK-STAT and Ras/Raf/MAPK pathways are considered as the chief signalling mechanisms that impart the biological effects. In addition, the activation of other kinases such as c-src and Fyn leads to the stimulation of intracellular mechanisms.

15.3.2.3.2 Biological Effects

Regulating development of mammary gland, maternal behaviour, initiation, and maintenance of lactation are the major biological effects of PRL. In addition, it acts as a luteotrophic factor in rodents and sheep. However, it is majorly concerned with regulating plumage and crop-milk secretion in birds.

15.3.2.3.2.1 Effects on Mammary Gland Development and Lactation

Prolactin drives the lobuloalveolar development during gestation and plays a considerable role in mammogenesis. It stimulates transcription and translation of casein gene associated with increase in amino acids uptake in alveolar cells. It triggers the synthesis of α-lactalbumin, a regulatory sub-unit for the lactose synthase system. The positive regulation α-lactalbumin activates the lactose synthase and triggers lactose synthesis. Therefore, PRL-dependent initiation of lactose and casein synthesis is responsible for lactogenesis and galactopoiesis.

15.3.2.3.2.2 Effects on Animal's Behaviour

The increased levels of PRL during the peri-parturient period impart maternal behaviour including nest-building behaviour, nursing, and cleaning of young ones. After parturition, the rise in PRL levels leads to an inhibition on GnRH secreting neurons with a subsequent delay in the onset of oestrous cycle. The rise in oestrogen levels during oestrous cycle stimulates the release of PRL and thought to have a positive effect on sexual receptivity in female animals.

15.3.2.3.2.3 Effects of PRL in Birds

It stimulates the secretion of crop milk for nourishing young ones. In addition, it has an effect on plumage pattern, broodiness in hens.

15.3.2.3.2.4 Regulation of Secretion

The activation of dopaminergic neuroendocrine cells due to a rise in PRL levels serves as the basic negative regulatory mechanism for its secretion. However, vasoactive intestinal polypeptide (VIP) positively regulates PRL secretion in birds. Further, regular milking has a positive effect on its secretion.

15.3.2.4 Adrenocorticotropic Hormone (ACTH)

ACTH targets different adrenocortical zones to stimulate the production of three different classes of steroid hormones collectively known as corticosteroids. Primarily, it is important for the secretion of glucocorticoids to alleviate the harmful effects of various kinds of stressors in animals.

15.3.2.4.1 Chemical Structure and Mechanism of Action

The release of hypothalamic CRH stimulates corticotropes in anterior lobe and neuroendocrine cells in intermediate lobe to synthesise a large precursor known as pro-opiomelanocortin (POMC). The POMC is cleaved in both anterior and intermediate lobes of pituitary, generating a single-chain polypeptide hormone of 39 amino acids length known as ACTH. ACTH binds to melanocortin 2-receptor (MC2-R), a membrane localised GPCR localised in adrenal cortical cells. It results in the production of cAMP, activating protein kinase A (PKA) with subsequent initiation of downstream cellular pathways.

15.3.2.4.2 Effect on Adrenal Cortex

The activation of PKA leads to generating an acute steroidogenic response characterised by increased de novo cholesterol synthesis by activating hormone-sensitive lipase (HSL), StAR, and CYP11A1 (p450scc). Their co-activation leads to increased conversion of cholesterol to pregnenolone, a rate-limiting step in the synthesis of corticosteroids. Additionally, the activation of zone-specific hydroxylases results in the production of distinct classes of corticosteroids.

15.3.2.4.3 Regulation of Secretion

The CRH has a positive effect on ACTH levels whereas it is negatively regulated by an increased level of corticosteroids especially cortisol.

15.3.2.5 Thyroid Stimulating Hormone (TSH)

With thyroid gland as the target organ, TSH plays a key role in stimulating it to produce thyroid hormones (T_3/T_4) and in regulating metabolism to cater the needs of animals during various physiological states.

15.3.2.5.1 Chemical Structure and Mechanism of Action

TSH is a heterodimeric glycoprotein with α and β chains, composed of 92 and 112 amino acids, respectively. It binds to specific GPCRs known as TSH receptors (TSHR) that are present on the cell membrane of thyroid follicular cells. The signal transduction involves the activation of adenylyl cyclase resulting in elevated cAMP levels with the subsequent triggering of PKA and downstream signalling pathways. They mainly effect the rate of transcription and translation of genes that are linked to the production of T3/T4.

15.3.2.5.2 Effect on Thyroid Gland

It positively affects all the steps involved in synthesis and secretion of T3/T4 from the thyroid gland. Briefly, it increases the synthesis of sodium-iodide symporter protein (NIS), thyroid peroxidase, and thyroglobulin. It also increases the endocytosis of stored colloid followed by iodination to produce and secrete thyroid hormones.

15.3.2.5.3 Regulation of Secretion

The TRH and GH are two principal stimulators for the secretion of TSH. Whereas, rise in T3/T4 circulatory levels inhibit its release.

15.3.2.6 Pituitary Gonadotropins: LH and FSH

Pituitary gonadotropins stimulate gonadal steroidogenesis and gametogenesis in post-pubertal animals. Therefore, normal secretion of gonadotropins helps in maintaining optimal reproduction, which is very essential in farm animals.

15.3.2.6.1 Chemical Structure and Mechanism of Action

Gonadotropins are heterodimeric glycoprotein hormones with a common α sub-unit (92 amino acids) and distinct β sub-units that determine their specific biological activity. The β sub-units of LH and FSH consist of 121 and 109 amino acids. The glycosylation of asparagine residues in α and β sub-units plays a vital role in determining specific biological effects and half-life of gonadotropins in circulation. Gonadotropins bind to their respective membrane-bound GPCRs, i.e. luteinising hormone receptor and follicle-stimulating hormone receptor (LHR and FSHR) to initiate signal transduction by activating adenylyl cyclase enzyme and followed by a rise in cAMP production. The increased cAMP levels stimulate PKA, thereby initiating downstream signalling pathways that stimulate cholesterol synthesis, cholesterol side chain cleavage by CYP11A1, StAR, and various hydroxylases. In addition, they also activate MAPK and AKT pathways that help in regulating cell cycle, proliferation, and apoptosis.

15.3.2.6.2 Biological Effects of LH

In female animals, LH acts on thecal cells of the growing follicles to stimulate synthesis and secretion of testosterone that later gets converted into oestrogen. The granulosa cells develop LHRs during the late phase of follicular growth and when stimulated with LH are responsible for developing critical changes that ensues ovulation. Moreover, LH essentially causes the transition of granulosa cells into luteal cells in most of the mammals and hence known as the chief luteotrophic factor. However, LH acts on the Leydig cells (hence it is also referred to as interstitial cell stimulating hormone (ICSH)) stimulating the production of testosterone, which is a principal regulator of male fertility. Therefore, the LH-dependent production of oestrogen, testosterone, and progesterone regulates key processes such as oestrous cycle, libido, and gestation.

15.3.2.6.3 Biological Effects of FSH

FSH stimulates the granulosa cells in ovarian follicles to produce oestrogen. The testosterone produced by theca cells is converted by FSH activated aromatase enzyme in granulosa cells. The dominant follicle that shows high sensitivity to FSH becomes the graafian follicle. In males, FSH acts on Sertoli cells to regulate their proliferation and differentiation. It also helps in sequestering testosterone in seminiferous tubules by producing androgen-binding protein (ABP) from Sertoli cells. Further, FSH dependent conversion of testosterone to estrogen in Sertoli cells is obligatory for the spermatogenesis and maintaining libido in male animals (Fig. 15.11).

15.3.2.6.4 Regulation of Secretion

GnRH is a chief stimulator for the secretion of pituitary gonadotropins. The rise in oestrogen levels has a negative effect on FSH and a positive effect on LH surge. Activin and inhibin are large proteins produced in gonads; respectively, they exert positive and negative regulation on the secretion of FSH. Higher levels of testosterone and progesterone impose a negative effect on the secretion of LH in male and females correspondingly.

Know More...

- All the glycoprotein hormones (TSH, LH, and FSH) of pituitary origin and placenta (eCG/hCG) are composed of a common α sub-unit composing 92 amino acids. Thus, their respective biological effects reside in the β sub-units.

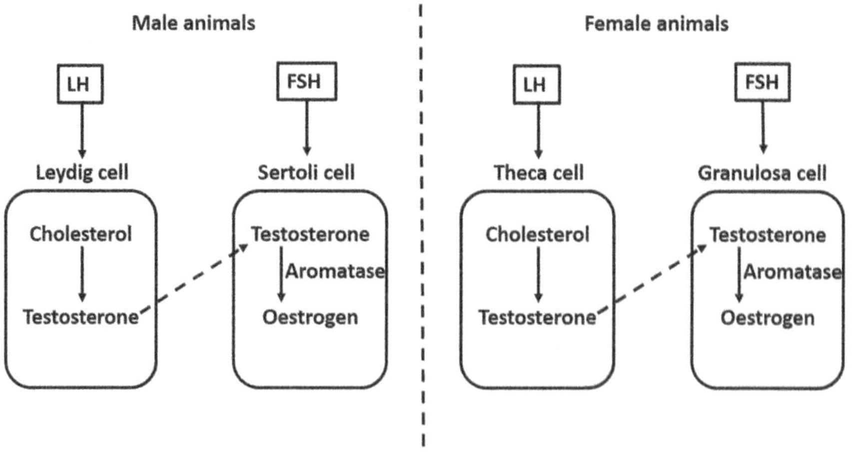

Fig. 15.11 Biological effects of gonadotropins on different somatic cells present in male and female animals. [Pituitary gonadotropins stimulate the production of gonadal steroids, i.e. testosterone and oestrogen in male and female animals, respectively. [*LH* luteinising hormone; *FSH* follicle-stimulating hormone]

15.3.3 Posterior Pituitary

Derived from neural ectoderm, posterior pituitary is composed of axons arising from magnocellular (MC) neurons present in PVN and SON of hypothalamus. Up on appropriate stimulus, these nerve terminals in the posterior pituitary secrete oxytocin (OT) and anti-diuretic hormone (ADH). In addition, glial cells present in the neurohypophysis commonly referred to as pituicytes.

15.3.3.1 Chemical Structure of Oxytocin and ADH

Both oxytocin and ADH are nonapeptides with minor differences in their amino acid composition. They are synthesised in MC neurons along with their specific precursor transport protein known as neurophysin and stored as secretory granules. However, during the axonal transport, the transport protein moiety is cleaved to produce the active hormone. The action potential propagated in response to an afferent neural stimulus on MC neurons results in the release of hormones from their nerve terminals.

15.3.3.2 Oxytocin

Widely known for its role in parturition, oxytocin also has an effect on milk ejection, sperm transport, social bonding, and ovulation in animals.

15.3.3.2.1 Mechanism of Action

It binds to specific GPCRs known as oxytocin receptors (OTR), and then activate PLC system to produce the secondary messengers DAG and IP_3. The secondary messengers in turn activates PKC and stimulates the release of Ca^{+2} from endoplasmic reticulum. The increased Ca^{+2} levels activate myosin-light chain kinase (MLCK), whereas activated PKC inhibit myosin-light chain phosphatase (MLCP) resulting in an increased formation of actin-myosin bridges. This leads to the initiation of smooth muscular contraction process.

15.3.3.2.2 Biological Effects

15.3.3.2.2.1 Effects on Uterus

The afferent neural stimuli due to the entry of foetus in to cervical region leads to the release of oxytocin, and this particular neuroendocrine reflex is known as Ferguson's reflex. This stimulates smooth muscle cells present in myometrium to initiate uterine contractions, which leads in the expulsion of foetus. Hence, it is also known as the "birth hormone". It is released while/after mating in male and female animals, aiding in the transport of spermatozoa in male and female reproductive tracts.

15.3.3.2.2.2 Effects on Mammary Gland

The contraction of myoepithelial cells around the alveoli leads to the ejection of milk from the mammary gland. In dairy animals, afferent neural stimuli due to tactile stimulation of udder, visual, or auditory stimuli leads to the secretion of oxytocin with subsequent ejection of milk.

15.3.3.2.2.3 Effects on Ovulation

Oxytocin stimulates the synthesis of PGF2α, which has a significant role in rupture of follicular membrane resulting in ovulation.

15.3.3.3 Anti-diuretic Hormone

Responsible for conserving body water during periods of extreme dehydration, ADH opposes water excretion (anti-diuresis) through urine and prevents increase in ECF osmolality.

15.3.3.3.1 Mechanism of Action

Anti-diuretic hormone binds to transmembrane GPCRs known as vasopressin receptors (VR). Three different types of vasopressin receptors are found, namely V_1, V_2, and V_3. The signal transduction pathway and effects of ADH depend on the type of receptors present on the target tissue.

15.3.3.3.2 Biological Effects

15.3.3.3.2.1 Effects on Kidney

It binds to V_2 receptors on tubular epithelial cells present in the distal collecting tubules and collecting ducts. This leads to the production of cAMP levels by activating the adenylyl cyclase enzyme, thereby initiating transcription, translation of the aquaporin gene. Consequent incorporation of aquaporins on tubular epithelial cells leads to the reabsorption of water resulting in the classic anti-diuretic effect of ADH.

15.3.3.3.2.2 Effects on the Vascular System

ADH binds to V_1 receptors present on the arteriolar smooth muscle cells to activate PLC resulting in the production of DAG and IP_3. The further activation of various kinases and rise in intracellular Ca^{+2} levels lead to the smooth muscle contraction. Due to its potent vasoconstrictor ability, ADH is also known as vasopressin. This particular vasoconstriction in reducing the GFR to conserve water and maintaining systemic blood pressure during periods of acute water loss or prolonged dehydration.

15.3.3.3.3 Regulation of Secretion

Residing in the supra-optic nucleus (SON), cells that detect osmolality of ECF are known as the osmoreceptors. An increase in the osmolality results in the shrinking of osmoreceptors and stimulates the secretion of ADH. While a decrease in osmolality of ECF has an opposite effect on its secretion. In addition to the osmoreceptors, the stretch receptors in heart stimulate the secretion of ADH in response to a 5–10% reduction of blood volume (Fig. 15.12).

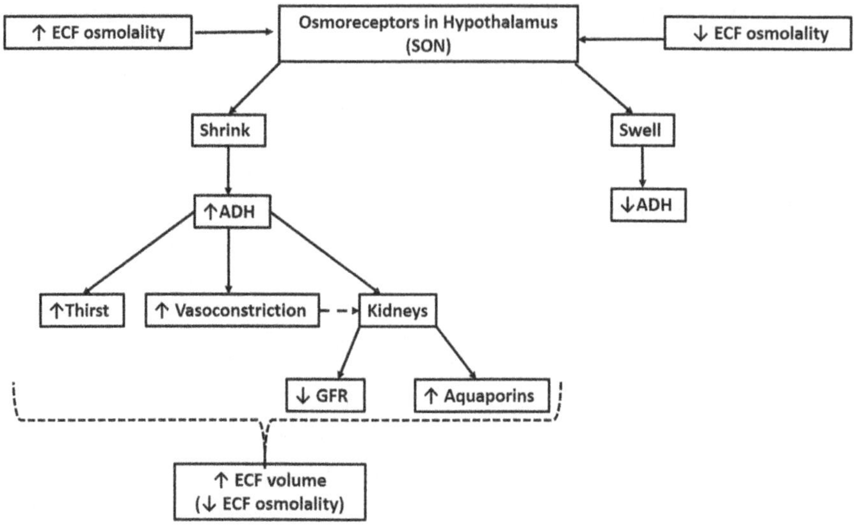

Fig. 15.12 Mechanism of secretion of ADH and its actions. [Osmoreceptors in the hypothalamus responds to the hyperosmolality of ECF by secreting ADH, which drives the thirst behaviour and conservation of body water by decreased excretion in animals. [*ADH* anti-diuretic hormone; *ECF* extracellular fluid; *SON* supra-optic nucleus; *GFR* glomerular filtration rate]

Know More...

- Neurophysin-I is the transport protein for the oxytocin, whereas the neurophysin-II is required for the axonal transport of ADH.
- All mammals have arginine vasopressin except in pigs, where lysine vasopressin is found.
- **Vasotocin:** Found in birds, a nonapeptide hormone possess both the biological activities of oxytocin and ADH.

15.4 Pineal Gland

Commonly referred as "biological clock", pineal gland is responsible for circadian rhythm in mammals by secreting melatonin. It is composed of glial cells and neurons known as pinealocytes.

15.4.1 Chemical Structure

Melatonin is derived from an indole ring containing amino acid tryptophan. The hydroxylation of tryptophan produces 5-hydroxytryptophan, followed by carboxylation reaction to yield serotonin (5-hydroxytryptamine). The conversion of serotonin to N-acetyl serotonin (NAS) is catalysed by the enzyme aralkylamine N-acetyltransferase (AANAT), considered as a rate-limiting step in melatonin synthesis. Finally, melatonin will be produced from N-acetyl serotonin by the action N-acetyl serotonin methyl transferase (ASMT).

15.4.2 Regulation of Secretion

Although pineal gland is widely credited as a prime factor in regulating circadian rhythm in mammals, the cellular and molecular mechanisms regulate it reside in the suprachiasmatic nucleus (SCN) of hypothalamus. The afferent sympathetic innervation to pineal gland is activated by the SCN during darkness. The adrenergic stimulation leads to synthesis of melatonin in pineal gland by activating the transcription and translation of AANAT and ASMT genes. In addition, the rhythmic expression of a certain group of genes known as clock genes (Per1-3, Cry1&2, Bmal1, and clock) especially Per1 stimulates the synthesis of melatonin. The increased synthesis of melatonin corresponds to its high levels observed during night hours. Melatonin acts on the hypothalamus and other peripheral tissues (especially ovary) to activate clock genes, in turn regulating their functioning.

15.4.3 Mechanism of Action

It binds to melatonin receptors, which belong to GPCRs family. MT1 and MT2 are two different types of melatonin receptors present in various tissues. Both of them inhibit the adenylyl cyclase, subsequent production of cAMP and affect various cellular signalling pathways.

15.4.4 Biological Effects

The role of melatonin in regulating circadian rhythm, sleep, and reproduction is very well documented. However, its

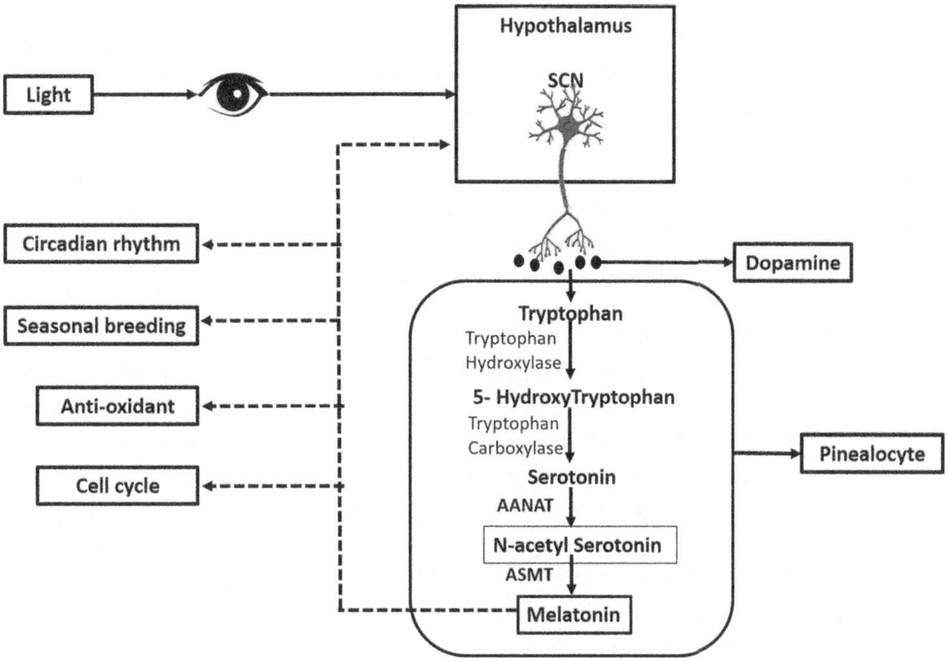

Fig. 15.13 Mechanism of synthesis of melatonin and its biological effects. [The supra-chiasmatic nucleus in the hypothalamus stimulates the synthesis and secretion of melatonin during dark period. Melatonin is the major humoral factor implicated in regulating cell cycle, seasonal breeding, and circadian rhythm in animals. [*SCN* supra-chiasmatic nucleus; *AANAT* aralkylamine N-acetyltransferase; *ASMT* N-acetyl serotonin methyl transferase]

other equally important properties are preventing oxidative stress, regulation of cell cycle and apoptosis.

15.4.4.1 Effects on Ovary

Melatonin by binding to its receptors controls some vital physiological processes such as growth of follicles, maturation of oocytes, and luteinisation. Much of them are attributed to the anti-apoptotic action of melatonin. In males, the anti-oxidative property of melatonin defends spermatozoa from oxidative damage. Hence, it ensures optimum fertility in animals by effecting steroidogenesis and ameliorating oxidative stress on gametes.

15.4.4.2 Effect on Seasonal Breeders

The changes in melatonin levels during different seasons act merely as a transducer in deciding a favourable period to reproduce. In case of short-day breeders like sheep, rise in melatonin levels with decreased duration of day light leads to the onset of oestrous cycle. Whereas in long-day breeders, onset of oestrous cycle corresponds with an increase in day light. This seasonal breeding effect of melatonin is mainly brought about by altering the secretory pattern of pituitary gonadotropins and thyrotropin.

15.4.4.3 Miscellaneous Effects

The stimulation of MAPK pathway and p53 genes by melatonin helps in regulating the apoptosis and cell cycle. Amelioration of oxidative stress by directly scavenging reactive oxygen species (ROS) or by production of anti-oxidant enzymes is a key property of melatonin (Fig. 15.13).

Learning Outcomes

- **Hormones:** Defined as chemical messengers that are secreted into blood in response to an appropriate stimulus, carried and act on a specific target organ or an organ system thereby bringing a well-defined biological response. Being secreted from the ductless glands, their ability to bind with specific receptors and initiation of signal transducing mechanisms that produce a specific target effect are the major defining properties of a hormone.
- **Signal transduction:** It involves the initiation of a cascade of enzymatic reactions that initiate signalling pathways that result in the production of secondary messengers or activation of enzymes or initiation of transcription and translation of genes.
- **Hypothalamus:** Situated in the diencephalon, it integrates nervous and endocrine system. Based on their effect on the secretory pattern of pituitary gland hormones, hypothalamic hormones are classified as releasing and inhibiting hormones. The neuroendocrine cells arranged in distinct nuclei are of two types, i.e. magnocellular and parvocellular neurons.
- **Hypophyseal portal system:** The hormones secreted from the magnocellular neurons of hypothalamus in median eminence are carried to the anterior pituitary gland by hypophyseal portal system. It is responsible for the hypothalamic regulation on pituitary hormones secretion.

(continued)

- **Pituitary gland:** Pituitary gland has three distinct regions: anterior lobe, intermediate lobe, and posterior lobe. The anterior lobe is functionally contiguous with hypothalamus, whereas the posterior lobe is structurally related to hypothalamus. The somatotropes, lactotropes, corticotropes, gonadotropes, and thyrotropes are different neuroendocrine cell types present in anterior pituitary and are responsible for the secretion of GH, PRL, ACTH, LH/FSH, and TSH. The posterior lobe is mainly composed of the axons arising from hypothalamic magnocellular neurons present in PVN and SON. The nerve terminals secrete oxytocin and ADH in to blood on appropriate stimulus.
- **Pineal gland:** The pineal resides near the third ventricle and regulates circadian rhythm in animals by secreting an effector hormone melatonin. The melatonin is derived from tryptophan with peak synthesis occurring during dark period. Melatonin plays a role in generating sleep-wake cycle, seasonal breeding, and fertility in farm animals. In addition, it mitigate the damage to DNA and other cellular organelles by scavenging ROS directly or indirectly by producing anti-oxidant enzymes.

Exercises

Objective Questions

Q1. What is the higher centre for the integration of endocrine and nervous systems?

Q2. The term hormone is derived from the Greek word ___________.

Q3. Which fatty acid acts as the precursor for the biosynthesis of prostaglandins?

Q4. What is the smallest peptide hormone?

Q5. The catecholamines are derived from the amino acid ______________.

Q6. The receptors for steroid hormones reside in the ______________.

Q7. The large precursor molecules produced by the rough endoplasmic reticulum during the synthesis of a peptide/protein hormone are known as ____________.

Q8. What are the secondary messengers produced due to the hormonal stimulation of G_q receptors?

Q9. What is the major neuroendocrine cell type present in the anterior pituitary gland?

Q10. The somatomedin type ________ is responsible for the post-natal growth of an animal.

Q11. The hormone responsible for the brooding behaviour in birds is _____________.

Q12. What are the glycoprotein hormones released from the anterior pituitary?

Q13. The hormone that has the biological activities of oxytocin and ADH found in birds is _____________.

Q14. The precursor molecule produced in the intermediate pituitary lobe for the synthesis of ACTH is _____________.

Q15. The specific region of genome to which a steroid hormone-receptor complex bind is known as ________________.

Q16. ______________ are the distinct neuroendocrine cells responsible for the production of posterior pituitary hormones.

Q17. The functional circuit that helps in the transport of hypothalamic hormones to act on the anterior pituitary hormones is called as ____________________.

Q18. Which neurotransmitter is also known as the prolactin-inhibiting hormone (PIH)?

Q19. Which hormones is responsible for seasonal breeding in animals?

Q20. The rate-limiting enzyme in the synthesis of melatonin is ___________.

Subjective Questions

Q1. Enlist the different chemical messengers that regulate physiological functions.

Q2. What are the general properties of hormones?

Q3. Describe the different classifications of hormones with examples.

Q4. What are the different types of receptors based on their cellular localisation?

Q5. Describe in detail the steps involved in the synthesis of polypeptide/protein hormones.

Q6. Describe in detail the mechanism of synthesis of steroid hormones.

Q7. Why post-translational modifications in protein/polypeptide hormones is important? Enlist the different types of post-translational modifications.

Q8. Describe in detail the steps in the production of cAMP as a secondary messenger by hormones. Give examples of hormones that act based on this mechanism.

Q9. Explain the downstream signalling pathways due to the activation of phospholipase C (PLC) system by hormone-dependent GPCRs.

Q10. What is the role of receptor tyrosine kinases in peptide signalling?

Q11. What are the different types of neuroendocrine cells in hypothalamus?

Q12. Enlist the different hypothalamic nuclei linked with endocrine activity.

Q13. Explain the structural and functional means by which hypothalamus and pituitary are connected.

Q14. What are the different types of neuroendocrine cells in adenohypophysis? Enlist their respective hormones along with their target effect.

Q15. Why GH is known as a diabetogenic and ketogenic hormone?
Q16. Describe the effects of GH on intermediary metabolism.
Q17. How PRL is responsible for lactogenesis? Describe the factors regulating its secretion.
Q18. Describe the mechanism of synthesis of ACTH and its biological effects.
Q19. Briefly describe the chemical structure of pituitary glycoproteins and explain the role of gonadotropins in regulating animal fertility.
Q20. Describe the steps involved in the synthesis of melatonin and its effect on seasonal breeding in animals.

Answer to Objective Questions

A1. Hypothalamus
A2. Ormao
A3. Arachidonic acid
A4. TRH
A5. Tyrosine
A6. Cytoplasm
A7. Preprohormones
A8. DAG, IP_3
A9. Somatotropes
A10. Somatomedin-C (IGF1)
A11. Prolactin
A12. LH, FSH, TSH
A13. Vasotocin
A14. Pro-opiomelanocortin (POMC)
A15. Hormone response element (HRE)/Steroid response element (SRE)
A16. Magnocellular neurons
A17. Hypothalamic-hypophyseal portal system
A18. Dopamine
A19. Melatonin
A20. Aralkylamine N-acetyltransferase (AANAT)

Keywords for the Answer to Subjective Questions

A1. Neurotransmitters, Hormones, Neurohormones, Paracrines, Autocrines, Cytokines
A2. High-affinity receptors, onset of action, signal transduction, and feedback mechanisms
A3. Based on the source of secretion, chemical nature, physiological action, and solubility
A4. Transmembrane, cytosolic, and nuclear receptors
A5. Pre-pro hormone, prohormone, and hormone
A6. Cholesterol, StAR, p450scc, pregnenolone, and tissue specific hydroxylase
A7. Biological activity, half-life, glycosylation, acetylation, sulfation, and amidation
A8. GPCRs, adenylyl cyclase, cAMP, and protein Kinase A
A9. GPCRs, phospholipase C, DAG, and IP_3
A10. Tyrosine kinase receptors, tyrosine kinase-associated receptors, insulin, GH, and PRL
A11. Magnocellular neurons and parvocellular neurons
A12. PVN, POA, DMN, ACN and SON
A13. Hypophyseal portal system, pituitary stalk
A14. Somatotropes-GH, lactotropes-PRL, Corticotropes (ACTH), gonadotropes (LH/FSH), and thyrotropes-TSH
A15. Peripheral utilisation of glucose, insulin resistance, gluconeogenesis in liver, and lipolysis
A16. Shift of metabolism from carbohydrates to fats, protein accumulation, and lipolysis
A17. Lobulo-alveolar growth, lactose, and casein synthesis
A18. Pre-POMC, POMC, and adrenal cortex
A19. α and β sub-units of TSH, LH and FSH, folliculogenesis, CL formation, spermatogenesis, and steroidogenesis,
A20. Tryptophan, serotonin, AANAT, ASMT, effect on thyrotropin and gonadotropin secretion

Further Reading

Textbooks

McDonald LE, Pineda MH, Dooley MP (2003) McDonald's veterinary endocrinology and reproduction. 5th ed./edited by M.H. Pineda, with the editorial assistance of Michael P. Dooley. Ames, Iowa: Iowa State Press. Print

Melmed S, Polonsky KS, Larsen PR, Kronenberg HM (2017) Williams textbook of endocrinology, 13th edn. Elsevier Inc., pp 1855–1916. https://doi.org/10.1016/C2013-0-15980-6

Nussey SS, Whitehead SA (2001) Endocrinology: an integrated approach. https://www.ncbi.nlm.nih.gov/books/NBK22

Articles

Bousfield GR (2019) Biosynthesis and posttranslational processing of peptide hormones. https://doi.org/10.1016/B978-0-12-801238-3.92822-8

Clarke IJ (2011) Hypothalamus as an endocrine organ. Compr Physiol 5(1):217–253. https://doi.org/10.1002/cphy.c140019

Devesa J, Almengló C, Devesa P (2016) Multiple effects of growth hormone in the body: is it really the hormone for growth? Clin Med Insights Endocrinol Diabetes 9:47–71. https://doi.org/10.4137/CMED.S38201

Hu J, Zhang Z, Shen WJ, Azhar S (2010) Cellular cholesterol delivery, intracellular processing and utilization for biosynthesis of steroid hormones. Nutr Metabol 7(1):1–25. https://doi.org/10.1186/1743-7075-7-47

Kim KH, Seong BL (2001) Peptide amidation: Production of peptide hormones in vivo and in vitro. Biotechnol Bioprocess Eng 6(4):244–251. https://doi.org/10.1007/BF02931985

Levin ER, Hammes SR (2016) Nuclear receptors outside the nucleus: extranuclear signalling by steroid receptors. Nat Rev. Mol Cell Biol 17(12):783–797. https://doi.org/10.1038/nrm.2016.122

Møller N, Jørgensen JOL (2009) Effects of growth hormone on glucose, lipid, and protein metabolism in human subjects. Endocr Rev 30(2):152–177. https://doi.org/10.1210/er.2008-0027

Posner BI, Laporte SA (2010) Cellular signalling: peptide hormones and growth factors. Prog Brain Res 181:1–16. https://doi.org/10.1016/S0079-6123(08)81001-1

Psarra AM, Solakidi S, Sekeris CE (2006) The mitochondrion as a primary site of action of steroid and thyroid hormones: presence and action of steroid and thyroid hormone receptors in mitochondria of animal cells. Molecular and cellular endocrinology 246(1-2):21–33. https://doi.org/10.1016/j.mce.2005.11.025

Scanes CG, Jeftinija S, Glavaski-Joksimovic A, Proudman J, Arámburo C, Anderson LL (2005) The anterior pituitary gland: lessons from livestock. Domest Anim Endocrinol 29(1):23–33. https://doi.org/10.1016/j.domaniend.2005.04.002

Talpur HS, Chandio IB, Brohi RD, Worku T, Rehman Z, Bhattarai D, Yang L (2018) Research progress on the role of melatonin and its receptors in animal reproduction: a comprehensive review. Reproduction in Domestic Animals 53(4):831–849. https://doi.org/10.1111/rda.13188

Weigel NL, Moore NL (2007) Kinases and protein phosphorylation as regulators of steroid hormone action. Nucl Recept Signal 5(1):nrs-05005. https://doi.org/10.1621/nrs.05005

Hormonal Regulation of Metabolism, Water, and Minerals

16

Balantrapu Achuta Anjani Sai Kumar

Abstract

The vital role of thyroid hormones, insulin, glucagon, corticosteroids, and catecholamines in regulating metabolism is a key for an animal's optimal growth and survival. The thyroid hormones (THs) are derived from tyrosine and have a permissive effect on growth hormone (GH) to produce growth, reproduction, and lactation in domestic animals. Apart from the THs, optimal secretion of insulin from the pancreas is responsible to mediate the growth-promoting effects of GH. As the only hypoglycemic hormone present in animals, the optimal secretion of insulin is a prerequisite for increasing glucose uptake in skeletal muscles and adipose tissue. Furthermore, the secretion of glucocorticoids and catecholamines from the adrenal gland alters the metabolic status of an animal to combat various stressors effectively. Whereas, the mineralocorticoids regulate the standard concentration of sodium and potassium ions in the blood, which is essential for the establishment of resting membrane potential in almost all cells. In addition to the aforementioned cations, the tight regulation of extracellular fluid (ECF) Ca^{2+} levels by hormones such as parathormone (PTH), calcitriol, and calcitonin is crucial for the excitation of neurons, skeletal muscles, and smooth muscles. Overall, the endocrine system warrants proper metabolic health by controlling energy homeostasis and mineral balance in animals.

B. A. A. Sai Kumar (✉)
Department of Veterinary Physiology, Rajiv Gandhi Institute of Veterinary Education and Research, Kurumbapet, Puducherry, India

© The Author(s), under exclusive license to Springer Nature Singapore Pte Ltd. 2023

P. K. Das et al. (eds.), *Textbook of Veterinary Physiology*, https://doi.org/10.1007/978-981-19-9410-4_16

Graphical Abstract

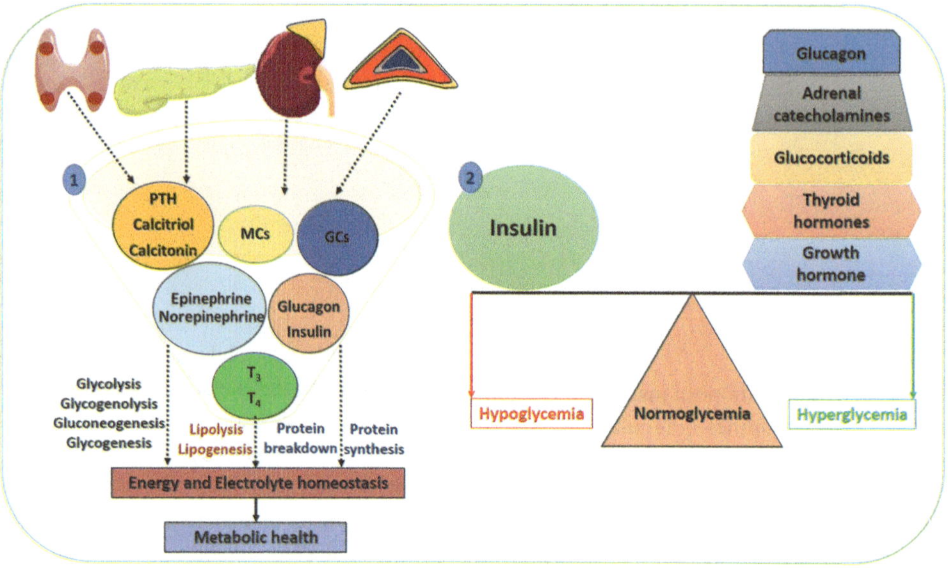

Description of the graphic: Role of the endocrine system in regulating metabolic health of animals (**1**). Hormones secreted from the thyroid gland, parathyroid gland, pancreas, and adrenal gland modulate a variety of metabolic processes to maintain optimal energy and mineral balance in animals (**2**). Hormonal regulation of blood glucose levels. [*MCs* mineralocorticoids; *GCs* glucocorticoids; T_3 triiodothyronine; T_4 tetraiodothyronine.]

Keywords

Thyroid hormones · Pancreas · Corticosteroids · Parathormone · Calcitriol

Learning Objectives

- Biosynthesis and target effects of thyroid hormones
- Endocrine functions of the pancreas
- Role of insulin and glucagon in regulating intermediary metabolism
- Regulation of Na^+ and K^+ in ECF by aldosterone
- Biological effects of cortisol on metabolism
- Calcium homeostasis by parathormone, calcitriol, and calcitonin

16.1 Thyroid Gland

Intermediary metabolism includes all the biochemical reactions that involve in the conversion of dietary nutrients into energy or cellular components. The energy derived through metabolism will be utilized to maintain homeostasis, growth, production, and reproduction in animals. It drives crucial processes such as cell proliferation, sustenance, and the synthesis of cellular components. The rate of metabolism varies according to different physiological states or stages in an animal's life. Hormones of the thyroid gland, pancreas, and adrenal gland play a major role in defining the plane of metabolism to support a specific physiological function in an animal and are hence known as metabolic hormones. They primarily affect metabolic processes such as glycogenolysis, glycogenesis, gluconeogenesis, lipogenesis, and lipolysis to meet the metabolic needs of an animal.

It is a bi-lobed gland present on either side of the trachea, derived from the foregut endoderm layer, and produces thyroid hormones (THs), i.e., triiodothyronine (T_3) and tetraiodothyronine (T_4). These hormones are responsible for multifaceted functions including organ development, growth, homeostasis, oxidative metabolism, reproduction, and production.

16.1.1 Histology

The endodermal-derived thyroid progenitor cells form thyroid follicles that constitute the thyroid gland. Thyroid follicles are often referred to as the manufacturing units of THs. Within these follicles are the specialized epithelial cells known as "thyrocytes" that serves as a site of synthesis, storage, and secretion of TH. The thyrocytes have polarized apical and basal membranes, which help in regulating specific bidirectional transport of substances back and forth from

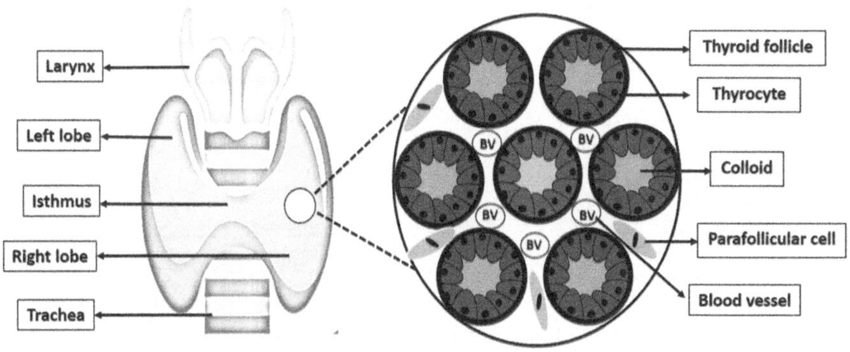

Fig. 16.1 Histology of the thyroid gland. [The thyroid gland is composed of follicles and parafollicular cells that are responsible for the manufacturing of thyroid hormones and calcitonin, respectively]

the lumen. Apart from thyrocytes, another type of neuroendocrine cell known as "parafollicular or C-cells" exists in close association with the thyroid follicles (Fig. 16.1).

16.1.2 Synthesis of Thyroid Hormones

16.1.2.1 Synthesis of Thyroglobulin

Thyroglobulin (TG) is a homodimeric (660 kDa) glycoprotein synthesized in thyrocytes and subsequently stored in the lumen. The thyroglobulin is secreted and stored in the follicular lumen and is commonly termed as "colloid". Each TG monomer has about 70 tyrosine residues and it acts as a scaffold for the synthesis of THs. The TSH-dependent stimulation of TSHR present on the basal membrane of thyrocytes increases the rate of TG gene expression and its subsequent translation. TG undergoes post-translational modifications that favor protein folding, trafficking, iodination, and hormonogenesis during its transit into the lumen.

16.1.2.2 Iodine Uptake

The dietary iodide absorbed from the GIT reaches thyroid gland via systemic circulation. A specialized "Sodium (Na^+)-Iodide (I^-) symporter (NIS)" present on the basolateral membrane helps in the secondary active transport of iodide into the cytoplasm of thyrocytes. Another apical membrane-bound iodide transport protein known as "Pendrin" helps in the rapid efflux of cytoplasmic I^- into the lumen. These carrier proteins confer the unique ability of thyrocytes to concentrate I^- by 30–60 fold within their cytoplasm and this exclusive phenomenon occurring is referred to as "iodide trapping." During the efflux of I^- into the follicular lumen, it is converted into Iodine (I) by the apical membrane-bound enzyme "thyroid peroxidase (TPO)" (Fig. 16.2).

16.1.2.3 Organification

In a process known as "Organification," the highly reactive iodine reacts with the tyrosine residues present on the TG to form monoiodotyrosine (MIT) and diiodotyrosine (DIT). While roughly half of the tyrosine residues present in each TG monomer can be iodinated depending on the availability of superficial tyrosine molecules.

16.1.2.4 Coupling

This refers to the combination of MIT and DIT molecules within TG to form triiodothyronine (T_3), tetraiodothyronine

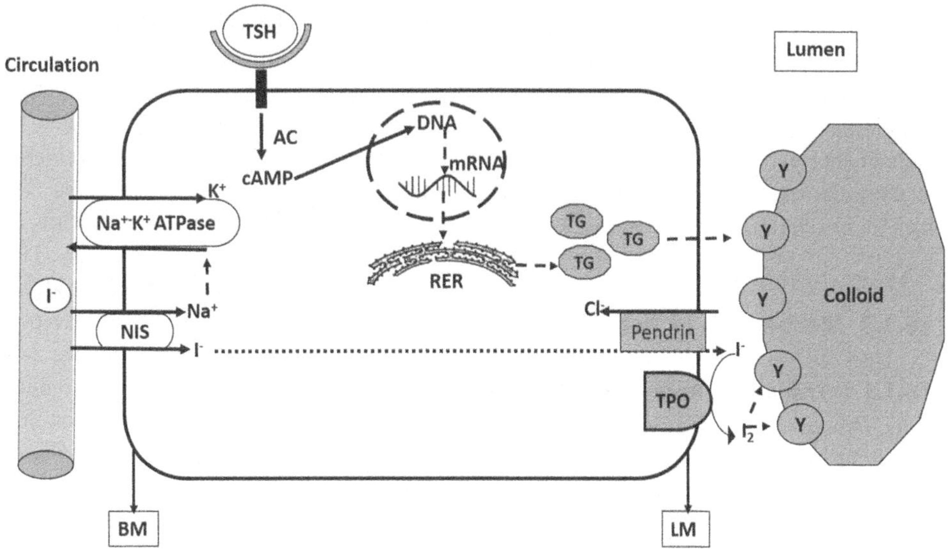

Fig. 16.2 Iodide trapping in thyrocytes [The iodide in the circulation is actively transported by the NIS protein and stored inside the thyrocytes or used for thyroid hormonogenesis. *TP* thyroid peroxidase; *TG* thyroglobulin; *AC* adenylyl cyclase; *NIS* sodium-Iodide symporter; *BM* basal membrane; *LM* luminal membrane; *RER* rough-endoplasmic reticulum]

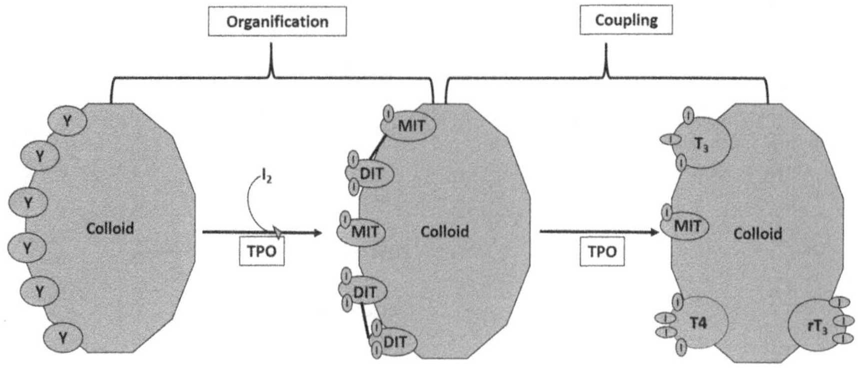

Fig. 16.3 Organification and coupling reactions in thyroid hormonogenesis [Iodination of tyrosine residues in the TG is catalyzed by TPO and results in the production of MIT and DIT, which will be further to yield THs. [*MIT* monoiodotyrosine; *DIT* di-iodo tyrosine; *T_3* triiodothyronine; *T_4* tetraiodothyronine reverse; *rT_3* triiodothyronine]

(T_4), and reverse-triiodothyronine (rT_3). Although they collectively comprise the thyroid gland secretions, only T_3 and T_4 are active hormonal forms that can elicit biological effects in the target tissues. Only a few out of the many MIT and DIT molecules embedded in TG undergo thyroid hormonogenic coupling reactions (Fig. 16.3).

16.1.2.5 Endocytosis and Lysosomal Degradation

The TSH-dependent stimulation of thyrocytes ensues endocytic reuptake of TG surrounding the apical membrane. Thus, the reinternalized TG is conveyed to lysosomes for enzymatic degradation resulting in the liberation of T_3, T_4, MIT, and DIT. The thyroid hormones are then released into the bloodstream through monocarboxylate transporter 8 (MCT8) present on the basal membrane. While the MIT, DIT, and TG molecules will be degraded to reutilize iodine for further synthesis of thyroid hormones. Principally, T_4 is the major form of TH produced by the thyroid gland.

16.1.2.6 Transport in Blood

THs are lipophilic and therefore require plasma proteins for their circulatory transport. More than 99% of THs in the circulation will be bound to the carrier proteins and liberated rapidly when required. The plasma proteins, such as thyroxine-binding globulin (TBG), thyroxine-binding prealbumin (TBPA or transthyretin), and serum albumin act as the carrier proteins of THs. The TBG is the major plasma protein that binds to THs, it has a greater affinity to T_4 than $T_3$and hence a long half-life (Table 16.1). They act mainly as an extra-thyroid pool of THs, preventing their rapid metabolism and excretion from the animal's body.

Table 16.1 Properties of thyroid hormones

S. No	Characteristic	T_3	T_4
1.	Proportion of secretory form	≈7%	≈93%
2.	Major binding protein	Serum albumin	Thyroxine-binding globulin (TBG)
3.	Affinity to thyroid receptors	Strong	Weak
4.	Potency	High (four times more than T_4)	Less
5.	Half-life	1 day	6–7 days

16.1.3 Mechanism of Action

16.1.3.1 Transport Across the Cell Membrane

Only free or unbound THs are carried through the MCT 8 present on the cell membranes of target cells. Since T_3 binds to the nuclear thyroid receptors (TRs) with a great affinity when compared to T_4, it is considered as the more potent and active form of TH. While T_4 is the major secretory product of the thyroid gland, subsequent conversion of T_4 into the active T_3 form by intracellular deiodinases takes place in target cells. 5′-deiodinase type 1 (D1), 5′-deiodinase type 2 (D2), and 5-deiodinase type 3 (D3) are different forms of deiodinases distributed in various tissues. The D1 and D2 enzymes are responsible for the conversion cum activation of T_4 to T_3. Whereas D3 catalyzes the conversion of T_3 to rT_3, causing its inactivation. Therefore, MCT 8 and deiodinases are essential factors in determining the magnitude of biological response in target tissues.

16.1.3.2 Intracellular Signaling

Produced either from the thyroid gland or by the conversion of prohormone T_4 in peripheral tissues, T_3 initiates a signaling cascade by binding to thyroid receptors (TRs) localized in the nucleus. TR α and TR β are the two major differentially expressed isoforms of TR found in various tissues, they determine the activation of specific regulatory pathways of metabolism. Generally, TR forms a heterodimer with retinoid X receptor (RXR) and binds to specific regions of DNA known as thyroid response element (TRE). When the heterodimeric complex is not bound to T_3, it is associated with corepressors like nuclear receptor corepressor (NCoR) or silencing mediator for retinoid and thyroid receptor (SMRT). These repressors recruit histone deacetylases (HDACs) that bind to the promoter regions of various genes and help in maintaining their repression. Subsequent

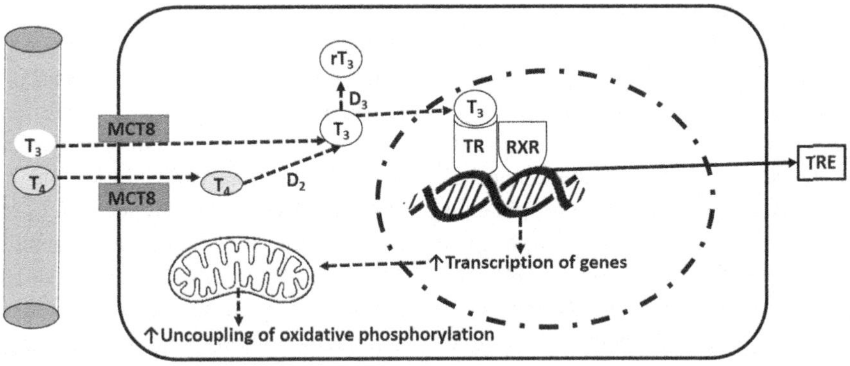

Fig. 16.4 Mechanism of thyroid signaling in the target cells [The thyroid hormones in the circulation are taken up by the target tissues and exert their biological effects by binding to the thyroid receptors residing in the nucleus. *T_3* triiodothyronine; *T_4* tetraiodothyronine reverse; *rT_3* triiodothyronine; *TRE* thyroid response element; *D2* 5′-deiodinase type 2; *D3* 5′-deiodinase type 3; *RXR* retinoid X receptor; *TR* thyroid receptors; *MCT 8* monocarboxylate transporter 8]

dissociation of corepressors and recruitment of coactivators take place when T_3 binds to the TR complex, resulting in the transcription of TH regulated genes (Fig. 16.4).

16.1.4 Biological Effects

16.1.4.1 Effect on Carbohydrate Metabolism

THs stimulate intestinal absorption, glycolysis, glycogenolysis, and gluconeogenesis in various tissues. The enhanced glucose production by the aforementioned pathways is necessary to maintain basal metabolic rate (BMR), thermogenesis, and animal growth (Fig. 16.5).

16.1.4.1.1 Effect on Intestinal Absorption

THs increase the absorption of glucose from the small intestine by upregulating the activities of Sodium (Na^+)-Glucose cotransporter 1 (SGLT1) and Na^+-K^+ ATPase pump.

16.1.4.1.2 Effect on Liver

The liver is a major organ regulating glucose homeostasis in animals. THs have direct effects on glucose uptake,

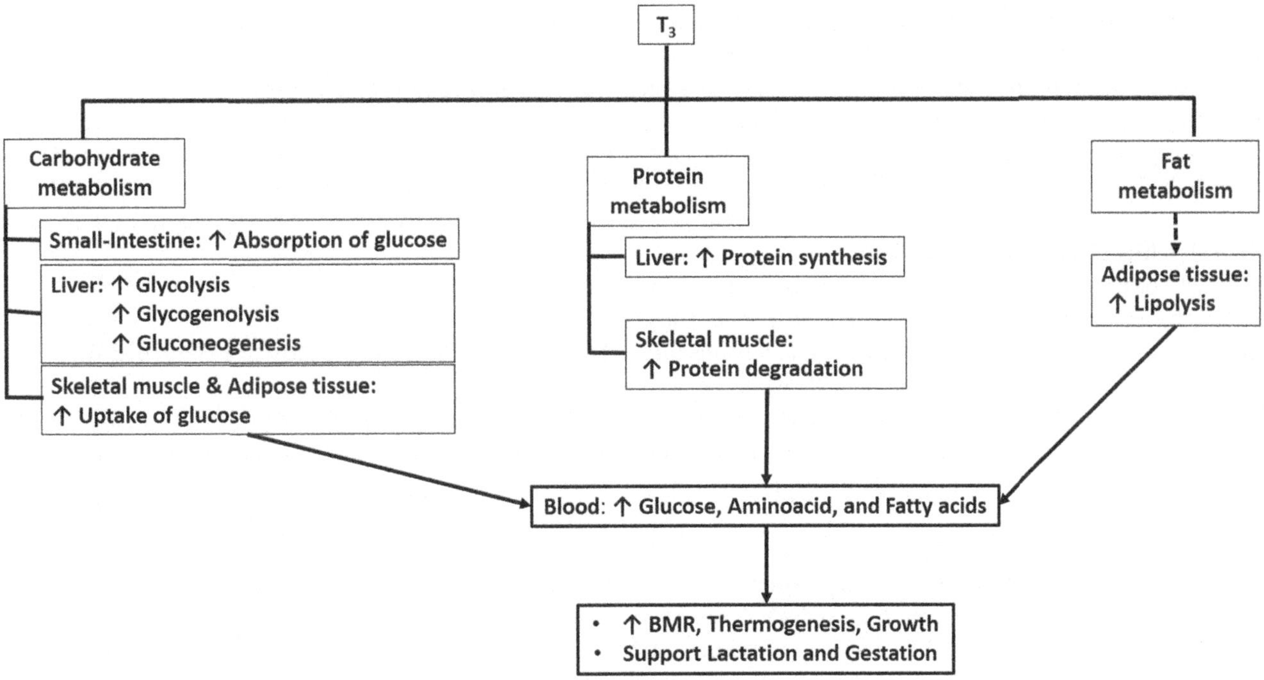

Fig. 16.5 Effect of thyroid hormone on metabolism. [The thyroid hormones increase the rate of glycogenolysis, gluconeogenesis, and lipolysis resulting in an elevated BMR to support the survival and production in animals. [↑ increase; *BMR* basal metabolic rate]

production, and oxidation in hepatocytes. They increase both the uptake and secretion of glucose from hepatocytes by stimulating the expression of glucose transporter-2 (GLUT2). Simultaneously, upregulation of glycolytic enzymes enhances subsequent oxidation of glucose via the glycolytic pathway. The key enzyme encoding genes of gluconeogenesis such as pyruvate carboxylase, phosphoenolpyruvate carboxykinase (PEPCK), and glucose-6-phosphatase are positively regulated by the THs. Moreover, THs also stimulates the rate of glycogenolysis in hepatocytes due to an increased rate of oxidation of glucose. Together, THs stimulate glycolysis, gluconeogenesis, and glycogenolysis in the liver leading to a concomitant rise in blood glucose levels.

16.1.4.1.3 Effect on Pancreas

THs play a crucial role in the development, maturation, and functioning of cells in the islets of Langerhans. They inhibit glucose-stimulated insulin release from the β cells leading to glucose intolerance.

16.1.4.1.4 Effect on Glucose Uptake in Skeletal and Adipose Tissue

The insulin-dependent upregulation of GLUT4 leads to increased glucose uptake in the skeletal muscles. In the same way, THs enhance the glucose uptake in adipocytes, which help in lipogenesis.

16.1.4.2 Effect on Protein Metabolism

Thyroid hormones stimulate both protein catabolism and anabolism in tissues. The degradation of proteins in skeletal muscles results in the elevation of plasma amino acid concentration to support gluconeogenesis in various tissues. In the liver, THs stimulate the synthesis of intracellular and secretory proteins. Altogether, THs stimulate protein turnover in the liver and skeletal muscle cells.

16.1.4.3 Effect on Fat Metabolism

Thyroid hormones stimulate the hepatic production of cholesterol by upregulating the HMG-CoA reductase gene. Thus, the increased amount of cholesterol is utilized to manufacture bile acids. They stimulate lipolysis in both white adipose tissue (WAT) and brown adipose tissue (BAT) to produce free fatty acids, which are used for thermogenesis. Further, THs also enhance lipogenesis to counter the depletion of lipid stores.

16.1.4.4 Effect on BMR and Thermogenesis

The altered levels of intracellular Na^+ and Ca^{2+} by THs augment the activity of Na^+-K^+ ATPase pump and sarcoplasmic/endoplasmic reticulum Ca^{2+}-dependent ATPase (SERCA) in skeletal muscle, heart, and other cells. The enhanced activity of the above-mentioned ion pumps corresponds to a proportional rise in the hydrolysis of ATP with subsequent generation of heat and elevation of BMR. Furthermore, the increased ATP requirement is ensured through the catabolism of glucose, fatty acids, and amino acids. In addition, exposure to cold invokes the activity of D2 and subsequent conversion of T_4 to T_3 in brown adipose tissue (BAT), initiating a thermogenic response. Altogether, THs increase metabolic heat production due to enhanced oxidation of glucose and fatty acids, a phenomenon that is commonly referred to as non-shivering thermogenesis.

16.1.4.5 Effect on Mitochondrial Functioning and Biogenesis

THs have a direct effect on mitochondrial biogenesis by binding to TRs localized in mitochondria leading to increased mtRNA and protein synthesis. In addition, the upregulation of nuclear transcription factors like nuclear respiratory factor 1 (NRF-1) and peroxisome proliferator-activated receptor gamma coactivator 1 alpha (PGC-1a) will further activate the transcription of nuclear genes that encode mitochondrial proteins. They increase ATP synthesis by stimulating ATP synthase needed for the functioning of Na^+-K^+ ATPase and SERCA. Furthermore, THs lead to the uncoupling of oxidative phosphorylation by stimulating proton leak from the inner mitochondrial membrane by uncoupling proteins 1/2/3 (UCP1/2/3) or by inhibiting the movement of reducing equivalents into mitochondria. This uncoupling leads to the generation of heat with a consequent decrease in ATP synthesis.

16.1.4.6 Miscellaneous Effects

The increased oxygen demand at the tissue level due to elevated mitochondrial respiration is met by increasing cardiac output, systemic blood pressure, and respiratory rate. THs have stimulatory effects on neuronal activity, GIT motility, sleep, and milk production.

16.1.5 Hypothalamic–Pituitary–Thyroid Axis

The secretion of THs is regulated primarily by TRH and TSH (Fig. 16.6) released from the hypothalamic–pituitary axis. TRH from the hypothalamus stimulates the secretion of TSH from the pituitary gland. Then, the TSH acts on the thyroid gland and stimulates the production of THs. Increased circulatory levels of THs exert negative feedback signals on the secretion of both TRH and TSH. Other hormones such as leptin, somatostatin, dopamine, and cortisol also can modulate their secretion.

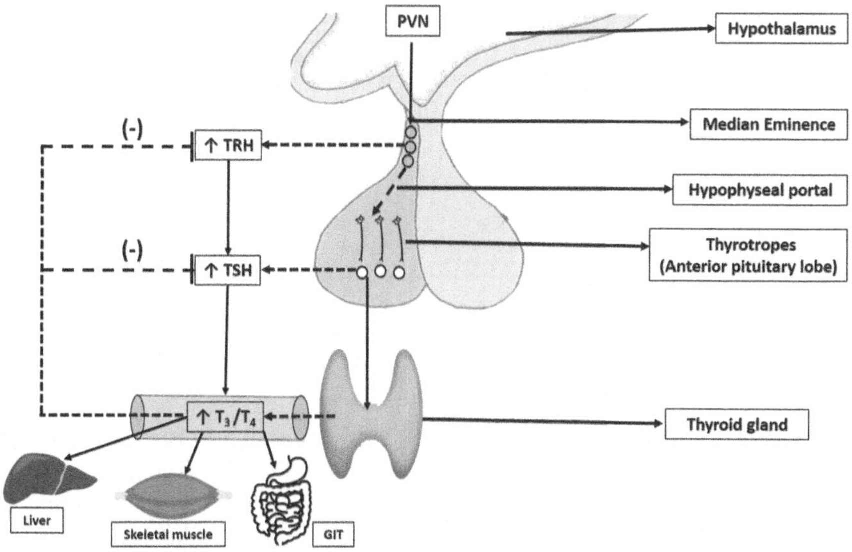

Fig. 16.6 Regulation of thyroid hormone secretion. [The increased circulatory levels of THs exert a feedback inhibition on the secretion of TRH and TSH from the hypothalamus and anterior pituitary, respectively. [↑ increase; *BMR* basal metabolic rate; *PVN* para-ventricular nucleus, *GIT* gastrointestinal tract; *TRH* thyrotropin releasing hormone; *TSH* thyroid-stimulating hormone]

Know More. . .

- Thyroid hormones have permissive effect on GH and absence of which leads to stunting of animal's growth.
- **Iodinated casein:** Resembles thyroid hormones and used in dairy cows to increase milk production.
- **Hypothyroidism:** Decreased circulatory levels of thyroid hormones. Seen in panhypopituitarism, iodine deficiency, and congenital deficiency of thyroid peroxidase. It results in "cretinism", characterized by impaired physical growth and neural development.
- **Hyperthyroidism:** Increased circulatory levels of thyroid hormones, often due to hyperactivity of thyroid gland seen in grave's disease and thyroid adenoma.
- **Goiter:** The pathological condition characterized by abnormal enlargement of thyroid gland. Goiter can be caused due to iodine deficiency (endemic goiter) and in grave's disease (toxic goiter).

16.2 Pancreas

The pancreas is an abdominal, mixed type of gland that significantly contributes to digestion and intermediary metabolism. The digestive enzymes secreted from exocrine pancreatic acini aid in intestinal digestion, whereas the different hormones from the endocrine pancreas are implicated in regulating various metabolic processes (Fig. 16.7). The deranged enzymatic and hormone secretory patterns of the pancreas will immediately affect animal metabolism, energy homeostasis, and production status.

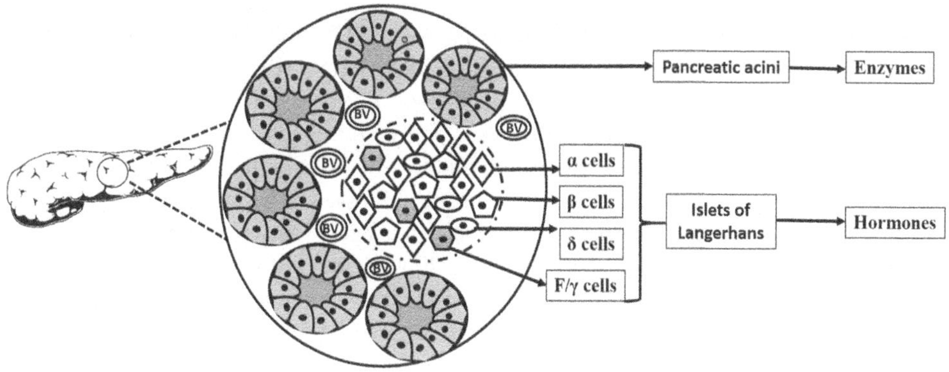

Fig. 16.7 Histology of pancreas. [The group of specialized endocrine cells dispersed amongst the pancreatic acini known as islets of Langerhans, which function to produce pancreatic hormones that are mainly concerned with regulating carbohydrate metabolism. [*BV* blood vessel]

Table 16.2 Different pancreatic hormones, their chemical nature, effect, and plasma half-life

S:No	Cell type	Hormones	Chemical structure	Effect	Half-life (in minutes)
1.	α cells	Glucagon	Peptide 29 a.a	Increase blood glucose levels (Hyperglycemia)	6–7
2.	β cells	Insulin	Polypeptide A Chain 21 a.a B Chain 30 a.a	Decrease blood glucose levels (Hypoglycemia)	3–5
3.	δ cells	Somatostatin	Polypeptide 32 a.a	Inhibit the secretion of insulin and glucagon	1–3
4.	F/γ cells	Pancreatic polypeptide	Polypeptide 36 a.a	Inhibits exocrine pancreatic secretions	6–7

16.2.1 Endocrine Pancreas

The endocrine function is imparted by islets of Langerhans, composed of specialized cells dispersed amidst the exocrine regions of the pancreas. They make up 2–3% of the pancreas, consists of four different endocrine cell types, i.e., α, β, γ, and δ. These distinct endocrine cells are responsible for the production of glucagon, insulin, somatostatin, and pancreatic polypeptide hormones, thereby regulating metabolic homeostasis (Table 16.2).

16.2.1.1 Glucagon

The islet α-cells secrete glucagon derived from the precursor proglucagon. The proglucagon is also expressed in other tissues such as the brain stem, hypothalamus, and enteroendocrine L cells. The prohormone convertases (PC1/2/3) further help in the process of conversion of proglucagon. The PC2 in α-cells converts proglucagon into glucagon, a polypeptide hormone with 29 amino acids. Whereas, the PC1 in the intestine and brain converts proglucagon to glucagon-like peptide 1 (GLP-1) and glucagon-like peptide 2 (GLP-2).

16.2.1.1.1 Mechanism of Secretion

Glucagon is primarily released in response to the decreased level of glucose in the blood. The presence of glucose transporter 1 (GLUT1) on the cell membrane of α-cells helps in the influx of glucose during normal conditions. Then, the glucose entered inside will be oxidized to produce ATP, which will be used to open ATP-sensitive potassium channels (K_{ATP} channels) and hence prevents depolarization by promoting the efflux of K^+ ions. During hypoglycemia, the consequent reduction in ATP production due to reduced glucose influx leads to the closure of K_{ATP} channels. Subsequently, the build-up of K^+ ions inside the cytoplasm triggers the opening of voltage-dependent Ca^{2+} channels allowing the influx of Ca^{2+} ions. The rise in the intracellular Ca^{2+} levels stimulates the exocytosis of glucagon stored in the form of vesicles (Fig. 16.8).

16.2.1.1.2 Mechanism of Action

The glucagon exerts its biological actions by binding to the glucagon receptors (GCGR) on the target cells, with the liver having more GCGR than any other tissue. Being a G-protein coupled receptor; activated GCGR primarily activates the adenylyl cyclase (AC) system resulting in the production of cAMP and followed by the activation of protein kinase A (PKA). Thus, activated PKA migrates to the nucleus, activates transcription factors such as cAMP response element-binding protein (CREB) to promote the transcription of genes that mediate specific biological effects.

16.2.1.1.3 Role of Glucagon on Intermediary Metabolism

Glucagon increases hepatic glucose production by stimulating glycogenolysis, gluconeogenesis, along with concomitant inhibition of glycolysis and glycogenesis (Fig. 16.9). It also stimulates the catabolism of lipids and amino acids. In addition, it stimulates a positive effect on heart rate and contractility and inhibits gastric acid secretion and appetite. Altogether, glucagon is a catabolic hormone that profoundly affects intermediary metabolism by stimulating hyperglycemia, ketosis, and ureagenesis.

16.2.1.1.4 Effect on Carbohydrate Metabolism

It stimulates the transcription of glucose 6-phosphatase (G6Pase) and phosphoenolpyruvate carboxykinase (PEPCK) genes that help in enhancing gluconeogenesis. In addition, activated PKA phosphorylates the pyruvate kinase (PK) to suppress glycolysis. Whereas, PKA-dependent activation of glycogen phosphorylase stimulates the release of glucose from the glycogen. The resulting increased glucose output from the liver acts to correct the hypoglycemic state. The increased glucose levels further stimulate the release of insulin from the pancreas, which will help in the mobilization of glucose in peripheral tissues. Henceforth, the hyperglycemic effect of glucagon is essential to ensure the proper functioning of absolute glucose-dependent tissues such as the brain and skeletal muscles.

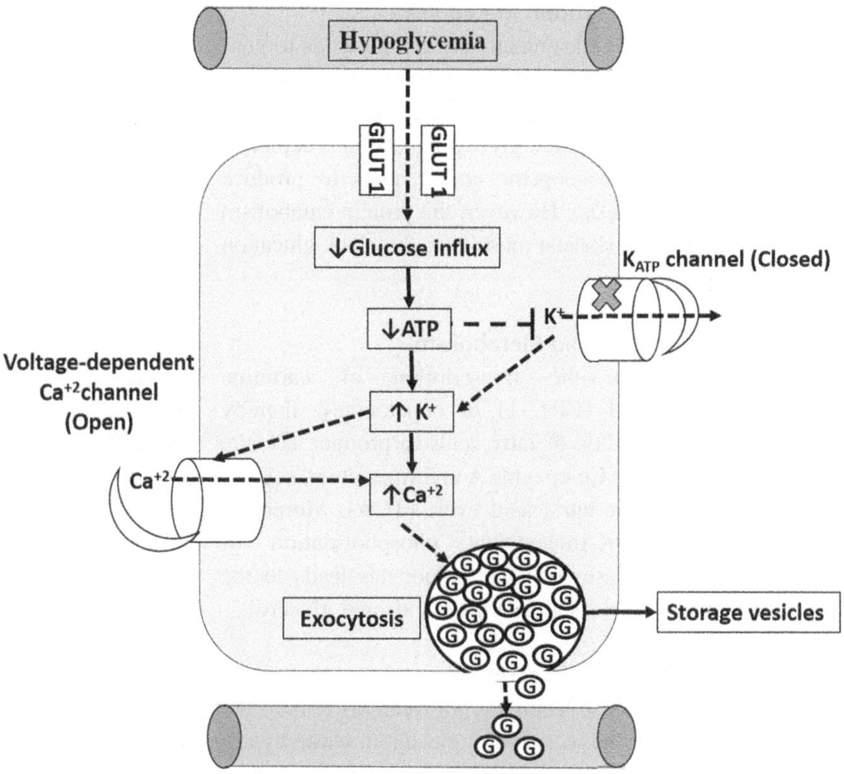

Fig. 16.8 Mechanism of glucagon secretion by α-cell. [The reduced ATP generation during hypoglycemia inhibits the outward movement of K^+, resulting in the depolarization of α-cells and exocytosis of glucagon. [*GLUT1* glucose transporter 1; *ATP* adenosine triphosphate; *G* glucagon; K^+ potassium; Ca^{2+} calcium; ↓ decrease; ↑ increase]

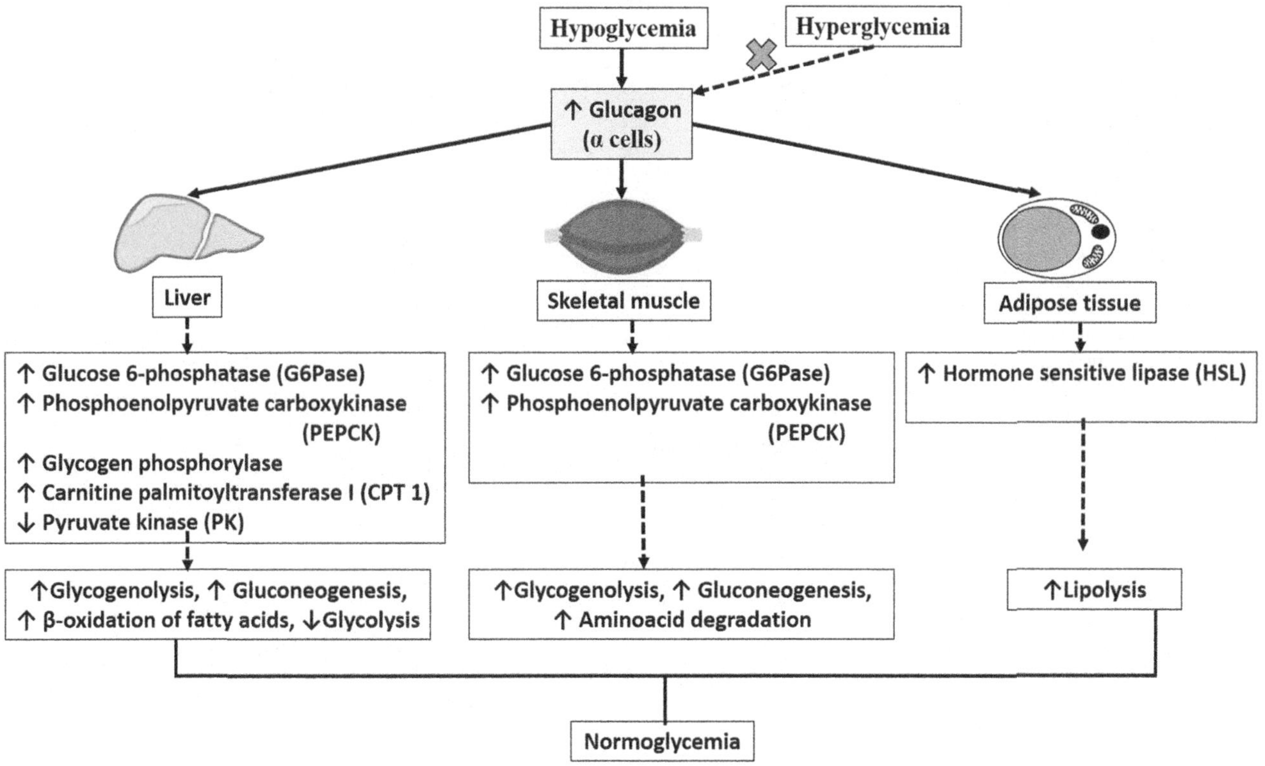

Fig. 16.9 Effects of glucagon on intermediary metabolism. [Glucagon increases circulatory levels of glucose by stimulating glycogenolysis, gluconeogenesis both in the liver and skeletal muscle. In addition, it prevents the utilization of glucose by increasing the lipolysis in adipose tissue and amino acid catabolism in skeletal muscles]

16.2.1.1.5 Effect on Protein Metabolism

It stimulates the uptake, deamination of amino acids to generate ATP in the liver and further facilitates the conversion of ammonia to urea by inducing the enzymes involved in ureagenesis. When the hepatic glycogen stores are depleted, glucagon recruits gluconeogenic amino acids to produce glucose (gluconeogenesis). However, the protein catabolism does not take place in skeletal muscles as they lack glucagon receptors.

16.2.1.1.6 Effect on Lipid Metabolism

Glucagon stimulates the transcription of carnitine palmitoyltransferase I (CPT 1) in hepatocytes, thereby activating the β-oxidation of fatty acids to produce acetate. The acetate reacts with Co-enzyme A to form acetyl-CoA and is metabolized via the citric acid cycle (TCA). Moreover, glucagon-induced PKA-dependent phosphorylation of hormone-sensitive lipase (HSL) in adipocytes leads to the catabolism of triglycerides to free fatty acids and glycerol.

16.2.1.1.7 Regulation of Secretion

Decreased blood glucose levels (hypoglycemia) remain the primary stimulus for the secretion of glucagon, while hyperglycemia has the opposite effect. In addition, ingestion of a protein-rich diet or increased levels of glutamine or alanine, cortisol, and β-adrenergic activity stimulates the release of glucagon. Other pancreatic hormones such as insulin and somatostatin act in a paracrine manner to inhibit the secretion of glucagon.

16.2.1.2 Insulin

Produced by the β-cells in islets of Langerhans, insulin is a heterodimeric polypeptide hormone consisting of A and B chains held together by two disulfide bridges. The A chain consists of 21 amino acids with an intra-chain disulfide bridge, whereas the B chain has 30 amino acid residues. The precursor molecule proinsulin will be acted upon by a trypsin-like enzyme to produce the mature insulin and C-peptide. The mature hormone along with C-peptide is stored as secretory vesicles in the cytoplasm and released when the need arises.

16.2.1.2.1 Mechanism of Secretion

Insulin is primarily secreted due to high circulatory glucose levels. The glucose passes through the cell membrane of β-cells through GLUT2, phosphorylated by glucokinase and subsequently metabolized to generate ATP. The energy thus produced decreases the efflux of K^+ by inhibiting the K_{ATP} channels. The accumulation of K^+ ions stimulates the ion gated Ca^{2+} channels, thereby depolarizing the cell and facilitating the exocytosis of insulin. The concomitant fall of ATP levels during hypoglycemia leads to the hyperpolarization of cells, thereby inhibiting the secretion of insulin (Fig. 16.10).

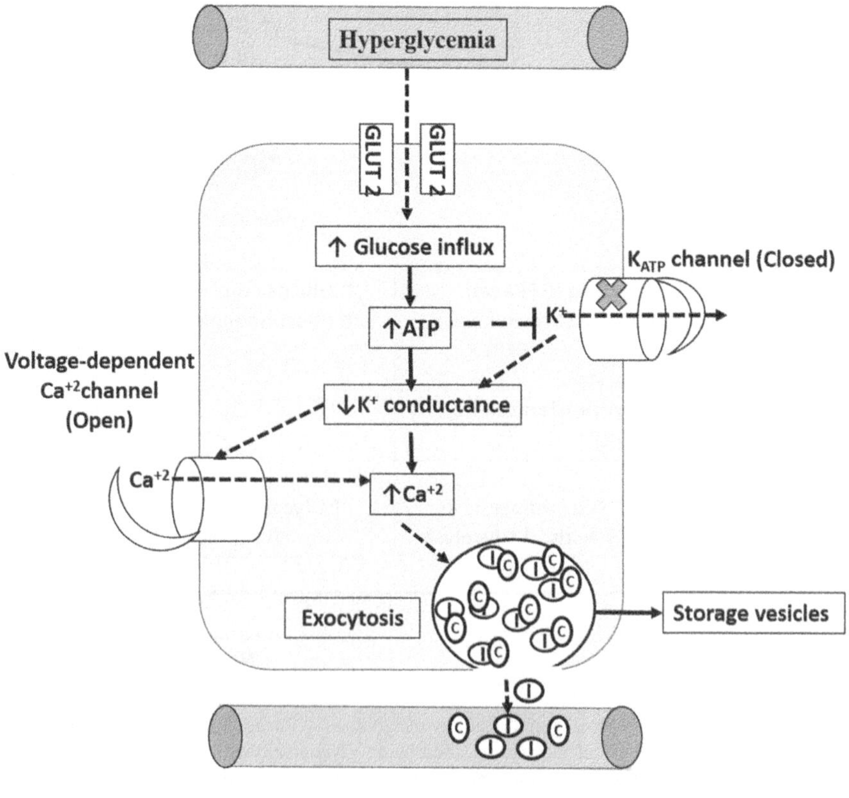

Fig. 16.10 Mechanism of insulin secretion by β-cells [Increased glucose entry during hyperglycemia prevents the conductance of K^+ ions, resulting in the opening of voltage-dependent Ca^{2+} channels and subsequent exteriorization of insulin from the storage vesicles. [*GLUT2* glucose transporter 2; *ATP* adenosine triphosphate; *I* insulin; *C* C-peptide; K^+ potassium; Ca^{2+} calcium; ↓ decrease; ↑ increase]

16.2.1.2.2 Mechanism of Action

The transmembrane insulin receptor (IR) is a disulfide-linked dimer that belongs to the receptor tyrosine kinase (RTK) family and plays a pivotal role in eliciting the downstream signaling pathways. The intracellular domain of IR has a tyrosine kinase activity, which will be activated when insulin binds to the extracellular region. The activated tyrosine kinase phosphorylates the tyrosine residues outside the kinase domain and these in turn act as sites for various docking proteins such as insulin receptor substrate 1-6 (IRS 1-6) and Shc (Src homology 2 domain containing). They further mediate the activation of the PI3K/AKT and Raf/Ras/MEK/MAPK pathways that are responsible for various biological effects.

16.2.1.2.3 Biological Effects

Insulin is the only hypoglycemic hormone acting against all other hyperglycemic hormones (GH, THs, cortisol, and catecholamines). Hence, it is widely considered as an important regulator of metabolism in animals. Principally, insulin regulates carbohydrate metabolism by affecting glycogenesis, glycogenolysis, gluconeogenesis, and glycolysis. In addition, it also regulates protein and lipid metabolism. With liver, skeletal muscles, adipose tissue, and endothelium as major target tissues, insulin has an overall anabolic effect on intermediary metabolism (Fig. 16.11). Moreover, the control on a wide range of metabolic pathways confers the permissive effect of insulin on the actions of GH.

16.2.1.2.3.1 Effect on Glucose Metabolism

Insulin-dependent regulation of glucose metabolism is mainly due to the activation of the PI3K/AKT pathway in target tissues. Activation of the PI3K/AKT pathway in skeletal muscles and adipose tissue enhances cellular uptake of glucose by stimulating the integration of GLUT4 on their cell membrane. GLUT4 mediated facilitated diffusion of glucose into these cells with its consequent entry into various metabolic pathways remains as the archetype effect of insulin. Thus, the increased glucose uptake increases glycolysis in both skeletal muscle and adipose tissue, while increased glycogen synthesis happens only in skeletal muscle. In the liver, insulin inhibits glycogen phosphorylase and side by side activates glycogen synthase resulting in decreased glycogenolysis with a simultaneous increase in glycogenesis, respectively. In addition, the insulin-dependent downregulation of PEPCK, G6Pase, and fructose-1, 6-bisphosphatase (FBP) genes inhibits gluconeogenesis in hepatocytes. Overall, insulin produces hypoglycemia by increasing glucose uptake in the liver and other peripheral tissues. Together with increased glycogenesis and decreased glycogenolysis, gluconeogenesis results in imparting the anabolic effect of insulin on glucose metabolism.

16.2.1.2.3.2 Effect on Amino Acid Metabolism

Insulin increases the skeletal muscle uptake of amino acids such as isoleucine, leucine, tyrosine, phenylalanine, and valine. Thus, the increased amino acid uptake facilitates the

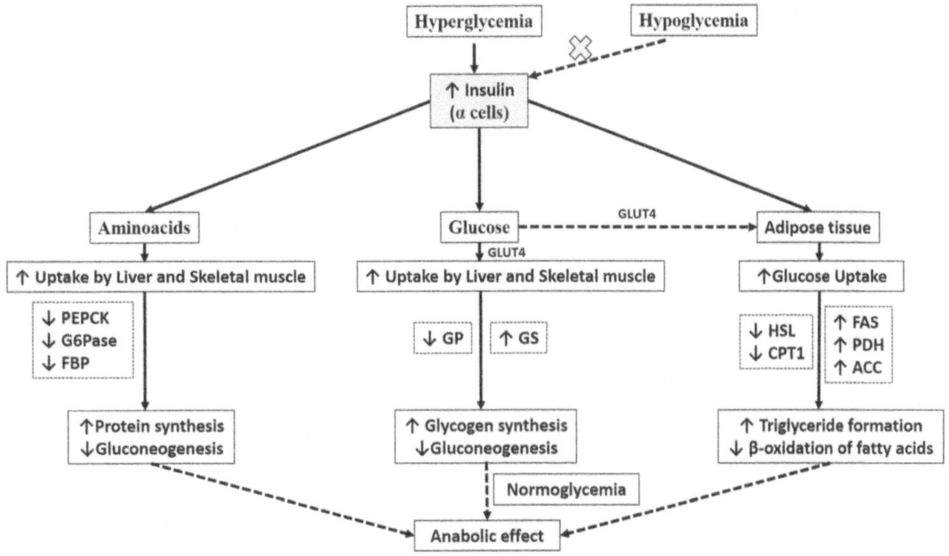

Fig. 16.11 Anabolic effects of Insulin on intermediary metabolism. [Insulin establishes normoglycemia by stimulating the glucose uptake, glycogenesis, and lipogenesis in liver, skeletal muscles, and adipose tissues. [*GLUT4* glucose transporter 4; *PEPCK* phosphoenolpyruvate carboxykinase; *G6Pase* glucose 6-phosphatase; *FBP* fructose-1, 6-bisphosphatase; *GP* glycogen phosphorylase; *GP* glycogen synthase; *HSL* hormone-sensitive lipase; *CPT1* carnitine O-palmitoyltransferase-1; *FAS* fatty acid synthase; *PDH* pyruvate dehydrogenase; *ACC* acetyl-CoA carboxylase (ACC); ↓ decrease; ↑ increase]

formation of new proteins along with a reduction in amino acid catabolism. Moreover, increased uptake of amino acids and inhibition of gluconeogenesis facilitate protein synthesis in the liver.

16.2.1.2.3.3 Effect on Lipid Metabolism

The insulin-mediated inhibition of hormone-sensitive lipase (HSL), carnitine O-palmitoyltransferase-1 (CPT1) reduces lipolysis and β-oxidation in adipocytes. Furthermore, increased glucose uptake along with upregulation of fatty acid synthase (FAS), pyruvate dehydrogenase (PDH), and acetyl-CoA carboxylase (ACC) genes helps in lipogenesis. Therefore, the formation of lipid stores along with a reduction in their breakdown results in the anabolism of lipids.

16.2.1.2.4 Regulation of Secretion

The rise in blood glucose levels (hyperglycemia) is the major metabolic stimulus for the secretion of insulin. However, an increase in circulatory levels of fatty acids, amino acids, GH, cortisol, gastrin, secretin, and cholecystokinin (CCK) positively regulates the secretion of insulin. However, hypoglycemia, somatostatin, and leptin inhibit the secretion of insulin.

16.2.1.3 Somatostatin

Produced from δ-cells in the pancreas, enteroendocrine D-cells, and in the brain (GHIH from the hypothalamus). Somatostatin secreted from the pancreas and intestine has 28 amino acids (SS-28) while the hypothalamic type has 14 amino acids (SS-14). SS-14, SS-28 were first isolated in the ovine brain and porcine gut, respectively.

16.2.1.3.1 Mechanism of Action

It binds to the somatostatin receptors (SSTRs) belonging to the GPCRs family. Activation of SSTR inhibits adenylyl cyclase (AC), leading to reduced intracellular cAMP and Ca^{2+} levels that further inhibit hormone secretion from target tissues.

Know More...

- Insulin is the first peptide hormone/protein to be sequenced by Fredrick Sanger in 1955.
- β-cell is the major cell type present in the islets of Langerhans.
- Diabetes mellitus (DM): A pathological condition due to a reduction in insulin production (Type-I DM) or in the number of insulin receptors (Type-II DM), characterized by hyperglycemia, ketosis, and skeletal muscle depletion.
- Bovine insulin and porcine insulin differ from human insulin only by 3 and 1 amino acids, respectively. Hence, they were used in treating diabetes mellitus in humans in the twentieth century.
- Although glucagon and insulin have antagonistic effects on various metabolic pathways, they both increase the cellular uptake of amino acids.

16.2.1.3.2 Biological Effects

Generally, it is a negative regulator of neuroendocrine, pancreatic, and GIT hormones. It inhibits the secretion of growth hormone (GH), thyroid-stimulating hormone (TSH), and prolactin (PRL) in the brain. It inhibits the secretion of insulin and glucagon in the pancreas in a paracrine manner by stimulating the efflux of K^+ with subsequent inhibition of Ca^{2+} influx. In addition, it inhibits the secretion of bile acids, gastric acid, pancreatic enzymes mainly by inhibiting the secretion of GIT hormones such as CCK, VIP, and gastrin.

16.2.1.3.3 Regulation of Secretion

GH, GHRH, and glucose regulate the secretion of the hypothalamic SS-14. The gastric SS-28 is primarily stimulated by the autonomic nervous system (ANS), CCK, and gastrin. In addition, Substance P produced in the intestine has a negative effect on the secretion of SS-28.

16.2.1.4 Pancreatic Polypeptide (PP)

Secreted from F-cells, PP is a peptide hormone with 36 amino acids and belongs to the neuropeptide Y (NPY) family of proteins. It binds to the Y4 receptor, a GPCR belonging to the NPY receptor family. When activated, it inhibits the AC system resulting in a reduction of cAMP levels. PP inhibits gastric motility, gall bladder contraction, and exocrine pancreatic secretion. In addition, it stimulates the secretion of gastric juice and suppresses anxiety. Its secretion is increased by a protein-rich diet, exercise, and fasting.

16.3 Adrenal Gland

16.3.1 Introduction

The adrenal gland located cranially on each kidney consists of an outer cortex and inner medulla, functions as two discrete endocrine glands with distinct embryological origins and endocrine activities. With a mesodermal origin, the adrenal cortex secretes cholesterol-derived hormones that are collectively known as corticosteroids. Whereas, the adrenal medulla derived from neural ectoderm is involved in the production of tyrosine-derived catecholamines. Together, corticosteroids and catecholamines help in regulating glucose

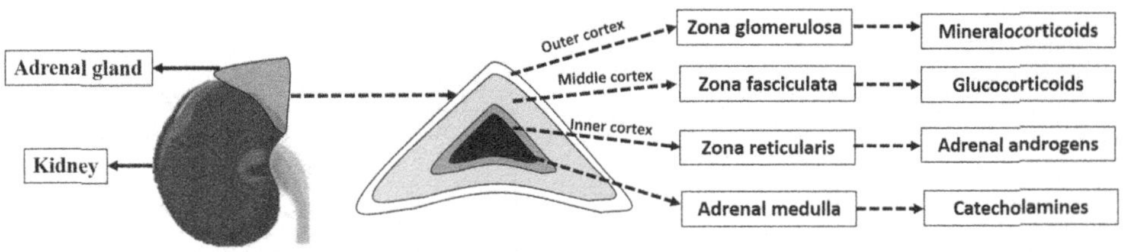

Fig. 16.12 Adrenal gland and its histology. [The adrenal gland comprises two distinct regions, i.e., cortex and medulla, attributed to secrete corticosteroids and catecholamines, respectively. The adrenal cortex has three layers, which are functionally and histologically different]

metabolism, electrolyte balance, and antagonize stressors. Hence, appropriate functioning of the adrenal gland is essential for an animal's survival.

16.3.2 Adrenal Cortex: Histology

The adrenal cortex has three different histological zones, namely: the outer zona glomerulosa, the central zona fasciculata, and the inner zona reticularis (Fig. 16.12). The presence of zone-specific hydroxylases in adreno-cortical cells helps in converting cholesterol to different classes of zone-specific steroid hormones. The corticosteroids are not stored in the cortical cells but their synthesis is rapidly stimulated in response to specific stimuli (Table 16.3).

16.3.3 Mechanism of Synthesis of Corticosteroids

The cholesterol required for the synthesis of corticosteroids is primarily derived from the circulation although a smaller proportion is derived from the de novo synthesis. The abundant lipid stores, mitochondria, and smooth endoplasmic reticulum are the major characteristics of cortical cells. The cholesterol influx or de novo synthesis from the cellular lipid stores depends on various stimuli like the adreno-corticotropin hormone (ACTH), altered ionic concentrations (K^+), etc. The cholesterol is transported into the inner mitochondrial membrane from the outer mitochondrial membrane by steroidogenic acute regulatory protein (StAR). Subsequently, the cholesterol is converted to pregnenolone by CYP11A1 (p450scc/Cholesterol desmolase). The formation of pregnenolone from cholesterol is a rate-limiting step that is primarily stimulated by ACTH. The pregnenolone will be further acted upon by zone-specific hydroxylases to be converted into mineralocorticoids in zona glomerulosa or glucocorticoids in zona fasciculata or sex steroids in zona reticularis.

16.3.4 Zona Glomerulosa: Site of Synthesis for Mineralocorticoids

The zona glomerulosa is a thin outermost layer with columnar cells arranged in irregular cords. They are responsible for the secretion of a class of hormones known as mineralocorticoids, which are implicated in regulating major electrolytes (Na^+, K^+) present in the blood. Aldosterone is the potent and major mineralocorticoid secreted across different species of animals. In addition, corticosterone and 11-deoxycorticosterone have slight mineralocorticoid activity. Within these cortical cells, pregnenolone is converted to progesterone by 3β-hydroxysteroid dehydrogenase (3βHSD). Then the subsequent conversion of progesterone by 21β-hydroxylase results in the production of 11-deoxycorticosterone. Further, the 11-deoxycorticosterone is converted to corticosterone by 11 β-hydroxylase. Finally, the aldosterone synthase that is present exclusively in the zona glomerulosa converts

Table 16.3 List of hormones secreted from the adrenal gland, their chemical nature, effect, and half-life in circulation

S. No	Part of adrenal	Hormones	Precursor	Effect	Half-life
1.	Zona glomerulosa (Cortex)	Mineralocorticoids (Aldosterone)	Cholesterol	Increase blood volume, hypokalemia	≈20 min
2.	Zona fasciculata (Cortex)	Glucocorticoids (Cortisol, corticosterone, 11-deoxy corticosterone)	Cholesterol	Increase blood glucose levels (Catabolic)	60–90 min
3	Zona reticularis (Cortex)	Androgens (DHEA, androstenedione)	Cholesterol	Anabolic and masculine effects	≈20 h
4.	Adrenal medulla	Catecholamines (Epinephrine, norepinephrine)	Tyrosine	Fight or flight response	2–3 min

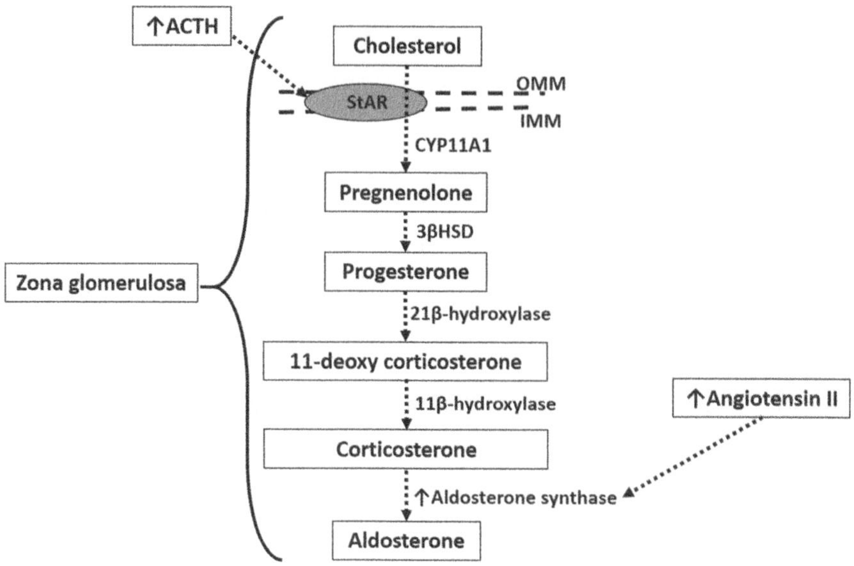

Fig. 16.13 Biosynthesis of aldosterone in zona glomerulosa [Produced from cholesterol, the corticosterone is converted exclusively in the zona glomerulosa by the enzyme aldosterone synthase to yield aldosterone. [*ACTH* adrenocorticotropic hormone; *StAR* steroidogenic acute regulatory protein; *OMM* outer mitochondrial membrane; *IMM* inner mitochondrial membrane; *CYP11A1* cholesterol side-chain cleavage enzyme; *3βHSD* 3β-hydroxysteroid dehydrogenase]

corticosterone to aldosterone (Fig. 16.13). The enzymatic activity of aldosterone synthase is regulated by angiotensin-II. Furthermore, the inability of zona glomerulosa cells to secrete cortisol or other androgens is due to the absence of 17 α-hydroxylase.

16.3.4.1 Mechanism of Action

Aldosterone exerts its biological effects mainly by binding to intracytoplasmic mineralocorticoid receptors (MR). Once the hormone-receptor complex is formed, it migrates to the nucleus and stimulates the transcription of Na^+-K^+ ATPase and epithelial sodium channels (ENaC) genes. Hence, the effects of aldosterone are not evident soon after its release and require a brief period. Most importantly, the principal cells (PC) and intercalated cells (IC) are recognized as the major cellular targets of aldosterone.

16.3.4.2 Biological Effects

The restoration of normal circulatory levels of Na^+ and K^+ by inhibiting natriuresis with a concomitant rise in potassium secretion from kidneys is regarded as the crucial biological effects. In addition, the rise in systemic circulatory volume and arterial blood pressure are secondary effects due to increased reabsorption of water from the renal tubules.

16.3.4.3 Effect on Principal Cells

The principal cells are present in the late distal tubule and effectively contribute to the reabsorption of Na^+ and secretion of K^+. The Na^+-K^+ ATPase located on the basolateral membrane pumps Na^+ ions in exchange for K^+ ions from blood, resulting in the establishment of low Na^+ and high K^+ concentrations inside. This resultant decrease in the intracellular Na^+ concentration facilitates its influx from the tubular filtrate through ENaC. During the reabsorption of Na^+, K^+ ions are secreted down the concentration gradient to maintain electrical neutrality. Furthermore, the reabsorption of Na^+ leads to the simultaneous movement of water and leads to a minor or no increase in the circulatory Na^+ levels. Whereas, the increased secretion of K^+ ions leads to decreased circulatory levels of K^+ (hypokalemia) (Fig. 16.14).

16.3.4.4 Effect on Intercalated Cells

Intercalated cells (IC) are the other type of distal tubular cells affected by aldosterone. It stimulates the H^+ ATPase/H^+-K^+ ATPase pumps present on the apical membrane to secrete H^+ ions and reabsorb K^+ ions. Thus, this particular activity of IC is critical for the excretion of H^+ ions and imparting a regulatory role in acid-base balance. In addition, there is an interdependency between the functioning of IC and PC.

16.3.4.5 Regulation of Secretion

Although ACTH is necessary for the production of aldosterone, the circulatory concentration of K^+ ions is by far the most potent stimulator for its secretion. The angiotensin-II is the second most potent stimulator of aldosterone production. It increases the secretion of aldosterone by directly acting on the zona glomerulosa cells and by stimulating the production of ACTH from the anterior pituitary. The initiation of the renin–angiotensin–aldosterone system (RAAS) plays an important role in regulating the circulatory volume and arterial pressure. Moreover, an increase in the circulatory concentration of Na^+ ions suppresses the secretion of aldosterone.

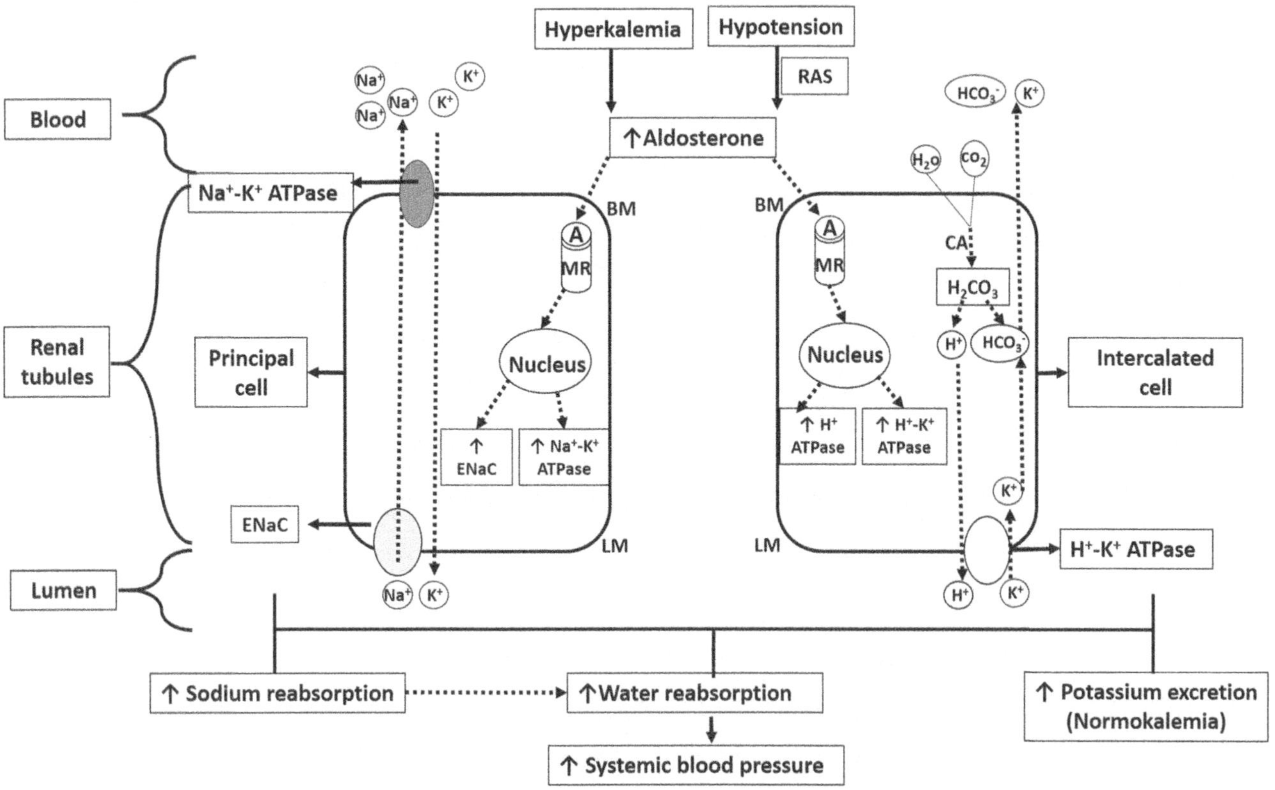

Fig. 16.14 Biological effects of aldosterone [Aldosterone acts on the principal and intercalated cells in the distal convoluted tubules to increase the tubular reabsorption of Na^+ along with water and tubular excretion of K^+ ions. [*A* aldosterone; *MR* mineralocorticoid receptor; *RAS* renin-Angiotensin system; Na^+ sodium ion; K^+ potassium ion; HCO_3 bicarbonate ion; *ENaC* epithelial sodium channels; H_2CO_3 carbonic acid; *CA* carbonic anhydrase; *BM* basal membrane; *LM* luminal membrane; ↓ decrease; ↑ increase]

16.3.5 Zona Fasciculata: Site of Synthesis for Glucocorticoids

The zona fasciculata is the middle and broadest layer of adrenal cortex comprising polyhedral cells. They secrete a group of hormones known as glucocorticoids, which are implicated in regulating metabolic pathways that enable the animals to endure various stressors. The presence of 17 α-hydroxylase in fascicular cells converts pregnenolone to 17-hydroxypregnenolone. Then, it is further converted to 17-hydroxyprogesterone by the enzyme 3βHSD. Consequent action of 21β-hydroxylase results in the conversion of 17 hydroxy-progesterone to 11-deoxycortisol. Finally, the 11-deoxycortisol is converted by 11 β-hydroxylase to produce cortisol, the predominant glucocorticoid in animals. Additionally, pregnenolone is converted to progesterone and further to 11-deoxycorticosterone and corticosterone, both of which are found in the circulation as minor glucocorticoids (Fig. 16.15). Cortisol then secreted will be bound to transcortin (corticosteroid-binding globulin) present in the circulation.

16.3.5.1 Mechanism of Action

Glucocorticoids exert their effects on the target tissues by binding to glucocorticoid receptors (GCR) localized in the cytoplasm. Generally, the GCRs in their inactive state are bound with heat shock protein 90 (Hsp90). Upon their interaction with glucocorticoids will result in the dissociation of Hsp90 with a simultaneous translocation of the hormone-receptor complex to the nucleus. In the nucleus, hormone-receptor complexes dimerize and bind to DNA (glucocorticoid response element) to regulate the transcription of genes.

16.3.5.2 Biological Effects of Glucocorticoids

The glucocorticoids have a prominent role in maintaining energy homeostasis by regulating metabolic pathways in the liver, adipose tissue, and skeletal muscles. Glucocorticoids increase gluconeogenesis, lipolysis, and glycogenolysis in various tissues resulting in hyperglycemia, mobilization of fat stores, and depletion of protein reserves. Moreover, it has a potent suppressive effect on the animal's immune system. In total, the secretion of glucocorticoids is obligatory for animal survival by regulating their metabolism to drive the functioning of various vital organs, especially the brain (Fig. 16.16).

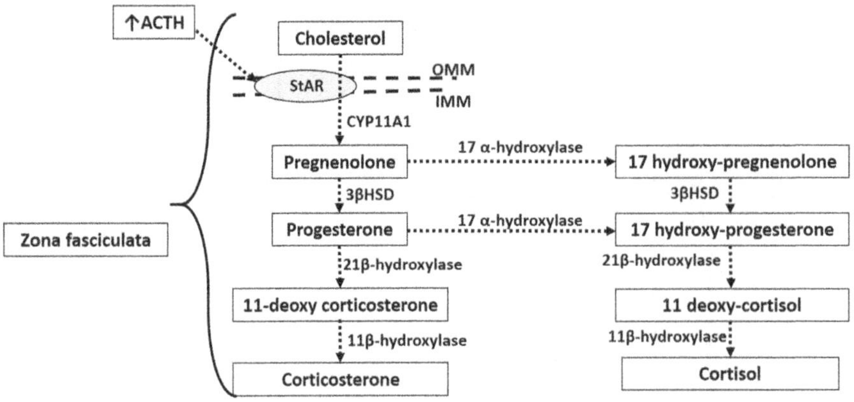

Fig. 16.15 Biosynthesis of glucocorticoids in zona fasciculata. [The cholesterol taken up by the fascicular cells is enzymatically converted to produce glucocorticoids such as corticosterone and cortisol. [*ACTH* adrenocorticotropic hormone; *StAR* steroidogenic acute regulatory protein; *OMM* outer mitochondrial membrane; *IMM* inner mitochondrial membrane; *CYP11A1* cholesterol side-chain cleavage enzyme; *3βHSD* 3β-hydroxysteroid dehydrogenase]

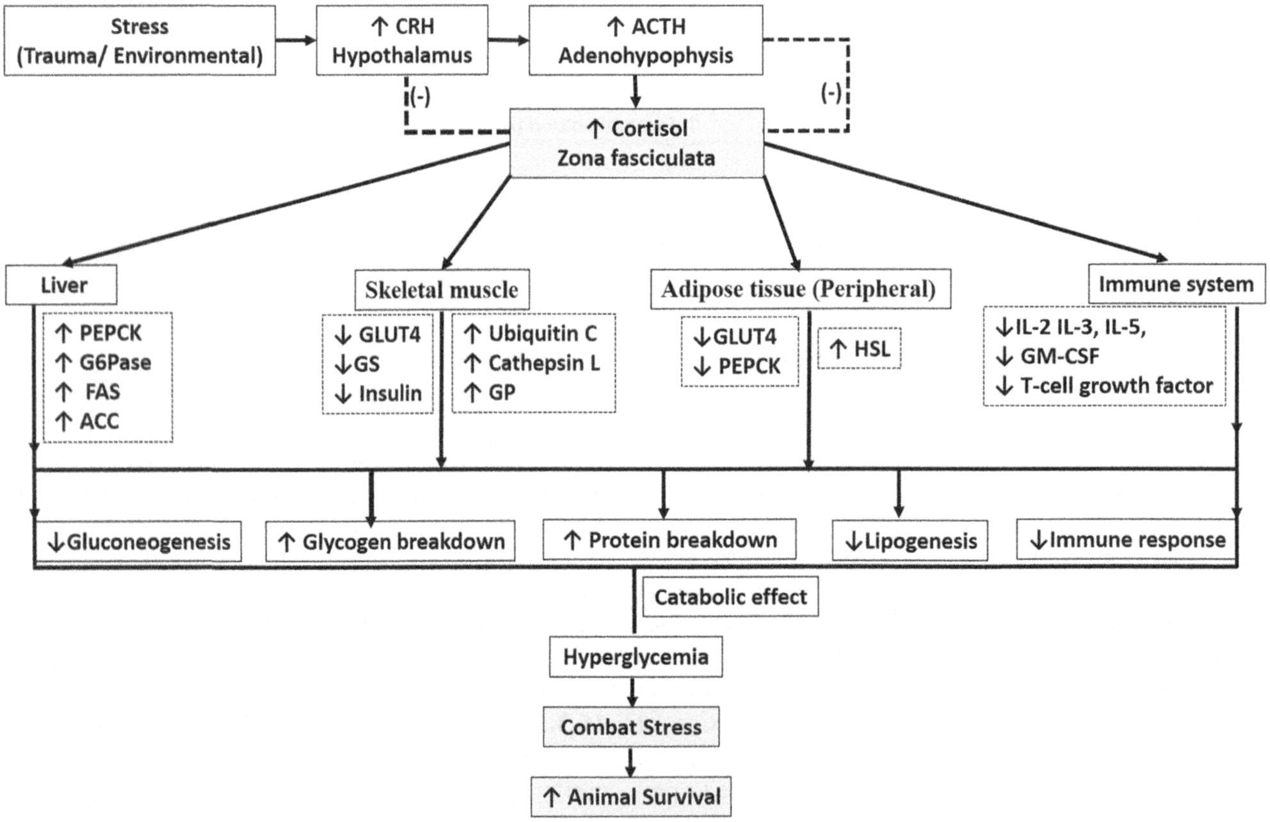

Fig. 16.16 Effects of cortisol on intermediary metabolism and its regulation of secretion [Glucocorticoids increase the catabolism of lipids and amino acids for cellular metabolism, thereby maintain a constant elevated blood glucose levels to combat stressful conditions. [*CRH* corticotropin-releasing hormone; *ACTH* adrenocorticotropic hormone; *PEPCK* phosphoenolpyruvate carboxykinase; *G6Pase* glucose 6-phosphatase; *GLUT4* glucose transporter 4; *FAS* fatty acid synthase; *ACC* acetyl-CoA carboxylase (ACC); *FBP* fructose-1, 6-bisphosphatase; *GP* glycogen phosphorylase; *GP* glycogen synthase; *HSL* hormone-sensitive lipase, *(−)* negative feedback inhibition; *IL-2* interleukin 2; *IL-3* interleukin 3; *IL-5* interleukin 5; *GM-CSF* granulocyte-macrophage colony-stimulating factor; ↓ decrease; ↑ increase]

16.3.5.2.1 Effect on Hepatic Metabolism

Glucocorticoids stimulate gluconeogenesis and glycogenolysis, respectively, by stimulating the transcription of key genes such as PEPCK and G6Pase. This helps in increased hepatic glucose production and released into the circulation. They also upregulate FAS and ACC genes, which leads to increased lipogenesis. Along with lipogenesis, simultaneous inhibition of the β-oxidation of fatty acids leads to an increase in hepatic lipid accumulation (hepatic steatosis).

16.3.5.2.2 Effect on Skeletal Muscle

Glucocorticoids inhibit the insulin secretion from β-cells, insulin-dependent uptake of glucose and amino acids by inhibiting the insulin-mediated PI3K/Akt signaling pathway. Thus, the inhibition of the PI3K/Akt pathway results in reduced glucose uptake and glycogenesis due to declined translocation of GLUT4 and downregulation of the glycogen synthase gene respectively. Furthermore, glucocorticoids increase the protein degradation by the proteasome, cathepsin-L, and ubiquitin C to produce free amino acids required for generating energy and glucose. Altogether, glucocorticoids produce a catabolic effect on skeletal muscles characterized by the exhaustion of glycogen and protein reserves.

16.3.5.2.3 Effect on Adipose Tissue

They activate hormone-sensitive lipase (HSL) in peripheral adipocytes, this leads to increased lipolysis and FFA production. In addition, they inhibit PEPCK and insulin-dependent glucose uptake to decrease triglyceride formation. Whereas, the same glucocorticoids stimulate the differentiation and hypertrophy of central adipocytes. Altogether, glucocorticoids redistribute lipids from peripherally located adipose tissues to central adipose depots, especially in the abdomen.

16.3.5.2.4 Effect on the Immune System

Glucocorticoids inhibit the production of cytokines like interleukin 3(IL-3), interleukin 5 (IL-5), GM-CSF which regulate maturation, differentiation, and survival of eosinophils. Their effect on inhibiting IL-2 and T-cell growth factor results in the inhibition of T cell proliferation and with a concurrent increase in T-cell apoptosis. Furthermore, glucocorticoids inhibit the migration of leucocytes by downregulating the expression of adhesion molecules and chemokines from the inflammatory site.

16.3.5.2.5 Hypothalamo–Pituitary–Adrenal (HPA Axis)

The secretion of glucocorticoids is primarily regulated through CRH and ACTH released from the hypothalamic–pituitary axis. The release of CRH is stimulated during physical stress, physiological stress, or behavioral stress. Further, the increased circulatory levels of glucocorticoids have negative feedback on the secretion of CRH and ACTH. The circadian rhythm also affects the secretion of cortisol, attaining peak secretion during the early morning.

16.3.6 Zona Reticularis and Adrenal Androgens

The dehydroepiandrosterone (DHEA) and androstenedione are the two adrenal androgens secreted from the innermost zona reticularis in response to ACTH (Fig. 16.17). The 17-hydroxypregnenolone produced from cholesterol is converted by the action of 17, 20 lyase to DHEA. Further, DHEA is converted to androstenedione by 3βHSD. They bind to albumin and sex hormone-binding globulin (SHBG) in circulation. These weak adrenal androgens cannot bind to androgen receptors and require transformation to potent forms such as testosterone and dihydrotestosterone to elicit

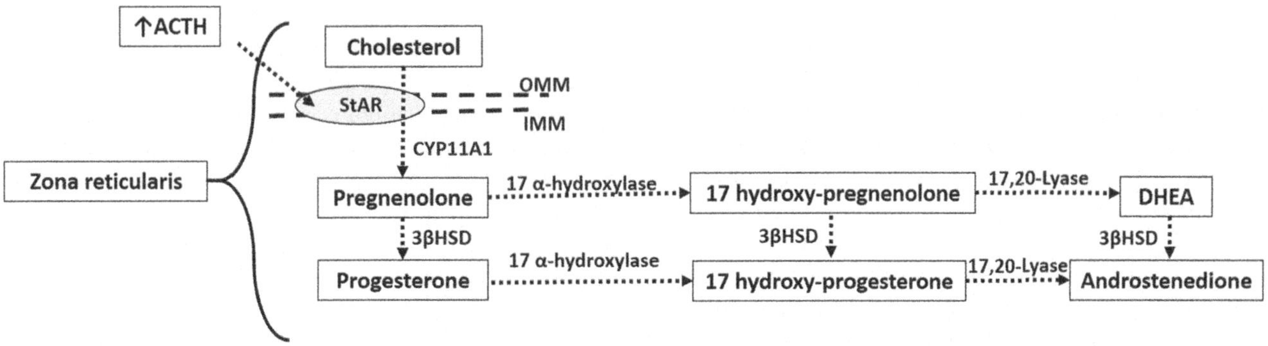

Fig. 16.17 Biosynthesis of adrenal androgens. [Zona reticularis is bestowed with the production of androgens from cholesterol, which are further converted in to active or more potent forms in the gonads. [*DHEA* dehydroepiandrosterone; *ACTH* adrenocorticotropic hormone; *StAR* steroidogenic acute regulatory protein; *OMM* outer mitochondrial membrane; *IMM* inner mitochondrial membrane; *CYP11A1* cholesterol side-chain cleavage enzyme; *3βHSD* 3β-hydroxysteroid dehydrogenase; ↑ increase]

target effects. Although the amount of adrenal androgens produced in male animals is negligible, they might play a role in producing an anabolic effect on muscle mass, bone density, and estrous behavior in female animals.

Know More...
- **Addison's disease:** Also known as hypoadrenocorticism, characterized by reduced secretion of corticosteroids.
- **Cushing's disease:** Pathological condition due to the hyperactivity of adrenal cortex.
- **Fetal cortisol** is the hormone that initiates the parturition reflex in animals.
- Circulatory cortisol levels are often used as a stress marker and useful assess well-being of animals.
- In birds, mice, and rats, corticosterone is the major glucocorticoid secreted from adrenal cortex.
- **Aldosterone escape:** The expansion of circulatory volume due to aldosterone triggers the release of atrial natriuretic peptide (ANP) from heart to induce natriuresis and diuresis.

16.3.7 Adrenal Medulla: Histology

The adrenal medulla is the innermost part of the adrenal gland, it forms one-fifth of the adrenal mass. Generally considered an extension of the sympathetic system, the adrenal medulla comprises postganglionic sympathetic cells that can secrete hormones. When stimulated by pre-ganglionic sympathetic neurons, they synthesize and secrete epinephrine (adrenaline) and norepinephrine (noradrenaline). Since these neuroendocrine cells display a high affinity toward chromium salts, they are also known as chromaffin cells.

16.3.7.1 Mechanism of Synthesis

Derived from the amino acid tyrosine, both epinephrine and norepinephrine are known as adrenal catecholamines. The tyrosine required for their synthesis is derived either from the diet or through the enzymatic conversion of phenylalanine. The further conversion of tyrosine to dihydroxyphenylalanine (DOPA) by tyrosine hydroxylase (TH) is the rate-limiting step in the biosynthesis of adrenal catecholamines. Subsequently, DOPA is formed from dopamine due to the enzymatic action of L-aromatic amino acid decarboxylase (AAAC). Later, norepinephrine is produced from dopamine under the influence of dopamine-β-hydroxylase (DBH). Additionally, 80% of chromaffin cells possess phenylethanolamine N-methyltransferase (PNMT) enzyme, which converts norepinephrine to form epinephrine. Therefore, epinephrine is the major catecholamine to be secreted from the adrenal medulla (Fig. 16.18).

16.3.7.2 Mechanism of Action

Epinephrine and norepinephrine bind to specific adrenergic receptors (AR), which are further classified into two major subtypes α and β. Epinephrine has an equal affinity toward both the receptor types, whereas norepinephrine predominantly excites β type of adrenergic receptors. The adrenergic receptors belong to the GPCRs family, upon activation they either stimulate or inhibit adenylyl cyclase (AC) and phospholipase-C (PLC) systems to produce biological effects in the target organs. Hence, the target effects of adrenal catecholamines depend on the type of receptor expressed on the target tissues.

16.3.7.3 Biological Effects

Even though the adrenal medulla is not essential for life, hormones secreted from it activate physiological and behavioral responses collectively known as "fight or flight" to overcome acute stress. The biological effects are predominant in the cardiovascular system, skeletal muscles, energy metabolism, GI tract, and kidneys.

16.3.7.3.1 Effects on the Cardiovascular System

Both epinephrine and norepinephrine directly stimulate the SA node, AV node, and Purkinje conduction system leading to an increased heart rate. In addition, they also increase the strength of myocardial contractions. Both these effects are mediated by the activation of β-ARs and downstream AC system. Furthermore, catecholamines produce α-AR-mediate vasoconstriction in the lungs, kidneys, and GIT. The concurrent vasodilation occurring in skeletal muscles due to the activation of β-ARs will lead to the redistribution of blood to them. The redistribution and vasoconstrictor effects of catecholamines lead to an increase in the systemic arterial pressure to maintain adequate blood supply to vital organs.

16.3.7.3.2 Effects on the Smooth Muscle System

Adrenal catecholamines have profound effects on smooth muscles present in various organs. The vasoconstrictor effect in different visceral organs is due to the contraction of vascular smooth muscles present in small arterioles and pre-capillary sphincter. They also act on smooth muscles present in bronchioles, GIT, and urinary bladder resulting in bronchodilation, inhibition of GIT motility, and urine retention, respectively.

16.3.7.3.3 Effects on Metabolism

The circulating catecholamines inhibit insulin and stimulate glucagon secretion by activating α-ARs and β-ARs, respectively. In addition, they stimulate glucose production through glycogenolysis in the liver and skeletal muscles. They also

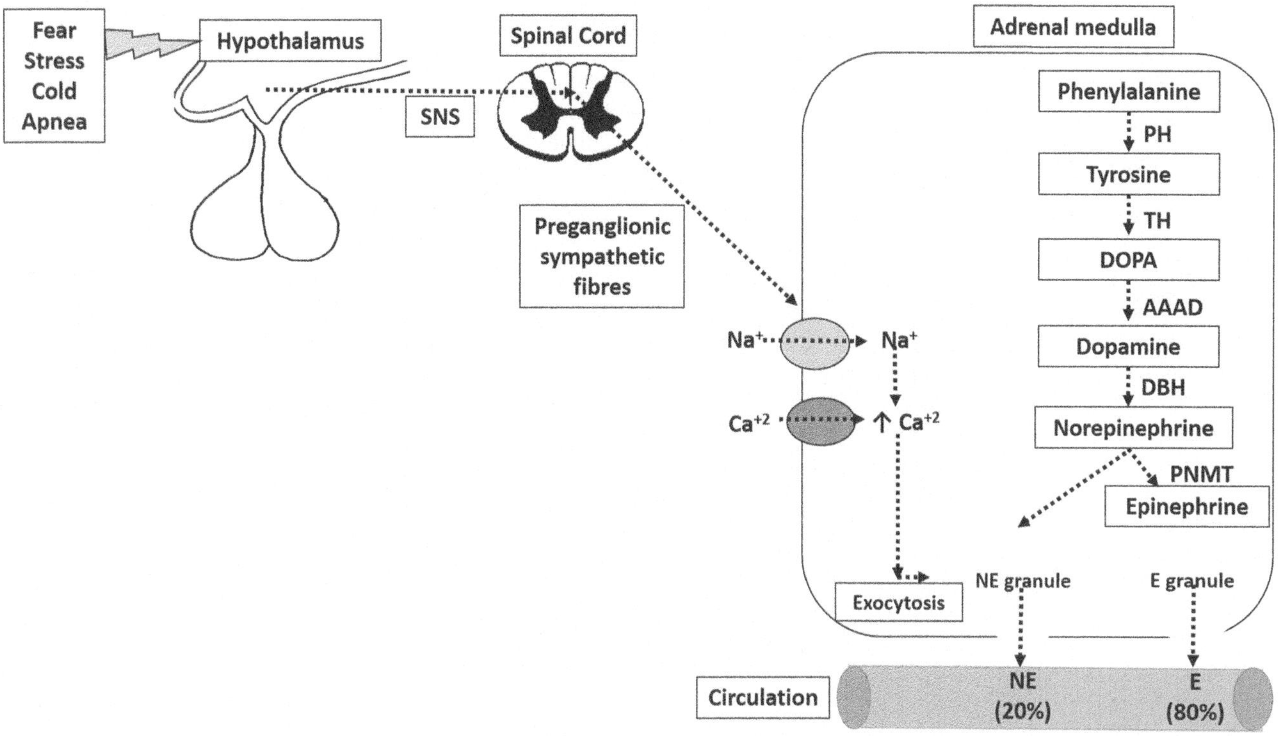

Fig. 16.18 Synthesis of adrenal catecholamines [Produced from tyrosine, the catecholamines are released into the circulation from the adrenal medulla in response to adverse situations such as fear, stress, cold, and apnea. [*SNS* sympathetic nervous system; *PH* phenylalanine hydroxylase; *TH* tyrosine hydroxylase; *AAAD* L-aromatic amino acid decarboxylase; *DBH* dopamine-β-hydroxylase; *PNMT* phenylethanolamine N-methyltransferase; *E* epinephrine; *NE* norepinephrine; Na^+ sodium ion; Ca^{2+} calcium ion; ↑ increase]

augment lipolysis by activating triglyceride lipase in adipose tissue to release free fatty acids. The aforementioned effects of adrenal medullary hormones will lead to an increased concentration of glucose and free fatty acids in the blood. The resultant changes in metabolism will help the skeletal muscles, heart, and brain to function normally even during adverse conditions.

16.3.7.3.4 Effect on Skeletal Muscles

The redistribution of blood to skeletal muscles from visceral organs will help in meeting the nutrients required for increased muscular activity seen during flight or fight response. They stimulate glycolysis and β-oxidation of fatty acids for deriving the energy required for muscular contraction (Fig. 16.19).

16.3.7.3.5 Miscellaneous Effects

Other effects of catecholamines include natriuresis, activation of the renin–angiotensin system, mydriasis, and inhibition of micturition. They also play a crucial in learning and memory consolidation.

16.3.7.4 Regulation of Secretion

The secretion of adrenal catecholamines is mainly due to the stimulation of the sympathetic nervous system (SNS) by cold, apnea, physical or environmental stress, and fear.

16.4 Calcium Homeostasis

16.4.1 Introduction

Nearly 99% of the total calcium is present in bones and teeth, with 1% inside cells and 0.1% in circulation. The ionized calcium (Ca^{2+}) present in circulation plays an indispensable role in blood coagulation, excitation of neurons, and contraction of smooth muscle/cardiac cells/skeletal muscles. Furthermore, a rise in the intracellular Ca^{2+} ions is a prerequisite for the exocytosis of enzymes and hormones. Hence, hormone-mediated calcium homeostasis of circulatory Ca^{2+} is critical for growth, reproduction, and production in animals. Parathormone (PTH), calcitriol, and calcitonin constitute the triad of hormones implicated in calcium homeostasis (Table 16.4).

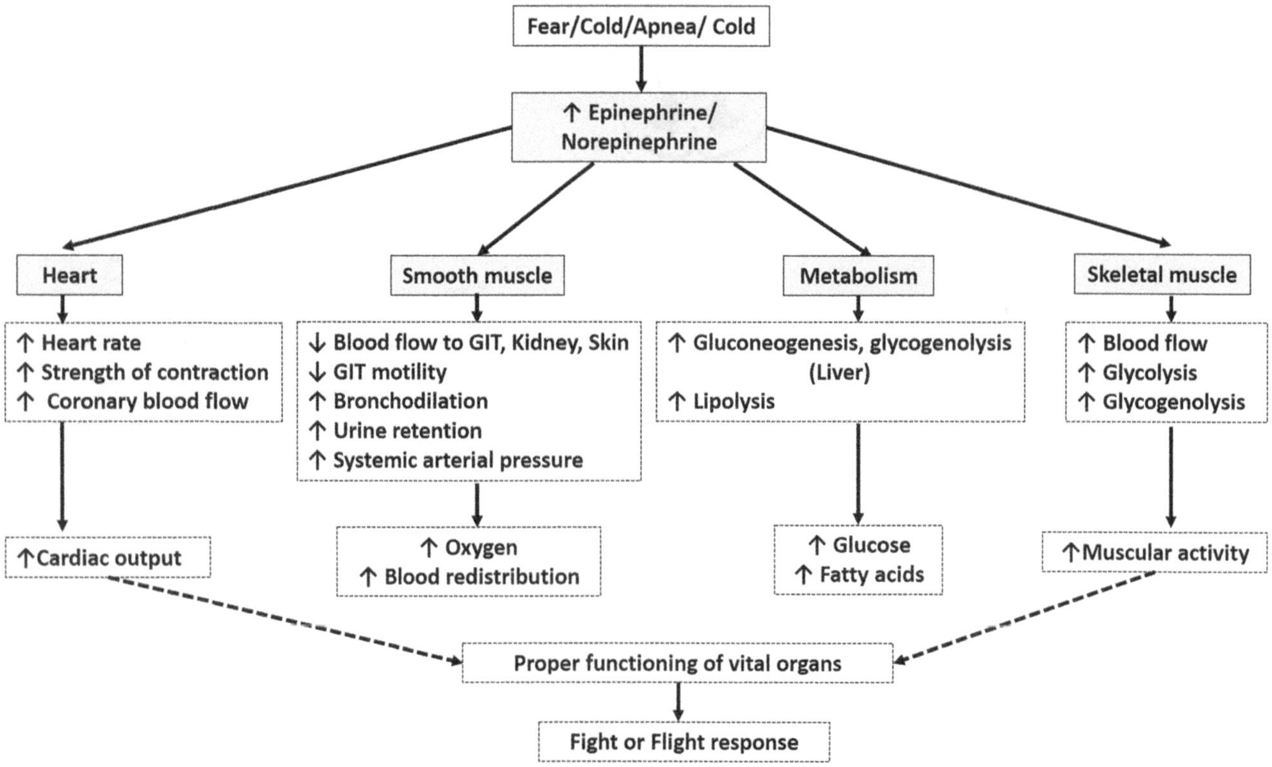

Fig. 16.19 Biological effects of catecholamines. [Catecholamines increase the heart rate, blood flow to skeletal muscles, and catabolism of glucose and fatty acids to support the increased skeletal muscular activity to either fight or avert a threat in animals. [↑ increase; ↓ decrease]

Table 16.4 Hormones involved in calcium (Ca2+) homeostasis

S. No	Source	Hormones	Chemical nature	Effect	Half-life
1.	Chief cells (Parathyroid)	Parathormone	Polypeptide 88 aa	Increase blood Ca^{2+} (hypercalcemic), phosphaturic	2–4 min
2.	Kidney	1,25-dihydroxy cholecalciferol (Calcitriol)	Cholesterol derivative	Increase absorption of Ca^{2+} in the intestine	3–6 h
3.	C-cells (Thyroid)	Calcitonin	Polypeptide 32 aa	Decrease blood Ca^{2+} (hypocalcemic)	10 min

16.4.2 Parathormone (PTH)

PTH is a single-chain polypeptide hormone isolated initially in the bovine parathyroid gland. It is synthesized and secreted from chief cells residing in the parathyroid gland. Oxyphil cells are an additional type of inactive cells present in the parathyroid gland. PTH is initially synthesized as a precursor polypeptide with 115 amino acids. Successive modifications in the precursor molecule yield an active form of PTH with 88 amino acid residues. The initial 34 amino acids confer biological activity of PTH on its target tissues.

16.4.2.1 Mechanism of Action

PTH binds to at least three specific types of GPCRs known as parathormone receptors 1/2/3 (PTH 1/2/3R). When bound to PTH, they trigger the production of second messengers such as cAMP and calcium to activate PKA and PKC, respectively. PTH activates PTH1R present on osteoblasts and tubular epithelial cells making them major target cells.

16.4.2.2 Biological Effects

The target effects of PTH on kidney and bone are aimed principally at increasing circulatory Ca^{2+} levels (hypercalcemia). Further, PTH facilitates the activation of Vitamin D_3 (Vit.D) in kidneys, thereby indirectly stimulating the intestinal absorption of dietary Ca^{2+}.

16.4.2.2.1 Effect on Bone

PTH acts on osteoblasts and stimulates the release of bone degrading proteases and cytokines that activate osteoclasts.

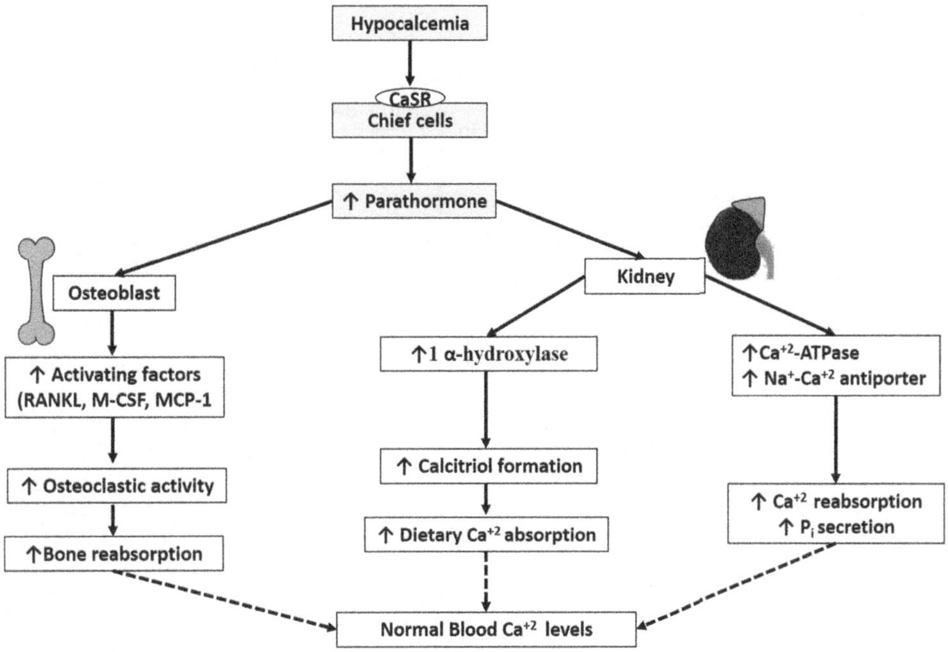

Fig. 16.20 Regulation of calcium by PTH [Parathormone increases the resorption of Ca^{2+} from the bones and kidneys to restore the circulatory Ca^{2+} levels to normalcy. *CaSR* calcium-sensing receptors; Ca^{2+} calcium ion; P_i inorganic phosphorus; *RANKL* receptor activator of nuclear factor kappa B ligand; *M-CSF* macrophage colony-stimulating factor; *MCP-1* monocyte chemoattractant protein-1; ↑ increase]

Two such molecules that mediate activation and differentiation of osteoclasts are the macrophage colony-stimulating factor (M-CSF) and receptor activator of nuclear factor kappa B ligand (RANKL). The production of monocyte chemoattractant protein-1 (MCP-1) from osteoblasts will stimulate the formation of new osteoclasts. The increased production, activation, and differentiation of osteoclasts result in the demineralization of bones. In addition, PTH stimulates osteocytes to redistribute Ca^{2+} from bone fluid to circulation, termed osteocytic osteolysis. Together, osteolysis and osteoclast-mediated bone resorption increase the mobilization of Ca^{2+} and P_i (phosphate) from bone into the blood.

16.4.2.2.2 Effect on Kidney

PTH stimulates the tubular reabsorption of Ca^{2+} by upregulating the Ca^{2+}-ATPase and Na^+-Ca^{2+} antiporter genes in ascending loop of Henle and distal convoluted tubule (DCT). It also inhibits the reabsorption of P_i in the proximal tubule, thus useful in excreting excess P_i (phosphaturic effect) accumulated during bone resorption. In addition, it stimulates the production of calcitriol (1, 25-dihydroxy cholecalciferol) by upregulating the 1-α-Hydroxylase gene in the kidney and increases the absorption of Ca^{2+} from GIT. Overall, increased mobilization of Ca^{2+} from bone with concurrent inhibition of its excretion produces the hypercalcemic effects of PTH (Fig. 16.20).

16.4.2.3 Regulation of Secretion

The calcium-sensing receptors (CaSR) present on chief cells will help to detect the minute-to-minute variations in circulatory Ca^{2+} levels. Increased binding of Ca^{2+} to CaSR results in the inhibition of PTH release from the chief cells, whereas vice versa is true for PTH release. Since the parathyroid is bestowed with an exceptional sensitivity toward circulatory concentration of Ca^{2+}, it is the foremost regulator of Ca^{2+} homeostasis.

16.4.3 Calcitriol (1, 25-Dihydorxy Cholecalciferol)

Vit.D_3 (Cholecalciferol) is a secosteroid derived from the precursor 7-dehydrocholesterol in the skin or from dietary supplements. Because of its lipophilic nature, Vit.D_3 binds to Vit.D binding protein (DBP) in circulation. Then, 25 α-hydroxylase present in the hepatocytes mediates the hydroxylation of Vit.D_3 to produce 25-hydroxycholecalciferol (25-hydroxy Vit.D_3). Finally, 25-hydroxycholecalciferol (Calcidiol) is further hydroxylated by 1α-hydroxylase in proximal tubular cells to produce the active form of Vit.D_3, i.e., 1, 25-dihydroxycholecalciferol, also known as calcitriol. The upregulation of 1α-hydroxylase is independently stimulated by a fall in Ca^{2+} and increased PTH in the blood (Fig. 16.21).

16.4.3.1 Mechanism of Action

Calcitriol binds to vitamin D receptor (VDR) localized in the nucleus, fundamentally a type of ligand-dependent transcriptional factor. Expressed in various tissues, VDR is abundant in bones, intestinal epithelium, parathyroid gland, skin, and even in germ tissues. Once activated, VDR forms a hetero

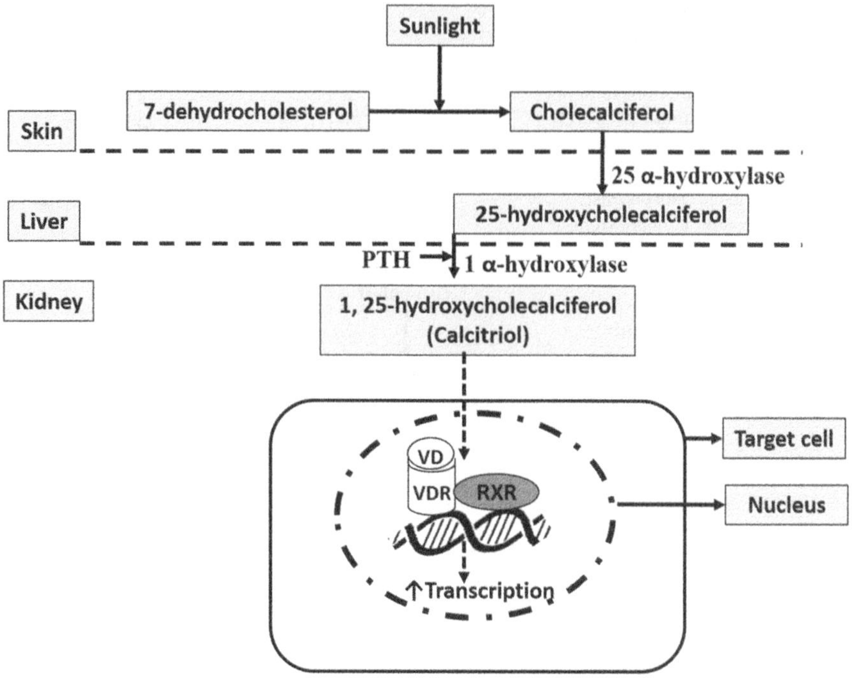

Fig. 16.21 Biosynthesis and mechanism of action of calcitriol [The cholecalciferol produced in the skin is sequentially hydroxylated in liver and kidneys to yield the active form calcitriol. The calcitriol produced binds to the VDR present in the nucleus to produce the target effects. [*PTH* parathormone; *VD* calcitriol; *VDR* Vit.D receptor; *RXR* retinoid X receptor; ↑ Increase]

dimer with RXR and binds to a specific DNA region known as vitamin D response element (VDRE). The binding of heterodimer results in the transcriptional activation of different genes that contribute to producing the biological effects.

16.4.3.2 Biological Effects

Calcitriol acts in concert with PTH to elevate the blood Ca^{2+} levels back to normal. Its major targets tissues are bones and the intestine epithelium. Apart from the hypercalcemic effect, calcitriol also regulates cellular proliferation, differentiation, and immune response. Together, increased absorption of calcium from the intestine and bones helps in increasing the blood calcium concentration.

16.4.3.2.1 Effects on the Intestinal Epithelium

Calcitriol stimulates enterocytes to produce more calcium-binding protein (CaBP), Na^+-Ca^{2+} pumps and increases the brush border permeability to Ca^{2+}. The Ca^{2+} forms a complex with CaBP and is subsequently absorbed into the blood. The presence of a Na^+-Ca^{2+} pump on the basolateral membrane helps in pumping out the Ca^{2+} accumulated in the enterocytes to the blood. The Na^+-K^+ ATPase pump is very much essential to maintain the activity of the Na^+-Ca^{2+} pump and hence this method of Ca^{2+}absorption is energy-dependent and considered an active process. Additionally, the Ca^{2+}-CaBP complex can be translocated from the gap junctions between the enterocytes or by fusing with lysosomes and exocytosed into the blood. Furthermore, it also increases the intestinal absorption of phosphate, thus increasing its concentration in blood.

16.4.3.2.2 Effects on the Bone

Calcitriol has a synergistic effect on the PTH-mediated resorption of Ca^{2+} from the bone. It activates osteoclasts by stimulating the paracrine signals from osteoclasts. In the absence of calcitriol, the effect of PTH on osteoclast activation is negligible.

16.4.3.3 Regulation of Secretion

The activation of Vit.D_3 depends on the circulatory levels of PTH and Ca^{2+}. Elevated PTH in the blood increases the calcitriol formation, whereas increased levels of Ca^{2+} result in the conversion of 25-hydroxycholecalciferol to inactive 24, 25-dihydroxy cholecalciferol.

16.4.4 Calcitonin

Calcitonin is a polypeptide hormone secreted from C-cells (also known as parafollicular cells) in the thyroid gland. The mature hormone consisting of 32 amino acids is derived from the precursor prohormone molecule with 136 amino acids.

16.4.4.1 Mechanism of Action

Calcitonin binds to calcitonin receptors (CTR) that belong to the GPCR superfamily. The activation of CTR initiates both AC and PLC systems to initiate downstream signaling pathways in the target cells, especially in renal tubular epithelial cells and osteoclasts.

16.4.4.2 Biological Effects

Released in response to hypercalcemia, calcitonin acts to bring the circulatory Ca^{2+}levels back to normal. It acts on the osteoclasts to suppress their release of acid phosphatase, motility, and differentiation resulting in the inhibition of bone resorption. In addition, calcitonin inhibits the renal tubular reabsorption of Ca^{2+} promoting its excretion. Together, a concurrent decrease in Ca^{2+} release from bone and its simultaneous excretion from kidneys produce a hypercalcemic effect.

16.4.4.3 Regulation of Secretion

Secretion of calcitonin is chiefly triggered by the rise of Ca^{2+} concentration in blood.

Know More...

Hyperparathyroidism: A pathological condition characterized by increased secretion of PTH.

Hypoparathyroidism: Condition characterized by decreased secretion of PTH.

Milk fever: A metabolic disease in post-parturient cows due to the decreased calcium levels in blood.

Rickets: Deficiency of Vit.D3 resulting in the abnormal bending of bones in young animals.

Osteomalacia: Deficiency of Vit.D3 in adult animals leading to an abnormal softening of bones.

Renal rickets: A pathological condition characterized by the absence of 1α-hydroxylase in kidneys and subsequent deficiency of calcitriol.

Learning Outcomes

- **Thyroid hormones:** Thyrocytes present in the thyroid follicles synthesize triiodothyronine (T_3) and tetraiodothyronine (T_4) from the amino acid tyrosine. While T_4 is the major secretory form (90%), biological effects are due to T_3. Hence, T_4 is converted to T_3 in the target cells by the action of deiodinases. It stimulates glycogenolysis, gluconeogenesis, and glycolysis in the liver resulting in the increased secretion of glucose into the circulation. T_3 stimulates glucose uptake, glycolysis, and protein degradation in skeletal muscle to generate more nutrients. Moreover, increased nutrients along with an increased mitochondrial number and activity to augment oxidative phosphorylation and subsequent thermogenesis.
- **Glucagon:** It is a polypeptide hormone released from the α-cells in response to hypoglycemia. Glucagon stimulates the rate of production of glucose from the liver by stimulating glycogenolysis, gluconeogenesis, and β-oxidation of fatty acids with concomitant inhibition of glycolysis. It also produces a marked degradation of skeletal muscle protein and lipolysis in adipose tissue to produce glucose via gluconeogenesis. The resultant increase in blood glucose levels supports the functioning of brain and other glucose-dependent cells present in the body.
- **Insulin:** Insulin produced from the β-cells is the only hypoglycemic hormone produced in animals. It is a heterodimeric polypeptide hormone released during hyperglycemia. It stimulates the glucose uptake in skeletal muscles and adipocytes to produce glycogen and triglycerides, respectively. It inhibits glucose production by inhibiting gluconeogenesis and stimulates protein synthesis in both the liver and skeletal muscles. The anabolic effects thus produced by insulin make it as one of the crucial mediator of growth due to GH.
- **Mineralocorticoids:** Mineralocorticoids are implicated in regulating the electrolyte balance and circulatory fluid volume. Aldosterone is the major mineralocorticoid synthesized exclusively in zona glomerulosa due to the presence of aldosterone synthase. Hyperkalemia is the most potent stimulator for aldosterone secretion, followed by angiotensin-II. Primarily, it affects principal cells and intercalated cells in distal tubules to increase Na^+ reabsorption, K^+ excretion, H^+ secretion, and water reabsorption. These biological effects result in bringing down the K^+ ion levels and restoring circulatory volume.
- **Glucocorticoids:** Released from zona fasciculata, they are responsible for regulating glucose levels in circulation. Cortisol is the major glucocorticoid present in most animals, whereas corticosterone is the primary glucocorticoid in birds, mice, and rats. The glucocorticoids have a catabolic effect on glycogen, protein, and adipose tissue stores present in an animal's body. Thus, the activation of various catabolic pathways results in hyperglycemia to support the functioning of glucose-dependent vital organs such as the brain during stress. In addition to its effects on metabolism, it causes potent inhibition of immune responses.
- **Ca^{2+} homeostasis:** Parathormone, calcitriol, and calcitonin constitute the triad of hormones that regulates calcium homeostasis. Parathormone and

(continued)

calcitriol are released from the parathyroid gland and kidneys, respectively, to elevate blood Ca^{2+} levels. During the period of hypocalcemia, they stimulate the reabsorption of Ca^{2+} from bone and kidney, increase the intestinal absorption of Ca^{2+} and renal excretion of phosphorus (P_i). Calcitonin released from the thyroid gland during hypercalcemia increases the renal excretion of Ca^{2+}, thereby decreasing the Ca^{2+}levels back to normal.

Exercises

Objective Questions

Q1. The glycoprotein secreted by the thyrocytes is known as ____________

Q2. The iodide trapping seen in the thyroid gland is an example of __________ type of active transport

Q3. ____________ is the enzyme required for the conversion of iodide into iodine in the thyroid gland

Q4. What is the major form of thyroid hormone produced by the thyroid gland?

Q5. What is the major plasma protein to which the thyroid hormones bind in the circulation?

Q6. The group of enzymes that metabolism of the thyroid hormones in the target tissue are known as __________

Q7. The increase in the metabolic heat production by the thyroid hormones in the target tissues is an example for the ____________ type of thermogenesis

Q8. What is the major endocrine cell type present in the islets of Langerhans?

Q9. What is the pancreatic hormone released during hypoglycemia?

Q10. The insulin receptor belongs to the ____________ family

Q11. ____________ is the widest layer in adrenal cortex

Q12. The embryological origin of the adrenal medulla is __________

Q13. The adrenal catecholamines are derived from ____________ amino acid

Q14. What is the major mineralocorticoid secreted from the adrenal cortex?

Q15. The major type of corticosteroid in birds and reptiles is __________

Q16. Which hormone is responsible for the aldosterone escape?

Q17. Which catecholamine preferentially binds to the β type of adrenergic receptors?

Q18. The hypercalcemic hormone is produced by __________ cells present in the parathyroid gland

Q19. The last step in the activation of calcidiol to form calcitriol is catalyzed by the enzyme ____________

Q20. The parafollicular cells in the thyroid gland are responsible for the production of ____________ hormone

Subjective Questions

Q1. Describe the process of iodide trapping in thyroid follicles.

Q2. Explain the sequential steps in thyroid hormone synthesis

Q3. How the thyroid hormones are transported in blood?

Q4. Describe the mechanism of action of thyroid hormones.

Q5. What are the biological effects of T_3?

Q6. Explain the regulation of the secretion of thyroid hormones.

Q7. What are the different cell types present in islets of Langerhans? Explain their endocrine function?

Q8. Describe the effects of glucagon on intermediary metabolism.

Q9. Describe the molecular mechanisms involved in the secretion of insulin

Q10. Briefly explain the intracellular signaling of insulin in target tissues.

Q11. Substantiate why insulin is known as an anabolic hormone.

Q12. Explain the histology of the adrenal gland.

Q13. Write a short note on the synthesis of mineralocorticoids

Q14. Describe the biological effects of aldosterone.

Q15. List out different glucocorticoids and their biosynthesis.

Q16. Why the secretion of glucocorticoids is considered essential for life?

Q17. What is the hypothalamus–pituitary–adrenal (HPA) axis?

Q18. Explain the steps involved in the synthesis of adrenal catecholamines.

Q19. Explain the events in the fight or flight response.

Q20. Explain the hormonal regulation of Ca^{2+} homeostasis

Answer to the Objective Questions

A1. Thyroglobulin

A2. Secondary

A3. Thyroid peroxidase (TPO)

A4. T_4(Thyroxine)

A5. Thyroxine-binding globulin (TBG)

A6. Deiodinases

A7. Non-shivering thermogenesis

A8. β-cells

A9. Glucagon

A10. Tyrosine kinase
A11. Zona fasciculata
A12. Neural ectoderm
A13. Tyrosine
A14. Aldosterone
A15. Corticosterone
A16. Atrial natriuretic peptide (ANP)
A17. Norepinephrine
A18. Chief cells
A19. 1α-hydroxylase
A20. Parathormone

Keywords for the Answer to Subjective Questions

A1. Sodium (Na^+)-Iodide (I^-) symporter (NIS), secondary active transport, pendrin
A2. Iodine uptake, organification¸ endocytosis
A3. Thyroxine-binding globulin (TBG), thyroxine-binding prealbumin (TBPA or Transthyretin), serum albumin
A4. MCT 8, 5′-deiodinase type 1 (D1), 5′-deiodinase type 2 (D2), thyroid receptors (TRs), retinoid X receptor (RXR), thyroid response element (TRE)
A5. Increases glucose production, basal metabolic rate (BMR), glycolysis, gluconeogenesis, glycogenolysis, lipolysis, basal metabolic rate (BMR).
A6. Hypothalamic–Pituitary–Thyroid axis, TRH, and TSH
A7. α cells-glucagon, β cells-Insulin, δ cells-somatostatin, and F/γ cells-pancreatic polypeptide
A8. Inhibition of glycolysis and glycogenesis, catabolism of lipids and amino acids, hyperglycemia, ureagenesis
A9. GLUT2, K_{ATP} channels, Ca^{2+} channels, and exocytosis
A10. Insulin receptor (IR), tyrosine kinase, insulin receptor substrate 1-6 (IRS 1-6), PI3K/AKT pathway, Raf/Ras/MEK/MAPK pathway
A11. Decreased glycogenolysis, increase in glycogenesis, inhibit gluconeogenesis, protein synthesis in the liver, decrease lipolysis, increase lipogenesis
A12. Zona glomerulosa, zona fasciculata, zona reticularis, and adrenal medulla
A13. Pregnenolone, progesterone, 11-deoxycorticosterone, corticosterone, aldosterone synthase, aldosterone
A14. Principal cells (PC), epithelial sodium channels (ENaC) genes, Na^+-K^+ ATPase, intercalated cells (IC), H^+ ATPase, H^+ ion secretion, normovolemia
A15. Cholesterol, pregnenolone, 11-deoxycorticosterone, corticosterone, cortisol
A16. Gluconeogenesis, glycogenolysis, lipolysis, hyperglycemia
A17. Hypothalamus-CRH, anterior pituitary gland-ACTH, adrenal-cortisol
A18. Phenylalanine, tyrosine, dihydroxy-phenylalanine (DOPA), dopamine, norepinephrine, epinephrine
A19. Increased cardia output, blood flow to skeletal muscle, increased muscular activity, and glycogenolysis
A20. Parathormone, calcitriol, and calcitonin

Further Reading

Textbooks

Feingold KR, Anawalt B, Boyce A, et al (eds) (2000) Endotext. South Dartmouth, MA: MDText.com, Inc. https://www.ncbi.nlm.nih.gov/books/NBK278943/

Hall JE, Hall ME (2020) Guyton and Hall textbook of medical physiology e-Book. Elsevier Health Sciences

McDonald LE, Pineda MH, Dooley MP (2003) McDonald's veterinary endocrinology and reproduction. 5th ed./edited by M.H. Pineda, with the editorial assistance of Michael P. Dooley. Ames, Iowa: Iowa State Press. Print

Nussey SS, Whitehead SA (2001) Endocrinology: an integrated approach. https://www.ncbi.nlm.nih.gov/books/NBK22

Reece WO, Erickson HH, Goff JP, Uemura EE (eds) (2015) Dukes' physiology of domestic animals. Wiley

Takei Y, Ando H, Tsutsui K (eds) (2015) Handbook of hormones: comparative endocrinology for basic and clinical research. Academic Press. https://doi.org/10.1016/C2013-0-15395-0

Articles

Braun D, Schweizer U (2018) Thyroid hormone transport and transporters. Vitam Horm 106:19–44. https://doi.org/10.1016/bs.vh.2017.04.005

Citterio CE, Targovnik HM, Arvan P (2019) The role of thyroglobulin in thyroid hormonogenesis. Nat Rev. Endocrinol 15(6):323–338. https://doi.org/10.1038/s41574-019-0184-8

Feldt-Rasmussen U, Rasmussen AK (2007) Thyroid hormone transport and actions. Pediatr Adolesc Med 11(R):80

Fleet JC (2017) The role of vitamin D in the endocrinology controlling calcium homeostasis. Mol Cell Endocrinol 453:36–45. https://doi.org/10.1016/j.mce.2017.04.008

Goltzman D (2018) Physiology of parathyroid hormone. Endocrinol Metab Clin 47(4):743–758. https://doi.org/10.1016/j.ecl.2018.07.003

Janah L, Kjeldsen S, Galsgaard KD, Winther-Sørensen M, Stojanovska E, Pedersen J, Wewer Albrechtsen NJ (2019) Glucagon receptor signaling and glucagon resistance. Int J Mol Sci 20(13):3314. https://doi.org/10.3390/ijms20133314

Kim K, Kopylov M, Bobe D, Kelley K, Eng ET, Arvan P, Clarke OB (2021) The structure of natively iodinated bovine thyroglobulin. Acta Crystallogr D Struct Biol 77(11):1451–1459. https://doi.org/10.1107/S2059798321010056

Mullur R, Liu YY, Brent GA (2014) Thyroid hormone regulation of metabolism. Physiol Rev 94(2):355–382. https://doi.org/10.1152/physrev.00030.2013

Newton R (2000) Molecular mechanisms of glucocorticoid action: what is important? Thorax 55(7):603–613. https://doi.org/10.1136/thorax.55.7.603

Vegiopoulos A, Herzig S (2007) Glucocorticoids, metabolism and metabolic diseases. Mol Cell Endocrinol 275(1–2):43–61. https://doi.org/10.1016/j.mce.2007.05.015

Veerappa S, McClure J (2020) Intermediary metabolism. Anaesth Intensive Care Med 21(3):162–167. https://doi.org/10.1016/j.mpaic.2020.01.003

William Tank A, Lee Wong D (2011) Peripheral and central effects of circulating catecholamines. Compr Physiol 5(1):1–15. https://doi.org/10.1002/cphy.c140007

Hormones of Gonads and Non-classical Endocrine Organs 17

Balantrapu Achuta Anjani Sai Kumar

Abstract

The non-classical endocrine organs are those that have crucial additional functions apart from secreting hormones. They are characterized by the presence of unspecialized hormone-secreting cells. The heart, kidney, thymus, placenta, and gastrointestinal tract (GIT) are prominent non-classical endocrine organs in animals. The natriuretic peptides secreted from the heart help in bringing back the circulatory volume to normal levels by promoting the renal excretion of sodium and water. The thymus too is implicated in the production of hormones such as thymulin, thymopoietin, and thymosins to potentiate the cell-mediated immunity in animals. The kidneys secrete a glycoprotein hormone known as erythropoietin (EPO), responsible for alleviating tissue hypoxia by stimulating erythropoiesis in the bone marrow. Likewise, enteroendocrine cells (EECs) in the GIT produce a wide range of peptide hormones that regulate digestion, motility, and appetite in animals. In addition to the nutrient transport function, the placenta in mammals evolved to produce key hormones such as progesterone, estrogen, and relaxin to maintain gestation and facilitate parturition. The gonads in animals are responsible for the secretion of steroid hormones implicated in sex-specific behavior, reproductive tract development, and reproductive phases such as the estrous cycle, gestation, and parturition.

B. A. A. Sai Kumar (✉)
Department of Veterinary Physiology, Rajiv Gandhi Institute of Veterinary Education and Research, Kurumbapet, Puducherry, India

© The Author(s), under exclusive license to Springer Nature Singapore Pte Ltd. 2023

P. K. Das et al. (eds.), *Textbook of Veterinary Physiology*, https://doi.org/10.1007/978-981-19-9410-4_17

Graphical Abstract

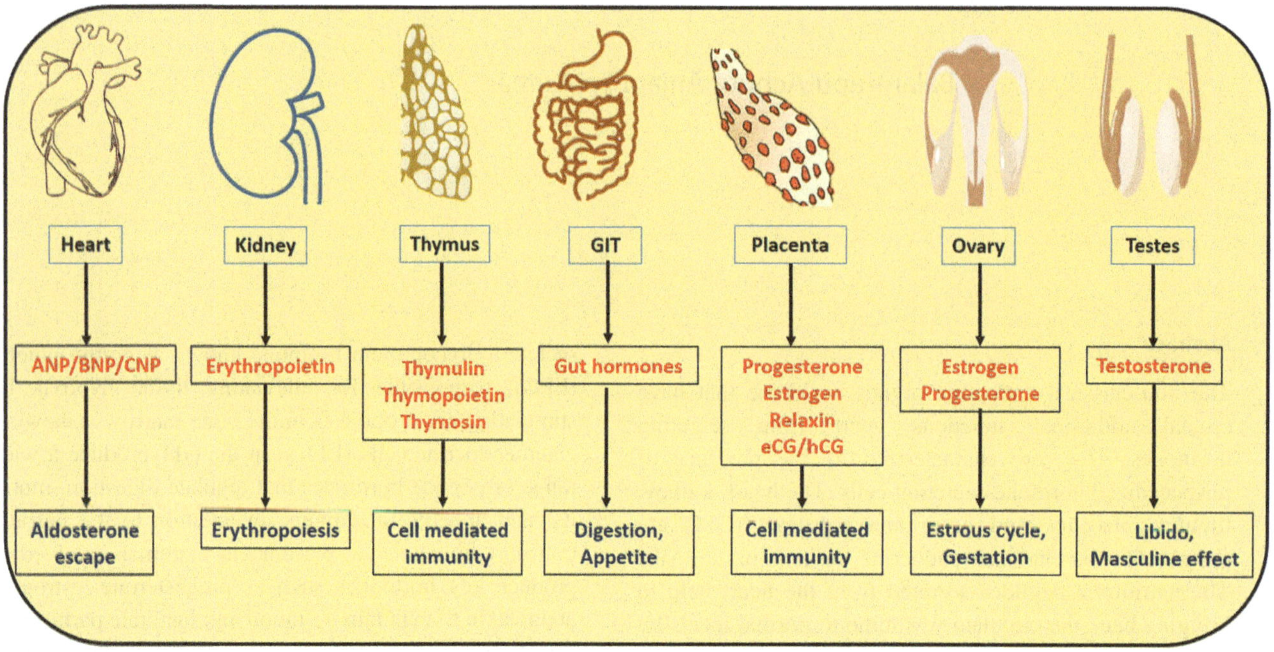

Description of the graphic: Gonadal hormones and other hormones secreted from the non-classical endocrine organs. [*GIT* gastrointestinal tract; *ANP* atrial natriuretic peptide; *BNP* brain/B-type natriuretic peptide; *CNP* C-type natriuretic peptide; *eCG* equine chorionic gonadotropin; *hCG* human chorionic gonadotropin.]

Keywords

Aldosterone escape · Erythropoietin · Thymic hormones · Placenta · Gonadal hormones

Learning Objectives

- Natriuretic peptides in blood volume regulation
- Role of erythropoietin on erythropoiesis
- Thymic hormones and their role as immunomodulators
- Different gastrointestinal hormones and their importance in GIT functioning
- Placental hormones
- Gonadal hormones

17.1 Heart as an Endocrine Organ

The endocrine activity of the heart comprises the release of various natriuretic peptide hormones from the cardiomyocytes. The atrial natriuretic peptide family includes three structurally similar peptides, i.e., atrial natriuretic peptide (ANP), brain or B-type natriuretic peptide (BNP), and C-type natriuretic peptide (CNP). The secretion of these hormones enables the heart to participate in regulating the circulatory volume and systemic blood pressure in animals.

17.1.1 Distribution in Tissues

The atrial cardiomyocytes are primarily responsible for the secretion of ANP although a small proportion is released from the ventricles. Originally isolated from the porcine brain tissue, BNP is produced chiefly from ventricles. These secretory types of cardiomyocytes are characterized by abundant rough endoplasmic reticulum (RER), golgi apparatus, and peptide storage granules. The CNP is a minor natriuretic peptide distributed in the cerebellum, hypothalamus, anterior pituitary, kidney, and endothelial cells.

17.1.2 Mechanism of Synthesis

The genes encoding different natriuretic peptides were evolved from the ancestral natriuretic peptide gene. Natriuretic peptide A precursor (NPPA), natriuretic peptide B precursor (NPPB), and natriuretic peptide B precursor (NPPC) genes, respectively, encode the ANP, BNP, and CNP. Initially those are synthesized as preprohormones (preproANP, preproBNP, and preproCNP). The proteolytic cleavage of signal peptides from preprohormones results in the production of prohormones that are stored as vesicles. Subsequent endoproteolytic action of serine proteases such as corin and furin acts on prohormones to produce mature hormones that are released through exocytosis. All three types of natriuretic peptides contain a 17-aminoacid residues ring linked by a disulfide bridge, which is essential for their biological activity.

Know More...

- Type C natriuretic peptide has a broken 17-aminoacid ring and hence cannot exert natriuretic effects.
- Birds and reptiles do not possess ANP.

17.1.3 Mechanism of Action

The target effects of natriuretic peptides are mediated by cell membrane-bound receptors coupled with guanylyl cyclase. Three different types of receptors were identified: NPR-A, NPR-B, and NPR-C. ANP and BNP exert their biological effects by binding to NPR-A. The binding of ANP or BNP to NPR-A results in a conformational change with consequent activation of the intracellular guanylyl cyclase domain. The cGMP formed from GTP in the presence of guanylyl cyclase is regarded as the secondary messenger in the signal transduction mechanism. The elevated cGMP levels affect various cGMP-dependent protein kinases (PKGs), cGMP gated ion channels, and cGMP-regulated phosphodiesterases to bring about specific target effects. Kidneys, blood vessels, and the adrenal cortex are regarded as the principal target tissues of natriuretic peptides.

17.1.4 Biological Effects

The natriuretic peptides (except CNP) are responsible for lowering the blood volume by enhancing the renal excretion of water (diuresis) and sodium (natriuresis). Apart from kidneys, they also cause the relaxation of vascular smooth muscles thereby bringing down the arterial pressure. They act on the pituitary gland, adrenal cortex, and small intestine to inhibit ADH, aldosterone secretion, and sodium absorption, respectively.

17.1.4.1 Effect on Kidneys: Aldosterone Escape

The increased intracellular levels of cGMP result in the activation of PKGs to cause a simultaneous dilation of afferent arterioles and constriction in efferent arterioles. The increased blood flow to the kidneys is correlated with an increase in glomerular filtration rate and filtration fraction. Moreover, the hampered reabsorption of sodium and water in proximal tubules and distal tubules hastens sodium and water excretion (Fig. 17.1). Further, the inhibitory role of ANP on renin secretion from the kidneys is well documented. Altogether, the natriuretic and diuretic effects of ANP/BNP override the effects of aldosterone on the circulatory volume, generally referred to as the aldosterone escape.

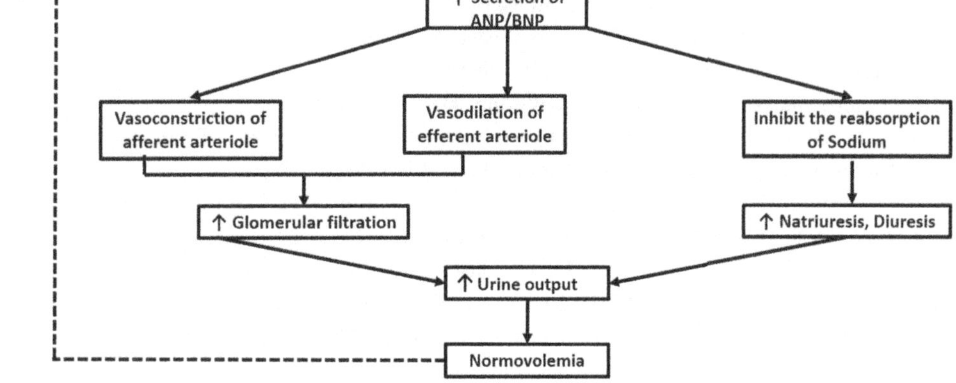

Fig. 17.1 Effects of natriuretic peptides on circulatory volume. [ANP/BNP released due to the stretch of heart musculature will lead to the increased excretion of sodium along with water and decreases the blood volume to normal level. ↑ = increase, (−) = Feedback inhibition]

17.1.4.2 Effects on the Vasculature

The increased intracellular levels of cGMP cause a decrease in the levels of Ca^{+2} ions. As calcium ions play a significant in smooth muscle contraction, a decrease in their intracellular levels leads to the relaxation of smooth muscles present in blood vessels and alters their permeability. The resultant vasodilation in the arterial tree and venous systems leads to a decrease in the peripheral resistance and preload, thereby reducing cardiac output. In addition, ANP has an anti-proliferative effect on vascular smooth muscles, fibroblasts, and cardiac cells.

17.1.5 Regulation of Secretion

The natriuretic peptides, especially ANP and BNP, are secreted in response to an increase in blood volume. The increased stretch of atrial muscle cells due to the rise in venous returns causes the secretion of ANP/BNP. Since the natriuretic and diuretic effects produced by natriuretic peptides are pressure induced, they are also known as pressure natriuresis and pressure diuresis. The vasoactive peptides such as endothelin 1 (ET-1), angiotensin II, and catecholamines are also implicated in increasing the expression of ANP in cardiomyocytes.

17.2 Erythropoietin

Erythropoietin (EPO) is a glycoprotein hormone (167 amino acids) released from the kidneys in response to hypoxia. The fetal liver and kidneys are major sites for the production of EPO. The interstitial peritubular cells in proximal tubules synthesize and secrete EPO during hypoxia and anemic stress.

17.2.1 Mechanism of Synthesis

Hypoxic inducible factor (HIF) is a transcriptional factor regulating the erythropoietin gene expression. It is a heterodimer made of HIF-α (HIF-1α/HIF-2α/HIF-3α) and HIF-1β subunits. Under normoxic conditions, HIF-1α will be hydroxylated by oxygen-sensitive hydroxylases such as prolyl hydroxylase (PHD) and factor inhibiting HIF (FIH). Further, the hydroxylated HIF-1α will undergo ubiquitination and proteasome degradation. During cellular hypoxia, the reduced activity of oxygen-sensitive hydroxylases leads to a reduction in the degradation of HIF-1α and stabilization of HIFα. The stabilized HIFα binds to a specific region in the genome known as the hypoxia-response element (HRE) to increase the transcription of the EPO gene.

17.2.2 Mechanism of Action

The activation of erythropoietin receptors (EPOR) when bound with EPO results in their dimerization and phosphorylation by Janus kinase 2 (JAK2). The transphosphorylation of EPOR initiates the phosphorylation of signal transducer and activator of transcription 5 (STAT5), PI3K/AKT, and SHC/MAPK pathways to regulate cellular proliferation and survival. The EPOR is distributed majorly in erythrocyte progenitor cells residing in the bone marrow, thus making them the primary target tissue.

17.2.3 Biological Effects

EPO is essential for the survivability of erythrocyte progenitor cells and their subsequent differentiation to colony-forming unit-erythrocyte (CFU-E) cells. It results in the swift passage of erythroblasts through different stages in erythropoiesis, thereby increasing the rate of erythropoiesis (Fig. 17.2). Apart from the erythrocyte progenitors, EPO is found to have angiogenic effects on the endothelial cells and neurotrophic effects in the brain and regulate bone mass and formation of skeletal muscle fibers.

17.2.4 Regulation of Secretion

Hypoxia is by far the potent stimulus for EPO secretion, with 90% of EPO formed in kidneys and the rest from the liver, brain, spleen, and lungs.

Know More...

- Since EPO is heavily glycosylated, it has a long circulatory half-life of 1 day.
- EPO increases the rate of erythropoiesis as high as ten times.
- Effects of EPO on erythrocyte numbers can only be seen after 5 days (time taken for erythropoiesis).

17.3 Endocrine Functions of the Thymus

The thymus is a primary lymphoid organ bestowed with the site of development, maturation, and differentiation of T-lymphocytes to direct adaptive immune responses. In addition, it produces hormones that modulate the function of T-lymphocytes present in the thymus and peripheral circulation. Thymulin, thymopoietin, and thymosin-α_1, β_4 are some of the important thymic hormones produced by thymic

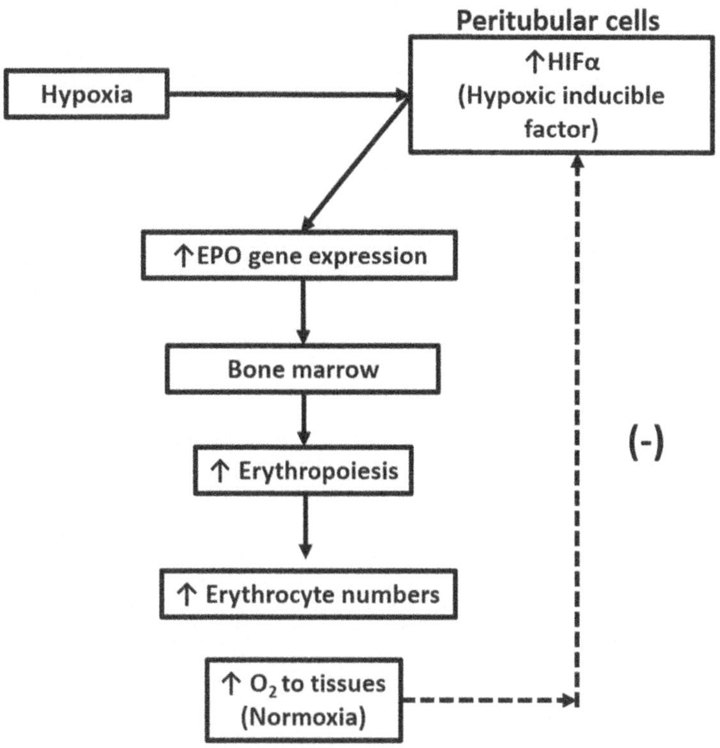

Fig. 17.2 Regulation of erythropoietin secretion and its effect on erythrocytes. [The peritubular cells in the kidney secrete erythropoietin in response to tissue hypoxia. It increases the rate of erythropoiesis and subsequently establishes normal oxygen levels in the tissue. ↑ = Increase, (−) = Feedback inhibition]

Table 17.1 List of thymic hormones and their function concerning the immune system

S. No	Hormone	Chemical nature	Function
1.	Thymulin (Facteur thymique serique)	Nonapeptide containing zinc	1. Increased generation of cytotoxic T-cells. 2. Intrathymic and extrathymic T-cell differentiation
2.	Thymopoietin	Polypeptide 49 amino acids	Induces the differentiation of T-lymphocytes
3.	Thymosin α_1, Thymosinβ_4	28 amino acids 43 amino acids	Stimulates lymphopoiesis and T-cell proliferation

epithelial cells (TECs), which have well-documented effects on the immune system (Table 17.1). Hormones such as GH, PRL, THs, and sex steroids influence the endocrine function of the thymus by acting on TECs. Nonetheless, the thymic hormones can also affect the production of hormones from the hypothalamic–pituitary axis and their target endocrine glands including the gonads.

17.4 Gastrointestinal Hormones

Considered as the only site for nutrient assimilation, the gastrointestinal tract (GIT) also produces hormones that are responsible for regulating secretions of exocrine glands, gastrointestinal motility, cellular proliferation, and differentiation in GIT. The first hormone to be discovered by Bayliss & Starling (1902) is secretin, which is a gut hormone. Considered the largest endocrine organ in the animal's body, GIT produces more than 20 peptide hormones.

17.4.1 Properties of the Gut Hormones

The hormone-producing cells are also known as the enteroendocrine cells (EECs), interspersed between gut mucosal cells in GIT. They are derived from the pluripotent intestinal stem cells in the intestinal crypts. Different types of EECs are present in the GIT, with each cell type having the ability to synthesize and secrete at least one kind of hormone (Table 17.2). The hormones thus produced can also act as autocrine or paracrine or neurocrine factors on the nearby target cells. Based on the structural homologies, the gut hormones are further classified into different families: the secretin family (secretin, glucagon-like peptides, VIP), the

Table 17.2 Source, chemical structure, receptors of different GIT hormones and their effect on feed intake in animals

S. No	Hormone	Cell type and location	Chemical structure	Receptor	Feed intake
1.	Gastrin	G cell (Stomach, duodenum)	34 aa—Big gastrin 17aa—Little gastrin 14aa—Small gastrin	CCK2R (G_q type)	↓
2.	CCK (Pancreozymin)	I/L cells (Small intestine)	33 aa	CCK1R (G_q type)	↓
3.	Ghrelin	A (X-like) cell (Stomach)	28 aa	GHS-R1a/ GHS-R1b (G_q type)	↑
4.	Motilin	M cell (Duodenum)	22 aa	MTLR	↑
5.	Secretin	S cells (Small intestine)	27 aa	SCTR (G_s type)	↓
6.	Vasoactive intestinal polypeptide (VIP)	Myenteric and submucosal neurons	28 aa	VPAC-I/ VPAC-II (G_s type)	↓
7.	Gastric inhibitory polypeptide (GIP)	K cell (Duodenum)	42 aa	GIPR (G_s type)	↓
8.	Glucagon-like peptide-1 (GLP-1)	L cell (Ileum, colon)	31 aa	GLP-1R (G_s type)	↓
9.	Glucagon-like peptide-2 (GLP-2)	L cell (Ileum, colon)	33 aa	GLP-2R (G_s type)	↓
10.	Oxyntomodulin (OXM)	Oxyntic cells (Fundus)	37 aa	GLP-1R (G_s type)	↓
11.	Gastrin-releasing peptide (GRP)/ Bombesin-like peptide	Stomach, small intestine (Neuropeptide)	27 aa	GRPR (G_q type)	↓
12.	Neuropeptide Y (NPY)	Neuropeptide (Expressed primarily in CNS, also released in GIT)	36 aa	Y1-6 (G_i type)	↑
13.	Peptide YY (PYY)	L cell (Jejunum, caecum, colon)	36 aa	Y1, Y2, Y5 (G_i type)	↓
14.	Somatostatin	D cell (Stomach, small intestine)	28 aa (SS-28) 14 aa (SS-14)	SSTR (G_i type)	↓
15.	Enteroglucagon	L cell	29 aa	GCGR (G_s & $G_q$type)	↓
16.	Leptin	Stomach, adipose tissue	167 aa	Lep-R	↓

G_q type G-protein coupled receptor that activates PLC; *G_s type* G-protein coupled receptor that activates AC; *G_i type* G-protein coupled receptor that inhibits AC

gastrin family (gastrin, CCK), and the PP-fold family (Neuropeptide Y (NPY), PYY, pancreatic polypeptide).

17.4.2 Biological Effects

The gut hormones have a unique property of activating at least one specific type of GPCR present on the target tissues. The signal transduction mechanisms include the activation of adenylyl cyclases, protein kinases, and membrane-bound ion channels. With overlapping target organs, multiple gut hormones are needed to produce the desired biological effects. The major biological effects of gut hormones include the regulation of ingestion, digestion, and metabolism of nutrients (Table 17.3).

17.4.2.1 Effects on Feed Intake

Based on their ability to stimulate or inhibit the intake of feed, they are classified as orexigenic and anorexigenic hormones, respectively. Orexigenic hormones like ghrelin, motilin, and insulin-like peptide 5 (INSL5) are secreted at a maximum rate before feeding to initiate hunger sensation. Whereas, presence of nutrients in the GIT induces the release of anorexigenic hormones such as cholecystokinin (CCK), glucagon-

Table 17.3 Biological effects and the regulation of secretion of gut hormones

S. No	Hormone	Biological effects	Stimulus	Inhibition
1.	Gastrin	• Stimulates gastric acid secretion • Antagonistic to GIP and Secretin • Growth and differentiation of gastric mucosa	• Gastrin-releasing peptide (GRP), • Amino acids	Somatostatin, ↑H^+ ion concentration
2.	CCK (Pancreozymin)	• Contraction of gall bladder • Relaxation of the sphincter of Oddi • Enzyme rich pancreatic secretion	• Dietary lipids and amino acids	Somatostatin, pancreatic peptide
3.	Ghrelin	• Increase gastric acid • Increase gastric emptying • Increase GH • Decrease insulin	• Feed deprivation	Substance P
4.	Motilin	• Increases intestinal motility • Migrating myoelectric complex	• Cyclical pattern of release (90 min interval)	Ingestion of feed
5.	Secretin	• Inhibits gastric acid release • Inhibits gastric emptying • Stimulates bicarbonate rich pancreatic secretion	• Presence of gastric acid in duodenum • Digestive by-products of fats and proteins	Somatostatin
6.	Vasoactive intestinal polypeptide (VIP)	• Inhibits gastric acid secretion • Relaxation of intestinal smooth muscle • Stimulates secretion of anions (Cl^-, HCO_3^-)in the intestine, insulin, and glucagon from the pancreas • Potent vasodilator	• Serotonin, • Acetylcholine (ACh), • Substance P (Enteric neurons)	Somatostatin
7.	Gastric inhibitory polypeptide (GIP)	• Stimulates insulin secretion • Induces satiety • Decreases gastric acid and gastric emptying	• Presence of fats and carbohydrates in the intestine	Somatostatin
8.	Glucagon-like peptide-1 (GLP-1)	• Induces satiety • Decreases gastric motility • Stimulates insulin secretion • Suppress glucagon secretion	• Carbohydrates and lipids in the ration • Leptin • GIP	Insulin
9.	Glucagon-like peptide-2 (GLP-2)	• Control growth and function of GIT by regulating epithelial integrity, secretion, blood flow, and motility	• Dietary carbohydrates and lipids, • GIP	Somatostatin
10.	Oxyntomodulin	• Stimulates insulin release • Increases satiety • Decreases gastric acid • Increases energy expenditure • Lipolysis	• Response to feed intake	–
11.	Gastrin-releasing peptide (GRP)/Bombesin-like peptide	• Increases gastrin secretion • Regulate food intake • Male sexual functions • Memory consolidation	• Gastric phase of digestion (Vagal stimulation)	–
12.	Neuropeptide Y (NPY)	• Induces food intake • Lowers energy expenditure • Circadian rhythm	• Low energy status	Leptin
13.	Peptide YY (PYY)	• Reduces gastric acid secretion, • Decreases gastric emptying • Inhibits appetite	• Dietary fats	Obesity
14.	Somatostatin	• Inhibits the release of GH, CCK, gastrin, motilin, secretin, GIP, VIP, etc.	• Increased uptake of nutrients	GH, IGF-1
15.	Enteroglucagon	• Increase enterocyte proliferation	• Feed intake	Increased blood glucose
16.	Leptin	• Decrease feed intake • Increase energy expenditure	• Increase adipose tissue (Positive energy balance)	Decreased fat tissue (Negative energy balance)

like peptide 1 (GLP1), peptide YY (PYY), and oxyntomodulin that stimulate the cessation of feed intake and cause satiety.

17.4.2.2 Effects on Digestion

The gut hormones modulate the secretion of bile, gastric acid, pancreatic enzymes, and bicarbonate ions to provide an ideal environment for the enzymatic breakdown of complex feeds ingested by the animal into simpler nutrients. In addition, they determine the rate of gastric emptying and motility patterns of the intestine, thereby providing appropriate time for digestion, absorption, and elimination of indigested feed.

17.4.2.3 Effects on Metabolism

Together the gut hormones regulate energy homeostasis by affecting the secretion of hormones and metabolic pathways The GLP1, oxyntomodulin, and gastric inhibitory peptide (GIP) prevent hyperglycemia by stimulating the secretion of insulin from the pancreas, hence known as incretins. On the other hand, ghrelin stimulates growth hormone secretion from the anterior pituitary and increases adipogenesis and circulatory glucose levels. In addition, hormones such as CCK and GLP1 cause meal-induced thermogenesis by stimulating lipolysis of brown adipose tissue (BAT).

17.4.2.4 The Gut–Brain Axis

The EECs play the role of nutrient sensing and produce hormones that help to relay the sensory information so accrued to various parts of the brain. Subsequently, the higher regulatory centers respond by regulating various physiological functions, especially the animal's appetite, thus forming a gut–brain axis.

17.5 Placental Hormones

The placenta acts as the site of attachment of the fetus, delivers nutrients and gases derived from maternal circulation, prevents the fetal allograft from the maternal immune system, and eliminates fetal metabolic-waste products. In addition, it acts as a transient endocrine organ by producing hormones that are vital for the maintenance of gestation, fetal growth, and parturition.

17.5.1 Progesterone

Progesterone stands as an absolute requirement for the maintenance of gestation in all mammals. Although corpus luteum is the primary source of progesterone in domestic animals, placental-derived progesterone also helps in the maintenance of gestation. The shift of progesterone production from the corpus luteum to the placenta is regarded as an essential phenomenon seen in the case of equines, sheep, and primates. This luteo-placental shift in progesterone synthesis has minor relevance in goats and pigs. Luteal-progesterone plays a prominent role in myometrial quiescence, endometrial growth and differentiation, immunosuppression to prevent fetal rejection, and cervical closure. However, the progesterone derived from the placenta is speculated to be useful for species with longer gestation.

17.5.2 Estrogen

Unlike follicle that produces 17β-estradiol, placenta in ungulate animals primarily secretes estrone in sulfate form. In primates, placental estrogen is implicated in trophoblast differentiation, mammary gland development, and uteroplacental blood flow. However, in cattle, maternal estrogen levels rise along with a concomitant decrease in progesterone levels at the end of gestation period due to an increased conversion of progesterone to estrogen by CYP17A1 (17-α-hydroxylase) in the placenta. Thus, the increased circulatory levels of estrogen lead to the ductular development of the mammary gland, remove the progesterone-mediated negative feedback on lactation, excite the myometrial tissue, and prepare the birth canal for parturition.

17.5.3 Chorionic Gonadotropins

As of now, only two chorionic gonadotropins: equine chorionic gonadotropin (eCG) and human chorionic gonadotropin (hCG) are reported in equines and primates, respectively. Like the pituitary gonadotropins (LH & FSH), the chorionic gonadotropins are made of a common α-glycoprotein chain. However, β-chains of chorionic gonadotropins (β_{CG}) are similar to the β_{LH} chain, except that they are heavily glycosylated. The hCG released by blastocyst in primates is responsible for the maternal recognition of pregnancy and maintenance of corpus luteum by binding to LH receptors. Whereas, distinct areas of the placenta known as endometrial cups secrete eCG to promote the formation of accessory corpora lutea, thereby helping in the maintenance of gestation.

17.5.4 Placental Lactogen (Chorionic Somatomammotropin)

The binucleate cells of trophectoderm in ruminants, primates, and rodents secrete placental lactogen, a hormone that has both somatotropic and lactogenic functions. Mature placental lactogen (PL) is a single polypeptide chain with 200 amino acids. In bovines, PL shares a structural homology of 50% and 23% with PRL and GH, respectively. The circulatory levels of PL start raising from Day 30 onwards, reaching a peak during the last trimester, and begin to fall when the animal approaches parturition. In ruminants, PL has luteotrophic actions and augments progesterone secretion from the corpus luteum. It plays a major role in the partitioning of maternal nutrients to support fetal growth by stimulating the uptake of maternal nutrients, and glycogenesis in fetal tissues. In addition, it stimulates lobuloalveolar growth in the mammary gland and exerts galactopoietic effects by stimulating dry matter intake.

17.5.5 Relaxins

They are peptide hormones belonging to the insulin family, produced primarily by the corpus luteum and placenta. Relaxin1 (RLN 1) plays a major role in reproductive functions in mammals. It is a polypeptide heterodimer hormone with 53 amino acids. The production of RLN1 rises at the end of gestation to promote cervical ripening, dilatation of pubic symphysis, and relaxation of the sacrosciatic ligament to aid the process of parturition.

Know More...

- The long half-life and potent FSH-like activity of eCG make it an ideal hormone used for multiple ovulation protocols/embryo transfer protocols in cattle and buffaloes.
- Since eCG is found in pregnant mares, it is also known as pregnant mare serum gonadotropin (PMSG).

17.6 Gonads

Gonads refer to the testes and ovaries present in male and female animals, respectively. They are responsible for the production of gametes and hormones, which are crucial elements governing sexual reproduction in animals (Table 17.4). The hypothalamic–pituitary axis secretes GnRH, LH, and FSH, which are essential for regulating the endocrine and gametogenic activities of gonads. In turn, the gonadal hormones produced exert negative inhibition on the release of hypothalamic GnRH and pituitary gonadotropins.

17.6.1 Ovary

The follicle and corpus luteum developed sequentially are the transient endocrine organs that secrete two major steroids: estrogen and progesterone, respectively. The granulosa cells present in the follicle produce estrogen from androgens synthesised in the theca cells. Estrogen plays a vital role in the estrous cycle, development of female secondary sexual characteristics, ductular growth in the mammary gland, and parturition. Apart from the steroidal hormones, peptide hormones such as inhibin, activin, follistatin, and anti-Mullerian hormone are produced from the ovarian follicle. Whereas progesterone secretion to support gestation and lobuloalveolar development in the mammary gland is the major endocrine function of the corpus luteum. In addition, it also secretes relaxin and oxytocin that are responsible for the widening of the birth canal and luteal demise in cyclical animals.

Table 17.4 List of gonadal hormones and their functions

S. No	Hormone	Chemical nature	Function
1.	Estrogen (Females: Granulosa cell, Males: Sertoli cell)	Steroid (18C)	• Estrus cycle • Female sexual behavior • Parturition • Ductular growth in mammary gland
2.	Inhibin (Females: Granulosa cell, Males: Sertoli cell)	Glycoprotein	• Inhibit the release of FSH from the anterior pituitary gland
3.	Activin (Follicular fluid and Rete testis fluid)	Glycoprotein	• Stimulates the secretion of FSH from the anterior pituitary gland
4.	Follistatin	Protein	• Inhibits FSH secretion • Neutralizes activin
5.	Testosterone (Females: Theca cell, Males: Leydig cell)	Steroid (19C)	• Masculine characters in male animals • Libido • Protein anabolism • Development of accessory sex glands
6.	Progesterone (Corpus luteum)	Steroid (21C)	• Maintenance of pregnancy • Lobuloalveolar growth of the mammary gland
7.	Anti-Mullerian hormone	Protein	• Regression of Mullerian duct

17.6.2 Testes

The Sertoli and Leydig cells are the somatic cell types with endocrine function in testes. The testosterone produced from the Leydig cells in response to LH imparts secondary sexual characteristics, anabolic effects on bone and muscle mass, and libido in male animals. Additionally, the Sertoli cells are responsible for the secretion of inhibin, which attenuates the secretion of FSH from the adenohypophysis.

Endocrinology to Rescue Wild Animals...

- Several wild animal species are staring at an impending threat of extinction more than ever in the history. The ex situ breeding and managemental practices play a huge role in reviving their population and curtail them from getting extinct. Endocrinological approaches such as measuring the plasma levels of hormones such as estrogen, progesterone, and cortisol help in assessing the reproductive status and well-being of different wild animals. They also greatly contribute in understanding the reproductive behavior of animals and aid in devising intervening strategies of breeding in captive wild animals.

Learning Outcomes

- **Natriuretic peptides:** ANP, BNP, and CNP are the major natriuretic peptides secreted from cardiac tissue and CNS. They are secreted due to the stretch of cardiac cells due to an increased blood volume. The natriuretic peptides (ANP and BNP) are responsible for the excretion of sodium and water from kidneys to regulate the circulatory volume to a normal level. They cause an increase in the glomerular filtration via their vasodilatory effect on afferent arteriole. Decreased tubular absorption of sodium and water leads to an increased excretion with a simultaneous decrease in the blood volume. The natriuretic and diuretic effects of natriuretic peptides antagonize the antidiuretic action of aldosterone, which is known as the aldosterone escape.
- **Erythropoietin:** It is a glycoprotein hormone released from the kidneys in response to tissue hypoxia. The bone marrow as its target organ, EPO acts on the erythropoietic stem cells to increase their proliferation, differentiation, and subsequent transformation into the erythrocytes.
- **Thymus:** A major organ determines the cell-mediated immunity in animals by acting as the site of differentiation and production of T-cells. Additionally, the thymic epithelial cells possess the endocrine activity and secrete hormones such as thymulin, thymopoietin, and thymosins. These hormones primarily act to regulate the activation and differentiation of T-cells and modulate the immune status of the animals.
- **Gut hormones:** The enteroendocrine cells present in the GIT produce a wide variety of peptide hormones. The enteroendocrine cells act as nutrient sensors and secrete the hormones that control the appetite, gastrointestinal motility, exocrine secretion of the pancreas, liver, and GIT. All these hormones work in concert to harvest appropriate amounts of nutrients by tightly regulating the process of digestion and metabolism.
- **Placenta:** Recognized as a crucial transient organ formed during the early gestation period and responsible for supplying the nutrients to the fetus. Furthermore, it also secretes important hormones such as progesterone, estrogen, chorionic gonadotropins, placental lactogen, and relaxin. Together, they ensure the maintenance of pregnancy, fetal growth, and progression of parturition.
- **Gonads:** The production of gametes and hormones that control sex-specific behaviors in animals are two important functions of gonads. The testes in male animals produce testosterone, responsible for libido, secondary sexual characteristic, anabolic effects on bone and muscles. Estrogen is the major circulatory steroid produced by the ovarian follicles responsible for reproductive cyclicity, increasing bone mass, growth of reproductive tract and mammary gland in female animals. In addition, corpus luteum formed after ovulation produces progesterone, which helps in the growth of endometrium and immunotolerance to support the fetal growth in the uterus.

Exercises

Objective Questions

Q1. __________ is a natriuretic peptide that cannot promote the excretion of sodium
Q2. Natriuretic peptides activate ______________ enzyme to initiate the signal transduction.
Q3. The antagonistic effects of natriuretic peptides on aldosterone are collectively known as ________________
Q4. The secondary messenger produced in the target cells when bound with natriuretic peptides is ________________
Q5. Erythropoietin is a __________ type of hormone, and released primarily due to ____________.
Q6. The thymic hormones are produced by______________ cells present in the thymus
Q7. Which thymic hormone requires zinc for its biological activity?
Q8. What is the first hormone to be discovered?
Q9. The cells in GIT that produce gut hormones are collectively called as ______________
Q10. Which hormone is responsible for migrating myoelectric complex?
Q11. The gut hormones that stimulate insulin secretion from the pancreas are known as _______
Q12. Which hormone is responsible for the relaxation of pelvic symphysis before parturition?
Q13. The specialized areas in equine placenta that secrete eCG are called as ____________
Q14. Which placental hormone has structural homology of both GH and PRL?
Q15. Which gonadal hormone is responsible for the lobuloalveolar development of the mammary gland?

Subjective Questions

Q1. Enlist the different types of natriuretic peptides and their major production sites.
Q2. Briefly describe the biological effects of ANP.
Q3. What is EPO? Describe its effect on the bone marrow.
Q4. Explain the mechanism of production of EPO from kidneys.
Q5. Describe briefly the endocrine role of the thymus
Q6. What are the orexigenic hormones produced by the GIT?
Q7. What are the biological effects of gut hormones?
Q8. Enlist various placental hormones and briefly describe their biological effects in animals
Q9. What are the cyclical endocrine structures developed in the ovary and the respective hormones secreted from them?
Q10. What are the different somatic cells present in the testes and list their hormones?

Answers to Objective Questions

A1. CNP
A2. Guanylyl cyclase
A3. Aldosterone escape
A4. cGMP
A5. Glycoprotein, tissue hypoxia
A6. Thymic epithelial cells
A7. Thymulin
A8. Secretin
A9. Enteroendocrine cells
A10. Motilin
A11. Incretins
A12. Relaxin
A13. Endometrial cups
A14. Placental lactogen
A15. Progesterone

Answers to Subjective Questions

A1. ANP, BNP, CNP, cardiac muscle, and CNS
A2. Natriuresis, diuresis, vasodilation in arterioles, and inhibit aldosterone secretion
A3. Erythropoietin, glycoprotein, kidney, and erythropoiesis
A4. Tissue hypoxia, HIF1α, oxygen-sensitive hydroxylases, and hypoxia-response element
A5. Thymic epithelial cells, thymulin, thymopoietin, and thymosins
A6. Ghrelin, motilin, and insulin-like peptide 5 (INSL5)
A7. Regulating GIT motility, exocrine pancreatic secretion, gastric acid secretion, and appetite
A8. Progesterone, estrogen, placental lactogen, chorionic gonadotropins, and relaxin
A9. Follicle, corpus luteum, estrogen, and progesterone
A10. Sertoli cells-estrogen, inhibin; Leydig cells-testosterone

Further Reading

Textbooks

Hafez ESE, Hafez B (eds) (2013) Reproduction in farm animals. Wiley
McDonald LE, Pineda MH, Dooley MP (2003) McDonald's veterinary endocrinology and reproduction. 5th ed./edited by M.H. Pineda, with the editorial assistance of Michael P. Dooley. Ames, Iowa: Iowa State Press. Print
Norman AW, Henry HL (2022) Hormones. Academic Press. https://doi.org/10.1016/B978-0-08-091906-5.00007-0

Takei Y, Ando H, Tsutsui K (eds) (2015) Handbook of hormones: comparative endocrinology for basic and clinical research. Academic Press. https://doi.org/10.1016/C2013-0-15395-0

Articles

Ahlman H, Nilsson O (2001) The gut as the largest endocrine organ in the body. Ann Oncol 12:S63–S68. https://doi.org/10.1093/annonc/12.suppl_2.S63

Alvarez-Oxiley AV, De Sousa NM, Beckers JF (2008) Native and recombinant bovine placental lactogens. Reprod Biol 8(2):85–106. https://doi.org/10.1016/S1642-431X(12)60006-0

Goetze JP, Bruneau BG, Ramos HR, Ogawa T, de Bold MK, de Bold AJ (2020) Cardiac natriuretic peptides. Nat Rev. Cardiol 17(11):698–717. https://doi.org/10.1038/s41569-020-0381-0

Hall CM, Glaser S, Alpini G (2017) Gastrointestinal Hormone (GI) Regulated Signal Transduction. https://doi.org/10.1016/B978-0-12-809324-5.03384-8

Latorre R, Sternini C, De Giorgio R, Greenwood-Van Meerveld B (2016) Enteroendocrine cells: a review of their role in brain-gut communication. Neurogastroenterol Motil 28(5):620–630. https://doi.org/10.1111/nmo.12754

Savino W, Dardenne M (2000) Neuroendocrine control of thymus physiology. Endocr Rev 21(4):412–443. https://doi.org/10.1210/edrv.21.4.0402

Schuler G, Fürbass R, Klisch K (2018) Placental contribution to the endocrinology of gestation and parturition. Anim Reprod 15(Suppl 1):822–842. https://doi.org/10.5167/uzh-167451

Part VII

Reproductive System

Development of Sex Organs

18

Pradip Kumar Das, Joydip Mukherjee, and Dipak Banerjee

Abstract

The development of sex organs is the consequence of cellular and morphological alterations triggered by molecular and endocrine mechanisms initiated during early embryonic life. In the first phase of sexual differentiation, the bipotent gonad is developed, followed by the differentiation of gonads, either testis or ovary. The differentiation of internal and external genital organs then undergoes a long maturation process until puberty. The primary determinants of male gonads are the *sry* gene in the sex (Y) chromosome and DMRT1 gene in homologous ZZ chromosomes, respectively, in mammals and birds. Later stages of gonadal development are mostly under endocrine control. The development of infertile or sterile animals results from either developmental anomaly during sexual differentiation or due to interspecies mammalian hybridisation. The detailed knowledge of sex determination and its modulation through assisted reproduction techniques is the key to cloning and transgenesis to generate economically competent farm animals, wildlife conservation and avoidance of developmental disorders like freemartins.

P. K. Das (✉) · J. Mukherjee · D. Banerjee
Department of Veterinary Physiology, West Bengal University of Animal & Fishery Sciences, Kolkata, West Bengal, India

© The Author(s), under exclusive license to Springer Nature Singapore Pte Ltd. 2023
P. K. Das et al. (eds.), *Textbook of Veterinary Physiology*, https://doi.org/10.1007/978-981-19-9410-4_18

Graphical Abstract

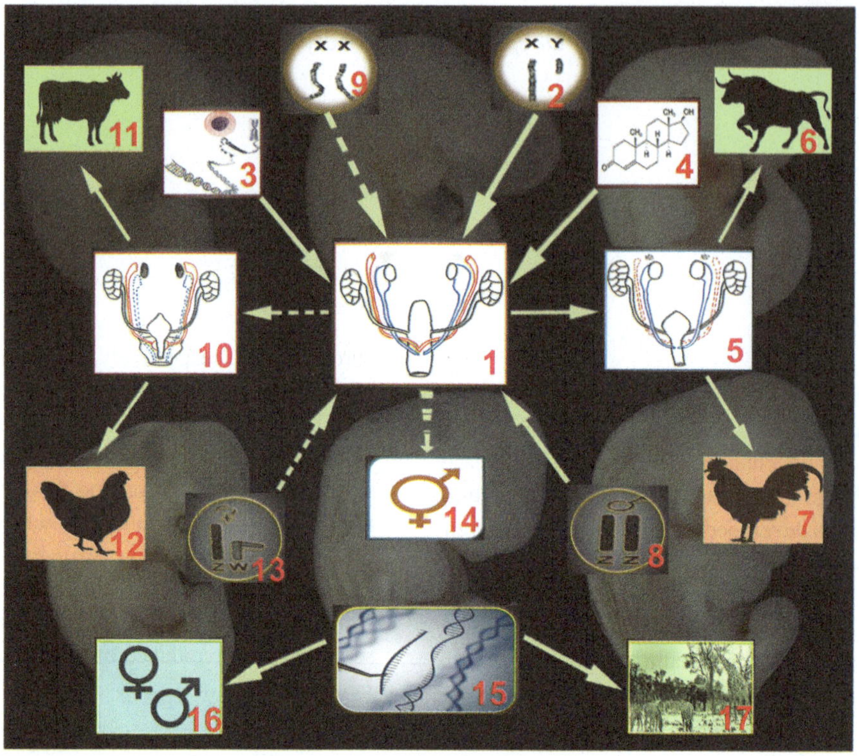

Description of the graphic: Sex development is initiated during the embryonic stage and continues entire foetal life. The bipotent gonad (**1**) is developed. In the presence of the Y chromosome (**2**) and with the influence of a group of genes (**3**), including the *sry* gene and testosterone (**4**), the male sex reproductive system (**5**) develops in mammals. The sex differentiation process continues with forming phenotypic male sex characters (**6**) until puberty. Similarly, in male birds (**7**), testes are developed in the presence of ZZ homologous chromosomes (**8**). In the absence of Y chromosome (**9**) (presented single dotted line) under the influence of ovarian determining factor (**3**), the female reproductive system (**10**) is developed and continued till puberty with various phenotypic female sex characters (**11**). Female bird (**12**) is developed due to the effect of the ZW chromosome (**13**). In coordination of the sex-determining factors leads to disorders of sex development (DSD) such as hermaphrodite (**14**). Various tools (**15**) are used to determine sex differentiation and to modulate the sex ratio (**16**) as well as to conserve the endangered animals (**17**)

Keywords

Bipotent gonad · Development of sex organs · Disorders of sex development · Sex determination in mammals and birds · Sex ratio

Learning Objectives

- The physiological phenomenon of development of bipotent gonad or bisexuality in early embryonic life
- Involvement of sex chromosome-specific genes, endocrines and biomolecules in the differentiation of sex in mammals and birds
- Principles of sex determination and its modulation through assisted reproduction techniques
- Factors and causes for the disorders of sex development (DSD) and impaired fertility (IF)
- Theories of sex ratio and its alteration to conserve the economic farm animals and wildlife

The central dogma of sexual differentiation states the sequential process of sex determination in three distinct steps. The chromosomal or genetic sex determined at fertilization triggers the development of gonadal sex (testis or ovary) from a bipotent gonad during embryogenesis. The gonadal sex determines the phenotypic sex (secondary sexual characteristics). In sexual differentiation, the male phenotype is induced, and the female phenotype is passively developed due to the absence of male determinants.

During the early embryonic stage, the development of sex organs is the consequence of cellular and morphological changes triggered by genetic and endocrine events. The embryonic development of the gonads can be categorized into two distinct phases. In the first indifferent phase (as it is common in both sexes) of gonadal development, the *bipotent genital ridge* comprises primordial germ cells (PGCs) and other supporting cells having characteristics of both male and female gonads are developed. The second phase is called the sex determination phase, in which the bipotent genital ridge undergoes differentiation to develop either testis or ovaries. The determination of phenotypic sex through the differentiation of internal and external genital organs commenced during embryonic life until puberty was influenced by genetic and endocrine control.

18.1 Bipotent Gonads

The indifferent or bipotent gonads arise, as paired structures, from the central region (mesonephros) within the intermediate mesoderm during the first half of embryonic development. The PGCs and supporting cells migrate to the genital ridge from the yolk sac situated at the base of the allantois at the posterior end of the primitive streak. Their site of migration determines the development of male and female gonads. The migration of PGCs at medullary and cortical regions of the gonadal ridge leads to the development of the Wolffian duct, which later transformed into male gonads (testes), and the Mullerian duct leads to the development of female gonads (ovaries), respectively (Fig. 18.1). Both the gonads appear on the ventromedial surface of the mesonephros. The *supporting cells* are developed into steroid-secreting cells, either testis-specific Sertoli cells or ovary-specific follicle (granulosa) cells.

18.2 Differentiation of Gonad

18.2.1 Differentiation of Sex in Domestic and Wild Mammals

The basic mechanism of sex determination is common in all mammals. It depends on the quantitative relationship between male and female-determining genes and their interface within the embryo. The females and males have XX and XY karyotypes, respectively. Hence, it was believed that the sex-determining factor was the presence of two X chromosomes for many decades. But, in the nineteenth century, it was proved that the determinant of maleness and femaleness was under the influence of the Y chromosome and the specific region of the Y chromosome that determines maleness was identified. The development of male gonads is triggered by the 'testis-determining factor' (TDF) encoded on the short arm of the Y chromosome. There are three candidates under TDF: H-Y antigen, zinc finger Y (ZFY) gene and 'sex-determining region on the chromosome Y'-gene or *sry* gene. The H-Y antigen is present only in males and classified as a minor histocompatibility antigen. The ZFY gene helps in spermatogenesis. The third candidate of TDF is *sry* gene which encodes a transcription factor that initiated the testes forming pathway at the medullary region of the developing bipotent gonad at the 7th week of embryonic life in humans. The testis starts producing two hormones during its development, namely testosterone and anti-Müllerian hormone (AMH). Testosterone and its metabolites (5α-dihydrotestosterone, 5α-DHT) promote the formation of accessory organs of the male reproductive system, whereas AMH regresses the Müllerian duct. It also stimulates Leydig cells to produce testosterone under the influence of placental human chorionic gonadotropin at about the 8th week of gestation.

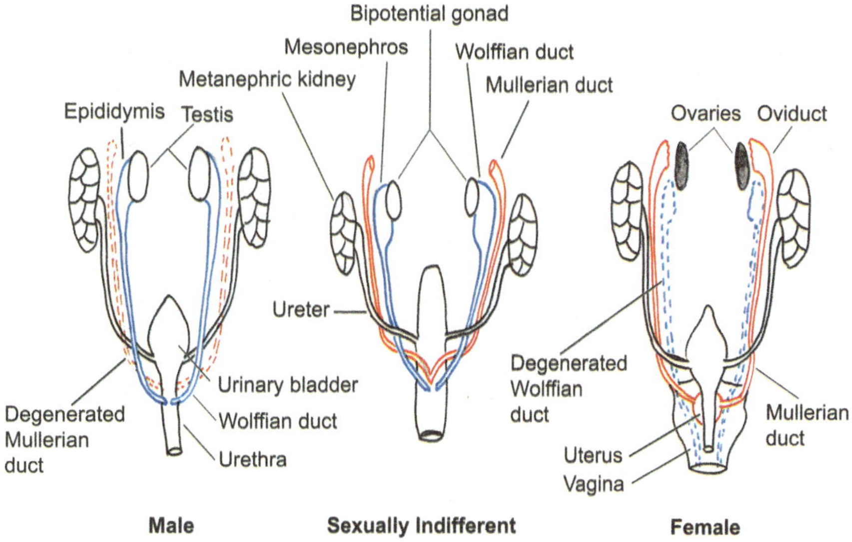

Fig. 18.1 Development of gonads in male and female from a bipotent foetus. [The sexually undifferentiated gonads (**middle**) comprising of both the male (Wolffian duct) and the female (Mullerian duct) gonads. The Wolffian duct leads to the development of male reproductive system (**left**), and the Mullerian duct leads to female reproductive system (**right**)]

The accessory sex organs of males are developed from the mesonephric duct under the influence of testosterone. The epididymis is developed from the proximal part of the mesonephric duct and connects with the rete testis via residual mesonephric tubules. The vas deferens is developed from the distal mesonephric duct. The seminal vesicle is developed as a diverticulum at the terminal portion of the epididymis. The testosterone stimulates the epithelium at the region of the prostatic urethra to form the ducts and stroma of the prostate gland. The bulbourethral glands are developed as an outward invagination from the anterior urethra.

The regression of mesonephros allows the testes to float free in the peritoneal cavity like with mesentery. From the caudal pole of the testes, a mesodermal band called gubernaculum originates and passes toward the inguinal region through the posterior abdominal wall. A peritoneal pouch named the processus vaginalis appears before the gubernaculum at the 6th month of gestation. This pouch develops into tunica vaginalis and holds the testes suspended after its descent around the 8th month of gestation.

The *sry* gene's counterpart in females for ovary determination is yet to be identified, and it was postulated that the development of the ovary is a default process. Recently, the 'Z' theory for ovarian determination has been proposed. The 'Z' factor, present in the XX karyotype, promotes ovarian development and suppresses testicular development. There are different candidate genes under the Z factor (ovary-determining genes). Dax1 (dosage-sensitive sex-reversal-adrenal hypoplasia congenital-critical region of the X chromosome gene1) was initially proposed as an ovary-determining gene due to its ovary-specific expression pattern, but later, it was found that Dax1 is not essential for ovarian development; rather, it has roles in testicular development. So, it was eliminated from ovarian determining factors. Wnt4, an ovarian specific gene, was reported to inhibit testis determining pathway and promote ovarian development. The most potential candidate of the Z factor is Foxl2, a transcription factor that plays a crucial role in normal ovarian development. Another candidate gene of Z factor, namely R-spondin family, member 1 (RSPO1), encodes a signalling molecule to regulate the WNT signalling pathway and promotes ovarian development. It also antagonizes testes formation. Under the influence of Z factor genes, the cortex of the ridge is activated and the *Mullerian* (paramesonephric) or female reproductive duct is developed as an invagination of the surface epithelium of the mesonephros (Fig. 18.1). The Mullerian ducts thus formed extend in cranial-caudal fashion and fuse at the caudal end to form a Y-shaped structure called uterovaginal primordium from which the uterus, the cervix, and the upper portion of the vagina are developed. Uterine development is initiated around the 8th week of gestation, and vaginal development continues until the 20th week in humans. Estrogenic receptor in the vaginal epithelium has been found during the 21st week. The fallopian tubes are developed from the unfused portions of Mullerian ducts. The caudal most uterovaginal primordium fuses the urogenital sinus to form the vagina and hymen.

Table 18.1 The key regulatory mammalian sexual development genes

Name of the genes	Major regulation/ activity
Wt1, Sf1, Lhx9, Emx2, L33	Development of bipotent gonad
Gata4/Fog2, Sry, Sox9, Sox8, Fgf9, Dax1, Pod1, Dhh, Pgdra, Pgds, Arx, Artx, Insl3, Lgr8, Hoxa10, Hoxa11, Amh, Misrl1, Pax2, Lim1, Dmrt1	Testis-determining pathway
Wnt4, FoxL2, Dax1, RSPO1	Ovary-determining pathway

The key regulatory mammalian sexual development genes and their probable roles in sex determination have been presented in Table 18.1 and Fig. 18.2.

18.2.2 Differentiation of Sex in Bird

In contrast to mammals, the males are homogametic (ZZ), and females are heterogametic (ZW) in birds. The avian species lack *sry* gene; instead, the testes determining factor for birds is Z chromosome-specific double-sex and Mab-3 related transcription factor #1(DMRT1). Another gene, namely, SOX9, is responsible for Sertoli cell differentiation. The ovary-determining factor of birds is two W-linked genes, avian sex-specific W-linked (ASW) and Female-Expressed Transcript 1 (FET1). Other ovary-determining factors like HINTW, FOXL2, WNT4 and RSPO1 are also present in W-chromosome. The avian embryos bearing ZZ chromosomes have a higher quantity, and a higher expression level of DMRT1 triggers testicular development. In females, a lower level of DMRT1 is insufficient to repress ovarian development genes, thus favouring ovarian development. The left gonad becomes an ovary with thickened cortex and vacuolated medulla during ovarian development, but the right one is regressed.

18.3 Disorders of Sex Development (DSD)

The disorders of sex development (DSD) and impaired fertility (IF) are serious concerns in animal breeding, particularly where artificial insemination is generally performed. There are three main categories of DSDs (originally adapted from humans) occurring in domestic and companion animals. The first category is the sex chromosome DSD found in freemartins and characterized by sex chromosome aneuploidies, sex chromosomal structural rearrangements

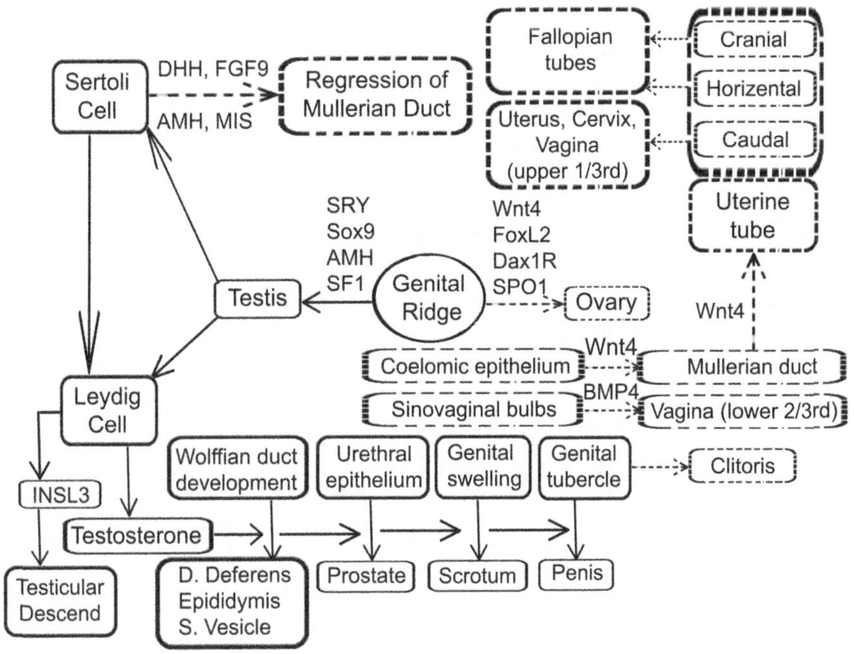

Fig. 18.2 Development of male and female reproductive organs under the influence of various genetic, endocrine and growth factors. [The bold lines showed the male gonad and duct development and dotted lines showed for females. Major genes involved in male reproductive organs development are ***Sry, Sox9, SF1,*** and the development of female reproductive organs is brought about by ***Wnt4, FoxL2, Dax1, RSPO1, BMP4***. **DHH** (Desert hedgehog) gene, **FGF9** (Fibroblast Growth Factor), **AMH** (Anti-Müllerian hormone) and **MIS** (Müllerian Inhibiting Substance) promote the regression of the Müllerian duct. Testosterone released from the Leydig cell further potentiates to develop the male reproductive system. **INSL3** (insulin-like peptide 3) is a relaxin peptide family member that helps descend the testis]

and lymphocyte chimerism (XX/XY). The XX DSDs are analogous to Turner's syndrome in humans which are observed in goats, pigs, horses and dogs. The XX DSDs are characterized by the presence of both male (testicles or ovotestes) and female (uterus, oviducts and ductus deferens) reproductive organs. The XY DSDs (analogous to Klinefelter's syndrome) include androgen insensitivity syndrome (AIS) and persistent Mullerian duct syndrome (PMDS) and are characterized by cryptorchidism and hypospadias. Under certain circumstances, animals with XX karyotype may express H-Y antigen and result in hermaphroditic females. It is reported in mares, mice, does, sows and Cocker Spaniel bitches.

18.3.1 Freemartins

One of the common examples of sex chromosome DSD is freemartinism. The freemartin is a sterile female calf result due to fusion of the chorioallantoic portions of the placenta occurring around days 28 and 30 of gestation in heterosexual twin births. The fusion of the placenta facilitates the common blood supply between twin foetuses. The male determining factors (testosterone, H-Y antigen, MIT) of the male foetus suppress the development of genitalia in the female foetus. The sterile female calf may have urogenital tract, vagina and external genitalia. The freemartins are common in cattle, goats and camelids.

18.3.2 Mammalian Hybrids

Interspecies *mammalian hybrids* are developed for various purposes. Some of them are sterile, and some are fertile depending on the chromosome numbers of breeding parents (Table 18.2). Some examples of such hybrids are mule, hinny, tigon and liger. The mule (male donkey X female horse) and hinny (male horse X female donkey) have 63 chromosomes. They are sterile due to chromosomal incompatibility despite their fertile parents (horse and donkey have 64 and 62 chromosomes, respectively). Mules can exhibit oestrus symptoms and can maintain horse embryos. Tigon is obtained from the breeding of a male tiger and a female lion. Liger is the result of breeding a male lion to a female tiger. Both lion and tiger have 38 chromosomes; hence, liger and tigon also have the same number of chromosomes and are fertile.

Table 18.2 Chromosome number (2n) in different domestic and wild animals and birds

Domestic animals and human	No.	Wild animals	No.	Birds	No.
Pig (*Sus*)	38	Kangaroo (Macropodidae)	16	Dove (Columbidae)	78
Domestic cat (*Felis catus*)	38	Giraffe (*Giraffa camelopardalis*)	30	Quail (*Coturnix coturnix*)	78
Mouse (*Mus musculus*)	40	Tiger (*Panthera tigris*)	38		
Rat (*Rattus norvegicus*)	42	Lion (*Panthera leo*)	38	Chicken (*Gallus gallus domesticus*)	78
Human (*Homo sapiens*)	46	Cat (*Felis silvestris catus*)	38		
Water buffalo (river type) (*Bubalus bubalis*)	48	Ferret (*Mustela putorius furo*) Hyena (Hyaenidae)	40 40	Turkey (*Meleagris*)	80
Water buffalo (swamp type) (*Bubalus bubalis*)	50	Rhesus monkey (*Macaca mulatta*) Dolphin (Delphinidae)	42 44	Pigeon (Columbidae) Domestic geese (*Anser anser domesticus*)	80 80
Sheep (Ovis aries)	54	Nilgai (Boselaphus tragocamelus)	46		
Yak (*Bosmutus*)	60	Chimpanzee (*Pan troglodytes*)	48	Parrot (Psittaciformes)	80
Goat (*Capra hircus*)	60	Elephant (Elephantidae)	56	Duck (*Anas platyrhynchos*)	80
Cattle (*Bos taurus*)	60	Red deer (*Cervus elaphus*)	68		
Donkey (*Equus asinus*) Guinea pig (*Cavia porcellus*)	62 64	White-tailed deer (*Odocoileus virginianus*)	70	Domestic pigeon (*Columba livia domestica*)	80
Horse (*Equus caballus*)	64	Bear (*Ursus* sps.)	74		
Dog (*Canis familiaris*)	78	Wolf (*Canis lupus*)	78		

18.4 Determination of Sex in Domestic and Wild Animals and Birds

18.4.1 Determination of Foetal Sex

A number of biotechnological tools and imaging techniques can be applied to determine foetal sex in animals during the various period of embryonic life (Table 18.3). The biotechnological tools identify sex-specific genes, amplicon, enzymes and proteins. Ultrasonography is the best imaging technique to determine foetal sex through phenotypic characteristics. The identification of males is more accessible than the females by molecular markers like *sry* gene using polymerase chain reaction (PCR) based sexing assays. The embryos can be sexed by DNA probes specific for the Y chromosome (H-Y antigen) or assayed for X-linked enzymes using the fluorescent antibody technique (FAT). The embryo sexing is useful in multiple ovulation embryo transfers (MOET) nucleus breeding schemes to generate more economic animals, avoid freemartins multiple births, develop transgenic and cloned animals, and conserve wildlife (discussed details in Chap. 24).

Table 18.3 Recognition of differentiation of foetal sex in different domestic animals and birds

Species	Duration from post-fertilization (week)	Mode of identification/genetic markers	Reference
Cattle	5–7	Y-specific sequences	da Cruz et al. (2012)
Cattle	4–34	sry gene; Y-specific amplicons	Xi et al. (2006), Wang et al. (2010); da Cruz et al. (2012)
Cattle	7–16	Ultrasonography	Quintela et al. (2011)
Cattle & Horse	9–10	Ultrasonography	Curran (1992)
Horse	4; 8–20	sry gene; sry& GAPDH	De Leon et al. (2012); Davoodian and Kadivar (2016)
Goat	5	Amelogenin gene	Chen et al. (2007)
Goat	6–17	Ultrasonography	Amer (2010)
Sheep	8–18	sry gene & GAPDH; Amelogenin gene	Kadivar et al. (2013); Davoodian and Kadivar (2016)
Pig	2	sry gene	Pomp et al. (1995)
Dog	4–7	sry gene & Ultrasonography	Prugnard et al. (2016)
Dog	3–4; 8	Ultrasonography	Orlandi et al. (2019); Khatti et al. (2018)
Cat	6	Ultrasonography	Zambelli and Prati (2006)
Mouse	2	sry gene	Larney et al. (2014)
Mouse	1	Ultrasonography	Pallares and Gonzalez-Bulnes (2009)
Bird	1	DMRT1 gene	Chue and Smith (2011)

Phenotypic characteristics are employed to determine postnatal sex in domestic and wild animals. But, it is difficult in birds, as most birds are monomorphic. Vent sexing is a common process in birds (discussed details in Chap. 19).

18.4.2 Determination of Sex in Birds

Different molecular techniques like karyotyping and PCR are generally used to differentiate sex in birds. Homologous ZZ chromosomes confirm that the male and ZW chromosomes validate females. The expression profile of the DMRT1 gene through PCR is used to identify male birds. Amplifying chromodomain helicase DNA-binding (CHD) gene, producing one amplicon fragment of the Z chromosome, confirms male, where two fragments for the Z and W chromosomes confirm the female birds.

Know More...

Sex Determination in Monotremes

The sex-determining region on the chromosome Y (sry or, *sry* gene) presents in all mammals, except the animals, viz. platypus, echidnas, steropodon, etc. under order monotremes. Monotremes are placental mammals and nurture the young by their milk but lay eggs. Monotremes comprise distinctive XY sex chromosome systems with several homologies with the avian Z chromosome.

Platypusus Echidnas Steropodon

18.5 Sex Ratio

The sex ratio is the ratio of males to females in a population. It is expressed as a percentage of males to females or males per 100 females. There are four types of sex ratios applied in animals during different stages of life: Primary sex ratio at fertilization; secondary sex ratio at birth, tertiary at puberty or, adult sex ratio (ASR) and quaternary sex ratio in post-reproductive stages. Theoretically, the primary sex ratio is 1: 1 or 50%. But, it depends upon many factors. Supplementation of omega-6 polyunsaturated fatty acids (PUFAs) to dam can increase the probability of male calves in cattle and sheep by influencing oocyte cellular functions and developmental potential. It is also evidenced in sheep and pigs that females with good physical conditions could produce more male offspring. Due to altered feeding patterns, the primary sex ratio can also be altered under the seasonal influence. Artificial insemination at the initiation of the oestrus period increased the probability of the female calves in cattle.

18.5.1 Manipulation of Sex Ratio

Assisted reproduction techniques can be applied to modulate primary sex ratio considering animal economic traits and ecological importance. The use of sex-sorted semen is now emerged as a promising technique to generate animals of the desired sex. This method is based on the sorting of X and Y chromosome bearing spermatozoa fluorescence-activated cell sorting (FACS) techniques. The molecular markers to identify X chromosome bearing spermatozoa are X-linked enzymes, like glucose-6-phosphate dehydrogenase (G6PD), hypoxanthine-guanine phosphoribosyltransferase (HPRT), phosphoglycerate kinase and alpha-galactosidase. The Y-chromosome bearing spermatozoa can be separated by the presence of *sex-specific proteins* (SSPs), which have distinct antigenic properties due to its H-Y antigen. The X- and Y-chromosome bearing spermatozoa are identified by detecting the chromosome-specific sperm protein constituents, like kinases, transmembrane proteins and chaperones. PCR, fluorescence in situ hybridization (FISH) and Raman spectroscopy techniques are also used to identify sexed spermatozoa through proteomics.

Learning Outcomes

- **Development of sex organ:** The embryonic development of the gonads occurs in two distinct phases. In the first indifferent phase, the bipotent genital ridge having characteristics of both male and female gonads is developed. The second phase is called the sex determination phase, in which the bipotent genital ridge undergoes differentiation to develop either testis or ovaries. The testis is developed under the genetic control of 'testes-determining factor' (TDF) encoded on the short arm of the Y chromosome. The 'Z' factor in the XX karyotype triggers ovarian development. The testis is developed from the medullary part of the genital ridge and ovaries from the cortex. In avian species, the males are homozygous (ZZ), females are heterozygous (ZW), and the testis-determining factor is DMRT1. The male accessory sex organs are developed under the influence of testosterone and its metabolites secreted from developing testes from the Wolffian duct, and the Mullerian duct differentiates into female accessory sex organs.

(continued)

- **Disorders of sex development (DSD):** The disorders of sex development (DSD) and impaired fertility (IF) are serious concerns in animal breeding, particularly where artificial insemination is generally performed. The sex chromosome DSD is seen in freemartins and is characterized by sex chromosome aneuploidies, sex chromosomal structural rearrangements and lymphocyte chimerism (XX/XY). The XX DSDs are analogous to Turner's syndrome and characterized by the presence of both male and female reproductive organs. The XY DSDs (analogous to Klinefelter's syndrome) include androgen insensitivity syndrome (AIS) and persistent Mullerian duct syndrome (PMDS) and characterized by cryptorchidism and hypospadias.
- **Sex differentiation in animals and birds:** A number of biotechnological tools and imaging techniques can be applied to determine foetal sex in animals during the various period of embryonic life by identifying sex-specific genes, amplicons, enzymes and proteins, whereas ultrasonography is the best imaging technique to determine foetal sex through phenotypic characteristics. It is effectively used in multiple ovulation embryo transfer (MOET) nucleus breeding programmes.
- **Sex ratio:** The ratio of males to females in a population is called the sex ratio. It can be expressed in four types according to different stages of life of the animals. The ratio can be altered naturally and by several manipulation techniques. These techniques are applied to conserve the wildlife and target the economic population for increasing production.

Exercises

Objective Questions

Q1. Which phase of the sex development is continued up to puberty?

Q2. What is the major role of the supporting cells of the bipotent gonad in the ovary?

Q3. Which gonad is developed from the medullary part of the genital ridge?

Q4. Which gene is primarily responsible for male gonad development in mammals?

Q5. What is the full name of DMRT1?

Q6. Why do mammals having Klinefelter's syndrome can be developed male gonad and in Turner's syndrome developed female gonad?

Q7. What is the fate of the embryonic genital tubercle in male and female mammals?

Q8. Why some interspecies mammalian hybrids are fertile, but some are sterile?

Q9. Which type of gonad is developed in the animals having a Y chromosome, but, sry gene is not fully expressed?

Q10. Which kind of sex differentiation can easily be recognized?

Q11. What is the sex ratio?

Q12. Which sex chromosome is larger and why?

Q13. Which chromosome carries the key candidate gene in the bird?

Q14. Which mammals lack the *sry* gene in the Y chromosome?

Subjective Questions

Q1. Write the developmental process of male gonad during embryonic life in mammals.

Q2. Write the developmental process of female gonad during embryonic life in mammals.

Q3. How are gonads developed in the bird?

Q4. Write the role of various endocrines and growth hormones in sex differentiation.

Q5. Explain: 'A tigon is fertile, but the mule is sterile'.

Q6. Why freemartins heifer can be sterile?

Q7. How does the sex differentiation can be recognized in mammals and birds and its importance?

Q8. What are the different sex ratios, and how can they be altered or manipulate?

Answer to Objective Questions

A1. The second phase or phase of differentiation of the gonad

A2. Steroidogenesis

A3. Testis

A4. sry gene

A5. Double-sex and Mab-3 related transcription factor #1

A6. Mammals having Klinefelter's syndrome belong Y chromosome, and Turner's syndrome do not belong Y chromosome

A7. In males, it forms the penis and in the female clitoris

A8. Fertile animals will occur when the involving species have compatible chromosome numbers, and sterility develops when they belong to the incompatible number of chromosomes.

A9. Ovary

A10. Male

A11. It is the ratio of males to females in a population and expressed as a percentage of males to females.

A12. The X chromosome, as it contains more DNA mass

A13. Z chromosome
A14. Mammals under order monotremes

Keywords for the Answer to Subjective Questions

A1. Y chromosome and sry gene, endocrines, growth factors
A2. XX karyotype, absence of sry gene, growth factors
A3. Karyotype ZZ or ZW, SOX9 gene, DMRT1
A4. Role of androgens, Mullerian inhibitors, TGF-β, FGF9, INSL3
A5. Hybrid, chromosome number, functional characters of the gonad
A6. Heterogeneous twin foetuses, masculinizing factors, suppression of ovarian development
A7. PCR, USG, MOET
A8. Four types, dietary manipulation, using ART

Acknowledgement We are grateful to *Prof. Sagar Sanyal*, Ex. Professor and Head, Department of Veterinary Physiology, West Bengal University of Animal and Fishery Sciences, Kolkata, West Bengal, India for his valuable guidance during the preparation of the chapter. We are also greatly indebted to the helps from *Dr. Parthib Roy*, Master Scholar, Department of Veterinary Gynaecology and Obstetrics for preparation of the figures.

Further Reading

Textbooks

Budik S (2017) Genetics in domestic animal reproduction. In: Abubakar M (ed) Trends and advances in veterinary genetics. https://doi.org/10.5772/67132

Pineda MH (2003) The biology of sex. In: Pineda MH, Dooley MP (eds) McDonald's veterinary endocrinology and reproduction, 5th edn. Iowa State Press, pp 201–238

Research Articles

Amer HA (2010) Ultrasonographic assessment of early pregnancy diagnosis, fetometry and sex determination in goats. Anim Reprod Sci 117(3–4):226–231. https://doi.org/10.1016/j.anireprosci.2009.05.015

Chen AQ, Xu ZR, Yu SD (2007) Sexing goat embryos by PCR amplification of X- and Y-chromosome specific sequence of the Amelogenin gene. Asian Australas J AnimSci 20(11):1689–1693

Chue J, Smith CA (2011) Sex determination and sexual differentiation in the avian model. FEBS J 278:1027–1034

Curran S (1992) Fetal sex determination in cattle and horses by ultrasonography. Theriogenology 37:17–21. https://doi.org/10.1016/0093-691X(92)90244-L

da Cruz AS, Silva DC, Costa EOA et al (2012) Cattle fetal sex determination by polymerase chain reaction using DNA isolated from maternal plasma. Anim Reprod Sci 131(1–2):49–53. https://doi.org/10.1016/j.anireprosci.2012.02.004

Davoodian N, Kadivar A (2016) Prenatal determination of farm animal fetal sex using free fetal DNA in maternal plasma. IOSR J Agric Vet Sci 9(11):38–45. https://doi.org/10.9790/2380-0911023845

De Canio M, Soggiu A, Piras C et al (2014) Differential protein profile in sexed bovine semen: shotgun proteomics investigation. Mol BioSyst 10:1264–1271

de Leon PMM, Campos VF, Dellagostin OA et al (2012) Equine fetal sex determination using circulating cell-free fetal DNA (ccffDNA). Theriogenology 77:694–698

Kadivar A, Hassanpour H, Mirshokraei P et al (2013) Detection and quantification of cell-free fetal DNA in ovine maternal plasma; use it to predict fetal sex. Theriogenology 79(6):995–1000. https://doi.org/10.1016/j.theriogenology.2013.01.027

Khatti A, Jena D, Singh SK et al (2018) Application of ultrasonography in canine pregnancy—an overview. Int J Livest 7(2):20–27. https://doi.org/10.5455/ijlr.20170204053145

Larney C, Bailey TL, Koopman P (2014) Switching on sex: transcriptional regulation of the testis-determining gene Sry. Development 141(11):2195–2205. https://doi.org/10.1242/dev.107052

Orlandi R, Vallesi E, Boiti C et al (2019) Contrast-enhanced ultrasonography of maternal and fetal blood flows in pregnant bitches. Theriogenology 125:129–134. https://doi.org/10.1016/j.theriogenology.2018.10.027

Pallares P, Gonzalez-Bulnes A (2009) Use of ultrasound imaging for early diagnosis of pregnancy and determination of litter size in the mouse. Lab Anim 43(1):91–95. https://doi.org/10.1258/la.2008.007139

Pomp D, Good BA, Geisert RD et al (1995) Sex identification in mammals with polymerase chain reaction and its use to examine sex effects on diameter of day-10 or −11 pig embryos. J Anim Sci 73(5):1408–1415. https://doi.org/10.2527/1995.7351408x

Prugnard C, Lamia AB, Cherel Y et al (2016) Early sex determination in the canine foetus by ultrasound and PCR. Anim Reprod Sci 165:56–68. https://doi.org/10.1016/j.anireprosci.2015.12.007

Quintela LA, Becerra JJ, Pérez-Marín CC et al (2011) Fetal gender determination by first-trimester ultrasound in dairy cows under routine herd management in Northwest Spain. Anim Reprod Sci 125(1–4):13–19. https://doi.org/10.1016/j.anireprosci.2011.02.022

Wang G, Cui Q, Cheng K et al (2010) Prediction of fetal sex by amplification of fetal DNA present in cow plasma. J Reprod Dev 56:639–642

Xi J, Jia B, Li R et al (2006) Rapid sex identification of embryos in different developmental stages by blood of pregnant dairy cattle. J Shihezi Univ (Nat Sci) 24:446–449

Zambelli D, Prati F (2006) Ultrasonography for pregnancy diagnosis and evaluation in queens. Theriogenology 66(1):135–144

Functional Morphology of the Male Reproductive System

19

Pradip Kumar Das, Joydip Mukherjee, and Dipak Banerjee

Abstract

The reproductive organs of males comprised of primary reproductive organs (testes), excurrent ducts composed of rete testis, vasa efferentia, epididymis, vas deferens, and urethra, accessory sex organs (seminal vesicle, ampulla, prostate and bulbourethral or Cowper's glands), and ancillary organs (penis and prepuce). The primary functions of the testes are spermatogenesis and steroidogenesis. The Sertoli cells of the testes aid nutritional support to spermatozoa. The spermatozoa are transported through the excurrent duct and stored in the epididymis for their maturation. The spermatozoa are suspended in the seminal plasma produced from accessory sex organs. The hypothalamic–hypophyseal–gonadal (HPG) axis controls the functions of the testes and other reproductive organs. The HPG axis becomes fully functional at puberty with the activation of the GnRH surge centre and subsequently stimulates the anterior pituitary for FSH and LH secretion. The LH stimulates Leydig cell steroidogenesis. The Leydig cells produce testosterone which facilitates spermatogenesis within the seminiferous tubules. This chapter highlights the functional morphology of the male reproductive system and the factors controlling the functions of each organ. Puberty and its associated events are also discussed. The entire chapter is divided into five sub-chapters to understand different aspects of the male reproductive system.

P. K. Das (✉) · J. Mukherjee · D. Banerjee
Department of Veterinary Physiology, West Bengal University of Animal & Fishery Sciences, Kolkata, West Bengal, India

© The Author(s), under exclusive license to Springer Nature Singapore Pte Ltd. 2023

P. K. Das et al. (eds.), *Textbook of Veterinary Physiology*, https://doi.org/10.1007/978-981-19-9410-4_19

Graphical Abstract

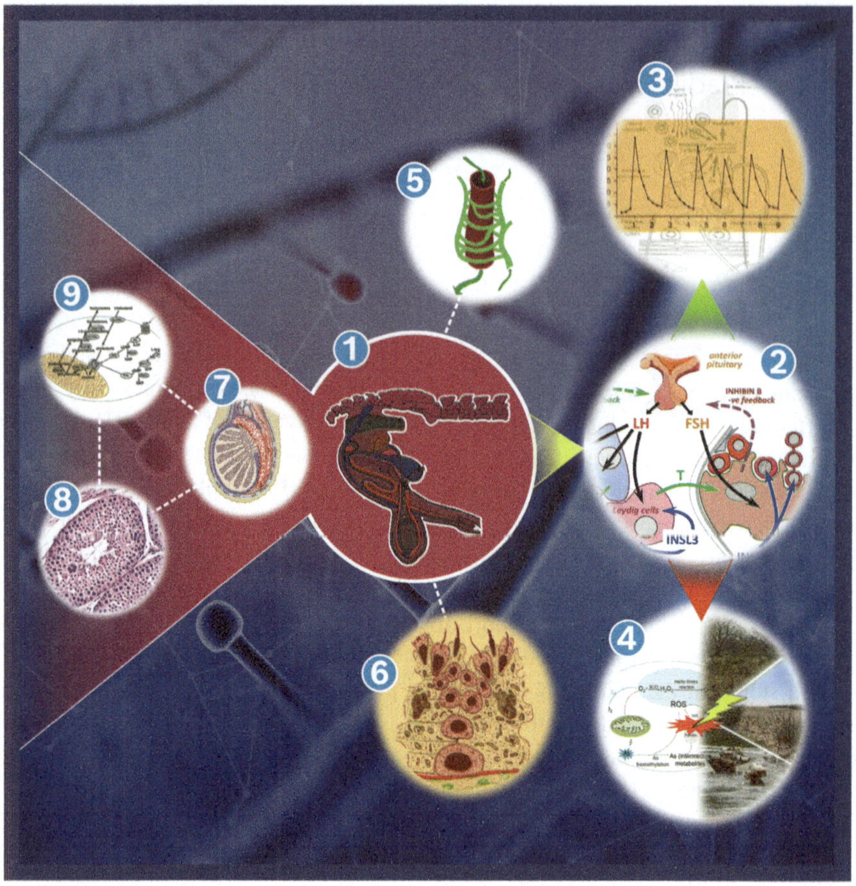

Description of the graphic: Male reproductive system (1) starts to develop from foetal life. Its functional role is primarily controlled by the Hypothalamus–Pituitary–Gonadal (HPG) axis (2), which triggers the onset of puberty characterized by luteinising hormone secretion at high amplitude (3). The activity of the HPG axis is also influenced by several other factors, viz. stressors (4). Special morphological arrangements, like the testicular thermoregulation mechanism (5) and the blood–testis barrier (6), provide a suitable microenvironment for spermatogenesis (7). The chapter elaborates the detailed internal morphological features of the testis (8) and other associate structures and their functions, viz. steroidogenesis (9). The entire functional characteristics of the system are species-specific and presented in the background of the graphic with the genomic structures.

Keywords

Excurrent tract · HPG axis · Puberty · Testes · Androgens

Learning Objectives

- Functional morphology of testes
- Role of excurrent tract and copulatory organs
- Accessory sex glands and their contribution to semen production
- Hypothalamus–Pituitary–Gonadal axis and endocrine regulation of male reproduction
- Biosynthesis of testicular androgens and their biological role
- Puberty in males and its determining factors

19.1 The Male Reproductive Organs

The male reproductive system consists of primary sex organ *testis* (testes in plural), excurrent tract, accessory sex glands, and ancillary organs. *The excurrent tract* begins from the rete testis, followed by efferent ducts (vasa efferentia), epididymis, vas deferens, and urethra. The accessory sex glands are the ampulla, seminal vesicles, prostate and bulbourethral glands (Cowper's glands). The ancillary organs are the penis and prepuce. There are species variations in the morphological features of male reproductive organs (Figs. 19.1 and 19.2).

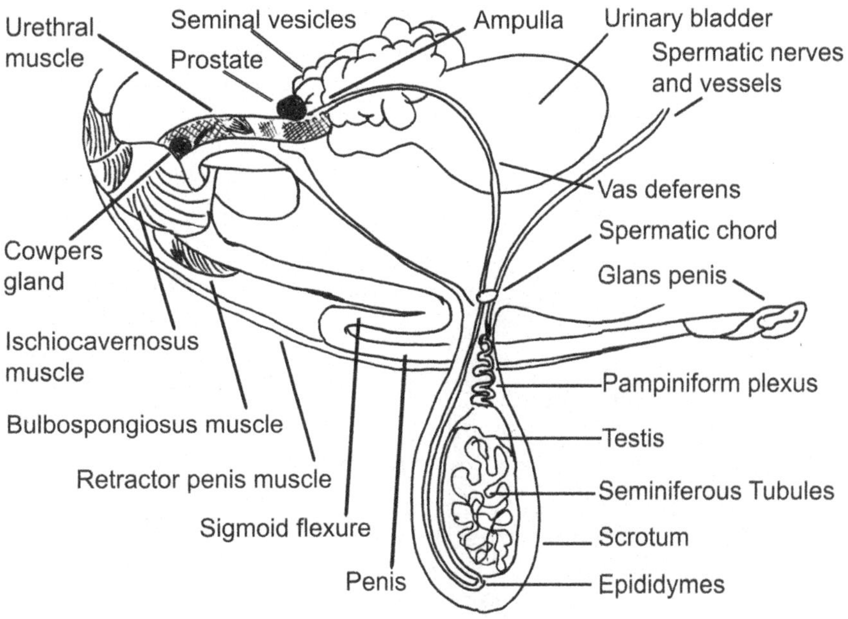

Fig. 19.1 Reproductive tract of bull

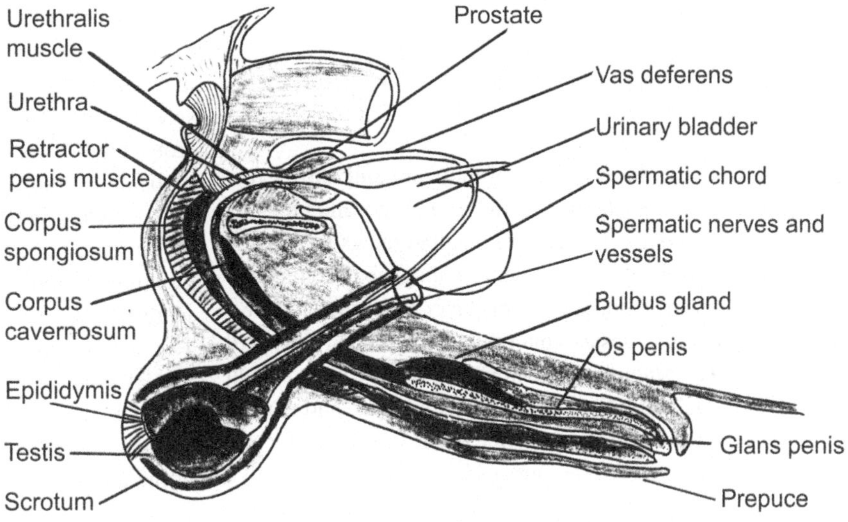

Fig. 19.2 Reproductive tract of a male dog

19.1.1 Testes

Testes are the paired structure situated outside the abdomen (in most species) in a purse-like structure called the scrotum made of skin and fascia. The spermatic cord attached with the superior pole of the testes helps to suspend the testis within the scrotal sac, and the distal end of the testes is attached with the scrotum by a scrotal ligament. The right and left testicles are separated by a muscular septum formed by dartos muscles. In some animals like mice, bat, the testes lie inside the body during their non-breeding season. But, it comes outside when breeding approaches. The testes are situated in rats, rabbits, and camels' perineal regions. Testes remain in the sub-anal region in the feline species like cat, tiger, lion, etc. The intra-abdominal testes are seen in monotremes, armadillos, sloths, elephants, rhinoceros, and birds. There is species variation in the morphological features of testes (Table 19.1). The testicular size varies throughout the year in seasonal breeders, like stallion, ram, and camel.

The testes are composed of testicular capsule, mediastinum, and parenchyma.

19.1.1.1 Testicular Capsule

Four layers encapsulate the testes. The innermost serous covering of the testes is called the vaginal process or tunica vaginalis. It is a double layer peritoneal process that envelopes the whole testes except the region where the epididymis and spermatic cord are attached. The outer and inner layers of tunica vaginalis are called parietal and visceral vaginal tunics, respectively (Fig. 19.3). The watery fluid

Table 19.1 Morphological features of the testes in different domestic animals

Characteristics	Bull	Stallion	Ram	Boar	Dog	Cat
Shape	Oval	Oval	Oval	Oval	Round to oval	Round to oval
Length (cm)	10–15	7.5–12.5	7.5–11.5	10–15	2–4	1.2–2
Diameter (cm)	5–8.5	4–7	3.8–6.8	5–9	1–2.5	0.7–1.5
Weight (gm)	200–500	200–300	200–400	600	7–15	0.7–1.5
Plane	Vertical	Horizontal	Vertical	Vertical	Oblique	Vertical

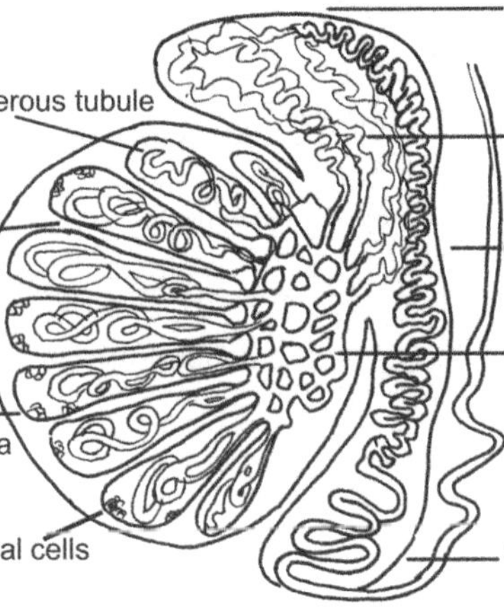

Fig. 19.3 Structure of testis. [The schematic diagram showing the longitudinal cross section of the testis along with the excurrent tract in bull]

between these two layers acts as a lubricant and allows free movement of the testicles within the scrotal sac, and prevents the testes from injury. The visceral vaginal tunics is attached with a fibrous white capsule of dense collagenous connective tissue, called tunica albuginea. This layer acts as an insulator and helps maintain testes' internal temperature. The third layer is called stratum vasculature, through which scrotal vasculature, including blood vessels, lymphatics, and nerves, enter into the testis. The fourth zone of the covering of testes is called septulae testis. It comprises loose areolar connective tissues, extends inward from the mediastinum testis, and separates the testicular parenchyma into many lobules.

19.1.1.2 Mediastinum Testes

It is a connective tissue sheet that runs through the centre of the testes from top to bottom (Fig. 19.3). The testicular mediastinum is divided into lobules by the trabeculae of the tunica albuginea. The lobules are filled with composed of seminiferous tubules. The mediastinum testes support the rete testis and allow the blood vessels and lymphatics.

19.1.1.3 Testicular Parenchyma

The testicular parenchyma comprises seminiferous tubules lined by different generations of maturing germ cells (spermatogonia, spermatocytes, spermatids spermatozoa) and Sertoli cells.

19.1.1.3.1 Seminiferous Tubules

These tubules are highly convoluted structures that converge at the apex of each testicular lobule (Fig. 19.3). In bull, each seminiferous tubule is 80 cm in length and about 110–250 μm in diameter. The seminiferous tubule cross section reveals three layers: the outer capsule, basement membrane, and testicular cells. The testicular cells comprise the germinal epithelium, Sertoli cells, and Leydig cells. The germinal epithelium and Sertoli cells are the lining cells of seminiferous tubules, whereas the Leydig cells are present in the interstitium outside the seminiferous tubule.

The Outer Capsule The basal lamina is composed of fibroblasts and *peritubular myoid* (PTM) cells. The PTM cells are spindle-shaped smooth muscle cells surrounding the seminiferous tubules and separating them from interstitial space. The PTM cells give structural support to the seminiferous tubules as it contains contractile proteins like actin, myosin, desmin, vimentin, and alpha-actinin. PTM cells also secrete ground substances for seminiferous tubule basement membrane. The growth factors (IGF-I, activin-A, peritubular modifying substance (PModS), and TGF-β) secreted from PTM control the function of the Sertoli cells. The PTM cells contain cellular retinol-binding protein (CRBP) involved in the internalization of retinol to support spermatogenesis. The PTM cells also assist sperm transport by increasing the contractility of excurrent tract with the help of prostaglandins and oxytocin.

The Basement Membrane Seminiferous tubule basement membrane (STBM) is composed of fibronectin, proteoglycans, and collagens (type I and IV) secreted from PTM cells. It gives structural support to the germinal epithelium and Sertoli cells.

Know More . . .

The left testis is generally larger in most mammals and birds, except in sharks, where the right one is larger. Intra-abdominal testes are found in birds, elephants, rhinoceros, sloths, whales, dolphins, and others in monotremes mammals. Testes remain in the sub-anal region in feline species, viz. cat, lion, and tiger. Testes

(continued)

appear in the perianal region in camel, rat, and rabbit. The total length of the convoluted seminiferous tubules in mice is about 2 m, in domestic cat 50 m, crab-eating fox 80 m, fowl (Nigerian indigenous chicken) 600 m, human 300–1000 m, goat 3.7 km, and bull 5.2 km.

19.1.1.4 Testicular Cells

The lining cells of the seminiferous tubules are of two types: the stratified germinal epithelium and Sertoli cells (named after Enrico Sertoli, Italian Physiologist who discovered the cells) (Fig. 19.4). The Leydig cells (named after German zoologist and anatomist Franz Leydig) are present at the interstitial space (between the tubules) and play a pivotal role in controlling the activities of seminiferous tubules. In crossbred bulls, the population of Sertoli cells are less due to small-sized seminiferous tubules, which affects their fertility.

19.1.1.4.1 Germinal Epithelium

The seminiferous epithelium is capable of producing spermatozoa hence called the germinal epithelium. The maturing germ cells are stratified, starting from spermatogonia at the base areas, followed by spermatocytes, spermatids, and spermatozoa towards the lumen.

19.1.1.4.2 Sertoli Cells

These cells are triangular or oval with a prominent nucleolus, fine chromatin, elongated mitochondria, small fibrils, and lipid droplet with glycogen (Fig. 19.4). The number of Sertoli cells varies with species. The number of Sertoli cells per gram of testes is about 40 x 10^6, 30 x 10^6 and 25 x 10^6 in bull, cat, and human, respectively. The numbers of Sertoli cells generally remain constant even after puberty. In seasonal breeders, the volume and functional activity increase during the breeding season.

Functions of Sertoli Cell Sertoli cells are involved in regulating spermatogenesis, formation of blood–testes barrier, and many other activities (Figs. 19.5 and 19.9). The proliferation and development of germ cells during foetal growth are regulated by Sertoli cells. The proliferation and differentiation of Sertoli cells cease after puberty. The cells then act as 'nurse cells' to feed the germ cells and aid the spermatogenesis by secreting stem cell factors (SCF). The maturation of spermatids into spermatozoa, i.e. spermiation occurs within Sertoli cells. The Sertoli cells provide nutrients to spermatozoa as the blood nutrients, such as glucose, are limited to spermatozoa due to the blood–testes barrier. Sertoli cells convert glucose into lactate, which the growing sperm cells utilize. They also help develop germ cells' iron transport by producing transferrin and ceruloplasmin. The androgen binding proteins (ABP) secreted from Sertoli cells bind with testosterone and transport it to the epididymis through seminiferous tubule fluid and aids epididymal transit, maturation of spermatozoa, and health of accessory sex glands. Sertoli cells also potentiate the action of testosterone after converting it to its active form 5α-dihydrotestosterone (DHT) by secreting the 5α-reductase enzyme. Sertoli cells also secrete some

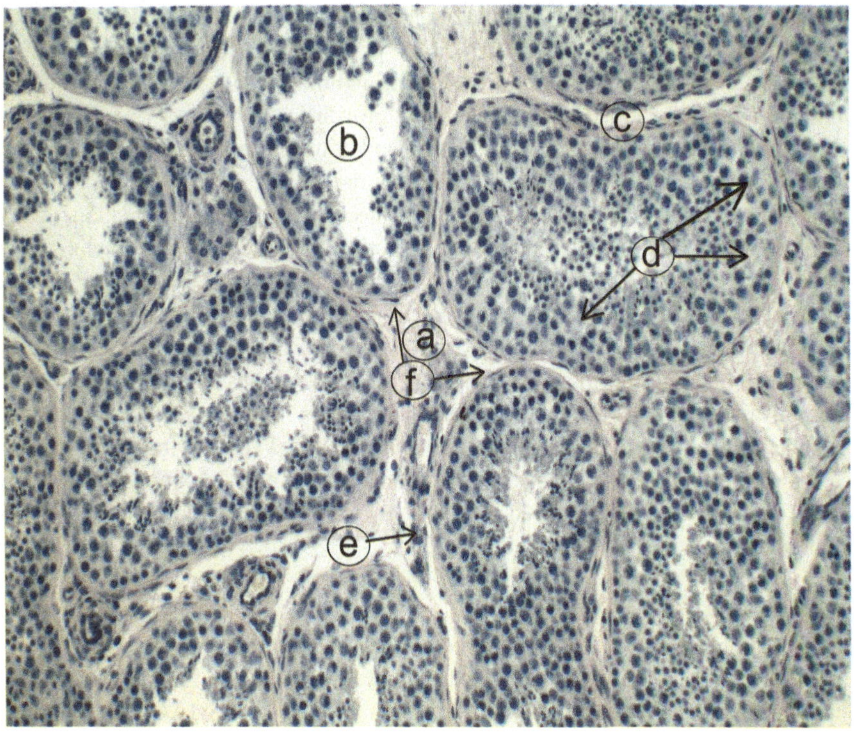

Fig. 19.4 Cross section of seminiferous tubules of rat. [The photomicrograph at 20X magnification showing the internal structure of the seminiferous tubules, contains **a** = interstitial space between the seminiferous tubules, **b** = lumen of the seminiferous tubule, **c** = blood–testis barrier (BBT), **d** = Sertoli cells, **e** = Leydig cells at the interstitial space, and **f** = peritubular myeloid cells surrounded the basement of the tubules]

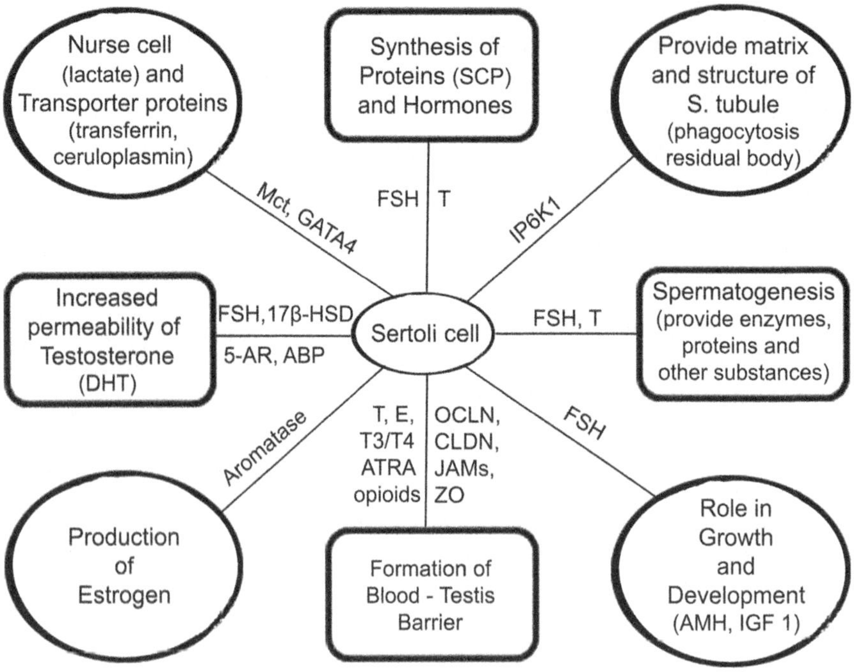

Fig. 19.5 Functions of the Sertoli cells. [The flow diagram showing the various function of the Sertoli cell with the involvement of different hormones, enzymes and growth factors. *FSH* follicle-stimulating hormone, mainly involved in hormone and various protein synthesis; *T* testosterone; *IP6K1* inositol hexakisphosphate kinase-1, an enzyme have role in cell metabolism and male fertility; *E* oestrogen; T_3/T_4 thyroid hormones; *ATRA* all trans retinoic acid, form of vitamin A; *OCLN* occluding; *CLDN* claudins; *JAMs* junctional adhesion molecules; *ZO* zonula occludens; *17β-HSD* 17 β-hydroxysteroid dehydrogenase; *5-AR* 5α-Reductase; *ABP* androgen binding protein; *Mct* monocarboxylate, lactate transporter; *GATA4* transcription factor GATA-4, a protein required for lactate synthesis; *DHT* 5-α-dihydrotestosterone; *SCP* Sertoli cell proteins; *AMH* anti-mullerian hormone; *IGF-1* insulin-like growth factor-1]

immune-related proteins (cytokines, chemokines, complement inhibitors, adhesion molecules) and provide immune protection to germ cells. However, Sertoli cells phagocytose, defective germ cells failed to complete spermatogenesis.

The Sertoli cells also produce activin, inhibin, anti-mullerian hormone (AMH), and insulin-like growth factor-1 (IGF-1). IGF-1 helps in spermatogenesis and testosterone production from Leydig cells. The liver borne IGF-1 is unable to pass the blood–testes barrier. Thus, the IGF-1 secreted from Sertoli cells is of great importance. Sertoli cells give structural support to the seminiferous tubule by adding the ground substances of the matrix. The proliferation and maturation of Sertoli cells are controlled primarily by follicle-stimulating hormone (FSH). But, the exogenous administration of FSH has a limited role in stimulating spermatogenesis as the numbers of Sertoli cells become fixed after puberty. In addition to this, several endocrine and paracrine factors are also involved in regulating Sertoli cell functions (discussed later in Sect. 19.2.7).

19.1.1.4.3 Leydig Cells

These are the large polygonal cells situated at the interstitial space of seminiferous tubules in extra-tubular connective tissue (Fig. 19.4). It is also known as an *interstitial cell*, and it is involved in the secretion of androgen (testosterone). Depending on their degree of maturation, LCs are of two types, foetal Leydig cells (FLCs) are present in male foetuses and matured LCs are called adult Leydig cells (ALCs). The numbers of ALCs decline with age, and testosterone production is subsequently decreased.

Foetal Leydig Cells (FLCs) The FLCs perform various functions, including determination of male sex and musculation, testicular descent and brain development together with the expression of the male behavioural pattern (Fig. 19.6). The androgens produced from FLC are responsible for developing the male genital tract during foetal life and testicular descent. The insulin-like peptide 3 (INSL3) secreted from FLC also potentiates testicular descent. The FLC derived testosterone helps in metabolic and neuroendocrine functions of the foetus, including brain development in conjugation with various growth and transcription factors. The action of FLC is gradually ceased after birth as its population diminished around puberty to support spermatogenesis but sufficient to develop male secondary sexual characteristics.

Adult Leydig Cells (ALCs) After puberty, the population of FLCs are gradually replaced by ALCs characterized by

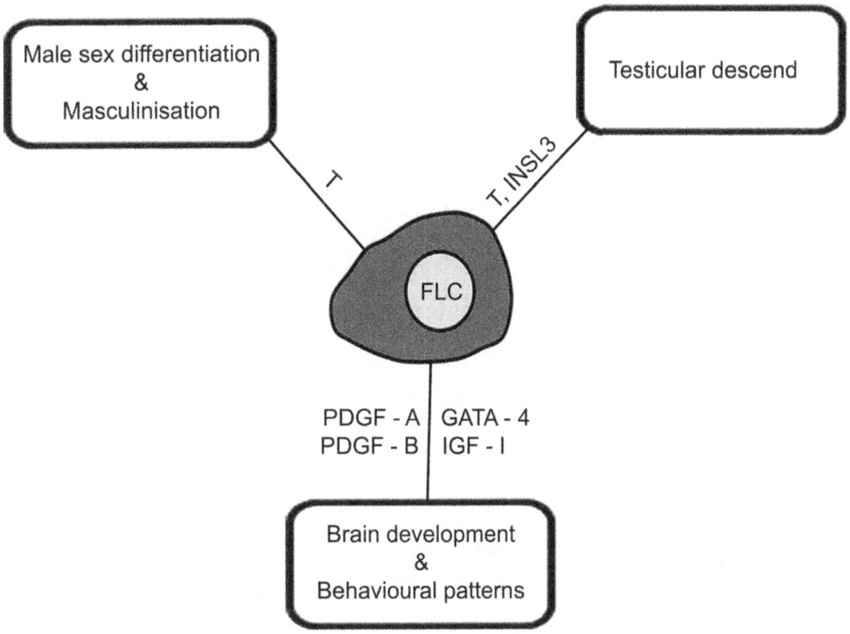

Fig. 19.6 Functional activity of foetal Leydig cells (FLC). [The diagram showing the functions of the FLC with the involvement of different hormones and growth factors. *T* testosterone, required for major activities of FLC; *INSL3* insulin-like-3, a peptide hormone responsible for descending of the testis; *PDGF-A* and *PDGF-B* are the platelet-derived growth factors; *GATA-4* is a transcription factor and *IGF-I* is insulin-like growth factor 1]

cellular hypertrophy (probably due to an increase in the nuclear volume, e.g. 5.6 pm in 4th week vs. about 7.4 μm in 75th week, in a bull) and hyperplasia. ALCs express LH/CG receptor (*LHCGR*), a G protein coupled receptor under the relaxin family peptide receptor 2 (*RXFP*$_2$). Luteinizing hormone or interstitial cell stimulating hormone (LH or ICSH) causes higher expression of the key steroidogenic enzymes like Hsd3β6 and Hsd17β3 to catalyse the conversion of androstenedione into testosterone (Fig. 19.7) with higher amplitude than FLCs. The testosterone produced from ALCs is either diffused into the seminiferous tubules and Sertoli cells by the paracrine process or enters into the systemic circulation. Sertoli cells further convert testosterone into its potent form and aid the spermatogenic process. The endocrine regulation of Leydig cell function is discussed in the next chapter.

19.1.1.4.4 The Cross-Talk Between Testicular Cells

All the cells are interrelated in their function (Fig. 19.8). The normal functioning of the testes depends upon the interaction between different testicular cells through endocrine and paracrine factors. The LH stimulates Leydig cell steroidogenesis and initiates a cascade of cellular cross-talk to promote spermatogenesis. The action of FSH on Sertoli cells facilitates cellular communications for germ cell development, differentiation of peritubular myoid cells, and Leydig cell function. The growth and differentiation of different testicular cells are regulated through several paracrine factors like IGF, TGF-α and β, and NGF. Some of these factors promote cell proliferation, and others regulate differentiation. The coordinated actions between these factors promote the rapid proliferation of germ cells and slower growth of peritubular and Leydig cells in adult testis.

19.1.1.5 Blood–Testis Barrier

A specialized multicellular barrier present between the vascular endothelium of the capillary blood vessels and the epithelium of the seminiferous tubules is called the blood–testis barrier (BTB) (Figs. 19.4 and 19.9). The tight junctions between the adjacent Sertoli cells form a barrier to dividing the seminiferous tubule into basal and adluminal compartments. The primary function of BTB is to isolate the germ cells from circulatory and lymphatic systems and local immune suppression to provide a biochemically and immunologically distinct microenvironment to support germ cell maturation. BTB restricts the transport of water, steroids, antibody-producing cells (B lymphocytes), electrolytes, paracrine factors, hormones, and toxic substances across the epithelium of Sertoli cells and allows FSH and testosterone to enter. The BTB also protects the developing germ cells from the exposure of germ cell-specific antigens, which otherwise leads to the production of anti-sperm antibodies to cause male infertility. The locally produced immune-suppressive substances (interleukins, interferons, and prostaglandins) from the Sertoli cells also provide an immune-suppressive microenvironment for germ cell maturation. The functional activity of BTB is mediated through several junctional proteins, namely *occludin, claudins*, junctional adhesion molecules (JAMs), and *zonula occludens* (ZO). The androgens, oestrogens, thyroid hormones, retinoic acid, and opioids influence the development of BTB.

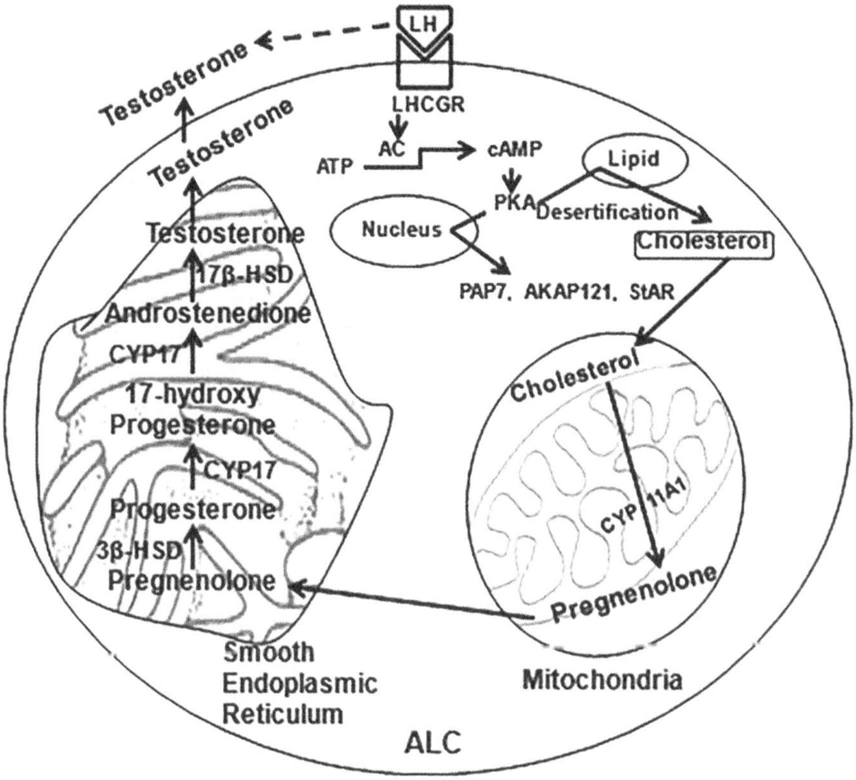

Fig. 19.7 Molecular mechanism of biosynthesis of testosterone in adult Leydig cell (ALC) by the influence of luteinising hormone (LH). [*LHCGR* G protein-coupled receptor for LH, activates AC; *AC* adenylate cyclase, the enzyme which converts ATP to cAMP, it activates PKA; *PKA* protein kinase, it de-esterifies the lipid droplet to cholesterol (free cholesterol); cAMP and PKA activate nucleus to produce steroidogenic proteins and enzymes like, PAP7, AKAP121, StAR along with voltage-dependent anion channel to import cholesterol into mitochondria; *PAP7* is A-kinase anchoring protein; *AKAP121* is A-kinase anchor protein 121; *StAR* is steroidogenic acute regulatory protein (transport protein); *CYP11A1* is an enzyme of the cytochrome P450 cholesterol side-chain cleavage ($P450_{SCC}$) superfamily member (family 11, subfamily A, polypeptide 1) converts cholesterol to pregnenolone within mitochondria; pregnenolone then migrate to smooth endoplasmic reticulum and converted to testosterone by involving enzymes 3β-HSD, CYP17, 17β-HSD; *3β-HSD* 3β-Hydroxysteroid dehydrogenase; *CYP17* cytochrome P450 17α-hydroxylase; *17β-HSD* 17β-Hydroxysteroid dehydrogenase]

19.1.1.6 Spermatic Cord

It is a tubular structure that suspends the testes into the scrotum. It originates from the inguinal ring, descends into the scrotum, and ends at the posterior margin of the testicle. The spermatic cord consists of the spermatic artery, spermatic vein, spermatic nerve, internal cremaster muscle, lymphatic vessels, vas deferens, and tunica vaginalis propria (Fig. 19.10). The spermatic cord provides blood and nerve supply to the testes.

19.1.1.7 Descent of Testes

The migration of testes from their intra-abdominal location to the scrotal sac is called testicular descent (Fig. 19.11). It starts around the 8th week of gestation in humans and is completed around the 35th week. Testicular descent is completed around the mid-gestation period in bull, ram, and buck. In boar, it occurs at the last quarter of pregnancy. The testicular descent in the stallion is completed just before birth or immediately after birth. Testicular descent in camels generally happens within a couple of days of postnatal life. The process needs 2–6 months after birth to complete the testicular descent in dogs. It may take about 2–3 years to complete descent. The descent of testes is a three-stage process. In the first stage of nephric displacement, the detachment of gonads from mesonephros occurs. In the second stage of transabdominal descent, the testes migrate towards the inguinal ring. The last stage is called inguinal descent when testicular migration is completed from abdominal cavity to scrotum. The descent of testes is guided by myxofibrous structure extending from testis to scrotum called gubernaculum testes. The testicular descent is mediated through the outgrowth and regression of gubernaculum testes. During the gubernacular outgrowth, rapid swelling of the gubernaculum dilates the inguinal canal and makes the way of testicular migration through the inguinal ring. During the regression of the gubernaculum, cellular remodelling occurs and it becomes a fibrous tissue rich in collagen and elastic fibre. The mechanical factor-like intra-abdominal pressure transmitting to the gubernaculum initiates the testicular descent, and protrusion of processus

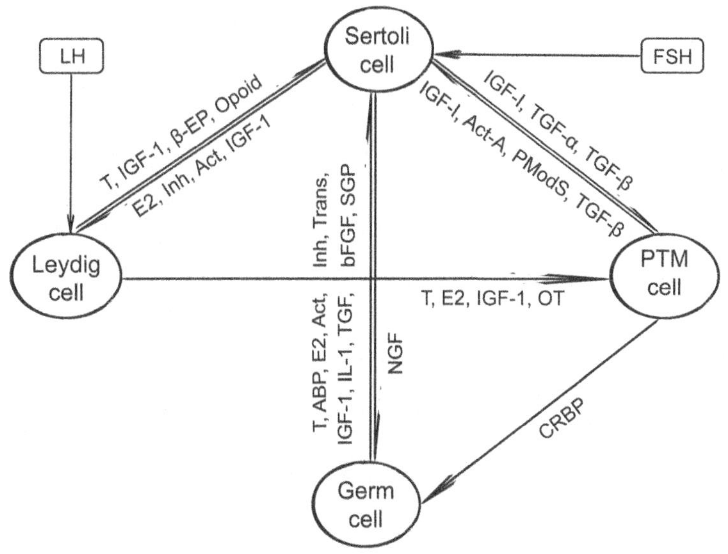

Fig. 19.8 Interactions among the testicular cells. [The involvement of different hormones and growth factors in testicular cells cross-talk is presented. *T* testosterone; *IGF-1* insulin growth factor 1; *β-EP* beta endorphin; *E2* oestradiol; *Inh* inhibin; *Act* activin; *PModS* peritubular modifying substance; *TGF* transforming growth factor; *OT* oxytocin; *ABP* androgen binding protein; *Trans* transferrin; *IL* interleukin; *bFGF* basic fibroblast growth factor; *SGP* sulphated glycoprotein; *NGF* nerve growth factor; *CRBP* cellular retinol-binding protein]

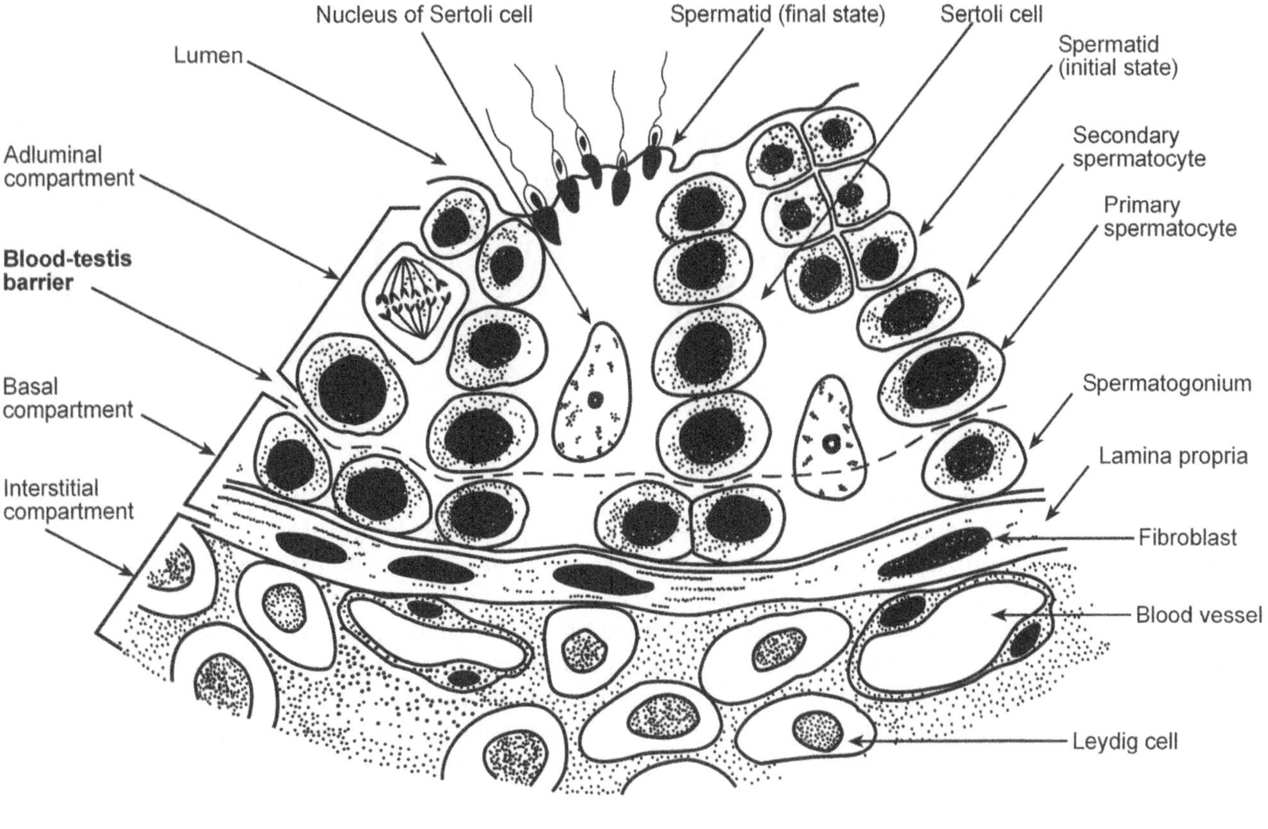

Fig. 19.9 Blood–testis barrier. [The cross section of seminiferous tubule depicting blood–testis barrier (BTB) in dotted line divides the testicular tissues in basal and adluminal compartments]

vaginal through the inguinal ring completes the descent process.

Many factors are involved in testicular descent. Androgen-independent insulin-like factor 3 (INSF-3) produced from Leydig cells plays a pivotal role in testicular descent. The mutation of this gene or antibody against INSF-3 leads to the failure of testicular descent. It causes the swelling of the gubernaculum and initiates testicular descent. The androgens have very little effect on the gubernaculum as gubernaculum doesn't possess androgen

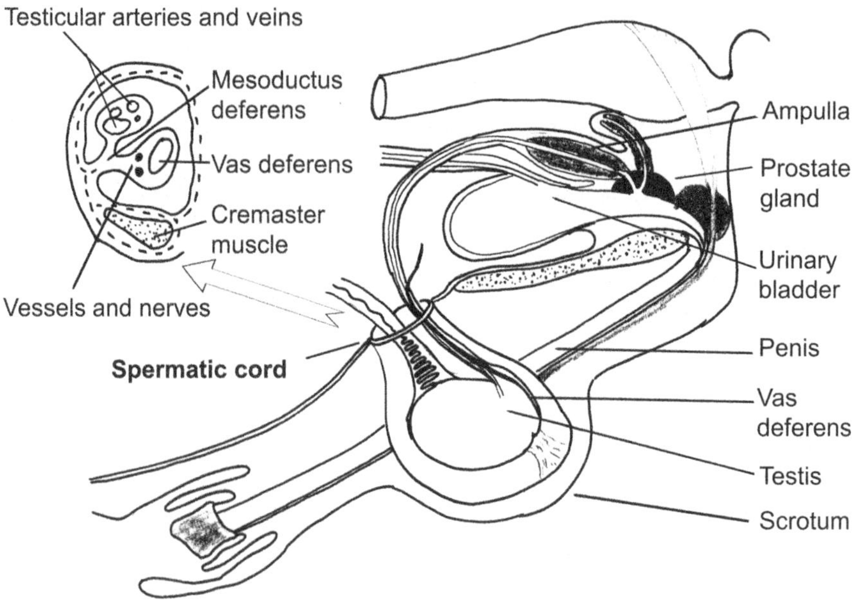

Fig. 19.10 Spermatic cord. [Diagram showing the spermatic cord in a dog. The cross section view of the spermatic cord depicting vas deferens, various blood vessels, nerves, and cremaster muscle is illustrated at the left top corner]

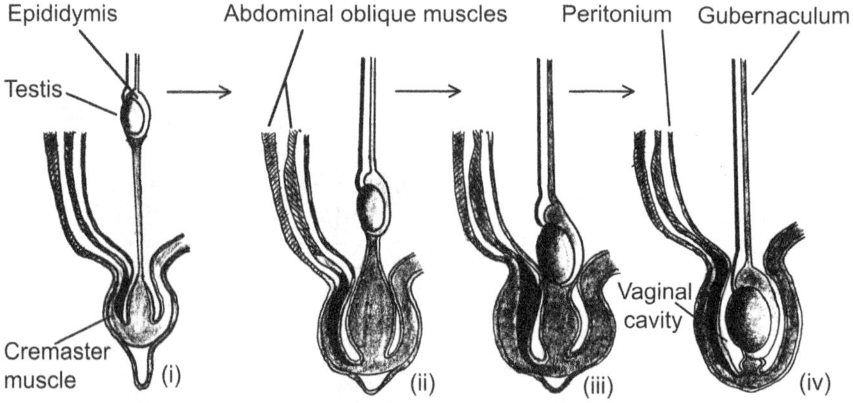

Fig. 19.11 Testicular descent. [Stages (i, ii, iii, and iv) of testicular descent from the foetal abdomen to scrotum through gubernaculum]

receptors; rather, androgen causes masculinization of genitofemoral nerves and stimulates to secrete calcitonin gene-related peptide (CGRP), causing rhythmic contraction of the gubernaculum testes. The other factors involved in testicular descent are oestrogen and epidermal growth factor (EGF). Oestrogen prevents the swelling of the gubernaculums, and EGF stimulates human chorionic gonadotropin (hCG) production from the placenta, which in turn stimulates androgen production from Leydig cells.

19.1.1.7.1 Disorders of Testicular Descent

Cryptorchidism and inguinal hernia are the two major disorders occurred due to disturbance in testicular descent.

Cryptorchidism The failure of testicular descent in the mature animal leads to a pathological condition called cryptorchidism. It may be unilateral (involving a single testicle, commonly right testicle) or bilateral (both testicles). The highest prevalence of cryptorchidism is seen in horses, followed by pigs. The companion animals also have a higher cryptorchidism incidence than cattle and sheep. The aetiology of cryptorchidism is multifactorial—the genetic, anatomical, and endocrine factors are involved in cryptorchidism. The hereditary predispositions of cryptorchidism are documented in pigs and horses. The candidate gene of cryptorchidism in pigs is SSC8, identified in Large White and Landrace. Thoroughbreds have less prevalence of cryptorchidism among the breeds of horses than American Quarter, Percherons, and American Saddlebreds. In sheep, cryptorchidism is associated with autosomal recessive mode inheritance, including the dominant gene with the incomplete penetrance. The endocrine factors involved in testicular descent are abnormal testosterone production or the absence of Müllerian inhibiting hormone. The foetus exposed to increased maternal oestrogen concentration may also develop cryptorchidism. The anatomical factors like torsion of the spermatic cord, scrotal hernia, and premature birth may interfere with testicular descent. The other miscellaneous causes of cryptorchidism include prolonged breech presentation, navel infections at the time of descent, exposure to anti-androgenic chemicals, and deficiency of maternal vitamin A. The cryptorchid animals can produce a subnormal amount of testosterone, and secondary sex characters are developed but fail to produce sperm due to elevated testicular temperature in the abdominal cavity.

Inguinal Hernia/Scrotal Hernia It results when the portion of the intestine drops into the scrotum with the testes due to enlargement of the inguinal canal. The abdominal viscera entered inside the scrotum may be located inside the vaginal process (indirect hernia) or within the vaginal process (direct hernia). Congenital inguinal hernia results from enlarged inguinal canal occurred due to abdominal compression during parturition. Acquired indirect hernias are common in stallions.

19.1.1.8 Thermoregulation of Testes

The efficient spermatogenesis requires an environment 2–6 ° C cooler than the core body temperature. The scrotum, therefore, has to provide the necessary thermal environment to support spermatogenesis. Five main anatomical features contribute to regulating the testicular temperature: tunica dartos smooth muscle, cremaster muscle, a countercurrent heat exchange system, absence of subcutaneous fat layer, and abundance of sweat gland. The tunica dartos surrounding the scrotum relaxes in response to higher environmental temperature holding the testis away from the body core. The reverse occurs during lower environmental temperatures. The relaxation of tunica dartos also increases the surface area of the scrotum for better heat exchange. The cremaster muscle also functions similarly to the dartos muscle. Still, the basic deference is that the contraction and relaxation of cremaster muscle lead to elevation and distension of testes, whereas dartos muscle controls scrotal skin. The testes receive comparatively cooler blood than other body organs due to a countercurrent exchange system. This testis system is facilitated by a network of small veins around the spermatic cord. This venous plexus is called the *pampiniform plexus*. The arterial blood carried through the testicular artery loses heat in exchange with the venous plexus, and a lower testicular temperature is achieved. In ram, the arteries and veins supplying the testes are superficial to the scrotum. It gives additional benefits to this species in terms of thermoregulation through evaporative heat loss. But, during hot-humid conditions, these evaporative heat exchange mechanisms are lost and may lead to '*summer sterility*'.

The absence of insulating covers in the subcutaneous fat layer in the scrotum facilitates conductive heat loss. A higher abundance of sweat glands in the scrotal skin promotes evaporative heat loss to maintain lower testicular temperature than the body core.

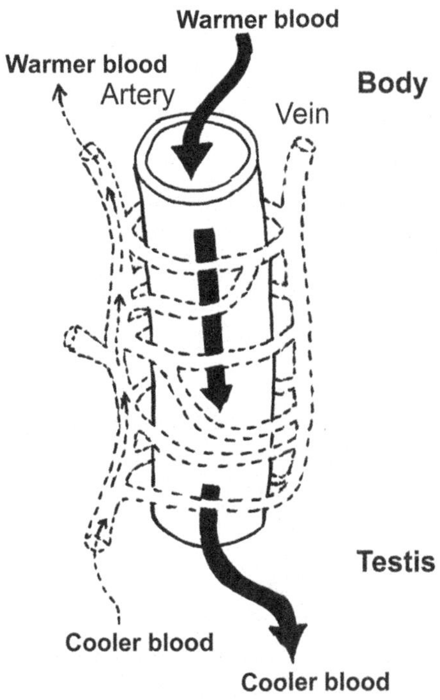

Fig. 19.12 Countercurrent heat exchange in testis. [The testicular artery (bold line) carries warmer blood towards the testes becomes cooler through the exchange of heat with the cooler blood of the testicular vein (dotted line)]

The testicular thermoregulation is less efficient in the boar due to the anatomical position of testes (less pendulous), and less sweat gland in scrotal skin resulted in more sperm abnormalities under extreme climatic conditions. The animals with intra-abdominal testes support spermatogenesis due to heat resistant spermatozoa or cooler core body temperature. The testicular temperature of sea mammals is generally lower than the abdominal temperature due to cooling employing the dorsal fin and associated vessels by *countercurrent heat exchange mechanism* (Fig. 19.12).

19.1.1.9 Testes of Birds

The avian testes are elliptical in shape and light yellow in colour, situated near the top of the kidneys. The testicular size shows seasonal variations and increases around the breeding season. The left testis is generally larger than the right. Birds have higher body temperatures than mammals. The testes lie adjacent to the air sac for an efficient heat exchange mechanism to maintain the desired testicular temperature for spermatogenesis. Spermatogenesis also occurs at night when the body temperature is comparatively lower. The avian spermatozoa are also heat resistant.

Testes can be considered a mixed gland in which the exocrine part, comprised of seminiferous tubules, is involved in spermatogenesis. The endocrine portion made of the

Table 19.2 The pH of fluid of various parts of the excurrent tract

Excurrent tract	pH
Testicular fluid	7.28–7.42
Rete testis	7.20–7.34
Efferent ducts	7.41–7.66
Epididymis	
(1) Caput	7.11–7.26
(2) Cauda	6.0–6.5
Vas deferens	
(1) Proximal part	6.85
(2) Distal part	7.39

Leydig cells produces male sex hormone through steroidogenesis. These functions are discussed in detail under subsequent sections/chapters.

19.1.2 Excurrent Tract

The excurrent tract of the male reproductive system extends from the rete testes up to the urethra and consists of rete testes, efferent tract, epididymis, vas deferens, and urethra (Figs. 19.1 and 19.2). Major functions of the tract are the maturation of spermatozoa, reabsorption of the excess fluid secreted from the seminiferous tubules, and providing passage for the expulsion of spermatozoa. The functional activity of the tract is controlled by the hormones in the paracrine and autocrine fashion with variable pH (Table 19.2).

19.1.2.1 Rete Testes

The highly convoluted seminiferous tubules are merged in straight tubules within testes are called rete testes (Fig. 19.3). In rats and mice, the rete testes are situated at the cranial pole of testes, but it is present at the centre of the testes in other species. The rete testis is lined by simple cuboidal or columnar epithelium but lacks germinal epithelium hence unable to produce spermatozoa. The columnar epithelium is either ciliated or non-ciliated. The ciliated cells are secretory and primarily involved in the movement of spermatozoa along with the fluid towards efferent ducts. The non-ciliated cells are mostly involved in selective reabsorption of the tubular fluid. The rete testes serve as collecting reservoirs of sperm. It acts as the first site for reabsorption of seminiferous tubular fluid and reabsorbs inhibin into the circulation. The junction between seminiferous tubule and rete testes is tubuli recti lined by Sertoli cells.

19.1.2.2 Efferent Ducts

Efferent ducts connect the rete testes with the epididymis (Fig. 19.3). These ducts are converged at the junction with epididymis. The lining cells are similar to the rete testes. The basement membrane of the efferent ducts is layered by smooth muscle and connective tissues. The smooth muscle cells are under sympathetic control and support the ciliary movement of the ducts. The lumen of the ducts is wider at the rete testes end and narrower at the epididymis end. In rats, mice, and guinea pigs, the ducts are merged and formed single duct before joining the epididymis. In other mammals, they join with the epididymis as parallel tubes. Before joining, the efferent ducts become tortuous and convoluted. Efferent ducts are mainly involved in the reabsorption of the testicular fluid and the movement of spermatozoa towards epididymis. Other functions of the ducts are ion transport, protein reabsorption, and steroid metabolism. The presence of Na^+/K^+ ATPase pump enables efferent ducts to absorb 70–95% of the testicular fluid, and around 50% of proteins released from the testes are reabsorbed from the efferent ducts. As a result, more viscous fluid and the spermatozoa reach the epididymis, which facilitates sperm maturation. The epithelium of the efferent duct also contains enzymes like acid phosphatase (endocytosis), carbonic anhydrase (bicarbonate absorption), glutathione S-transferase (GST) (cellular detoxification), and sulphated glycoprotein-1 (SGP1) (endocytosis), as well as the receptors for testosterone, oestrogen, vitamin D_3, oxytocin, inhibin, opioid, and proenkephalin to control the activity of the duct.

19.1.2.3 Epididymis

It is a highly convoluted single tube that remains in close contact with testes (Fig. 19.3) and descends into the scrotum through the spermatic cord. Length of epididymis varies with species. It is 40–50 m in bull, nearly 50 m in boar and ram, 70–80 m in the stallion, 3 m in rat, 2 m in cat, and 6–7 m in human. Morphologically it is divided into three parts, head (caput), a body (corpus), and a tail (cauda). The epididymis is lined by ciliated and non-ciliated columnar epithelium cells with long microvilli, which increase absorptive surface. There are four different cell types in the epididymal epithelium: principal, basal (*narrow cells*), clear, and halo cells. The most abundant cells are principal cells involved in secretion and absorption. The tight junctions between two adjacent principal cells form a blood–epididymis barrier that continues with BTB and provides immune protection of post-pubertal germ cells. The basal cells are flat and elongated act like macrophages involved in detoxification due to glutathione S-transferases (GST) and lysosomal enzymes. The clear cells are involved in the disposal of cytoplasmic droplets of the spermatozoa. Halo cells are believed to be lymphocytes and monocytes involved in the immune protection of the male reproductive tract. The cauda epididymis is connected to a highly muscular duct called vas deferens. The spermatozoa undergo final maturation during their transit through the epididymis, acquiring their motility and fertilizing capabilities (the role of epididymal factors in sperm maturation has been discussed in detail in the spermatogenesis chapter). The cauda epididymis is the principal storage site of mature spermatozoa. The epididymis is also involved in the reabsorption of epididymal and testicular fluid to the tune of 95% and 5–30%, respectively, to increase the osmolality of testicular fluid with the help of Na^+/K^+-ATPase pump (10–20 mosM of the seminiferous tubules in contrast to 200 mosM at the epididymis, in rat). The concentration of the spermatozoa is also more in the cauda than rete testis (10^9/mL from 10^4/mL, in rat). The epididymal secretions help acidify testicular fluid during its transit from caput (7.11–7.26) to cauda (pH 6.0–6.5) favoured by zinc and various proteins to promote sperm maturation. However, environmental pollutants, viz. heavy metals, oppose acidification. Epididymis also protects the spermatozoa from xenobiotics and oxygen radicals. The functions of the epididymis are under the control of testosterone. Enzymes like steroid 5α-reductase and neurotrophins (nerve growth factor, NT-3) are also reported to control epididymal functions. The epididymal functions are impaired by certain immunosuppressive drugs (cyclophosphamide), fungicides (benzimidazole), industrial gas (methyl chloride), sulfonates (ethyl dimethyl sulfonate) that may lead to infertility.

19.1.2.4 Vas Deferens

The vas deferens or *ductus deferens* is a tubular structure that originates from the caudal epididymis at the posterior border of each testis (Fig. 19.3) and enters the pelvic cavity through the inguinal canal to join the ejaculatory duct (union of the vas deferens and the duct of seminal vesicle) (Figs. 19.1 and 19.2). The distal portion of the vas deferens is dilated to form a sac-like structure called the ampulla. The length of vas deferens is about 30 cm in bull, 15 cm in buck, 29 cm in ram, and 35 cm in camel. The vas deferens is lined by pseudostratified columnar epithelium made of columnar cells and basal cells. The columnar cells are lined with cilia at their luminal surface. Beneath the epithelial layer, a tissue stroma is made of elastic fibres. The boarders of vas deferens are lined by circular and longitudinal smooth muscles with sympathetic innervations. The contraction of these muscles during ejaculation causes slow peristaltic waves that facilitate sperm transport epididymis to the urethra. Vas deferens act as temporary storage of spermatozoa before ejaculation. The sperm released from the cauda epididymis into the vas deferens will never return to the epididymis again if failed to expel out from the body. In a vasectomy, the passage of vas deferens is blocked through surgical interventions. It prevents the movement of spermatozoa from the epididymis to the urethra and causes permanent sterilization of males. The vasectomy does not assure to inhibit testosterone production and spermatogenesis. The spermatogenesis in vasectomized animals continued and stored at the cauda epididymis, resulting in the rupture of the epididymis, leading to sterility. Reversible contraception can also be possible by inserting an obstructive device into the vas deferens to block spermatozoa's flow temporarily.

19.1.2.5 Urethra

The urethra is a fibromuscular tube of the urogenital system as it carries urine from the bladder and semen from the excurrent tract to the exterior of the body (Figs. 19.1 and 19.2). It is broadly divided into two segments, namely the pelvic urethra and penile urethra. The pelvic urethra extends from the internal opening of the bladder up to the prostate. The pelvic urethra can be subdivided into the pre-prostatic and prostatic urethra. The pre-prostatic urethra is extended up to the prostate, and the prostatic urethra continues through the prostate joins with the penile urethra. Two ejaculatory ducts, each from vas deferens and seminal vesicle, drains sperm and ejaculatory fluids to the prostatic urethra. The prostatic ducts also open in this portion through the prostatic sinus to contribute prostatic secretions into the ejaculates. These openings are called *colliculus seminalis*. The pelvic urethra is lined with transitional epithelium. The penile urethra is composed of membranous urethra located in the deep perineal pouch extending through the external urethral sphincter. It has a bend at the ischial arch. It is the narrowest portion of the penile urethra. The lower and longest part of the penile urethra remains within the penis up to the urinary meatus at glance penis. It is covered with spongy elastic fibres hence called the *spongy urethra*. The bulbourethral glands open in the proximal portion of the spongy urethra.

Like pelvic urethra, the inner lining of the penile urethra is made of transitional epithelium, but in some species, it may change to stratified squamous epithelium before termination. The penile urethra is covered with erectile tissue called the *cavernous body* and *cavernous spongiosum* to control fluid flow. The penile urethra in ruminants is 'S'-shaped along with the sigmoid flexure of the penis (Fig. 19.1). Hypospadias is an anomaly of urethral opening at the body or head of the penis instead of the tip, making difficulty in urination and ejaculation. The aetiology may be congenital or endocrinal. There are two urethral sphincters, the internal urethral sphincter situated at the junction between the urinary bladder and urethra and the external urethral sphincter surrounds the membranous urethra. The internal urethral sphincter is made of smooth muscle and under involuntary (autonomic) control, and the external urethral sphincter is formed by skeletal muscles (*urethralis muscle*) and under voluntary control. The internal sphincter regulates the involuntary flow of urine from the bladder to the urethra, and the external sphincter controls the voluntary urine flow from the bladder to the urethra. The internal sphincter blocks the retrograde flow of semen into the urinary bladder at the time of ejaculation by relaxing and contracting the internal urethral sphincter during urination and ejaculation, respectively.

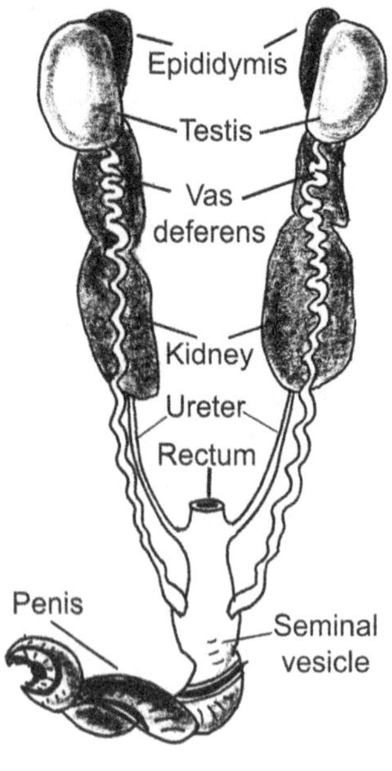

Fig. 19.13 Reproductive system of bird. [Long vas deferens and characteristic copulatory organ (penis) along with various components of the male reproductive system of bird are illustrated]

19.1.3 Excurrent Tract of the Bird

There are two major parts of the excurrent tract of the birds (Fig. 19.13). These are (1) vas deferens and (2) copulatory organs.

19.1.3.1 Vas Deferens

The vas deferens originates from each testis leading to the cloaca. At the opening (from the testis), it has a small and flattened area and resembles mammals' epididymis. The vas deferens is relatively narrow at the proximal part and gradually widens towards the cloaca. It has several bends and twists in its passage. The spermatozoa become mature, transported, and stored into it. The vas deferens are terminated at the swollen seminal vesicle (*glomus*) in the cloaca wall. Unlike mammals, the seminal vesicles of birds store the spermatozoa for a limited period. The sperm of the bird can be collected for artificial insemination by pressing these vesicles.

19.1.3.2 Copulatory Organs

Most birds do not have a penis, like mammals, and they have a small quantity of erectile tissue, known as *papilla*. The papilla is a small bump-shaped structure situated on the posterior wall of the cloaca. It is the rudimentary copulatory organ, generally not used in mating, but mostly the sexing of birds can be done by identifying it as early as day-old (Fig. 19.14). During mating, the birds of two opposite sex positioned their cloaca openings opposite each other to transfer the spermatozoa from male to female. The spermatozoa

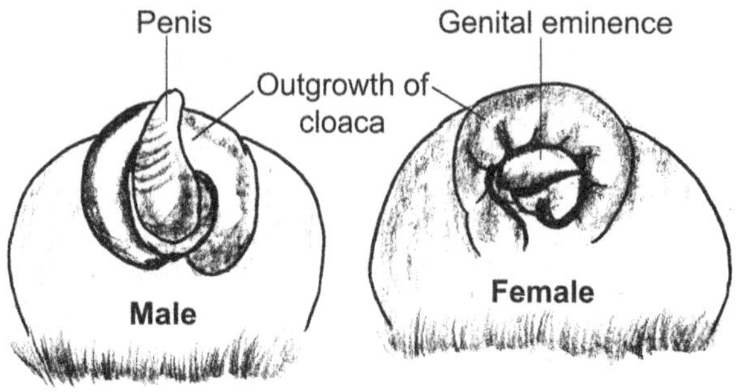

Fig. 19.14 Characteristic feature of the copulatory organ of male and female bird. [The enlarged penis of male (left) and genital eminence of female (right) are illustrated, which can be used to identify their sex at day-old birds]

are expelled from the body through the cloaca. The cloaca is a common body opening used for the expulsion of urinary and digestive waste also. The developed copulatory organ, like a penis, is present in male swans, ducks, geese, and ostriches.

In birds, the secretion of the epithelium cells of the excurrent tract is the major source of seminal plasma as they lack accessory sex glands. In addition, some lymph-like secretions are found during ejaculation originate from paired paracloacal vascular bodies.

19.1.4 Accessory Sex Glands

The male accessory sex glands include the ampulla, seminal vesicles, prostate, and bulbourethral (Cowper's) glands (Figs. 19.1 and 19.2). The secretions from these glands are the main contributor to seminal plasma. All the glands drain their secretion into the urethra, except the bulbourethral gland, which opens into the *penile urethra* at the base of the penis. There are huge species variability regarding the functional morphology and nature of secretions of the accessory sex glands (Table 19.3). Animals that possess both seminal vesicles and the bulbourethral gland have seminal alkaline fluid, as secretion of both the glands is alkaline. The seminal alkaline pH favours the neutralization of the acidic secretions of the female genital tract. The seminal plasma acts as a vehicle to carry the spermatozoa to the site of fertilization. Moreover, the secretory products of the accessory glands promote the metabolic activity and potentiality of the spermatozoa by providing nutrients, buffering constituents, and bioactive substances. Androgens control the functions of the accessory sex glands. Most accessory sex glands contain the 5α-reductase enzyme, which converts testosterone to dihydrotestosterone.

19.1.4.1 Ampulla

Ampulla (Ampullae in plural) is the glandular swelling of the vas deferens above the urinary bladder. The ampulla is lined

Table 19.3 Presence of accessory sex glands in different male animals and pH of its secretion

Animal	Ampulla	S. vesicle	Prostate	Bulbourethral	Seminal pH
Ruminant	++	+	++	+	6.0–7.0
Stallion	++	+	+	++	7.2–7.7
Boar	–	+	++	++	7.3–7.5
Canine	rud	–	++	–	6.4–7.0
Feline	rud	–	++	++	6.0–7.0
pH	Slightly acidic	Alkaline	Slightly acidic to alkaline	Alkaline	

$^{+}$present, $^{++}$large size, $^{-}$absent, ^{rud}rudimentary

by simple columnar epithelium without excretory ducts. The ampulla is present in ruminants, horses, camels, rodents (rat, squirrel, etc.), and bats. It is well developed in camels and some species of bats. The primary function of the ampulla is to store sperm like reservoirs. The ampullary secretion is serous in nature and yellowish-white in colour. But, the secretion of ampulla contributes little in seminal plasma except for bull (0.5–2.0 mL). But, in the stallion, ampulla secretes ergothioneine, an antioxidant and cytoprotective substance of the seminal plasma. The ampullary contraction depends on its surrounding smooth muscle, which is stimulated along with vas deferens.

19.1.4.2 Seminal Vesicles

The seminal vesicles or the *vesicular glands* are paired, coiled, and extended fibromuscular glands present at the dorso-cranial aspect of the pelvic urethra lateral to the ampulla. It opens directly into the prostatic urethra, except in bull, where it opens at vas deferens. It is absent in almost all carnivores, camel, and domesticated rabbits but well developed in bull, boar, ram, stallion, rat, and guinea pig. The seminal vesicles are lobulated in bull, ram, and buck; and sac-like in stallion and boar. The glands and their ducts are lined by pseudostratified columnar epithelium, and the wall is composed of smooth muscles innervated with sympathetic and parasympathetic nerve fibre. The parasympathetic nerves control the secretion of glandular tissues, whereas the smooth muscles are under the control of both sympathetic and parasympathetic nerves. Several neuropeptides are involved in seminal vesicle secretion, like neuropeptide Y (rats, guineapig, and humans); substance P (guineapig), vasoactive intestinal polypeptide (VIP) (mice), and gastrin-releasing peptide (rabbit).

The seminal vesicles contribute more than 50% of the ejaculate. The vesicular fluid is yellowish in colour, viscid, and alkaline. The predominant component of the vesicular secretion is the citric acid controlling the pH of the seminal plasma for its affinity towards divalent cations (viz. calcium, magnesium, and zinc). The seminal vesicle can synthesize fructose from either blood glucose or sorbitol, which acts as the chief energy source in seminal plasma. The concentration of fructose is inversely related to sperm motility as the spermatozoa utilize fructose as their major energy source. Therefore, the concentration of fructose in semen indicates the functional index of seminal vesicles. Other than citric acid and fructose, various biologically active substances are secreted by seminal vesicles (Table 19.4). MHS-5 protein (antigen) is the marker to identify seminal vesicular secretions in chimpanzee, gorilla, orangutan, and human semen. It is also used to distinguish the ejaculatory sperm (semen) from the epididymal sperm. The epithelium of the seminal vesicle is infiltrated with macrophages and T lymphocytes (CD4 and CD8) as the gland is prone to infection.

19.1.4.3 Prostate

The prostate is a tubule-alveolar gland that surrounds the urethra at the base of the urinary bladder and opens into the urethra. It is present in almost all domestic species and the only accessory sex gland in carnivores and cetaceans (whales and dolphins). The prostate is a heterogeneous gland divided into two parts, the pars propria (body of the prostate) and pars disseminata (disseminated portion). In bulls, both parts are distinguishable. In rams, only pars disseminata is present. In stallions, the prostate has two lateral lobes connected through the isthmus. Pars disseminata is well developed in boars. Large-sized prostates exist in dogs and cats. The prostate of a cat has four lobes. The secretion of the prostatic is apocrine. It is thin, milky, alkaline, with low protein contents. The prostatic fluid is serous in dogs and mucous in bulls. The prostate contributes 30% of seminal plasma in human, but secretory volume is comparatively low in bulls and dogs. A proteolytic substance called prostate-specific antigen (PSA) of prostatic secretion helps to liquefy the vaginal plug for free passage of spermatozoa. PSA is used as a marker of prostate cancer. In dogs, *Kallikrein-2* (KLK-2), a serine protease enzyme, induces PSA production. The prostate gland contains large quantities of chloride ions, calcium, zinc, citrate, and polyamines. Prostatin or prostatic binding protein has been identified in rats. Prostatic acid phosphatase (PAP) and prostate-specific protein (PSP) are identified in human prostatic secretions. The prostate is capable of secreting oxytocin that exerts a paracrine effect on the prostate to stimulate the growth of the prostate and its contraction during ejaculation. In some species, like a rat, mouse, guinea pig, hamster, rodents, and bats, the secretion of this gland helps to

Table 19.4 Major contribution of seminal vesicle

Components		Functions
Electrolytes	Potassium	Activation of ATPase
	Bicarbonate	Regulates cAMP level and thereby sperm motility
	Phosphate	Semen coagulation property
Proteins	Insulin-like peptide	Promotes the growth of seminal vesicles and prostate glands
	Protein C inhibitor (PCI)	Protects the sperms and other seminal plasma proteins from proteolytic damage
	Sperm-coated antigens (lactoferrin, ferriplan, MHS-5 antigen)	Sperm-zona pellucida recognition
	Semenogelin and vesiculase	Form a coagulated gel-like matrix/vaginal plug
	Fibrinogen Sperm motility inhibitor (SPMI) Trophoblast lymphocyte cross-reactive (TLX) antigen	Clotting of semen Inhibits sperm motility (prostatic proteases degrade SMPI immediately after ejaculation) Prevent female immune response against spermatozoa and embryo
Prostaglandins	Prostaglandin $F_2\alpha$ (PGF2α), E_1 and E_2	Promotes the transport of sperm
Reducing substance	Vitamin C	Prevents sperm agglutination
Enzyme	5α-reductase	Converts testosterone to dihydrotestosterone
Mucin		Acts as a sticky agent to hold the spermatozoa in the female genital tract for a long time

coagulate the semen immediately after intromission of the penis to form a *vaginal plug*; hence, it is called *coagulating gland*. The prostatic secretion in dogs is excreted through urine, other than ejaculate, and used to mark their territory. The growth and functions of the prostate are dependent on androgens, and glandular dysfunction is associated with endocrine deregulations. Benign prostatic hyperplasia (BPH) is one such pathological condition occurred in old age dogs and humans due to over production of dihydrotestosterone (DHT). Prolactin is also reported to induce BPH. BPH is associated with sperm abnormality and reduced motility. Bacterial prostatitis is also common in sexually mature males. The swollen prostate reduced the passage of the urethra, resulting in difficulty in urination. Anti-androgen drugs (to restrict 5α–reductase), *progesterone* therapy, or *castration* are recommended to control the BPH.

Know More ...

The human prostate can secrete unique, organic substance spermine synthesized from putrescine (precursor is *arginine*) under the influence of testosterone. The spermine formed *spermine phosphate crystals* together with phosphate, and it has bacteriostatic properties. This is used as a marker and used to identify the existence of human semen in veterolegal cases.

19.1.4.4 Bulbourethral (Cowper's) Glands

The gland was named after English surgeon; William Cowper discovered the gland in the seventeenth century. It is analogous to Bartholin's glands in females. The two Cowper's glands are situated beneath the prostate gland in the urogenital diaphragm covered by skeletal (bulbospongiosus, bulbocavernosus, and urethral) and smooth muscles. It is well developed in boar, camels, cats, rodents, and elephants, whereas small in bull, ram, stallion, and human. It is absent in dogs and most carnivores. Ducts of the gland Cowper's glands open into the penile urethra and are lined by pseudostratified epithelium. Cowper's gland's transparent, alkaline, and viscous secretion is released upon sexual arousal before coitus. It has five major roles:

1. The secretion helps to flush out the residual drops of urine and unwanted foreign substances from the urethra and penis for clean passage of semen.
2. The secretion contains many mucin or gelatinous substances that lubricate the passage of the penis and vagina for smooth ejaculation and transportation of spermatozoa.
3. The alkaline secretion buffers the acidic environment of the female genital tract and protects the sperm from the harsh environment.
4. The mucous facilitates the retention of the semen for a long time in the female genital tract. The secretion of Cowper's gland of boar is more viscous, having high sialomucin (sialomucoprotein) concentration, and holding the semen for nearly 4 days in the female genital tract.
5. The secretory products also provide energy to the spermatozoa.

The gland is highly responsive to oestrogens and estrogenic chemicals. Hence, the animals offered with a rich source of phytoestrogens (like clover pastures) may have metaplastic or cystic lesions in the bulbourethral gland.

19.1.5 Ancillary Organs

19.1.5.1 Penis

Morphology The penis is a fibroelastic muscular urogenital structure that assists in urination and ejaculation. It is made up of erectile tissue (contains several elastic fibres, sinuses, and large free space), dense connective tissues, lymphatics, blood vessels, smooth and skeletal muscle innervated with the autonomic and central nervous system. The penis contains two types of tissue, a pair of corpora cavernosa and a single corpus spongiosum. The corpora cavernosa (*cavernous body*) is a kind of smooth muscle with a resting tone. The corpus cavernosum consists of cavernous tissue that form is paired columns and surrounded by connective tissue. The corpus spongiosum surrounds the urethra and is made up of vascular tissues. It appears as a spongy tissue enlargement of the pelvic urethra. The corpus spongiosum extends over the corpus cavernosum and forms a glans penis at the penile extremity. A considerable variation in the morphology of the glance penis corresponds to the morphology of the female genital tract. The penis has three segments, the root, body, and glans penis. The root is connected with the ischial arch and attached with corpora cavernosa and ischiocavernosus muscle. The body of the penis is surrounded by corpora cavernosa and corpus spongiosum with a thick connective tissue covering called *tunica albuginea*. The body of the penis is made up of erectile tissues, which promote the rigidity of the penis. The rigidity of corpora cavernosa is more in comparison to the corpus spongiosum. The corpus spongiosum thus facilitates the reduction of the pressure on the urethra for the passage of semen during ejaculation. The last part of the penis is the glans, also called the head of the penis. It contains touch and pressure receptors (*Pacinian corpuscles*) with several sensitive non-myelinated nerve endings.

Morphological Variations In dogs, the distal portion of the corpus cavernosum transforms into bone called *os penis*. The penis of bull, boar, ram, and deer is fibroelastic, containing large connective tissue, elastic fibre, and small erectile tissues. In contrast, the musculovascular penis of stallions, dogs, cats, and humans have more erectile tissue and less connective tissue. The penis of ruminants and boar is 'S' shaped supported by a fibroelastic *sigmoid flexure* attached with *retractor penis muscle* (Fig. 19.1). During an erection, the retractor penis muscle is relaxed and facilitates the protrusion of the penis to the female genital tract. The rigidity of the penis is achieved after profuse blood supply in the venous sinuses through *helicine arteries.* The penis of dogs, bats, bears, seals, rodents, and certain primates have os penis (Fig. 19.2).

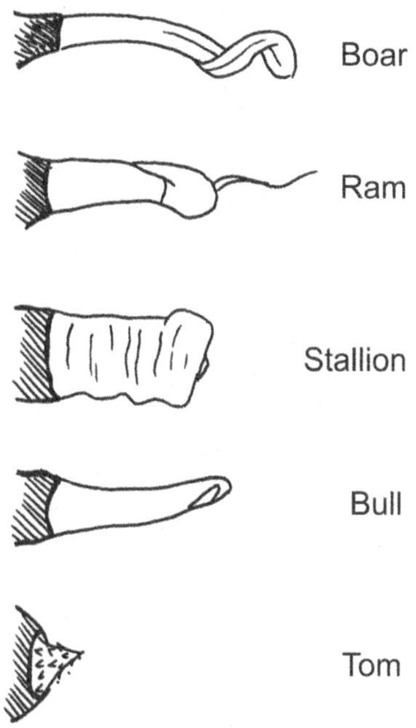

Fig. 19.15 Spiral deviation in penis of different domestic male

Functional Variations The *bulbus glandis* is present at the glans penis of the dog enlarging after intromission and helps to form copulatory tie during mating, which subsequently relaxes after ejaculation, and the penis comes out. The glans enlarges in ram, buck, and stallion. Os penis of the cat has cornified spines, and it stimulates the ovulatory response in the queen. In some species, the free end of the penis is spiraled (Fig. 19.15) after intromission. It depends upon the arrangement of adjacent supported lamellae or collagen fibres of the penis. After completion of erection, the penis returns to its normal shape by the *dorsal epical ligament.* If the penis's spiralling fails to return to its normal state, the penis can appear as a '*corkscrew penis*'.

Penile Erection Penile erection is a coordinated process involving circulatory, nervous, and muscular systems. Increased blood flows into cavernosal sinuses through helicine arteries facilitate the erection process—increased arterial pressure and venous occlusion help trap the blood, resulting in penile engorgement and rigidity. The somatic nerves attached with the ischiocavernosus, ischio-urethralis, and bulbocavernosus muscles are contracted and support the rigid penis. The parasympathetic nerves, originating from the sacral region, are connected with these muscles and erectile tissues. In ruminant and boar, the retractor penis muscle relaxes, facilitating the penis to become elongated. After ejaculation, the sacral-originated sympathetic nerves return to their normal tone, causing contraction of helicine arteries and retractor penis muscle. Thus the pressure around the veins is reduced along with increased outflow and restoration of normal blood flow. The penis also returns to its normal position under the prepuce. The role of penis in ejaculation is discussed in detail in spermatogenesis chapter. The disorder of penile erection is called erectile dysfunction or *incompetence* that occurs due to structural or morphological defects without psychological influences.

Know More . . .
Most male marsupials comprise an 'S'-shaped bifurcated penis used only during copulation. Its penis bifurcates into two columns corresponding with two lateral vaginas of the females.

19.1.5.2 Prepuce

The invagination of the abdominal skin covering the penis is called the prepuce. At birth, the epithelial lining of the penis and sheath of the prepuce are fused to form balanopreputial fold that obstructs the penile erection through prepuce. The fold disrupts at puberty under the influence of testosterone and allows free passage of the penis through prepuce during erection. In boar, this fold splits and enables the penis to move freely but remains as a ridge-like structure that holds urine and other wastes. In bull and dog, the fold may persist as a fibrous band causing incomplete protrusion. In a dog, the band is called the *frenulum*. The dog's prepuce is very loose, which assists in holding the glans penis when the bulbus glandis bulge. The prepuce is absent in the cat, and its penis moves backwards and downward from the ischial arch.

19.2 Endocrinology of Male Reproduction

Reproduction and fertility are regulated by coordinated synchrony of the endocrine orchestra involving the hypothalamus, pituitary, and gonads called the hypothalamic–pituitary–gonadal (HPG) axis controlling the stimulation and inhibition of sex steroid secretion and gonadal functions. The gonadal sex steroids also exert negative and positive feedback loops to regulate the HPG circuit. Impairment in the HPG axis may lead to infertility.

19.2.1 Hypothalamic–Pituitary–Gonadal (HPG) Axis

HPG axis comprises three main components, namely hypothalamus, pituitary, and gonads. The gonadotropin-releasing hormone (GnRH) of the hypothalamus is the key regulator of the HPG axis in vertebrates. GnRH secreting neuronal cell bodies are clustered around the medial preoptic area (POA), arcuate nucleus, and their projections terminate at median eminence. GnRH is carried to the anterior pituitary via the portal circulation. Under the influence of GnRH, the gonadotrophs of anterior pituitary secret luteinizing hormone (LH) and follicle-stimulating hormone (FSH), which facilitates gametogenesis and steroidogenesis in the gonads. The gonads produce both sex steroids and gametes under the control of LH and FSH. The sex steroids, namely oestrogen, progesterone, and androgen, exert negative feedback on GnRH release to control gametogenesis and reproductive cyclicity (Fig. 19.16).

19.2.1.1 Hypothalamic GnRH Secretion

The secretion of GnRH from the hypothalamus occurred in two modes, pulsatile and surge modes. In pulsatile mode, GnRH releases in an episodic manner during childhood, whereas surge mode causes transient and copious GnRH release around puberty. Both these pulsatile and surge modes of GnRH secretion are under neuroendocrine control. Pulsatile GnRH is regulated by a GnRH pulse generator at the mediobasal hypothalamic area (MBH). Sex steroids exert negative feedback over this pulse generator via opioid neurons.

In contrast, the GnRH surge generator at the preoptic area (POA) is influenced by GABA and arginine vasopressin (AVP) neurons. The functionality of GnRH pulse and surge generator depends on complex interactions between glutamatergic cells, GnRH, and other neurons together with certain neuropeptides. Among these neuropeptides, kisspeptin and RFamide-related peptide-3 (RFRP-3) have emerged as stimulatory and inhibitory neuroendocrine integrators, respectively, to control the HPG axis. The Kiss1 gene encodes kisspeptin is a potent stimulatory neuropeptide of GnRH secretion. The inhibitory neuropeptide of GnRH secretion is RFRP-3, encoded by the Rfrp gene. The avian orthologue of RFRP-3 is called gonadotropin-inhibiting hormone (GnIH).

19.2.1.2 Pituitary Gonadotropin Secretion

Two glycoprotein hormones, namely LH and FSH, secreted from the gonadotrophs of the anterior pituitary (adenohypophysis) control spermatogenesis (LH/FSH) and testosterone production (LH). Secretion of FSH or LH from gonadotropes depends upon the character of GnRH pulse. Low amplitude and irregular pulse of GnRH facilitate FSH secretion. In contrast, the release of LH is stimulated by a high-frequency GnRH pulse. After the synthesis, LH and FSH are stored in different secretory granules to release upon GnRH stimulation. Measurable quantities of LH and FSH in the peripheral blood are seen around the 12th week of gestation in humans. FSH is predominant over LH during foetal life, and the female foetus has a higher FSH/LH ratio, which gradually changes during development.

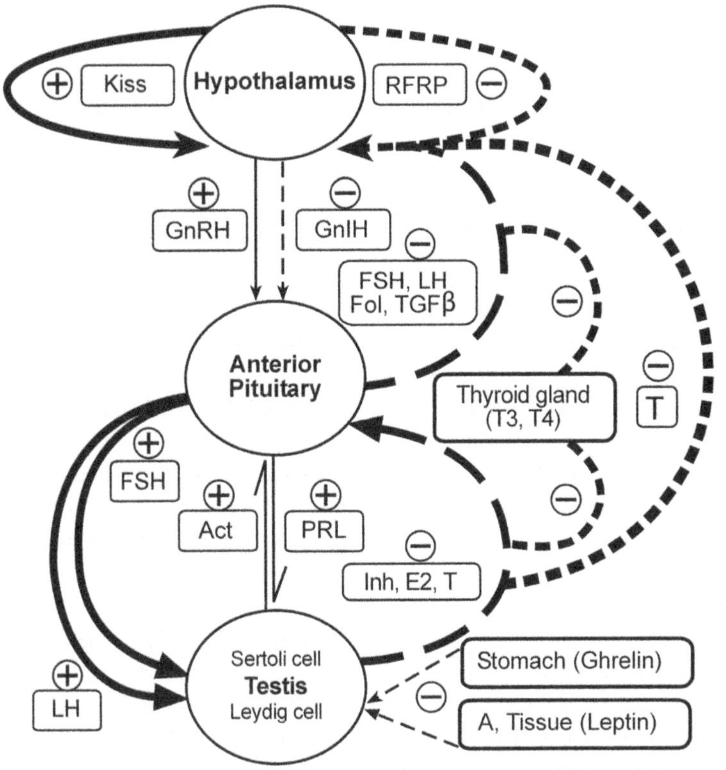

Fig. 19.16 Hypothalamus–Pituitary–Gonadal (HPG) (HPT or hypothalamic–hypophysis–Testis HHT) axis. [Up-regulations (marked as a **bold arrow** on the left side) are controlled by the *GnRH* gonadotropin-releasing hormone, secreted from the hypothalamus; *FSH* follicle-stimulating hormone and *LH* luteinising hormone, secreted from anterior pituitary (AP) act over Sertoli cells and Leydig cells, respectively; *PRL* prolactin, release from AP and act over Leydig cell for testosterone synthesis; *Act* activin, released from Sertoli cells and acts over the AP for FSH synthesis; *Kiss* kisspeptin, released from the preoptic areas of hypothalamus and promote GnRH release. **Down-regulations** (marked as **dotted lines** in the right side) are controlled by the *T* testosterone, produced by the Leydig cell and get potency from Sertoli cells, act over hypothalamus during its higher concentration to reduce secretion of GnRH and directly over AP to reduce synthesis of FSH and LH; *Inh* inhibin and *E2* oestrogens, released from Sertoli cells and acts over AP to inhibit FSH and LH respectively; *Fol* follistatin and *TGFβ* transforming growth factor(s) beta, secreted from AP during higher FSH level to reduce GnRH secretion; excess FSH and LH also inhibited hypothalamus to reduce the secretion of GnRH; *RFRP* RFamide-related peptide acts like GnIH; *GnIH* gonadotropin-inhibiting hormone, over the hypothalamus/AP to reduce GnRH secretion; **ghrelin** secretes mainly from stomach and **leptin** from adipose tissue act on the Leydig cells to reduce testosterone biosynthesis; *T3* tri-iodothyronine and *T4* thyroxin, released form thyroid gland, act over hypothalamus to reduce secretion of GnRH and directly over AP to inhibit secretion of LH]

19.2.1.3 Gonadal Steroidogenesis

Pituitary LH acts over the Leydig cells to stimulate testosterone synthesis. LH acts through LH/chorionic gonadotropin receptors (*LHCGR*). The receptor-ligand binding activates adenylyl cyclase and increases cAMP production. cAMP stimulates the transcription of steroidogenic acute regulatory protein (StAR), which helps in the cholesterol transportation to the inner mitochondrial membrane to initiate steroidogenesis to produce testosterone. Testosterone is transported in the plasma in conjugation with sex hormone-binding globulin (SHBG) of hepatocytes or androgen binding protein (ABP) of testes. A considerable amount of testosterone returns to the seminiferous tubule through the testicular countercurrent exchange mechanism to ensure a steady supply of testosterone in the tubule to support spermatogenesis. A high testosterone level inhibits LH secretion and thus exerts negative feedback on the HPG axis. Testosterone helps in the development and function of the testes, development of secondary sexual characteristics, and stimulation of spermatogenesis and libido.

In addition to testosterone, Sertoli cells also produce two protein hormones, activin and inhibin, to regulate the HPG axis. Activin stimulates GnRH secretion, and inhibin inhibits FSH secretion.

19.2.2 Development of Hypothalamic–Pituitary–Gonadal (HPG) Axis

Among the domestic animal species, the ontogeny of HPG axis development is mainly studied in ovine species. In sheep, the first GnRH neurons appeared in the medial portion of the nasal placode around the 26th day of embryonic life. These neurons are migrated along with the nasal septum towards cribriform plate during 26–35th days of embryonic life and ultimately lodged into their final site in the preoptic area during 35–45th days of gestation. The axonal projections of GnRH neurons reach the median eminence during 45–60th embryonic days. The GnRH neurons become functional from embryonic days 80–120 as indicated by the expression of the β chain of LH receptors, and pulsatile release of LH starts. LH pulses facilitate the release of testosterone in male foetuses. The LH secretion ceases from 120th day to gestation up to postnatal day 60, which reappears around 70–140, but the frequency is low. From 140 to 210 postnatal days, the LH pulses increase in frequency but at a lower amplitude.

19.2.3 Endocrinology of Male Sex Determination

The initial gonadal differentiation is genetically controlled by male and female determining factors around 6 weeks of gestation (discussed in earlier Chap. 18). The post gonadal sex determination is under endocrine control. Anti-mullerian hormone (AMH), testosterone, and insulin-like factor three produced from foetal testes are the key regulator of male sex determination. LH controls the foetal testicular steroidogenesis, and FSH stimulates Sertoli cells to produce AMH. The initial gonadal steroidogenesis is gonadotropin independent; then placental hCG controls the steroidogenesis up to 10–20 weeks, then gradually declines in humans. The expression of LH in male foetus initiates around 10 weeks of gestation and reaches a peak at 20 weeks and controls foetal steroidogenesis. Dihydrotestosterone, the active metabolite of testosterone, is formed by the enzyme 5-α reductase 2 (SRD5A2) stimulates the development of the prostate, penis, and scrotum. Androstenedione is the predominant foetal androgen as 17β-hydroxysteroid dehydrogenase type 3 (HSD17B3) is not expressed in FLC. The foetal androgens promote the development of Wolffian duct derivatives to form external male genitalia. The AMH causes regression of Müllerian ducts around 8–10 weeks of gestation in humans. Insulin-like 3 (INSL3) mediates testicular descent.

19.2.4 Pre-pubertal Suppression of HPG Axis

GnRH pulsatility is more during the infantile period but diminishes during the juvenile period. This juvenile period of sexual quiescent known as neurobiological brake occurred by two different mechanisms.

19.2.4.1 Steroid-Dependent Mechanism (Gonadostat Hypothesis)

According to the *gonadostat hypothesis*, gonadal steroids negatively affect GnRH neuronal activity. The GnRH secreting neurons show higher sensitivity to the gonadal steroids, and a small amount of androgens can suppress GnRH secretion. The steroid sensitivity of GnRH neurons gradually decreases during the pre-pubertal periods, which leads to increased secretion of GnRH and HPG activation. This pre-pubertal HPG axis suppression mechanism is validated in sheep, hamsters, ferrets, and cattle.

19.2.4.2 Steroid-Independent Mechanism

The 'gonadostat' is failed to explain the pre-pubertal HPG axis suppression in rats and monkeys as neonatal castration leads to lower gonadotropins levels during the infantile period, which increases progressively during the juvenile phase. Further, the GnRH neurons lack oestrogen receptor α (ERα) in these species. Therefore, a steroid-independent mechanism involving neuronal pathways was proposed to explain pre-pubertal sexual quiescent in these species. There are some inhibitory neurotransmitters like γ-amino butyric acid (GABA), neuropeptide Y (NPY), dopamine (DA), and endogenous opioid peptides (EOP) are thought to involve in this process. The inhibin produced from Sertoli cells also has a negative effect on FSH secretion. More recently, the neuropeptide RFRP-3 (gonadotropic inhibitory hormone, GnIH) was reported to inhibit GnRH synthesis or secretion from the hypothalamus.

19.2.5 Mini Puberty

The neonatal activation of the HPG axis is called mini puberty, occurred in some mammals with an increased level of testosterone around mid-gestation and early postnatal life. In cattle and sheep, this testosterone peak is seen during the last third of gestation and diminished at birth, but in rodents, the event occurs during the early postnatal period. Mini puberty helps to develop the genital organs.

19.2.6 Endocrine Regulation of Puberty

Puberty is triggered when the neurobiological brake is removed and the GnRH pulse generator is reactivated. Kisspeptin plays a significant role in controlling the onset of puberty through activating the GnRH pulse generator for intermittent release of GnRH. The suppressive action of oestrogen and GABA on kisspeptin synthesis is diminished around puberty, and kisspeptin reactivates GnRH pulse generation. The kisspeptin induced GnRH pulse generation is mediated through the kisspeptin/neurokinin B/dynorphin A (KNDy) neuronal pathway. KNDy neuronal cluster is situated at the arcuate nucleus of the hypothalamus comprising kisspeptin, neurokinin B, and dynorphin A. KNDy neurons secrete neurokinin B, which binds with tachykinin NK3 receptor (NK3R) and triggers Ca^{2+} influx into KNDy neurons in an autocrine/paracrine manner. Increased intracellular calcium stimulates KNDy neurons for pulsatile kisspeptin secretion, which in turn controls pulsatile GnRH secretion from the median eminence of hypothalamus through a G protein-coupled receptor named (GPR54/KISS1R).

The exact cues that initiate the timing of puberty are yet to be elucidated. But, several sensory inputs from the internal and external environment like growth, body fat/composition, photoperiod, and olfactory signals are thought to be involved in the timing of puberty. Out of these factors, energy metabolism and photoperiod emerge as important determinants of the HPG axis activation around puberty.

19.2.6.1 Energy Metabolism and HPG Axis Activation

The timing of puberty coincides with the optimum body reserve, and leptin, an adipocytokine (cytokine synthesized from adipose tissue), is thought to be involved in this integration. Serum leptin levels are positively correlated with the fat mass, and leptin concentration is decreased upon starvation. Therefore, leptin signals the hypothalamus for the optimal energy level of an individual. In the arcuate nucleus of hypothalamus, KNDy neuronal clusters coexist with the leptin receptor (leptin Receptor Long Isoform, LepRb). The leptin signalling is mediated through the activation of KNDy pathways. But the optimum leptin level is required to stimulate the HPG axis. Too low leptin during starvation and high leptin in obesity can suppress the HPG axis. The malnutrition suppresses the HPG axis through neuropeptide Y (discussed later). Overnutrition leads to central leptin resistance and decreases the expression of Kiss1 or NKB and its receptor to induce reproductive dysfunctions.

19.2.6.2 Photoperiod and HPG Axis Activation

In seasonal breeders, the day length (photoperiod) regulates the activation of the HPG axis around puberty to ensure successful birth during the favourable time of the year. The effect of photoperiod on the HPG axis is controlled by the photo-neuroendocrine circuit, where melatonin secreted from the pineal gland is the central player (Fig. 19.17). The dark and light information is perceived through photoreceptors at the retina. The photoreceptors transmit the dark light signal to the suprachiasmatic nuclei (SCN) of the hypothalamus through the retinohypothalamic tract (RHT). SCN controls the melatonin synthesis from the pineal gland via the polysynaptic pathway. The synthesis and release of melatonin are restricted during the night-time. So, the short melatonin peaks are seen during long days (LD) in summer and short days (SD) during winter which are associated with longer melatonin peaks. Melatonin affects the GnRH secreting neurons through the stimulatory signal of Kisspeptin (Kiss) and the inhibitory signal of RFRP. But melatonin peaks and associated neuroendocrine effects (kisspeptin and RFRP signalling) are reversed in short-day and long-day breeders. In long-day breeders (Syrian hamsters), the short melatonin peaks during summer stimulate the release of kisspeptin and inhibit RFRP, which initiates puberty during summer. In contrast, the longer melatonin peaks during short days in winter facilitate puberty in short-day breeders (sheep) by stimulating kisspeptin and inhibiting RFRP.

19.2.7 Endocrine Regulation of Sertoli Cell Functions

The activity of Sertoli cells after puberty is controlled primarily by FSH through the inhibin-activin-follistatin axis. Various growth factors also regulate Sertoli cell function in an autocrine and paracrine manner.

19.2.7.1 Role of Inhibin, Activin, and Follistatin

Activin and inhibin are secreted from Sertoli cells. Activin exerts stimulatory and inhibin inhibitory effects on FSH secretion. Activin is a cytokine, a member of the TGF-β protein superfamily. It enhances FSH biosynthesis and secretion from the anterior pituitary. But, excess activin-A production reduces the release of FSH. The concentration of the spermatozoa in the tubule controls the secretion of activin and inhibin. At low sperm concentration, the inhibin secretion is decreased, and activin triggers FSH release from the pituitary to stimulate spermatogenesis. Follistatin is secreted from the anterior pituitary and negatively stimulates FSH secretion. The inhibin feedback on the pituitary FSH is set up during early postnatal life, but the maximum sensitivity is seen at the age of puberty. The FSH stimulates the metabolic activity of Sertoli cells and spermatogenesis after the attainment of puberty. But, when Sertoli cells are exhaustive and sperm counts are too high in tubules, a glycoprotein hormone *inhibin* (*inhibin B*) suppresses the synthesis and release of FSH from the anterior pituitary via a negative feedba*ck mechanism*. It also reduces the secretion of extracellular matrix components from Sertoli cells. Follistatin is an autocrine glycoprotein secreted from the anterior pituitary when the FSH level is more. Follistatin is an activin-binding protein (or FSH-suppressing protein, FSP) that binds and neutralizes activin, and hence FSH is decreased at the peripheral circulation or the pituitary level.

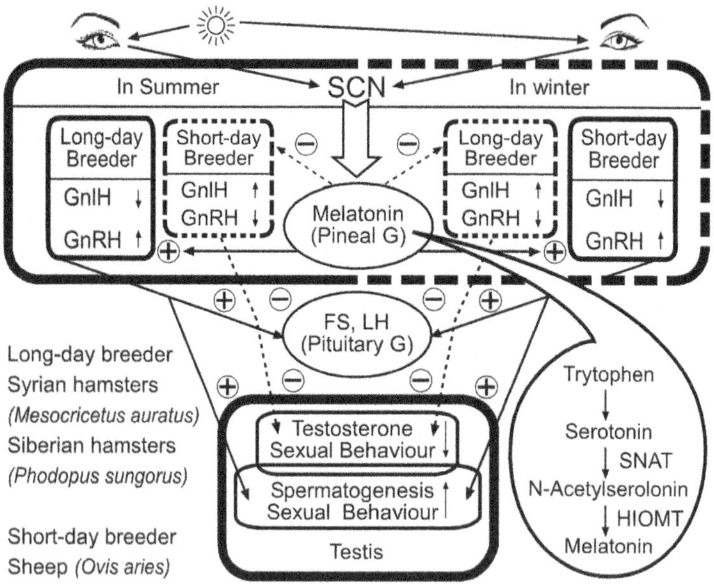

Fig. 19.17 Effect of photoperiod on HPG axis [The action of melatonin to control the reproductive function in males is summarized; melatonin biosynthesis is explained stepwise in the inset. The light stimulates photoreceptors in the retina of the eye and sends impulses to the SCN of hypothalamus. **SCN** (suprachiasmatic nucleus) stimulates the pineal gland for melatonin biosynthesis. During short-day photoperiod (in winter), at long dark phase (night), the **SNAT** (serotonin-N-acetyltransferase) and **HIMOT** (hydroxy indole-O-methyltransferase) enzymes are activated, and melatonin synthesis increases. The reverse mechanism is occurred in long-day breeders (during summer). Increased melatonin suppresses **GnIH** (gonadotropin-inhibitory hormone). It activates **GnRH** (gonadotropin-releasing hormone) pulse generator in hypothalamus resulting in the secretion of **FSH** (follicle-stimulating hormone) and **LH** (luteinising hormone) from anterior pituitary gland to promote spermatogenesis and sexual behaviour. The entire pathway is explained with the bold line with a positive (+) sign. Decreased melatonin level increases GnIH and suppresses GnRH pulse generator to reduce the secretion of FSH and LH, resulting in inhibition of testosterone secretion, marked by a dotted line with a negative sign (−)]

19.2.7.2 Role of Other Proteins

Several growth factors like insulin-like growth factor (IGF), fibroblast growth factor (FGF), epidermal growth factor (EGF), and transforming growth factor α (TGFα) are involved in the Sertoli cell functions, but IGF is primarily involved in regulating Sertoli cell number and testicular size. The secretion of IGF is FSH dependent. Two crucial enzymes involved in Sertoli cell activity are 5′-adenosine monophosphate-activated protein kinase (AMPK) and silent mating type information regulation two homolog 1 (Sirtuin 1 or SIRT1). The major gene responsible for Sertoli cell proliferation is *c-Myc* (*cellular Myelocytomatosis*). The c-Mycis an oncogene expressed under the influence of testosterone. Certain cyclin-dependent kinase inhibitors inhibit the Sertoli cell proliferation, viz. p21Cip1, p27Kip, p19INK4, and the gap junction protein *connexin 43* (Cx43). Some xenobiotic agents like ethane di-methanesulphonate have a cytotoxic effect on Sertoli cells.

19.2.8 Endocrine Regulation of Leydig Cell Functions

The differentiation of ALC and steroidogenesis is controlled by pituitary LH and some locally produced factors. The LH receptors (LHR) in the foetal Leydig cells (FLCs) start expressing from embryonic day 16 in mice, and LH binds with its receptor (LHR) on Leydig cells and initiates intracellular signalling mechanisms for steroidogenesis (details in Sect. 19.2.1.3). Androgens secreted from FLCs cause foetal masculinization, the development of male external genitalia, the accessory sex organs, and the male-specific neuronal network in the brain. *Desert hedgehog* (DHH) protein and *platelet-derived growth factor* (PDGF, PDGF-A, and PDGF-B), produced from the Sertoli cells, induce FLC differentiation. Its number increases throughout the foetal period. The differentiation of Sertoli cells is also influenced by transcription factors *GATA-4* and *IGF-1*. The FLCs decline after birth, and the number of ALCs rapidly increases to occupy the interstitial space of the adult testis during the pubertal period. Androgens, along with other proteins such as nuclear receptor subfamily 5 group A member 1 (NR5A1), Wilms tumor protein (WT1) and DHH protein, promote the differentiation of FLCs to form ALCs. The steroidogenesis is regulated within a precise range as excess or overproduction is detrimental.

Along with LH, several growth factors and cytokines act in an autocrine/paracrine manner to regulate steroidogenesis. Insulin growth factor I (IGF-I) and fibroblast growth factor 9 (FGF9) promote steroidogenesis, whereas transforming growth factor-beta (TGFβ), AMH, tumour necrosis factor-alpha (TNFα), IL-6 and IL-1 have inhibitory effects on steroidogenesis. Epidermal growth factor receptor (EGFR) also inhibits LH-induced steroidogenesis. 5' AMP-activated protein kinase (AMPK) is an enzyme involving cellular energy homeostasis that negatively regulates steroidogenesis. The activity of ALCs and testosterone production gradually diminished with age due to the reduced activity of steroidogenic enzymes, like Cyp11a1, Hsd3b, Cyp17a1, 17-ketoreductase, and Hsd11b2. The decreased activity of *antioxidants enzymes*, like copper and zinc superoxide dismutase (Sod1), manganese superoxide dismutase (Sod2), and glutathione peroxidase (Gpx), also affect the Leydig cell function in old age. Due to the failure of antioxidant activity, ROS production is increased. It interferes with cholesterol transport to the mitochondria and its conversion to pregnenolone in the steroidogenic pathway. Ghrelin and leptin reduce the activity of key steroidogenic enzymes, hence reducing testosterone release. INSL3 is the major secreted product of the ALC and is used as a biomarker of LC function and the onset of puberty. Prolactin increases the affinity of LH receptors to its ligand on Leydig cells to increase testosterone production. Thus, excess testosterone production in hyperprolactinemia causes downregulation of the H-P-G axis, leading to hypogonadism. Oestradiol suppresses the enzyme 17β-hydroxysteroid dehydrogenase and decreases testosterone synthesis. In boar and stallion, the production of oestradiol from testosterone by aromatase can reach the interstitial space and binds with its receptor at LC to affect steroidogenesis.

The Relationship Between LH and Testosterone Release Generally, every LH pulse is followed by a GnRH pulse (Fig. 19.18). But, every LH pulse does not yield testosterone release (Fig. 19.19). Normally the testosterone pulses are followed by a few narrowly spaced LH pulses. In mice, the LH pulse and testosterone pulse ratio is about 2:1. The testosterone release occurs within 10 min after the LH pulse appearance, which peaked within 20–30 min and gradually declined to reach baseline within 60–80 min (Table 19.3).

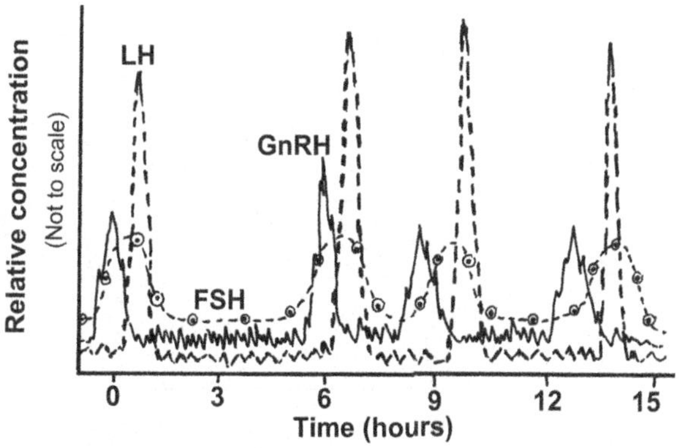

Fig. 19.18 Relative blood concentration and associations among GnRH, FSH, and LH in adult male mammals. [*GnRH* gonadotropin-releasing hormone (solid line); *FSH* follicle-stimulating hormone (dotted line with a circle); *LH* luteinizing hormone (dotted line). Secretion of FSH occurs immediately after the GnRH pulse, and LH secretion starts at the end of the GnRH pulse. The LH pulses are sharp and decline within10–20 min due to less half-life, whereas FSH pulses are flat, indicating more persistency than LH (five times more) than LH secretion. The concentration of FSH is comparatively less than LH due to the continuous release of inhibin]

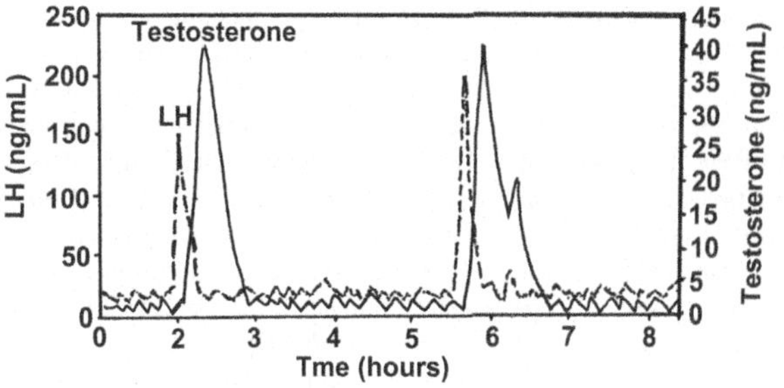

Fig. 19.19 Relationship between luteinizing hormone and testosterone in male mouse. [Luteinizing hormone presented (dotted line) and testosterone (solid line)]

19.2.9 Stress-Induced HPG Axis Suppression

19.2.9.1 Role of Hypothalamic–Hypophyseal–Adrenal (HPA) Axis

Stress is a well-known suppressor of gonadal functions in males and females by inhibiting the GnRH pulse generator. During stress, the hypothalamic–hypophyseal–adrenal (HPA) axis is activated to release the secretion of corticotropin-releasing hormone (CRH) from the hypothalamus, adrenocorticotropin-releasing hormone (ACTH) from the pituitary, and glucocorticoids (corticosterone) from the adrenal gland. CRH suppresses the synthesis and secretion of GnRH by changes in the expression of GnRH and its receptor (GnRHR) genes in the hypothalamic-hypophyseal unit. But, evidence has suggested that stress hormones (corticotrophin-releasing hormone and corticosterone) cannot suppress the HPG axis alone. Good numbers of other endocrine and neural signalling are implicated in stress-induced downregulation of the HPG axis. Some authors postulated that stress-induced suppression of the HPG axis is mediated through prostaglandins (PGs) as the stressors induce the activity of brain cyclooxygenase-2 (COX-2), an enzyme required for PG-synthesis. Recently the role of RFRP-3 GnRH and LH inhibition during stress has been documented.

19.2.9.2 Role of Anorexigenic Peptides

Orexigenic and anorexigenic peptides are released to convey information about the nutritional status of an animal. The

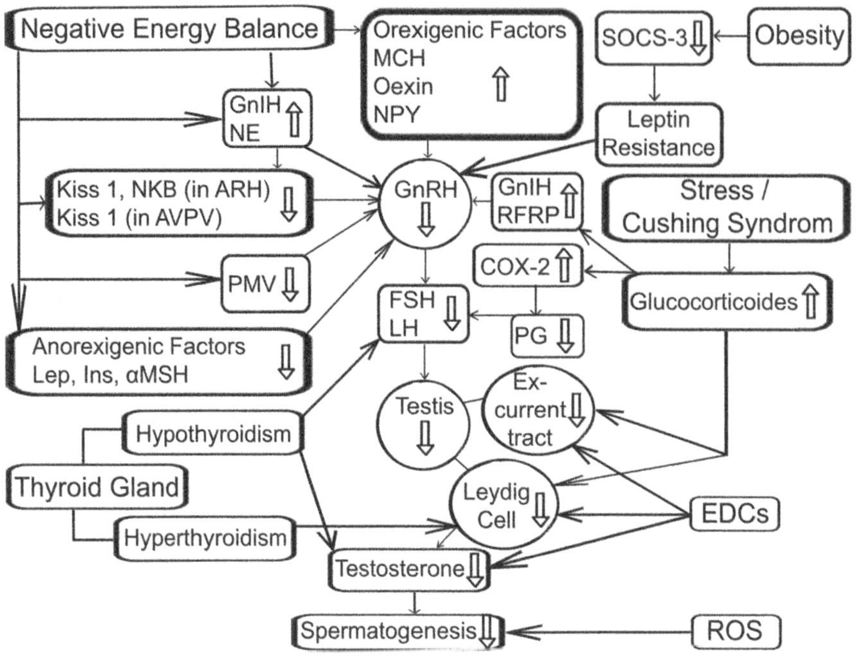

Fig. 19.20 Stress-induced suppression of the H-P-G axis. [The downregulation of the H-P-G axis under negative energy balance (**NEB**) through orexigenic factors like **MCH** (melanin-concentrating hormone), **orexin** (from lateral hypothalamus), and **NPY** (neuropeptide Y, from the arcuate nucleus, **ARH**). These orexigenic peptides directly suppress the neurokinin B(NKBB)-stimulated firing of kisspeptin (**Kiss 1**) neurons in the anteroventral periventricular nucleus (**AVPV**), the **PMV** (ventral pre-mammillary nucleus). The decreased secretion of anorexigenic peptides like **Ins** (insulin), Leptin (**Lep**), and **αMSH** (melanocyte-stimulating hormone) during negative energy balance also suppresses the HPG axis. The optimum nutrition favours the release of Leptin (**Lep**) which stimulates GnRH release through the kisspeptin/neurokinin B/dynorphin A (**KNDy**) pathway. In *obesity*, **SOCS-3** (suppressor of cytokine signalling) is inhibited, causing leptin resistance followed by downregulation of GnRH. Secretion of glucocorticoids (*stress* or *Cushing's syndrome*) suppresses the H-P-G axis through gonadotropin-inhibiting hormone (GnIH) or **RFRP** (RFamide-Related Peptide) secretion. Glucocorticoids also affect the activity of **Cox-2** (cyclooxygenase inhibitor-2), which decreases **PG** (prostaglandin) synthesis resulting in inhibition of LH secretion; glucocorticoids suppress the testosterone synthesis in the Leydig cell (**LC**) of the testis by inducing LC apoptosis and inhibiting the activity of LH receptor, P450SCC, StAR, CYP17, and 11β-HSD enzymes cause the suppression of spermatogenesis; testosterone action in the epididymis, vas deferens, the prostate is also affected by glucocorticoids. The function of LC is followed by testosterone production. The thyroid dysfunctions (hyperthyroidism and hypothyroidism) also affect spermatogenesis. **EDCs** (endocrine disruptors) and **ROS** (reactive oxygen species)] also affect the testicular functions]

orexigenic peptides like neuropeptide Y (NPY) originate during calorie restriction or hypoglycaemia and secrete anorexigenic peptides (leptin, insulin, and αMSH) during optimum body growth as well as obesity to control the appetite and feeding behaviour. These anorexigenic and orexigenic peptides modulate the HPG axis (Fig. 19.20). The stimulatory action of anorexigenic peptides, particularly leptin, has been discussed in 'energy metabolism and HPG axis activation' (Sect. 19.2.6.1). The orexigenic peptides like NPY and αMSH directly suppress the neurokinin B-stimulated firing of kisspeptin neurons and hence the HPG axis during malnutrition.

19.2.9.3 Role of Endocrine Disruptors Chemicals (EDCs) on the H-P-G Axis

Various phytoestrogen, persistent organic pollutants (POPs), and organochlorine insecticides negatively affect the HPG axis. They are collectively called *Endocrine disruptors* (EDCs). They exert their actions by modulating synthesis, release, transport, metabolism, and eliminating other neuro-endocrine factors involved in the HPG axis. The potential EDCs are agricultural waste, like triazoles, Imidazoles, and triazines; industrial products, like nonyl-phenols, octyl-phenols, bisphenol A, phthalates, organotins, perfluorooctane sulphonate, parabens, cadmium, etc.; metals, like cadmium. The EDCs damage the germ cells, Sertoli cells, and Leydig cells, alter various enzymatic activity, disrupt the antioxidant system, and lower the level of FSH and LH, which ultimately results in reduced testosterone synthesis, decreased sperm concentration, increased production of abnormal spermatozoa with hypomotility.

The role of different neuroendocrine factors in the H-P-G axis regulation is summarized in Table 19.5.

Table 19.5 Various neuroendocrine factors involved in the H-H-G axis

Hormones-like substances	Nature	Source	Major role
GnRH	Hypothalamic neuropeptide	Preoptic area/infundibular nucleus of hypothalamus	Activator of gonadotroph—secretes FSH, LH—H-H-G axis
FSH	Glycoprotein hormone	Gonadotroph of the anterior pituitary	Activator of Sertoli cell—spermatogenesis—puberty/H-H-G axis
LH	Glycoprotein hormone	Gonadotroph of the anterior pituitary	Activator of Leydig cell—Testosterone—Sertoli cell—spermatogenesis—puberty/H-H-G axis
Testosterone	Steroid hormone	Leydig cell	Activator of Sertoli cell—spermatogenesis—puberty/H-H-G axis
Activin	Homodimeric protein hormone	Sertoli cell, Leydig cells, prostate	Stimulator of GnRH, GnRHR, biosynthesis of FSH/LH—H-H-G axis
Inhibin	Heterodimeric glycoprotein hormone	Sertoli cell, Leydig cells, prostate	Suppressor of activin, FSH—H-H-G axis
Follistatin	Monomeric glycoprotein hormone	Folliculostellate (FS) cells of the anterior pituitary	Suppressor of activin, FSH—H-H-G axis
GnIH	Hypothalamic neuropeptide	Paraventricular/dorsomedial hypothalamic nuclei	Suppressor of GnRH—H-H-G axis
Prolactin	Polypeptide hormone	Lactotrophs of the anterior pituitary	Enhance LH receptor activity—testosterone production—spermatogenesis—puberty
Oestrogen	Steroid hormone	Sertoli cell	Suppressor of GnRH/H-H-G axis
Kisspeptin	Hypothalamic neuropeptide	Anteroventral periventricular nucleus (AVPV)	Activator of GnRH—LH—H-H-G axis—puberty
Leptin	Polypeptide hormone	Adipose tissue (also from skeletal muscle and stomach fundus)	Activator of GnRH/H-H-G axis, advance puberty
Ghrelin	Polypeptide hormone	Epsilon cells of the stomach fundus	Suppressor of GnRH, LH, Leydig cell, testosterone synthesis—H-H-G axis
IGF-1 (Somatomedin C)	Polypeptide hormone	Multiple mesenchymal cell types of liver, also peripheral tissues of bone	Activator of GnRH, gonadotroph (LH), Leydig cell, Sertoli cell—H-H-G axis
Melatonin	Neurohormone (amino acid-derived hormone)	Pineal gland	Suppressor of GnRH/H-H-G axis, delayed puberty
GABA (in stress)	Gamma-aminobutyric acid	Hypothalamic preoptic area	Suppressor of GnRH, LH—H-H-G axis and activator of acrosome reaction (sperm maturation)—puberty/maturation
Neuropeptide Y (NPY) (in chronic release)	Neuropeptide	Arcuate nuclei of the preoptic area	Suppressor of GnRH—H-H-G axis/puberty/maturation
RFamide-related peptides-3 (RFRP-3)	Neuropeptide	The dorsomedial nucleus of the hypothalamus (DMH) or paraventricular nucleus (PVN)	Suppressor of GnRH, biosynthesis of testosterone—H-H-G axis/puberty/maturation
MKRN-3	Makorin ring finger protein 3 (Probable E3 ubiquitin-protein ligase makorin-3)	Arcuate nucleus of the preoptic area	Suppressor of GnRH—FSH, LH—H-H-G axis
Neuroestradiol	Steroid hormone	Median eminence of the hypothalamus	Stimulus sexual behaviour (at low level), act as GnIH—suppressor of GnRH/H-H-G axis (at extreme level)
Glucocorticoid/cortisol (in excess, Cushing's syndrome	Steroid hormone	Zona fasciculata of adrenal cortex	Suppressor of GnRH, FS,H & LH and biosynthesis of testosterone—H-H-G axis
Thyroid (hyperthyroidism)	Tyrosine-based amino acid-derived hormone	Epithelial/follicular cells of the thyroid gland	Highly inducer of FSH/LH release (through testosterone potentiality using SHBG)—Suppressor of GnRH/H-H-G axis
Insulin (in diabetic state/insulin resistance)	Protein hormone	Pancreatic β-cell	Suppressor of SHBG—testosterone—H-H-G axis

19.3 Testicular Androgens

The term androgen is derived from the Greek word *andr—meaning 'man'*. Androgens are the natural or synthetic steroid hormones that regulate the development and functions of testes, expression of male secondary sexual characteristics, masculinization, libido, and spermatogenesis. The major androgens are testosterone, dihydrotestosterone (DHT), and androstenedione, out of which testosterone is the most abundant androgen in blood.

19.3.1 The Site of Androgen Production

In the male, the testis is the principal site of androgen production and constitutes about 95% of the total androgen secreted in the body. Ovaries are also able to produce a minimal amount of androgens. The Leydig cell of the interstitial space of the seminiferous tubule of testes is the major site of testosterone production. The predominant extragonadal source of androgens is the zona reticularis of adrenal cortex. But, the extragonadal androgens are less potent and contribute only 5% of total androgen production in mammals. The androgen production starts during foetal life and supports the differentiation of male sex organs. The peak androgen production under the influence of LH around puberty supports spermatogenesis.

19.3.2 Chemistry of Androgens

Androgens belong to the group of steroid hormones containing 19 carbon atoms; hence, they are called C-19 steroids. Testosterone ($C_{19}H_{28}O_2$) has 17β-hydroxy and 3-oxo groups and unsaturation at C-4-C-5 (Fig. 19.21). Dihydrotestosterone (DHT, 5α-DHT) ($C_{19}H_{30}O_2$) is a metabolite of testosterone formed by the action of the enzyme 5-α-reductase. Androstenedione ($C_{19}H_{26}O_2$) is the direct precursor of testosterone. The structure of testosterone is similar in all mammals, reptiles, birds, and fish.

19.3.3 Biosynthesis of Androgens (Steroidogenesis in Testis)

The synthesis of androgens and other associated intermediates (progesterone, oestrogens, cortisol) has occurred in the testes through a series of enzymatic reactions called steroidogenesis (Fig. 19.22). Cholesterol is the precursor molecule of androgen biosynthesis. Cholesterol is incorporated in the Leydig cells from low-density lipoproteins either by receptor-mediated endocytosis or can be synthesized de novo from acetyl-coenzyme A. In the beginning, cholesterol is transferred from the outer to the inner mitochondrial membrane of Leydig cells by the steroidogenic acute regulatory protein (StAR). Cholesterol (C-27) then undergoes oxidative cleavage at its side-chain to produce pregnenolone (C-21) by the enzyme cytochrome P450 oxidase(s) (P450scc, CYP11A1). Pregnenolone production is the rate-limiting step of androgen biosynthesis. Pregnenolone has several fates. It either converts to the C-21 group of steroids like progesterone, cortisol, etc. or gives rise to C-19 dehydroepiandrosterone (DHEA). The conversion of pregnenolone to DHEA is mediated by the enzyme cytochrome P450 17α-monooxygenase (CYP17). The DHEA is converted to androstenedione by the enzyme 3β-hydroxysteroid dehydrogenase (3β-HSD) at the endoplasmic reticulum. The androstenedione is converted to testosterone (C-19) by the enzyme 17β-HSD reducing the C-17 keto group. The enzyme 5α-reductase can convert testosterone into its potent form (two to threefold), dihydrotestosterone (DHT or 5α-DHT). In males, 5α-reductase is highly expressed in the prostate gland, seminal vesicles, epididymis, skin, hair follicles, and brain.

19.3.3.1 Biosynthesis of Oestrogen in Testes

The testosterone may be converted to oestradiol (C-18 steroid) by the enzyme aromatase (Fig. 19.22). The aromatase can convert the androstenedione into estrone (another C-18 steroid). The oestradiol can be transformed into estrone by 17β-HSD. The enzyme aromatase is highly expressed in the Sertoli cells, adipose tissue, bone, and the brain.

19.3.3.2 Biosynthesis of Progesterone and Cortisol in Testes

The enzyme 3β-HSD converts pregnenolone into progesterone (C-21 steroid). The progesterone can be further metabolized into corticosterone (C-21) with the help of CYP21 and CYP11 (Fig. 19.22). Progesterone can further be converted into two other intermediates of C-21 groups, viz. cortisol and cortisone, with the help of CYP17.

19.3.4 Transport of Androgens

The majority of the androgens are bound with the plasma proteins, and only a small fraction is available in free form to act on seminal vesicles, bone, muscle, and prostate gland. Globulin has a higher affinity for testosterone compared to albumin. The predominant plasma proteins responsible for androgen transport are sex hormone-binding globulin (SHBG) (also known as Testosterone Estradiol Binding Globulin, TeBG or steroid-binding protein, SBP) and androgen binding protein (ABP). SHBG is a β-globulin mainly produced from hepatocytes and contains single androgen

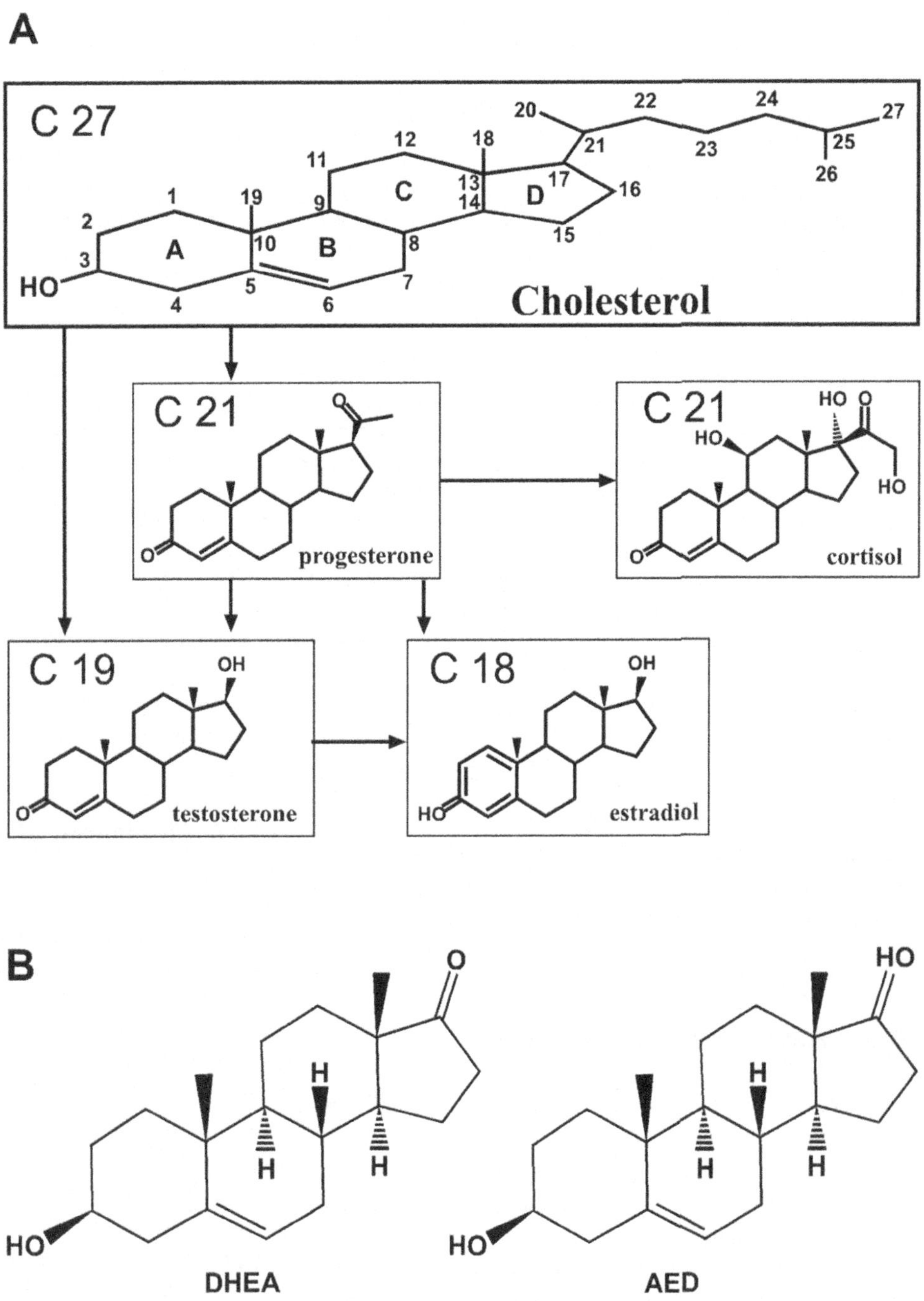

Fig. 19.21 Steroid hormones androgens and their precursor (cholesterol). [**a**: Depicted the chemical structures of the precursor of testosterone, the C-19 and **b**: depicted the chemical structures of two major forms of androgen, the DHEA (dehydroepiandrosterone) and AED (androstene)]

binding site per molecule. ABP is mainly produced from Sertoli cells in the testes. Albumin can also bind with androgens but dissociates quickly due to less affinity. Dehydroepiandrosterone and androstenedione remain primarily in an albumin-bound form in the human serum in contrast to testosterone and dihydrotestosterone bound principally with SHBG. The other proteins capable of binding with androgens are prostate binding protein (PBP), uteroglobin, and a1 acid glycoprotein. The SHBG also binds with oestrogens, and other steroid hormones, like progesterone and cortisol.

The distribution of androgens in free and bound forms in normal human serum is presented in Table 19.6. The testosterone dissociates from its carrier proteins in the capillaries, and the endothelial glycocalyx causes structural modifications of the androgen binding sites and reduces its affinity towards androgens. Spermatic veins are mainly responsible for androgen transport in general circulation. Androgens diffuse into the capillaries directly from Leydig cells or are transported via interstitial fluid.

Fig. 19.22 Steroidogenesis of the testes. [Reactions under dotted lines occurred in foetal Leydig cells (**FLC**), whereas the entire reactions took place place in adult Leydig cells (**ALC**). Major enzymes involved in the process are cytochrome P450 oxidase (s) (P450scc, CYP11A1, CYP11, CYP17, CYP21), hydroxysteroid dehydrogenase (HSD, 3β-HSD, 17β-HSD), aromatase and 5α-reductase]

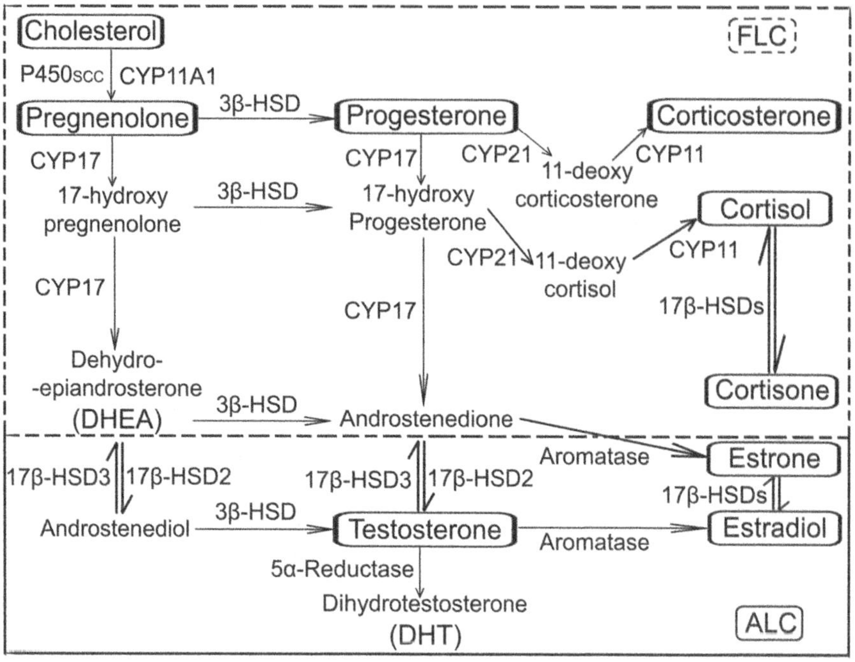

Table 19.6 Distribution of androgens in free and bound form in normal serum of human

Androgens	Free (%)	Albumin bound (%)	SHBG bound (%)
Testosterone	1	30	69
Dehydroepiandrosterone	4	88	8
Androstenedione	7	85	8
Dihydrotestosterone	1	21	78

(Ref: Dunn et al. 1981; Mendel 1989)

19.3.5 Plasma Levels of Androgens

The concentration of androgens in the plasma is dynamic and influenced by species (Table 19.7), general and testicular health, circadian rhythms, age, and environment. The diurnal variation of testosterone concentration with morning peak has also been found. In older age, the baseline level is decreased due to the reduction of ALC. Increased androgens synthesis is seen in seminiferous tubules hypertrophy due to increased gonadal temperature. Impaired function of androgens may occur due to its hypo- and hyper-secretion, metabolic disturbances, and receptor insensitivity. Muscular exercise, sleep (REM), vitamin D, zinc, and calcium in the ration can increase androgens production. In contrast, stress, high cortisol, obesity, dietary fat, estrogenic substances in feed and unilateral cryptorchidism reduces testosterone production.

19.3.6 Mechanism of Action of Androgens

Androgens enter the target cells either through direct diffusion or receptor-mediated endocytosis. The receptor-mediated endocytosis is facilitated by LDL receptor, steroid carrier, or steroid channels. In LDL receptor-mediated endocytosis, the lipoprotein bound androgens bind with LDL receptors at the plasma membrane and are taken up. The LDL is degraded in the lysosomes, and androgens enter different metabolic pathways. The hormone and its carrier proteins internalize endocytosis of androgens into the target tissues through steroid carrier or steroid channels, and the carrier is degraded intracellularly. Recently a low-density lipoprotein named megalin has been involved in the receptor-mediated endocytosis of androgen-SHBG complexes across the plasma membrane. Androgens act through intracellular nuclear receptors. The receptor has several domains, namely a ligand-binding domain (binds with hormone), a constitutionally activating function domain (leads to the activation of the receptors), a nuclear localization signal domain (helps in translocation of the hormone-receptor complex into the nucleus), and a highly conserved

Table 19.7 Normal circulatory level of androgens in adult male animals

Animal	Testosterone (ng/mL)	Reference	DHEA	Reference	AE	Reference
Bull	6.00	Woźniak et al. (2016)	1.24 ± 0.14 pmol/mL	Fustini et al. (2017)	80–100 pg/mL	Kanchev and Dobson (1976)
Buffalo bull	2.07 ± 0.10	Malfatti et al. (2006)	–		106 ± 35 pg/mL	Hemeida et al. (1985)
Buck	4.60 ± 0.80	Georgie et al. (1985)	–		–	
Ram	2.20 ± 6.20	Roselli et al. (2002)	–		–	
Stallion	2.00 ± 0.85	Inoue et al. (1993)	52.0 ± 43.8 ng/mL	Knych et al. (2014)	–	
Boar	4.50	Bonneau et al. (1987)	352.56 ± 47.9 pg/mL	Tagliaferro and Ronan (2001)	–	
Dog	2.30 ± 0.20	DePalatis et al. (1978)	5.2–30.5 ng/mL	Frank et al. (2003)	2.7–48.8 ng/mL	Frank et al. (2003)
Tom cat	1.93 ± 0.64	Villaverde et al. (2010)	–		–	
Rat buck	1.54 ± 0.27	Takikawa and Wakabayashi (1994)	–		–	
Human	2.50–4.00	Liverman and Blazer (2004)	2.6 ± 0.9 ng/dL	Vecchione et al. (2018)	4.5 ± 0.9 ng/dL	Vecchione et al. (2018)

DHEA dehydroepiandrosterone; *AE* androstenedione

DNA-binding domain (binds with genomic DNA and induce transcription).

19.3.7 Biological Roles of Androgens

Androgen receptors are present in various tissues, such as testes, epididymis, seminal vesicles, prostate, muscles, kidney, heart, spleen, salivary glands, pituitary, and hypothalamus, exerting a variety of functions (Fig. 19.23 and Table 19.8).

19.3.7.1 Role in Reproduction

Spermatogenesis Testosterone is the predominant androgen that regulates spermatogenesis. After secreted from Leydig cells, testosterone diffuses into the seminiferous tubule and acts as a paracrine factor to promote spermatogenesis. Testosterone regulates four critical spermatogenesis processes, viz. maintenance of blood testes barriers, meiosis, spermatid adhesion with Sertoli cells, and sperm release.

Testicular Descent Transinguinal migration of testes is mediated by the androgens. Androgens, particularly

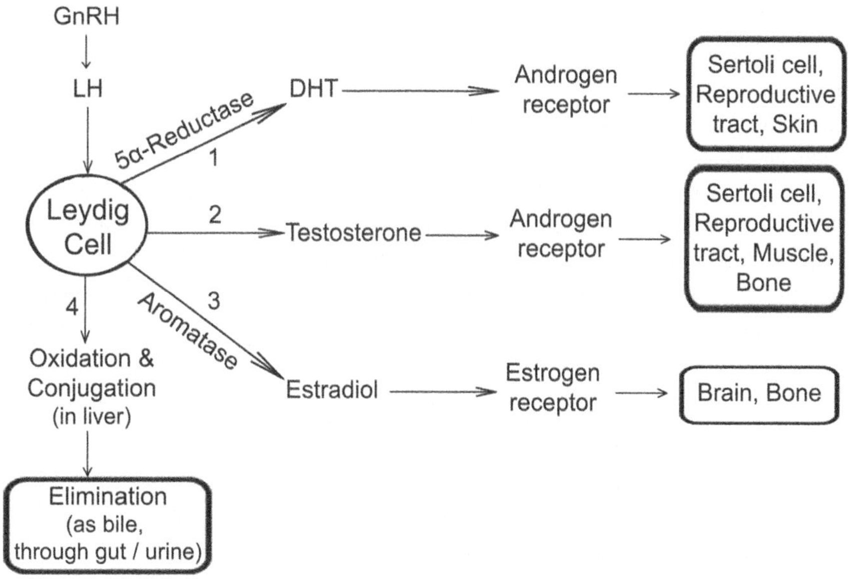

Fig. 19.23 Major site of production, functions, and metabolism of testosterone. [1, 2, and 3 are the major functional ways and 4 is the metabolism and elimination route]

Table 19.8 Function of testosterone

Stage of life	Type of function		
	Androgenic		Anabolic
	In reproduction	Other than reproduction	Systemic
In foetal life by FLC	Sex differentiation; masculinization of the male genital tract and external genitalia; testicular descent	Sex characteristics; development of brain	Protein anabolism
Before puberty (also during the non-breeding season)	Control GnRH release; attainment of puberty	Protein synthesis; development of body structure and skin; growth of hair and feather; change of voice	Bone growth
After puberty (also in breeding season)	Maintenance of structure and function of reproductive organs as well as H-H-G axis; attainment of puberty and sexual maturity; sex desires; spermatogenesis	Protein synthesis; stimulates erythropoiesis, body structure maintenance, hair and feather growth, and voice. Closing of the epiphysis, effect on adipose tissue (by converting oestrogens) Sexual behaviour	Sexual behaviour

testosterone, cause the reduction of the gubernaculum by altering its viscoelastic properties and helps in testicular descent.

Growth of Male Genitalia Both testosterone and DHT favour the growth of male genitalia. The masculinization of the Wolffian duct is under the influence of testosterone, and DHT controls the differentiation of external genitalia. The growth of the penis is directly correlated with testosterone concentration after puberty.

Growth of Accessory Sex Glands The main androgen that acts over epididymis, vas deferens, and seminal vesicles is *DHT,* produced from testosterone by 5α-reductase. DHT acts over the epididymis, seminal and prostate to control these glands' growth, development, maturation, and secretion. The growth of the prostate gland is also mediated via oestrogen produced after the aromatization of testosterone.

Development of Secondary Sexual Characteristics Testosterone plays a vital role in sexual maturity. The androgens promote masculinization through their anabolic effects on smooth and striated muscles. Testosterone causes glycogen synthesis in the striated muscles and hypertrophy of the fibres. Androgens stimulate bone mineralization, linear growth, prevent osteoporosis and increase bone density. The DHT acts on the skin to increase its thickness and texture, the quantity of melanin, sweat glands, and sebaceous glands. It also promotes sebum production. In humans, the development of pubic hairs and beards is under androgens' influence. The change in the voice around puberty in humans is mediated by testosterone, and it causes the growth of the larynx and the vocal cords.

19.3.7.2 Role in Brain Development

Androgens play a significant role in the sex-specific brain development of males during intrauterine life. Testosterone and DHT promote the cognition skill of males during adolescence by stimulating the cortico-limbic system.

19.3.7.3 Role in Sexual Behaviour

Androgens regulate a variety of sexual dimorphic behavioural patterns. Androgens are responsible for the expression of aggressive behaviour in male animals, and the copulatory patterns of the males are under androgenic control. The specialized sexual behaviours (musth in elephants, male dominance in mice) are also regulated by androgens.

19.3.7.4 Role in the Haematopoietic System

Androgens stimulate erythropoiesis by increasing erythropoietin production. The action of androgens on haematopoietic stem cells promotes haemoglobin synthesis. The androgens are associated with blood coagulation and fibrinolytic mechanism, and androgen deficiency leads to decreased fibrinolysis.

19.3.7.5 Blood Pressure Regulation

Testosterone causes vasoconstriction by upregulating thromboxane A2, angiotensin II, endothelin-1, and norepinephrine. Testosterone also helps in vascular remodelling and affects atherosclerosis, and it promotes sodium and water retention in the body.

19.3.7.6 Role of Androgens in Birds

In birds, testosterone promotes the development of comb, wattle, and plumage. It regulates bird's vocalization (call and song) around mating season. Testosterone increases the bioavailability of carotenoids in birds. The carotenoids are responsible for the red, yellow, and orange colours of the skin and feathers. Commercially the rooster is castrated (caponization) to eliminate the effect of testosterone. The caponised birds accumulate fat and can produce tenderer, juicier and flavoured meat.

Know More...

Measurement of glucuronidated or sulphated products of testosterone in urine and faeces (faecal testosterone, fT) using high-performance liquid chromatography (HPLC), radioimmunoassay (RIA), and enzyme-immunoassay (EIA) is considered as one of the best non-invasive processes for assessment of male sex steroid in the vertebrate mammals. The rate of metabolism is significantly variable with species, gut passage time, and circadian rhythm (more in the dark phase). *Conservation Physiology* extensively uses this procedure to study and monitor the reproduction, welfare, and ecological balance of wild lives and laboratory animals.

19.3.8 Metabolism and Fate of Testosterone

The half-life of testosterone is only 10–100 min. Both testosterone and DHT are metabolized mainly in the liver. Most testosterone is metabolized either by glucuronidated or sulphated conjugating with glucuronide by glucuronosyltransferases and joining with sulphate by sulfotransferases. The 17-ketosteroids androsterone and etiocholanolone are the two other forms of metabolism of testosterone with the involvement of 5α- and 5β-reductases, 3α-hydroxysteroid dehydrogenase, and 17β-HSD enzymes. The 17-ketosteroids androsterone and etiocholanolone are further than glucuronidated or sulphated. The testosterone conjugated glucuronidated or sulphated products are finally released from the liver in the bile and excreted through urine or gut. Only a small fraction is excreted unchanged form of testosterone in the urine. Hence, glucuronidated or sulphated products of testosterone in urine and faeces are used to measure testosterone concentration in wild lives.

19.4 Puberty in Males

Puberty can be defined as the age at which the male gonads are capable of releasing the gametes. In males, it is characterized by the development of secondary sexual characteristics ability to copulate and produce sperm. The onset of puberty requires complex and integrated biological sequences that lead to progressive sexual maturation and full reproductive capacity. The term puberty and sexual maturity are not synonymous as sexual maturity is the state of full reproductive capacity when the males can produce a sufficient number of viable spermatozoa capable enough to fertilize the ovum. If the animals are bred at puberty, when they may not sexually mature may attain conception failure. For example, the ejaculates of bull, ram, boar, and stallion contain about 50 million spermatozoa with more than 10% mortality around puberty. Still, this concentration of spermatozoa is not sufficient to establish a pregnancy. They required more time to gain adequate spermatozoa for successful fertilization. This interval between puberty and sexual maturity is called adolescence. It is about about 50 days in bulls. Puberty in males can be detected through sexual behaviour such as mounting and erection. But these behavioural signs do not guarantee the presence of spermatozoa in the ejaculates. The age of puberty in different animals has been presented in Table 19.9.

Table 19.9 Age for the onset of puberty and sexual maturity of male domestic and wild animals

Animal	Age of puberty (month)	Age of maturity (month)
Bull	12–20	14–24
Buffalo bull (*B. bubalis*)	14–36	24–44
Buck	7–8	10–12
Ram	7–20	10–24
Boar	4–8	5.5–7.5
Stallion	10–24	12–26
Camel	3–4 year	4–5 year
Dog	4–6	12–16
Tomcat	5–10	6–12
Rabbit buck	3.5–6.0	4–7
Rat buck	1.5–2.0	2.0–2.5
Mouse buck	1.0–1.5	1.2–2.0
Rooster (Cock)	6–7	7.5–8.0
Human (Male)	11–13 year	14–18 year
Lion (Male) (*Panthera leo*)	1–2.5 year[a]	3–4 year
Tiger (Male) (*Panthera tigris*)	–	4–5 year
Cheetah (Male)[b] (*Acinonyx jubatus*)	1.5–2 year	–
Deer buck (Cervidae)	9–15	15–18
Bear (Ursidae)	–	2–4 year
Giraffe (*Giraffa*)	–	9–10 year
White Rhinoceros (*Ceratotherium simum*)	–	10–12 year
African Elephants (*Loxodonta africana*)	10–12 year	10–20 year
Asian Elephants (*Elephas maximus*)	12–14 year	10–20 year

[a] Zoo animals attain puberty earlier than wild
[b] Male cheetahs attain puberty earlier than their females

19.4.1 Endocrine Basis of Puberty

The attainment of puberty is a consequence of HPG activation in response to external and internal factors. The activation of the HPG axis results in the secretion of a large quantity of testosterone for spermatogenesis and the development of characteristics sexual behavioural patterns. The concentration of testosterone from pre-pubertal to pubertal

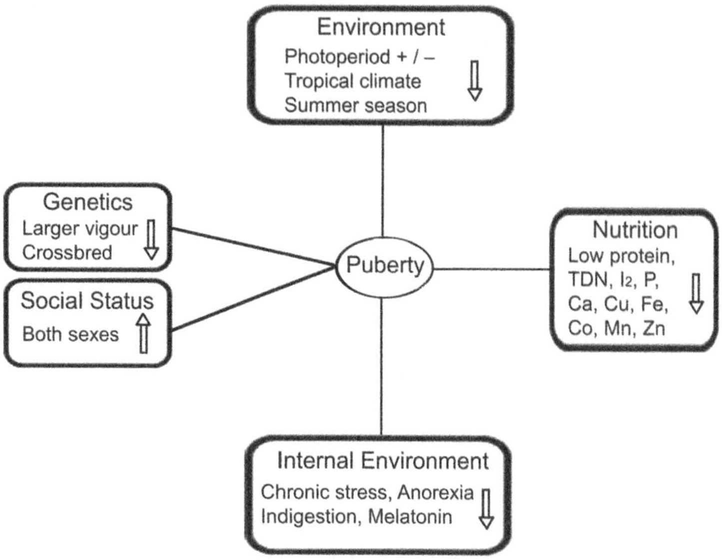

Fig. 19.24 Effect of various factors on puberty

period can be increased to the tune of 70–890 pg/mL in bulls. The detailed mechanism of HPG activation around puberty has been discussed (see Sect. 19.2.6 Endocrine regulation of puberty).

19.4.2 Factors Affecting the Puberty

The physiological events of puberty are the result of the integration between the environment and the HPG axis. Therefore, it depends on an individual's seasonal rhythms, genetic makeup, body growth, nutritional status, social status, and stress response (Fig. 19.24 and Table 19.9).

19.4.2.1 Genetic Factors

The onset of puberty is a multigenic trait with high heritability (0.2–0.48 in heifers). Generally, species of smaller size experience puberty at an early age compared to larger-sized species, particularly in cattle and horses. In rabbits, the miniature breeds reach sexual maturity at 4–5 months; medium breeds at 5–6 months; large breeds at 7–8 months; and giant breeds after 8 months. Crossbred animals reach puberty and maturity earlier. There are 10,650 genetic makers within 60 QTL regions of the X chromosome associated with puberty in cattle irrespective of breeds identified through genome-wide association studies (GWAS) using microsatellites. The most important candidate gene for the early puberty phenotype in cattle is IGF1. Studies have shown that seven genes from the IGF1 signalling pathway (IGF1R, IGFBP2, IGFBP4, EIF2AK3, PIK3R1, GSK3B, and IRS1) were associated with the onset of puberty in Brahman cattle under tropical climate.

19.4.2.2 Nutritional Factors

The role of nutrition and body growth in puberty attainment has been established for a long time. According to the 'critical fat mass hypothesis', critical fat is essentially required to attain puberty. As a thumb rule, the onset of puberty generally occurs when an animal reaches its 55–65% of adult body weight, depending on breed. The animals with a low protein and low total digestible nitrogen (TDN) diet experience delayed puberty. The deficiency of some specific elements like iodine, phosphorus, calcium, copper, iron, cobalt, manganese, zinc, vitamin A, and vitamin E also suppresses the pubertal onset. Metabolic cues such as glucose, insulin, and leptin have been shown to regulate the HPG axis directly or indirectly through a complex neuronal network (see Sect. 19.2.6.1 in Energy metabolism and HPG axis activation).

Know More...

The Evolutionary Advantage of Seasonal Breeder

Wild animals have various physiological adaptations and uniqueness in reproductive performance. In male animals, puberty generally attains later than in females as the spermatogenesis process is relatively longer, comparing the follicular maturation and ovulation in its complementary female. Male also required more energy than females for courtship, competing with others in the herd. Animals with short life expectancy are very 'opportunist' in sexual performance throughout the year, while animals with more extended life expectancy become 'strategic' in reproduction to give birth in the favourable season with plenty of food availability to facilitate the growth of

(continued)

the young. Thus, the animals, including wildlife, who have longer life expectancies are mostly seasonal breeders. The wild animals have high vigour and a narrow spectrum of food habits, are mostly dependent on regional food sources for maintaining their energy balance, and become seasonal breeders.

19.4.2.3 Environmental Factors

The geographical location, season, and photoperiod are the important environmental determinants to regulate the onset of puberty. Animal of tropical region attains puberty earlier. Photoperiod is an important environmental cue that determines puberty onset in seasonal breeders (see Sect. 19.2.6.2 Photoperiod and HPG axis activation). The onset of puberty is determined by the birth timing in the seasonal breeders. The lambs born during spring attain puberty in the autumn of next breeding season, but the lambs born during autumn reach puberty 10–12 months later in the autumn of the next breeding season. The delay may be due to prolonged steroid feedback over the hypothalamus. In cattle and buffaloes, the long day of photoperiodic exposure increases body weight gain but hastens the pubertal onset. Animal of tropical region attains puberty later. The animals exposed to heat stress exhibit delayed puberty. Increasing day length (including artificial light) reduces the age of puberty and sexual maturity in birds.

19.4.2.4 Interaction with the Opposite Sex

The interaction with the opposite sex causes early puberty and maturity due to bio-stimulation by pheromone. Male pheromone can induce early puberty in cows, sheep, goats, and pigs. The introduction of males can induce LH surge and ovulation in a flock of anestrus sheep and goats during the non-breeding season. This phenomenon is called the male effect and is used widely to induce oestrus in these species.

19.4.2.5 Stress

Prolonged or chronic stress has a negative effect on the HPG axis, thus delaying the onset of puberty. Stress-induced HPG axis suppression is mediated through the HPA axis detailed in the previous chapter (Sect. 19.2.9 Stress-induced HPG axis suppression). Different disease conditions viz. TB, Johne's diseases, and FMD delay puberty.

19.4.3 Manipulation of Puberty

Age at puberty and sexual maturity are important economic traits in farm animal practices. Delayed puberty causes huge economic losses. The age of puberty in farm animals can be manipulated through genetic selection, nutritional management, improvement of the microenvironment through housing and other managemental practices, and hormonal interventions. The scrotal circumference is an important selection marker of males to get offspring with lower age of puberty. Crossbreeding between Zebu and exotic cattle is also recommended for the same reason. Balanced nutrition together with good husbandry practices helps to achieve better growth and early sexual maturity. Proteins and energy are the most important nutrients that influence the growth of animals. Adequate mineral supplementation is also required for optimum metabolic process of the body and improves the pubertal onset.

Learning Outcomes

- **Testes:** The primary sex organ of the male reproductive system is the testes which serve two important functions, viz. spermatogenesis and steroidogenesis. Spermatogenesis occurs within seminiferous tubules of the testes with the active support of Sertoli cells and Leydig cells. The Leydig cells are primarily concerned with steroidogenesis by modulating its function by containing steroidogenic enzymes such as cytochrome P450 oxidase (s) (P450scc). Functional features of both the cells are unique in an individual animal's pre-pubertal, adult and senile state. The specialized thermoregulatory system and blood–testis barriers help to create a congenial environment for spermatogenesis.
- **Excurrent tract:** The excurrent tract consists of rete testis, efferent ducts (vasa efferentia), epididymis, vas deferens, and urethra. It involves in maturation, storage, and passage of spermatozoa. The spermatozoa gain their maturity during their passage through the excurrent tract, particularly epididymis. The secretion of the excurrent tract contains different bioactive compounds that aid the maturation process. There are species differences concerning the morphological features of the excurrent tract. The avian excurrent tract is typically different from the mammals.
- **Accessory sex glands:** The secretions of accessory sex glands, viz. ampulla, seminal vesicles, prostate, and bulbourethral glands (Cowper's glands) are collectively called seminal plasma that acts as a vehicle for spermatozoa during its transport. The seminal plasma acts as a buffer and protects the spermatozoa from the harsh acidic environment of the female genital tract. The functional morphology of accessory sex glands varies between species and results in

(continued)

distinct properties of semen and spermatozoa between species.

- **Ancillary organs:** The penis and prepuce, two accessory structures of the male reproductive system involved in the ejaculation of semen into the female genital tract. The penis is a fibroelastic structure covered by a mucocutaneous tissue called the prepuce. The morphological features of the penis and prepuce vary between species, resulting in different ejaculatory features between species.
- **H-P-G axis:** The gonadotropin-releasing hormone (GnRH) of the hypothalamus stimulates the anterior pituitary to release follicle-stimulating hormone (FSH) and luteinising hormone (LH). The FSH and LH, in turn, control spermatogenesis and steroidogenesis. The H-P-G axis remains quiescent till puberty. The pubertal activation of the H-P-G axis is mediated by the nutritional factors and photoperiod with the involvement of neuroendocrine factors. The H-P-G axis is influenced by age, stress, environmental factors (season), and circadian rhythm.
- **Androgens:** The major androgens are testosterone, dihydrotestosterone (DHT), and androstenedione, out of which testosterone is most abundant in blood. The testosterone is synthesized in the Leydig cells and transported in conjugation with plasma proteins. After metabolism, it is excreted through faeces and urine. Testosterone is primarily involved in spermatogenesis, and it also helps to develop the secondary sex characteristics of a male.
- **Puberty:** Puberty is the age at which the male gonads are capable of releasing the gametes. In males, it is characterized by the development of secondary sexual characteristics ability to copulate and produce spermatozoa. The onset of puberty requires complex and integrated biological sequences that lead to progressive sexual maturation and full reproductive capacity through the activation of the H-P-G axis.

Exercises

Objective Questions

Q1. Which part of the testis shows exocrine activity, and which part has an endocrine role?

Q2. Extreme ambient temperature causes temporary infertility in certain species, termed as ______?

Q3. Which cell population has a direct physiological role in the seasonal breeding activity of males?

Q4. Which endocrine axis has 'auto-control' in the Sertoli cell function?

Q5. Which cells of the male reproductive system utilize vitamin A (retinol) in the spermatogenesis process?

Q6. What is the basis of the functional relationship between GnRH secretions with FSH and LH from gonadotropes?

Q7. In the hypothalamus, which neuroendocrine bio-molecule acts like the RFamide-Related Peptide (RFRP)?

Q8. When a male can produce its offspring successfully, it is called ____?

Q9. In any stress, which hormone reduces the responsiveness of the receptors for Luteinizing hormone in Leydig cell?

Q10. Which is the biologically most potent form of testosterone?

Q11. Which enzyme is responsible for converting testosterone to oestradiol?

Q12. Which is the major site of production of sex steroid-binding globulin (SHBG)?

Q13. Which forms of testosterone are mostly available in urine and faeces?

Q14. In which form of testosterone facilitates brain development?

Q15. Which peptide hormones have a role in epididymal migration?

Q16. Which part of the epididymis is mostly responsible for morphological and ionic changes of spermatozoa?

Q17. Fertility-associated proteins are released mostly from which part of the male reproductive system?

Q18. Which part of the excurrent tract acts as a common pathway for transmission of both urine and semen?

Q19. Which glandular secretions clean the passage of the reproductive tract before ejaculation of semen?

Q20. In a dog, which part of the penis favours locking during mating?

Subjective

Q1. Why ablations of the gene for INSL3 cause cryptorchidism?

Q2. Write the adoptive characteristics features of avian species in testicular thermoregulation.

Q3. Why is the blood–testis barrier significant to maintaining fertility?

Q4. Why can't the exogenous administration of testosterone alone influence the spermatogenesis process?

Q5. Write the functional differences between FLC and ALC.

Q6. Write the various factors that can control the HHG axis.

Q7. Write the factors affecting puberty.

Q8. How is obesity related to spermatogenesis?

Q9. Why does spermatogenesis initiate only after puberty?

Q10. Write the biological functions of testosterone.

Q11. Write the role of the epididymis in sperm maturation.
Q12. Write the role of seminal vesicles in semen production in a bull.
Q13. Write the role of specific biomarkers to access the prostate gland activity.
Q14. Describe the functional and morphological features of the penis during erection?
Q15. Write the events of erection and ejaculation in the bull?

Answer to Objective Questions

A1. The exocrine part is the seminiferous tubules, and the endocrine part is the interstitial or Leydig cells
A2. Summer sterility
A3. Sertoli cell
A4. Inhibin-activin-follistatin axis
A5. Peritubular myoid (PTM) cells
A6. Characteristics of pulsatile release of GnRH
A7. Gonadotropin-inhibiting hormone (GnIH)
A8. Sexual maturity
A9. Glucocorticoids
A10. Dihydrotestosterone (DHT or 5α-DHT)
A11. Aromatase
A12. Liver
A13. Glucuronidated or sulphated products of testosterone
A14. Oestradiol
A15. Oxytocin and vasopressin
A16. Body or corpus of the epididymis
A17. Epididymis
A18. Urethra
A19. Bulbourethral (Cowper's) glands
A20. Bulbus glands

Keywords for the Answer to Subjective Questions

A1. Testicular descent, Role of genetic factors, Cryptorchidism
A2. Testis position, period of spermatogenesis, genetical adaptability of avian spermatozoa
A3. Morphological structure of blood–testis barrier, micro-environment, damage of barrier, and development of immune response
A4. Mechanism of action of testosterone, role of Sertoli cells in spermatogenesis, the functional relationship between testosterone and Sertoli cells
A5. Steroidogenesis, organogenesis, relation with puberty
A6. Axis activators, axis suppressors, physiological modulation in various rhythms, including seasonal breeding
A7. Genetical variation, environmental variation, hormones involves in the HHG axis
A8. Functional relationship with obesity and HHG axis, role of oestrogen, leptin, and testosterone in spermatogenesis, leptin and puberty relationship
A9. Role of adult Leydig cells, activation of cytochrome P450 oxidases, role of steroidogenic acute regulatory protein (StAR)
A10. In foetal life, before puberty, after puberty
A11. Biochemical changes of seminal plasma, structural and functional changes in spermatozoa, secretory role of the epididymis
A12. Major secretory role of seminal vesicles, secretion of proteins, modulation of glandular function
A13. Relationship with various hormones and benign prostatic hyperplasia (BPH), prostate-specific antigen (PSA), prostatic acid phosphatase (PAP)
A14. Erectile tissues, penile erection, sigmoid flexure, and species variations
A15. The rigidity of the penis, emission, and expulsion, spiralling of the penis

Acknowledgement We are grateful to *Prof. Sagar Sanyal*, Ex. Professor and Head, and *Dr. Prabal Ranjan Ghosh*, Associate Professor, Department of Veterinary Physiology, West Bengal University of Animal and Fishery Sciences, Kolkata, West Bengal, India, for their valuable guidance and assistance during the preparation of the chapter. We are also greatly indebted to the help of *Dr. Parthib Roy*, Master Scholar, Department of Veterinary Gynaecology and Obstetrics, for the preparation of the figures.

Further Reading

Textbooks

Liverman CT, Blazer DG (2004) In: Testosterone and Aging: Clinical Research Directions. Institute of Medicine (US) Committee on Assessing the Need for Clinical Trials of Testosterone Replacement Therapy. Washington (DC): National Academies Press (US); pp 1–240. https://www.ncbi.nlm.nih.gov/books/NBK216175/

Melmed S, Polonsky KS, Larsen PR, Kronenberg HM (2017) Williams textbook of endocrinology, 13th edn. Elsevier, pp 1855–1916. https://doi.org/10.1016/C2013-0-15980-6

Noakes DE, Parkinson TJ, England GCW (2019) Veterinary reproduction and obstetrics, 10th edn. Elsevier, pp 1–837. https://doi.org/10.1016/C2014-0-04782-X

Research Articles

Bonneau M, Carrié-Lemoine J, Prunier A, Garnierm DH, Terqui M (1987) Age-related changes in plasma LH and testosterone concentration profiles and fat 5α-androstenone content in the young boar. Anim Reprod Sci 15(3–4):241–258

Byrne CJ, Fair S, English AM, Urh C, Sauerwein H, Crowe MA, Lonergan P, Kenny DA (2018) Plane of nutrition before and after 6 months of age in Holstein-Friesian bulls: II. Effects on metabolic and reproductive endocrinology and identification of physiological markers of puberty and sexual maturation. J Dairy Sci 101(4):3460–3475

DePalatis L, Moore J, Falvo RE (1978) Plasma concentrations of testosterone and LH in the male dog. J Reprod Fertil 52(2):201–207. https://doi.org/10.1530/jrf.0.0520201

Dunn JF, Nisula BC, Rodbard D (1981) Transport of steroid hormones: Binding of 21 endogenous steroids to both testosterone-binding globulin and corticosteroid-binding globulin in human plasma. J Clin Endocrinol Metab 53:58

Farooqui MM, Chandrapal A, Prakash A (2011) Anatomical study on the descent of testis in prenatal goat (*Capra hircus*). Int J Morphol 29(2):318–324

Frank LA, Rohrbach BW, Bailey EM, West JR, Oliver JW (2003) Steroid hormone concentration profiles in healthy intact and neutered dogs before and after cosyntropin administration. Domest Anim Endocrinol 24:43–57

Fustini M, Galeati G, Gabai G, Mammi LE, Bucci D, Baratta M, Accorsi PA, Formigoni A (2017) Overstocking dairy cows during the dry period affects dehydroepiandrosterone and cortisol secretion. J Dairy Sci 100(1):620–628

Georgie GC, Mehta SN, Dixit VP, Sengupta BP, Kanaujia AS (1985) Peripheral plasma testosterone levels in two Indian breeds of goats and their reciprocal crosses. Anim Reprod Sci 9(1):95–98. https://doi.org/10.1016/0378-4320(85)90045-4

Gervasi MG, Visconti PE (2017) Molecular changes and signaling events occurring in spermatozoa during epididymal maturation. Andrology 5(2):204–218. https://doi.org/10.1111/andr.12320

Hemeida NA, El-Baghdady YR, El-Fadaly MA (1985) Serum profiles of androstenedione, testosterone and LH from birth through puberty in buffalo bull calves. Reproduction 74(2):311–316. https://doi.org/10.1530/jrf.0.0740311

Inoue J, Cerbito WA, Oguri N, Matsuzawa T, Sato K (1993) Serum levels of testosterone and oestrogens in normal and infertile stallions. Int J Androl 16:155–158

Ivell R, Heng K, Anand-Ivell R (2014) Insulin-like factor 3 and the HPG axis in the male. Front Endocrinol 5(article 6):1–6. https://doi.org/10.3389/fendo.2014.00006

Kanchev LN, Dobson H (1976) Plasma concentration of androstenedione during the bovine oestrous cycle. J Androl 71(3):351–354. https://doi.org/10.1677/joe.0.0710351

Knych HK, Arthur RM, Stanley SD, McKemie DS (2014) Disposition and metabolic profile of the weak androgen Dehydroepiandrosterone (DHEA) following administration as part of a nutritional supplement to exercised horses. Drug Test Anal 7(1):39–47. https://doi.org/10.1002/dta.1716

Kotini M, Barriga EH, Leslie J, Gentzel M, Rauschenberger V, Schambony A, Mayor R (2018) Gap junction protein Connexin-43 is a direct transcriptional regulator of N-cadherin in vivo. Nat Commun 9:3846. https://doi.org/10.1038/s41467-018-06368-x

Malfatti A, Barbato O, Todini L, Terzano GM, Debenedetti A, Borghese A (2006) Blood testosterone levels in Italian Mediterranean buffalo bulls managed in two different breeding conditions. Theriogenology 65(6):1137–1144. https://doi.org/10.1016/j.theriogenology.2005.07.012

Mendel CM (1989) The free hormone hypothesis: A physiologically based mathematical model. Endocr Rev 10:232

Meroni SB, Galardo MN, Rindone G, Gorga A, Riera MF, Cigorraga SB (2019) Molecular mechanisms and signalling pathways involved in Sertoli cell proliferation. Front Endocrinol 10:224. https://doi.org/10.3389/fendo.2019.00224

Nicolas D, Thierry DC, Christian L, d'Hellencourt DC, Vance LT, Joel NC, Olivier M, Olivier K, Elisabeth P (2018) Steroid transport, local synthesis, and signaling within the brain: roles in neurogenesis, neuroprotection, and sexual behaviors. Front Neurol 12:84. https://doi.org/10.3389/fnins.2018.00084

Roselli CE, Stormshak F, Stellflug JN, Resko JA (2002) Relationship of serum testosterone concentrations to mate preferences in rams. Biol Reprod 67:263–268

Tagliaferro AR, Ronan AM (2001) Physiological levels and action of dehydroepiandrosterone in Yucatan miniature swine. Am J Phys Regul Integr Comp Phys 281(1):R1–R9

Takikawa M, Wakabayashi K (1994) Quantitative analysis of hypothalamic-hypophyseal-testicular system: Why testosterone can act under negative feedback control. Endocr J 41:257–265

Vecchione MB, Eiras J, Suarez GV, Angerami MT, Marquez C, Sued O, Ben G, Pérez HM et al (2018) Determination of dehydroepiandrosterone and its biologically active oxygenated metabolites in human plasma evinces a hormonal imbalance during HIV-TB coinfection. Sci Rep 8:6692. https://doi.org/10.1038/s41598-018-24771-8

Villaverde AISB, Fioratti EG, Ramos RS, Ferreira JCP, Balieiro JCC, Lopes MD (2010) Relationship of testosterone concentration in blood and seminal plasma to sperm production in domestic cat. Biol Reprod 83(Suppl. 1):536. https://doi.org/10.1093/biolreprod/83.s1.536

Woźniak B, Witek S, Matraszek-Żuchowska I, Kłopot A, Posyniak A (2016) Levels of the natural hormones 17β-oestradiol and testosterone in serum of cattle: results from population studies in Poland. J Vet Res 60:461–466. https://doi.org/10.1515/jvetres-2016-0055

Spermatogenesis and Semen

20

Pradip Kumar Das, Joydip Mukherjee, and Dipak Banerjee

Abstract

The spermatozoa are the male gamete produced in the seminiferous tubule through spermatogenesis. The spermatozoa are derived from germinal epitheliums that undergo a series of mitotic divisions followed by meiotic divisions to produce early stages of haploid gamete called spermatids (spermatocytogenesis). The spermatids undergo differentiation to transform spermatozoa (spermiogenesis). The Sertoli cells play a pivotal role in spermiogenesis by providing nutrients to the spermatozoa. The spermatozoa are released from the Sertoli cells through spermiation. The spermatozoa released from the Sertoli cells are rarely motile without any fertilizing capability. During epididymal transit, the spermatozoa acquire some proteins required for protection against reactive oxygen species (ROS), gaining progressive motility and fertilizing ability. The spermatozoa are then mixed with the secretion of the accessory sex glands known as seminal plasma at the pelvic urethra and expelled out through a synchronized physiological event called ejaculation. The chapter encompasses the process of spermatogenesis and sperm maturation during epididymal transit, the formation of seminal plasma and its composition, the neuro-endocrine regulation of ejaculation and physio-biochemical properties of semen, including different biomarkers of sperm maturity divided into two subchapters, spermatogenesis and semen.

P. K. Das (✉) · J. Mukherjee · D. Banerjee
Department of Veterinary Physiology, West Bengal University of Animal & Fishery Sciences, Kolkata, West Bengal, India

© The Author(s), under exclusive license to Springer Nature Singapore Pte Ltd. 2023

P. K. Das et al. (eds.), *Textbook of Veterinary Physiology*, https://doi.org/10.1007/978-981-19-9410-4_20

Graphical Abstract

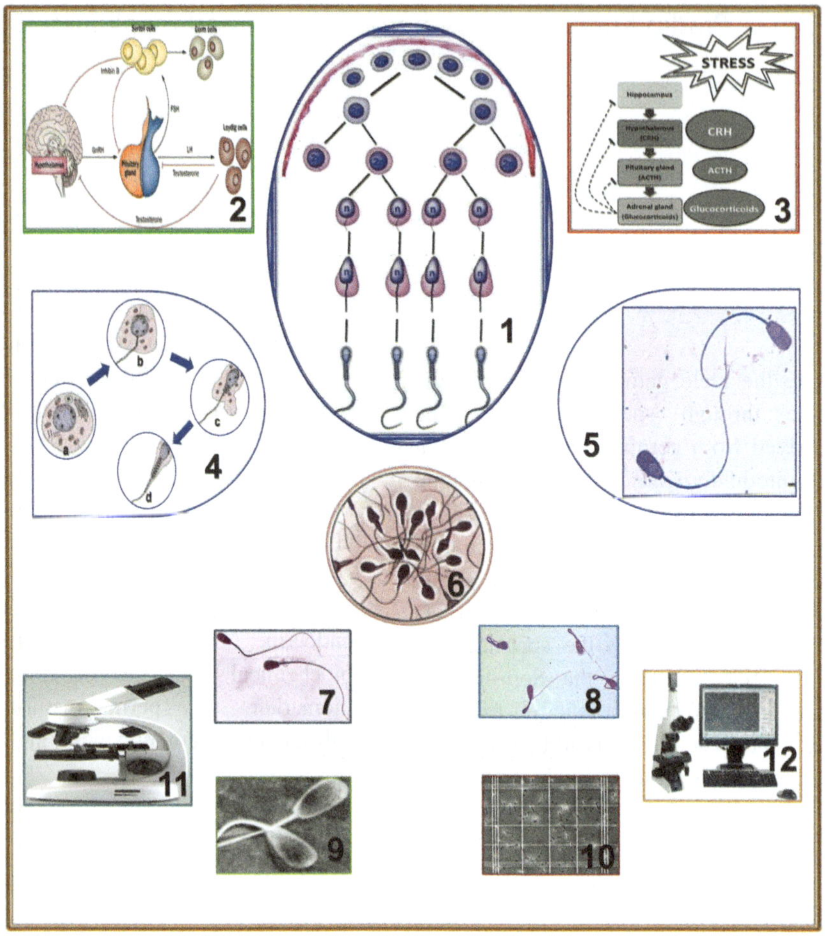

Description of the graphic: The spermatogenesis involves the synchronized process of cell division (**1**) through which the spermatogonium (stem cells) is developed into a haploid male gamete, spermatozoa. The entire process is controlled by endocrine factors associated with the hypothalamic–pituitary–gonadal (HPG) axis (**2**). Various stressors affect spermatogenesis mainly through the hypothalamic–pituitary–adrenal axis (**3**). Through spermiogenesis (**4**), mature spermatozoa have distinct head, neck, mid-piece, and tail (**5**) with progressive motility. The evaluation of spermatozoa viz. sperm motility (**6**), the proportion of morphologically normal (**7**) and abnormal (**8**) spermatozoa, and sperm concentration (**10**) are the essential pre-requisites to screen the semen for successful fertilization using various conventional techniques as well as (**11**) advanced computer-assisted semen analyser (**12**).

Keywords

Spermatogenesis · Epididymal transit · Ejaculation · Semen · Spermatogenic efficacy

Learning Objectives

- Mechanism of spermatogenesis and the factors affecting it
- Epididymal transit of spermatozoa and its physio-biochemical alterations to gain motility and fertilizing capability
- Formation of seminal plasma
- Ejaculation and its associated events
- Physio-biochemical composition of semen

20.1 Spermatogenesis

Spermatogenesis is a complex and synchronized process of cell division and differentiation of germ cells resulting in potential motile haploid male gamete, spermatozoa, in the seminiferous tubule (Fig. 20.1). The diploid germ cells (spermatogonia) undergo mitotic and meiotic divisions to

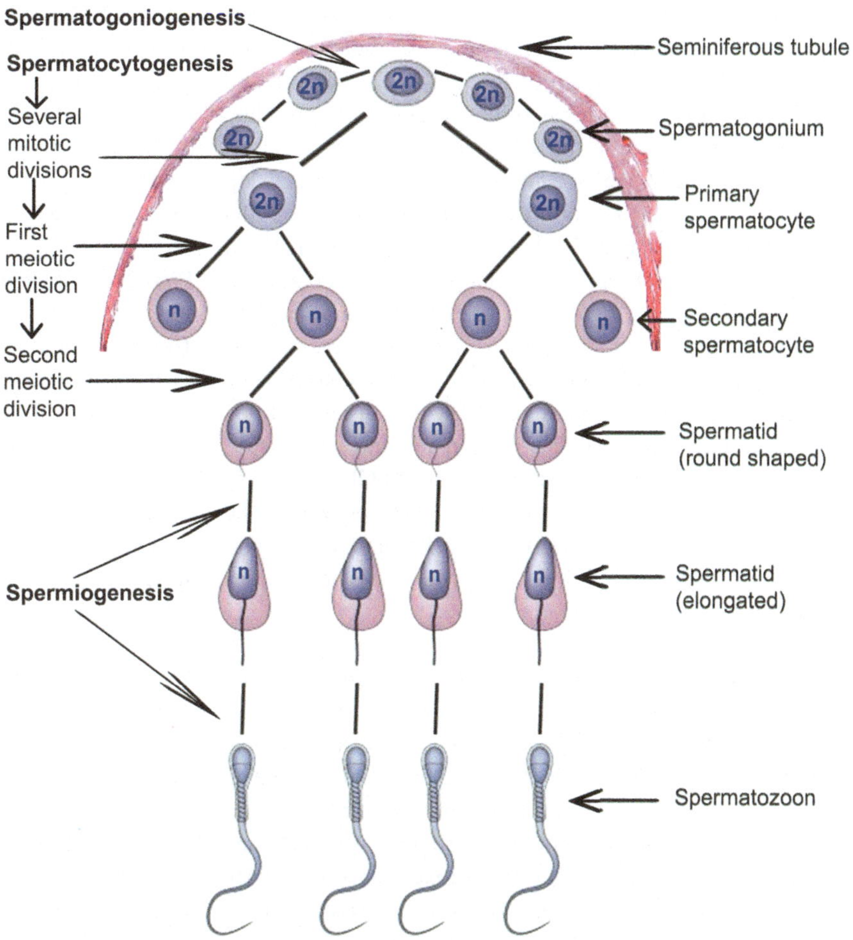

Fig. 20.1 Spermatogenesis. [The **spermatocytogenesis** involves both mitotic and meiotic divisions. In mitotic divisions, the **spermatogonium** is transformed into **primary spermatocytes** through **spermatogenesis**. The formation of **secondary spermatocytes** and **spermatids** occurs through **first** and **second meiotic divisions** under spermatocytogenesis. In **spermiogenesis**, the **rounded** spermatids are transformed into **elongated** spermatids and, finally, spermatozoa (Sing: **spermatozoon**) through morphological changes. The diploid cells are presented as '2n' and haploid as 'n']

form spermatids. The spermatids undergo metamorphosis to become mature spermatozoa and release into the tubular lumen.

Spermatogenesis can be divided into three main stages: *spermatocytogenesis*, *spermiogenesis*, and *spermiation*. Spermatocytogenesis involves the cycles of both mitotic and meiotic cell divisions, resulting in the formation of spermatids from spermatogonia. The mitotic divisions of spermatocytogenesis have a dual role; firstly, it helps the renew stem cells to produce spermatogonia and maintain a steady pool. Secondly, the spermatogonia become primary spermatocytes to produce mature spermatozoa. The meiotic cycles of spermatocytogenesis involve two cell divisions that reduce the chromosome number to yield haploid round spermatids. The spermatids then undergo metamorphic changes to become mature spermatozoa through spermiogenesis. In spermiation, the mature spermatozoa are released into the lumen of seminiferous tubules from the germinal epithelium.

The seminiferous tubule contains a variety of cells involving different stages of spermatogenesis. Thin cytoplasmic bridges connect all these cells until spermatozoa are formed. The spermatogenesis is under the endocrine control of testosterone, follicle-stimulating hormone (FSH), and oestrogen.

20.1.1 Spermatocytogenesis

Spermatocytogenesis is a proliferative phase in which many germinal epithelial cells are multiplied through a series of mitotic divisions followed by meiotic divisions to produce early stages of haploid gamete (Fig. 20.1). The spermatocytogenesis starts before puberty in most animals and birds, and the time required to complete the mitotic division is generally 21 days in the bull.

20.1.1.1 Mitotic Divisions

Development of Spermatogonial Stem Cells (SSCs) The SSCs are the undifferentiated spermatogonia developed from primordial germ cells (PGCs). SSCs have the capabilities of self-renewal and can convert into pluripotent stem cells. Further, they can transfer the genome from one generation to the next, and they are the only adult stem cells of this kind. The PGCs migrate to the genital ridge and are differentiated into gonocytes during 10.5 days post-fertilization in mice. The replications of gonocytes are arrested at the G0/G1 stage of the cell cycle. The mitotic capability of the gonocytes is resumed after birth, and they start migrating centre to the periphery of the seminiferous tubule and reach the basement membrane to develop spermatogonia. The transformation of gonocytes into the spermatogonia required 2 months after birth in pigs and goats, 3 months in sheep, and 4 months in cattle. A particular microenvironment is essential for the survival and development of SSCs. It is called 'niche', made by Sertoli cells, peritubular myoid cells and Leydig cells. Sertoli cells provide nutrients to the SSCs and germ cells. The Sertoli cells derived growth factors (glial cell line-derived neurotrophic factor, GDNF and basic fibroblast growth factor, bFGF) are essentially required for the self-renewal property of SSCs. The Leydig cells and peritubular myoid cells produce colony-stimulating factor 1 (CSF1), which potentiates the actions of GDNF.

Proliferation of Spermatogonia The spermatogonia are classified into two broad categories, type A and type B spermatogonia. The type A spermatogonia has a prominent round nucleus with condensed chromatin and peripheral nucleoli. It may also contain a nuclear vacuole. Cytologically, type A spermatogonium has two types, Ad (dark) and Ap (pale) spermatogonia. The Ad spermatogonia don't have proliferative capabilities under normal circumstances. They are considered the testicular stem cells, but when the spermatogonial concentration is severely reduced (as in radiation), the Ad spermatozoa may show proliferative capability.

In contrast, the Ap spermatogonia have the proliferative and self-renewal capabilities. Several types A spermatogonia remain as resting type A spermatogonia (intermediate spermatogonia), which later differentiated into type B spermatogonia by mitotic cell division. The type B spermatogonia contain central nucleoli with dispersed chromatin and no nuclear vacuole. The type B spermatogonia undergo mitotic division to form primary spermatocytes. Generally, one type A spermatogonium can produce 12 primary spermatocytes. The primary spermatocytes then enter meiotic divisions to develop spermatozoa.

The spermatogonial subpopulations vary with species and breeding seasons. Three types of spermatogonium have been found in bull, type A spermatogonium, intermediate spermatogonium, and type B spermatogonium. In stallions, A_1, A_2, A_3, B_1, and B_2 spermatogonia are found. Usually, type A spermatogonia are more available during the non-breeding season, whereas type B spermatogonia are abundant in the breeding season.

20.1.1.2 Meiotic Divisions

The diploid primary spermatocytes are larger than spermatogonia and contain highly condensed chromosomes with coarse chromatin. The primary spermatocytes then cross the blood–testis barrier and enter the meiotic divisions. They undergo two meiotic divisions to form spermatids and require 23 days in the bull.

First Meiotic Division In the first meiotic division, primary spermatocytes generate haploid secondary spermatocytes. The first mitotic division is also called reductional division, as there is the reduction of chromosome number and the separation of homologous chromosomes. The reshuffling of genetic material also occurs during this stage. The first meiotic division is testosterone dependent. Hence, before puberty, secondary spermatocytes are not developed. In hypophysectomized animals, the presence of secondary spermatocytes is also limited. Secondary spermatocytes are spherical cells with interphase nuclei.

Second Meiotic Division It is called equational division, in which the separation of the daughter chromatids is occurred to form haploid spermatids from secondary spermatocytes. The spermatids are round and mitotically inactive. The second meiotic division is testosterone independent and requires less time. Spermatids are transformed into spermatozoa through metamorphic changes.

20.1.1.3 Immune Protection of Spermatids

The spermatids are autoimmunogenic as recombination of genetic materials occurs during their production through meiosis. Due to the recombination of genetic materials, the haploid gamete may differ from the parent somatic cells and are susceptible to auto-immune attack. The blood–testis barrier (BTB) provides an immune-privileged environment to the spermatids for their survival by eliminating the somatic parent cells' humoral or cellular immune components. The temperature of the testes is also an essential criterion for meiosis as DNA polymerase and recombinase enzymes require a particular temperature for their action. Therefore, the descent of descending testes and its thermoregulation favour spermatogenesis.

20.1.2 Spermiogenesis

The transformation process of spermatids into spermatozoa is called spermiogenesis (Figs. 20.1 and 20.2). It is a metamorphic process characterized by nuclear condensation and structural shaping, flagellum formation, and cytoplasm expulsion. The entire process occurs within the cytoplasm

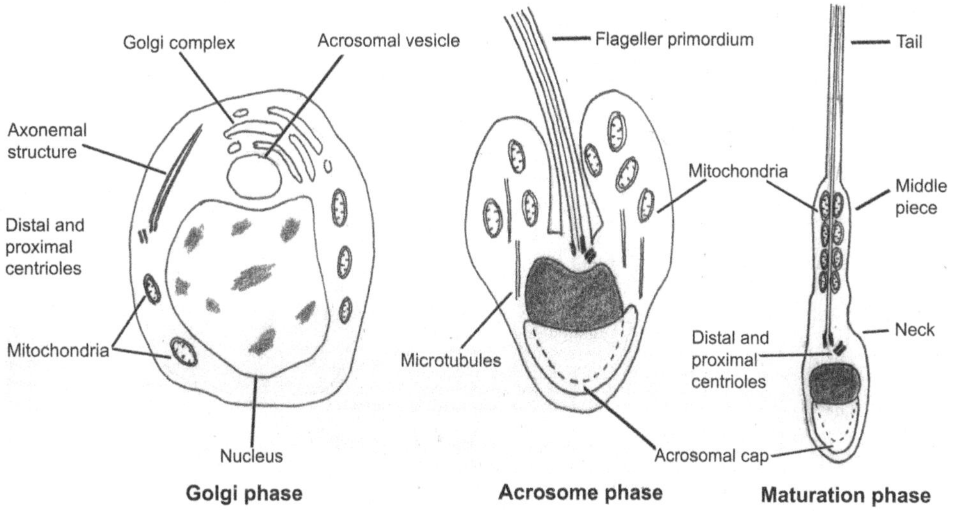

Fig. 20.2 Spermiogenesis. [Three distinct phases of spermiogenesis, viz. **Golgi phase**, **Acrosomal phase** or cap phase, and **Maturation phase** are presented where structural changes of various organelles occur]

of the Sertoli cells under the influence of FSH. The Sertoli cells provide the nutrients, enzymes, hormones, and other substances required for spermiogenesis (the role of the Sertoli cells in spermatogenesis has been discussed in detail in an earlier chapter). It commonly takes 17 days in the bull. The entire spermiogenesis process is usually divided into four phases: Golgi phase, cap phase, acrosomal phase, and maturation phase.

20.1.2.1 Golgi Phase

In this phase, the nucleus is compressed with tightly packed chromatin. The nucleus is transcriptionally inactive. The structure of the nucleus occurs with one side oval and the opposite one narrower. In birds, the nucleus is long or cylindrical. It is spiraled in the passerine group of birds. The granules of the Golgi vesicles of spermatids are merged at the oval side to form a pro-acrosomal vesicle. The centrioles and mitochondria migrate to a position opposite to acrosomal vesicles. The nucleus and acrosome together form the head of the sperm. The proximal centriole gives rise to the attachment point for the tail, and the distal centriole gives rise to the developing axoneme. The mitochondria combine to form a mid-piece. The structure axoneme, composed of a group of microtubules, is initiated to develop. The spermatid becomes extended, and the elongation process begins. At the end of this phase, the elongated part is turned into the mid-piece of the sperm. The mitochondria are helical and more elongated than non-passerine birds in passerine birds.

20.1.2.2 Cap Phase

The acrosomal vesicle flattens to form a distinct cap to cover almost half of the nucleus. It is conical or tapering shaped in the bird. The acrosomal vesicle is made of an outer acrosomal membrane and an inner acrosomal membrane. It contains lysosomal enzymes like hyaluronidase and proteases that help in fertilization. The distal centriole forms the axoneme or flagellum that projects away from the nucleus to the lumen of the seminiferous tubule (Fig. 20.2).

The acrosomal vesicle covers almost half of the nucleus, resulting in the head cap of the cell (spermatozoa). The acrosome acts like a lysosome and contains lysosomal enzymes, like hyaluronidase and proteases, which have a vital role in fertilization.

20.1.2.3 Acrosomal Phase

The nucleus of the spermatid begins to elongate, and the acrosome covers the majority of the anterior nucleus. A unique microtubular system called *manchette* is extended from the posterior portion of the nucleus. It contains a bunch of nine peripheral double microtubules and two single tubules in the centre. Some of the microtubules are developed into the post-nuclear cap. The tail is covered with the extension of the cell membrane. During the acrosomal phase, spermatids are deeply embedded in the Sertoli cells, with their tails protruding towards the lumen of the seminiferous tubule.

20.1.2.4 Maturation Phase

In this phase, the manchette migrates towards the tail and begins to disappear. Mitochondria are assembled around the flagellum to form mid-piece. The mid-piece portion contains a large amount of adenosine triphosphate (ATP) to provide energy to the sperm. The manchette microtubules form the post-nuclear cap. The dense outer fibres cover the flagellum. It contains a bunch of nine peripheral double microtubules and two single tubules in the centre. The junction between the middle piece and principal piece annulus is formed. The spermatids are elongated and extrude the rest of the cytoplasm in the form of a spheroidal lobule called residual body. The residual bodies are phagocytosed by the Sertoli cells. The secretory activity of the Sertoli cells (inhibin, ABP and interleukin-1 and 6) depends upon the elongation of

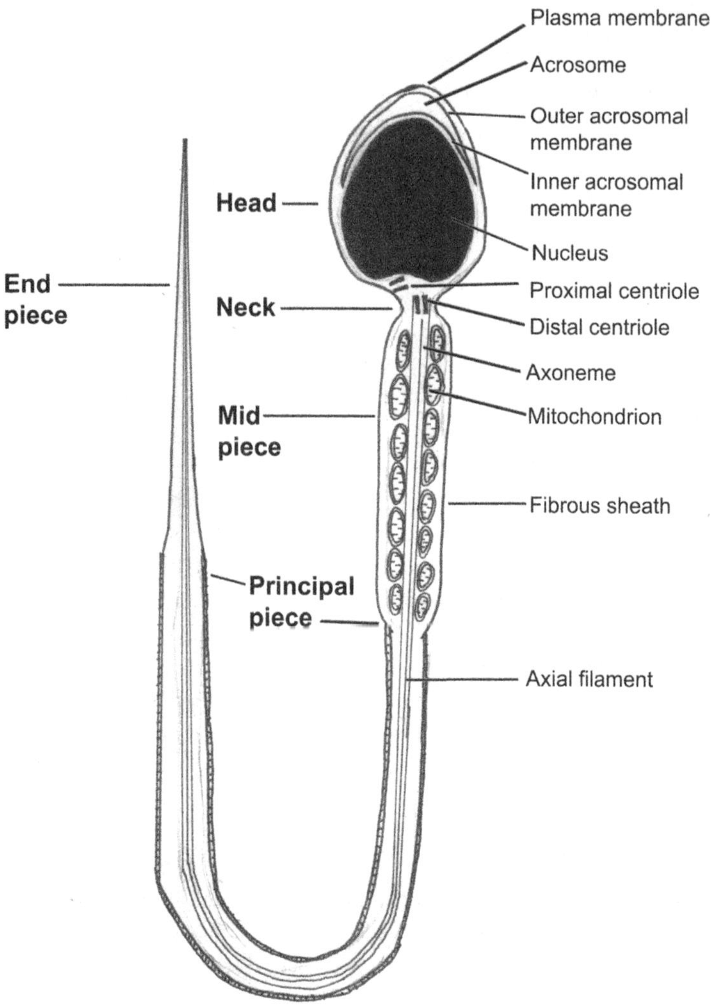

Fig. 20.3 Structure of fully developed spermatozoa

spermatids and residual bodies. A new spermatogenic cycle begins after the degradation of residual bodies.

The remnant cytoplasm (after phagocytosed by the Sertoli cells) adheres at the neck region of the elongating spermatid in the form of a cytoplasmic droplet (CD). The mammalian CD is about 2 μm in diameter and composed of organelle-derived membranes and cytosol. The CDs are removed from the spermatozoa during its passage through epididymis. They are generally seen at the neck in the immature sperm of caput epididymis and migrate further, and retain at the end of mid-piece in the sperm in cauda epididymis. Their migration along the mid-piece indicates sperm maturation, and the retention of CD at the mid-piece is one of the vital sperm abnormalities. Phospholipid binding protein (PBP) present in the ampulla and seminal vesicles of the bull helps in the removal of CD. PBP is generally absent in accessory sex glands of boar; hence the spermatozoa of porcine ejaculate contain more CD than other species. The enzyme 15-lipoxygenase (15LOX) and the ubiquitin-proteasome help degrade cytoplasmic droplets. The CD also helps in sperm volume adaptation and protects the spermatozoa from the hypotonic challenge. There is a relationship between the presence of CD and the progressive motility of the spermatozoa. The sperm possess CD display progressive motility, and the sperm without CDs are mostly non-motile. The relationship between CD and motility can explain because CDs act as mitochondrial modulators and potentiate mitochondrial activity and membrane potential. Despite several beneficial roles, the sperm cytoplasmic droplet remains an enigma as the retention of CD in the ejaculated sperm is associated with infertility. The sperm with CD have poor binding capabilities with zona pellucida and a reduced pregnancy rate. More recently, the CDs are considered the normal morphological occurrence in the human spermatozoa and the terminology 'excess residual cytoplasm' is designated as abnormal. The spermatozoa with 'excess residual cytoplasm' can now be considered immature spermatozoa without terminal differentiation.

The fully matured spermatozoa (Fig. 20.3) morphologically contain five major components with various cellular organelles (Table 20.1). The head of spermatozoa has a shape characteristic of different species. The spermatozoa of

Table 20.1 Spermatozoa and its organelles with major functions

Parts of spermatozoa	Organelles	Major function
Head	Acrosome (lysosome) and nucleus	Acrosomal enzymes involve in fertilization, and nuclear chromatin contains the genetic materials
Neck	Centrioles (2 Nos.)	Attach the tail and mid-piece with head
Mid-piece	A sheath of ring-shaped mitochondria wrapped the axoneme	Provide the energy for the flagellar movement
Principal piece	A sheath of ring fibbers enveloped the axoneme	Give support to the tail for movement
Tail	The 9 + 2 microtubules structure of the axoneme, covered with the plasma membrane	Provide thrust for forwarding movement

bulls and humans have paddle-shaped heads. The spermatozoa of rodents have hook-shaped heads. The head contains an oval and flattened nucleus in which a nuclear membrane surrounds compact chromatin. The acrosome covers the anterior 2/3rd of the nucleus. The acrosome has an outer and inner acrosomal membrane. Hydrolytic enzymes like acrosin, hyaluronidase, zonalysin, esterase, and acid hydrolases are present within the acrosome. These enzymes are resealed during the acrosomal reaction and help to penetrate the zona pellucida of the ovum during fertilization. The tail is made of the capitulum, mid-piece, principal, and terminal pieces. The capitulum lies at a depression in the posterior nucleus called the implantation socket. The anterior portion of the tail contains laminated columns that aid the flexibility of the neck. The axonemal components of the tail are made of 9 pairs of microtubules arranged radially around the central filaments. There are nine dense fibres surrounding the microtubular arrangements at the flagellum of spermatozoa. The mid-piece is composed of a mitochondrial sheath arranged in a helical pattern. The mid-piece contains a large amount of adenosine triphosphate (ATP) to provide energy to the sperm. The annulus is the junction between mid-piece and principal piece. The principal piece makes the major portion of the tail and connects with the terminal piece.

20.1.3 Spermiation

After the completion of spermiogenesis, the spermatozoa are released from the Sertoli cell into the lumen of the seminiferous tubules through a process called spermiation. The spermatozoa are arranged perpendicularly to the tubular wall and gradually expelled out into the lumen of the tubule. The release of spermatozoa is facilitated due to the attenuation of the slender cytoplasmic stalk connecting the spermatids with the residual body. Once the spermatids separate from the residual body, it becomes spermatozoa. The breakage of the stalk results in proximal CDs in the neck region. The residual bodies are degraded by Sertoli cells.

20.1.4 The Final Maturation of Spermatozoa

The testicular spermatozoa are immature and non-motile. They gain motility and fertilizing capability during their transit through the epididymis. The epididymis cells have high metabolic, secretory, and endocytic activity regulated by androgens which facilitate the absorption of fluids, and secretion of ions, antioxidants, and proteins.

20.1.4.1 Epididymal Transit of Spermatozoa and Storage

The sperm move from the head of the epididymis towards the tail by the hydrostatic pressure gradient. Most mammals generally use transit time for 1–2 weeks (Table 20.2). The peristaltic contraction of the epididymis facilitates the sperm transit, which is controlled by testosterone, oxytocin, vasopressin, and PGF2α. Frequent ejaculation favours sperm transit. In bull, daily ejaculation can reduce the transit time by up to 3 days and increase sperm concentration. The rate of movement is highest at the head, followed by the body and

Table 20.2 Transit time of the spermatozoa from caput to cauda in epididymis

Animal	Epididymal migration time (day)	Reference
Bull	8–14	Robaire et al. (2006)
Buffalo bull	6–8	Bhakat et al. (2015)
Ram	12.5; 14	Lino (1972); Robaire et al. (2006)
Boar	9–14	Briz et al. (1995)
Stallion	8–10	Varner (2015)
Dog	10	Olar et al. (1983)
Rat buck	4	Kempinas and Klinefelter (2015)
Mice buck	9–10	Robaire et al. (2006)
Rabbit buck	9–10	Swierstra and Foote (1965)
Human (male)	10–12	Robaire et al. (2006)

tail. In bull, it is 420 mm/2 h at the head, 64 mm/2 h at the body, and 25 mm/2 h in the cauda epididymis and vas deferens. The lowest speed at the tail of the epididymis increases the transit time and favours sperm storage. In the tail of epididymis, 50–80% of total spermatozoa are stored, which can be sufficient for ten successive ejaculations in stallions and bulls. At the cauda of the epididymis, the spermatozoa remain quiescent, and three to fivefolds can increase their metabolic activity upon ejaculation. The exact mechanism of this metabolic quiescence is unknown; however, the presence of specific enzymes, proteins, and luminal pH may be the contributing factors. The epididymal spermatozoa can be viable up to 2–3 weeks in most mammals, and the unutilized spermatozoa are released into the urethra and excreted through urine.

20.1.4.2 Morphological and Biochemical Changes of Spermatozoa During Epididymal Transit

The spermatozoa undergo many morphological, biochemical, and functional changes during epididymal transit necessary for successful fertilization. The major changes are summarized in Table 20.3.

The sperm released from the head of the epididymis are rarely motile and gain a progressive motility pattern during their epididymal transit (Table 20.4). The factors that promote this progressive motility are forward motility protein (FMP), the elevation of cyclic adenosine monophosphate (cAMP), and increased sperm pH. FMP alters the permeability of the sperm plasma membrane and allows the influx of Ca^{++} inside the sperm. Cyclic adenosine monophosphate (cAMP) activates cAMP-dependent protein kinases. These cAMP-dependent protein kinases, in turn, activate phosphoprotein phosphatase to phosphorylate multiple intra sperm phosphoproteins. The phosphorylation of intra-sperm proteins initiates flagellar movement, and sperm gain their motility.

20.1.4.3 Epididymal Proteins and Their Role in Sperm Maturation

During epididymal transit, the spermatozoa acquire some proteins required for protection against reactive oxygen species (ROS), gaining progressive motility and fertilizing capability. The proteins are thought to be transferred from the epididymis through some membranous vesicular structure called epididymosomes. These epididymosomes transfer

Table 20.3 Morphological and biochemical changes of spermatozoa during epididymal transit

Maturational changes		Effect
Morphological	Narrowing of the sperm acrosome	Favours the sperm motility
	Migration of cytoplasmic droplets	
	Changes in the cytoskeletal structure	
	Alterations in membrane fluidity	Protects the sperm from hypoosmotic stress
Biochemical	Formation of disulphide cross-links between protamine molecules	Chromatin condensation
	Increases in negative surface charges	Prevents sperm aggregation during storage and non-specific binding to the female reproductive tract
	Reduction in the membrane lipids	Utilization as an energy source
	Relocalization of surface antigens	Acquisition of forwarding motility Facilitate sperm-egg binding and improves fertilizing capability
	Increase in intra-sperm cAMP	Activates cAMP-dependent protein kinases to regulate flagellar motility
	Increase in sperm pH	Favours motility
	Activation of phosphoprotein phosphatase	Phosphorylation of intra-sperm phosphoproteins to regulate sperm motility
	Increase in intracellular calcium	Favours motility
	Alterations in the ionic composition (Na+, K+, Cl-),	Favours motility

Table 20.4 Spermatozoal characteristics in different parts of the epididymis in bulls

Parts of epididymis	Sperm concentration	Sperm characteristics
Head (caput)	8–25 × 10^9	Non-motile, non-fertile, proximal cytoplasmic droplet, low disulphide linking, high cholesterol to phospholipid ratio, increase in total surface negative charges, increase in sialic acid residues, increase in membrane fluidity, increase tRNA fragments
Body (corpus)	8–25 × 10^9	Negligible expression of motility, little fertilizing capability, translocation of the cytoplasmic droplet, moderate to high disulphide linking, ability to bind with oocytes
Tail (cauda)	10–50 × 10^9	Expression of normal motility, expression of fertilizing capability, high disulphide linking, distal cytoplasmic droplet, ability to bind with oocytes

Table 20.5 Role of epididymal proteins in sperm maturation

Name of the proteins	Functions
Oestrogen sulfotransferase (EST)	Helps in the sperm cholesterol metabolism during the maturation process
Macrophage inhibitor factor (MIF)	Modulation of the beating of sperm flagella and motility
Murine sperm adhesion molecule 1 (SPAM1)/PH-20	Helps in sperm–egg interaction
Glutathione peroxidase (GPX5)	Protects the sperm from ROS
P26h (hamster), P25b (bovine)	Helps in sperm–zona pellucida binding
Forward motility protein (FMP)	Alters the permeability of sperm plasma membrane and favours calcium influx
Anti-sticking factor (ASF)	Sperm surface glycoprotein helps to prevent sperm aggregation
Quiescence factor (QF)	Helps to maintain the quiescence state of spermatozoa to save energy
Immobilin	Immobilizes the spermatozoa till ejaculation by creating a viscoelastic environment
Taurine	Protects the spermatozoa from the harmful xenobiotics and reactive oxygen species (ROS)

selected proteins into the spermatozoa in an apocrine manner without fusing the sperm plasma membrane. Different proteins present in the epididymosomes and their roles are summarized in Table 20.5.

20.1.5 Spermatogenic Cycle

Upon examining the cross-sections of a seminiferous tubule, it may find definite cellular associations among the sperm forming cells as the seminiferous germinal epithelium cells (type A spermatogonia, the stem cells) are continuously regenerated by mitotic division to replenish the parental cell in a cyclic process. Each such spermatogenic association is designated as a stage of the seminiferous epithelial cycle. The time required to reappear in the same stage within a given seminiferous tubular segment is called the spermatogenic cycle (Fig. 20.4).

The cycle has two key features: firstly, the new spermatogonia start their divisions at a constant time interval

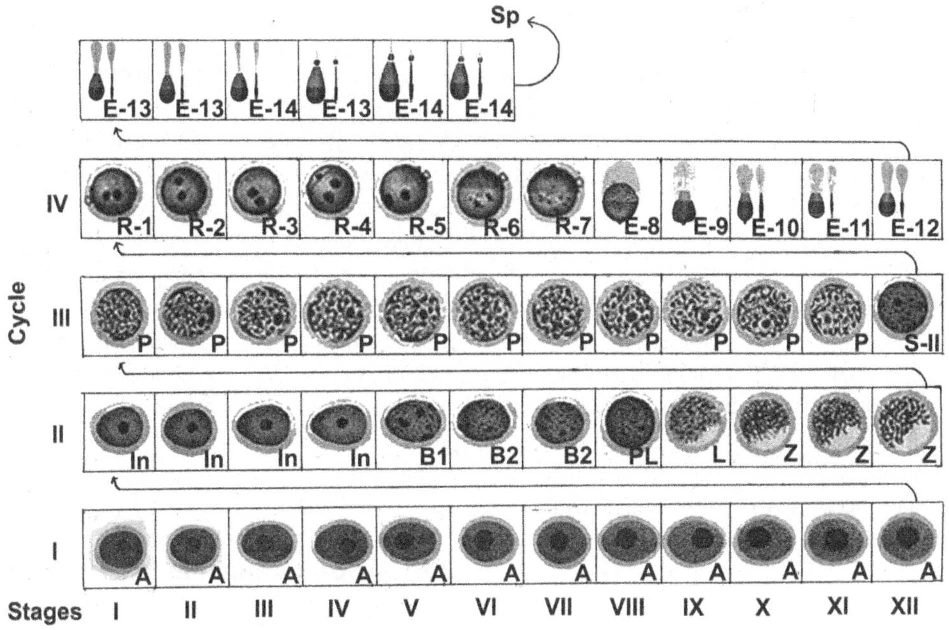

Fig. 20.4 Spermatogenic cycle of bull. [Figure showing the germ cell differentiation and progression from type A spermatogonia to spermatozoa in the 12 stages (written horizontally in roman I to XII) of the seminiferous epithelium cycle (presented vertically in roman I to IV) based on the development of the acrosome during spermiogenesis. Each stage can be found within the same transverse section of the seminiferous tubule. The duration of each cycle is about 13.5 days. Around 4.5 cycles are required to develop a spermatozoon from a spermatogonium. So the total duration of spermatogenesis is about 4.5 x 13.5 days = 61 days. The cell progression is continued from the right side in each row (cycle). The specific cell type in each cycle for a particular stage (column) is presented by alphabet(s) like, A = type A spermatogonia, In = intermediate spermatogonia, B1 = type B1 spermatogonium, B2 = type B2spermatogonium, PL = preleptotene spermatocyte, L = leptotene spermatocyte, Z = zygotene spermatocyte, P = pachytene spermatocyte, S-II = secondary spermatocyte, R-1 to R-7 = round spermatids and E-8 to E-14 = elongating/elongated spermatids, and Sp = Spermatozoon. Duration of mitotic proliferation continued for 21 days (all the stages of A to P), meiosis (S-II) continued for 23 days, and spermiogenesis (R-1 to E-14) for 17 days.] [**Source:** Segatelli et al. 2013; Staub and Johnson 2018]

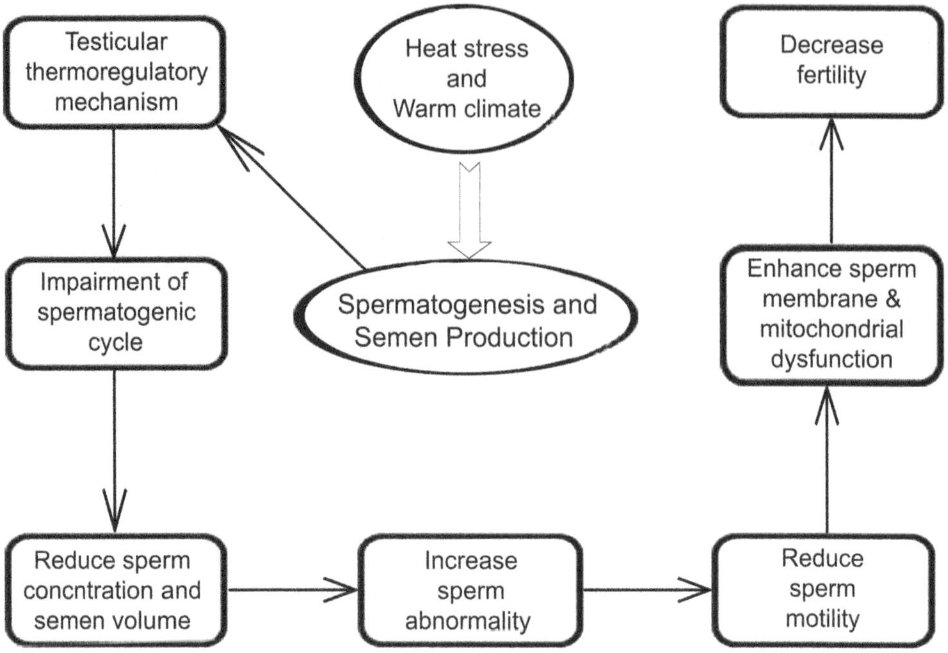

Fig. 20.5 Effect of heat stress or warm climate in spermatogenesis and semen production

at one point of the tubule. Secondly, the rate of germinal cell differentiation is always the same with a fixed duration. Therefore, the duration of a cycle is fixed and species-specific (Table 20.5). In a cycle, diverse developmental stages of spermatozoa are found in different regions of the seminiferous tubule. In a particular segment of the tubule, the recurrence of the same stage of the stem cell occurs. Various stages of spermatozoa are progressed cyclically. This process is continued longitudinally from the base of the seminiferous tubule towards the lumen.

20.1.6 Spermatogenic Wave

The spermatogenic stages are highly coordinated not only with time but also in space. The serial transverse section of the seminiferous tubule exhibits that stage I is followed by II and stage III is by stage IV, and so on. The spermatogenic wave is the distance between the same stages within the seminiferous tubule, and one tubule may have several complete waves.

20.1.7 Duration of Spermatogenesis

The time to complete a spermatogenic cycle varies between species and depends upon the number of spermatogenic stages in a cycle. In bull and mouse, one cycle consists of 12 different stages of cells; a rat has 14 different stages of cells in a cycle. To complete the entire spermatogenesis process for generating a single spermatozoon, a $4^1/_2$ cycle (3.9–4.7) is usually required in almost all domesticated mammals. The duration of spermatogenesis varies within 30–75 days in all domestic mammals (Table 20.6).

20.1.8 Germ Cell Degeneration

All the cells that are generated during various stages of spermatogenesis are not differentiated into spermatozoa. Few cells degenerate at different stages of spermatogenesis, and the numbers of degenerated cells are species-specific. In bull, about of 30% cells degenerate between spermatogonia A to intermediate spermatogonia. Additional 30% degeneration occurs during the formation of type B spermatogonia. In stallion, mostly degeneration occurs during the formation of type B spermatogonia. Some cells also degenerate at the end of meiosis. In humans, 30–40% of cells degenerate at the end of meiosis.

20.1.9 Spermatogenic Efficiency

Spermatogenic efficiency denotes daily sperm production (DSP) per gram of testes. It can be measured qualitatively by studying histological sections of seminiferous tubules and quantitatively by cell counting. The spermatogenic efficiency is effective for species comparison (Table 20.5). The spermatogenic efficiency depends upon the number of SSCs and Sertoli cells. Domestic animals can categorize into three groups based on their spermatogenic efficiency. Animals like boar, stallion, buck, ram, rabbit, and mice have high

Table 20.6 Duration of a spermatogenic cycle and spermatogenesis and daily production of spermatozoa with testicular weight in different mammals

Species	Duration of a spermatogenic cycle (days)	Duration of spermatogenesis (days)	Daily production (10^9 Nos.)	Gross weight of testes (g)
Bull	13.50	61	6.00–7.50	500–700
Buffalo bull	8.60	38	1.94–2.20	250–400
Buck	10.60	48	2.90–3.30	70–100
Ram	10	47	10.00–11.60	320–550
Stallion	12	57	5.00–6.00	340–500
Boar	8.60	39	16.20–23.00	750–1500
Dog	14	60	0.37–0.50	15–31
Tomcat	10.40	49	0.30–16.00	5–21
Rabbit buck	11–12	50	0.20–0.30	2.80–6.00
Rat buck	13	60	0.09–0.27	1.64–3.70
Mouse buck	9	35	0.30–0.40	0.20–0.25
Cock	3	12–13	0.80–2.50	15–25
Human	16	74	0.10–0.13	35–50

Data compiled from various sources

spermatogenic efficiency and can produce 20–30 million sperm per gram of testis. The bull, buffalo bull, and cats have average spermatogenic efficiency and can produce 10-–20 million sperm per gram of testis. Humans have low spermatogenic efficiency and can produce less than ten million sperm per gram of testis. The lower spermatogenic efficiency in humans can be explained by their fewer germ cells, longer duration of spermatogenesis, and longer cycle length. In addition, the proportion of seminiferous tubules and the seminiferous epithelium is lower in humans compared to stallions, bulls, and rats. The spermatogenic efficiency can also explain puberty. As a thumb rule, 50 million spermatozoa with 10% motility denote puberty in the bull at about 42 weeks (38–46 weeks), depending on breed.

20.1.10 Factors Affecting Spermatogenesis

Several factors alter spermatogenic efficiency, and they can classify into physical factors, chemical factors, nutritional factors, endocrine factors, genetic factors, age, pathological factors, and miscellaneous factors. The detailed mechanisms of these factors affecting spermatogenesis are summarized in Table 20.7.

20.1.11 Special Characteristics in Avian Spermatogenesis

Avian spermatogenesis divides into spermatocytogenesis and spermiogenesis. But, the duration of spermatogenesis is generally four times faster than mammals. Spermatogenic cycle time is also less with less mitotic division. There is no stem cell pool in avian testes. Hence, more spermatozoa produce within a short time. The epididymal transit time is also less in avian spermatozoa. Spermatogenic efficiency is around four times more than mammals (four times more spermatozoa per gram of testis). The spermatozoa of chicken have spindle-shaped heads and are challenging to differentiate from mid-piece. The spermatozoa are stored in extragonadal ducts, and their survivability is poor. Therefore, more mating is required to ensure more viable spermatozoa fertilize the ovum when the hens are in a clutch.

Know More . . . .
The efficiency of the Sertoli cells for spermatozoa production is considered the major factor in assessing the spermatogenic efficiency of any animal. The wild boar (*Sus scrofa scrofa*) is one of the species that contain the highest number of Sertoli cells (42 million per gram of testis) and Leydig cells (157 million per gram of testis) among the mammals. But, the efficiency of sperm production by the Sertoli cells is less in wild boar, nearly 50% than the domestic pig; the size of the Leydig cells (400 μm^3) is about fivefold smaller than the domestic pig. Thus, the spermatogenic efficiency of wild boar is less.

20.2 Semen

The semen is a biological fluid containing spermatozoa and seminal plasma. Seminal plasma is the cumulative secretions of accessory sex glands. The mixing of seminal plasma and spermatozoa occurs in the urethra during ejaculation. Seminal plasma acts as a vehicle to carry, nourish, and modulate

Table 20.7 The factors affecting the spermatogenesis

Factors				Mechanism
Physical	Irradiation			Injury to spermatogonia, spermatocytes, and spermatids. Spermatocytes are most susceptible
	Hyperthermia (Fig. 20.5)			Affecting testicular thermoregulation (environmental temperature more the 41 °C has a serious effect on spermatogenesis), Poor energy metabolism of spermatozoa. Decreased plasma membrane integrity. Loss of motility of spermatozoa, DNA fragmentation. Primary spermatocytes and round spermatids are most susceptible
	Hypothermia			Damage to spermatogenic functions occurs at −25 °F. Stagnation of blood and hypoxia of the testicular tissue
	Light			Short photoperiod (below 12 h) decreases the gonadotropin and affects spermatogenesis
	Low oxygen tension			Testicular ischemia—causes germ cell apoptosis. Low oxygen tension at high altitudes impairs spermatogenesis in experimental models
Chemical	Antispermatogenic drugs	Cadmium		Necrosis of testicular tissue. Increases the permeability of BTB
		Alkaline agents (Busulfan)		Destruction of spermatogonia
		Diamines		Causes maturation deletion of spermatozoa
		Nitrogen-containing compounds		Arrest of spermatogenesis at primary spermatocyte stage
		Drugs affecting cell division (hydroxyurea)		Interfere with DNA synthesis
	Environmental contaminants	Organochlorine compounds	Dichlorodiphenyltrichloroethane (DDT)	Spermatogenic cell degeneration and loss of germinal epithelium
			Cyclodines (dieldrin, aldrin)	Germ cell damage. Decreases plasma testosterone. Decreases prostatic secretion
			Benzene hexachloride	Degenerative changes in the spermatozoa. Damage to the seminiferous tubule
			Misc. organochlorine compounds (Kepone, polychloroprene used as insecticides)	Damage to the seminiferous tubule
		Organophosphates	Dichlorvos	Damage to the seminiferous tubule
			Carbamates	Increases testosterone hydroxylation
		Feed additives	Diethyl stilbesterol	Atrophy of seminiferous tubule
			Colouring agents (metanil)	Vascular damage
			Alcohol (a most important cause of male feminization in humans)	Atrophy of seminiferous tubule. Decreased testosterone secretion. Decreases testicular retinal dehydrogenase activity (interfere with vit-A metabolism)

(continued)

Table 20.7 (continued)

Factors				Mechanism
		Industrial chemicals	Organopolysiloxane, Ethylene oxide cyclic tetramer	Premature release of spermatozoa
Nutritional	Low plane of nutrition			Delayed puberty, hypoplasia of the testes. Reduced secretion of FSH and LH
	Low plane of nutrition			Fat deposition around the testes in the scrotum; insulates the testes to affect thermoregulation
	Vitamin deficiencies	Hypovitaminosis A		Vit-A is epitheliotropic, and its deficiency leads to degeneration of seminiferous tubule epithelium, testicular atrophy, and hypoplasia
		Hypovitaminosis E		Testicular damage in rats. Role in domestic animals is non-significant
Endocrine	Exogenous steroids			A small dose of testosterone impairs spermatogenesis by inhibiting the HPG axis
	Anterior pituitary hormones			The deficiency of LH and FSH causes testicular atrophy
Genetic	Heredity			Testicular hypoplasia may be congenital due to a single recessive autosomal gene. Acrosomal abnormality of spermatozoa can also be heritable
	Inbreeding			Abnormalities of the seminiferous tubule
	Cytogenetic disturbances	Stickiness of chromosome		Fail to separate during anaphase
		Pyknotic nucleus and multiple spindle formation		Formation of giant cells with multiple nuclei due to extrachromosomal division
	Hybridization			The pairing of the homologous chromosome is impossible due to the odd chromosome number of the hybrid (e.g. mule)
Pathological	Congenital	Testicular hypoplasia		Lack of mitotic division and block in spermatogenesis
		Cryptorchidism		Thermoregulation of testes is affected
		Scrotal or inguinal hernia		Interfere with testicular thermoregulation
	Acquired factors	Testicular degeneration (due to trauma, cold, heat, castration, irradiation, toxaemia, FMD, vaccination against FMD and Rinderpest)		The seminiferous tubule is reduced in size
		Testicular fibrosis		Degenerative changes in germinal epithelium
Age				Increased germ cell degeneration (bulls are able to generate fertile spermatozoa till 19 years of age)

Table 20.8 Composition, source, and function of seminal plasma

Components	Source	Function
Water	Epididymis (minor), prostate gland, bulbourethral gland, seminal vesicles	Liquid vehicle for the sperm
Mucus	Bulbourethral gland	Acts as a lubricant for the passage of semen
Bicarbonate buffers	The prostate gland and bulbourethral gland	Neutralize the acidic secretions of the vagina
Carnitine	Epididymis	Metabolism of fatty acids to provide nutrition to the spermatozoa
Glycerylphosphocholine	Epididymis	Nutrition to the spermatozoa
Fructose	Seminal vesicles	Major energy source to the spermatozoa
Fibrinogen	Seminal vesicles and prostate gland	Clots semen
Ascorbic acid	Seminal vesicles	Nutrition to the spermatozoa
Prostaglandins	Seminal vesicles and prostate gland (little)	Contraction of the vas deferens
Fibrinolytic enzyme	Prostate gland	Liquefies semen
Citric acid	Prostate gland	Nutrition to the spermatozoa
Zinc	Prostate gland	Stabilizes the DNA-containing chromatin in the spermatozoa

the functions of spermatozoa. Therefore, good quality semen production is one of the main contributors to male fertility and depends upon several factors like breed, age, nutritional status, season, and endocrine factors. Some chemicals and drugs can also modulate semen character.

20.2.1 Formation of Seminal Plasma

Different components of seminal plasma are mainly produced from the accessory sex glands. These components mix at the time of ejaculation. The initial ejaculates comprise the secretions of Cowper (bulbourethral) glands that constitute around 5% of total ejaculate. The second fraction of the ejaculate derived from the prostate consists of about 15–30% of the total ejaculates. A small portion of the secretions from the ampulla and epididymis occurs following the second fraction. Finally, the remaining amount of the ejaculate secretes from the seminal vesicle. The composition of seminal plasma summarizes in Table 20.8. The biochemical composition of the semen in different species presents in Table 20.9. Various biochemical tests are generally performed to assess seminal plasma's biochemical constituents to evaluate the functions of reproductive organs. Fructose is the biomarker for assessing seminal vesicles; citric acid, zinc, and acid phosphatase for the prostate gland; free L-carnitine, glycerophosphocholine, and alfa-glucosidase for the epididymis.

20.2.2 Sperm Concentration in the Semen

The concentration of spermatozoa in the semen mainly varies between species, breed, and ejaculatory fractions (Table 20.10 and 20.11). The effective sperm concentration required for fertilization depends on the initial progressive motility and morphological features. In domestic animals, the progressive motility and morphologically normal spermatozoa are generally 70–90%, which is reversed in wild animals. The fresh semen may contain a more morphologically abnormal percentage of spermatozoa with less progressive motility in wild species. The clinical terminologies related to spermatozoa concentration and motility summarize in Table 20.12. The morphological abnormalities of spermatozoa are of three types. Primary abnormality or structural deformity is related to hereditary or developmental origin. Secondary abnormalities are associated with functional alterations of the reproductive tract. Tertiary abnormalities are due to semen handling and processing. In ruminant and canine species, the sperm motility and viability in the semen are nearly 70–80% and morphological abnormalities are lower than 30%. Morphological abnormalities are evaluated microscopically. In recent years, a *computer-assisted integrated sperm analysis system* (ISAS) has been used to assess sperm motility, viability, and structural configurations. The DNA integrity of spermatozoa can be evaluated through *Comet assay* or *sperm chromatin structure assay* (SCSA). Sperm *fluorescence in situ hybridization analysis* (FISH) is also performed to assess fertilizing capability of the spermatozoa. The sperm membrane integrity can evaluate through the *Hypoosmotic sperm swelling test* (HOST), and *acrosome integrity* can analyse through fluorescent staining.

20.2.3 Sperm Quality Biomarkers

Protamines (PRM), nuclear proteins, are considered the biomarkers of sperm quality. Mammalian spermatozoa have two types of protamines, PRM1 and PRM2. Their

Table 20.9 Composition of seminal plasma in some domestic animals (in mg/dL unless otherwise stated)

Constituents	Bull[a]	Ram[1]	Goat[1]	Buffalo[1]	Dog[2]	Stallion[3]	Boar[4]
Fructose, mg/dL	150–900	150–600	875	368–815	4	4.1–7.8[8]	20–40
Glucose, mg/dL	300	0.90–1.60	4.80–8.80	13–52	406–902	20–24	600–725 (inositol)
Citric acid, mg/dL	340–1150	110–260	–	440–444	2–14	29.8–41.6[8]	110–260
Total proteins, g/dL	3.80	2.30–2.50	0.77–1.48	–	4.6–11.0	0.80–4.13	1.84
Total lipids, mg/dL	29	254–396	–	150–175	–	62–134	97.3
Phospholipids, mg/dL	149.1	–	57	6.9–59.4	–	–	86.8
Cholesterol, mg/dL	312.16	–	–	117.83	–	5.50–21.00	–
Glutamic acid, mg/dL	1.0–8.0	4.5–5.2	–	4.28	–	–	–
Sodium (Na), mg/dL	140–280	120–258	60–183	260–278	940–1000	–	290–850
Potassium (K), mg/dL	80–210	50–140	76–255	192–205	107–163	–	90–440
Calcium (Ca), mg/dL	35–60	6–15	5–15	30	3.96–5.98	12–20	2–6
Phosphorus (P), mg/dL	9	4.8–12.0	–	8–9	126–234	8.73–4.02	2
Chlorine (Cl), mg/dL	110–290	86	82–215	303–347	1438–1502	–	150–430
Magnesium (Mg), mg/dL	7–12	2–13	1–4	4.3–5.7	0.9–2.1	–	5–15
Zinc (Zn), mg/dL	2.6–3.7	56–179	–	0.80–1.17	0.10–0.26 mg/g[5]	–	1.69–2.17[9]
Testosterone, pg/mL	210–1310	25–375	–	970	–	–	–
Oestrogen, pg/mL	20–166	–	–	43.67	–	–	–
Prostaglandins, ng/mL	5–10	500–20,000	–	–	–	–	–
Alkaline phosphatise (ALP)	246 BU/dL	14,895–40,818 mU/mL	–	315 BU/dL	5758–8767 IU/L[6]	11.84–18.03 U/L	37,945–54,870 U/L[9]
Aspartate amino transferase (AST)	345–623 SFU/mL	190–256 mU/mL	–	166 U/mL	88.4–200.0 U/mL[7]	158–220 U/L	3.00–26.30 U/L[9]
Alanine aminotransferase (ALT)	15.0–18.3 SFU/mL	39–148 mU/mL	–	34 U/mL	–	11–21 U/L	–
Lactate dehydrogenase (LDH)	1909 U/mL	968–1697 mU/mL	–	1621 BBU/mL	1374–1636 U/mL[7]	782–2082 U/L	–

BBU Berger-Broida units; *BU* Bodansky units; *LDH* lactate dehydrogenase; *SFU* Sigma Frankel units; *UI* international units

Source: [1] Juyena and Stelletta (2012), [2] Wales and White (1965), [3] Talluri et al. (2017), [4] Frunză et al. (2008), [5] Hidiroglou and Knipfel (1984), [6] Shalini and Antoine (2018), [7] Murdoch and White (1967), [8] Carluccio et al. (2006), [9] Rodríguez et al. (2013)

Table 20.10 Production and requirement of spermatozoa for fertilization in domestic animals

Species	Volume ejaculate (mL)	Concentration of sperm (million/mL)	Total sperm in an ejaculation (million)	No. of sperm required for successful insemination[a] (million)	No. of possible females can be bred from an ejaculate[a]
Bull	5.0–8.0	1200	9600	8	800–1500
Buffalo	3.0–8.0	500–1000	3000	10	250–350
Buck	0.7–1.4	2500	2500	25–35	75
Ram	0.8–1.2	2000	2000	50–60	40
Boar	150–250	270	58,000	2500	15–17
Stallion	60–100	200	10,000	500–600	20–25
Dog	2.0–10.0	60–500	600–1500	100–200	3–7
Tomcat	0.1–0.5	1700	570	5–50	5–10

[a] The number of spermatozoa may be doubled, and possible females to be bred would halve when semen is used in frozen semen technology; however, the potentiality of the animal and characteristics of semen can influence the number of required spermatozoa

Table 20.11 Seminal volume and characteristics of spermatozoa in some free-living wildlife

Animal	Seminal volume (mL)	Sperm concentration (million/mL)	Initial progressive motility (%)	Morphological abnormality (%)	Reference
African elephant (*Loxodonta africana*)	56 (18–94)	818 (68–1568)	81 (52–100)	55 (41–69)	Luther (2016)
Asian elephant (*Elephas maximus*)	20 (15–26)	1502 (920–1905)	27 (19–33)	27 (14–37)	Thongtip et al. (2008)
Southern white rhinoceros (*Ceratotherium simum simum*)	24 (0–48)	83 (0–179)	82 (74–90)	62 (48–76)	Luther (2016)
Alpaca (*Vicugna pacos*)	2	17.6	15–64	49	Flores et al. (2002); Juyena et al. (2012)
Kangaroo (*Macropus giganteus*);**Macropus fuliginosus*	25	31.2 (24–38.5)	77	17.9*	Johnston et al. (1997); Lane (2014)
Spotted deer (*Axis axis*)	0.2–7	4–4000	35–80	30–50	Shivaji et al. (2003)
Brown bear (*Ursus arctos*)	2.55 (2.11–2.78)	207 (158–256); 472	77 (71–81)	22	Ishikawa et al. (1998); Anel-López et al. (2017)
Tiger (*Panthera tigris*)	1.4 (0.3–3.7)	42.1 (12–84)	47 (25–80)	25.2	Shivaji et al. (2003)
Lion (*Panthera leo*)	3.9 (1.3–9)	52.1 (20–95)	63 (35–90)	22.9	Shivaji et al. (2003)
Indian Leopard(*Panthera pardus*)	1.6 (0.5–4)	55.8 (10–142)	57 (20–90)	28.1	Shivaji et al. (2003); Jayaprakash et al. (2001)
Namibian cheetahs (*Acinonyx jubatus*)	3.7 (0.6–6.8)	20.4 (3.5–66)	78 (70–90)	78.3 (55–95)	Crosier et al. (2006)
Jaguars (*Panthera onca*)	4.1 (3.4–4.8)	152 (64–240)	73 (67–79)	26.5 (23–30)	Morato et al. (2001)
Ferret (*Mustela putorius furo*)	0.2	88.8 (73–105)	38 (32–42)	47.5	Toledano-Díaz et al. (2021); Strickler (2010)
Rhesus monkey (*Macaca mulatta*)	0.06–0.6	51 (5–500)	43–93	93 (90–96)	Dong et al. (2008); Liu et al. (2016)
Indian White-Backed Vulture (*Gyps bengalensis*)	0.37 (0.05–1.4)	58.4 (7–143)	46.8 (0–67)	27.8 (10–35); 55–65	Umapathy et al. (2005); Santiago-Moreno et al. (2016)
Brazilian rattlesnakes (*Crotalus durissus terrificus*)	18.5 (3–70) μL	1380 (940–2230)	64	NA	Zacariotti et al. (2007)

NA not available

Table 20.12 Some common terminology used in semen evaluation

Terminology	Indication
Normozoospermia	Normal ejaculate
Oligozoospermia (oligospermia)	Very less sperm concentration
Asthenozoospermia	Less than 50% sperm concentration with forwarding motility
Teratozoospermia	Less than 30% of spermatozoa with normal morphology
Oligoasthenoteratozoospermia	Disturbance of the above three variables
Azoospermia	No spermatozoa in the ejaculate
Aspermia	No ejaculate (volume)
Hyperspermia	Profuse ejaculate (volume)
Hypospermia	Too less than normal ejaculate (volume)
Pyospermia	Abnormal presence of leukocytes

proportions vary between species, and altered PRM1 and PRM2 ratios may lead to infertility.

20.2.4 Properties of Semen

20.2.4.1 Volume

The semen volume varies with species (Tables 20.9 and 20.10), age, body size, and between ejaculates. The semen volume is fairly constant with species but may differ between ejaculates. The semen volume increases with age, body size, and vigour. Teasing the bulls is practised to increase semen volume. A decrease in semen volume is seen at a young age, excessive use of males for breeding, incomplete ejaculation, and bilateral seminal vesiculitis. Domestic animals that contain lower semen volume have higher sperm concentrations.

20.2.4.2 Colour

The semen of bull and buck is creamy, milky white, or opaque, and buffalo bull is whitish compared to bull semen. The semen colour of stallions, boar, and dogs is pearly white to grey and translucent. The deviation from the normal colour indicates pathological conditions. Orchitis or haemorrhage in the male reproductive system leads to brown, dark red, or pinkish coloured semen. Semen can turn into yellow-green in *Pseudomonas aeruginosa* infection. An increased number of spermatogenic cells in the semen may cause dull and dirty white semen. Urine contamination leads to yellowish semen colour. Semen appears curdy with the presence of clots in genital tract infections.

20.2.4.3 Odour

The typical odour of the semen is due to the presence of basic amines, like putrescine, spermine, spermidine, and cadaverine. Semen may have an unusual smell like strong, fishy due to reproductive tract infections.

20.2.4.4 Viscosity

The consistencies of semen may vary from watery, milky, or creamy, depending on the spermatozoa concentration. The specific gravity of the semen of the bull is 1.0361. There is a positive correlation between spermatozoa's viscosity and sperm concentration in semen. Various pathological conditions will lead to alteration in the viscosity of semen, viz. less milky (pathology of the epididymis), purulent flocculi (in seminal vesiculitis), and thick viscous (catarrhal conditions of accessory sex glands).

20.2.4.5 pH and Buffering Capacity

The semen has a high buffering capacity (higher than most of the body fluids), which helps maintain its pH near neutrality in an acidic vaginal environment to protect the DNA of spermatozoa. The components contributing to the buffering action of the semen are HCO_3/CO_2 (contribute 25% of buffering activity), proteins (29% of buffering capacity), low–molecular weight components, viz. phosphate, citrate, and pyruvate (47% of buffering capacity). The pH of the ruminant semen is slightly acidic, as the bulbourethral gland is smaller in size (Table 20.13). The bulbourethral gland is absent in canines; hence, the pH of semen is also slightly acidic.

20.2.4.6 Osmolarity

The osmolarity of the semen is higher than plasma. The presence of sugars and ionic salts contributes to this high osmolarity. The osmolarity of human semen varies from 250 to 400 mOsm. The osmolality of stallion semen ranges from 331 to 336 mOsm.

Table 20.13 The semen pH varies with species

Species	Semen pH
Bull	6.4–7.8
Ram	5.9–7.3
Boar	7.3–7.5
Stallion	7.2–7.8
Cock	7.2–7.6

20.2.5 Ejaculation

Ejaculation is the biological process by which the seminal plasma is forcefully expelled out of the body. The ejaculation occurs through synchronized physiological events and is divided into two phases, emission and expulsion. The spermatozoa moved from the epididymis to the prostatic urethra and mixed with the seminal plasma during emission. During the emission phase, the neck of the bladder is closed to prevent retrograde movement of semen into the bladder. After the bladder neck closure, the spermatozoa coming through the vas deferens mix with prostatic secretions in the prostatic urethra. The spermatozoa containing prostatic secretion then mix with seminal vesicle fluid. The secretions from the Cowper's glands contribute a little to the seminal fluid. The expulsion phase initiates once the emission phase is over. It mediates through rhythmic contraction of striated pelvic perineal muscles, and the urethral sphincter facilitates the ejection of semen through the penis via the urethral meatus. There are species variations regarding the mixing of spermatozoa with seminal plasma. In bull, ram, and buck, the spermatozoa are mixed instantaneously with the seminal plasma. In contrast, spermatozoa are mixed sequentially with the fluid of the accessory sex glands in the stallion, boar, and dog; hence the ejaculates from these species contain three first 'sperm free', followed by 'sperm rich' and 'sperm poor' fractions of semen.

20.2.5.1 Neural Control of Ejaculation

The process of ejaculation is mediated through complex neurovascular mechanisms. There are several central and peripheral neurological factors involved in ejaculation. The encapsulated nerve endings termed Krause-Finger corpuscles at glance penis and free nerve endings in other portions of the penis are temperature and pressure-sensitive. The tactile receptors in bull and ram are more sensitive to warmth, and the receptors in stallion and boar are more sensitive to slipperiness or pressure within the vagina. During the natural coitus or in an artificial vagina stimulates, these tactile receptors of the penis, and the signals are transmitted through two afferent sensory inputs. The main sensory input comes through the dorsal nerve of the penis. The other sensory afferent input comes through the hypogastric nerve of the paravertebral sympathetic chain. These two afferent inputs terminate at the medial dorsal horn and the dorsal grey commissure of the spinal cord. The efferent nervous system of ejaculation constitutes sympathetic, parasympathetic, and motor nerves. The locations of preganglionic sympathetic and parasympathetic fibres are thoracolumbar segments of the spinal cord and sacral parasympathetic nucleus, respectively. The motor neurons are located in the sacral spinal cord. The sympathetic nerves innervate to the smooth muscle of epididymis, vas deferens, seminal vesicles, and prostate

Table 20.14 Role of various neurotransmitters and hormones in the ejaculation process

Agents		Functions
Neurotransmitters	Dopamine	Stimulates ejaculation
	Serotonin	Inhibits ejaculation
	Nitric oxide	Inhibits ejaculation
Hormones	Oxytocin	Helps in epididymal contractions and sperm motility, Stimulates CNS for ejaculation
	Prolactin	Inhibits GnRH and dopamine production, thus exerting an inhibitory on male sexual desire.
	Thyroid hormones	Delayed and premature ejaculation is associated with both hypo- and hyperthyroidism
	Glucocorticoids	Elevated cortisol levels are seen during ejaculation in animals; cortisol replacement in Addison disease improves ejaculation
	Oestrogens	Controls the emission phase of ejaculation by altering epididymal contractility, luminal fluid reabsorption, and sperm concentration
	Androgens	Low and high levels of androgens are associated with delayed and premature ejaculation, respectively.

and facilitate the emission process. The somatic nerves are innervated in ischiocavernosus, bulbocavernosus, and ischiourethralis muscles. The parasympathetic nerves are connected with erectile tissues. Various neurotransmitters and hormones are involved in the ejaculation process. They are summarized in Table 20.14.

Know More . . . .

Seminal Characteristics in Wild Animals

Testosterone level is reduced in wild animals due to inbreeding, affecting semen quality. The testosterone concentration of outbred Asiatic lions demonstrated (1.8 ng/mL) is higher than the inbred Indian lions (less than 1 ng/mL), as found in the Gir forest of India, which leads to the production of 60% abnormal with pleomorphic spermatozoa. The high proportions of pleiomorphic spermatozoa have also been reported in wild-born free-living cheetahs (*Acinonyx jubatus*) in Namibia of Africa, and other sanctuaries or reserves. The captive wild animals may donate more volume of semen, but the concentration of spermatozoa is more in a free-living wild animal.

Learning Outcomes

- **Spermatogenesis:** Spermatogenesis is a series of complex synchronized processes of mitotic and meiotic cell divisions, which occur in the seminiferous tubules of the testis. It has two distinct phases, spermatocytogenesis and spermiogenesis. Various hormones like testosterone, FSH and oestrogens have been involved in spermatogenesis. The time required to complete the entire process differs between species, and birds need very little time to complete the spermatogenesis.
- **Spermatocytogenesis:** Spermatocytogenesis is initiated by mitotic divisions and ends with the early stages of meiotic divisions in a cyclic process called the spermatogenic cycle. The spermatogonia are the germinal epithelium cell layers, considered stem cells. It continuously generates type A spermatogonia and is stored as an intermediate stage. Later, it differentiated into type B spermatogonia followed by primary spermatocytes. The primary spermatocytes generate secondary spermatocytes by first meiotic division with the influence of testosterone after puberty. Secondary spermatocytes undergo second meiotic division and form spermatids through a testosterone-independent process.
- **Spermiogenesis:** The haploid spermatid transformed into spermatozoa by metamorphic change through spermiogenesis under the influence of FSH. Sertoli cells play a major role in this process. It has five distinguished steps, Golgi phase, cap phase, acrosomal phase, and maturation stage. The mature spermatozoa contain a distinct head, neck, mid-piece and tail and various cellular organelles. At the end of this phase, the spermatozoa become elongated and leave the Sertoli cells through spermiation.
  Nucleus and cellular organelles are typically shaped with the spermatozoa acquiring various proteins and enzymes required for fertilization.
- **Spermatogenic efficacy:** All the spermatogonia are not converted to spermatozoa; some degenerate during different stages of spermatogenesis. The effectiveness of spermatozoa production depends upon the testicular mass, including the number of stem cells and Sertoli cells and the diameter and length of

(continued)

the seminiferous tubules. Different species have different spermatogenic efficacy and can be categorized as high, average, and low. Various stressors, mainly heat stress, reduce the effectiveness.

- **Semen:** Spermatozoa, along with seminal plasma, constitute the semen. Its characteristics vary in different species. The quality of the semen of a particular species also depends upon the breed, age, nutritional status, season, and hormonal status of the animal. Various tests are performed to evaluate semen before using it for breeding purposes.
- **Spermatozoa:** Morphologically normal spermatozoa with progressive motility are required for fertilization. Species-specific sperm morphological features and concentration, along with its viability and motility, can be assessed by routine semen evaluation techniques. The integrity of its acrosome, DNA fragment, and some other cellular characteristics can be evaluated by advanced tests.
- **Seminal plasma:** Different species have different physical and biochemical characteristics. Seminal plasma contains various enzymes, proteins, phospholipids, and other biochemical components to support the spermatozoa in increasing its fertilizing ability and viability.
- **Ejaculation:** The semen is forcefully expelled out of the body through synchronized physiological events into two phases, emission and expulsion. Several neurohormones and hormones are involved in the complex neurovascular mechanisms of ejaculation.

Exercises

Objective Questions

Q1. Which cells are considered the stem cells of the seminiferous tubule?

Q2. Which type of cell division occurs in spermatocytogenesis?

Q3. Are spermatids haploid or diploid cells?

Q4. What are intermediate spermatogonia?

Q5. Which stage of spermatocytogenesis is testosterone dependent?

Q6. Which part of the seminiferous tubules holds spermatocytogenesis?

Q7. Why is thermoregulation significant in spermatogenesis?

Q8. Which phase of spermatogenesis occurs within Sertoli cells?

Q9. Which structure of the spermatozoa contains lysosomal enzymes?

Q10. Significant structural changes occur in which phase of spermiogenesis?

Q11. Which structure is formed in combination with the group of microtubules?

Q12. Citric acid is evaluated to assess the functional activity of which accessory sex gland?

Q13. Which part of the spermatozoa provides energy to the spermatozoa for its movement?

Q14. Spermiation is occurred at which part of the seminiferous tubule?

Q15. What is the common duration of the spermatogenic cycle in mammals?

Q16. What is the major cause of the production of pleomorphic spermatozoa in wild animals?

Q17. In which process the permeability of the sperm membrane to bicarbonate is increased?

Q18. Write the name of the following animals serially, according to the concentration of spermatozoa in ejaculate—bull, buck, boar, and tomcat.

Q19. What do you mean by oligozoospermia?

Q20. What is the major source of energy in seminal plasma?

Subjective

Q1. Why is spermatocytogenesis considered a proliferative process?

Q2. Write the different steps of spermiogenesis.

Q3. Describe the spermatogenic cycle.

Q4. What is spermatogenic efficiency?

Q5. How heat stress affects sperm characteristics?

Q6. Write down the unique features of avian spermatogenesis.

Q7. Describe the factors that determine the fertilizing capabilities of spermatozoa.

Answer to Objective Questions

A1. Spermatogonia.

A2. Both the mitotic and meiotic.

A3. Haploid cells.

A4. The resting type A spermatogonia.

A5. First meiotic division.

A6. At the adluminal compartment of the tubule between Sertoli cell and basal layer.

A7. The DNA polymerase and recombinase activities are altered with the fluctuation of testicular temperature.

A8. Spermiogenesis.

A9. Acrosome.

A10. Golgi phase.

A11. Axoneme.

A12. Prostate gland.
A13. Mid-piece.
A14. The lumen of the seminiferous tubules.
A15. 30–75 days.
A16. Inbreeding.
A17. Hyperpolarization.
A18. Buck, tomcat, bull, and boar.
A19. Very less sperm concentration in the fresh ejaculate.
A20. Fructose.

Keywords for the Answer to Subjective Questions

A1. Mitotic divisions, meiotic divisions, number of cells generated from spermatogonia.
A2. Five stages of spermiogenesis, structural changes of the spermatids, spermiation.
A3. Various developmental stages of spermatozoa, longitudinal movement, cyclic process.
A4. Degeneration of germ cells, testicular mass, factors alter the spermatogenic efficiency.
A5. Testicular thermoregulation, effect on sperm morphology, structural deformity.
A6. No stem cell, longitudinal structure, less survivability.
A7. Routine semen evaluation, advance test, morphology and biomarker of sperm quality.

Further Reading

Textbooks

Deviche P, Hurley LL, Fokidis HB (2011) Avian testicular structure, function, and regulation, Ch: 2. In: Hormones and reproduction of vertebrates, Volume 4d (Birds). Elsevier, pp 27–70

Research Articles

Aire TA (2014) Spermiogenesis in birds. Spermatogenesis 4(3): e959392. https://doi.org/10.4161/21565554.2014.959392

Anel-López L, Ortega-Ferrusola C, Martínez-Rodríguez C et al (2017) Analysis of seminal plasma from brown bear (*Ursus arctos*) during the breeding season: its relationship with testosterone levels. PLoS One 12(8):e0181776. https://doi.org/10.1371/journal.pone.0181776

Bhakat M, Mohanty T, Gupta AK, Prasad S, Chakravarty AK, Khan H (2015) Effect of season on semen quality parameters in Murrah Buffalo Bulls. Buffalo Bull 34:100–112

Briz MD, Bonet S, Pinart B, Egozcue J, Camps R (1995) Comparative study of boar sperm coming from the caput, corpus, and cauda regions of the epididymis. J Androl 16(2):175–188

Carluccio A, Tosi U, Battocchio M, Veronesi MC, De Amicis I, Contri A (2006) Citric acid and fructose seminal plasma concentrations and semen characteristics in the stallion. Ippologia Cremona 17(1):29–32

Crosier AE, Pukazhenthi BS, Henghali JN et al (2006) Cryopreservation of spermatozoa from wild-born Namibian cheetahs (*Acinonyx jubatus*) and influence of glycerol on cryosurvival. Cryobiology 52:169–181. https://doi.org/10.1016/j.cryobiol.2005.10.011

Dong Q, Rodenburg SE, Huang C et al (2008) Effect of pre-freezing conditions on semen cryopreservation of rhesus monkey. Theriogenology 70(1):61–69. https://doi.org/10.1016/j.theriogenology.2008.02.008

Flores P, García-Huidobro J, Muñoz C et al (2002) Alpaca semen characteristics previous to a mating period. Anim Reprod Sci 72(3–4):259–266. https://doi.org/10.1016/s0378-4320(02)00095-7

Frunză I, Cernescu H, Korodi G (2008) Physical and chemical parameters of boar sperm. Lucrări Ştiinlifice Medicină Veterinară, Timişoara, vol XLI, pp 634–640. https://www.usab-tm.ro/vol8MV/101_vol8.pdf

Gemeda AE, Workalemahu K (2017) Body weight and scrotal-testicular Biometry in three indigenous breeds of bucks in arid and semiarid agroecologies. Ethiopia J Vet Med 2017:5276106. https://doi.org/10.1155/2017/5276106

Hidiroglou M, Knipfel JE (1984) Zinc in mammalian sperm: a review. J Dairy Sci 67:1147–1156

Ickowicz D, Finkelstein M, Breitbart H (2012) Mechanism of sperm capacitation and the acrosome reaction: role of protein kinases. Asian J Androl 14(6):816–821. https://doi.org/10.1038/aja.2012.81

Ishikawa A, Matsui M, Tsuruga H et al (1998) Electroejaculation and semen characteristics of the captive Hokkaido brown bear (*Ursus arctos yesoensis*). J Vet Med Sci 60(8):965–968. https://doi.org/10.1292/jvms.60.965

Jayaprakash D, Patil SB, Kumar MN (2001) Semen characteristics of the Captive Indian Leopard, *Pantherapardus*. J Androl 22:25–33

Johnston SD, Blyde D, Gamble J et al (1997) Collection and short-term preservation of semen from free-ranging eastern grey kangaroos (*Macropus giganteus*: Macropodidae). Aust Vet J 75(9):648–651. https://doi.org/10.1111/j.1751-0813.1997.tb15362.x

Juyena NS, Stelletta C (2012) Seminal plasma: an essential Attribute to spermatozoa. J Androl 33(4):536–551. https://doi.org/10.2164/jandrol.110.012583

Juyena NS, Vencat J, Pasin G (2012) Alpaca semen quality in relation to different diets. Reprod Fertil Dev 1-8. http://dx.doi.org/https://doi.org/10.1071/RD12050

Kempinas WDG, Klinefelter GR (2015) Interpreting histopathology in the epididymis. Spermatogenesis 4(2):00–00. https://doi.org/10.4161/21565562.2014.979114

Lane ML (2014) Sperm competition and sexual selection in western grey kangaroos *Macropus fuliginosus*. Thesis for the Bachelor of Science in Conservation and Wildlife Biology with Honours, Murdoch University, pp 1–92. https://researchrepository.murdoch.edu.au/id/eprint/25342/1/Sperm_competition_and_sexual____selection_in_western_grey_kangaroos_Macropus_fuliginosus.pdf

Lebelo SL, Horst GVD (2016) Ultrastructural changes occurring during spermiogenesis of the Vervet monkey, Chlorocebus aethiops. Asian J Anim Sci 10:247–254. https://doi.org/10.3923/ajas.2016.247.254

Lino BF (1972) The output of spermatozoa in rams II. Relationship to scrotal circumference, testis weight, and the number of spermatozoa in different parts of the urogenital tract. Aust J Biol Sci 25:359–366

Liu Z, Nie YH, Zhang CC, Cai YJ, Wang Y, Lu HP, Li YZ, Cheng C, Qiz ZL, Sun Q (2016) Generation of macaques with sperm derived from juvenile monkey testicular xenografts. Cell Res 26:139–142. https://doi.org/10.1038/cr.2015.112

Ming-Wen L, Meyers S, Tollner TL et al (2013) Damage to chromosomes and DNA of rhesus monkey sperm following cryopreservation. Androl 28(4):493–501. https://doi.org/10.2164/jandrol.106.000869

Luther I (2016) Semen characteristics of free-ranging African elephants (*Loxodonta africana*) and Southern white rhinoceros (*Ceratotherium simum simum*) using Computer-aided sperm analysis, Electron microscopy and Genomics as diagnostic tools. PhD thesis,

Department of Medical Bioscience, University of the Western Cape, pp 1–301. https://core.ac.uk/download/pdf/84116393.pdf
Morato RG, Conforti VA, Azevedo FC et al (2001) Comparative analyses of semen and endocrine characteristics of free-living versus captive jaguars (*Panthera onca*). Reproduction 122(5):745–751. https://doi.org/10.1530/rep.0.1220745
Murdoch RN, White IG (1967) Studies of the distribution and source of enzymes in mammalian semen. Aust J Biol Sci 21:483–490
Olar TT, Amann RP, Pickett BW (1983) Relationships among testicular size, daily production and output of spermatozoa and extragonadal spermatozoa reserves of the dog. Biol Reprod 29(5):1114–1120. https://doi.org/10.1095/biolreprod29.5.1114
Robaire B, Hinton BT, Orgebin-Crist MC (2006) The epididymis. In: Neill JD (ed) Knobil and Neill's physiology of reproduction, 3rd edn. Elsevier, New York, pp 1071–1148
Rodríguez AL, Rijsselaere T, Beek J, Vyt P, Soom AV, Maes D (2013) Boar seminal plasma components and their relation with semen quality. Syst Biol Reprod Med 59(1):5–12. https://doi.org/10.3109/19396368.2012.725120
Santiago-Moreno J, Esteso MC, Villaverde-Morcillo S et al (2016) Recent advances in bird sperm morphometric analysis and its role in male gamete characterization and reproduction technologies. Asian J Androl 18(6):882–888. http://dx.doi.org/https://doi.org/10.4103/1008-682X.188660
Segatelli TM, França LR, Pinheiro PFF et al (2013) Spermatogenic cycle length and spermatogenic efficiency in the gerbil (*Meriones unguiculatus*). J Androl 25(6):872–880. https://doi.org/10.1002/j.1939-4640.2004.tb03156.x
Shalini I, Antoine D (2018) Semen characteristics in German Shepherd dogs. Int J Curr Microbiol Appl Sci 7(3):2304–2312. https://doi.org/10.20546/ijcmas.2018.703.270
Shivaji S, Kholkute SD, Verma SK et al (2003) Conservation of wild animals by assisted reproduction and their molecular marker technology. Indian J Exp Biol 41:710–723
Staub C, Johnson L (2018) Review: spermatogenesis in the bull. Animal 12(S1):s27–s35. https://doi.org/10.1017/S1751731118000435
Strickler TL (2010) Improving assisted reproductive technologies in the endangered Black-Footed Ferret: artificial insemination and sperm cryopreservation. MSc thesis, Graduate School, Ohio State University, pp. 1–136. https://etd.ohiolink.edu/apexprod/rws_etd/send_file/send?accession=osu1267057806&disposition=inline
Swierstra EE, Foote RH (1965) Duration of spermatogenesis and spermatozoan transport in the rabbit based on cytological changes, DNA synthesis and labeling with tritiated thymidine. Am J Anat 116(2):401–411. https://doi.org/10.1002/aja.1001160206
Talluri TR, Mal G, Ravi SK (2017) Biochemical components of seminal plasma and their correlation to the fresh seminal characteristics in Marwari stallions and Poitou jacks. Vet World 10(2):214–220. https://doi.org/10.14202/vetworld.2017.214-220
Thongtip N, Saikhun J, Mahasawangkul S et al (2008) Potential factors affecting semen quality in the Asian elephant (*Elephas maximus*). Reprod Biol Endocrinol 6:9. https://doi.org/10.1186/1477-7827-6-9
Toledano-Díaz A, Castaño C, Velázquez R (2021) Cryopreservation of ferret (*Mustela putorius furo*) sperm collected by rectal massage and electroejaculation: comparison of a decelerating and an accelerating freezing rate protocol. Vet Med Sci 7(1):256–263. https://doi.org/10.1002/vms3.362
Umapathy G, Sontakke S, Reddy A et al (2005) Semen characteristics of the captive Indian white-backed vulture (*Gyps bengalensis*). Biol Reprod 73(5):1039–1045. https://doi.org/10.1095/biolreprod.105.043430
Varner DD (2015) Odyssey of the spermatozoon. Asian J Androl 17:522–528. https://www.ajandrology.com/text.asp?2015/17/4/522/153544
Wales RG, White IG (1965) Some observations on the chemistry of dog semen. J Reprod Infertil 9:69–77
Zacariotti RL, Grego KF, Fernandes W et al (2007) Semen collection and evaluation in free-ranging Brazilian rattlesnakes (*Crotalus durissus terrificus*). Zoo Biol 26(2):155–160. https://doi.org/10.1002/zoo.20126

Functional Morphology of the Female Reproductive System

21

Pradip Kumar Das, Joydip Mukherjee, and Dipak Banerjee

Abstract

The female reproductive system consists of the primary sex organ, the ovary, the reproductive tract, and the secondary sex organs. The ovaries' size, shape, and anatomical positions show huge variability among mammals. The reproductive tract includes oviducts, uterus, cervix, vagina, and vulva and remains in the pelvic cavity suspended with the ligaments. The ovaries are responsible for oogenesis and hormone production. The reproductive tract performs the functions like gamete transport, fertilisation, and maintenance of pregnancy. The functional morphology of the female reproductive tract varies with the stages of the reproductive cycle under the influence of endocrine factors and is used as a tool to predict the stages of the reproductive cycle in animals. Birds have a morphologically distinct female reproductive system.

P. K. Das (✉) · J. Mukherjee · D. Banerjee
Department of Veterinary Physiology, West Bengal University of Animal & Fishery Sciences, Kolkata, West Bengal, India

© The Author(s), under exclusive license to Springer Nature Singapore Pte Ltd. 2023

P. K. Das et al. (eds.), *Textbook of Veterinary Physiology*, https://doi.org/10.1007/978-981-19-9410-4_21

Graphical Abstract

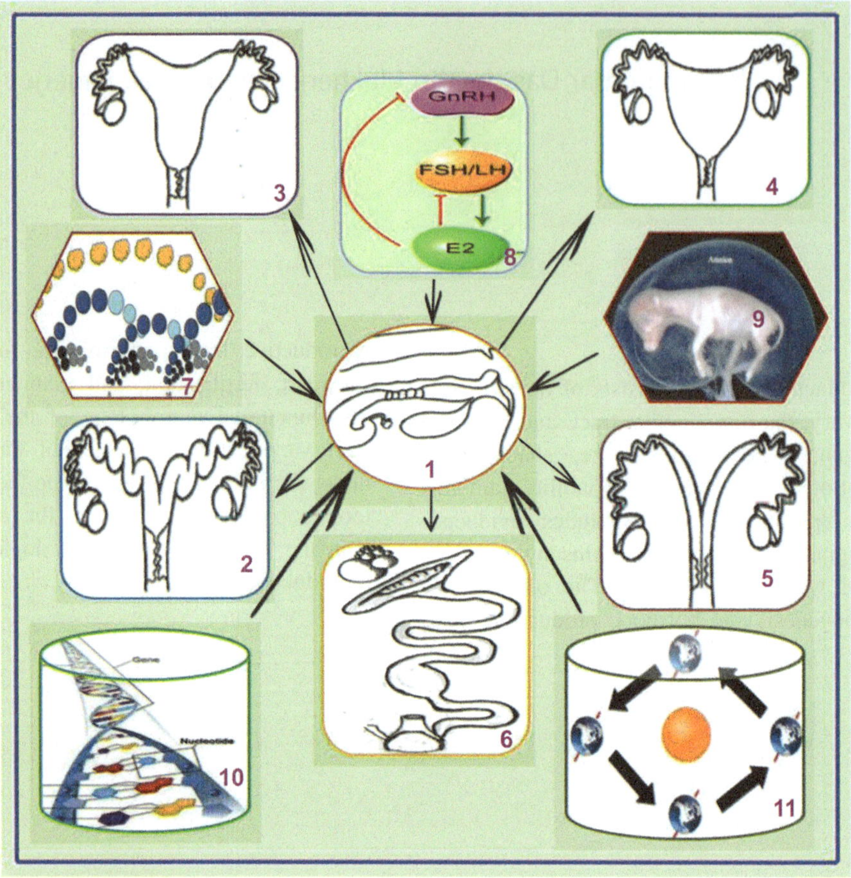

Description of the graphic: Female reproductive tract lying in the pelvic cavity below the rectum (**1**) can be categorised into bicornuate (**2**), bipartite (**3**), simplex (**4**) and duplex (**5**), according to morphological features of the uterus and cervix. The female reproductive system of birds (**6**) is morphologically different from mammals. The physiological activities of the reproductive system alter during different reproductive states (presented by reverse arrow), such as ovarian follicular dynamic and oestrous cycle (**7**), the influence of endocrines, e.g. oestrogens (**8**), pregnancy (**9**), other factors like breed (**10**) and environment (**11**)

Keywords

Morphology · Primary female reproductive organs · Secondary female reproductive organs · Avian female reproductive system

Learning Objectives

- Functional morphology of ovary.
- Functional morphology of reproductive tract consisting of oviduct, uterus, vagina and vulva in female reproduction.
- Characteristics of accessory sex organs, viz. vestibular glands, vestibular bulbs, hymen and clitoris.
- Uniqueness in secondary sex organs in avian.

The female reproductive system consists of the primary sex organ (ovary), reproductive tract consisting of the oviduct, uterus, vagina and vulva and accessory sex organs, viz. vestibular glands, vestibular bulbs, hymen and clitoris (Fig. 21.1). Ovaries are paired organs divided into left and right ovaries responsible for oogenesis and the production of hormones. A pair of tortuous oviducts remains close to the ovaries and is divided into a funnel-like opening, the infundibulum, followed by a convoluted ampulla and isthmus tube. The uterus consists of two uterine horns fused with the isthmus of the oviduct, a body and a neck continued towards the exterior of the body through the cervix. Through suspensory ligaments, the entire reproductive tract and

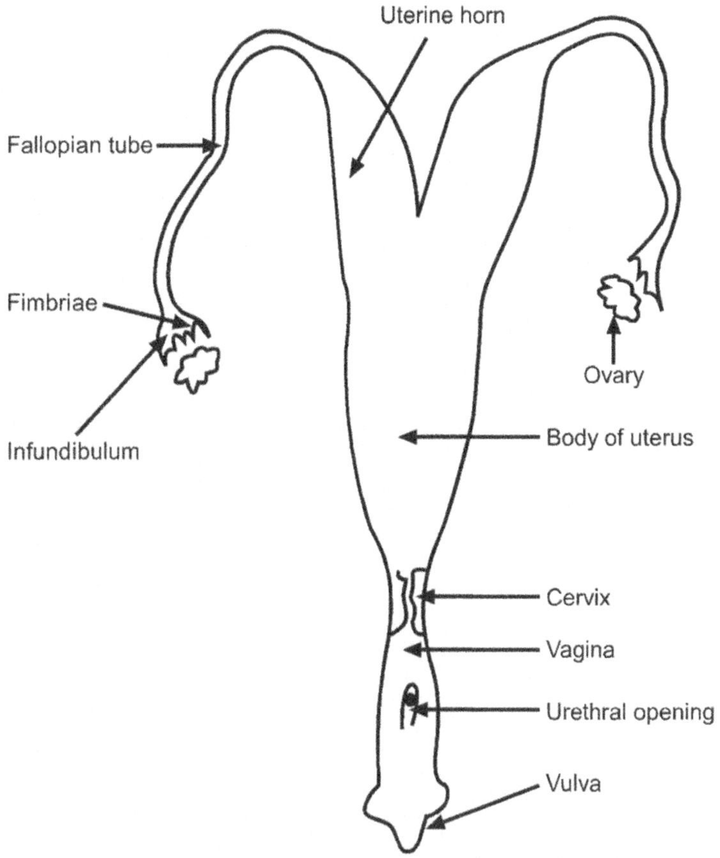

Fig. 21.1 Reproductive system of ewe. Figure shows the major parts of the female reproductive system with one **body of the uterus**, and both sides horn of uterus (**uterine horn**), **fallopian tube** ended with **infundibulum** having **fimbriae** at the edge to receive the ovum from the **ovary**. The single body of the uterus is continued with the single **cervix**, **vagina** and **vulva**. The urethra opens in the vagina at the **urethral opening**

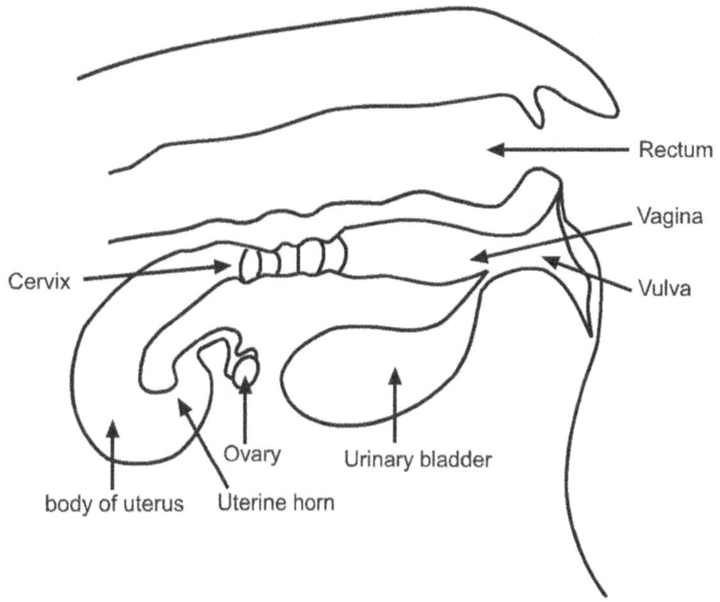

Fig. 21.2 Position of cow's reproductive tract. Figure shows the lateral view of the cow's reproductive tract, including the **ovary**, **uterine horn**, **body of uterus**, **cervix**, **vagina** and **vulva** with **urinary bladder**. It can palpate through the **rectum**

ovaries suspend within the pelvic cavity. These ligaments support the various parts of the reproductive system and provide the channel for nerves and blood vessels to the reproductive system. The urinary bladder is located below the reproductive tract and opens at the vagina. The rectum is positioned above the reproductive tract (Fig. 21.2), and the tract can be palpated and manipulated per rectal examination. There is variability in the morphological features of the female reproductive tract between species (Table 21.1) during different stages of the reproductive cycle.

Table 21.1 The morphometric features (in cm) of the different parts of the female reproductive system of animals

	Cow	Buffalo	Ewe	Doe	Mare	Sow	Bitch	Queen
Rt ovary ($L \times W \times T$)	3 × 2 × 1.5	2.5 × 1.7 × 1.9	1.2 × 0.9 × 0.9	(L × W) 1.5 × 0.7; 1.3–1.7 × 0.9[a]	(Each ovary) (L × W) 3 × 2.5 (NS) 7.5 × 3.5 (S)	2.5 × 1.2 × 0.3	($L \times W$) 2 × 1[b]	(Each ovary) 0.9 × 0.5 × 0.4[c]
Lt ovary ($L \times W \times T$)	2.5 × 1.6 × 1.3	2.4 × 1.6 × 1.9	1.2 × 0.9 × 0.8	($L \times W$) 1.4 × 0.6; 1.3–1.7 × 0.9[a]		2.5 × 1.2 × 0.2	($L \times W$) 1.4 × 1.1[b]	
Weight of ovary (g) (Rt, Lt)	(3.8, 3.1); (6.3, 5.4)[d]; 10–20[e]	(3.8, 3.4)	(0.9, 0.8); (3–4)[e]	(0.61, 0.60); (1.8–3.5)[e]	120[f] (each); 40–80[e] (each)	(2.3, 1.3); 3–10[e] (each)	0.1–1.5 (each)	0.1–0.3 (each)
Oviduct (L)	21; 16.5[d]; 25[g]	20.7	15[g]	10.2; 15[g]	25–30[g]	15–30	5–9[g]; 4–7	3–5
Horn (L)	27.5–31.5; 23[d]; 35–40[e]	27 × 2.6 (W)[h]; 26 × 2.1 (W)[i]	11.8 × 4 (W)[j]	12.3 × 2.8 (W)	20–25	33–104 × 0.2 (W)	10–14	6–10
Shape of uterine horn	Round parallel to each other	Round parallel to each other	Round	Round	Straight and divergent ('T') shaped. Uterine body equal to uterine horn	Intestinal loop appearance	Straight and divergent	Straight and divergent
Body of uterus ($L \times W$)	3.5 × 3	2 × 3.3[h]; 1.4 × 2.7[i]	8.7 × 6.3[j]; 1–2[e]	2.5 × 2.7	18–20	3.7 × 2.4 × 0.5 (T)	1.4–2 (L)	1.5–2 (L)
Cervix ($L \times W$)	5.6 × 4.9	5.2 × 2.9[h]; 5.9 × 2.5[i]	6.7 × 2.2[j]	3.4 × 1.8	6 (L); 7–8 × 3.5–4[e]	15 × 2.5 × 0.8 (T)	1.5–2 × 0.5 × 1.5	1–1.5 × 0.4–0.6
Vagina (L)	25 × 6 (W); 30[g]	29.5[k]	10.6–11.4[l]	7.1 × 4 (W)	15–20	10 × 2.5 (W) × 1.1 (T)	35[m]	–
Vulva (L)	9 × 5 (W)	–	–	2.7	12–15[n]	5.5 × 2 (W) × 0.5 (T)	–	–
Reference	Khaton et al. (2015)	Alwan et al. (2005)	Hoque et al. (2016)	Gupta et al. (2011)	Morel (2015)	Vicencio et al. (2017)	Pineda and Dooley (2003)	

Rt = right, Lt = left, L = length, W = width, T = thick, NS = non-season, S = season

[a] Haque et al. (2016)
[b] Eker and Salmanoglu (2006)
[c] Conze and Wehrend (2017)
[d] Drennan and Macpherson (1966)
[e] Pineda and Dooley (2003)
[f] McAuliffe (2013)
[g] Vicencio et al. (2017)
[h] Lohachit et al. (1981)
[i] Devkota and Singh (2017)
[j] Jaji et al. (2013)
[k] Carvalho et al. (2010)
[l] Knight et al. (2016)
[m] Lévy (2016)
[n] Bergfelt (2016)

21.1 Suspensory Ligaments

Three suspensory ligaments help to suspend the genital tract in the pelvic cavity and hold the genital tract in place. These are *broad ligament*, *utero-ovarian ligaments* and other supporting ligaments. The broad ligament, the suspensory connective tissue, originates from the peritoneum and holds the ovaries, oviduct and uterus. The parts of the broad ligament attached to the ovaries are called *mesovarium* suspended from the dorsolateral wall to the abdomen and attached to the hilum of the ovary. It is mainly involved in neurovascular supply to the ovaries. *Mesosalpinx* is the part of the broad ligament that supports the oviduct and helps orient the infundibulum around the ovary to direct oocytes into the oviduct. In the bitch, the mesosalpinx encloses the ovary forming a bursa around the ovary. The largest part of the broad ligament is the *mesometrium*, which supports the uterine horns and (or) uterine body. It is continuous with the dorsal peritoneum and hangs from the dorsal body wall. The utero-ovarian ligament attaches the ovaries to the uterus. This ligament is called the proper ligament of the ovary. The mesovarium and utero-ovarian ligament form a bursa like a pouch called *ovarian bursa*; other ligaments hold the uterus and cervix on the pelvic floor.

21.2 Ovary

Ovaries are paired organs suspended in the abdomen with the mesovarium in most animals. In ruminants, it remains close to the ventral abdominal wall, slightly above to mid of the pelvic inlet; in sows, positioned at the ventrolateral side of the pelvic inlet and generally remains within the intestines; in dogs and cats, located at the dorsal part of the abdomen behind the kidneys below 3–4 lumbar vertebrae; in mare, below the 4–5 lumbar vertebrae. The actual potion of the ovaries is variable depending on the animal's age, parity and breeding season. The shape of ovaries is generally oval or almond-shaped in almost all mammals and bean-shaped in horses, and berry shaped in the sow. The size of the ovary depends upon several factors, like, the number and size of the ovum, seasons (in the case of the seasonal breeder) and age. Size enlarges during the breeding season and the early stage of life, immediate to puberty. Most domestic females have a larger and heavier right ovary than the left ovary (Table 21.1). The free surface of the ovary is physiologically active where follicles and corpus luteum develop.

21.2.1 Structure

The outmost layer of the ovary is made of a single layer of cuboidal cells called germinal epithelium, which is continuous with the peritoneum lining the body cavity. The type of germinal epithelium cells is changed from squamous to cuboidal with the advancement of age. Immediately below the epithelium cells, the dense irregular connective tissue layer or tunica albuginea is situated, made of fewer cells and closely packed fibres. Beneath the tunica albuginea is the ovarian cortex that contains follicle and corpus luteum in various stages of development and regression. Fibroblasts, collagen, reticular fibres, blood vessels, nerves, lymphatic vessels, and smooth muscle form the connective tissue of the ovarian cortex. The central part of the ovary is called the medulla. The ovarian cortex is concerned with ova and hormone production; hence, it is considered the functional unit of the ovary. The ova are released from the entire surface of the ovary in most species, except the horse. The ova are released at one edge of the ovary over its surface, having a long groove in the horse called the *ovulation fossa*. The medulla comprises blood vessels, lymphatics, nerves, loose areolar connective tissue and various elastic and reticular fibres. Its structure remains static during various phases of the reproductive cycle. It has a narrow depression along the mesovarian edge, the *hilum*. It acts as the passage of blood vessels, lymphatics and nerves to the ovary. Small masses made of some solid tubules or cords, the *rete ovarii*, are found near the hilum.

The right ovary is rudimentary in birds and usually contains only medullary tissue, and the left ovary is functional. The developing follicles appear like a cluster of tiny grapes. The weight of the ovary of chickens (*Gallus domesticus*) is about 0.3–0.5 g in young and about 60 g in adults, and the same in quail (*Coturnix coturnix japonica*) is 0.06–0.09 g and 10–12 g, respectively. The ovaries of amphibians, reptiles, birds and monotremes have large medullary spaces with fluid-filled cavities called *lacunate*.

21.2.2 Function

Primary functions of the ovaries are gametogenic and hormone production. It produces the female gamete (ova or oocyte, in singular ovum) and various hormones, including oestrogen, progesterone, relaxin, activin and inhibin, under the influence of follicle-stimulating hormone (FSH) and luteinising hormone (LH). The ovarian hormones control the reproductive cycle, the activity of the reproductive tract and secondary sex characteristics.

21.3 Oviduct or Fallopian Tube (Synonym: Salpinx or Tunnel of Love)

The oviducts are the paired, narrow, tortuous tubes that connect the ovary to the uterus and suspend in the mesosalpinx. It is the site of fertilisation in many species.

Oviducts are about 20–30 cm long and 3 mm thick. The oviduct divides into three regions, namely the infundibulum, ampulla and isthmus. The infundibulum is the funnel-shaped structure opening adjacent to the ovary. The opening of the infundibulum (*ostium tubae abdominale*) contains numerous tiny fingers like projections called *fimbriae*. They *are* positioned laterally very close to the ovary and increase the surface area of the infundibulum. The gliding movement of the fimbriae over the ovary's surface helps capture the ovum into the fallopian tube at the time of ovulation. The *ovarian bursa* is the additional structure that ensures ovum capture into the infundibulum. It is made up of connective tissues and the ligaments adjoining the ovary. It is not part of the oviduct, instead developed from mesovarium, mesosalpinx, fimbria and infundibulum regions. It is prominent in rodent and sow and poorly developed in cow and ewe. The infundibulum directly terminates at the ampulla, which constitutes about one-half of the oviductal length and merges with the isthmus of the oviduct. The junction between ampulla and isthmus is called the *ampullary-isthmus junction*, thc site of fertilisation in most mammals. The ampulla is highly convoluted; mucosal cells develop profoundly and feel soft in touch. The diameter of the isthmus is smaller than the ampulla with well-developed muscular layer. The opening of the isthmus into the uterus is called *ostium tubae uterinum*. The junction between the uterus and the isthmus is called the *utero-tubal junction*. In cows, the utero-tubal junction forms a kink, thus blocking the movement of embryos. The kink straightens with decreasing cortisol levels and allows the embryo to enter the uterus. In pigs, the utero-tubal junction is guarded by a finger-like mucosal process and in cats as a barrier to sperm entry and prevents polyspermy.

21.3.1 Histological Features

The oviduct wall has three major layers: mucosa, muscularis and outer serous coat. The oviductal mucosa is made of primary, secondary and tertiary folds, which are highest at the infundibulum and gradually decrease towards the isthmus. The mucosal epithelium contains a single layer of non-ciliated secretory and ciliated columnar epithelial cells together with '*peg cells*' *(depleted secretary cells)*, more abundant at the ampulla and isthmus. The number and activities of the cells vary during different phases of the reproductive cycle. The proportion of ciliated cells is highest at the fimbriae and declines gradually towards the isthmus. The second muscular layer is composed of inner circular and outer longitudinal smooth muscles. Oviductal muscularis thickness increases from the ovarian to the uterine end. These muscles facilitate peristaltic movement of the oviduct and help in gamete transport. The serosa consists of connective tissues, and an external coat originated from the peritoneum. The oviductal musculature at the isthmus is innervated by adrenergic nerves, making the isthmus act like a sphincter to regulate egg transport.

21.3.2 Functions

The oviduct captures and transports the ovum and spermatozoa to the site of fertilisation in opposite directions almost simultaneously. The oviduct also provides a favourable environment for sperm capacitation, fertilisation and cleavage of the embryo. The oviduct transports the early embryo to the uterus, and it gives a protective and nourishing environment to the sperm, oocyte and early embryo.

21.4 Uterus

The uterus is the tubular hollow structure that connects the oviduct to the cervix. It is called the organ of pregnancy as the implantation and development of the foetus occur in the uterus. The uterus consists of a corpus (body), cervix (neck) and two horns (cornua). The mammalian uteri can be classified into four types based on the development of uterine horns, viz. *bicornuate, bipartite, simplex and duplex* (Fig. 21.3). The bicornuate type uterus is characterised by two well-developed uterine horns and a small uterine body connected with a single cervix. The small uterine body results from the poor fusion of paramesonephric ducts during embryonic development. Bicornuate type uterus is found in pigs, dogs, cats, elephants, whales and dolphins. The uterus of ruminants and mare are characterised by moderate to poorly developed horns with a relatively large uterine body. This type of uterus is called the bipartite uterus. In the uteri of the mare, the uterine horn is shorter with a large uterine body than the ruminants due to a high degree of paramesonephric duct fusion. The simplex type of uterus is seen in primates and humans. A single uterus and cervix characterise it without a uterine horn. Two uterine horns form two different compartments in the duplex uterus that open into two cervical canals. The duplex uterus is of two types. In the first category, two cervical canals terminate into two vaginas. This type of uterus is seen in marsupials (opossum). The males of this species have characteristics forked penis that simultaneously deposits semen in each of the cervical canals during mating.

The second category of the duplex uterus is characterised by two cervical canals with a single vaginal canal, as seen in the rabbits, rats, mice and guinea pigs. Generally, the length of the uterine horns is related to litter size. The large uterine horns are characteristic of polytocous species like pigs and bitch to accommodate large litters. The uterus of sow, mare, bitch and queen has longitudinal folds. The morphometric features of the uterus of different animals have been

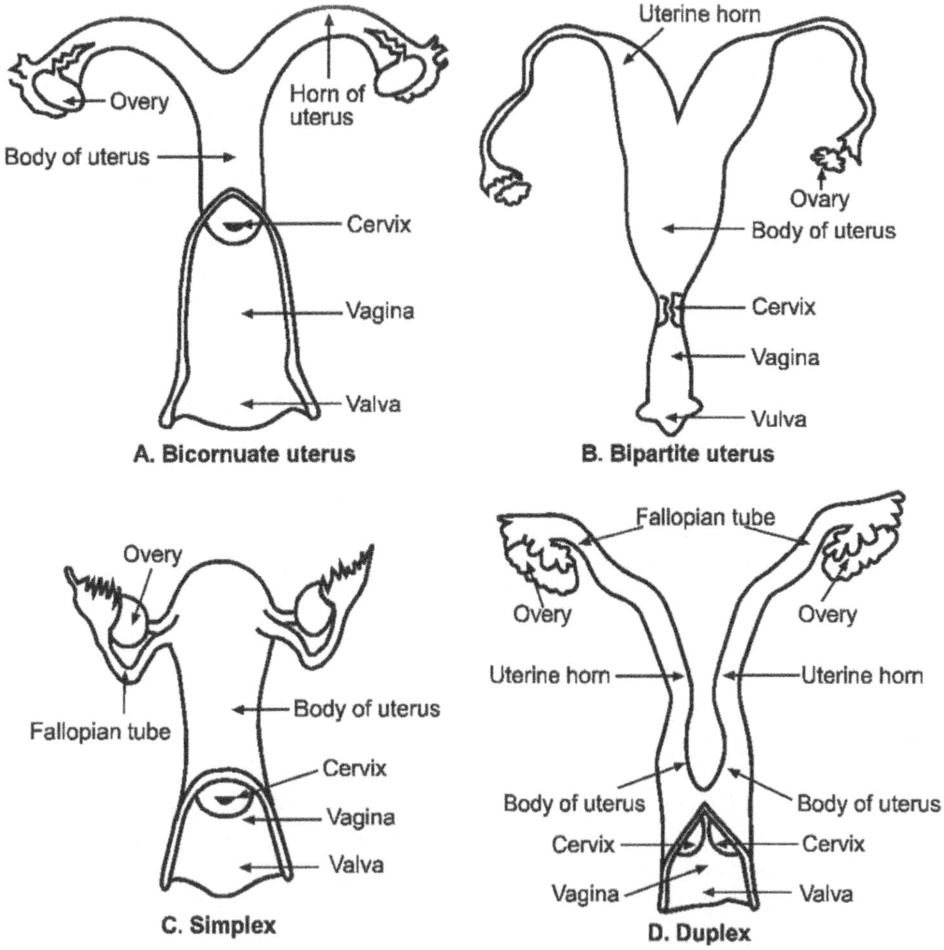

Fig. 21.3 Types of the uterus in mammals. Figures show four types of the uterus, viz. (**a**) **Bicornuate uterus** having two well-developed horns of the uterus, a single body of uterus and cervix; (**b**) **Bipartite uterus** comprising of two moderate size uterine horns and single well-developed uterus and cervix; (**c**) **Simplex** without horns, a single body of uterus and cervix and (**d**) **Duplex** having two horns of the uterus, body of uterus and cervixes

presented in Table 21.1. In artificial insemination, semen is deposited in the body of the uterus. In natural mating, the semen is generally deposited either in the body of the uterus or the cervical canal in dogs, pigs and horses.

21.4.1 Histological Features

The histological features of the uterus reveal a thin outer layer, perimetrium, thick myometrium and inner endometrium. The *perimetrium* is continued with the suspensory ligaments and supports the uterus to remain in position. The *myometrium* is made up of a thinner outer longitudinal and thick inner circular layer of smooth muscles. The muscles exhibit hypertrophy and hyperplasia during pregnancy to support the foetus. The endometrium is composed of an epithelial lining of the lumen followed by a glandular layer and connective tissue layer. The epithelium lining of the lumen is columnar and exhibits structural modification under the influence of hormones during different phases of the reproductive cycle. The endometrial glands are simple, branched and tubular, which coiled progressively towards their ends. These glands are lined by ciliated columnar epithelium. In ruminants, some mushroom-shaped fleshy vascular growths are present over the endometrium, called *caruncles*, where the lobules of the foetal membrane, the *cotyledons*, are attached. The mother's nutrients and the foetus's metabolites are transported through these caruncles. The secretion of the uterine glands is called uterine milk or histotrophs, which supports the survival of the peri-implantation embryo. The secretion of uterine milk is under the influence of progesterone.

21.4.2 Functions

The uterus has the following major functions.

Sperm transport—The myometrial contraction helps transport the spermatozoa from the site of ejaculation to the site of fertilisation.

Luteolysis and control of ovarian cyclicity—The uterus produces $PGF_2\alpha$, which helps to regress the corpus luteum and control the ovarian cyclicity.

Nourishment of the peri-implantation embryo—Endometrial glands secrete uterine milk to provide nutrition to free-living zygote before implantation.

Contribution to the placenta—The uterus helps form the placenta and aids in nutrition, excretion and exchange of CO_2 and O_2.

Expulsion of the foetus and foetal membrane—Strong myometrial contraction at the time of parturition helps expel the foetus and foetal membrane.

21.5 Cervix

The cervix is a thick-walled marrow projection between the uterus and vagina. It acts as a physical barrier to protect the uterus from the invasion of foreign materials, particularly during pregnancy under the influence of progesterone. It considers the point of orientation while examining the genital tract through the rectum in cows. The spindle-shaped flattened passage within the cervix is called the cervical canal (*canalis cervicis*) that opens into the anterior vagina through the external os (*os externum*) and into the uterus through the internal os (*os uterinum*). The finger-like projections at the internal and external os are termed *plicae palmatae*. The os is generally opened during the oestrus phase to allow sperm and during parturition to expel the foetus and remains tightly closed during other phases of the cycle. In cows, the penis touches the external os. The semen is deposited at the os, but in mares, the semen is directly deposited into the uterus due to a dilated cervix. During the opening, the cervix becomes flaccid. The cervix secretes mucus. The consistency of cervical mucus alters during different phases of the reproductive cycle. During oestrus, the cervical mucus is clear and watery, but it becomes thick and sticky during pregnancy and forms a plug to prevent the passage of spermatozoa and microorganisms. The portion of the cervix that projects into the vagina is called *portio vaginalis*.

The junction of the cervix and vagina has a blind pouch called the *fornix* and is prominent in the mare and absent in sow. The spindle-shaped flattened passage within the cervix is called the *cervical canal*, characterised by transverse or spirally interlocking ridges called annular folds. The folds are prominent in cows and ewe and close the cervix securely. The number of annular folds is 5–6 in cows and 3–4 in buffalo. In sow, the fold is arranged as interdigitated pads and appears like a corkscrew adapted to the spiral twisting of the tip of the boar's penis. Several longitudinal folds of mucous membrane characterise the mare's cervix that projects into the vagina.

21.5.1 Histological Features

The cervix consists of mucosa, muscularis and serosa. The mucosal lining of the cervix is composed of tall columnar epithelium. There are goblet cells between the cervical epithelium responsible for cervical mucous secretion. The cervical wall is made of fibroelastic collagenous and muscular tissue responsible for the high tensile strength of the cervix. The cervix of the mare is more dilated due to less cartilaginous tissue.

21.5.2 Functions

The cervix acts as a physiological barrier against external pathogens. It facilitates sperm transport and acts as a sperm reservoir. The cervix is also responsible for selecting viable sperm and prevents the entry of non-viable and abnormal sperm. The dilation of the cervix at the time of parturition helps expel the foetus and foetal membrane. The cervical plug at the time of pregnancy protects the foetus from infections.

21.6 Vagina

The vagina is a female's copulatory organ and also serves as the urination organ. This thin-walled tube is highly elastic and can expand considerably during the expulsion of the foetus; hence, it is also called the birth canal. The muscle of the vagina can contract and expand during *sexual arousal* or *courtship* with the secretion of lubricating mucous.

Histological Features The vagina consists of mucosa, muscularis and serosa layer. The mucosal epithelium varies considerably in different regions. It is columnar and secretary in nature near the cervix becomes stratified squamous towards the posterior vagina. Some touch and pressure-sensitive *chemotactic receptors* are present in the vagina, which generates sexual stimulation and mucous secretion in response to coitus.

Function Vaginal is the copulatory organ where the semen deposits. During the natural service, the semen is deposited in the anterior vagina in most ruminants. The spermatozoa are stored in a vaginal pouch during the non-breeding season in bats. The vagina acts as an excretory duct for cervical, uterine and oviductal secretions. The vaginal environment protects the upper reproductive tract from external pathogens.

21.6.1 Suburethral Diverticulum

The urinary bladder is opened at the floor of the vagina in the form of a blind sac called the *suburethral diverticulum*. It is prominent in cow and sow and short and broad in mare. The suburethral diverticulum has two fossae in bitch. The diverticulum guides during catheterisation but must be avoided during intrauterine infusion and artificial insemination; otherwise leads to injury.

21.7 Vulva

The tract's external portion that extends from the vagina to the exterior opening is called the *vulva*. It is the highly vascular and thick fold of skin comprising two lips, namely *labia majora* and *labia minora* (singular labium). The labia majora is analogous to the scrotum of the male. It protects the reproductive tract from the external environment and contains many sebaceous and sweat glands. Two labia are joined at the dorsal and ventral commissure. The dorsal commissure is generally rounded, and the ventral is pointed in all domestic animals, except in horses, where the dorsal commissure is pointed and the ventral commissure is rounded. In small ruminants, the labia are not developed prominently. In ewe, ventral commissure is longer. In sow, the dorsal commissure is wrinkled and has a small number of hairs. In carnivores, the labia are mostly round in shape and generally pigmented with the covering of dense hair. A transverse cutaneous fold appeared at the dorsal commissure, and the ventral one remains as pointed in carnivores. It usually remains as wrinkled and dry. But, under the influence of oestrogen, it became congested with blood and turned swelled and pinkish, a typical oestrus symptom of oestrus. Vulva remains open during mating, oestrus and parturition. It also helps to void the urine.

21.7.1 Vestibule

This is a tube-like structure present between the labia and the muscles of the vagina. It extends from the suburethral diverticulum to the external vulva; hence, it performs reproductive and urinary functions. It slops ventrally and opens into the vulva. It is about 10–12 cm long in cow and mare, 5–6 cm in sow, 2.5–3 cm in the ewe, doe and bitch and 0.5–1.5 cm in queen.

21.7.2 Vestibular Glands

There are many glands called vestibular glands or *bulbovestibular* (*Bartholin's*) glands at the vestibule wall. They are accessory sex glands of the female and resemble the bulbourethral glands of males. The secretions of these glands act as a lubricant and ease mating and parturition. The vestibular glands can also secret pheromone, which stimulate the males. The vestibular glands are well developed in cows, queens and ewe that open through a single duct on each side of the vestibule. The vestibular glands are smaller in bitch and mare, which open as linear series. Vestibular glands are partially developed in rodents. On the vestibule floor, the remnants of the Wolffian duct are present in the form of a blind pouch called *Gartner's ducts*. The prostate gland is partially developed in rodents, bitch, mare and ewe with smaller vestibular glands.

21.7.3 Vestibular Bulbs

It is present on the vestibule wall in mare and bitch. It contains profuse veins (venules) with erectile tissues and is homologous to the *cavernous* structure of the penis. After intromission, the bulb becomes engorged and erects with the trapping of blood that holds the glans penis, resulting in the *locking of the penis*.

21.7.4 Hymen

In horses, llamas, elephants, chimpanzees, manatees, whales, guinea pigs and women, a well-developed transverse fold of membranous covering is present at the junction between the vagina and vestibule that partly covers the external vaginal opening, called the *hymen*. It separates the vagina from the vulva. In domestic animals, it is poorly developed and appears as small oblique folds at the vagino-vestibular junction. It seals the opening of the vagina during the non-breeding season in guinea pigs.

21.7.5 Clitoris

It is a small mass of erectile tissue suspended by a ligament from the pubic bone and found within the ventral commissure of the vulva. It is homologous to the glans penis of the male. Apart from erectile tissue, the clitoris contains tiny blood vessels and sensory nerve endings. The clitoris is very small in small ruminants and sows, and large in carnivores and mares. The body (corpus) and glans (*glans clitoridis*) have two parts. The body is analogous to the prepuce of a male. The erectile tissues are elongated and extended internally, called *crura*. They attach to the ischiatic arch through its left and right branches. Two crura are closely associated with forming the body placed in a fossa, *fossa clitoridis*, covered with the mucosal fold. The contraction of the

Table 21.2 Major characteristics differences of the functional morphology of different female reproductive systems

Parts of the system	Species	Characteristics
Ovary	Cow, sow, ewe, human	Cortex at outside and medulla at inside; ovulation over the entire surface
	Mare	Cortex and medulla innervated; ovulation occurs only at the ovulation fossa that may result in deep-seated corpus luteum
	Fowl	Only left ovary functional; large medullary space with fluid-filled cavities (lacunate); hierarchical follicular growth
Uterus, cervix and vagina	Elephants, rabbits, rodents, opossums, aardvarks	Duplex: two uterine horns, no uterine body, two cervixes, one or two vaginas
	Cow, ewe, mare, sow, bitch, queen	Bicornuate: two uterine horns, one uterine body, one cervix, one vagina
	Primates, monkeys, human	Simplex: no uterine horns, one uterine body, one cervix, one vagina
	Fowl	The secondary sex organs are infundibulum, magnum, isthmus, uterus (shell gland), vagina and cloaca
Cervix	Cow, buffalo, ewe	Annular rings—folding of the small smooth musculature
	Mare	Longitudinal folds
	Sow	Interdigitating pads, corkscrew like, no fornix
	Bitch, queen	Irregular
Vulva	Cow, ewe, sow, mare	Labia majora
	Human	Labia majora and minora
Hymen	Ewe, horses, elephants, humans, chimpanzees, manatees, whales, guinea pigs	Well developed
	Domestic animals	Poorly developed

vestibular and vulval sphincter muscles elevates the clitoris, and the clitoris protrudes between the vulval lips. It is called the winking of the clitoris. The clitoris becomes swollen and erect during coitus and has a role in tactile sexual stimulation. The stimulation of clitoris following the artificial insemination has increased the conception rate by promoting LH surge and ovulation.

Major characteristics of the functional morphology of female reproductive organs are summarised in Table 21.2.

> Know More.........
> **Skene's glands**, also known as *periurethral glands*, are paired glands situated at the anterior wall of the vagina near the lower urethra, secretes a fluid homologous to plasma ultrafiltrate. These glands are the source of female ejaculate and are considered the 'female prostate'.

21.8 Secondary Sex Organs of Birds

The secondary sex organs of birds consist of the infundibulum, magnum, isthmus, uterus (shell gland), vagina and cloaca. The size of the entire reproductive tract is variable, depending on the species. It is about 70–80 cm in fowl (Fig. 21.4) and about 60 cm in guinea fowl. The oviduct is divides into three parts. The first part of the avian oviduct is *infundibulum* (*ostium*). It captures the ovum (*yolk* of the egg) and is also considered the site of fertilisation in avian species. The next part of the oviduct is the magnum. It constitutes the

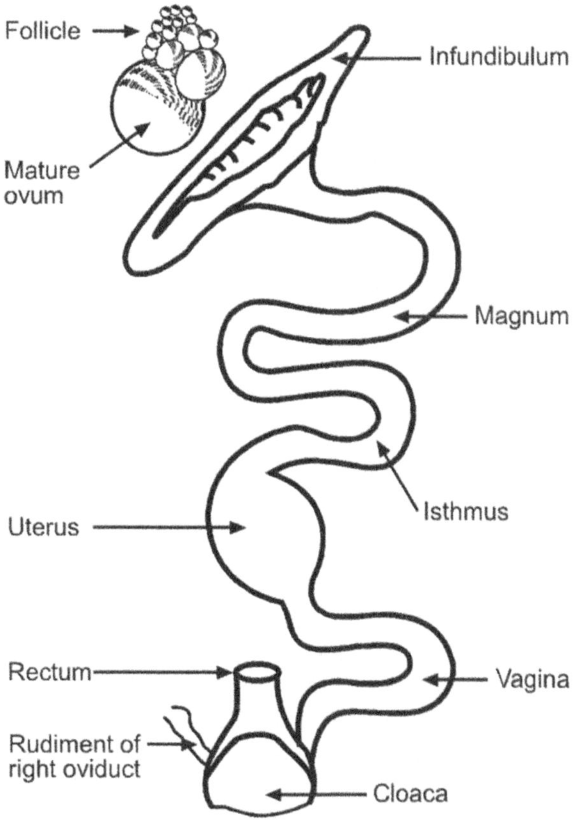

Fig. 21.4 Female reproductive system of birds. Figure shows the cluster of **follicles** with a **mature ovum** and with reproductive tract comprising **infundibulum** (10%), **magnum** or ampulla (50%) and **isthmus**, **uterus** and **vagina**, each containing 14% of the entire length of the tract. The position of the **rectum**, **cloaca** and **rudiment of the right oviduct** is also shown in the figure

Table 21.3 Time spent in the oviduct to form an egg in different domestic birds

Part of oviduct	Fowl	Duck	Quail	Turkey	Guinea fowl
Infundibulum	15–30 min	15–30 min	15–30 min	15–30 min	15–30 min
Magnum	3 h	2 h 45 min	2 h 15 min	2 h 45 min	2 h 30 min
Isthmus	1 h 15 min	2 h 15 min	1 h 45 min	1 h 15 min	1 h
Uterus	19–21 h	18–19 h	19–20 h	22–24 h	20–21 h
Vagina	Few min	Few min	Few min	Few min	Few min
Total	24–25 h	23–25 h	24–25 h	26–29 h	24–25 h

Data compiled from various sources

largest portion of the avian oviduct. The magnum has thick muscular walls containing an albumin secreting gland that helps form the albumin layer over the yolk. The *chalazae* (in singular, chalaza) form during albumin secretion. Chalazae are the albuminous spring-like cords that originate from the two opposite ends of the yolk and extend towards the extremities of the egg to hold the yolk in position. The magnum proceeds to a narrow part of the tract called the *isthmus*. The soft shell membrane of the egg forms, and water and minerals add to the isthmus. The isthmus terminates at the uterus. It has secretory glands that form a hardy egg shell. The white colour of the egg is due to the secretion of calcium carbonate from the shell glands. Various colours or mosaic patterns of eggs in birds, lizards and reptiles are due to the secretion of different pigments from these glands that are controlled genetically to camouflage and protection from the predators. At the last phase of the shell formation, a gelatinous protein-rich fluid, called *bloom*, is formed over the shell. Bloom can protect the egg against the invasion of microorganisms. The next part is a narrow muscular walled *vagina* followed by a *cloaca*. The muscle of the vagina helps to expel the egg through the cloaca. Just before laying, the pores of the egg shell are sealed by the mucous secreted by the vaginal glands. The special vaginal tubules (*spermatheca*) are found in the vagina of birds, lizards and reptiles, which can store the spermatozoa. Spermatozoa are stored in storage glands (*sperm nests*) found in birds' infundibulum and uterovaginal junction. Spermatozoa can be stored in the *seminal receptacles* located in the infundibulum in some snakes. An ovum takes at least 24 h to form an egg (Table 21.3). Hence, a bird can't lay two eggs a day.

21.9 Some Developmental Disorders in the Female Reproductive System

Some defects may cause abnormal development of the female reproductive system. The formation of ovotestes and/or hypoplastic gonads is seen due to chromosomal disorders. Instead of a bipartite uterus, hypoplastic uterine horns and a single uterine body may form in cattle, horses and other animals. With hypoplastic vesicular glands, freemartinism is common in cattle, goats and horses. Ovotestes can also be developed in the bursa of dogs and pigs, where hypoplastic testes with large epididymis develop nearer to the inguinal canal or scrotal sac and are attached to the uterine horns. Agenesis in one or both ovaries may occur in ruminants, swine and dogs, but the occurrence is rare.

Learning Outcomes

The female reproductive system comprises of two ovaries with a reproductive tract constituting oviducts, uterus, cervix, vagina and vulva. The histomorphology of ovaries and reproductive tract are variable in different species, breeds, seasons, breeding times and phases of the reproductive cycle. The broad ligament is the main supporting ligament that holds the different parts of the female reproductive tract in position.

- **Ovary**: Ovary is generally oval with pointed ends. Normally, the right ovary is more active in almost all mammals. In birds, the left ovary is functional. The position of the ovary is variable in different species. The ovary is divided into the outer cortex and inner medulla, and the blood vessels, lymphatics and nerves are entered through the hilum. Cortex is involved in ova and hormone production.
- **Oviduct or fallopian tube**: Fallopian tube captures the ovum through its finger-like projection called fimbriae at the infundibulum. The remaining portion of the oviduct consists of the ampulla and isthmus. The ampullary-isthmic junction is the site of fertilisation in mammals, and the isthmus joins the uterus through utero-tubal junction. The number and activities of oviductal cells are varied during different phases of the reproductive cycle. Oviduct helps in gamete transport and acts as the site of fertilisation.
- **Uterus**: The uterus consists of two uterine horns, the neck and body of the uterus, which communicates with the vagina through the cervix. Generally, four

(continued)

types of uterine structure are found in various mammals. The uterus is morphologically classified as bipartite, bicornuate, simplex and duplex uterus. It supports the pregnancy with its varied patterns of cellular structures and fleshy vascular growths over the endometrium, caruncles and lobules of the foetal membrane, the cotyledons.

- **Cervix**: The cervix is a thick-walled heavy sphincter-like configuration that divides the internal environment of the uterus from the external environment of the vagina. The cervical canal is characterised by transverse or spirally interlocking ridges called annular folds. It acts as a physical barrier to protect the uterus from the invasion of foreign materials, particularly during pregnancy.
- **Vagina**: Vagina is a female's copulatory organ and also serves as the organ of urination. This thin-walled tube is highly elastic and able to expand considerably during the expulsion of the foetus; hence, it is also called the birth canal. The opening of urethra at the vagina form a blind sac is called the suburethral diverticulum, which assists during catheterisation.
- **Vulva**: The exterior opening of the female reproductive tract is the vulva, consisting of two lips, the labia majora and labia minora. The secretion of vestibular glands moistens the passage during copulation and parturition and secretes pheromone during oestrus. There are two additional structures, the hymen and clitoris, which play a role during coitus and tactile sexual stimulation.
- **Secondary sex organs of birds**: Birds have infundibulum, magnum, isthmus, uterus (shell gland), vagina and cloaca as secondary sex organs. Different parts of this tract have individual contributions to the formation of an egg. The ovum has to spend at least 24 h in different parts of the tract to form a whole egg, and thus it is impossible to two eggs in a single day.

Exercises

Objective Questions

Q1. Which part of the broad ligament is attached to the fallopian tubes?
Q2. Which part of the ovary (cortex or medulla) exhibits gametogenic function?
Q3. Which ovary is functional in almost all mammals?
Q4. Ovulation fossa is present in which animal species?
Q5. Which is the site of fertilisation in most mammals?
Q6. The ovarian bursa is well developed in which of the following animals—cow, ewe and mice?
Q7. Whether the appearance of fimbriae endocrine dependent?
Q8. Which type of uterine configuration is found in the dog?
Q9. What are the major features of the bipartite uterus?
Q10. How many cervixes are found in duplexes?
Q11. Which parts of the female reproductive tract are involved in the formation of bloom in the bird?
Q12. Which gland of the female reproductive system secretes pheromone during oestrus?
Q13. Which physiological configuration of the cervix blocks the passage of spermatozoa?
Q14. In which particular phase(s) of the reproductive cycle does the vagina become open in animals?
Q15. In which animal the spermatozoa may be stored in a vaginal pouch during the non-breeding season?
Q16. Which structure of the vagina assists catheterisation?
Q17. Which organ of the female reproductive system is analogous to the male's scrotum?
Q18. Which parts of the female reproductive tract of birds secrete albumin?
Q19. Which structure of the female reproductive system is homologous to the cavernous structure of the penis?
Q20. Name the special vaginal tubules of birds that hold the spermatozoa.

Subjective Questions

Q1. Describe the various pattern of the uterus according to its structural configuration.
Q2. Write the various functional roles of the cervix in different phases of the reproductive cycle.
Q3. Describe the physiological role of the vagina in reproduction.
Q4. Describe the role of the vulva and its accessory structures in mating.
Q5. Write the role of various parts of secondary sex organs in birds in egg formation.

Answer to Objective Questions

A1. Mesosalpinx
A2. Cortex
A3. Right
A4. Horse
A5. Ampullary-isthmic junction
A6. Mice
A7. Yes
A8. Bicornuate
A9. Two moderate to poorly developed horns with a moderate to well-developed single uterus and cervix

A10. Two
A11. Uterus
A12. Vestibular glands
A13. Cervical plug
A14. During oestrus and parturition
A15. Bat
A16. Suburethral diverticulum
A17. Labia majora of vulva
A18. Magnum
A19. Vestibular bulbs
A20. Spermatheca

Keywords for the Answer to Subjective Questions

A1. Bicornuate uterus, bipartite uterus, simplex and duplex
A2. Physical barrier, role of progesterone, cervical plug
A3. Copulatory organ, birth canal, chemotactic receptors
A4. Influence of oestrogens on vulva, vestibule and its glands and bulbs, hymen and clitoris
A5. Role of secretion of various parts of the tract, formation of chalazae, bloom

Further Reading

Textbooks

McAuliffe SB (2013) Reproductive disorders (Chapter 12). In: Knottenbelt and Pascoe's color atlas of diseases and disorders of the horse, 2nd edn. Elsevier Ltd., Philadelphia, PA, pp 443–513. https://doi.org/10.1016/C2010-0-66008-1

Morel MCGD (2015) Equine reproductive physiology, breeding and stud management, 4th edn. CABI, Wallingford, pp 1–411. ISBN: 978-1-78064-442-4.

Pineda MH, Dooley MP (2003) McDonald's veterinary endocrinology and reproduction, 5th edn. Iowa State Press, Ames, IA, p 597

Research Articles

Abiaezute CN, Nwaogu IC, Okoye CN (2017) Morphological features of the uterus during postnatal development in the West African Dwarf goat (Capra hircus). Anim Reprod 14(4):1062–1071. https://doi.org/10.21451/1984-3143-AR886

Alwan AF, Al-Saffar HE, Khammas DJ (2005) Biometry of genital organs in Iraqi female buffalo. Iraqi J Vet Sci 19(1):77–81. https://doi.org/10.33899/ijvs.2005.37421

Bergfelt DR (2016) Anatomy and physiology of the mare. In: Equine medicine (11 chapter). Available at: https://veteriankey.com/anatomy-and-physiology-of-the-mare/

Carvalho NAT, Gimenes LU, Reis EL, Cavalcante AKS, Mello JE, Nichi M, Nicacio AC, Nasser LFT, Rezende LFC, Wisnesck CA, Moura CEB, Benedicto HB, Bombonato PP, Baruselli PS (2010) Biometry of genital system from buffalo (Murrah) and bovine (Nelore) females. In: Anatomy and fisiology, Proceedings 9th World Buffalo Congress, pp 276–290

Conze T, Wehrend A (2017) Sonographische Darstellung der physiologischen Ovarienbei der Katze. Tierarztl Prax Ausg K 45(3):199–203. https://doi.org/10.15654/TPK-160164

Devkota B, Singh AK (2017) Morphometry of the female reproductive organs of the Murrahcross buffaloes in Chitwan. Nepal J Agric Forest Univ 1:113–117

Drennan WG, Macpherson JW (1966) The reproductive tract of bovine slaughter heifers (a biometrical study). Can J Comp Med Vet Sci 30: 224–227

Eker K, Salmanoglu MR (2006) Ultrasonographic monitoring of follicular development, ovulation and corpora lutea formation in a bitch. Turk J Vet Anim Sci 30:589–592

Gupta MD, Akter MM, Gupta AD, Das A (2011) Biometry of female genital organs of back Bengal goat. Int J Nat Sci 1(1):12–17. https://doi.org/10.3329/ijns.v1i1.8609

Haque Z, Haque A, Quasem MA (2016) Morphologic and morphometric analysis of the ovary of black Bengal goat (Capra hircus). Int J Morphol 34(1):13–16

Hewitt SC, Winuthayanon W, Korach KS (2016) What's new in estrogen receptor action in the female reproductive tract. J Mol Endocrinol 56(2):R55–R71. https://doi.org/10.1530/JME-15-0254

Hoque MSA, Islam Md M, Selim Md AS (2016) Interspecies differences on ovarian parameters between black Bengal goat and indigenous Bengal sheep in view of in vitro maturation. Adv Life Sci 6(3):54–60. https://doi.org/10.5923/j.als.20160603.02

Jaji AZ, Buduwara RA, Akanmu AI, Zachariah M, Luka J, Gambo B (2012) Pregnancy related biometric changes in the ovaries and uterus of the Sahelian goat. Sokoto J Vet Sci 10(1):17–21

Jaji AZ, Akanmu AI, Buduwara RA, Elelu N, Kigir E, Mahre MB, Gambo B (2013) Pregnancy related biometrical changes in the ovaries and uterus of the Balami sheep. Glob J Med Res Vet Sci Vet Med 13(2):1–7

Jones SM, Troxel TR (2019) Understanding reproductive physiology and anatomy of the mare. In: Horses. National Institute of Food and Agriculture, United States Department of Agriculture, Corp Ext. Available at: https://horses.extension.org/2019/07/understanding-reproductive-physiology-and-anatomy-of-the-mare/. 3.10.2019

Khaton R, Sarder MJU, Gofur MR (2015) Biometrical studies of reproductive organs of dairy cows of different genotypes in Bangladesh. Asian J Anim Sci 9:388–395. https://doi.org/10.3923/ajas.2015.388.395

Knight KM, Moalli PA, Nolfi A, Palcsey S, Barone WR, Abramowitch SD (2016) Impact of parity on ewe vaginal mechanical properties relative to the nonhuman primate and rodent. Int Urogynecol J 27(8):1255–1263. https://doi.org/10.1007/s00192-016-2963-2

Lévy X (2016) Videovaginoscopy of the canine vagina. Reprod Domest Anim 51(Suppl. 1):31–36. https://doi.org/10.1111/rda.12785

Lohachit C, Bodhipaksha P, Tesprateep T (1981) Studies on the biometry of the reproductive tract in Thai female swamp buffalo. Food and Agricultural Organisation of the United States, AGRIS. Available at: http://agris.fao.org/agris-search/search.do?recordID=XB8121585

Ono M, Akuzawa H, Nambo Y, Hirano Y, Kimura J, Takemoto S, Nakamura S, YokotaH HR, Higuchi T, Ohtaki T, Tsumagari S (2016) Analysis of the equine ovarian structure during the first twelve months of life by three-dimensional internal structure microscopy. J Vet Med Sci 77(12):1599–1603. https://doi.org/10.1292/jvms.14-0539

Vicencio NG, Viernes VD Jr, Ocampo LC, Ocampo MB (2017) Gross anatomy of the female reproductive organs of Philippine Native pig (Susscrofa L). Int J Agric Technol 13(7.2):1627–1638

Female Reproductive Physiology 22

Pradip Kumar Das, Joydip Mukherjee, and Dipak Banerjee

Abstract

The hypothalamic-pituitary-ovarian (HPO) axis is the key regulator of female reproductive functions. HPO axis integrates internal and external signals to coordinate reproductive functions through the neuroendocrine network. Gonadotropin-releasing hormone (GnRH) plays a central role in the HPO axis to control pituitary gonadotropin secretion and ovarian steroidogenesis. The ovarian steroids, particularly estradiol regulate GnRH secretion through the negative feedback mechanism. At the initiation of puberty, the negative feedback of estradiol decreases, leading to activation of the GnRH surge centre and commencement of the estrous cycle, growth of the follicles, and ovulation. The estrous cycle divides into four distinct stages: estrus, proestrus, metestrus, and diestrus. The follicles' development started during the foetal life and continued through a dynamic process during various phases of post-natal periods called folliculogenesis. Rhythmic alterations of follicular dynamics occur during each estrous cycle after puberty. The maturation of the female gamete, the ovum, is completed after puberty which is subsequently released from the matured or Graafian follicles at the end of estrus, followed by the formation of corpus luteum (CL). The metestrus and diestrus stage is under the control of progesterone secreted from CL. The CL undergo lysis if the animals fail to conceive by prostaglandin F2α (PGF2α) secreted from the endometrium. Successful conception prevents the release of PGF2α, and the CL is sustained throughout the pregnancy. The reproductive cycles of females are species specific and are controlled by several factors like environmental cues, photoperiod, nutritional state, stress, and diseases. This chapter encompasses the integrated neuroendocrine and molecular mechanisms involved in different aspects of female reproduction, such as initiation of puberty, ovarian steroidogenesis, oogenesis, and folliculogenesis, including ovulation in domestic, wild animals, and birds.

P. K. Das (✉) · J. Mukherjee · D. Banerjee
Department of Veterinary Physiology, West Bengal University of Animal & Fishery Sciences, Kolkata, West Bengal, India

© The Author(s), under exclusive license to Springer Nature Singapore Pte Ltd. 2023

P. K. Das et al. (eds.), *Textbook of Veterinary Physiology*, https://doi.org/10.1007/978-981-19-9410-4_22

Graphical Abstract

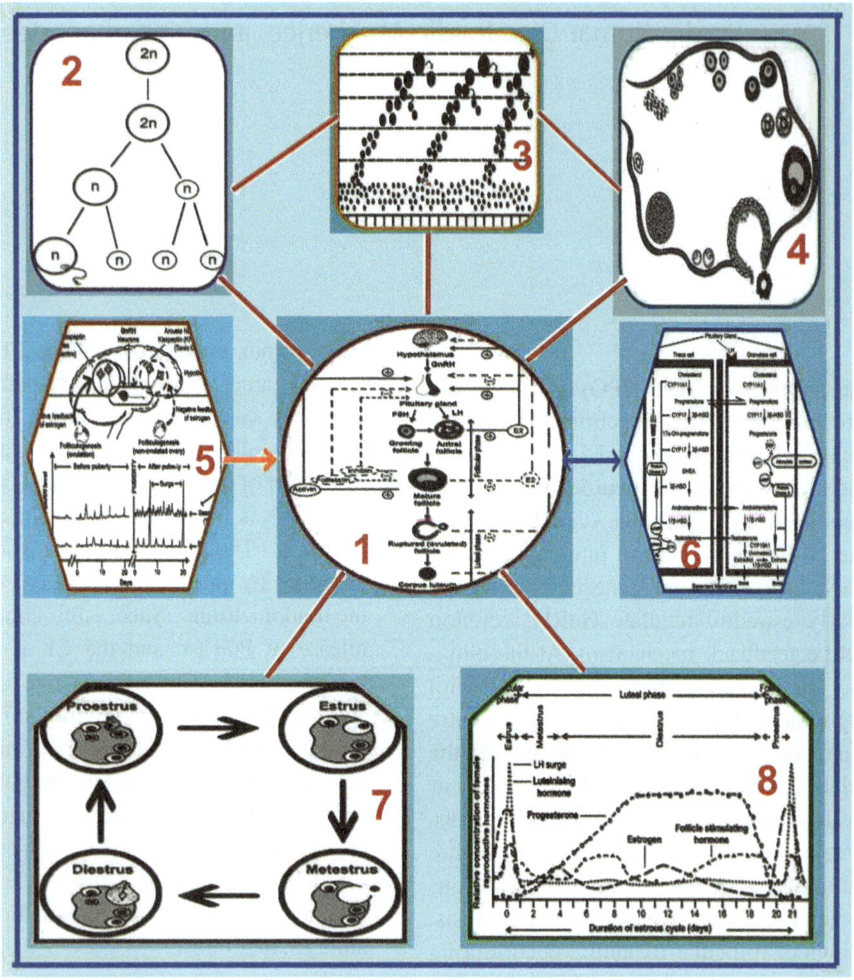

Description of the graphic: All the events of female reproduction control by the Hypothalamic-Pituitary-Ovarian (HPO) axis (**1**). The oogenesis and folliculogenesis (**2**) initiate during foetal life and undergo rhythmic alterations during each estrus cycle to become dominant follicles through follicular waves (**3**). The matured follicle undergoes ovulation and the formation of the corpus luteum (**4**). The activation of the HPO axis leads to the attainment of puberty (**5**), controls ovarian steroidogenesis (**6**), rhythmic sexual behavioural patterns, the estrous cycle (**7**), and altered endocrine milieu (**8**)

Keywords

HPO axis · Puberty and estrous cycle · Oogenesis and folliculogenesis · Ovulation in mammals and birds · Corpus luteum

Learning Objectives

- Role of various hormones, growth factors, and cytokines in female reproduction.
- The neuroendocrine control of puberty and the estrous cycle.
- Ovarian gametogenesis and steroidogenesis, including its endocrine and molecular mechanisms.
- Follicular dynamics, ovulation, and corpus luteum formation and regression.
- Ovulation and egg formation in birds.

22.1 Endocrinology of Female Reproduction

22.1.1 Hypothalamic-Pituitary-Ovarian Axis

The reproductive activity of females is regulated by the hypothalamic-pituitary-ovarian (HPO) axis comprising the hypothalamus, the anterior pituitary, and the ovaries (Fig. 22.1). This HPO axis is responsible for regulating both centrally and peripherally produced reproductive

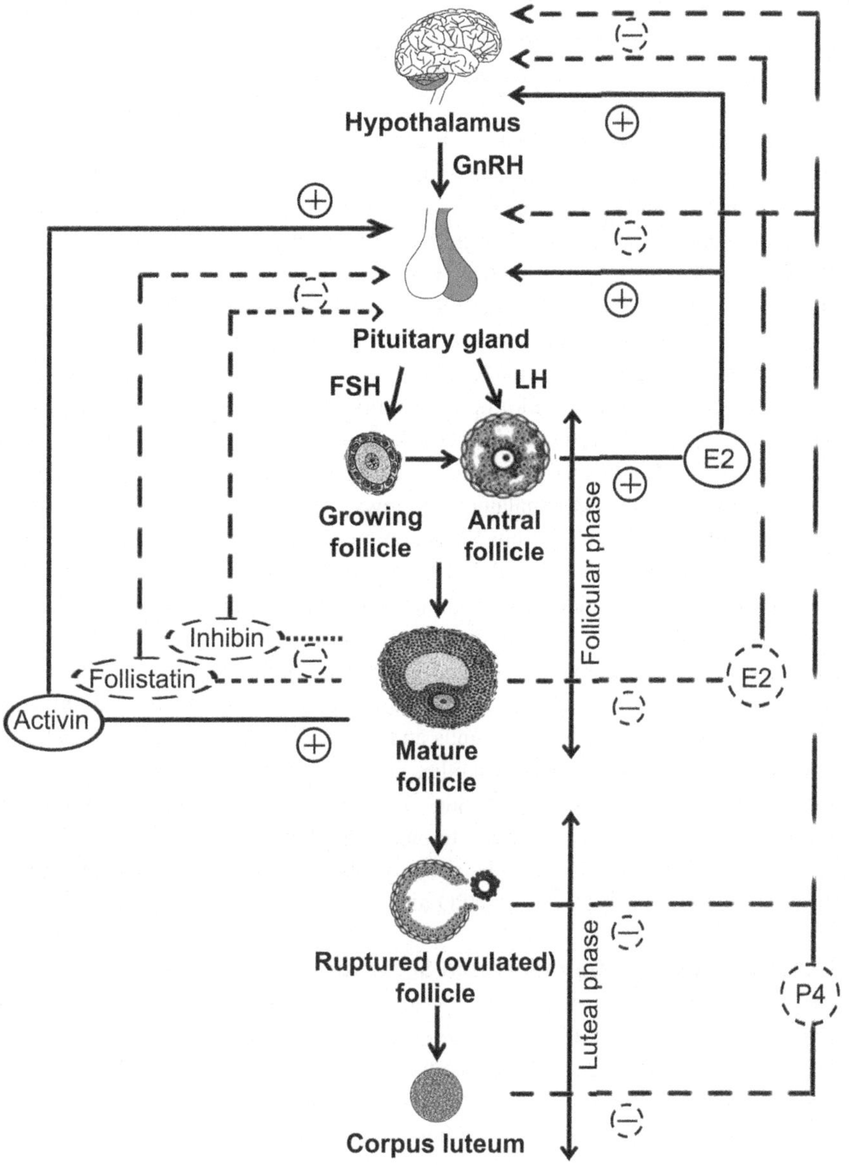

Fig. 22.1 Hypothalamus-Pituitary-Ovary (HPO) axis. Figure shows the interrelationship among the **hypothalamus**, **pituitary gland,** and **ovary**. The hypothalamus releases **GnRH** (gonadotropin-releasing hormone), which acts on the pituitary gland to secrete gonadotropins, **FSH** (follicle-stimulating hormone), and **LH** (luteinising hormone). The FSH acts on the **growing follicles** and stimulates their growth to become **antral follicles**. LH acts on the mature follicle and causes **rupturing of the follicle** or ovulation, followed by the formation of the **corpus luteum**. The first two steps, the follicular development and development of mature follicle, are considered the **follicular phase**, where released **E2** (estrogen) and **activin** positively stimulate the axis to release gonadotropins (illustrated by the solid positive (+) lines). The release of FSH from the pituitary gland is suppressed in the follicular phase (shown by the dotted negative (−) lines) by the **inhibin**, **follistatin**. The FSH secretion is also negatively controlled by excess secretion of E2 in the follicular phase, and secretion of **P4** (progesterone) in the **luteal phase**, comprising ovulation and corpus luteum formation

hormones. The central part of the axis includes GnRH, secreted from the hypothalamus, and gonadotropins, viz. LH and FSH are released from the anterior pituitary.

22.1.1.1 Hypothalamic GnRH Secretion

GnRH is a decapeptide produced in the hypothalamus to regulate the release of two gonadotropins, namely LH and FSH, from the anterior pituitary. Two different isoforms of GnRH have been identified for the vertebrate species. GnRH-I (hypothalamic origin) acts as a neurohormone to release pituitary gonadotropin, and GnRH-II (secreted from mid-brain regions) has neuromodulatory roles. GnRH is released from the hypothalamus in two fashions, pulsatile or tonic and surge. GnRH is continuously secreted at low pulses from the ventromedial and arcuate nucleus in tonic or pulsatile secretion. The surge centre or pre-ovulatory GnRH centre is a pre-optic nucleus, the anterior hypothalamic area, and the suprachiasmatic nucleus from which GnRH is secreted at high pulses followed by rapid declining at the time of ovulation. The surge centre is sensitive to estrogens. Steroid hormones, along with inhibins and activins produced by gonads, modulate the secretion of gonadotropins by negative and positive feedback control.

The secretion of GnRH is controlled by some small peptides called RFamides. They are so named due to the Arg-Phe-NH2 motif at the C-terminus. Two groups of RFamide peptides, namely kisspeptins and Gonadotropin-Inhibiting Hormone (GnIH), have stimulatory and inhibitory roles in GnRH secretion, respectively. The GnRH-secreting neurones are co-localised with the kisspeptin–neurokinin B–

dynorphin (KNDy) neuronal network. The kisspeptin-secreting neurones in the hypothalamus are identified in the anteroventral periventricular nucleus (AVPV), the rostral periventricular region of the third ventricle (RP3V), periventricular nucleus (AVPV), and arcuate nucleus in case of rodents. The predominant site for kisspeptin-secreting neurones in ruminants is the arcuate nucleus. In humans, the kisspeptin neurones are localised at the rostral pre-optic area (POA) and in the infundibular nucleus. Kisspeptin is a potent stimulator of the HPO axis. It stimulates GnRH-secreting neurones by activating protein kinase C involving inositol triphosphate and diacylglycerol pathway. The RFamides were found to decrease gonadotropin secretion in a dose-dependent manner, so designated as GnIH. Two RFamide-related peptides (RFRPs), namely RFRP-1 and RFRP-3, have been identified in the hypothalamus of cows, rat, rhesus macaque, and humans that inhibits the secretion of LH from the anterior pituitary. The RFRP-3-secreting neurones are localised in the dorsomedial hypothalamus, and their axonal projections extend up to the pre-optic area and come in contact with GnRH neurons. GnIH and RFRPs act through their receptors GPR-74 and GPR-147. In humans, GPR-147 has been identified in the hypothalamus and pituitary. Other than kisspeptins and RFRPs, several other factors exert stimulatory (norepinephrine, neuropeptide Y, dopamine) or inhibitory (beta-endorphin, progesterone, interleukin-1) roles in controlling the HPG axis. Estradiol has both a stimulatory and inhibitory role over the HPO axis.

22.1.1.2 Pituitary Gonadotropin Secretion

The anterior pituitary secretes two gonadotropins from its gonadotroph, FSH and LH, glycoproteins in nature. The frequencies of GnRH pulses determine the secretion of FSH and LH. Slow GnRH pulse frequency ($<$1 pulse per 2–3 h) increases FSH secretion by augmenting FSH-β gene transcription. In contrast, a rapid GnRH pulse ($>$1 pulse per h) increases LH-α and LH-β gene transcription causing the release of LH. The LH pulse frequency increases during the follicular phase, particularly in the pre-ovulatory period and gradually declines during the luteal phase. The FSH stimulates follicular growth and estradiol formation by inducing the aromatase enzyme. At the late stages of the follicular phase, activins and estradiol enhance the actions of FSH. LH facilitates ovarian steroidogenesis in the pre-ovulatory follicles. The theca and granulosa cells of the ovary are stimulated independently by LH and FSH. The ovarian steroidogenesis in the pre-ovulatory follicle is mediated through LH receptors on theca and FSH receptors on granulosa cells (see 'two cell two gonadotropin theories'). The steroidogenic acute regulatory protein (StAR protein) facilitates androstenedione production, which serves as an estrogen precursor after its diffusion to the granulosa cells.

22.1.1.3 Ovarian Hormones

Besides producing mature oocytes, ovaries act as dynamic endocrine glands that secret steroid and peptide hormones and play pivotal roles in female reproduction. The ovaries have two important steroid hormones: *estrogens* (*estradiol, estrone,* and *estriol*) and *progesterone* (*progestin*). *Activins, inhibins*, and *follistatin* are the peptide hormones produced in the ovaries.

22.1.1.3.1 Steroidogenesis

Cholesterol is the prime precursor of all steroid hormones. Steroidogenesis in the ovaries requires the delivery and uptake of cholesterol precursors and their conversion into estrogen and progesterone through a series of enzymatic reactions catalysed by steroid cytochrome P450 (CYP) hydroxylases and hydroxysteroid dehydrogenases.

22.1.1.3.1.1 Cholesterol Uptake

Cholesterol can't dissolve in the body fluid due to its hydrophobic nature and is carried through lipoproteins. In cattle, pigs, and humans, cholesterol is incorporated by the steroidogenic cells in low-density lipoprotein (LDL). The follicular and luteal cells of the ovaries have lipoprotein receptors, namely scavenger receptor class B member 1 (SR-BI). It binds with LDL and incorporates cholesterol inside the cells as lipid droplets in the form of cholesterol esters. The cholesterol ester hydrolase enzyme converts the cholesterol esters to free cholesterol in the cytoplasm of these cells. This free cholesterol carries from the outer to the inner mitochondrial membrane through steroidogenic acute regulatory protein (StAR). This step is the rate-limiting step of steroidogenesis.

22.1.1.3.1.2 Steroidogenesis

In the mitochondria of the ovarian cells, free cholesterol is converted to pregnenolone by the enzyme cytochrome P450 (a cholesterol side-chain cleavage enzyme, P450scc, CYP11A1). The pregnenolone then diffuses into the smooth endoplasmic reticulum, where it converts to progesterone by the enzyme 3β-hydroxysteroid dehydrogenase (3β-HSD).

22.1.1.3.1.3 Steroidogenesis in Corpus Luteum

The luteal cells have the receptors for LH, responsible for upregulation of StAR and LDL in the luteal cells along with stimulation of P450scc. Thus, progesterone is produced in luteal cells. The lutein cells lack P45017-OH, 17-α-hydroxylase and aromatase; hence, androstenedione and estrogens are not synthesised in CL.

22.1.1.3.1.4 Steroidogenesis in the Placenta

In sheep, horses, cats, guinea pigs and humans, the placenta, apart from the ovaries, plays a pivotal role in maintaining the pregnancy by producing progesterone. The trophoblast cells are mainly responsible for steroidogenesis (androgens and

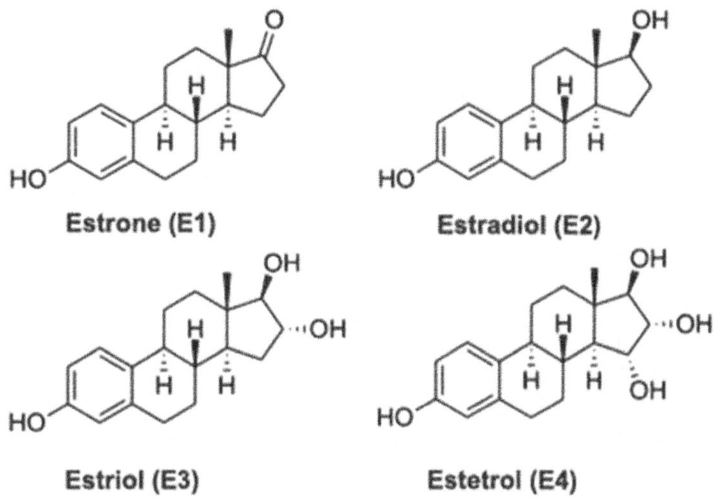

Fig. 22.2 The chemical structures of estrogens. Estrone (E1) has single hydroxyl (–OH), Estradiol (E2) has two hydroxyls, Estriol (E3) has three hydroxyls, and Estetrol (E4) has four hydroxyls. Estrone is synthesised from androstenedione by the enzyme aromatase. It is also produced reversibly from estradiol by the enzyme 17β-hydroxysteroid dehydrogenase (17β-HSD) in various tissues like the ovary, liver, uterus, and mammary gland. Estradiol is chiefly available in females produced from androstenedione/testosterone by aromatisation in the granulosa cells of the ovarian follicle; beyond that, it can be produced in adrenal glands, fat, liver, the breasts, and brain. Estriol is mostly found in pregnant females produced in the placenta. Estetrol is an intermediate product, synthesised during pregnancy only in the foetal liver from estradiol (E2) and estriol (E3) by the actions of two enzymes, 15α-hydroxylase, and 16α-hydroxylase

estrogens) in cows, sheep, pigs, and rats. In these species, the enzyme 17α-hydroxylase (Cytochrome P450 17A1, CYP17A1) mainly controls the steroidogenesis process. In species such as horses, rhesus monkeys, and humans, the placenta is devoid of these enzymes; hence, the gonadal interstitial cells and adrenal gland provide the androgen for the production of placental estrogens in horses and primates, respectively. The placenta of the horse and monkey secretes both estrogens and progesterone, but the placenta of mouse and rabbit secretes only estrogen. In the rat, ovarian estrogens suppress the activity of 17α-hydroxylase in the placenta, but placental androgens act as the precursors of ovarian estrogens for aromatisation.

22.1.1.3.2 Estrogens

22.1.1.3.2.1 Chemical Structure

All estrogens are chemically made of estrane skeleton (C-18), known as C-18 steroids (Fig. 22.2). The aromatisation of the estrogens' A-ring leads to a planar structure where all carbons are in the same plane. The estrogens have hydroxyl groups at positions 3, 15, and 16, and a combination of these groups yield different forms of estrogens like estrone, estradiol, and estriol. Estradiol is the most potent estrogen and ten times more potent than estrone and 80 times more potent than estriol. In cyclic females, estradiol is the major form of estrogen produced from the aromatisation of androgens. In pregnant females, estrone is the common estrogen produced from androstenedione. Estriol is generally found in primates during pregnancy.

22.1.1.3.2.2 Source

Estrogen is mainly synthesised in the granulosa cells of the ovary (Fig. 22.3). The extragonadal sources of estrogens are the placenta, adrenal glands, and brain. Adipocytes also secret a small amount of estrogens that originate from the conversion of peripheral androstenedione to estrone. Phytoestrogens are the estrogenic compounds available in various plants and feed sources such as alfalfa, red clover, white clover, subterranean clover, berseem clove, and the seed of soybean, sunflower, sesame, etc. The estrogens are subjected to gastrointestinal and hepatic inactivation; hence, difficult to apply in the oral route. Estrogens are also synthesised in bulls, boars, stallions, dogs, and men testes. Equilin and equilenin are the two estrogenic compounds produced in the mare placenta.

22.1.1.3.2.3 Transportation and Activation

Estradiol, like other steroids, is hydrophobic and circulates in conjugation with plasma proteins. Only 1–2% estradiol is circulated in free form. Estradiol has a high affinity to binding with sex hormone-binding globulin (SHBG) synthesised in the liver. But, due to less availability of globulin, nearly 40% of estradiol is transported in conjugation with SHBG and the remaining portion is circulated after binding with plasma albumin. However, albumin has less affinity for estradiol.

22.1.1.3.2.4 Mechanism of Action

Estrogens act through estrogen receptors, namely ER-α and ER-β. Both these receptors are present in the ovaries. But ER-β is predominant in the granulosa cells, and ERa-α is

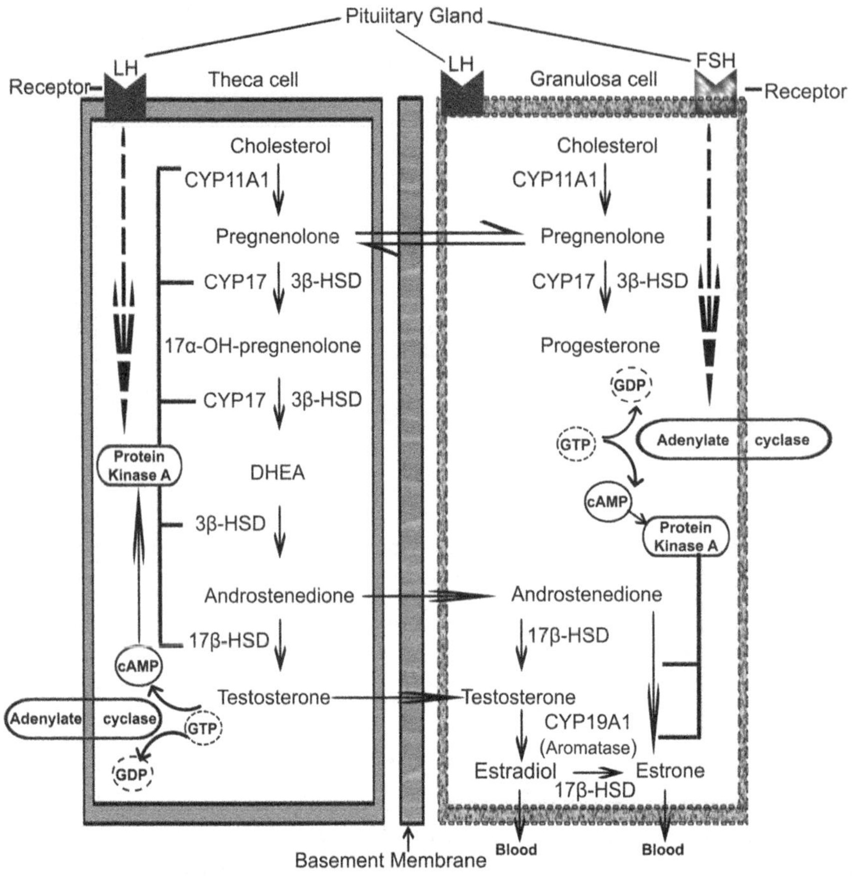

Fig. 22.3 Two cell two gonadotropins. Figure shows steroidogenesis in the ovary. Two gonadotropins, the **FSH** (follicle-stimulating hormone) and **LH** (luteinising hormone), are released from the **pituitary gland**. The **receptor** for FSH presents in granulosa cells, and receptors for the LH are present in both **granulosa** and **theca cells**. The FSH and LH stimulate **adenylate cyclase** and produce **cAMP** (cyclic adenosine monophosphate) to activate **protein kinase A** utilising adenosine triphosphate (ATP), **GTP** (guanosine triphosphate) and **GDP** (guanosine diphosphate). The protein kinase A, in turn, stimulates the various enzymes required for steroidogenesis (shaded arrow). The **cholesterol** is utilised by the **CYP11A1** (cytochrome P450, a cholesterol side-chain cleavage, P450scc) to produce **pregnenolone** that can be diffuse to both **thecal** and **granulosa cells** (reversible arrow) and converted to **progesterone** or **DHEA** (dehydroepiandrosterone) as well as **androstenedione** by the enzymes **3β-HSD** (3β-hydroxysteroid dehydrogenase) and **CYP17** (cytochrome P450). The androstenedione is further converted into **testosterone** by **17β-HSD** (17β-hydroxysteroid dehydrogenase). The end products of cholesterol in the theca cell are androstenedione, and testosterone can transfer to the granulosa cell through the **basement membrane**, where the influence of **aromatase** or **CYP19A1** produces **estradiol** and **estrone.** Thus, the availability of estrogens in the **blood** is regulated in granulosa cells with the influence of FSH, and its precursors are regulated in the theca cell by the LH

situated in the thecal and luteal cells. There are three proposed pathways of estradiol signalling. In the 'classical pathway', estradiol binds with its cytosolic receptors (ER-α and ER-β), and the conformational changes (receptor dimerisation) occur. The hormone-receptor complex translocates to the nucleus and binds with estrogen-responsive elements (EREs) in the regulatory regions of estrogen-responsive genes. In a 'tethered signalling pathway', estrogen receptors interact with non-ERE response elements to regulate gene expression after binding with transcription factors (activating protein-1, AP-1 or the stimulating protein-1, Sp1) other than estrogen. In the 'nongenomic pathway', estrogens interact with plasma membrane-bound receptors, initiate cytoplasmic signalling pathways, and activate MAPKs.

22.1.1.3.2.5 Functions of Estrogens

Estrogens have various roles in female reproduction.

Ovarian effects: Estradiol facilitates granulosa cell proliferation by upregulating the expressions of several genes required for granulosa cell differentiation together with FSH and LH in an autocrine or paracrine fashion. Estradiol also enhances FSH-induced aromatase expression. Estrogens trigger an increased blood flow to the ovaries, thus favours ovulation.

Effects on other reproductive organs: Estradiol regulates the contractility and secretory activity of the cervix, uterus, and fallopian tube by epithelialisation, vascularisation, and fat deposition together with the oxytocin and prostaglandins. Estrogens enhance the rate of protein synthesis and uptake of glucose and water to support the growth of the lining epithelium and underlying muscular tissue (endometrium) of the uterus.

Expression of behavioural estrus and sex desire: It is responsible for sexual receptivity or sex desire, i.e. onset of heat (discussed in the estrous cycle).

Roles in gamete transport: Estradiol promotes gamete transport (both sperm and ovum) by increasing the contraction of oviductal smooth muscle and ciliary beat frequency (CBF) of the oviduct through the phosphorylation of protein kinase C and A (PKC and PKA) and production of cAMP. Estradiol stimulates the release of antioxidants in the oviductal fluid and reduces sperm stress during sperm transport.

Role in implantation: Estradiol favours the embryo transport from the fertilisation site to the implantation site after stimulating smooth muscle contraction, inducing fluid production and flow, and increasing the CBF.

Immunoprotection of the embryo: Estradiol protects the embryo from the maternal immune system via the activation of ER-α in the oviductal epithelial cells.

Role in parturition: Estrogens cause rhythmic contractions of the uterus during parturition. The E2 and estrone sulphate are mainly available in early pregnancy and late gestation. A higher level of E2, around 24 h before the parturition, helps remove the progesterone block and initiates the parturition (discussed details in parturition).

Development of secondary sexual characteristics: Estrogens are responsible for expressing secondary sexual characteristics in females, viz. hair growth, skin texture, body configuration, voice, etc.

Effects on mammary gland: Estrogen stimulates ductal growth of the mammary gland through ER-α expressed in the mammary epithelium and stroma.

22.1.1.3.2.6 Non-reproductive Roles of Estradiol

Other than reproductive effects, estradiol is also involved in the functioning of different systems like the heart, skin, muscle, bone, brain, and liver.

Estrogens help modulate total cholesterol levels by decreasing LDL or increasing HDL cholesterol through nuclear and extranuclear ER-α and ER-βb by altering the HMG-CoA reductase gene promoter as it contains an estrogen-responsive estrogen element-like sequence (Red-ERE). Decreased level of estrogens leads to atherosclerosis and the risk of a heart attack in postmenopausal women.

Estrogens accelerate bones' linear growth and epiphyseal closure and increase bone density and strength. Hence, skeletal growth is arrested after puberty and osteoporosis after menopause in females.

Estrogen affects the structure and function of muscle, tendon, and ligaments. Muscle weakness is evident in postmenopausal women due to lower estrogen levels.

Estrogen receptors are present in several brain regions like the hippocampus, cerebral cortex, claustrum, hypothalamus, subthalamic nucleus, amygdala, and thalamus. Estrogens promote cerebral blood flow, neuronal activity, and anti-inflammatory effects in the CNS, thus acting as neuroprotective and neurotrophic agents. The estrogens control sex-specific brain activity. Estrogens also increase serotonin levels to influence sexual desire in males and females. In the male, the testosterone in the brain is converted to estrogen by aromatisation. Estrogens influence the sensory inputs of vision, audition, and olfaction and their integration with the motor neurons for muscles of the genital tract to exhibit lordship, mounting, and other sexual behaviour.

Estrogens play a pivotal role in metabolisms such as food intake, glucose homeostasis, lipolysis/lipogenesis, and osmoregulation. Most of these functions exhibit conjugation with other peptides involved in energy expenditure like leptin, neuropeptide Y, pro-opiomelanocortin, or melanin-concentrating hormone). Estrogens regulate water and salt balance stimulating reabsorption by inducing arginine vasopressin (AVP), atrial natriuretic peptide (ANP), renin, and aldosterone. Estrogens stimulate lipid and lipoprotein metabolism in the liver and upregulate some serum proteins like thrombin and fibrinogen. Therefore, estrogen therapy may lead to venous thromboembolism.

Estradiol indirectly regulates the expression of cardiac heat shock proteins 70 and 72 (HSP70 and 72) and reduces apoptosis by stabilising the mitochondrial membrane and averting apoptosome formation.

Estrogens have anti-inflammatory properties. Estrogen regulates the activity of immune cells through the regulation of cellular metabolism. ERR-α controls the metabolic activity of T cells and influences T cell activation.

22.1.1.3.3 Progesterone

Progesterone is essentially required to establish and maintain pregnancy (pro-gestational) and the reproductive cycle.

22.1.1.3.3.1 Chemical Structure

Progesterone belongs to the class 'progestogen', and the molecule has a 21-carbon skeleton called the pregnane skeleton (C-21) (Fig. 19.21). Both natural and synthetic forms of progesterone are available, and synthetic progestogens are usually referred to as progestins.

22.1.1.3.3.2 Source

Major source of progesterone (P4) is the corpus luteum of the ovary in all mammals. The luteal cells contain the enzymes required to synthesise progesterone from the cholesterol (Fig. 22.3). It can also produce in adrenal glands and placenta during pregnancy, particularly in mare, ewe, queen, and human. But the placental progesterone is not sufficient to maintain the pregnancy in cattle, goats, pigs, dogs, and rats. In the case of sheep, horses, cats, and humans, the placental progesterone from the mid-pregnancy is sufficient to support the pregnancy independent of CL. In camels, a large quantity of placental progesterone produces from the multinucleate giant cells by day 30–35 of gestation. The placenta of mare and ewe produces 5α-pregnane instead of progesterone. In cows, progesterone produces from the placenta during pregnancy's latter half (6–8 months). Placental progesterone is absent in doe, sow, bitch, camel, and rabbit. The growing follicles can produce a small quantity of progesterone in bitch.

The plasma progesterone concentration is highest in the sow, lowest in the ewe and intermediate in the cow. Progesterone secretion sustains throughout the luteal phase of cyclic females. The LH primarily controls progesterone synthesis, and the PGF2α generally destroys the progesterone-producing cells. Hence, progesterone activity is governed by the pulsatile release of pituitary LH and the PGF2α of the uterine endometrium. Phyto-progesterone is available in the *Juglans regia*. The *Dioscorea mexicana* contains progesterone-like steroids (diosgenin), which can act as the precursor of progesterone. Progesterone can also produce in the nervous system. Progesterone may also be available as the chief transitional substance for circulating androgens and estrogens.

22.1.1.3.3.3 Transportation

About 98–99% of progesterone is transported in protein-bound form. Nearly 80% of progesterone is combined with albumin, 18% with corticosteroid-binding globulin (CBG), or transcortin and less than 1% with SHBG.

22.1.1.3.3.4 Mechanism of Action

The progesterone receptor (PR) has two major isoforms, PR-A and PR-B. After ligand binding, the hormone-receptor complex translocates to the nucleus and binds hormone-responsive elements (HRE) at regulatory regions hormone-responsive genes to initiate new protein synthesis. The receptors for androgens, mineralocorticoids, and glucocorticoids can also recognise the HRE of progesterone. Recently, it has been identified that the actions of progesterone are mediated by membrane-localised progestin receptors other than classical PRs (PR-A and PR-B). The membrane-bound progesterone receptors are called the progesterone membrane component (PGRMC1 and 2) and membrane progestin receptors (mPs). The actions of progesterone occur through these receptors and are mediated after the initiation of intracellular signalling pathways and subsequent cellular responses. They can also modulate the genomic actions of progesterone.

22.1.1.3.3.5 Function of Progesterone

Regulation of uterine functions/maintenance of pregnancy: Progesterone plays a central role in maintaining pregnancy by modulating uterine physiology, such as endometrial maturation, reduction of uterine contractility, and favours uteroplacental circulation. It also modulates maternal immune response by suppressing the inflammatory mediators and preventing the foetal allograft's rejection for supporting pregnancy. Thus, susceptibility to metritis may occur during the postpartum period when progesterone level is diminished suddenly. Progesterone helps differentiate endometrial stromal fibroblasts into specialised secretory decidual cells that secret uterine milk to provide nutritional support to the embryo before implantation. This process is called *decidualisation* of the uterus and potentiates placental development.

Regulation of ovarian functions: Progesterone plays pivotal roles in oocyte meiosis, ovulation, luteinisation, and maintenance of CL. Progesterone helps in meiosis resumption by disrupting gap junctions between cumulus cells. Progesterone inhibits FSH release, thus suppressing follicular growth, ovulation, and estrogen level, which favours the gestation.

Effect on the oviduct: Progesterone stimulates the morphology and function of the luminal epithelium of the fallopian tube to regulate the volume and composition of the oviductal fluid. It also regulates the muscular activity of the oviduct.

Role in lactation: During pregnancy, progesterone in combination with prolactin favours epithelial proliferation leading to the formation of alveoli. But progesterone inhibits lactation during pregnancy. Copious milk secretion occurs immediately after parturition only when progesterone levels markedly decrease.

Inhibition of sexual behaviour: Progesterone inhibits sexual behaviour. It acts in the ventromedial hypothalamus or the brain's pre-optic area to block the LH surge and estrogen-induced sexual behaviour. Progesterone inhibits the ion channels and vomeronasal sensory neurons and suppresses pheromone-induced sexual behaviour.

Non-reproductive roles: Progesterone metabolites like di-hydro progesterone (DHP) and 3α, 5α-tetra hydro progesterone (allopregnanolone) are available in the central and peripheral nervous system. They can modulate neuronal and astroglial plasticity, enhance neural survivability, support the myelination process, increase neurogenesis in adulthood, and inhibit lipid peroxidation and anti-inflammatory properties. Thus, progesterone therapy is effective in brain injury, ischaemia, and peripheral neuropathy.

Progesterone is reported to increase appetite to facilitate a positive energy balance during pregnancy. A high progesterone level reduces the natriuresis (lack of sodium retaining aldosterone activity). Thus, extracellular volume is reduced.

Progesterone as contraceptive and application in assisted reproductive technologies (ART): Progesterone is used as a contraceptive as it suppresses follicular development and ovulation by inhibiting FSH secretion. Exogenous progesterone can be administered through implants (Intrauterine device, IUD, or intravaginal) or oral and parenteral routes. But the most preferred route is vaginal implants. It increases the bioavailability of progesterone by 40-fold more than oral progesterone. The follicular development is suppressed till the withdrawal of progesterone. Upon withdrawal, it results in immediate secretion of FSH followed by follicular development and ovulation within 2–3 days. This functional activity of progesterone is extensively applied to synchronise the ovulation in integrated artificial insemination programme, superovulation, embryo transfer technology (ETT), and assisted reproductive technology (ART).

Metabolism and Excretion of Sex Steroids

The half-life of ovarian steroids is less (Table 22.1). Estrogens are metabolised into estrogenically inactive metabolites like estrone and estriol by cytochrome P450 (CYP) enzymes, mainly in the liver. Cytochrome P450 oxidase enzymes (CYP3A4 and CYP1A1) cause the oxidation of the 17β-hydroxyl group. The inactive metabolites undergo sulphate and glucuronide conjugation in the liver. Estradiol is excreted via the urine (nearly 75%) or faeces (25%). Estriol is the main estrogen metabolite found in the urine. Progesterone lacks a hydroxyl group; hence, it can't be easily sulphated and esterified. Thus both the solubility and half-life of progesterone are more than estrogen. Nearly 60–65% of progesterone is metabolised by 5α-reductase and 5-β-reductase to form allopregnanolone and pregnanolone, respectively. A small amount of progesterone is also metabolised into 11-deoxycorticosterone by 21-hydroxylase. Most progesterone is metabolised in the liver, GI tract (particularly in the duodenum), and kidney. Inactivated progesterones are excreted by the kidney in conjugated form (through glucuronidation or sulphation).

Table 22.1 Biological half-life of steroid hormones

Steroids	Half-life
Cortisol	60–100 min
Aldosterone, DHEA, androstenedione, estradiol, and testosterone	Less than 20 min
Progesterone	3–90 min

The urinary concentrations of the inactivated steroid compound provide an essential clinical index of reproductive function and determinate the stage of ovarian dynamics. In wild animals, estimation of the inactivated form of steroids from urine and faeces are estimated to evaluate the reproductive status. Some inactive metabolites of estrogens are estrone sulphate, estradiol glucuronide, 2-hydroxy estrone, 2-hydroxy estradiol, 16α-hydroxy estrone, and 16α-hydroxy estradiol. Estrone is also considered a less active form of estradiol. Inactivated form of progesterone is pregnenolone sulphate.

22.1.1.3.4 Control of Ovarian Steroidogenesis

22.1.1.3.4.1 Endocrine Control (Two Cell Two Gonadotropin Theory)

Ovarian steroidogenesis is controlled by two gonadotropins (FSH and LH) acting on the thecal/luteal cells and granulosa cells, respectively. Hence, it is called the 'two cell, two gonadotropin theory'. Figure 22.3 depicts the mechanism of the 'two cell, two gonadotropin theory'. The receptors for FSH are present in the granulosa cell, and receptors for LH are abundant in both granulosa and theca cells. The synthesis of estradiol requires a synergistic relationship between theca cells which produce androgens (i.e. dehydroepiandrosterone (DHEA), androstenediol, androstenedione, testosterone under the influence of LH, and these androgens then diffuse into granulosa cells and converted to estrogens (i.e. estrone, estradiol) by the action of cytochrome P450 aromatase (CYP19A1) stimulated by FSH. LH acts through G protein-coupled receptors after stimulating adenylate cyclase to produce cAMP, which stimulates protein kinase A (PKA). The PKA then vigorously phosphorylates the StAR, transports cholesterol from the outer mitochondrial membrane to the inner mitochondrial membrane, and initiates steroidogenesis of androgens (DHAE).

On the other hand, FSH stimulates adenylate cyclase via G protein-coupled receptors. The cyclic adenosine monophosphate (cAMP) generated from adenosine triphosphate (ATP) activates protein kinase A to stimulate the expression of the respective steroidogenic enzymes such as NADPH cytochrome P450 reductase, which transfers electrons to aromatase; HSD3B2 which converts DHEA to androstenedione, and type 1 17β-hydroxysteroid dehydrogenase (HSD17B1), the 'estrogenic' 17β-HSD that reduces estrone to estradiol. The estrone is produced by the action of cytochrome P450 aromatase (P450arom or aromatase) enzyme and the estrone and subsequently converted to estradiol by 17β-HSD (CYP19A1) (Fig. 22.3). Hence, granulosa cells are an abundant source of estrogens. Progesterone production is limited in the granulosa cells due to the lack of the 3β-HSD enzyme. But, in theca cells, pregnenolone can be converted to 17α-hydroxypregnenolone by P45017-OH under the influence of LH. Thus, in the ovary, progesterone can be synthesised with the influence of LH only, but the involvement of both LH and FSH are required for estrogen production. The ovarian theca cells can be compared with the testicular Leydig cells and granulosa cells with the Sertoli cells, considering their steroidogenesis activity.

Activins and inhibins have paracrine effects on ovarian steroidogenesis. The granulosa cell-derived inhibin is stored in antral fluid, diffuses to the adjacent thecal cell layer, and positively stimulates androgen synthesis in LH-stimulated theca cells. Activin stimulates estradiol synthesis by upregulating the aromatase enzyme and FSH receptors.

22.1.1.3.4.2 Molecular Control

Several LH-induced signalling molecules regulate the activity of StAR and control the steroidogenesis (Table 22.2), viz. cAMP and PKA, insulin-like growth factors (IGFs), etc.

22.1.1.3.5 Ovarian Peptide Hormones

22.1.1.3.5.1 Activin

Activins are transforming the growth factor-beta (TGF-β) superfamily of cytokines secreted from the granulosa cells of the ovary. Structurally activins are dimers of inhibin β subunits that act through the classical TGF-β signalling pathway. There are several forms of activins based on the types of β subunits. They are activin A ($\beta_A\beta_A$), activin B ($\beta_B\beta_B$), activin C ($\beta_C\beta_C$), activin D ($\beta_D\beta_D$), and activin E ($\beta_E\beta_E$). Activin controls ovarian and testicular development. It stimulates FSH secretion, promoting oocyte maturation and granulosa cell steroidogenesis. Activin has a negative paracrine effect over LH-induced theca cells for androgen production. It also helps in folliculogenesis by preventing luteinisation of the premature antral follicle. It is also involved in endometrial repair, decidualisation, and pregnancy maintenance. Other functional features of activins include morphogenesis of the embryo, particularly the development of the limb's nervous system and the development of facial and dental structures. In the testis, activins modulate germ cell development and Sertoli cell proliferation.

Table 22.2 LH-induced signalling molecules to regulate steroidogenesis

Name of the signalling molecules	Functions
cAMP and PKA	Promotes steroidogenesis by phosphorylation of StAR cAMP regulates CYP19 and CYP17 expression
Phospholipase C (PLC)	Activates PKA
Src and extracellular-regulated kinases (ERKs)	Increases StAR expression
Insulin-like growth factors (IGFs)	Increases CYP19 and CYP17 expression, increases transcription of LH receptor
Epidermal growth factor (EGF) receptor	Promotes phosphorylation of StAR

22.1.1.3.5.2 Inhibin

Inhibins are the glycoproteins composed of α-subunit and β-subunit. Based on the β-subunit, inhibin can be of two types, inhibin A (βA subunit) and inhibin B (βB-subunit). Inhibin B is more biologically active compared to inhibin A. They are secreted from granulosa cells of the ovary and Sertoli cells of the testes. Inhibin is also synthesised from the placenta and can be present in the foetus. Unlike activin, it inhibits the synthesis and release of the FSH. Apart from FSH inhibition, inhibin exerts paracrine effects on the gonads. Inhibin functions as a regulatory hormone in mares during the follicular phase of estrous cycle. It inhibits progesterone synthesis in the ovary. The concentration of inhibin decreases with a declining ovarian follicular reservoir; hence, it can be used as a potential marker for ovarian function. Inhibin is three times more potent than follistatin.

22.1.1.3.5.3 Relaxin

The predominant source of relaxin is the corpus luteum in both pregnant and non-pregnant animals. It is also synthesised in the placenta (horse). In males, relaxin is produced from the prostate and released in seminal fluid. In recent years, the heart's atria have been identified as an extragonadal source of relaxin. The functions of relaxin include inhibition of uterine contractility and relaxation of the uterine muscles and ligaments during pregnancy in synergy with progesterone, inhibition of the collagen synthesis in the estrogen-primed cells of the cervix, vagina, and pubic symphysis to soften the birth canal. It promotes angiogenesis in the endometrium, especially during the implantation of mares. In horses and rats, relaxin has an important role in ovulation to guiding the ova into the fallopian tube. It causes proteolysis in the follicular walls by influencing the secretion of gelatinases and tissue inhibitors of metalloproteinases. The receptors for relaxin are present in granulosa and theca cells of follicles; thus, it facilitates follicular development in pigs and humans. It is also involved in mammogenesis.

22.1.1.3.5.4 Follistatin

Follistatin is a single chain glycoprotein of the ovarian follicular fluid that acts as an activin-binding protein. It neutralises activin and indirectly suppresses FSH secretion from the anterior pituitary. Follistatin can be neutralised by activin.

22.1.2 Role of HPO Axis in Female Reproduction

The events of the female reproductive biology are regulated by a complex interplay between the nervous and endocrine systems. The hypothalamus, the central component of the HPO axis, secrets GnRH to stimulate pituitary gonadotrophs for secreting FSH and LH. The gonadotropins (FSH, LH), in turn, regulate the gonadal functions. Together with inhibins and activins, ovarian steroids influence gonadotropin secretion in a feedback manner. The FSH acts on the ovarian follicle's granulosa cells and controls folliculogenesis and estrogen synthesis. A high level of estrogen, low level of LH, and absence of inhibin initiate follicular recruitment. The follicular dominance is achieved in a milieu with low FSH and high LH and inhibin. Inhibin suppresses the FSH secretion from the pituitary and restricts further follicular growth. The estrogen is produced from the matured follicles under the influence of FSH and LH (details in two cell two gonadotropin mechanism). When the estrogen level reaches a threshold, it helps to manifest the behavioural estrus. It stimulates the GnRH surge centre at the pre-optic and suprachiasmatic centre of the hypothalamus to release GnRH at high pulses, which facilitates pre-ovulatory LH surge and ovulation.

The theca and granulosa cells undergo luteinisation after ovulation, and the corpus luteum is formed. The large luteal tissue secrets progesterone and oxytocin, whereas the small luteal tissue secrets progesterone. The progesterone supports pregnancy by causing endometrial hypertrophy, secretion of uterine milk and blocking uterine contraction (details in the endocrine control of pregnancy). A higher progesterone level exerts strong negative feedback over the hypothalamus and prevents pre-ovulatory follicular growth. If an animal fails to conceive, the corpus luteum undergoes luteolysis by the action of PGF2α. The PGF2α is secreted from the endometrium by the action of oxytocin. After binding with its receptor at the endometrium, oxytocin stimulates the enzymes for the synthesis of PGF2α from cholesterol. Estrogens upregulate the expression of oxytocin receptors in the uterus. The synergistic and agonistic activities of the gonadotropins and sex steroids to control the HPO axis have been summarised in Table 22.3 and Fig. 22.1.

22.1.2.1 Factors Affecting the HPO Axis

Several factors regulate the HPO axis.

Table 22.3 Interrelationship between the gonadotropins and sex steroids

Production level	Estrogens (E2)	Progesterone (P4)	FSH	LH
Nil	–	FSH ↑	–	–
Low	LH ↑	LH ↑	LH ↑, E2 ↑	–
Moderate	LH ↑↑, FSH ↓	LH ↑↑	LH ↑, E2 ↑, (also Inhibin ↑)	FSH ↓
High	FSH ↓↓	LH ↑↑	LH ↑↑, E2 ↑↑, FSH ↓↓	FSH ↓↓

22.1.2.1.1 Genetical and Congenital Factors

Some animals may have inherited HPO axis insufficiency due to genetic mutations of the HPO axis components. Turner syndrome, Kallmann's syndrome (in humans) and chromosomal aberrations may directly affect the HPO axis. Hypothalamic–pituitary signal dysfunctions may occur due to genetic disorders that cause acute ischemia or compression and autoimmunity.

22.1.2.1.2 Pathological or Physiological Dysfunction

Obesity, hyperprolactinemia and hypothyroidism affect the HPO axis by reducing GnRH and gonadotropins. The extragonadal sources of estrogens may occasionally disturb the ovarian function in conditions like Cushing's syndrome, tumours or cysts in the adrenal or ovary and congenital adrenal hyperplasia. In these conditions, conversion of peripheral androstenedione to estrone in adipose tissue and skin is increased, and this estrogen causes cyclic irregularities. This phenomenon is common in polycystic ovarian syndrome. Chronic parasitism causes prolonged hypothyroidism, leading to reduced GH-dependent IGF-I synthesis in the liver, resulting in inhibition of the HPO axis. Due to regional acute ischemia or compression and autoimmunity, panhypopituitarism also affects the HPO axis. Phytoestrogenic compounds negatively affect the HPO axis. Metabolic disturbances, such as ketosis in high yielding cows, reduce the secretion of both FSH and LH. The metabolites like high GH, non-esterified fatty acid (NEFA) and beta-hydroxybutyrate (BHB), and lower glucose and IGF-I are thought to modulate the HPO axis under these conditions.

22.1.2.1.3 Stress

Stress-induced suppression of the HPO axis occurs under the influence of gonadotropin-inhibitory hormone (GnIH) and cortical hormones (Fig. 22.4). The GnIH is also called RFamide-related peptide 3 (RFRP) in mammals. Its concentration is increased during stress and leads to inhibition of GnRH secretion. Synthesis of cortical-releasing factor (CRF) during stress reduces GnRH secretion as both CRH and GnRH neurones are localised in the pre-optic area at the

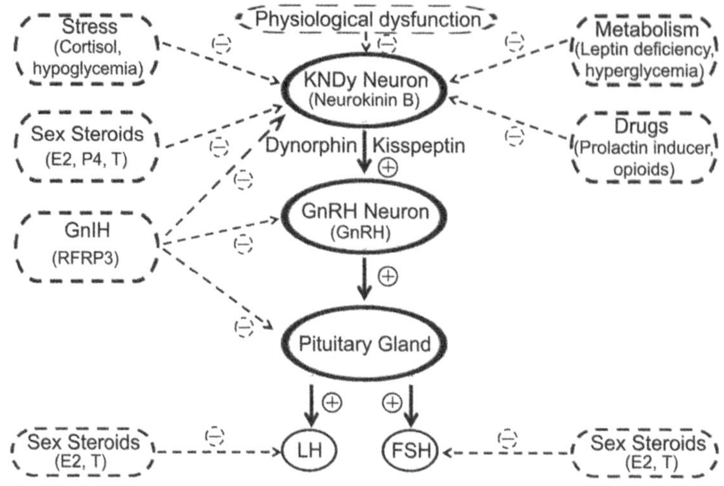

Fig. 22.4 Factors control the release of gonadotropins. The sketch shows various factors that are controlled the release of gonadotropins, the **LH** (luteotropic hormone) and **FSH** (follicle-stimulating hormone) from the **KNDy (kisspeptin–neurokinin B–dynorphin) Neuron** or neuronal network and **GnRH** (gonadotropin) **Neuron** of the hypothalamus followed by the **pituitary gland**. The network systems are depicted by bold circles, bold arrows and circled positive (+) sign. Dotted circles and bold arrows illustrate the negative or inhibitory factors and a circled negative (−) sign. Most factors are affected by the KNDy Neuron, viz. **GnIH** (gonadotropin inhibitory hormone), an RFamide-related peptide 3 (**RFRP3**); various **sex steroids** in variable concentrations (e.g. **E2** = estrogens, **P4** = progesterone and **T** = testosterone); in the **stress**-causing release of **cortisol**, **hypoglycaemia** and similar conditions; occurrence of **physiological dysfunctions** like cyst, acute and chronic illness; disturbances in **metabolism** due to **leptin deficiency**, **hyperglycaemia**, and other conditions; and various **drugs** like **prolactin inducer** or stimulating, **opioids,** etc. Among such factors, GnIH can also directly affect GnRH neurons and the pituitary gland. The sex steroids, viz. E2 and T in various concentrations can directly affect LH and FSH's functional activity

hypothalamus. CRF down-regulates GnRH gene expression in the hypothalamus (Fig. 22.5). FSH secretion is more affected in response to stress than LH secretion. In nutritional and environmental stress, HPO axis is affected through specific mediators like adipokines, cytokines, and adipose tissue-derived factors, fatty acids and (Table 22.4). Long-term exposure to corticosteroids, ACTH and stress, cause an inhibitory effect on the HPO axis. But, acute release or administration of ACTH and some corticosteroids to the estrogen-primed animals cause stimulation of the HPO axis and pre-ovulatory LH surge. Malnutrition causes decreased adipose tissue-derived leptin levels, and the stomach originates ghrelin. These two peptides result in a decrease in GnRH and LH pulse.

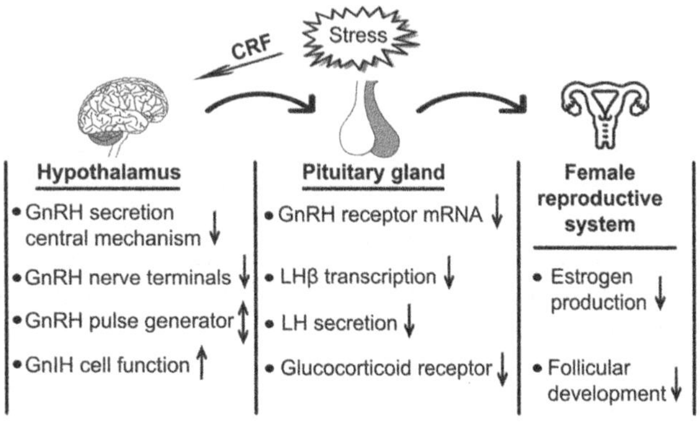

Fig. 22.5 Effect of stress on the female reproductive system. Figure shows the stress caused by to release of **CRF** (cortical-releasing factor) over the **hypothalamus**, followed by the **pituitary gland** and **female reproductive system**. It affects **GnRH** (gonadotropin-releasing hormone) and **GnIH** (gonadotropin-inhibitory hormone) neurons in the hypothalamus. Receptors for GnRH, **LH** (luteinising hormone) and glucocorticoid hormone in the pituitary gland are affected as a consequence of it. Ultimately, estrogen production and follicular development have been arrested. The down arrow denotes the various modulations of the mechanism of actions for inhibitory action, the upward arrow by influencing action and both side arrows indicate the together actions

22.1.2.1.4 Environment

Season and photoperiod affect the HPO axis, particularly in the temperate region, through melatonin production in the pineal gland (discussed in detail in the puberty section). Seasonal influence on the reproductive cycles in seasonal breeders also operated through the action of melatonin over the HPO axis. The HPO axis is affected in cattle when the temperature-humidity index (THI) exceeds 70, and high THI affects the feed intake, growth, and hormonal imbalances to suppress the HPO axis.

22.1.2.1.5 Physiological Factors

Inter-estrus interval is variable and can be modulated in different species due to genetic variation and in the animals' specific physiological state like the postpartum period. High

Table 22.4 Effects of various adipokines, cytokines and adipose tissue-derived factors, fatty acids and peptides on HPO axis

Bio-molecules	Major effector cells/tissues	Other effector cells/tissues	Mediated by	Effect	Impact
Adipokines					
Leptin	Hypothalamus, pituitary, ovaries	Adipocytes, somatotrophs, gonadotrophs, thyreoidotrophs	Circadian rhythm during the reproductive cycle	Increase—GnRH, LH, FSH, E2	Early puberty, seasonal breeding
Adiponectin	Pituitary	Adipocytes, theca cells, cumulus cells, oocyte, uterus, oviduct	GnRH, LH, FSH, estrous cycle	Increase—FSH, P4, insulin-induced LH, IGF-I induced P4, E2	Regulation of HPO axis, early gestation, reproductive cycle
Visfatin	Ovaries	Adipocytes, primary granulosa cells, cumulus cells, oocytes	Obesity, type-2 diabetes	Increase LH	Ovary functional irregularities, PCOS
Resistin	Ovaries	Adipocytes	Gonadotropins, gonadal steroids, IGF-I	Decrease steroids Increase—FSH and LH	Ovarian steroidogenesis
Chemerin	Ovaries	Adipocytes, granulosa cells, theca cells, corpus luteum, oocyte	TNFα, insulin, androgen	Decrease—antral follicular growth, steroidogenesis, FSH-induced aromatase activity	Apoptosis of granulosa cells
Cytokines					
Interleukin-6 (IL-6)	Immune cells, uterus, placenta	Adipocytes, gestational tissues	Adipose tissue, T cell differentiation	Increase immune response, MIF in ovarian function	Embryo implantation, placental development
Tumour necrosis factor (TNFα)	Ovaries	Adipocytes, macrophages, uterine cells, trophoblast	Fibroblastic growth, collagenase stimulation	Decrease—FSH-induced LH receptor, LH secretion Increase—prostaglandin synthesis	Follicle development, ovulation, formation and regression of CL, endometrium function, pregnancy
Fatty acids					
Non-esterified fatty acid (NEFA)	Ovaries, embryo	Adipocytes, immune cells	TNFα, LH	Decrease—follicle and oocyte growth, LH Increase—TNFα in macrophages	Follicular growth, embryo quality
Free fatty acid (FFA)	Ovaries	Adipocytes, oocytes	Prostaglandins	Increase—granulosa cell apoptosis	Cumulus oocyte complex (poor quality)
Peptides					
Kisspeptin	Hypothalamus	Hypothalamic neurone, adipocytes	Sex hormones, seasonal breeding, leptin, food intake	Increase—GnRH, FSH, LH	Influence HPO axis
RFamide-related peptides (RFRPs)	Hypothalamus (at DMN)	Hypothalamus (at POA in primates; AVPV and INF in humans; ARC in non-humans)	Kisspeptin neurons, GnRH neurons, and the pituitary	Decrease—LH release	Inhibit HPO axis

Source: Calejja-Agius et al. (2009), Prins et al. (2012), Tsatsanis et al. (2015), Dobrzyn et al. (2018), Hu et al. (2019)
GnRH gonadotropin-releasing hormone, *LH* luteinising hormone, *FSH* follicle-stimulating hormone, *E2* estrogens, *P4* progesterone, *IGF* insulin-like growth factor, *PCOS* poly cystic ovarian syndrome, *MIF* Mullerian inhibiting factor, *CL* corpus luteum, *DMN* dorsomedial hypothalamic nucleus, *POA* pre-optic area, *AVPV* periventricular nucleus, *INF* infundibulum, *ARC* arcuate nucleus

yielding cows undergo ketosis immediately after the parturition due to high GH, non-esterified fatty acid (NEFA), beta-hydroxybutyrate (BHB) and low insulin, glucose, and IGF-I. This condition reduces the secretion of both FSH and LH.

In the postpartum period, high metabolic disturbances (ketosis in high yielding cows) occur. All the factors that affect the puberty and estrous cycle are mediated through the various components of the HPO axis.

22.1.2.1.6 Endocrine Factors

22.1.2.1.6.1 Prolactin

Estrogens influence the lactotroph cells of the anterior pituitary to secrete prolactin during puberty and late pregnancy. Prolactin inhibits gonadotropin secretion. In birds, prolactin plays a significant role in brooding behaviour and is associated with metabolic changes that occur during brooding.

22.1.2.1.6.2 Oxytocin

Oxytocin, a hypothalamic neuropeptide stored in the posterior pituitary, stimulates the uterine smooth muscle activity. It acts synergistically with estrogens during parturition to make strong myometrial contractions.

22.1.2.1.6.3 PGE2

PGE2 produces in the ovary, uterus, and embryonic membranes. It helps to soften the cervix, augments uterine contraction, and prepares the tract during parturition, particularly in horses and humans. It is also involved in ovulation and progesterone secretion from the corpus luteum.

22.1.2.1.6.4 Human Chorionic Gonadotropin (hCG)

The hCG is a glycoprotein synthesised in the trophoblast cells of a blastocyst in primates and humans. Its function is similar to LH, so it helps to secrete progesterone and estrogens. It can protect the embryo from the maternal immune system during the first phase of pregnancy. To confirm pregnancy, the presence of this hormone can use in the maternal blood.

22.1.2.1.6.5 Equine Chorionic Gonadotropin (eCG)

The eCG, a glycoprotein, is released from the chorionic tissues of the placenta in horses and primates. It acts like FSH and helps with follicular growth and ovulation. Thus, it helps form accessory corpora lutea to ensure progesterone production during pregnancy. The eCG uses to induce superovulation in other species.

22.1.2.1.6.6 Placental Lactogen

Placental lactogen is a polypeptide hormone of placental origin. It has structural and functional similarities with growth hormones. Hence, it is called chorionic somatomammotropin. It helps to provide energy to the developing foetus during pregnancy by altering the metabolic status of the mother in humans. It has an important role in lactation and maintenance of corpus luteum in the rat. Placental lactogen supports the pregnancy by the steady progesterone production from the corpus luteum. There are two forms of placental lactogen-I and II. The placental lactogen-I can bind with the same receptor as prolactin due to structural similarity and can mimic the activity of prolactin.

22.1.2.1.6.7 Summary of Endocrine Factors Involved in Female Reproduction

According to the involvement of female reproduction, reproductive hormones can be classified into three types. These are primary, secondary, and tertiary hormones (Table 22.5). Primary reproductive hormones have a direct role in reproduction and secondary hormones are involved indirectly by influencing the growth and functional activity of reproductive organs or related endocrine glands. Tertiary hormones are the neurohormones that engage in the secretion of other reproductive hormones.

22.1.3 Pheromones and Pheromone-Induced Sexual Behaviour

The chemical substance synthesised and released by the animals into the surroundings affects the physiology and behaviour of the other animals of the same species called *pheromone*. Pheromones are an important medium of communication in both mammals and non-mammalian species. Pheromones play crucial roles in searching for mates, searching for foods and other interactions like alarming predators. Pheromones can be either male-specific, female-specific, or combined types. In mammals, pheromones are released through urine, faeces, vaginal secretion, saliva, and modified scent (cutaneous) glands, including hair and wool. The receiving animals perceive pheromones through their olfactory system or in combination with other sensory systems (auditory, visual, or tactile system) and combine with odorant-binding proteins (OBPs), major urinary proteins (MUPs), and/or similar soluble pheromone carrier proteins for their transport through biological fluids. The MUPs are produced in the liver and excreted through urine. The OBPs are mostly found in nasal tissues. The 1-octen-3-ol is one of the soluble pheromone carrier proteins. It also increases the bioavailability of the pheromones by protecting them from metabolism. Sex pheromones are exclusively involved in the socio-sexual communications between the male–female interactions (Table 22.6). Sex pheromone modulates acceleration of puberty, induction of estrus, minimises the postpartum anestrus, exhibition of estrus symptoms, synchronisation of ovulation, influences courtship and maternal behaviour, including reduced aggressiveness to young by influencing synchronise hormonal activities (Table 22.7). Sex pheromones are of three types. (a) Releaser pheromones—Releaser pheromones are the pheromones which cause immediate response like immobilisation reflex in the sow. (b) Signaller pheromones—Signaller pheromones indicate the identity or presence of the pheromone producer (sender) like a mother–neonatal relationship. (c) Primer pheromones—Primer pheromones make slow and long-acting responses like the advancement of puberty and reduction of postpartum anestrus duration.

22.1.3.1 Bio-stimulation (Interaction with the Opposite Sex)

The stimulus induced by the presence of males to modulate estrus and ovulation through pheromones, genital

Table 22.5 Hormones and growth factors involved in female reproduction with their sources

Primary	Secondary	Tertiary
Hypothalamus	*Hypothalamus*	*Hypothalamus*
Gonadotropin-releasing hormone (GnRH)	Thyrotropin-releasing hormone (TRH)	Prolactin-releasing hormone (PRH)
Gonadotropin-inhibiting hormone (GnIH or RFamide-related peptide 3, RFRP) Kisspeptin	Corticotropin-releasing hormone/ factor (CRH/CRF) or corticoliberin	Prolactin-inhibiting hormone (PIH) or Dopamine
Growth hormone-releasing hormone (GHRH)	*Anterior pituitary*	Norepinephrine
Somatostatin or growth hormone inhibiting hormone (GHIH)	Growth hormone (GH)	Gamma-aminobutyric acid (GABA)
Prolactin-inhibiting factor (PIF)	Thyroid-stimulating hormone (TSH)	*Paraventricular supraoptic nucleus*
Anterior pituitary Follicle-stimulating hormone (FSH)	Adrenocorticotropic hormone (ACTH)	Oxytocin Endorphin
Luteinising hormone (LH)	*Posterior pituitary*	Vasopressin
Prolactin (PRL)	Vasopressin	*Brain and pituitary*
Posterior pituitary	*Liver*	Fibroblast growth factor (FGF)
Oxytocin	Insulin-like growth factor 1 (IGF-I or somatomedin C)	*Smooth muscle cells, activated macrophages, endothelial cells, and mesenchymal cells*
Ovary	*Male embryo*	Platelet-derived growth factor (PDGF)
Estrogen Progesterone	Mullerian inhibiting factor (MIF or AMH)	*Macrophages*
Inhibin	*Submaxillary and other glands*	Transforming growth factor-β (TGF-β)
Activin	Epidermal growth factor (EGF)	Tumour necrosis factor (TNFα)
Follistatin	*Thyroid gland*	*Progenitors (in bone marrow)*
Oocyte maturation factor	Thyroxin (T4)	Haematopoietic growth factor (HGF)
Relaxin	Tri-iodothyronine (T3)	*Adipocytes*
Placenta	Thyro-calcitonin	Adipokines
Placental lactogen (PL or chorionic somatomammotropin)	*Parathyroid gland*	
Prostaglandin (PGF2α)	Parathormone	
Estrogen	*Adrenal cortex*	
Progesterone	Sex steroids	
Equine chorionic gonadotropin (eCG, PMSG)	Corticoids	
Human chorionic gonadotropin (hCG)	*Pancreas* Insulin	
Placental protein 13	*Pineal gland*	
Cervix	Melatonin	
Relaxin		
Uterus		
Prostaglandin (PGF2α)		

stimulation, or other external cues (including olfactory, visual, and auditory signals) is called bio-stimulation. Different animals exhibit sex behaviours through bio-stimulation.

22.1.3.1.1 Flehmen Response (Flehmen Reaction or Flehming or Flehmening)

The *flehmen response* is a typical behavioural posture of the animal after receiving the pheromone from the environment. The behavioural posture includes twisting the upper lip to expose the front teeth and gum with the extension of the head for better inhalation through the nostril with deep breathing for a few seconds. This response is mainly exhibited by the males, often in females, of various ungulate mammals, including cattle, buffalo, sheep, goats, horses, and cats (like a domestic cat, tiger, etc.). It is also displayed by rodents like guinea pigs and mice. The chemicals (pheromone) or scents that the animals perceive are non-volatile organic compounds (non-VOCs) and secreted mainly through the urine and or genital organs of the females during proestrus and estrus, the time of sexual receptivity. The perceptive organs of these pheromones are the vomeronasal organ (VNO) for the pheromones in liquid form and main olfactory system (MOS) for aerosol molecules. During the time of inspiration, the lumen of VNO increases and blood pressure decreases to facilitate proper mixing of pheromones with mucus-binding protein, followed by increases in blood pressure and decreases in VNO lumen, which expel the pheromones from the VNO organ. In buffalo, rapid tongue strokes at the rostral and medial palate generate a separate route from the prompt response. This response attracts the male for

Table 22.6 Source and chemical nature of some sex pheromones in domestic animals

Animal	Compounds	Sources	Functions
Cow	6-Methyl-1-heptanol, 2-methyl-7-hydroxy-3-4 and heptene	Vaginal secretions	Estrus identification
	Acetaldehyde	Milk and blood	Sex desire
	Trimethylamine, acetic acid, phenol 4-propyl, pentanoic acid and propionic acid	Saliva	Sex desire
	Acetic acid, 1-iodo undecane and propionic acid	Faeces	Sex desire
	Trimethylamine, acetic acid, phenol, propionic acid, and 3-hexanol	Vaginal fluid	Sex desire and mounting
	1-Iodo undecane	Urine	Sex desire
Buffalo	4-Methyl phenol, 9-octa decenoic acid, and 1-chlorooctane	Urine	Sex desire and mounting
	9-Octadecenoic acid	Vaginal fluid	Mounting
	4-Methyl phenol and trans-verbenol	Faeces	Estrus identification
Sheep	Fatty acid I and II (1, 2-hexadecanediol and 1, 2-octadecanediol)	Wool or hairs (ram)	LH stimulation
	C16 and diols	Fleece	LH stimulation and ovulation
	Amniotic fluid	Placenta	Accelerate the maternal response
Goat	4-Ethyl octanoic acid, octanoic acid, and 2,6-di-t-buthyl-4-methyl phenol	Fleece	LH stimulation
Pig	16-Androstene steroid	Saliva (boar)	Puberty attainment and fertility in sow
	Group of fatty acids (in definite proportion): hexadecanoic acid, cis-9-octa decenoic acid, 9,12-ctyladecanoic acid, dodecanoic acid, tetradecanoic acid, and decanoic acid	Skin (lactating sow)	Reducing stress in piglets
	5α-Androsterone-16-en-3-one, 5α-androsterone-16-en-3α-ol, and quinoline	Saliva (boar, estrus sow induced)	Standing posture
Horse	*p*-Cresol and *m*-Cresol	Urine	Ovulation marker
	p-Cresol	Urine	Augment erection
Cat			
Feline facial pheromones	F1: Oleic acid, caproic acid, trimethylamine 5-aminovaleric acid, *n*-butyric acid, α-methyl butyric acid	Unknown	Unknown
	F2: oleic acid, palmitic acid, propanoic acid, and *p*-hydroxy phenylacetic acid	Face (male, sebaceous secretions)	Facial marking
	F3: Oleic acid, azelaic acid, pimelic acid, palmitic acid	Face (male, sebaceous secretions)	Reduce negative scent-marking behaviours, induce grooming, reduce anxiety, and improve feeding, activity and playing behaviour
	F4: 5β-cholestan acid 3β-ol, oleic acid, pimelic acid, *n*-butyric acid	Face (male, sebaceous secretions)	Sexual marking
	F5: Palmitic acid, isobutyric acid, 5-aminovaleric acid, *n*-butyric acid, α-methyl butyric acid, trimethylamine, azelaic acid, *p*-hydroxyphenyl acetic acid	Unknown	Unknown
Feline-appeasing pheromone (FAP)	Lauric acid, myristic acid, stearic acid, linoleic acid, oleic acid, valeric acid, azelaic acid, pimelic acid, palmitic acid	Mammary sebaceous glands during lactation	Establish bonds and positive relationships between kittens and queen
Feline interdigital semiochemical (FIS)	Two fractions 1. Linoleic acid, valeric acid, lactic acid 2. Propionic acid, cyclohexylacetic acid, cyclopentylpropionic acid	Sebaceous gland secretions	Induce scratching behaviour
Dog	Tri-methyl amine and methyl-dihydroxy benzoate	Sebaceous secretions (female)	Attract male
Mouse	2-*Sec*-butyl-dihydrothiazole (SBT) and dehydro-*exo*-brevicomin (DHB)	Urine (male)	Advance puberty, estrus synchronise

Source: Wani et al. (2013) and Mucignat-Caretta (2014)

Table 22.7 Role of some sex pheromones in different species

Role of sex pheromones	Species
Pheromone of males	
Increase the onset of puberty	Cattle, sheep, goat, pig, mouse, and rodents
Induction and synchronisation of estrus during anestrus (by influencing LH secretion—*male effect*)	Sheep, goat, pig, mouse, and wild ungulates
Reduction of postpartum anestrus	Cattle, buffalo, and pig
Influence the standing/mating posture	Pig, cattle, buffalo, dog, and cat
Pheromone of females	
Estrus indication	Cattle, buffalo, and horse
Enhance sperm quality and libido (erection of the penis)	Cattle, buffalo, horse, and dog
Maternal pheromones	
Reduce agonistic behaviour	Pig
Reduce anxiety	Dog, cats, and rabbit
Neonatal pheromones	
Maternal responsiveness	Sheep, horse, elephant

courtship and improves the sperm quantity and libido in males. The characteristics of Flehmen response are used to identify the estrus in females to synchronise reproduction and breeding management in wild ungulates. Animals also exhibit flehmen response after parturition to recognise the neonates, where females perceived the non-VOCs from the newborn and the amniotic fluids. It often occurs in sheep, horses, and elephants. Goat shows the flehmen reaction in response to the pheromone synthesised from the modified sex hormone (androgen) and excreted through urine.

22.1.3.1.2 Ram Effect

The presence of ram in the flock of ewe can promote the occurrence of estrus. This phenomenon is called the *ram effect*. It is widely practised in sheep farming to synchronise the estrus and reduce the silent estrus incident. Introduction of the ram for at least 9–13 days before the end of the anestrus augments ovarian activity and fertility. Bio-stimulation for the ram effect mediates through the pheromones secreted through the wool and wax of the ram (Table 22.6), which induces the LH secretion in anestrus ewes. The characteristic behavioural response of the ram effect includes cessation of movements of ewes during estrus after the physical contact with the rams. This typical sheep behaviour is called *tupping* and is used to detect the estrus in ewes.

Besides the ram effect, pheromone-induced behavioural responses are manifested by the ewes for offspring recognition, the development of filial attachment in ewes, and the development of suckling behaviour in lambs. The maternal-offspring passion also develops in response to the pheromones in amniotic fluid other than oxytocin secreted during parturition. The pheromones also assist in developing the perpetual olfactory memory in the brain.

22.1.3.1.3 Self-Enurination (SE) in Buck (Scent-Urination, Urine-Marking)

Domesticated bucks show a typical behavioural response during their breeding season termed *SE*. The spreading of urine characterises it onto the face, beard, and front legs to display olfactory signals to attract estrus females. The buck turns its head and shoulders downwards and emits urine from the erect penis to disperse the urine onto the face and beards. The manifestation of SE occurs when a buck approaches estrus females before mating to induce estrus and ovulation. The vocalisation of bucks also causes a similar effect.

22.1.3.1.4 Bull Effect

The resumption of postpartum ovarian activity in cows is seen when the bulls (even vasectomised) are introduced to the farm. This phenomenon is called the *bull effect*. The onset of puberty in cows can accelerate through bio-stimulation. The cervical mucus of estrus females can promote ovarian activity in heifers and postpartum cows. The pheromones secreted from cattle are amended in Table 22.6. The pheromones such as *1-iodo undecane*, secreted through urine and faeces of females during estrus, can induce increased *libido* in bulls. Hence, the 1-iodo undecane is considered a biochemical marker to identify bovine estrus.

22.1.3.1.5 Vandenbergh Effect

John Vandenbergh first described the sexual behavioural response due to bio-stimulation in mice. He reported that puberty could be accelerated in female mice exposed to the urine of adult males. Vandenbergh effect is also seen in pigs. Boar pheromones (Table 22.6) are responsible for the Vandenbergh effect in females to stimulate reproductive behaviour and performance in gilts and sows. The male pheromones (5α-androsterone-16-en-3-one and 5-α-androsterone-16-en-3α-ol) are produced in testes and sub-maxillary salivary glands that are released through saliva. The sow comes close to the boar and makes the nose to nose contact with frequent movement of the tongue to receive the pheromones through VNO. This posture of sows is called *chomping behaviour*. The pheromone causes *immobilise* (or *standing*) *reflex* in estrus sow to hasten the copulation process. The saliva of boars is often used to augment reproductive activity in females. A group of fatty acids (Table 22.6) secreted from the skin of lactating sow reduce the agonistic behaviour in piglets for feed and house space after weaning.

22.1.3.1.6 Pheromone-Induced Sexual Behaviour in Dog and Cat

Pheromones secreted from sebaceous glands of male dogs and cats enhance the sexual in females. The supra-caudal glands in the peri-anal region of the cat and circumanal glands of the dog around the anus become active during spermatogenesis and estrus, respectively, to promote courtship behaviour in both the sexes. Methyl-dihydroxy benzoate secreted from the sebaceous gland of bitches during estrus can modulate sexual behaviour in males. A significant quantity of pheromones secretes in cats (Table 22.6) with specific functions. The tomcats rub their face to deposit facial pheromones (F2 type) in an object near sexually active females as a sexual display that improves courtship.

22.1.3.1.7 Whitten Effect

In mice, the presence of a male in a group of females can initiate ovarian activity and stimulate gonadotropin secretion. This phenomenon is called the *Whitten effect*. The pheromone of male mice is secreted through urine, and the females perceive it through smell. It causes synchronisation of estrus in females with irregular sexual cycles. The pheromones responsible for the Whitten effect are 2-sec-butyl-dihydrothiazole (SBT) and dehydro-*exo*-brevicomin (DHB). The same compounds also cause puberty augmentation in mice called the Vandenbergh effect.

22.1.3.1.8 Bruce Effect

The presence of an unknown male terminates the pregnancy in female rodents called the *Bruce effect*. It was named after Hilda Margaret Bruce, who identified the behaviour in 1959. The major histocompatibility complex (MHC) class 1 protein released from the males through their urine is perceived through VNO. The females learn the specific MHC of the males during mating. Still, when an unfamiliar male is introduced into the pregnant flock, the scent released by the male activates the neuro-endocrine pathway via the cortico-medial amygdala, accessory olfactory tract, and stria terminalis, which in turn stimulates the release of dopamine. Dopamine prevents the secretion of prolactin and causes pregnancy termination. The Bruce effect can also be seen in pigs and domestic horses. The mouse can sense estrogen through nasal ingestion; hence, excess estrogen in pen may cause the Bruce effect. The female rodents perceive MHC molecules through VNO. Oxytocin helps to recognise the MCH molecules of the breeding partner and does not terminate the pregnancy.

Know More ……

Musk

The organic aromatic substance secreted from the preputial gland of a kind of male deer (*Moschus* spp.) inhabiting China, India, Pakistan, Afghanistan, Tibet, Siberia, Mongolia, and North Vietnam is termed 'Musk'. The deer is named musk deer. *Muscone* is the main active ingredient having several androstane derivatives with specific proteins. The deer secrets musk to define the territory and attract females. Unmated male musk deer produce a more significant amount of muskcone than mated males due to certain bacteria in higher concentrations that control the metabolic pathways of androstane derivatives. Certain plants and animals produce similar kinds of strong odorous substances called *musk plants* and *musk animals* (muskrat, musk duck, musk turtle). Musk is one of the most expensive animal products commercially used as a perfume fixative in the world.

22.1.3.1.9 Interspecies Communication

Pheromones secreted from the urine, vaginal fluid, saliva, faeces, and milk during different phases of the cow's estrous cycle can attract other species like dogs, mice, and rats. These animals can identify estrus cows, and dogs can sense the luteal phase of cows.

22.1.3.2 Pheromone like Chemicals

Some bio-chemicals produce by one animal influence the activity of others, but effects are not similar to pheromones like allomones, kairomones, synomones, and interomones. These are generally released from insects. Allomones are the semiochemicals released by one organism to affect the behaviour of the other species to favour the sender, not the receiver. Many insects (stink bugs, bombardier beetles, blister beetles) use the allomones to defend the predators by emitting pungent smells or repugnant chemicals. Kairamones are the semiochemicals that benefit the receiver, not the sender. Kairamones are used by parasites, predators, parasitoids, omnivores, and herbivores to search for food. Synomones regulate interspecific communications benefitting both the sender and the receiver and can control both intraspecies, and some are interspecies communications. Interomones are semiochemicals which affect the physiological response of other species.

22.1.3.2.1 Endocrine Disruptors

The chemicals that interfere with the functions of the endocrine system by either mimicking the hormonal activity or acting as anti-hormone are called endocrine disruptors (EDs) or endocrine-disrupting chemicals (EDCs). They can generate inside the body due to disturbance in the synthesis process or can introduce from the environment as synthetic chemicals. The synthetic EDs that affect reproductive functions are diethylstilbestrol (DES) or stilbestrol (nonsteroidal estrogen); industrial chemicals like phthalates, bisphenol A; pesticides (DDT) and organochlorine pollutants like polychlorinated biphenyl (PCBs). They cause DNA methylation or reprogramming of the genes in germ cells and affect the follicular dynamics. Long-term exposure to phytoestrogens like clover and fungal toxin contaminated feed cause reproductive disturbances in sheep.

Know More ....

EDCs and Infertility

Exposure to endocrine-disrupting chemicals (EDCs) and persistent organic pollutants (POPs) causes infertility in the animals. Female mammals are more vulnerable to EDCs and POPs, which can be transmitted through the placenta and milk and may cause congenital anomalies in the endocrine system. The wild animals are also susceptible to EDCs and POPs-induced infertility through the milk and food web. The animals of the polar regions of the globe are at higher risk due to more concentrations of the POPs in those regions.

22.2 Puberty in Females and Estrous Cycle

Puberty is the ability of the animal to produce gamete, i.e. ovum in females. The onset of puberty results from integrated sequences of biological events that lead to progressive maturation of sexual characteristics to attain full reproductive capacity. The onset of puberty is related to activating the hypothalamic-pituitary-ovarian (HPO) axis. Before puberty, the tonic release of hypothalamic GnRH and subsequently pituitary LH cause a small amount of estrogen secretion in the growing follicles. The estradiol at low concentration exerts negative feedback over the pituitary and hypothalamus (see *gonadostat hypothesis* in male puberty section). At the initiation of puberty, the negative feedback of estradiol decreases as the hypothalamus becomes less sensitive to estradiol, leading to activation of the surge centre and commencement of the estrous cycle and ovulation (Fig. 22.6). Ovulation occurs, and plasma progesterone concentrations reach above 1 ng/mL in most domestic animals. Attainment of puberty in the animal did not assure sexual maturity. A considerable period after puberty is required for the animals to have regular ovarian cycles, acquire fertility, and get the uterus capable of supporting pregnancy. Sexual maturity is more closely related to proper growth or bodyweight than the animal's age.

22.2.1 Factors Affecting Puberty

The attainment of puberty depends on the maturation of the hypothalamic-pituitary-ovarian axis, and the HPO axis maturation requires adequate body weight and age. Thus, the factors which can alter the maturation of the hypothalamic-pituitary-ovarian axis are the genetics of the animals, nutritional level, energy metabolism, appropriate body size, body weight environment, and photoperiod.

Generally, females attain puberty earlier than males of the same species, with minor exceptions. In the male rabbit, the onset of puberty is earlier than in males. Early attainment of puberty occurs in small-sized breeds than in heavy breeds within a species. Long-lived animals attain puberty earlier. The external modulators of the HPO axis are photoperiod (particularly in goats, sheep, and horses in tropical areas), contact with the male (cattle and goat) and hard size or size of social groups (cattle and pig). The pheromones secreted from males influence the HPO axis to control LH and estrogen secretion. But the pheromones hasten puberty in sheep and goats.

22.2.1.1 Breed, Age, and Body Weight

The functionality of the HPO axis is highly related to the adequate somatic growth of the animal. Hence, age and body weight play a significant role in achieving reproductive efficiency. However, it is species specific and varies between breeds of a particular species (Table 22.8). The age of puberty is moderately heritable in all species. In cattle and buffaloes, it ranges between 0.16–0.57 and 0.24, respectively. In general, the puberty occurs in bovine species at 11 months (male 7–18 and female 9–24); in ovine, 7 months (male 6–9 and female 4–14); in porcine, 7 months (male 5–8 and female 5–7); in equine 14 months (male 10–24 and female 12–19); canine 9 months (male 5–12 and female 6–24) and feline 9 months (male 8–10 and female 4–12). Usually, puberty occurs when an animal reaches 55–60% of its adult body weight in beef cattle, 30–40% in dairy cattle, 60–65% in buffalo, 40–65% in sheep, 40–60% in goat, 40–50% in pig, 60–65% in horse, and 40–60% in the dog.

22.2.1.2 Nutrition

Plane of nutrition, particularly protein and energy and calcium and phosphorus in the ration, directly influence the HPO axis. Lack of optimum level of nutrition inhibits ovarian follicular development and steroidogenesis and reduces the synthesis and release of GnRH, FSH, LH, and GH. Several endocrine factors link energy metabolism and reproduction. Leptin, produced from adipose tissue, skeletal muscle, stomach, mammary tissue, placenta and pituitary, activates the GnRH pulse generator in heifers, rodents, and other non-ruminants to activate the HPO axis. GH and IGF-I affect the concentration and amplitude of LH pulses, and insulin acts through IGF-I and increases LHRH expression. Thus, leptin and IGF-I are the indicators of adequate nutritional status and HPO axis maturation. Neuropeptide Y (NPY) acts between interneurons of GnRH and LH under stress,

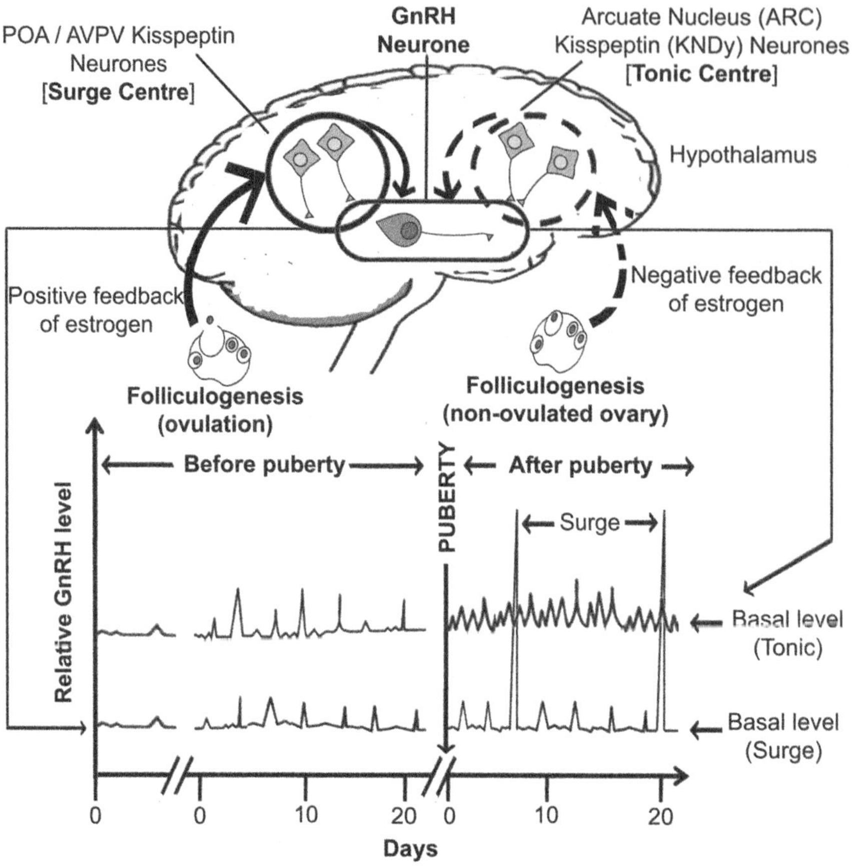

Fig. 22.6 Control of gonadotropin-releasing hormone secretion and commencement of puberty in cattle. The upper part of the graph shows the estrogenic control on **GnRH** secretion of (gonadotropin-releasing hormone) in the **hypothalamus** with the occurrence of **puberty,** and the lower part of the figure illustrates the secretory pattern of GnRH. **During pre-pubertal periods**, the **tonic GnRH centre** comprises an **arcuate nucleus (ARC)** with **kisspeptin** (**KNDy** = kisspeptin/neurokinin B/dynorphin) **neurones** is under the negative control of estrogen released from the **non-ovulated follicles of the ovary** during folliculogenesis process. It facilitates the secretion of low-frequency GnRH from the **GnRH neurone** (**negative feedback of estrogen**, depicted by a dotted curved arrow). **After puberty**, the tonic centre becomes non-sensitive to estrogen feedback, and GnRH is secreted at a high frequency (shown as a bold-faced curve-linear graph in the lower part of the figure). At the same time, the **surge centre** comprising **POA/AVPV** (pre-optic area/anteroventral periventricular nucleus) **Kisspeptin neurones** is sensitised with the estrogens (**positive feedback of estrogens**, depicted with a bold-face curved arrow) at the onset of puberty. It causes sudden spikes (**surges**) of GnRH secretion from the GnRH neurone (shown in the lower part of the figure). The positive feedback of estrogens causes activation of the hypothalamic-pituitary-ovarian axis and the profuse release of estrogen results in **ovulation**. The events of pulsatile release of GnRH inform the tonic and surge centres before and after puberty and are presented at a scale of **20 days** on the *X*-axis of the graph, considering one estrous cycle of cattle. The *Y*-axis depicts the **relative amplitude (level) of GnRH** secretion

and it increases food intake and storage of energy as fat and reduces anxiety and stress.

22.2.1.3 Environment

Season, particularly the season of birth, and photoperiods are two major factors that modulate puberty and sexual maturity.

22.2.1.3.1 Season of Birth

The young born during autumn and winter attains puberty earlier than the young born during spring. It is due to the long photoperiod and fodder availability in autumn and winter, favouring growth. The high environmental temperature in summer reduces the activity of gonads. Sheep are usually bred in autumn, and young are born in spring. Lambs born in spring and autumn will attain puberty earlier than summer-born lambs. The autumn and winter (September to February) are the buffalo's breeding season, and the calving season is spring (July to November). Puberty occurs earlier in spring-born buffalo calves than in their summer-born mates. Horse favours long-day period for breeding and foaling as well. In the horse, the frequency of FSH and LH pulses is more in early spring and decreases gradually and becomes lowest during winter.

22.2.1.3.2 Photoperiod

Photoperiod means daylight exposure, affecting puberty onset in seasonal breeders. Long-day photoperiod consists of a maximum of 16 h of day length, and short-day

Table 22.8 Physiological age and body weight at puberty of various breeds of domestic animals

Species	Breed	Physiological age of puberty (month)	Bodyweight at puberty (kg)
Cattle	Jersey	8–10	160–180
	Holstein Frisian (in the USA)	12–13	265–289
	Holstein Frisian (in Australian)	8–12	200–230
	Brown Swiss	10–11	280–300
	Sahiwal	30–46	225–250
	Gir, Red Sindhi (in India)	36–40	240–250
	Indigenous (tropical region)	27–40	160–210
	Indigenous (in India)	20–40	80–200
	Jersey × Indigenous (in India)	15–18	160–180
	Holstein Frisian × Indigenous (in India)	15–18	180–210
Buffalo	NilliRavi	30–33	450–520
	Murrah	33–36	320–360
	Surti	30–46	280–330
	River buffalo	15–18	250–450
	Asian Swamp	21–25	300–330
Goat	Florina, Mountain Black	6–12	20–30
	Angora, Black Bengal	6–8	8–10
	Jamunapari	8–9	12–14
Sheep	Bergamacia	7–9	20–30
	Dorset, Rambouillet	6–8	40–45
	Merino	7–8	30–40
	Deccani	9–11	20–22
	Garole	5–12	7–10
	Munjal	10–12	22–27
Pig	Large White Yorkshire, Landrace	6–8	100–140
	Ghoongroo	7–9	50–60
	Zovawk	2.5–3	4.5–6
Horse	Thoroughbred	7–13	270–410
	Morgan	12–15	340–370
	Spiti	20–35	175–225
Dog	German Shepherd	5–8	23–30
	Labrador Retriever	6–12	15–30
	Golden Retriever	9–11	20–25
	Beagles	9–10	3–6
	Dachshund	7–18	6–14
	Pug	4–6	3–6
Cat	Domestic cat	4–5	2.5
Rabbit	Miniature and medium breeds	3–6	3
Rat	Standard (Laboratory rat)	30–40 day	50–125 g

Source: Data compiled from various sources

photoperiod denotes a maximum of 8 h of daylight exposure. Photoperiod modulates the reproduction through the secretion of melatonin hormone from the pineal gland. Melatonin influences the hypothalamus to secrete more GnRH and LH (Fig. 19.17). Melatonin secretes during the dark phase of the day. Thus, a short-day photoperiod (long dark-phase) stimulates melatonin synthesis. The exposure to short photoperiod during pre-pubertal life (animals born in autumn and winter) facilitates the attainment of puberty by secreting more melatonin. Thus, attain early puberty. But, in the horse, increased melatonin decreases GnRH activity. Hence, they have more GnRH activity during summer (long-day photoperiod), when melatonin secretion is less.

Seasonal breeders are mostly found in the temperate zone (60–70° N/S), where photoperiod alters the reproduction in animals. The effect of photoperiod on reproduction is less evident in the domestic animals closer to the equator, where the daylight and dark phases are almost equal.

22.2.1.3.3 Seasonal Breeder

A seasonal breeder shows reproductive activity during a particular season and maintains reproductive quiescence during the rest of the year. Seasonal breeders are generally *short-day* and *long-day breeders* (Table 22.9). In short-day breeders, decreasing day length (in autumn and winter) influences the onset of estrus (sheep and goat). In contrast, increasing day length during summer and spring stimulates estrus in long-day breeders (mares). Generally, the long-day breeders give birth within the same breeding season or same season of the following year. Seasonal breeding is a kind of adaptive strategy for the survival of the offspring and mothers. The changes in the food resources available that control the animal's energy expenditure plays a central role in reproductive seasonality through integrated endocrine and genetic mechanisms. Therefore, apart from photoperiod and melatonin, the seasonal breeding is controlled by neuropeptide Y (NPY), Kisspeptin, GnIH, gonadotropins, estrogen, GH, IGF-I, leptin, and thyroid hormones that are related to the energy metabolism of an animal. Generally, small mammals and birds show sexual activity during spring and summer, and they have short gestation or incubation period and give birth within the same season.

Table 22.9 Seasonal breeders with their breeding season

Seasonal breeder	Breeding season	Animals
Short-day breeder	Autumn and winter	Sheep, goat, buffalo, fox, deer, and elk
Long-day breeder	Spring and summer	Horse, cat, hamster, groundhogs, and mink

Cattle, pigs, and rabbits are the non-seasonal breeders. But, ovulation rate, conception rate, and litter size are less in summer in pigs. Parity one is more susceptible to reproductive infertility in the sow. There are another group of mammals competent to be fertile at any time or can resume fertility at a short period in a favourable environment, called *opportunistic breeders*. Human is an example of opportunistic breeder and can mate throughout the year. Golden spiny mouse temporary halts its reproduction in high salinity in drying desert; it can resume fertility during rainfall. Small rodents and tree kangaroo are also opportunistic breeders. The dog considers a seasonal breeder, but the exact breeding season(s) are difficult to determine. They show breeding activity after 6 months. The rainy season is one of the preferred breeding seasons for the dog may be due to higher humidity and less environmental temperature that favours the pheromone signals for the activation of sexual response.

22.2.2 Estrous Cycle

The rhythmic sexual behavioural pattern exhibited by the female animals after the attainment of puberty is called the *estrous cycle*. The sign of sexual receptivity is called *estrus* or *heat*, which denotes the initiation of the cycle. Estrous cycle can also be defined as the duration between two successive estruses. The estrous cycle is classified into four distinct phases: estrus, metestrus, diestrus and proestrus. During the proestrus and estrus, follicular development or generation of follicular wave(s) occur, followed by ovulation. Hence, these two phases are collectively called the follicular phase. The luteal phase is characterised by the formation of the corpus luteum and its lysis. The luteal phase consists of metestrus and diestrus.

> Know More ....
> **Estrus: Etymology**
> The name *estrus* was derived from the Greek word 'oistros'. In ancient Greece (1690–1700), the *oistros* was a *gadfly*, a biting fly under the genus Oestrus that hurt animals and forced them to react. The reaction of the animals attacked by such flies is similar to the behaviour during the estrus. In 1850, the word *oistros* was used as the Latin word 'oestrus'; later, in 1890 as 'estrus'.

The events and duration of these phases (Fig. 22.7) are species-specific (Table 22.10 and 22.11). The events of the estrous cycle are regulated mainly by two gonadotropins (FSH and LH) and ovarian sex steroids (estrogens and progesterone) hormones that change dramatically during different phases of estrous cycle (Fig. 22.8). The animals showing a regular estrous cycle are called cyclic animals. Pregnancy causes physiological cessation of the cycle, and it resumes after the parturition as postpartum estrus. The failure to express the estrous cycle without any physiological reasons (such as pregnancy) is termed 'anestrus' (Fig. 22.7). In the majority of the cases, anestrus is temporary due to some pathological reasons, but prolonged anestrus leads to infertility, and persistent infertility leads to sterility.

22.2.2.1 Classification of Estrous Cycle

Estrous cycle can classify in several ways. Based on the occurrence of the estrous cycle, animals can be classified into monoestrous, polyoestrous, and seasonally polyestrous. Monoestrous animals exhibit only one estrous cycle per year. The true *monoestrous* animals are foxes, wolves, and bears. Animals like cows and sows show frequent, periodic estrous cycles throughout the year are called *polyestrous* animals. Animals that exhibit periodic estrous cycles only during a particular season are called *seasonally polyestrous* animals. They can be classified further into short-day breeders and long-day breeders. Animals that come in estrus during the decreasing day length are called *short-day breeders*, for example, buffaloes, deers, and ewes of non-tropical regions. The animals that show estrus during the increasing day length are called *long-day breeders* (mare and queen). Based on ovarian activity, the estrous cycle can be classified into regular, spontaneous, and induced ovulatory estrous cycles. In the animals showing a *regular estrous cycle*, ovulation and formation of the corpus luteum (CL) occur regularly during a cycle. The corpus luteum formation is independent of mating, for example, cow, sheep, mare, and sow. In the *spontaneous ovulatory cycle*, ovulation occurs spontaneously, but CL will not be functional until mating, for example, rat and mouse. In the *induced ovulatory cycle*, ovulation and CL formation depends on whether mating has occurred or not, for example cat, rabbit, mink, and camel. There are sensory neurones in the female genital tract of induced ovulators which are sensitised during copulation and stimulate the hypothalamus for LH surge, which causes ovulation and subsequent CL formation. The classification of the estrous cycle based on ovarian activity is presented in Table 22.10.

22.2.2.2 Estrus

It is a well-defined period characterised by sexual receptivity or acceptance of males. The most distinguishable feature of this phase is behavioural change. The estrus begins with the first acceptance and ends with the last acceptance of the male. A large and matured Graafian follicle protrudes above the ovarian surface. Oviducts are tonic with mature epithelia and active cilia. Contractions of the oviduct occur that facilitate sperm transportation after insemination. The fimbriated end of the oviduct comes close affinity to the Graafian follicle to

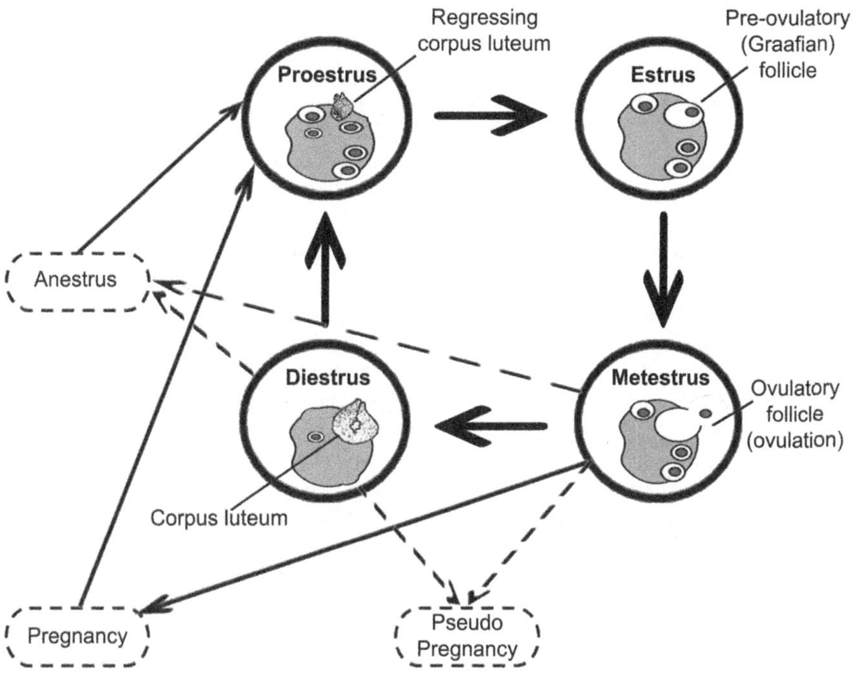

Fig. 22.7 Various phases of cow's estrous cycle. Figure depicts the four distinct phases of estrous cycle with associated ovarian changes. The cycle is initiated with **estrus**, followed by **metestrus**, **diestrus,** and **proestrus** (boldface circle). The estrus is characterised by **pre-ovulatory** or **Graafian follicle's** presence, and **ovulation** is occurred during metestrus in the cow; **corpus luteum** is formed in diestrus and **regressed** in proestrus. **Pregnancy** leads to prolonged diestrus, and the cycle resumption occurs from the proestrus after the termination of pregnancy (straight arrow). The pseudo pregnancy also causes prolonged anestrus (dotted arrow), and the cycle is restored from proestrus after the termination of pseudopregnancy

Table 22.10 Characteristic pattern of various types of estrous cycle

	Characteristics			
Species	Cycle	Follicular development	Ovulation and CL formation	Function of CL
Cow, doe, ewe, mare, sow	Long	Spontaneous (FSH)	Spontaneous (LH surge)	Spontaneous
Rat, hamster, mice	Short	Spontaneous (FSH)	Spontaneous (LH surge)	Induced (prolactin)
Cat, rabbit, ferret, mink	Induced	Spontaneous (FSH)	Induced (LH surge)	Induced

capture the ova during ovulation. The secretion of oviductal fluid increases. The uterus appears tonic and becomes turgid. There is increased blood supply to the uterus. Uterine mucosal growth is evident, and there is increased mucous secretion. Vaginal mucosa becomes pale pink due to increased vascularity. In dogs and cats, thickening of vaginal mucosa and desquamation of cornified epithelial cells occurs. The cervix is relaxed and edematous. Stringy mucous hangs from the vulva. There is increased neutrophilic infiltration into the uterine lumen, and the stroma becomes edematous. Glandular ducts secrete a thin serous fluid which flushes the tract and assists sperm transportation. In most domestic species, ovulation occurs during estrus, but in cows, ovulation takes place 12 h after the end of estrus. In induced ovulators, the estrus may be prolonged up to 7–10 days in the absence of males.

During estrus, the level of LH, as well as estrogens, is increased gradually. It causes the selection of the dominant follicle and its maturation. LH surge occurs at the end of this phase with the profuse synthesis of estrogens favouring ovulation. Secretion of FSH is also reduced due to high estrogen levels, which favours follicular maturation. The synthesis of PGF2α has also occurred at the time of ovulation. The morphological alterations of the female reproductive tract are manifested mainly by the estrogens with progesterone priming. The absence of progesterone priming results in poor estrus manifestation (silent estrus), as seen in the first estrus after puberty and the first postpartum estrus. Silent estrus is common in cow, ewe, and sow and less frequent in mare and bitch; as progesterone starts secreting from theca externa of the Graafian follicle before estrus in these animals. The behavioural signs of estrus are also manifested by estrogens acting over the pre-optic area, ventromedial hypothalamus (VMH), amygdale, midbrain, and pituitary. The sexual receptivity is manifested through estrogen receptor α at the ventromedial hypothalamus (VMH), the centre for sexual receptivity. Progesterone has an inhibitory role on

Table 22.11 Duration of estrous cycle and its various phases

Species	Estrous cycle (day)	Estrus[a] (h)	Ovulation time (from the onset of estrus, h)	Insemination time[b] (from the onset of standing estrus, h)	Metestrus (day)	Diestrus (day)	Proestrus (day)
Cow	21–22	18–19	25–32	10–11 (from the end of s. estrus)	3	16	2
Buffalo	17–24 (21)	12–30 (19–21)	24–48 (34)	24	2–3	11–15	2–3
Ewe	16–17	24–36	30–36	12–18 (twice)	2	11	2
Doe	21	32–40	30–36	Every 12-h interval (thrice)	2	11	2
Mare	19–25	4–8 days	1–2 days (before the end of estrus)	1 day before ovulation or every day from day 3 to the end of estrus	2	13–14	2
Sow	19–20	48–72	35–42	Day of estrus to next day morning	1–2	14	2
Bitch	70–100	4–13 days	Day 4	2–5 days after ovulation	---	60–80	5–9
Queen (domestic cat)	7–42 (21)	3–17 (6) days	24–60 (after mating[c]) or induction by eCG and hCG; after 25–30 h of hCG administration	Immediate before ovulation (confirmed by cytological test), after 20–22 h of hCG administration	---	35–40[d] (if ovulated but not pregnant)	1–2
Guinea pig	13–21 (16)	8–11	1–2 day	During ovulation[e]	3	11–12	1–1.5
Rat	4–5[f]	12/25–27	Night on the day of estrus	End of proestrus to entire estrus	21 h/6–8 h	57 h/10^2	12–14 h

Source: Data collected from various sources
[a] Estrus is considered for the standing estrus period
[b] Repeated mating is preferred in the natural breeding programme in polytocous species
[c] For successful ovulation 3–4 mating is required within 24 h for ovulation
[d] After formation of active corpus luteum in pseudopregnancy stage
[e] Other than the intravaginal route, intraperitoneal route of insemination can be used in guinea pig as there is an opening of the ventral part of the ovarian bursa
[f] Up to ovulation

sexual receptivity in all induced ovulators (rabbit, ferret), guinea pig, and hamster as it antagonises the estrogenic effects on the CNS. Some neuronal networks involved in the expression of sexual receptivity are GnRH neurons (at the medial pre-optic area), noradrenergic neurons (at the medial pre-optic area and the VMH-secreting norepinephrine), and dopaminergic neurons (at ventral tegmental areas and the substantia nigra of midbrain-secreting dopamine).

22.2.2.2.1 Behavioural Estrus

The estrus is characterised by the appearance of a series of visible psychological behaviour under the influence of estrogens. The most predominant sign of estrus is standing to be mounted. The animals under estrus remain standing and allow other (cow or bull) animals to be mounted with the elevation of hindquarters, called *lordosis* reflex. It is also called *standing heat* and may continue up to 18–19 h in a cow (the total estrus phase may be up to 20–30 h). The female on standing heat may allow a maximum of 50 times mounting with a few seconds duration in a cow. The other primary signs include bellowing, red and swollen vulva, and clear cervical mucus that appears as viscous elastic strands hanging from the vulva and sometimes spread over the tail, flanks, and perineal region. Upon microscopic examination, the cervical mucus shows characteristics 'ferning pattern' due to high chloride content. In some cows and mostly in heifers, bloody mucus may discharge 1–3 h after the estrus. It is called *metestrus bleeding*. This sign confirms that the animal was in estrus. The metestrus bleeding is due to dripping from the uterine blood vessels caused by excess estrogens during estrus. The secondary signs of estrus are restlessness, sniffing the genitalia, licking the vulva of other animals, lip curling, frequent micturition, rubbing the chin on the back or rump of the other animal (chin-resting), loss of appetite and reduced milk yield. The characteristic female sexual behaviour manifests only during the estrous cycle's estrus phase in all mammalian species, except humans. This distinguishable external expression helps identify the reproductive state of the animals, and hence, the period of estrus is considered day 1 (the day of initiation) of the cycle.

22.2.2.2.1.1 The Ideal Time for Insemination

The time of ovulation is species specific (Table 22.11); hence, the timing of insemination also varies between species. The ideal time of insemination can predict based on several factors, such as the duration of standing heat, ovulation time, the viability of ovum and spermatozoa, and the duration taken by the sperm to achieve fertilising capability

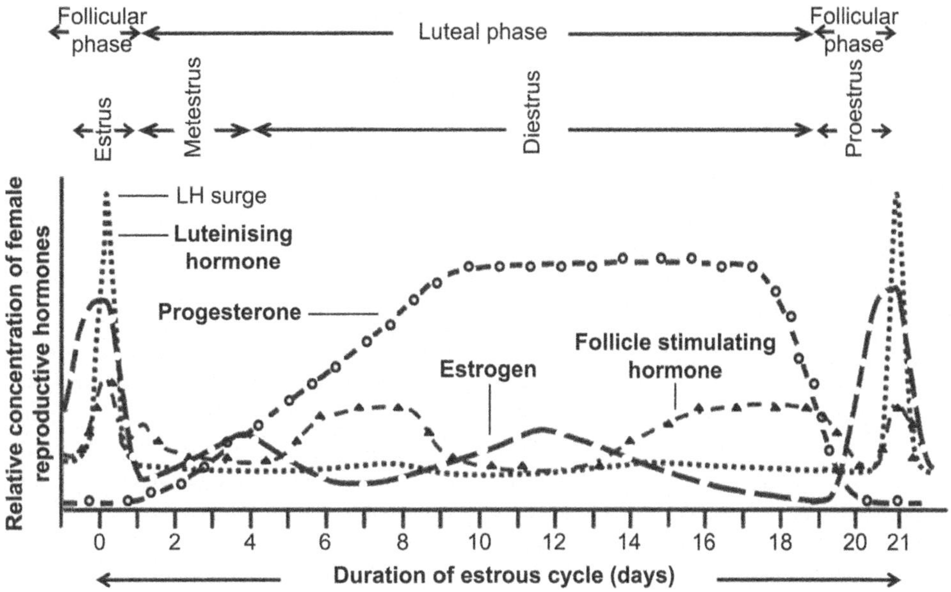

Fig. 22.8 Relative concentration of various female reproductive hormones in different phases of estrous cycle in cow. Figure shows the relative changes of the concentrations of luteinising hormone (LH), **follicle-stimulating hormone** (FSH), **estrogen** and **progesterone** (***Y*-axis**) during different phases of the estrous cycle of the cow (***X*-axis**). Different phases of the cycle, viz. **estrus** (0–1 day), **metestrus** (1–4 days), **diestrus** (4–19 days), and **proestrus** (19–21 days), are illustrated above with the hormonal profile (line arrow). The cycle initiated at estrus with the **LH surge.** The concentration of estrogen also rises at the initiation of estrus and continues up to the entire estrus phase. The progesterone level gradually increases during the metestrus, and the concentrations of other hormones decrease. In diestrus, the progesterone level peaked on day ten and sustained up to day 17, followed by a sudden fall to reach a base level after day 19. The concentration of FSH and estrogens follow characteristics of wavy patterns during diestrus. Proestrus (day 20–21) is characterised by a higher level of FSH and estrogen. The proestrus and estrus consider follicular phase, and metestrus and diestrus are called the luteal phase; marked at the top of the figure and demarketed with a line arrow

Table 22.12 The duration of different events to predict the ideal time of insemination in cow

Events	Duration
The length of standing heat	8–9 h
Ovulation	24–32 h after the start of standing heat (avg 28 h)
Fertile life of ova after ovulation	12 h
Sperm viability after insemination	18–24 h
Sperm transport (capacitation)	6 h

in the female reproductive tract (capacitation). The average duration of such events in cattle is presented in Table 22.12.

Considering the above facts, the a.m./p.m. or p.m./a.m. rule is practised in artificial insemination in cows. A cow exhibited in standing heat in the morning should inseminate in the afternoon of that day. A cow shown in standing heat in the afternoon or evening should be bred the following morning.

22.2.2.2.2 Heat Detection Techniques

Appropriate detection of estrus or heat is the essential prerequisite for a successful breeding programme. Visual observation is the best method to detect the standing heat, but it is of limited use during silent heat and repeat breeding animals. There are several aids to detect the standing heat in animals, such as (1) Teaser bull: Vasectomised bulls are routinely used to detect heat in farms with an efficiency of 84%. But sexually transmitted diseases are the major demerits of this method. (2) Milk progesterone analysis. During proestrus and estrus, the level of progesterone will decrease. In cows, low progesterone levels may continue for nearly 6 days. But, it is not an ideal tool for detecting the standing heat. (3) Mount detector device is an electronic gadget that can detect mounting and is very useful for identifying the postpartum estrus. (4) Animal wise action lists can sometimes assist in detecting the heat in the herd and encaged animals. (5) Videography of the animals' behaviour is more accurate, but it is costly. (6) Vaginal electrical resistance (ER) probe can be used to detect heat, where the resistance value is minimum during standing estrus. The maintenance cost of the device and transmission of diseases are the major demerits of this tool. (7) Vaginal cytology is extensively used to detect heat in laboratory animals. (8) Periodic evaluation of estrogens and progesterone metabolites in urine and faeces are used to detect heat in wild animals during the breeding season, followed by confirmation through the cytological test.

22.2.2.2.3 Postpartum Estrus

The immediate estrus with subsequent occurrence of ovulation and formation of corpus luteum after parturition is called postpartum estrus. The uterus returns to its normal tone with proper myometrial activity to support the successive gestations. Generally, the postpartum estrus is anovulatory, and the animal may not be able to fertilise after insemination. During the postpartum period, the progesterone level is minimised due to luteolysis. The regressed corpus luteum transformed into scar tissue. Follicular wave is initiated under the influence of FSH. Lactation, particularly in high yielding animals, causes increased prolactin secretion, which inhibits gonadotropin release and delays the occurrence of postpartum estrus (lactational anestrus), whereas weaning advances the period. The occurrence of postpartum estrus varies between species. It is nearly 45–60 days in cow, 30–90 days in buffalo, 33–90 days (depending upon season and latitudinal location) in doe and ewe, 6–15 days in mare, 90 days or more in bitch, 45–60 days in queen (domestic cat), 3–18 h in rat, 2–14 h in guinea pig, 14–24 h in the mouse.

22.2.2.3 Metestrus

It is the transition period between ovulation and the formation of CL. It is a poorly developed period. In cows, ovulation occurs in the first part of this phase. It is mainly under the influence of progesterone produced by CL; hence, this period is considered under the luteal phase. Progesterone inhibits FSH secretion and restricts the further growth of Graafian follicles. The level of LH remains at a moderate level with declining estrogens. In cows, the growth of endometrial glands begins, and capillary haemorrhage occurs that leads to menstrual bleeding. The secretion of mucus is decreased. The uterus becomes less tonic, and the superficial epithelium becomes hypertrophic and pseudostratified cells are modified to tall columnar epithelium cells. Uterine contraction is reduced. In most animals, fertilisation takes place in this phase.

22.2.2.4 Diestrus

It is the most prolonged phase of the estrous cycle. The corpus luteum matures and secrets progesterone. Several small follicles of successive wave(s) will appear in the ovary. Due to the effect of progesterone, there is increased hypertrophy of endometrial glands, and the uterine muscles become flaccid. The basal cells of the endometrial glands become active and secrete thick *uterine milk* into the lumen of the uterus. The cervix is closed, and vaginal mucus is scanty. The vaginal mucous membrane becomes pale. Estrogen priming is essential to get the optimum effect of progesterone during diestrus as the estradiol activates progesterone receptors in the endometrium.

On the other hand, the progesterone antagonises estradiol and causes negative feedback on the GnRH for the secretion of FSH. Thus, the estrogen and FSH level is minimised with the increasing level of progesterone (Fig. 22.8). In cows, the corpus luteum attains its maximum size within 7–8 days after ovulation, i.e. within the first 4 days of the diestrus and sustains up to day 16. If the animal becomes pregnant, the peak size of the corpus luteum will persist throughout pregnancy to provide progesterone. The corpus luteum regresses in non-pregnant females by PGF2α at the end of this phase, and the progesterone level is dropped suddenly.

22.2.2.5 Proestrus

It is an ill-defined period that usually lasts for 2 days in domestic animals. Growth of the follicles occurs under the influence of FSH with increased production of estradiol. Estrogens cause the formation of the antrum and also cause endometrial hypertrophy. Estrogens also stimulate the expression of pre-ovulatory sexual behaviour in some animals like rats. Increased vascularity of uterine mucosa causes hyperplasia of endometrial cells. The level of progesterone is minimum during this phase. The ovaries are characterised by remnants of CL (corpus albicans) of the earlier cycles. There is increased growth of oviductal cells and cilia. The vaginal epithelium thickens due to increased vascularity and is cornified in dogs and cats. The LH level rises at the end of the proestrus.

The primary follicles are primarily involved in the wave. The size of the follicles is species specific. Ovary and the entire reproductive tract become hyperaemic. The uterus and vagina are distended. Non-cornified nucleated epithelial cells are available on vaginal cytology. This is the first phase of the follicular stage.

22.2.2.6 Anestrus

Anestrus is the period of complete reproductive incompetence due to ovarian inactivity when the animals fail to express estrus behaviour. When estrus behaviour is less intense, it is called *sub-estrus*. It is the period between the diestrus of a cycle and the proestrus of the next cycle. During anestrus, the temporary inactivation of the HPO axis during anestrus restricts follicular development and ovulation. The secretion of estrogen is decreased and results in weak estrus behaviour. The inactivation of the HPO axis is due to reducing GnRH secretions from the GnRH pulse generator. Anestrus is of two types. In *anovulatory anestrus*, follicle maturation is affected due to reduced FSH, and no ovulation occurs due to lack of LH. In *persistent luteal anestrus*, the CL does not regress, and the progesterone secreted from this CL inhibits FSH secretion. The ovary of anestrus animals becomes oval and somewhat flattened in shape with large numbers of small follicles, and the endometrium becomes atrophic with dense stroma and small endometrial glands.

In bitch, two types of persistent anestrus are seen, primary persistent anestrus and secondary persistent anestrus. Anestrus without a previous estrous cycle is called primary persistent anestrus, whereas secondary persistent anestrus occurs after the first estrous cycle. The primary persistent anestrus develops due to genetic and chromosomal disorders, autoimmune diseases (oophoritis), ovarian aplasia, ovariectomy, or ovariohysterectomy at a young age and progesterone-releasing ovarian cysts. Management problems, like over-exercise and nutritional deficiency, also act as predisposing factors for primary persistent anestrus in bitches. The regular interval between two estruses in bitch is about 4–10 months; hence, if a bitch fails to return heat after a gap of 10–18 months from its previous estrus. It can be designated as secondary persistent anestrus and seen in an aged animal with an irregular estrous cycle. Infections in the reproductive tract, metabolic diseases, endocrine disturbances (Cushing's disease, dysfunction of the thyroid gland), systemic diseases, and altering reproductive hormonal balance may cause secondary persistent anestrus. Luteal cyst and the ovarian tumour may cause both primary and secondary anestrus in bitch.

22.2.2.6.1 Cause of Anestrus

Anestrus may occur due to several reasons (Table 22.13). Physiological anestrus occurs due to pregnancy, lactation, and breeding seasons. Sometimes congenital defects, nutritional deficiencies, and stress factors also suppress the HPO axis and cause anestrus. The anestrus period may also vary between species, breed, parity, and age.

22.2.2.6.1.1 Cystic Ovarian Diseases

Cyst is a fluid-filled structure. The follicular and luteal cyst may develop in the ovary, particularly in cattle. A *follicular cyst* is formed within unovulated follicle, caused due to lack of LH surge required for ovulation. Prolonged exposure to phyto-estrogen and administration of estrogen and progesterone may reduce the optimum LH level for ovulation, causing the development of a follicular cyst. The pathological state builds up the cystic ovarian disease (COD). A thin layer of theca cells covers a follicular cyst. The theca cells continuously produce progesterone. But it does not have the receptors for PGF2α. Hence, luteolysis is not occurred, causing anestrus. Animals having follicular cyst with anestrus may exhibit constant estrus (heat) due to estrogen produced by granulosa cells. When the cyst develops in the corpus luteum, it is called a *luteal cyst*. The luteal cyst contains a thick luteal cell layer, which will produce continuous progesterone and causes anestrus. The thick layer of luteal cells can respond to exogenous PGF2α.

22.2.2.6.2 Management of Anestrus

The anestrus can be treated after removing the underline causes with proper management and specific hormonal therapy. PGF2α is frequently used to treat luteal cysts. First, FSH and LH are applied to induce ovulation and CL formation for the follicular cysts. Then, PGF2α is administered 7 days apart to induce luteolysis. Estrus can be induced using hCG and eCG in wild cats (like Amur leopard cat) during the non-breeding season. The presence of males can increase ovulation rate and reduce the period of anestrus in the non-breeding season of seasonal breeders, like goats and sheep, along with gonadotropin administration. In bitches, serotonin-antagonists and dopamine-agonists can reduce the

Table 22.13 Factors causing anestrus in animals

Factors		Mechanism
Physiological	Pregnancy	High progesterone suppresses FSH and LH.
	Lactation (suckling anestrus)	Prolactin suppresses GnRH secretion.
	Season	The reproductive quiescence during the non-breeding season in seasonal breeders is due to melatonin that suppresses the HPO axis.
Nutritional deficiency	Negative energy balance	Continuous energy deficiency in high yielding animals leads to negative energy balance and blocks the pulsatile release of GnRH from the hypothalamus (Fig. 22.9).
	Deficiency of vitamins (carotene) and minerals (Cu, Co, Mn, P)	The deficiency of specific vitamins and minerals like carotene, copper, cobalt, manganese, and phosphorus may cause anestrus as they act as cofactors in the activity of various enzymes required for the synthesis of reproductive hormones.
Congenital defects	Freemartin, uterus unicornis, ovarian aplasia, and ovarian hypoplasia	Dysfunction of ovaries and other reproductive organs.
Drugs	Chronic supplementation of anabolic steroids (glucocorticoid)	Negative feedback on the HPO axis
	Prolong exposure to phyto-estrogen and administration of estrogen and progesterone	
Pathological conditions	Ovarian cyst	Follicular cysts: Animals exhibit constant estrus (heat) due to estrogen produced by granulosa cells. Luteal cysts: Progesterone suppresses the HPO axis.
	Pyometra, foetal mummification	Persistent CL causes suppression of FSH and LH.
Stress		Hypothalamic-hypophyseal-adrenal axis suppresses the HPO axis.

anestrus period by increasing FSH levels and decreasing the prolactin levels in lactating animals.

22.2.2.6.3 Silent Heat

The lack of behavioural estrus symptoms is termed silent heat. It is completely different from anestrus as reproductive organs undergo normal cyclic changes in silent heat. The predominant cause of silent heat is the lack of estrogen receptors. The silent heat is commonly seen in the animals experiencing the first estrus due to the absence of progesterone priming. Silent heat can also occur during the first postpartum estrus. Seasonal breeders also exhibit silent heat at the beginning of every breeding season. Persisted silent heat may cause anestrus due to insufficient estrogen production, nutritional deficiency, and various pathological conditions. In persisted silent heat, ovulation may occur.

22.2.3 Factors Affecting Estrous Cycle

22.2.3.1 Species and Breed

Species variations in the lengths of estrous cycles among species (Table 22.11) are determined primarily by the duration of the luteal phase. Large mammals have a longer luteal phase than animals with small body stature. The evolutionary advantages of having shorter luteal phases in small mammals are short gestation periods, absence of lactational anestrus, and early attainment of puberty in young. Small mammals are susceptible to predation, and a lengthy non-pregnant cycle is not advantageous. The animals with a longer estrous cycle are prone to extinction. The duration of cycle length and occurrence of estrus cycle varied between breeds of certain species. In sheep, more duration with fewer cycles in a season is seen in the Santa Ines breed than Romney Marsh breed. Generally, the breeds of the tropical region, Dorsets and the fine-wool breeds like Rambouillet and Merino have short anestrus than the other breeds like Hampshire, Suffolk, and Border Leicester. The breeding season starts in Rambouillet breed around July, whereas Landrace initiates their breeding during September. A short estrous cycle and proestrus often occur in German shepherds than Pointers in dogs. German shepherd breeds also exhibit long estrus.

22.2.3.2 Seasons and Photoperiod

The photoperiod, latitude, and season cause HPO axis deviation and seasonal reproductive activity in certain species. Estrus duration is increased during summer with less mounting activity in cattle. Total daily light exposure is important to control the onset of estrous cycle. The duration of seasonal anestrus in long-day breeders may reduce by increasing photoperiod duration with artificial light. In sheep and goats, decreased day length influences the onset of the estrous cycle. In mares, increased total daily hours of light hasten the onset of estrous cycle. Rapid alterations in the exposure to artificial light can extend the cyclic activity in short-day breeders. Melatonin implantation also has a similar effect.

22.2.3.2.1 Temperature

The secondary sign of estrus is displayed more in the early morning and late evening due to favourable ambient temperatures ranging between 75 and 85 °F in the cow. Excessive heat during summer directly reduces reproductive efficiency. In cattle, the estrous cycle length can be increased up to 25 days during summer compared to 20–22 days in cool weather. The reduced reproductive activity around summer is due to reduced thyroid activity, high progesterone from gonadal and non-gonadal (adrenal gland) sources, oxidative stress, and reduced LH secretion. Heat stress also affects follicular development with diminished aromatase activity and lower production of 17β-estradiol (Fig. 22.9). Reduced estradiol and poor LH concentration cause silent heat around summer. The poor quality oocytes with reduced follicular fluid and less viable granulosa cells produced under thermal stress may affect fertilisation rates. It is designated as summer sterility usually occurs in high yielding cows and buffaloes of the tropical and sub-subtropical region.

In sow, summer stress (beyond 35–40 °C ambient temperature) negatively modulates the HPO axis causing less ovulation rate, reduced litter size, increased anestrus, and duration of postpartum estrous occurrence. The effect of high ambient temperature and heat stress on reproductive cycles can be minimised by increasing antioxidant levels, hormonal manipulation, and nutritional and housing management. In rats, cold exposure enhances progesterone levels and diminishes estrogen levels. In an extremely cold environment, LH level increases and results in impaired ovarian function with irregular estrous cycle, decreased ovulation rate, and even induced follicular cysts.

22.2.3.3 Nutrition

Nutritional deficiency impairs the secretions of gonadotropic hormones from the pituitary gland. Negative energy balance has a detrimental impact on the reproductive activity of dairy animals. Insufficient energy intake during the growth period results in delayed onset of estrous and reduced fertility in females. Malnutrition also causes poor estrus manifestations and inadequate follicular maturation. Dietary fat, calcium salt, and long-chain fatty acids often influence the follicular growth under energy-deficient conditions, particularly in lactating bovines and swine. Feeding phytoestrogens like clover (red clover), barley and oat grains, alfalfa, and some growing legumes (mature and dry legumes have less estrogen) may cause delayed estrus.

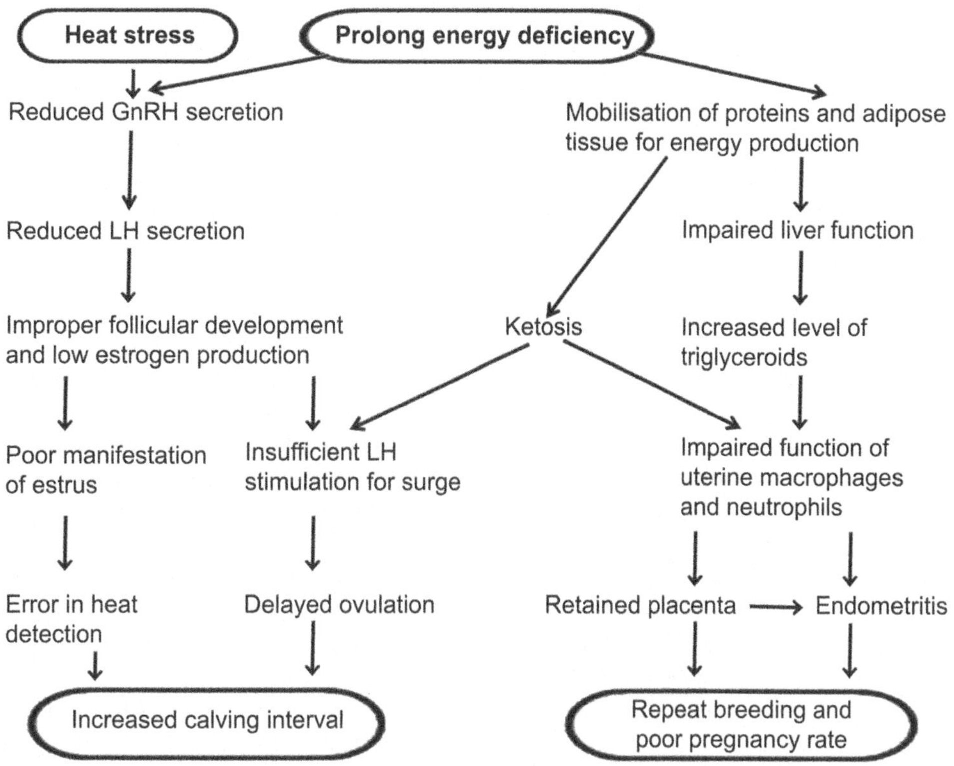

Fig. 22.9 Effect of malnutrition on the female reproductive system. Figure shows the effect of **heat stress** cause to insufficient hormone secretion. Constant scarcity of sufficient nutrition leads to **prolonging energy deficiency.** The consequence of it results in various hormonal insufficiency and changes in the metabolic or systemic status of the animal. It affects the cellular and physiological function of the female reproductive system directly with an **increased calving interval**, **poor pregnancy rate**, and **repeat breeding**

22.2.3.3.1 Nutritional Flushing

Increasing nutrient intake suddenly from sub-optimum to the optimum level to improve body condition before and during the breeding time to augment ovulation and conception are called flushing. It is generally practised in ewes but can be applied to sows, does or even cows. Flush-feeding positively impacts the GnRH pulse generator and hypophyseal sensitivity to GnRH. Thus, flushing induced gonadotropin release increases the follicular wave and ovulation rate. Nutritional flushing also influences the release of some metabolic hormones and growth factors like insulin, growth hormone, IGF-I, and leptin, which affect the HPO axis and stimulate gonadotropin release. Flushing is generally practised during pre-estrus, 1–2 weeks earlier than estrus during breeding time. Flushing is more beneficial for lean animals than fatty animals. Flushing is helpful in sows, as feed intake is generally reduced in post-weaning conditions.

22.2.3.4 Lactation

Suckling anestrus is common in high yielding cattle and buffaloes due to its high prolactin and oxytocin. Both these two hormones have a negative impact on the HPO axis. In hyperprolactinemia, dopamine secretion is increased, which inhibits the GnRH release from the hypothalamus and subsequently FSH and LH release from the anterior pituitary. Prolactin suppresses the estrogen-induced LH release, and oxytocin suppresses the GnRH pulse generator. All these factors cause impaired estrus manifestation and ovulation. Poor follicular development leads to low estradiol production. Hence, the occurrence of silent heat is increased. Some high yielding animals also experienced negative energy balance during the lactation period, which adversely affects the HPO axis's functionality.

22.2.3.4.1 Interaction with the Opposite Sex

The presence of males can influence the occurrence of estrus through bio-stimulation. The bio-stimulation is mediated mainly through pheromones. Flehmen response is a common phenomenon in a wide range of ungulate mammals. The other examples of bio-stimulation are the bull effect in cattle, ram effect in sheep, self-enurination in goats, Vandenbergh effect in mice and boar, Whitten effect in mice, and Bruce effect in rodents.

22.2.4 Distinguishable Attributes of Estrous Cycle in Various Species

22.2.4.1 Cattle

The cow is a non-seasonal polyestrous animal, and the duration of estrous cycle is 19–21 days. Different attributes of the estrous cycle in cows are discussed in different sections as a model animal.

22.2.4.2 Buffalo

Buffaloes are known as *shy breeders* due to their less intense estrus behaviour. The occurrence of silent heat is very common in buffaloes, particularly during summer, due to heat stress. Buffaloes exhibit estrus behaviour from evening to night; hence, estrus detection is challenging for buffaloes. The major causes of silent estrus in buffaloes are the small ovarian size and less primordial follicles (20,000 vs 150,000 in cattle) that produce a small quantity of estrogen (17-β-estradiol). The follicular atresia is also more in buffaloes than in cattle with fewer antral follicles. The vasectomised teaser can detect standing estrus; milk progesterone estimation and ultrasonography are used to confirm the estrous cyclic stages.

22.2.4.3 Ewe

Sheep are the seasonally polyestrous short-day breeder. The estrous cycle length varies from 16 to 17 days, depending on the breed. The duration of breeding seasons is also diverse. Breeds like Leceister, Scottish Blackface, Texel, and Shetland have a short breeding season (less than 4 months). The breeds of the medium breeding season (4–6 months) are Hampshire, Suffolk, and Charollais. The long breeding season (6–8 months) is seen in Rambouillet, Dorset, Finn, and Romanov. The tropical breeds of sheep (Nellore, Bellary, Malpura, Yankasa, and Uda) are generally considered non-seasonal breeders. Progesterone priming is important for follicular wave and sexual receptivity in ewe; hence, the silent estrus usually occurs of the first cycle in each breeding season. Estrus symptoms are less prominent in sheep than in other domestic animals like cattle, pigs, etc. The presence of ram (*ram effect*), even with the teaser and androgen-treated whether, can be used to detect estrus. In general, the induction of estrus and ovulation are seen within 48 h of the ram effect. But, the first estrus induced by the ram effect is of shorter duration (with 5–6 days of diestrus). It is followed by a true estrus, seen around 25 days after the ram effect. This process is effectively used to synchronise estrus in ewes. Some animals exhibit long estrus duration with sexual receptivity and high ovulation rate due to high gonadotropin. Flushing before breeding also creates a higher ovulation rate and increases litter size in ewes.

22.2.4.4 Doe

The reproductive character of goats depends upon the habitat. The breeds of the temperate region are seasonal breeders. Autumn and winter are their breeding seasons. The breeds of the tropical area are non-seasonal breeders. The tropical breeds of goats exhibit prolonged anestrus with reduced litter size, probably due to malnutrition. Flushing, manipulation of photoperiod and hormonal treatment can reduce the duration of anestrus in seasonal breeders. The does exhibit a short estrous cycle followed by a regular cycle. The short estrous cycle is related with less ovulation rate. In tropical regions, does are often exhibited a short estrous cycle during high rainfall. In does, generally, four or more follicular waves are seen in one cycle, and ovulation occurs from the follicle(s) of the last wave. The ovulation rate is usually 1–3 eggs per doe, with a maximum of 5 depending upon the breed and management of the animal. Application of exogenous progesterone is generally used to synchronise the estrous, and eCG and FSH or hCG are used for superovulation in does. The corpus luteum maintained its peak luteal activity from day 7 to 10 of the cycle and regressed from day 11 in non-pregnancy. In pregnancy, luteal progesterone is the primary source of progesterone to maintain the gestation.

22.2.4.5 Sow

The onset of puberty occurs in most pig breeds at 5–7 months. The pig is a non-seasonal polyestrous polytocous animal. Photoperiod has little or no influence on the estrous cycle, but long darkness exposure may delay the estrus. Standing estrus is coincides with ovulation and generally seen in the middle of the estrus phase and continues up to 8–36 h. Duration of standing estrus is usually more in sows than gilts. Typical behavioural estrus symptoms are restlessness, decreased appetite, increased vocalisation, pricked ears, and swollen and reddish or pinkish coloured vulva. Back-pressure test (BPT) is used to detect estrus in the sow to see the standing reflex during the heat. The sow gets immobilised for 10–15 s after applying back pressure on both sides. The onset of estrus can be accelerated by the presence of boar (Vandenbergh effect, details in *sex pheromone*). The application of boar saliva is often used to synchronise the estrus in the sow. The boar presence also stimulates oxytocin secretion in the sow to facilitate sperm transport after the insemination. Hormonal manipulation with the combinations of PMSG and hCG to the gilt, aged 7 months or more, can also accelerate the onset of puberty. But, the results are questionable as cyclic gilt would not respond appropriately due to suppression of progesterone by PMSG (when the gilt is in the luteal phase) or FSH, LH, and estrogen will reach too high upon PMSG treatment (when the gilt is in follicular phase). Multiple breeding during the standing estrus period can enhance the litter size. The visual symptoms of ovulation are the reddening of the vulva before ovulation, which subsides after ovulation. Hence, artificial insemination should practice when the reddening of the vulva is subsided. About 15–30 follicles can ovulate from both ovaries in one cycle, depending upon mainly the breed and nutritional status. Gilts can be bred on the second or third cycle to achieve better litter size. Exposure to unfamiliar boar during pregnancy may cause abortion, compared with the Bruce effect in mice. Lactational anestrus is very common after farrowing in the sow. Nutritional factor influences the duration of postpartum estrus. Early weaning at 21 days can reduce the period of postpartum anestrus with the appearance of estrus within 4–7 days. Proper nutrition during the postpartum period is essential for uterine involution and the growth of the piglets. Hence, the low weaning weight of piglets during the first lactation causes reduced litter size in second (or subsequent) lactations. It is called second litter syndrome.

22.2.4.6 Mare

Mare is a seasonally polyestrous animal. In non-tropical regions, summer is the breeding time; autumn considers a transition time and winter is characterised by anestrus in mare. In the northern hemisphere, the duration of estrus is shorter than in winter due to rapid follicular development under favourable photoperiod. But the duration of diestrus does not alter due to season. Mares exhibit prolonged estrus of 4–8 days, comprising 20–30% of the total duration of estrous cycle. Sometimes a period (1–2 days) of decreased behavioural signs is seen in the mares during the estrus stage, known as *split estrus* or *slit heat*. It is because follicles of one follicular wave fail to develop fully, and a second follicle develops to secret estrogen leading to the second strong phase of estrus behaviour. Normal estrus behaviours of mares include squealing at other horses and frequent urination with small amounts. After approaching the stallion, estrus-mares lowered the pelvis and spread her hind limbs. The mare exposes her perineal region with the elevation of the tail. The rhythmic eversion of the clitoris, called *clitoral winking,* is also documented in estrus mares.

In some cases, unique facial expressions are seen characterised by relaxed facial muscles and lowered head with ears turned to the side. The characteristics of Flehmen's response are also common in mares. Mares exhibit estrus behaviour during the anovulatory period due to estrogenic steroids of adrenal gland origin and unique among other ungulates. Another distinguishing feature of the mare estrous cycle is the postpartum estrus called 'foal heat'. It is generally seen during the first week after birth, and the mare can be conceived at this time. But, breeding is usually preferred 2–-3 months after the foaling. The ovulation can occur before LH surge in the mare; hence, progesterone starts to release from the primary corpus luteum.

22.2.4.7 Rat

The commencement of puberty and appearance of estrous cycle were seen around the 33rd day after birth in female rats. The beginning of estrous cycle is indicated by vaginal opening. High estradiol concentration associated with vaginal unfolding was the main reason for sexual receptivity and first ovulation. From the onset of sexual maturity up to the age of 12 months, the mean cycle length in the female rat is 4 days. In some individuals, 6-day is seen. The estrous cycle in rats was characterised by four stages: proestrus, estrus, metestrus, and diestrus. Sometimes the cycle may consist of five phases having two metestrus, metestrus-I (15–18 h) and metestrus-II (6 h) (Table 22.11). Ovulation occurs from the beginning of proestrus to the end of estrus. A persistent estrus with constant sexual receptivity was seen in females aged 6–18 months.

A state of pseudopregnancy of 10–14 days' duration was seen in aged females who failed to breed after persistent estrus. The rat estrous cycle is characterised by frequent and abrupt changes in the concentration of hormones. The FSH, LH, and prolactin levels increase in proestrus's afternoon, and a surge usually occurs in the late afternoon. The estrous cycle in rats is influenced by light and age, nutrition, dark and light cycle, temperature, stress and social relationships. The female rodents cycle occurs more regularly when males are present in the room (i.e. the Whitten effect) although mice are more sensitive to this phenomenon than rats. Period of anestrus has also occurred in rats. It may be due to low progesterone in the absence of corpus luteum along with high prolactin, cause of pituitary tumours or pituitary dysfunction of other reasons, which suppress LH secretion. The cyclic rats contain three or more generations of variable sets of corpora lutea. The first half of pregnancy is maintained by the progesterone secreted from the corpus luteum. In the second half, the progesterone is released from the placenta.

22.2.4.7.1 Heat Detection by Vaginal Cytology in the Rat

Rodents' estrous cycles were characterised by histomorphological alterations in ovaries, the uterus and the vagina. There are several methods to determine the stages of the estrous cycle in rats. Vaginal cytology is one of the most reliable tools to classify estrous cycle in rats. The basic principle behind this method was the rapid alterations in the vaginal mucosa during different stages of estrous cycle under the influence of sex steroids, particularly estradiol, which facilitated the cornification of the vaginal epithelium. The desquamation of mucosal epithelium occurred under estradiol withdrawal. The vaginal smear can obtain after inserting a cotton-tipped swab moist with physiological saline into the vagina, followed by subsequent rolling against the vaginal wall. The stains visualised exfoliated vaginal epithelial cells were 0.1% crystal violet, Wright's Giemsa, Toluidine blue O, and a single differential stain. There were three types of cells: leucocytes, cornified epithelial cells, and nucleated epithelial cells. Their relative proportions in vaginal smear evaluated the estrous cycle stage (Table 22.14).

22.2.4.8 Guinea Pig

Female guinea pigs attain puberty at 2 months of age (55–70 days). The estrous cycle length of the guinea pig is 16–19 days having the estrus period of 6–11 h. Different phases of the estrous cycle can detect by identifying cell types of the vaginal smear. Mainly three kinds of epithelial cells are seen during different cycle phases (Fig. 22.10). The cells are the parabasal (immature or smallest squamous epithelium having a large nucleus compared to cytoplasm),

Table 22.14 Gross and microscopic features of rat reproductive organs during estrous cycle

Phase of cycle	State of phase		Histology/gross examination		
		Vaginal smear	Vagina	Uterus	Ovary
Estrus	Entire	Non-nucleated **cornified cells** (less gradually), **infiltration of WBC,** and **large basophilic epithelial cells**	(1) Loss of exterior mucoid and cornified layers, (2) reduced thickness of epithelium cell layers, (3) infiltration of WBC, (4) **detachment of the cornified epithelium**	**Degeneration of endometrial glandular cells and lining cells**, infiltration of WBC	More degenerated and small newly formed corpora lutea, basophilic cytoplasmic cells, centrally fluid-filled cavity
Metestrus	Entire	Infiltration of WBC, less cornified and basophilic epithelium cells; all the **three types of cells are the almost same proportion**	(1) Squames at lumen, (2) **cornified epithelium—completely detached,** (3) S. granulosum and S. spinosum—gradual loss, (4) **thin epithelium layer,** (5) infiltration of WBC	Vacuolar degeneration of endometrial epithelium. **Infiltration of WBC**	Cytoplasm of new corpora lutea—less basophilic than estrus and smaller than diestrus. No fibrous tissue
Diestrus	Beginning	Infiltration of WBC; **predominated WBC** among the three types of cells	(1) S. basale—1 layer (columnar epithelium cell), (2) S. spinosum—multiple layers (polyhedral cells), (3) **S. granulosum—absent,** (4) **infiltration of WBC**	(1) Small and inactive, (2) cuboidal/columnar epithelium cells—thin layer, (3) often degenerating cells	
	End	Less mucus and WBC, **nucleated basophilic cells,** and often vacuolated cells	(1) S. basale—multiple layers (columnar epithelium cell), (2) **S. granulosum—newly formed**, (3) Less infiltration of WBC	Edema in the stroma to the nearer endometrial epithelium cells	Maximum size corpora lutea—few (generated from last ovulation) with vacuoles in centrally positioned cells
Proestrus	Beginning	**WBC absent, predominated nucleated epithelial cells**	(1) S. granulosum—flattened with **keratohyalin granules**, (2) S. corneum—cornified cells (newly formed), (3) often infiltration of WBC	Large cuboidal to the columnar epithelium, **no degenerated cells**	Degenerated corpora lutea with central fibrosis and cytoplasmic vacuoles
	End	WBC absent, **nucleated acidophilic epithelial cells,** and **cornified cells**	(1) S. granulosum—cuboidal with mucin containing cytoplasmic vacuoles, (2) **S. corneum—fully cornified with the outward mucoid cell layer**	Lumen dilated, edema in the stroma, prominent endometrial vasculature, and **more degenerated endometrial epithelium and few inflammatory cells**	Noticeable degenerated corpora lutea with central fibrosis and cytoplasmic vacuoles

Note: Bold words are the major distinguishable features. *S. basale* stratum basale, *S. spinosum* stratum spinosum, *S. granulosum* stratum granulosum, *S. corneum* stratum corneum

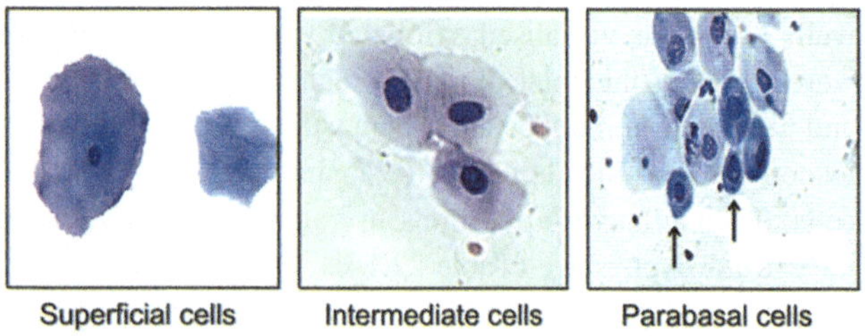

Fig. 22.10 Vaginal cytological changes during different phases of estrous cycle. Figure depicts the appearance of three distinct types of exfoliated vaginal epithelial cells viz. **superficial cells**, **intermediate cells,** and **parabasal cells** in various phases of estrous cycle

intermediate and superficial cells, and RBC and WBC (neutrophil). Proestrus is characterised by the presence of superficial nucleated cells and RBCs. Superficial cells are the largest epithelial cells appearing as polygonal flat shapes, transparent, eosinophilic, with very small and dark nucleus or enucleated. Enucleated superficial cells are prominent during the proestrus-estrus transition phase. In estrus, superficial cells with a pyknotic nucleus are dominated along with RBC infiltration. No parabasal cells are found in estrus, but their presences, together with neutrophils, are more in metestrus. The diestrus phase is characterised by similar cell types and large and small intermediate cells. Intermediate cells are larger than parabasal cells having an almost round or oval shape with large and prominent nuclei (small intermediate) and polygonal shape with small nucleus (large intermediate). Multichromatic (multicoloured) stains like Papanicolaou stain or Pap stain are generally used in cytological staining techniques.

22.2.4.9 Mouse

Mice are polyestrous non-seasonal breeders and spontaneous ovulator. The estrous cycle of the mouse is very short, 4–6 days. The age of maturity is 6 weeks, and the reproductive cycle continues throughout the life span (2–3 years in laboratory mice). Lactational anestrus is also common in mice due to the luteotropic action of prolactin. The postpartum estrous period is also very less, 12–20 h. The phase of proestrus and estrus lasts nearly 1–2 days, whereas sexual receptivity lasts for about 13 h. The estrogen level gradually increases in proestrus and remains higher in the morning until ovulation and slowly reaches baseline in the afternoon. Estrogen and progesterone levels gradually increase in diestrus within 1–2 days. Thus, ovulation generally occurs during the night due to a surge of FSH and LH. However, the FSH and LH levels remain low in estrus, metestrus, and diestrus. Ovulation occurs around 10 h after the beginning of estrus. Generally, 10–20 ova can be ovulated in a cycle.

Various cytological changes occur in the vagina, uterus, and ovary during different phases of the estrous cycle due to estrogen's action. The visual assessment of the vagina and vulva is one of the most accepted, non-invasive and fastest techniques for estrous cycle assessment. The vaginal opening is swollen, broad, wrinkled, and reddish-pink during proestrus, while in estrus, it is less swollen, less moist, and less pink. Metestrus is characterised by the presence of white cellular debris with a narrow vaginal opening. In diestrus, the vaginal opening is closed, and no swelling occurs. Vaginal cytology is also a reliable method of estrus detection in mice. The presence of numerous leukocytes, nucleated epithelial cells, and cornified cells is evident during different cycle phases.

The cycle continues throughout the life span after maturity (6 weeks), except during lactation due to prolactin. Hence, within about 25 days, a mouse can give birth to two successive litters. Thus, mice are used as an animal model in the experimental research study. The Whitten effect can synchronise the estrous cycle, and puberty can be enhanced through the Vandenbergh effect. The presence of an unfamiliar male (other than a mating partner) may cause abortion due to the Bruce effect.

22.2.4.10 Bitch

The ovarian cycle of dogs occurs twice a year except for Basenji breed (cycles once a year). Bitches enter their first heat between 6–10 months, but it may occur at 18–24 months in some breeds. The proestrus begins with vaginal bleeding and ends when a bitch allows a male to mate. The duration of proestrus is about 9 days (ranges from 3–25 days). The bitches are playful and attractive compared to males but refuse to mount. In late proestrus, the bloody discharge fades and becomes transparent to straw colour. The proestrus is under the influence of estrogen synthesised from developing follicles. The estrogen level peaks in late proestrus and then declines at the basal levels at the beginning of estrus. The progesterone concentration increases from basal levels at late proestrus and increases at the onset of estrus and ovulation. The source of progesterone is partially luteinised follicles before ovulation and developing CL. Estrus starts with the first acceptance of males and ends with refusal. The duration of estrus is about 9 days but may range from 2 to 18 days. The standing heat in bitches is characterised by a declining estrogen level and increasing progesterone level. The bitch attracts males from a distance due to pheromones. The vaginal discharge becomes straw-coloured or pink at the time of estrus. Ovulation occurs 24–48 h after the LH surge at 24–72 h. The eggs are released as primary oocytes and become fertile after 24–72 h. The mature ova have a life span of 2–4 days. The diestrus begins with the end of standing heat and ends when the progesterone reaches its basal level. The unique feature of CL in bitches is that the CL remains functional whether it is pregnant or not. The luteal phase ends in a pregnant bitch after the parturition. But, in non-pregnant bitch, the CL functioned for 75–100 days and regressed due to ageing.

LH and prolactin are the two luteotropic factors. The luteotropic action of prolactin is seen in the second half of the luteal phase. The uterine involution occurs during anestrus and extends up to 4.5 months. Vaginal cytology is used to identify the estrus in bitches. The proestrus is characterised by increased cornified cells with RBC and leukocytes. In estrus, mostly cornified cells are evident in the vaginal smear. There is an abrupt change in the vaginal cells from superficial to basal cells with increased leukocyte infiltration in diestrus. The anestrus is characterised by foam cells and leukocytes. The best time for mating is between the tenth and 14th day of estrous cycle. The breeding time can also be predicted by progesterone assay (Table 22.15).

Long proestrus is one of the common disturbances in the estrous cycle of bitch due to a lack of estrogenic peak during proestrus with higher progesterone levels. Split estrus is also another problem in bitch. In split estrus, the proestrus starts due to incomplete luteolysis (in chronic premature luteolysis or hypothyroidism), when the progesterone level falls

Table 22.15 Prediction of breeding time based on progesterone level

Events	Progesterone level	Time of mating
LH surge (stimulus for ovulation)	1.5–2.0 ng/mL (4.5–6 nmol/L)	Four days later
Ovulation (2 days after LH)	5.0–5.5 ng/mL (15–18 nmol/L)	Two days later
Fertile period	10–30 ng/mL (30–90 nmol/L)	Immediately
Ideal mating time is around 15–24 ng (48–75 nmol)		
Two matings, 48 h apart		

slightly. It results in the initiation of proestrus, but proper follicular development will not achieve due to progesterone. Thus, proestrus is prolonged, followed by either failure to estrus or exhibiting a short estrus but no ovulation. Vaginal bleeding and vulvar swelling during this time may be confused with estrus. After about a month, the normal cycle will start. This incidence is called split estrus, wherein the first phase is a pseudo or short non-ovulatory estrus followed by a true estrus. This incidence is mostly common in young experiencing the first estrus in each breeding season but can be seen in adults.

22.2.4.11 Cat (Queen)

The onset of puberty varies from 3 to 12 months of age. The Burmese cats mature earlier than the Persian and free-ranging cats. Environmental factors influence the onset of puberty, particularly in the regions away from the equator. Cats are seasonally polyestrous animals. Short-haired breeds come to cycle throughout the year. Increasing day length makes the females cycle all round the year. Being an induced ovulatory, queens have three possibilities of estrous cycle

1. *Proestrus, estrus (non-bred), inter-estrus*
2. *Proestrus, estrus (sterile mating), diestrus, anestrus*
3. *Proestrus, estrus (fertile mating), pregnancy*

Proestrus is a short phase of 1–3 days characterised by rubbing against objects by the queens, followed by rolling with purring, rhythmic opening and closing of claws, squirming and scratching. The queens exhibit a monotone howling (heat cry) for 3 min at a time during proestrus and estrus. Queens are spraying urine and sebaceous secretions to attract males. The estrus behaviour includes crunching lordosis and a copulatory stance. Tails are laterally displaced, and serosanguinous discharge from the vulva is observed. The estrus lasts for 4–6 days, and the sexual receptivity is seen on the 3rd and 4th. The queens do not ovulate unless mating occurs. Ovulation can be induced by stimulation with male penile spines or artificially by glass rods (several insertions of 10 s for 5–10 min after 48 h). The period between two successive estrous cycles of a cyclic queen is called the inter-estrus phase. It is 2–17 (average 10) days. The copulation before estrus will not induce ovulation, and hormonal applications such as eCG (for multiple follicular developments) and hCG (for ovulation) can be applied to augment the ovulation process. The heat symptoms will subside within 1–2 days after ovulation in the absence of fertilisation and the diestrus period continues for 35–40 days, termed pseudopregnancy. If the queen gets pregnant, she will not return to estrus till the seasonal peak or the following year. About 10% of pregnant queens display estrus behaviour at the 3–6th week of gestation. In the absence of tomcats, females are in estrus for 10–14 days, and the estrous cycle is 29 days. If the breeding does not occur, metestrus is an inter-estrus period between two successive estruses.

22.2.5 Menstrual Cycle

The term 'menstrual cycle' is derived from the Latin word 'mensis', meaning months, and the cycle lasts for approximately 28 days. The periodic discharge of blood and other substances (sloughed lining of endometrial cells of the uterus and mucus), including an unfertilised ovum from the female genital tract of human and non-human primates (monkey and ape), is called menstruation. The repetition of such events from one menstruation to the next is called the *menstrual cycle*. It is generally a cycle of 28 days, with 21–35 days. It starts after puberty, with the development of the HPO axis, at the age of 10–11 years in girls with irregular cycle duration, whereas regular cyclicity develops after 17 years after maturity.

22.2.5.1 Phases of the Menstruation Cycle

The menstrual cycle can broadly classify into pre-ovulatory and post-ovulatory phases. The pre-ovulatory phase divides into bleeding (menstruation, day 1–day 4) and the proliferative phase (day 5–day 14). The post-ovulatory phase is also called secretory phage (day 14–day 28). The first day of bleeding is considered day 1 of the cycle. The normal amount of blood discharged during menstruation is about 4–40 mL of arterial origin, and the blood is unclotted due to fibrinolysin content. Ovulation generally occurs on days 12–14, and the ovum can be viable up to 24 h after ovulation. The changes in the ovaries, endometrium, and endocrine alterations during different menstrual cycle days are summarised in Table 22.16.

Table 22.16 Characteristics of the menstruation cycle

Day	Ovarian, endometrial, and endocrine changes
1–5	The menstrual flow starts along with endometrial shedding
6–7	Increased secretion of FSH and LH under the influence of GnRH; FSH stimulates the follicular growth and estrogen secretion; menstrual flow ceases along with endometrial repair
8–12	Estrogen secretion increases which exert negative feedback for FSH and LH secretion; endometrial growth starts
13–14	LH surge on day 14 causes ovulation
15–24	CL secrets progesterone; progesterone alters the endometrium to accept fertilised ovum if the fertilisation takes place
25–28	Degeneration of CL leads to decreased progesterone and estrogen secretion; lower progesterone causes degenerative changes in the endometrium leading to endometrial shedding and menstrual flow

Know More ....

Mittelschmerz

The word *Mittelschmerz* comes from the German words 'middle' and 'pain'. It means the pelvic and lower *abdominal pain* that occurs during ovulation halfway through the menstrual cycle. Hence, it is also called ovulation pain. Pain is generally one-sided, corresponding with the ovary involved in ovulation. It is a physiological process different from the abdominal pain related to some endometriosis and inflammation due to sexually transmitted infections.

22.2.5.2 Menstruation Cycle and Estrous Cycle

Other than the duration of the cycle and its occurrence, the estrous cycle and menstrual cycle have some striking differences. The major difference is in sexual receptivity. The animals are sexually receptive only during the estrus phase of the cycle, whereas females are receptive throughout the cycle in the menstrual cycle. The corpus luteum in the menstruation cycle can synthesise a small quantity of estrogen, but CL in the estrous cycle cannot produce estrogen. Sometimes, menstruation may be confused with metestrus bleeding in cows, but these two are different phenomena. Menstrual bleeding is due to the imbalance between estrogen and progesterone regulates the endometrial lining. The metestrus bleeding in cows is due to high estrogen levels during estrus, leading to blood leakage from vessels on the surface of the uterus. The metestrus bleeding can be called *pseudomenstruation* and is seen in bitches during proestrus and estrus.

22.3 Ovarian Dynamics

The ovary performs two major functions, gametogenesis or oogenesis and steroidogenesis or production of steroid hormones. These two activities are started during the foetal life and continued through a dynamic process during various phases of post-natal periods. Rhythmic alterations of these functions are seen during each estrous cycle after puberty.

The term oogenesis is used to describe the developmental process of the ova (singular ovum) or the female gamete. The oogenesis depends upon the interaction between the oocyte and follicular cells surrounding it. The ovarian follicles are the functional unit of the ovary that appears as a cluster of somatic cells to protect and nourish the germ cell or oocyte. The follicular cells are developed through a process termed *folliculogenesis*.

22.3.1 Oogenesis

Oogenesis is a complex differentiation process leading to the production of functional oocytes. Oogenesis is initiated at the embryonic stage and completed after puberty. The process of oogenesis is commenced with the migration of PGCs into the gonadal ridge and serves the function of stem cells. It proliferates to form oogonia, remains in ovarian nests and is connected by intercellular cytoplasmic bridges. The somatic cells are the pregranulosa cells, or granulosal cells originated from mesonephros cells or ovarian surface epithelium (species-specific). Oogenesis occurs in the outmost layers of the ovaries and is divided into three different phases. These are *oocytogenesis, ootidogenesis,* and the maturation phase *(oogenesis proper).*

22.3.1.1 Oocytogenesis

The primordial germ cells (PGC) of the foetus in the ovary are transformed into the oocyte by a series of mitotic processes called *oocytogenesis*, or the *prenatal phase* of oogenesis. It has two stages. In the first stage, the PGC is transformed into *oogonia* (singular oogonium). Later, the oogonia are developed into an *oocyte* or *primary oocyte* (Fig. 22.11).

Oogonia: Oogonium is a diploid cell that resembles spermatogonium. An oogonium is generally spherical and larger in shape with a prominent nucleus compared to

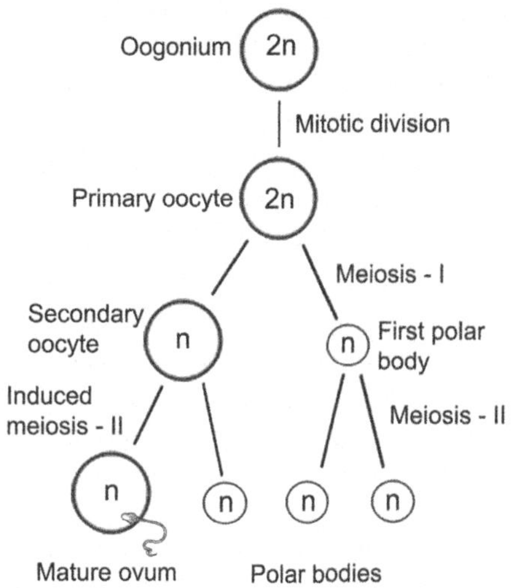

Fig. 22.11 Oogenesis. Figure shows the formation of a mature ovum from oogonium. The **oogonium** is developed into a **primary oocyte** through ***mitotic division***. The haploid **secondary oocyte** and a **(first) polar body** are formed by ***meiosis-I*** division. ***Meiosis-II*** is continued, and ovulation occurs with the resumption of this cell division, in the presence of spermatozoa (after penetration, during fertilisation) as ***induced meiosis-II*** and **mature ovum** and second **polar bodies** are formed

the surrounding somatic cells. It can be differentiated microscopically by randomly scattered fibrillar and granular material from the somatic cell with a more condensed nucleus with a darker outline.

Primary oocyte: The oogonium is enlarged and transformed into primary oocytes. This process is completed before birth or immediately after birth. Hence, there will be no scope to develop a primary oocyte after birth. In contrast, spermatogenesis continues after birth and is a transient stage of ova that generates haploid secondary oocytes. The oocytogenesis ends with the development of a primary oocyte.

22.3.1.2 Ootidogenesis

The process of transforming the primary oocytes into *ootids* by *meiotic cell division* is called *ootidogenesis* (Fig. 22.11). The meiotic cell division has two distinct stages, meiosis I and meiosis II.

Meiosis I: It starts during embryonic life but is arrested at the diplotene of prophase I until puberty. After puberty, the cell division is completed under follicle-stimulating hormone (FSH). At the end of meiosis I, the *synapsis* occurs, but each chromosome has still two chromatids, and the primary oocyte is divided asymmetrically into two daughter haploid cells. The cell having more cytoplasm is called a *secondary oocyte*, and the other has less cytoplasm called the *first polar body*. Depending on the species, the secondary oocyte remains in this *dictyate* stage until puberty or ovulation or fertilisation.

Meiosis II: Immediately after meiosis I, the two haploid daughter cells undergo meiosis II. But, for the secondary oocytes, meiosis II is again halted at metaphase II; however, the first polar body completes its meiosis and generates two second polar bodies. The secondary oocyte is called an *ootid*, and its nucleus is called a germinal vesicle (GV). The ootids are devoid of fertilising capabilities and remain in a dictyate state. This arrest is to inactivate DNA to protect it so that it is not vulnerable to possible damage during its lifetime for proper fertilisation.

22.3.1.2.1 Maturation Phase

The ootids mature into the ovum with fertilising capabilities in the maturation process. The maturation phase is completed before ovulation or immediately after ovulation at the fallopian tube, depending on species (details in Chap. 23). The luteinising hormone (LH) surge controls the meiosis process during ovulation. LH surge facilitates completing the meiosis I and extrusion of the first polar body, followed by halting meiosis II at metaphase II.

22.3.2 Folliculogenesis

The synchronised transformation of ovarian follicles holds the primary oocyte encompasses the growth and development or atresia of follicles through morphological and functional changes is termed *folliculogenesis*. There are two types of follicular pools within the ovaries: the non-growing pool and the growing pool. The non-growing pool contains primordial follicles, and the growing pool contains primary, secondary, and tertiary follicles. The primordial follicles enter into the growth phase throughout the reproductive life of an animal. The primordial follicle and primary follicle are developed during the *prenatal stage*. Secondary follicles are found in the antral stage, and the tertiary follicles are formed during the *pre-ovulatory stage* of follicular development. The primordial follicles are transformed into mature follicles, leading to ovulation and corpus luteum formation in a sequential pattern within the ovaries called the *ovarian cycle* (Fig. 22.12).

22.3.2.1 Ovigerous Cords and Primordial Follicles

The pregranulosa cells-oogonia complexes are gradually fused and form a tube-like structure called ovigerous cords. The wall of the cord is made up of pregranulosa cells. The formation of somatic cell-germ cell complexes (pregranulosa cells-oogonia complexes) are mediated through the chemotactic attraction between stem cell factor (SCF, or steel factor), a cytokine secreted from the granulosal cells and the expression of the c-kit mRNA/protein receptors (CD 117, a receptor tyrosine kinase) on the oogonial cell surface. The basic fibroblast growth factor (bFGF), released from the oogonia, increases the production of kit ligand in pregranulosa cells and facilitates the interaction. The oogonia that fail to establish this interaction are prone to apoptosis, and the granulosa cells become free to attach with other active oogonia. Ovigerous cords secrets the basal lamina and form the primordial follicle. It is more prominent in rodents, not ruminants, due to long foetal life and slow ovarian development. The ovigerous cords disappear before birth, and the oogonia either transform to the active form (within the primordial follicles) or undergo apoptosis. The primordial follicle consists of an oocyte surrounded by flattened (squamous) pregranulosa cells or follicle cells (Fig. 22.13a). These primordial follicles then enter the growing pool and are transformed into primary follicles. The protease enzymes break down the cytoplasmic bridge between the oocytes during follicle formation. Incomplete breakage of cytoplasmic bridges leads to multi-oocyte follicles (MOFs) or polyovular follicles as they contain more than one oocyte.

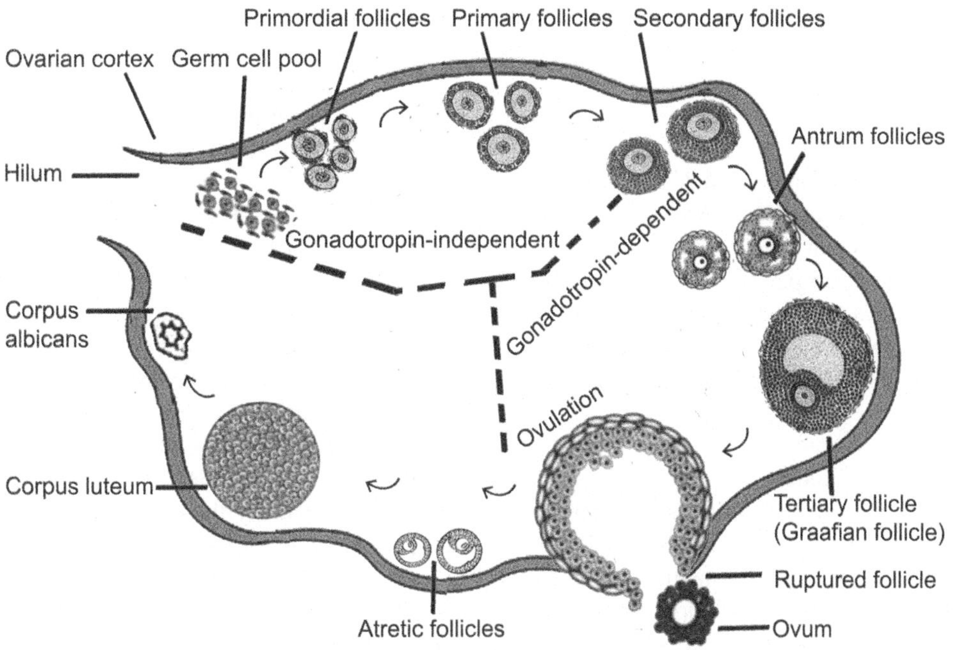

Fig. 22.12 Ovarian cycle. Figure showing the follicular development (clockwise) in two phases, **gonadotropin-independent** or pre-antral phase and **gonadotropin-dependent** or antral phase. The follicles from the **germ cell pool** of the **hilum**, a part of the **ovarian cortex**, developed into **primordial follicles**, followed by **primary follicles** in the gonadotropin-independent phase. The **secondary follicles** are developed from primary follicles, followed by **antrum follicles** under the influence of follicle-stimulating hormone (FSH). The **tertiary follicle** or **Graafian follicle** is developed under the influence of luteinising hormone that leads to **ovulation** with the release of **ovum** from the **ruptured follicle** with subsequent formation of **corpus luteum**. The ovarian surface also contains the scar tissue bearing **corpus albicans** of early ovulated follicles. The ovarian cortex also has **atretic follicles** that fail to form Graafian follicles and don't undergo ovulation

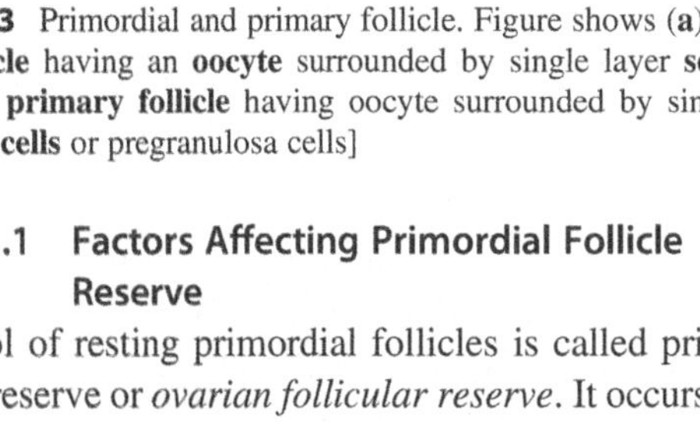

Fig. 22.13 Primordial and primary follicle. Figure shows (**a**) **primordial follicle** having an **oocyte** surrounded by single layer **squamous cells**, (**b**) **primary follicle** having oocyte surrounded by single layer **cuboidal cells** or pregranulosa cells]

22.3.2.1.1 Factors Affecting Primordial Follicle Reserve

The pool of resting primordial follicles is called primordial follicle reserve or *ovarian follicular reserve*. It occurs mainly during 90–140 days of foetal life in cows. The size of follicle reserve is related to the animal's fertility, and the animal that has more reserve has longer reproductive life. Animals with poor follicular reserve have a weak response to *superovulation* protocol. Some internal and external factors reduce ovarian follicular reserve by altering hormonal steroid levels and reducing the responsiveness of the receptors of these hormones (androgen receptor, AR and estrogen receptor, ER) to the follicular cells. Internal factors such as *anti-Mullerian hormone* (AMH), *growth and differentiation factor 9* (GDF9), *bone morphogenetic protein* (BMP15), and maternal nutrition and health of dam are reported to alter the primordial follicular reserve. Anti-Müllerian hormone (AMH) is a glycoprotein secreted by granulosa cells of small follicles.

Serum AMH is undetectable during infancy and rapidly increases with the onset of puberty, reflecting the follicular recruitment. AMH secretion declines when the dominant follicle separates from the antral follicle. The AMH is used as a marker to evaluate ovarian follicular reserve. GFD9 causes growth and differentiation of granulosa cells and expands cumulus cells. The polymorphism of the GDR9 genes is associated with infertility in sheep. Lower expression of GFD9 is reported in women with the polycystic ovarian syndrome (PCOS). BMP 15 causes proliferation and differentiation of granulosa cells and inhibits cumulus cell apoptosis. In ewe, the DEAD (Asp-Glu-Ala-Asp) box polypeptide-4 as a factor with the expression of DDX4, a protein-coding gene, is used as a molecular marker to recognise the size of primordial follicle reserve. The level of BMP 15 in the follicular fluid is correlated with oocyte quality. External factors like *phytoestrogens and bisphenol A* (BPA) or *endocrine-disrupting chemicals* (EDCs) from plastics, pesticides, and industrial chemicals reduce the follicular reserve.

22.3.2.2 Primary Follicle

The oocyte within the primordial follicle increases its size. The squamous pregranulosa cells are transformed into cuboidal and form a layer around the growing oocyte to form primary follicles (Fig. 22.13b). These follicles are generally located under the tunica albuginea at the deepest part of the ovarian cortex. In cows, the primordial follicle and primary follicles are usually formed during 90 and 140 days of foetal life, respectively.

22.3.2.3 Secondary Follicle

The secondary follicle has an oocyte (transformed into oogonium), called a secondary oocyte, surrounded by zona pellucida and two or more layers (multilayer) of somatic follicular granulosa cells. These granulosa cells originated from pregranulosa cells. The granulosa cells nearer to the basement membrane are called mural granulosa cells, and the cells closer to the follicular antrum are known as antral granulosa cells. Theca cells are situated over the granulosa cell layer (Fig. 22.14). These follicles are in a growing stage close to the ovarian epithelium. The glycoproteinous substances released from the granulosa cells formed the zona pellucida to cover the oocyte for its protection (Fig. 22.14). The granulosa cells are responsive to FSH and theca cells to LH. The secondary follicles generally start to form during 210 days of foetal life in cattle.

22.3.2.4 Tertiary Follicle or Vesicular Follicle

The secretion of proteoglycans (hyaluronan and chondroitin sulphate) creates an osmotic gradient that drives fluid from thecal vasculature. This fluid (liquor folliculi) accumulates, results in the splitting of granulosa cell layers, and gradually forms a central fluid-filled space called the antrum. In the species with the large follicle, additional alpha-trypsin inhibitors in the follicular fluid cross-link with hyaluronan to facilitate the retention of these molecules within the antrum to maintain the osmotic gradient.

22.3.2.5 Graafian Follicle (*Named for the Dutch Anatomist Regnier de Graaf, 1641–73*)

Graafian follicles are a heterogeneous family of relatively large follicles with 400 μ (0.4 mm) diameter and 150–200 μ (0.2 mm) in large and small mammals. The cell type of tertiary follicle is similar to the secondary follicle with more granulosa and theca cells and a large antrum; as such, the term antral follicle is synonymous with the Graafian follicle. But, to be more precise, the matured form of antral follicles is called Graafian follicles (Fig. 22.15). FSH and LH trigger the maturation process along with estrogen.

Histologically, the Graafian follicle contains six distinct components, namely the theca externa, theca interna, basal lamina, granulosa cells, oocyte, and follicular fluid. The morphological features of the Graafian follicle do not alter as growth proceeds. Theca externa is the outermost layer characterised by the presence of smooth muscle cells innervated by autonomic nerves. The contractile activity of these smooth muscles facilitates the ovulation process. The theca interna, composed of loose connective tissue and blood vessels, is situated just beneath the theca externa. The thecal cells and granulose cells are separated by a thin layer of extracellular matrix called basal lamina or follicular basement membrane. The granulose cells of the Graafian follicles have four different layers. The layer beneath the basal lamina is called membrana granulosa, composed of a pseudostratified columnar epithelium anchored to the basal lamina. The periantral granulosa cell layer is situated just after the membrana granulosa. The cumulus oophorus (discus proligerus) is a cluster of granulosa cell layers surrounding the oocyte. The innermost layers of the cumulus oophorus adjacent to the zona pellucida are called *corona radiate* towards the antrum. The cumulus cells are essential for oocyte survival as they provide metabolic and nutritional support to the oocyte. The cumulus cells perform glycolysis

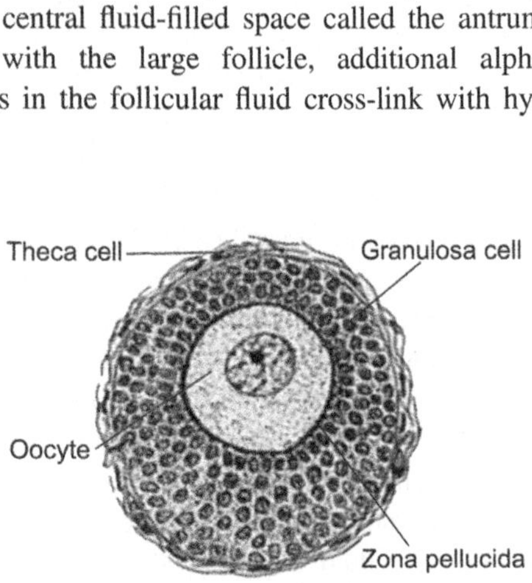

Fig. 22.14 Secondary follicle. Figure shows **oocyte, zona pellucida, granulosa cells**, and **theca cells**

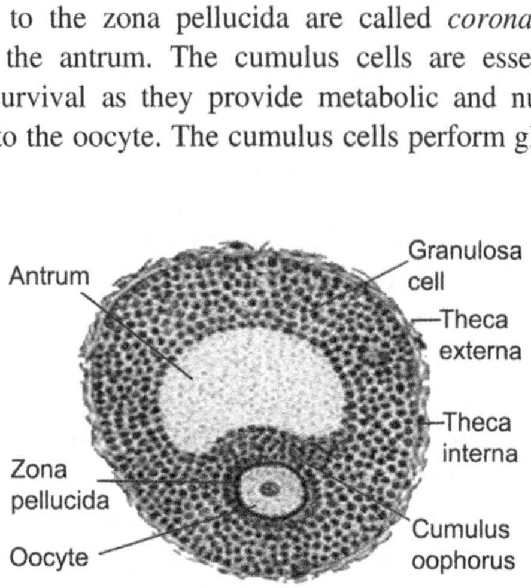

Fig. 22.15 Tertiary follicle. Figure shows follicle having **antrum** and **oocyte** covered with **zona pellucida**, followed by **cumulus oophorus** with multilayer **granulosa cells** and **theca interna** as well as **theca externa**

and transport pyruvate to the oocytes for energy production as oocytes cannot metabolise glucose independently. The cumulous cells also transport endogenous molecules like ATP, cAMP, and cyclic GMP to maintain meiotic arrest.

The major functions of the follicular cells are to promote the maturation of the follicle and the oocyte and to release the oocyte from the ovary at the time of ovulation to form the corpus luteum and production of steroid hormones. The other functions of the follicles are to protect the oocyte (building zona pellucida, discussed earlier), meiotic arrest of the oocyte and secretion of yolk-forming materials. The granulosa cells of some hibernating bats can store glycogen to provide the energy to the oocyte.

The first growing follicles are generally available in the innermost part of the cortex, near the medulla, where the proliferating oogonia mainly exist nearer to the ovary's epithelium. Major characteristic differences of various stages of oocyte or premature ovum of mammals and its corresponding follicle can easily be distinguished microscopically (Figs. 22.13, 22.14, and 22.15) with certain features (Table 22.17). The size of oocytes and follicles of different stages are species specific (Table 22.18) and can be identified by ultrasonography.

22.3.2.6 Period of Folliculogenesis

Folliculogenesis is initiated with the commencement of oogenesis and the formation of oogonia. Time taken to start the oogenesis and complete the folliculogenesis is presented in Table 22.19. The primordial follicle reserve is initiated during foetal life, and most of the primordial follicles are formed before birth in cows, ewe, mare, and human. Primordial follicle formation is continued after birth in sows, queens, and mice. In the rabbit, both oogenesis and folliculogenesis are initiated after birth.

22.3.2.7 Follicular Waves

The earlier stages of follicular growth (follicles with a diameter <9 mm) are independent of gonadotropins. The follicles beyond 9 mm diameter required FSH and LH for antral formation (gonadotropin-dependent follilcular growth). The wave-like pattern of gonadotropin-dependent follilcular growth is called follicular waves. The dynamics of antral follicles consist of recruitment, selection, dominance, and atresia. The follicular waves can be seen during the prepubertal period, estrous cycles, and postpartum periods. During one inter-ovulatory interval, two, three, or four waves can be observed in cattle.

22.3.2.7.1 Recruitment

The term 'recruitment' denotes the stage when the follicles enter the growing pool from the non-growing pool. In this stage, a group of small antral follicles start growing and produce estradiol. Some of these recruited follicles are selected, and the rest of the follicles undergo *atresia*.

22.3.2.7.2 Selection

It involves the emergence of dominant follicles from recruited pre-ovulatory follicles. After selection, some follicles are destined for dominant follicles. The remaining follicles are called *subordinate follicles* and undergo atresia.

22.3.2.7.3 Dominance

At this stage, the follicles are functionally dominant and are capable of ovulation. The dominant follicles are capable of producing estradiol and inhibin.

22.3.2.7.4 Atresia

Atresia denotes degeneration and resorption of follicles. It can occur at any stage of a follicle's primary or secondary

Table 22.17 Distinguished characteristic features of the various stages of follicles in mammals

Type of germ cell	Nature of germ cell	Chromosome character	Corresponding follicle	Type of somatic cell available	Character of the somatic cells	Character of follicle
Oogonium	Dormant, Mitotic	Diploid (2N)	Primordial follicle	Pregranulosa cells	Single layer of flattened cells	Small size
Primary oocyte	Meiosis I (dictyate in prophase I)	Diploid (2N)	Primary follicle	Pregranulosa cells	Mitotic, single layer of cuboidal cells	Larger size, have zona pellucida, corona radiate, no antrum
Secondary oocyte	Meiosis II (arrest in metaphase II)	Haploid (1N)	Secondary (growing) follicle	Granulosa cells, theca cells	Mitotic, Multiple granulosa cells layers; receptor for FSH and LH	Follicle larger than the primary one, have antrum
Ootid	Meiosis II	Haploid (1N)	Tertiary follicle or, Graafian follicle	Granulosa cells, theca cells	Granulosa cells make cumulus oophorus; theca cells in—interna and externa layers; receptors for FSH, LH, and estrogens	Follicle and antrum grow more and mature with the largest size before ovulation, steroidogenic
Ovum		Haploid (1N)				

Table 22.18 Diameter (in μm) of the various follicles available in the ovary of domestic animals

Follicle types	Cow	Buffalo	Ewe	Doe	Sow	Bitch	Queen	Human
Primordial follicles	30–40	25–40	20–25	35–45	30–40	20–30	35–60	30–40
Primary growing follicles	140	40–60	44–48	54–150	60–70	30–90	60–110	40–50
Secondary growing follicles	230–250	50–70	270–275	230–250	340–360	80–210	90–150	300–350
Tertiary follicle (cell layers)								
Granulosa cell	40–50	35–40	45–50	40–50	360–700 (three layers)	320–650 (three layers)	160–200 (three layers)	45–100
Theca interna	90–100 (both)	60–65	103 (both)	70–80				350–600 (both)
Theca externa		55–60		120–140				
Matured oocyte	110–130	65–80	95–140	100–110	100–110 (105)	70–120 (113)	80–120	110–130 (120)
Cumulus–oocyte complex (COC) (mm)	20	16	6	22	8	6	4	23

References: Gougeon (2004), Alwan et al. (2005), Griffin et al. (2006), Ptak et al. (2006), Reynaud et al. (2009), Songsasen et al. (2009), Bächler et al. (2014), Conti and Chang (2016), Haque et al. (2016), Hoque et al. (2016), Sasan et al. (2016)

Table 22.19 Period of folliculogenesis in mammals

Species	Formation of reserve primordial follicle and initiation of meiosis-I (days after conception)	Completion of most of the primordial follicles formation (days after conception)	Period of gestation (days after conception)	Period of dictyate stage of the primary follicle or first occurrence of ovulation (age of puberty, months)	Maximum period of folliculogenesis continued and ovulation occurred (years)
Cow	80	130	280	15	20
Ewe	52	110	150	6	16
Sow	48	139	114	6	12
Mare	70	200	340	12	30
Queen	45	100	60	6	15
Rabbit	2 (after birth)	17 (after birth)	30	4	7
Mouse	12.5	21	19	40 (days)	18 (months)
Human	60	210	270	12	50

Source: Monniaux et al. (2014)

growth. The follicular atresia facilitates by hormone-mediated apoptosis of granulosa cells. The ovigerous cords break down during atresia, resulting in apoptosis of pregranulosa cells, followed by replacement of fibrous materials. The oocytes and the oogonia, which are not transformed into the primary oocyte, are degraded and become part of the ovarian stroma. Atresia of the follicle can occur at any stage of development. In cows and humans, rapid follicular atresia is seen at birth, puberty, and certain age of adult life. The last rapid atresia occurs during cows' 8–10 years of age and 35–40 years in humans. Generally, the follicle number reduces by about 1/20th between birth and puberty (Table 22.20). The maximum number of follicles is lost before birth in bovines than in other domestic animals.

Table 22.20 Follicular atresia during foetal life in various domestic animals

Animals	Maximum follicles developed in foetal life	Follicles during birth	Loos of follicle (%)
Bovine	28,00,000	1,50,000	95
Ovine	8,50,000	80,000	91
Murine	70,000	12,000	83
Porcine	12,00,000	5,00,000	58

22.3.2.7.5 Duration of Follicular Wave

The duration of a follicular wave depends upon the selection of dominant follicle(s) and is species specific. Other factors like the stage of lactation, milk yield, nutritional status, postpartum period, and luteal phase duration also influence the duration of the follicular wave. Successive two follicular waves at an interval of 7–10 days are common in cow. Hence, within a span of one estrous cycle (19–21), 2–3 waves can occur (Fig. 22.16). Similarly, two waves are common in buffalo, 1–2 in the mare and four or more in doe. In humans, inter-wave intervals (IWI) generally vary from 6 to 11 (for 3 wave patterns) and 14 to 15 days (2 wave patterns) in one menstruation cycle (28 days interval).

22.3.2.7.6 Major Wave and Minor Wave

In major waves (ovulatory wave), the divergence of follicles occur, and one follicle becomes dominant and destined for ovulation, whereas others become subordinate. But, no divergence occurs in a minor wave (anovulatory follicular wave), and no dominant follicle develops. The loss of follicular

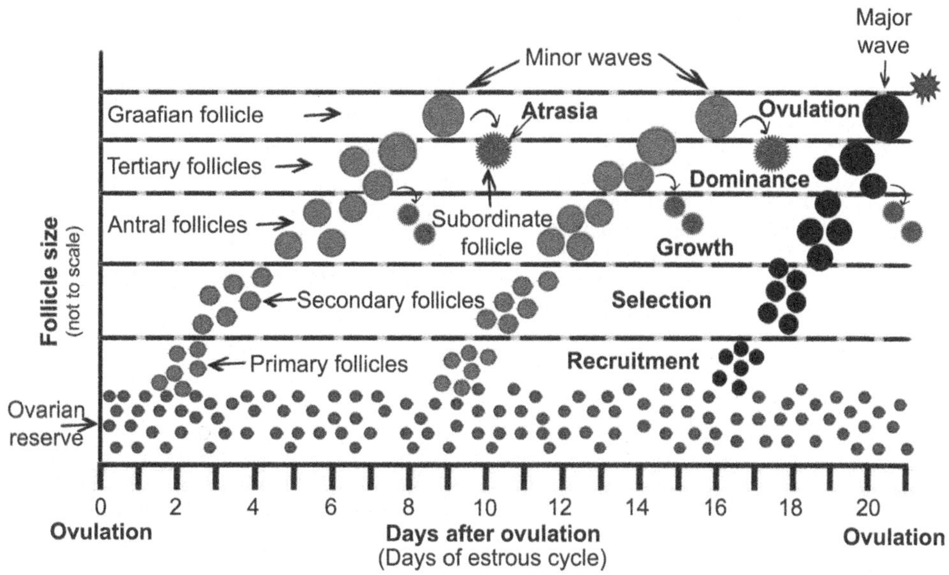

Fig. 22.16 Follicular waves in cattle. Figure shows the three follicular waves in a span of one estrous cycle (21 days) in cattle, comprising one ovulated **major wave** and two non-ovulated **minor waves**. The days between two successive ovulations or an **estrous cycle** period are shown on the '*X*'-axis. The major wave is shown with **dark black** coloured follicles, and the minor waves are shown with **shaded black** coloured follicles. Two minor waves originated on day 1 and day 9, respectively, after the day of last ovulation. In contrast, the major wave originated on day 16, when the progesterone level minimises. The events of the follicular wave, separated by horizontal dotted lines, are shown in boldface, expressing follicular **recruitment**, followed by **selection**, **growth,** and **dominance** in the form of **primary follicles**, **secondary follicles**, **antral follicles**, **tertiary follicles**, **Graafian follicle**. **Ovulation** occurred only from the Graafian follicle of the major wave. The Graafian follicles of the minor waves and all the non-dominated follicles become **atretic** or regressed. Few primary follicles are involved in the event of recruitment, with the influence of FSH from the primordial follicle reserve or **ovarian reserve**. Gradual increase of **follicle size** is represented in the '*Y*'-axis (not to scale)

divergence in minor waves is due to insufficient gonadotropins and a high progesterone level that restricts the follicular growth below 10 mm in diameter in cattle, 7–10 mm in sow, 22–23 mm in mare. The major wave is seen in the follicular phase immediately after the luteal phase when the FSH level is more, and the minor wave(s) mainly occurs in the luteal phase. The major and minor follicular waves are well recognised in mares due to their large follicular diameter compared to other species (Table 22.21). In a single estrous cycle, 5–12 follicles have more than 10 mm in diameter, and the largest follicle has a diameter of 35–55 mm. In mares, the major waves are further classified into primary waves in which (one follicle ovulates with oestrus and secondary waves that comprise a dominant follicle that is unable to ovulate or may ovulate after the primary wave by secondary ovulation. Minor waves are more frequent in spring.

The characteristics of follicular waves in different species have been summarised in Table 22.21.

Table 22.21 Characteristics of follicular waves in different species

Species	Number of follicular waves per cycle	Maximum follicle diameter (mm)	
		Non-ovulatory	Ovulatory
Cattle	2 or 3	10–15	12–20
Sheep	2–4	5–7	6–7
Goat	3 or 4	5–9	6–9
Horse (Major wave)	1 or 2	30–45	40–55
Pig	0	6–7	7–10
Buffalo	2 or 3	10–16	13–18
Dog	Few subliminal waves	1–2	7–11
Cat	1 every 8–24 days	0.1	3–4
Llama	1 every 11–20 days	9–16	9–12
Camel	1 every 17–19 days	17–64	9–19
Chicken	0	6–8	Up to 40

22.3.3 Ovulation

Ovulation is a biological process in which the oocyte is released from the mature Graafian follicle. It is an inflammatory process sequentially controlled by the neuroendocrine system. Two subsequent events occur at the oocytes and surrounding follicles during the ovulation process. In oocytes, the resumption of meiosis and the structural remodelling of the follicles are happened to release the maturing oocyte (Fig. 22.12).

Site of ovulation: The ovulation can occur at any point on the ovarian surface, but the site is restricted to the *ovulation fossa* in mares. It is probably due to the unique ovarian

Table 22.22 Factors involved in oogenesis, folliculogenesis, and ovulation

Name	Source	Functions
Bone morphogenetic proteins (BMPs), BMP4, and BMP7	Theca/ stromal cells	Formation of PGC and regulation of gene expression
OCT4, NANOG	Germ cells	PGC survival
Factor in the germline alpha (FIGα)	Germ cells	Expression of the glycoproteins to form zona pellucida
GATA4, FOXL2, LHX9, WT1, WNT4, and SF1	Somatic/ granulosa cells	Ovarian determination
BCL2, BCLX	Somatic/ granulosa cells	Anti-apoptotic factor helps in follicle survival
BAX	Somatic/ granulosa cells	Pro-apoptotic factor that promotes cell death
CASP2	Both germ cell and granulosa cell	Regulation of apoptosis
TATA-binding protein 2 (TBP2)	Germ cells	Progression of follicular development
Folliculogenesis-specific basic helix-loop-helix (Figla)	Germ cell	Upregulation of oocyte-specific gene (Pou5f1, Zp2, Ivns1abp, Vbp1, Padi6, and Rbpms2) and down-regulation of testes specific gene (Sp3, Hdac2, and Ogt) expression
Forkhead box transcription factor (Foxl2), newborn ovary homeobox (NOBOX)	Germ cell	Regulation of oocyte-specific genes
A disintegrin and metalloproteinase with thrombospondin motifs 1 (*Adamts1*) and prostaglandin-endoperoxide synthase 2 (*Ptgs2*)		Follicular rupture
Pentraxin 3 (Ptx3) and TNF alpha-induced protein 6 (Tnfaip6)		Cumulus expansion
Amphiregulin (Areg)		Oocyte maturation
Steroidogenic acute regulatory protein (Star) and cytochrome P450 family 11 subfamilies a member 1 (Cyp11a1)		Luteinisation

Reference: Schuermann et al. (2018)

structure in mares (kidney bean-shaped). The ovulation fossa is situated as a thin, pointed edged surface (wedged-shaped) at the concave side of the ovary. Before ovulation, the large size mature Graafian follicles reach the fossa. In some animals (whales), the ovulation predominates in one ovary; but it alternates between the ovaries in most mammals. The presence of previous corpus luteum restricts the growth of the follicles in rhesus monkeys. So, the ovulation alternates between the ovaries. In the case of cows, sheep, and horses, the ovulation is independent of previous CL's presence and can occur at random between the ovaries.

22.3.3.1 Theories of Ovulation

There are several theories to explain the mechanism of ovulation, and neither of these theories can explain the ovulation process alone.

Follicular pressure theory: During the growth of the follicles, the amount of liquor folliculi increases; this exerts pressure on the follicular wall to rupture it.

High osmotic pressure theory: The electrolytes present in the liquor folliculi (Na, K) increase the osmotic pressure, leading to the follicular rupture.

But, these two theories did not accept the same events in cystic ovaries without follicular rupture.

Ischemic theory: The accumulation of follicular fluid exerts pressure on the follicular wall. As a result, ischemia occurs at a point that leads to stigma formation and rupture of the follicle. This theory is partly accepted.

22.3.3.2 Mechanism of Ovulation

A recent theory explains that ovulation is a combination of physiological, biochemical, and biophysical mechanisms triggered by pre-ovulatory LH surge, the involvement of other endocrine and paracrine factors.

The steroidogenesis of pre-ovulatory follicles is regulated through LH receptors present in thecal cells and FSH receptors on granulosa cells. LH stimulates steroidogenic acute regulatory protein (StAR protein) to produce androstenedione when it binds with the thecal cell receptors. The androstenedione is diffused into the granulosa cells and converted to estradiol by the action of aromatase (CYP19) under the influence of FSH. Here two cells (theca and granulosa) are involved in estradiol production under the influence of two gonadotropins (LH and FSH). Hence, this theory is called the 'two cell two gonadotropin theory' (Fig. 22.3). There is more estradiol production in the late follicular phase, and it achieved its highest critical value. There is a dramatic change in the action of estradiol from negative to positive feedback for the secretion of GnRH and LH, respectively, at the pituitary and hypothalamic levels. As a result, the LH-secreting cells of the anterior pituitary become highly sensitive to GnRH. LH surge controls several key events of ovulation, such as the resumption of oocyte meiosis and expansion of the cumulus-oocyte complex, and follicular remodelling.

22.3.3.2.1 Induction of the Resumption of Oocyte Meiotic Maturation

The mitotic arrest in the oocyte is maintained by a high concentration of cyclic adenosine monophosphate (cAMP). Adenylyl cyclase 3 (AC3) helps in cAMP synthesis, whereas oocyte-specific phosphodiesterase (PDE3A) causes cAMP breakdown. The steady level of high cAMP throughout mitotic arrest is maintained after the inhibition of PDE3A with the production of cyclic guanosine monophosphate (cGMP) by NPR2 guanylyl cyclase of granulose cells (mural and cumulus). A high level of LH at the time of ovulation reduces the intra-oocyte cAMP level by downregulating NPR2 guanylyl cyclase. The high level of LH also minimises the gap junctions between the oocyte and CGCs and facilitates the diffusion of cGMP within the follicles. The reduced cAMP levels within the oocytes further activate the maturation promoting factor (MPF), a kinase that helps germinal vesicle breakdown (GVBD). The germinal vesicle (GV) is the spherical nucleus of the oocyte, which contains chromatin (DNA) and the nucleolus. GVBD refers to the dissolution of the oocyte nucleus. The MPF also favours the spindle assembly and chromosome segregation to complete the first meiotic division and the formation of the first polar body. The diameter of both follicle and oocyte governs the initiation of meiotic resumption. The antral follicle and oocyte size are essential parameters to evaluate oocyte in vitro maturation (IVM) in Assisted Reproductive Technology (ART). In cow, ewe and doe, resumption of meiosis occurs when the diameter of the oocyte, including zona, is generally more than 110 μm as well as the follicle reaches 2–3 mm in diameter. But, follicles having more than 5 mm in diameter of these species are shown to perform better in embryo production. However, in mare, COC compactness is considered a better criterion than the diameter of oocyte and follicle.

22.3.3.2.2 Cumulus Oocyte Complex (COC) Expansion

Pre-ovulatory LH surge induces the expression of epidermal growth factor-like ligands (EGF-Ls), such as amphiregulin, epiregulin, and betacellulin, and several transcription factors like CCAAT enhancer-binding protein (C/EBP) α/β and progesterone receptor (PGR). EGF-Ls also stimulate the production of pentraxin 3 (PTX3) and hyaluronan synthase 2 (HAS2). These are the cumulus matrix proteins that cause the expansion of COC.

22.3.3.2.3 Follicular Remodelling and Rupture

LH surge causes an increase in PGF2α and PGE2 levels. PGE2 stimulates the production of plasminogen activators and increases plasminogen activity, enhancing the follicular wall elasticity and cell migration for mixing theca and granulosa cells during the formation of CL. PGF2α causes the rupture of epithelial cell lysosome at the follicular epithelium. The hydrolase's from the lysosome cause hydrolysis of albuginea cells and theca cells. Combining these enzymes and cellular apoptosis leads to thinning extracellular matrix (ECM) layers after hydrolysing laminin, collagen type-IV, fibronectin, and proteoglycans. Thecal fibroblasts are migrated, the surface epithelial cells are sloughed from the follicle at the apex region, and a thin and a circumscribed area is formed called *stigma*. Rapid dissolution of these ECM components results in apical puncture and channel for the COC. This perforation makes a passage for the vascular constituents during the formation of the corpus luteum immediately after ovulation. PGF2α causes contraction of smooth muscles of ovarian stroma and theca externa leading to ovarian and follicular contractions. Ovarian contractions lead to follicular rupture, and follicular contractions lead to the expulsion of the oocyte. After the eviction, the ovum and surrounding cells in a gelatinous mass protrude at the ovarian surface and are swept into the ostium by mobile kinocilia of the fimbriae.

In some cows, acute inflammatory reactions occurred during collagenolysis and the release of histamine and prostaglandins with leukocyte recruitment. Under this circumstance, the follicle becomes luteinisation without ovulation and forms a cyst (cystic corpora lutea). To hasten this inflammatory process, some immunological factors, like Intercellular Adhesion Molecule 1 (ICAM-1), are secreted in response to LH surge to recruit leukocytes, mainly macrophages and neutrophils in follicular and perifollicular thecal layers, respectively.

22.3.3.3 Types of Ovulation

Ovulation is of two types based on the underlying neuroendocrine mechanisms. In spontaneous ovulators, like cattle, sheep, goats, pigs, horses, rats, mice, monkeys, and humans, the LH surge is induced by Graafian follicles' ovarian steroids (estradiol). In contrast, the induced ovulator, like rabbits, ferrets, cats, and camels, mating is the main stimulus to cause LH surge instead of 'spontaneous' steroid-induced LH. The somatosensory stimuli originate from the sensory neurones of the female genital tract of these species and activate the noradrenergic neurons of the midbrain and brainstem. These noradrenergic neurons have their projections in the MBH and promote GnRH release, which in turn causes a pre-ovulatory LH surge. A goat is a spontaneous ovulator, but the presence of a male in the flock of goats (male effect) induces an increase in the frequency of LH that gives rise to a pre-ovulatory LH surge. The evolutionary origin of induced ovulation can well explain through male-induced activation of the GnRH neuronal system in females of several species.

A single oocyte is released during ovulation in monotocous species like cows and mares. In polytocous animals, like pigs, dogs, cats, and rats, more than one oocyte is released from different Graafian follicles during ovulation. The small breeds of bitch can ovulate 2–10 oocytes, and the large breed can 5–20 oocytes in one ovulation process. The ovulation can occur from both ovaries at a time in polytocous animals, whereas a single ovary is involved in monotocous animals.

In rare conditions, when two growing follicles reach nearly 10 mm in size with similar sensitivity to FSH, they can transform into two simultaneous dominant follicles. It causes the ovulation of two Graafian follicles at a time and leads to dizygotic or non-identical twins. In superovulation protocol, more numbers of growing follicles become equally responsive to FSH, causing ovulation of many follicles in each wave.

22.3.4 Control of Oogenesis, Folliculogenesis, and Ovulation

22.3.4.1 Endocrine Control

The early stages of follicular development are gonadotropin-independent, as the single layer of cuboidal granulosa cells is non-responsive to gonadotropins. FSH and LH essentially require antrum formation and follicular growth beyond 9 mm diameter. The granulosa cells of secondary follicles are responsive to FSH. High levels of FSH, low LH, and no inhibin initiate follicular recruitment. The gonadotropin stimulates mitosis of granulosa cells and antral fluid formation. FSH also increases the sensitivity of granulosa cells for LH by increasing LH receptors. LH receptors only stimulate the thecal cells from the beginning of thecal cell formation.

When the follicles enter the selection stage, inhibin is produced from the follicles and inhibits FSH release. At the stage of follicular dominance, the larger follicles produce more estrogen and inhibin. A higher estrogen level stimulates a pre-ovulatory LH surge. Inhibin reduces FSH secretion and causes the antral follicles to undergo atresia. Insulin-like growth factors (IGFs) are also involved in follicular growth. IGF stimulates the PI3K pathway that mediates primordial follicle activation. They also sensitise the granulosa cells for FSH action during the terminal phase of follicular development. The actions of IGFs are controlled by a series of IGF-binding proteins (IGFBP), namely IGFBP-2 and IGFBP-4, secreted from blood or synthesised locally within the follicles. The decreased concentration of IGFBP-2 and IGFBP-4 during the follicular growth leads to increased bioavailability of IGF. Therefore, the low amount of IGFBP4 facilitates dominant follicles to attain FSH independence, whereas higher levels of IGFBP2 in subordinate follicles lead to follicular atresia. The follicular cells synthesised steroid hormones that regulate the follicular maturation processes. Larger follicles produce more estrogen. The corpus luteum produced after ovulation produces progesterone to maintain pregnancy.

Follistatin synthesised from granulosa cells competes with activin for its receptor; thus, it counteracts the activin. Follistatin inhibits folliculogenesis by inhibiting bone morphogenetic protein-15 (BMP-15), a stimulatory factor for granulosa cell proliferation. Recently, an intraovarian factor, C-type natriuretic peptide (CNP), has been involved in the preantral and antral follicle growth and the inhibition of oocyte maturation. CNP is secreted from the granulosa cells in response to FSH and acts to its receptor NPRB expressed in the secondary follicles to stimulate follicular development through cGMP production. But, in the cumulus cells, CNP inhibited phosphodiesterase 3A (PDE3A) enzyme and increased intra-oocyte cAMP levels, suppressing oocyte maturation. The CNP level decreases in response to pre-ovulatory LH surge to complete meiosis within pre-ovulatory oocytes. In human, α-fetoprotein, a steroid-binding protein, blocks estrogens and progesterone in the ovary; thus, it controls the folliculogenesis process.

The ovulation is induced by an LH surge (generated by a higher estradiol level) that triggers a biochemical cascade. PGE2 causes the activation of tPA and uPA to augment follicular remodelling along with PGF2α that causes the rupture of epithelial cell lysosome at the follicular epithelium. LH surge also upregulates progesterone receptors in the granulosa layer of bovine pre-ovulatory follicles. The progesterone stimulates collagenase activity and helps in follicular wall thinning. PGF2α induces ovarian and follicular contractions before ovulation. The GnRH pulse happens to be very rapid with a constantly higher level throughout the LH surge. But this constant exposure leads to desensitisation of LH secretion, resulting in termination of LH surge. Immediately after the ovulation, the ruptured follicles are transformed into corpus luteum and release progesterone. In the absence of conception, the corpus luteum regressed leads to decreased progesterone and inhibin. It favours pulsatile GnRH secretion and FSH to promote the growth of the new follicles.

22.3.4.2 Molecular Control

There are nearly 70 genes involved in folliculogenesis in mammals. Several factors and hormones control them. The primordial follicle activation is mediated through phosphoinositide 3-kinase/phosphatase and tensin homolog deleted on chromosome 10 (PI3K/PTEN) pathway regulating cell proliferation and apoptosis. The activation of the PI3K pathway leads to the activation of AKT, a protein kinase that mediates cell proliferation and survival by regulating the phosphorylation of transcription factor forkhead box O-3 (FoxO3). Usually, this FoxO3 induces the transcription of cell cycle arrest genes. It becomes phosphorylated, translocates from the nucleus to the cytoplasm, and is inactivated to allow follicular development. PTEN is a negative regulator of PI3K pathway. The arrest of primordial follicles is controlled through a separate signalling pathway involving tumour suppressor tuberous sclerosis complex 1 (TSC1) and the mammalian target of rapamycin complex 1 (mTORC1). Tsc1 negatively regulates mTORC1 to maintain follicular quiescence. The activation of AKT also phosphorylates and inactivates Tsc1 to begin follicular awakening. Genes that regulate the meiotic arrest of oocytes are species specific. In cows, the YBX2 gene is involved in meiotic arrest. The gene is expressed around 140 days of foetal life and can be used as a marker for the diplotene oocytes. The meiotic arrest is triggered by cyclic GMP, which diffuses into the primary oocyte through the gap junctions and stimulates the phosphodiesterase (PDE3) to promote the degradation of cyclic AMP and subsequent meiotic arrest.

Apoptosis of the germ cells (oogonia) occurs due to lower cKIT gene expression in germ cells. The atresia of the follicles is mediated through time-dependent changes in pro-apoptotic factors and anti-apoptotic factors that regulate cell death. Lower expression of anti-apoptotic B-cell lymphoma/leukaemia-2 (Bcl-2) family of proteins (helps in cell survival) and higher expression of pro-apoptotic BAX trigger follicular atresia. CASP2, released from both germ cells and granulosa cells, regulates apoptosis. Several other factors in oogenesis, folliculogenesis, and ovulation are summarised in Table 22.22.

22.3.4.3 Oocyte-Derived Paracrine Factors

The oocyte secreting different paracrine factors regulates the differentiation of somatic granulosa cells and reproductive hormones (Fig. 22.17). R-spondin-2, growth differentiation factor-9 (GDF9), and bone morphogenetic protein-15 (BMP15) are the three important paracrine factors that regulate granulosa cells' growth and differentiation. These three factors are transforming growth factor β (TGF-β) superfamily proteins. The R-spondin-2 transcripts are identified only in the oocytes of primary and larger follicles. R-spondin-2 promotes the development of primary follicles to the secondary stage and functions like FSH. GDF9 promotes the growth of follicles beyond the primary stage. BMP15 is involved in follicular development. The mutation of bone morphogenetic protein (BMP15, in mouse and human) and growth differentiation factor (GDF9, in mouse and human) genes reduce the protease activity and induce multi-oocyte follicles (MOFs) formation.

22.3.4.3.1 Postpartum Ovulation

IGF-I and insulin promote estradiol production. In high-yielding dairy cows, its role has significantly been identified in resuming ovarian function and ovulation in postpartum conditions. The role of IGF-I signifies 'metabolic signals' for ovulation. Poor nutritional status like lesser body fat, glucose, non-esterified fatty acids (NEFA), total cholesterol and aspartate aminotransferase (AST), as well as low body condition score (BCS) with high GH, is the primary reason for the anovulatory condition in postpartum state. In cows, at least 3 weeks postpartum is required for ovulation. However, after 5 days of calving, the medium-sized follicles may grow into large follicles but cannot ovulate instead of becoming atretic.

22.3.5 Corpus Luteum

The temporary heterogeneous endocrine structure consists of steroidogenic luteal cells, fibroblasts, endothelial, pericytes, and immune cells formed in the ovum-free follicle on the ovarian surface called corpus luteum (in plural corpora lutea).

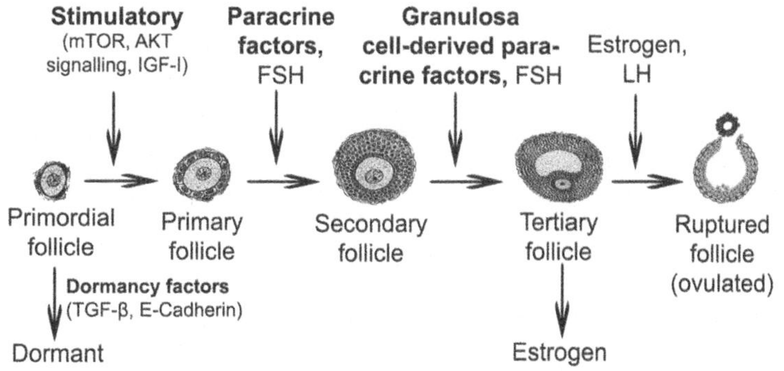

Fig. 22.17 Role of paracrine and endocrine factors in folliculogenesis. In the **primordial follicle** reserve, some primordial follicles are in a **dormant** state induced by dormancy factors such as **TGF-β** and **E-cadherin**. Some of the primary follicles are activated by **mTOR** and **AKT signalling** and give rise to the primary follicle. Insulin-like growth factor-I (**IGF-I**) also promotes the growth of primordial follicles. Follicle-stimulating hormone (**FSH**) and **oocyte-derived paracrine factors** like growth differentiation factor 9 (GDF9), bone morphogenetic protein (BMP6, BMP15), and fibroblast growth factors (FGFs) regulate the growth and differentiation of **primary and secondary follicles**. FSH induces aromatase expression for **estrogen** biosynthesis, and a high estrogen level stimulates **LH** surge for **ovulation**

Table 22.23 Cellular constituents of corpus luteum and their endocrine products

Follicle	Corpus luteum	Endocrine products
Theca cells	Theca lutein cells	Androgens, progesterone
Granulosa cells	Granulosa lutein cells	Progesterone, estrogen, relaxin, inhibin A, and oxytocin

It plays a central role in regulating the reproductive cycle and maintenance of pregnancy. The process of formation of corpora lutea is known as luteinisation. During luteinisation, the granulosa and theca cells transform into lutein cells that can produce significant amounts of progesterone and moderate estradiol and inhibin A (Table 22.23). In Latin, 'lutin' means yellow. The cow's corpus luteum (CL) is yellow due to the large quantity of lutein pigment (carotenoids). But, it is mostly red in other species due to high vascularisation. The cells of CL have distinguished morphological, endocrine, and biochemical features that enable them to secret a wide range of endocrine products (Table 22.24).

22.3.5.1 Formation of CL

The formation of the CL begins with the expression of genes to regulate cell cycles in theca and granulosa cells, followed by the breakdown of the follicular basal membrane to allow the migration of endothelial cells, fibroblasts, and theca cells into the avascular granulosa layer. Many tiny blood vessels rupture at ovulation, leading to haemorrhage, and it appears as a blood clot penetrates at the centre of the former follicle. It is called *corpus haemorrhagicum*. Locally released PGE2 activates the plasmin to absorb the clotted materials and induce cellular differentiation. Corpus haemorrhagicum subsequently becomes functional CL by tissue remodelling and vascularisation.

22.3.5.1.1 Upregulation of Genes for Luteinisation

The binding of LH with its receptors activates the signalling pathway through the stimulatory guanine nucleotide-binding protein Gs and adenylyl cyclase (AC) to increase cAMP and activate cAMP-dependent protein kinase (PKA). Upon activation, the catalytic unit of PKA translocates to the nucleus and phosphorylates several transcription factors required for luteinisation. They are mitogen-activated protein kinase3 (MAPK3), matrix metalloproteinases (MMPs), tissue inhibitor of metalloproteinases (TIMP), JunD, Frizzled Class Receptor 4 (FZD4), estrogen receptor (ERα/ERβ), prolactin receptor (PRL-R), steroidogenic acute regulatory protein (stAR), vascular endothelial growth factor (VEGF), basic fibroblast growth factor (bFGF) and endocrine gland-derived VEGF (EGVEGF), cholesterol side-chain cleavage enzyme (P450scc), 3β-hydroxysteroid dehydrogenase (3βHSD), and FSH receptor (FSH-R).

22.3.5.1.2 Tissue Remodelling

LH-induced expression of several MMPs and TIMPs facilitates tissue remodelling by disrupting the extracellular matrix (ECM), followed by cell migration and neovascularisation. MMP-2 (gelatinase A), expressed in luteal and endothelial cells, and MMP-9 (gelatinase B), expressed in stromal cells, are two important MMPs involved in tissue remodelling by cleaving type IV collagen. Alpha

Table 22.24 Functional characteristics of corpus luteum (CL) in different species

Species	Peak luteal activity (days)	Major luteotropic substance	Diameter of peak luteal cell (μm) [CL (mm)]	CL regressed in non-pregnancy/luteal phase (days)	Peak level continued in pregnancy (days)	Major luteolytic substance
Cow	8–9	LH	15–40 [25–29]	18–19	275–290	PGF2α
Ewe	6	LH, PRL	31–34 [5–14]	12–15	50–70	PGF2α
Sow	7–8	LH	20–28 [9–10]	13–16	114–115	PGF2α
Mare	12–14	LH	10–15 [29–32]	14–15	25–30 (pri. CL) 130–150 (Sec. CL)	PGF2α
Bitch	15–25	PGE2, PRL	30–40 [17–18]	30–90 (45)	58–68	PGF2α (preg), passive degeneration (Non-Preg)
Queen	12-16, 21	PGE2, LH, PRL	17–20 [22–24]	21-50 (37)	45–50	PGF2α
Human	9	LH, hCG	15–25 [2–5]	11–17 (14.2)	70	PGF2α
Rat	2	E2, LH, PRL	12–20 [1.2–2]	2–3 generation	21–24	PRL

Source: Compiled from various sources
LH luteinising hormone, *PRL* prolactin hormone, *hCG* human chorionic gonadotropin, *E2* estrogen, *PGE2* prostaglandin E2, *Pri* primary, *PGF2α* prostaglandin F2α, *Preg* pregnancy

2 macroglobulin (α2M) and TIMPs inhibit MMPs. Prolactin stimulates the expression of α2 macroglobulin. The ECM acts as a 'scaffold' protein to hold the luteal cells. The theca and granulosa cells undergo extensive hypertrophy and differentiate into steroidogenic luteal cells. Granulosa cells are differentiated into large luteal cells, and theca cells are transformed into small luteal cells. Large luteal cells synthesised 2–40 folds more progesterone than small cells in ruminants and rodents. In several species, a considerable mixing between these cell types occurred during the luteinisation process, except in primates, where two cell populations remain relatively separate and designated as granulosa-lutein cells and theca-lutein cells, respectively. At the initial stage of luteinisation, all the lutein cells proliferate and contain large quantities of fat droplets (cholesterol as the precursor for progesterone) within their cytoplasm. The structure gradually becomes a solid mass over the surface of the ovary. Later, the cells become hypertrophied, and the vascular network regenerates throughout the structure with the presence of fibroblast. The large lutein cells have abundant rough endoplasmic reticulum, prominent Golgi bodies, and secretory granules, making them endocrine cells and synthesising progesterone.

22.3.5.1.3 Vascularisation

Neovascularisation is essential for the formation and maintenance of CL as it requires profuse blood flow to transport nutrients, hormones, and lipoprotein-bound cholesterol. Immediately after ovulation and LH surge, theca-derived pericytes (the first vascular cell) start invading luteal parenchyma and proliferates rapidly to give rise to many blood vessels in the CL. Vascular endothelial growth factor (VEGF), basic fibroblast growth factor (bFGF), and endocrine gland-derived VEGF (EGVEGF) help in CL angiogenesis. These growth factors are expressed in response to LH.

The formation of the corpus luteum takes 8–9 days in cow and sow. In non-pregnant animals, the functional CL regresses after 17–18 days and completely regressed within 19 days. Upon regression, it becomes an avascular scar tissue called corpus albicans. The functional CL persists throughout the gestation period in pregnancy and is called the corpus luteum graviditatis. LH controls the development and maintenance of CL function. In pregnant women, the functionality of CL is also maintained by the hormones like human chorionic gonadotropin (hCG, like LH) secreted from a blastocyst (trophoblast) starts at day 9 post-fertilisation.

22.3.5.2 Functions of CL

22.3.5.2.1 Luteal Steroidogenesis

The expression of key proteins involved in the uptake, transport, and processing of cholesterol to progesterone enables the corpus luteum to act as a transient endocrine organ during and after luteinisation. The large and small luteal cells contribute around 85% and 15% progesterone synthesis in the cow. The progesterone secreted from CL helps in the embryo's implantation and maintenance of the pregnancy. The luteal steroidogenesis begins with the uptake of cholesterol. CL can incorporate cholesterol from both HDL and LDL. Still, HDL is the main source of cholesterol for the CL due to scavenger receptor class B type I (SR-BI) and HDL receptor for selective uptake of HDL-derived cholesterol ester. The expression of SR-BI is increased several folds in response to LH/hCG during the development of the CL. But, due to its hydrophobic nature, cholesterol requires a carrier protein named sterol carrier protein-2 (SCP-2) for its intracellular movement to the mitochondria where the steroidogenic enzymes are located. The mitochondrial P450scc transforms cholesterol into progesterone. But the amount of progesterone secretion not only depends on the amount of cholesterol or the expression of P450scc. Another enzyme, 20α-HSD, that catabolises progesterone into the inactive 20α-DHP, also controls the progesterone secretion. PRL, LH, and estradiol regulate the proteins involved in luteal steroidogenesis.

In humans and rodents, CL also synthesises androgens and estrogens in addition to progesterone. The androgen produced from the CL is androstenedione. The enzyme P45017 α-hydroxylase/C17–20 lyase (P450c17 or CYP17) helps converts progesterone into androstenedione. But, in pre-ovulatory follicles, P450c17 is expressed only in theca and interstitial cells, not in granulosa cells.

22.3.5.2.2 Synthesis of Protein Hormones

The CL also produces relaxin responsible for softening the pubic symphysis and uterine cervix and the relaxation of the myometrium during parturition. Large luteal cells produce oxytocin, which helps synthesise prostaglandin from the uterus in ruminants and pigs.

22.3.5.3 The Fate of CL

The corpus luteum has two fates depending on the occurrence of fertilisation. If fertilisation and implantation occur, the CL persists throughout the pregnancy to secret progesterone and maintains the pregnancy as corpus luteum graviditatis. The CL is regressed to form corpus albicans if fertilisation doesn't occur in the alternate fate. The life span of the corpus luteum is governed by the synchronised activity of the pituitary, ovary, uterus, and the embryo through LH, progesterone, oxytocin, and prostaglandin. The substances that support CL to sustain are called luteotropic, and those that terminate the CL are termed luteolytic. The actions of luteotropic and luteolytic substances on CL are different in various species. The LH, progesterone, and prostaglandin E2 (PGE2) are the luteotropic, whereas oxytocin and prostaglandin F2α (PGF2α) are luteolytic in most of the domestic species. In rodents and carnivores, prolactin plays a vital role in CL formation. In many, but not all, animal species, luteolysis is mediated by uterine prostaglandin F2α (PGF2α); hence, called primary luteolysin. Luteolysis occurs through the induction of multiple biochemical pathways that inhibit progesterone secretion and apoptosis. Uterine PGF2α is the major substance that causes luteolysis in ruminants, pigs, horses, guinea pigs, hamsters, rabbits, and rats. Placental PGF2α is the major luteolytic substance in canine or feline species (discussed later). Estrogens are considered luteotropic in some species like pigs, rats, and rabbits. LH also helps to sustain the life of CL. During the persistence of CL, no FSH is synthesised from the pituitary due to the high secretion of progesterone from the CL. But, immediate after regression of CL, the FSH can be synthesised in the pituitary, and a new follicular wave or a new cycle can start. Hence, luteolysis is the key factor in initiating the cycle for ovulation of follicles, and synthetic luteolytic drugs are administered to augment fertility and estrus synchronisation and the superovulation process.

22.3.5.3.1 Synthesis of PGF2α

During the middle of the diestrus, prolonged exposure to progesterone down-regulates its receptor at the uterine endometrium, leading to higher estrogen receptor expression. It generally occurs around day 14 of the estrous cycle in cattle. After binding with its receptor, the estrogens upregulate the oxytocin receptor at the endometrium. The oxytocin facilitates the conversion of arachidonic acid to prostaglandins (PGF2α) (Fig. 22.18). Pulsatile secretion of PGF2α causes luteolysis. The pulsatile release of PGF2α and the appearance of the local luteolytic factors are species specific (Table 22.7). It takes 18–19 in cow, 14–15 in mare, 13–16 in sow, and 11–12 days in women unable to conceive.

In pregnant animals, the IFN-τ released from the conceptus suppresses estrogen receptor expression at endometrium vis-à-vis synthesis of PGF2α to sustain the CL. Thus, foetus acts as a luteotropic substance. In the case of polytocous species, the number of foetuses determines the PGF2α synthesis. In pigs, less than four foetuses may cause abortion due to luteolysis with the action of PGF2α. A high progesterone level can also inhibit the synthesis of oxytocin and PGF2α.

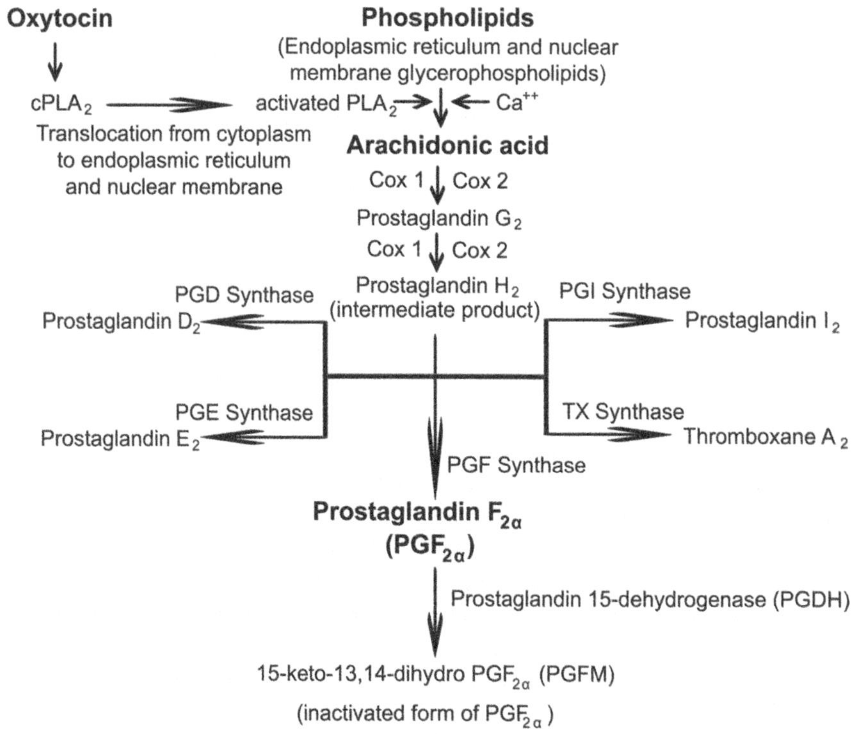

Fig. 22.18 Biosynthesis and metabolism of Prostaglandin F2α (PGF2α). **Oxytocin** activates cytoplasmic Phospholipase A2 (**cPLA2**) and facilitates its translocation from cytoplasm to the membrane of endoplasmic reticulum and nucleus where the **activated PLA2** converts membrane glycerophospholipids (**Phospholipids**) into **Arachidonic acid** in the presence of calcium ion (Ca^{2+}). Arachidonic acid is converted to **Prostaglandin F2α (PGF2α)** by the actions of **Cox1, Cox2,** and **prostaglandin (PG) F Synthase** with the intermediate products like **Prostaglandin G2** (PGG2) and **Prostaglandin H2** (PGH2). PGH2 can be transformed into **Prostaglandin D2** by Prostaglandin (PG) D Synthase. Similarly, **Prostaglandin E2, Prostaglandin I2,** and **Thromboxane A2** can also be synthesised by their **synthases.** The PGF2α is metabolised to its inactivated forms like **15-keto-13, 14-dihydro PGF2α (PGFM)**, under the influence of **Prostaglandin 15-dehydrogenase (PGDH)**

Thus, small-sized CL is regressed earlier due to its low progesterone secretion.

Due to its short biological half-life, PGF2α cannot be measured directly in serum. Instead, its plasma metabolites 15-keto-13,14-dihydro-PGF2α (PGFM) can be estimated and obtain an indirect measurement of PGF2α.

22.3.5.3.2 Transport of Uterine PGF2α to the Ovary

The PGF2α is synthesised and secreted from the endometrium in a pulsatile manner and directly reaches the ovary through a unique vascular utero-ovarian plexus (UOP), bypassing the systemic circulation. Uterine PGF2α enters the uterine vein, and the uterine vein forms the utero-ovarian vein after joining the ovarian vein. From this utero-ovarian vein, PGF2α enters the ovarian artery through the utero-ovarian plexus (UOP) and ultimately reaches the ovary for luteolysis. In ruminants, local synthesis of PGF2α occurs through an autocrine-signalling loop. The uterine PGF2α upregulates the expression of prostaglandin-endoperoxide synthase 2 (PGHS-2) or cyclooxygenase-2 (COX-2) in luteal cells to synthesise PGF2α locally.

22.3.5.4 The Regression of CL (Luteolysis)

The regression of the CL occurs in two phases, functional regression associated with a decrease in progesterone synthesis and structural regression, where the luteal cells undergo programmed cell death. The structural regression initiates after the initial decline in progesterone. In rats and some other species, the CL persists even after functional luteolysis and produces some inactive metabolites of progesterone. In most domestic species, both types of luteolysis coincide. In cows, the concentration of progesterone below1 ng/mL denotes functional luteolysis.

22.3.5.4.1 The Role of PGF2α on Functional Regression

PGF2α doesn't inhibit progesterone synthesis, but it causes the metabolism of progesterone to 20-αDHP through the activation of the enzyme 20-α-hydroxysteroid dehydrogenase (20αHSD). In addition to this, PGF2α is also shown to reduce cholesterol transport by decreasing SCP-2 and StAR expression. PGF2α blocks LH-induced cAMP accumulation and thus mediates anti-LH action on luteal cells. Several other transcription factors are reported to be involved in functional luteolysis, such as nur77 and endothelial cell-derived peptide endothelin-1 (ET-1). Transcription factor nur77 induces 20αHSD, whereas ET-1 directly inhibits progesterone secretion.

22.3.5.4.2 The Role of PGF2α on Structural Regression

PGF2α are involved in several cellular and molecular events that lead to structural luteolysis, such as vasoconstriction, apoptosis, infiltration of immune cells, increased metalloproteinase activity, and induction of oxidative stress.

22.3.5.4.2.1 Vasoconstriction

PGF2α induces the expression of several vasoactive factors, such as endothelin-1 (ET-1), endothelin converting enzyme (ECE), endothelin type A and B receptors (ETA-R/ETB-R), angiotensin II (Ang II) and angiotensin-converting enzyme (ACE). During the initial stage of luteolysis, there is increased blood flow to the CL through nitric oxide (NO) production. PGF2α stimulates this NO production by activating endothelial nitric oxide synthase (eNOS). Due to increased blood flow and secrets ET-1 and Ang II, the capillaries are subjected to high shear stress. These two local factors induce chronic vasoconstriction of the CL arterioles.

22.3.5.4.2.2 Apoptosis

PGF2α, along with other factors like PRL, TNF, and Fas ligand, triggers cell death signals. The apoptosis is mediated by both the extrinsic or death receptor-mediated pathway and the intrinsic or extrinsic mitochondrial pathway through the activation of caspases. In the extrinsic pathway, the Fas ligand (FasL), after binding with the TNF receptor (TNFR-3, -4, and -5), activates caspase-8. On the other hand, the intrinsic pathway is activated in response to stress stimuli that alter the mitochondrial membrane permeability to release cytochrome c. The cytochrome c, in turn, combines with protease-activating factor-1 and procaspase-9 to form an apoptosome that ultimately activates caspase-9. These caspases, in turn, cleave some essential intracellular peptides like actin, poly (ADP-ribose), polymerase (parp) (involved in DNA repair), and protein kinases to facilitate cell death.

22.3.5.4.2.3 Immune Cell Infiltration

PGF2α induces the expression of intercellular adhesion molecule-1 (ICAM-1) and monocyte chemoattractant protein-1 (MCP-1) in the luteal cells, resulting in monocyte/macrophage infiltration into the CL and phagocytose the luteal cells.

22.3.5.4.2.4 Increased Metalloproteinase Activity

The expression of luteal cell-specific tissue inhibitor of metalloproteinase-1 (TIMP-1) is induced by PGF2α. TIMP-1 increases the metalloproteinase activity to cause structural regression.

22.3.5.4.2.5 Induction of Oxidative Stress

PGF2α decreases the activity of the antioxidant machinery of the CL by reducing protective enzymes and antioxidant vitamins (ascorbic acid). It also causes the down-regulation of genes necessary to eliminate free radicals.

22.3.5.5 Persistent CL

Persistent CL is a pathological condition where the CL is failed to regress beyond the day of structural luteolysis (day 19 in cow and day 16 in mare). It may be due to delayed ovulation, embryonic loss after the maternal recognition of pregnancy, and chronic uterine infections. Within 4–5 days post ovulation, uterine infections damage the receptors for progesterone and oxytocin in the endometrium, interfering with PGF2α synthesis and persistent CL. The administration of nonsteroidal anti-inflammatory drugs interferes with the endometrial PGF2α synthesis and may cause persistent CL. In persistent CL, growing follicles of the existing follicular waves become dominant; but fail to ovulate. The animals are not in heat and can be mistakenly considered pregnant. There are some distinguished features of persistent CL in contrast to cyclic CL and CL during pregnancy. In mare, persistent corpus luteum appears as a well-demarcated hyperechoic structure on the ovary, with better uterine tone and pale, tight, and dry cervix examined on speculum. The expressions of immune tolerance-related factors (PGES and forkhead/winged-helix transcription factor 3) were upregulated in pregnancy CL but not in persistent CL.

The upregulation of genes involved in lymphangiogenesis, inflammation, and apoptosis were in the same proportion in persistent and CL of pregnancy, but not in the case of cyclic CL. The diagnosis of persistent CL can be made upon history, rectal palpation, progesterone assay, a transvaginal ultrasound, USG-guided ovarian, and endometrial biopsy. The level of serum progesterone concentration higher than 1.0 ng/mL is reported in mare exhibiting persistent CL. The treatment protocol for persistent CL in cows includes the injection of PGF2α alone or in combination with GnRH (48–56 h after PGF2α injection). However, a single PGF2α injection is reported to give better results in the treatment of persistent CL. Sometimes, uterine wash or messaging stimulates the synthesis of the PGF2α and is applied to induce luteolysis. Some contraceptives, like the intrauterine device (IUD), are used to diminish the PGF2α synthesis from the endometrium and induce transient sterility in animals.

> Know More ………
>
> **Longest CL**
>
> The CL with the most extended life span among mammals is found in the lynx (animal under the Felidae family). Physiologically persistent corpora lutea exist after pregnancy and pseudo-pregnancy for up to 2 years and can secret progesterone. Thus, folliculogenesis is inhibited during the non-breeding season(s). The progesterone level is temporarily reduced during the period of estrus and parturition in the breeding season and induces this polyestrous animal as a monoestrous. The presence of a potential higher level of superoxide anions causing apoptosis of the growing follicles has been identified in lynx and is a physiological adaptation.

22.3.5.6 CL in Dogs and Cats

The corpus luteum of bitch remains functional for a particular period, whether the bitch is pregnant or not. The secretion of progesterone (P4) is similar in pregnant and non-pregnant bitches throughout the entire luteal life span before the prepartum luteolysis. The luteal phage ends with the onset of parturition (around 65 days) in pregnant bitches; but in non-pregnant bitches functional CL exists for a more extended period (75–100 days). The major luteotropic factors in bitches are PGE2 (initial phase) and prolactin (later phase, 25 days from the CL formation) in synergy with LH. The prostaglandin E2 (PGE2) as a luteotropic factor (at the initial luteal phase) is identified that acting through an autocrine/paracrine fashion. PGE2 is mediated by the upregulation of steroidogenic acute regulatory (StAR) protein. The prolactin acts as a luteotropic factor from 25 days of the luteal phase. There are several other factors, like lymphocytes (CD4 and CD8), cytokines (IL8, IL10, IL12, TNFα, or TGFβ), tropic factors (IGF and VEGF), and some glucose transporters and insulin-sensing receptors also act as luteotropic factors in these species. Another uniqueness of canine CL is that the uterus has no role in CL regression. It is probably due to anatomic independence between the uterine vein and the ovarian artery. Progesterone and estrogen also have a luteotropic role. The ageing of CL brings about the regression of CL. A passive degenerative process in association with apoptotic signals is evident during the time of luteolysis in ferret and mink due to reduced expression of PRLR and LHR in CL. However, the exogenous PGF2α can induce luteolysis in non-pregnant (after day 30 of the luteal phase) and pregnant (in the second half of pregnancy). In queen, the CL may continue a maximum of 40–50 days in the non-pregnant luteal phase (also called pseudo-pregnancy) or approximately 65 days in pregnancy. In dogs, the CL is the only source of progesterone, unlike in cats, where placental progesterone can support the pregnancy from day 40 to 45 of gestation. Diminished progesterone levels from day 45 of pregnancy induce utero/placental PGF2α release, which stimulates StAR expression and the appearance of vacuoles and connective tissue elements, collagen fibres, and apoptotic factors in luteal cells. These factors lead to luteolysis in pregnant bitch during the time of parturition. In the case of induced ovulator like cats, rats, and rabbits, pseudo-pregnancy is developed in non-fertilised ovulation.

22.3.5.7 CL in Mares

In mares, two corpora lutea form during gestation. The primary is formed due to ovulation and exists up to the 35th day of pregnancy. It is the primary source of progesterone during this period. The primary CL is regressed around 25–30 days of gestation, and the secondary CL is formed as a result of ovulation or luteinisation of follicles under the influence of equine chorionic gonadotropin (eCG), formerly called pregnant mares' serum gonadotropin (PMSG) secreted from a specialised endometrial cups. The eCG acts like FSH and LH can induce follicular development and ovulation. The endometrial cup formation occurs around the 35th day of gestation as a girdle-like band of specialised cells developed from the foetal trophoblast and embedded within the uterus after its detachment from the foetal trophoblast. The secondary CL secretes progesterone from day 80 to nearly mid of pregnancy, day 130–150 of gestation. The placenta is the source of progesterone for the remainder of gestation in these species. Oxytocin and PGF2α induce the regression of CL in mares only when they have functional luteal tissue.

22.3.6 Ovulation in Birds

Female chicken that attains puberty is called a pullet. The sexually matured pullet capable of forming eggs or ova is called an adult or hen. A pullet or hen contains nearly 4000 ova or oocytes in their left ovary, and a few of them are surrounded by different nutrients and gradually become a yolk in their sac or *follicle*. The follicles are continuously migrated towards the outer circumference of the ovary. The yolk and ovum rupture from the follicles and dissociates from the stalk after acquiring sufficient yolk or nutrients for the chick. This process is called ovulation. Only one ovum can be ovulated at a time. The diameter of a matured pre-ovulating follicle is about 40 mm in fowl. Each follicle requires nearly 10 days time duration in fowl to become mature before ovulation when it reaches about 30–31% weight of the whole egg.

In birds, the follicles of various developmental stages and blood vessels and nerves are suspended together from the ovary. It is called a stalk or pedicle. The domestic fowl generally takes 9 days to develop a fully matured oocyte. The funnel-like infundibulum of the oviduct captures the yolk after ovulation. Then it passes through different parts of the oviduct before laying. The term oviposition refers to the process of laying or shedding completely developed eggs. Various egg layers and shells are formed over the yolk in different parts of the oviduct to become a complete egg (Table 22.25) having species-specific shape and colour (discussed in detail in an earlier chapter). Time taken from ovulation to oviposition depends upon the species (discussed in detail in an earlier chapter). It takes nearly 24–25 h for all domestic birds. Thus a bird can't lay two successive eggs in a day. Two consecutive ovulations can occur in fowl, turkeys and Japanese quail with an interval of 15–30 min and duck and guinea fowl slightly earlier (15 min). The post-ovulatory follicles may remain up to 24 h in fowl and then regress without forming corpus luteum.

Table 22.25 Role of different parts of the oviduct in egg formation

Parts of the oviduct	Contribution
Ovary	Yolk containing ovum
Magnum	White or albumin
Isthmus	Shell membrane
Uterus or shell gland	Shell

22.3.6.1 Control of Ovulation

22.3.6.1.1 Neuro-endocrine Axes

Avian reproductive and ovulation are regulated by the synchronised interactions of different hormones (Table 22.26) and peptides. The central neuroendocrine axis is the hypothalamic-pituitary-gonadal (HPG) axis. The hypothalamus integrates the upregulation of the axis and the secretion of GnRH–FSH–LH, followed by ovarian steroids and controls the reproduction. The GnRH neurones are stimulated by environmental factors like photoperiod, seasons, food availability, and specific reproductive signals like courtship, sound or song behaviour. Most birds are seasonal breeders, and the photoperiod influences their

Table 22.26 Role of various hormones in avian female reproduction

Hormone	Function
FSH	Follicular development, follicular hierarchy, yolk development and steroidogenesis
LH	Steroidogenesis, ovulation and prostaglandin secretion
Inhibin	Inhibit FSH secretion
Estrogen	Development of oviduct, in flow of nutrients to the ovary and oviduct for supplying to egg, growth of the plumage, mating and nesting behaviour
Progesterone	LH secretion and surge, formation of albumen
Androgen	Development of oviduct, formation of albumen and growth of comb
Corticosterone	Down-regulate HPO axis
Prolactin	Reduces ovulation, promotes broodiness, nesting behaviour and caring for the young
IGF	Activation of FSH and LH, follicular hierarchy
Prostaglandin	Ovulation, shell formation and oviposition
Thyroid hormones	Inhibit ovulation, atresia of the hierarchial follicle, reduced LH and estrogen, and promote progesterone
Parathyroid hormone	Shell formation and oviposition
Vastocin	Oviposition, uterine contraction, reproductive behaviour

breeding. Long photoperiod and favourable seasons upregulate the HPG axis as it facilitates access to food, favouring gaining energy. Stress-induced suppression of reproductive activity is mediated through GnIH after activating the hypothalamic-pituitary-adrenal (HPA) axis. Avian reproduction is also regulated by the hypothalamic-pituitary-thyroid (HPT) axis.

22.3.6.1.2 FSH and LH

The FSH and LH secreted from the anterior pituitary gland under the influence of GnRH play a predominant role in the ovulation process. The FSH regulates follicular growth and maintains the follicular hierarchy. A single ovum is transformed into a pre-ovulatory follicle from 7 to 10 hierarchical follicles after a rapid growth phase of a small follicular pool. The LH stimulates both the hierarchial and non-hierarchical follicles for steroidogenesis. Androgen and estrogens are produced from the thecal layers of the small follicles and progesterone from the granulosa cells of the pre ovulatory follicles and small follicles. But, progesterone is converted into androgen and or estrogen when it reaches the theca layer. The production of androgen and estrogen gradually decreased in the pre-ovulatory follicle with the progression of follicular size due to reduced receptivity of FSH receptors in the thecal cells. Thus, the matured pre-ovulatory follicle is capable of only production of progesterone. Progesterone causes a positive feedback loop to stimulate the profuse release of LH from the anterior pituitary, resulting in the LH surge and ovulation. The highest inhibin secretion occurs in the large pre-ovulatory follicle and inhibits the next ovulation by down-regulating FSH secretion. All the steroids are involved in follicular development and the oviduct development along with up-and down-regulation of the HPO axis.

22.3.6.1.3 Prolactin

Prolactin promotes broodiness characteristics as well as nesting behaviour. Its level is increased immediately after oviposition and thus reduces the next ovulation. Prolactin is induced to increase vasoactive intestinal polypeptide (VIP) and reduce gonadotropin-releasing hormone (GnRH) from the pre-optic area (POA) of the hypothalamus. It results in decreased secretion of LH from the anterior pituitary. The level reduces when chicks are self-dependent on food. The secretion of prolactin is controlled genetically. Its level is more in indigenous or non-descript breeds of fowl, and mostly all wild birds show more broodiness characteristics than layer varieties of breeds.

22.3.6.1.4 Thyroid Hormones and Parathyroid Hormone

The nuclear receptors (TRα and TRβ0) and plasma membrane receptors (integrin, αVβ3) of both thyroxine (T4) and triiodothyronine (T3) are present in the ovarian follicular cells. The T3 inhibits the ovulation process by inducing atresia of pre-ovulatory follicles, decreasing estrogen levels by reducing the activity of thecal cells of the pre-ovulatory follicles and non-hierarchical follicles, and diminishing the function of LH. But it can increase progesterone levels by influencing the granulose cells of the pre-ovulatory follicle. Thyroid hormones induce the moulting and hatching process; thus, referred to as the *hatching hormone*. The parathyroid hormones and parathyroid hormone-related proteins promote the relaxation of the proximal oviducts. They also facilitate increased blood flow at the oviduct and surrounding gland to release various egg-forming materials, including calcium.

22.3.6.1.5 IGF

The insulin-like growth factors (IGF-I and IGF-II), synthesis in the ovary, enhance FSH and LH's receptivity over the thecal and granulosa cells. However, IGF-II augments the recruitment of follicles, and the follicles cannot respond to IGF-II stimulation atretic. Conversely, the urokinase enzyme is active only in small rapid, growing follicles and expressed little in larger follicles. Thus, urokinase is also considered one of the follicular hierarchy determinants.

22.3.6.1.6 Prostaglandins

Prostaglandins involve in the ovulation and egg formation process. It helps to rupture the stigma and facilitates ovulation. The PGF2α acts over the shell gland to secrete materials for forming the eggshell. It also helps contract the oviduct to move the ova downward. In reverse, the PGE2 acts over the uterovaginal sphincter to relax the oviduct, resulting in ease of laying.

22.3.6.1.7 Vasotocin

Vastocin belongs to the vasopressin family peptide, involves uterine contraction and regulates oviposition. It also influences social and reproductive behaviours, osmoregulation and glycogenolysis.

22.3.6.2 Factors of Ovulation

Genetics (species and breed) (Table 22.27) and nutritional state are the major factors affecting ovulation.

The indigenous or non-descript birds lay less number of eggs in a year. Layer birds are genetically capable of producing more eggs, and various genetic manipulations developed for commercial layer birds. About 2021 and 2623 genes identified in low egg-producing hens (LEPH) and high egg-producing hens (HEPH) affect the HPO axis and regulate the laying process.

Ovulation is delayed when ova takes a long time to receive its optimum nutrients from the surroundings. Thus, poor nutrition causes delayed ovulation. A good hen having good genetic make-up with the optimum level of nutrition

Table 22.27 Occurrence of ovulation in a year of some domestic pure breed birds

Type of birds	Breeds	Ovulation (lay egg) (in a year)
Layer chicken	Leg Horn, Golden Comet, Australorp	250–280
Indigenous chicken	Assel, Baladim, Kampung	30–100
Layer duck	Campbell, Runner, Buff	250–340
Indigenous duck	Nageswari, Pati	60–150
Quail (dual type)	Coturnix (Japanese quail), Bobwhite	200–300
Layer guinea fowl	Grey and White type guinea fowl	180–200
Indigenous guinea fowl	Pearl, Lavender, White	40–100
Geese	Zie	70–100
Turkey	Beltsville small white	70–100

Data compiled from various sources

may produce about 300–310 eggs in a year. The size and weight of the ova or eggs are also genetically altered. The size and weight of quail egg is about 3.5 cm (length) × 2.7 cm (diameter) and 10 g; while 22–11.5 cm (length) × 18 cm (diameter) having about 140 g weight found in goose egg. A chicken egg's average size and weight is about 6.2 cm (length) × 4.3 cm (diameter) and 55 g.

22.3.6.3 Clutch

The birds generally lay eggs in a group, successively for a few days, called a clutch, following an asynchronous gap or pause (break). Clutch is comparable with the litter size of mammals. Length of clutch and pause varies from 1 or 2 to 100 depending on species, breed and age. Habitat, latitude, food availability, and the presence of predators influence the clutch size. The nest supports the brood of the eggs; hence, nesting birds have a comparatively larger clutch size. Wild migratory birds have low clutch size (1–2 eggs) compared to non-passerines like ducks and geese as many as 20 eggs. Birds having a long pre-pubertal period (4–5 years), like marine birds, have a low clutch. Marine birds that scavenge far from their habitat have smaller clutches (one egg) than those which explore near the colony (2–3 eggs). Higher latitudes favour larger clutches as the birds can gather more food per unit of time. Birds of tropical non-seasonal rain-forest and nearer to the equator contain smaller and uniform clutch sizes throughout the year due to a shortage of food compared with the same species and breeds of the polar zone during the spring and summer. Extreme weather and global warming reduce the eggs' viability, thus reducing the clutch size. More extended day length increases clutch size as extended photoperiod facilitates to search for more food. Daily total exposure to light (including artificial light) requires about 14–16 h in fowl and about 14–18 h in quail to maintain optimum clutch size. Older birds have larger clutch sizes than young ones. Younger birds generally take 1–2 weeks after their first laying to gain a stable clutch size. Peak clutch size often occurs within 3–5 weeks from the day of 1stlaying or attainment of puberty in fowl. Puberty may attain in fowl at about 4–6 months, and they can lay eggs for up to 3 years or more; but, after 1.5 years, the clutch size is reduced, and the break period becomes longer.

22.3.6.4 Formation of Defective Eggs

Various defective eggs may occur due to defects in ovulation and egg formation processes (Table 22.28). Simultaneous ovulation of two ova results in a double-yolk egg. Blood spots within the egg may occur due to the rupturing of blood vessels in the ovary and oviduct. Deficiency of vitamin A and vitamin K in feed, presence of lucerne and fungal toxin in feed and use of specific drugs, like sulphaquinoxaline, may cause such spots. A lack of carotenoids may lead to pale yolk formation. Watery whites or albumin may be occurred due to fungal toxins in the feed, ammonia in the shed and certain diseases like infectious bronchitis and egg drop syndrome. Greenish coloured albumin may form due to the excess use of riboflavin and cyclopropene fatty acids in cottonseed. Any stress develops various grooves over the eggshell, called the misshapen egg. The deficiency of vitamin-D3 and calcium and salt-rich drinking water will cause thin shell eggs and different deformed eggs. Young birds may produce such deformed eggshells due to immature shell glands.

Table 22.28 Various egg deformities with their origin

Major sources	Deformities in egg
Young and aged layers	Double yolk egg, pale yolk, meat spots in the egg, thin shell, rough shell, misshapen eggs, various shell deformities
Dietary origin	Blood spots in the egg, pale yolk, discoloured yolk, watery whites, discoloured white, thin shell, gross crack, pimple egg, off odours, and flavours
Drug residue	Blood spots in the egg, discoloured yolk
Faulty lighting programme	Blood spots in the egg, rough shell, various shell deformities
Noise/stress in the shed	Blood spots in the egg, thin shell, rough shell, misshapen eggs, various shell deformities
Improper handling and storage	Discoloured yolk, watery whites, discoloured white, gross crack, various cracks and marks in eggshell, rotten eggs, off odours, and flavours
Occurrence of diseases	Blood spots in the egg, watery whites, thin shell, gross crack, rough shell, misshapen eggs, mottled shell, and other various shell deformities
Genetical factor	Double yolk egg, meat spots in the egg, watery whites

Source: Das and Roy (2016)

Learning Outcomes

- **Endocrinology of female reproduction**: The reproductive activity of females is controlled by the hypothalamic-pituitary-ovarian (HPO) axis comprising of hypothalamus, the anterior pituitary, and the ovaries. This HPO axis is responsible for regulating both centrally and peripherally produced reproductive hormones. The central part of the axis includes GnRH from the hypothalamus and gonadotropins from the anterior pituitary, LH and FSH. Ovaries act as dynamic endocrine glands that secret steroid and peptide hormones. The ovaries have two important steroid hormones: estrogens (estradiol, estrone, and estriol) and progesterone (progestin). Activins, inhibins, and follistatin are the peptide hormones produced in the ovaries. All these hormones act in a coordinated fashion to control different reproduction events in animals. The secretion of GnRH is controlled by neuropeptides such as kisspeptin and RFRP3. HPO axis is influenced by several factors like nutrition, photoperiod, and stress. Besides these reproductive hormones, pheromones are also involved in regulating sexual behaviour. In mammals, pheromones are released through urine, faeces, vaginal secretion, saliva and modified scent (cutaneous) glands, including hair and wool. Animals perceive the pheromones through vomeronasal organs and express characteristics of sexual behavioural patterns.
- **Puberty and estrous cycle**: Puberty is the ability of the animal to produce gamete, i.e. ovum in females. The onset of puberty results from integrated sequences of biological events that lead to progressive maturation of sexual characteristics to attain full reproductive capacity. Puberty results due to the activation of the HPO axis. Estradiol suppresses GnRH secretion through a negative feedback mechanism. At the initiation of puberty, the negative feedback of estradiol is decreased, leading to activation of the GnRH surge centre and commencement of the estrous cycle, growth of the follicles, and ovulation. The rhythmic sexual behavioural pattern exhibited by the female animals after the attainment of puberty is called the estrous cycle. The estrous cycle classifies into four distinct phases: estrus, metestrus, diestrus, and proestrus. During the proestrus and estrus, follicular development or generation of follicular wave(s) occurs. Hence, these two phases are collectively called the follicular phase. The metestrus and diestrus phases are characterised by the formation of the corpus luteum and are called the luteal phase. The length of estrous cycle, duration of estrus, and time of ovulation are species specific. The estrus cycle is characteristically different from the menstruation cycle found in primates and humans.
- **Oogenesis and folliculogenesis**: The developmental process of the female gamete or ovum is called oogenesis. Oogenesis is initiated at the embryonic stage and completed after puberty in three different phases: oocytogenesis, ootidogenesis, and maturation (oogenesis proper). The ovarian follicles are the functional units of the ovary that appear as a cluster of somatic cells to protect and nourish the oocytes. The follicular cells are developed through a process termed folliculogenesis. The folliculogenesis process encompasses the growth and development or atresia of follicles through morphological and functional changes. Several hormone and growth factors are involved in the oogenesis and folliculogenesis process.
- **Ovulation and corpus luteum formation**: Ovulation is the biological process that involves the shedding of the oocyte from the mature Graafian follicle. It is an inflammatory process sequentially controlled by the neuroendocrine system. Two subsequent events occur at the oocytes and surrounding follicles during the ovulation process. In oocytes, the resumption of meiosis and the structural remodelling of the follicles release the maturing oocyte. The temporary heterogeneous endocrine structure comprises steroidogenic luteal cells, fibroblasts, endothelial, pericytes, and immune cells formed in the ovum-free follicle on the ovarian surface called corpus luteum. It plays a central role in regulating the reproductive cycle and maintenance of pregnancy by secreting progesterone. The process of formation of corpora lutea is known as luteinisation. The corpus luteum has two fates depending on the occurrence of fertilisation. If fertilisation and implantation occur, the CL persists throughout the pregnancy as corpus luteum graviditatis. The CL is regressed to form corpus albicans if fertilisation doesn't happen. The life span of the corpus luteum is governed by the synchronised activity of the pituitary, ovary, uterus, and the embryo through LH, progesterone, oxytocin, and prostaglandin. Prostaglandin $F2\alpha$ is the principal luteolytic substance in domestic animals.
- **Ovulation in birds**: The term oviposition refers to the process of laying or shedding completely

(continued)

developed eggs. The yolk and ovum rupture from the follicles after acquiring sufficient yolk or nutrients for the chick through the ovulation process and captured by the infundibulum. Various egg layers and shells are formed over the yolk in different parts of the oviduct to develop a complete egg. Several neuroendocrine factors are involved in the ovulation process of birds. The birds generally lay eggs in a group, successively for a few days, called a clutch, following an asynchronous gap or pause (break).

Exercises

Objective Questions

Q1. What are the products of meiosis-I?
Q2. Which gonadotropins are mainly responsive to which kind of cells of the oocyte?
Q3. Which kind of antral follicles secrete gonadotropin surge-attenuating factor?
Q4. Why does ovulation occur after mating in induced ovulation?
Q5. Which ovary is more functional in ruminants?
Q6. How does follistatin acts on FSH?
Q7. Which follicles are termed subordinate follicles?
Q8. Which hormone plays the major role in meiotic resumption?
Q9. What is the role of StAR in steroidogenesis?
Q10. How do estrogens perform their anti-inflammatory role?
Q11. How many carbons are in the progesterone skeleton?
Q12. Which physiological phenomenon will occur when the stimulatory mechanism between the kisspeptin and NKB networks with the GnRH surge centre is developed?
Q13. Write the role of resistin in steroidogenesis.
Q14. How many days generally a follicle will take to mature before ovulation in fowl?
Q15. Write the difference between diestrous and diestrus.
Q16. Which reflex cause the occurrence of standing estrus behaviour?
Q17. Why does follicular wave not occur during pregnancy?
Q18. Anestrus persisted without occurring estrous cycle is termed as _______.
Q19. Write a few major physiological reproductive adaptations in females of small animals.
Q20. Write the name of the specific protein that causes the Bruce effect.

Subjective Questions

Q1. Write the major differences between spermatogenesis and oogenesis.
Q2. Write the factors to control oogenesis and folliculogenesis.
Q3. Write the interrelationship between gonadotropins and ovarian steroid hormones.
Q4. Write the mechanism of ovulation.
Q5. Describe the events in the follicular wave.
Q6. Write the various steps of luteolysis in a cow.
Q7. Why granulosa cells can produce estrogens, but theca cells can't?
Q8. Write the role of hormones and growth factors in female reproduction.
Q9. Write the various factors affecting puberty.
Q10. Write in brief the role of gonadotropins and ovarian steroids in estrous cycle of an ewe.

Answer to Objective Questions

A1. Secondary oocyte and first polar body
A2. Granulosa cells are responsive to FSH and theca cells to LH
A3. Small antral follicles
A4. LH surge is mating-induced
A5. Right ovary
A6. By reducing the responsiveness of the receptors for activin
A7. The unsuccessful antral follicles become dominant follicles, followed by regression
A8. LH (Surge)
A9. Transport cholesterol to the inner mitochondrial membrane from the outside
A10. By mobilising the polymorphonuclear leukocytes or neutrophils
A11. 21
A12. Puberty
A13. Decrease steroidogenesis
A14. 10 days
A15. Diestrous means occurring of two estrous cycles in a year, and diestrus is a phase of estrous cycle
A16. Lordosis reflex
A17. Due to the lack of FSH
A18. Primary persistent anestrus
A19. Earlier puberty, more litter size, short gestation, very less duration of lactation anestrus
A20. Major histocompatibility complex (MHC) class 1 protein

Keywords for the Answer to Subjective Questions

A1. Oogonium and spermatogonium, duration, stage of gamete production
A2. Gonadotropins, sex steroid hormones, several proteins
A3. Relationship between FSH and LH with estrogens and progesterone
A4. LH surge, role of proteolytic substances, production of local paracrine effectors

A5. Changes of follicles up to dominant follicle from the ovarian pool, atresia, hormonal changes
A6. Role of PGF2α, oxytocin, various tissue degenerating substances, and cytokines
A7. Two cells two gonadotropins, presence of specific enzymes in both the cells
A8. Role of primary, secondary, and tertiary hormones
A9. Breed, age and body weight, nutrition, endocrines, and growth factors and environment
A10. Role of FSH, LH, estrogen, and progesterone in estrus, metestrus, diestrus, and proestrus

Further Reading

Textbook

Conti M, Chang RJ (2016) Endocrinology: adult and pediatric, vol II, 7th edn. Elsevier, Philadelphia, PA, pp 1–77. ISBN: 978-0-323-18907-1.

Das PK, Roy B (2016) Eggopedia. Parul Prakashani Pvt Ltd., pp 1–150. ISBN: 9789385555961.

Gougeon A (2004) Chapter 2: Dynamics of human follicular growth: morphologic, dynamic, and functional aspects. In: The ovary, 2nd edn. Academic Press, pp 25–43. https://doi.org/10.1016/B978-012444562-8/50003-3

Mucignat-Caretta C (2014) Chapter 16: Cattle pheromones, Table 16-1: Cattle pheromones: identification, source, and functions. In: Neurobiology of chemical communication. CRC Press/Taylor & Francis, Boca Raton, FL

Research Articles

Oogenosis and Folliculogensis

Alwan AF, Al-Saffar HE, Khammas DJ (2005) Biometry of genital organs in Iraqi female buffalo. Iraqi J Vet Sci 19(1):77–81. https://doi.org/10.33899/ijvs.2005.37421

Bächler M, Menshykau D, ChDe G et al (2014) Species-specific differences in follicular antral sizes result from diffusion-based limitations on the thickness of the granulosa cell layer. Mol Hum Reprod 20(3):208–221. https://doi.org/10.1093/molehr/gat078

Brady K, Liu HC, Hicks JA et al (2020) Transcriptome analysis of the hypothalamus and pituitary of turkey hens with low and high egg production. BMC Genomics 21(1):647. https://doi.org/10.1186/s12864-020-07075-y

Griffin J, Emery BR, Huang I et al (2006) Comparative analysis of follicle morphology and oocyte diameter in four mammalian species (mouse, hamster, pig, and human). J Exp Clin Assist Reprod 3(2). https://doi.org/10.1186/1743-1050-3-2

Haque Z, Haque A, Quasem MA (2016) Morphologic and morphometric analysis of the ovary of black Bengal goat (*Capra hircus*). Int J Morphol 34(1):13–16

Hoque MSA, Islam MM, Selim MAS (2016) Interspecies Differences on ovarian parameters between black Bengal goat and indigenous Bengal sheep in view of *In vitro* maturation. Adv Life Sci 6(3):54–60. https://doi.org/10.5923/j.als.20160603.02

Monniaux D, Clément F, Dalbiès-Tran R et al (2014) Theovarian reserve of primordial follicles and the dynamic reserve of antral growing follicles: what is the link? Biol Reprod 90(4):85, 1–11. https://doi.org/10.1095/biolreprod.113.117077

Ptak G, Matsukawa K, Palmieri C et al (2006) Developmental and functional evidence of nuclear immaturity in prepubertal oocytes. Hum Reprod 21(9):2228–2237. https://doi.org/10.1093/humrep/del184

Reynaud K, Gicquel C, Thoumire S et al (2009) Folliculogenesis and morphometry of oocyte and follicle growth in the feline ovary. Reprod Domest Anim 44(2):174–179. https://doi.org/10.1111/j.1439-0531.2007.01012.x

Sasan JS, Uppal V, Bansal N et al (2016) Histological exploration of Graafian and atretic follicles of buffalo ovary: a seasonal study. Buffalo Bull 35(1):135–146

Songsasen N, Fickes A, Pukazhenthi BS et al (2009) Follicular morphology, oocyte diameter and localisation of fibroblast growth factors in the domestic dog ovary. Reprod Domest Anim 44(Suppl 2):65–70. https://doi.org/10.1111/j.1439-0531.2009.01424.x

Corpus Luteum, Puberty—Maturity, Steriodogenesis and Estrous Cycle

Calejja-Agius J, Muttukrishna S, Jauniaux E (2009) Role of TNF-α in human female reproduction. Expert Rev Endocrinol Metab 4(3):273–282. https://doi.org/10.1586/eem.09.4

Casarini L, Crépieux P (2019) Molecular mechanisms of action of FSH. Front Endocrinol 10:305. https://doi.org/10.3389/fendo.2019.00305

Dobrzyn K, Smolinska N, Kiezun M et al (2018) Adiponectin: a new regulator of female reproductive system. Int J Endocrinol 2018: 7965071. https://doi.org/10.1155/2018/7965071

Hansen PJ (2019) Reproductive physiology of the heat-stressed dairy cow: implications for fertility and assisted reproduction. Anim Reprod 16(3):497–507. Epub 28 Nov 2019. https://doi.org/10.21451/1984-3143-ar2019-0053

Hu KL, Chang HM, Li R et al (2019) Regulation of LH secretion by RFRP-3—from the hypothalamus to the pituitary. Front Endocrinol 52:12–21. https://doi.org/10.1016/j.yfrne.2018.03.005

Jones ASK, Shikanov A (2019) Follicle development as an orchestrated signaling network in a 3D organoid. J Biol Eng 13:2. https://doi.org/10.1186/s13036-018-0134-3

Khan A, Khan MZ, Umer S et al (2020) Cellular and molecular adaptation of bovine granulosa cells and oocytes under heat stress. Animals (Basel) 10(1):110. https://doi.org/10.3390/ani10010110

Kowalewski MP (2012) Endocrine and molecular control of luteal and placental function in dogs: a review. Reprod Domest Anim 47 (Suppl. 6):19–24. https://doi.org/10.1111/rda.12036

Kowalewski MP (2014) Luteal regression vs. prepartum luteolysis: regulatory mechanisms governing canine corpus luteum function. Reprod Biol 14:89–102. https://doi.org/10.1016/j.repbio.2013.11.004

Lucy MC (2019) Stress, strain, and pregnancy outcome in postpartum cows. Anim Reprod 16(3):455–464. https://doi.org/10.21451/1984-3143-ar2019-0063

Prins JR, Gomez-Lopez N, Robertson SA (2012) Interleukin-6 in pregnancy and gestational disorders. J Reprod Immunol 95(1–2):1–14. https://doi.org/10.1016/j.jri.2012.05.004

Schuermann Y, Rovani MT, Gasperin B, Ferreira R, Ferst J, Madogwe E, Gonçalves PB, Bordignon V, Duggavathi R (2018) ERK1/2-dependent gene expression in the bovine ovulating follicle. Sci Rep 8:16170. https://doi.org/10.1038/s41598-018-34015-4

Tsatsanis C, Dermitzaki E, Avgoustinaki P et al (2015) The impact of adipose tissue-derived factors on the hypothalamic-pituitary-gonadal (HPG) axis. Hormones 14(4):549–562. https://doi.org/10.14310/horm.2002.1649

Wani A, Dhindsa SS, Tawheed AS, Chowdhary SRA, Balwinder K (2013) The role of pheromones in animal reproduction – a review. Progress Res 8(1):14–18

Zubeldia-Brenner L, Roselli CE, Recabarren SE et al (2016) Developmental and functional effects of steroid hormones on the neuroendocrine axis and spinal cord. J Neuroendocrinol 28(7):10. https://doi.org/10.1111/jne.12401

Fertilisation, Gestation and Parturition 23

Pradip Kumar Das, Joydip Mukherjee, and Dipak Banerjee

Abstract

The fusion of the cytoplasm and pronuclei of the male and female gametes along with the intermingling of the chromosomes of maternal and paternal origin to form a diploid zygote is known as fertilisation. The fusion of male and female pronuclei is achieved through a series of events like gamete transportation, the interaction of male and female gametes through receptor–ligand interactions and zygotic genome activation. The egg cortical reaction facilitates the block of polyspermy. The zygote undergoes a series of mitotic cell divisions (cleavage) to form 2-cell, 4-cell, 8-cell, 16-cell, morula and blastocyst. Morula proceeds towards the uterus and gradually transforms into blastocyst after developing a cavity called blastocoele. The blastocyst then attaches to the endometrial surface through a process called implantation. The intrauterine period of embryonic and foetal development in viviparous mammals is called the gestation or pregnancy period. The physiological exchange between mother and foetus is mediated through a transient structure called the placenta. The structure and type of placenta are species specific. After completing intrauterine life, the foetuses and foetal membranes are expelled from the uterus to the external environment through a biological process called parturition. Mother regains its reproductive cycles after a considerable period from the parturition termed the postpartum interval. The uterus is transformed into a virgin state by uterine involution. All these processes are regulated by a synchronised endocrine orchestra together with biochemical and molecular mechanisms. Any disturbance in the process above may lead to fertility-related anomalies.

P. K. Das (✉) · J. Mukherjee · D. Banerjee
Department of Veterinary Physiology, West Bengal University of Animal & Fishery Sciences, Kolkata, West Bengal, India

© The Author(s), under exclusive license to Springer Nature Singapore Pte Ltd. 2023
P. K. Das et al. (eds.), *Textbook of Veterinary Physiology*, https://doi.org/10.1007/978-981-19-9410-4_23

Graphical Abstract

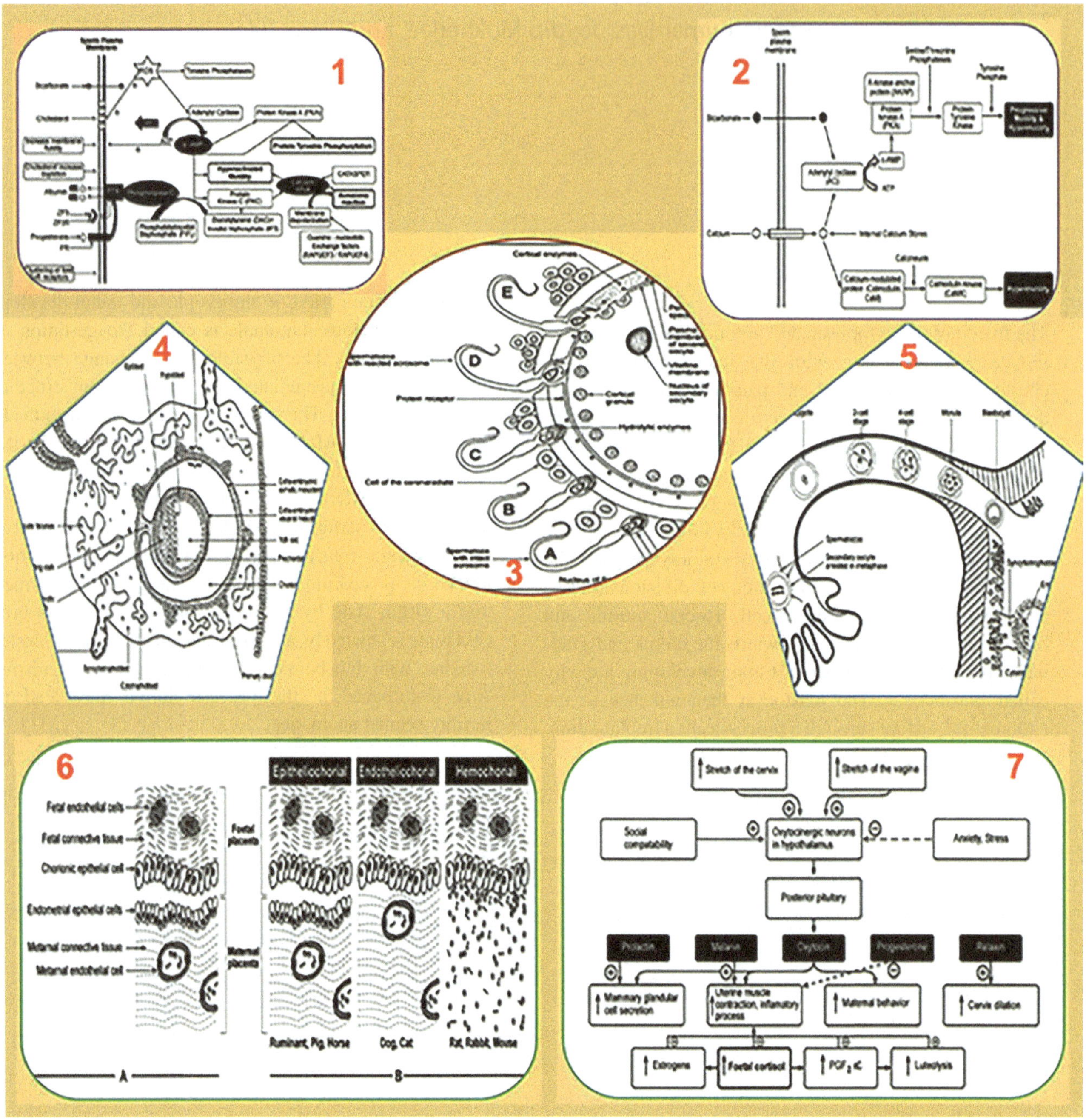

Description of the graphic: Fertilisation occurs through a series of biochemical and biophysical alterations in the spermatozoa, viz. capacitation (**1**) and hyperactivation (**2**) and the ovum (**3**). Fusing male and female gametes develops a single-cell zygote that undergoes cell divisions, differentiation and gastrulation (**4**). It consequently anchors with the endometrium through implantation (**5**). The maternal and foetal tissues coalesce from the placenta (**6**) to support the growth and development of the foetus(es). The foetus(es) expel out through a complex physiological phenomenon (**7**) called parturition

Keywords

Fertilisation · Implantation and gestation · Placenta · Parturition · Fertility-associated anomalies

Learning Objectives

- The mechanisms of transportation of sperm and ovum to the site of fertilisation.
- The species-specific biochemical and molecular mechanisms occur in sperm and ovum before and after fertilisation.
- Process of implantation and formation of the placenta.
- Gestation and postpartum physiology.
- Role of different factors in controlling gestation.

23.1 Fertilisation

Fusing the pronuclei of the male and female gametes and intermingling the chromosomes of maternal and paternal origin to form a diploid zygote is known as fertilisation. In mammals, the fertilisation occurs inside the body of females and is hence called internal fertilisation. The introduction of sperm into a female's reproductive system is called insemination, which can be done naturally and artificially. In natural mating, the semen is introduced into the vagina or close to the cervix, whereas during artificial insemination, semen is mostly deposited into the cervix or uterus. Soon after the semen deposition, the sperm has to travel through the cervix towards the uterus to reach the fallopian tubes, where it interacts with the ovum. In most mammals, the fertilisation usually occurs in the oviduct's ampullary-isthmus junction. The period between insemination and sperm–ovum interaction is crucial for fertilisation as both sperm and ovum have a fixed fertilisable life span within which they have to interact (Table 23.1). The sperm has to reside in the female reproductive tract for a considerable period to achieve maximum fertility, during which the spermatozoa undergo some changes to become fertile. This process is called capacitation and the time required for capacitation varies between species (Table 23.1). The fertilisation process is a series of species-specific biochemical and molecular changes that include receptor–ligand interactions, activation of signalling cascades and nuclear transformations.

23.1.1 Sperm Transportation

The movement of sperm in the female genital tract from the site of semen deposition to the site of fertilisation (ampulla) is called sperm transportation. The transport of spermatozoa in the female reproductive tract occurs in three phases: initial rapid transport, colonisation and sustained transport. During natural copulation, the seminal fluid is usually deposited in the anterior vagina. Then the spermatozoa are colonised inside the crypts of the cervix and released in two modes. In the first mode, the spermatozoa exhibit initial rapid transport and reach the distal end within 15–30 min in most species (Table 23.1). Still, most of these spermatozoa cannot fertilise the ovum as the capacitation is not completed. The initial rapid transport is facilitated by the contractility of the female reproductive tract rather than sperm motility. In the

Table 23.1 Preparatory time for the gametes for fertilisation in female genitalia

Species	Site of insemination[a]	Time for initial rapid transport (min)	Time for sperm capacitation (h)	Time for ova migration/ maturation (h)
Cattle	Cervix (vagina)	2–15	2–4 (6 max)	8–10 (72 max)
Buffalo	Cervix (vagina)	5–15	4–8	>4
Sheep	Cervix (vagina)	6–10	1–2	10–25 (66 max)
Goat	Cervix (vagina)	10–15	4	98 (max)
Horse	Uterus (cervix or uterus)	15–30	3–24	30–36[b] (144 max)
Pig	Uterus (cervix or, uterus)	15–30	2–6	12–48
Dog	Uterus (uterus)	Several minutes	2–4	4–7 (days)[c]
D. cat	Vagina or uterus (vagina or cervix)	20	3–4	18–36 (144–168 max)
G. pig	Peritoneum (uterus)	1–30	16–18	5–10 min (48 max)
Rat	Uterine horn (uterus)	15–30	1–4	88 (max)
Rabbit	Vagina (vagina)	Few minutes	5–11	55 (max)
Mice	Vagina (vagina)	15	2	72 (max)
Human	Vagina (vagina)	5–30	5–7	24 (80 max)

D. cat Domestic cat, *G. pig* Guinea pig, *max* maximum

[a] Insemination considered as artificial insemination; and the site of natural insemination mentioned in the bracket

[b] In in vitro

[c] Longer duration is required due to completion of meiosis-I and II (up to metaphase-II)

second mode of transportation, the maximum spermatozoa undergo slow transport by swimming through the cervical mucus. Sperm transport depends upon the structure and activity of cervical, uterine and fallopian tube epithelial cells and the contractile activity of smooth muscles of these organs. Ovarian hormones mainly control the epithelium's secretory activity, favouring sperm transport, whereas progesterone suppresses it. The contraction of the uterus and fallopian tube is controlled by oxytocin released during intercourse. The contractility of the uterus and fallopian tubes is also affected by prostaglandins.

23.1.1.1 Site of Semen Deposition

The site of semen deposition varies considerably among species and the mode of insemination (Table 23.1). The semen is deposited in the anterior vagina during natural mating in most domestic animals. Some species, such as pigs and horses, deposit the semen directly into the uterus. The murine rodents deposit semen directly into the uterus, but some remain in the vagina and coagulate to form a copulatory plug promoting sperm transport. Coagulation also occurs in human semen, but it forms a loose gel, unlike murine rodents. The gel's structural proteins comprise semenogelin I and semenogelin II, and their glycosylated form is secreted from seminal vesicles. The gel is degraded by prostate-specific antigen (PSA). The intention of forming the gel is to hold the spermatozoa at the cervical os and to give protection from the vaginal environment. The semen of some primates also coagulate to form a soft gel. Female chimpanzee mates with more than one male, and a compact gel produced from the semen prevents further mating. The copulatory tie forms during the mating in some carnivores where the penis serves as a copulatory plug for sustained insemination.

23.1.1.2 Transportation Through Vagina

The spermatozoa have to overcome the immune defence machinery of the vagina that acts as a natural barrier against external infections. These include acidic pH (pH 4) of the vagina and other immunological responses. The acidic pH of the vaginal secretion is due to the lactic acid produced by the action of anaerobic lactobacilli over the vaginal epithelium from glycogen. The acidic vaginal pH is detrimental to spermatozoa. The alkaline seminal plasma neutralises the acidic vaginal environment. The secretions of seminal vesicles, prostate gland and bulbourethral glands are alkaline (details in accessory sex gland section), which neutralises the vaginal acidic medium (pH 7.2) within a few minutes of post insemination and maintains this pH for up to 2 h. The spermatozoa become more motile in an alkaline medium. The immunosuppressive agents present in the seminal plasma protect the spermatozoa from the immune protective components of vaginal secretions. The immunosuppressive agents of the seminal plasma include prostasomes, prostaglandin E, polyamines, immunoregulatory cytokines and lymphocyte suppressing proteins. Immediately after the insemination, the semen coagulates in the acidic medium to form a clot. The proteolytic substances of the seminal plasma dissolve the clots and allow the spermatozoa to move within 30 min. Spermatozoa that cannot move within 2 h remain in the vagina even up to 12 h, depending on species, but almost lose their motility. At first, a rapid sperm transport is occurred within seconds after ejaculation, followed by slow transport. The majority of spermatozoa move through a slow transport process.

23.1.1.3 Transportation Through Cervix

The spermatozoa reach the internal os and cervix within 1.5–3 min of insemination by sperm motility and contractility of the myometrium. In most domestic animals, the spermatozoa are trapped within the mucosal folds of the cervical crypts that act as sperm reservoirs. The spermatozoa are released slowly by the female reproductive tract's motility and contractile activity. The cervix provides a favourable environment to spermatozoa and protects them from phagocytosis. Cervix also aids the sperm energy metabolism and helps the capacitation process. Cervix also helps in sperm selection and restricts the entry of non-motile and abnormal spermatozoa. The cervix allows the sperm migration nearer to ovulation time and blocks the sperm entry during other phases of the reproductive cycle by forming a cervical plug.

The cervical mucous acts as a barrier to sperm transport. The cervical mucous comprises flexible linear glycoprotein molecules named mucins. The long mucin molecules are aligned themselves to form a hydrogel of the 3D network. The mucin molecules are glycosylated during the oestrus under the influence of oestrogen. The glycosylated mucins have a water holding capacity, and the mucous become highly hydrated during oestrus. The cervical plug liquefies during oestrus under oestrogen and highly hydrated cervical mucous, and the cervix appears to widen to allow the sperm to move through the cervix. The spermatozoa penetrate the cervical mucous through their motility and rheological properties. The spermatozoa swim in a straighter path guided by the secretory flow of cervical mucous with the orientation along the long axis of mucin threads. The non-motile spermatozoa are unable to penetrate the cervical mucous. Thus, cervix helps to select the non-motile spermatozoa. The penetration of cervical mucous by the spermatozoa is facilitated by a glycoprotein called beta-defensin 126 (DEFB126) identified in primates. The DEFB126 provides a highly negative surface charge to the spermatozoa essential for cervical mucous penetration. The sperm has to escape the immune responses of the cervix as the insemination stimulates the migration of neutrophils and macrophages into the cervix. The normal and motile spermatozoa can avoid phagocytosis, but neutrophils phagocytose the non-motile and abnormal spermatozoa due to complement-fixing anti-sperm antibodies. The neutrophils interact with the spermatozoa through L-selectin that binds with the sialic acid on the sperm surface. Human sperm moves at a speed of 1.5–5 mm/min in the female genital tract and reach the uterus within an hour. The pig, horse and dog deposit the semen directly to the uterus or end of the cervix; hence, the spermatozoa cross the cervix or uterus rapidly.

23.1.1.4 Transportation Through Uterus

Contractile activity of the uterine smooth muscles along the length of the uterus propels the spermatozoa and watery cervical mucus from the cervix into the uterus. Strong contractile activity of myometrium is seen during oestrus compared to the luteal phase in cows and ewes. The contraction of uterine smooth muscle is stimulated by oestrogen that stimulates the secretion of PGF2α. The oestrogen from the pre-ovulatory follicles reaches the endometrium through a counter-current exchange mechanism between the ovarian vein and ovarian artery, then transported through the uterine artery to act over the endometrium. In boars, seminal plasma contains a considerable amount of oestrogens directly deposited into the uterine cavity and aids the uterine contraction. In certain animals, like cow and rabbit, mating or copulation induce uterine contraction probably by releasing oxytocin. Stimulation of vulva and per-rectum uterine messages during artificial insemination may have a similar effect. Fright or fear may inhibit sperm transportation due to the release of epinephrine causes vasoconstriction.

23.1.1.5 Transportation Through Utero-Tubal Junction (UTJ)

The UTJ acts as an anatomical and physiological barrier to sperm from the uterus to the oviduct. This UTJ barrier resembles a filter that allows only healthy motile sperm to the isthmus. The abnormal and poor-quality sperm are arrested here. The structure of UTJ is simple in humans, but in cows, pigs, rabbits and many other species, the structure is complicated due to numerous mucosal folds. Cul-de-sacs characterise the bovine UTJ originated from the mucosal folds that entrap the spermatozoa. The lumen of UTJ is squeezed during the oestrus by the contraction of thick, smooth muscle. The presence of viscous mucus in the lumen of UTJ further restricts the passage of sperm. The spermatozoa penetrate the utero-tubal junction by their linear progressive motility. After crossing the UTJ, the viable spermatozoa reside in the isthmus, attaching to its wall for capacitation. Different proteins, namely fertilin β, calmegin or testis-specific angiotensin-converting enzyme (ACE), help sperm migration through UTJ. Infertility may result due to deficiency of these proteins.

23.1.1.6 Sperm Transport in the Oviduct

From the millions of ejaculated spermatozoa, nearly 10,000 or fewer spermatozoa enter the fallopian tube. The spermatozoa can enter any side of the tubes, but those that enter ipsilaterally to the ovulation side are capable of fertilisation. The oviduct provides a favourable environment for the spermatozoa as the oviduct does not induce any immune reactions, unlike the vagina, cervix and uterus. The isthmus is considered a sperm reservoir in cattle, sheep, pigs, rabbits, mice and other species. In addition to the cervix and UTJ, the oviduct itself acts as a barrier in sperm selection. Only viable spermatozoa bind with the wall of the fallopian tube epithelium till the fertilisation and rest are eliminated. The interactions of sperm with oviductal epithelial cells (OEC) are mediated through receptor–ligand interactions, and this binding enables the sperms to become viable for at least 48 h. The bovine seminal vesicle proteins like PDC109 (seminal plasma A1orA2, Binder of SPerm (BSP), viz. BSPA3 and BSP30K present in the sperm heads interact with the annexin family of proteins of OEC. The BSP also contains two heparin-binding domains. BSP's binding with heparin blocks BSP–annexin interaction and detachment of spermatozoa from the oviductal epithelium. Hence, heparin is extensively used in in vitro capacitation process. Two proteins, namely Hsp60 and GRP78 of the apical membrane of the bovine and human OEC, are capable of binding with the sperm membrane.

23.1.1.7 Sperm Loss During Transportation and Its Manipulation

The numbers of spermatozoa are gradually reduced during the transportation from the deposition site to the site of fertilisation. In sheep, its number reduces from 1000 million to only 1000 from the site of deposition to the site of fertilisation, respectively, in pigs the number varies from 8000 million to 1000, in rats 60 million to 20–100, in guinea pig 80 million to 25–50, and in humans 200 million to 10–1000. The insufficient sperm numbers at the site of fertilisation lead to fertilisation failure, particularly in sheep and pigs.

The sperms are susceptible to phagocytic attack by the neutrophils and macrophages infiltrated into the uterine lumen after coitus. This process is called spermophagy and is mostly seen in mice, rats and rabbits. The damaged sperms and seminal debris are mainly vulnerable to phagocytic attack, and seminal plasma constituents generally protect the normal sperms. But the normal sperms are deposited into the anterior vagina and are also susceptible to phagocytic attack as their immune protections are when they reach into the uterus. In dogs, endometrial glands act as sperm reservoirs. In some species with an extensive ovarian bursa (mice), the non-fertilised sperm can enter the peritoneal cavity.

The induction of oestrus by progestogen or prostaglandin F2α sometimes causes disturbances in sperm transport in the ewe. Prolonged exposure to phytoestrogen and abrupt luteolysis may disturb sperm transportation at the subsequent oestrus. All these disturbances in sperm transport are due to a lack of optimum oestrogen level. The sperm transportation can be enhanced by adding some compounds to the semen or applying them to the female reproductive tract. Majorities of these compounds are species-specific such as prostaglandin E1 in rabbits, sheep and goats, estradiol-17β in sheep, carbacholine in pigs and amylase or glucuronidase in cattle.

The spermatozoa undergo two important pre-fertilisation events for its activation at the isthmus: the capacitation and hyperactivation.

23.1.2 Capacitation

The capacitation is the physio-biochemical alterations in the sperm plasma membrane that enables them to become fertile (Table 23.2). The site and time required for capacitation vary between species (Table 23.1). In species where sperm are deposited at the anterior vagina, capacitation begins when the sperm migrates through the cervix. In species where semen is deposited in the uterus, capacitation is initiated at the uterus and completed in the isthmus of the oviduct. The process of capacitation includes activating multiple signal transduction mechanisms that lead to modification of surface molecules, alterations in the intracellular ionic concentration and activation of the enzymatic system (Fig. 23.1). The capacitation includes both membrane and cytoplasmic events. The membrane remodelling is the most pronounced event of the capacitation that leads to increased fluidity of the phospholipid bilayer and promotes acrosomal reaction. The albumin and high-density lipoproteins of female genital tract secretions remove cholesterol from the sperm plasma membrane and alter its fluidity. The modifications in the sperm plasma membrane expose the ligands that bind with the zona pellucida. The activation of membrane-bound enzymes helps the sperm to penetrate the cumulus oophorus. The

Table 23.2 Major biochemical reactions and the role of some specific bio-molecules involved in the capacitation

Bio-molecules	Effects	Probable mechanism
1. Bicarbonate	Membrane flexibility for ion transport	Influence adenylyl cyclase (AC)/cyclic adenosine 3′,5′-monophosphate (cAMP)/ protein kinase A (PKA) signalling pathway
2. Calcium	Alkaline cytoplasm and phosphorylation	Activate (hyperpolarisation) membrane voltage-dependent channels and increase cAMP
3. AC and cAMP	Second messenger and PKA lead to membrane remodelling, ATP synthesis	AC activate cAMP
4. PKA	Hyperpolarisation of the sperm plasma membrane, intracellular alkalinisation and calcium ion trigger an acrosomal reaction (AR)	Tyrosine phosphorylation, a series of biochemical reaction, activate phospholipase D (PLD)
5. PLD	Actin formation	Actin polymerisation
6–8. Adenosine, fertilisation-promoting peptide (FPP), calcitonin	G proteins stimulate capacitation	Regulates AC/cAMP (initially stimulate followed by inhibiting), acts as first messengers
9. Angiotensin II (AII)	Intracellular calcium ion	Indirectly regulates AC/cAMP
10. Albumin	Membrane flexibility for ion transport, membrane phospholipid and lipid reorganisation	Cholesterol depletion
11. Glucose	ATP synthesis	Major precursor for the glycolytic pathway
12–13. Pyruvate and lactate	Glycolysis and mitochondrial oxidative phosphorylation (OXPHOS), ATP synthesis	Regenerate cytosolic NAD^+
14. Platelet-activating factor (PAF)	On capacitation by—intracellular calcium ion followed by sperm motility and on AR and embryo implantation	Signalling phospholipid acts by activating phospholipase that converts diacylglycerol (DAG) to inositol triphosphate (IP3)
15. Progesterone	On capacitation and hyperactivation through PAF-mediated action, intracellular calcium, efflux of chloride and cholesterol and on AR—phospholipases and tyrosine phosphorylation of sperm proteins	Influences PAF, activates acrosome surface receptors
16–18. Oestrogen, genistein and 4-tert-octylphenol (OP)	Ion channels, calcium fluxes, cyclic nucleotides (cAMP), various kinases	Acts on acrosome surface receptors for oestrogen (ESR1 and ESR2) in low concentration
19–20. Seminal antioxidants (like catalase, superoxide dismutase and ergothioneine) and mitochondrial aldehyde dehydrogenase 2	Protect sperm cell degeneration during capacitation	Reduce cytoplasmic excess concentrations of reactive oxygen species (ROS), like superoxide anions and hydrogen peroxide

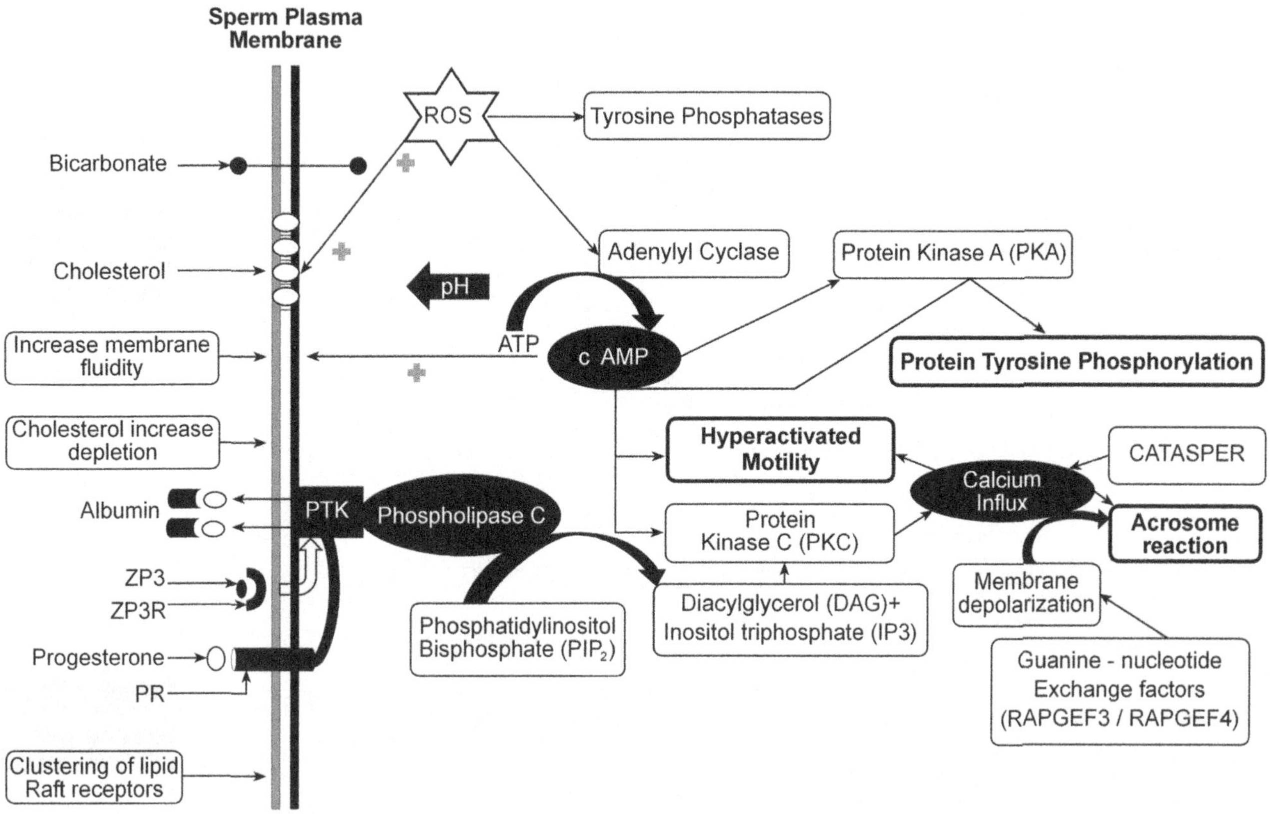

Fig. 23.1 Mechanism of capacitation and acrosomal reaction. Capacitation process is initiated with the increased plasma membrane fluidity and permeability for bicarbonate, depletion of cholesterol and clustering of lipid raft receptors. Albumin acts as a cholesterol acceptor. The PTK (= protein tyrosine kinase) is activated by the progesterone (PR = progesterone receptor) and ZP3 (= zona pellucida 3 proteins, ZP3R = ZP3 receptor). The activation (+) of cAMP-PKA-(protein) dependent tyrosine phosphorylation and ROS (= reactive oxygen species) production cause calcium influx within the spermatozoa followed by increased motility (hyperactivated motility) as well as the occurrence of the acrosomal reaction. (*Source*: Leemans et al. 2019)

alterations in the membrane protein conformation increase the permeability of the plasma membrane for calcium and bicarbonate (HCO_3^-) by activating T-type (voltage-sensitive) calcium channels in the sperm membrane. Increased HCO_3^- concentration leads to higher intracellular pH and activation of adenylyl cyclase, which in turn causes cAMP production and cAMP-dependent PKA activation. The PKA phosphorylates protein tyrosine which facilitates sperm zona binding. The γ-aminobutyric acid (GABA) acts as an inducer of both capacitator and hyperactivation in sheep, rats and humans but plays an inhibitory role in hamsters. Progesterone and oestrogen of follicular fluid act as inducers for capacitation in many other species like cattle, sheep, goats, pigs, horses, dogs, mice, golden hamsters and humans. These steroids facilitated the cholesterol and chloride ions efflux and increased intracellular calcium ions. Progesterone can also stimulate the capacitation process by inducing the follicular fluid's platelet-activating factor (PAF). The capacitation is considered an oxidative process in humans, where superoxide anion (O^-), nitric oxide (NO) and hydrogen peroxide (H_2O_2) are produced. The time requirement for capacitation to achieve sufficient spermatozoa for fertilisation is species specific (Table 23.2).

23.1.3 Hyperactivation

The hyperactivation of the spermatozoa denotes a state in which the sperm exhibit vigorous motility due to increased flagellar beating amplitude and asymmetrical beating pattern. Hyperactivation facilitates the detachment of spermatozoa from the isthmus epithelium. Hyperactivation also aids the movement of spermatozoa through the viscoelastic mucus-filled tortuous bending of the ampulla due to the flexibility of the sperm head at a wider angle. The viscoelastic substances are generally made by long-chain polyacrylamide or methylcellulose present in the tubular lumen and at the extracellular matrix of the cumulus oophorus. Hyperactivation also enables spermatozoa to penetrate the zona pellucida. The capacitated spermatozoa can bind with the zona, but only hyperactivated sperms can penetrate it. The peristaltic movement of the oviduct and the cilia-directed ductal fluid

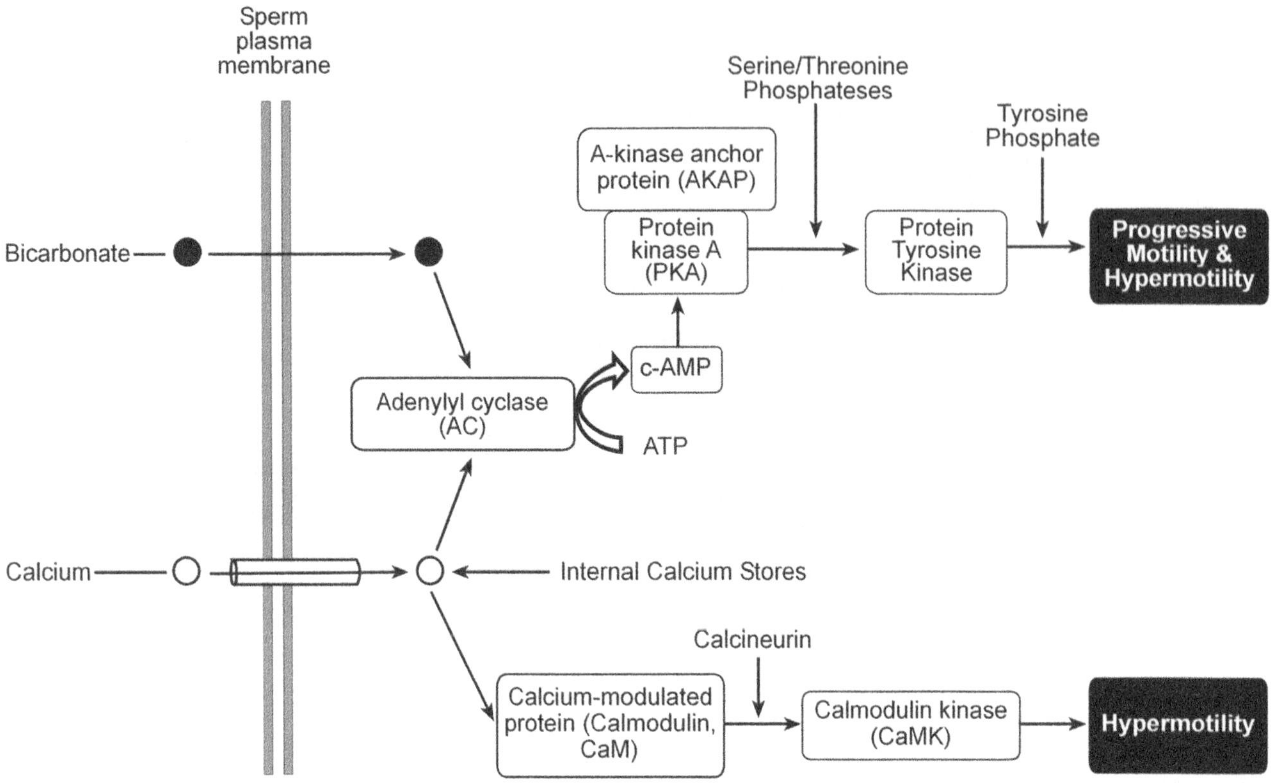

Fig. 23.2 Molecular mechanism of hyperactivation. Schematic representation of signalling pathways involved in the activation of the mammalian sperm (hyper). Progressive motility is regulated by AC/cAMP/PKA pathway. Hyper motility induces when this pathway is activated together with the calmodulin kinase (CaMK) activation. (Image modified from Turner 2006)

movement also facilitate the sperm migration through the ampulla. It has also been observed that the little (about 2 °C) warmer temperature in ampulla than isthmus favours the hyperactivation of rabbit sperm.

The hyperactivation of spermatozoa is induced by calcium either directly binds with plasma membrane phospholipids or through intracellular calcium receptor calmodulin (CaM). The Ca^{2+}-CaM complex stimulates the activity of various enzymes like adenylyl cyclases, phosphatases, phosphodiesterases and protein kinases (PK). Activation of these enzymes, particularly PKA, causes the phosphorylation of axonemal dynein and increases ATP consumption, leading to hyper motility (Fig. 23.2).

The equine spermatozoa are unique in terms of premature natural acrosome reaction. The spermatozoa lose their acrosomal integrity immediately after the semen collection and poorly bind with zona pellucida, which hinders semen collection and preservation. The use of capacitated sperm with appropriate inducers, like progesterone, calcium ionophore and heparin, improves the sperm binding and fertilisation rate.

The two processes, capacitation and hyperactivation, make the spermatozoa completely mature and prepare them ready for fertilisation process.

23.1.3.1 Process of Sperm Release from Oviductal Epithelium

The capacitation and hyperactivation enable the sperm to detach from the oviductal epithelium near ovulation. The capacitation causes the shedding of epithelium binding proteins from the sperm plasma membrane flowed by increased flagellar movement acquired through hyperactivation. Endocrine changes that trigger ovulation have a minimal role in sperm detachment in many species. The BSP proteins have heparin-binding sites. The glycosaminoglycans released during late oestrus bind with these BSP proteins and facilitate sperm detachment. Heparin-like compounds of the oviductal fluid also bind with annexin, the oviductal receptors for BSP and cause sperm detachment. The hyperactivation of the spermatozoa is characterised by asymmetrical flagellar beating with an increased amplitude which generates sufficient force to detach spermatozoa from the oviductal epithelium.

23.1.4 Migration of Ova and Completion of Oocyte Maturation

During ovulation, the oocyte is captured by the infundibulum. This process is called *ova pick-up* or *egg pick-up*. The infundibulum surrounds the ovary, which facilitates receiving the oocyte (cumulus-oocyte complex, COC) within the oviduct. The COC moves towards the ampulla by the rhythmic movements of hair-like fimbriae and kinocilia of the fallopian tube. Kinocilia is the particular type of motile cilia abundant in the fallopian tube that assists in propelling the fluid. The time required for COC to reach the site of fertilisation is species specific (Table 23.1). The state of maturity of the oocyte is not the same in all species. In most species, the haploid secondary oocyte is ovulated in the form of an ootid with a germinal vesicle (nucleus) at its arrested meiosis II stage. In dogs and horses, the meiosis I is completed after ovulation, but the cell division continues up to meiosis II, like in other animals, before fertilisation.

About 48–96 h is required in the dog after ovulation to complete meiosis I, followed by 60–108 h for meiosis II (metaphase–II). Hence, approximately 4–7 days after LH surge, the oocyte of a dog is capable of fertilisation. Meiosis II completes when the spermatozoa interact with the ovum.

23.1.5 Process of Fertilisation

The sperm and oocyte encounter requires three critical steps, namely sperm migration through cumulus cells, sperm attachment and migration through zona pellucida, gamete fusion or karyogamy or amphimixis and block of polyspermy (Fig. 23.3). An optimum environment is required for the process of fertilisation. Different ions, namely calcium, bicarbonate, sodium and magnesium, maintain the optimum pH necessary for fertilisation.

23.1.5.1 Sperm Migration Through Cumulus Cells (If Present)

Immediately after the ovulation, the ovum is surrounded by loosely packed follicle cells called cumulus oophorus in certain species like rodents. The intercellular matrix of cumulus oophorus contains a cementing substance made of mucopolysaccharide substances hyaluronic acid. Hyaluronidase, present on the outer surface of the sperm acrosome of the

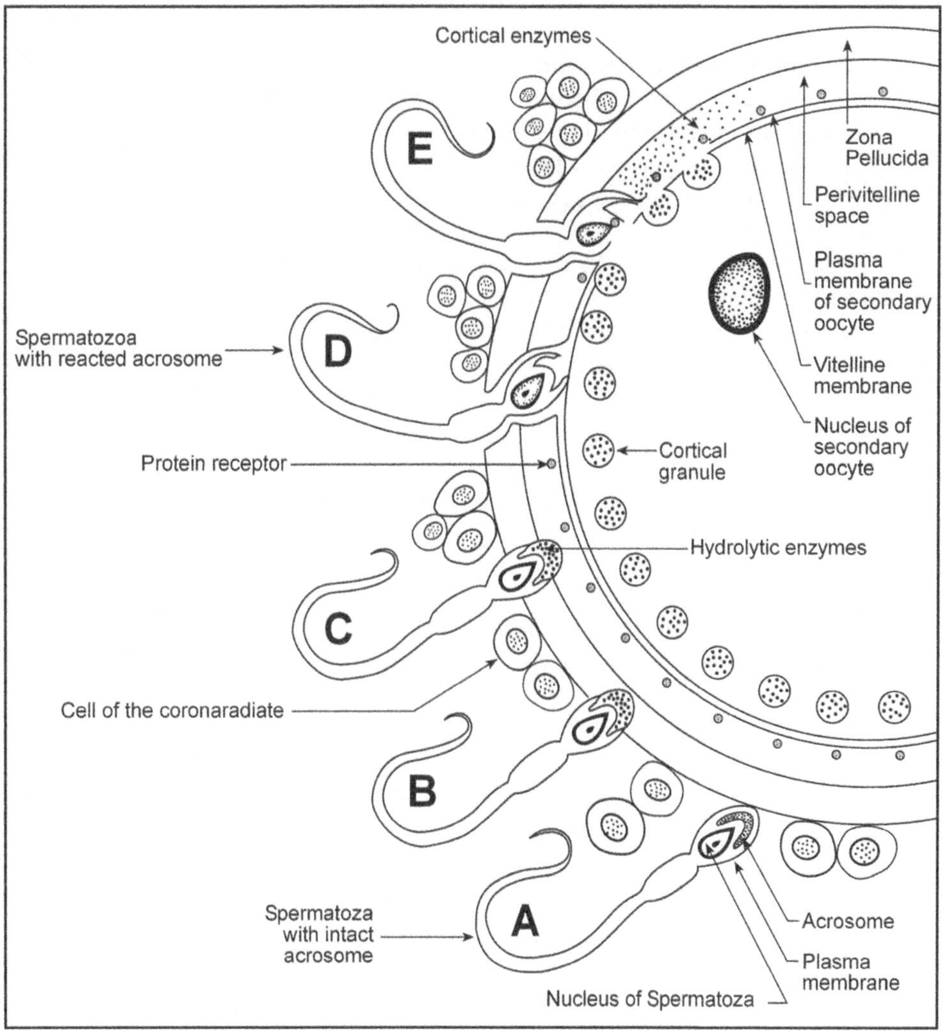

Fig. 23.3 Process of fertilisation. The events of the acrosomal reaction in the fertilisation are elaborated in steps A to E; where A = the spermatozoa come in contact with the zona pellucida of the oocyte, B = spermatozoa start a reaction with the zona pellucida, C = spermatozoa reached the perivitelline space, D = spermatozoa (with reacted acrosome) fused with the plasma membrane of the secondary oocyte and E = nucleus of the spermatozoa enters into the secondary oocyte rupturing the vitelline membrane. The occurrence of cortical reaction is depicted with the fusion of cortical granules involving its enzymes in the perivitelline space, causing impermeable further spermatozoa after the acrosomal reaction. (*Source*: Monroy 2020)

Table 23.3 Egg-binding proteins (EBPs) on sperm membrane and their complement sperm receptor on zona pellucida of oocytes

Egg-binding proteins (EBPs)	Complement sperm receptor on ZP
β-Galactosyltransferase	*N*-acetylglucosamine (GlcNAc) residues on ZP3
Sperm protein-56	ZP3 oligosaccharides
Zonadhesins	Zona proteins
Spermadhesins	Carbohydrate residues of ZP3
Zona receptor kinase	ZP3
Mannose-binding protein	mannose residues of ZP
Sperm protein-17 (Sperm-specific autoantigen)	ZP3
Sperm agglutination antigen-1	Surface antigen of human sperm

spermatozoon, helps to digest the hyaluronic acid and allows the sperm to migrate through the cumulus oophorus. The penetration of cumulus cells is of little importance in cattle as the cumulus oophorus is usually absent 3–4 h after ovulation. However, hyaluronidase is found in the bull spermatozoa. Arylsulfatase in the boar spermatozoa helps penetrate cumulus cells in this species.

23.1.5.2 Sperm Attachment and Migration Through Zona Pellucida

The oocyte's zona pellucida consists of three glycoproteins, namely zona proteins 1, 2 and 3 (ZP1, ZP2 and ZP3). Zp1 and ZP2 are the structural proteins to maintain the integrity of zona pellucida. ZP 3 acts as the sperm binding receptor. The sperm plasma membrane also contains two zona binding sites (1) the primary zona binding region, responsible for sperm binding with zona pellucida and (2) acrosomal reaction-promoting region (ARPR), which binds with ZP3 and initiates acrosomal reactions. Several species-specific egg-binding proteins (EBPs) on the sperm membrane bind with zona pellucida (Table 23.3) of the corresponding species. Hence, the sperm of any particular species cannot fuse with the oocyte of other species.

23.1.5.2.1 Acrosomal Reaction

Acrosomal reaction is the multiple fusions between the plasma membrane and the spermatozoa's outer acrosomal membrane to form vesicles to release the acrosome contents by exocytosis.

23.1.5.2.2 Signal Transduction

The binding of EBPs with the receptors of ZP3 activates multiple signalling cascades (Table 23.4). Two different types of receptors are present in the sperm plasma membrane that binds with ZP3. One is G protein-coupled receptor, and another is a tyrosine kinase (TK) receptor. These receptors have different second messenger systems to augment intracellular calcium levels. G protein-coupled receptor acts through cAMP and phospholipase C (PLC) second messenger system, whereas receptor tyrosine kinase is coupled with PLC.

Elevated calcium triggers the depolymerisation of the inter-membrane actin network and activation of phospholipases for exocytosis of the acrosomal contents. The acrosomal reaction begins with the multiple fusions between the sperm plasma membrane and outer acrosomal membrane that leads to the formation of many vesicles through vesiculation. The acrosomal enzymes are released through tiny pores created during vesiculation. These vesicles are sloughed after the acrosomal reaction, leaving the inner acrosomal membrane and equatorial segment intact. The secretory product of acrosome is called sperm lysine, which contains (1) hyaluronidase—helps to dissolve cumulus cells, (2) corona penetrating enzyme—it breaks the corona radiata of the cumulus-oocyte complex (COC) and (3) acrosin or zona lysine—it is a zymogen present within the acrosomal region and converted to acrosin that digests the zona pellucida. After penetration of cumulus, corona radiate and zona pellucida, a single spermatozoon fuses with the plasma membrane of the secondary oocyte. Progesterone also induces acrosomal reactions after binding with sperm surface receptors. Progesterone increases the pH of the sperm head cytosol and intracellular calcium level.

Assessment of the acrosomal integrity of sperm is one of the essential evaluation criteria for semen analysis and can do immediately after semen collection. The hypoosmotic swelling (HOS) test is routinely used to assess acrosomal integrity. Monoclonal antibodies and indirect

Table 23.4 Signal transductions to initiate the acrosomal reaction

Receptor	Second messenger system	Signal transduction
G protein-coupled receptor	cAMP	Activation of adenylyl cyclase (AC) leads to cAMP production and protein kinase A (PKA) activation. The PKA activates a voltage-gated Ca^{2+} channel at the outer acrosomal membrane to allow the entry of Ca^{2+} from the acrosome to the cytosol.
G protein-coupled receptor	Phospholipase C (PLC)	PLC hydrolyses phosphatidyl-inositol bisphosphate (PIP2) to diacylglycerol (DAG) and inositol-trisphosphate (IP3). DAG helps in the translocation of protein kinase C (PKC) at the plasma membrane. PKC stimulates voltage-gated Ca^{2+} channel (L) in the plasma membrane, leading to more intracellular calcium.
Receptor tyrosine kinase	Phospholipase C (PLC)	

Table 23.5 Sperm proteins and their receptors on oocyte surface to facilitate gamete fusion

Sperm proteins	Receptors on oocyte surface
Fertilin α (ADAM1)	CD9
Fertilin β (ADAM2)	α6β1 integrin, α9β1 integrin
Cyritestin (ADAM2)	CD9
CD46	β1 integrin
Izumo	Juno

immunolabeling techniques are also used to evaluate the same.

23.1.5.3 Gamete Fusion

Once the sperm has traversed through zona pellucida, the head moves into vitelline space to interact with the vitelline membrane. The penetration of the vitelline membrane activates the ovum, and the resumption of meiosis occurs. The female pronucleus is formed after the completion of meiosis. The equatorial region of the sperm is incorporated into the plasma membrane of the ovum. Several molecules have been identified in sperm and ovum that facilitate gamete fusion (Table 23.5).

Other molecules responsible for gamete fusion are spermosin, HYAL5T, angiotensin-converting enzyme 3 (ACE3), trypsin-like acrosin and SPAM1. The beating of sperm tails stops immediately after the sperm-egg fusion. The fusion results in actin polymerisation and the extension of microvilli. The cytoplasm of the oocyte starts swelling and forms a fertilisation cone. Then the sperm is drawn by the microvilli of the egg, and the sperm nucleus, together with other organelles, is incorporated into the cytoplasm of the oocyte. All the cellular organelle of the spermatozoon, except periacrosomal materials, are engulfed by the fertilisation cone. The periacrosomal materials are infused into the oocyte cytoplasm, which favours oocyte activation for resumption of meiosis II. The nucleus and acrosomal tubules containing centrioles and mitochondria of the mid-piece of spermatozoon remain in the cone. Only the nucleus and centrioles are involved in the fertilisation process, and other structures do not have any role. The acrosomal tubule dissolves, and the centriole divides into two halves which form the mitotic spindle. The mammalian oocyte does not have any centriole. The nuclear envelope of sperm is disintegrated, followed by chromatin decondensation.

The factor responsible for chromatin decondensation is called the male pronucleus growth factor. A new nuclear envelope of sperm develops within the oocyte cytoplasm, forming a male pronucleus. The male and female pronuclei migrate to the centre of the ovum. Then nuclear membranes of both male and female pronuclei disperse, and chromosomal intermixing occurs. This process is called karyogamy or amphimixis. It occurs within 24 h of ovulation in cattle. A diploid zygote is formed due to the karyogamy process. The formation of zygotes denotes the end of fertilisation. The chromosomes aggregate in the prophase of first cleavage division, leading to zygote formation and the restoration of the diploid state. The fusion of male and female gamete is called syngamy.

In polytocous animals, more than one oocyte is ovulated and simultaneously fertilised. Sometimes, two embryos can develop in monotocous animals due to specific fertilisation abnormalities (discussed in detail in the fertility-related abnormalities section). In the case of polyspermy (more than one sperm involved in fertilisation), four cells are developed with improper numbers and types of chromosomes. These cells may die or undergo abnormal development.

23.1.5.4 Block to Polyspermy/Egg Cortical Reaction

Polyspermy is the process of fertilisation of an oocyte by more than one spermatozoon. Polyspermy usually leads to embryo death. Immediately after the entries of spermatozoa through the zona pellucida, the surface of the ovum changes continuously to prevent further sperm binding, called block to polyspermy. The process is mediated by egg cortical reaction, where the exocytosis of the cortical granules (CG) present below the oolemma causes zona block. The block of polyspermy usually occurs at the zona pellucida in most species. But, in rabbits, a secondary block at the vitelline membrane also occurs. The block of polyspermy occurs in two steps fast and a slow block.

23.1.5.4.1 Fast Block

Within a second after the fertilisation, the membrane potential of the oolemma changes to the depolarised state by a massive influx of Na^+ ions. It is called the fast block to polyspermy as the sperm cannot penetrate a membrane where the potential is more than −70 mV. The fast block of polyspermy is intended to prevent sperm attachment to the oocytes. The oolemma undergoes rapid repolarisation within a minute through K^+ leakage.

23.1.5.4.2 Slow Block/Cortical Reaction

In this process, the cortical granules release their content to modify the extracellular matrix of the zona pellucida. It acts as a permanent barrier to prevent further entry of spermatozoa. In cortical reaction, secretory vesicles are fused with the oolemma and release their contents (Fig. 23.3). Two classes of proteins, namely soluble NSF-attachment protein receptors (SNAREs), are involved in the translocation of cortical granules and membrane fusion. SNAREs have two components, vesicular (v) and target membrane (t), found in cortical granules and oolemma. Vesicle-associated membrane protein (VAMP) and

Table 23.6 Cortical granular contents and their role in block to polyspermy

Cortical granule contents		Role
Proteinases	Tissue-type plasminogen activator (tPA)	Zona hardening and blocks sperm penetration by converting plasminogen to plasmin
	ZP2 proteinase	ZP2 (120 kDa) is converted to ZP2f (90 kDa) by proteolysis. ZP2f causes zona hardening.
Ovoperoxidase		Catalyses the tyrosines cross-linking in the zona resulting zona pellucida hardening
Calreticulin		Blocks the carbohydrate moieties of glycoproteins required for sperm–oocyte interaction
N-Acetylglucosaminidase		Cleaves the terminal *N*-acetylglucosamine residues of zona protein to prevent sperm binding
p32		Prevents sperm binding
Peptidyl arginine deiminase (PAD/ABL2antigen/p75)		Forms an extracellular matrix in the perivitelline space called cortical granule envelope
Glycosaminoglycans		Attracts water into the perivitelline space and allows it to expand and form the hyaline layer

synaptotagmin are v-SNAREs situated at the vesicle membranes. Syntaxin and synaptosome-associated protein of 25 kDa (SNAP-25) is t-SNARE found in oolemma. The interaction between v and t SNAREs results in membrane fusion. The cortical granules contain several proteins and enzymes that alter the zona pellucida and vitelline membrane to facilitate block to polyspermy (Table 23.6).

23.1.5.4.3 Molecular Mechanism of Egg Cortical Reaction

Cortical granules' exocytosis involves a calcium-dependent pathway. The binding of sperm with ZP3 initiates G protein-coupled receptor signalling and activation of PLC, which cleaves phosphatidylinositol 4,5-biphosphate (PIP2) into inositol 1,4,5-triphosphate (IP3) and diacylglycerol (DAG). PIP2 and DAG act as second messengers. IP3 binds to its endoplasmic reticulum receptors and facilitates the release of Ca^{2+} from the endoplasmic reticulum. The Ca^{2+} binds its receptors at the endoplasmic reticulum around the cortical granules and results in more Ca^{2+} releases (Ca^{2+}-induced Ca^{2+} release, CICR process), and a wave-like calcium spreading occurs to rupture the granules. DAG induces PKC activation, which phosphorylates ion-exchange proteins for Na^+ to H^+ and increases Na^+ and H^+ output. The pH of the ovum rises from 6.8 to 7.3, favouring the awakening of oocytes from metabolic inertia. The CG exocytosis process involves two important regulatory proteins, calmodulin (calcium-binding) and gelsolin (involved in restructuring cortical F-actin). Due to insufficient CG, polyspermy occurs in particular mammals, like pigs and marsupials. IP3-mediated granular exocytosis is also absent in these species.

23.1.5.5 Activation of Zygote and Initiation of Mitosis

The zygote becomes activated immediately after fertilisation. A zygote is totipotent and can form all cells, including an extraembryonic membrane. The fertilised ovum undergoes a series of events, like degradation of maternal products (RNA and protein), post-translational regulation and epigenetic reprogramming. All these events are collectively called maternal-to-zygotic transition. Ca^{2+} plays a central role in activating the oocyte from meiotic arrest and triggers the embryonic development programme. High intracellular Ca^{2+} inside the oocyte coincides with the fertilisation through Ca^{2+}-induced Ca^{2+} release (CICR process). Mature oocytes are in M II block due to the action of the M-phase-promoting factor (MPF) that forms a complex with cyclin B and cyclin-dependent kinase p34cdc2. High MPF activity leads to stabilisation of the meiotic spindle and chromatin condensation. The rise of Ca^{2+} leads to proteolysis of cyclin B and inactivation of MPF to resume meiosis. The amplitude and duration of Ca^{2+} spikes required for oocyte activation are species specific. In mice, Ca^{2+} spike occurs at an interval of 10 min. In humans, pigs and cows, Ca^{2+} spike occurs every 30–60 min.

23.1.5.5.1 Zygotic Genome Activation (ZGA)

The initiation of gene expression after fertilisation is called zygotic genome activation (ZGA). Before the activation of the oocytes, the transcription arrests as the transcription factors are unable to bind with their motifs due to chromatin condensation. Several *pioneer factors* facilitate the binding of transcription factors with their motifs to induce ZGA. The first identified essential pioneer factor in influencing ZGA is Zelda. It generates the transcription of hundreds of genes by histone acetylation and nucleosome remodelling that facilitates the binding of other transcription factors. Other pioneer factors involved in ZGA include Nanog, Oct4 and SoxB1. Nanog and SoxB1 are involved in nucleosome destabilisation, and Oct4 is the key factor in inducing pluripotency (the ability of the individual cell to form all tissue lineages).

23.1.6 Fertility-Associated Proteins

Several species-specific proteins are involved in different aspects of acrosomal reaction and gamete fusion. They are called fertility associated proteins and are used as a biomarker to assess fertility in animals. The sperm acrosome-associated 1 (SPACA1), a tyrosine-phosphorylated protein, is required for acrosomal reaction in bull, boar and mice. Another acrosomal protein, Izumo sperm-egg fusion 1 (IZUMO1), involves an acrosomal reaction and gamete fusion in bull and mouse. The sperm nuclear protein protamine 1 in bull and protamine 2 in primates and rodents are essential in protein synthesis during early embryogenesis. Some sperm transmembrane proteins like the adenylyl cyclase 10 (ADCY10, a bicarbonate sensor), osteopontin (Ca ion-binder) and Na^+/K^+-ATPase found in bull and humans involves in acrosomal reaction for ionic exchange.

23.1.7 Failure of Fertilisation

Improper ovulation, obstruction in the oviduct, abnormal oocyte and ovarian adhesions are the major inter factors of fertilisation failure in animals. In assisted reproductive technology, inappropriate prediction of ovulation time and insemination techniques are the major causes of fertilisation failure. Ovulation-related disorders include delayed ovulation, silent heat, anovulatory oestrus, poor managemental practices, malnutrition and environmental factors, like heat stress. Pathological conditions like ovarian cyst and endometritis may also lead to fertilisation failure. The animal can be considered a repeat breeder when it fails to conceive after repeated insemination attempts despite its normal reproductive cycle. Dietary supplementation of omega-6-rich polyunsaturated fatty acids (PUFAs) as calcium salt directly affects the oocyte to increase male sex offspring in cattle.

23.1.7.1 Development of Twin

The development of twins in monotocous species is considered abnormalities. When two offspring are born at the exact birth, they are called twins. There may be three types of twin, viz. identical twin, fraternal twin and semi-identical twin.

23.1.7.1.1 Identical Twin

The identical twins are also called monozygotic twins, where a single oocyte is fertilised by a single spermatozoon leading to the formation of a single zygote. The single zygote develops up to blastocyst, but the inner cell mass splits into two parts to develop two separate foetuses. Embryo-splitting techniques in in vitro fertilisation intends to develop such type of twins. The monozygotic twins are phenotypically identical, and hence, they are called identical twins.

23.1.7.1.2 Fraternal Twin

In fraternal twins, two separate oocytes fertilise independently, followed by the formation of two zygotes. They undergo a separate implantation process in the uterus. Hence, it is called a dizygotic twin. The term fraternal twin describes two different spermatozoa used in fertilisation. Phenotypically they may or may not be identical but may have diverse sequences on each chromosome and other sex; hence, it is called non-identical twins. Such incidence is due to hyperovulation, often seen in advanced age. Genetics and nutrition may be predisposing factors for fraternal twin development.

23.1.7.1.3 Semi-identical Twin

The occurrence of semi-identical twins is rare. A semi-identical twin develops when an unfertilised oocyte is mitotically divided into two oocytes and fertilised separately by different spermatozoa. The offspring may be of another sex with non-identical genetic and phenotypic characteristics; hence, it is termed semi-identical twins and occurs in various mammals like cattle, sheep, dolphins, elephants and humans.

23.2 Cleavage and Implantation

The mammalian embryo undergoes a series of mitotic cell divisions to form 2-cell, 4-cell, 8-cell, 16-cell, morula and blastocyst stages called cleavage. The zygote is termed an embryo when it starts mitotic division. The embryo is known as morula once it reaches 16-cells and proceeds towards the uterus during this stage (Table 23.7). Morula transforms into blastocyst after developing a cavity called blastocoele (Fig. 23.4). This process is called embryogenesis. Embryogenesis in mammals occurs through three stages: (1) cleavage, followed by the formation of blastula or blastocyst stage, (2) gastrula stage and (3) organogenesis stage. The last two stages, gastrula and organogenesis, result in the development of the body and the shape of an organisation; hence, these two stages are collectively called morphogenesis. After the morphogenesis, an embryo is termed a foetus.

23.2.1 Cleavage

The first rapid series of mitotic cell division of the zygote is called cleavage (Table 23.7). The first cleavage results in two-cell embryo, and the daughter cells are called blastomeres. The cells become progressively smaller throughout the cleavage with no net increase in the size up to successive three divisions (up to the eight-cell stage). It resulted in the decreased weight of the embryo from the single-cell zygote and is called a negative growth. The

Table 23.7 Stage of early embryonic development before implantation (period considered after fertilisation)

Developmental stage and number of cells	Major characteristics	Bovine	Ovine	Porcine	Equine	Mouse	Human
1-cell	Zygote	0–1 d	0–24 h	0–24 h	0–24 h	0–20 h	24 h
2-cell	Cleavage	1–2 d	24 h	140–16 h	24 h	20–38 h	48 h
4-cell	Cleavage	1–2 d	1.3 d	1 d	1.5 d	38–50 h	60 h
8-cell	Moves for implantation	2–4 d	1.5 d	2.5 d	3 d	50–62 h	72 h
16-cell	Moves for implantation, totipotent	3–4 d	–	–	–	60–74 h	3.5 d
Early morula (16–32-cell)	Pluripotent, trophoblast and ICM	4–5 d	3–4 d (morula)	3–4 d (morula)	4–5 d (morula)	60–74 h	4 d
Tight morula (32–64-cell)	Compaction	4–6 d				3.5 d	4 d
Early blastocyst (64–128-cell)	Small blastocoele	6–7 d	6–7 d (blastocyst)	5–6 d (blastocyst)	6 d (blastocyst)	4 d	4.5 d
Blastocyst (64–128-cell)	Large blastocoele	6–8 d				4.5 d	5 d
Expanded blastocyst (128–256-cell)	Thinning ZP	7–9 d	–	–	–	5 d	5 d
Hatching blastocyst (128–256-cell)	ZP degeneration, elongation of embryo	8–10 d	7–8 d	6 d	8 d	5 d	5.5 d

Source: Lopes and Mummery (2014), Soom et al. (1997), Seidel and Seidel (1991)
ICM inner cell mass, *ZP* zona pellucida, *d* day, *h* hour

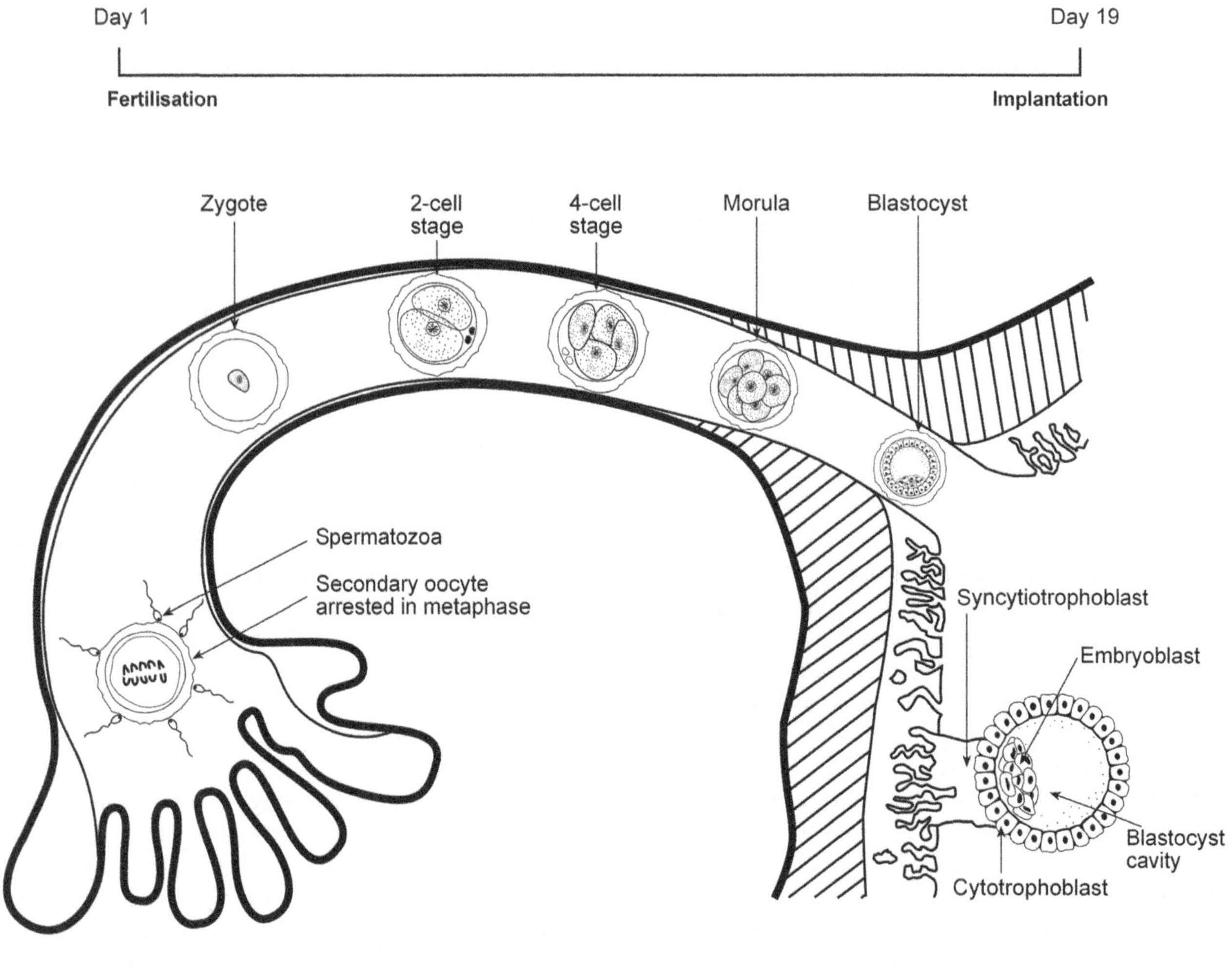

Fig. 23.4 Fertilisation to implantation. The fertilisation occurs at the **ampullary-isthmus junction**. The **zygote** undergoes a series of mitotic cell divisions (cleavage) to form 2-cell, 4-cell, 8-cell, 16-cell, morula and **blastocyst**. Morula proceeds towards the uterus and gradually transforms into blastocyst after developing a cavity called blastocoele. The blastocyst then attaches to the **endometrial surface** through implantation

proportion of decrease in the cellular mass is nearly 20% and 40% in cows and sheep, respectively. The cell's nuclei size increases to maintain the appropriate amount of nucleic acid. The first stage of cell division is occurred longitudinally, followed by the second longitudinal cell division at 90° to the plane of the first. The third division occurs in the perpendicular plane of the first two divisions. The rates of cleavage vary among species. It takes about 144 h (6 days), 96 h (4 days) and 120–140 h (5 days) in cows, sheep and goats, respectively, to reach the morula stage. The cleavage process continues until a solid mass of cells called morula is formed.

23.2.1.1 Morula

The 16-cell stage of an embryo is called *morula* (singular morulae), which develops after four rounds of cleavage divisions. The weight of the embryo starts increasing from the morula stage compared to the one-cell stage. Initially, the cells of the morula are *totipotent*. It becomes *pluripotent* and gives rise to two layers of cells. The outer layer is called the trophoblast. The inner cell mass remains in a cluster. The cells of the morulae undergo compaction due to cell-to-cell adhesion, and the shape of the cells turns spherical to polygonal. This stage is called a *tight morula*, and the size is reduced compared to the early morula stage. The cells of tight morula could be as many as 32. Less compaction results in poor quality embryos. The embryo migrates from the ampulla to the uterus in this stage by the smooth muscle contraction and ciliary movement of the oviduct. In bovines, it occurs within 3–4 days after fertilisation. The morula contains little yolk (except for pigs and horses), and hence, rely on the mother for their nutrition. It is provided by oviductal and uterine glands (histotrophs).

Most of the embryos are generally evaluated during this stage. The evaluation criteria include the number and shape of the cells, the degree of compaction, appearance of excluded blastomeres, character of peri-vitelline space and regions of degeneration. Embryos are graded based on these criteria as excellent, good, fair, poor, degenerative and unfertilised (details in Chap. 24). Excellent and good quality embryos are generally used for embryo transfer technology (ETT). The embryo with an appropriate category may also preserve, but their sustainability is poor.

23.2.1.2 Blastocyst Formation or Blastulation

The tight morula gradually acquires fluids to form a cavity (*blastocoele* or *exocœlome*) in between an outer layer of the blastocyst (*trophoblast* or *trophectoderm*) and inner cell mass (the *embryoblast*) at one pole. The inner cell layer of the trophectoderm is known as *cytotrophoblast*, and the outer one is called *syncytiotrophoblast*. The structure thus formed is called the *blastocyst*. The time of blastulation varies between species. It takes 6 days in pigs and horses, 7 days in sheep and 8 days in cattle after fertilisation. The trophoblast becomes flat and makes the epithelial wall of the blastocyst. The internal secretion of fluid forms the blastocoele by the blastomeres. This process is called *cavitation*.

The blastocoele pumps the fluid between the cells, and the blastocyst is turned into an *expanded blastocyst*. The blastocyst is still under the covering of zona pellucida. The sperm receptors at the zona pellucida are lost after fertilisation, and the gelatinous capsule protects the embryo from the invasion of pathogens and the maternal immune system. Any degeneration of the zona pellucida before the blastulation affects blastocoele formation. The thickness of zona pellucida (10 μm) and perivitelline space is reduced during the development of blastocoele. It acts as an indicator to evaluate good quality blastocyst. Usually, the zona pellucida is harder, and the blastocyst is smaller in size in vivo than in vitro embryo production. In cattle, the size of the blastocyst is about 200–203 μm in vivo and 217–221 μm in vitro.

23.2.2 Maternal Recognition of Pregnancy

Maternal recognition of pregnancy (MRP) is the biological process by which conceptus (elongated blastocyst) signals its presence to the mother and prevents the luteolytic mechanism from sustaining the life span of the corpus luteum. The ultimate goal of maternal recognition of pregnancy is to ensure the continuous release of progesterone to maintain the pregnancy. The major signalling agents that cause the MRP are generally species specific (Table 23.8).

23.2.2.1 Interferon Tau (IFNτ) and Inhibition of Luteolytic Mechanism

Trophoblast-derived interferon Tau (IFNτ) acts as an anti-luteolytic factor in domestic ruminants. IFNτ is a Type I IFN family member that acts through Type I IFN receptors (IFNAR) in the LE, GE and stroma. Upon binding with its receptor, IFNτ inhibits the transcription of the ESR1 gene

Table 23.8 Pregnancy recognition signals in mammals

Animal	Agents for MRP	Day of production	Maternal recognition of pregnancy (days after conception)
Cow	Bovine Interferon Tau (bIFNτ)	12–38	16–17
Ewe	Ovine Interferon Tau (oIFNτ)	9–21	12–13
Sow	Estradiol (E_2)	11–30	12
Mare	Equine chorionic gonadotropin (eCG)	14–16	14–16
Human	Human chorionic gonadotropin (hCG)	11	

Table 23.9 IFN-stimulated genes and their role in implantation

Name of the ISG	Functions
HIF2A (transcription factor)	Induces angiogenesis and glucose transport by promoting the expressions of VEGF and SLC2A1, respectively
SLC2A1	Glucose transport
Wingless-type mouse mammary tumour virus integration site family, member 7A	Promotes uterine–conceptus interactions
CTSL (cathepsin L)	Cysteine proteinase
CST3 (cystatin C)	Proteinase inhibitor
LGALS15 (galectin 15)	Promotes trophectoderm cell migration and adhesion
GRP (gastrin-releasing polypeptide)	Affects morphogenesis and angiogenesis
IGFBP1 (insulin-like growth factor binding protein 1)	Induces mitogenic response

through a signalling pathway via IFN regulatory factor (IRF) 2. The inhibition of the *ESR1* gene prevents the action of oestrogen from inducing the expression of oxytocin receptor (OXTR). Ultimately, the action of oxytocin to synthesise luteolytic PGF2α is blocked. In addition to its anti-luteolytic action, IFNT regulates the expression of several IFN-stimulated genes (ISGs) induced by the progesterone for endometrial differentiation and implantation of the conceptus (Table 23.9).

23.2.2.2 Estradiol and MRP in Pigs

The pregnancy recognition signal in pigs is the oestrogen secreted by the conceptuses. The release of oestrogen by the conceptus is biphasic. The oestrogen appears first at 11–12 of pregnancy, followed by a sustained release at days 15–30 for conceptus attachment and placental development. The anti-luteolytic action of oestrogen in pigs can be explained by the endocrine/exocrine hypothesis. According to this model, the oestrogen secreted from the conceptus directed endometrial-derived PGF2α away from uterine vasculature (endocrine) and sequestrated into the uterine lumen (exocrine). In the uterine lumen, PGF2α inactivates into its inactive 13,14-dihydro-15-keto prostaglandin F2α metabolite. In addition to the anti-luteolytic mechanism, oestrogen also helps in the migration and spacing of blastocysts. To establish a pregnancy, the presence of at least two conceptuses in each uterine horn is mandatory initially to establish a pregnancy. Oestrogen also increases fibroblast growth factor 7 (FGF-7) expression in the endometrium for proliferation and differentiation of trophectoderm.

23.2.2.3 Chorionic Gonadotropin (CG) and MRP in Horses and Primates

CG is the key maternal recognition signal in horses and primates. It is a glycoprotein hormone secreted from the trophoblast. CG acts like an LH agonist and performs anti-luteolytic actions together with induction of steroidogenesis in CL. The anti-luteolytic actions of CG include (1) maintaining stable luteal blood flow, (2) increases in the Bcl2/Bax ratio to prevent apoptosis and (3) preventing tissue remodelling by modulating matrix metalloproteinase-2 (MMP-2) functions and recruitment of macrophages. CG increases steroidogenesis by inducing the expression of the steroidogenic acute regulatory protein (StAR), cytochrome P450 cholesterol side-chain cleavage (P450scc) and 3-β-hydroxysteroid dehydrogenase (3β-HSD).

23.2.2.4 Pregnancy Recognition in Rodents

Semicircadian prolactin surges due to cervical stimulation are required for pregnancy in rodents. These surges are responsible for converting the corpus luteum (CL) of the cycle into the corpus luteum of pregnancy (CLP).

23.2.2.5 Pregnancy Recognition in Dogs and Cats

In dogs and cats, MRP is not essentially required to establish pregnancy as the life span of CL usually is about 60 days, irrespective of whether conception occurred or not.

23.2.3 Implantation

Implantation is a biological process of attachment of the embryo to the endometrial surface followed by the invasion of the epithelium to form the placenta (Fig. 23.4, Tables 23.7 and 23.10). Both the embryo and the uterus undergo a series of physiological process that ultimately favours implantation. Implantation requires crosstalk between a receptive uterus and the conceptus for a limited period called the "window of implantation". The time for initiation of the implantation process is species-specific. In sow, the process starts after 2 days of fertilisation; in sheep, 2.5–3 days, cattle, 3–4 days; and in horses, 5.5–6 days. In humans, it starts the 8–10 days after ovulation and continues till the second week. Endometrial differentiation is an essential prerequisite for the initiation of the implantation process. Oestrogen induces endometrial differentiation. Progesterone also acts over the oestrogen primed endometrium to reinforce further differentiation to make a suitable environment for embryo implantation. The process of implantation can be divided into six distinct phases: (1) shedding of the zona pellucida or zona hatching, (2) blastocyst elongation orientation and spacing, (3) intrauterine migration and spacing, (4) apposition, (5) adhesion and (6) endometrial invasion.

23.2.3.1 Zona Hatching

The release of the blastocyst from the zona pellucida is called zona hatching, and the expanded blastocyst is transformed into a hatched blastocyst. Zona pellucida prevents premature

Table 23.10 Period of implantation and gestation period of some mammals

Domestic animals	Implantation[a] (days, stage)	Gestation (days)	Wild animals	Gestation (days)
Cattle	19–35 (post gastrula)	279–292 (286)	Bison (American)	217
Buffalo	19–30 (post gastrula)	281–334 (283)	Chimpanzee	230–292 (286)
Sheep	12–18 (post gastrula)	142–152 (147)	Deer (white-tailed)	201
Goat	14–25 (post gastrula)	145–155 (150)	Elephant (African)	645
Pig	13–20 (gastrula)	112–115 (113)	Fox	52
Horse	30–38 (organogenesis)	330–342 (336)	Giraffe	420–450 (430)
Camel	25	360–420 (390)	Gorilla	255–260 (257)
Dog	18–20	58–65 (61)	Hippopotamus	225–250 (237)
Cat	13 (12–14[b])	58–67 (64)	Kangaroo	42
Rat	5.5	21–23 (22)	Leopard	92–95 (93)
Guinea pig	6	56–74 (65)	Lion	108
Mouse	4.5 (blastocyst)	19–21 (20)	Monkey (rhesus)	164
Rabbit	6.5 (gastrula)	28–35 (31)	Rhinoceros	450
Hamster	4	16–23 (20)	Seal	330
Ferret (domestic)		41–42 (41)	Squirrel	30–40 (35)
Human	6–7 (blastocyst)	259–375 (270)	Tiger	105–113 (109)
Semi-domestic animals		**Period (days)**	Whale	480–590 (535)
Elephant (Asian)		617	Wolf	60–68 (64)
Donkey		365	Zebra	361–390 (375)

Compiled from various sources
[a] Days count after standing oestrus
[b] Days after mating

implantation of the embryo. Usually, the zona pellucida collapses by some enzymatic reaction. It usually occurs 9–11 days post ovulation in the cow. The blastocyst generally comes out from its embryonic pole, the side opposite the inner cell mass (ICM). The zona hatching occurs by two forces. The mechanical pressure exerted by the growing blastocyst and the enzymatic lysis of the zona pellucida. Depending on the species, several proteases are involved in the zona lysis, such as serine proteases, cysteine proteases and metalloproteinases. Cathepsins are a cysteine protease actively engaged in zona hatching. Ovastacin also helps in the zona hatching process by removing cortical granules. The zona hatching is regulated by hormones, growth factors, cytokines and transcription factors. The predominant growth factors involved in zona hatching are heparin-binding epidermal growth factor (HB-EGF), transforming growth factor-beta (TGF-β) and leukaemia inhibitory factor (LIF). The cyclooxygenase-2 (COX-2) inhibitors, prostaglandins (PGs, E2 or I2), plasmin and trypsin play an essential role in the hatching process. Calcium is also required for zona hatching as some of the mechanisms of zona hatching are calcium dependent.

23.2.3.2 Blastocyst Elongation and Orientation

The shedding of zona pellucida is followed by the rapid growth of the blastocyst. The hatched blastocyst gradually becomes elongated and moves from the oviducts to the uterine horns for implantation. In cow, the spherical blastocyst (3 cm) is transformed into a filamentous thread-like structure (25 cm) called conceptus from day 13 to day 25 post fertilisation. This elongation occurs through continual hyperplasia of the trophectoderm and entire endoderm. The blastocyst does not elongate in the horse but somewhat increases in diameter by 2–3 mm/day to become a spherical baseball-like form. At this stage, the trophoblast secretes a hormone called pregnancy serum protein B (PSPB or PAG). It influences the corpus luteum survivability and helps to secret progesterone for embryo development and maintenance of gestation. The level of PSPB is gradually increased during the gestation period and reaches a maximum during the day of parturition.

The conceptus develops a specific orientation concerning the uterus. In most domestic species, the early conceptus is arranged so that the yolk sac is found on the endometrial side of the uterine lumen and the embryonic disc lies on the anti-mesenteric side.

23.2.3.3 Intrauterine Migration and Spacing

In polytocous species, intrauterine migration and equidistant spacing are essentially required for embryo survival. The embryos of polytocous species (rabbit, pig, rat and mouse) enter the uterine horn at late morula and early blastocyst. They orient themselves at the longitudinal axis so that the inner cell mass (ICM) is situated at the mesometrial side of the uterus. In the ruminant embryo, the migration is limited, and the embryo rarely passes through the body of the uterus into the contralateral horn. Sheep embryo tends to migrate when multiple ovulations occur in the same ovary. There is no correlation between the side of ovulation and the side of

Table 23.11 Stages of trans-uterine migration and associated mechanisms

Stages of transuterine migration	Mechanism	Factors responsible
Stage-I: Embryo is floating in the uterine lumen	Oestrogen is secreted from the blastocyst that the embryo for their orientation	Oestrogen from blastocyst or some unknown factors of endometrium that sense the embryo
Stage-II: Individual separation of embryo	Synchronised myometrial contractility	1. Prostaglandin (PG): • Relaxation: PGD_2 and PGI_2 • Contraction: $PGF_{2\alpha}$ and TXA_2 • Both contraction and relaxation; PGE_2 2. Ovarian steroids: A balance between oestrogen and progesterone is required for synchronised myometrial contraction. 3. Adrenergic signalling
	Reabsorption of luminal fluid	1. Aquaporins (AQP) water channels (AQP2, AQP5 and AQP8 in humans) 2. Ion channels: • The cystic fibrosis transmembrane conductance regulator (CFTR) • CAMP-activated Cl channel • Epithelial Na^+ channel (ENaC) 3. Wnt/b-catenin signalling: Activation of circular smooth muscles
Stage-III: Stromal oedema and immobilisation of embryo	Immobilisation of embryo	1. Steroid hormone 2. Inflammatory signals by PGs, histamine and nitric oxide (NO)

embryo attachment in the horse. Intrauterine migration and spacing occur in three stages (Table 23.11).

23.2.3.4 Apposition

The trophectoderm adheres with the adhesive receptors of the endometrial luminal epithelium. The presence of an anti-adhesive substance such as mucin 1 (MUC1) prevents the blastocyst adhesion. The progesterone facilitates the declining of MUC1 from the endometrial luminal epithelium, and the adhesive receptors such as integrins are exposed to trophoblast for initial apposition. The endometrial glandular epithelium secretes histotroph under the influence of progesterone that nourishes the developing blastocysts.

23.2.3.5 Adhesion

The adhesion between blastocyst trophectoderm and endometrial luminal epithelial is achieved through the interaction between cell adhesion molecules, such as glycosylated cell adhesion molecule (GLYCAM1) 1, galectin 15 (LGALS15) and secreted phosphoprotein 1 (SPP1 or osteopontin) with their receptors (integrins and glycoconjugates). The cell adhesion molecules are expressed on the trophoblast apical surface and interact with their receptors at the luminal epithelium.

23.2.3.6 Endometrial Invasion

In this phase, the giant binucleate cells (BNC) of the trophoblast fuse with the LE to form multinucleated syncytial plaques. The BNC develops from the mononuclear trophectoderm cells through mitotic polyploidy (nuclear divisions without cytokinesis). The giant BNC migrate to the trophoblast and fuse with the individual luminal epithelium to form trinucleated foetomaternal hybrid cells. The remaining luminal epithelium, unable to form hybrid cells, undergoes apoptosis. The migration and fusion of BNC are continued till the syncytial plaques are limited in size to 20–25 nuclei. Then no further nuclear divisions occur, and syncytial plaques and linked with tight junctions to form caruncular syncytia. The caruncular syncytia expand from cotyledons. The giant BNC serve two essential functions: (1) formation of foetomaternal syncytial plaques that give rise to cotyledon of placentome and (2) synthesis of chorionic somatomammotropin hormone 1 (placental lactogen), pregnancy-associated glycoproteins (PAGs) and progesterone that facilitates the growth of endometrial glands and differentiation of endometrium. In ruminants, the endogenous retroviruses (ERV) are involved in the fusion of BNC and LE.

In pigs, the trophoblast and uterine epithelium undergo loose apposition immediately after the blastocyst elongation. There is interdigitation between the microvilli of two epithelial surfaces (trophectoderm and LE), and later the trophoblastic surface becomes modified to form an absorptive surface called areolae. The nutrient uptake by the developing conceptus is facilitated through these areolae. In this species, the attachment begins on day 13 and is completed around days 18–24.

In ruminants, an initial transitory attachment occurs between trophoblast and LE. Trophoblast develops finger-like villi that penetrate the lumen of uterine glands and act as a temporary anchor. The centres of caruncles become depressed and cytoplasmic protrusions are developed from the trophoblast epithelium. The permeability of the caruncular capillaries increases on day 15. Between days

16–19, the effective attachment occurs through the interpenetration of uterine microvilli and cytoplasmic projections of trophectoderm (Fig. 23.4).

In the mare, the attachment occurs between the surface epithelium of the embryonic vesicle and uterine lining through interdigitations.

23.2.3.7 Types of Implantation

Implantation can classify as invasive or non-invasive based on the degree of invasion or penetration. In primates and rodents, the blastocyst penetrates the uterine mucosa, followed by the phagocytosis of uterine LE. The blastocyst migrates the uterine stroma. In this type of implantation, the endometrial stromal cells and endothelial cells of the blood vessels undergo decidualisation in the presence of leukocytes under the influence of progesterone to form a special kind of tissue called decidua. It suppresses the mother's immune response to prevent the immune rejection of foetuses. The decidua also secretes various growth factors, cytokines, insulin-like growth factor binding protein 1 (IGFBP1), prolactin and different extracellular matrix proteins like fibronectin and laminin, favouring the invasion process. The decidua becomes the part of placenta with the advancement of pregnancy.

In contrast, the implantation in domestic ruminants, carnivores, pigs and horses is non-invasive. The conceptus remains within the uterine lumen and is embedded in the uterine wall.

Based on blastocyst orientation, implantation can be of three types centric, eccentric and interstitial. In the centric type of implantation, the embryo(s) remains at the centre of the uterus and its size increases before implantation. It is non-invasive implantation seen in all domestic ruminants, carnivores, pigs and horses. In an eccentric pattern, the small-sized blastocyst(s) invades one side (generally the reverse side of the mesometrium) of the uterus. This implantation pattern is generally invasive and occurs in some rodents like rats and mice. In guinea pigs, humans and other primates, the small-sized blastocyst(s) is entered deep into the endometrial epithelium and attached to the subepithelial connective tissue of the endometrium. It is called interstitial or nidation or nest making pattern of implantation. It is also of invasive type.

In humans, a cellular structure called pinopode or uterodome is thought to be involved in uterine receptivity. Pinopode is a large cellular protrusion on the uterine epithelial surface under the influence of hormones. Other than humans, pinopodes are seen in mice and rats. Pinopods act as a clinical marker of endometrial receptivity in such species. Pinopods are involved in regulating uterine luminal contents and regulation of implantation associated proteins.

23.2.4 Gastrulation

Gastrulation is the formation of three germ layers, viz. ectoderm, endoderm and mesoderm. In blastocyst, the embryonic cells differentiate into outer trophoblast and inner cell mass (embryoblast) (Fig. 23.5). Dramatic changes occur at the embryoblast, giving rise to epiblast (outer layer) and hypoblast (inner layer). The hypoblast is small and cuboidal in shape. It provides the nutrients to all the embryo cells by forming a yolk sac until the development of the placenta is functional. In the next stage, the inner cell mass cells are organised into a sheet of columnar epithelium called an embryonic disc. The trophoblast and embryonic disc are collectively called ectoderm, the first primary germ layer. Delamination of the cells from the inner surface leads to forming a second layer. The third and final germ layer forms between the first and second layers as a wave of cellular emigration from the embryonic disc called mesoderm. The parts of primary germ layers beyond the embryonic disc are extraembryonic parts that give rise to the extraembryonic membrane. Blastulation ends with the formation of three primary germ layers.

23.2.5 Organogenesis

A thick structure develops along the midline of the embryonic disk, called a primitive streak. The primitive streak determines the major axis, and accordingly, the embryo's left, right, cranial and caudal regions are designated. Thus, three body axes are developed in the embryo anterior-posterior (head-tail), left-right (lateral-medial) and dorsal-ventral (back-belly). A node (primitive node) is extended from cranial to caudal end through its midline, called a primitive groove. The outer layer cells start to fold up towards the inside and gradually separate along with the primitive streak. This process is called invagination. The invagination causes the replacement of the outer cells with new cells and forms definitive endoderm. Similarly, the internalised cells form definitive ectoderm, and the cells that remain between definitive ectoderm and endoderm are called definitive mesoderm. In cattle, it occurs within the third week after fertilisation.

Three different germ layers have the potency to give rise to all tissues and organs of the organism (Table 23.12). The heart is the first organ that develops from the endoderm. The process by which the germ layers of the embryo into tissues and organs is called organogenesis. The embryo is termed as foetus when organogenesis is initiated and continued. In cattle, it occurs around the end of the fourth week.

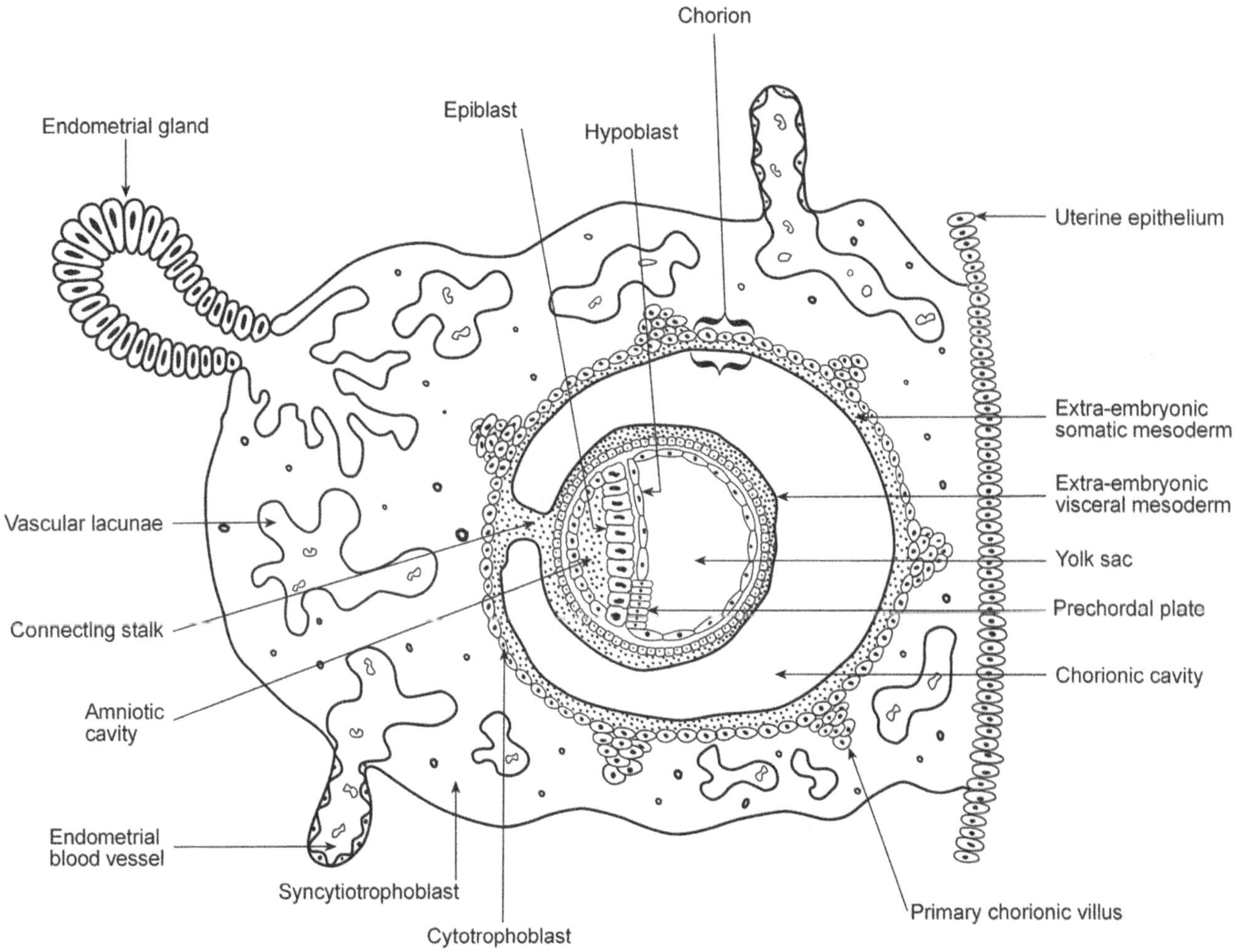

Fig. 23.5 Gastrulation. The blastocyst consists of outer **trophoblast** and **inner cell mass** (ICM). At the time of implantation, the trophoblast differentiates into two distinct layers. The outer **syncytiotrophoblast** assists in the implantation by releasing hydrolytic enzymes and hCG. The inner **cytotrophoblast** surrounds the somatic mesoderm. The inner cell mass gives rise to **hypoblast** and **epiblast**. Hypoblast is differentiated from **extraembryonic mesoderm,** which gives rise to amnion, allantois, chorion, and visceral **yolk sac**. The epiblast forms three primary germ layers: ectoderm, mesoderm, and endoderm

Table 23.12 Organs developed from germ layers

Germ layers	Organs
Endoderm	Digestive system, respiratory system, thymus, thyroid, parathyroid, the epithelial lining of the gastrointestinal tract, respiratory tract, excretory tract, auditory duct and some endocrine glands
Mesoderm	Uro-genital system, muscular system, skeletal system and vascular system
Ectoderm	The nervous system, organs of special senses, pituitary glands, hair, sebaceous glands, facial cartilage, tooth dentin

23.2.5.1 Formation of Extraembryonic Membrane

The formation of an extraembryonic membrane starts with separating the mesoderm into an inner and outer layer by a narrow cavity. The outer mesoderm is called somatic mesoderm, and the inner layer is called splanchnic mesoderm. The intervening cleft between the outer and inner mesoderm is called embryonic coelom. The splanchnic mesoderm is associated with viscera formation, and the somatic mesoderm develops connective tissue of the body wall. The somatic mesoderm fuses with the underlying ectoderm to form somatopleure, and the splanchnic mesoderm fuses with the underlining ectoderm to form splanchnopleure. The formation of amnion, chorion and yolk sac develops due to the folding of somotopluerae and splanchnicplurae. Amnion and chorion are developed from somatopleure, whereas allantois and yolk sac are derived from splanchnopleure. The amnion and chorion are fused to form an amniochorion.

Further invaginations of splanchnicplureae subdivide the blastocoel into primitive gut and yolk sac. The yolk sac comprises extraembryonic endoderm on its inner surface and splanchnic mesoderm on its outer surface. The splanchnic mesoderm is the site of primitive haematopoiesis in the developing embryo. In domestic animals, the yolk sac is small and absorptive function during early pregnancy in mares and carnivores.

The allantois develops from the ventral part of the hindgut. The internal layer of the allantois forms from the endoderm, and the external layer originates from the splanchnic mesoderm. Allantois serves as temporary storage of urine in developing foetuses. The fusion of the outer layer of allantois with overlaying chorion leads to the formation of allantochorion. The formation of blood vessels at the allantochorion acts as a transient organ for gas exchange.

The internal epithelium cell lining of the chorion develops from vascularised mesenchyme tissue, which covers the exocelema, amnion, allantois and yolk sac. The amnion is made up of squamous epithelium cells and makes a membranous layer containing mesenchyme connective tissue. It guards the foetus against mechanical pressure along with the nutrient exchange process.

The allantois comprises a squamous epithelial cell layer with a basement membrane and extraembryonic mesenchyme developed from the embryonic intestine as an extraembryonic urinary bladder. The yolk sac is involved in the exchange process at the early embryonic stage; later, it merges with chorion when the placenta develops in most mammals. It is rudimentary in pigs, guinea pigs and humans.

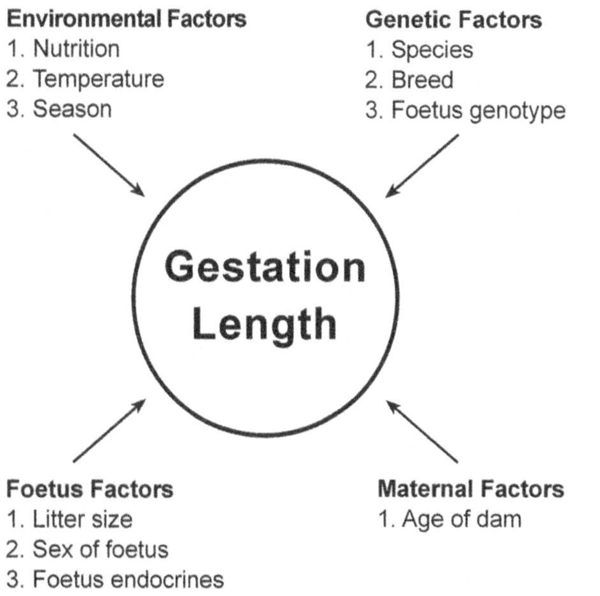

Fig. 23.6 Factors affecting gestation. The pregnancy success is influenced by **environmental** (nutrition, environmental temperature and season), **genetic** (species, breed and foetal genotype), **maternal** (maternal age, number of births, maternal nutrition) and **foetal** factors (litter size, sex of foetus and foetal endocrine factors)

23.3 The Physiology of Gestation

The intrauterine period of embryonic and foetal development in viviparous mammals is the gestation or pregnancy period. The gestation period starts after fertilisation, and the length of the gestation period is species specific (Table 23.10). The rate of pregnancy in single insemination is called fertility. Among the domestic animals, pigs have the highest fertility rate (90%), followed by sheep (85%), beef cattle (45%), dairy cattle (35%) and humans (25%). The first 2–3 weeks of pregnancy are the most vulnerable period for developing an embryo as most embryonic losses occur during this period. Pregnancy success is influenced by nutrition, environment, disease, genetics, and management (Fig. 23.6).

Nutrition: The lactation imposes a tremendous nutritional challenge on dairy cattle, and the cows undergo negative energy balance (NEB). The body reserves are mobilised to meet the metabolic need for milk production. Severe NEB leads to poor reproductive performances, and the first postpartum ovulation is delayed. The resumption of cyclicity within 4 weeks postpartum is economically beneficial for dairy cattle, and optimum nutrition during the immediate post-calving period is essentially required to achieve this goal with high embryonic survival.

Diseases: The incidence of diseases in livestock affects embryonic mortality. In cattle, fertility is compromised in mastitis, retained placenta, uterine infections, displaced abomasum, hypocalcaemia and ketosis. Common uterine infections are metritis, which occurs in 40% of dairy cows within a week of parturition, followed by subclinical endometritis (30%) and endometritis (20%) that persist beyond 3 weeks postpartum. All these infections hamper the embryo survival rate.

Environment: Environmental stressors cause early embryonic mortality due to poor oocyte quality and depressed HPO axis.

Genetics: Chromosomal defects, mutation in the individual gene and genetic interactions lead to poor embryonic mortality. The genetic predisposition to early embryonic loss is implicated due to inbreeding. Brachyspina syndrome, a rare recessive genetic disorder that develops in the cattle due to deletion in the bovine FANCI, has been identified in Holstein dairy cattle and causes embryonic mortality. Its incidence is low (<1/100,000), but carrier frequency is high (7.4%) in the Holstein cattle. In Jersey cattle, a nonsense mutation in CWC15 spliceosome-associated protein homolog (*S. cerevisiae*) leads to embryonic mortality and repeated abortions.

23.3.1 Physiological Changes During Gestation

23.3.1.1 Vagina and Vulva

During gestation, the vulva becomes oedematous and vascular. The vagina remains dry and pale throughout the gestation period, but it becomes swollen and pliable at the end of pregnancy.

23.3.1.2 Cervix

The cervical mucous becomes viscous and forms a plug to protect against infections. This cervical plug is formed under the influence of progesterone.

23.3.1.3 Uterus

The embryo spends over 98% of its gestation life in the uterus. The uterus undergoes gradual changes to accommodate the foetus. Under the influence of high progesterone and low estradiol, the uterus becomes soft and flaccid to make a congenial environment to accommodate the foetus.

23.3.1.4 Ovary

The most prominent gestational change in the ovary is the presence of the corpus luteum that secrets progesterone. The oestrus is suspended during gestation under the influence of progesterone. In mares, the secondary CL develops between 35th and 150th day of pregnancy, which regresses along with primary CL around the seventh month of pregnancy.

23.3.1.5 Pelvic Ligaments

The pelvic ligaments become relaxed during the time of pregnancy. The most pronounced changes occur around parturition under the influence of estradiol and relaxin. The relaxation of pelvic ligaments is common in cows and ewe compared to the mare.

23.3.2 Maternal Adaptions During Pregnancy

Pregnancy imposes tremendous metabolic challenges to the mother for optimum foetal growth and nutrition. An adequate supply of nutrients is essentially required for the survivability of foetuses. The mother has to undergo physiological adaptations to ensure a continuous supply of nutrients for the foetuses and to maintain the mother itself. The physiological adaption strategies during pregnancy are summarised in Table 23.13.

23.3.3 Control of Gestation

Endocrines control gestation. Nerves have no role in maintaining gestation. Progesterone is the chief hormone for maintaining gestation. Early pregnancy is controlled by the luteal progesterone in all domestic mammals. The presence of embryo/conceptus influences the hypothalamus-hypophyseal tract for constant releasing of gonadotropin (luteotropin). It results in persistent corpus luteum followed by inhibition of luteolytic PGF2α synthesis and assures continual luteal progesterone. The later part of the pregnancy is controlled by the placental progesterone in mare and ewe. The luteal progesterone is continued to sustain the pregnancy in other mammals. Various placental, ovarian and foetal hormones and growth factors control the different phases of gestation (discussed in detail in the following areas). Progesterone reduces or blocks the muscular tone of the female reproductive tract, referred to as progesterone block. Progesterone favours endometrial growth, influences to secrete endometrial gland secretion (uterine milk) and support the placentation in farm animals. Oestrogens are synthesised both in the ovary and in the placenta in low quantity during gestation to support the action of progesterone as well as udder development. At the last stage of pregnancy, with the influence of foetal pituitary and adrenal hormones, the level of oestrogens is increased, which influences the parturition process by relaxing pelvic structures, cervical dilation and promoting oxytocin (discussed details in following areas).

23.3.4 Placenta

The transient structure of mammals that connects the developing foetus with the uterine wall through the umbilical cord for the physiological exchange between mother and foetus is called the placenta. It develops soon after the blastocyst implantation and expels during parturition and the foetuses. Placenta provides nutrients and eliminates waste, helping in gaseous exchange and thermoregulation. It also acts as a temporary endocrine gland and defends against infection. The placenta's shape, structure, and configuration differences are species specific (Table 23.14) and depend primarily on the uterine structure and litter size. The mammals having placenta are also termed eutherian mammals.

23.3.4.1 Structure of Placenta

The placenta has two basic parts; the foetal part develops from the chorion of the blastocyst, and the maternal part develops from the endometrium of the uterus or maternal tissues. The placenta's foetal parts comprise three layers: the endothelium lining of allantoic capillaries, the connective tissue of chorioallantoic mesoderm and the chorionic epithelium of chorioallantoic mesoderm and chorionic trophoblast. Similarly, the maternal part of the placenta also has three layers: the endothelium lining of the blood vessels, the epithelial cells and the connective tissue of the endometrium (Fig. 23.7a). In the chorioallantoic placenta, all three layers of the foetus are involved. Still involvements of different

Table 23.13 Maternal adaptation during pregnancy

Adaptation	Process
Cardiovascular adaptations	• Increase in blood volume facilitates the exchange of nutrients and gases between mother and foetus • Increased cardiac output by more than 50% • Increased uterine blood flow • Increased stroke volume • Decreased systemic and pulmonary vascular resistance • Decreased arterial blood pressure
Haematological adaptations	• Decreased erythrocyte count, Hb, PCV and MCV • Increased procoagulant factors (I, V, VII, VIII, IX, X), adhesive proteins (vwf), fibrinolytic proteins (plasminogen activator inhibitors)
Respiratory adaptations	• Increased minute ventilation and alveolar ventilation • Vital capacity remains unchanged • Increased tidal volume • Decreased functional residual capacity • Decreased total lung capacity • Increased dead space
Renal adaptations	• Increased GFR and renal plasma flow • Increased reabsorption of sodium and water • Decreased BUN and creatinine • Increased renin secretion • Decreased glucose reabsorption
Metabolic adaptations	• During early pregnancy, the mother is in an anabolic state with increased lipogenesis • During the last part of gestation, the anabolic state shifts towards a catabolic state • Increased gluconeogenic activity • Glycerol is used as a primary gluconeogenic precursor • Increased lipolysis at late gestation
Immunological adaptions	• Increased proportion of natural killer (CD335) cells, cytotoxic T cells (CD8), and macrophages and dendritic cells. • Th1–Th2 shift: Reduced expression of inflammatory cytokines (Th1/Th17) and increased expression of anti-inflammatory cytokines (Th2) • Increased expression of proteins for immune tolerance such as programmed cell death ligand-1 (CD274), lymphocyte activationgene-3 (CD223) and cytotoxic T-lymphocyte-associated protein-4 (CD152) • Higher expression of indoleamine 2,3 dioxygenase (IDO) converts tryptophan to kynurenine and converts naïve T cells into FoxP3+ regulatory T cells • Pregnancy-associated glycoprotein (PAG) PAG reduces endometrial T-cell proliferation • IFNT causes the induction of regulatory T cells that promote immunosuppressive functions

Table 23.14 Different types of the placenta in domestic mammals

			Involvement of endometrial tissue(s)		
Mammals	According to shape and attachment	According to layer involvement	Endothelium	Epithelium	Connective
Ruminants	Cotyledonary	Epitheliochorial (Synepitheliochorial)	+	+ (−)	+
Horses	Diffuse	Epitheliochorial	+	+	+
Pigs	Diffuse	Epitheliochorial	+	+	+
Dogs	Zonary	Endotheliochorial	+	−	−
Cats	Zonary	Endotheliochorial	+	−	−
Rats	Discoid	Hemochorial	−	−	−
Rabbits	Discoid	Hemochorial	−	−	−
Guinea pigs	Discoid	Hemochorial	−	−	−
Mouse	Discoid	Hemochorial	−	−	−
Human	Discoid	Hemochorial	−	−	−

Note: Synepitheliochorial (earlier termed as Syndesmochorial) is a lack of endometrial epithelium after implantation

endometrium tissues are occurred by the influence of progesterone and species specific, which develop various types of the placenta.

Chorion is the double-layered foetal membrane that originates from the trophoblast and extraembryonic mesoderm of the foetus that emerges into the endometrium of the uterus during pregnancy. The chorion, along with its villi together called chorion frondosum. Another structure of foetus involved in placentation is alantois (in plural allantoides or allantoises). Allantois (in plural allantoides or allantoises) is a hollow pouch composed of an extraembryonic tissue. It connects with the urinary organs of the foetus to act as a

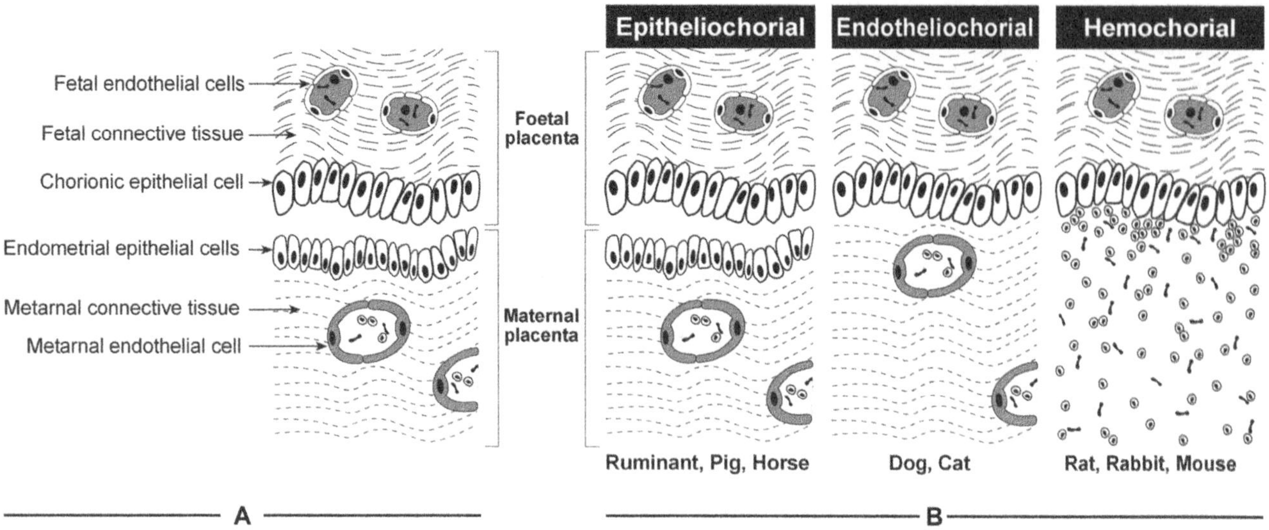

Fig. 23.7 Typical features of the placenta in domestic animals. (**a**) Types of tissue involvement in the placenta; (**b**) Types of placenta according to tissue involvement. The placenta's foetal parts comprise three layers: **endothelium** lining of allantoic capillaries, the **connective tissue** of chorioallantoic mesoderm and the **chorionic epithelium** of trophoblast. Similarly, the maternal part of the placenta also has three layers: the **endothelium** lining of the blood vessels, the **epithelial cells** and the endometrium **connective tissue**. The placenta can classify based on the retention of maternal layers. All three maternal layers retain in the **epitheliochorial placenta**. Only uterine endothelium remains in the **endotheliochorial placenta**. But no layer of maternal lining exists in the **hemochorial placenta**

waste reservoir. In reptiles, birds, and marsupials, it helps in gaseous exchange and removal of liquid wastes of the foetus. The fusion of chorion and allantois leads to the formation of chorioallantoic placenta. In some mammals (mostly in ruminants), the chorioallantoic surface is attached to the endometrium in the form of a band-like structure called placentome. The placentome contains caruncles derived from the endometrium and cotyledon from the chorion of the foetus. The number of placentomes varies between species. In cattle, sheep and goats, the number ranges between 75 and 125; in deer, 4–6 and in giraffes, 150. The villi of the chorion remain inside the oval or circular patches inform of cotyledon for gaseous and nutrient exchange. Six layers, three from foetal or chronic and three form maternal tissues, made the placenta.

23.3.4.1.1 Paraplacental Structures

Additional or accessory structures can also form in certain species to support the general exchange processes, called paraplacental structures. These structures are generally developed immediately after implantation and continue before the formation of the placenta. Still, it will continue till the end of gestation together with the placenta in some animals. The yolk sac or vitelline sac is considered one of the paraplacental structures. It is found in almost all mammals in early embryonic life but becomes non-functional within the first trimester of gestation except in rabbits and many other rodents. It attaches to the embryo's midgut through a layer of endometrial epithelium and vascularised foetal mesenchyme. The yolk sac has a vital role in exchanging iron molecules in rabbits; iron and calcium molecules in rats. Subplacenta is a paraplacental structure found in some rodents. In guinea pigs, it develops from mesenchyme and is present at the roof of the central excavation in between the placental disc and decidua. The cluster of multinuclear giant cells involves in nutrient exchange. Hematoma or hemophagous organ is another type of paraplacental structure that develops in carnivores (dogs and cats) and remains up to three-fourths of the gestational period. Hematoma or hemophagous organ has a vital role in exchanging iron-rich substances. It helps to exchange the extravasation materials of maternal blood and trophoblast cells digested maternal erythrocytes. The pigments from the degraded haemoglobin after erythrophagocytosis cause in colouration of the hematoma or hemophagous organs with a green border in dogs and a brown border in cats. The placental hematoma or hemophagous areas are also involved in iron transfer in bovines, ovines and caprines. The areola is a dome-shaped paraplacental structure present in different ungulates, mainly pigs and horses, for its high absorptive capacity due to a capillary network and a cavity. The areola plays an essential role in absorbing calcium and iron (in the form of ferritin) from maternal blood. The chorionic girdle is a paraplacental structure present in horses. It starts to develop from trophoblast around day 15 and lasts about day 57 of gestation. One of the major functions of the chorionic girdle is to develop equine chorionic gonadotrophin (eCG) from the endometrial cups other than nutrient exchange.

23.3.4.2 Types of the Placenta

The placenta can classify histologically, morphologically and structurally. Histologically, the placenta can be classified into three types: endotheliochorial, epitheliochorial and hemochorial (Fig. 23.7b). In the endotheliochorial placenta, the chorionic villi of the trophoblast are attached to the endothelium of maternal or endometrial blood vessels. It is mostly seen in dogs, cats and other carnivores. In the epitheliochorial placenta, the chorionic villi of the trophoblast are attached to the epithelial lining of the endometrial glands. It is found mainly in ruminants, horses and the lower group of primates. Synepitheliochorial (previously termed Syndesmochorial) is another type of epitheliochorial placenta where the endometrial epithelium is disintegrated after implantation by some foetus-derived mediator substances. It directly contacts the foetal trophoblast with the maternal connective tissue found in ruminants. The ruminant placenta contains many binucleated cells originating from the trophoblast, which combine with caruncular epithelial cells to make small syncytia. Placental lactogen is synthesised from these binucleated cells. In the hemochorial placenta, the chorionic villi are attached to the mucosal lining of the endometrium, and the maternal blood vessels remain in direct contact with the chorionic epithelium of the foetus. It can directly transfer the nutrients but may lead to immune reactions in the foetus. It is present in rats, rabbits, guinea pigs, mice, humans and other higher primates. In the hemochorial placenta, particularly in humans, decidua is formed by decidualisation. Decidua is the modified part of the mucosal lining of the endometrium that attaches with the trophoblast of the foetus and becomes a part of the placenta. The additional or auxiliary structure of the placenta may occasionally develop in humans to provide the blood vessels in different parts (lobes) is called succenturiate placenta. The number of lobes may be two (bilobed or bipartite), three (trilobed or tripartite) or more. It causes a risk of profuse bleeding during parturition.

The placenta can also be classified into four categories according to its morphology, i.e. shape and point of attachments between foetal and endometrial tissues. These are diffuse, cotyledonary, zonary and discoid (Fig. 23.8). In diffuse types of the placenta, nearly all the allantochorionic surface villi are involved in the placenta present in horses and pigs. In cotyledonary types, cotyledons from the chorioallantoic surface are attached discretely to the endometrium, which is common in ruminants. The alantochorionic part of the placenta of some mammals appears like a band of tissue with the endometrium is called the zonary type of placenta found in dogs, cats and other carnivores, also in elephants. In most rodents and primates, including humans, the placenta is flat and circular disc-shaped, called discoid type placenta. In the discoid placenta, the villi of the foetal surface are attached to the maternal tissues throughout the circular plate.

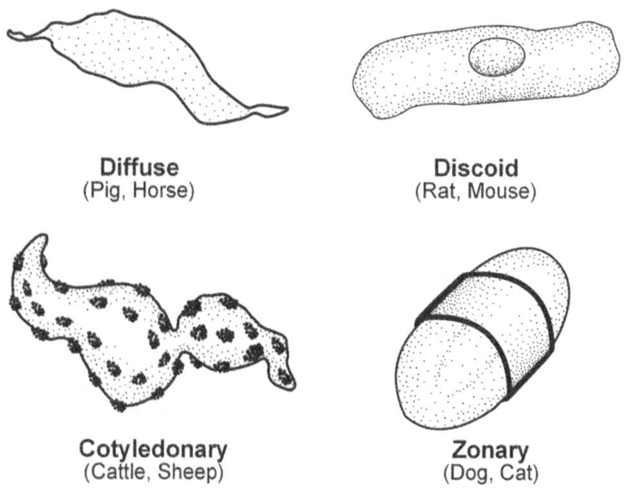

Fig. 23.8 Types of placenta according to shape and attachment. On the basis of shape and attachment points, the placenta classifies into **diffuse** (involving the entire surface of the allantochorion), **cotyledonary** (multiple attachment points), **zonary** (placenta appears as a band surrounding the foetus) and **discoid** (single placenta of discoid shape)

According to the structure (architecture) of the chorionic membrane and its orientation with the endometrium, the placenta can be classified into three types. The crumpled surface of the endometrium (uterine epithelium), when attached to the chorionic folds, it is called the folded placenta. It is found in the pig. In the villous placenta, three types of villi are attached in chorion; primary villi are found in trophoblastic columns. It branches in extraembryonic mesoderm as secondary villi and contains secondary villi tertiary villi involved in the nutrient exchange. It is found in ruminants, horses and primates. When chorionic villi are surrounded by the maternal blood vessels and form a network at the junction of the foeto-maternal junctional space, it is called the labyrinth zone. The labyrinthine placenta occurs in carnivores and rodents.

Based on the loss of maternal tissue during parturition, the placenta can be classified into two types deciduate and non-deciduate. In the deciduate kind of placenta, the maternal epithelium, submucosa, deciduate cells and foetal placenta are shaded during parturition. The endothelial part remains in the mother's uterus during parturition. It generally occurs in humans and monkeys. In dogs and cats, similar types of maternal tissues are lost. Non-deciduate type of placenta is found in ruminants, pigs and mares where the foetal membrane is shaded, no maternal tissues are expelled like the deciduate type.

23.3.4.3 Placental Exchange

The placenta is involved in the physiological exchange of various substances from mother to foetus and vice versa. It provides nutrients, oxygen, multiple bio-molecules, hormones and immune modulators from the maternal end required for foetal growth. Placenta also eliminates the metabolic waste products of the foetus. The placenta allows a majority of substances to pass through it. However, some substances are required to metabolise before placental exchange.

23.3.4.4 Functions of Placenta

23.3.4.4.1 Foetal Nutrition

The nutrients required for foetal growth can be categorised into histotrophs and hemotrophs. Histotrophs are the secretory products of the endometrial glands and the extravasation materials. Hemotrophs are the substances directly transferred from the maternal blood to the foetus through the placenta. The histotrophic and hemotrophic substances are physiologically exchanged through the chorioallantoic placenta. The chorion plays a major role in exchange processes. The hemotrophic substances are generally transported to the non-areolar region of the epitheliochorial placenta.

There are several transporter proteins involved in the nutrient exchange process. Uteroferrin, a transporter protein synthesised in uterine glands, assists in exchanging the iron molecule from mother blood in pig and horse through the areola. A transporter glycoprotein, transferrin, transfers the iron molecule in the hemochorial type of the placenta like a rat, rabbit, guinea pig, mouse and human.

Various essential nutrients or bio-molecules are transferred through the placenta from mother to foetus carried by the blood. During the advanced stage of pregnancy, more than 80% of the uterine blood flows restricted to the cotyledonary areas, where nearly 15% flows to the endometrial and the rest to the myometrium.

Glucose is the major energy source for foetuses concerned with the endotheliochorial and hemochorial placenta. Glucose can be converted to fructose in the trophectoderm cells of epitheliochorial and synepitheliochorial placenta to provide additional energy. Glucose can also be utilised for the biosynthesis of some essential substances like glycosaminoglycans, phospholipids, nucleic acids and other substances. Glucose is transported by passive diffusion due to concentration gradient and facilitated diffusion. The rate of glucose transport depends upon the foetal glucose requirement, types of the placenta and availability of glucose transporters. Maximum glucose transportation occurs in the hemochorial placenta. The insulin-independent six hexose transporters, GLUT1, GLUT2, GLUT3, GLUT4, GLUT5 and GLUT7, play a vital role in facilitating glucose transport through the mammalian placenta. The expression of glucose transporters is species specific.

In rodents and humans, GLUT1 and GLUT3 transporters; in cattle and sheep, GLUT1, and in mice GLUT1 and GLUT2 are predominant. GLUT1 and GLUT3 are also involved in water transport and glucose in different animals. The connexin 26 glucose transporters are seen in rats and mice. Glucose can also be transported through sodium-glucose co-transporters using a sodium/potassium pump system. The sodium-glucose transporter family (SGLT) like SGLT1, SGLT2, SGLT4 and SGLT5 are involved in glucose transport. Progesterone, interferon tau, growth hormone (in late pregnancy), oestrogen (in rats) and maternal obesity influence the transport and uptake of glucose and fructose. The maternal glucose may transport to the foetal blood in the form of lactate, and the foetus utilises it as an energy source. Lactates are transported by the lactate-hydrogen ion co-transport system (the monocarboxylate transporter). The other monocarboxylates, like pyruvate and β-hydroxybutyrate, are also transported by the same system. Some dicarboxylates, like succinate, malate, fumarate, α-oxoglutarate and citrate (the intermediate products of some tricarboxylates), are transported by dicarboxylate transporters through transmembrane electrochemical sodium ion gradient.

The concentration of amino acids is generally more in foetal blood than in maternal blood. Hence, an energy-dependent active transport process is required to transport amino acids against the concentration gradient. The amino acid transporters are classified according to the type of amino acids (acidic or anionic, basic or cationic and neutral). Mainly the sodium-dependent amino acid transporters are involved in this process. The EAAC1, GLAST1 and GLT1 are the sodium-dependent transporters involved in acidic or anionic amino acids (like aspartate and glutamate) transportation in the rat. The 4F2HC is another sodium-dependent transporter involved in lysine, arginine and histidine (the basic or cationic amino acids) in rats and humans. The same amino acids can be transported by sodium-independent transporters like MCAT1 in rats and CAT-4 in humans. Some amino acids (mainly the anionic amino acids) need to be converted into other amino acids before placental transport like serine is converted to glycine. The urea and ethanol are diffused passively through the placental membrane.

Various fatty acids and triglycerides and their metabolites like choline, cholesterol, steroids hormones and fat-soluble vitamins are transported through the placenta. The lipase enzyme initially hydrolyses the triglycerides into free fatty acids on the maternal surface of the syncytiotrophoblast. Some essential fatty acid transporters are fatty acid translocase (FAT), plasma membrane fatty acid-binding protein (FABPpm), fatty acid transporter protein (FATP) and cytoplasmic fatty acid-binding protein family (FABP) in rats and humans. Cholesterols are transported in lipoproteins, and the energy-dependent active ABC transporters are used to transport the cholesterols. The triglycerides can be stored in the trophoblast and used for energy through β-oxidation or can transform into different polyunsaturated fatty acids.

Various nucleosides are transported by two types of placental nucleoside transporters, the equilibrative and concentrative. The equilibrative nucleoside transporters transfer both purine and pyrimidine bases. The concentrative nucleoside transporters involve dipyridamole and nitrobenzylthioinosine transfer.

The lipid-soluble vitamins are transferred by diffusion and carrier-mediated transport process. Vitamin A is transported by retinol-binding protein. Vitamin D ($1,25(OH)_2D_3$) is transported faster than $25(OH)D_3$. Vitamin E or α-tocopherol is transferred in small quantities, and vitamin K is mainly impermeable to the placental barrier. The water-soluble vitamins are either transported directly or form metabolites in the syncytiotrophoblast. Ascorbic acid (vitamin C) is diffused in its two metabolised form, the dehydroascorbate (oxidised state), which diffuse faster or is reduced as ascorbate. The riboflavin is transported in the form of flavin adenine dinucleotide and flavin mononucleotide transporter. The folate is transported by sodium-independent transporter, and thiamine transportation favours by Ca-dependent transporters.

Major macro minerals like sodium, potassium, chlorine, calcium, iron, magnesium, phosphorus and iodine and micro minerals like zinc, aluminium, sulphur, chromium and molybdenum transfer through the placenta. Some heavy metals like cadmium, mercury and lead can also transport through the placenta. Contaminated drinking water and food, exposure to polluted soils, industrial waste and pesticides are the major sources of these heavy metals. Sodium is transported by sodium ion–hydrogen ion exchange process, as conductance and as a co-transport system with other molecules like organic solutes and inorganic anions. Sodium acts as a co-transporter for essential nutrients like amino acids, dicarboxylates, serotonin, some vitamins and phosphates. Sodium can also be transported by a Na^+–K^+ pump. Calcium transports through the placenta like the intestinal absorption process, involving both the energy-dependent active process and facilitated diffusion. In rats and rodents, the yolk sac contains calcium transporter proteins. A magnesium pump is involved in the transportation of magnesium. Phosphates and iodine are transported against the concentration gradient by a sodium-dependent active transportation process. Parathyroid hormone reduces phosphate transportation. Its transportation is modulated by sodium and amino acid concentration. Oxytocin, hCG, prolactin and 17β-estradiol facilitate the iodine uptake by the trophoblast cells in different species. Iron is mainly transported in the form of ferritin with the help of specific protein transporters like uteroferrin and transferring (discussed in detail in an earlier section).

Transporter proteins transport zinc in the placenta, and the placenta generally serves as a partial barrier to aluminium transportation. Hydrogen ion transportation helps maintain the acid-base balance in the foetus and placenta. The hydrogen ions are transported mainly by sodium ion–hydrogen ion exchange process, proton pump, co-transport with organic ions like lactic acid (monocarboxylate) transportation and protein-mediated transportation (active transportation by carrier proteins) process. Chloride ion is transported by anion exchanger, as HCO_3^- and competing with Br^-, I^-, NO_3^- and SCN^-. It is also transported by co-transporter with organic substances like taurine and serotonin. Sulphates (SO_4^{2-}), selenium (SeO_4^{2-}), chromium (CrO_4^{2-}) and molybdenum (MoO_4^{2-}) are mainly transported as anion exchangers, with HCO_3^-. Sulphates (SO_4^{2-}) may also act as anion exchangers to transport other substances.

The movement of immunoglobulin (Ig, antibody) depends upon the types of the placenta. In hemochorial types of the placenta, the IgG can transport to the foetus through Ig-binding proteins through endocytosis. No Igs are generally transported to the foetus in the epitheliochorial and endotheliochorial placenta. The offspring of the mammals having epitheliochorial and endotheliochorial placenta are received the Igs through colostrums. In carnivores (dogs and cats), some Igs can be transferred in a minimum quantity from mother to foetus in the last trimester of gestation. It moves through a hemochorial placental structure.

Bilirubin and various drugs are generally lipophilic and transferred from the foetus to maternal blood without metabolism as an unconjugated form. Such substances may be metabolised in the foetal liver and develop a conjugated form (like conjugated-bilirubin). The metabolised conjugated substances are transported inadequately due to their water-soluble property. Hence, the inability of the foetal liver for its improper metabolism facilitates the excretory substances from the foetus.

23.3.4.4.2 Gaseous Exchange

The gaseous exchange usually takes place through the chorion. Fully oxygenated blood is entered into the placenta from maternal circulation through uterine arteries. The uterine artery gives rise to numerous spiral arterioles that open into the intervillous space. Deoxygenated blood from the foetus is first carried to the placental circulation and then communicates with uterine veins. The chorionic villi and umbilical arteries are situated in the intervillous space. At this site, both the oxygenated and deoxygenated blood comes. Still, due to placental oxygen uptake, a partial pressure gradient is established where the intervillous space has a lower partial pressure of oxygen than the maternal blood. The endothelin and prostanoids have a vasoconstriction role, whereas nitric oxide causes vasodilatation. The foetoplacental circulation is susceptible to hypoxia, which leads to excessive free radicals resulting in pre-eclampsia and other pregnancy complications. Melatonin has an antioxidant role in the placenta.

Further, the oxygen-carrying capacity of foetal haemoglobin is more than mother haemoglobin in most domestic mammals. Foetal haemoglobin is absent in horses, dogs, rabbits and mice. The carbon dioxide is exchanged between maternal and foetal tissues through the placenta due to partial pressure differences.

23.3.4.4.3 Endocrine Role of the Placenta

Placenta produces several steroid and glycoprotein hormones that promote foetal growth and modulate the maternal physiological system in terms of ovarian activity, growth of the mammary gland and parturition. The placenta produces several cytokines, but placental hormones are generally used as a biological marker for confirmation of pregnancy. Placental hormones are categorised into four types according to their production and action, steroid hormones, prolactin-growth hormone-related hormones, neuroactive hormones and other groups of hormones (Table 23.15). The interaction of placental hormones and growth factors and maternal

Table 23.15 Interaction of placental hormones and growth factors during pregnancy in rodents and primates, including human

				Impact on mother physiology					
Hormones/ growth factors	Expression profile	Up-regulation	Down-regulation	Energy production	Bone metabolism	Immune system	Lactation	Cardiovascular system	Maternal behaviour
Steroid hormones									
E2	1T↑, 2T↑↑, 3T↑↑↑	LEP, OXY	PRL	↑↑		↑↑	↑↑	↑↑	↑↑
P4	1T↑, 2T↑↑, 3T↑↑↑		PL, OXY	↑↑		↑↑	↑↑	↑↑	↑↑
Prolactin-growth hormone-related hormones									
PRL	2T↑↑, 3T↑↑↑	PL, IGF2	ACT	↑↑			↑↑		↑↑
PL	1T↑, 2T↑↑, 3T↑↑↑	IGF2, E2	ACT	↑↑			↑↑		↑↑
GH	2T↑↑	P4		↑↑			↑↑		↑↑
GHRH	2T↑↑, 3T↑↑	GH							
IGF2	1T↑, 2T↑↑, 3T↑↑↑	PLF	LEP	↑↑					
PLF	2T↑↑, 3T↑↑								
PRP	3T↑↑								
Neuroactive hormones									
OXY	3T↑↑	PRL		↑↑	↑↑		↑↑	↑↑	↑↑
MEL	1T↑, 2T↑↑, 3T↑↑↑	E2		↑↑			↑↑		↑↑
SER	1T↑, 2T↑↑, 3T↑↑↑	E2, PRL, GH		↑↑			↑↑		↑↑
KISS	1T↑, 2T↑↑, 3T↑↑↑	CG		↑↑				↑↑	
TRH	1T↑, 2T↑↑, 3T↑	PRL, CG		↑↑					↑↑
Other hormones									
REL	1T↑↑↑, 2T↑↑, 3T↑↑	PRL, GH				↑↑	↑↑	↑↑	
CG	1T↑↑↑, 2T↑↑, 3T↑↑	P4	LEP, E2			↑↑			
ACT	1T↑, 2T↑↑, 3T↑↑↑	E2, P4, GHRH			↑↑			↑↑	
LEP	1T↑, 2T↑↑↑, 3T↑↑	PL, CG		↑↑			↑↑	↑↑	↑↑
PTHrP	3T↑↑	P4		↑↑	↑↑		↑↑		

References: Petraglia et al. (2010), Feng et al. (2016), Napso et al. (2018)

Note: The expression profile of the hormones is demonstrated considering the total gestational period in three trimesters (first trimester, second trimester and third trimester as 1T, 2T and 3T, respectively). The upward arrow marking (↑) indicates up-regulation. Single, double and triple arrow (s) denote(s) the intensity of responsiveness. *E2* oestrogens, *P4* progesterone, *PRL* prolactin, *PL* placental lactogen, *GH* growth hormone, *GHRH* growth hormone releasing hormone, *IGF2* insulin-like growth factor 2, *PLF* proliferins, *PRP* proliferin-related proteins, *OXY* oxytocin, *MEL* melatonin, *SER* serotonin, *KISS* kisspeptin, *TRH* thyrotropin-releasing hormone, *REL* relaxin, *CG* chorionic gonadotropin, *ACT* activin, *LEP* leptin, *PTHrP* parathyroid hormone-related protein physiological adjustment is important for maintaining pregnancy and optimum foetal growth.

23.3.4.4.3.1 Placental Steroids

Two major steroid hormones, progesterone and oestrogen, are produced from the placenta and modulate the functions of the endometrium. Pregnenolone is the precursor of both oestrogen and progesterone produced from cholesterol in uninucleated trophoblast epithelial cells (UTC), and progesterone is synthesised either in binucleated dominated trophoblast giant cells (TGC) or foetal cotyledons with the help of cytochrome P450SCC in UTCs and 3β-hydroxysteroid dehydrogenase in the TGCs or, foetal cotyledons or, maternal caruncles. The maternal caruncles have more steroidogenic activity during the latter half of gestation, mainly in cattle. The progesterone thus produced is transported to both maternal and foetal systems. The permeability of progesterone in maternal tissue and foetus is almost the same in mammals. Still, the progesterone level is generally less in the foetus than in maternal tissue due to more metabolism. The progesterone is converted into 20α-hydroxy-progesterone by the 20-α-hydroxysteroid dehydrogenase enzyme (20α-HSD). It is mainly expressed in foetal cotyledons during late pregnancy and before parturition. Hence, it is metabolised before the diffusion of progesterone to maternal caruncles. In sheep, placental progesterone synthesis is biphasic, once in 50–70 days and another in 90–120 days of gestation. The placenta can produce sufficient progesterone after 150–200 days of pregnancy in horses and after 6 weeks in women independent of CL. Equine placental steroidogenesis is unique. Initially, the luteal progesterone produced by the influence of eCG supports placental and foetal growth during the first phase of pregnancy. Later, foetal pregnenolone and dehydroepiandrosterone (androgen) appear in the placenta and act as the major substrates to synthesise the 5-α-pregnanes and oestrogen-like compounds, respectively. In rats and mice, three types of steroids, progesterone, testosterone and oestrogens, are produced from the placenta in different phases. Placental testosterone is used as the precursor of progesterone and oestrogens. The level of placental progesterone production peaked during half of pregnancy (around 12 days of gestation), followed by a decline from day 14 to base within day 16. In humans, estriol and a tiny amount of equilin are synthesised. For some animals, like a dog, no additional steroid is required for maintaining pregnancy.

Progesterone acts as muscle relaxant and local immunosuppressive agent. It also involves endometrial differentiation and closure of the cervix. Progesterone also influences the HPO axis to reduce the secretion of FSH, followed by inhibition of oestrogen, LH surge and ovulation. Progesterone suppresses glucocorticoids and inhibits the prolactin receptor in mammary glands.

Placental oestrogen is oestrone in nature in contrast to ovarian estradiol in most of the higher groups of mammals. Mare can also synthesise other oestrogenic compounds like equilin and equilenin from the placenta. The horse's placenta has a deficiency of 17α-hydroxylase enzyme and a higher level of 3βHSD and CYP11A1; hence, foetal androgen is utilised to produce oestrogenic compounds like equilin and equilenin during pregnancy. The level of the oestrogenic compound during pregnancy is found to occur maximum in mares among the various domestic mammals. In cattle, the oestrone is generally produced as an inactivated form of oestrone sulphate, converted to active oestrogen by an enzyme steroid sulfatase (StS) in the caruncular epithelium. The foetal adrenal gland controls the rate of production and metabolism of placental oestrogens. The foetal adrenal is responsible for the activation of 17α-hydroxylase (a placental cytochrome P450) required in the oestrogen production pathway. Under the influence of 17α-hydroxylase and aromatase, foetal androgen, placental progesterone and other steroidogenic precursors are converted into oestrogenic compounds during steroidogenesis. Hence, the production rate of oestrogens in the placenta increases when the foetal pituitary-adrenal axis is matured at the late stage of gestation. Oestrogen is mainly responsible for the growth of endometrium and myometrium along. It also controls uterine blood flow during the pre-implantation period and at the time of parturition to increase the myometrial tone. During pregnancy, oestrogen also facilitates mammogenesis in some animals like horses, rats and mice. It enhances progesterone sensitivity, the production of prolactin, IGF-1 and phospholipids, regulates salt and water retention and has a crucial role in foetal neuroendocrine development, digestion, energy storage and other chemical homeostatic mechanisms. It influences maternal behaviour and controls the parturition process.

In domestic eutherian mammals, uteroplacental tissues can produce prostaglandin (PGF2α) immediately before parturition at the last stage of pregnancy. It is due to progesterone and oestrogen ratio alteration and activation of prostaglandin-endoperoxide synthase 2 (PGHS-2). The PGHS-2 is one of the major inducers for the synthesis of PGF2α, which is suppressed during gestation. The major roles of PGF2α are to increase contractility of the myometrium, destruction of the chorioallantois, partition of the placenta from the uterine attachment, and uterine involution and cervical dilatation.

Hence, the abundant oestrogenic and progestin compounds are formed simultaneously from the equine "foeto-placental unit", a rare phenomenon in domestic mammals. The increased level of pregnane from mid-pregnancy, followed by a reduced level immediately before parturition, is controlled by the presence of the receptor for pregnane.

Testosterone is mainly available in the first half, mainly from the ovarian source, where the placenta becomes the major source in the second half. As a result, testosterone level gradually increased and attained a peak a few days earlier to parturition (about day 18), then declined. Low levels of oestrogens are also produced in the placenta; these are 20 alpha-hydroxy-4-pregen-3-one and 17 beta-estradiol.

23.3.4.4.3.2 Placental Protein Hormones

The placenta produces several protein hormones. Major placental protein hormones are placental lactogens, chorionic gonadotropins and relaxin.

Placental lactogens (PL) are structurally similar to growth hormones and prolactin. It is synthesised in the trophoblasts of ruminants, rodents and primates. The terminology of placental lactogen based on species like bovine placental lactogen is known as bPL, sheep (ovine) as oPL, rat as rPL, mouse as mPL, hamster placental lactogen haPL and human as hPL. The bPL generally appears from 4 months of pregnancy. The oPL secrets from day 50 of gestation and continue to produce large quantities throughout the pregnancy. The rate of production of rPL and mPL is correlated with the litter size. The mPL is generally three types mPL-I (Prl3d1, produced at mid-pregnancy), mPL-II (Prl3b1, produced at latter half of pregnancy) and mouse PRL (prolactin). The rat can also make a variety of rPL, like PLI (Prl3d4, produced in the latter half of pregnancy). The PL is produced under the influence of prolactin and somatotropin-linked genes in ruminants and primates, respectively.

In humans, the chorionic somatomammotropin genes (ICSH-1 and ICSH-2) are involved. Several growth factors, depending on species, influence the synthesis of PL. It is influenced by IGF-1, angiotensin II, phospholipase A2 and epidermal growth factor in humans.

The PL is considered to involve in mammary gland development (mammogenesis) together with IGF-1, epidermal growth factor (EGF) and transforming growth factor-α (TGF-α). It influences maternal angiogenesis. It can also affect the mobilisation of the nutrients to the foetus during pregnancy by inhibiting maternal insulin. Gestational diabetes is the consequence of this insulin resistance due to PL. The glucose, amino acids, free fatty acids and ketone bodies are mainly transferred to the foetus by the influence of PL. The PL, along with prolactin, influences the choroid plexus and brain to induce maternal behaviour during pregnancy by altering the neural processes during the pregnancy and postpartum period. The PL is also involved in modulating autoimmune reactions and cell-mediated immunity and is associated with thymus and bone marrow functions. The PL, in combination with progesterone and prolactin, stimulates erythropoiesis. The PLI and PLII enhance progesterone synthesis by reducing the expression of 20-α-hydroxysteroid dehydrogenase (20α-HSD).

Chorionic gonadotropin (CG) is produced from trophoblasts and has functional similarities with pituitary gonadotropins. It is secreted in the horse (eCG, previously regarded as pregnant mare serum gonadotropin, PMSG) and primates, including humans (hCG). The mouse can also produce chorionic gonadotrophic with functional similarities with hCG, and its production rate is relatively increased with the litter size. The level of CG is increased from the stage of implantation. Hence, the presence of CG in urine or blood confirms the pregnancy at a very initial stage. The hCG acts over the LH-R in the corpus luteum, supporting its life span and preventing regression. Hence, hCG provides maternal recognition of pregnancy in humans, like interferon tau in ruminants. The eCG is structurally and functionally similar to LH. It is synthesised from the chorionic girdle, a special glandular structure and endometrial cups of the mare. The eCG appears at the end first month (from around day 25) of gestation and continues up to 3–4 months. It supports the primary corpus luteum to generate progesterone. It also influences the development of supplementary corpora lutea in mare.

The placenta is the major source of a polypeptide insulin-superfamily hormone, relaxin, produced from the trophoblast of the foetus. It appears from 12 weeks in the mare and reaches a peak at mid-pregnancy that corroborated with the levels of PL of cats and dogs where relaxin appears after third (reached peak at the 36th day) and fourth weeks of pregnancy, respectively. The concentration of relaxin rapidly declines at parturition. The Corpus luteum is the major source of relaxin in other mammals. In cattle and sheep, relaxin-like hormone, the insulin-like peptide 3 (INSL3), is produced in the corpus luteum. In the horse, trophoblast has enzyme furin which converts preprorelaxin into relaxin. The major functions of relaxin include endometrial angiogenesis through vascular endothelial growth factor (VEGF), relaxation of pelvic ligaments and pubic symphysis and cervical dilatation to facilitate parturition. It also involves the remodelling of the extracellular matrix. Relaxin can influence the growth of the uterus, vagina and cervix in late pregnancy in the pig.

23.3.4.4.3.3 Placental Proteins

The placenta of rodents can produce some luteotrophic PRL-like proteins, such as proliferin (Prl2c2) and proliferin-related protein (Prl7d1) from giant cells and cytotrophoblasts, respectively. Ruminants and rodents can produce several *prolactin-related proteins* (*PRP*) from the placenta. The placenta can secrete about six types PRPs in bovine (bPRP) and sheep (oPRP), 22 types in rat (rPRP) and mice (mPRP). The goat also secretes PRPs (cPRP) from the placenta. PRP helps in implantation, placentation and maintenance of pregnancy.

The trophoblastic cells and syncytium of sheep and goats and trophoblast of humans can produce placental growth hormone. It secretes from day 27 to 75 of pregnancy in sheep, with a peak at day 40–45. The human placental growth hormone is generally produced up to the second trimester, and then its activity is replaced by pituitary growth hormone. In early pregnancy, the placenta can produce some pregnancy-associated glycoproteins (PAG) in cattle (bovine placental specific protein, BPSP), sheep, goats, pigs, horses and mice. The PAGs belong to the pepsin-like protein family but have no enzymatic role. It is secreted from the trophectoderm in ruminants and involved in immunoregulation at maternal–foetal interaction during pregnancy. It is also used to identify the pregnancy in ruminants. The rodents and human placenta can synthesise an enzyme 11-β-hydroxysteroid dehydrogenase (11β-HSD2), which can protect the foetus from the high level of glucocorticoid by inactivating cortisol. The rodents produce it from mid-pregnancy, and in humans, it continues throughout the gestation. Hence, stress can affect pregnancy in rodents and humans. The human trophoblast cells can produce unique microRNAs (miRNAs), known as trophomiRs that suppress the maternal immune response against the developing foetus. In humans, it is usually used as a biomarker for some pregnancy disorders like pre-eclampsia and foetal trisomy 21.

The placenta of different animals secretes various types of peptide hormones, like inhibin A, activin A, adipokines and TGF-β superfamily proteins. Some neuropeptides like a gonadotropin-releasing hormone (GnRH), corticotropin-releasing hormone (CRH), thyrotropin-releasing hormone (TRH), somatostatin and ghrelin are also synthesised from the placenta. The placenta is also involved in the synthesis of many growth factors, like IGFs (from endometrial cells) and vascular endothelial growth factor (VEGF) under the influence of CG. The growth factors facilitate angiogenesis and placental exchange, and CRH involves steroidogenesis. Placenta produces cytokines like interleukin-8 (IL-8) and transcription factors (NF-B) from epithelial cells, cytotrophoblast cells and fibroblasts interacting with the immune cells. All are involved in the regulatory network for foetal growth, placental development and parturition. It is considered that cytokines derived from maternal adipose tissue and placental cells are originated from common molecules and involved in similar inflammatory events.

23.3.4.5 Placenta in Different Animals

23.3.4.5.1 Pig

The chorion and allantois grow rapidly during days 18–30 of pregnancy and fuse together within day 60. The exponential growth of the placenta continues up to day 70. The angiogenesis and the capillary bed are increased in the last 40 days of gestation. The allantochorion surface is highly folded to make ridge-like structures that adhere to the endometrium's grooves. The fold of the chorion builds a special structure called areolas made by trophoblasts cells that receive the endometrial secretions and are involved in erythrophagocytic activity and iron transportation from the degraded haemoglobin.

23.3.4.5.2 Horse

Structural characteristic of the horse's placenta is similar to a pig. The trophoblast of the horse forms a narrow band-like structure (later, chorionic girdle) at the junction between allantois and yolk sac. Later, it develops into ridges followed by glands from where the eCG secrets. The glands transformed into a decidua-like cell special structure, the endometrial cup. The glands hold the chorionic girdle in position, and the girdle starts to increase rapidly. The hyperplastic girdle cells gradually invade the endometrial epithelium, basement membrane and uterine stroma within day 40 of pregnancy and form eCG-secreting endometrial cup cells. The endometrial cups reach their maximum size within days 55–70 of gestation, having round shaped binucleated cells with maximum secreting capability. The leukocyte accumulation progressively occurred immediately after the maturation of endometrial cups from days 70 to 80. The B cells, T cells and various macrophages destroy the cups within days 100–140 of gestation. The major histocompatibility (MHC) class I antigen is potentiated by the immunological destruction process. Failure of development of endometrial cups within day 70–80 causes abortion. Mare bred with donkey develops one supplementary corpus luteum, whereas breeding with horse results in 2–3 corpora lutea. The supplementary corpus luteum formation generally starts around days 20–25 of gestation in horses. Both the primary and supplementary corpora lutea are regressed around 26 weeks, but placental progestagen (5-α-pregnanes, metabolite of progesterone) production is continued.

23.3.4.5.3 Dogs and Cats

Implantation is centric and anti-mesometrial, and the placenta is zonary-endotheliochorial in dogs and cats. The endometrial epithelium degenerates during implantation to form syncytiotrophoblast or syncytium at the centre of chorioallantois. The syncytium enters the endometrial epithelium and attaches to the endothelium of the maternal capillaries. The invading villi of the foetus later unite together to form a labyrinthine-type placenta. The dog placenta has a labyrinth zone and the junctional zone, sponge zone, glandular zone and hemophagous zone. The region between the labyrinth and gland zones is called a junctional zone. The trophoblast of this transitional area invades the endometrial gland cavity and contains a single layer of tall columnar cells with microvilli. The deep part of the junctional zone is termed as sponge zone. The placenta in dogs has hemophagous zones located on both maternal and foetal ends of the placenta, having high columnar trophectoderm involved in active phagocytosis, digestion of erythrocytes and iron transportation.

Both the dogs and cats have paraplacental structures, hematomas for its nutrients exchange. The pregnancy is maintained in the dog by the luteal progesterone and depends upon the pituitary gonadotropin (LH) prolactin. Hence, confirmation of pregnancy by measuring progesterone level is difficult in the dog, where corpus luteum is persisted more than during the pregnancy period in pseudopregnancy. The progesterone level can also confuse to confirm the pregnancy or pseudopregnancy in cats. The cat placenta can start to produce progesterone at around 3 weeks of ovulation, which may be perplexed when a luteal cyst exists. However, progesterone level is reduced initially at 10–12 days of ovulation in a pseudopregnant cat due to low StAR and 3βHSD mRNA expression though its expression is further increased from mid-pregnancy. Relaxin is synthesised in the placenta from about 4 weeks in dogs and cats. Hence, relaxin can be used to determine the pregnancy in dogs and cats.

23.3.4.5.4 Rodents and Humans

Both the rodents and humans have histologically hemochorial and morphologically discoid type placenta, but structurally humans and primates have a villous type, whereas rodents contain labyrinthine type placenta. The maternal blood comes into direct contact with foetal chorion without fluid exchange due to low blood pressure, resulting in backflows of deoxygenated blood through endometrial veins. The placental exchange is similar mainly in rodents and humans because of their histological similarity. It has three layers between maternal and foetal blood vessels, which resemble the maternal-foetal counter-current arrangement. But the trophoblastic epithelial cell layer does not remain uniform throughout the gestational period in rodents and humans. The trophoblastic layer is gradually reduced during different stages of gestation in humans. Three layers exist during the entire pregnancy period in rats and mice, termed haemotrichorial whereas two layers are persisted in rabbits and the first trimester of human pregnancy, regarded as haemodichorial; and a single layer is present in guinea pigs and the last two trimesters of human pregnancy, known as haemomonochorial. Hence, placental exchange in humans during the first trimester of pregnancy can be compared with rabbits, and the last two trimesters can be compared with guinea pigs. Thus, in various research trials of drugs and other bio-molecules, the rabbit and guinea pig model is used instead of the human, according to the stage of gestation. Different immunoglobulins are transferred through the chorio-allantoic placenta in humans, whereas it occurs in rodents through the yolk sac. Within rodents, the trophoblastic invasion of the maternal arteries is also variable. More invasion of trophoblast occurs in rats and guinea pigs than the mouse.

23.3.4.6 Foetal Sex-Specific Placental Activity

Presence or absence of sex-bearing Y chromosome in foetus influences the placental responses in some animals, including humans. It is due to the presence of some specific coding genes that demonstrates sexually dimorphic differences and influence the growth and development of the placenta and foetus. The specific coding genes influence the production of placental and manipulate the function of steroids, neuropeptides (serotonin, melatonin and oxytocin), prolactin, growth hormone and various growth factors (including placental lactogen and IGF). It results in up- and-down-regulation of the flow of different nutrients and other essential proteins through the placenta.

In cattle, buffalo, sheep, goats and other ungulates, the presence of a female sex-bearing foetus influences the greater expression of interferon tau (IFN_T) than in males. Hence, the signal for maternal recognition of pregnancy is better expressed in the female sex-bearing foetus, caused to influence more anti-luteolytic effects. The male sex-bearing foetus grows faster than the female foetus by enhancing metabolism and amino acid transportation in cattle, like all other eutherian mammals. In mice, breed-specific characteristics changes occur. The placental growth and metabolic rate are up-regulated in male spiny mouse foetus than in female foetus by influencing glucose transporter, nutrient supply, cell growth and other systems with the dynamic changing of specific genes like glucose transporter protein type 1 (*Slc2a1* or *Glut1*), insulin-like growth factor 1(*Igf1r*), mitogen-activated protein kinase kinase 1 (*Map2k1*) where the expression of said genes are fixed pattern in the female foetus. The olfactory sense is up-regulated more in female NIH Swiss mice with a higher expression of olfactory receptor 154 (*Olfr154*). The same mouse can express more steroid receptors (*Esr1* and *Ar*) in the female foetus.

In contrast, the male foetus-bearing placenta up-regulated more *Prl* gene for luteotrophic prolactin functioning resulting in better maternal recognition of pregnancy in mice (opposite nature to ungulates). The female foetus-bearing rabbit up-regulates the *Lxra* gene, causing more mobilisation of fat, whereas the fats (triglycerols) are accumulated more in the male foetus-bearing placenta. In humans, female foetus-bearing placenta expressed more hcg and interferon tau (IFN_T) gene causing better maternal recognition of pregnancy by more luteotrophic action than male foetus-bearing placenta. The male foetus-bearing placenta is more susceptible to uterine infection and maternal stress. It is due to the up-regulation of glucocorticoid receptor alpha (*GRαD2*) causes more metabolisms of glucocorticoids by increasing the 1β-hydroxysteroid dehydrogenase (11β-HSD2) enzyme in male foetus-bearing placenta. In addition, the female foetus-bearing placenta can also up-regulate specific genes to generate more immune responses against infection, like *JAK1, IL2RB, Clusterin, LTBP, CXCL1* and *IL1RL1*, where the similar genes are down-regulated in male foetus-bearing placenta. The renin-angiotensin system is up-regulated, due to *ACE1* and *ACE2* gene expression in deciduas, causing alteration in blood pressure and sodium reabsorption, in both the foetus and mother, having a female foetus compared to the male foetus. Thus, the male foetus-bearing mother is more prone to hypertension (diastolic pressure) than the female foetus-bearing mother.

23.3.4.7 Foetal Growth

Foetal growth has a tremendous influence on animal production. Low birth weight is associated with reduced energy reserve and low thermoregulatory activity, which leads to neonatal mortality. However, the greater bodyweight of calves also creates a parturition hindrance. Foetal growth is influenced by the number of foetuses (birth weight is inversely related to the number of foetuses), sex of foetuses (male foetuses have higher growth than females), age and parity of the cow (birth weight and parity are directly proportional). Foetal growth is decreased in heat and increased in the cold. Inadequate nutrition of the mother also negatively affects foetal growth. The foetal length is measured by crown-rump length (CRL), which indicates the normal growth of the foetuses. It is the straight length from the occipital magnum to first caudal vertebra. The age of the foetuses (in a month) can be calculated from CRL (in cm) by a formula.

$$\text{Gestational age (month)} = \sqrt{}(\text{CRL} + 1) - 1.$$

Different events of foetal growth in bovines with gestational age are depicted in Table 23.16.

Table 23.16 Events of foetal growth in bovines

Events of foetal growth	Age in days (considering ovulation as day 1)
Placentation and start of organogenesis	42
Gonadal development	45–60
Bone ossification starts	50–60
Completion of rumen differentiation	70
Increased caruncular vascularisation and blood flow	120
Completion of caruncular arterial vascularisation	150
The appearance of brown fat	190
Further cellular differentiation and growth of all tissues	Last third of gestation

Source: Pohler et al. (2020)

23.3.4.8 Foetal Fluids

23.3.4.8.1 Amniotic Fluid

It is a hypotonic, clear, colourless mucoid fluid situated within the amniotic cavity. The predominant source of amniotic fluid is the urine through foetal swallowing, foetal lung secretions and foetal nasal and buccal are also added to amniotic fluid. The components of amniotic fluid in a full-term human foetus are foetal urine: 800–1200 mL/day, foetal lung liquid: 170 mL/day, oral-nasal secretions: 25 mL/day, intramembranous flow: 200–400 mL/day, and foetal swallowing: 500–1000 mL/day. In the first trimester, AF is isotonic to maternal plasma. AF composition changes as foetal urine are added to amniotic fluid in humans during the second half of pregnancy. In cattle, the protein content of AF increases during the second and third trimester of pregnancy. The level of creatinine and urea increased as the gestation progressed. Excess AF accumulation leads to a pathological condition called polyhydramnios or hydramnios. The volume of amniotic fluid varies with species. The volume of AF is 5–6 L in cattle, 3–7 L in mare, 350–700 mL in sheep, 400–1200 mL in goats, 40–200 mL in sows and 80–100 mL in dogs. Amniotic fluid protects the foetus from external shock, prevents adhesions, and aids in parturition by providing lubrication to the birth canal. The removal of amniotic fluid is done by amniocentesis to evaluate foetal pathology.

23.3.4.8.2 Allantoic Fluid

It is a clear, viscous and amber coloured fluid. It derives from foetal urine and secretions from the amniotic membrane. It contains a low Na, Cl and glucose level and a high K, Mg, fructose, creatinine, urea and uric acid. The volume of amniotic fluid increases with gestation age. Allantoic fluid volume is 4–5 L in cow, 8–20 L in mare, 500–1500 mL in the ewe, 100–200 mL in sow, 10–50 mL in bitch and 3–15 mL in the cat. Allantoic fluid stores the foetus's excretory products and helps maintain the osmotic pressure of foetal plasma.

23.3.4.9 Foetal Growth Restriction (FGR) or Intrauterine Growth Restriction (IUGR)

Foetal growth restriction (FGR) occurs when intrauterine foetal growth is restricted or retarded due to congenital abnormalities or pathophysiological alterations. FGR can be developed for several reasons like hypoxia, hyperthermia and other pathophysiological conditions that interfere with maternal nutrient transfer. Certain hormones and growth factors control the transportation of nutrients. Hence, by assessing these hormones and growth factors, the FGR can be evaluated along with the direct measurement of the foetuses. The genetic abnormalities to cause GFR can be evaluated through the expression of certain species-specific genes.

In mice, the extracellular signal-regulated kinase 3 (*ERK3*) genes are responsible for visceral growth. In humans, placental protein regulators *CSH1* (chorionic somatomammotropin hormone 1, placental lactogen) of syncytiotrophoblasts origin and *KISS1* (kisspeptin 1, metastasis suppressor); of cytotrophoblasts, origin, and *PEG10* (paternally expressed 10) are used to predict FGR.

23.3.4.10 Factors Affecting Gestation Length

Major influencing factors on gestation can be categorised as genetic, maternal, foetal and environmental (Fig. 23.6). Gestation length is variable in different species (Table 23.10), breeds and foetal genotypes. Generally, the larger species have a longer gestation length. Breed variation is also noticed within a species, like the gestation length of the medium-wool and meat-type breeds of sheep ordinarily have a shorter gestation period than the fine-wool breeds. The ewes bred to white-faced, wool-breed rams generally exhibited a little longer gestation period than those bred to black-faced and meat-type rams. Season of breeding may affect the gestation length in the same species, particularly in the horse (Marwari mares). Different genotypes of the foetus affect the gestation length. The gestation period of mare X stallion is 320–350 days, whereas in mare X jack donkey (mule), it is 360–380 days, and jenny donkey X stallion (hinny) has a shorter gestation length than the horse.

Age of dam is an important maternal factor that causes variation in gestation length. Generally, the young have a slightly shorter gestation length than the older. The parity also has an effect. Litter size, foetal sex and endocrine disturbances also modulate the gestation length. A large litter size reduces the gestation length. Thus, polytocous animals have a shorter gestation length than monotocous animals. The occurrence of twin foetuses in monotocous animals causes to generally 5–6 days shorter gestation length. Male foetuses usually have 1–2 days more gestation length than female foetuses. Anterior pituitary and adrenal hormones of the foetus are involved in the initiation of parturition at the end of gestation. Nutrition and season are the two important environmental factors influencing the gestation length. Season causes severely affect the gestation length in seasonal breeders like the mare, when conceived in late summer or autumn, reduced the gestation length than those conceived in early spring. Malnutrition in the dam can reduce the gestation length. Poor nutrition, particularly during the last half of gestation, affects birth weight and survivability. Deficiency of Vit A and iodine increase the gestation length.

23.4 Parturition and Postpartum Physiology

The physiological process of delivering foetuses and foetal membranes from the uterus to the external environment is called parturition or *eutokia*. It occurs in three stages with species-specific time duration (Table 23.17).

Table 23.17 Duration of various events of parturition (in hour)

	Stage-1	Stage-2	Stage-3	
Species	Cervical dilation and initiation of contractions	Foetal expulsion	Placental expulsion	Parturition terminology
Cattle	2–6 (24 max)	2–5	8–12	Calving
Buffalo	2–6	0.5–1	6–12	Calving
Sheep	2–5	0.5–2	0.5–8	Lambing
Goat	4–8 (12 max)	0.5–4	1–8	Kidding
Pig	2–12	2.5–3	1–4	Farrowing
Horse	1–4	5–40 min	1	Foaling
Dog	2–12	1 (24 max)	6–12	Whelping
Cat	5–30 min (2 max)		24 (max)	Queening
Rat	1–4.5 (2.5 avg)	5–40 min	1–2.5 (1.5 avg)	Parturition
Rabbit	5–30 min (10 min, avg)		–	Kindling
Human	5–6	30–45 min	5–30 min	Childbirth

Source: Compiled from various sources
Max maximum, *avg* average

23.4.1 Foetal changes before parturition

The foetus undergoes physiological and structural changes before the parturition for the extrauterine life. The physiological changes include

1. Lung maturation and its expansion by the secretion of surfactants that reduce the alveolar surface tension.
2. Development of glycogen stored in the liver for energy supply till the initiation of suckling.
3. Increased output secretion of catecholamines and tri-iodothyronine for metabolic activity and thermoregulation.

The structural changes of the foetuses that occur before the parturition include

1. Closure of the ductus arteriosus.
2. Closure of foramen ovale within a few hours of birth.

23.4.2 Factors Responsible for the Initiation of Parturition

Throughout the gestation, the uterus remains quiescent under the influence of progesterone secreted from CL and placental. Uterine contraction initiates at the time of parturition due to decreased progesterone levels and increased oestrogen levels. Several other bio-molecules like prostacyclin, relaxin, nitric oxide and catecholamines begin the parturition process. The foetal glucocorticoid surge is the universal signal for the parturition in most domestic species. The physical,

Table 23.18 Factors responsible for the initiation of parturition

Factors		Effects
Physiological factors	Increased foetal size	Uterine irritability
	Uterine distension	Reversal of progesterone block
	Placental fatty degeneration including the presence of infarcts	Interfere foetal nutrition and foetal detachment from the uterus
Biochemical factors	Increase in CO_2 tension in maternal blood due to foetal activity	Increase uterine contractility
	Release of foetal serotonin	Induces collagenase activity and stasis of cotyledonary blood supply
Endocrine factors	*Foetal*	
	1. Increased foetal corticotrophin-releasing hormone and ACTH	Increased foetal cortisol secretion
	2. Increased cortisol secretion	Converts P4 to E2 and release of PG
	3. Increased endogenous opioids	Induces ACTH secretion
	Maternal	
	1. Reversal of progesterone block	Increased uterine contraction
	2. Secretion of relaxin	
	3. Increased placental oestrogens	
	4. Release of pro-inflammatory cytokines	Dilation of birth canal and release of PG
	5. Release of PG	Dilation of pubic symphysis and relaxation of sacro-sciatic ligaments Cervical softening of the cervix Stimulates smooth muscle contractility
	6. Release of oxytocin	Induces myometrial contractility

biochemical and hormonal signal factors to initiate the parturition process are presented in Table 23.18.

23.4.3 Mechanism of Parturition

At the end of gestation, the foetal hypothalamic–pituitary–adrenal (HPA) matures and increases cortisol secretion (Fig. 23.9). Foetal cortisol influences the functional maturation of foetal physiological systems, particularly the lungs and the cardiovascular system requires for the newborn immediately after birth. Increased foetal cortisol up-regulates cyclooxygenase 2 (COX2) and steroidogenic key enzyme 17a-hydroxylase-C17,20-lyase (CYP17) in the trophoblast cells. COX2 increases prostaglandin production. Activation of CYP17 leads to the metabolism of pregnenolone to dehydroepiandrosterone, and the level of progesterone falls along with increased oestrogen level. The altered endocrine milieu (decreased progesterone, increased oestrogen) induces contraction associated proteins in the myometrium. PG promotes inflammatory processes and increases blood flow to the uterus and placenta and myometrial contraction initiation. Relaxin causes the dilation of the cervix.

23.4.3.1 Ferguson Reflex

Ferguson reflex is a neuroendocrine reflex induced by the distension of the cervix caused by the foetus that stimulates a series of neuroendocrine responses to expel the foetus. The uterine distension signals the nucleus tractus solitarii (NTS) through spinal and vagal afferent nerves. NTS neurons project into the magnocellular oxytocin neurons at para ventricular neurone (PVN). The neurones of NTS release nor-adrenaline that acts over the $\alpha 1$ receptors n PVN to activate the oxytocin neurons to release oxytocin that initiates myometrial contraction.

Melanin of the pineal gland controls the circadian rhythm through the suprachiasmatic nucleus (SCN) and stimulates the uterine contraction, determining the time of parturition. Melanin generally affects the uterus during the resting phase of the day; hence, parturition usually occurs in the daytime in the nocturnal animals like rodents (rabbits) and during late night or early morning hours in humans.

23.4.4 Inflammatory Changes During Parturition

The inflammatory process before and during the parturition is termed cervical ripening. It starts a few days before parturition with a slow progression rate but immediately before, the inflammatory reaction accelerates rapidly. The types of cytokines involved in this process are species specific. In cows, the IL-1β, IL-8 and IL-10 increase, whereas TNF-α decreases. Granulocyte colony-stimulating factor (G-CSF), leukaemia inhibitory factor (LIF) and prostaglandin (PGE_2)

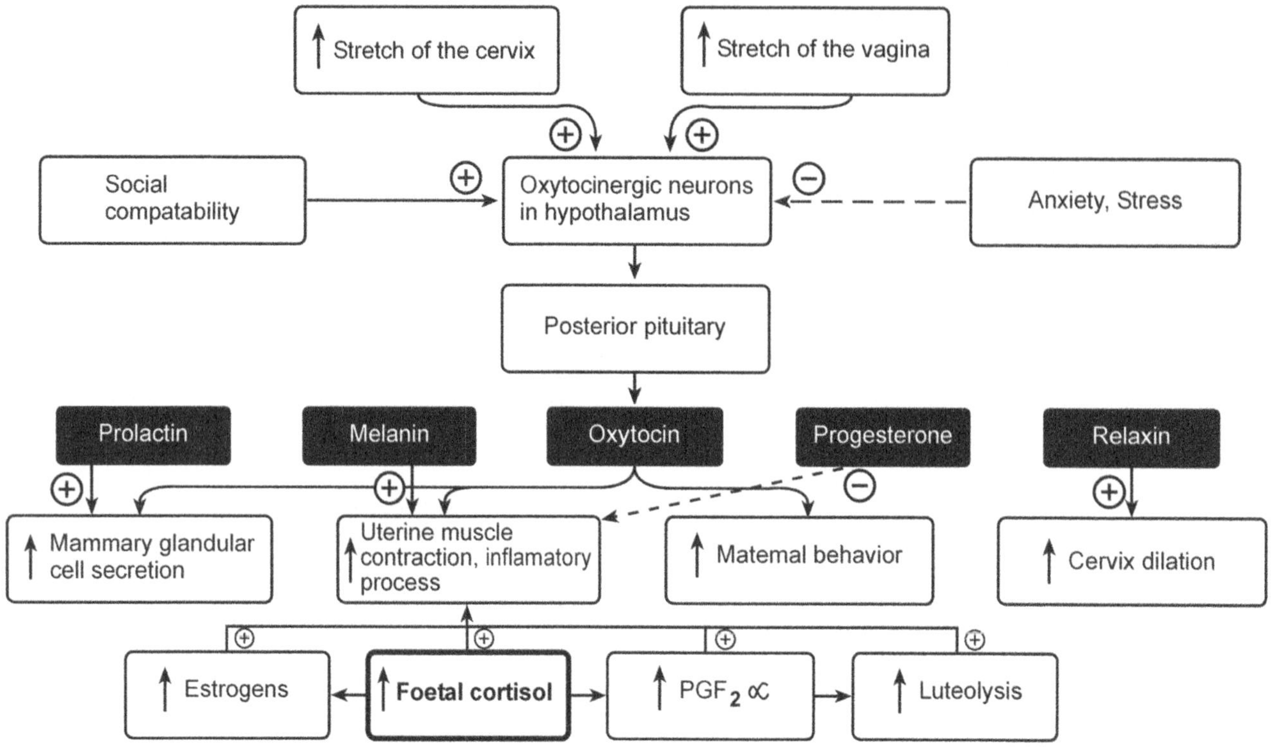

Fig. 23.9 Neuroendocrine mechanism of parturition. The cervical stretching induces **oxytocin** secretion from the **posterior pituitary**. Oxytocin initiates myometrial contraction. Higher secretion of PGF2α causes luteolysis and decreased **progesterone** secretion. The myometrium overcomes the progesterone block to initiate contraction. **Relaxin** causes cervical dilation. Other associated endocrine factors of parturition are **prolactin** (induces maternal behaviour) and **melanin** (causes uterine contraction). The **foetal cortisol** also promotes myometrial contraction

are increased during stage-1 of parturition. All these bio-molecules cause neutrophils into the cervical tissues and induce the release of matrix metalloproteinases (MMPs) from the stromal cells, fibroblast and smooth muscle cells. Some of the important MMPs are MMP-1 (Fibroblast collagenase), MMP-2 (Gelatinase-A), MMP-8 (leukocyte collagenase) and MMP-9 (Gelatinase-B). MMPs cause depletion of the collagen network and loosen the interactions between the bio-molecules present in the extracellular matrix of cervical connective tissue to decrease cervical rigidity and soften it.

23.4.5 Events of Parturition

There are three events of parturition that occurs sequentially. Stage-1 of parturition is the dilation of the cervix. It is also referred to as the onset of labour. The dilation of the cervix characterises it. Initially, the dilation is occurred slowly, called the latent phase, followed by an active phase. Stage-2 or the expulsion or delivery of the newborn commences after full dilation of the cervix and continues till the expulsion of foetuses. The first phase of expulsion is passive. The foetus moves down through the vagina, followed by a short duration of active phase with the contraction of abdominal and uterine muscle contraction that push the foetus to expel out. Stage-3 is the placental shedding stage when the placenta is expelled out after delivery of the young. Pluriparous animals require less time to complete each stage than primiparous animals. The species-specific characterised events of parturition are depicted in Table 23.19.

23.4.5.1 Stage-1

The cervix remains firmly closed during the entire gestation period in the form of a cervical plug. In stage-1, this plug is dissolved completely by the inflammatory process. The cervix becomes relaxed. The uterine peristalsis starts at the apex of the horn after getting free from the progesterone block. The foetus's orientation changes and it rotates along the axis. During this stage, animals usually exhibit some characteristic behavioural changes, considered parturition approaching symptoms. Absolute cervical dilation denotes the completion of this stage and commencement of stage-2.

Behavioural changes before parturition: Immediately before parturition is uneasiness or discomfort, anorexia, rising tail and mucous discharge from the oedematous and flaccid vulva. The softening and relaxation of the pelvic ligaments near the pin bones causing depression in the area generally occurs in domestic mammals. In a mare, the sinking

Table 23.19 Characteristic features of parturition in different animals

Species and acts of parturition	Signs of parturition	Mechanism of initiation of parturition	Parturition process
Cow (calving)	1. Distended and swollen mammary gland 2. Swollen and prominent teats with viscoid and clear secretions 3. Subcutaneous oedema of udder surrounding tissues ***Immediate signs*** 1. Kicking the abdomen, switching the tail, and frequent laying down and rising	Foetal cortisol stimulates the release of PGF2α and alterations in progesterone and oestrogen ratio	1. Stage-I (dilation of the cervix) 2. Stage-II (expulsion or delivery of the newborn) 3. Stage-III (expulsion of placenta)
Mare (foaling)	1. Oozing of colostrums from the teats 2. The secretion dries off and results in teat sealing called waxy seal 3. Swollen flange region behind the elbow ***Immediate signs*** 1. Colicky symptoms, switching of tail 2. Sweat patches on the flank a few hours before foaling	Increased oxytocin at the end of pregnancy induces the synthesis of PGF2α, and the combined effect of these two hormones facilitates uterine contraction	Stage-I: Frequency of uterine contraction increases and pushes the foetus into the cervix and pelvic canal. The rotation of the foetus occurs from a dorso-pubic to a dorso-sacral position. Mares may roll to facilitate the rotation of the foetus. Stage-II: The period between the rupturing of chorioallantois and the expulsion usually lasts 15–30 min. The cervical distension caused by the foetus initiates Ferguson's reflex for abdominal contractions. Stage-III: It is the expulsion of the foetal membranes within 3 h after foaling
Bitch (whelping)	1. Enlarged mammary gland ***Immediate signs*** 1. Digging and scratching of the floor 2. Chewing, panting 3. Copious green vaginal discharge before, during and after parturition	Foetal cortisol stimulates the release of PGF2α and alterations in progesterone and oestrogen ratio	1. *First stage of labour*: It usually lasts 1–12 h. And may extend up to 36 h in primiparous bitches. Cervical dilation is completed by the end of this stage, along with the appearance of a water bag at the cervix, but it will not be visible from the outside. 2. *Second stage of labour*: Characterised by involuntary uterine and voluntary abdominal contractions. This stage usually lasts for 3–12 h. The first water bag appears in the vulva, and the bitch can be seen lying, standing, or crouching. The bitch breaks it by licking or biting to release allantoic fluid. Each puppy contains an intact second sac (amnion), and the bitch breaks it open and stimulates the puppy. 3. *Third stage of labour*: There are continuous uterine and abdominal contractions that facilitate placentas' expulsion.
Queen (kittening/ kindling)	1. Seeking of dark and dry area 2. Irritable and defensive (some queens may become hysterical) ***Immediate signs*** 1. Digging of floor 2. Defection posture 3. Vocalisation	Foetal cortisol stimulates the release of PGF2α and alterations in progesterone and oestrogen ratio	1. *Contraction phase*: Abdominal and uterine muscles contract 2. *Emergence phase*: Litter passes through the birth canal. The amniotic sac is broken by contraction of the uterus 3. *Delivery phase*: Expulsion of the foetus from the vulva. The licking stimulus initiates foetal respiration 4. *Placental phase*: Placenta is expelled from the genital tract. Queens often eat the placental tissue.

(continued)

Table 23.19 (continued)

Species and acts of parturition	Signs of parturition	Mechanism of initiation of parturition	Parturition process
Sow (farrowing)	Isolation and nest building (48–24 h before farrowing)	The release of PGF2α from the foetoplacental unit in response to foetal cortisol triggers farrowing	1. Stage 1 (The pre-farrowing period): It starts from 10 to 14 days before the date of farrowing and is characterised by the development of the mammary glands and swelling of the vulva 2. Stage 2 (The farrowing process): It ranges from 3 to 8 h, and each piglet is usually delivered every 10 to 20 min. In most pigs, the head is born first, but more pigs are presented backwards towards the end of the farrowing period 3. Stage 3 (Delivery of the placenta): It usually takes 1–4 h, indicating the completion of farrowing
Ewe (lambing) and doe (kidding)	Little changes in the mammary gland	Foetal cortisol stimulates the release of PGF2α and alterations in progesterone and oestrogen ratio	Same as cow

of croup muscles is not so profound. Udder enlarges, becomes oedematous, and even secretion may appear from a teat in cow and litter bearing animals like the sow, bitch and queen. It occurs 2–4 weeks earlier in litter bearing animals and fourth months in the calf. Bitch and cat usually remain calm and quiet before parturition.

23.4.5.2 Stage-2

This stage begins with the entry of the foetus and foetal membrane into the pelvic canal. The allantochorion (commonly known as the water bag) is expelled first, followed by the amnion, which contains the foetus. The rapid abdominal and uterine contractions expel the foetus through the vulva. The maximum contraction occurs during the expulsion of the foetal head. This maximum straining is followed by rest, but soon after that, the foetal thorax passes the vulva. The hips and hindlimbs are expelled simultaneously. The mare and sow usually lie in lateral recumbency while cow, bitch and ewe prefer to lie on their sternum. The offspring are generally born with the intact umbilical cord, but the movement of offspring and mother ruptures the cord. It takes longer to expel the foetuses in polytocous animals, depending on the number of foetuses. The duration between two successive piglets is about 10–20 min in the pig.

A longer time to complete the delivery process occurs in some abnormal conditions like twins, monster calf and abnormal posture of the foetus. During delivery, lateral recumbency is the usual posture in most domestic animals but standing posture may also occur in pluriparous dams. The behavioural changes at this stage include licking and grooming. The licking in the nasal area facilitates the removal of mucous and placental tissues for the ease of respiration. The grooming behaviour is absent in sows, and the piglets often die due to respiratory obstruction. Bitches and queen nurse their offspring in recumbency while cow, ewe and doe nurse their young in standing. A strong bond between mother and foetus is developed primarily due to pheromone perceived by olfaction.

Table 23.20 Duration of parturition in different species

Species	Stage-I (h)	Stage-II (h)	Stage-III (h)
Mare	1–4	0.2–0.5	1
Cow and buffalo	2–6	0.5–1	6–12
Ewe	2–6	0.5–2	0.5–8
Sow	2–12	2.5–3	1–4
Bitch	6–12	3–6	

23.4.5.3 Stage-3

Occasionally, the placenta is expelled within 3–8 h after the delivery of the foetus (Table 23.20). The expulsion of the foetal membrane is facilitated by the loosening of chorionic villi from maternal crypts. The expulsion of the foetal membrane occurs in two stages. The failure of placental expulsion within 24 h after parturition leads to retained placenta.

23.4.6 Foetal Presentation During Parturition

The foetus rotates from its position throughout the entire gestation period and orients itself to the normal parturition posture for smooth delivery at stage-1. Generally, the normal foetal presentations are of two types in domestic animals. These are either anterior presentations or posterior presentations. In the anterior presentation, two forelimbs and the head are present towards the vulva parallel to the animal's spine. In the posterior presentation, one or both of the hind limbs and tail remain towards the opening of the birth canal parallel to the animal's spine. In polytocous animals like pigs, the foetuses are positioned about half-and-half between facing forwards and facing backwards. Different abnormal presentation leads to parturition complications.

23.4.7 Parturition Complications

Difficulty during parturition is termed dystocia. Dystocia and retained placenta are the two major complications that arise during parturition. Each foetus has an individual placenta; hence, increased litter size may lead to dystocia. Positioning of the head between forelimbs in multiple foetuses causes dystocia on many occasions. Dystocia is mainly developed due to inappropriate cervical dilation and mostly occurs in cattle, goats, sheep and humans. The cervix of cattle is generally more cartilaginous; hence, incomplete cervical dilation is more. Several gestational anomalies like twins and uterine torsion may cause incomplete cervical dilation in cattle. The weak myometrial contraction due to hypomagnesaemia, hypocalcaemia, hyposelenemia and old age predisposes to incomplete cervical dilation and dystocia. Insufficient production of certain enzymes like 5α–reductase, and increased tissue inhibitor of matrix metalloproteinase-1 (TIMP-1) fail cervical ripening. The deficiency of 5-α-reductase causes sustained progesterone production. A large size foetus may cause a hip lock. If the placenta is not driven out within the period of stage-3 (Table 23.20), the state is termed as retention of the placenta. Low oxytocin is considered one of the causes of retention of the placenta.

23.4.8 Induction of Parturition

In some cases, parturition is induced through therapeutic and managemental interventions to release the mother from several conditions like traumatic reticulo pericarditis, broncho-pneumonia, pre-partum cervical prolapse, downer cow syndrome and pregnancy toxaemia. Prolonged gestation is also required for parturition induction. The manipulation of the time of parturition is also desirable as calving in daylight facilitates better care for the newborn and the mother. The parturition can induce either through therapeutic interventions or managemental procedures.

23.4.8.1 Therapeutic Interventions for Parturition Induction

Several drugs like glucocorticoids, oxytocin and PGF2α are used alone or in combinations to induce parturitions in domestic and pet animals. The drugs' choice and doses vary considerably among different species (Table 23.21).

23.4.8.2 Managemental Interventions for Parturition Induction

Most parturition in cows happens in the hour of darkness. The easiest and most effective method for avoiding night calving is feeding cows at night. The exact mechanism is unknown, but it is postulated that feeding alters intra-ruminal

Table 23.21 Therapeutic intervention to induce parturition in different species

Species	Drugs	Remarks
Cow	Dexamethasone Betamethasone Flumethasone	• Occurrence of retained placenta in 90% of cases • Slower onset of milk production • Delayed uterine involution
	PGF2α	• Can be used as early as 275 days of gestation • Parturition can be used within 2–3 days after administration
Mare	Corticoids	• Can be applied at 24 h interval • Average induction time is 4 ± 1 days • Can be started at 321 days of gestation • Prolonged corticosteroid administration may cause immune suppression in foal
	PGF2α	• Only synthetic forms are recommended as natural prostaglandins cause strong contraction and lead to early placental separation and foetal weakness • Foaling occurs within 4 h
	Oxytocin	• Doses vary with the degree of cervical relaxation • Cervical relaxation needs to be monitored
Goat/sheep	Corticoids	• Induction time varies between 30 and 35 h after administration
	PGF2α	
Swine	PGF2α	• Should not be applied till 111 days of gestation • Induction time 24–30 h after administration
	Corticoids	• Can be applied on days 101–104 of gestation
Canine	Dopamine agonist/anti-prolactin compounds (cabergoline)	• May cause hypotension and emesis

pressure. The frequency of rumen contraction falls a few hours before parturition. Feeding increases the rumen contraction, and hence the parturition can delay night feeding.

23.4.9 Postpartum Physiology

The time between the parturition and the occurrence of the first oestrous is called postpartum period (interval) or puerperal period. The duration depends upon species, breed, nutritional status, age, parity and environmental conditions. Each species required a considerable time to exhibit the postpartum oestrus. It varies widely, only 24 h (rabbit) to 60 days (cow). The major physiological changes occur in the ovary, uterus, and mammary gland, together with behaviour changes to nourish the young.

Failure to exhibit oestrus and ovulate beyond such period after parturition is considered postpartum anestrus.

23.4.9.1 Postpartum Ovarian Changes

Pregnancy interrupts the HPO axis by secreting large quantities of the placental steroids, and resumption of ovarian cyclicity depends upon the interaction between hypothalamus-pituitary. The CL inhibits the growth of the antral follicles even 20 days after the parturition in cows and reduces the frequency of ovulation. The resumption of ovarian cyclicity occurs in three distinct phases. In the first phase, which occurs within 2–4 weeks postpartum, the pituitary LH secretion is increased. In the second phase, the sensitivity of the hypothalamus to the positive feedback of estradiol is restored. In the third phase, the HPO axis is recovered from the suckling induced suppression due to prolactin.

23.4.9.2 Uterine Involution

The restoration of uterine volume to its non-pregnant state is called uterine involution. It is characterised by decreased uterine volume, elimination of bacterial infections and regeneration of uterine epithelium. In cows, the diameter and length of the gravid uterus are reduced to half by 5 days postpartum and 15 days postpartum, respectively, and completed with 40–50 days postpartum (Table 23.22). The constriction and occlusion of caruncular blood vessels lead to necrotic changes in the caruncle within 48 h of parturition. The necrotic layer is spread over the stratum compactum by day 5 postpartum. Some of the necrotic materials slough off through lochia (the uterine discharge secreted during early postpartum, composed of blood, mucus, remnants of foetal membranes, and maternal tissue together with foetal fluids). The shedding of necrotic caruncles is completed by day 15 postpartum, and the smooth surface of the uterus is restored by day 19 postpartum. Other than the intercaruncular areas, the endometrial tissues start regenerating immediately after the parturition and completely regenerate by day 8 postpartum in the cow. The caruncular re-epithelialisation starts from 25 days postpartum and heals entirely by day 40–60. The constriction of the cervix begins within 10–12 h of parturition in cow and undergoes atrophy and shrinkage. PGF2α helps in uterine involution.

Table 23.22 Uterine involution and occurrence of postpartum oestrus in different species

Species	Uterine involution time	Occurrence of postpartum oestrus
Cow	50–60 days postpartum	–
Doe	20–25 days postpartum	25–45 days postpartum
Ewe	30–25 days postpartum	30–40 days postpartum
Mare	13–25 days postpartum	7–9 days postpartum (foal heat)
Bitch	28–35 days postpartum	–
Sow	20–25 days postpartum	45–50 days postpartum

The important events in the ovary during this period are regression of corpus luteum, no follicular growth, decreased ovary size and appearance as anestrus. Uterine involution and repairing of the endometrium occur. Nursing to the young is common in most mammals after parturition, except mares and rats. Copious milk secretion with a positive energy balance has been found in the mammary epithelium cells. Suckling increases opioid peptides (β-endorphin) from the hypothalamus, which inhibits gonadotropin secretion, causing to increase postpartum period. Thus, weaning favours to re-establishment of the HPO axis. It results in oestrous, and ovulation and the animal can conceive further. The oestrous after parturition is termed postpartum oestrus and can affect by light cycles, temperature and humidity. Any kind of stress negatively affects the HPO axis through cortisol, followed by a delayed postpartum period. Hot, dry and extreme weather cause delays in the postpartum period. Parturition in the non-favourable season to the seasonal breeder also delays the period. The development of anomalies during pregnancy and parturition may also cause delays in the postpartum period.

23.4.10 Initial Care of Newborn

23.4.10.1 The Onset of Respiration

During the parturition, PO_2 and blood pH is decreased, and PCO_2 is increased due to placental separation. High PCO_2 stimulates chemoreceptors at the carotid sinus and initiates respiration. Tactile and thermal stimulation is also required to initiate respiration. After birth, the fluid and mucous from the upper respiration tract need to be cleared first. Brisk rubbing the chest and scrapping of nostrils are the tactile stimuli to induce sneezing and clear the upper airways. Doxapram hydrochloride can also be applied to induce breathing. In extreme cases, oxygen therapy is required.

23.4.10.2 Thermoregulation

The newborn has inadequate sub cutaneous fat insulation and poor thermoregulatory control. The initial thermoregulation is achieved by producing metabolic heat and reduction in heat loss. For adequate metabolic heat production, sufficient nutrition is essentially required. The newborn should be placed in a warm environment to minimise heat loss. Application of body coat can also reduce heat loss. The ideal temperature for the newborn puppy is 95–100 °F during the first week, 85 °F during the second week and 70–85 °F during the third week.

23.4.10.3 Care of Umbilicus

The hemostatic clamp must be removed from the umbilical cord and must ligate to prevent further haemorrhage. Cutting the umbilical cord close to the abdomen is advisable, along with proper disinfection with antiseptic.

23.4.10.4 Feeding of Colostrum

Feeding of colostrums is required to boost the newborn's immune system by transferring immunoglobulins. Nutrients and antibodies increase the immune system. Increased level of oxytocin induced by suckling stimulus facilitates colostrum ejection. Cleaning of teats and physical assistants to a newborn is required for initial colostrum feeding. As a thumb rule, colostrum at the rate of 10% of a calf's body weight can be fed to the newborn. Colostrums replacement therapy or colostrum from other animals should give in case of death of the dam.

23.4.11 Postpartum Reproductive Disorders

Due to stress and immune suppression around the peripartum, the animals are prone to infections. Some postpartum disorders like mastitis, metritis, retained placenta, and lameness may also occur from digestive or nutritional malfunctions directly affecting uterine development.

23.4.11.1 Mastitis

Mastitis has a direct effect on reproduction. Inflammation due to mastitis leads to stress and hyperthermia because of increased cortisol, reactive oxygen species (like NO) and cytokine (TNF-α). It affects locally and in the HPO axis, resulting in abnormalities in the oestrous cycle, ovulation, decreased oocyte competence, and failure of fertilisation. In addition, the pathogen-derived endotoxin and altered immune system cause endocrine imbalance and increased production of PGF2α, followed by luteolysis and an adverse uterine environment. It causes early embryonic mortality or abortions in dairy cattle. Mastitis affects reproduction in dogs also.

23.4.11.2 Metritis

Metritis and endometritis are uterine infections. They cause poor follicular growth and postpartum anestrus due to altering GnRH secretion, decreased responsiveness of LH to the follicular cells and lower aromatase expression followed by less oestrogen secretion. The inflammatory response in the endometrial epithelium generates TNF-α, nitric oxide synthase (NOS) and prostaglandin-endoperoxide synthase 2 (PTGS2 or COX-2). It influences more PGE production than PGF (PGF2α). Thus, luteolysis is disturbed. Endometrial immunity is also compromised through altered steroid hormone concentration, somatotrophins and local regulatory proteins due to LPS of the pathogens. Higher progesterone and IGF I, released in metritis, play an immunosuppressive role through the serine proteinase inhibitors, glycan-binding proteins (galectins) and galactose-β1,4-*N*-acetylglucosamine. *Systemic inflammation* and metabolic syndrome also affect reproductive function by a similar mechanism. Hence, metritis, endometritis and systemic inflammation directly affect the reproductive function to cause impaired fertilisation, twins, stillbirth, dystocia and retained placenta. It often occurs in cattle, sheep and dogs.

23.4.11.3 Retained Placenta

Retention of the placenta may occur due to certain venereal diseases and managemental and nutritional factors. It may cause chronic infection and alteration in hormonal balance and immune compromisation of the mother. It results in twin, abortion, stillbirth, dystocia, labour problem and shortened gestation. *Subinvolution* of the placenta after the first parturition within 3 years of age is a common problem in the dog.

23.4.11.4 Lameness

Lameness is considered chronic stress to the animal. Hence, the HPO axis is inhibited through cortisol and other stress bio-molecules. So, LH pulse frequency and oestrogen level are reduced. It causes poor oestrus signs, failure of ovulation, delayed ovulation and low progesterone production from the immature corpus luteum. Thus, oocyte maturation and the process of fertilisation are disturbed. Infertility may develop due to the formation of an anovulatory cyst. Delay postpartum oestrus has also occurred.

23.4.11.5 Digestive or Nutritional Factors

Reproductive anomalies may also occur due to nutritional deficiency and toxicity of certain chemicals during in postpartum state. It causes improper placental growth and abortion with other reproductive hazards. The deficiency of vitamin A results in thickening and erosion of the placenta, followed by abortion in late gestation. It also causes an irregular oestrous cycle. The lack of vitamin D develops silent oestrus and delays ovulation. Iodine deficiency is also a cause of abortion. Deficiency of iron, copper, cobalt and protein leads to anaemia and energy-deficient condition leads to anestrus and irregular oestrus due to disturbance in the HPO axis and ovarian activity. Magnesium and phosphorus deficiency, mostly during postpartum of high yielder animals, also causes a similar effect. Selenium and vitamin E deficiency causes metritis. Excess blood urea nitrogen (BUN) due to protein-rich ration reduces fertilisation. Toxicity of nitrites and nitrates from fertiliser and some plants during drought, excess phosphorus (from the plant's seed), and mycotoxins from fungus caused abortion and retained placenta. Diarrhoea, bloating, bleaching, constipation, nausea, poor gut health, chronic irritation and cramping cause anxiety and stress, lead to deviation in the HPO axis and ovarian dysfunction. Gluten (a protein found in wheat, rye and barley)-rich diet may develop an autoimmune disorder called celiac disease in the genetically predisposed dog, cats, rats and mice. It causes ovulatory problems along with osteoporosis, depression and anaemia. Various oestrogenic compounds like bisphenol A, DDT, polychlorinated biphenyls, polybrominated diphenyl ethers, petroleum by-products, pesticides, herbicides and plastics disrupt the steroidogenesis process during foetal growth.

23.4.12 Diseases Related to Embryonic Mortality and Foetal Abnormality

23.4.12.1 Congenital Disorders

Congenital deformity in the reproductive tract, like malformation of the Mullerian duct, results in abnormalities in the fallopian tubes and uterine cavity. Like the Boxers breed of dog, some small animals have a small-sized uterus, causing dystocia. The ovarian bursal adhesion may occur in specific conditions due to improper ovarian manipulation by rectal palpation. It appears in large ruminants and results in inappropriate capturing of ova during ovulation and fertilisation anomalies. Other placental abnormalities include small placenta, premature placental separation and umbilical cord complications. Common defects in dog foetal developments are canine congenital sensorineural deafness (CCSD, degeneration of inner ear), canine-dilated cardiomyopathy and hypocalcaemia originated canine eclampsia (mostly in small-breed accompanying with more litter size).

Certain diseases like bovine viral diarrhoea (BVD), vibriosis, brucellosis, chronic leptospirosis and trichomoniasis cause early embryonic mortality and infertility. Different pathogens affect different modes, but in general, adhesions in the region of the ovaries, obstruction in the oviducts, and inflammation in the uterus, placenta and cervix occur. Hence, the normal reproductive function is affected.

23.4.12.2 Bovine Viral Diarrhoea (BVD)

BVD causes early embryonic mortality or foetal death or birth of a malformed calf in cattle according to severity and time of infection. The infection also causes ovaritis and leads to the malfunctioning of follicular cells resulting harmful environment for the embryo or foetal development. It may also cause retention of the placenta.

23.4.12.3 Vibriosis

Vibriosis is caused by the bacterium *Campylobacter foetus* (earlier *Vibrio foetus*). It is common in cattle and sheep and also in humans. It is a venereal disease and spreads by infected bulls. The bacteria cause inflammation in the trophoblasts and chronic villi, resulting in necrotic placentitis and uterine infection. Infection may continue 1–8 months of pregnancy. The consequence is abortion and infertility.

23.4.12.4 Brucellosis

Brucellosis is a venereal disease caused by the bacterium *Brucella abortus*. It causes infection in the endothelial cells of the capillaries and sub-chorionic tissues. The oedematous growth results in the premature termination of pregnancy and retention of the placenta. It affects cattle, sheep, goats, pigs, dogs and humans.

23.4.12.5 Leptospirosis and Trichomoniasis

Leptospirosis caused by *Leptospira* genus of bacteria. Transmission occurs during natural mating, where young are more susceptible and cause early embryonic mortality. Cattle, sheep, goats, pigs and dogs are affected more. Trichomoniasis is caused by a protozoon, *Trichomonas foetus*. The infective organism transmits through the infected animal, semen and instrument used for insemination. It causes uterine infection and abortion. It is common in cattle.

23.4.12.6 Other Pathogens

There are other organisms like streptococci, staphylococci, corynebacteria, diplococci, micrococci, and moulds that cause local infection in the reproductive tract; the consequence of this result in impaired gestation and infertility may arise in the chronic stage.

Learning Outcomes

- **Fertilisation**: Internal fertilisation occurs in all the primates and non-primates domestic mammals. Semen is deposited at the vagina by natural insemination and around the cervix by artificial insemination. The time of insemination depends upon the time of ovulation and it is species specific. After insemination, spermatozoa are transported to the ampulla, the site of fertilisation, through a uterine tube, and subjected to species-specific biochemicals and molecular changes known as capacitation and hyperactivation for final maturation. In most mammals, the secondary oocyte completes its maturation after ovulation in the oviduct immediately before fertilisation with the induction of spermatozoa. A series of biochemical reactions, like acrosomal reaction in the spermatozoa, cortical reaction in the oocyte, entry of sperm and initiation of meiosis and karyogamy occur to complete the fertilisation process. Several proteins and enzymes are involved in successful fertilisation or conception.
- **Gestation**: The period between conception and giving birth to young is the period of pregnancy or gestation. Its duration depends upon the species. Several factors like genetic, maternal, foetal, endocrines and environment control the gestation. The embryo is implanted in the uterus during its blastocyst stage and then transforms into the foetus. During gastrulation, the foetus produces three germ layers for proper attachment with the uterine wall through the placenta. Various organs are developed at the organogenesis stage. The gestation is maintained by several endocrines and growth factors like interferon Tau (IFNτ). The presence of these species-specific bio-molecules in the animal confirms the maternal recognition of pregnancy.
- **Placenta**: The placental types are species specific. According to layer involvement, these are epitheliochorial, endotheliochorial and hemochorial. These are cotyledonary, diffuse, discoid and zonary according to shape and attachment. The placenta supports the foetus to exchange nutrients and metabolites with the mother, physical and immunological protection, thermoregulation and proper development with the influence of several endocrines and bio-molecules. Placenta itself secretes some endocrines. Foetal sex-specific placental activity has also occurred.
- **Parturition**: The species-specific complex physiological phenomenon occurs during the termination of pregnancy, the parturition. Interactions of several endocrines initiate the event involving various cellular and muscular activities in the female reproductive tract to generate pressure and inflammatory processes for expelling the young and the placenta. Faulty foetal presentation causes the occurrence of dystocia. Several congenital and anatomical deformities may also occur.

Exercises

Objective Questions

Q1. Why do spermatozoa reach rapidly at the site of fertilisation immediately after insemination cannot fertilise?
Q2. Which part of the female reproductive tract is considered a sperm reservoir in the cattle?
Q3. Which biological activity provides the zona penetrating ability to the spermatozoa?
Q4. Why is secondary oocyte not completely maturated before ovulation in cows?
Q5. Which stage of fertilisation causes the permanent blockage of polyspermy?
Q6. What is progesterone block?
Q7. The embryo is termed as a foetus from which morphological stage?
Q8. What is the significance of zona hatching?
Q9. What is maternal recognition of pregnancy?
Q10. Eccentric type of implantation occurs mostly in which animals?
Q11. Which kind of placenta does not involve any endometrial tissue?
Q12. What is hemotroph?
Q13. When eCG start to secret in a mare?
Q14. Why can relaxin determine pregnancy in dogs and cats, not progesterone?
Q15. In which sex the interferon tau is better expressed?
Q16. What is cervical ripening?
Q17. Manipulation of which hormone is advised to correct retention of the placenta?
Q18. How many oocyte(s) are involved in the occurrence of fraternal twins?
Q19. Why dystocia mainly occurs in the Boxer breed of dog?
Q20. Why vitamin A deficiency causes abortion in late pregnancy?

Subjective Questions

Q1. Why is the concentration of spermatozoa reduced from the insemination site to the fertilisation site?
Q2. Describe the role of some bio-molecules involved in the capacitation.
Q3. Describe the process of fertilisation diagrammatically.
Q4. Describe the various factors affecting gestation.
Q5. Classify the mammalian placenta according to histological and morphological structure, shape and attachment.
Q6. Describe the endocrine role of the placenta.
Q7. Write the process of parturition in cow.
Q8. Neuroendocrine mechanism of parturition.

Answer to Objective Questions

A1. Lack of capacitation process
A2. Isthmus
A3. Hyperactivation
A4. Meiosis II is completed with the induction of spermatozoa at the oviduct
A5. Cortical reaction
A6. Reduce or block the muscular tone of the female reproductive tract in favour of gestation
A7. Gastrula
A8. The blastocyst can expand and proceed to the implantation process
A9. The state of the mother, caused by the conceptus, promotes the gestation
A10. Rat and mouse
A11. Hemochorial
A12. Substances that are directly transferred from the maternal blood by blood vessels through the placenta
A13. Around 25 days of gestation
A14. Corpus luteum persists more than during the pregnancy period in dogs, and placental progesterone can be synthesised in the cat with a luteal cyst and low progesterone in the first 2 weeks. In contrast, placental relaxin is synthesised from about 4 weeks in the dog and 3 weeks in the cat.
A15. Female
A16. The inflammatory process in parturition
A17. Oxytocin
A18. Two
A19. Small-sized uterus
A20. Due to thickening and erosion of the placenta

Keywords for the Answer to Subjective Questions

A1. Vaginal toxic environment, cervical plug, transportation through the uterus, morphological and mucus barrier at UTJ
A2. Role of ions like bicarbonate and calcium; hormones like oestrogens and progesterone; various proteins, glucose and metabolites
A3. Acrosomal reaction in the spermatozoa, cortical reaction in the oocyte, entry of sperm and initiation of meiosis, karyogamy
A4. Genetic, maternal, foetal, endocrines and environmental factors
A5. Involvement of endometrium; shape and point of attachments between foetal and endometrial tissues; structure of the chorionic membrane and contact with the endometrium; loss of maternal tissue during parturition
A6. Placental steroid, placental protein hormones, placental proteins, placental prostaglandin
A7. Stage1, stage 2 and stage 3
A8. Role of progesterone, oestrogens, oxytocin, relaxin, melanin and foetal cortisol

Further Reading

Books

Carlson BM (2019) The reproductive system (Ch. 14). In: The human body. Andre Gerhard Wolff, London, pp 373–396. https://doi.org/10.1016/B978-0-12-804254-0.00014-4

Das PK, Mukherjee JM (2019) Immune responses of mammary gland (Ch. 2). In: Sar TK (ed) Mastitis: symptoms, triggers and treatment. Nova Science Publisher, Inc., New York, pp 89–118. ISBN: 978-1-53616-124-3

Lopes SMCS, Mummery CL (2014) Differentiation in early development (Ch. 10). In: Essentials of stem cell biology, 3rd edn. Academic Press, pp 121–139

Research Articles

Ardon F, Markello RD, Hu L et al (2016) Dynamics of bovine sperm interaction with epithelium differ between oviductal isthmus and ampulla. Biol Reprod 95(4):90. https://doi.org/10.1095/biolreprod.116.140632

Borghesi J, Mario LC, Rodrigues MN et al (2014) Immunoglobulin transport during gestation in domestic animals and humans—a review. J Anim Sci 4:323–336. https://doi.org/10.4236/ojas.2014.45041

Chavatte-Palmer P, Tarrade A (2016) Placentation in different mammalian species. Ann Endocrinol (Paris) 77:67–74. https://doi.org/10.1016/j.ando.2016.04.006

Conley AJ, Ball BA (2019) Steroids in the establishment and maintenance of pregnancy and at parturition in the mare. Reproduction 158(6):R197–R208. https://doi.org/10.1530/REP-19-0179

Croxatto HB (2002) Physiology of gamete and embryo transport through the fallopian tube. Reprod Biomed Online 4(2):160–169. www.rbmonline.com/Article/228

Feng Y, Zhang P, Zhang Z et al (2016) Endocrine disrupting effects of triclosan on the placenta in pregnant rats. PLoS One 11(5): e0154758. https://doi.org/10.1371/journal.pone.0154758

Ferramosca A, Zara V (2014) Bioenergetics of mammalian sperm capacitation. Biomed Res Int 2014:902953, 1–8. https://doi.org/10.1155/2014/902953

Gahlot SC, Kumaresan A, Kumar S et al (2017) Incomplete cervical dilatation in animals—an update. Int J Sci Environ Technol 6(2):1036–1048

García-Vázquez FA, Gadea J, Matás C et al (2016) Importance of sperm morphology during sperm transport and fertilisation in mammals. Asian J Androl 18(6):844–850. https://doi.org/10.4103/1008-682X.186880

Harayama H, Minami K, Kishida K et al (2017) Protein biomarkers for male artificial insemination subfertility in bovine spermatozoa. Reprod Med Biol 16(2):89–98. https://doi.org/10.1002/rmb2.12021

Hung PH, Suarez SS (2010) Regulation of sperm storage and movement in the ruminant oviduct. Reprod Fertil Dev 67(1):257–266. https://doi.org/10.5661/RDR-VII-257

Ickowicz D, Finkelstein M, Breitbart H (2012) Mechanism of sperm capacitation and the acrosome reaction: role of protein kinases. Asian J Androl 14(6):816–821. https://doi.org/10.1038/aja.2012.81

Jin SK, Yang WX (2017) Factors and pathways involved in capacitation: how are they regulated? Oncotarget 8(2):3600–3627. https://doi.org/10.18632/oncotarget.12274

Kim SM, Kim JS (2017) A review of mechanisms of implantation. Dev Reprod 21(4):351–359. https://doi.org/10.12717/DR.2017.21.4.351

Kota SK, Gayatri K, Jammula S et al (2013) Endocrinology of parturition. Indian J Endocrinol Metab 17(1):50–59. https://doi.org/10.4103/2230-8210.107841

Leemans B, Stout TAE, Schauwer CD, Heras S, Nelis H, Hoogewijs M, Van Soom A, Gadella BM (2019) Update on mammalian sperm capacitation: how much does the horse differ from other species? Reprod 157(5):R181–R197. https://doi.org/10.1530/REP-18-0541

Markkandan K, Ahn K, Lee DJ et al (2018) Profiling and identification of pregnancy-associated circulating microRNAs in dairy cattle. Genes Genom 40(10). https://doi.org/10.1007/s13258-018-0668-2

Monroy A (2020) Fertilization. Encyclopedia Britannica. https://www.britannica.com/science/fertilization-reproduction

Napso T, Yong H, Lopez-Tello J et al (2018) The role of placental hormones in mediating maternal adaptations to support pregnancy and lactation. Front Physiol 9:1091. https://doi.org/10.3389/fphys.2018.01091

Petraglia F, Imperatore A, Challis JRG (2010) Neuroendocrine mechanisms in pregnancy and parturition. Endocr Rev 31(6):783–816. https://doi.org/10.1210/er.2009-0019

Pohler G, Franco GA, Reese ST, Smith MF (2020) Physiology and pregnancy of beef cattle. In: Bazer FW, Lamb GC, Wu G (eds) Animal agriculture: sustainability, challenges and innovations. Elsevier Inc.. ISBN: 978-0-12-817052-6

Ponraj P, Chang S, Rajesh NV et al (2017) Prevalence of various pathological conditions in female buffaloes (*Bubalus bubalis*). Asian Pac J Reprod 6(2):58–67. https://doi.org/10.12980/apjr.6.20170203

Rijsselaere T, England GCW, Freeman SL et al (2014) Current knowledge on the transport and fate of spermatozoa in the reproductive tract of the bitch. Reprod Domest Anim 49(Suppl. 2):2–7. https://doi.org/10.1111/rda.12299

Rosenfeld CS (2015) Sex-specific placental responses in foetal development. Endocrinology 156(10):3422–3434. https://doi.org/10.1210/en.2015-1227

Schuler G, Fürbass R, Klisch K (2018) Placental contribution to the endocrinology of gestation and parturition. Proceedings of the 10th International Ruminant Reproduction Symposium (IRRS 2018); Foz do Iguaçu, PR, Brazil, September 16th to 20th, 2018. Anim Reprod 15(Suppl. 1):822–842. https://doi.org/10.21451/1984-3143-AR2018-0015

Seidel GE Jr, Seidel SM (1991) Evaluation of embryos (Ch. 7). In: Training manual for embryo transfer in cattle. FAO Animal Production and Health Paper 77. ISBN: 92-5-102804-4. http://www.fao.org/3/t0117e/T0117E07.htm

Siemieniuch MJ, Jursza E, Szostek AZ et al (2012) Steroidogenic capacity of the placenta as a supplemental source of progesterone during pregnancy in domestic cats. Reprod Biol Endocrinol 10:89. https://doi.org/10.1186/1477-7827-10-89

Soom AV, Boerjan ML, Bols PEJ et al (1997) Timing of compaction and inner cell allocation in bovine embryos produced *in vivo* after superovulation. Biol Reprod 57:1041–1049

Sumigama S, Mansell S, Miller M et al (2015) Progesterone accelerates the completion of sperm capacitation and activates CatSper channel in spermatozoa from the rhesus macaque, Biol Reprod 93(6). https://doi.org/10.1095/biolreprod.115.129783

Turner RM (2006) Moving to the beat: a review of mammalian sperm motility regulation. Reprod Fertil Dev 18(2):25–38. https://doi.org/10.1071/RD05120

24 Assisted Reproductive Technologies in Farm Animals

Ayan Mukherjee, Pradip Kumar Das, Dipak Banerjee, and Joydip Mukherjee

Abstract

The global human population is anticipated to rise from 7.9 billion today to 9.7 billion by 2050 and a sense of Malthusian crisis has set in: how can all these extra mouths be fed? With the scarcity of arable land, clean water, and energy, increased production of food is the major concern in the twenty-first century for achieving global food security. Foods of animal origin will play an important role to meet the daunting challenge of feeding a huge population because animal products are a significant source of high-quality protein and other vital micronutrients. Also, farm animals are essential for a sustainable agricultural system, particularly for smallholder farmers. However, increased animal disease burden, abrupt changes in global climatic conditions, and poor reproductive efficiency are some of the hindrances in front of the production of farm animals. To meet the escalating demand of animal food products and sustainable agricultural production, the adoption of modern reproductive biotechnologies are of great relevance. All these frontiers technologies related to animal reproduction ranging from manipulation of male and female gametes, transgenesis, and stem cell technology are collectively referred as assisted reproductive technologies (ART). This chapter is aimed to highlight the various ART used worldwide to augment the reproductive efficiency in farm animals. Some of them are conventional technologies like artificial insemination (AI), multiple ovulation and embryo transfer (MOET), superovulation, etc. Besides those, light will be shed on advanced technologies involving gamete and embryo level manipulation such as in vitro embryo production (IVEP), transgenesis, and sexing of spermatozoa and embryos.

A. Mukherjee (✉)
Department of Animal Biotechnology, West Bengal University of Animal & Fishery Sciences, Kolkata, West Bengal, India

P. K. Das · D. Banerjee · J. Mukherjee
Department of Veterinary Physiology, West Bengal University of Animal & Fishery Sciences, Kolkata, West Bengal, India

© The Author(s), under exclusive license to Springer Nature Singapore Pte Ltd. 2023

P. K. Das et al. (eds.), *Textbook of Veterinary Physiology*, https://doi.org/10.1007/978-981-19-9410-4_24

Graphical Abstract

Description of the graphic: Assisted reproductive technologies in farm animals include in vitro embryo production and embryo transfer (**1**), sex sorting of spermatozoa by fluorescence-activated cell sorter (**2**), somatic cell nuclear transfer or cloning developed by Ian Wilmut with the cloned sheep "Dolly" (**3**), precise editing of genome by gene transfer (**4**), transgenic animal production methods vis-à-vis pronuclear microinjection, embryonic stem cell, and spermatogonial stem cell-mediated transgenesis and sperm-mediated gene transfer (**5**)

Keywords

Sperm · Oocyte · Embryo · Animal cloning · Transgenesis

Learning Objectives

- Knowledge of traditional method of assisted reproductive technologies.
- Familiarity with in vitro embryo production by different techniques.
- Method to produce offspring of desired sex by sperm sexing.
- Methods and application of genome level manipulation of livestock by transgenesis.

24.1 Artificial Insemination

Artificial insemination (AI) is the most widely utilized ART in farm animal species, having revolutionized the sector over the twentieth century.AI is the most common method of breeding intensively reared domestic livestock, such as dairy cattle, buffalo, and pigs. AI is increasing in horses, beef cattle, sheep, goat, and dog. It has also been employed on rare or endangered animals such as monkeys, elephants, deer, and wild felids in conservation breeding.

AI is a process by which sperm is collected from the male, processed, stored, and artificially introduced into the female reproductive tract for the purpose of conception by using means other than sexual intercourse or natural insemination.

24.1.1 Milestones

1780	Lazzaro Spallanzani reported the first successful use of AI in dog
1922	E.I. Ivanoff successfully conducted AI in cattle and sheep
1940	The scientists of Denmark introduced the straw method for packing of liquid semen.
1940	Philips and Lardy developed egg yolk phosphate diluter for preserving the fertility and motility of refrigerated bull spermatozoa.
1941	Salisbury et al. developed egg yolk citrate diluter
1949	Polge, Smith, and Parkes discovered cryoprotective effect of glycerol in frozen semen technology
1960	Adler first froze the semen packed in straws by using liquid nitrogen vapor
1964	Cassou developed the straws by reducing their size and named it as "medium French straw" (135 mm long and 2.8 mm diameter with 0.5 mL semen capacity)
1968	Cassou further developed the straws by reducing their diameter for better freezing and named it as "mini French straw" (135 mm long and 2.8 mm diameter with 0.5 mL semen capacity)
1972	Simmet introduced a new straw called "mini tube" or "German straws" or "Landshut system" in Germany

24.1.2 Procedure

24.1.2.1 Collection of Semen

Artificial vagina method is commonly used for collection of semen in most domestic animals bull, ram, and stallion. Artificial vagina consists of a greased liner placed between two outer jackets, with warm water filling the area between them. To create an environment like natural vagina, air is forced via the nozzle into the water jacket to build pressure in it, and the rubber liner is subjected to the same pressure. Male animals are allowed to mount either an estrous female or a phantom. The ejaculate is deposited into an insulated collecting vessel attached to one end of the artificial vagina liner. Also, electro-stimulation and manual stimulation of ampullae of the ductus deferens through rectal wall are performed in boar or dog or in case of injured bull of high quality.

24.1.2.2 Semen Processing

Despite its importance in stimulating spermatozoa in the female reproductive tract, seminal plasma is harmful to long-term sperm survival outside the body. If the semen is used for preservation spermatozoa become exposed to seminal plasma for a long time. So, semen extenders are added to the semen to dilute noxious elements in seminal plasma, to provide nutrients for the spermatozoa during in vitro storage and to buffer their metabolic by-products. Also, the use of extender allows the semen to be split into numerous semen doses, each carrying a definite quantity of spermatozoa optimum for a successful fertilization process.

24.1.2.3 Preservation of Semen

Cryopreservation of semen is routinely used in livestock breeding sector especially in cases of cattle and buffalo bull semen. Sperm cryopreservation helps in the propagation of animals with better genetic features and species conservation. The spermatozoa are combined with a cryoprotectant such as glycerol and a protective solution including lipoproteins, carbohydrates, and a cryoprotectant. These components aid in the preservation of membrane integrity throughout the chilling and rewarming processes. However, sperm motility must be maintained so that frozen spermatozoa can reach and fertilize the oocytes following insemination. In some species, such as stallion, boar, and goat, semen, the seminal plasma, is extracted by centrifugation before combining with the cryoextender. The extended semen is packed in straws and frozen in liquid nitrogen vapor before being stored in liquid nitrogen.

24.1.2.4 Estrus Detection and Ovulation

Estrus detection is critical if the female is to be inseminated at the correct moment, as a successful outcome of AI depends on the deposition of spermatozoa at a suitable time relative to ovulation. Males of the same species are naturally good at recognizing estrus females, but because many livestock breeding units do not have male animals nearby, it is critical that husbandry staff learn to recognize estrous behavior. Although certain domestic animals, such as dairy cows, have well-developed estrous behavior, others do not. Restlessness or increased activity, vocalization, chin resting, vulva swelling, vaginal discharge, and mounting other cows are all indicators of estrus in cows. However, these behavior vary between individual animal and breed.

24.1.3 Insemination

24.1.3.1 Recto-Vaginal Method

The AI gun is inserted through the vulva to the vagina and cervix with one hand in the rectum to guide the gun in proper place. The semen is deposited by pressing the piston of the gun. The gun is removed after depositing the semen in the cervix.

24.1.3.2 Vaginal Speculum Method

Vaginal speculum is placed in the vagina of the cow. Then inseminating tube is passed through the speculum and semen is deposited at the cervix.

24.2 Cryopreservation of Male Gamete

Semen cryopreservation is useful to preserve the superior and threatened genetic resources, genetic exchanges between captive as well as wildlife conservation centers, and semen having low sperm concentration and motility. Cryopreservation is one of the possible ways to establish germ-plasm bank. It is effective in those wild species where narrow genetic variations exist like cheetah. Semen of wild animals are to be preserved precisely as most (usually more than 60%) of the spermatozoa are pleomorphic and teratospermic in nature. The chemicals and processes are to be accurate during cryopreservation.

24.2.1 Collection of Male Gamete

Artificial vaginas or vaginal condoms are generally used to collect the male gamete. Male gamete along with semen can be collected by electro-ejaculation method. Hazards may arise to restrain the animals by this process. Post-coital sperm recovery and epididymal spermatozoa from recently poached or dead animals are the alternate processes of semen collection from wildlife. C ryopreservation of testicular tissue is also used successfully in domestic bovines, porcines, cats, and mice and some wild species like monkeys, blackbuck (*Antilope cervicapra*), jungle cat (*Felis chaus*), lion (*Panthera leo*), leopard (*Panthera pardus*), Timor deer (*Rusa timorensis*), Tenasserim muntjac (*Muntiacus feae*), and Sumatran serow (*Capricornis sumatraensis*).

24.2.2 Cryopreservation of Epididymal Spermatozoa

The quality of epididymal spermatozoa remains almost unchanged if the collection occurs within 4 h after death. The rate of fertility is vigorously altered after 24 h. The rate of abnormality is reduced if the epididymis is preserved at 5 °C immediately after death and the cryopreservation process starts within 24 h. The epididymal spermatozoa are not fully matured and have certain abnormal features, like the presence of nonmotile spermatozoa of about 15%, cytoplasmic droplets of nearly 14%, dead and less intact plasma membrane of 13% in bull. But, epididymal spermatozoa exhibited effective results following species-specific protocol of cryopreservation. Successful cryopreservation was done from the epididymal spermatozoa of a gaur (*Bos gaurus*) bull and pregnancy was confirmed after 41 days by ultrasonography inseminating to the Holstein cow twice with 80×10^6 spermatozoa. Cryopreservation of epididymal spermatozoa followed by successful pregnancy was demonstrated in certain semi-domesticated and wild animals like alpaca (4×10^6 sperm per insemination). Pregnancy has been achieved by inseminating 4×10^6 sperm per dose in wild cheetahs, wolves, and other animals where higher litter size (three) was obtained with a higher dose at $6–16 \times 10^6$ sperm. Cryopreserved post-thaw spermatozoa having 40–50% motility and 50–60% intact acrosomes are biologically viable. Isolation of effective motile spermatozoa: Certain species-specific substances sometimes are used to clear the unwanted substance from the semen samples before cryopreservation to get better post-thawed motile spermatozoa. Papain, a protease used in alpaca to dissolve the protein mucin 5B of the semen cause to reduce the viscosity without disturbing the motility, acrosome integrity, viability, and DNA integrity of the spermatozoa. The epididymal sperm can be preserved after separating the contaminated cells and cellular debris by single-layer centrifugation using species-specific colloid formulations. Androcoll-C, a glycidoxypropyltrimethoxysilane-coated silica optimized for dog spermatozoa (where "C" stands for canine,

Androcoll-E is a similar substance used for equine species), may be used at 30% to separate the carcasses of the epididymal spermatozoa of canine wildlife like wolf, etc.

24.2.3 Cryopreservatives and Cryoprotectants

Cryopreservatives and cryoprotectants are used in the semen during cryopreservation to protect the spermatozoa from thermal shock, formation of ice crystals within spermatozoa, and reduce viscosity. The Cryopreservatives comprised a combination of various substances that are used as diluents or extenders during cryopreservation. The effective extender has the characteristics of cryoprotectiveness, energy sources. Egg yolk with varying concentrations from 5% to 20% is used in different species. Egg yolk is used along with glycerol, dimethylsulfoxide (DMSO), and ethylene glycol are the common cryoprotectants, where glycerol is mostly effective but has some toxic effects on certain species. These substances act as intracellular cryoprotectants. Egg yolk and glycerol at 4–6% are mostly used in cryopreservation of domestic mammalian male gamete. This combination is also used to preserve the semen of some wild species like Asian elephants (*Elephas maximus*), rhinos (*Ceratotherium simum, C. simum cottoni*), leopards (*Neofelis nebulosa*), bears (*Ursus arctos*), rhesus monkeys, felids, and marine mammals. DMSO and dimethylacetamide are effective in saltwater crocodiles (*Crocodylus porosus*) and kangaroos (*Macropus giganteus*). Glycerol is to be added at a particular temperature and specific step of cryopreservation, otherwise it may drag the water causing to shrinkage of the sperm. In cheetah sperm, glycerol is generally added slowly for a period of 60 min at ambient temperature just before cryopreservation. Some monosaccharide, disaccharide, and polysaccharide sugars like lactose, maltose, glucose, fructose, and citrate- or tris-based (SHOTOR), skimmed milk is usually used to provide the energy and act as diluter as well as extracellular cryoprotectant. Sugars trap the salt in the unfrozen water caused to inhibit the eutectic freezing and to increase the viscosity resulting to prevent the fast cooling, pattern of crystallization, and protect the sperm membrane. Glucose and fructose are permeable to sperm membrane and provide energy. Hence, monosaccharides are comparatively less effective than lactose in respect of its cryoprotective role. Lactose is larger in size than glucose and fructose and cannot easily enter inside the sperm, rather protect from osmotic change during dilution resulting in better survivability of the spermatozoa. Thus, lactose is effective than other diluents. It is most suitable for epididymal spermatozoa cryopreservation and for those species which have less fructose in semen like alpaca. Lactose is also effectively used in buck and garut ram semen. Some efficient extenders are 5% egg yolk extender with glycerol diluted with lactose at 1:3. It is effective in alpaca. Glucose or fructose is routinely used as a diluter in many species like cattle, sheep, and pig. Successful results were also obtained by using raffinose (trisaccharide) in Etawah buck, trehalose (non-reducing sugar having di-glucose) in garut ram, trehalose and sucrose in bull, and maltose in garut ram semen. Better result has been obtained in lactose-based semen in camel. Skim milk is effective in Asian elephant (*Elephas maximus indicus*) and the Indian rhinoceros (*Rhinoceros unicornis*).

24.2.4 Form of Freezing

Semen can be frozen in straw or in pellet form and can be preserved in liquid nitrogen at −196 °C. Preservation can also be possible on a dry ice block in the laboratory. Straws are more secure and easy to handle. It has a more surface-to-volume ratio, resulting in better post-thaw motility.

24.2.5 Rate of freezing

The rate of freezing should follow the surface-to-volume ratio of the spermatozoa and the permeability of the membrane. Thus, small-headed spermatozoa (like alpaca) require a faster freezing rate than large-headed (like bull). In general, slow freezing at −0.5 °C/min demonstrated optimum result in most of the mammalian species.

24.3 Sexed Semen Technology

The goal of sexed semen technology is to generate a calf of a specific sex. Because of this, sex preselection has a significant impact on the profitability of livestock-based industry. Females are required for milk production and the production of calves, while male calves are usually chosen for meat production because of their better feed conversion efficiency and lean-to-fat ratio. Furthermore, males of superior genetic quality are utilized as sires in artificial insemination programs.

The principle behind the sexed semen technology is the inherent differences between X and Y chromosome-bearing spermatozoa in mass and motility, surface charges, swimming patterns, volumetric differences, centrifugal countercurrent distribution, and immunological properties (Table 24.1).

24.3.1 Procedure

The flow cytometry method of sperm sorting is the only approach that has shown to be economically feasible and

Table 24.1 Differences between X- and Y-bearing spermatozoa

Properties	X-bearing sperm	Y-bearing sperm	Methods of separation
Size	Larger	Smaller	Percoll gradient
Motility	Slower	Faster	Swim-up technique
Surface charge	Faster migration to cathode	Slower migration to cathode	Free flow electrophoresis
Sperm surface antigen	Absence of HY surface antigen	Presence of HY surface antigen	Immunological sexing
DNA content	More DNA	Less DNA	Flow cytometry

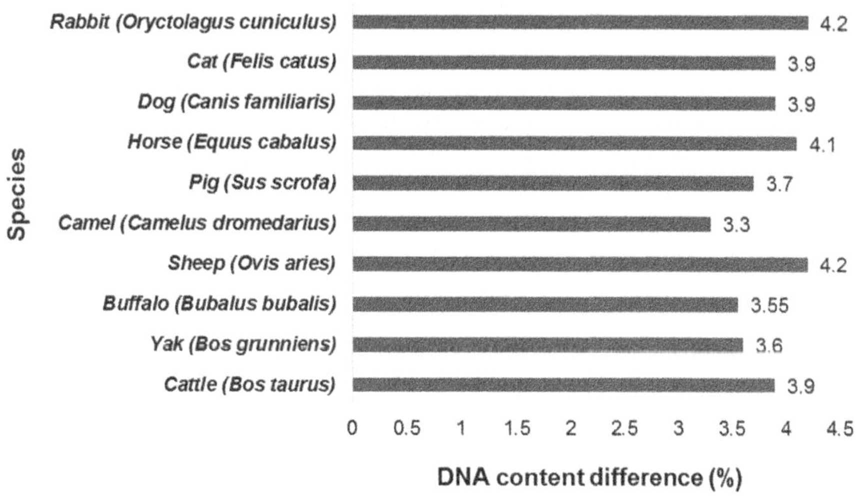

Fig. 24.1 DNA content differences of X and Y chromosome-bearing spermatozoa in different domestic animal species. (*Graphical representation drawn from Garner and Siedel* 2003)

yielded encouraging results thus far. The difference in DNA content between X and Y chromosome-bearing spermatozoa is utilized for separating two types of spermatozoa. Figure 24.1 represents differences in DNA content in domestic animal species.

Semen sample to be sorted is first diluted with bis-benzimide (Hoechst 33342) dye in standard dilution. The dye Hoechst 33342 diffuses across an intact sperm membrane and binds to the A/T base pairs in the minor groove of DNA. Hoechst 33342 has a 350/460 nm absorption and fluorescence emission spectra, which makes it a highly useful marker for determining the precise quantity of DNA in sperm cells. The flow cytometer measures the DNA content difference between two types of sperm using two fluorescence detectors that measure the strength of the signal from the Hoechst 33342 attached to the sperm DNA when excited by a laser. The diluted DNA is allowed to pass through the flow cytometer channel and a vibrating crystal breaks the stream into droplets. An argon laser beam of 351 and 364 nm wavelength illuminates the stained sperm. X chromosome-bearing sperm with more DNA content glows brighter than the Y chromosome-bearing sperm. The two separated populations are deflected into opposite streams for collection due to the presence of charged plates at the discharge point. On the flow cytometer, fluorescence histograms differentiate the sorted populations, and the software also allows for the gating out of dead and moribund sperm. An enriched population of flow-sorted sperm is collected by relative gating of the individual population (Fig. 24.2).

Due to multiple processes involved with the conventional method of sperm sorting there occurs irreversible alteration in sperm which ultimately leads to the compromised fertility potential of sex-sorted semen. Insemination with sex-sorted semen shows 10% less fertility compared to unsorted semen and increasing the number of sex-sorted sperm per inseminate cannot bridge this gap. The fertility of sex-sorted semen is compromised by irrevocable biochemical changes occurring due to additional subprocesses during sorting procedure including a prolonged holding time prior to staining, exposure to a laser beam, separation into X- and Y-sperm and finally exposure to an electrical field for enrichment of sorted population. Drastic changes in the sperm environment in every subprocess are caused due to mechanical, physical, and biochemical stresses to the sperm cells.

24.3.2 Next Generation Sperm Sorting Technologies

Advanced methods of sperm sorting technologies involve several modifications of different steps in the original technique. Pretreatment of sperm, optimization of the medium in which sperms are maintained, sperm staining medium, sheath fluid, and freezing medium are such modifications that have been integrated to maintain the physiological pH of the sperm

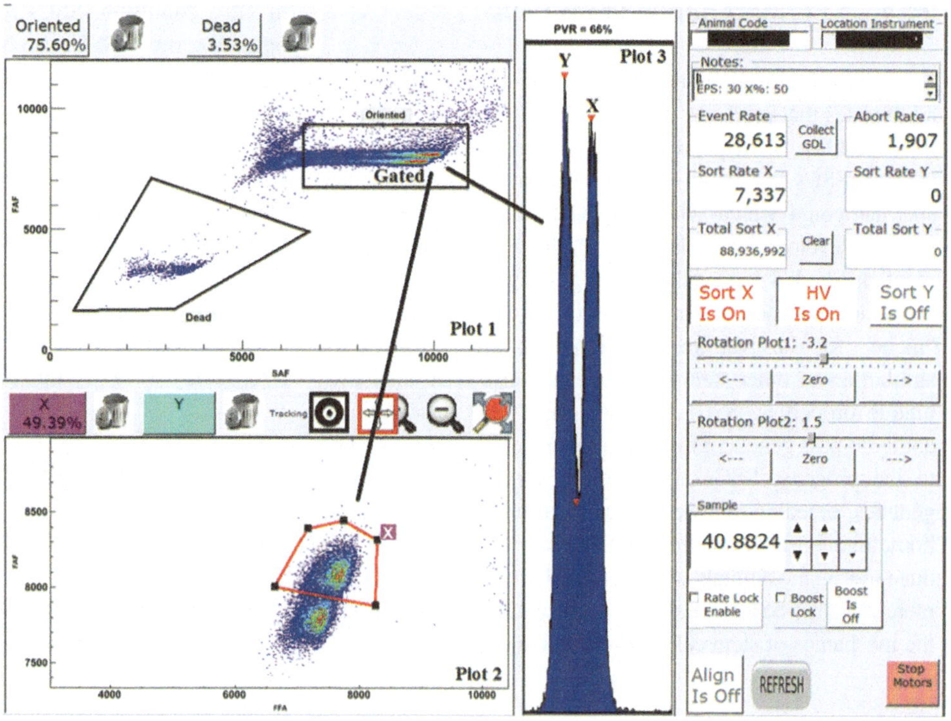

Fig. 24.2 Flow cytometric separation of X- and Y-bearing spermatozoa. Plot 1 identifies live/dead sperm populations and gates only the cells within the oriented region to plot 2 and plot 3. All the dead and moribund sperm are removed from the sorting process. Plot 2 allows for the gating of the desired sex (X or Y or both) and Plot 3 evaluates resolution by measuring peak to valley ratio (PVR). (*Source: Vishwanath and Moreno* 2018)

throughout the procedure and ensure low-dose freezing after sorting. Furthermore, the introduction of microfluidics- and nanotechnology-based tools has aided in the rapid progress of sperm sorting technologies recently.

24.3.2.1 Microfluidics Dielectrophoretic Chip-Based Sperm Sorting

Microfluidics is a technique for controlling tiny volumes of fluid on the micro- and nanoscale through channels that are less than a thousand micrometers in diameter. It may be used to separate a variety of cells, including sperm. Dielectrophoresis (DEP) is a noninvasive cell separation technology that uses nonuniform electric fields in a suitable solution to regulate cell mobility. Cells with different electrical surface properties are separated by being attracted toward (positive DEP; pDEP) or repelled from (negative DEP; nDEP) the location of the greatest electric field. DEP is frequently amalgamated with the microfluidic chip method to achieve high performance for sorting applications. The flagellum pattern and velocity of human X and Y sperm differ, and this difference is depending on the dielectrophoretic field and medium employed. Based on this difference the combined technology, the microfluidic dielectrophoretic system (MF-DEP) has been preliminarily applied for the enrichment of X chromosome-bearing sperms in bull. The technique is safe for sperm and the efficiency of sorting depends upon several factors like electrical voltage applied, sorting cycle, sorting buffer, flow rate, and frequency. Further efforts are much required for large-scale application of this technology in livestock sector.

24.3.2.2 Magnetic Nanoparticle (MNP)-Based Sperm Sorting

MNP-based sperm sorting relies on the difference in zeta potential between X and Y chromosome-bearing sperm. Zeta potential is a negative electro-kinetic potential of about −16–−20 mV acquired by the sperm membrane during spermatogenesis and epididymal maturation as a result of sperm surface coating with sialic acids. Semen sample is mixed with negatively charged nanosize magnetic microbeads (about 50 nm in diameter), incubated for 10 min, and then exposed to a magnetic field for around 20 min to isolate X- or Y-bearing sperm. The Zeta potential of Y-bearing spermatozoa is −16 mV, whereas the Zeta potential of X-bearing spermatozoa is −20 mV; hence, the Y-bearing sperm population will form complexes more easily by binding to the MNP. The complexes will stick to the inner wall of the test tube if a magnetic force is applied to it, while the X-bearing population will remain suspended in the medium and may be recovered by gentle aspiration. Preliminary studies in equines have revealed that MNP-based sperm sorting does not alter the sperm qualitative attributes like viability, motility, and chromatin integrity, and, thus, maybe a good alternative to traditional flow cytometry-based sorting.

24.4 Spermatogonial Stem Cells

Spermatogenesis, an intricate and male-specific process in adult testes that produces a steady amount of spermatozoa for male fertility, is dependent on spermatogonial stem cells (SSCs). SSCs comprise a subpopulation of undifferentiated spermatogonia which when removed from the testes show broader developmental plasticity. During embryonic development, primordial germ cells (PGCs) give rise to spermatogonia. SSCs depend upon a specific microenvironment known as a "niche" for survival and development. SSCs have some similarities to other stem cells, but they also have their own unique properties. SSCs are adult stem cells that have the ability to pass genetic information from the paternal generation to the progeny. SSCs are therefore very useful in animal genetics, breeding, and reproduction if gene transfection and homologous transplantation of SSCs are synthesized to produce transgenic animals with increased productivity and commercial value. SSCs are also an excellent model for studying the mechanics of stem cell self-renewal and differentiation.

24.4.1 Isolation and Enrichment of Spermatogonia from Domestic Animals

In domestic animal testes, SCs are extremely rare. As a result, for continued culture or manipulation of these cells, it is important to identify and enrich SSCs with high viability and purity.

Fluorescence-activated cell sorting (FACS) and magnetic-activated cell sorting (MACS) are routinely used for the isolation and enrichment of SSCs as they can differentiate SSCs from diverse testicular cell population with the help of several SSC-specific biological markers (Table 24.2).

24.4.2 In Vitro Culture of SSCs from Domestic Animals

SSCs remain in the basement membrane of seminiferous tubules in a three-dimensional (3D) microenvironment. In in vitro condition, SSCs require a similar environment for optimal growth. SSCs can be identified, cultured in vitro, and sperm generated from them can be used in IVF or ICSI of oocytes. In order to stimulate the proliferation of germ cells in culture, the right environment is required. Various culture conditions and growth agents, such as glial cell line-derived neurotrophic factor (GDNF), leukemia inhibitory factor (LIF), epidermal growth factor (EGF), or fibroblast growth factor 2 (FGF2), are supplemented in the cell culture media to keep SSCs growing and multiplying. Porcine SSCs are cultured with or without feeder cells in agarose-based 3D hydrogels and 2D culture plates.

Table 24.2 Spermatogonial stem cell markers used for isolation of SSCs in domestic animals

Species	SSC markers
Cattle	*UCHL1*(PGP9.5), *ZBTB16, DBA, THY1*(CD90), *NANOG2, POU5F1, Claudin-8*
Buffalo	*UCHL1*(PGP9.5), *DBA, POU5F1*
Goat	*UCHL1*(PGP9.5), *ZBTB16, THY1(CD90)*
Sheep	*ZBTB16, DBA*
Pig	*UCHL1*(PGP9.5), *ZBTB16, DBA, GFRα1 THY1*(CD90), *NANOG2, POU5F1, SSEA1*
Equid	*ZBTB16, GFRα1, CSF1R*

24.4.3 Applications of SSCs

24.4.3.1 Genome Editing Via SSCs

SSCs edited with TALEN or CRISPR/Cas9 and implanted into mice's seminiferous tubules created spermatogenetic cell colonies that produced genetically modified sperm. As a result, gene editing can be used to create genetically altered model animals.

24.4.3.2 Generating Transgenic Animals

SSCs can be taken from postnatal animals, grown in vitro, and genetically targeted. SSCs that have been genetically modified can be differentiated into sperm in vitro or in vivo, which can be used to create transgenic embryos, ESCs, and progeny.

24.4.3.3 Restoration of Fertility

The use of SSC autotransplantation to treat infertility is a feasible approach for restoring fertility in male animals. Ablation of the endogenous germline is required for use as a breeding tool in animals; otherwise, the mixing of donor and recipient sperm production would occur after SSC transplantation. However, the process is the initial stage of development in livestock species.

24.5 Multiple Ovulation and Embryo Transfer Technology (MOET)

The success of frontier areas of reproductive technologies depends upon the fruitful transfer of embryos in domestic animals. The major application of ETT is to take advantage of female reproductive potential, such as having more offspring from valuable donors and having offspring from infertile donors. In addition, ETT is a very useful technique for the conservation of elite genetic resources. The approach is also used to develop new breeding strategies such as reducing generation intervals and evaluating desirable genes in a short amount of time.

24.5.1 Milestones

1890	Walter Heape transferred two 4-cell Angora rabbit embryos into an inseminated Belgian doe, which subsequently gave birth to four Belgian and two Angora young.
1949	Birth of four piglets after the application of nine embryos into the oviduct of a recipient sow reported by A.V. Kvasnicki.
1951	Successful embryo transfer in a cow and calf born (Willet et al. 1951)
1972	W.R. Allen and L.E.A. Rawson reported equine offspring production by ET

24.5.2 Procedure

24.5.2.1 Donor Animal Selection

The most significant factors to consider when choosing a donor animal are the donor animal's genetic superiority, the purity of the breed from which the donor animal is being chosen, normal physiology and health conditions, normal reproductive status, age, and the economic value of the possible offspring.

24.5.2.2 Superovulation

Superovulation, also known as super stimulation, is a treatment that aims to boost the donor animal's ovulation rate and thus the number of available oocytes without interfering with the physiological and endocrinological processes involved in oocyte maturation, ovulation, and fertilization, as well as embryonic and fetal development.

Almost 150,000–200,000 primordial follicles are there in the ovaries of a newborn female calf. At 4 years of age of a heifer or cow, the total number of follicles including primordial, primary, secondary, and Graffian follicles become reduced to about 77,000. During each ovulation cycle single mature oocyte from the Graffian follicle is released with the effect of Luteinizing Hormone (LH). In superovulation follicular stimulation with FSH or alike hormone is done that allows release of more oocytes during ovulation. In each estrus, ten or more live oocytes can be retrieved from superovulated cows and heifers. On 80–85% of superovulated normal viable donors, around five transferrable embryos can be retrieved.

24.5.3 Application of Superovulation Protocols in Different Farm Animals

Different protocols are followed for superovulation in different animals. The protocol varies according to the species and breed of the animal. Figure 24.3 depicts some common protocols of superovulation in farm animals.

24.5.3.1 Embryo Recovery

ET methods for embryo recovery or flushing are typically performed 7 days after insemination using nonsurgical techniques. To prevent straining, the donor is administered an epidural anesthesia near the tail head. A flexible rubber tube catheter is inserted into the body of the uterus through the cervix. To keep the catheter in position, the cuff is inflated with saline solution, which is pumped into the uterine horns through perforations in the catheter tip that precede the cuff. The solution-filled uterine horn is gently massaged and the fluid carrying the embryos is pulled back out through the catheter.

24.5.3.2 Evaluation of Embryos

Viable embryos with characteristic morphological features are selected for transfer. Table 24.3 and Fig. 24.4 show the morphological features of transferable embryos.

24.5.3.3 Recipient Animal Selection

Recipient animals are selected on the basis of several criteria like normal physiological and health status, good reproductive condition, lack of any reproductive abnormalities, compatibility with the donor in terms of the size of the fetus, and ease of synchronizing the estrus.

24.5.3.4 Embryo Transfer

In this step, embryos are placed in the uterine horn without causing damage to the endometrial lining of the uterus. The process of embryo transfer may be done either nonsurgically or surgically. In sheep, goats, and pigs, the abdomen is opened and the embryo is implanted into the tip of the uterine horn. In cattle and buffalo, embryos are transferred nonsurgically using a special type of catheter.

24.5.3.5 Nonsurgical Method

Viable, good-quality embryos are loaded into a 0.25 mL straw and loaded into the AI gun. The recipient animal is administered an epidural anesthesia to reduce rectal contractions. The insemination gun is inserted through the cervix and into the uterus corresponding to the ovary having a corpus luteum. The embryos are lodged as far as possible into the uterine horn without applying force. If twin calves are desired, embryos are placed in both uterine horns.

24.5.3.6 Surgical Method

Maintaining all the aseptic conditions a 2 in. incision is made with a scalpel on a pre-shaved 6 in. square located some 6 in. in front of the hip joint. By gripping the uterus with the fingers of one hand, the uterus and ovaries are pulled close to the incision opening. A blunt needle is used to make a small incision in the uterine horn. The embryo is placed in the uterus using a 0.25 mL straw coupled to a tiny syringe. A few

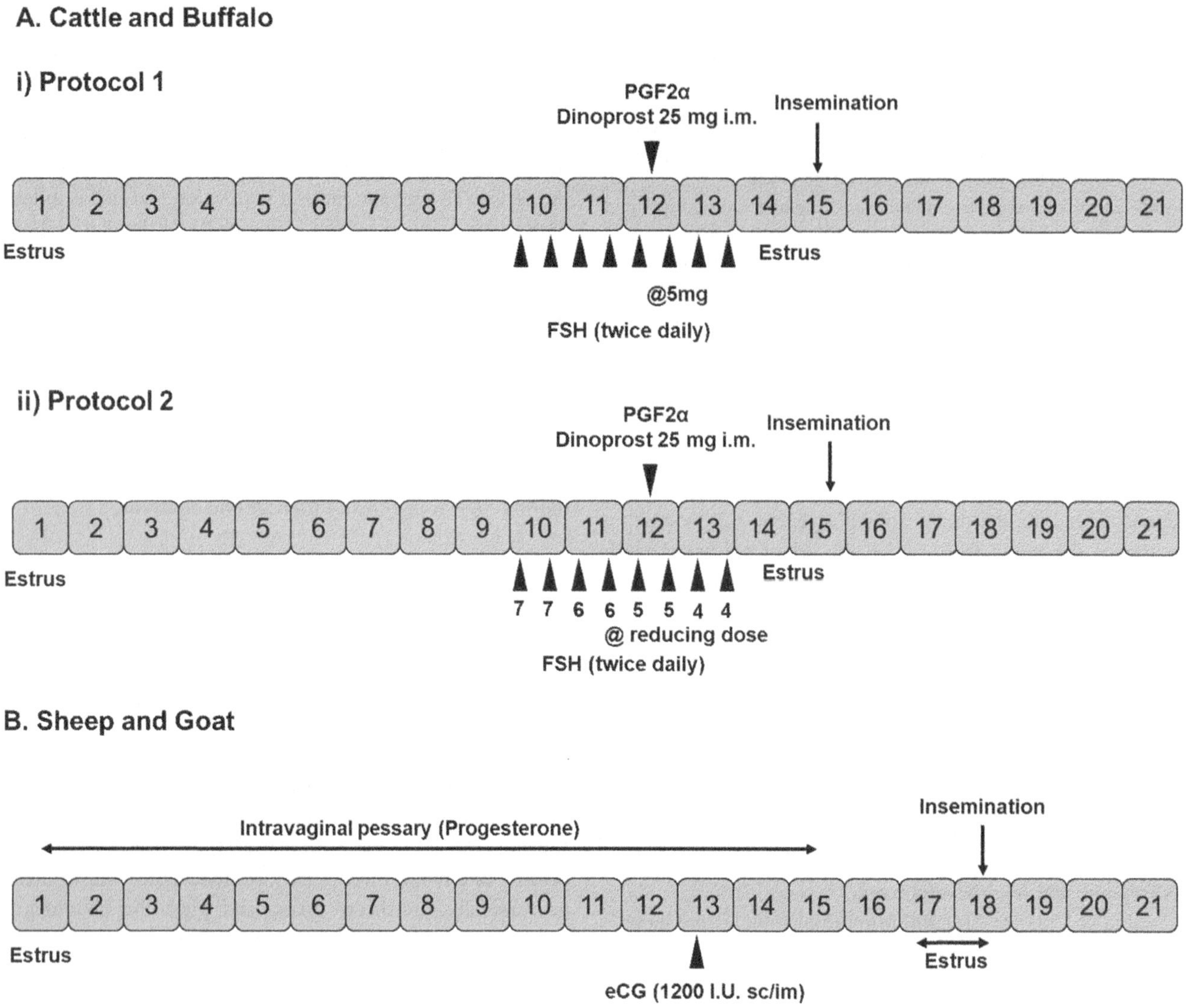

Fig. 24.3 Different superovulation protocols followed in livestock species. (**a**) (i) Protocol 1 with a uniform dose of PGF2α in cattle and buffalo, (ii) Protocol 2 with reducing dose of PGF2α in cattle and buffalo. (**b**) Superovulation protocol in sheep and goat

Table 24.3 Different embryonic stages and their morphological characteristics

Embryonic stages	Morphological characteristics
Morula	Individual blastomeres are difficult to distinguish, and the embryo's cellular mass takes up the majority of the perivitelline space.
Compact morula	Individual blastomeres have merged into a compact mass, and the embryo mass occupies 60–70% of the perivitelline space.
Early blastocyst	An embryo forms a fluid-filled cavity called a blastocoel, which resembles a signet ring and occupies 70–80% of the perivitelline space while maintaining trophoblast and inner cell mass differentiation.
Mid blastocyst	There is substantial differentiation of the outer trophoblast layer and the more compact inner cell mass. The blastocoel is prominent, occupying the majority of the perivitelline space.
Expanded blastocyst	The overall diameter of the embryo is drastically increased to 1.2–1.5×, with the zona pellucida thinning to approximately one-third of its original thickness.
Hatched blastocyst	Embryo completely shed zona pellucida

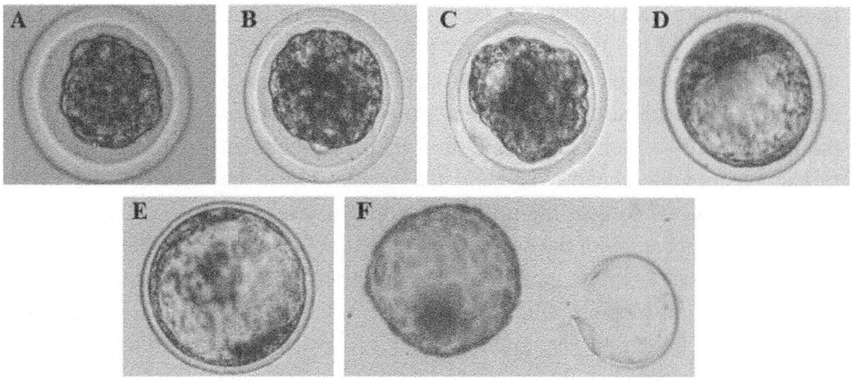

Fig. 24.4 Different stages of transferrable embryos. (**a**) Morula, (**b**) compact morula, (**c**) early blastocyst, (**d**) blastocyst, (**e**) expanded blastocyst, (**f**) hatched blastocyst. (*Source:* https://veteriankey.com/evaluation-of-in-vivo-derived-bovine-embryos/)

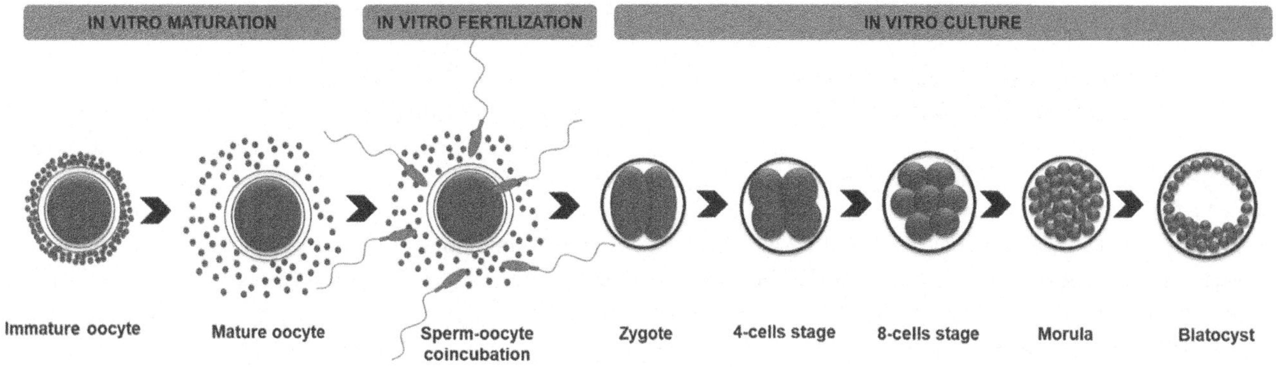

Fig. 24.5 Steps of in vitro embryo production. Immature oocyte is incubated for 24 h in maturation medium. Mature oocyte is co-incubated with sperm for in vitro fertilization and presumptive zygotes are cultured in vitro

stitches are used to close the incision, and an antibiotic solution is put on the stitch region to prevent infection.

24.6 In Vitro Embryo Production

In vitro embryo production (IVEP) refers to the processes of in vitro oocyte maturation (IVM), in vitro fertilization (IVF), and the early days of in vitro embryo culture (IVC). Usually, immature oocytes are collected from the ovaries of common domestic animals, through transvaginal ultrasound-guided oocyte retrieval (OPU: Ovum Pick-Up) procedures in live animals, or after the animal's death after slaughter. A schematic diagram of various steps of in vitro embryo production is depicted in Fig. 24.5.

24.6.1 Transvaginal Ultrasound-Guided Oocyte Retrieval (OPU: Ovum Pick-Up)

Transvaginal Ultrasound-Guided Oocyte Retrieval or OPU is an ART program that has been successfully used in cow and mare for the last three decades. Originally, the technique was introduced in human reproductive medicine. Subsequently, this minimally invasive technique became a popular method for collecting immature oocytes repeatedly from valuable donor animals of high genetic merit. A modified version of a transvaginal ovum pick-up technique, originally developed for use in human reproduction, was applied in cattle.

Animals may be sedated with an anesthetic and intestinal relaxant; however, this is not mandatory. Fecal matter is removed from the rectum and 2% lidocaine is used for epidural anesthesia for easy transrectal manipulation. Vulvar lips and adjacent areas are washed and disinfected thoroughly to lessen the chance of infection.

There are three major parts in the OPU machinery system: an ultrasonographic scanner with an appropriate transducer (probe), an aspiration pump, and a needle guidance system coupled to an oocyte collecting tube. OPU device, comprising of the transducer and the needle guidance system, is introduced into the vagina. Depending on the left or right ovary for oocyte retrieval process, the head of the ultrasound transducer is directed craniodorsally to the left or right of the cervix. With the other hand inserted per rectum, the ovary is held against the head of the transducer so that ovary and follicles become visible on the ultrasound screen. A special biopsy line programmed within the ultrasound scanner's software guides the precise positioning of the follicle for fruitful puncture. OPU handle can be operated with one hand from outside and the operator inserts the needle gradually until it pierces the vaginal wall and the needle is seen entering the ultrasound field. Simultaneous hand-guided movement of the needle and per rectal manipulation of the ovary helps in proper positioning of the ovary and the needle. Once the needle pierces the follicle the aspiration pump is triggered with the foot pedal. With the suction pressure, the follicular fluid and cumulus–oocyte complexes are collected into the oocyte collection medium within the embryo filter. Subsequently, the content of the embryo filter is washed and transferred into petridish. Oocytes are identified with a stereomicroscope, graded and transferred to in vitro maturation medium.

24.6.2 Follicular Wave Synchronization of Donor Animals Prior to OPU

The success of OPU is governed by a plethora of factors, one of which is the quality of oocytes. In bovine, oocytes are retrieved from the ovarian follicles of at least 2 mm size on a random day of estrous cycle. But most of the follicles from which oocyte aspiration is performed are atretic in nature. IVEP with oocytes aspirated from the growth phase of the follicle results in better production of blastocysts compared to the dominance phase. Different hormonal treatment protocols have been applied for follicular wave synchronization of donor animals. Multiple FSH injections (40, 80, and 80 mg) at 24 h interval increased the number of follicles for aspirations. Another factor that determines the outcome of OPU technique is coasting, the period between the last hormonal treatment and OPU. Along with multiple FSH injections, administration of a LH injection 6 h prior to OPU procedure increases the blastocyst production. A combination treatment of progesterone and estradiol benzoate or GnRH application on a random day of the estrous cycle synchronizes follicular wave after 2–4 days.

24.6.3 Collection of Oocytes from Slaughterhouse Ovaries

Immature oocytes are obtained from ovaries collected from freshly slaughtered animals in a slaughter house. The ovaries are transported to the laboratory in a lukewarm normal saline solution (35–37 °C) containing antibiotic. Subsequently, follicles present on the ovarian surface are aspirated to obtain the oocytes. The oocytes are then searched and graded on the basis of the homogeneity of their ooplasm and the mass of cumulus cells surrounding the oocytes. Cumulus–oocyte complexes (COCs) with an unexpanded cumulus mass of >2 layers of cumulus cells, and homogeneous, evenly granular ooplasm is selected for IVM.

24.6.4 In Vitro Maturation

Usable quality oocytes are washed in maturation medium several times and cultured in a group of 15–20 oocytes per droplet of the same medium for 24 h, overlaid with mineral oil in small petridish. Maturation medium is supplemented with LH and FSH and in some cases estradiol, growth hormones, and insulin.

24.6.5 In Vitro Fertilization

Frozen straws of semen samples are thawed and processed in a fertilization medium containing compounds like caffeine, heparin that promotes capacitation and acrosomal reaction. Mature oocytes are also washed in fertilization medium, placed in a small droplet in petridish and processed semen samples containing the required amount of spermatozoa are added to it. The fertilization is carried out in a CO_2 incubator for 18 h.

24.6.6 In Vitro Culture

After 18 h of sperm-oocyte co-incubation, the cumulus cells are washed off the oocytes by gentle pipetting. For culture of bovine embryos modified Charles Rosenkrans medium with amino acids (mCR2aa) containing bovine serum albumin (BSA) and fetal bovine serum (FBS) are used. After the oocytes are washed several times in IVC medium they are transferred to the original granulosa bed and cultured for 8–9 days. Cleavage rate is observed on day 2 and subsequent stages of embryonic development are recorded at regular intervals.

24.6.7 Application

- Exploitation of female reproductive capacity (more offspring from valuable donors).
- Significant facilitation of import and export of valuable genetic material.
- Development of new breeding strategies.
- Twin production (Embryo splitting).
- Introduction of new genes into closed herds.
- Manipulation of embryos.
- Transgenesis.

24.7 Embryo Sexing

Embryo sexing technique is used to predetermine the sex of offspring. Widespread use of embryo transfer technique has created the scope to control the sex of an offspring. The technique is of high demand for the conservation of endangered species and also in various laboratories and pet animals. Embryo sexing is an alternative to sexed semen technology as sorted semen is expensive. There are two commonly used procedures, invasive and noninvasive methods.

24.7.1 Invasive method

24.7.1.1 Karyotyping/Cytogenetic Analysis

A limited number of cells are taken from a biopsy and grown in a medium containing colcemid, a mitotic cell division stopping drug. The cells are then made to swell, causing the chromosomes to scatter. After being fixed and stained with a permanent DNA dye, such as Giemsa, the slides are viewed under a microscope. In metaphase, cells produce a spread of chromosomes that may be distinguished by their distinct banding patterns. The tiny size of the Y chromosome makes it easy to spot. The accuracy of sexing is 100% in this procedure but the process is labor-intensive and time-consuming.

24.7.1.2 Y-Specific DNA Probe

In this technique, a tiny number of cells from the embryo are biopsied, and perfect cellular DNA from these cells is hybridized to a tagged sequence of DNA that is unique to the Y chromosome and hence the embryo's male sex. Probes against several Y chromosomal genes like *SRY*, *ZFY*, and *DDX3Y* have been generated in livestock species like cattle, sheep, and embryonic sex has been determined.

24.7.1.3 Fluorescent In Situ Hybridization (FISH)

Y chromosome-specific DNA probes are labeled with DNA labeling materials like digoxigenin or avidin. The probe is hybridized to prefixed embryonic biopsy material on a slide with a suitable hybridization buffer and condition. Finally, washing is done with appropriate buffer and bound digoxigenin- or avidins-labeled probes are detected with fluorescin-conjugated anti-digoxigenin or anti-avidin antibody (Fig. 24.6).

24.7.2 Noninvasive method

Circulating cell-free fetal DNA (ccff DNA): Analysis of free fetal DNA analysis in maternal circulation is a noninvasive and effective method for determining fetal sex in animals.

24.8 Animal Cloning by Somatic Cell Nuclear Transfer

Animal cloning refers to the production of genetically identical copies of individual animals by means of nuclear transfer. Animal cloning involves advanced methods of microsurgery, embryo culture, and transfer into recipients (surrogate mothers). Animal cloning is a technique by which a nucleus from a cell of the donor animal is inserted into an enucleated oocyte, and reconstructed oocyte is allowed to grow upto embryonic development and embryos are transferred into a recipient animal. The method is therefore also known as somatic cell nuclear transfer (SCNT).

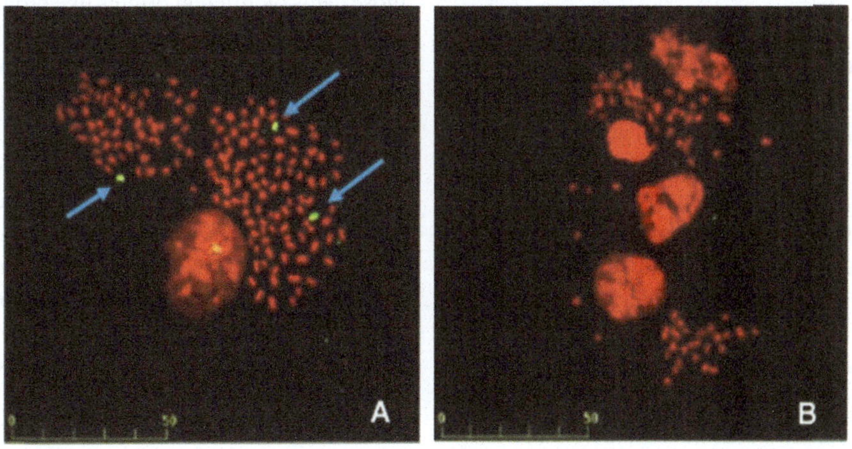

Fig. 24.6 Sexing of in vitro fertilized bovine embryos by FISH. Y chromosomal probe of *Bos taurus* (BtY2-L1) was detected by FITC-conjugated anti-avidin antibody (yellow). Propidium iodide (PI) was used for counterstaining the spread. (**a**) Metaphase nuclei of male-putative blastomeres (arrow showing binding of Y-specific probe). (**b**) Metaphase nuclei of female-putative blastomeres (Y-specific probe is not bound). (*Source: Lee et al.* 2004)

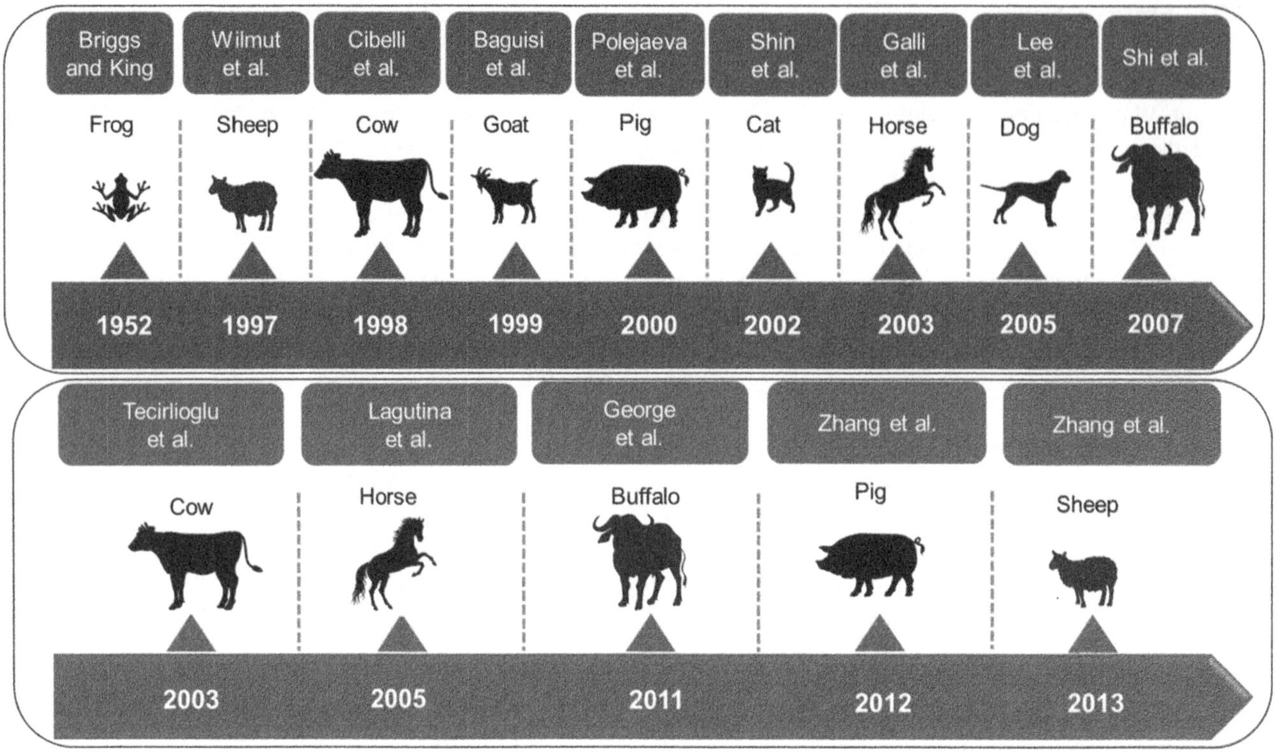

Fig. 24.7 Timeline of the development of live offspring of farm animals. (**a**) Traditional cloning method and (**b**) hand-guided cloning method

Based on the method of enucleation process animal cloning may be either traditional cloning (TC) or hand-made cloning (HMC). TC uses micromanipulator apparatus for enucleation event, whereas HMC is a more advanced procedure. In HMC, enucleation of zona-free mammalian oocytes is performed with sharp metal blades for bisection of zona-free oocytes under stereomicroscope or density gradient centrifugation or chemicals.

24.8.1 Milestones

Briggs and King transferred nuclei from early cleavage stage embryos to enucleated eggs of the North American leopard frog *Rana pipiens* and tadpoles developed. Comparatively large size of the eggs makes micromanipulation easier in amphibians. Figure 24.7a, b depicts the timeline of the development of SCNT animals both by TC and HMC methods.

24.8.1.1 Hand-Made Cloning (HMC) Method

Figure 24.8 represents schematically the different steps of HMC method.

24.8.1.1.1 Preparation of Donor Somatic Cell or Nuclei

Sources of donor cells in case of HMC vary from species to species and conditions of the nuclear reprograming procedure. Depending on the type of donor cells used in HMC technique blastocyst production rate varies. Adult and fetal fibroblast cells are commonly used as donor cells in farm animals like cow, buffaloes, sheep, and goat. Apart from those, cells of different origins like pronuclear stage embryos, embryonic blastomeres, cumulus cells, granulosa cells, embryonic stem cells, lymphocytes, milk somatic cells, and urine epithelial cells have been used efficiently during HMC process in different species. The donor cells must be quiescent at G0 or arrested at G1 phase for optimum efficiency of cloning process. HMC with donor cells of S and G2/M phases cause premature chromosome condensation leading to chromosome pulverization and chromosomal aneuploidy, respectively which ultimately lead to poor cloning efficiency. Synchronization of cell cycle stage of donor cells is performed by treatment with chemicals like cycloheximide, DMSO, and roscovitine or serum starvation or contact inhibition by cell confluence.

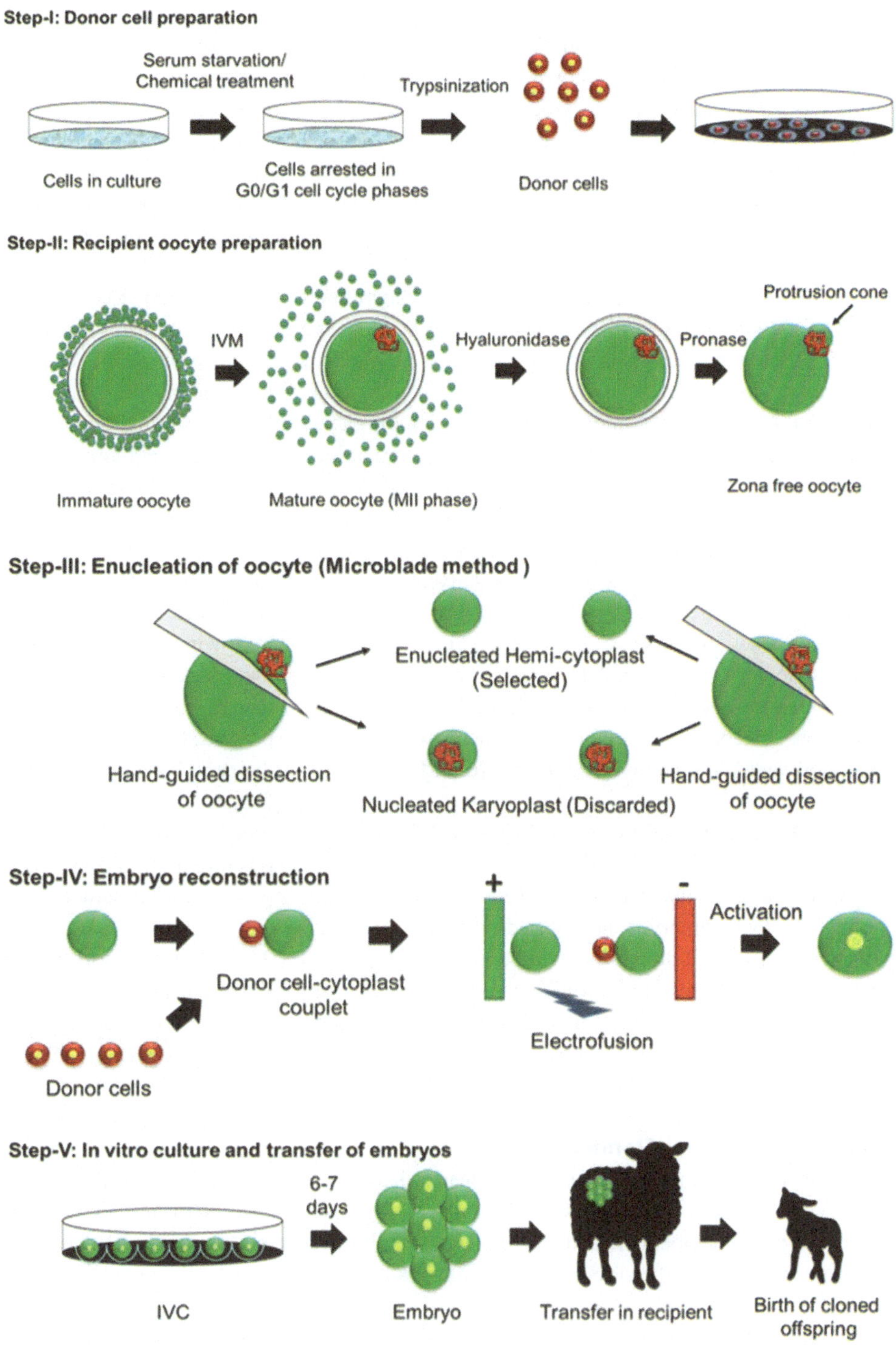

Fig. 24.8 Steps of hand-guided cloning. Recipient and donor cells are prepared. Protrusion cone containing the oocyte nucleus is enucleated. Donor cells and enucleated oocytes are fused by electrical pulse. Finally, the reconstructed embryos are cultured for 6–7 days before transfer to the recipient animals

24.8.1.1.2 Recipient Oocyte Selection, In Vitro Maturation, and Enucleation

Various sources of oocytes are used for HMC method. Oocytes may be obtained from live animals by ultrasound-guided ovum pick-up (OPU) or laparoscopic technique following hormonal stimulation of animals. Also, abattoir ovaries from nonstimulated females are also used for HMC. Good quality oocytes are selected on the basis of compact cumulus–oocyte complex (COC), and homogeneous ooplasm. In vitro maturation of oocytes is carried out for 22–24 h in a maturation medium containing the hormones like FSH, LH, and estradiol. After that, hyaluronidase is used to remove cumulus cells, pronase is used to remove zona, and zona-free oocytes are processed for enucleation. Enucleation of oocytes is done by cutting the polar body (protrusion cone) with the help of a microblade or chemically by the use of demecolcine, nocodazole, etoposide, caffeine, MG132, etc., or density gradient centrifugation. After bisection, a portion of oocyte with chromatin is known as karyoplast and that without chromatin is known as cytoplasts or hemi-cytoplasts (also known as demioocyte). Cytoplasts are screened by fluorescent staining and selected for embryo reconstruction purpose.

24.8.1.1.3 Reconstruction of Embryos

In this step, demioocytes are exposed to phytohemagglutinin (PHA) for a few seconds. PHA makes a sticky layer on the surface of the demioocytes. This helps in easy attachment of the donor cell on its surface to form a couplet. As there occurs substantial loss of cytoplasm during the dissection of oocyte, another demioocyte is attached to the couplet to compensate for the loss of cytoplasm. A demioocyte-donor cell couplet and a single demioocyte are aligned in a BTX fusion chamber connected to an electrofusion device and a single low-voltage AC pulse in an electrofusion medium is delivered. Alternatively, two separate electrofusion events may be performed in the reconstruction process, initially during fusion of demioocyte-donor cell couplet fusion and second during couplet-demioocyte fusion. After electrofusion, chemical activation of rebuilt clone embryos is done with calcium ionophore and *N*-6 dimethylaminopurine (6-DMAP).

24.8.1.1.4 In Vitro Culture of Embryos and Transfer

HMC embryos are cultured in vitro individually in microwells using the well-of-the-well (WOW) system. Activated HMC embryos are placed in each microwell which ensures three-dimensional blastomere arrangements in zona-free embryos. Embryonic quality and stages are assessed at definite interval of upto 7–8 days. Finally, good quality embryos are selected and transferred aseptically in the uterus of animals after 7–8 days.

24.8.1.2 Traditional Cloning method

Traditional cloning methods employ micromanipulator apparatus which consists of a contrast optical system, stereomicroscope, micromanipulator, and microinjectors. Following steps are followed for the classical method of SCNT:

24.8.1.2.1 Enucleation

Enucleation of oocytes is done when the cell cycle phase of the oocyte is stalled at Metaphase II stage. During this stage, chromosomes remain condensed in the form of metaphase plate or meiotic spindle near the protruded polar body. Enucleation is performed in a small cohort of oocytes, usually with 10–20 oocytes at a time. The oocyte, polar body, and enucleation pipette are perfectly aligned for the removal of the metaphase chromosome. For optimum alignment, the oocyte is held gently by the holding pipette and the enucleation pipette is used to rotate the egg appropriately to bring the inner zona surface, polar body, and pipette tip in the same focal plane.

After securing the proper alignment, the holding pipette is made tighter and the enucleation pipette is introduced into the zona pellucida, piercing the oocyte membrane, adjacent to the metaphase plate.

A small amount of cytoplasm and polar body is gently sucked into the enucleation pipette. Enucleated oocyte is transferred from the micromanipulation chamber to the culture medium. Epi-illumination with Hoechst 33342 dye is done to confirm the successful enucleation.

Hoechst 33342 dye is done to confirm the successful enucleation and finally transferred to the culture medium. For preventing damage of the oocyte membrane during the process of piercing by enucleation pipette Cytochalasin B is used in the micromanipulation drop that destabilizes the actin cytoskeleton.

24.8.1.2.2 Donor Cell Preparation

Donor cells are harvested and prepared as described in HMC. Following harvesting cells are transferred into a fresh manipulation drop containing enucleated oocytes. Several cells are loaded into the cell transfer pipette at a time to hasten the process. Cells are lodged in the perivitelline space by placing the oocyte in such a manner that the slit in the zona pellucida formed during enucleation is close to the tip of the cell transfer pipette. Within the zona, the cell is transferred by pushing it out of the pipette with the microinjector, ensuring proper contact between the cell and the oocyte membrane.

24.8.1.2.3 Fusion of Donor Cell-Oocyte Couplet

Electrical stimulation is the most popular method of fusion of donor cell-oocyte couplet. Other methods like the use of polyethylene glycol (PEG), or Sendai virus have also been used by several researchers in different species. Electrofusion medium is a non-ionic, slightly hypotonic medium. A common electrofusion medium consists of 0.3 M mannitol, 0.1 mM $MgSO_4$, and 0.05 mM $CaCl_2$ which vary from species to species. Mannitol determines the osmolarity of the medium. Divalent magnesium cation helps to maintain membrane contact between the cell and oocyte calcium ion provides activation stimulus at the time of fusion. Fusion chambers of BTX instruments and temperature of the fusion medium are key factors that influence the efficiency of cloning experiment. A four-well plate is used for fusion purpose. Well 1 contains manipulation medium, Well 2 contains a 1:1 mixture of manipulation medium and fusion medium, Well 3 contains fusion medium, and Well 4 contains manipulation medium. From the manipulation, microscope couplets are moved to Well 1, then to Wells 2 and 3, where they settle to the bottom of each well. Once fused, they are placed into Well 4. Finally, they are placed in a culture medium and kept in the incubator. Chemical activation of reconstructed clone embryos is done with calcium ionophore and *N*-6 dimethylaminopurine (6-DMAP).

Know More. . . .

Improving Animal Cloning Efficiency by Epigenetic Reprogramming

Since the birth of Dolly 25 years ago significant improvement has occurred in the efficiency of cloning procedure. Still the proportion of cloned embryos that develop to full term remains very low, greatly limiting the application of SCNT technology. The major reason behind that is incomplete epigenetic reprogramming during cloned embryo development. Epigenetic modifications are heritable changes in gene expression without alterations in genomic DNA sequences. Several epigenetic changes like DNA methylation, histone modification, genomic imprinting, and X chromosome inactivation (XCI) occur during the progression from fertilized oocyte to differentiated embryo and they also play a key role in embryo development following SCNT. Present research on SCNT focuses to improve epigenetic reconstruction in cloned embryos and several strategies have been devised so far. Improving DNA methylation reprogramming by DNA demethylation reagents and Dnmts knockdown has successfully ameliorated genome DNA methylation and histone modification in cloned embryos. Another method for improving the development competence of cloned embryos is to modify histone marks. Trichostatin A (a class I and II Histone deacetylase or Hdac inhibitor), scriptaid (a low-toxicity synthetic Hdac inhibitor), and valproic acid all increase histone acetylation, particularly H3K9ac and H3K14ac, boost gene expression levels in cloned embryos, and hence promote SCNT-mediated nuclear reprograming.

24.9 Animal Transgeneis

Transgenic animals are genetically engineered animals that have a foreign gene or genes of interest introduced into their genome, which are transmitted and expressed by their progeny. The creation of novel ways of gene delivery in mammals was aided by advances in assisted reproductive technologies (ART) and molecular biology procedures, which hastened the development of transgenic technology. The idea of genetic manipulation of animal by introducing genes into fertilized eggs became real during the 1980s. Gradually transgenesis has become a mighty technique for analyzing the function of a plethora of genes, for generating animal models of human diseases, and for producing pharmaceutically important proteins secreted through milk. The process of expressing a gene of pharmaceutical proteins in the mammary gland is termed as "pharming." Mammary gland is chosen for expressing the proteins because milk can be collected from the animal without any harm to the animal.

There are several steps in the process of animal transgenesis.

24.9.1 Pronuclear Microinjection

In this technique, exogenous DNA is injected into the male pronucleus of zygotes with the help of a pulled glass needles. Foreign DNA molecules become inserted into their genomes at random sites. Embryos produced are subsequently transferred into a surrogate mother where viable offspring grow.

24.9.2 Embryonic Stem Cell Engineering

Embryonic stem (ES) cells are pluripotent stem cells derived from the inner cell mass of the blastocyst. These cells have the ability to divide indefinitely under ideal culture conditions. Because of this property, ES cells can be easily propagated and manipulated by inserting a DNA construct encoding the desired gene or genes. The procedure entails isolating and cultivating ES cells in vitro prior to inserting the transgene. The transgenic ES cells are then separated from non-transgenic cells and allowed to multiply to generate transgenic ES cell colonies. The chimeric animals are then evaluated for germline transmission, and pure transgenic animals are created using various breeding procedures. The use of ES cells for transgenesis has a number of advantages, as transgene integration can be tested using selectable markers.

24.9.3 Sperm-Mediated Transgenesis

Spermatozoa can naturally uptake exogenous DNA by using a simple incubation process. DNA attaches to the plasma membrane of sperm cells via a DNA-binding protein found in sperm. The mechanism mediated by CD4 molecules internalizes about 15–20% of the DNA attached to sperm. In order to insert the gene of interest within sperm various techniques are used to transfer sperm carrying a gene of interest into oocytes, with variable degrees of success. Electroporation or lipofection can help enhance DNA uptake by sperm. Exogenous DNA attaches to the subacrosomal region of the sperm head of different species. Exogenous DNA interacts with DNA-binding protein(s) (DBP) of 30–35 kDa on the sperm cell surface, forming a DNA/DBP complex that is triggered by CD4-mediated internalization and is dependent on MHC class II expression. DNA/DBP/CD4 enters the

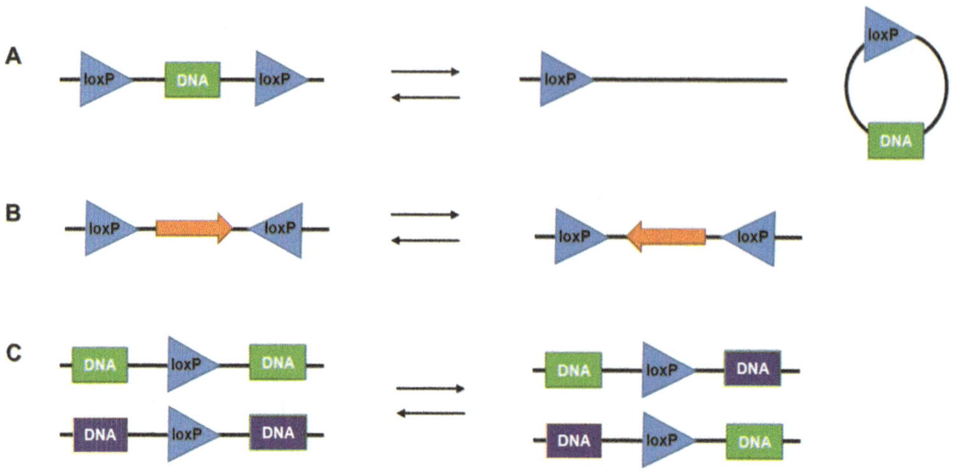

Fig. 24.9 Cre-lox system of recombination. (**a**) Excision—*cis* placement of *loxP* sites in same direction. (**b**) Inversion—*cis* placement of *loxP* sites in opposite direction. (**c**) Translocation—*trans* placement of *loxP* sites

nucleus and reaches the nuclear matrix. A small percentage of foreign DNA recombines with the sperm chromosomal genome at a few "accessible" chromatin locations.

24.9.4 Virus-Based Transgenesis

Viral vectors have been used to mediate foreign gene transfer by delivering and integrating transgenes into the host genome. Viral vectors can be divided into non-integrating viral vectors (e.g., adenoviral vectors), and integrating viral vectors that are mostly derived from a retrovirus, lentivirus, and adeno-associated virus (AAV). Viral vectors can be employed to perform transgenesis into zygotes in two ways: zona-free embryo viral transduction and perivitelline space (subzonal) injection.

24.9.5 Recombinase-Mediated Transgenesis

Recombinases are natural enzymes that facilitate genetic recombination at specified sites. Recombinases can accomplish deletions, insertions, and inversions in DNA sequences by interacting with their own recognition sites. For a variety of reasons, site-specific recombinases have been incorporated into genome editing initiatives. Cre recombinase, Flp recombinase, and PhiC31 integrase have all been used to manipulate livestock genomes.

The basis of the Cre-lox recombinase system is the ability of the P1 bacteriophage cyclization recombination (Cre) recombinase gene (cre) to effect recombination between pairs of 34 base pair *loxP* sites. This 34-bp sequence consists of two 13-bp inverted or palindromic repeats separated by an 8-bp spacer region (Fig. 24.9). The Flp/FRT system, which is derived from the yeast *Saccharomyces cerevisiae*, works similarly to the Cre/loxP system in that flippase (Flp) detects and cleaves two FRT recognition sites. The PhiC31 integrase from *Streptomyces* bacteriophage has been used to mediate homologous recombination between attB and corresponding pseudo attP sites.

24.9.6 Transposon-Mediated Transgenesis

DNA transposons are mobile DNA elements that can integrate into the chromosomes of host cells, allowing them to function as gene transfer vectors. These systems are used as binary tools consisting of a transposon vector and a supply of the transposase enzyme. Mechanism of transposition involves excision of a transgenic cassette of interest flanked by transposon-inverted repeat sequences from a plasmid vector, followed by genomic integration (Fig. 24.10).

Know More. . . .

- *Gene knock-in*: Gene knock-in technology modifies the genetic locus of interest by replacing DNA sequence information one-for-one or by adding sequence information not found on the genetic locus.
- *Gene knockout*: Gene knockout is the total removal or permanent deactivation of a gene through genetic engineering.
- *Gene knock down*: Gene knock down is the deactivation or suppression of genes, rather than complete deletion.
- *Conditional knockout*: Conditional knockout is an approach to knockdown studies in genes that would be lethal if they were completely knocked out.

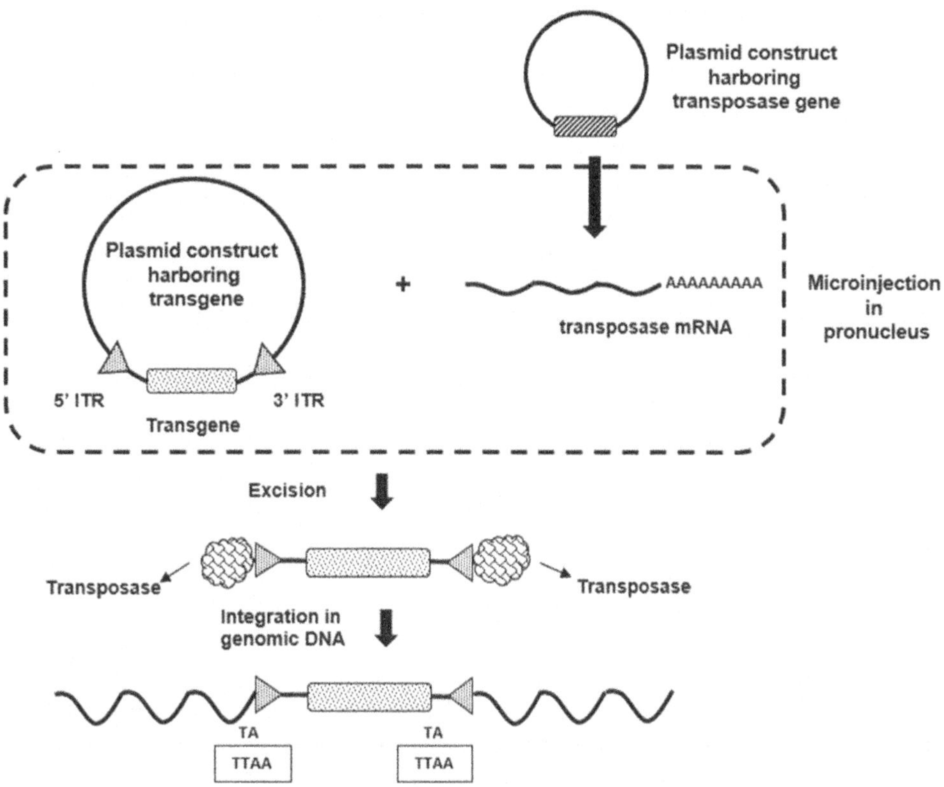

Fig. 24.10 The plasmid-based binary transposon system. Binary transposon platform is made up of the transposon construct harboring the gene of interest flanked by inverted terminal repeats (ITR) and the transposase protein. After the components are delivered to the cells, the transposase protein is translated and attaches to the ITRs flanking the transgene. Transposase catalyzes the transposon's excision and subsequent genomic integration. Integrations happen at sequence motifs TA or TTAA. Duplication of these motifs occurs at the insertion site to flank the transposon

24.9.7 Application

24.9.7.1 Improving Milk Production and Composition

The creation of transgenic animals that produce more milk, yield milk with better nutrient content, or yield milk with beneficial protein, is required to increase livestock growth or survivability through milk composition alteration. Increased expression of a number of these proteins in milk may benefit the developing offspring's growth, development, health, and survival. Bioactive substances like insulin-like growth factor (IGF), transforming growth factor (TGF-α), and lactoferrin in the milk play important roles in the development and maturation of the gut, the immune system, and the endocrine organs of neonates. Transgenic animal-secreting proteins with physiological roles within the mammary gland itself such as lactalbumin, lysozyme, lysostaphin, or other antimicrobial peptides have been produced.

24.9.7.2 Improving Growth and Carcass Composition

Porcine growth hormone (PGH) genes were introduced into the pig genome, which boosted the growth rate without causing arthritis or aberrant skeletal growth. Transgenic pigs with a human metallothionein promoter/porcine growth hormone gene construct showed significant improvements in economically important traits like growth rate, feed conversion, and body fat/muscle ratio while avoiding the pathological phenotype seen in previous growth hormone constructs. Transgenic pigs for human insulin-like growth factor-I produce 30% more loin mass, 10% more carcass lean tissue, and 20% less total carcass fat. The generation of the first pigs transgenic with a spinach *desaturase* gene, which produces greater levels of unsaturated fatty acids, was a significant step toward the production of healthier pork recently. In striated muscle, these pigs have a greater ratio of unsaturated to saturated fatty acids, indicating that they eat a diet rich in unsaturated fatty acids.

24.9.7.3 Generating Disease-Resistant Animals

Disease resistance in animal is a polygenic trait. In an attempt to boost infection resistance, transgenic constructs containing the immunoglobulin-A (IgA) gene have been successfully introduced into pigs, sheep, and mice. The development of cattle devoid of the prion protein, which prevents infection and transmission of spongiform encephalopathies, such as scrapie and bovine spongiform encephalopathy, was a noteworthy feat. To combat mastitis, transgenic dairy cows that exude lysostaphin in their milk have been developed. Lysostaphin is an antimicrobial peptide that protects the mammary gland against infection by killing *Staphylococcus aureus* bacteria in a dose-dependent manner.

24.9.7.4 Improving Hair and Fiber Production

Transgenesis using the sheep wool keratin and keratin-associated protein (*KAP*) gene may alter the protein composition of wool fiber, resulting in fiber types with superior processing and wearing properties.

24.9.7.5 Modification of Digestion

Phytase is an enzyme that converts inorganic phosphorous to organic phosphorous, increasing the amount of phosphorous available to animals. This enzyme is generally found in ruminants; however, it is not found in monogastric animals. The phytase enzyme is expressed in the saliva of transgenic pigs, which improves digestion and lowers feed costs in pig production.

24.9.7.6 Transgenic Animal as Bioreactor

The use of transgenic animals as bioreactors was initially focused on the mammary glands, but blood, bladder, eggs, and male accessory glands are now all considered bioreactors for medicinal proteins.

24.9.7.7 Transgenic Mammary Glands

Human milk lysozyme is a vital protein for innate immunity however commercial manufacturing of this enzyme is difficult to come by. Human *lysozyme* expressed in a cow's mammary gland produces "value added" milk which is beneficial for orphan children. α1-protease inhibitor expressed in transgenic sheep (Tracy) mammary gland is purified and used for the treatment of lung emphysema. Nexia biotechnologies (Canada) has successfully transplanted the spider silk gene into goats, and the goats' milk now contains the protein that makes up spider silk. This is known as *Biosteel*, the strongest fiber on the earth. This is used for manufacturing bullet-proof vests and suture silk for closing up of wounds.

24.9.7.8 Transgenic Blood

Human hemoglobin was extracted from transgenic pig blood and used to make a blood substitute for human patients.

Application in xenotransplantation: Pigs are considered as a popular source of organs for transplantation in humans since their physiology and other characteristics, such as organ size, indicate that they are among the best non-primate potential donors. The major antigen responsible for hyperacute rejection is the $\alpha_{1,3}$-galactose ($\alpha_{1,3}$-Gal) epitope. The enzyme $\alpha_{1,3}$-galactosyl transferase ($\alpha_{1,3}$-GT) produces this epitope, although the enzyme is inactive in humans. The gene has been knocked out in pig used as organ donor by transgenesis technology.

Application as disease model: The transgenic pigs were employed as a large animal model for *retinitis pigmentosa*, a human eye disease. Transgenic rabbits are used as an animal model for human cardiovascular disease and atherosclerosis.

Know More. . . .

Precise Genome Editing by Engineered Nuclease

Site-specific double-strand break is the key to precise genome editing. Several distinct classes of nucleases have been discovered and bioengineered for this purpose. These are the Zinc finger nucleases (ZFNs), transcription-activator like effector nucleases (TALEN), meganucleases, and the clustered regularly interspaced short palindromic repeats (CRISPR/Cas9) system.

Learning Outcomes

- **Artificial insemination**: AI is the most common method of breeding intensively reared domestic livestock, such as dairy cattle, buffalo, and pigs. AI is a process by which sperm is collected from the male, processed, stored, and artificially introduced into the female reproductive tract for the purpose of conception by using means other than sexual intercourse or natural insemination.
- **Sperm cryopreservation**: Sperm cryopreservation helps in the propagation of animals with better genetic features and species conservation. The spermatozoa are combined with a cryoprotectant such as glycerol and a protective solution including lipoproteins, carbohydrates, and a cryoprotectant. These components aid in the preservation of membrane integrity throughout the chilling and rewarming processes.
- **Sexed semen technology**: Sexed semen technology is used to generate a calf of a specific sex. Sex preselection has a significant impact on the profitability of livestock-based industry. The inherent differences between features of X and Y chromosome-bearing spermatozoa are utilized for sorting of sperm.
- **Spermatogonial stem cells (SSCs)**: SSCs comprise a subpopulation of undifferentiated spermatogonia in testis that controls the process of spermatogenesis. SSCs depend upon a specific microenvironment known as a "niche" for survival and development. SSCs are useful in animal genetics, breeding, and restoration of male fertility.
- **Multiple ovulation and embryo transfer (MOET)**: MOET is a widely used technique for the conservation of elite genetic resources. The animal is induced for multiple ovulation through various hormonal treatment protocols and inseminated

(continued)

subsequently. Embryos are flushed, graded, and transferred to the recipient animal's uterus.

- **In vitro embryo production (IVEP)**: In vitro embryo production (IVEP) refers to the processes of in vitro oocyte maturation (IVM), in vitro fertilization (IVF), and the early days of in vitro embryo culture (IVC). Immature oocytes are collected by ovum pick-up (OPU) or from slaughterhouse ovaries, and cultured in maturation medium. In vitro fertilization is carried out and zygotes generated are cultured for 7–8 days in vitro upto the stage of blastocyst.
- **Somatic cell nuclear transfer (SCNT)**: is a technique by which a nucleus from a cell of the donor animal is inserted into an enucleated oocyte, and the reconstructed oocyte is allowed to grow upto embryonic development, and embryos are transferred into a recipient animal.
- **Transgenesis**: Transgenesis is the process of introducing foreign gene or genes of interest introduced into the animal genome. The gene (s) are transmitted and expressed by their progeny.

Exercises

Objective Questions

Q1. What are the major cryoprotectants used for the cryopreservation of sperm?
Q2. What are the sperm parameters to be considered for determining freezing rate of sperm?
Q3. Which dye is used for flow cytometry-based sex sorting of spermatozoa?
Q4. Which property of sperm cell is used for Magnetic nanoparticle (MNP)-based sperm sorting?
Q5. Which techniques are used for isolation and enrichment of spermatogonial stem cells?
Q6. How many primordial follicles are present in the ovary at the time of birth of a calf?
Q7. Which chemical is used for attachment of two demioocytes to form a couplet during hand-guided cloning?
Q8. What is coasting?
Q9. In which year cloned sheep "Dolly" was born?
Q10. What are the epigenetic modification agents used for increasing animal cloning efficiency?
Q11. Which protein is required for binding of foreign DNA with sperm in sperm-mediated transgenesis?
Q12. Give an example of non-integrating viral vector used in transgenesis.
Q13. Which antimicrobial peptide gene is transferred to mammary gland of animal for mastitis prevention?
Q14. What are the nucleases required for precise editing of genome?
Q15. Which gene is transferred for Biosteel production?
Q16. Which gene is knocked out from pig genome for using the animal as organ donor?
Q17. Which hormones are supplemented in in vitro maturation medium of oocyte?
Q18. What is the recommended post-thaw motility of cryopreserved spermatozoa for performing AI?
Q19. What is the freezing rate for slow freezing of sperm?
Q20. What are the key components of plasmid-based binary transposon system?

Subjective Questions

Q1. What are the symptoms of estrus in animals?
Q2. What are the components used as cryoprotectant? Mention their functions.
Q3. What are the major differences between X- and Y-bearing spermatozoa?
Q4. Describe the culture method of spermatogonial stem cells?
Q5. How restoration of male animal fertility can be done using spermatogonial stem cells?
Q6. Write the steps of hand-guided somatic cell nuclear transfer method.
Q7. What is "pharming"? How is it done?
Q8. How does epigenetic reprogramming help to increase the cloning efficiency?
Q9. What is the usefulness of embryo transfer in animal?
Q10. How transposon-mediated transgenesis is performed?
Q11. Describe the protocol of superovulation protocol in small ruminants.
Q12. Elucidate the mechanism of Cre-lox recombination system.
Q13. Describe the advanced methods of sperm sexing.
Q14. What are the applications of spermatogonial stem cells?
Q15. How transgenic technology can be applied for improving milk production and composition?

Answer to Objective Questions

A1. Glycerol, dimethylsulfoxide (DMSO), and ethylene glycol
A2. The surface-to-volume ratio of the spermatozoa and the permeability of the membrane
A3. Bis-benzimide (Hoechst 33342)
A4. Difference in zeta potential between X and Y chromosome-bearing sperm
A5. Fluorescence-activated cell sorting (FACS) or magnetic-activated cell sorting (MACS)
A6. 150,000–200,000

A7. Phytohemagglutinin
A8. The period between the last hormonal treatment and OPU
A9. 1997
A10. Trichostatin A, scriptaid, and valproic acid
A11. DNA-binding protein (DBP)
A12. Adenoviral vectors
A13. Lysotaphin
A14. Zinc finger nucleases (ZFNs), transcription-activator-like effector nucleases (TALEN), meganucleases, and the clustered regularly interspaced short palindromic repeats (CRISPR/Cas9) system
A15. Spider silk gene
A16. $\alpha_{1,3}$-Galactosyl transferase ($\alpha_{1,3}$-GT)
A17. LH and FSH
A18. 40–50%
A19. −0.5 °C/min
A20. Transposon construct harboring the gene of interest flanked by inverted terminal repeats (ITR) and the transposase protein

Keywords for the Answer to Subjective Questions

A1. Restlessness or increased activity, vocalization, chin resting, vulva swelling, vaginal discharge, and mounting other animals
A2. Glycerol, dimethylsulfoxide (DMSO), and ethylene glycol
A3. Motility, surface charge, sperm surface antigen, DNA content
A4. Glial cell line-derived neurotrophic factor (GDNF), leukemia inhibitory factor (LIF), epidermal growth factor (EGF), or fibroblast growth factor 2 (FGF2)
A5. Three-dimensional microenvironment, supplementation of various growth factors, 3D-agarose gel
A6. Preparation of donor somatic cell or nuclei, recipient oocyte selection, in vitro maturation and enucleation, reconstruction of embryos, in vitro culture of embryos, and transfer
A7. The process of expressing a gene of pharmaceutical proteins in the mammary gland is termed as "pharming"
A8. DNA methylation, histone modification, genomic imprinting, and X chromosome inactivation
A9. Exploitation of female reproductive capacity (more offspring from valuable donors), import and export for valuable genetic material, development of new breeding strategies, embryo splitting, introduction of new genes into closed herds, manipulation of embryos, transgenesis
A10. Plasmid-based binary transposon system
A11. Day 1 to Day 15 intravaginal pessary (progesterone), Day 13 eCG (1200 IU), Day 17 estrus, Day 18 insemination
A12. Excision, inversion, translocation
A13. SexedULTRA™ technology of sperm sorting, Microfluidics dielectrophoretic chip-based sperm sorting, Magnetic nanoparticle (MNP)-based sperm sorting
A14. Genome editing, generating transgenic animals, restoration of fertility in male animals
A15. Creation of transgenic animal producing lactalbumin, lysozyme, lysostaphin, or other antimicrobial peptides

Further Reading

Textbooks

Gordon I (2017) Reproductive technologies in farm animals, 2nd edn. CABI, Wallingford

Pursley JR, Cibelli J (2020) Reproductive technologies in cattle. In: Presicce GA (ed) Reproductive technologies in animals. Elsevier, London

Research or Review Articles

Bihon A, Assefa A (2021) Prostaglandin based estrus synchronization in cattle: a review. Cogent Food Agric 7(1):1932051

Galli C, Duchi R, Colleoni S, Lagutina I, Lazzari G (2014) Ovum pick up, intracytoplasmic sperm injection and somatic cell nuclear transfer in cattle, buffalo and horses: from the research laboratory to clinical practice. Theriogenology 81(1):138–151

Garner DL, Seidel GE Jr (2003) Past, present and future perspectives on sexing sperm. Can J Anim Sci 83(3):375–384

Hansen PJ (2020) Implications of assisted reproductive technologies for pregnancy outcomes in mammals. Annu Rev Anim Biosci 8:395–413

Jahnke MM, West JK, Youngs CR. Evaluation of in vivo-derived bovine embryos. Veterian Key. https://veteriankey.com/evaluation-of-in-vivo-derived-bovine-embryos/

Lamas-Toranzo I, Guerrero-Sanchez J, Miralles-Bover H, Alegre-Cid G, Pericuesta E, Bermejo-Alvarez P (2017) CRISPR is knocking on barn door. Reprod Domest Anim 52:39–47

Lee JH, Park JH, Lee SH, Park CS, Jin DI (2004) Sexing using single blastomere derived from IVF bovine embryos by fluorescence in situ hybridization (FISH). Theriogenology 62(8):1452–1458

Monzani PS, Adona PR, Ohashi OM, Meirelles FV, Wheeler MB (2016) Transgenic bovine as bioreactors: challenges and perspectives. Bioengineered 7(3):123–131

Moore SG, Hasler JF (2017) A 100-year review: reproductive technologies in dairy science. J Dairy Sci 100(12):10314–10331

Pancarci SM, Jordan ER, Risco CA, Schouten MJ, Lopes FL, Moreira F, Thatcher WW (2002) Use of estradiol cypionate in a presynchronized timed artificial insemination program for lactating dairy cattle. J Dairy Sci 85(1):122–131

Steele H, Makri D, Maalouf WE, Reese S, Kölle S (2020) Bovine sperm sexing alters sperm morphokinetics and subsequent early embryonic development. Sci Rep 10(1):1–3

Vishwanath R, Moreno JF (2018) Review: Semen sexing-current state of the art with emphasis on bovine species. Animal 12:s85–s96

Willet EL, Black WG, Casida LE, Stone WH, Buckner PJ (1951) Successful transplantation of a fertilized bovine ovum. Science 113(2931):247. https://doi.org/10.1126/science.113.2931.247

Part VIII

Lactation Physiology

Lactation Physiology

25

Joydip Mukherjee, Pradip Kumar Das, and Dipak Banerjee

Abstract

Lactation is a complete process of milk synthesis and secretion including development of mammary gland (mammogenesis), initiation of milk synthesis and secretion (lactogenesis), maintenance of milk secretion (galactopoiesis), and restoration of virgin-like mammary gland (involution). Mammogenesis starts around 32 days of embryonic life but a series of structural and functional development, differentiation, and involution are noticed during different physiological state of animals. Lactogenesis is the biological process of onset of milk secretion which includes the enzymatic and cytological differentiation of mammary alveolar cells in early pregnancy to full lactation after parturition. The mammary gland utilized blood precursors to synthesize various milk components. Galactopoiesis is the maintenance of established lactation which follows a dynamic curve with a rapid accelerating phase, peak yield followed by a declining phage till the end of lactation. The suckling stimulus causes milk ejection from mammary gland by neuro-endocrine reflex. Mammary involution is a biological process to restore mammary gland into a virgin-like state through a series of tissue remodeling processes after the cessation of milk secretion. The interactions of different hormones and growth factors are essentially required in each stage of lactation process.

J. Mukherjee (✉) · P. K. Das · D. Banerjee
Department of Veterinary Physiology, West Bengal University of Animal & Fishery Sciences, Kolkata, West Bengal, India

© The Author(s), under exclusive license to Springer Nature Singapore Pte Ltd. 2023

P. K. Das et al. (eds.), *Textbook of Veterinary Physiology*, https://doi.org/10.1007/978-981-19-9410-4_25

Graphical Abstract

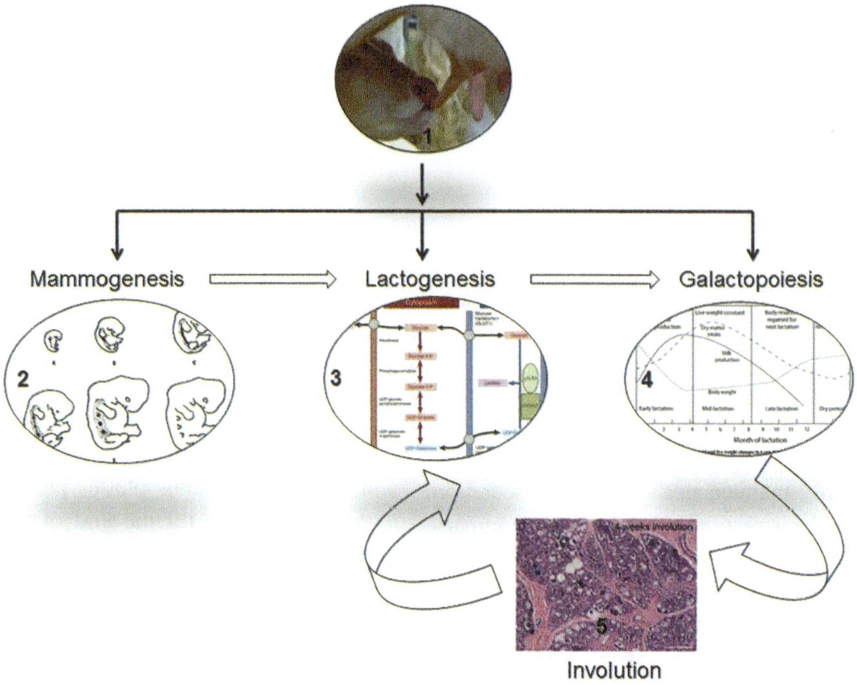

Description of the graphic: Lactation (**1**) is a complete process of mammogenesis (**2**), lactogenesis (**3**), and galactopoiesis (**4**). The mammary gland is developed from mesoderm through a series of structural and functional differentiation called mammogenesis (**2**). Lactogenesis is the biological process of onset of milk secretion which includes the enzymatic and cytological differentiation of mammary alveolar cells together with the synthesis of milk constituents (**3**). Galactopoiesis is the maintenance of established lactation which follows a dynamic curve with a rapid accelerating phase, peak yield followed by a declining phage till the end of lactation (**4**). Mammary involution is a biological process to restore mammary gland into a virgin-like state through a series of tissue remodeling processes (**5**) after cessation of milk secretion

Keywords

Mammogenesis · Lactogenesis · Galactopoiesis · Involution · Hormones · Growth factors

Learning Objectives

- The external and internal structure of the mammary gland.
- Development of mammary glands during different physiological states.
- Biosynthesis of milk constituents and milk secretion.
- The composition of milk.
- Factors affecting milk yield and composition.

25.1 The Mammary Gland

Mammary glands are the distinguishing characteristics of all mammals that have evolved to nourish the newborn offspring for a certain period of post-natal life. It is an exocrine gland modified from sweat (sudoriferous) gland. From the evolutionary point of view, the mammary gland of bovines is the most advanced form which can produce far more milk than required for a calf. Genetic selection, nutritional strategies, and advanced managemental practices make the mammary gland capable of producing an increased volume of milk.

25.1.1 External Anatomy of Mammary Gland

The development of mammary gland is almost similar in different species but there are striking differences among species in regards to the general anatomy of the mammary gland (Table 25.1).

Mature mammary gland consists of udder, teat or nipple, associated ducts, and alveoli composed of epithelial secretory cells and supporting tissues.

Know More.....

The largest mammary gland is seen in blue whales. They have two mammary glands like human and each weighted around 250 pounds and 5 ft in length. The glands produce around 37,500 pounds of milk during a single lactation cycle.

Table 25.1 Comparative features of mammary glands in different species

Species	Teat position			Total glands	Openings per teat
	Thoracic	Abdominal	Inguinal		
Human	2	–	–	2	15–25
Cattle	–	–	4	2	1
Goat	–	–	2	2	1
Sheep	–	–	2	2	1
Pig	4	6	2	12	2
Dog	2	6	2	10	8–14
Cat	2	6	–	8	3–7
Horse	–	–	2	2	2
Elephant	2	–	–	2	10–11

Source: Nickerson and Akers (2011)

25.1.1.1 Udder

In bovines, the udder is composed of four mammary glands. Being a skin gland, the udder is covered with hair except for the teats. The appearance of udder of bovines is square or more or less rounded and saccular. The right and left halves of udder are separated by a longitudinal intermammary groove or sulcus intermammaricus. The separation of the front and rear udder quarters is not well demarcated externally but internally a thin connective tissue provides anatomical barriers between the front and rear glands (quarters) on either side of the udder and there are no direct connections between front and rear udder quarters. Further, the rear quarter produces 60% of total milk. The mammary glands are directly connected with abdominal cavity via inguinal canal by a pair of narrow oblique passages that allow blood and lymph vessels and nerves. In animals, the weight and capacity of the udder increase with age up to 6 years, with the greatest increase occurring between first and second lactation.

The udder of a good milch cow should have the following characteristics:

- The udder should be large enough to produce large volume of milk and should have a good udder attachment.
- The desirable shape of the udder should be long, wide, and moderate in depth with evenly balanced and symmetrical udder quarters.
- The udder should extend well forward.
- It should have a reasonable height from floor level.
- The rear attachment should be high and wide.
- It should be soft to touch and have good mammary vein.

25.1.1.2 Teats

Milk from each gland is emptied through cylindrical or conical-shaped teats (Papilla mammae). Usually, only one teat drains one gland hence bovines have four teats. No sebaceous or sweat glands are found in the teat wall. Supernumerary teats or extra teats are found in 50% of cows which are generally removed before 1 year of age. The pseudo-teats are having no internal streak canal hence no connection to the internal structures of the gland. A good teat should be of moderate size, having proper placement and enough tension on the sphincter muscle around the teat orifice to allow easy milking and prevent leakage of milk.

25.1.2 Supporting Structures of the Udder

Supporting or suspensory structure of the udder is required to maintain proper udder attachment with the body as the weight of lactating mammary gland of a Holstein cow is around 50 kg (>100 lb). The mammary suspensory system is composed of six different types of tissues.

Skin (Tissue-I): The suspension capacity of skin is very low. Nevertheless, it plays a vital role in covering the mammary gland.

Fine areolar subcutaneous tissue or Superficial fascia (Tissue-II): It attaches the skin to the underlying tissue. It has poor suspension capacity.

Coarse areolar subcutaneous tissue (Tissue-III): This tissue is required to attach the front quarter with the body wall after forming a loose bond between abdominal wall and dorsal surface of front quarter. The weakening of this tissue causes the udder to loose from abdominal wall. However, it has also poor suspension ability.

Two parallel lateral suspensory ligaments (right and left lateral suspensory ligament) originate at the sub-pelvic tendons and travel vertically to cover exterior of the mammary gland. It is attached to the lateral surface of the udder with numerous lamellae and inserted into the glandular tissue and the connective tissue stroma to support lobules and lobes in the parenchyma. The lateral suspensory ligaments are made of fibrous tissues mainly collagen which is non-stretchable. Therefore, it provides main suspensory support to the udder without elasticity. There are two layers of lateral suspensory ligaments.

Superficial layer of lateral suspensory ligaments (Tissue-IV): It originates from sub-pelvic tendon and moves downward and forward to reach the udder. It covers the udder below the skin.

Deep layer of lateral suspensory ligament (Tissue-V): It is the inner part of lateral suspensory ligaments. It also originates from sub-pelvic tendon and is thicker than the superficial layer.

Median Suspensory Ligament (Tissue-VI): With the greatest tensile strength and being located in the body's center of gravity, the median suspensory ligament is the main supporting system of the udder in cow. It is composed of two heavy yellow elastic sheets attached to the medial surface of the udder. Unlike lateral suspensory ligaments, it has elastic property and allows the mammary gland to stretch when fills with milk. The loosening of the ligament leads to pendulous udder.

Sub-pelvic tendon (Tissue-VII): It is not a part of udder suspensory system but it gives rise to superficial and deep lateral suspensory ligaments.

25.1.3 Internal Structure of the Udder

The udder is composed of two types of tissues. The parenchymal or secretory tissues are composed of alveoli and ducts. The connective tissue supports the secretory tissue.

25.1.3.1 Alveoli

The primary secretory units of the mammary gland are alveoli. The alveoli are globular structure having a diameter of 50–250 mm depending upon the accumulated milk volume. A number of alveoli join together into a common duct that is surrounded by connective tissues to form a lobule. Many lobules are again surrounded by connective tissues to form lobes.

The secretory epithelium of the mammary gland is cuboidal to columnar in nature that forms the peripheral lining of each alveolus. The secretory epithelium is responsible for the synthesis and secretion of milk. They absorb the blood precursors for milk constituents from adjacent capillaries and the milk is released in the lumen of the alveolus.

The mammary epithelial cell is composed of cytoplasm together with the organelles (nucleus, rough endoplasmic reticulum, mitochondria, and Golgi apparatus) covered by plasma membrane. Rough endoplasmic reticulum is situated adjacent to the basement membrane near nucleus. The Golgi apparatus is situated between the nucleus and the apical membrane. Secretory vesicles are present in Golgi apparatus as terminal swellings and contain casein. Fat is synthesized in the rough endoplasmic reticulum stored in fat droplets. Both, the fat droplets and the secretory vesicles are channelized towards the apical membrane guided by microtubules (situated perpendicular to plasma membrane) and secreted at the lumen. The cytoplasm is also populated with mitochondria and ribosomes.

Myoepithelial cells: These are modified smooth muscle with long cytoplasmic projections that surrounds each alveolus and small ducts. Due to their long cytoplasmic projections, they are also termed as basket cells. Myoepithelial cells have the ability to contract in response to oxytocin which allows the squeezing of epithelial cells and facilitates milk secretion into the lumen of alveoli.

25.1.3.2 Duct System

The duct system carries the milk from the secretory tissue to the teats. The ducts are composed of a double-layered epithelium surrounded by myoepithelial cells. Individual ducts originating from each alveolus connect together to form inter-lobular ducts which join together to form inter-lobar ducts. The branching points of ducts have a narrow opening and then form a sinus-like enlargement before it narrows again. This constriction at the point of branching prevents milk leakage by the gravity of the teat and gland cisterns. Interlobular ducts join to form large ducts which drain into cistern, the collecting spaces. Usually, 5–20 large ducts join with the gland cistern of the udder.

The arrangement of the ducts varies between species depending on the number of openings per teat. Cattle, buffalo, sheep, and goats have one opening per teat hence they have only one final duct per gland. Mare and sows have two main ducts and associated openings. Dogs and cats have 10 or more openings per teat containing 10 or more final ducts.

25.1.3.3 Gland Cistern (Sinus Lactiferous or Udder Cistern)

The major ducts open in the gland cistern. It serves as the storehouse of milk between inter-milking periods. Generally, it is circular in shape but sometimes it may appear as pockets of various sizes at the terminal of major ducts. The capacity of the gland cistern varies from 100 to 400 mL.

25.1.3.4 Teat Cistern (Sinus Papillaris)

The cavities within the teat are termed as teat cistern. Teat cistern originates from the gland cistern and is continuous with the gland cistern. At the junction between the gland and teat cistern, there is a constriction called an annular fold composed of dense connective tissue with a thickness of 2–6 mm. Sometimes a horizontal septum forms at the annular fold and blocks the milk flow from gland cistern to teat cistern leading to a blind quarter and requires surgical interventions to correct it. The teat cistern is layered by double-layered epithelium consisting of columnar (luminal) and cuboidal (basal) cells. The storage capacity of teat cistern is 10–50 mL.

25.1.3.5 Streak Canal (Ductus Papillaris)

The distal opening of the teat cistern is called streak canal or teat canal through which milk is let down. The length of the streak canal is 8.5 mm (5–13 mm) and the diameter is 0.46 mm (0.4–1.63 mm). The sphincter at the streak canal is composed of circular smooth muscles. The contraction of these muscles facilitates tight closure of the streak canal between milking and prevents the leakage of milk. The incompetency of these muscles leads to leaky teats and increases the chance of mastitis. In contrast, cows with tight sphincter are called "hard milkers" due to poor milk flow and increased milking time. Just above the streak canal, there are a series of 6–10 longitudinal folds known as Furstenberg's rosette. The tissue folds do not have any role in preventing milk leakage rather they increase the surface area of the epithelium and stroma and provide local defense against the pathogen after recruiting leukocytes, especially lymphocytes and plasma cells. The epithelial lining of the Furstenberg's rosette abruptly changes from double layered to single layer of stratified squamous epithelium which is continuous with the outer skin. The desquamation at this site forms keratin which occludes the lumen of the teat canal between milking and prevents bacterial entry.

25.1.4 Blood Vascular System of the Mammary Gland

The mammary gland derives the precursors of milk from blood. In a high-yielding cow, 400–500 units of blood are required to synthesize one unit of milk with an average blood flow rate of 280 mL/s. The lactating udder receives 8% of the total blood volume. Thus, to meet this huge demand, the mammary gland must have a well-developed blood vascular system.

25.1.4.1 Arterial Supply

Arterial blood from the heart flows initially through posterior dorsal aorta which after entering into the abdominal cavity becomes abdominal posterior dorsal aorta. This aorta runs parallel to vertebral column and at the level of sixth lumber vertebrae, it diverges to form right and left iliac arteries and later into the internal and external iliac arteries. The external iliac artery gives rise to the external pudendal artery or mammary artery and reaches the dorsal surface of the udder via the inguinal canal. After emerging from the inguinal canal, mammary artery and associated milk vein follow a S-shaped route which allows the lengthening of the blood vessels during the stretching of median suspensory ligaments during distension of the udder. The mammary artery forms subcutaneous abdominal artery which in turn divides into anterior and posterior mammary arteries. The subcutaneous abdominal artery supplies to the anterior dorsal portion of each side of the udder. The anterior and posterior mammary arteries vertically enter into the parenchyma of the fore and rear quarters of each side and terminate in a capillaries network surrounding the alveoli. The teats receive blood supply from papillary arteries which arise from mammary arteries. At the teat, papillary artery and venous plexus form corpus cavernosum.

25.1.4.2 Venous Drainage

Veins leaving the mammary gland run anti-parallel to the arteries. There are three veins to carry blood away from the mammary gland.

External pudic vein (middle mammary vein) runs parallel to the external pudic artery and joins the external iliac vein after ascending through the inguinal canal. The external pudic vein follows a route similar to that of the artery but the flow of blood is in the reverse direction.

Subcutaneous abdominal vein (milk vein/anterior mammary vein) leaves the mammary gland at the anterior end of the front quarters and runs along the abdominal wall. This vein is visible under the skin on the abdomen of the cow. It enters the body cavity through a depression at the xiphoid process called as "milk wells" and empties into internal thoracic vein.

Perineal vein (posterior mammary vein) drains the rear halves of the gland parallel to the perineal artery in an upward and backward direction. After turning the ischial arch it joins the internal pudic vein.

A venous circle is formed by anterior and posterior mammary veins to prevent venous outflow when the cow is lying down.

Papillary veins drain the teat and communicate with mammary veins upon the venous circle at the base of the udder.

25.1.4.3 Lymphatic System of the Udder

The main function of the lymphatic system is to circulate interstitial fluids originating from the capillaries of mammary parenchyma and to carry waste products away from the udder. Majority of the afferent lymphatic ducts converge to form larger lymphatics and run toward the dorsal portions of the udder. They terminate at the supra-mammary lymph nodes situated on the right and left halves of the udder. Around 1–7 numbers of supra mammary lymph nodes are situated above the caudal border of the mammary gland and their sizes range from 4 to 10 cm. The branches of the lymphatic vessels then pass through inguinal, iliac, and pre-femoral lymph nodes that join the lumbar lymph trunk and thereby continue to the thoracic duct to drain their content at the anterior vena cava.

High-yielding cows often experience udder edema due to the accumulation of fluid between skin and glandular tissue during the periparturient periods. The etiology of udder edema is due to the imbalance of hydrostatic and osmotic pressures which result in fluid flow out of the capillaries into the interstitial spaces. Increased capillary permeability due to the damage of the capillary wall and obstruction of the lymphatic system is also the predisposing factors behind udder edema. Massage of the udder in the direction of supra-mammary lymph nodes is used to correct udder edema which allows the movement of lymph towards supra-mammary lymph nodes (as the movement of lymph is always in dorsal direction towards supra-mammary lymph node) and prevents the accumulation of lymphatic fluids.

25.1.4.4 Nerve Supply to the Udder

The sensory or the afferent nerves of the udder carry the information from the udder to the brain and involve in the initiation of the milk ejection reflex. The motor supply to the udder is entirely autonomic or sympathetic and composed of motor fiber to the smooth muscles of arterial walls and teat sphincter. They control the blood flow to the udder by altering the diameter of the blood vessels and involved in the inhibition of the milk ejection reflex.

The main spinal nerves to the udder are the first, second, third, and fourth lumbar nerves and the external spermatic nerves. The first lumber nerve supplies the anterior portion of the udder without parenchyma. The second, third, and fourth lumber nerves fuse together to form inguinal nerve. The caudal portion of the udder is supplied with perineal nerve which is composed of the nerve fiber from second, third, and fourth sacral spinal nerves. Innervation of the udder is highest in the dermis of the teats. The terminal ending of these innervations are sensitive to physical stimuli like pressure, touch, and stretching.

The neurons of the sympathetic nerve fibers are located in the lateral horns of the spinal cord. The circular smooth muscles undergo continuous rhythmic contraction between milking relax at the time of milking allowing dilatation of teat canal and for milk flow in response to sympathetic nervous control.

25.1.5 Mammary Gland Immunity

Mammary gland protects itself from invading pathogens by means of anatomical, cellular, and soluble defense factors that act in coordination with each other. The defense mechanism of the udder is one of the prime area of research in the field of lactation physiology as mastitis or udder infection is the costliest disease in dairy cows. In addition, detailed knowledge on udder immunity will help to develop early detection kits for mastitis diagnosis and to develop appropriate therapeutic interventions. The immune components of mammary glands are either blood origin or locally synthesized factors together with udder morphology.

25.1.5.1 Anatomical Defenses

The first line of mammary gland defense is mediated through teat and associated structures. The sphincter present in the streak canal restricts the entry of pathogens during the inter-milking periods. The Fürstenberg rosette at the internal end of the streak canal acts as a mechanical sealant against the entry of pathogens as well as local immune defense due to the localization of neutrophils around the rosette. Keratin secreted from the squamous epithelium at the streak canal also has bacteriocidal properties. The teat canal also acts as the physical barrier against pathogens by the peristaltic action of the smooth muscle lining of the teat canal. Rounder and pointed teats were reported to be more resistant to intramammary pathogens compared to flat, funnel-shaped, and cylindrical teats.

25.1.5.2 Cellular Defense

The cell populations in the mammary glands are collectively called as "milk somatic cells" (SCC) which are basically body-derived cells composed of mammary epithelial cells (70–75%) and leukocytes (25–30%). The leukocytes mainly comprise of polymorphonuclear neutrophils (PMN) (15–17% of total SCC), macrophages (20–21% of total milk SCC), and lymphocytes (46–60% of total milk SCC). Milk somatic cells are used as an indicator of udder inflammation and milk quality throughout the world. In a normal healthy quarter, the total somatic cell populations are below 100×10^3 cells/mL of milk. But udder injury or inflammation leads to the migration of leukocytes to the site of infections and the leukocytes escape from capillary to alveolar epithelium and penetrates the basement membrane. During their penetration through basement membrane, epithelial cells are also sloughed off. Thus, during intramammary infection, SCC may increase up to several folds.

The roles of different cell populations are summarized in Table 25.2.

25.1.5.2.1 Factors Affecting Milk SCC

Mastitis or intramammary infections: Mastitis or intramammary infections are the main factors that lead to increased milk SCC. Milk SCC from an individual quarter depends upon the infection status of the quarter and SCC ≤100,000 cells/mL could be considered as threshold or negative for the California mastitis test. The increased milk SCC during mastitis is due to the efflux of PMN from blood.

Species: The buffaloes are reported to have low milk SCC compared to cattle whereas goats have higher milk SCC. The values of milk SCC in different species have been presented in Table 25.3.

Stage of lactation: It has been reported that SCC increases with progressing lactation regardless of whether the cow is infected or not.

Breed: SCC variation has been noted between breeds of dairy animals. Generally, SCC is higher in high-yielding cows compared to low yielders. The milk SCC in different breeds of cattle has been presented in Fig. 25.1.

Age and parity: Various researchers have reported that SCC increases with increasing age and is primarily due to an increased prevalence of infection in older cows. Young primiparous cows have less SCC due to less milk production compared to multiparous cows.

Milk yield: There is an inverse relationship between milk yield and milk SCC. High-producing cows have lower

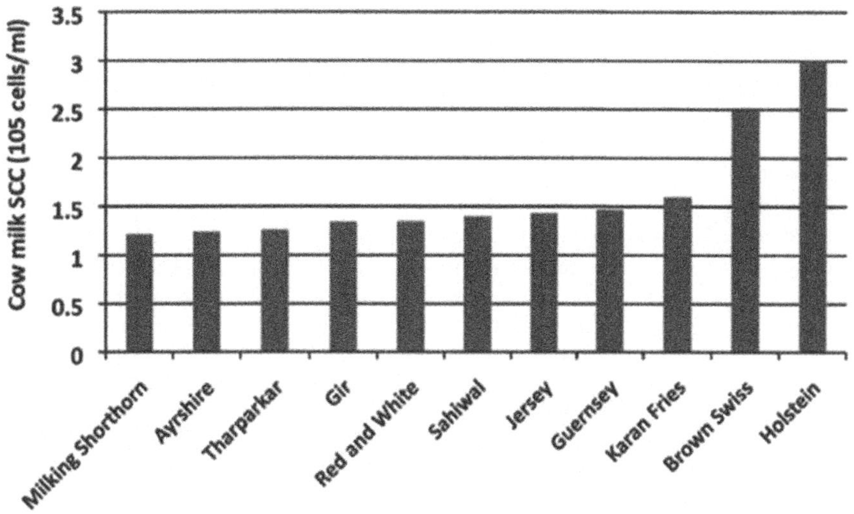

Fig. 25.1 Milk SCC in different breeds of cattle. (Source: Alhussien and Dang 2018)

Table 25.3 Milk SCC in different species

Species	Milk SCC ($\times 10^5$ cells/mL)	References
Cattle	1–3	Alhussien and Dang (2018)
Buffalo	0.8–1.2	De et al. (2011)
Sheep	<4	Alhussien and Dang (2018)
Goat	7–10	Alhussien and Dang (2018)
Pigs	0.2–0.3	Skrzypczak et al. (2012)

immunity due to their production stress and the immunity is compromised leading to more milk SCC.

Season: Milk SCC are generally highest during the hot humid season and then decreased in summer and winter respectively in tropical climate due to higher stress during extreme weather and poor quality fodder which causes reduced feed intake and compromises the immunity of the animals.

Udder quarter: The rear quarter produces more milk compared to the front quarter. The milk collected from rear quarter showed higher milk SCC compared to the fore quarter as milk SCC is inversely related to milk SCC. It was also found that mastitis scores might be greater on the side of the udder on which the cow preferred to lie.

Body condition score (BCS) and body weight: An increase in the BCS at calving was associated with reduced milk SCC in first and second parity cows. But higher BCS around calving in cows with more than third parity showed higher SCC.

Milk fractions: Milk SCC varies with various fractions of milk collected during milking. It has been reported that early and mid-strips of milk had lower levels of SCC which increased significantly in the late strip.

Infectious agents: It has been reported that cows and buffaloes infected with *S. aureus* are having higher milk

Table 25.2 Functions of milk somatic cells

Cell types	Functions	References
Polymorphonuclear neutrophils (PMN)	• Phagocytosis of invading pathogens by generation of reactive oxygen species (ROS)	Rainard and Riollet (2014), Mukherjee and Das (2019), Mukherjee et al. (2013)
Macrophages	• Bacterial phagocytosis like neutrophils • Expresses class II MHC antigen and helps in antigen processing and presentation • Releases chemical messengers or chemoattractants like cytokines, leukotrienes, etc., and facilitates the migration of neutrophils • Convert plasminogen to plasmin, and stimulate urokinase-plasminogen activator during involution	
Lymphocytes	• Recognizes antigens through specific membrane receptors • Produces various immunoregulatory cytokines after Ag-recognition with the help of MHC class II molecules (T lymphocytes) • Humoral immune responses by the production of antibodies	
Mammary epithelial cells	• Synthesis and processing of dimeric of IgA in monogastric animals • Helps in leukocytes migration during inflammation through the production of cytokines and chemokines such as interleukin-8 (IL-8) • Production of arachidonic acid metabolites β-defensins, lactoferrin, and acute phase proteins	

Table 25.4 Soluble/humoral defense factors of mammary gland

Factors		Roles
Complement system		Opsonization and neutrophil migration
Lactoferrin		Reduces iron availability to microorganisms and acts as a bacteriostatic agent.
Transferrin		Bactericidal agent act by altering membrane permeability
Lysozymes		Cleaves bacterial cell wall proteoglycans
Lactoperoxidase and myeloperoxidase		Bactericidal agent
Cytokines	Interleukin-1 (IL-1)	Helps in neutrophil recruitment at the site of inflammation and increases its phagocytic potential
	Interleukin-6 (IL-6)	• Helps in the influx of monocytes in the mammary gland • Causes differentiation of B cells including Ab production • Inhibits TNF-α (anti-inflammatory action)
	Interleukin-6 (IL-8)	• It is a potent neutrophil-chemoattractant factor • Helps in neutrophil degranulation during inflammation
	Interleukin-10 (IL-10)	Prevents the other pro-inflammatory cytokines, chemokines, and eicosanoids (anti-inflammatory cytokines)
	Tumor necrosis factor-α (TNF-α)	Helps in the accumulation of phagocytes 1 in the mammary gland
	Granulocyte-macrophage colony-stimulating factor (GM-CSF)	Stimulates neutrophil migration and phagocytosis
Chemokines		Helps in leukocyte migration
Host Defense Peptides (HDPs)		Antimicrobial

Source: Mukherjee and Das (2019)

SCC followed by *Escherichia coli* and *Streptococcus agalactiae*.

Diurnal variations: both diurnal and infradian rhythmicity in milk SCC have been studied in cows. There was no significant diurnal variation in milk SCC in the cows with less than fourth parity. However, a significant diurnal variation in the DLC was recorded in the cows above fourth parity.

Managemental factors: There are many management factors that play the most important role in the development of contagious disease like mastitis in dairy animals. Among these, unhygienic conditions are more important in increasing the chances of intramammary infection (IMI) and resulting in high SCC. Teat injuries and leakers commonly develop because of stall and platform design raising the incidence of mastitis and causing higher SCC. The supplementation of antioxidant vitamins (Vitamin A, C, and E and β-carotene) and minerals (selenium, zinc, and copper) were reported to reduce milk SCC in cows. Cows treated with melatonin have been reported to possess improved milk quality and enhanced immunity. A vaccine against *S. aureus* named MASTIVAC I was reported to improve udder immunity and reduce milk SCC.

25.1.5.3 Soluble/Humoral Defense Factors

Soluble defense mediated by secreted antibodies, complement proteins, and certain antimicrobial peptides. Majority of these defense mechanisms are originated from blood and extracellular fluids. Important humoral defense factors concerned with mammary gland immunity are described in Table 25.4.

25.2 Mammogenesis

The term "mammogenesis" can be defined as the growth and development of the mammary gland. Mammogenesis occurs through a series of structural and functional development, differentiation, and involution associated with the growth and reproductive stages of animals and regulated by hormones and growth factors. Mammary secretory tissues are developed from the ectoderm; however, the blood and lymph vessels, connective tissue, fat pad, and smooth muscles are derived from the mesoderm. The mammogenesis in female can be broadly classified into five stages namely prenatal, prepubertal, postpubertal, pregnancy, and early lactation.

25.2.1 Prenatal Development of Mammary Gland

The prenatal development of the mammary gland of bovines starts around 32 days of embryonic life. The fetal ectoderms on either side of inguinal region give rise to mammary band which is the first developmental stage of prenatal mammogenesis. Different stages of mammogenesis are as follows.

Mammary band: Mammary band developed from the ectoderm of the inguinal region differentiated from mesenchyma. It is made of a single layer of flattened cuboidal cells that appeared as an abroad band on either side of trunk from upper limb to lower limb. The formation of mammary bands occurs on 32nd day of embryonic life.

Mammary streak: Further layering of ectodermal cells over mammary band develops around 34th day of embryonic life.

Mammary lines: Further the transient mammary line develops from lower ectodermal layer (Malpighian or germinal layer) which is composed of several layers of cells around 35th day of embryonic life.

Mammary crest: Around 37th day of embryonic life, the ectodermal cells of mammary lines are begun to divide into the mesenchymal cell layer leading to the formation of mammary crest.

Mammary hillock: Continual inward growth of the ectodermal cells into the mesenchymal layer forms a dome of ectodermal cells into the mesenchyma called mammary hillock occurred around 40th day of embryonic life.

Mammary bud: It is the spherical or globular-shaped ectodermal layer into the mesenchymal layer normally seen around the 43rd embryonic day. The ectodermal layer sinks entirely into the mesenchyma with a small depression at the outer pole called mammary pit. Mammogenesis is identical in both sexes up to mammary bud stage. The structure of mammary buds is different in male and female. In females, mammary buds are ovoid and do not form as deep a mammary pit as in the males. In males, mammary buds are spherical. The mammogenesis is faster in females compared to males after the mammary bud stage.

Early teat formation: Rapid growth of the mesenchyme surrounding the mammary buds forces the mammary buds to rise above the epithelium with a slight opening at the distal end of the buds. It starts around the 65th day of embryonic life.

Formation of primary sprout: The solid core of the lower ectodermal layer (Malpighian or germinal layer) invaginates into the mammary buds following the least resistant path which pushes the mesenchymal layer aside and leads to the formation of primary sprouts at the 80th day of embryonic life. It gives rise to gland and teat cistern.

Formation of secondary sprout: At about 90th day of embryonic life, when the primary sprouts reach its maximum growth (up to 16 cm in cattle), many secondary sprouts emerge as branching of primary sprouts. The upward growth of secondary sprouts in various angles into the mesenchymal layer leads to the formation of tertiary sprouts which in turn convert to the duct system of the udder. However, during fetal mammogenesis very limited growth of primary sprouts is evident.

Canalization of primary and secondary sprout: Formation of lumen in the solid core of epithelial cells in the primary and secondary sprouts is called canalization normally seen around 100th day of embryonic life. It is mediated by the separation of cells in the primary and secondary sprouts.

Formation of gland cistern: Canalization of the inner end of the primary sprout in both directions leads to the formation of the gland cistern at 110th day of embryonic life. It can be well recognized at 4 months of fetal life and during that time the layers surrounding the cavities are reduced.

Formation of teat cistern: From 130th day of embryonic life, the progression of the canalization primary sprouts without disintegration towards the distal end leads to the formation of teat cistern. During the growth of the teat, the tip of the mammary bud is opened. Initially, it looks like a duct but horizontal movements of the cells increase the cavity of teat cistern and the layers are reduced. At the distal end, streak canal is formed when the duct becomes narrow at its distal end.

The dermis of the skin surrounding the udder is developed from the mesenchymal tissue below the Malpighian layer. The fibrous tissues are formed as threads or bundles perpendicular to the base of the udder. The connective tissues appear as whorls (aggregation of cells in the center with a circular periphery). These connective tissues are subsequently replaced by secretory tissues during the development of mammary alveoli.

All these aforesaid developmental stages of mammary gland are completed within the first 6 months of fetal age and no further developments are seen prior to birth.

Prenatal mammogenesis in doe and ewe is similar to cow. However, in goat fetus, hair anlagens are evident on the teat skin. In mouse, prenatal mammogenesis starts around 11th day of embryonic life on each side of the trunk similar to cow. Mammary buds are developed around 12th–14th day of embryonic life. Differentiation of mammary gland in females begins around 15th day which is characterized by sinking of mammary buds into the mesenchyme tissue and leads to the formation of primary mammary cord. The distal end of the primary mammary cord develops lobulo-alveolar system of mouse mammary gland. The prenatal mammogenesis in sow is occurred in the same direction as in mouse but the teat has two ducts.

25.2.2 Postnatal Development of Mammary Gland

25.2.2.1 Mammogenesis from Birth to Puberty

Mammogenesis from birth to puberty is characterized by the appearance of connective tissue and fats in the mammary gland. A substantial amount of secretory tissue growth also takes place during this time. During the initial phage of mammogenesis, the growth of mammary gland is proportionate with the body growth which is termed as *isometric growth* of mammary gland. This isometric growth persists for 3 months in cow and until 22nd to 23rd day in rats. Beyond these periods, mammary gland grows three times faster than body growth which is termed as *allometric growth.* The allometric growth of mammary gland is characterized by an increase in mammary gland DNA content which is around 1.96 times faster than body growth. During the isometric phage of mammary growth, the udder size is a result of the continued increases in fat pad and connective tissue. The duct system and mammary parenchyma grow a little and the growth of secretory alveoli is not appreciated during this phase. Extensive growth and development of the ductal network is evident during the allometric growth phase which invades the surrounding adipose tissue or mammary fat pad which will determine the extent of lobulo-alveolar development during gestation. Another characteristic feature of mammary development during allometric growth phase is terminal ductile lobular units (TDLU) which develop when elongation and branching of the ducts occurred. Estrogen is responsible for cell multiplication at the tip of TDLU and enlargement of the ducts. Mammogenesis from birth to 6 months of age is often used to predict milk production in mature animals. However, litter correlation exists between glandular growth and milk production.

25.2.2.2 Mammogenesis During Estrous Cycle

The development of mammary gland during the estrous cycle is characterized by the branching of secondary and tertiary sprouts, the growth of buds from these branches, and the growth of mammary ducts. The growth of mammary gland occurring in each estrous cycle is appeared to be lost by the process of regression in the next cycles. However, a small amount of positive growth takes place during each cycle as the regression is comparatively less compared to the growth. Mammary duct system appeared as one layer lining like alveoli. The amount of DNA content was studied as an indicator of cell proliferation in mammary gland of rats and it was reported that DNA content/100 g body weight increased during the first four cycles and no further development occurred after that. As per the individual phage of estrous cycle is concerned, there was 8% higher DNA content in estrus compared to proestrus in rats and heifer and maximum growth occurred during the estrogenic phage of the estrous cycle. Histological changes occurring during estrous are characterized by large alveolar lamina filled with secretions which were shrunken during diestrus. The cuboidal-shaped epithelium in estrus phase is turned into columnar during diestrus.

25.2.2.3 Mammogenesis During Pregnancy

In cattle, exponential growth of mammary gland occurs throughout gestation. It has been reported that 48–94% growth of mammary gland and 60–65% growth of mammary parenchyma occur during gestation. A marked increase in the gland cistern occurs during the fifth to sixth months of pregnancy. The proportion of secretory tissues of the mammary gland increases by branching of the ducts and end buds during fourth month of gestation and appears as found during lactation. These secretory tissues replace the fat tissues and form very small lobules. These lobules are joined together to form the lobes. There is a little increase in the duct length (maximum up to 3 cm). The large ducts appear with two layered linings whereas the ductules and the alveoli are having single-layered cuboidal cells. The connective tissues that separate the lobules and lobes contain numerous blood capillaries. The secretory activity of the alveoli begins from ninth month of pregnancy and the alveolar epithelium becomes distended with granular cytoplasm and appearance of fat droplets. The development of mammary gland in cow during pregnancy is summarized in Table 25.5.

It has also been reported that mammary growth during gestation was proportional to the litter size in sheep, goats, and pigs except in cow.

Table 25.5 Summary of mammary development during pregnancy in cow

First trimester	Second trimester	Third trimester
• Most of the duct growth • Little increase in the secretory tissue	• Growth of lobulo-alveolar system • Glandular proliferation increase near large ducts entering the gland cistern • Further branching of small ducts • Formation of end buds • Secretory tissue replaces adipose tissue and forms small lobules • Alveoli differentiate at the end of terminal ducts (smallest ducts)	• Marked increase in growth of duct secretory tissue, vascular system, and lymphatic system • Alveoli initiate some secretory activity • Epithelial cells become distended • Fat droplets are present in the lumen of alveoli

25.2.2.4 Mammogenesis During Lactation

Around 10% mammary growth occurs during lactation which is characterized by increased number of secretory tissues due to the mitotic proliferation of cells before and after parturition. Around 65% increase in mammary DNA has been estimated from 10 days before parturition to 10 days post-calving which is maximum at peak lactation. After peak lactation, very little cell proliferation occurs and the destroyed cells are not replaced by newly formed cells. In sows, mammary volume due to hypertrophy and hyperplasia of the secretory cells occur around 28 days of lactation period which was associated with increased DNA content around this period. In sheep and goats, udder volume was exponentially increased during up to the last third of pregnancy which gradually declined in lactation.

25.2.3 Hormones and Growth Factors in Mammogenesis

A series of investigations on hypophysectomized and ovariectomized animals established the participation of both ovarian hormones (estrogen and progesterone) and hypophyseal hormones (prolactin and growth hormone) in mammogenesis. It is now well established that estrogen stimulates duct growth and estrogen and progesterone in combination stimulate lobulo-alveolar growth and the mammogenic action of estrogen and progesterone is mediated only in presence of prolactin and growth hormone. Other mammogenic hormones and growth factors are placental lactogen, glucocorticoids, oxytocin, and insulin like growth factors.

Estrogen: Estrogen induces ductal growth of mammary gland. The effect of estrogen on mammogenesis is species-specific. In laboratory animals (mouse, rat, and rabbit) and cats, physiological dose of estrogen induces duct growth and prolonged administration of a high dose of estrogen causes alveolar growth. In these animals, estrogen has the same mammogenic potential for both male and females. In ruminants and guinea pigs physiological dose of estrogen can induce extensive lobulo-alveolar growth including duct growth but the mammogenic potential of estrogen in these animals are more in females compared to males. In the third category animals which include bitches and ferrets, estrogen alone has little or no role in mammary development. The ductal mammogenesis in response to estrogen occurs when the animals attain puberty. Estrogen induces synthesis of IGF-1 from stromal cells of mammary gland which induce epithelial cell proliferation. In another way estrogen in paracrine manner, induces the release of amphiregulin (AREG), an epidermal growth factor family which binds its stromal cell receptors induce the release of FGFs to stimulate the proliferation of luminal cells. The net effect of estrogen on mammogenesis is the elongation of ducts, side branching, formation of terminal end bud (TEB) together with alveologenesis.

Progesterone: Progesterone in combination with estrogen induces lobule-alveolar growth of mammary gland. Progesterone is responsible to produce a lactation-competent mammary gland by causing alveologenesis and side branching of ducts. Progesterone also promotes the differentiation of alveoli during lactogenesis together with prolactin. Progesterone induces the synthesis of tumor necrosis factor ligand superfamily, member 11 (TNFSF11), also known as RANKL (receptor activator of NFKB1 ligand) which in turn initiates cell proliferation.

Growth hormone (GH): Growth hormone can induce mammary growth in hypophysectomized, adrenalectomized, and ovariectomized rats. It was also established that estrogen and/or progesterone were unable to induce mammary growth without growth hormone or prolactin. According to the modified somatomedin hypothesis, growth hormone acts in two ways in the mammary gland. Firstly, GH directly stimulates the growth and proliferation of mammary parenchyma. Secondly, the indirect mammogenic effect of growth hormone is mediated by the secretion of IGF-1 from either liver or mammary stromal cell which acts via autocrine, endocrine, and paracrine mechanisms.

Prolactin: The role of prolactin on mammary growth and differentiation was well established in laboratory species specifically in rabbits where prolactin helps to stimulate lobulo-alveolar system during pregnancy. However, the same effect was questionable in rats and cattle as there was no elevation of prolactin level in these species during pregnancy. Prolactin stimulated the branching of ducts and regression of end buds in virgin animals whereas around pregnancy it stimulates lobulo-alveolar growth. After binding with its specific receptors on the mammary gland epithelium, prolactin induces the expression of whey acidic protein (WAP) gene. WAP is one of the major whey proteins of milk which regulates the proliferation of mammary epithelial cells.

Glucocorticoids: Cortisol causes differentiation of mammary lobulo-alveolar growth in cattle and cortisol-primed differentiation was essential for the action of prolactin to induce milk protein synthesis during lactogenesis.

Insulin: The mammogenic actions of estrogen and progesterone in hypophysectomized animals are enhanced with the action of exogenous insulin. Higher dose of insulin was proved to be mammogenic when studied in vitro. The action of insulin may mimic the action of IGF-1 as insulin exerts its effect through IGF-1 receptors.

Placental lactogen: Placental lactogen is mammogenic in rodents but its effect on mammary gland growth and development in cattle is questionable as the concentration of placental lactogen was very low in dam compared to fetus and exogenous administration of placental lactogen had little effect on metabolism in lactating cows.

Insulin like growth factors (IGF): The insulin like growth factor family comprises of ligands (such as IGF-I, IGF-II, and insulin), receptors (IGF-IR, IGF-IIR, and insulin R), and IGF-binding proteins (IGFBPs). The role of IGF on mammogenesis in terms of cell proliferation, migration, and apoptosis has been well established in cattle, goats, sheep, pigs, and mice till puberty. The predominant source of circulating IGFs is liver though some extra-hepatic sources of IGFs have also been identified in uterus and mammary gland as well. The expression of IGF in the mammary gland is low and thus it was postulated that IGF can be transported from liver to mammary gland. The IGF helps in DNA synthesis and stimulates the cells in late G_1 to enter into S phase thus increasing cell proliferation.

Epidermal growth factor (EGF) family: EGF family includes four members including ligands (EGF, ErbB2, ErbB3, and ErbB4) and their receptors. The other growth factors like transforming growth factor (TGF)-α, heparin-binding EGF (HB-EGF), amphiregulin (AR), and neuregulins (NRGs) are also included in this group as they also act through receptor tyrosine kinase (RTK). EGF receptors have been identified in nonpregnant, pregnant, and lactating cow and sheep. But the level of expression was similar in all these stages. EGF family exerts stage-specific effects on the mammary gland. They affect both mammogenesis in terms of ductal growth including alveolar differentiation during early puberty (EGFR, ErbB2, and ErbB3) whereas ErbB4 helps in mammogenesis during late pregnancy and lactation.

Other growth factors: There are several other growth factors either stimulatory or inhibitory to mammogenesis. Platelet-derived growth factor (PDGF) was found to be stimulatory to the proliferation of myoepithelial cells by increasing DNA synthesis.

Transforming growth factor-β (TGF-β) has been identified as a mammary gland-derived growth inhibitor having the primary function to induce differentiation. In vitro experimentation showed that TGF-β inhibits ductal growth, lobulo-alveolar development, and suppresses casein synthesis in pregnant mice.

Vitamin A in mammogenesis: Retinoic acid was involved in mammary epithelium development during embryogenesis even after birth. The role of retinoic acid during mammary gland involution was also well established.

25.2.4 Mammary Gland Involution

Involution is a biological process by which the lactating mammary gland undergoes a series of tissue remodeling processes after cessation of milk secretion to restore into a virgin-like state. The mechanisms of involution have been well studied in mice after inducing the involution by weaning at peak lactation and teat sealing but there are notable differences in the mechanisms of mammary involution among species. In ruminants, lactation overlaps pregnancy and at the time of milk stasis, cows are at their last trimester of pregnancy while goats are at their first days to the second trimester of pregnancy. Therefore, the stimuli for mammary involution is opposed by the pregnancy-induced mammogenic and lactogenic stimuli. A brief non-lactating period prior to lactation in those species is required to maximize the production. This non-lactation period with limited involution is termed as dry period.

The mammary gland involution occurs in two phases namely

Reversible phase: In this phase, the events of involution can be stopped if the suckling stimulus is reintroduced and the gland can revert to a state of milk production.

Irreversible phase: *The* mammary gland is unable to return to a state of milking without being restimulated by mammogenic and lactogenic stimuli.

25.2.4.1 The Events of Mammary Gland Involution

The involution process (both reversible and irreversible) is mediated through a series of events involving apoptosis and tissue remodeling.

Inhibition of milk secretion: The stasis of milk is an essential prerequisite for the initiation of involution. The transition of mammary epithelial cells from secretory columnar epithelium to a non-secretory squamous cell is noticed upon milk stasis. Cessation of milking or suckling stimulus facilities the engorgement of mammary alveoli with milk which exerts some pressure to cause distension of udder that favors the release of some local inhibitory factors for milk stasis. One of these factors, "feedback inhibition of lactation," has been discussed in earlier section (see factors affecting milk yield and composition). The other notable factors are mentioned below (Table 25.6).

Cell death and regression of epithelial cells: The regression of epithelial tissues and their shedding in the alveolar lumen are evident in both reversible and irreversible phases of involution. The regression of epithelial cells is seen in mouse mammary gland as early as 12 h of weaning and peak around day 2 and day 3. During the secretory phase, the epithelial architecture is well maintained by lactogenic factors like prolactin (PRL), glucocorticoids

Table 25.6 Local factors and their role in milk stasis

Factor	Probable mechanism	Species
Serotonin (5-HT)	• Inhibits milk protein gene expression • Blocks serotonin-specific reuptake transporter (SERT) and prevents the reuptake of serotonin by epithelial cells • Disrupts epithelial tight junctions required for maintaining the columnar secretory epithelial cells	Mouse, bovine, and human
Lactoferrin (LTF).	• Suppresses casein expression	Guinea pig, mouse, pig, and human
Interleukin (IL)-6	• Decreases the sensitivity of the epithelial cells to lactogenic hormones	Mouse

Table 25.7 The factors responsible for regression and death of epithelial cells

Mechanism	Factors	Probable mechanism
Inhibition of permissive conditions for epithelial regression	IGF-binding proteins (IGFBPs)	Prevents the permissive action of IGF-I in mammary gland
	Suckling stimulus	Decreased secretion of PRL and GC
Local pro-apoptotic factors	Serotonin (5-HT)	Induces apoptosis in the suprabasal cells causing regressive and irreversible changes like pyknotic fragmented nuclei
	Transforming growth factor β (TGFβ)	Acts locally in an autocrine manner in inducing epithelial cell death
	leukemia inhibitory factor (LIF)	Induces epithelial cell death in mouse mammary glands
Anoikis (is apoptosis induced by lack of correct cell/ECM attachment)	Fibronectin, laminin, Vit-D receptor, oncostatin-M	Impaired cell and extracellular matrix attachment

(GC), and IGF-I. But due to the absence of suckling stimuli, the mammary epithelial cells become refractory to these lactogenic hormones and initiate the release of pro-apoptotic factors which mediate the cell regression. The factors responsible for the regression and death of epithelial cells are discussed in Table 25.7.

Involution-associated Immune response and clearance of cell debris: The stasis of milk in the mammary gland results in the accumulation of residual milk and cell death results in shedding of epithelial cells along with cellular debris. The removal of these waste materials is crucial for maintaining the normal health of the mammary gland. The induction of immune response followed by clearance of cell debris are thus two important mechanisms of involution.

Activation of neutrophils: Increased leukocyte counts during the time of mammary involution along with higher expressions of pro-inflammatory cytokines and their receptors (e.g., IL-1α, IL-1β, and IL-13) in mice suggests the elucidation of immune responses in the mammary gland. Involvement of 5-HT as an immune mediator during early involution may induce immune response either directly by activating the immune cells or indirectly after the release of pro-inflammatory cytokines.

Activation of acute phase response (APR): Around 12 acute phase proteins for activation of macrophage (e.g., CD68) and B cell chemokine (e.g., CXCL14) have been identified during different phages of involution in mouse mammary gland. Induction of APR also initiates the phagocytic ability of non-preferential phagocytes like mammary epithelial cells.

B lymphocyte activation: Activation of B lymphocytes are responsible for the local synthesis of immunoglobulins (IgA, IgM, and IgG) underlying the epithelium in both mouse and ruminant mammary gland.

Induction of an immune response recruits both preferential (neutrophils and macrophages) and non-preferential (epithelial cells) phagocytes to clear cell and tissue debris. Mammary epithelial cells usually respond during the early phase of involution and mostly ingest apoptotic cells, casein micelles, and milk fat globules. In contrast, professional phagocytes (macrophages) respond late and their role in clearing cell debris is limited compared to epithelial cells.

Disruption of epithelial tight junctions (TJ): Mammary TJs provide an apical seal and holds the mammary epithelial cells together. It also provides a barrier for the paracellular transport of fluids and ions between the lumen and the interstitial space. Disruption of mammary epithelial TJs facilitates the progression of mammary involution. The disruption of TJs are highly reversible and can be restored by the initiation of suckling stimulus within 18–24 h in goats and cows by positive actions of PRL and GC. But the continual absence of suckling decreases the permissive actions of PRL and GC on TGs along with the release of only locally produced disrupting factor 5-HT.

Remodeling of extracellular matrix (ECM) and basement membrane (BM): The stroma of the mammary gland is composed of a variety of tissues like adipose tissue, connective tissue, fibroblast, vascular components, and ECM.

Table 25.8 The components of the plasminogen system and matrix metalloproteinase (MMP) system in mammary gland ECM remodeling

System	Components	Source	Mechanism
Plasminogen system	Plasminogen (Plg)	Synthesized in the liver and released into circulation	Plasminogen is converted to plasmin by uPA. Plasmin in turn results ECM/BM breakdown.
	Plg activators Urokinase type Plg activator (uPA) (main activator operating in the mammary gland), Tissue type Plg activator (tPA)	Stromal fibroblast	
	Plg inhibitors		
Matrix metalloproteinase (MMP) system	Maspin	Myoepithelial cells	Prevents ECM/BM breakdown
	Tissue Inhibitor of MMP (TIMP)	Stromal fibroblast	

Table 25.9 Factors regulating vascular remodeling during mammary gland involution

Factors	Source	Functions
Vascular endothelial growth factor (VEGF)	professional and non-professional phagocytes, mammary stromal cells, especially adipocytes	Favors vascular regression and angiogenesis
Prolactin (PRL)	Systemic	Favors the generation of vaso-inhibitory 16K PRL and facilitates vascular regression
Cleaved PRL (16K)	Local (adipocyte)	Anti-angiogenic
5-HT	Mammary epithelial cells	Acts as vasoactive and mitogenic agent for endothelial and vascular smooth muscle cells

ECM or basement membrane (BM) is a cementing substance composed of laminin, collagen, fibronectin, and integrins synthesized from myoepithelial cells. The main function of ECM is to anchor epithelial and myoepithelial cells. The breakdown of BM via proteolysis to become thick, folded, and discontinuous are the characteristic features of ECM remodeling during mammary gland involution. This part of involution is irreversible.

The remodeling of ECM is mediated by two systems namely plasminogen system and matrix metalloproteinase (MMP) system. The components of these two systems and their mechanism are depicted in Table 25.8.

Vascular remodeling: Vascular remodeling consists of vascular regression and angiogenesis. The mammary gland vasculature is composed of a capillary network spreading as a honeycomb structure surrounding each alveolus during lactation. The alveolar engorgement due to milk stasis leads to an increase in the perivascular capillary volume seen in the initiation of involution. The vascular regression begins around day 6 of involution in mouse mammary gland characterized by clusters of capillaries and the vascular network becomes similar to the virgin gland by day 10 of involution. The angiogenesis occurs simultaneously with vascular regression in adipose tissue. There are several local and systemic factors that regulate vascular remodeling during mammary gland involution as summarized in Table 25.9.

25.3 Lactogenesis

"Lactogenesis" was initially defined as the action of lactogenic hormones on mammary gland and histological alterations in the tissue. Later, lactogenesis was used to describe the onset of copious milk at the time of parturition and the appearance of organelles related to milk synthesis and secretion. Lactogenesis comprises of two-stage process; appearance of pre-colostrum (stage-I) and onset of copious milk secretion at parturition (stage-II). Considering the above two facts, a more precise definition of lactogenesis can be obtained. Lactogenesis is the biological process of onset of milk secretion which includes the enzymatic and cytological differentiation of mammary alveolar cells in early pregnancy to full lactation after parturition.

25.3.1 Enzymatic and Cytological Differentiation of Alveolar Cells Before the Onset of Lactation

The stage-I of lactogenesis is characterized by enzymatic and cytological differentiation of alveolar cells divided into four phases namely proliferative phase (early pregnancy), secretory differentiation phase (mid-pregnancy), secretory activation (parturition), and lactation.

The proliferative phase initiates immediately after conception and is characterized by extensive proliferation of mammary epithelial cells as indicated by increased DNA content. Around 25% of proliferative cells are evident till day 5 of pregnancy. Alveolar buds were developed from proliferating epithelial cells that progressively become milk-secreting lobules. The proliferation decreases during mid-pregnancy and a network of capillaries encompasses each alveolus.

The secretory differentiation starts around second half of pregnancy and is characterized by several biochemical changes required for initiation of milk synthesis such as the increased activity of lipid synthetic enzymes along with the expression of adipophilin protein and incorporation of ^{14}C-acetate in the mammary glands of pregnant rabbits and rats. The activity of alkaline phosphatase was reported to be increased by tenfolds in rat mammary gland by the end of pregnancy. The endogenous respiration in the mammary gland gradually increased from pregnancy to postpartum. Increased expression of β-casein RNA has also been observed during this phase.

The phase of secretory activation is characterized by the onset of copious milk secretion and coincides with the drop of plasma progesterone level and higher prolactin level after parturition. A higher level of prolactin initiates transcription of milk protein genes. The cytological alterations of mammary epithelium during this phase are characterized by an increased number of Golgi and endoplasmic reticulum which acts as a machinery for the synthesis of various milk components. The probable markers of secretory activation phase are lactose, protein, citrate, and sodium.

The last phase of lactogenesis is characterized by continuous production of milk. It is again subdivided into two sub-phases namely colostral phase (milk contains a large amount of immunoglobulins and immune defense proteins) and mature secretion phase (production of a large volume of milk to support newborn). Rapid proliferation of mammary epithelium together with the development of active enzymatic machinery for the synthesis of milk constituents are the characteristic features of this phase. Activation of enzymatic machinery required for lactose, fat, and protein synthesis during the initiation of lactation has been well documented in several species like pig, cow, rat, mouse, guinea pig, and rabbit and showed huge inter-species variation in the enzymatic development of mammary gland. In rats and guinea pigs, the activities of these enzymes were increased during late pregnancy and early lactation and achieved maximal activity around the second or third day of lactation. Whereas, in cows, no significant increase in enzyme activities occurred during late pregnancy and early lactation. The probable explanation for this was the fact that the growth of mammary gland in cows required more time compared to rats and guinea pigs and during this proliferative phase only a small proportion of newly formed secretory cells are available in mammary tissue.

25.3.2 Milk Synthesis and Secretion

25.3.2.1 Metabolic Adaptations During Early Lactation/Nutrient Partitioning

During the initiation of lactation, the majority of the cows are in a state of negative energy balance (NEB) which indicates the energy required for maintenance and milk production are greater than the energy intake. NEB is not only due to limited feed intake but also due to high metabolic priority for milk production. In the dairy cow, energy requirements for lactation can reach 80% of net energy from intake and lactose production can utilize 85% of circulating glucose. The energy requirement for milk production is obtained by mobilization of body fat and muscle proteins. Moreover, there is excessive partitioning of nutrients towards mammary gland is the fundamental consequence of NEB which is most severe during the first week of postpartum. The partitioning of nutrients is regulated by homeostatic and homeorhetic mechanisms and under genetic control. The metabolic adaptations during peripartum period were aimed to increase endogenous glucose production together with the delivery of glucose and nonesterified fatty acids (NEFA) to the mammary gland for milk production which is facilitated by endocrine factors such as decreased insulin, leptin insulin like growth factors, and thyroid hormones together with the elevated level of GH, cortisol catecholamines, and glucagon. The major metabolic adaptions to cope with NEB are summarized in Table 25.10.

25.3.2.2 Biosynthesis of Lactose

Lactose is the major milk sugar in most species. However, sialyllactose is the predominant sugar present in rat and mouse milk. Lactose is a disaccharide composed of D-glucose and D-galactose, joined through ß-1,4-glycosidic linkage whereas sialyllactose is an oligosaccharide. Lactose is the main contributor to milk osmolality and it is responsible for drawing water into the milk. It can also be noted that lactose is one of the least variable component of milk.

25.3.2.2.1 Precursors of Lactose

Blood glucose is the precursor of milk lactose and two molecules of glucose are required for the synthesis of one molecule of lactose. The predominant sources of glucose are dietary or endogenous production. In ruminants, the dietary carbohydrates are degraded in the rumen by microbial fermentation to volatile fatty acids and only 10% of the glucose requirement is filled up by dietary glucose and the animal has to produce the remaining 90% of required glucose through endogenous production by gluconeogenesis and glycogenolysis in liver. Gluconeogenesis and glycogenolysis contribute around 85% of glucose requirement in the mammary gland. Propionate is the main precursor for neoglucogenesis (45–60% of endogenous glucose is synthesized from propionate) together with lactate and amino acids, particularly alanine. Glycerol, produced from lipolysis of adipose tissue,

Table 25.10 The major metabolic adaptions during the state of NEB

Function	Metabolic changes	Tissues involved
Feed intake and digestion	Increase in food and water intake	Central nervous system
	Hypertrophy of digestive tract	All segments of GI tract
	Increased nutrient absorption	
Lipid metabolism	Increased lipolysis Decreased lipogenesis	Adipose tissue
Glucose metabolism	Increased gluconeogenesis and glycogenolysis	Liver
	Utilization of acetate for energy (ruminanta)	Mammary gland
Protein metabolism	Mobilization of protein reserve	Muscle and other body tissue
Mineral metabolism	Increased absorption and mobilization of reserve	Gut, bone, kidney, and liver
Water metabolism	Increased absorption and expansion of plasma volume	Gut, kidney, CNS

also aids a little contribution to neoglucogenesis, particularly around parturition. The glycogenolysis reached a peak around 5 days postpartum. Together with higher production of endogenous glucose, the animal must ensure less peripheral utilization of glucose in tissues like muscle and fat which is mediated through transient insulin resistance around parturition in cows. However, the animals have to spare glucose for milk synthesis.

25.3.2.2.2 Glucose Sparing Action in the Mammary Gland of Cattle

In ruminants, blood glucose level is lower compared to non-ruminants (40–80 vs 80–120 mg/dL). During the initiation of lactation, the majority of glucose is channelized for milk lactose production and little glucose is available for maintenance energy requirement. So, under this high energy-demanding stage, the ruminants must have a well-developed glucose-sparing mechanism in mammary gland to ensure that glucose carbon is not used in synthetic and oxidative reactions whereas other substrates like acetate can effectively be used. One such example is seen in fatty acid synthesis in the mammary gland where the exclusion of glucose is linked with the absence of citrate cleavage enzymes essential for the generation of cytoplasmic acetyl CoA from glucose.

25.3.2.2.3 Utilization of Glucose in the Mammary Gland

The majority of glucose (60–70%) in the mammary gland is utilized for milk lactose synthesis and the remaining (20–30%) goes through the pentose phosphate shunt to generate NADPH (used as reducing equivalents in milk fatty acid synthesis). A small amount of glucose is also utilized for the generation of ATP, ribose sugar (DNA and RNA), and synthesis of glycerol (used for milk triglyceride synthesis).

25.3.2.2.4 Glucose Uptake in the Mammary Gland

Glucose uptake in the mammary epithelial cells is mediated by two processes namely passive facilitative diffusion process and sodium-dependent glucose transport. The concentration gradient established across plasma membrane favors this diffusion process. Two distinct classes of glucose transporters are involved in the glucose uptake process namely, facilitative glucose transporters (GLUT) and sodium-linked glucose cotransporters (SGLT), for facilitative diffusion and sodium-dependent glucose transport, respectively. GLUT has 14 known isoforms of which GLUT1 is the predominant glucose transporter for rat, mouse, human, and bovines. Bovine mammary gland also expresses GLUT3, GLUT4, GLUT5, GLUT8, and GLUT12. GLUT2 is expressed in human breast tissue but absent in rat or bovine. Sodium linked glucose transporters (SGLT1) are also expressed in human, rat, and bovine mammary gland during both lactating and non-lactating state. Bovine mammary tissue also expresses SGLT2 at low level. Another group of newly identified glucose transporters SWEET1 was found in the Golgi apparatus of mouse mammary gland and thought to be involved in glucose uptake by the Golgi during lactose synthesis.

The expression of these glucose transporters is lactation stage-specific, lower expression was reported during virgin state which was reported to be increased around 40-folds during mid-lactation and declined sharply around involution.

25.3.2.2.5 Biochemical Pathway of Lactose Synthesis

Glucose is phosphorylated at its sixth position to form glucose-6-phosphate with the help of hexokinase. Phosphoglucomutase then transfers phosphate group from position 6 to 1 to form glucose-1-phosphate which combines with uridine triphosphate to form uridine diphosphate glucose and liberates pyrophosphate catalyzed by the enzyme UDP-glucose pyrophosphorylase. UDP-glucose is converted to UDP-galactose by UDP-galactose 4-epimerase. Finally, UDP-galactose combines with glucose to form lactose with the help of lactose synthetase enzyme.

The enzymatic machinery required for lactose synthesis particularly UDP-glucose pyrophosphorylase and UDP-galactose 4-epimerase increase markedly after the onset of mammary gland in rat and guineapig but not in cattle suggested that the enzymes already present in the mammary gland of cattle during pregnancy.

The enzyme lactose synthetase is a complex of two proteins that combines reversibly, in 1:1 stoichiometry. The protein A of this enzyme complex is galactosyltransferase which transfers galactose from UDP-galactose to terminal non-reducing *N*-acetylglucosamine residues of glycoproteins as follows:

$$\text{UDP} - \text{galactose} + N - \text{acetyl} - \text{D} - \text{glucosamine} \rightarrow N - \text{acetyllactosamine} + \text{UDP}$$

The protein B of lactose synthetase complex, i.e., α-lactalbumin inhibits this reaction and allows UDP-galactose to combine with glucose to form milk lactose.

$$\text{UDP} - \text{D} - \text{galactose} + \text{D} - \text{glucose} \rightarrow \text{Lactose} + \text{UDP}$$

25.3.2.2.6 Secretion of Milk Lactose

After synthesis, lactose is transported to Golgi complex and encapsulated in Golgi membrane to form secretory vesicles. These vesicles move towards apical surface and fused with the apical membrane and discharge their contents by exocytosis.

25.3.2.3 Biosynthesis of Milk Fat

The predominant constituents of milk fat in bovines are triglycerides (98%), diacylglycerol (1%), cholesterol (<0.5%), phospholipids (1%), and free fatty acids (0.1%). The other minor lipid components are ether lipids, hydrocarbons, and fat-soluble vitamins. The major fatty acids in bovine milk are saturated fatty acids (65%), monounsaturated fatty acids (25%), and polyunsaturated fatty acids (10%).

25.3.2.3.1 Precursors of Milk Fat and Their Synthesis

The precursors of milk fat are fatty acids and glycerol. The fatty acids of milk originate from two major sources, circulatory uptake and de novo synthesis in mammary epithelium. Long-chain fatty acids (>16 C) are obtained from circulating lipids. The short- (4–8 C) and medium-chain fatty acids (10–14 C) arise from de novo synthesis. Fatty acids with 16 C are produced from both sources.

25.3.2.3.2 Sources of Circulating Fatty Acids

Mammary gland obtains circulating fatty acids from circulating lipoproteins and nonesterified fatty acids (NEFA). There are two important sources of circulating fatty acids, viz., absorbed lipids from the digestive tract and the lipids mobilized from fat reserve of the body. In ruminants, the predominant source of circulating fatty acids are dietary though a small proportion (<10%) of circulating fatty acids are obtained from fat mobilization. The latter is increased during the state of negative energy balance. Whales undergo starvation during the entire lactation period and the fatty acids are mobilized from body fat to mammary gland which is reflected in their milk fat composition. Fat mobilization is also important in humans and around 60% of fatty acids in milk are derived from body fat mobilization.

25.3.2.3.3 De Novo Synthesis of Fatty Acids in the Mammary Gland

The generation of fatty acids by de novo synthesis varies between species, in elephant de novo synthesis is the principal source of fatty acids synthesis and in the seal, and it is derived from circulation. About half of the fatty acids in the milk of ruminants are derived from de novo synthesis from acetate. However, butyrate contributes the first four carbons of fatty acids originating from de novo synthesis. Reducing equivalent like NADPH is also required for fatty acid synthesis which is derived either from pentose phosphate cycle and the isocitrate cycle (ruminants) or from pentose phosphate cycle and the malate transhydrogenation cycle (non-ruminants). The other required cofactors are Mn^{2+}, Biotin, and HCO^{3-}. The de novo fatty acid synthesis occurs in the cytosol and the basic building block is acetyl CoA. The reaction is catalyzed by fatty acid synthase (FASN) and acetyl CoA carboxylase (ACC). Fatty acid synthase is a multi-enzyme complex consisting of acyl carrier protein (ACP) and 6 different enzymes namely b-ketoacyl synthase (KS), acetyl/malonyl-CoA transferase (MAT), b-hydroxyacyl dehydratase (DH), enoyl reductase (ER), b-ketoacyl reductase (KR), and thioesterase I (TE I). The biochemical pathways of fatty acid synthesis are depicted below.

The enzyme thioesterase is the key regulator of chain length of fatty acids. It has two subtypes. TE I is a part of FASN whereas TE II is a tissue-specific enzyme observed only in nonruminants that is independent of FASN. Both TE I and II can interact with each other and FASN to produce all kinds of fatty acids (C8, C10, C12).

Glycerol is synthesized from glucose or circulating glycerol. The primary source of glycerol is glycerol-3-P derived either from glycolytic pathway or lipolysis of triglycerides.

25.3.2.3.4 Esterification of Fatty Acids

Synthesis of triglycerides through the esterification of glycerol-3-phosphate and fatty acids is evident in most cell types including mammary gland. C_{12}–C_{16} fatty acids are concentrated on C-2 atom of glycerol whereas short-chain fatty acids (C_4–C_5) are concentrated on C-1 and C-3 atom of glycerol. ATP, Mg^{2+}, and CoA are the cofactors required for etherification process. The proposed model of esterification of fatty acids is given in Fig. 25.2.

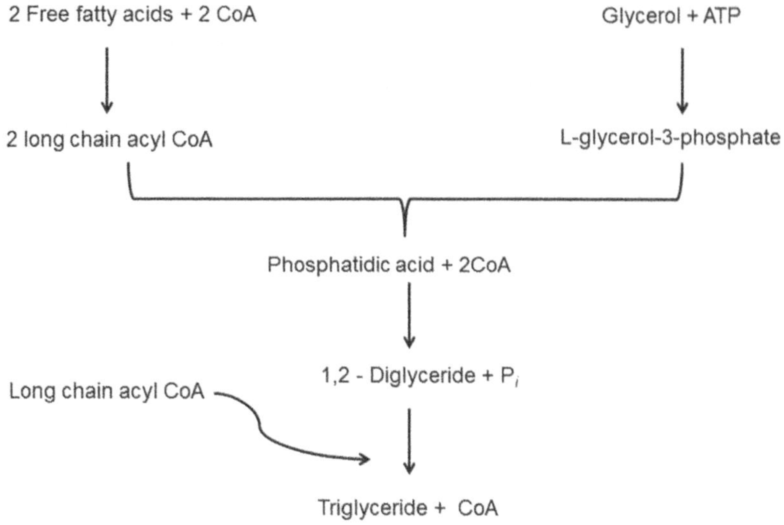

Fig. 25.2 Esterification of fatty acids

Christiesomes are the cell fragments that contain endoplasmic reticulum, mitochondria, and lipid droplets, identified in goat milk. These are involved in triglyceride synthesis.

25.3.2.3.5 Secretion of Milk Lipids

Milk lipids are secreted as droplets of various sizes. These lipid droplets are mainly composed of triglycerides (95%) with lesser amounts of sterols, partial glycerides, phospholipids, glycolipids, and hydrocarbons. The droplets are covered by a membrane composed of polar lipids and proteins known as milk fat-globule membrane (MFGM). The membrane originates either from apical plasma membrane (primary membrane) or from endoplasmic reticulum and other intracellular compartments. Based on the size there are two types of milk droplets namely cytoplasmic lipid droplets (CLDs) (1–5 μm) and microlipid droplets (MLDs) (0.5 μm or less). There are two proposed routes of milk lipid secretions.

Apical vesicle route: Lipid droplets originate as MLDs in the rough endoplasmic reticulum and transit towards apical membrane alone or in combinations to form CLDs (MLDs are fused during their transit towards apical membrane).

Secretory vesicle route: In this mechanism, CLDs are surrounded by secretory vesicles which in turn fuse together to form intracytoplasmic vacuoles and transported towards apical membrane. In both cases, the contents of apical and secretory vesicles are released by exocytosis.

A combination of both apical and secretory vesicle routes may also help to release milk fat droplets.

Lipid droplet proteins called as adipophilin is involved in the production of CLDs during secretory differentiation of mammary alveolar epithelium and play a pivotal role in both formation and secretion of milk lipids.

25.3.2.4 Milk Protein Synthesis

25.3.2.4.1 Precursors of Milk Proteins

The predominant precursors of milk proteins are the free amino acids of the plasma free amino acids. It has been estimated that 70% of amino acids perfusing mammary gland are derived from blood. Some nonessential amino acids were synthesized in the mammary gland after utilizing the nitrogen of certain amino acids, particularly arginine and ornithine. Other minor sources of amino acids are the glutathione of erythrocytes and plasma oligopeptides.

Amino acid uptake by the mammary gland: The substrate, amino acids from blood, is transported through the basolateral membrane to mammary secretory cell. Different groups of amino acids require different transporting system. Based on the substrate specificity and transport mechanism, amino acid transporters have been classified into

- Na^+-dependent transporters utilize the electrochemical gradients of Na^+ and other ions to actively transport amino acids. These transporters are also energy-dependent and facilitate unidirectional amino transport. Based on the amino acid specificity Na^+-dependent transporters may be of two types (Table 25.11).
- Na^+-independent transporters are non-energy dependent and facilitate bidirectional transport of amino acids. The details of Na^+-independent transporters have been presented in Table 25.12.

Genes for milk protein synthesis: Majority of the milk proteins are synthesized from transcription of tissue-specific genes under the influence of hormones.

Table 25.11 Na^+-dependent transporters in the mammary gland in different species

Na^+-dependent transporters	Types	Amino acids involved	Species
Transporters for basic and acidic AA	L-type amino acid transporter-2 (LAT-1)	Lys, Arg, His, Leu, Ile, Met, Ala, Ser, Thr, Val	Pig, human, rat
	L-type amino acid transporter-2 (LAT-1)	Lys, Arg, His, Gln, Leu, Met, Ala, Cys	Pig
Neutral amino acid transporters	Sodium-coupled neutral amino acid transporter (SNAT)	Ala, Ser, Gly, Pro, Cys, Gln	Rat, mouse, cow
	Alanine, serine, cysteine amino acid transport protein (ASCT)	Ala, Ser, Gly, Val, Thr, Cys, Gln	Pig, human, rat, cow

Table 25.12 Na^+-independent transporters in the mammary gland in different species

Na^+-independent transporters	Types	Amino acids involved	Species
Transporters for basic and acidic AA	Cationic amino acid transport protein (CAT)	Lys, Arg, His, Ornithine	Pig, human, rat, cow
	Amino acid transporter rB (rBAT)	Lys, Arg, Ornithine, Cystine, Leu	Pig
	Excitatory amino acid transporter (EAAT)	Asp, Glu	Mouse, rat
Neutral amino acid transporters	L-type amino acid transporter (LAT)	Ala, Ser, Val, Thr, Leu, Ile, Met, Phe, Tyr, Trp, His	Mouse, rat, cow
	T-type amino acid transporter (TAT)	Phe, Tyr, Trp	Mouse

The localization of casein gene locus has been identified in chromosome 6 in cattle, sheep, and goat, chromosome 5 in mice, and chromosome 4 in human. β-lactoglobulin-encoding gene locus has been identified in chromosome 3 in sheep and chromosome 11 in cow and goat. The α-lactalbumin-encoding gene has been linked to chromosome 12 in man, chromosome 3 in sheep, chromosomes 5 in bovine and goat, and chromosome 5 in pig.

25.3.2.4.2 Mechanism of Milk Protein Synthesis

Hormone-induced transcription factors initiate the expression of aforesaid genes and the protein biosynthesis is initiated by following steps.

Transcription: In this step, messenger RNAs are formed in the nucleus which carries the codes of specific proteins. Non-coding sequences (introns) are removed from coding sequences (exons) by mRNA splicing. The formed mRNAs are localized in ribosomes in the rough endoplasmic reticulum.

Amino acid activation: Amino acids in the cytoplasm are activated by reaction with ATP and attachment to transfer RNA (tRNA). The tRNAs are specific for each amino acid.

Translation: The mRNA contains codes for amino acids (codon). The anti-codons in the tRNA recognize codon to form appropriate amino acid–tRNA complex. This complex moves and appropriate amino acid–tRNA complex is added to form peptide chain.

25.3.2.4.3 Intracellular Transport and Processing of Milk Proteins

Transport in the endoplasmic reticulum: Proteins are synthesized in the ribosomes of rough endoplasmic reticulum as a long polypeptide chain. A short peptide of 16–30 amino acids present at the N-terminus of newly synthesized proteins is released by the action of a protease. This peptide acts as an open reading frame encoding the rest of the protein called signal sequence. These signal sequences are recognized by receptors on the ER and the proteins are translocated in the endoplasmic reticulum. After translocation in the endoplasmic reticulum, cysteine-rich peptide residues undergo disulfide bond formation that allows the folding of nascent polypeptides in the lumen of the endoplasmic reticulum. The oxidizing environment inside the lumen of the endoplasmic reticulum favors this bond formation and protein disulfide isomerase (PDI) catalyzes this reaction. Appropriate folding of these proteins is an essential prerequisite for the transport of these proteins into Golgi complex. The other co-translation modifications are *N*-glycosylation of α-lactalbumin and oligomerization of caseins.

Transport in the Golgi and trans Golgi network: The protein cargo emerging from the endoplasmic reticulum is transported to tubule vesicular network between endoplasmic reticulum and Golgi complex known as ER-Golgi intermediate compartment (ERGIC). These ERGIC are responsible for the protein trafficking between endoplasmic reticulum and Golgi. Inside the Golgi complex and trans Golgi network lots of posttranslational modifications of the proteins occur (Table 25.13).

Table 25.13 Posttranslational modifications of major milk proteins in Golgi and trans Golgi network

Posttranslational modification	Enzymes involved	Proteins involved
Glycosylation	Glycosyltransferases	Ƙ-casein
Phosphorylation	Kinases	Caseins
Sulfation	Tyrosylprotein sulfotransferase	No sulfation of major milk proteins however sulfation of proteoglycans occurs in mammary epithelial cells
Proteolytic processing	Endoproteases	No direct evidence of cleavage of major milk proteins

25.3.2.4.4 Transport in Secretory Vesicles and Secretion

After phosphorylation of caseins in the Golgi apparatus, calcium ions are attached with casein ester phosphate groups in the terminal Golgi to form calcium caseinate phosphate and micelles formation occurs. Two types of secretory vesicles are formed. One type of vesicle is composed of chains of small spherical particles and filamentous structures. The other honeycomb-type vesicle contains densely packed casein micelles. Two types of secretory pathways of proteins arc proposed.

Exocytosis: The major milk proteins (Caseins, α-lactalbumin, and β-lactoglobulin) are secreted by exocytosis. The secretory vesicles are channelized to the apical pole guided by microtubules. At the apical pole, secretory vesicles are fused with the apical membrane and emptied their contents in alveolar lumen.

Transcytosis: Immunoglobulins, transferrin, and serum albumin are secreted by transcytotic pathways. It involves a series of complex sorting events. Transcytosis begins with the uptake of material by the basal membrane of mammary epithelial cells and enters into basolateral early endosome (BEE) compartment. The substances within BEE undergo a rapid recycling pathway to a common endosome recycling (CER) compartment. Within the CER there may be further sorting and the vesicles are channelized either apical or back to basolateral membranes for secretion.

25.3.2.5 Fluid and Electrolyte Transport Across the Mammary Gland

Transport of water and small solutes across the epithelial cells is mediated by a family of 28 kDa membrane proteins with high permeability for water called aquaporins (AQP). AQP are activated by arginine-vasopressin via V2-R receptors and translocated to the cellular membrane for water transport. Various studies have confirmed the presence of different AQP in mammary gland of several species as summarized in Table 25.14.

The concentration of different ions between extracellular fluid, mammary epithelial cells, and milk has been depicted in Table 25.15.

Table 25.14 Aquaporins in mammary gland of different species

Aquaporin	Species
AQP1, AQP2, AQP3, AQP4, AQP5, AQP6, and AQP7	Rat
AQP1 and AQP3	Human
AQP1, AQP3, AQP4, and AQP7	Bovines

Table 25.15 The concentrations of different ions between extracellular fluid, mammary epithelial cells, and in guinea pig

Ions	Extracellular fluid (mM)	Mammary epithelial cell (mM)	Milk (mM)
Sodium (Na^+)	150	43	8
Potassium (K^+)	4.5	143	24
Chloride (Cl^-)	116	62	12

Source: Linzell and Peaker (1971)

The electrochemical gradients of these ions keep the inside of the epithelial cell electrically negative compared to milk (−41 vs 3 mV). In the basolateral membrane, Na^+ and K^+ gradients are maintained by Na^+–K^+-ATPse pump and in the apical membrane Na^+ and K^+ are passively distributed.

Na^+–K^+–Cl^- cotransport system have also been identified in the apical surface of lactating rat mammary epithelium which confirmed that the uptake of K^+ by the mammary tissue is dependent on both sodium and chloride.

Na^+/H^+ exchange and Na^+/HCO_3^- cotransport are involved in the regulation of mammary cell pH.

Rat mammary tissue expresses a Na^+-dependent phosphate transport system which helps to absorb free inorganic phosphate from blood and utilizes this inorganic phosphate for casein micelle formation.

Mammary secretory cells are having less quantities of calcium (<0.1 μM) compared to blood and milk. It suggests that there may be an active calcium pump across the basolateral membrane which pumped calcium back to the intersitium.

25.3.2.6 Secretion of Vitamins in Milk

Vitamins are generally taken up by the mammary gland from blood. The supplementation of vitamins in the diet enhances the vitamin concentration in the milk except vitamin C.

Fat-soluble Vitamins (A, D, E, K) come from feed along with exposure to sunlight. In ruminant animals, Vitamin A is synthesized from β-carotene in the intestinal mucosa. Vitamin A in the milk is present as esterified retinol in contrast to unesterified retinol which is predominant in plasma.

Ruminant microflora synthesizes B vitamins in the rumen and the content of B vitamins in the milk cannot be altered through dietary manipulation.

The ergosterol in the feed is the precursor of vitamin D. It can also be obtained from the skin in response to exposure of sunlight that activates 7-dehydrocholesterol in the skin of the animal.

Vitamin D in the milk of cows comes from activation of ergosterol in feed or from the animal's exposure to sunlight. This activates 7-dehydrocholesterol in the skin of the animal. Milk contains vitamins E and K.

25.3.2.7 Hormonal Control of Lactogenesis

Prolactin: Prolactin plays a pivotal role in lactogenesis and was considered as the part of lactogenic hormone complex together with growth hormone and insulin in different species including cattle. But prolactin alone can initiate milk secretion in rabbit. Prolactin binds with its specific membrane plasma membrane receptors of mammary secretory alveolar cells and leads to transcriptional activation of major milk protein genes.

Growth hormone: Earlier, the growth hormone was considered as a member of lactogenic hormone complex as administration of exogenous GH in hypophysectomized mice was proved to be lactogenic. But in cattle administration of exogenous GH during late pregnancy had no role in lactogenesis. A surge of GH secretion was reported at the time of parturition in cow but the first stage of lactogenesis begins before that. So it was logical to conclude that GH was not lactogenic, especially in cattle.

Insulin: Insulin is involved in partitioning of nutrients during the initiation of lactation. Insulin increases the utilization of glucose and lipids in mammary gland of rats and decreased glucose uptake in adipocytes. But the uptake of acetate, β-hydroxybutyrate, triglycerides, amino acids, and glucose are independent of insulin and plasma insulin concentration was negatively correlated with milk yield in bovines. However, infusion of glucose at the time of high insulin concentration increased milk protein in cow.

Estrogen: The concentration of estrogen is increased with impending parturition and reported to be involved during lactogenesis. The initiation of lactation by estrogen is mediated either by the release of prolactin or by increasing the prolactin receptor in the mammary gland. However, estrogen had no or very little role in milk yield during an established lactation and exogenous administration of estrogen in multiparous animals was reported to suppress the lactation by interfering milk ejection reflex.

Progesterone: Progesterone was reported to suppress the lactation and removal of progesterone during pregnancy-initiated lactation and decreased secretion of progesterone during peripartum coincides with copious milk secretion. Progesterone also suppresses the synthesis of normal milk constituents like casein, lactose, and α-lactalbumin. The proposed mechanism by which progesterone suppresses lactogenesis is either by blocking the prolactin receptors or glucocorticoid receptors in the mammary gland.

A combination of estradiol-17β and progesterone at high doses (as around calving) for 7 days was reported to induce lactation in sterile cows (Smith and Schanbacher 1973). The success rate was 70% and the induced animals had a production potential of 70% of their normal production.

Glucocorticoids: Administration of glucocorticoids and adrenocorticotropic hormones in therapeutic doses suppressed lactation in cattle. Uptake of glucocorticoids in the mammary gland and its subsequent binding with glucocorticoid receptors were positively correlated with glucose uptake in mammary tissue in cattle. Glucocorticoids are lactogenic in rats and regulate the secretion of α-lactalbumin and β-casein.

Thyroid hormones: Administration of thyroxine or thyrotropin-releasing hormone increases milk production in cows. The lactogenic effect of thyroid hormones is probably due to their influence on mammary gland metabolism and nutrient uptake in the mammary gland. During lactation, conversion of thyroxine to triiodothyronine (active form of thyroid hormones) in the mammary gland was increased compared to other body tissue and the mammary gland was in euthyroid state compared to the rest of the body (hypothyroid).

Prostaglandins: Prostaglandin $F_{2\alpha}$ ($PGF_{2\alpha}$) is reported to act as local autocrine lactogenic inhibitor during late pregnancy. Just before parturition, the concentration of $PGF_{2\alpha}$ venous blood declines rapidly due to its metabolism into 13,14-dihydro-15, keto-$PGF_{2\alpha}$ (DHKPGF$2_{2\alpha}$) by the mammary epithelium and the mammary gland regains its secretory activity. $PGF_{2\alpha}$ is also responsible for controlling the permeability of mammary epithelial cells and secretory rate as well.

25.4 Milk and Its Composition

Production of milk is the distinguishing feature of mammals to nourish their young. It is a whitish liquid produced by mammary secretory cells of the udder of the female produced by a series of physio-biochemical mechanisms. Milk secreted in the first few days after parturition is termed as colostrum. Milk can also be defined as a complex chemical substance containing fat as emulsion, proteins, and some minerals in colloidal suspension and lactose, soluble protein together with some minerals as true solution. The International Congress of Food held in 1909 in France defined milk as "milk is the product of the total, full, and uninterrupted milking of a dairy female in good health, also nourished and not overworked." It must be collected properly and not contain colostrum.

Milk exists in four physical phases

- *Gas phase*: It is essentially composed of dissolved gases (5% by volume), mainly CO_2, N_2, and oxygen O_2 at the time of milking.
- *Fatty phase*: It is composed of cells, fat (2–5 m in diameter) comprises of lipids and fat-soluble elements. The fat globules are surrounded by phospholipids and protein membranes.
- *Colloidal phase*: It contains casein micelles together with calcium and magnesium in the form of phosphates and citrates.
- *Aqueous phase*: It is the aqueous phase that comprises of lactose, soluble proteins (whey protein), and minerals. To keep the milk isotonic to blood plasma the lactose and minerals maintain an inverse relationship.

25.4.1 Colostrum

Colostrum is the mammary gland secretion immediately after parturition usually up to 3 days postpartum. It contains less lactose and high fat, protein, and minerals compared to milk. Another striking feature of colostrum is that it contains a large amount of immunoglobulins, particularly IgG. The gut of the neonates allows the passage of these immunoglobulins and facilitates the transfer of passive immunity from mother to offspring.

Synthesis and secretion of colostrum: The level of progesterone decreases very fast around parturition, while the level of estrogen together with prolactin and glucocorticoids is increased. The alteration of the endocrine milieu allows the differentiation of mammary secretory epithelium. The synthesis of lactose and fatty acids starts as early as 30 and 15 days prepartum, respectively. Prolactin and glucocorticoids induce the onset of "copious mileticulum." The appearance of IgG receptors (especially IgG1) in mammary secretory epithelia allows the uptake of IgG from blood immediately after calving. Synthesis of all these constituents cause increased mammary gland volume and colostrum secretion commences. The amount of colostrum secreted by an individual Holstein cow varies from 23.1 ± 2.5 to 36.4 ± 2.1 L (pooled first four milking).

Composition of colostrum: The components of bovine colostrum have been summarized in Table 25.16.

Functions of colostrum: Colostrum is an excellent energy-rich nutritional supplement. The predominant function of the colostrum is to transfer passive immunity in form of antibodies. IgG-1 is the principal immunoglobulin type in colostrum followed by IgM, IgA, and IgG-2. Feeding of colostrum significantly reduces calf mortality. The cellular components (polymorphonuclear leukocytes and macrophages) present in the colostrum produce lysozyme, complement components, and interferon which protects the newborn against enterocolitis. Lactoferrin, an iron-binding protein present in the colostrum is also having antibacterial and antiviral properties. Intake of colostrum affects metabolism, endocrine systems, and the nutritional state of the calves including the development and function of the gastrointestinal tract. The growth facts of the colos-

Table 25.16 Composition of bovine colostrum

Factors	Constituents	Amount	References
Nutritional factor	Energy (kcal/100 mL)	130	Guthrie (1989)
	Protein (g/100 mL)	14.9	
	Lactose (g/100 mL)	2.6	
	Fat (g/100 mL)	6.7	
Immune factors	Lactoferrin (mg/mL)	100	Stelwagen et al. (2009)
	IgA (mg/mL)	3.9	
	IgG (mg/mL)	47.6	
	IgG2 (mg/mL)	2.9	
	IgM (mg/mL)	4.2	
Growth factors	Epidermal growth factor (EGF) (μg/L)	30–50	Stelwagen et al. (2009)
	Transforming growth factor (TGF α) (μg/L)	2.2–7.2	
	TGF β (mg/L)	1–2	
	Insulin like growth factor (IGF) (mg/L)	10	
	Vascular endothelial growth factor (VEGF)	NA	
	Growth hormone (GH) (ng/L)	<0.03	
Cellular components	Somatic cells (leucocytes and epithelial cells) (cells/mL)	1,479,000	Ontsouka et al. (2003)

trum favor cell growth and tissue repair. Colostrum is also having a mild laxative effect which allows the passage of the first stool (meconium) along with the excretion of bilirubin.

25.4.1.1 Transition of Colostrum into Milk

The composition of colostrum changes with each hour with a declining trend in its biological and nutritive values. The ability to absorb immunoglobulins by the calves also decreases by 1/3 after 6 h, 2/3 after 12 h. Therefore, it is recommended to feed colostrum immediately after birth (0.5–1 h). Table 25.17 depicts the changes in the composition of bovine colostrum over time.

25.4.2 Composition of Milk

The mammary gland can be considered as a complex bio-machine that is capable of synthesizing new components from those existing in plasma. The compositional differences (folds) between milk and plasma are depicted in Table 25.18.

The composition of milk varies greatly among different species. The gross composition of milk in different species has been presented in Table 25.19.

Table 25.17 Changes in the composition of bovine colostrum over time

Hours after calving	Protein (%)	Fat (%)	Lactose (%)
0	16.8	6.7	2.9
6	11.7	6.1	3.5
12	6.3	4.4	3.9
24	5.5	4.1	4.1
48	4.8	3.9	4.2
120	3.6	3.8	4.5
Milk	3.2	3.8	4.6

Source: Puppel et al. (2019)

Table 25.18 The compositional differences (folds) between milk and plasma

Components	Milk	Blood plasma
Fat	35	3
Lactose	49	0
Caseins	27	0
α-Lactalbumin and β-lactoglobulin	4	0
Albumin and globulin	1.5	75
Citric acid	2	0
Chlorides	1.6	6
Phosphates	2.5	0.3

Source: Alais et al. (1985)

Table 25.19 The gross composition of milk in different species

	Average composition (%)				
Species	Water	Fat	Protein	Lactose	Ash
Buffalo	84.20	6.6	3.9	5.0	0.7
Cow	86.30	4.9	3.4	4.1	0.7
Sheep	83.70	6.0	4.8	4.9	0.8
Goat	86.50	4.5	3.5	4.7	0.8
Pigs	80.6	8.2	5.8	4.8	0.63
Horse	89.1	1.6	2.7	6.1	0.51
Camel	87.61	5.38	2.98	3.20	0.7
Human	87.43	3.75	1.63	6.98	0.21

Source: Mehta (2015)

25.4.2.1 Milk Fat

Fat is one of the important milk constituents ranging from below 3.0% to more than 6.0%. The economic value of milk is determined by its fat content. Fat serves as the source of energy (9 kcal/g) and carrier of fat-soluble vitamins. The desirable flavor of milk products is largely depending on milk fat. The fat in milk is present as milk fat globule (MFG) with a diameter of 0.1 to more than 22 μm. The surface of this spherical globule known as milk fat-globule membrane (MFGM) protects the nonpolar core of fat globule and stabilizes the MFG in an aqueous environment. Fat has a lower density compared to surrounding aqueous phase and hence can be separated by centrifugation.

The milk fat comprises of triacylglycerols (95%), diacylglycerol (2%), cholesterol (<0.5%), phospholipids (1%), and free fatty acids (FFA) (<0.5%).

The components of milk fat can also be classified as saponifiable fat (99–99.5%) comprises of glyceride (monoglycerides, diglycerides, and triglycerides), phospholipids, cholesterol ester, and FFAs and non-saponifiable fats (0.30–0.45%) includes cholesterol, carotenoids (β-carotene), fat-soluble vitamins, traces of squalene (hydrocarbon), and waxes. The yellow color of milk fat is due to β-carotene.

Triglycerides determine the properties of milk fat. Triglycerides act as a solvent for sterols, carotenoids, and tocopherol as they are apolar in nature and not surface active. Diglycerides are also apolar in nature and their properties are similar to triglycerides. The monoglycerides are polar in nature.

The majorities of milk fatty acids are statured and constitute more than half of the milk fatty acids amounting to 19 g/L whole milk. The predominant saturated fatty acids of milk are butyric acid, capric acid, caprylic acids, lauric acid, and stearic acid. These fatty acids are having some potential health benefits. Butyric, capric, and caprylic acids are reported to have anticancer activities. Lauric acid may have antiviral and antibacterial properties and is able to kill *Helicobacter pylori*. Both capric and lauric acids are able to inhibit cyclooxygenase (COX) and can be used as anti-inflammatory agents.

The predominant monounsaturated fatty acid present in the milk is oleic acid which accounts for about 8 g/L of whole milk. Oleic acid has the ability to lower cholesterol and triglyceride concentration in blood.

Linoleic-(18:2 omega-6) and α-linolenic (18:3 omega-3) acids are the main polyunsaturated fatty acids (PUFA) in milk. They can be converted into eicosanoids via arachidonic acid (20:4 omega-6) and eicosapentaenoic acid (EPA) (20:5 omega-3). The eicosanoids produced from linoleic acid via arachidonic acid facilitate platelet aggregation and increase the risk of coronary heart diseases. EPA partially blocks the conversion of arachidonic acid to eicosanoids. PUFA also have some positive role in signal transduction. The ratio between omega-6 and omega-3-fatty acids in milk is low and favorable compared to most other non-marine products. The conjugated linoleic acid present in the milk in the form of cis 9, trans 11 isomer also have anti-carcinogenic effect.

Phospholipids and glycosphingolipids constitute 1% of total lipids. They involve in stabilizing the emulsion, cell–cell interactions, and immune recognition. They also act as receptors for certain hormones and growth factors.

Know More……

The fattiest (fat percentage 61%) milk is produced by hooded seals. Their lactation period is also the shortest among mammals (4–5 days) and within this short period they try to provide as much as calorie to their pups for their weight gain before the mother seal return to the sea in search of food.

25.4.2.2 Milk Protein

Milk contains more than 200 types of proteins constituting around 95% of nitrogen in milk. The biological value of milk protein is high and it is a good source of essential amino acids. The milk proteins are classified into two major groups namely casein and whey protein at the ratio of 80% and 20%, respectively. The properties of different milk proteins have been presented in Table 25.20.

25.4.2.2.1 Casein

Casein in milk exists in form of colloidal calcium phosphate as micelle and precipitates upon decreasing pH. The casein micelle is composed of 94% protein and 6% mineral (calcium and phosphorus). Caseins are considered as naturally denatured proteins due to high proline content. Casein has four different fractions namely αs1-, αs2-, β-, and κ-casein. αs- and β-caseins remain as phosphoproteins whereas the κ-casein molecules are glycosylated.

The genes responsible for β-casein production are highly polymorphic in cow and 13 allelic variants have been identified. The most common allelic variant of bovine β-casein is A1 and A2 variants. The A1/A2 variants are determined by a pair of genes at sixth chromosome with two major alleles A1 and A2 β-casein alleles. Homozygous A2A2 or A1A1 cow will produce only A1 and A2 β-casein, respectively whereas heterozygous A1A2 cow will produce A1 and A2 β-casein in equal amounts. A1 β-casein variant is commonly found in milk from crossbred and European breeds of cattle and A2 milk is found in indigenous zebu cattle and buffaloes of India. Based on this milk β-casein variations, milk can be classified as A1 or A2 type. A1 milk contains only A1 or A1A2 β-casein but A2 milk contains only the A2 type of β-casein. The major difference between A1 and A2 variants of bovine β-casein is at 67th amino acid which is histidine in A1 and proline in A2 variants.

The digestion of A1 β-casein milk yields a bioactive peptide composed of 7 amino acids called beta-casomorphin 7 (BCM-7) which causes some potential health hazards like diabetes and cardiac diseases. But the presence of proline in A2 milk at 67th position prevents the formation of BCM-7 thus not related to any of such health issues.

A hydrolytic product of β-caseins is called γ-caseins representing 29–209, 106–209, and 108–209 residues of β-caseins.

κ-Casein is the only glycosylated milk protein consisting of different amounts of carbohydrates such as galactose (1%), galactosamine (1.2%), and *N*-acetylneuraminic acid (2.4%).

Table 25.20 Properties of different milk proteins

			(Residues/molecule)			
Property	Average concentration (%)	Molar mass	Amino acid	Phosphoserine	Cysteine	Isoelectric pH
αs1-Casein	1.1	23,614	199	7–9	0	4.5
αs2-Casein	0.3	25,230	207	10–13	2	5.0
β-Casein	0.9	23,983	209	5	0	4.8
κ-Casein	0.3	19,023	169	1	2	5.6
β-Lactoglobulin	0.32	18,283	162	0	5	5.2
α-Lactalbumin	0.12	4176	123	0	8	4.3
Serum albumin	0.04	66,267	582	0	35	4.8

Source: Mehta (2015)

25.4.2.2.2 Whey Proteins

These are the soluble milk proteins comprising around 20% of total milk protein. The whey proteins are globular in nature and remain in secondary and tertiary structures. There are four major classes of whey proteins like β-lactoglobulin (50%), α-lactalbumin (25%), immunoglobulins (9%), and bovine serum albumin (6%). α-Lactalbumin is the smallest among the whey proteins and helps in lactose synthesis. The bovine serum albumin is identical to blood albumin. Immunoglobulins are the most heterogeneous group of whey proteins that exist as monomers or polymers made up of two light and two heavy chains. There are five classes of immunoglobulins namely IgA, IgG, IgM, IgE, and IgD. Colostrum contains a large amount of immunoglobulins of which IgG is the highest in concentration.

Lactoferrin (LF) is an iron-binding glycoprotein composed of a single polypeptide chain comprising of 689 amino acid residues. It has a molecular weight between 76 and 80 kDa. Bovine LF can act as antibacterial, antiviral, immune modulator antioxidant, anticancer, and anti-allergic agent.

Lactoperoxidase present in the milk also has an antibacterial property. It is present in both milk (13–30 mg/L) and colostrum (11–45 mg/L) of bovines.

It has been reported that milk proteins have antihypertensive effects as it can inhibit angiotensin-converting enzyme, opoid-like activities, and antithrombin properties.

Milk is rich in essential and branched-chain amino acids that act as a substrate for protein synthesis and gluconeogenesis.

Taurine, one of the most abundant intracellular amino acid in humans present in breast milk (18 mg/L) and bovine colostrum. It has a role in the conjugation of bile acids, neuromodulation, and retinal development. It also has antiarrhythmic and antioxidant effect. Taurine concentration is low in bovine milk (1 mg/L) but it is higher in goat milk (46–91 mg/L).

Glutathione, a tripeptide composed of cysteine, glycine, and glutamic acid is also present in milk which acts as an antioxidant.

25.4.2.3 Milk Sugars

The main sugar in milk is lactose, a disaccharide that comprises of α-D-glucose and β-D-galactose joined by β-1,4-glycosylic linkage. But in rat and mouse milk is sialyllactose an oligosaccharide. Lactose in milk may exist as both α- and β-lactose, with an equilibrium ratio of β/α = 1.68 at 20 °C. Lactose is normally found in dairy products in two forms namely

Crystalline forms: α-hydrate or α-lactose monohydrate and anhydrous β-lactose exist as crystalline forms.
Amorphous forms: It is a mixture of alpha and beta lactose.

Lactose is only 25% as sweet as sucrose. Lactose is responsible for colligative properties of milk like depression of osmotic pressure and freezing point and elevation of boiling point.

There are some other sugars in milk like glucose, fructose, glucosamine, galactosamine, *N*-acetylneuraminic acid, and oligosaccharides in addition to lactose.

25.4.2.4 Minerals

The major minerals present in milk are calcium, magnesium, potassium, and sodium exist as bicarbonates, chlorides, citrate, and bicarbonates forms. The minerals are distributed between soluble and colloidal phase. The principal mineral of soluble phase is calcium (66%) whereas phosphorous (55%) is the main mineral of colloidal phase. Milk is a rich source of dietary calcium. It is associated with casein which improves its absorption in the GI tract. Milk also contains some trace minerals. The mineral and trace elements in milk have been presented in Table 25.21.

25.4.2.5 Vitamins

Milk contains both fat-soluble (A, D, E, and K) and water-soluble (B1, B2, B6, B12, pantothenic acid, niacin, biotin, folic acid, and vitamin C) vitamins. During milk processing fat-soluble vitamins are present in cream whereas water-soluble vitamins are retained in the whey. Table 25.22 depicts the amount of different vitamins present in milk.

Table 25.21 The mineral content in milk

Minerals	Amount (g/100 mL)
Sodium	0.048
Potassium	0.143
Calcium	0.117
Magnesium	0.011
Chloride	0.110
Phosphate	0.230
Citrate	0.175
Sulfate	0.0100
Trace elements	**Amount (μg/L)**
Zinc	4000
Aluminum	500
Iron	400
Copper	120
Molybdenum	60
Manganese	30
Nickel	25
Silicon	1500
Bromine	1000
Boron	200
Fluorine	150
Iodine	60

Source: Mehta (2015)

Table 25.22 Vitamins present in milk

Vitamins	Concentration (mg/L)
A (retinol)	0.4
D (calciferol)	0.001
E (tocopherol)	1.0
B1 (thiamine)	0.4
B2 (riboflavin	1.7
B6 (pyridoxine)	0.6
B12 (cyanocobalamin)	0.005
Nicotinamide	1
Pantothenic acid	3.5
Biotin	0.03
Folic acid	0.05
C (ascorbic acid)	20

Source: Belitz et al. (2009)

25.4.2.6 Enzymes

Milk contains around 60 different types of enzymes. They are synthesized either in mammary secretory cells or derived from blood. Some of the enzymes are of microbial origin. The enzymes are having both beneficial and undesirable effects. Enzymes like plasmin, catalase, and *N*-acetyl-β-D-glucosaminidase are secreted during mammary gland infection (mastitis) and thus acts as markers of udder infection. The enzymes like alkaline phosphatase, lactoperoxidase, and γ-glutamyl transpeptidase are indicators for the thermal stability of milk. The enzymes present in milk and their functions are presented in Table 25.23.

25.4.3 Intolerance to Milk Components

Milk is nature's most nutritious food but there are several myths regarding the potential health hazards like respiratory problems and asthma after milk consumption. Though there are few reports on the asthma-like symptoms after milk consumption, inflammatory reactions and increase in mucus production have not been confirmed after milk consumption. There are some health problems associated with intolerance to milk components

Milk allergy: Milk proteins may act as potential allergens and mediate allergic manifestations particularly in children (0–3 years) with an incidence rate of 2–5%. It is no longer a problem after the age of 3 years. The allergic reaction induced by milk may be rapid with the symptoms like vomiting, anaphylaxis, and wheezing. The slower onset of milk allergic reaction is most common and characterized by loose stool, vomit, and reduced weight gain. Milk allergy can be treated by avoiding milk proteins.

Lactose intolerance: It occurs due to the deficiency of lactase enzyme which hydrolyzes lactose into glucose and galactose. Lactase activity is developed in infants which disappeared after weaning. Human milk is having a high amount of lactose and the lactose intake may reach 30–40 g/day after breastfeeding. If the lactase activity is not sufficient the excess intake of lactose may lead to lactose malabsorption including diarrhea, bloating and flatulence, abdominal pain, and gaseous accumulation in the intestine. Congenital lactase deficiency is a rare genetic condition where lactase activity is decreased or absent at birth and remains low throughout life. Primary lactose intolerance occurs due to decreased lactase activity after weaning. Secondary lactose intolerance or acquired hypolactasia can be caused by low lactase activity due to damage of the intestinal lining in intestinal surgery or diarrheal diseases such as gastroenteritis, inflammatory bowel disease, cow's milk intolerance, or AIDS.

Galactosaemia: Digestion of lactose in milk yields increased concentrations of galactose. The galactose is catabolized by phosphorylation at position 1, and then converted to UDP-galactose and glucose-1-phosphate in Leloir pathway (Fig. 25.3). One of the important enzyme of Leloir pathway is galactose-1-phosphate uridyltransferase. Galactosemia develops due to deficiency of enzyme galactose-1-phosphate uridyltransferase which leads to

Table 25.23 Enzymes present in milk and their functions

Enzymes	Origin	Remarks
Lipases	Casein micelles	Catalyzes the hydrolysis of triglycerides to form fatty acids and causes rancidity in milk and milk products
Esterases	Serum	Activity is high in colostrum and during udder infections
Lactoperoxidase (LPO)	Serum	LPO is the most abundant enzyme in milk. It has bactericidal activity
Catalase	Leucocytes	Antioxidant enzymes. Indicator of mastitis
Plasmin	Casein micelle	Contributes to primary proteolysis in cheese
Alkaline phosphatase	Fat-globule membrane	Used as an index of pasteurization
Acid phosphatase	Fat-globule membrane	Used as an index of heat stability of milk
Lysozyme	Serum	Breakdown of peptidoglycan polymers of bacterial cell wall
γ-Glutamyltransferase	Fat-globule membrane	Indicator for heat stability of milk

Source: Belitz et al. (2009) and Mehta (2015)

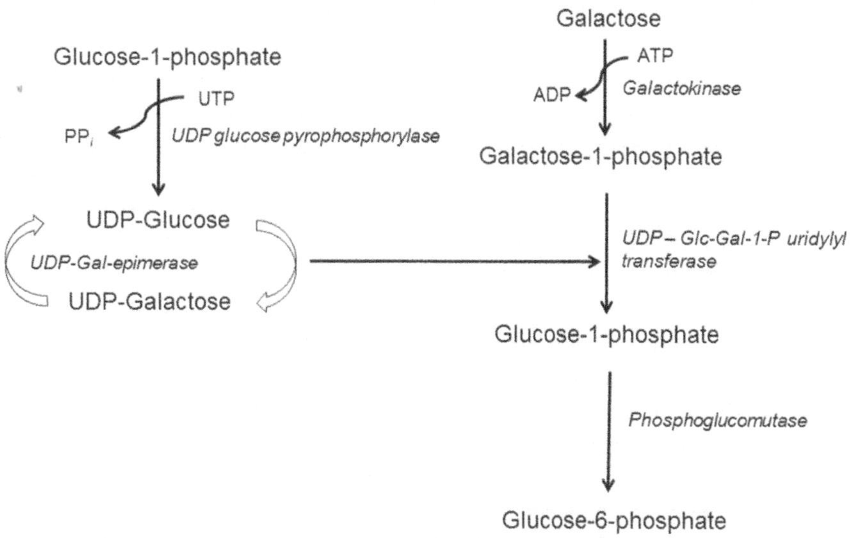

Fig. 25.3 Leloir pathway of galactose utilization

accumulation of galactose-1-phosphate in eye lens thereby resulting in cataract. The other symptoms of galactosemia are hepatomegaly and splenomegaly, ascites, and feeble mindedness.

25.5 Galactopoiesis

Galactopoiesis can be defined as the maintenance of established lactation. The term may often be used to describe the enhancement of established lactation. In cattle, the milk production follows a dynamic curve with a rapid accelerating phase and reached peak around 6 weeks then decline till the end of lactation. The typical lactation curve of bovines can be split into early (day to 100 days), mid (101–200 days), and late (201–300 days) followed by a dry period of 65 days.

Due to high energy requirement and nutrient partitioning, the animal exhibits negative energy balance (NEB) during early and peak lactation and the mobilization of body reserves leads to decreasing body weight during this critical period. During NEB state, the health and fertility of the animals may be compromised.

The ability of the animals to sustain the peak yield with a lesser decrease is termed as production persistency. An animal with higher production persistency has increased overall gain per lactation alternatively high persistency may also have a negative impact on the animal's health and fertility.

Galactopoiesis depends upon milk synthesis and milk removal. Mammary gland has to ensure a sufficient number of secretory cell population by increasing cell proliferation and decreasing cell loss.

The key components that contribute to galactopoiesis and lactation persistency are discussed below.

25.5.1 Activity of Milk Secretory Cells

During the initiation of lactation, the numberof secretory cells are highest with less activity and milk production per cell was lowest. During peak lactation, the secretory cell differentiation leads to an increase in average yield per cell which is sustained throughout the lactation period. Decline in milk yield after peak lactation is therefore due to loss in secretory cell number due to apoptosis rather than decreased secretory activity. Apoptosis is the key mechanism which maintained the secretory cell dynamics of the mammary gland. Hormones and growth factors are important determinants for secretory cell dynamics. However, nutrition, oxidative stress, and frequency of milking also regulate secretory cell of apoptosis. Thus, better lactation persistency can be achieved by slowing down the apoptosis by manipulating the aforesaid determinants.

25.5.2 Milk Secretion and Milking Frequency

After being secreted from secretory epithelium, the milk is distributed between two main components. Some portion of milk remains in the lumen of mammary alveoli and small ducts (alveolar milk) and some portion descend to cistern (cisternal milk). The cisternal milk can be obtained without milk ejection whereas the alveolar milk can only be obtained after milk ejection reflex. Huge species differences exist with respect to the storage of milk in different compartments. In goat and sheep, cisternal milk filling initiates immediately and 75% milk can be stored following a milking interval of 12–14 h. In contrast, there is no milk in the cistern up to 2 h post milking and only 20% cisternal storage of milk is achieved between normal milking interval. To obtain maximum persistency milk has to be removed from the mammary gland. Incomplete milking hastens lactation persistency. Absence of milk removal leads to increased intramammary pressure and decreased blood flow to mammary gland. Some local autocrine inhibitors of milk secretion such as feedback inhibitor of lactation (FIL) result in the accumulation of milk in the alveolar ducts and cause partial inhibition of milk synthesis and secretion. If long-term stasis of milk occurs, it can lead to complete cessation of lactation and mammary involution begins.

Local factors: *Feedback inhibitor of lactation (FIL)*: The rate of milk secretion in response to the frequency of milking is regulated by a local inhibitory glycoprotein termed as a Feedback inhibitor of lactation (FIL). It is an active whey protein identified in cows, goats, marsupials, and humans. FIL prevents the differentiation of mammary secretory epithelial cells and negative feedback on the synthesis of milk protein and lactose. Another hypothesis stated that FIL induces apoptosis by reducing the number of prolactin in the mammary gland.

Beside FIL, there are several other local factors thought to be involved in autocrine regulation of milk secretion (see mammary gland involution). One such inhibitory factor in the form of proteolytic casein fragment has been identified in the udder of lactating goat. The enzyme carbonic anhydrase was identified in the capillary endothelium which is related to mammary metabolism, milk composition, and milk flow rate in goat.

Table 25.24 The roles of somatotropin in different body tissues

Tissue	Role in lactation
Mammary gland	Increases the milk synthesis with normal composition Increases the nutrient uptake Increases the number and activity of secretory cells Increases blood flow consistent with an increase in milk yield
Liver	Increases gluconeogenesis Decreases the ability of insulin to suppress gluconeogenesis
Adipose tissue	Increases lipogenesis during positive energy balance and lipolysis during negative energy balance Decreases the ability of insulin to stimulate lipogenesis Decreases the ability of adenosine to inhibit lipolysis Increases the ability of catecholamines to stimulate lipolysis
Muscle	Decreases glucose uptake
Kidney and intestine	Increases the production of 1.25-vitamin D3 Increases the sbsorption of Ca, P, and other minerals required for milk Increases the ability of 1,25-vitamin D3 to stimulate Ca-binding protein
Whole body	Decreases glucose oxidation Increases oxidation of nonesterified fatty acids in negative energy balance Increases cardiac output consistent Increases productive efficiency (milk per unit of energy intake)

25.5.3 Hormonal Control of Galactopoiesis

25.5.3.1 Somatotropin

Growth hormone or somatotropin is involved in tissue metabolism and nutrient partitioning during lactation. The main roles of somatotropin are to promote lipolysis in adipocytes and gluconeogenesis in the liver during the state of negative energy balance. Exogenous administration of growth hormone for 10–12-week treatment period resulted in 40% increase in milk yield of dairy cows and genetic engineering explored the potential of using recombinantly derived bovine somatotropin (rbST) in the biology of lactation and its commercial use in augmentation of lactation in bovines. The review of Bauman and Vernon (1993) on the roles of bovine somatotropin in different body tissues is summarized below in Table 25.24.

The direct role of bST on the mammary tissue is a matter of debate as it was reported that growth hormone did not bind to receptors in the bovine mammary gland. Rather an alternate hypothesis was proposed on the indirect effects of ST in association with the IGF complex which binds to its receptor on mammary epithelial cells and mediates the action of growth hormone. The nutritional status of the animals plays a crucial role in regulating IGF axis in cattle. In animals with adequate nutritional status, administration of bST causes an increase in circulating IGF-I levels but this effect of bST was abolished in undernourished animals.

25.5.3.2 Prolactin

The role of prolactin in monogastric animals on galactopoiesis is well-established. In rat, suppression of prolactin resulted in decreased milk secretion and administration of prolactin increases milk secretion during the early phase of lactation in rabbits. But the galactopoietic role of prolactin in ruminant was questionable as neither administration of prolactin nor its inhibition was correlated with milk yield in cows. But recent reports showed that prolactin is well involved in the galactopoiesis in bovines which is released during milking and nursing in response to mammary gland stimulation. The galactopoietic role of prolactin in bovines is further supported by the fact that a long-day photoperiod increases prolactin concentration and milk production and administration of melatonin (which secrets at night) for 12 weeks decreased prolactin and milk production. Prolactin stimulates the synthesis of milk constituents such as caseins and lipids. Prolactin may induce secretory activity of mammary epithelium as decreased milk production by prolactin antagonists appeared to be the result of a reduction in cell activity.

25.5.3.3 Thyroid Hormones

Thyroid hormones are galactopoietic in nature. Administration of T_3 and T_4 or feeding of thyroprotein increases milk production for 2–4 months along with increased butterfat percentage. Long-term supplementation of thyroid hormones or thyroproteins increased milk production during early lactation but there was a rapid decline in later lactation. Supplement thyroid hormones during the negative energy balance state of early lactation has little effect on galactopoiesis as the galactopoietic effect of thyroid hormones are mediated through increased metabolism. Thyroid hormones potentiate the action of other galactopoietic hormone such as prolactin for lactose and casein synthesis.

25.5.3.4 Glucocorticoids

Evidence on glactopoietic effect of glucocorticoids in ruminants are controversial. Studies reported that adrenalectomy reduces milk yield in cows. In contrast, the administration of dexamethasone inhibits milk production in dairy cows. This negative effect of glucocorticoids on galactopoiesis is thought to be induced by the inhibition of prolactin release or the reduction of mammary responsiveness to prolactin. Glucocorticoids also decrease plasma IGF-I in cows and goats.

25.5.3.5 Insulin

Insulin concentration is negatively correlated with milk production. But administration of insulin together with supplementation of extra glucose stimulates lactation. Insulin has no role in mammary uptake of acetate, β-hydroxybutyrate, triglycerides, amino acids, and glucose.

25.5.3.6 Ovarian Steroids

Exogenous administration of 17β-estradiol reported to decrease milk production in dairy cows. Ovariectomy in nonpregnant dairy cows improves persistency of lactation. Oral contraceptives containing estrogen also reduces milk production in woman. The mechanism by which estrogen affects milk production is yet to be discovered. But the inhibition of lactogenic effect of prolaction and GH by estradiol could be the reason for decreased milk production upon estrogen administration.

Progesterone has a negative effect on lactogenesis but during galactopoiesis or established lactation progesterone has no effect on milk yield. It may be due to the fact that mammary tissue lacks progesterone receptor during lactation and progesterone has a greater affinity for milk fat rather its own intracellular receptor.

25.5.4 Factors Affecting Milk Yield and Composition

In mammals, milk is primarily produced to nourish the young. But the cow possesses most advanced type of mammary gland from the evolutionary point of view which can able to produce far more milk than the calf can consume. Milk production can be augmented further through genetic selection, nutritional manipulation or supplementation, better managemental strategies, and advanced milking technologies. The quantity and quality of milk are dependent upon the availability of secretory tissues in the mammary gland and efficiency of these tissues to synthesize milk components together with the availability of suitable nutrients as the precursors of milk components. The selection of cows for increased production may compromise their fertility and disease resistance. Production stress also compromises the host immunity and makes the animals more prone to metabolic and systemic diseases which in turn affect milk production and composition.

There are several genetic and non-genetic factors that affect milk yield and compositions

25.5.4.1 Breed

Milk yield varies considerably among different breeds of cattle. Generally, heavier breeds produce more milk compared to lighter breeds. Zebu cattle have lower production potential compared to exotic cows. Table 25.25 depicts the comparison of production performances of zebu, exotic, and cross-bred cattle along with different buffalo breeds.

In zebu cattle, the milk composition varied significantly among different breeds except for lactose. The fat percentage was higher in Jerseys and Guernseys, and more compared to Holstein Friesians and Ayrshires. The highest intrabreed variability was seen in fat percentage followed by solids-not-fat (SNF), protein, and lactose. The composition of milk constituents among different breeds of cattle and buffalo has been presented in Table 25.26.

25.5.4.2 Genetics

Economical traits such as milk yield were dependent on genetics. Milk yield has heritabilities around 0.25 whereas percentage milk fat, protein, and lactose have heritabilities around 0.5 (Table 25.27). The heritability of lactation persistence, peak yield, and milking rate is moderate while that of mastitis resistance is low.

The repeatability of milk yield and different milk production traits are presented in Table 25.28. Milk yield is highly repeatable (repeatability 0.95). The percentage of protein and SNF are also highly repeatable (0.85 and 0.90) but the repeatability of fat percentage is low (0.60). The characters with moderate to high repeatability can be improved through genetic selection.

The correlation coefficients between milk yield and various milk production traits are presented in Table 25.29. The fat, protein, solid not fat, and total solids are negatively correlated with milk yield.

Table 25.25 Milk yield of different cattle and buffalo breeds

Species	Types	Breeds	Milk yield (kg/lactation)	References
Cattle	Zebu	Sahiwal	2000–4000	Department of Animal Husbandry and Dairying. Govt. of India, New Delhi
		Red Sindhi	2000–4000	
		Tharparkar	1800–3500	
		Gir	2000–6000	
		Haryana	1000–2000	
	Exotic	Holstein	6000–8000	TANU Agritech Portal, Tamil Nadu Veterinary and Animal Science University, Chennai, India
		Ayrshire	4000–6000	
		Jersey	3000–5000	
		Brown Swiss	4000–6000	
		Guernsey	3000–5000	
	Crossbred	Karan Swiss (Sahiwal and Red Sindhi × Brown Swiss)	5000–6000	Thiagarajan (2014)
		Karan Fries (Tharparkar × Holstein Friesian)	3000–4000	
Buffalo		Murrah	1500–2500	TANU Agritech Portal, Tamil Nadu Veterinary and Animal Science University, Chennai, India
		Nili ravi	1500–1850	
		Jaffarabadi	100–1200	
		Surti	900–1300	
		Mehsana	1200–1500	
		Godavari	1200–1500	

Table 25.26 Milk composition of different cattle and buffalo breeds

Species	Breeds	Fat (%)	Protein (%)	Lactose (%)	Total solid (%)	References
Zebu cattle	Sahiwal	4.23 ± 0.18	3.60 ± 0.05	5.38 ± 0.07	13.99 ± 0.23	Sarkar et al. (2006)
	Red Sindhi					
	Tharparkar	4.37 ± 0.20	3.92 ± 0.05	5.35 ± 0.08	14.22 ± 0.25	Sarkar et al. (2006)
Exotic cattle	Holstein	3.56	3.01	4.61	11.91	
	Ayrshire	3.97	3.28	4.63	12.69	
	Jersey	4.97	3.65	4.78	15.15	
	Brown Swiss	3.8	3.18	4.8	12.69	
	Guernsey	4.58	3.49	4.78	13.69	
Crossbred cattle	Karan Swiss					
	Karan Fries	3.91 ± 0.14	3.58 ± 0.04	5.39 ± 0.5	13.69 ±:0.7	Sarkar et al. (2006)
Buffalo	Murrah	7.53 ± 0.19	4.03 ± 0.05	--	16.53 ± 0.20	Misra et al. (2008)
	Bhadawari	7.43 ± 0.26	3.92 ± 0.07	--	17.70 ± 0.28	
	Mehsana	6.46 ± 0.17	3.87 ± 0.05	--	5.59 ± 0.18	
	Surti	6.17 ± 0.20	3.93 ± 0.05	--	4.96 ± 0.21	

Table 25.27 Heritabilities of different milk production traits in cow

Traits	Heritability
Milk yield	0.25
Fat %	0.50
Protein %	0.50
Peak yields	0.30
Milking rate	0.40
Persistency	0.40
Mastitis resistance	0.10

Source: Wilcox (1992)

Table 25.28 Repeatabilities of different milk production traits in cow

Traits	Repeatability
Milk yield	0.95
Fat %	0.60
Protein %	0.85
SNF %	0.90

Source: Wilcox (1992)

Table 25.29 The correlation coefficients between milk yield and various milk constituents

Traits (%)	Correlation coefficient
Fat	−0.3
Protein	−0.3
SNF	−0.2
Total solid	−0.3

Source: Wilcox (1992)

25.5.4.3 Environment

Lactating cows are more cold-resistant due to the production of metabolic heat as a result of increased feed intake. The dairy cows may sustain their production until the temperature fall below −5 °C. *Bos indicus* of tropics are more heat tolerant compared to *Bos Taurus* of European origin due to low BMR and lower feed intake. But simultaneously they also have lower milk production.

Lactating cows are more susceptible to heat stress which decreases their feed intake. A significant decline in milk production was found at a temperature humidity index (THI) of 77. The critical values for minimum, mean, and maximum THI for milk production were 64, 72, and 76, respectively for dairy cows. There was 0.32 kg decrease in milk yield with the increase per unit THI. Milk composition traits except milk protein are highest in hot humid season compared to other seasons. The milk of cows calved during winter exhibits more milk fat and SNF compared to the cows calved during summer.

25.5.4.4 Nutrition

The mammary secretory epithelium requires a constant supply of nutrient precursors to produce milk. Therefore, both the milk yield and composition are affected by dietary manipulation. Through dietary manipulation, fat and protein content have been altered up to a range of 0.3–0.6%, respectively but the lactose was reported to be unchanged with respect to dietary manipulation. Acetate and butyrate are the main precursors of milk fat. Lower intake of roughage to carbohydrate decreases milk fat by decreasing the production of acetate and butyrate. The concentration of propionate is negatively correlated with milk fat but it promotes milk protein synthesis by increasing the availability of glutamate. On a dry matter basis, the minimum forage-to-concentrate ratio of 40: 60 is required to ensure milk fat percentage above 3.6. In contrast, finely chopped forage increases the amount of propionate hence the milk protein percentage. Starch is essentially required for optimum microbial protein synthesis and its inclusion in the diet positively influences milk yield and the percentage of protein in milk. Milk fat percentage can also be increased by adding a minimum of 28% neutral detergent fiber. Supplementation of saturated fatty acids in the diet was reported to increase milk fat percentage, in contrast, unsaturated fatty acids decrease fat percentage. But the rumen microbes are unable to utilize lipids as an energy source thus supplementation of fatty acids in the diet may decrease milk protein percentage. Therefore, it is recommended that fatty acid supplemented diets should be enriched with amino acids to maintain optimum milk protein percentage. There are certain feed additives that promote milk fat synthesis such as sodium bicarbonate, magnesium oxide, and methionine hydroxy analog with high-energy diets. They facilitate the transfer of blood lipids to mammary gland.

25.5.4.5 Parity and Stage of Lactation

It was reported that the total milk production of a cow reaches a peak around fifth lactation when the cow is 7–8 years old with maximum skeletal size which is around 30% more compared to first lactation. The recurring increments of milk production are 13% from first to second, 9% from second to third, 5% from third to fourth, and 3% from fourth to fifth lactation. There is a plateau in milk production after fifth lactation after which milk production declines beyond 12 years of age. Mammary glands also increase in size on subsequent lactations till fifth which is due to skeletal maturation and increase in body weight to accommodate a larger udder.

Milk constituents are also influenced by the lactation stage. The fat and SNF content are decreased by 0.05% and 0.1%, respectively in each successive lactation. Parity has no significant effect on fat and SNF content in some Zebu (Tharparkar, Red Sindhi) and crossbred (Karan swiss cow).

The transition from colostrum to milk occurs within the first few days postpartum with abrupt changes in the composition. Colostrum contains more protein and minerals but less lactose than milk. Colostrum also has 25% total solid which is much higher than milk. Colostrum is also having more calcium, magnesium, sodium, phosphorus, and chloride but less potassium compared to milk. However, these changes are still continuing up to 6–8 weeks postpartum at a slower rate. An increase in the protein content decreases lactose and vice versa and the SNF content remains fairly constant till the 8th month of lactation and then increases under the influence of hormones of pregnancy but declines gradually in unbred cows over the remaining lactation period.

25.5.4.6 Milking Management

Milking at unequal intervals results in less milk production compared to those milked at regular intervals. This reduction in milk production is more pronounced in high-yielding cows compared to low yielders. Milking twice a day yields 40% more milk production than once a day which can be further increased to 5–20% thrice a day and 5–10% when milked four times a day. Increasing the frequency of milking to three times a day can increase milk yield by up to 20%. The probable reasons for increased milk yield in response to milking frequency are less intramammary pressure with frequent milking, increased stimulation for oxytocin release, and less negative feedback on secretory cell results due to accumulation of milk. Around 10–20% of total milk is left in the udder after milking termed as residual milk. Poor milking procedure decreases the milk yield by increasing the fractions of residual milk.

25.5.4.7 Pregnancy

Milk production of cow is compromised during gestation especially after 4 or 5 months due to nutrient partitioning for the growth and maintenance of the fetus. The proportion of milk constituents is increased with the advancement of gestation after fourth month of pregnancy. Milk SNF, protein, and lactose contents are altered during pregnancy but fat and mineral content are not affected in HF crossbred. The alterations in the milk composition during gestation can be explained by increased estrogen levels in the maternal circulation.

25.5.4.8 Dry Period

A dry period of 42–60 days is common practice that facilitates the replacement of old and damaged mammary epithelial cells and increases milk production in the next lactation. The milk production was reported to be decreased by 4.5% for a short dry period (4–5 weeks) and by 19.1% for no dry period. But after conventional dry period the cow exhibits negative energy balance during early lactation due to reduced feed intake and high milk yield which may continue for several months associated with metabolic disorders and impaired fertility. Therefore, shortening of dry period is recommended to improve the energy balance in early lactation through decreased milk yield after calving. The milk protein, lactose, and SNF percentage were reported to be increased by a shorter dry period but milk fat content was not affected.

25.5.4.9 Body Condition Score (BCS)

The state of NEB during the transition period and early lactation mobilizes body adipose for milk production. Therefore, BCS is negatively correlated with milk yield. Superior milk-producing cows genetically have lower BCS throughout lactation compared to those of lower genetic merit. Data have suggested that 20% of the increase in milk production is due to increased body weight. BCS is slightly negatively correlated to milk fat, lactose, and SNF content and positively correlated with protein.

25.5.4.10 Photoperiod

Increased photoperiod has a positive effect on milk production in many species including cattle. Increased milk production at a tune of 25 kg/cow has been reported with increasing light exposure from 12 to 16–18 h. The exposure of light in the retina stimulates photoperiod which transmits inhibitory signal to pineal gland through retino-hypothalamic tract. In pineal gland, this inhibitory signal decreases melatonin synthesis by inhibiting the enzyme *N*-acetyltransferase. Decreased melatonin stimulates prolactin release which mediates galactopoietic effect. But prolactin is only galactopoietic in rodents and it had a limited role in milk yield in an established lactation in cattle. Therefore, this hypothesis did not hold true for cows, instead this photoperiodic induction of galactopoiesis may be mediated through increased IGF-I secretion. Increasing day length exposure did not affect milk composition but a minor decrease in milk fat percentage has been reported. Photoperiodic induction of milk yield can also be mediated through increased dry matter intake in cattle.

25.5.5 Milk Ejection Reflex

It is a neuroendocrine reflex leading to passive withdrawal of milk from alveoli, cistern, and ducts of mammary glands. As discussed in previous chapter, after secretion from secretory alveoli, milk is stored in the udder in the form of cisternal and alveolar fractions. The cisternal milk fractions can be removed after loosening the teat sphincter barrier. The alveolar milk fraction that locates in the alveoli and small ducts are fixed by capillary force. The removal of alveolar milk requires its forceful expulsion into the cistern. Milk ejection reflex facilitates this forceful milk expulsion.

25.5.5.1 Pathways of Milk Ejection Reflex

Neuroendocrine reflex, as the name implies, consists of neural (afferent) and endocrine (efferent) pathways.

Neural (afferent) pathway: The neural component starts after the stimulation of pressure-sensitive nerve receptors located at the tip of teats. The nerve impulse travels through the spinothalamic nerve tract to the brain, especially to the supraoptic nuclei (SON) and paraventricular (PVN) nuclei of the hypothalamus. These hypothalamic nuclei secrete oxytocin stored in the secretory terminals of the neurohypophysis. The suckling-induced oxytocin is released into bloodstream which initiates the endocrine (efferent pathway).

Endocrine (efferent) pathways: Once oxytocin is released, it reaches the mammary gland through systemic circulation. Oxytocin binds with its specific receptors in the mammary gland and contracts the myoepithelial cells. These cells are located between the basement membrane and epithelial cells of alveoli. They are also termed as basket cells and are having long cytoplasmic processes which cover epithelial cells. Due to the contraction of myoepithelial cells, the interalveolar pressure increases leading to the expulsion of milk from the alveoli into the cisternal system.

Mechanism of action of oxytocin on myoepithelial cell contraction: The receptors of oxytocin belong to rhodopsin-type (Class 1) of the G-protein coupled receptor superfamily. Ligand binding triggers intracellular signaling pathways that activates Phospholipase-C (PLC) which converts phosphatidyl inositol-bis-phosphate (PI2) into diacylglycerol (DAG) and phosphatidyl inositol-tri-phosphate (PI3). PI3 mediates the release of calcium from sarcoplasmic reticulum. Calcium after binding with calmodulin (CaM) activates myosin light chain kinase (MLCK) and phosphorylation of myosin. Phosphorylation of myosin initiates the contraction of myoepithelial cells.

25.5.5.2 Inhibition of Milk Ejection Reflex

Inhibition of milk ejection is brought about by either central (brain) or peripheral (mammary gland) level. In central inhibition, the release of oxytocin from the neurohypophysis is inhibited whereas in peripheral inhibition the secretion of oxytocin is normal but oxytocin is unable to exert its effect on mammary gland.

25.5.5.2.1 Central Inhibition

A number of stimuli are associated with the central inhibitory pathway of the milk ejection reflex. But the possible mechanism responsible for this inhibition was not clearly understood in cows. Two possible mechanisms of central inhibition are proposed.

Opioid control: The endogenous opioid system (EOP) is thought to be the main regulator of central inhibition of oxytocin release. The EOP system inhibits oxytocin release at three different levels, viz., neuronal terminals in the neurohypophysis, supraoptic, and paraventricular nuclei of the hypothalamus and inputs to oxytocin neurons. Pro-opiomelanocortin is the common precursor of both β-endorphin and cortisol. Therefore, it seems that cortisol may be a mediator of central inhibition. But the administration of cortisol does not have any role in milk ejection in cattle. Therefore, the involvement of cortisol in this inhibitory control may be ruled out.

Noradrenergic control: The noradrenergic cells (A2 cell group) of nucleus tractus solitarii in medullary structures and dopaminergic cells of the posterior and periventricular hypothalamus also involve in the oxytocin release. A2 cells secret noradrenalin which mediates excitatory control over oxytocin release via α 1-adrenergic receptors in PVN and SON whereas adrenaline released from adrenal medulla inhibits oxytocin release via β-adrenergic receptors. However, the role of catecholamines in the central inhibition of milk ejection reflex is questionable as peripheral catecholamine concentration was unaltered during central inhibition in cows and adrenergic blocking agents failed to abolish central inhibition of milk ejection reflex in cows.

25.5.5.2.2 Peripheral Inhibition

The failure of oxytocin to stimulate milk ejection at the mammary gland level under normal oxytocin secretion is known as peripheral inhibition of milk ejection. It is brought about either through the blocking of oxytocin receptor in the mammary gland or by the ability of oxytocin to reach mammary gland through the systemic circulation. Catecholamines are important mediators of peripheral inhibition of milk ejection. There is a close association between oxytocin-containing neuron and sympathetic nervous system in brain and the effect of oxytocin sympathetic nervous system in the mammary gland. The arteries, arterioles, milk duct system, and smooth muscles are having sympathetic innervations. α-adrenoceptor agonists stimulate the contraction of these muscle whereas β-adrenoceptor agonists and α-adrenergic antagonists cause the relaxation of these muscles. Earlier it was believed that sympathetic stimulation inhibits blood flow in the udder by vasoconstriction and oxytocin hardly reaches myoepithelial cells. Administration of an α-adrenergic agonist increased the teat length without reducing the milk flow in the cistern indicating that the contraction of the vascular smooth muscles of the udder are not responsible for the inhibition of milk flow. Another hypothesis suggested that the inhibition of milk ejection by α-adrenergic agonists is possibly due to constriction of teat wall and mammary ducts and enhanced milk flow by β-adrenergic agonists is possibly due to relaxation of the teat sphincter, teat wall, and large mammary ducts.

25.5.5.3 Factors Affecting Milk Ejection Reflex

Suckling stimulus: Both suckling stimulus and machine milking have a positive effect on the milk ejection reflex; however, suckling stimulus is stronger compared to machine milking. The suckling stimulus has a potent effect on milk ejection in cows with previous suckling experience.

Presence of calf: The release of oxytocin is more when the milking is performed in presence of calf. But immediate removal of calf after postpartum before initiation of milking did not influence oxytocin release. The separation of own calf and suckling by an alien calf also decreases oxytocin secretion.

Relocation: Change in the milking surroundings causes inhibition of milk ejection reflex. Relocation within familiar surroundings also inhibits milk secretion and oxytocin release. The relocation causes emotional stress to the animals and releases β-endorphin which in turn inhibits oxytocin release by the central inhibitory pathway.

Methods of milking: Hand milking induces more oxytocin release compared to machine milking. However, mechanical stimulation is sufficient enough to induce normal oxytocin secretion and milk ejection under normal conditions.

Stress: Stress in the animals by any means results in inhibition of milk ejection. Stress-induced inhibition of milk ejection is thought to be mediated by adrenaline. Some authors suggested that inhibition of milk ejection upon stress is mediated through the release of ACTH and cortisol release in rat but machine and hand milking or even suckling stimulate the release of cortisol in cows. Thus, its inhibitory role in milk ejection is not well established.

Learning Outcomes

- **Mammary gland**: Mammary glands are the exocrine glands modified from sweat (sudoriferous) gland. From the evolutionary point of view, the mammary gland of bovines is the most advanced form which can produce far more milk than required for a calf. Mature mammary gland consists of udder, teat or nipple, associated ducts, alveoli composed of epithelial secretory cells and supporting tissues. The primary secretory units of the mammary gland are alveoli.
- **Mammogenesis**: The growth of the mammary gland is termed as mammogenesis and occurs through a series of structural and functional development, differentiation, and involution regulated by hormones and growth factors. Mammary secretory tissues are developed from the ectoderm; however, the blood and lymph vessels, connective tissue, fat pad, and smooth muscles are derived from the mesoderm. Mammogenesis in female can be broadly classified into five stages namely prenatal, prepubertal, postpubertal, pregnancy, and early lactation. Majority of mammary growth occurs around gestation and lactation. Hormones like estrogen, progesterone, placental lactogen, glucocorticoids, oxytocin, and insulin like growth factors play pivotal role in mammogenesis.
- **Lactogenesis**: Lactogenesis is the biological process of onset of milk secretion which includes the enzymatic and cytological differentiation of mammary alveolar cells in early pregnancy to full lactation after parturition. Lactogenesis comprises of two stage process; appearance of pre-colostrum (stage-I) and onset of copious milk secretion at parturition (stage-II). The precursors for milk synthesis are either derived directly from blood or synthesized de novo. Prolactin, growth hormone, insulin, estrogen, progesterone, glucocorticoids, thyroid hormones, and prostaglandins regulate galactopoiesis. It is a neuroendocrine reflex leading to passive withdrawal of milk from alveoli, cistern, and ducts of mammary glands called milk ejection reflex.
- **Galactopoiesis**: Galactopoiesis can be defined as the maintenance of established lactation. In cattle, milk production follows a dynamic curve with a rapid accelerating phase and reached peak around 6 weeks then decline till the end of lactation. The typical lactation curve of bovines can be split into early (day 1 to 100 days), mid (101–200 days), and late (201–300 days) followed by a dry period of 65 days. Milk yield and composition vary with species, breed, stages of lactation, and nutritional status of the animals.
- **Mammary gland involution**: Involution is a biological process by which the lactating mammary gland undergoes a series of tissue remodeling processes after cessation of milk secretion to restore into a virgin like state mediated through a series of events involving apoptosis and tissue remodeling.

Exercises

Objective Questions

Q1. The main supporting system of the udder in cow is ___________.

Q2. Gland cistern of a cow can store ___________________ mL of milk.

Q3. What is the primary function of Furstenberg's rosette?

Q4. What is the common name of subcutaneous abdominal vein?

Q5. Synthesis and processing of dimeric of IgA in monogastric animals is facilitated by which cellular component of mammary gland?

Q6. The prenatal development of mammary gland of bovines starts around ________ days of embryonic life.

Q7. ___________________ hormone helps in mammary ductal growth and ___________________ hormone helps in mammary lobulo-alveolar growth.

Q8. What facilitates the glucose sparing action in the mammary gland of cattle?

Q9. Specialized cell fragments involved in triglyceride synthesis in goat mammary gland is known as _________________.

Q10. A2 milk is found in _________________ cattle.

Subjective Questions

Q1. "Fear and excitement inhibit milk ejection reflex" justify the statement.
Q2. What is the difference between allometric and isometric growth of the mammary gland?
Q3. What is feedback inhibition of lactation?
Q4. How does milk SCC help to indicate intramammary infections?
Q5. How does the exposure of light facilitate galactopoietic effect?
Q6. Briefly describe the mechanism of milk fat secretion.
Q7. Why milk ejection reflex is called neuroendocrine reflex?
Q8. What is the role of myoepithelial cells?
Q9. What is lactogenic hormone complex?
Q10. Briefly describe the partitioning nutrients around lactogenesis?

Answer to Objective Questions

A1. Median suspensory ligament
A2. 100–400 mL
A3. To provide local defense against pathogen by recruiting leukocytes especially lymphocytes and plasma cells
A4. Milk vein
A5. Mammary epithelial cells
A6. 32 days
A7. Estrogen, progesterone
A8. Absence of citrate cleavage enzymes essential for generation of cytoplasmic acetyl CoA from glucose
A9. Christiesomes
A10. Zebu cattle

Keywords for Subjective Questions

A1. Catecholamines, peripheral inhibition
A2. Body growth, mammary growth, relative proportion
A3. Local inhibitory glycoprotein, milk secretion
A4. Infection, leukocyte recruitment, higher SCC
A5. Photoperiod, IGF, mammary epithelial growth
A6. Cytoplasmic lipid droplets, microlipid droplets, apical vesicle route, secretory vesicle route
A7. Neural stimulation, oxytocin release, milk secretion
A8. Modified smooth muscle, contraction, squeezing of mammary epithelium
A9. Growth hormone, prolactin, insulin
A10. Metabolic adaptation, negative energy balance, glucose-sparing action

Further Reading

Books

Akers RM (2001) Lactation and the mammary gland. Iowa State Press, Ames, IA, p 278
Alais C, Alais C, Godina AL (1985) Ciencia de la leche: principios de técnica lechera. Editorial Reverté, Barcelona
Banerjee GC (1998) Lactation. In: Textbook of animal husbandry, 8th ed. IBH Publishing Co. Pvt. Ltd., New Delhi, pp 340–380
Bauman DE, Davis CL (1974) Biosynthesis of milk fat. In: Larson BL, Smith VR (eds) Lactation: a comprehensive treatise, vol 2. Academic, New York, pp 31–75
Cowie AT (1984) Lactation. In: Austin CR, Short RV (eds) Reproduction in mammals, vol 3, 2nd edn. Cambridge University Press, Cambridge
Cowie AT, Tindal JS (1972) The physiology of lactation, Monographs of the Physiological Society No. 22. Edward Arnold Ltd., London
Dang AK, Singh M (2002) Somatic cell counts of milk. National Dairy Research Institute, Karnal, Haryana
Folley SJ (1956) The physiology and biochemistry of lactation. Oliver and Boyd, Edinburg
Guthrie AH (1989) Introductory nutrition. Times Mirror/Mosby College Publishing, St. Louis, MO
Mehta BM (2015) Chemical composition of milk and milk products. In: Cheung PCK, Mehta BM (eds) Handbook of food chemistry. Springer-Verlag, Berlin
Mukherjee J, Das PK (2019) Immune responses of mammary gland. In: Sar TK (ed) Mastitis: symptoms, triggers and treatment, 1st edn. Nova Science Publishers, Inc., New York, p 89
Nickerson SC, Akers RM (2011) Mammary gland | anatomy. In: Fuquay JW, Fox PF, McSweeney PLH (eds) Encyclopedia of dairy sciences, vol 3, 2nd edn. Academic Press, San Diego, pp 328–337
Schmidt GH (1971) Biology of lactation. W. H. Freeman, San Francisco
Thiagarajan R (2014) Textbook of animal breeding. Satish Serial Publishing House, New Delhi
Turner CW (1952) The mammary gland, 1st edn. Missouri, CO, Lucas Brothers Publishers, p 383
Wilcox CJ (1992) Genetics: basic concept. In: Van Horn HH, Wilcox CJ (eds) Large dairy herd management. ADSA, Champaign, IL, pp 1–8

Research Articles

Alhussien MN, Dang AK (2018) Milk somatic cells, factors influencing their release, future prospects, and practical utility in dairy animals: an overview. Vet World 11(5):562–577
Bauman DE, Vernon RG (1993) Effects of exogenous bovine somatotropin on lactation. Annu Rev Nutr 13:437–461
Belitz HD, Grosch W, Schieberle P (2009) Food chemistry, 4th revised and extended edn. Springer, Berlin, pp 498–545
De K, Mukherjee J, Prasad S, Dang AK (2011) Effect of different physiological stages and managemental practices on milk somatic cell counts of Murrah buffaloes. Buffalo Bull 30(1):72–75
Department of Animal Husbandry and Dairying. Govt. of India, New Delhi, India.
Linzell JL, Peaker M (1971) Mechanism of milk secretion. Physiol Rev 51(3):564–592

Masson PL, Heremans JF (1971) Lactoferrin in milk from different species. Comp Biochem Physiol B 39(1):119–129

Misra SS, Sharma A, Bhattacharya TK, Kumar P, Saha RS (2008) Association of breed and polymorphism of α-s1and α-s2casein genes with milk quality and daily milk and constituent yield traits of buffaloes (*Bubalus bubalis*). Buffalo Bull 27:294–301

Mukherjee J, Varshney N, Chaudhury M, Mohanty AK, Dang AK (2013) Immune response of the mammary gland during different stages of lactation cycle in high versus low yielding Karan Fries crossbred cows. Livestock Sci 154:215–223

Ontsouka CE, Bruckmaier RM, Blum JW (2003) Fractionized milk composition during removal of colostrum and mature milk. J Dairy Sci 86:2005–2011

Puppel K, Gołębiewski M, Grodkowski G, Slósarz J, Kunowska-Slósarz M, Solarczyk P, Łukasiewicz M, Balcerak M, Przysucha T (2019) Composition and factors affecting quality of bovine colostrum: a review. Animals 9:1070–1084

Rainard P, Riollet C (2014) Innate immunity of the bovine mammary gland. Vet Res 37:369–400

Sarkar U, Gupta AK, Sarkar V, Mohanty TK, Raina VS, Prasad S (2006) Factors affecting test day milk yield and milk composition in dairy animals. J Dairy Foods Home Sci 25(2):129–132

Skrzypczak E, Babicz M, Szulc K, Marcisz M, Buczyński JT (2012) The analysis of variability of pH level and somatic cell count (SCC) in the colostrum and milk of Złotnicka white sows. Afr J Biotechnol 11(20):4687–4692

Smith KL, Schanbacher FL (1973) Hormone induced lactation in the bovine. I. Lactational performance following injections of 17-β estradiol and progesterone. J Dairy Sci 56:738–743

Stelwagen K, Carpenter E, Haigh B, Hodgkinson A, Wheeler TT (2009) Immune components of bovine colostrum and milk. J Anim Sci 87(13 Suppl):3–9

TANU Agritech Portal. Tamil Nadu Veterinary and Animal Science University, Chennai, India.

Tucker HA (2000) Hormones, mammary growth, and lactation: a 41-year perspective: symposium: hormonal regulation of milk synthesis. J Dairy Sci 83:874–884

Tucker HA, Petitclerc D, Zinn SA (1984) The influence of photoperiod on body weight gain, body composition, nutrient intake and hormone secretion. J Anim Sci 59:1610–1620

Part IX

Physiology of Growth and Behaviour

Physiology of Growth

26

C. Devaraj, M. R. Reshma Nair, and V. Sejian

Abstract

Growth is a complex biological process which occurs through a series of cellular proliferation, cellular differentiation, and morphogenesis resulting in a progressive increase in the size and weight of an animal. Growth takes place through the processes of increase of cell numbers and increases of cell size. Growth and development of cells, tissues, and organs occur both during prenatal and postnatal life of an organism. Several factors that regulate growth during prenatal periods include heredity, endocrine factors, maternal age and nutrition, sex of the fetus, litter size, and environment. Factors which influence postnatal growth include genetic factors, nutritional, climatic, managemental, and health status. Thus, the knowledge of biological process and factors that influences the growth of meat-producing animals and poultry species is very much essential to maximize the growth, meat and egg production efficiency, and quality of the product yield by the farm animals. However, the growth process may be manipulated by using various growth enhancers, feed additives, and supplementation of growth factors to improve the growth of animals. This chapter provides an overview of the physiological process of growth during various developmental stages of meat-producing animals and details of growth manipulation, growth anomalies, aging, and senescence in domestic farm animals.

C. Devaraj (✉) · M. R. R. Nair · V. Sejian
ICAR-National Institute of Animal Nutrition and Physiology,
Bangalore, Karnataka, India

© The Author(s), under exclusive license to Springer Nature Singapore Pte Ltd. 2023
P. K. Das et al. (eds.), *Textbook of Veterinary Physiology*, https://doi.org/10.1007/978-981-19-9410-4_26

Graphical Abstract

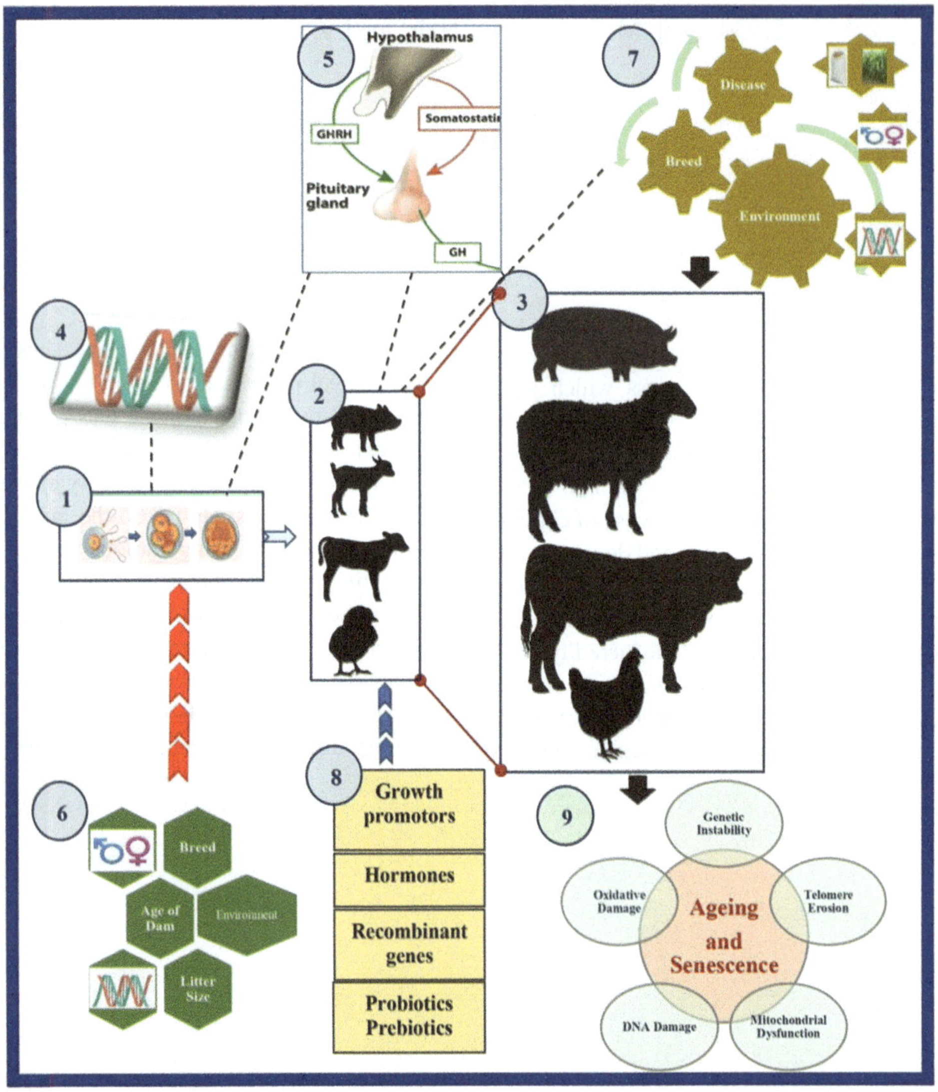

Description of the graphic: Prenatal growth and development process in farm animals (**1**), postnatal growth process in farm animals (**2**), growth of adult farm animals (**3**), both pre and postnatal growth is controlled by various genes (**4**). Growth during prenatal, postnatal, and adult life is regulated by the Hypothalamus-Pituitary-Growth hormone axis (**5**). Growth and developmental process during prenatal life is affected by various factors related to the mother, fetus, and environment (**6**), similarly postnatal and adult growth also affected by various factors (**7**). The growth of meat-producing animals and poultry is being manipulated using various strategies in order to enhance production efficiency (**8**) and finally due to the gradual decline of all the functional characteristics, animals undergoes aging and senescence (**9**). All aspects of the growth of farm animals are elaborated in this chapter

Keywords

Environment · Growth · Hormones · Prenatal growth · Postnatal growth · Senescence

Learning Objectives

- Understanding the growth process in domestic animals.
- Knowing the different factors affecting prenatal and postnatal growth.
- Application of different techniques to enhance growth in meat-producing animals and poultry.
- Information about growth anomalies, aging, and senescence in domestic animals.

26.1 Growth and Development

Growth is defined as a progressive increase in the size and weight of an animal due to the production and accretion of new biochemical units during a specific period of time. It is a fundamental process of all living organisms and is characterized by combination of increase in total cell size and number with protein deposition and differentiation. An increase in the size of a cell, tissue, or organ is called as hypertrophy whereas an increase in the number of cells in a tissue or an organ is known as hyperplasia. Any decrease in the size of a cell in an organ or tissue is called as atrophy. Generally, structural tissues and organs increase in size via cellular hypertrophy and hyperplasia. During prenatal growth period, all cells in tissues or an organ grow by hyperplasia whereas during postnatal period tissues having mitotic ability continue their multiplication throughout life; however, certain cells lose their mitotic ability and grow only by enlargement or incorporation of satellite cells (e.g., nerve and muscle cells).

Development is the process where the fertilized egg undergoes a series of diverse processes (cell division and differentiation) to form a fully functional new organism/animal. Due to the developmental process body shape and form are changed. It is regulated by the genetic makeup of the animal. Growth occurs in three distinct processes, viz., (1) Cell division (mitosis) where cell numbers increased, e.g., fertilized egg into daughter cells, formation of blood cells, hair follicles, and ectoderm cells. (2) Incorporation of substances taken from the environment leading to cell expansion or enlargement, e.g., proteins and minerals. (3) Cell enlargement and differentiation: increase in the size of a cell and changes in cell shape and form, e.g., skeletal tissue and nervous tissue.

When cell numbers increase (hyperplasia) their DNA content also increases and most of the cells contain only one nucleus and each nucleus contains the same amount of DNA. Therefore, the DNA content of a tissue could be used as an index of number of cells found in the tissue. Whereas in hypertrophy, cell size increases but DNA content remains the same. In most of the cells, hypertrophy arises due to increased protein content. Therefore, the protein/DNA ratio is considered as an index of cell size and growth. RNA/protein ratio indicates capacity for protein synthesis. The protein/DNA ratio differs with species, age, tissues, and other factors. Under optimal nutrition conditions protein/DNA ratio is maximum. During a short period of starvation, tissue protein level decreases while no changes in DNA would be observed. When feeding is resumed, cellular protein level increases rapidly to reach the previous protein/DNA ratio. This rapid rate of growth after a period of starvation is called as compensatory growth. It is a process where growth is lesser than normal for some period due to undernutrition, after which the animal gains rapid live weight with the availability of good nutrition.

26.1.1 Growth Pattern in Animals

Growth can be divided into two periods based on the occurrence,

1. Prenatal growth
2. Postnatal growth

26.1.2 Prenatal Growth

The period of growth between conception and birth is called prenatal growth. During this period, fertilized ovum fully transforms into a new organism with defined tissues, organs, and nervous system at birth.

The prenatal period is divided into three periods:

26.1.2.1 Pre-embryonic/Ovum Period

The period from fertilization to implantation is referred as pre-embryonic period or period of ovum. Figure 26.1 describes the various stages of pre-implantation embryonic development. Subsequently, a sequence of coordinated events takes place as follows:

- Fertilization
- Cleavage
- Formation of blastocyst
- Implantation

Fertilization is a process involving fusion of a sperm and ovum leading to the formation of zygote or embryo with a diploid number of chromosomes. After fertilization, the zygote undergoes differentiation and forms different kinds of cells, referred as cytogenesis or histogenesis. This development is followed by a course of transformations to form different organs, which culminates the differentiation. The organization of different cells into different organs, each with a particular structure and function is known as morphogenesis or organogenesis.

Cleavage is the process by which the zygote undergoes repeated mitotic cell divisions and produces progressively smaller daughter cells known as blastomeres. Each blastomere will develop into specific tissues based on their location in the blastocyst. This event occurs when it passes down the oviduct and there will be an increase in total size. The zygote is divided into 16 or more blastomeres which then enter the uterus and form a compact ball of cells called as morula. Fluids accumulate between the cells and form a hollow, ball-shaped cellular structure called a blastocyst. Fluid-filled space is called the blastocoele or blastocystic cavity. Inner cells of the blastocyst constitute the inner cell mass that gives rise to the embryoblast which will become the proper embryo. Whereas, surrounding cells form an outer cell mass that gives trophoblast, which will become the fetal membranes. The trophoblast and part of the inner cell mass form the fetal part of the placenta, amnion, and chorion.

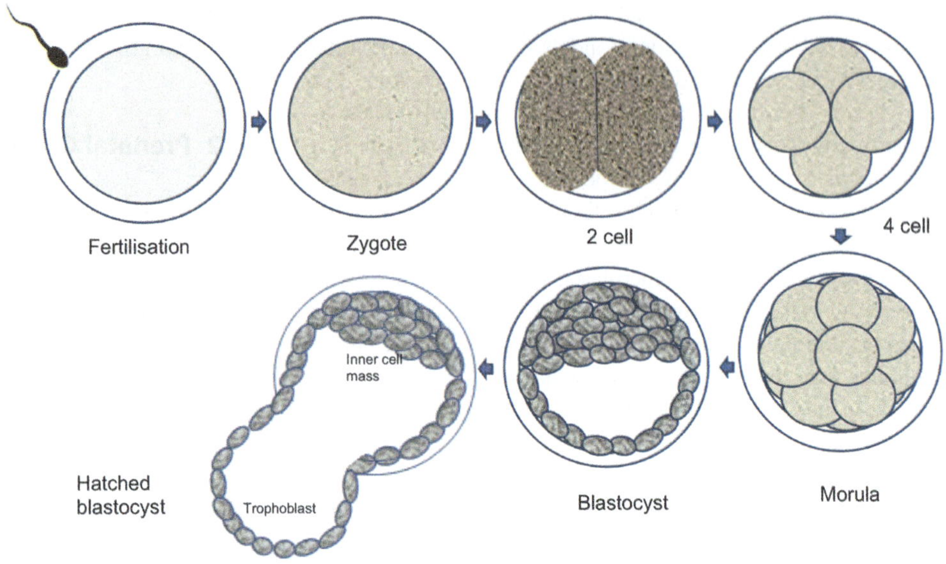

Fig. 26.1 Pre-implantation embryonic development. Fertilized eggs undergo differentiation and form different type of cells and organs. Zygote becomes blastomere through repeated mitotic cell division then forming blastocyst; finally, embryo hatches and grows continuously in the uterus until implantation

Uterine glands are important for implantation and embryo survival because secretions from the uterine glands nourish the blastocyst.

Implantation: The embryo hatches out of the zona pellucida (zona hatching) and continues to grow, freely floating in the uterus until the implantation begins. Trophoblast cells start to pierce between the epithelial cells of the uterus. Attachment of the blastocyst to the wall of the uterus is known as implantation, in which the embryo is called a fetus, that will develop into a full-term young one at the end of the pregnancy period. Implantation that occurs outside the uterus is called as ectopic pregnancy.

26.1.2.2 Embryonic Period

This period is usually considered as the most important stage as major tissues, organs, and their system are developed during this phase. This period is characterized by rapid cell division, cell determination, and differentiation. Cell determination is the process by which cells follow a specific pathway to form a particular tissue. Thus, it moves to a specific region of the embryo where they undergo further transformation and are differentiated into particular tissues or organs. Cell differentiation is defined as the transformation of unspecialized cells into a specific cell type. During this period embryo undergoes a series of successive changes for the ultimate formation of the fetus at birth.

After implantation, the embryo undergoes rapid elongation and the trophoblast differentiates into cytotrophoblast (mitotically active single nucleus cells) and syncytiotrophoblast (a rapidly growing multinucleated mass). The growing embryo is surrounded by the amnion. The blastomeres of the blastula are rearranged and invaginated at one end resulting in the formation of a three-layered blastocyst called the gastrula. Three germ layers namely, ectoderm (outer), endoderm (inner), and mesoderm (middle) layers develop and give rise to the tissues and organs of the embryo.

The ectoderm give rise to the structures and organ such as skin, epithelium of the oral and nasal cavities, central nervous system, peripheral nervous system, mammary glands, pituitary gland, sweat glands, and teeth. Mesodermal layers give rise to muscle, bones, connective tissue, and the circulatory, urinary and genital systems. Endoderm becomes the gastrointestinal tract, respiratory systems, and bladder. It also forms the mucosal lining and glands of the digestive and respiratory systems.

26.1.2.3 Fetal Period

This period starts when cell proliferation and organ systems of the embryo complete its development in rudimentary state and ends at birth. In bovine fetal period starts from 9th week of pregnancy and ends at birth. This period is characterized by a period of fast growth and development, and maturation of all organs and systems. Initially, fetal length increases rapidly than its weight, in later stages length increases slowly but weight increases rapidly. During this period size of the fetus increased dramatically which is determined by various factors such as genetic factors, environmental factors, age of the mother, nutrition, and management. Muscle fiber number is fixed by the time of birth. During this period all the organs and systems are fully functional to support itself and for its survival after birth.

26.1.3 Postnatal Growth Period

The period of growth from birth to death is referred as postnatal growth. This period is categorized as young and

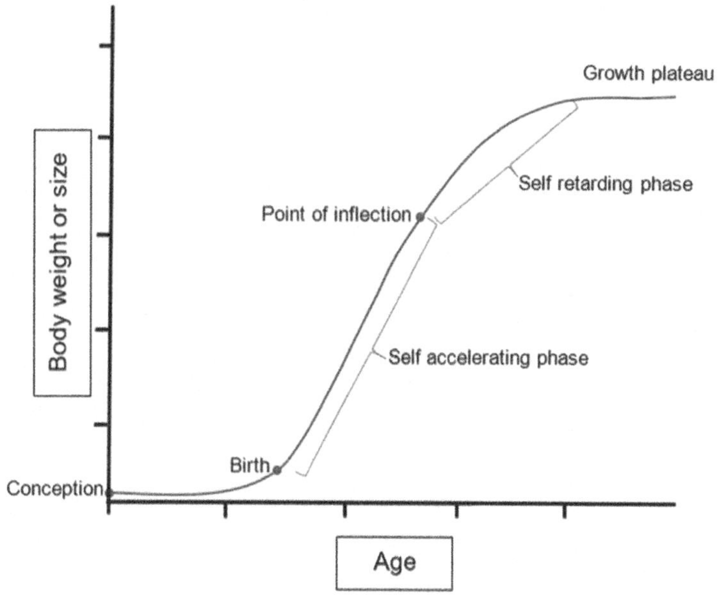

Fig. 26.2 Growth curve of farm animals. Sigmoid growth curve consists of rapid self-accelerating phase and slow retarding phase. Once animal reaches the point of inflection growth rate starts to slow down

adult period. Further young stage is classified as preweaning, weaning, and postweaning and adult period is divided into prepubertal, pubertal, and postpubertal period. Duration of this period greatly varies depending on the species.

Ex. mouse—2 years
Sheep/goat—15 years
Cattle—30 years
Elephants and humans—60 years

During this period, the growth of mainly muscle, bone, and fat tissues takes place. The rate of growth depends on the age of the animal, type of tissue, genetic makeup, and plane of nutrition.

26.1.4 Growth Curve

Initially, the animal grows slowly, then undergo a period of rapid growth followed by slow or stagnant growth when adult size is attained. Plotting the live body weight of an animal against the age yields growth curve. Generally, the growth curve of all the species follows a similar pattern of sigmoid or S-shape. Figure 26.2 describes the growth curve in farm animals.

The growth curve contains two phases:

26.1.4.1 Self-Accelerating Phase

During this phase, the growth rate accelerated to the maximum until mature weight is reached. It is the steep slope of the curve. There are two forces acting on the growth rate of an organism which determine the shape of the growth curve.

26.1.4.1.1 Growth Accelerating Force

Growth accelerating force exists in all cells in the body and is due to the increase in the number of mitotic cells (hyperplasia), increase in size of the cells (hypertrophy), and inclusion of substances like proteins, vitamins, and minerals.

26.1.4.1.2 Growth Retarding or Decelerating Force

Due to lack of space or food supply for tissue growth, the growth rate slowdown, and from this point onwards growth pattern shows stagnation or reduction. The growth retarding force is found in the surrounding cells' environment. During the initial linear phase of growth, the two opposing forces are more or less in balance.

26.1.4.2 Self-Decelerating Phase/Retarding Phase

It is the decreasing slope of the curve. During the retarding phase, the rate of growth decreases and stops. It is the final phase of growth curve and starts when the animal attains maturity. There is an in-built mechanism that controls further growth by means of reducing feed intake. There is a regulation of dry matter intake and reduction in the body weight gain until the maintenance requirement is balanced due to the action of somatostatin.

A point at which the accelerating and decelerating phase meet, the growth rate starts to decrease. This point is known as the **point of inflection**. At this point, puberty occurs in all species and is also known as the **point of pubertal inflection**. This point specifies the time of maximal growth, pubertal age, starts increased mortality rate and point of reference to decide the age equivalents of various domestic animals. This point starts at 14 years in humans, 9–12 months in cattle, and 6–7 months in sheep and goat.

26.1.4.3 Negative Growth Phase

In old age, all the parts of the body start to degenerate and this phase is known as negative growth phase.

> **Know More**............
> - Growth decides the size and shape of the animals.
> - Animal growth is a complex biological process determined by a variety of factors.
> - Growth curve is sigmoid shape in all domestic animals.
> - Marbling is the intramuscular fat deposition that completes when animals reach maturity.

26.2 Regulation of Growth

The complex process of growth is controlled by the interaction between genetic factors, endocrine system, and nutritional and environmental factors. Among these regulatory factors, endocrine system is considered as the primary system that controls the growth in animals. Somatotroph axis (growth hormone, GH releasing hormone, somatostatin) and thyrotroph axis (thyrotropin releasing hormone, thyroid hormone stimulating hormone) regulate the growth in farm animals. During early embryonic period, most of the skeletal muscles develop from mesodermal progenitor cells and are controlled by a range of positive (e.g., Insulin-like growth factor-1 (IGF-1)) and negative (myostatin) signals. Growth hormone (GH), thyroid hormones, glucocorticoids, insulin, prolactin, and gonadal steroids influence growth in animals during various stages of life. Growth rate and feed conversion ratio increased in calves, lambs, and pigs subjected to exogenous growth hormone (GH) injection. In pigs, giving GH during early pregnancy increases fetal IGF and increases myofiber number at birth. During postnatal period, the growth rate is regulated by the genetic makeup of the animal and also by the environment. Hypothalamus controls the secretion of pituitary hormones which affects growth such as GH, thyroid stimulating hormone, gonadotropins, prolactin, and adrenocorticotropic hormone (ACTH). Hypothalamus is also a major regulator of feed intake which will influence the growth of an animal by increasing the intake of available nutrients.

The GH is the major hormone regulating growth and its release is controlled by GH releasing factor and somatostatin, both are secreted from the hypothalamus. GH exerts its effects on the growth of muscles and connective tissues through IGF from the liver. The IGF-1 acts on muscle growth by stimulating the differentiation and proliferation of myoblasts and increase amino acid uptake and protein synthesis with the help of thyroid hormones, insulin, and anabolic steroids. GH also exerts its direct metabolic effects on adipose tissues leading to lipolysis. Injection of purified chicken GH in the chicken increased the body weight in the treatment groups compared to the control; however, the increase in the body weight was transient and the chicken GH's effect diminished at the end of the study. Chicken subjected to exogenous intravenous administration of human pancreatic GH releasing factor showed a significant increase in the body weight during early stages; however, the bodyweight diminished similar to the GH. On the contrary, the effect of thyroid hormones (T3 and T4) impact on the body weight of chicken was not transient and showed significant body weight gain. GH injection in pigs increased the GH receptors in the liver indicating the presence of higher GH in blood.

Insulin increases growth hormone receptor in the adipose and hepatic tissue in dairy cows during periparturient stage. Insulin stimulates protein synthesis in the skeletal muscle of pigs, during the early postnatal period. Thus, insulin has an important role in protein deposition in the skeletal muscle of growing animals. Insulin exerts its action in the in vitro muscle cells by binding with IGF-I receptors. Infusion of balanced amino acid, which is the precursor of protein synthesis in the fast growing pigs showed increased protein synthesis in the skeletal muscle of pigs. Similarly, protein diet feeding also increased protein synthesis in the skeletal muscle of suckling lambs. In the skeletal muscle, GH treatment increased protein deposition, and thus enhancing the growth of muscles.

Under in vitro conditions in different cell lines GH and IGF-I increase the differentiation, proliferation, replication, and lipogenic enzyme activity of pre-adipocytes. Similarly, the T3 hormones under in vitro conditions increase lipogenic enzyme activity, fatty acid synthase synthesis, and increased pre-adipocyte proliferation and differentiation. Glucocorticoids accelerate the differentiation of adipocyte tissues.

Fetal growth rate is not affected by the size of uterus during early stages of pregnancy. However, there is a correlation between fetal growth and placental size during later stages of pregnancy. Placenta facilitates the transport of nutrients, gases, and metabolic waste products between mother and fetal circulations. Apart from this, placenta provides a large surface area for nutrient exchange and syntheses growth factors, nutrients, and immune modulators which are necessary for blood flow and delivery of nutrients to the fetus. Maternal nutrition is the most important factor that immensely contributes to growth and development of the fetus during pregnancy and maternal health. Nutrient availability during pregnancy decides the weight and size of the fetus thus affecting the growth of fetus. It is well known that maternal nutrient restriction during pregnancy leads to fetus that are smaller at birth than fetuses from normally fed animals.

26.2.1 Biochemical and Genetic Determinants of Growth

Normal body growth requires the integrated function of the endocrine system, metabolic, and other growth factors involved in the hypothalamo-pituitary growth axis. The growth of animals is mainly determined by genetic factors as well as non-genetic factors. The genetic factors include breed, feed conversion efficiency, and disease resistance. Several genes that control the growth and development of animals. Growth and body composition differ between breeds of all farm animals. The rate of maturation and body weight gain varies between different breeds. Feed conversion efficiency of individual animals influences their energy utilization and growth. Disease resistance traits of an animal reduce the impact of disease by preventing infection or by reducing the further growth of the pathogen in the host.

An adequate and continuous supply of protein is required for normal growth and maintenance of the animals. Proteins are made of various amino acids, including which are not synthesized in the animal body called as essential amino acids. Essential amino acids must be provided in diets for optimal growth and development. Inadequate supply of protein causes retardation of growth whereas excessive protein supply cannot increase protein synthesis beyond the genetic potential of the animal. Dietary fats are a good source of energy for animals. Some fatty acids are essential for growth. Dietary fats are digested into fatty acids, absorbed and deposited as body fat. Body fat composition varies between breeds and species.

26.3 Factors Affecting Growth

Several factors control the growth of an animal during prenatal and postnatal life periods that may influence their productive performances. A few of these factors are listed below. Figure 26.3 describes the various factors that influence prenatal growth in farm animals.

26.3.1 During Prenatal Period

26.3.1.1 Heredity

Fetal growth is mostly determined by its own genotype during the early and mid-pregnancy periods whereas maternal genotype influences fetal growth during late pregnancy. There are variations among the species, breeds, and individuals. The contribution of the maternal genotype is greater than the paternal genotype to the genetic variability.

26.3.1.2 Hormones

Fetus secretes several hormones and its functional display is based on the receptors present in the target system.

26.3.1.2.1 Pituitary Hormones

In some species growth hormone is biologically active where the young ones born with their eyes open and stand up immediately after birth are known as precocial young ones, e.g., cow, mare, sheep, and goats.

The growth hormone is physiologically inactive in some species and the immature young ones are born with closed eyes, e.g., dog, cat, and lab animals.

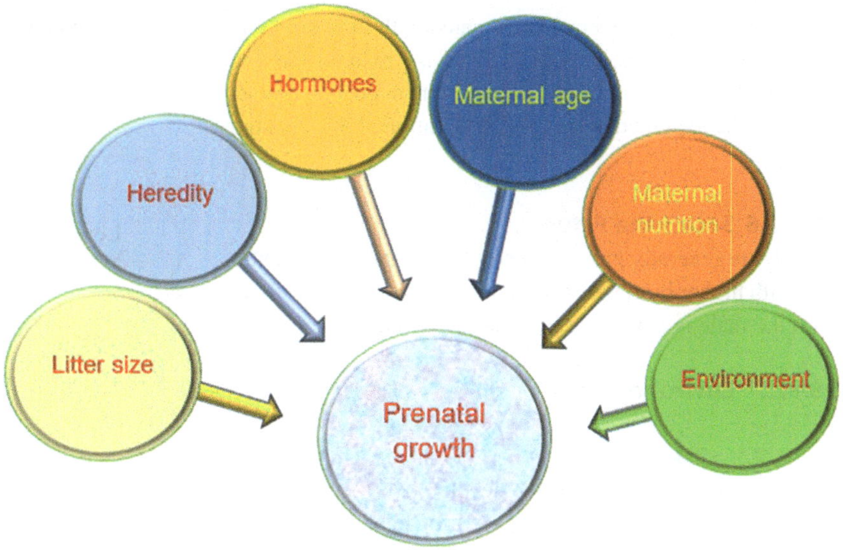

Fig. 26.3 Factors affecting prenatal growth in farm animals. Various factors such as genetic background, maternal age, nutrition, endocrine, and environmental factors affect the growth of the fetus during prenatal period

26.3.1.2.2 Thyroid Hormones

The fetal thyroid hormones such as T3 and T4 are necessary for morphogenesis, differentiation, and growth of the fetus.

26.3.1.2.3 Gonadal Hormones

The fetal gonadal hormones (estrogen and progesterone) are necessary for the development of the sexes. In males, androgen (testosterone) synthesized by the fetal testes is required for testicular descent into the scrotum.

26.3.1.3 Maternal Age

Fetal growth is directly associated with mother size. The growth of the fetus is faster in a large dam than a smaller dam. Maternal age also impacts the size of the fetus. As the age of the mother increases, the size of the fetus also increases. But in the older animals with obesity, the young ones born are smaller because of restricted space in the gravid uterus. Fetal size differs with the length of the gestation period. During the long gestation period, chances are more for fetal growth than in the short gestation period.

26.3.1.4 Maternal Nutrition

Fetal growth depends on the nutritional status of the dam. Fetal growth rate is reduced when the dam nutrition is poor. Although the fetus will grow slowly and have a less birth weight, nutritional deficiency of vitamins and minerals will cause developmental abnormalities in the fetus.

Nutrient restriction affects organ development differently in the fetus. The nervous system, skeletal, and heart growth are least affected, whereas the development of lungs, kidneys, and muscles are moderately affected. Poor nutrition severely affects the development of the skin, spleen, thymus, and liver.

26.3.1.5 Litter Size

Fetal birth weight is inversely associated with litter size, e.g., in polytocous species, increased litter size reduces the fetal growth rate due to variations in the nutrient supply, mass, and action of the fetal membrane and the duration of the gestation period.

26.3.1.6 Sex of the Fetus

Male fetus generally grows at a rapid rate than their counterpart with higher birth weight than the female one. If twins are born the male is always larger than the female.

26.3.1.7 Environment

26.3.1.7.1 Placenta

The placenta decides fetal development since it is transporting all the nutrients required for fetal growth. Fetal growth is positively correlated to the placenta size. Fetus size is smaller in the small size placenta because of retarded fetal growth. Nutrient supply by the placenta to multiple litters is lesser compared to single litter which affects the growth of the fetus.

26.3.1.7.2 Temperature

Temperature directly affects the fetus growth since it is influencing the metabolic activity of the dam as well as the fetus. Either heat or cold temperature will decrease the fetus growth depending on the intensity of heat stress. Especially chronic heat exposure restricts fetal growth due to impaired placental development during the early to mid-gestation period.

26.3.2 During Postnatal Period

Figure 26.4 describes the various factors that influence postnatal growth in farm animals.

26.3.2.1 Sex

Differential rate of growth and development as well as tissue composition are associated with the sex of animals. Generally, male grow faster and attain better weight gain than female of equivalent ages even if both the sexes are provided with same nutrition. Male animals are more muscular and have less fat than female.

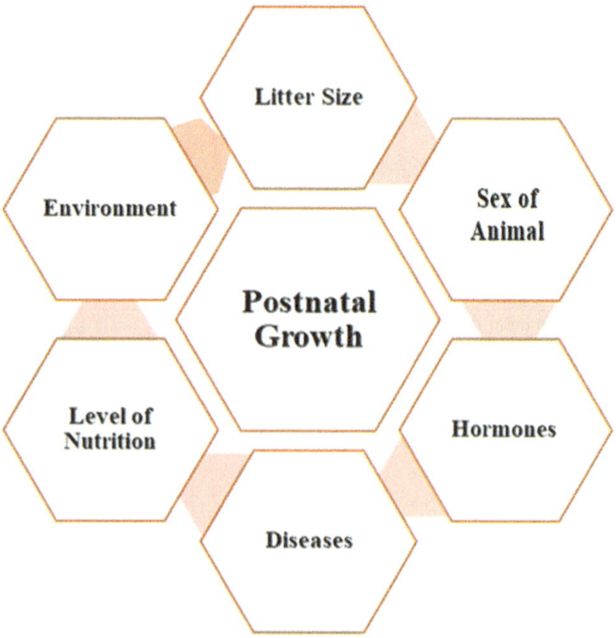

Fig. 26.4 Factors influencing postnatal growth in farm animals. Various factors such as plane of nutrition, sex of animal, litter size, disease, endocrine, and environmental factors affect the growth of animal during postnatal period

26.3.2.2 Litter Size

Larger litter size leads to low birth weight with less growth performance of the newborns. Single litter has better birth weight than twins and multiple litters. There is a chance of shortage of milk supply by the dam in larger litter size.

26.3.2.3 Level of Nutrition

The influence of plane of nutrition is very important because of its relation to the economics of milk and meat production. All animals require a level of nutrition for the normal growth and development of muscle and fat tissues. The carcass composition of meat animals are significantly affected by the quality and quantity of nutrient supplied to them. Poor nutrition during the initial phase of growth leads to a stunted growth. Adequate quantities of protein, vitamins especially vitamin B complex, A, and C are required for the optimal growth of an animal.

26.3.2.4 Hormones

The endocrine system regulates the growth of various tissues and organs, and partition the energy required for various physiological process related to growth and development. Both endogenous and exogenous hormones influence the growth rate or body composition by promoting the translation, transcription, and amino acid uptake for protein synthesis.

26.3.2.4.1 Growth Hormone (GH)

It stimulates the uptake of amino acids, synthesis of protein, and decreases the degradation of protein in all tissues. It produces lean tissue growth in animals and plays a role in hormone-stimulated protein deposit, nutrient partitions, and the utilization of lean tissue growth and also from fat deposition. Exogenous administration of GH stimulates growth and feed conversion ratio, and decreases the fat content in carcasses of pigs, cattle, and lambs.

26.3.2.4.2 Insulin-Like Growth Factor 1 (IGF-1)

It is necessary for bone growth since it stimulates chondrocyte growth. It is also playing an important role in protein, glucose, and fat metabolism.

26.3.2.4.3 Thyroid Hormones

It is required by all animals for their normal growth. Its deficiency causes retarded growth.

26.3.2.4.4 Insulin

It is a very important hormone involved in the growth and development of muscle tissues. It stimulates the uptake of glucose and transport of amino acids required for normal growth. It also reduces protein degradation.

The gonadal hormones play an important role in the growth and development of the animal. Androgens stimulate muscle growth and development by increasing protein synthesis and reducing fat deposition. Some tissues are more sensitive than others to androgen, depending on their action in reproduction.

26.3.2.5 Environmental Factors

26.3.2.5.1 Environment

Rearing environmental conditions have a marked influence on growth rate and even on body composition. The subject of heat regulation in farm animals has a wide economic significance. Climate influences growth indirectly through the availability of fodder and water resource and directly have adverse effects on the animal itself. In tropical countries, mostly poor quality and quantity of fodder are available during summer that causes a reduced growth rate. However, in the temperate zone, fodder scarcity due to winter leads to decreased growth rate.

26.3.2.5.2 Photoperiod

Impact of photoperiod on animal growth varies among the species. Long photoperiods stimulate the lean body growth in calves, lambs, and kids and also hasten the onset of puberty.

26.3.2.6 Diseases

Disease condition hinders the growth rate of animals since it affects the dry matter intake, digestion, and absorption of nutrients.

26.4 Growth in Meat-Producing Animals and Poultry

26.4.1 Growth of Organs

In general, organs and tissues grow at different rates relative to each other as well as to the whole body. Even though their growth rate is genetically determined, several other factors, viz., nutrients supply, endocrine and growth factors also control their growth. Generally, growth pattern of all tissues and organs follows the sigmoid or S-shaped growth curves.

26.4.2 Isometric Growth

Any organ or part of the body that grows at a same rate as that of the general body growth is known as isometric growth.

26.4.3 Allometric Growth

Any organ or body part that grows at a different rate than the general body growth is known as allometric growth.

The growth of tissues or organs in the body follows both isometric and allometric growth patterns during various stages of life. During the prenatal growth period, some systems developed more rapidly than others depending on the functional needs of the body and the developmental stage of the animal. For example, during prenatal phase nervous system developed at maximum rate followed by the circulatory system, connective tissues, bones, muscle, and finally the adipose tissues. However, during the postnatal growth period, initially loins grow rapidly followed by the pelvis, thorax, head, and finally the legs.

In general, growth and development of meat animals consist both increase in total body cells and differentiation of these cells. Due to differentiation, some tissues, organs, or parts of the body grow at different rates and mature at different time from that of the whole body which results in a change in shape and size of the meat-producing animals. For example, the head grows faster in early life and proportionally larger, while other parts like limbs make rapid growth later and form a larger proportion of the body weight. The shape, size, and composition of animals vary continuously during the growth period. The major body tissues show differential growth patterns in the order as follows: the skeleton, muscle, and fat. Rate of growth of different tissues in animals is as follows: nervous system followed by bone then fat. Fat is the last tissue to mature in the body. There is a difference in maturation exists within the fat tissues based on their location of deposition. Orders of maturation of various fat tissues in animals are as follows: perinephric fat, intermuscular fat followed by subcutaneous fat, and intramuscular fat.

Different sequence of maturity of tissues or organs has major implications for the growth of meat animals with desirable carcass quality. For lean meat production, animals may be killed before much fat deposition has been taking place. There is a variation in the composition of animals as it grows due to the differential growth rate of different parts of the body. Muscle, bone, and fat are the major tissues of carcass and they show differential growth rate during the development of meat animals. Differential growth of these tissues determines carcass composition and quality. Age and weight at slaughter are also some of the important factors that influence the carcass composition. As age and weight increase, muscle and bone percentage decrease and fat percentage increases.

26.4.4 Muscle Growth

Muscle is the major body tissue by weight basis that increases in size after birth by an increase in the size of myoblast cells. Muscle fibers are formed by the fusion of single nucleated cells called myoblasts which in turn form multinucleated myotubes. These are immature muscle fibers. They elongate into mature fibers. Only a few cells have retained their replication potential called satellite cells. The muscle fibers constitute more than three-fourth of the muscle mass and also its size and number determine the muscle mass of the animal. Muscle fibers number are genetically determined and fixed at the time of birth. Hence, the postnatal muscle fiber weight is increased due to increase in its diameter and length. The size of the muscle fiber is affected by both age and nutrition. During early stages of growth, muscle fiber diameter increases significantly then progressively slow down till mature size is reached. Animals having larger number of muscle fiber at birth grow faster than animals with smaller numbers. Once the muscle fiber reaches its maximum growth further weight gain must be due to fat deposition. Therefore, for those animals that attained maturity, the increase in muscle weight increases without an increase in muscle fiber size could be partially clarified on the basis of intramuscular fat deposition.

26.4.5 Protein Deposition

Maintenance of protein concentration in the body involves controlled regulation of protein turnover. The protein deposition in farm animals depends on the balance between the rates of protein synthesis and degradation in tissues. This process is affected by various factors such as nutritional status, tissue types, environmental conditions, and general physiological status. Growth rate have been correlated with different rate of protein synthesis. Protein synthesis in the body is an energy-consuming process and a considerable amount of energy is used to form the activated amino acids to be linked together. Protein-containing tissues are being continuously turned over. For one unit of net protein deposition, about 5 units of protein are being synthesized. About 15–33% of energy needed for maintenance is being diverted towards the maintenance of the existing protein and synthesis of new protein. Protein deposition in animals has negatively correlated with growth and body weight gain. Various factors regulate protein deposition in animals. Endocrine status of an animal is a major factor that governs the growth rate and the relative rates of protein and fat deposition. Normally the rate of protein deposition is positively linked with the rate of protein degradation due to the consequence of the need for remodeling of the tissues as it grows. The balance between the synthesis and degradation of protein gives fine control to the overall rate of protein deposition. The rate of protein deposition also depends on the protein quality, energy content of the diet, and protein intake by the animals. Dietary protein and energy intake increases protein deposition since it requires energy and also dietary protein can contribute to the energy demands of the animal. Supplementation of additional energy resulted in an increase in protein synthesis and decrease in protein degradation that ultimately leads to enhanced net protein deposition. Additional protein supplementation with constant energy caused an increase in both protein synthesis and protein degradation which resulted in a smaller net increment of protein deposition. Protein metabolism is influenced by the hormonal status of the animal. Growth hormone concentration positively correlates to protein deposition. In contrast, catabolic hormones corticosteroid decreases the protein synthesis and deposition in muscle tissues and increases the degradation of muscle protein. The concentration of insulin favors the protein synthesis and amino acids uptake by the cells lead to protein deposition. Male hormone androgen tends to stimulate protein deposition in animals. Estrogen tends to stimulate protein synthesis and deposition indirectly by stimulating the secretion of growth hormones or thyroid hormones.

26.4.6 Fat Deposition

Fat cells are called adipocytes. Its production and deposition are normal during growth and development of the meat animals. The rate of fat deposition has shown great variation in different parts of the body. Their size differs in different fat depots and different stages of growth. Fat tissues are widely distributed throughout the body and vary between species. Based on their deposition in adult meat animals it is classified as

(a) Subcutaneous (between the skin and muscle) fat
(b) Intermuscular fat (between muscles)
(c) Perinephric fat (surrounding the kidney)
(d) Intramuscular fat (within the muscle/between the muscle fiber) also called marbling
(e) Omental fat (surrounding the stomach)
(f) Mesenteric fat (surrounding the mesentery of intestines)

Carcass quality is determined by the amount and distribution of fat. Intramuscular fat in the carcass is desirable since it gives flavor and juiciness to the meat whereas fat accumulation in the body cavity and surrounding the organs decreases the carcass quality.

26.4.7 Growth in Birds

In general, growth is a complicated process that is regulated by various genes, hormones, nutrients, and environmental factors. Growth in avian species is divided into five phases such as embryonic, post-hatch, prepubertal, pubertal, and adult. Time required for an egg to hatch differs among the species. Fertilized chicken egg takes 21 days for hatching. However, the sequence of developmental process is very similar to all species of birds. Growth patterns are typically s-shaped like other livestock species. After chicks are hatched, they required supply of water and protein-rich feed, and most importantly warm environment. Visible growth changes in growing chicks will be evident during the 5–6 weeks of age. During this period the teenage (prepubertal) female is referred as pullet whereas male is called as cockerel. Gender differences are more evident between the weeks 7 and 15. During the prepubertal stage birds require adequate amount of protein, vitamins, minerals, and electrolytes for optimal growth and production. Around 15–18th weeks, birds start to lay eggs. Once the pullet started laying egg, female birds become hen and male birds become rooster. During the egg-laying stage from 18 to 72 weeks, birds requires essential amino acids, vitamins, and minerals, particularly at right ratios of calcium and phosphorous for improved growth performance, good quality egg, and strong bones. Growth rates vary among the species and also within the species depending on the feed availability and surrounding environment. Bigger size birds grow more slowly than smaller birds. Around 72 weeks of age, molting may begin. During molting stage, protein supplement is very much required for the regrowth of feathers.

26.5 Growth Manipulation

Enhancement of growth rate is possible by the use of various growth promotors, feed additives, and growth factors which will reduce the production cost of the animals.

26.5.1 Growth Promotors

Growth promoters are substances that are used in feed as additives/supplements or injections as a medicine to increase animal production by increasing body weight and product output. There are different types of growth promotors used in the animal and poultry industry to augment growth and production. These can be classified based on their mode of action and by the nature of the agents. These growth promotors are divided into gut flora modifiers (antibiotics, antibacterial, probiotics, prebiotics, nutraceuticals, organic acids), nutrient modifiers (enzymes, vitamins, and minerals), and physiological regulators (hormones) that have been widely used to promote health and production of animals.

Antibiotics are given at sub-therapeutic dosage to stabilize the gut microbes for enhancing growth performance, feed conversion, and prevention of some disease occurrence. Frequently used as growth stimulants in domestic animals include bacitracin, penicillin, flavomycin, chlortetracycline, oxyteracycline, doxycycline, colistin sulfate, avilamycin, erythromycin, aureomycin, tiamulin, lincomycin, furazolidone, enrofloxacin, and neomycin sulfate.

26.5.2 Hormones

Hormonal growth promotors are widely used in meat-producing animals for better performance in growth and enhancing feed conversion efficiency. Natural (estradiol, progesterone, and testosterone) as well as synthetic hormones (melengestrol acetate, trenbolone acetate, and zeranol) are used as implants and feed additives. The main aim of the use of these growth promotors are (1) to improve meat quality; (2) to enhance feed conversion efficiency; and (3) to increase milk production. Hormonal growth promotors increase growth by stimulating appetite, improving feed efficiency, and increase rate of weight gain.

Major hormones that regulate growth and development during prenatal and postnatal life are growth hormone, Insulin-like growth factors (IGFs), insulin, thyroid hormones, glucocorticoids, prolactin, and gonadal steroids (androgens and estrogens). Generally, the role of gonadal hormones in growth and development has been studied by evaluating the effect of hormone deficiency after endocrine tissue ablation as well as effect of excess hormones administered to animals in vivo. Fetus secretes several hormones and its functional display based on the receptors present in the target tissue of fetus. Fetal hormones may affect its own growth and also regulates maternal function to assure suitable environmental conditions for its continued growth and development till the parturition. These hormones also stimulate mammary gland growth (mammogenesis), lactogenesis, and galactopoieses either directly or indirectly by interacting with maternal endocrine system for its survival during the postnatal period.

26.5.2.1 Growth Hormone (GH)

GH is the most important hormone affecting the growth and development. It is biologically active in some species and their young ones are born with opened eyes and stand up immediately after birth. These young ones are known as precocial young ones, e.g., cow, mare, sheep, and goats. Growth hormone physiologically inactive in some species and their immature young ones are born with their eyes closed, e.g., dog, cat, and laboratory animals. Exogenous administration of GH has been reported to enhance the growth rate and feed conversion efficiency of calves, pigs, and lambs; however, it is not producing any effects on fetal growth during prenatal period since a smaller number of GH receptors in fetal life and increases rapidly after delivery. In dairy animals, administration of GH reportedly increases milk production whereas in pigs and lambs carcass composition shifted towards lean meat quality (less fat and more protein).

26.5.2.2 Insulin-Like Growth Factors (IGF)

The insulin-like growth factors system (IGF-I and IGF-II) plays a major role in regulation of fetal and placental growth during the pregnancy period. IGF-I stimulates fetal growth when adequate nutrients are available and IGF-I synthesis is sensitive to undernutrition. IGF-II regulates placental growth and nutrient transfer. IGF-I and II are expressed in fetal tissues from early pregnancy (pre-implantation) to before birth. IGF-I primary growth factor supports fetal growth whereas IGF-I plays role in the later stage of pregnancy. After birth, IGF-I plays a predominant role and its production becomes GH-dependant to ensure appropriate postnatal growth in the new nutritional environmental conditions. Deficiency of IGF-I, II, or undernutrition retards fetal growth.

26.5.2.3 Thyroid Hormones

The thyroid hormones (T3 and T4) are traceable in fetal circulation from early gestation and play important role in fetal growth and development. These hormones are important for morphogenesis, metabolic, tissue accretion, and differentiation in the fetus in all species including humans. These hormones' bioavailability in fetal plasma and tissues is regulated developmentally and also varies with species, pregnancy stage, nutrient availability, and uterine environment. Its deficiency during prenatal life retards the intrauterine growth, is associated with developmental abnormalities of some individual tissues, and compromises its adaptation to the external environment during postnatal life.

26.5.2.4 Insulin

Insulin is an important hormone that regulates fetal glucose metabolism and is essential for optimal growth, development, and metabolism of fetus. Insulin stimulates glucose uptake in insulin-sensitive tissues and also the conversion of glucose into glycogen in the fetal liver. It also has an effect on protein metabolism for optimal fetal growth.

26.5.2.5 Gonadal Hormones

The gonadal hormones of the fetus are necessary for sexual differentiation. In male, testosterone produced by the fetal testes is necessary for the testicular descent into the scrotum.

26.5.2.6 Glucocorticoid

Cortisol, a major stress hormone secreted from fetal adrenal gland that prepares the fetus for delivery and also supports the maturation of organs (lungs, thyroid gland, and GI tract) and metabolic pathways (glucose metabolic pathway in liver) during the transition from intrauterine to extrauterine life. After birth, a high concentration of cortisol in the newborn animal due to birth-related stress has a high impact on the adequate adaptation of the animal to extra-uterine life. Cortisol enhances the maturation of thyroid axis and somatotropic axis around birth leading to increased conversion of T4 to T3 and prenatal fetal growth (growth independent of GH) to postnatal growth, respectively.

26.5.3 Recombinant Gene Transfer Technology—Advantages and Limitations

Farm animals provide high-quality protein in terms of meat, milk, and egg for human consumption. The production performance of animals mainly depends on the genes, environment, and their interactions. Apart from that they are utilized for various research purposes which help us to improve our understanding of the basic and applied physiology of animals. Recent advances in genetic engineering technologies open a new era for scientific and industrial applications in the field of whole genome sequencing and genetic manipulation for improving meat, milk, and egg production through recombinant DNA (rDNA) transfer technology. Recombinant DNA is a DNA molecule produced through laboratory methods from DNA materials derived from two or more sources to produce a hybrid. Gene transfer technology: it is defined as an introduction or transfer of genes, i.e., DNA sequences into an organism by using genetic engineering technologies to obtain desirable traits. The complex process of animal growth is regulated by the endocrine system which also mediates the effects of nutritional, environmental, and genetic factors in animals (NRC 1988). The main hormones that affect growth in animals are growth hormone, thyroid hormone, glucocorticoids, prolactin, and gonadal hormones. Among these hormones, growth hormone (also known as somatotropin) is a protein hormone secreted from the pituitary gland that stimulates growth. It is generally considered as the most important hormone governing the growth and development of skeletal muscle and bone, lipolysis, and milk production. Many studies reported that recombinant bovine growth hormone (bGH) is physiologically active and possess the same properties as native pituitary bGH. Based on the functions of growth hormones, it is a drug of interest in animal production field since it has been proven as the efficient growth promotor in ruminants and pigs through better feed conversion, and also a production enhancer to improve lactation performance in dairy cows. Recombinant growth hormones (rGHs) are synthesized in large quantities in several forms corresponding to various animal species (bovine, equine, porcine, fish, etc.) and widely used across the country to stimulate milk production and as a general growth promoter in meat producing animals.

Advantage

1. Production of transgenic animals that are of great value to basic research and medicine.
2. Synthesis of proteins, peptides, vaccine, amino acids, and enzymes of interest.
3. Improving disease resistance.
4. Enhancing the production of meat, milk, and egg with higher nutrient content.

Disadvantages

1. Concern about biosafety of proteins or transgenic animals or by-products generated by rDNA technology.
2. Gene transfer affects the functionality and stability of existing gene in animals.

26.5.4 Probiotics

Probiotics are free-dried cultures of friendly bacteria such as *Lactobacillus* sp. or Bifida bacteria and yeasts. Live microbial dietary supplements are widely produced and used for harmonizing the microbial population in the gut by the production of different compounds, competitive elimination and displacement of pathogens from enterocytes, maintenance of gut pH and thereby enhancing the health and immune status of animals.

26.5.5 Prebiotics

Prebiotics are defined as non-digestible food ingredient/supplement that selectively stimulates the growth of some or all of the non-pathogenic favorable organisms (bacteria) in the gut of the host. Prebiotics enhance the immune system thereby improving health and production. The commonly used prebiotic includes oligosaccharides of galactose, fructose or mannose, lactulose, inulin, galacto-oligosaccharides (GOS), mannan oligosaccharides (MOS), and fructo-oligosaccharides (FOS). Prebiotics shows their beneficial action on the host by selectively feeding the harmless microbes at the cost of the harmful ones.

26.5.6 Organic Acids

Organic acids have been commonly used in commercial compound feeds which include lactic, formic, acetic, propionic, tartaric, fumaric, and citric acids. Due to their antimicrobial effect, they have been proven effective in maintaining the growth performance of animals. It also lowers the pH which leads to the reduction of many pathogenic organisms in the gut.

26.5.7 Exogenous Enzymes

Exogenous enzymes are commonly added in feeds as additives which are highly specific proteins, promoting a particular biochemical process (hydrolysis). These include non-starch polysaccharides degrading enzymes (glucanase, cellulase, arabinose, xylanase, arabino-xylanase), proteases, and phytase. These enzymes have the potential to aid many digestive processes either by hydrolyzing the nutrients or making them more available or by hydrolyzing the anti-nutritional factors like lectins, gels, phytates, or polyphenols.

26.6 Growth Anomalies

Generally, the incidence of growth anomalies in domestic animals is low. However, developmental abnormalities are due to both genetic and environmental factors that can be single and multiple in nature.

Developmental anomalies in farm animals are the impact of genetic and environmental factors. Congenital abnormalities are abnormalities of structure and function present at birth. Causes of congenital defects are unknown however they may be due to genetic (inherited by autosomal recessive genes) or environmental factors or a combination of both. Genetic defects occur due to genes missing or present in excess or gene mutation or gene translocation. Environmental factors that cause abnormalities are nutritional deficiencies (vitamins and mineral) in dam, teratogenic drugs or chemical exposure, viral infection, radiations, mechanical interference of the fetus, rectal palpation for pregnancy diagnosis, and dam exposure to toxic agents during organogenesis.

Examples: Hypotrichosis, Dwarfism, Osteopetrosis, spider lamb syndrome, split wing, complex vertebral malformation, etc.

Major impact of abnormal development is the loss of the nonviable fetus leads to financial loss and maintenance cost of the dam retained for another year. To control these abnormalities, it is better to avoid animals that carry these genes. A better way for prevention is selective breeding. Use bulls or semen from reputed breeders where bulls are tested for genetic disorders.

26.6.1 Congenital and Inherited Anomalies

Developmental anomalies in farm animals are the impact of genetic and environmental factors. Studying the etiology of the abnormalities will help to overcome and prevent the problem associated.

(a) Congenital malformations or congenital deformities—Structural abnormalities present at birth.
(b) Developmental or congenital abnormalities—Any defects and anomalies, in both the functional and morphological imperfections.
(c) Monsters are generally deformed fetuses.
(d) Teratology is the branch of embryology and prenatal pathology dealing with abnormal development and congenital defects.

26.6.2 Inherited Defects

Inherited defects occur due to mutation in the genes and the impact of which may vary. The occurrence might be due to the impact of single gene or by polygenetic effect as in case of cattle hypertrichosis, where the different genes at two interacting loci control the trait. Breeding of both the horses with Overo color pattern had led to the failure of intestinal tract innervation secondary to ileocolonic aganglionosis due to congenital anomalies. Similarly, breeding of dairy goats with dominant polled traits resulted in polled intersex goats. Such heritable defects have been reported to occur due to different embryonic anomalies like embryonic death, stillbirth, and birth of compromised offspring.

Mating or artificial insemination of monophosphate synthase (DUMPS) deficient Holstein cattle results in lethal autosomal recessive traits, where normal fertilization and embryonic development will occur: however, the fetus will be born dead. Similarly, homozygosity results in arginosuccinate synthetase deficiency thus affecting the urea cycle where the animals are healthy at birth while the ammonia concentration increases in body eventually leading to death after a few days of birth. This condition is known as citrullinemia. Canine X-linked muscular dystrophy is seen in Golden Retrievers, Labrador Retrievers, and other canine breeds, which occurs due to the presence of a single copy of a defective allele on an X chromosome the defect arises due to a defect on the X chromosome.

26.6.2.1 Infectious Agents

Infectious agents esp. viral infection affects fetal or embryonic development. Bovine viral diarrhea virus (BVDV) during prenatal infection causes congenital disorders like cerebellar hypoplasia, alopecia, brachygnathia, internal hydrocephalus, ocular defects, and impaired immunocompetence. Similarly, infection in pregnant ewes with BVDV infection also resulted in congenital anomalies. Pestivirus infections also cause congenital defects. Pregnant ewes infected with the border disease virus exhibit embryonic and fetus mortality or congenital defects involving the skin, skeletal, nervous, endocrine, and immune systems.

26.6.2.2 Environmental Factors

Normal fetal growth is influenced both directly by the placental (amniotic) environment and indirectly by the maternal environment. Maternal health, correct placentation, maternal nutrition, excess concentration of drugs and chemicals, xenobiotics, and physical agents are the factors causing developmental anomalies.

Table 26.1 Vitamin and mineral deficiencies causing development defects in animals

S. no.	Compound	Defects
1	Iodine deficiency	Congenital goiter or cretinism
2	Copper deficiency	Enzootic ataxia
3	Manganese deficiency	Congenital limb
4	Vitamin D deficiency	Neonatal rickets
5	Vitamin A	Eye defects or harelip
6	Deficiencies of choline, riboflavin, pantothenic acid, cobalamin, and folic acid	Teratogenic effects
7	Hypervitaminosis A	Teratogenic effects

26.6.2.3 Nutritional Factors

Both vitamin and micro-mineral deficiencies are implicated in a variety of developmental defects (Table 26.1).

26.6.2.4 Physical Agents

During twinning, large fetal size causes congenital joint contracture due to uterine crowding. Transverse or caudal presentation results in torticollis, scoliosis, and limb abnormalities in foals. Similarly, umbilical cord twisting in foals is associated with urachus. Aggressive transrectal palpation of amnionic vesicle before 42 days of gestation induced atresia coli due to disruption of vascular supply to the intestinal tract in cattle.

26.7 Aging and Senescence

Aging is defined as the gradual decline of physiological functions and organism fitness, which leads to age-dependent fitness loss, diseases eventually mortality.

26.7.1 Oxidative Damage

Oxidative stress refers to elevated reactive oxygen species (ROS) at intracellular level where 2–3% of oxygen atoms are reduced insufficiently to ROS by the mitochondria. The ROS oxidizes and damages cell membranes, proteins, and even nucleic acids. Over-expression of ROS-destroying enzymes in drosophila had 30–40% longer life than the control. Flies with mutant the *methuselah* gene live 35% longer than the wild-type files, where the mutants have developed resistance to paraquat, a poison that generates ROS within cells. Thus, genetic control can enhance aging process. Similarly, *Caenorhabditis elegans* mutants with the higher secretion of ROS-degrading enzymes live much longer than the wild ones. Mice lacking protein $p66^{shc}$ give resistance to ROS, leading to an increase in the life span of mice compared to the control. The protein $p66^{shc}$ is involved in apoptosis signaling pathway, their absence might play a role in the increase of the animal's life span. Calorie restriction aids in slowing the aging process by preventing ROS synthesis. Use of ROS inhibitors like vitamins E, C, and E in the diet increases the longevity of flies and nematodes.

26.7.2 Genetic Instability

Mutation in the protein synthesis process produces more altered protein; likewise mutation in DNA-synthesizing enzymes, DNA repair enzymes will affect the aging process.

Mitochondrial genome damage

Mitochondrial mutation is 10–20 times faster than nuclear DNA mutation. Mitochondrial function declines due to mutation in the mitochondrial genome which is associated with the aging process. Impacts of mitochondrial mutation are

1. Defects in energy production
2. Production of ROS by faulty electron transport
3. Induce apoptosis

All these eventually enhance the aging process.

26.7.3 Telomere and Aging

Telomeres are non-coding specialized repetitive DNA sequences located at the ends of chromosomes. Its functions are to maintain chromosome stability. When telomeres size reaches a particular smaller size, it inhibits cell division and initiates senescence processes. The size of telomeres shortens with aging. The length of the chromosomes is maintained by telomerase which prevents the shortening by adding the telomere onto the chromosome at each cell division. Telomerase may not be a factor in determining the differences in the aging rate among species. Telomere shortening might help in determining some age-related properties of organs. However, the activation of telomerase hinders a barrier to the growth of developing cancers while lack of telomerase activity provides a tumor suppressor function.

Learning Outcomes

- Growth is a complex biological process by which animals become larger over a period of time.
- Animal growth pattern is categorized into prenatal and postnatal growth which is regulated by several factors, viz., genetic, nutritional, and environmental factors.

(continued)

- Growth process might be manipulated by using various growth enhancers, feed additives, and supplementation of growth factors to enhance the meat, milk, and egg production in domestic animals.
- Growth anomalies observed in domestic animals might be due to genetic and environmental factors.
- Animals undergo aging and senescence once attain maturity.

Exercises

Objective Questions

Q1. Progressive increase in size and weight of an animal during a specific period of time is known as ______?
Q2. What is hypertrophy?
Q3. What is hyperplasia?
Q4. What are periods of growth in farm animals?
Q5. What is prenatal growth?
Q6. What are the phases of prenatal growth?
Q7. Which period is considered as the most important during prenatal growth?
Q8. Gastrointestinal tract developed from which germ layer?
Q9. What is the shape of the growth curve in farm animals?
Q10. What are the phases of growth curve?
Q11. Which force is responsible for rapid growth of a cell?
Q12. Which point is the indicator maximum growth rate?
Q13. At which point the growth rate started to decrease?
Q14. What are the main factors regulating the growth of farm animals?
Q15. What are the axes governing the growth process in animals?
Q16. Which is the hormone mediating the GH effects in muscle and connective tissue growth?
Q17. Where is IGF produced?
Q18. What is the correlation between birth weight and litter size?
Q19. Growth of organs or parts of the body follows which type of growth?
Q20. What are the different phases of growth observed in birds?
Q21. Gradual loss of physiological functions and organism fitness known as ______?
Q22. Which enzyme maintains the length of the chromosomes?

Subjective Questions

Q1. Describe the prenatal growth patterns in farm animals?
Q2. Write the events takes place during the pre-embryonic period of growth?
Q3. Write the characteristics of the growth curve in farm animals?
Q4. Describe biochemical and genetic determinants of growth.
Q5. Write the various factors regulating prenatal growth?
Q6. Write the various factors regulating postnatal growth?
Q7. Describe the growth in meat-producing animals and poultry?
Q8. Describe the growth manipulation in farm animals?
Q9. Describe the growth anomalies in farm animals?
Q10. Write about aging and senescence?
Q11. Describe the growth pattern in birds?

Answer to Objective Questions

A1. Growth
A2. Increase in size of a cell, tissue, or organ
A3. Increase in number of cells
A4. Prenatal and postnatal growth period
A5. Growth period between conception and birth
A6. Pre-embryonic/ovum period, embryonic, and Fetal period
A7. Embryonic period
A8. Endoderm
A9. Sigmoid or S-shape
A10. Self-accelerating phase and self-retarding phase
A11. Growth accelerating force
A12. Point of pubertal inflection
A13. Point of inflection
A14. Genetic, endocrine, nutritional, and environment
A15. Somatotroph axis and thyrotroph axis
A16. Insulin-like growth factor 1 (IGF-1)
A17. Liver
A18. Negative
A19. Urethra
A20. Isometric and allometric
A21. Embryonic, post-hatch, prepubertal, pubertal, and adult
A22. Aging
A23. Telomerase

Keywords for Answer to Subjective Questions

A1. Pre-embryonic period of growth, embryonic period, fetal growth
A2. Fertilization, cleavage, formation of blastocyst, implantation
A3. Phases of the growth curve, self-accelerating phase, self-retarding phase, negative growth phase
A4. Genetic factors, role of amino acids, role of fatty acids
A5. Heredity, hormones, maternal age, maternal nutrition, litter size, fetus sex, environment
A6. Sex of animal, litter size, plane of nutrition, hormones, environmental factors, diseases

A7. Growth of organs, muscle growth, protein deposition, fat deposition

A8. Use of different growth promotors, hormones, recombinant gene transfer, prebiotics, probiotics, organic acids, exogenous enzymes

A9. Congenital anomalies, inherited defects, developmental defects-infectious agents, environmental factors, nutritional factors, physical agents

A10. Oxidative damage, genetic instability-mitochondrial genome damage, role of telomere

A11. Growth phases, embryonic, post-hatch, prepubertal, pubertal, adult

Further Reading

Brody S (1945) Bioenergetics and growth. Hafner, New York

Buttery PJ (1983) Protein deposition in animals. Outlook Agric 12(4):172–178

Christopherson RJ (2010) Physiology of growth and reproduction in livestock. In: Hudson RJ (ed) Animal and plant productivity, Encyclopedia of life support systems. EOLSS Publications, Oxford

Gibson GR, Roberfroid MB (1995) Dietary modulation of the human colonic microbiota: Introducing the concept of prebiotics. J Nutr 125:1401–1412

Gicquel C, Le Bouc Y (2006) Hormonal regulation of fetal growth. Horm Res Paediatr 65(Suppl. 3):28–33

Gilbert SF (2000) Aging: the biology of senescence. In: Developmental biology, 6th edn. Sinauer, Sunderland, MA

Kwon H, Ford SP, Bazer FW, Spencer TE, Nathanielsz PW, Nijland MJ, Hess BW, Wu G (2004) Maternal nutrient restriction reduces concentrations of amino acids and polyamines in ovine maternal and fetal plasma and fetal fluids. Biol Reprod 71(3):901–908

Lawrence TLJ, Fowler VR, Novakofski JE (2012) Growth of farm animals. CABI, Wallingford

Miller EL (2004) Protein nutrition requirements of farmed livestock and dietary supply. In: Protein sources for the animal feed industry, expert consultation and workshop, Bangkok. Food and Agriculture Organization of the United Nations, Rome, pp 29–76

Nagashima H, Matsunari H, Nakano K, Watanabe M, Umeyama K, Nagaya M (2012) Advancing pig cloning technologies towards application in regenerative medicine. Reprod Domest Anim 47: 120–126

Owens FN, Dubeski P, Hanson CF (1993) Factors that alter the growth and development of ruminants. J Anim Sci 71(11):3138–3150

Rousseaux CG (1988) Developmental anomalies in farm animals: I. Theoretical considerations. Can Vet J 29(1):23–29

Scheaffer AN, Caton JS, Redmer DA, Reynolds LP (2004) The effect of dietary restriction, pregnancy, and fetal type in different ewe types on fetal weight, maternal body weight, and visceral organ mass in ewes. J Anim Sci 82(6):1826–1838

Wu G, Bazer FW, Wallace JM, Spencer TE (2006) Board-invited review: intrauterine growth retardation: implications for the animal sciences. J Anim Sci 84(9):2316–2337

Physiology of Animal Behavior

27

P. Visha and V. Sejian

Abstract

Animal behavior has emerged as a multi and interdisciplinary science, encompassing such sciences as physiology, pathology, immunology, endocrinology, ecology, genetics, and neuroscience. Neuroethology merges into cellular biochemistry, physiology, and molecular genetics and also at the other end, the study of animal groups extends from the evolutionary theory to social ecology. External events as well as genetic factors affect the way the nervous system develops and ultimately the way the individual behaves later on as an adult. In most animals, behavior patterns reflect the complexity and functional organization of the nervous system. Behavior is a complex response mediated by the nervous system and is modulated by the endocrine system. The behavior of animals has been broadly classified and studied extensively under categories such as learning, feeding, sexual, maternal, and social behaviors. Understanding in depth these various types of animal behavior can help the researchers to identify various indicators which could serve as potential guidelines for assessing their well-being. Behavior is closely related to the welfare of the animal. Hence, behavioral studies help the veterinarians and animal scientists to identify the cause of any abnormal behavior and address it. Also as many clinical problems, animal training methods have a behavioral component, understanding the physiological basis of behavior will help the veterinarians and animal handlers to advocate better approaches towards the animals.

P. Visha (✉)
Veterinary College and Research Institute, Tamil Nadu Veterinary and Animal Sciences University, Salem, Tamil Nadu, India

V. Sejian
ICAR-National Institute of Animal Nutrition and Physiology, Bangalore, Karnataka, India

© The Author(s), under exclusive license to Springer Nature Singapore Pte Ltd. 2023
P. K. Das et al. (eds.), *Textbook of Veterinary Physiology*, https://doi.org/10.1007/978-981-19-9410-4_27

Graphical Abstract

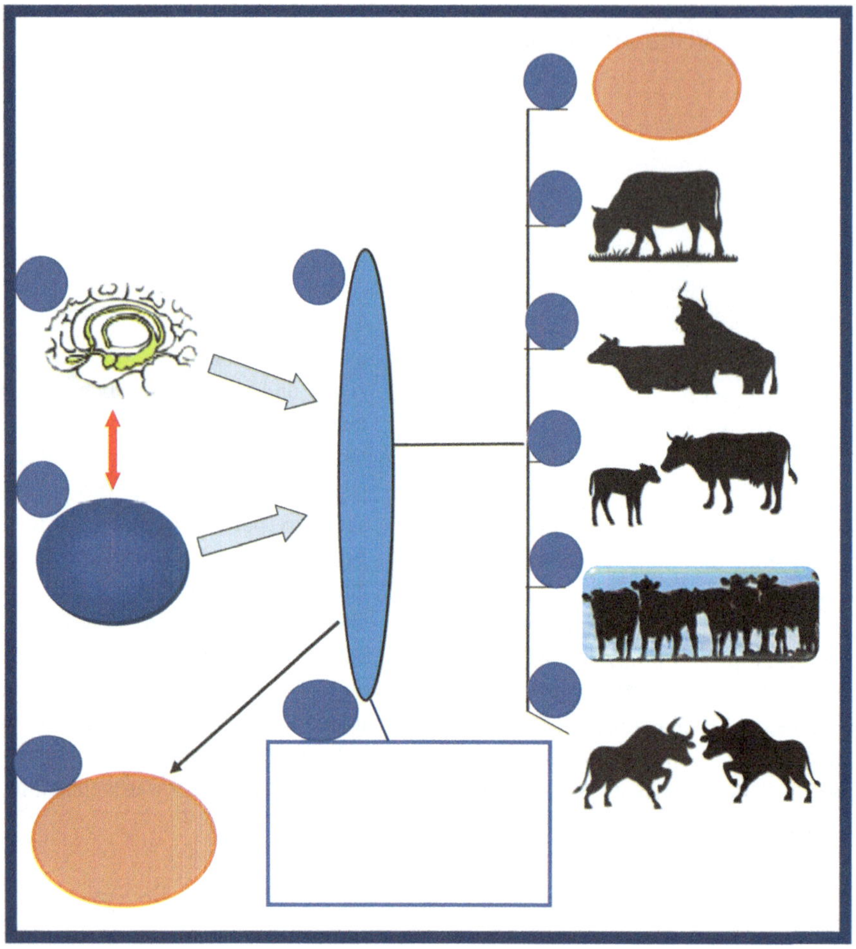

Description of the graphic: Nervous system (limbic) (**1**) and endocrine system (**2**) control the different types of behavior in animals (**3**). Learning behavior where individual animals acquire new responses and new capacities (**4**). Feeding behavior differs with each animal and it is a sign of health influenced by various factors (**5**). Sexual behavior is important in all species of animals including proceptive and receptive behavior by the female; courting and mate guarding by the male, as well as actual copulation (**6**). Maternal behavior exhibited by the mother in response to neonate plus prior experience of being a mother and it is under control of both hormonal and neural (**7**). Animals in groups exhibit various social behaviors such as dominance hierarchy, peck order, teaching, cooperation, and playing (**8**). Aggressive behavior is usually displayed to threaten or attack to resolve competitive disputes over limited resources (feed, territory), to increase their reproductive potential, or to escape from threatening situations (**9**). Animals communicate not only by auditory signals but also by visual and olfactory signals (**10**). Various behavioral disorders also showed in each species (**11**). All aspects of the behavior of all livestock species are elaborated in this chapter

Keywords

Ethology · Neuroendocrine · Communication · Learning · Behavioral problems

Learning Objectives

- To understand the importance of ethology in relation to the veterinary science and animal welfare.
- To recognize the basic physiological and genetic mechanisms involved in the development and evolution of behavior.
- To appreciate the intricate interplay of neuroendocrine system in exhibition of a behavior.
- To differentiate the various types of learning and behavioral conditioning.
- To understand the differences in the feeding, sexual, maternal, and social behaviors in different species and the manifestation of clinical behavioral disorders.

- To learn the various modes of communication in different animals.
- To understand the training responses in dogs and horses.
- To know the significance of wild life ethology.

27.1 Introduction to Ethology and Its Importance in Veterinary Science

Animal behavior is fascinating and interesting to common man as well as to the animal scientists and veterinarians. Presently, it has emerged into a multi and interdisciplinary science, encompassing such sciences as physiology, pathology, immunology, endocrinology, ecology, genetics, and neuroscience. Veterinary ethology enables the veterinarian to study the animal not only as an entity but also as an open system interacting with the surrounding environment by exhibiting different behaviors. The application of ethology has wide implications in areas of research, farm management, wildlife management, and also in clinical practice. Therefore, any animal management system needs to be evaluated from an ethological point of view, before problems of illness and trauma can be addressed. Furthermore, preventive diagnosis and treatment of behavioral problems form part of the clinical approach. With the insight of ethology, veterinarians can guide the animal owners regarding the animal's specific behavior and thus help in preventing any behavioral abnormalities exhibited by the animal.

27.2 Ethology: Definition and Its Importance in Animal Welfare

Ethology can be defined as the study of animal behavior, its causation, and biological function. Behavior in turn can be defined as the sum of all observable or measurable and overall responses or even lack of response in some cases that occur in response to changes in an animal's internal and external environment. Behavior as a whole has a genetic basis, and hence is subjected to natural selection, and can be modified through experience and learning. Behavior may involve one individual reacting to a stimulus or a physiological change or it may also involve two individuals or a group reacting and responding to the activities of one another.

Behavior is closely related to the welfare of the animal. Hence, behavioral studies help veterinarians and animal scientists to identify the cause of any abnormal behavior and address it. The behavior of the animals is bound to vary depending on the species, breed, age, sex, physiological status of the animal, and surrounding environment. An understanding of the behavior of animals will guide animal scientists to prevent and manage various behavioral disorders thus ensuring the welfare of the animal.

Behavioral issues are often encountered in various situations ranging from,

- rearing a newly weaned pup or calf
- milking reflex inhibition due to loss of calf
- transporting animals especially horses in trucks over long distances
- managing a newly adopted pet
- anxious behavior shown by a deserted pet or when it is temporally separated from the owner
- managing a large herd or group of animals in a farm
- managing animals in restricted enclosures such as zoos
- managing an anxious pet/farm animal suffering from fear/pain presented to the veterinarian for treatment
- feeding behavior disorders leading to obesity or starvation, all of which have to be dealt with carefully

Further, it has been estimated that dogs, cats, and sometimes even horses are euthanized for behavioral problems than for any other single cause. Hence, behavioral studies help to guide the veterinarians to solve the problems before the animal becomes a potential danger to the owner or the public or before the owner can no longer live with the animal.

27.3 Physiological Concept of Behavior, Neuroendocrine Integration for Behavioral Manifestation

Behavior is an integrated manifestation or interplay of the neuroendocrine system.

27.3.1 Neurobiology of Behavior

Behavior is a complex biological phenomenon mediated through the nervous system consisting of brain, spinal cord, and sensory and motor neurons. The cerebral cortex, especially the neocortex, in the brain contains the centers for the integration of sensory stimuli, conscious reasoning, memory, and reflection. The limbic system in the cerebral cortex consisting of hippocampus, olfactory cortex, parts of thalamus and hypothalamus, is involved in the immediate control of basic behavioral programs related to feeding, aggression, and sexual behavior. The limbic system also connects to sensory areas in the neocortex and is responsible for attaching emotions and feelings to behavior.

Stimuli from the surroundings constantly flow across the individual. All the sensory stimuli (visual, olfactory, auditory, and mechanic) perceived by the sense organs, are transmitted to the various areas of the brain where they are interpreted in a meaningful and purposeful way to the animal. The hippocampus, the parietal, and prefrontal cortex areas of the brain also play a vital role in the spatial learning process, and in particular the number of dendritic spines in these regions of the brain is correlated with learning ability. Key stimuli are usually linked to specific behavioral responses. Motor activity is the central aspect of all behavior occurring as a result of synchronized pathways of muscle contraction which causes the animal to move in a functional manner in response to the stimuli.

27.3.2 Neurophysiology of Behavior

Behavior is primarily controlled by the central nervous system involving brain and spinal cord and is also regulated by endocrine factors. The cerebral cortex involving all four cortical lobes (parietal, temporal, frontal, and occipital lobes) exerts control over cranial and spinal regulation of motor actions and controls all kinds of behavior. The main cortical areas controlling the behavior involve hypothalamus, amygdala, limbic system, basal nuclei, and hippocampus. Prefrontal cortex (PFC) (anterior part of the frontal lobe) is the most important cortical part involved in the control of executive, social, emotional, or instinctive behaviors.

The hypothalamus has one of the most complex circuitries of any brain region. There are both neural interconnections and also extensive non-neural communication pathways between the hypothalamus and other brain regions and the periphery. Hypothalamus mediates the control of behavior through several circuitries and pathways. Hypothalamic inputs to various motor pattern generators may increase the probability of specific behaviors. For example, when hungry, most animals need to forage for food, then explore it by sniffing and licking, and then finally consume it. The hypothalamus may reduce the threshold for activating motor pattern generators for locomotion, sniffing, and oral behaviors that are involved in the ingestion of food. Descending outputs from the hypothalamus to the sensory system may sensitize or desensitize them. Finally, hypothalamic control of autonomic responses may cause signals that reach higher cognitive regions to elicit the appropriate behavior. Similarly, hypothalamic regulation of endocrine systems is also interlinked and exhibits feedback control on the brain centers. For example, many neurons in the brain have receptors for steroid hormones involved in stress responses, reproduction, or salt depletion, and changes in these hormones may alter the likelihood of expression of various complex behaviors regulated by those neuronal systems.

The amygdala is a central processing area which accords the level of priority to a given stimulus. It then sends projections to the hypothalamus for further integration and coordination with the brain stem areas to initiate the body's response including the "fight or flight" responses (e.g., increase in respiratory rate, blood pressure, etc.)

The limbic system of the brain controls a variety of behaviors that are essential for survival. The limbic system predominantly regulates appropriate responses to stimuli having emotional, motivational, or social importance, which includes innate behaviors such as mating, aggression, and defense. The innate behaviors which ensure the survival of the individual or its offspring and species propagation include the establishment of social hierarchy, mating, maternal care, and defense. These behaviors are regulated and modulated by sensory stimuli such as sound, touch, and, most importantly, smell in rodents.

The neural circuitry that regulates innate behaviors are intricately influenced by endocrine factors, primarily sex hormones such as testosterone and estrogen. Both circulating and local brain levels of estrogen and testosterone are expressed in a sex-dependent manner and they refine the neural circuits involved in sexually dimorphic behaviors. Most of the limbic circuitry structures express estrogen receptors in both females and males. Estrogen is the primary hormone in the induction of maternal care in females. Virgin female rats inhibit their aversion and show attraction to pups after supplementation with estradiol, behaving more like nursing females. While estrogens shape the programming of sexually dimorphic circuits, testosterone acts directly through the androgen receptor and is essentially required for the activation and modulation of male-typical displays such as territorial aggression, urine marking, and mating.

Basal nuclei are a set of subcortical gray matter collections located in the vicinity of the diencephalon having an essential role in rewarding value-guided or motivated behavior by deciding the choice of behavior that will be more rewarding to follow. Hippocampus involved in the processing and retrieval of memory also exerts its influence on behavior. Autonomic nervous system plays a key role in the expression of innate as well as learned behavior.

27.3.3 Endocrine Moderation of Behavior

Behavior is a complex response mediated by the nervous system and is modulated by the endocrine system. Hormones determine and influence the probability that a specific sensory input leads to a specific behavioral response. Hormonal changes might modify some ongoing behavior by increasing or decreasing the frequency or duration of that behavior, or they might trigger the onset or end of a behavior or behavioral sequence. Hormones do not initiate or inhibit any behavior by themselves; however, they influence the sensitivity of the neural circuits involved in exhibiting the various behaviors. In addition, hormones might prime animals so that they are more or less likely to behave in a specific way in a specific environment. For example, when baseline levels of testosterone are high, males are primed for aggressive behavior and display aggression when encountering another male. Testosterone shows a hormonal-behavioral feedback loop. Also if an animal wins a fight, partly as a result of behaviors resulting from high baseline levels of testosterone, the act of winning may in turn increases the probability of winning future fights by further increasing testosterone levels or by lowering the level of stress hormones such as cortisol. Similarly, oxytocin has a direct behavioral effect by calming an aggressive animal.

Hormones also affect the organization of behavior systems during embryonic development. For example, female mice gestate many fetuses at the same time. If a developing male mouse fetus is surrounded by female fetuses, it is often exposed to low levels of circulating testosterone and high levels of estrogen in the uterus. When such males mature, they tend to be less aggressive and less sexually active than males that were surrounded by male fetuses in utero. This shows how a specific behavior could be fundamentally altered by hormonal influence early in development.

Similarly, in rodents, the sex of the two siblings surrounding an individual in utero can have dramatic effects on testosterone levels and an individual's behavior after birth. Males that were surrounded by two other males in utero are more aggressive and mark their territories by scent and mount more females than males that were surrounded by two females. Also, as a result of high testosterone levels, these males tend to exhibit less parental care. This relationship between in utero testosterone exposure and subsequent effects on brain activity especially on the preoptic area exhibits the tight connection between the hormonal and neurobiological basis of behavior.

The fight or flight response is another example showing the influence of hormone on the animal's behavior. When an individual confronts stress or a predator, the hypothalamus initiates adrenaline and cortisol release. Adrenaline has a wide spread effect on all parts of the body including, an increase in cardiopulmonary activity thus delivering increased glucose and oxygen to the brain, skeletal muscles, and heart. These effects enable the animal to quickly flee from a predator or perhaps to fight against the danger.

27.3.4 Endocrine Influence on Social Behavior

The social behavior in animals is influenced by hormones which in turn are genetically controlled. Vasopressin and oxytocin are the two related hormones which play a central role in reproduction and parental care in mammals. For example, Prairie voles are monogamous, both males and females have a single mate for a given breeding season and males often display parental care and guard their mates. Meadow voles have a polygynous mating system, wherein males mate with multiple females during a breeding season, and males display very little parental care behavior toward their mates. One of the major differences in the male behavior towards their offspring and their mate is centered on the vasopressin system in these two species. The Prairie voles had higher numbers of vasopressin receptors (which are controlled by a gene known as avpr1a) in the ventral pallidium area of their brain than in the Meadow voles and this is considered to be responsible for the difference in male social behavior in Prairie versus meadow voles.

The relationship between behavior, hormones, and the nervous system is very intricate. One such dramatic example is the "Bruce effect," named after Hilda Bruce. Pregnant female mice abort their litters and reabsorb the embryos if a strange male mouse (not the father of the litter) comes in contact with them. Just the smell of a strange male is sufficient to induce abortion; it happens even when females come into contact with the litter soiled by another one. The pregnancy-blocking effect is brought about by inhibition of prolactin secretion and consequent reduction in the progesterone which is needed to maintain pregnancy.

27.3.5 Behavioral Plasticity

The plasticity of behavior is an array of behavioral responses to varying environmental conditions. Animals respond to environmental change by dispersing/migrating, adjusting through phenotypic plasticity, or adapting through genetic changes. Phenotypic plasticity involves the tendency of a particular genotype to produce different phenotypes under altered environmental conditions. It allows an animal to adjust its behavior to suit the conditions of its immediate environment and, in so doing, increase its fitness. Phenotypic plasticity is the ability of an organism with a given genotype to change its phenotype in response to changes in the environment. Plastic responses can have adaptive significance for organisms in unpredictable environments, migratory species, and organisms in novel environments. The ability of individuals, populations, or species to switch between behaviors across situations can have significant ecological and evolutionary implications. For example, phenotypic plasticity can play a role in the diversification process and species range-expansion.

Broadly, three types of behavioral plasticity can be identified: differences in ontogenetic development, adjustments through learning, and the innate ability to respond to a variety of stimuli. Differences in the ontogeny of behavioral patterns may be due to varying social environments. The ontogeny of behavior is similar to that of morphological plasticity because it does not represent an immediate response and is not usually reversible. Learning which is defined as the modification of behavior by experience, results in behavioral plasticity, with an immediate and reversible response. Plasticity is expected to evolve by means of natural selection when it provides plastic individuals with a fitness advantage over less plastic individuals. For example, natural selection favors individuals that adjust their activity levels in response to changes in predation risk.

Behavior induces plastic changes in the brain components part, e.g., increase in size, connections, dendritic arborization, axonal sprouting, spine density, synaptic organization and formation, and neurogenesis (in hippocampus).

Know More . . . . .

Behavioral ecology refers to the study of how animal behavior develops its evolution and its contribution to reproductive success and survival. It deals with the individual animal's behavior in relation to the environment and its ecological pressures. Young animals can be considered as miniature adults, gradually growing in size, but their behavioral responses must change as well to keep pace with their growth and the ever-changing environment. The behavioral and morphological changes occur in young animals which may be totally different from that of adults. Similarly, some specialized infantile behavior patterns do not always disappear but may return in a slightly different context. For example, Baby Meerkats (an African mongoose, *Suricata suricatta*) limp and behave passively when an adult seizes them by the scruff of the neck. This reflex facilitates them to be moved without injury. Similarly, during copulation, the adult female Meerkats relax when seized in a neck bite by the male. The neural basis for the reflex remains beyond infancy and even if they are not used again, some juvenile reflexes remain during the latter part of their life also.

27.4 Concept of Learning, Instinct, Habituation, Imprinting, Reinforcement, Conditioning, and Temperament Scoring

27.4.1 Learning

Learning is a process whereby an individual acquires new responses and new capacities. Learning results in behavioral changes within an animal's lifetime which can introduce a new dimension into behavioral evolution. Behavior can have a genetic basis and so change occurs by natural selection operating on inherited mutations, favoring those changes in which the interactions with the environment lead to greater success. An animal which grows up in a persistent social group or has frequent contact with others is influenced by them and it will profoundly affect the way its behavior develops. The parent and the other adults in a social group form a constantly changing set of "models" from which young animals learn. Learning from others and copying can be a potential factor for change in behavior, which is both rapid and extensive. The famous long-term studies on Japanese macaques (*Macaca fuscata*) have recorded the development of a "food-washing" culture in some groups. It might have originated in some monkeys copying humans who were providing them with sweet potatoes after washing the earth sticking to sweet potatoes. Later on, it could be observed that due to social learning, some coastal groups of monkeys also began to wash food in streams.

Learning as a whole helps the animal to survive better in ever-changing environment. Animals learn about potential mates, familial relationships, aggression, and about predators, e.g., prey often in areas that contain both predatory and nonpredatory species. Encounters with predators are not the same always, as at any given time, some predators might be in hunting mode while others might not be actively hunting the prey. If prey can learn to distinguish between dangerous and benign encounters with potential predators, they may free up that time for their other vital activities such as foraging or mating.

Learning and memory formation are important to both wild and domesticated animals. Dogs and horses are often valued based on their ability to learn. In farms, the cows learn to approach the milking area, feeders and also remember which cow will butt her. Pigs and poultry learn to operate electronically controlled feeders and drinkers. Many animals learn to fear veterinarians who are easily recognizable by their white coats and their shiny examination tables.

Learning occurs in two distinct phases: short-term memory which is a synaptic event wherein impulses pass from one neuron to another due to sustained neurotransmitter release and long-term memory which requires new protein (cAMP response element binding protein) synthesis. These proteins form new synaptic connections which are the basis for long-term memory. The cellular and molecular mechanisms for learning are similar throughout various species ranging from fruit flies to large mammals. The hippocampus appears to be essential for explicit spatial learning, but the striatum is necessary for tasks that require incremental learning of associations. For learning simpler implicit tasks such as the classical conditioning of blinking when a puff of air is directed at the eye, only the cerebellum is necessary.

27.4.2 Instinct

Instinct is a stereotyped, species-typical behavior that is fully functional for the first time it is performed, without the process of learning. Such behaviors are usually triggered by a particular stimulus or cue and are not readily modified by subsequent experience. For instance, a kangaroo rat instantly performs an automatic escape jump maneuver when it hears the sound of a striking rattlesnake, even if it has never encountered a snake before. Instinctive behaviors play an essential role in survival. Instinct develops along with the developing nervous system and gradually evolves over the generations to match an animal's behavior to its environment. Instinct behavior does not essentially require learning or practice, but it is needed for those who appear appropriately the first time. Animals may not have any pre-set responsiveness but tend to modify their behavior on the basis of their individual experiences. They learn how to behave appropriately and perhaps practice or even copy from others to produce the best response. Pre-set behavior which requires no learning or practice is especially advantageous for animals having short life span and no parental care.

27.4.3 Habituation

Habituation is an extremely simple form of learning, wherein an animal, after a period of exposure to a stimulus, stops responding and ignores it. Habituation occurs after repeated exposure. Habituation is more if the stimulus is not offensive or harmless. Lack of continued response to strong odors is a common example of sensory habituation. Habituation to complex stimuli may occur at the level of hippocampus in the brain. Habituation helps in filtering the large amounts of information received from the surrounding external environment. By habituating to less important signals, an animal learns to focus its attention on the most important features of its environment. For example, birds very soon learn to ignore the scarecrow which had scared and made them fly away when it was initially placed in a field and later the birds get habituated to it. Thus, habituation can be considered as a simple form of learning. It is relatively long-lasting and is "stimulus-specific," i.e., only the stimuli which are repeated without reinforcement are affected whereas the animal remains alert to others.

Another good example may be observed in species that depend on alarm calls to convey information about approaching predators. As the animal gets habituated, the animals stop giving alarm calls when they become familiar with other species in their environment that turn to be non-predators.

27.4.4 Imprinting

"Imprinting" refers to the various behavioral changes wherein a young animal becomes attached to a "mother figure" and/or a future mating partner. This type of response to a mother figure is usually called "filial imprinting." Imprinting is distinctive because it occurs soon after hatching or birth and commonly results in an attachment which is difficult to change. Imprinting is an early form of learning having major effects on how young birds and mammals attach to a parent figure which offers protection, and interestingly, rather later may influence the young animal's choice of a sexual partner. Imprinting-like phenomena are also associated with the social development of mammals. For example, the popularly known example is of baby ducks and geese, which are observed to closely follow their mother on their early forays away from the nest. The baby birds form an attachment to an individual that was present as they hatched out and moved about after hatching.

27.4.5 Operant Conditioning

Learning in which behavior is affected by its consequences is called as operant conditioning. This type of learning is seen in animals as they associate performing a specific behavior with some sort of benefit or detriment. Operant conditioning is classified as a type of associative learning. When behavior is followed by something the animal wants the behavior gets strengthened. Conversely, if the behavior is followed by something the animal fears or dislikes the behavior is likely to weaken also in cases where nothing follows the behavior. Operant conditioning is useful when the animals are trained to perform behaviors, as it primarily deals with voluntary behavior.

Operant conditioning continues throughout the animal's life. They learn through trial and error about the behavior that brings them stimuli they want and which brings them stimuli they dislike. The stimuli affecting the animal's behavior can be a variety of foods, touches, sounds, activities, and odors. It depends on the individual animal whether a stimulus is a reinforcement or punishment. Both reinforcement and punishment must occur within 2 s of the behavior and before the performance of the behavior or else their effect on the behavior is likely to be weakened tremendously.

27.4.6 Reinforcement

The treat given for exhibiting behavior during operant conditioning is called as reinforcement. There are two types of reinforcements used in operant conditioning: positive and negative reinforcements. Positive reinforcement involves adding a stimulus, the animal wants immediately after or during the behavior it is performing. Adding this stimulus strengthens the behavior that precedes it. Rewarding the dog a piece of bone whenever it obeys the owner's command would serve as positive reinforcement. Negative reinforcement takes away a stimulus when a behavior happens, causing the behavior to strengthen. For example, when a dog lunges at the approaching person causing the person to back up, the animal learns lunging behavior removes the scary person. Consequently, the behavior of lunging at the person will strengthen.

27.4.7 Classical Conditioning

This type of associative learning is the one in which an animal associates a novel stimulus with another stimulus that has already caused a reaction. After repeated pairings, the novel stimulus will also cause the same reaction. The typical pattern for classical conditioning is the animal associating a conditioned stimulus with an unconditioned stimulus to elicit an unconditioned response. Again, after many pairings, the conditioned stimulus will elicit a conditioned response. A conditioned stimulus is often defined as a stimulus that initially fails to elicit a particular response but then comes to do so when it becomes associated with a second (unconditioned) stimulus.

The conditioned stimulus must occur before an unconditioned stimulus to elicit the unconditioned response as the animal is learning about the conditioned stimulus. The unconditioned response occurs naturally after the unconditioned stimulus but the conditioned response is learned when the conditioned stimulus repeatedly precedes the unconditioned stimulus. This type of learning was studied by Russian Ivan Pavlov in the late 1800s and the early 1900s. His experiments with dogs involved the reflex of salivation. Most of the behaviors involved are involuntary; they include reflexes, emotions, and secretions. Classical conditioning is powerful as learning takes place within the unconscious mind. For example, a cat learns being put in its carrier signifies it is taken to the veterinary hospital. Once when it is placed in the carrier, it starts to feel as if in the hospital. In a way, the carrier becomes the hospital for this cat. We can effectively use classical conditioning to our advantage by teaching animals to associate the hospital and/or the carrier with stimuli they enjoy.

27.4.8 Spatial Learning

Spatial learning is the ability to recognize surroundings and memorize a route. In contrast to other forms of learning, it is probably explicit, i.e., the animal is aware that it is learning.

27.4.9 Temperament scoring

The temperament of an animal influences the animal's judgemental capabilities about the surrounding environment. If the animal is highly reactive and fearful, it will judge more stimuli to be dangerous and thus will become sensitized to them. Although the animal's temperament is largely present at birth, it can and is modified throughout life by experience. During an animal's socialization period, they can be exposed to many stimuli, creating a board basis of exposure to many objects, experiences, sounds, and living things. This allows them to have the best chance of being able to habituate to the stimuli commonly found in their environment throughout their life.

Hormones also influence learning and temperament. Glucocorticoids have a major role in mediating stress responses and learning in animals. In adult animals, glucocorticoids easily cross the "blood-brain" boundary and bind with the receptors in the hippocampus, where they can modulate the emotional state and cognitive abilities of the individual. Glucocorticoids not only influence learning and memory in adult animals but also have an influence on fetus. For example, when pregnant female rats are stressed and level of glucocorticoid rises, the offspring of such females show high levels of anxiety and perform sub-optimally in learning tests indicating that stressing pregnant mothers affects anxiety levels in their offspring.

27.4.10 Culture

Extensive observation of social groups in nature commonly reveals enduring differences in their behavior. Individual patterns have arisen in one group or another and persist because young animals acquire them from the adults with whom they grow up. These persistent behavioral features are referred as "culture" and the changes as cultural evolution. They are acquired and passed on across generations/between individuals. Cultural evolution is possible only among animals living in contact with others and having the ability to modify their behavior by copying and practicing. For example, rats have been shown to acquire the tendency to dig for hidden food by watching others doing so. Having learnt this, they themselves can become models from whom others learn thus initiating a chain of cultural transmission. Cultural transmission may occur via vertical (information is transmitted across generations from parent(s) to offspring), oblique (transfer of information across generations via unrelated individuals, but not through parent/offspring interactions), or horizontal (transfer of information between individuals of same age groups) transmission.

27.5 Types of Animal Behavior

The behavior of animals has been broadly classified and studied extensively under categories such as feeding, sexual, maternal, and social behaviors. These studies have helped us to better understand the animals and address their behavioral problems.

27.5.1 Feeding Behavior

Feeding behavior is a circadian rhythm and an important sign of health which is influenced by a variety of internal and external factors. The factors can act on a short-term or long-term basis or be operational only in emergency situations.

Short-Term Control
Taste can either stimulate or inhibit intake. Gastric factors, particularly gastric fill, can suppress feeding whereas the hormone ghrelin can stimulate it. The changes within the gastrointestinal tract such as an increase in osmotic pressure and release of the hormone cholecystokinin cause satiety thus inhibiting intake.

Long-Term Control
The long-term control of feeding in which intake is controlled as part of the regulation of body weight or body fats is especially important in pigs and other domestic animals. The feedback from fat cells to the brain is by leptin which acts on the brain, and as a result, feeding signals are inhibited and feed intake decreases. All these signals are integrated into the brain, involving a variety of neurotransmitters and anatomical sites. The central nervous system controls of feeding are complex. Neuropeptide Y (NPY) and agouti-related protein (ARP) stimulate feeding, whereas cocaine and amphetamine-related transcript (CART) and melanocyte stimulating hormone (MSH) suppress feeding by their actions in the hypothalamus. In addition, when the sympathetic nervous system is stimulated, lipolysis occurs and decreases fat stores. Another feedback system is peptide YY (PYY), a peptide released from intestinal cells when ingested fat enters the cells, and suppresses feeding.

Emergency Control
Lack of metabolizable glucose induces eating in response as an emergency mechanism; however, increasing blood levels of glucose does not suppress feeding.

27.5.1.1 Feeding Behavior in Ruminants

27.5.1.1.1 Ingestive Behavior in Cattle
Cattle eat most of their meals during daylight but may graze at night in hot weather. Grazing time is 400–650 min/day at a bite (prehension) rate of 60 per min. Cattle will also eat more in groups than they will individually, and first-calf heifers will eat more if grouped with older cows. Another effect of the social hierarchy during group feeding of grains is that subordinate cows eat faster than dominant ones, probably because they have less time to eat as they are more likely to be displaced by a dominant cow.

27.5.1.1.2 Grazing Behavior
The preferences of ruminants for plant species vary and depend on various factors including the growth stage of the plant (ruminants prefer fast-growing, succulent species). Cattle wrap their tongues around grass and pull to prehend it. This method of foraging limits them to plants that are higher than 10 mm. Cattle eat less when environmental temperatures are high. In addition, higher temperature also effects plant growth that may render them less palatable. Food intake falls in estrous cattle.

27.5.1.1.3 Ruminating Behavior
The cows spend nearly 7–10 h ruminating. The composition of the diet especially neutral digestive fiber and the particle size influence the rumination behavior. Apart from the dietary factors, rumination behavior can be altered by the state of the cow and management strategy. For example, the onset of estrus reduces daily rumination times and the diurnal pattern of rumination.

One of the most common abnormal behavior is the fat cow syndrome wherein the dry cows may overeat, gain weight, and later at calving or during lactation become ill or even die.

27.5.1.2 Feeding Behavior in Sheep
Sheep grazes up to 12 h/day, and most of this time is spent biting. Sheep masticate 60–70 times/min. Sheep prehends grass by breaking it between their lower incisors and upper dental pad. They select leaves rather than stems. Sheep eat more when they are cold and less when they are hot. For example, sheared sheep eat 50% more after shearing as a result of the loss of insulation of their wool. Sheep are much more sensitive to unpleasant taste than other ruminants.

27.5.1.3 Feeding Behavior in Goats
Goats eat 12 meals a day with most of it during the daytime. Goats are more selective feeders than sheep and have a longer vertical reach than sheep of the same weight. Browsing behavior is a skill that they learn. Goats learn to break twigs off the plants rather than to chew them off.

27.5.1.4 Feeding Behavior in Dogs
Dogs that have free access to food throughout the day eat many small meals a day, mainly during daylight hours and once-a-day feeding is not "natural," or not much preferred, by dogs. Addition of another animal to the household increases the original pet's interest in food. During estrus, bitches tend to eat less. Both cats and dogs have a lower metabolic rate following castration or spaying, which also accounts for their tendency to gain weight.

The common behavioural problems encountered are obesity, pica, grass eating, coprophagia, and psychogenic polyphagia.

27.5.1.5 Feeding Behavior in Cats

Similar to dogs, cats eat many small meals (12) per day when given free access to food, but unlike dogs, cats eat both in the light and in the dark. A feral cat with good hunting skills might easily catch 12 mice (or 3 rats) per day. Cats have neither been shown to increase their food intake when housed in groups rather than individually, nor does the sight of one cat eating stimulate other cats to eat when food is offered.

The common behavioral problems reported are wood chewing, pica, obesity, anorexia, and plant eating.

27.5.1.6 Feeding Behavior in Pigs

Pigs are essentially diurnal animals; therefore, most feeding takes place during the day. They tend to eat every 45 min during the day and every two-and-half hours at night. The pigs were observed to eat 8–12 meals/day, with the number of meals decreasing as the pig grew larger. Being a social animal, when one pig goes to the feeder, all the pigs follow it to the feeder. Pigs show a marked preference for sweet substances. Food intake is inversely related to environmental temperature; therefore, in hot weather, they eat less and increase food intake in lower temperatures.

27.5.1.7 Feeding Behavior in Horses

In free range, horses normally graze continuously for several hours and then rest, depending on the weather conditions and distances that have to be traveled to obtain water and sufficient forage. Grazing behavior is selective to the species it prefers to ingest. Horses show dynamic averaging in which they return to the last patch they grazed after a short absence. Horses prehend at a rate of 25 bites/min. Intake rate depends on bite size and rate. Horses sometimes exhibit pica, including geophagia (especially the soil rich in iron and copper) and ingestion of stones.

27.5.2 Sexual Behavior

Sexual behavior is important in all species of animals. Sexual behavior includes proceptive and receptive behavior by the female and courting and mate guarding by the male, as well as actual copulation.

27.5.2.1 Physiological Bases of Sexual Behavior

Adult sexual behavior depends on a variety of factors for its expression: These factors are the sex of the animal, perinatal organization, action of hormones, past social and sexual experience, the attractiveness of the potential mate, and the external environment. Both in males and females, the hypothalamic–pituitary axis is involved in the exhibition of sexual behavior. Testosterone acts upon the anterior hypothalamus–preoptic area in conjunction with appropriate stimuli from an estrous female to produce male sexual behavior.

27.5.2.2 Role of Pheromones in Sexual Behavior Manifestation

Pheromones are volatile chemicals secreted/excreted outside the animal's body to trigger a social response in the members of the same species. There are alarm pheromones, food trail pheromones, sex pheromones, and appeasing pheromones. They are found in the vaginal secretions/urine and are of value in breeding activity, reducing aggression and fear, and encouraging feeding in animals. The vomeronasal organ is a paired tubular organ located between the hard palate and the nasal cavity into which pheromones are aspirated. Receptor neurons in the lining of the organ detect pheromones and send information more directly to the hypothalamus than neurons in the main olfactory system. In ruminants and horses, flehmen or lip curl accomplishes this by closing the nostril while the animal breathes deeply. Cats gape and dogs tongue using their tongue to move material into the opening of the incisive ducts that open into the vomeronasal organ. In mice, the vomeronasal organ allows the recognition of sex.

27.5.2.3 Cattle

27.5.2.3.1 Cow

Cow is a nonseasonal, continuously cycling breeder. The onset of estrus occurs more often in the evening and ceases in the morning. Actual sexual receptivity lasts 13–14 h. The estrous cow exhibits increased motor activity and a decrease in feed intake. The more active a cow is, the higher her fertility. Investigative behavior such as flehmen, sniffing, rubbing, and licking increase as does premounting behavior such as standing behind the cow and resting the chin on the back of another, usually another estrous cow. The cow bellows frequently, switches her tail and raises or deviates it to one side, and urinates frequently. If a bull is available, the cow tends to approach it. Estrogen levels peak when the cow stands to be mounted. The cow that mounts is usually preovulatory. The mounting cows are also usually dominant over the mounted cows. This represents an interaction between hormonally mediated behavior and social influences.

Visual cues such as homosexual mounting are received over greater distances than olfactory cues and may attract bulls which are separated from the cows. Furthermore, bulls choose cows that are mounting and being mounted in preference to those who are not. Aggressive behavior also increases markedly during estrus.

Silent heats and Nymphomania are commonly reported sexual behavior problems.

27.5.2.3.2 Courtship Behavior in Cattle

The courtship sequence is actually a series of reciprocal interactions between male and female. Starting late in the cow's proestrus, the bull begins to graze beside the cow, guarding her from any other cattle. His attempts to mount will be repulsed by the cow. During proestrus, most females are attractive to the male and attracted to him but not yet receptive. The bull may attempt to herd the cow away from the rest of the herd. Periodically, the bull will smell and lick the cow's vulva, often followed by the flehmen response. Some, but not all, dairy bulls flehmen to estrous urine. Flehmen is followed by an increase in circulating LH. When the cow is in full estrus, the bull will have a partial erection while guarding her, and accessory gland fluid or precoital discharge will drip from the penis. The bull frequently nudges the female's flanks and may rest his head across the cow's back while they stand in a T-position. He makes several mounting attempts with a partial erection before the female will stand for him. When she is ready, the cow remains immobile and the bull mounts immediately. He fixes his forelegs just cranial to the pelvis of the female as he straddles her. Ejaculation occurs within seconds of intromission and is noted by a marked, generalized muscular contraction. The bull's rear legs may be brought off the ground during this act. Dismounting and retraction of the penis follow immediately. The stimulation of sexual behavior caused by a new female is called the Coolidge effect. The Coolidge effect occurs in bulls so that the mounts with intromission and decreased mounting intervals occur when a novel female is provided hourly.

Commonly noted problems among bulls include masturbation and impotence.

27.5.2.4 Sheep

27.5.2.4.1 Ewe

Sheep are short-day breeders, showing breeding activity in the winter months as the photoperiod decreases. The ewe is polyoestrous and will cycle several times during one breeding season if not bred; the average cycle for the ewe is 16 days (range 14–20). The actual period of estrous receptivity is 30–36 h in the ewe. The duration of estrus in lambs and in the first estrus of the season is shorter than that of the normal estrus.

27.5.2.4.2 Ram Effect

Most sheep breeding is still done naturally, with the rams pastured with the ewes all year or introduced in the late summer. The introduction of a ram, when the ewes have not been with one tends to synchronize estrus in a high proportion of the ewes 15–17 days later the process is mediated through LH, a pulse of which is released within minutes of exposure to a ram. If contact with the male is maintained, a preovulatory surge of LH at around 36 h occurs, accompanied by a rise in FSH. Ovulation occurs, but no estrous behavior (silent heat). Sexual behavior with subsequent ovulation appears 18 or 25 days after the introduction of the ram. The ram effect occurs only in sexually experienced ewes.

27.5.2.4.3 Courtship Behavior

The estrous ewe follows or seeks out a ram, may circle, sniff the male's body and genitals and then thrust her head against his flanks, fans, or wags her tail, and stands to be mounted. An estrous ewe calls frequently with nonspecific bleats and becomes more active. Standing occurs when the female is receptive, and the ram investigates and nudges her. Like cows, most ewes are in standing estrus at night. Perception via the vomeronasal organ is important. Ewes do differ in their sexual attractiveness to rams and are stable from estrus to estrus. Wooly ewes are preferred to short ones. Further, the attractiveness of estrous ewes for rams depends, in part on the bacterial flora present in their vagina. The bacteria may react with pheromones or themselves contributing to attractivity.

A ritualized kicking with a foreleg is performed as the ram orients himself behind the ewe; the leg is raised and lowered in a stiff-legged striking manner. Tongue flicking accompanies the nudging. The head is tilted and lowered while the shoulder is brought into contact with or oriented toward the flank of the ewe; simultaneously, the ram utters low-pitched vocalizations or gargling or courting grunts. Several abortive mounts may be made with pelvic thrusting but without intromission. When the tip of the glans penis contacts the vulvar mucosa, a strong pelvic thrust accomplishes intromission and ejaculation occurs immediately.

27.5.2.5 Goats

27.5.2.5.1 Doe

Goats, like sheep, are short-day breeders. Does in estrus show frequent wagging of the tail, vocalization, urination, and mounting of other females. This is probably proceptive behavior in that it does not occur as frequently if males are in the same enclosure. Does are in estrus for 39 h once every 20–21 days. There is a male effect in goats similar to that in sheep. Contact with bucks induces an immediate increase of luteinizing hormone followed by ovulation in anestrous goats. In contrast to ewes, a large proportion of does exhibits sexual behavior with the first ovulation.

27.5.2.5.2 Buck

In general, billy goat or buck behavior is similar to ram behavior, kicking at the doe with the front legs and stretching his neck toward her and emitting the gobbling vocalization. The buck holds his tail straight up during courtship. A component of the mating sequence unique to the goat is the urination by bucks on their own forelegs and beards during courtship. This behavior is termed enurination. Although various functions have been attributed to enurination, including increasing the intensity of the buck's odor and advertising his nutritional fitness, it occurs most frequently in a situation of sexual frustration when the buck is restrained from mating. Observation of does mounting one another or copulation by a male and female stimulates ejaculation sooner and with increased frequency.

Silent heat is one of the notable clinical problems in does.

27.5.2.6 Horses

27.5.2.6.1 Mare

27.5.2.6.1.1 Estrous Cycle

Mares are long-day breeders and cycle in the spring. Foaling occurs mostly in the late winter and early spring. Mares show individual preferences for stallions, and the preferences are influenced by the stallion's vocal behavior. The more the stallion neighs, the more likely the mare is to approach him.

The commonly observed problems in mares include anestrus, split, and prolonged estrus.

27.5.2.6.2 Stallion

Stallions exhibit libido throughout the year but show peak sexual behavior in the spring season. Seasonal changes are also seen in sperm number and testosterone levels. Stallions with a harem have a higher level of testosterone than those in a bachelor herd, which in turn have a higher level than stalled stallions.

27.5.2.6.3 Courtship Behavior in Horses

Courtship behavior will vary with the management practices involved. Driving, herding, or snaking with a distinctive head-down position is behavior usually elicited by the presence of another stallion. An adequate period of sexual foreplay is essential and males may tend to a female for several days before she is fully receptive. Nipping and nuzzling begin at the mare's head and proceed gradually along the body of the mare to the perineal area. During this testing phase, he exhibits the flehmen response. As sexual excitement increases, the male calls with neighs and roars. Full erection usually develops over several minutes in the mature stallion. Several mounts are usually made before intromission and ejaculation. During copulation, the stallion rests his sternum on the mare's croup and may reach forward to bite her neck. Ejaculation occurs around 15 s after intromission and after approximately seven thrusts, and intromission lasts less than 45 s.

Some common problems of sexual behavior in stallions are lack of sexual interest towards receptive mares, self-mutilation, lack of ejaculation, masturbation, and tendency to injure handlers or mares.

27.5.2.7 Pigs

27.5.2.7.1 Sow

Sows are nonseasonal breeders. After regular cycling commences, the sow will cycle every 18–24 days (average—21 days) until bred. Puberty occurs at 5–8 months. The presence of a boar leads to the occurrence of estrus at an earlier age.

27.5.2.7.1.1 Estrous Cycle

Sow shows an increase in activity nearing the estrus. Urination is frequent, as is calling to the male. The call is a soft, rhythmic grunt. An estrous female approaches the boar and sniffs him around the head and genitals. Estrous sows attempt to mount other estrous females, but subordinate sows rarely mount dominant ones.

27.5.2.7.2 Boar

Pigs are unusual in that defeminization occurs well after birth and is under the control of estrogenic metabolites that act as late as 3 months postnatally.

27.5.2.7.3 Courtship Behavior in Pigs

After contact with an estrous female has been made, the boar will pursue the female, attempting to nose her sides, flanks, and vulva. Unique to the pig is the boar's "courting song," which is used during this phase of courtship. This is a series of soft, guttural grunts, about 6–8/s. Tactile stimulation of the female continues and increases in intensity as the boar's sexual excitement increases. The boar usually emits urine rhythmically; Pheromones in the urine may further increase the female's willingness to stand. Several mounting attempts might be made until the female becomes immobile, after which mounting and intromission follow rapidly. Ejaculation occurs within 3–20 min, with an average of 4–5 min.

27.5.2.8 Dogs

27.5.2.8.1 Bitch

27.5.2.8.1.1 Estrous Cycle

The domestic dog, unlike most of its canid relatives, is a nonseasonal breeder. The length of each estrous cycle is extremely variable from individual to individual and sometimes from one heat to the next in the same bitch. Usually, one to four cycles yearly may be seen, with two being the most usual.

27.5.2.8.1.2 Courtship Behavior in Bitches

The hormonal levels and sexual behavior is very highly correlated in dogs. The first proestrus and estrus of a bitch's life are shorter than subsequent ones and the levels of LH and estradiol are lower. She is less attractive to the males and less proceptive. Courtship behavior is marked by play behavior in the proestrous part of the cycle, but this play behavior decreases during estrus. Urination becomes more frequent as estrus approaches and the posture used will frequently be the squat raise. The attraction of males and proceptivity appear in proestrus, but receptivity occurs later, during estrus. During estrus, she stands more quietly to allow male investigation and eventually intromission toward the end of estrus. When the male touches her vulva, she will flex her body laterally; while he is thrusting, she will move her perineum so as to increase the probability of intromission. After the lock or the copulatory tie has been established, she may roll or twist and turn.

27.5.2.8.2 Male Dogs

Sexual behaviors may appear in 5-week-old male pups, and mounting behavior becomes an important part of the male's social repertoire as it matures. As with many other mammals, mounting is used as a sign of dominance; a submissive animal will stand for a more dominant male.

27.5.2.8.2.1 Courtship Behavior in Dogs

Male dogs are attracted to estrous bitches. The urine of the estrous bitch appears to be more attractive to the dog than vaginal secretions, but a component of the vaginal secretions, methyl *p*-hydroxybenzoate, induces male sexual behavior and is considered as pheromone acting as a "releaser" of sexual behavior in the male.

Males may show extreme interest or indifference to females, although mating may occur successfully in either case. Play behavior may be marked or absent. The male sniffs the female's head and vulva; he may lick her ears. While canids do not show the classic flehmen response of ungulates, it is possible that the "tonguing" response seen during this olfactory investigation accomplishes the transport of pheromones to the vomeronasal organ in a manner similar to that postulated for ungulates. The male mounts in response to female immobility, thrusts his pelvis and when intromission has been achieved, the rate of thrusting increases. Engorgement of the bulbus gland and contraction of the vaginal muscles following intromission result in the copulatory lock or tie, a phenomenon most closely associated with canids but not restricted to them. The male will usually dismount and turn around so that male and female are facing opposite directions while ejaculation occurs. The lock may last 10–30 min, after which the bulbus decreases in size and the pair separates.

Impotence and lack of socialization are common sexual behavioral problems in dogs.

27.5.2.9 Cats

27.5.2.9.1 Queen

27.5.2.9.1.1 Estrous Cycle

The queen is seasonally polyestrous, and most cats exhibit cycle at least twice yearly if not bred. If unbred, the cat will cycle every 3 weeks for several months. Actual estrus lasts 9–10 days without copulation and around 4 days if the cat is bred. Most felids, including domestic cats, are induced ovulators, and thus breeding may be accomplished whenever the female shows receptivity.

27.5.2.9.1.2 Courtship Behavior in Female Cats

Anestrous females will call and purr, will be restless, and show increased general motor activity. If she is a house cat she may run from one room to the next, stopping to call at each door or window. She may be very affectionate towards the owners. Urination occurs frequently, and she may spray. She rubs her head and flanks on furniture; glands in these areas may produce pheromones indicative of estrous. She crouches, elevates her perineal region, and treads with her back legs. During proestrus, she will roll and solicit the male's attention but act aggressively if he mounts. This may be termed postural acceptance and affective rejection. When fully receptive, she becomes immobile and stands crouched in lumbar lordosis with her tail deviated to one side. An estrous female may show a darting behavior in the presence of several tomcats. She will repeatedly run a short distance from the toms, and this may be her means of assessing the relative strength of the males as they chase her and try to displace one another.

27.5.2.9.2 Tom

27.5.2.9.2.1 Courtship Behavior in Male Cats

The male probably locates an estrous female via olfactory cues deposited as pheromones in the urine and by sebaceous gland secretions. A male placed with a female for mating will spend some time investigating and marking the area with urine and anal gland secretions before mating. The cat shows a flehmen response, or gape, similar to that of ungulates. He calls to the female, circles her, and sniffs her genitalia. A nonreceptive female will actively, even violently, rebuff a male. When a female is receptive, the male approaches her from the side and behind and then mounts with the front legs. Intromission follows a forward stepping with arched back and pelvic thrusts. Ejaculation occurs seconds after intromission, and intromission usually lasts less than 10 s. The penis is covered with numerous small spines that apparently cause an intense stimulation as evidenced by the loud copulatory cry of the female with intromission. Copulation may occur every 10–15 min for several hours.

27.5.3 Maternal Behavior: Formation of Bond Between Mother and Fetus, Concept of Critical Period, Vocalization

Maternal behavior is influenced by hereditary, experiential, neural, and hormonal factors. The combination of the proper hormonal milieu and the stimulus for maternal behavior, the neonate, plus prior experience of being a mother can elicit and influence maternal behavior. The stimulation of maternal behavior appears to be under both hormonal and neural control. E.g., in ewes, estrogen rises and progesterone falls at parturition and hormonal priming by estrogen and progesterone, plus vaginocervical stimulation, is necessary in order to reduce aggression towards, or withdrawal from, alien lambs by ewes. Experience is also necessary for full expression of maternal behavior because only multiparous ewes would show positive maternal behavior. The fact that primiparous ewes routinely reject their lambs if they have been delivered by cesarean section also indicates the importance of neural stimulation by the passage of the lamb through the vaginal canal. The fact that multiparous ewes will readily accept their lambs even if they have been delivered by cesarean section indicates the importance of prior experience in ovine maternal behavior. The presence of neonates can induce maternal behavior in virgin females and also in males. This phenomenon is called concaveation.

27.5.3.1 Cattle

27.5.3.1.1 Parturition

A greater proportion of births will occur during the day if the cows are fed late at night. Signs of nearing parturition are relaxation of the sacrosciatic ligament, distention of the udder and teats, slackening of the tissue of the perineum and vulva, and mucous discharge from the vulva. Cows will alternate standing and lying much more frequently in the hours before parturition. Arching of the back and tail elevation occurs for 1–3 h before the chorioallantoic membrane ruptures. When the membrane ruptures, the cow often licks the fluid. A periparturient cow will sniff and lick other calves, especially if she is within 24 h of parturition and the other calf has just been delivered, whereas, after parturition, all her activities are directed toward her own calf. Licking of alien calves does neither cause rejection of the cow's own calf nor does suckling of an alien mother cause a calf to fail to suckle its own dam. Other cows will sometimes push or butt a newborn calf. Placentophagia occurs in approximately 82% of cattle.

27.5.3.1.2 Postpartum Behavior

Contact between the cow and her calf for as brief a period as 5 min postpartum results in the formation of a strong, specific maternal bond. Cows groom their calves during the early postpartum period, especially on the back and abdomen. Licking the calf occupies up to half the cow's time during the first hour postpartum; heifers lick less. Licking not only dries and stimulates the calf but also results in analgesia in the cow.

As soon as the calf's shoulders are free of the mother's vulva, the newborn calf begins to shake its head, snuffle, and sneeze. Some calves remain motionless for up to 30 min after birth, but within an hour most calves can stand. For the first time, the calf may take 30 min to an hour to locate the teats, but it will be able to locate it more quickly in subsequent nursings. Sucking problems and Cross-fostering are common abnormal maternal behavior.

27.5.3.2 Sheep

27.5.3.2.1 Parturition

Maternal behavior in sheep has an important clinical relevance, as most lamb mortality occurs within the first week of life in range-reared sheep. Lambs may be born at any time of the day or night, with peak frequencies being noted between 9 and 12 in the morning and between 3 and 6 in the evening. A few days before parturition, the ewe withdraws from the flock and seeks shelter. Shelter seeking by the ewe improves the environment into which the lamb is born so that its chances of survival are better. Nearing parturition, the ewe will withdraw and seek a corner in the pen. She will show restlessness, circle, vocalize, rub her head on her flanks, lick herself, and paw at her bedding 60–90 min before parturition. Grazing and ruminating cease. The interval between the onset of labor and the appearance of the lamb is usually 30–60 min but may vary. Even before parturition, nearly 20% of ewes show maternal behavior toward other lambs. This prepartum maternal behavior results in lamb stealing.

27.5.3.2.2 Postpartum Behavior

When the lamb is born, the ewe begins to lick it during most of the first hour after parturition. Primiparous ewes bleat more. Licking of the lamb is very important especially in cold or windy weather because it serves to dry the neonate and it additionally serves to stimulate the lamb. While the lamb is recumbent, the ewe licks its head, even restraining the lamb with a front leg to prevent it from standing. Licking of the perineal area stimulates the lamb to rise. Usually within the first 30 min, the lamb tries to stand and the ewe continues to lick the lamb mostly on the hindquarters. If the ewe stops licking, the lamb gives distress calls. Finally, licking of the lamb by the ewe establishes the maternal–offspring bond, which enables to identify her lamb by smell and taste. Usually, the fetal membranes are licked off the lamb and ingested, but the placenta is not eaten. The lamb raises and shakes its head and bleats. Although lambs are able to stand within 30 min or an hour of birth, it may take 2–3 h before they find the udder.

The "critical period" during which an ewe will accept a lamb is the first several hours after parturition. Normally, a ewe will stay within two meters of her lamb for most of the first day. If a ewe's lamb is removed immediately after birth, before she has licked it, the ewe will accept any lamb presented to her. After the ewe has spent 30 min to 2 h with a lamb, her own or a substitute, she will not accept any other lamb. Mutual recognition by the ewe and lamb depends on olfaction, audition, and visual cues. Common maternal behavioral problems include Cross-fostering, Mismothering, and poor mothering.

27.5.3.3 Goats

As parturition approaches, does, especially multiparous ones, leave the herd and seek a sheltered place, almost always near a vertical object. The does will defend this area both before and for the first day after the kid is born. A doe is usually very agitated vocalizing, urinating, and moving, but this response disappears just before parturition. As parturition approaches, they grunt, paw the ground, kick, and lick their backs. Does usually lie down for kidding. After parturition, the kid is sniffed and immediately licked; the head is the primary target. The licking continues for 2–4 h.

Goat kids should stand within 20 min and suckle within an hour. Within 4 h after the kid's birth, the doe can recognize it by sight and sound. Two-day-old kids can identify their mothers visually. Olfaction seems to be important for selective maternal behavior. Kid rejection is one of the most common maternal behavioral problems in goats.

27.5.3.4 Horses

27.5.3.4.1 Parturition

The onset of parturition in mares is heralded by waxing of the udder, but the length of time between the appearance of udder waxing and the appearance of the foal might be quite variable, up to 21 days. Body temperature is lower on the day prior to parturition. The mare will walk more and stand less on the evening of parturition. In the first stage of labor, which lasts for about 4 h, the mare is restless and will crouch, straddle, and urinate. The smell of fetal fluids is attractive to parturient mares. The mare will exhibit the flehmen response in response to the amniotic fluid that is expelled. Sweat will appear on her elbows and flanks. During the second stage of labor, the mare will lie in lateral recumbency. This second stage is very violent and very short in horses, lasting less than a half hour.

27.5.3.4.2 Post-Parturient Behavior

Ordinarily, the foal is delivered in such a way that the mare needs only turn her head to meet her foal muzzle to muzzle. Licking, as well as sniffing, is concentrated first on the head of the foal and later on the hindquarters, particularly the perianal area.

Standing and suckling occur within 1–2 after birth for pony foals. Foals suckle four times per hour at 1 week of age and gradually decrease the frequency to once per hour by 5 months. Mares are very protective of their foals and are usually very aggressive towards other horses and sometimes towards people for the first day or two after foaling. The first hour is probably critical for the mare to learn to recognize her foal selectively. The foal appears to take much longer, perhaps as long as a week, to recognize the mare. The neighs (or whinnies) of the separated mare and foal are impressive, and horses make use of these calls to locate one another. Mismothering can occur in equids, although it is much rarer than in sheep.

27.5.3.5 Pigs

27.5.3.5.1 Parturition

Most of the farrowing take place in the afternoon or night. After labor begins, most sows lie in lateral recumbency. The sow will swish her tail violently as abdominal straining takes place. Parturition usually takes 3–4 h but varies considerably with litter size and the condition of the gilt.

27.5.3.5.2 Behavior of the Sow

Maternal behavior in sows is exhibited as calmness and care when lying down to avoid piglets and remaining in the nursing posture following milk ejection and protectiveness in response to piglet squeals and human approach to piglets. When not confined, the sow will eat the placenta. It may be recycling of nutrients or a form of defense against predators by removing odors or as a means of inducing analgesia. Sows do not often lick their newborns.

27.5.3.5.3 Behavior of the Neonatal Piglet

Piglets make a most startling transition from fetal to independent existence. They may be apnoeic for 5–10 s after birth. Then, they give a few gasps before beginning to breathe regularly. Their eyes and ears are open, and they are able to walk immediately, although their gait is staggering for the first few hours. Most piglets are nursed within 30 min of birth. Piglets are attracted to soft, warm surfaces, pig vocalizations, and the sow's odors and they move in the direction of the sow's hair growth. The firstborn piglet appears to use thermal, tactile, and olfactory cues to find the udder, whereas the later-born probably respond to the suckling sounds of their older littermates and walk straight to it because social facilitation is strong in pigs at birth.

27.5.3.5.4 Mutual Recognition

Sows and piglets apparently use olfaction to identify one another but need more than a day and possibly as long as a week to learn. The piglets can identify their dam's feces, milk, and urine odors, as well as their vocalizations. Sows respond to playback of piglet separation calls by vocalizing, but cannot discriminate their own from other piglets on the basis of their voices. However, she learns to identify her own piglets by the time they are a week old on the basis of olfaction. Piglets can easily be fostered by another sow when they are less than a day old. After that, the fostered piglets walk around, and are reluctant to suckle, as they have already formed a bond with their dam.

Cannibalism occasionally occurs in sows, especially in nervous primiparous gilts and sows suffering from mastitis will refuse to nurse.

27.5.3.6 Dogs

27.5.3.6.1 Parturition

During late pregnancy, the activity of the animal decreases and its appetite increases. The animal will be anorectic as abdominal pressure increases as it nears whelping and takes only small quantities of feed. The animal will be usually restless, shows nesting behavior, and grunts each time it sits, especially if it is carrying a large litter. At parturition, the bitch licks itself, any soiled bedding, and, almost incidentally, the puppy. The pup's head, umbilicus, and perineum are the area's most frequently licked. A pup is delivered every 30–60 min on average. The bitch cuts the umbilical cord with her molars and then licks the stump. The placenta is expelled 5–15 min after each pup is born. If the placenta has not been delivered, it may extract it by pulling on the cord. The placenta is usually eaten by the bitch. As the labor progresses the animal usually becomes silent and less active unless dystocia occurs. Labor can be interrupted if the bitch is disturbed. Depending upon the temperament of the dog, 15 min to an hour is required for normal parturition to proceed. Amniotic fluid is very important because if pups have been cleaned the bitch will not accept them.

27.5.3.6.2 Postpartum Behavior

The puppies move forward by paddling and after they encounter the mother's body, they nuzzle through the fur, usually attempting to burrow beneath her. Puppies have two vocal signals, the whine and the grunt. The whine is emitted whenever the puppy is cold, hungry, or separated from its litter or mother. The grunt is apparently a pleasure communication that occurs when sought-after warmth or reunion occurs.

Bitches lick their puppies a great deal. The licking arouses the pups to eat, and, when it is directed at the anogenital area, it stimulates urination and defecation that would otherwise not occur spontaneously. The bitch keeps the nest area clean by consuming the urine and feces of her puppies for at least the first 3 weeks of their lives. Licking helps in retrieval which is unique to dogs. Bitches usually do not carry their puppies. Instead, they lead them back to the nest by licking the pup's head. The common problems encountered are pseudopregnancy, maternal rejection, and repeatedly moving the pups.

27.5.3.7 Cats

27.5.3.7.1 Parturition

Gestation period in cats is 63–66 days. Births have a seasonal distribution, with the greatest number of litters being born in the summer and the least in autumn and early winter.

Most cats will choose a cave-like place, such as a closet or a linen cabinet. Parturition in the cat is characterized by a great amount of licking by the queen: self-licking mostly directed at the belly and genital area, licking of the fetal fluids from her body or the floor, and licking the kittens. The queen is typically very restless and other behaviors include frequent lying down, sitting up, squatting, bracing lordosis, licking of the vulva, circling, and walking around the cage.

The queen will eat the placenta as it is delivered. In the process of eating the placenta, the queen stretches the umbilical cord so that little bleeding occurs when the cord is severed. Most kittens are suckling within an hour or two of birth.

27.5.3.7.2 Postpartum Behavior

Even before the opening of their eyes, kittens can return to their nest from several feet away by using olfactory cues. Queens retrieve their kittens in response to auditory, rather than visual signals. The more the kitten vocalizes, the more apt the queen is to retrieve it. The queen usually picks the kitten up by the scruff of the neck, though occasionally it grasps the skin at the back of the head or even the kitten's whole head.

Grooming plays an important part in feline maternal behavior as in most other species. Queens lick their kittens frequently and in particular lick the perineum to stimulate urination and defecation for the first 2–3 weeks of life. In common with most carnivores, female cats ingest the kittens "urine and feces" for several weeks postpartum thereby keeping the nest clean. Cats will accept alien kittens that are not too much older than their own at the time of parturition. Infanticide, mismothering, and overgrooming are the common problems observed.

27.5.3.7.3 Parental Care and Its Influence on Behavior

Parental influence and parental contacts vary among different species. For example, marsupial infants are born after a very brief gestation period and they are fused to the mother's nipples and are no more than externalized fetuses. Their sensory capacities are very much limited at this stage and mother–infant interactions is only through nursing. But even when they become independent, they are constantly with the mother, carried in the pouch thus representing a far higher degree of contact than the part-time attention received by baby rats and mice.

In contrast, the mother rabbits (Oryctolagus) visit their litter only once a day initially. and the milk is pumped into the babies. The young ones are briefly groomed and then the nest is covered and left; the mother herself has another nest chamber elsewhere. Probably this behavior helps to minimize the chances of predators finding the nest by picking up the mother's scent.

A similar explanation is suggested for the yet more extreme maternal behavior of tree shrews (Tupaidae), a rather distinct group of small insectivore-like creatures that have a distant linkage to the primates. They build two separate nests, in one of which the adults sleep while the female produces the litter in the other. The young are visited briefly every second day; they suckle rapidly and become distended with milk. The mother then leaves them with the very minimum of grooming and typically does not return for 48 h. The young are capable of grooming and licking themselves from the day of birth. They stay alone in the nest until they emerge at 33 days of age.

The primates, have some of the longest and closest associations between parents and offspring, lasting for 6 or 7 years and beyond in the great apes. The relationship between a primate mother and its offspring is absolutely crucial for almost every aspect of behavioral development. Separation, particularly in the early months, can have very severe effects on infants.

27.5.4 Milking Behavior and Factors Influencing Milking Behavior

Milk let-down is a neuroendocrine reflex initiated by oxytocin release following stimulation of the teat by a suckling calf. However, the cows with calves soon start to release oxytocin when other stimuli from the calf are detected, and cows milked using a milking machine may respond to other cues, such as the sounds, in a similar manner. Milk let-down by dairy cows in response to the typical sounds of a milking parlor is the best example of classical conditioning on farms. Milk let-down is a conditioned response to the conditioned stimulus of clanking noises, made by the washing of milking vessels, etc., in the milking parlor.

Know More ......

Rhesus monkeys are more restrictive, rarely allowing their babies to leave their side for several weeks and control outside contact for a much longer period. However, the Barbary macaque (Macacasylvanus) of North Africa which is a close relative to the rhesus monkeys behaves quite differently. Mothers allow other individuals, males as well as females, to hold their babies even within a few days of birth. In all primates, it is commonly observed that newborns and young generally are objects of intense interest to other members of a group, particularly young females. "Aunting" behavior has frequently been reported from primates and is also an important part of parental behavior within the matriarchal groups of African elephants. They too have a long infancy and a rich social environment. Elephant calves do their best if they grow up in a group where they have not just their mother's attention but of other females and juveniles as well.

There is a breed variation among cattle with respect to how readily such conditioning can occur, for example, breeds such as the Salers in France do not readily let-down milk to stimuli other than those emanating from real calves as compared to Friesians or Holsteins. Farmers need to be aware of the fact that milk let-down in a parlor is a conditioned response and that such learning depends upon adequate experience and training.

27.5.4.1 Factors Influencing Milking Behavior

- Animals should not be handled roughly or beaten while leading to the milking parlor. They should not be coerced into the milking area.
- A frightened or stressed animal may show inhibition in the conditioned milk let-down. Experiments have shown better milk let-down in response to music played in the milking area.
- Strangers should not be allowed in the milking area during milking.
- Presence of disturbing stimuli in the milking parlor may prevent the young animal from learning and the older animal that is conditioned may be inhibited from exhibiting the response.
- Veterinary operation involving discomfort for the cow should be avoided out in the milking parlor and carried out in a separate facility.
- Providing concentrates during milking may strengthen the conditioned milk let-down.
- Barking dogs may be avoided in the milking area.

Nervous cows that are difficult to get into the bail or those not settling down for easy milking will not be suitable as good milking cows. These may be replaced with cows which have a quiet and relaxed temperament.

27.5.5 Social/Gregarious Behavior

In wild, animals often tend to prefer living in social groups as it provides protection against predator attacks, enhance hunting prospective, mating opportunities, helps in learning, finding feed, grasslands, and water resources. Animals in groups exhibit various behaviors such as dominance hierarchy, peck order, teaching, cooperation, and playing.

27.5.5.1 Dominance Hierarchy

It is a type of social hierarchy that arises when members of a social group interact with one another, to create a ranking system. In social living animals especially in wild, members compete for limited feed resources and mating opportunities. Rather than fighting they confront; a relative rank order is established between the members of the same sex in a species. This order is thus created based on repetitive social interactions and is subject to change whenever the dominant animal is challenged by his/her subordinate. Hence, these hierarchies are not fixed and depend on a number of changing factors such as age, sex, body size, intelligence, and aggressiveness.

Several brain areas are linked to hierarchical behavior in animals. One of the important areas is prefrontal cortex, a region involved in decision-making and social behavior. High-ranking macaques have a larger rostral prefrontal cortex in larger social groups. Amygdala, thalamus, and dorsal raphae nucleus are also associated with social ranking behavior.

27.5.5.2 Peck Order in Chickens

The social rank order established in chicken is referred as peck order. The peck order influences the feeding, drinking, egg laying, mating, and crowding activities. The stronger members of a flock are at the top of peck order while weaker or submissive birds are placed lower in the order. Hens high in the pecking order chase out other hens out of nest boxes they favor. There is a reduction in aggression and a lower incidence of conflicts when a peck order is established.

27.5.5.3 Teaching in Animals

Teaching is a behavior often seen in animals and it helps in imparting some new information to offsprings/individuals faster than they would otherwise receive it. Mother cheetah captures live prey and allows its young to interact with this prey, making sure that the prey does not escape along the way.

Another example can be observed wherein the meerkats young pups begin following groups of foragers, and the pups are assisted in their own foraging attempts by older groupmates called "helpers." Helpers will often incapacitate scorpions by removing their stingers and presenting them to pups as food. Very young pups were either fed dead or incapacitated scorpions, but as the pups got older, the helpers presented them more and more often with live scorpions.

Similarly, the key element of elephant society is the matriarchal group led by the oldest and largest female with her daughters and their offsprings. A baby born into a group is known to all others, it is nourished and closely protected by its mother and other relatives for several years. Slowly it acquires the adult feeding behavior, learning how to select food and migrate according to the seasonal changes of vegetation and water supply.

27.5.5.4 Cooperation

Cooperation refers to an outcome in which two or more interacting individuals receive a net benefit from their joint actions, despite the costs they may have to pay for undertaking such actions. For example, jointly hunting prey may provide each of the two hunters with food, even though there are costs (possible injury, energy expended) associated with hunting. In addition to looking at outcomes (i.e., successfully capturing prey), it is also important to examine cooperation in terms of individual action.

Know More . . . . .

An interesting case of cooperation is observed in the Rodriquez fruit bat (*Pteropus rodricensis*). In this bat, during the birthing process, unrelated female helper continues providing assistance by grasping the wings of the pregnant female, providing both protection and warmth, and subsequently cleaning (licking) newborn pups upon their emergence. Once pups are born, helpers guide the newborns into a suckling position, where they can obtain milk from their mother. These bat "midwives" cooperate with pregnant mothers during virtually every stage of the birthing process. Assistance during birth has been recorded in marmosets, Indian elephants, African hunting dogs, raccoon dogs, and bottle-nosed dolphins.

27.5.5.5 Social Grooming

Many animals groom, or clean, themselves in an attempt to remove external parasites. This sort of grooming might involve scratching an area of skin or licking it. Social grooming, or allogrooming, in which one individual grooms another, is one of the most obvious and frequently observed cooperative behaviors. While social grooming often serves the function of removing parasites from the body of a partner, it may also have many other functions as well. Social grooming has been considered a large part of the glue that holds primate troops together. One benefit linked to social grooming in primate groups is "tension reduction." One way in which primates of a group reduce the chances of escalating violence is through social grooming, which has the effect of lowering the level of tension between putative combatants. From a proximate perspective, one means by which social grooming lowers tension levels is by increasing the circulating levels of hormones such as endorphins and opioids. This sort of tension reduction through social grooming may play a role in nonprimate species as well. Primates also appear to exchange social grooming for access to scarce resources such as water or food, entrance into new groups, aid in chasing potential predators away, and future association with individuals who possess "special skills" that they themselves do not.

Other examples of specialized behavioral learning include food-storing behavior, i.e., when the chance arises, a number of bird and mammal species collect more food than they can eat immediately and hide it. In this way, they can take advantage of a rich food source while it is there and reduce the degree to which they have to share it with others. Some animals make a single larder to which they return regularly, interesting some species hide food items singly or in caches dispersed around their territory.

27.5.5.6 Play

Animals play with each other and with objects spending long periods. Although the adults may continue to play intermittently, it tends to fade out with age. It is commonly suggested that play has a role in the development of adult behavior. Play is an energy-consuming activity and 5–20% of the "surplus" energy available is expended while playing. The cost is not just in energy; young desert bighorn sheep (*Ovis canadensis*) cavorting in play through their harsh environment are often pierced by cactus thorns; young gelada baboons (*Theropithecus gelada*) playing on the steep cliffs of Ethiopian gorges have been noticed to fall 5–10 m and limp painfully away. Another significant cost arises because vigilance is likely to be reduced when young animals play, making it potentially dangerous.

The play behavior is very conspicuous among the ungulates, carnivores, whales, elephants, and primates. Each group has its own typical play repertoire; ungulates butt each other, mock fights and chases are common among primates, while carnivores show elaborate stalking, leaping, and prey-catching behavior. The mock fights among kittens or the amazing high-speed chases and wrestling bouts of young monkeys play a vital part in the development of skills which are going to become vital for survival during adult life.

Apart from the obvious physical skills, numerous other functions have been suggested for play such as gaining knowledge about potential prey species, gaining knowledge of the social group and one's position within it, exploration of the environment, and so on. It prepares animals for the unexpected. Mothers often take part in play sessions and their influence can be detected indirectly as well. Play builds up the flexibility of responsiveness and hence improves survival. Play in young mammals takes place at a time when crucial maturation events are occurring in the brain and muscles which lead to the differentiation of synapses in the cerebellum and the control of complex movements.

With some members of the cat family, there is a very conspicuous type of play that seems particularly directed towards the teaching of prey-catching. Domestic cats bring home live prey which they "play" with, letting them loose only to pounce on them again, often repeatedly. Cheetah mothers may bring back a crippled gazelle fawn to her young ones.

27.5.6 Agonistic (Combat or Aggressive) Behavior

In animals, aggression is a threat or harmful action directed against another individual and includes snarling, growling, and biting. Aggressive behavior is usually displayed to threaten or attack to resolve competitive disputes over limited resources (feed, territory) or to increase their reproductive potential, or to escape from threatening situations. However, aggression can have multiple motivations, especially in pet animals.

Aggression is a normal communication modality and is not necessarily pathological or abnormal behavior. Normal aggression is demonstrated in situations that warrant aggression. The intensity of the aggression by the animal is modified, depending on the situation and the relative level of the threat. For example, a normal aggressive display would be if a stranger broke into the home unannounced and was bitten.

27.5.6.1 Conflict-Induced Aggression

This aggression typically occurs when the pet seeks out the pet owner, perhaps even solicits attention, then shows conflict behaviors (freezes, shifts eyes) and aggresses. The pet could leave the situation but chooses not to because of conflicting motivations (desire to be with the pet owner and uncertainty of what will happen next.)

27.5.6.2 Possessive Aggression

Possessive aggression is described as guarding of an item (food) from people.

27.5.6.3 Alliance-Induced Aggression

Alliance-induced aggression refers to aggression between dogs in the same household that fight only in the proximity of the pet owner. Usually, the more fearful and most owner-dependent dog becomes anxious when the other dog approaches the owner and consequently the less confident dog aggresses.

27.5.6.4 Idiopathic Aggression

Idiopathic aggression refers to an unexplained and intense aggression in which no identifiable aggression-provoking stimuli can be determined on a complete medical investigation.

27.5.6.5 Maternal/Hormonal-Induced Aggression

Maternal aggression occurs when a bitch defends her puppies from people or other animals. It is considered to be normal behavior to a certain degree.

27.5.6.6 Play-Induced Aggression

Canine mouthing and play biting are considered normal forms of interactions and communication for dogs but can become problematic when directed toward humans. Although not considered as a true aggression, over-exuberant play can be damaging to the human–animal bond.

27.5.6.7 Redirected Aggression

Redirected aggression is incidental to another form of aggression or emotional arousal. It occurs when a dog or cat cannot reach the target of its aggression and out of frustration changes its focus of aggression to an object, person, or other animal that is not the stimulus for the aggressive arousal.

27.5.6.8 Territorial Aggression

Dogs are genetically programmed to be territorial. Territorial aggression refers to the aggression directed towards people or animals that are not part of the dog's immediate social group or are unfamiliar to them when they enter the dog's perceived territory. The territory may be a fixed location (i.e., house, room, yard, and bed) or a mobile territory around the dog (i.e., leash, proximity to owner). Territorial aggression is influenced by breed, age, sex, and socialization experience.

27.5.7 Communicating Behavior and Modes of Communication in Different Species

Animals communicate not only by auditory signals but also by visual and olfactory signals. Communication in animals depends on their capacity to perceive messages. Many of the stimuli animals encounter come from other animals and these form the basis of communication. Animals communicate with each other using signals, which often closely match the sensitivities of the sense organs of the animal receiving them.

27.5.7.1 Visual Signals

Every animal communicates with one another, with animals of other species through various visual cues. The body posture movements and overall behavior differ among species.

27.5.7.1.1 Dogs

A dog's emotional state can be identified by observation of its ears, mouth, facial expression, tail, and overall body position and posture. A calm dog stands with ears and tail hanging down. When it becomes alert, its tail and ears are pointed upward. The dog may point with one front foot and as the dog becomes more aggressive, the hair on the shoulders and the rump rise and the lips are drawn back. The ears remain forward and the tail wags slowly. On increasing aggression, the lips get retracted and the teeth are exposed in a snarl. The dog stands straight. As the dog gets frightened, the ears go back and are flattened against the head and the tail is tucked between the legs.

Dogs greet their owners as they do with their mothers: by licking their faces. Puppies lick their mothers' faces begging for regurgitated feed.

27.5.7.1.2 Cats

A cat carries its tail high when greeting, investigating, or frustrated. During stalking, the tail is depressed and the tip is wagged. A calm and relaxed cat usually stands with its tail hanging; its ears are usually kept forward.

27.5.7.1.3 Pigs

The tail, particularly in piglets, is a good index of their general well-being in most breeds. Although Vietnamese mini pigs do not curl their tails, a tightly curled tail indicates a healthy pig in most breeds, and a straight one indicates some sort of distress.

27.5.7.1.4 Horses

The horse's ears are probably the best indicator of its emotions. The alert horse looks directly at the object of interest and holds its ears forward. During aggression, the ears are pointed back, and the flatter the ears are against the head, the more aggressive the horse is.

27.5.7.2 Vocalizations

The animals vocalize with one another to communicate.

27.5.7.2.1 Cattle, Sheep, and Goats

Vocal communication in prey species such as cattle may be the most important in disseminating information about general safety or danger. It might have been more important for cattle to be alert and ready to flee than to communicate more precise information in their calls. The moo is low-pitched and most common. The other common vocalizations are higher pitched—hoot or roar consisting of repeated brief calls, usually by a distressed cow. A threatening bull gives a roar of high amplitude. A very hungry calf will give a high-intensity "menh" call. Grunting sounds are heard during copulation. Vocal communication in sheep consists of bleating in distress or to initiate contact. Ewes rumble to their newborn lambs and rams make a similar call while courting. The snort is an aggressive communication in sheep.

27.5.7.2.2 Dogs

The common vocal communications of dogs are bark, whine, howl, and growl.

Barking is a territorial call of dogs. It is used to guard a territory and to demarcate its boundaries. Whining is a care-soliciting call of the dog used by puppies to communicate with the mother, who provides warmth and nourishment. Mature dogs whine when they need relief from pain or are in even a mildly frustrating situation, such as when they want to escape outdoors or reach a rabbit for which they are digging. Howling is a canine call that occurs more frequently in wild canids, coyotes, and wolves and in some breeds of dogs, such as huskies, malamutes, and to a lesser extent in hounds. Growling is an aggressive or distance-increasing call in dogs.

27.5.7.2.3 Cats

The common vocalization call include Murmur, Purr, Growl, Squeak, Hiss, Shriek, Spit, Estrus call, and Chatter. Howl and yowl of an aggressive cat, Mowl, or caterwaul, Mew, Moan, and Meow.

27.5.7.2.4 Pigs

Vocal signals are probably the most important means of communication in pigs. Nearly, 20 calls have been identified, Bark, Grunt, and squeal being the common ones.

Bark is given by a startled pig. A shorter grunt is given by an excited or investigating pig. Continuous short grunts are given by a threatening sow and may precede an attack on anyone who disturbs her litter. In a milder form, it can be a greeting. A long grunt (0.4–1.2 s) might be a contact call and is often associated with pleasurable stimuli, especially tactile ones. The squeal is a more intense vocalization indicating arousal and an injured pig screams.

27.5.7.2.5 Horses

Neigh and nicker are common vocal calls which evoke a reply. The neigh (or whinny) is a greeting or separation call that is most often heard when adult horses or a mare and foal are separated. The soft nicker is a care-giving or care-soliciting call given by a mare to her foal upon reunion and probably is mutually recognized. A horse may also nicker to its caretaker and a stallion to a mare in estrus.

27.5.7.3 Olfaction

Olfactory acuity is probably the most important sense in the domestic animals because individual odor recognition and pheromonal release are an important part of their communication and especially during breeding. Olfaction to animals is what writing is to humans, a message that can be transmitted in the absence of the sender. The sender must be present for auditory or visual signals to be sent, but an odor persists for minutes (or days) after the sender has gone. Dogs probably have the greatest olfactory acuity and this macrosmatic species hence is the one most investigated. They can distinguish between the odors of identical twins and detect the odors of fingerprints 6 weeks after the fingerprints were placed on the glass. Bloodhounds are indeed the most sensitive of all dog breeds. Dogs frequently are trained to sniff out drugs and natural gas leaks, and bloodhounds have been used for centuries to track people; apparently, no modern invention is as reliable as the canine olfactory mucosa. A large dog has over 15 times the area of the epithelium of a human and 100 times the density of sense cells per unit area. Pigs and bears use their sense of smell to detect roots and bulbs or burrowing rodents well below ground. Males can also determine whether a female is ready to mate from her scent.

27.5.7.3.1 Olfactory Signals in Cats

Cats use several olfactory signals such as scent marking and rubbing with the secretions such as anal secretion, urine, and facial gland secretion.

Male cats scent marks, that is, spray urine, more than females, but both sexes do it. They spray trees along their most frequently traveled path. Spraying is also done by cats that are subjects of aggression. The smell of tomcat urine is due to the sulfur-containing amino acid, felinine, which is an important olfactory component in territorial spraying. Cats are well known for their fastidious covering of their feces, but in some situations, such as outside their core living area, cats may leave their feces uncovered. Cats probably use fecal and anal sac odor for communication. Cheek rubbing (bunting) behavior may also be a form of olfactory communication in that glandular secretion from the cat's face is deposited on the object bunted.

27.5.7.3.2 Olfactory Signals in Boars

Boars may use behavioral signs more than pheromones to determine the sexual receptivity of the sow. Boars are the only male ungulates that do not exhibit flehmen. Instead, they gape like a cat does when they encounter sow urine. Females can identify intact males, by the strong boar odor produced by the androgen metabolites present in both the saliva and in preputial secretions.

27.5.7.3.3 Other Signals

Duck-billed platypuses, sharks, and rays have the capacity to detect the electric fields generated by muscular activity in animals such as the crayfish and shrimps on which they feed. Electric fish go one stage further and create their own electrical environment by using specially modified muscle tissue which helps in social communication and to find prey.

27.5.8 Behavioral Disorders

27.5.8.1 Dogs and Cats

Failing to respond when called, jumping up on owners or visitors, and running away are minor behavior problems which may strain the owner–dog bond. Tail chasing, light chasing, circling, and digging for imaginary prey are compulsive problems. Phobias, especially the fear of storms, are often a serious problem. Destructive chewing behavior caused by oral exploration, escape attempts, or separation anxiety are common complaints reported by the owners. Signs of separation anxiety include destruction at doors, general destruction in the house, house soiling, excessive salivation, distress vocalization, or self-trauma. Other problems include urine marking, hyperexcitability, and acute conflict and stereotypical behaviors.

27.5.8.2 Horses

Stall walking in horses may occur as a stereotypy, a repetitive, functionless behavior seen mostly in confinement. Pawing is a response to frustration, a displacement activity that might have originated from the activity of uncovering food buried under snow. Most horses kick the stall walls with their hooves and a few horses may knock their hocks against the wall. Cribbing is an oral behavior wherein the horse grasps a horizontal surface, such as the rim of a bucket or the rail of a fence, with its incisors, flexes its neck, and aspirates air into its pharynx. Some horses may aspirate air without grasping an object. This is called aerophagia or windsucking or pneumovaginitis.

As horses are generally neophobic, that is, they are afraid of new or strange things many of them exhibit failure to load when transported on the ground by trailers. Both innate behaviors and learning contribute to loading problems. Dark interior of the trailer; the hollow sound of hooves on the ramp, and the instability of the ramp and vehicle, and unpleasant experiences with loading or riding in a trailer may exaggerate horse's innate fears. Hence, a horse that is reluctant to enter a trailer may be willing to follow another. Horses are less stressed when traveling with a companion horse.

27.5.9 Training of Dogs and Horses

Specially trained working dogs have a variety of roles in our society, ranging from livestock handling and care in farms, to military and law-enforcement dogs who detect bombs, fugitives, or drugs; to search and rescue dogs who save people trapped by natural disasters such as earthquakes or avalanches. Guide and service dogs contribute invaluable assistance to the visually and hearing impaired and people with limited mobility and in recent years they are being trained as therapy dogs to help autistic children, the aged, and people in depression to overcome their health issues and interact with the society in a better way. Human–animal interactions documented to have positive health effects in the areas of social behavior, interpersonal interactions, and mood; improvement in stress reduction scores, heart rate, and blood pressure; and reduction in fear and anxiety.

Training becomes a necessary component in rearing animals, especially in pets and we often need to train them for basic daily life skills and specific purposes-oriented skills. Being highly intelligent social animals, they have proven themselves capable of learning many extraordinary tasks to assist us. Dogs know when we are paying attention to them and understand our perspective. They can convey information to people as well as read our signals to them. Dogs can signal the location of a hidden object by using "showing behaviors," like alternating their gaze from the object to their human companion repeatedly until the target is acquired. Studies report that dogs use multiple facial features of humans such as the eyes, midface, and mouth to determine emotional meanings and respond accordingly. Thus, training the animals involves intelligent interaction between the trainer and the animal demanding a lot of patience and understanding.

27.5.9.1 Training in Dogs

There are two types of associative learning that are particularly useful in training and working with dogs: classical conditioning and operant conditioning. Classical conditioning involves reflex-like, involuntary, or emotional responses, and occurs when a stimulus that was previously meaningless ("neutral") to the dog takes on the power to elicit such a reflex-like response. Practically, we can use classical conditioning to build a positive association between the dog and something or someone (i.e., some stimulus or trigger) in the dog's environment. For example, if we want the dog to be friendly with children, veterinarians and other dogs, and cats, we can achieve it by some classical conditioning, i.e., we may ensure that whatever the trigger is, it becomes a predictor of good things. So, each time the puppy sees a child or veterinarian approaching on a walk speaks in a happy voice and offer him a delicious treat essentially, we are trying to teach him that "children/veterinarian mean good things are going to happen."

Another very important form of associative learning is operant conditioning. Behaviors that are followed by positive outcomes (i.e., are rewarded or reinforced in some way) are more likely to be repeated; behaviors that are followed by negative consequences (i.e., result in some form of punishment) are less likely to be repeated. One important aspect of using rewards in dog training is the speed at which the dog is rewarded for successfully performing the behavior. If the reward takes longer than a second or two to reach him, he may not understand exactly why he has received the reward (or, he may have performed a different, less desirable behavior in the meantime). Provision of immediate reward and reinforcement to the dog can greatly improve the effectiveness of the training.

Shaping a way to train a new behavior, especially a fairly complex one, is to break down the desired behavior into tiny increments or steps, and reinforce the dog at each incremental step until we get the full behavior from the dog.

Continuous reinforcement means that every time the dog performs the behavior, he earns a reward and is the best schedule to use when teaching a dog a new behavior. Intermittent reinforcement, on the other hand, means that the dog does not receive a reward every time he does the behavior, but only periodically, and either on a fixed schedule or on a variable schedule. Intermittent reinforcement should only be used once the dog has successfully learned the behavior, but it is often the best way to maintain the acquired response (i.e., the dog will continue to work in anticipation of the reward, even when the reward does not happen each and every time).

Today, rewards-based training is considered as the science in dog training. Older methods that rely heavily on force and positive punishment are not in vogue. Scientific literature on dog training methods and veterinary medicine highlight and emphasize the dangers of positive punishment (increased fear, high risk of increased aggression, disruption to human–animal bond).

27.5.9.2 Training in Horses

While human–horse interaction has existed for many millennia through hunting, it is only relatively recently that horses have been used for agriculture, transport, war, and, more recently, for sport and leisure. Since the beginning of domestication, various techniques for horse training have been developed and passed on to subsequent generations orally or through the literature. However, all these techniques are constrained by the biology of the horse. When it comes to getting the perfect efficient output from the horses in sport and work, we need to be well acquainted with their behavior. Effective and humane training always takes account of the animal's ethology, even though the training systems differ.

Horse training differs fundamentally from the reward-based training methods used for marine mammals, exotic carnivores, and most companion animals because it largely relies on negative reinforcement. During their early training, horses learn that the correct responses result in the reduction of pressure from the reins when they stop or slow. Pressure from the rider's legs is reduced when the horse moves forward. To be effective and humane, the application of pressure must be subtle and its removal immediately once the horse complies.

Training a New Behavior in Horse

When a horse is taught a new behavior, the behavior should be reinforced after every occurrence initially and then the reinforcement can be gradually spaced after occurrence. In this manner, the behavior is more likely to be rapidly learned and efficiently remembered well. For this, the horse must achieve the goal in order for it to be rewarded.

An effective trainer must maintain his position as controller and apply the three Cs:

1. Communicate—the voice command tells him what he should do
2. Coordinate—apply the aids correctly to reinforce this command
3. Cohere—if you reward him appropriately, he will want to respond to you

As the horse becomes more confident and able, he will learn to discriminate more subtle signals than a voice command. He will prepare to jump when we want him to even before we think we have asked him.

27.5.9.2.1 Behavior Chaining

Some tasks are more complicated and can be learned more efficiently by using a technique known as behavior chaining. It is most often used to teach complex behavior sequences, for example, in performing animals in the circus or in films. In these circumstances, we do not want to give a series of commands but want the animal to do a number of actions one after the other.

27.5.9.2.2 Punishment and Its Problems

Punishment is an integral part of the learning process but there are many problems with its use as a training aid. As a result, it tends to be misused and so causing a lot of concern. However, punishment is not a particularly efficient training tool, as it does not specifically tell the horse what it should be doing. It just signals that one specific activity in a given context should not be performed. Repeated use of punishment reduces its effectiveness and habituation occurs. Physical punishment causes discomfort, inhibits learning, and if used excessively, emotional changes might occur which cause serious concerns such as aggression and learned helplessness. Punishment makes a fearful horse worse and increases timidity.

27.5.10 Ethology of Wild Animals

It refers to studying the behavior of wild animals. Heini Hediger was the Swiss biologist and first scientist known as the "Father of Zoo Biology," who stressed the importance of ethology in captive wild animal management. The survival of a wild animal depends upon getting food, protection from other animals and adverse environment, and getting resistance to diseases, parasites, and predators. Wild animals' behavior is species-specific and like other mammals, wild animals also exhibit specific pattern of behavior such as learning, feeding, sexual, social, communication, and agonistic behavior.

27.5.10.1 Communication Behavior in Wild Animals

Animals respond to external stimuli that come from signals given by other animals of their own species—from their parents or from their own offspring, for example, or from rivals or potential mates. Mammals use hearing, sight, and smell at a distance but for many of the more social ones that spend a good deal of time in physical contact, tactile communication is also important. Chimpanzees touch and kiss each other's hands in gestures of reconciliation after a fight. Animals that rely on vision for finding their way about and locating prey will also tend to communicate through visual signals. Sound has one big advantage in that it can go around corners and through some barriers, e.g., high-frequency (50–100 kHz) sounds are used by bats for echolocation. Animals can also help sound to travel further over longer distances by elevating themselves above ground. Crickets that sing from trees or shrubs can spread their signal over 14 times the area of those that sing from the ground and consequently attract more females. The territorial songs of birds are usually delivered from a raised song post which also increases their effective range, while grassland birds such as meadowlarks (Sturnella) and pipits (Anthus) sing while they fly.

In the animal kingdom, alarm calls are quite unique, e.g., monkeys give different alarm calls when they see different predators and chickens and ground squirrels also have different alarm calls for aerial and ground predators. Predator-specific alarm calls are particularly well-developed in vervets. When a vervet monkey sees a leopard (Panthera pardus) or other large cats, it gives a loud barking alarm call. When the other vervets hear this, they run up into trees so that the leopard cannot catch them. On the other hand, if the vervets catch sight of an eagle that prey on them they give a completely different alarm call—a sort of double-syllable cough. Since both these eagles are highly skilled at taking monkeys both from the ground and from the trees, a monkey on the ground hearing the eagle alarm call immediately looks up into the air and then runs into the thickest bush available. Yet another type of alarm call is given when the monkeys encounter snakes such as pythons or cobras. When a vervet hears a snake alarm call, it stands up on its hind legs and looks down into the grass.

Whales and elephants both use infra-sound to communicate over long distances. Elephants use low-frequency "rumbles" (10–35 Hz) that are very powerful and clearly effective over distances of 1–2 km to keep in touch with other members of their group. Some bats and owls also hunt in complete darkness not by emitting sounds of their own but by being extremely sensitive to sound made by their prey. The barn owl's hearing is so good, for example, that it can locate a mouse in complete darkness simply by homing in on the sound of the mouse rustling through leaves or even just chewing.

In response to environmental stimuli such as visual, auditory, mechanical (touch), and chemicals (scent, taste, and pheromones) wild animals exhibit different type of behaviors. Among these behaviors feeding/eating, fly/run away, sexual, and maternal are considered as major behavioral responses. Aggressive behavior is more prevalent in wild animals. The major causes are competition of males for successful selection of females, fight over food resources, disturbance of social organization, incompatibility in the composition of groups, and distortion of social role.

Learning Outcomes

Significance of studying animal behavior: Behavior is closely related to the welfare of the animal. Hence, behavioral studies help the veterinarians and animal scientists to identify the cause of any abnormal behavior and address it. An understanding of the behavior of animals will guide the animal scientists to prevent and manage various behavioral disorders thus ensuring the welfare of the animal.

Endocrine moderation of behavior

Behavior is a complex response mediated by the nervous system and is modulated by the endocrine system. Hormones determine and influence the probability that a specific sensory input leads to a specific behavioral response. Hormonal changes might modify some ongoing behavior by increasing or decreasing the frequency or duration of that behavior, or they might trigger the onset or end of a behavior or behavioral sequence.

Behavioral plasticity

The plasticity of behavior is an array of behavioral responses to varying environmental conditions. This determines the inherent potential of an animal to adapt to different environmental conditions. Broadly, three types of behavioral plasticity can be identified: differences in ontogenetic development, adjustments through learning, and the innate ability to respond to a variety of stimuli.

(continued)

Types of animal behavior

The behavior of animals has been broadly classified and studied extensively under categories such as learning, feeding, sexual, maternal, and social behaviors. ***Learning***: Learning is a process whereby an individual acquires new responses and new capacities. Learning results in behavioral changes within an animal's lifetime which can introduce a new dimension into behavioral evolution; ***Feeding behavior***: is a circadian rhythm and an important sign of health which is influenced by a variety of internal and external factors; ***Sexual behavior***: Sexual behavior is important in all species of animals. Sexual behavior includes proceptive and receptive behavior by the female and courting and mate guarding by the male, as well as actual copulation. ***Maternal behavior***: The combination of the proper hormonal milieu and the stimulus for maternal behavior, the neonate, plus prior experience of being a mother can elicit maternal behavior. The stimulation of maternal behavior appears to be under both hormonal and neural control.

Wild animal behavior

It refers the studying the behavior of wild animals. Wild animals' behavior are species-specific and like other mammals, wild animals also exhibit specific pattern of behavior such as learning, feeding, sexual, social, communication, and agonistic behavior.

Exercises

Objective Questions

Q1. The phenomenon wherein a young animal develops attraction to an object or an animal is called as ___________.

Q2. The type of communication which involves touch in behaviors such as social bonding, infant care, grooming, courtship, and mating is ___________.

Q3. The organization of a group of animals in such a way that some members of the group have greater access to resources like food or mates than others is called as _______.

Q4. The area in the brain that accodes the level of priority to a given stimulus is ______.

Q5. The decrease in response to repeated or continuous stimulation is called as ______.

Q6. Pavlov conditioned dogs to salivate upon hearing a ringing bell by repeatedly presenting them with food while ringing a bell. This is an example of ______.

Q7. Coprophagy is seen in ________.

Q8. The fixed behavioral responses that animals exhibit even without learning them are known as ____________.

Q9. Stealing of young ones of other animals of their species by prepartum animals is seen in ___________.

Q10. Vomeronasal organs are involved in detection of __________.

Q11. Brooding behavior of the birds is influenced by ___________.

Q12. Group feeding animals tend to increase ________.

Q13. Punishing the animals during training leads to ____________.

Q14. Enurination is a unique sexual behavior seen in __________.

Q15. Hormone that depresses feeding is _________.

Q16. Ensuing of abortion of the fetus when a strange male mouse comes in contact with thepregnant female is known as ___________.

Q17. One of the major hormones associated with maternal behavior is ___________.

Q18. Peck order is a social ranking established in __________.

Q19. One of the plastic changes induced by behavior in the brain components is ____.

Q20. In prey species such as cattle the most important mode of communication in disseminating information about general safety or danger is __________.

Subjective Questions

Q1. Discuss the role of nervous system in development and exhibition of behavior.

Q2. Write briefly on the role of pheromones in animals.

Q3. Explain the control of feeding intake and the differences in feeding behaviors in animals.

Q4. Discuss the role of hormones on influencing the behavior.

Q5. Write a note on appeasing hormones.

Q6. Describe in detail the significance of social behavior.

Q7. Discuss in detail about the agnostic behavior observed in domesticated animals.

Q8. Write briefly on the various modes on social communication in animals.

Q9. Write note on the types of learning.

Q10. Discuss in brief on the milking behavior and its factors controlling in cow.

Q11. Discuss in detail about the maternal behavior in animals.

Q12. Describe the various behavioral disorders related to sexual behavior.

Q13. Write briefly on the significance of wild animal ethology.

Q14. Discuss in detail the importance of ethology and role in animal welfare.
Q15. Describe the importance of training in domesticated animals and the commonly employed techniques.

Answer for the Objective Questions

A1. Imprinting
A2. Tactile
A3. Hierarchy
A4. Amygdala
A5. Habituation
A6. Operant conditioning
A7. Rabbits
A8. Instincts
A9. Ewes
A10. Pheromones
A11. Prolactin
A12. Feed intake
A13. Habitutation
A14. Goats
A15. Cholecyctokinin
A16. Bruce effect
A17. Oxytocin
A18. Chicken
A19. Synaptic organization and formation
A20. Vocal

Keywords for the Answers to the Subjective Questions

A1. Cerebral cortex, limbic system, hypothalamus, neural circuits
A2. Volatile chemicals secreted/excreted, vomeronasal organ, breeding activity, reducing aggression and fear, and encouraging feeding
A3. Short-term, long-term and emergency control, hormones, grazing, and ruminating behavior in cattle
A4. Influence during embryonic development, modify/increase/decrease an ongoing behavior, influence social behavior
A5. Dams, maternal care, reduce aggressiveness, clinically used for behavioral problems
A6. Increased learning, and species-specific, survivability
A7. Threaten, attack to access limited resources, feed, territory, and mating partner
A8. Vocal, olfactory, visual, and tactile communication
A9. Habituation, reinforcement, conditioning operant conditioning
A10. Neuroendocrine reflex, classical conditioning, oxytocin, stress, feeding concentrates
A11. Influence by hormone, hereditary, neural and experiential factors, nursing, grooming
A12. Lack of sexual interest towards receptive mares, self-mutilation, lack of ejaculation, masturbation and tendency to injure handlers, lack of socialization
A13. Species-specific, learning to communicate, alarm calls, searching for feed and water resources, social learning
A14. Understand the animal's need, helps in better management, treatment, transport, and training
A15. Training for the basic daily life skills, used for livestock handling, rescue dogs, assist visually impaired, therapy dogs, stress reduction purpose. Reinforcement, operant conditioning, reward

Further Reading

Atkinson T (2018) Practical feline behaviour understanding cat behaviour and improving welfare. CABI Publishing, Oxfordshire

Aubrey M, Dawkins M (2012) An introduction to animal behavior, 6th edn. Cambridge University Press, Cambridge

Barnard C (2004) Animal behaviour: mechanism, development, function and evolution. Prentice-Hall, Harlow

Beaver BVG (2009) Canine behavior: insights and answers. Elsevier Health Sciences, St. Louis, MO

Broom DM, Fraser AF (2007) Domestic animal behaviour and welfare, 4th edn. CABI Publishing, Wallingford

Case LP (2010) Canine and Feline behavior and training: a complete guide to understanding our two best friends. Delmar, Cengage Learning, Clifton Park, NY

Dugatkin LA (2014) Principles of animal behavior, 3rd edn. W. W. Norton and Company, New York

Fraser AF (2010) The behaviour and welfare of the horse, 2nd edn. CABI Publishing, Wallingford

Houpt KA (2011) Domestic animal behavior for veterinarians and animal scientists. Wiley-Blackwell, Ames, IA

Jensen P (2007) Behavioural biology of dogs. CABI Publishing, Wallingford

Lethbridge E (2009) Knowing your horse: a guide to equine learning, training and behaviour. Wiley-Blackwell, Chichester

Phillips C (2002) Cattle behaviour and welfare, 2nd edn. Blackwell Science, Oxford

Price EO (2002) Animal domestication and behavior. CABI Publishing, Wallingford

Shaw JK, Martin M (2014) Canine and Feline behavior for veterinary technicians and nurses, 1st edn. Wiley-Blackwell, Chichester

Part X

Environment and Thermoregulation

Environmental Physiology and Thermoregulation in Farm Animals 28

G. Krishnan, M. V. Silpa, and V. Sejian

Abstract

Environment is surroundings of a living organism consisting of natural forces, living things that facilitate circumstances for the progress and improvement along with risk and impairment. Environmental physiology facilitates the understanding of the interrelationship between the environmental factors and animal. The environmental factors that influence animals' performance are ambient temperature (T_a), relative humidity (RH), radiation, precipitation, atmospheric pressure and wind velocity. Homeothermic animals sustain a consistent inner body temperature (T_c) by adjusting the heat produced or increased by metabolism and heat loss to the environment. Thermoregulation is accomplished by physiological, morphological and behavioural mechanisms. Chronic heat stress compromises the animal production. Therefore, animals combat the heat load and reduce its adverse impact on production, reproduction through physiological process. When the animal experiences the heat stress, it stimulates sensible and insensible heat loss through conduction, convection, radiation and evaporation. The energy demand increases during winter to maintain the T_c by shivering or non-shivering thermogenesis. The body temperature regulates chemical reactions in the body, with an increase in temperature accelerating such reactions and a decrease in temperature slowing them down. Further, heat stress associated physiological responses of animals activate endocrine, autonomic and central nervous systems along with redistribution of blood flow. The different systems act in a coordination in relation to the level of stress so as to maintain the homeostasis by stimulating physiological mechanisms to reduce the adverse impacts. The stress controlling systems differ between entities in association with their past exposure, physiological status, genetic predisposition, the extent and severity of stress.

G. Krishnan (✉) · V. Sejian
Centre for Climate Resilient Animal Adaptation Studies, ICAR-National Institute of Animal Nutrition and Physiology, Bangalore, Karnataka, India

M. V. Silpa
Institute of Animal Breeding and Genetics, Justus-Liebig-Universität Gießen, Gießen, Germany

© The Author(s), under exclusive license to Springer Nature Singapore Pte Ltd. 2023

P. K. Das et al. (eds.), *Textbook of Veterinary Physiology*, https://doi.org/10.1007/978-981-19-9410-4_28

Graphical Abstract

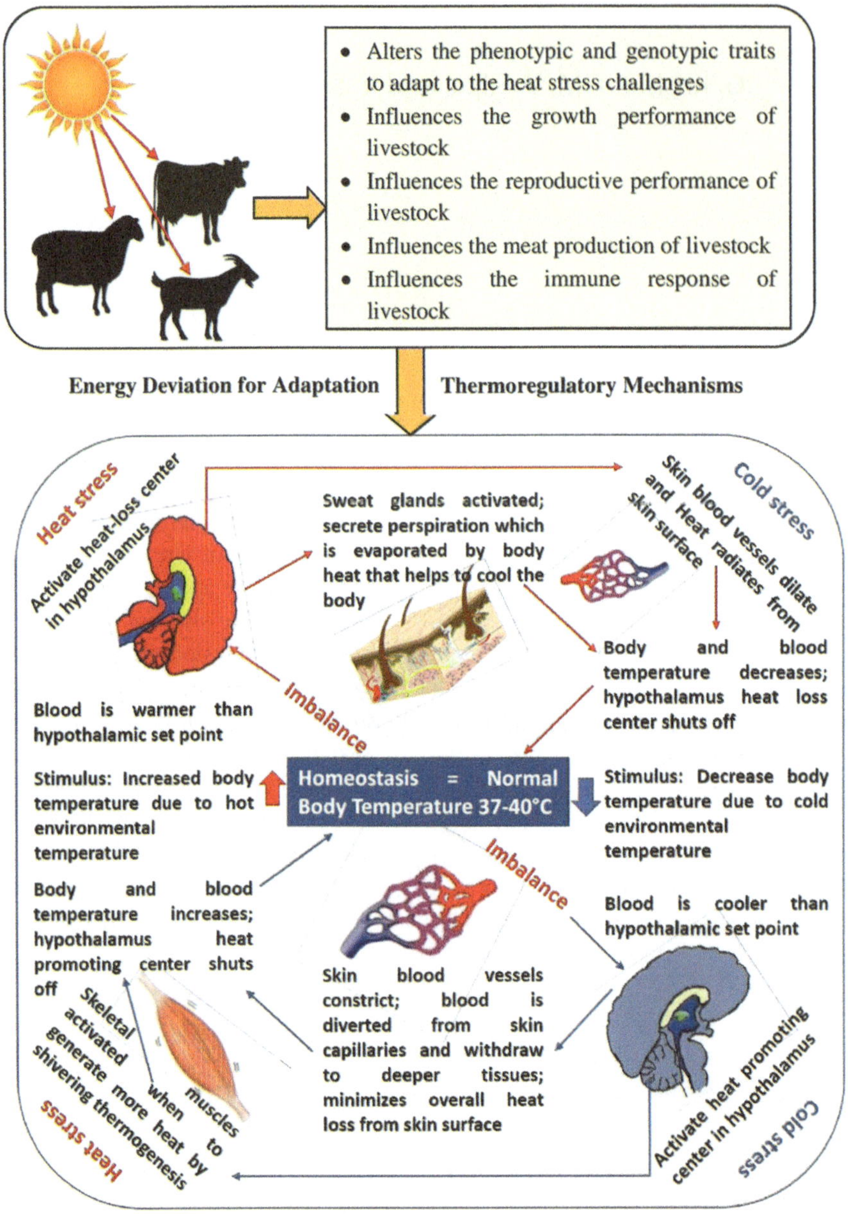

Homeothermic animals maintain their body temperature relatively constant with narrow range. The decrease in body temperature stimulates the hypothalamus and initiates the shivering and non-shivering thermogenesis to enhance heat increment. Similarly, the increase in body temperature also activates the hypothalamus to enhance heat dissipation mechanisms to reduce the core body temperature to its normal range

Keywords

Animal · Environment · Heat and cold stress · Thermoregulation · Adaptation

Learning Objectives

- The chapter aims to understand all important definitions and terms pertaining to environmental physiology of farm animals.
- The chapter highlights the various cardinal weather variables which could influence animal responses to its environment.
- The chapter elaborates the different effects of heat stress on the productive response in farm animals.
- The chapter defines the thermoneutral zone for animals and its significance.
- The chapter describes the thermoregulatory mechanisms of domestic animals.
- The chapter also provides details on various indices to evaluate heat stress response in farm animals.

- The chapter also highlights in brief the different approaches to maintain livestock productivity during heat stress.

28.1 Introduction

Environment is critical for all ecosystems to survive on earth. Therefore, any living species surviving on earth must possess inherent abilities to cope with the ever-changing environment. Recent climate change projections are quite alarming for any sector including livestock. The animals possess inherent abilities to cope with environmental stress but while doing so they mostly compromise their productive responses. Hence, it is very important to study the animal physiology associated with environmental sciences as this will help us to understand the intervening point to improve livestock production in the prospective climate change scenario.

Ecology is the division of biology that deals with the interactions between living things and their biophysical environment which comprised of living and non-living components. Ecology comprises many aspects of biodiversity, production and biomass, distribution of population in addition to competition between them within and among ecosystems. Whereas, animal ecology refers to the association of individuals with their environments, in addition to physical elements and living things, and the importance of these associations for evolution, population growth and regulation, interactions between species, the composition of biological communities, and energy flow and nutrient cycling through the ecosystem. In the event of domestic animals, ecology comprises their association with human society and economy, and especially concern with their productivity. Ecology is divided into two branches as autecology and synecology, where autecology, also known as species ecology, describes the association of a single species with the living and non-living aspects of its environment. Autecology is principally dealing with the quantifiable variables like light, humidity, and available nutrients so as to study the necessities, life history and behaviour of the organism or species at experimental conditions. The current advancement in laboratory techniques and technological developments has empowered physiologists to learn physiological variables in free living animals, in addition to laboratory experiments. Whole-animal physiology/physiological ecology/environmental physiology is the study of animals within their natural habitat. Environmental physiologists primarily focus on studying and understanding the animal's function along with their response to environment at different stages of life. Overall development in the life sciences with recent molecular biological techniques facilitates the environmental physiologists to study substantial differences in animal physiology within and between species to understand the mechanisms and their gene regulations.

Environment refers to the kind of habitat or surroundings of a living organism which include both the biotic and abiotic factors, which leads to the development and growth of the organism along with risk and impairment. Each species has a precisely defined environment within a biome in the totality of all the external factors it experiences. The behavioural choices opted by an organism, along with its absolute presence, modify its environment. The microenvironment or microhabitats or microclimate is dealt with environmental and evolutionary physiology. The individual animal selects a range of its environment where to spend its time, feed and rest. These opportunities are changed on a magnificent spatial scale with a quick temporal scale for small animals and it takes much more coarsely for large animals. Climate can be referred as a long-term (over 30 years) average condition of the meteorological variables of a particular region. Thirty years is the standard time period for well execution of statistics which was suggested by the world meteorological organization. Therefore, climate change refers to variations in the climate for a long course of time which reappears with regular time interval, but is not strictly periodic. The existence of a pattern without any real precision in the recurrence of events is termed as climatic cycle (climatic oscillation). Macroclimate refers to the climatic condition over a larger area. Mesoclimate refers to the climatic conditions over smaller areas, which may not be representative of the general climate while the microclimate refers to the climatic conditions directly surrounding the animal.

28.2 Important Terms and Concepts Associated with Environmental Physiology of Animals

Bioclimatology is the section of climatology which describes the association of climate with living organisms such as animals and plants.

Biometeorology is an interdisciplinary science studying the interactions between atmospheric processes and living organisms—plants, animals and humans.

Climate: The long-term (some 30 years) average condition of the meteorological variables in a given region. The periodic alterations of weather conditions such as temperature, wind, rain, snowfall, humidity and clouds in a broader locality moderated for years.

Weather: The short-term day-to-day fluctuations of the metrology variables, as distinguished from climate, which is the long-term manifestation of the weather.

Microclimate: The climate condition directly surrounding the animal.

Climate change: It refers to changes in measures of climate such as temperature and rainfall for a given area persisting for an extended period, typically decades or longer. Climate change can involve cooling or warming. Climate change may result from natural factors (e.g. changes in the sun's energy, volcanic eruption), natural processes within the climate system (e.g. changes in ocean circulation) and human activities (e.g. burning of fossil fuels, agriculture).

Greenhouse effect: The greenhouse effect is the process by which radiative energy leaving a planetary surface, like the earth, is absorbed by some atmospheric gases, called greenhouse gases (GHGs). The GHGs transfer this energy to other components of the atmosphere, and it is re-emitted in all directions, including back down towards the surface.

Global warming: Global warming refers to an increase in average global temperature which in turn causes climate change solar radiation in the form of light waves passes through the atmosphere, most of this radiation is absorbed by the earth and radiated back into space in the form of infrared waves. Part of the infrared waves is trapped by the atmosphere making the earth just warm enough to be livable. But because of too much GHGs that thicken the atmosphere, most if not all of the infrared waves are now trapped making the earth warmer.

Basic Weather Parameters

Weather is referring to numerous elements of environment such as ambient temperature, atmospheric pressure, humidity, precipitation, solar radiation and wind which can be computed to evaluate the condition of atmosphere.

Animal Adaptation

Animal adaptation is the developmental changes that favour greater survival of an animal in its territory.

Adaptation (biological): The morphological, anatomical, physiological, biochemical and behavioural attributes that encourage the animal well-being and support its activity in distinct surroundings.

Adaptation (genetic): The genetical changes in the animal's characteristics which benefits the persistence of a population in a specific environment. This may be established by developmental transformations over many generations (selected by nature) or inheriting definite genetic properties (selection by man).

Adaptation (physiological): The competence and mechanism of improvement of the animal to itself, to other living material and to its external physical environment.

Acclimatization: A long-term adaptive physiological modifications that enhanced the resistance to uninterrupted or frequent experience to complex climatic stressors (normally produced under field conditions). **Acclimation** represents the physiological or behavioural modifications that develop within the lifetime of an organism and which lessen or augment the resilience of a strain influenced by experimentally induced stressful changes specifically, climatic factors.

28.2.1 Natural Adaptation of Animals to Environment

Any animal cutting across species has their own inherent inability to cope with external environment. They acquire this potential from generation after generation in a specific environment which brings in alterations in both phenotypic and genotypic traits in these animals. Through these several generational changes they acquire the capacity to live and produce maximum in a particular region.

28.2.1.1 Morphological and Anatomical Adaptation

Morphological adaptation is a structural change which gives an animal a greater chance of survival in its habitat. Short and thin hair with light colour, light pigmented skin, large numbers of sweat glands, slender legs and less subcutaneous fat are the morphological adaptations in farm animals. **Anatomical adaptations** are the physical features of the animals which help them to survive in specific environment. The illustration of anatomical adaptations includes cat's claw, rabbit's back legs and shape of an animal.

28.2.1.2 Theories of Morphological Adaptation

Gloger's (1833) ecogeographical *rule refers* ***animal*** colouration with climatic diversity. Skin pigments protect animals from UV radiation. Animals in cooler and dry regions have light-coloured skin while darker in high temperature and humidity. **Bergman's rule** (1847) describes body size to in relation climate where, the animals at higher latitudes are larger and thicker in comparison to those near the equator for efficient conservation of heat. The heavy cattle breeds (Holstein) are well adapted to cooler temperature while smaller breeds (Jersey) prefer warmer. **Wilson's rule** (1854) interrelates insulating capacity to climate, and animals in cold region have a dense, heavy, thick external coat, whereas those are in warm climate have short, glossy and thin hair. **Allen's rule** (1877) interrelates animal appearance with temperature; animals living in cold zones have shorter extremities and bodily appendages than those in warm climate.

28.2.1.3 Characters of Well-Adapted Animals

Adaptations are any behavioural or physical characteristics of an animal that help it to survive in its environment. The modifications in body parts, body coverings and behaviours favour the survival of an animal to a specific environment.

The better adaptive attributes to warm climates include a wide range of physiological functions, behavioural and morphological alterations. On the physiological perspectives, the typical characteristics of well-adapted farm animals include their capability to cope with periodic water and feed scarcity. In addition, they have the ability to walk long distances in exploration of water and feed, digest low quality feeds, be resilient to heat stress and highly immune against ticks, and tick-borne diseases and other tropical diseases. The well adapted animals demostrate resilience behaviorally to graze without seeking shade, not reducing their feed intake, and drinking water at normal levels. Further, with increased standing time and rumination time as well these animals do not concentrate their urine and faeces. Morphologically, the well-adapted animals possess light/white and thin coat colour which will facilitate them to reflect solar radiation and induce effective evaporative cooling mechanisms.

28.2.1.4 Energy Exchange Processes Between Animal and Environment

Animals interchange heat load with their immediate surroundings through conduction, convection, radiation and evaporation. These processes were described in subsequent subsections.

28.3 Impact of Climatic Variables on Animal Production

The primary environmental factors that influence animals are ambient temperature (T_a), relative humidity (RH), radiation, precipitation, atmospheric pressure and wind velocity. The high T_a and RH for long duration results in heat stress. Therefore, animals try to adjust their physiological means to conquer such stressful situations by compromising their production and reproduction. The T_a is among the primary factors associated with thermal stress and it is altered by wind, precipitation, RH and solar radiation. Animals adopt the adaptive mechanism to compensate within their limits on exposure to varied T_a by adjusting feed intake, metabolism and improving the heat loss mechanisms. The animals are able to maintain a relatively constant body temperature in a range of temperature limit by modifying behavioural and physiological responses. The escalation in the ambient temperature causes heat stress as a result of inefficiency of the animal to eliminate excessive heat load. The enhanced environmental temperature impacts production performances of farm animals like reduction in milk yield as well as changes in milk composition and reproduction. Hence, both the extremities of climatic conditions impact negatively on livestock welfare, performance and health.

28.3.1 Temperature

Temperature is the degree of hotness or coldness measured on definite scale. The high ambient temperature beyond the animal's physiological limit may impact its biochemical processes. The temperature alterations may affect the enzyme activities, protein synthesis, and degradation at the cellular level. The cellular components are directly affected by temperature changes, whereas the cell elements are negatively influenced. Ultimately, as a consequence of the cellular damage, the synthesis and production of heat shock proteins, molecular chaperones that protect cells during cold or heat stress, get activated. At the same time, some species have the innate mechanism of surviving in extremely cold regions with low temperatures that often touch below freezing levels. For example marine species contain anti-freezing components within themselves that enables their survival.

28.3.2 Solar Radiation

The animals receive solar radiation directly as are flection from clouds and surrounding surfaces and indirectly from terrestrial or long-wave radiation emitted by surroundings surfaces. The total impact of solar radiation on an animal depends upon the variance between the collective solar and long-wave radiation received and the long-wave radiation emitted by the animal. There are some factors that influence the total effect of solar radiation such as shade, ground cover, clouds, hair coat and insulation characteristics of animal. Generally, the net gain of heat by solar radiation exceeds the effective T_a during hot climatic conditions by 3–5 °C culminating in heat stress, whereas the increase in effective T_a in winter is highly beneficial. Thus, increase in T_a directly impairs the animal performance by disturbing the heat balance. In addition, environmental temperature hampers the animal's heat exchange by convection and evaporation. As the temperature increases, the rate of evaporation is enhanced which develops as a vital means of heat dissipation. The reduction in feed intake due to elevated environmental temperature leads to a decrease in the secretory pattern of calorigenic hormones, particularly growth hormone, catecholamines and glucocorticoids which reduces the thermogenic processes of digestion and metabolism. The scientific committee on animal health and welfare set the upper critical temperature in calves at 30 °C and RH below 60%, and it should be below 27 °C when RH exceeds 80%. Further, temperatures between 15 and 29 °C influence growth performance; however, temperatures above 30 °C negatively impact daily weight gain.

28.3.3 Relative Humidity

Humidity is an essential element of environment and routinely expressed as relative humidity (RH). It is the ratio of the current absolute humidity relative to the maximum humidity at a specific temperature, representing the amount of water vapour in the air at that temperature. The moisture content of air impacts an animal's heat balance, especially during warmer environmental conditions where the evaporative heat loss is critical for homeothermy. Higher environmental vapour pressure depresses the vapour pressure gradient from the skin and respiratory tract to the external environment, thereby decreasing rate of evaporation. The occurrence and prevalence of infectious diseases are sensitive to RH which favours growth of pathogenic organisms. Lower RH leads to dryness of the ocular mucosa and the stratum corneum of the skin along with reduced skin temperature. The suitable percentage of RH for mammals varies in the range of 30–70%.

28.3.4 Wind Velocity

The wind speed influences the amount of convective and evaporative heat dissipation mechanisms. The enhancement of heat loss or gain is maximum at low speed of air since air movement disturbs the closed air over the body. The effective air velocity suggested for dairy cattle is 1.8–2.8 m/s during hot environmental conditions. Further, wind velocity above 1.0 m/s cools animal very effectively during hot periods wherein the RH is high. However, increasing air velocity above 1.67 m/s has small extra benefit in the convective heat loss.

28.3.5 Precipitation

Animals occasionally encounter an extreme weather in combination of low temperature, wind and rain or wet snow which adversely affects its body heat balance. The logging of water in an animal's fur or hair transfers still air, which decreases the external insulation. Further, precipitation also smooths the fur and decreases its depth, which results in reduced insulative value.

28.3.6 High-Altitude Environment

The high-altitude environment is characterized with extreme cold, mountainous terrain, reduced oxygen in the air, high solar radiation and short vegetations and/or herbage growing seasons. The reduced partial pressure of oxygen at high altitude leads to decreased oxygen loading in lungs followed by insufficient oxygen supply to the tissues via the circulating blood. The reduced level of tissue oxygenation limits aerobic metabolism which may impact an animal's feed and water requirements, locomotor activities and impair the internal heat production. However, in mammals, the modifications of haemoglobin function mediate an adaptive response to high-altitude hypoxia. The partial pressure of the oxygen is high when the atmospheric air enters into the alveoli of the lungs which facilitate the diffusion of oxygen across the respiratory membrane and arterial bloodstream. The oxygen in the arterial blood binds to haemoglobin and transported to the tissues wherein the haemoglobin discharges oxygen which disseminated into the cells across the capillary walls. At the same time, the carbon dioxide and other metabolic end-products get into the venous bloodstream which is delivered at the lungs.

28.4 Effects of Heat Stress on Production

28.4.1 Impact of Heat Stress on Livestock Growth

Growth, the increase in the live body mass or cell multiplication, is controlled genetically and environmentally. The higher environmental temperature influences average daily body weight gain. The elevated ambient temperature reduces growth rate by decreasing the anabolic activity with enhanced tissue catabolism as an effect of increased catecholamines and glucocorticoids during heat stress in livestock.

28.4.2 Impact of Heat Stress on Livestock Milk Production

The increased environmental temperatures and humidity enhance the animal's body temperatures that resulted in reduced feed intake, decreased milk yield and disrupted reproductive functions. Heat stress causes a reduction in feed intake and growth rate which ultimately affects milk production and reproduction performance in dairy cows. Further, the cellular and molecular reactions to heat stress influence mammary gland metabolism, energy distribution and health status of animal. The elevated temperature with high heat load negatively impacts ovarian activity exclusively in buffalo and crossbred cows that are inefficient in eliminating heat from the skin. The water scarcity further impacts production and reproduction in farm animals. Heat stress negatively affects the secretory functions of the udder that results in decreased milk production during high environmental conditions. The lactating cows are incompetent to cope with heat stress particularly during early lactation which adversely affects the total milk production. The high-producing dairy animals are more susceptible to heat stress due to greater metabolic heat production and eventually decreased milk yield with lower level of fat, solids, lactose and protein. Hence, high-producing animals are more affected by heat stress than low producers. Heat stress affects milk production during the first 60 days of lactation when the high-producing cows are in negative energy balance. The catabolic processes which are initiated to mobilize energy to meet the lactation demands further add on to the heat load by increasing the metabolic heat production.

The impact of climatic stress on milk production of dairy animals is predicted to be 1.8 million tonnes at present. The models-based prediction of loss of milk production on different climatic scenarios suggested to be 1.6 million tonnes by 2020 and more than 15 million tonnes by 2050. The northern part of India is predicted to experience greater climate-related reduction in milk production of cows and buffalos.

28.4.3 Impact of Heat Stress on Meat Production

There are significant research reports on impact of heat stress on meat quality and composition in cattle, sheep, goat, pig and broilers. The increased temperature with relative humidity resulted in higher meat pH, loss of juiciness, high cooking and drip loss. Heat stress reduces the energy utilization and enhances the energy expenditure for thermoregulation which deteriorates the meat quality by decreasing the muscle glycogen that increases the muscle pH. The primary functional properties of meat such as colour, water holding capacity and myofibrillar fragmentation index were also negatively influenced during heat stress in ruminants. In addition, the animal management practices during heat stress indirectly affect the meat quality. The rearing of heat-tolerant *Bos indicus* cattle is an effective adaptation strategy against the prevailing harsh climatic conditions which results in tougher and less juicy beef. Besides the qualitative alterations driven by the heat load on the animals, carcass weight losses in heat stressed animals also have economic significance. Ante-mortem temperature stress is a major determinant for live carcass weight losses, hot carcass weight and retail meat yield. The diversion of energy towards thermoregulation in combination with reduced feed intake during heat stress resulted in live weight losses. Therefore, the above information confirms that heat stress deteriorates the qualitative and quantitative characteristics of meat depending upon thermotolerance and animal origin.

28.4.4 Impact of Heat Stress on Livestock Reproduction

28.4.4.1 Female Reproduction

The environmental temperature influences the reproductive performance of female animals at different stages of pubertal development, conception and embryo development. Stress suppresses reproductive efficiency of animals through activating the hypothalamic–pituitary–adrenal (HPA) axis which enhances the secretion of adrenocorticotropic hormone (ATCH) from the pituitary gland. ACTH acts on the adrenal and secretes glucocorticoids and catecholamines which are primarily stress relievers. Heat stress reduces the extent and magnitude of estrous cycle and affects follicular development with higher incidence of apoptosis in the antral and pre-antral follicles. Heat stress suppresses growth and development of follicles, thereby altering the oocyte function. The prolonged secretion of ACTH prevents ovulation and follicular development by modifying potency of follicular selection, dominance and follicular steroidogenesis. The higher level of glucocorticoids prevents the meiotic maturation of oocytes and corticotropic releasing hormone which consequently suppress the ovarian steroidogenesis. The hot environmental condition delays the commencement of puberty in female animals. In addition, high temperature decreases production of gonadotropin-releasing hormone (GnRH) which in turn results in delayed estrus. The reduction in GnRH also decreases luteinizing hormone (LH) and estradiol that lowers length and intensity of estrus in heat stressed animals. The comprehensive negative impact of heat stress modifies the reproductive behaviour of higher prevalence of anestrus and silent heat in farm animals. The higher level of ACTH and cortisol during heat stress decreases the activity of granulosa cells aromatase that results in lower estradiol secretion resulting in reduction of estradiol-induced sexual behaviour. The reduction in the concentration of estradiol depresses the estrus signs, gonadotropin surge, ovulation, transport of gametes with eventual reduction in fertilization rate. Heat stress impairs the follicular dynamics by altering the development of follicles with reduced follicular dominance and encouraging numerous large follicles with prolonged dominance which disrupts the normal estrus cycle. In addition, lower level LH and negative energy balance of animals in hot environmental conditions inhibit the maturation and ovulation of dominant follicles. The extended follicular dominance alters the normal oocyte maturation and reduces their developmental competence. The development of such small dominant follicle results in ovulation of defective oocyte. Heat stress also decrease the oocyte developmental competence by inhibiting growth and maturation with higher oxidative damage and apoptotic cell death that results in irreversible alterations on cytoskeleton and meiotic spindle. The heat stress reduces fertilization capacity of oocyte and further having a negative impact on its embryo development stages, thereby reducing the fertility in animals. The embryonic death during heat stress could be due to impaired protein synthesis, significant oxidative cell damage, decreased pregnancy recognition and lower progesterone levels.

28.4.4.2 Male Reproduction

Bull is known as half of the herd where bull's fertility is equally or more important for fertilization of oocyte to produce a good, viable and genetically potential conceptus. Males possess an exceptional physiological mechanism of testicular thermoregulation to sustain its reproductive competence during hot environmental conditions. Further, presence of numerous sweat glands in scrotum of male animals is efficiently involved in the local thermoregulation. The testicular temperature is 4–5 °C lower than the rectal temperature which is prerequisite for the optimum semen production. The male animals are vulnerable to heat stress that directly affects the sperm quantity and quality and results in low fertility. Therefore, preferable environmental temperature for efficient semen production varies between 15 and 20 °C. The high environmental temperatures enhance the oxidative metabolism of glucose in spermatic cells due to mitochondrial dysfunctions and increased the reactive oxygen species. Therefore, the testicular temperature must be 2–6 °C cooler than core body temperature for the desired production of fertile semen with good quality and fertile sperm.

Heat stress reduces the level of testosterone, which has an impact on libido and results in low sperm concentration and motility with increased dead and abnormal sperm. The higher testicular temperature also enhances spermatogonial germ cell apoptosis, degradation of Sertoli and Leydig cells, DNA damage specially in pachytene spermatocytes and spermatids. The sexual behavioural alteration of bulls is attributed to the reduced levels of testosterone. Heat stress impairs the semen production and quality not only on day of collection but also during epididymal maturation or spermatogenesis which can go up to 70 days before collection. Further, biochemical elements of semen including fructose, sodium and potassium cation, citric acid, total phosphorus and calcium levels are decreased in heat stressed bulls. Further, higher level of lipid peroxidation during heat stress due to oxidative stress deteriorates the semen characters in bulls.

28.5 Heat Shock Proteins and Its Regulation

Heat stress leads to severe cellular damage resulting in production of cellular toxins and impaired protein synthesis. During such a stressful situation, the heat shock proteins (HSPs) play a vital role to maintain cellular homeostasis and thus are the ideal biomarker for heat stress. This group of proteins are well known as the 'molecular chaperones' for their vital cytoprotective functions like folding of proteins in various intracellular compartments, maintenance of structural proteins, refolding of misfolded proteins, translocation of proteins across membranes and into various cellular compartments, prevention of protein aggregation and degradation of unstable proteins. HSPs also regulate the apoptotic signalling pathways where HSP27, HSP70 and HSP90 proteins are predominantly antiapoptotic, and HSP60 is proapoptotic. The principal HSPs that have chaperonic activity belong to five conserved classes, HSP33, HSP60, HSP70, HSP90, HSP100 and the small HSPs. Heat shock factors (HSFs) present in the cytosol are attached with HSPs and maintained in an inactive state. Heat stress or stressors activate HSFs and detach from HSPs, and phosphorylation of HSFs by protein kinases leads to formation of trimers in the cytosol. These HSF trimer complexes enter into the nucleus and bind with heat shock elements (HSE) in the promoter region of the HSP genes. HSP mRNA is transcribed and exits the nucleus to enter in the cytosol, where new HSPs are synthesized. The primary activity of HSPs is to sustain translation and protein integrity in the cells. Cells that become thermo-tolerant also produced less HSP during a second challenge compared to the previous heat stress exposure, suggesting HSP synthesis is also regulated depending on the levels of these proteins within the cell. HSPs are highly important in all the organisms which protect the cells against varied stressful circumstances and critically essential for normal cellular function.

28.6 Impact of Heat Stress on Muscle Function

The muscular performance decreases with increased incidence of intense work load and varies with the intensity, duration and mechanisms (shortening, isometric, stretching) involved in the contractions. The muscle fatigue also enhances the higher accumulation of lactic acid in muscle. The heat generated during skeletal muscle contraction increases body and tissue temperatures. The increased temperatures (>40 °C) of muscle, both due to muscular function and primarily rise in environmental temperature, cause an early onset of fatigue. Additionally, the increased temperatures cause protein unfolding, denaturation and aggregation that further affect the muscle enzyme activity and contractile protein functions. The high temperature and protein denaturation activate the synthesis of heat shock proteins. The increased HSP level supports the skeletal muscle contractions during heat stress and protects cardiac tissue against ischemic stressors. Particularly, the higher expression of Hsp72 ensures protection of skeletal muscle damage in heat stress. In addition, elevated HSPs levels provide a cross-tolerance effect to cardiac and skeletal muscle.

28.7 Impact of Heat Stress on Immune System

The immune system is foremost important component of body for the defense mechanism against the invading pathogens. The invading pathogens are acted upon by the innate and adaptive immune systems. The equilibrium between the innate and the adaptive immune systems is regulated by Hypothalamic Pituitary Adrenal (HPA) and Sympathetic-Adreno-Medullar (SAM) axis. In addition, toll-like receptors (TLRs), HSPs, NOD-like receptors (NLRs), C-type lectin receptors (CLRs), AIM2-like receptors (ALRs), RIG-I like receptors (RLRs) and cytokines are also involved in the maintenance of immune system. Innate immunity is non-specific which is the first line of defense; whereas, adaptive immune system is induced by the entry of the pathogen which is more specific to act on the invading antigen. The cell-mediated immunity (CMI) and humoral mediated immunity (HMI) are the components of adaptive immune system. The T-lymphocytes are responsible for CMI, and B-lymphocytes constitute HMI. The mast cells, natural killer cells, eosinophils, basophils and phagocytic cells are fundamentals of the immune system. The balance between CMI and HMI is highly essential for effective functioning due to their interdependent actions. The heat stress impairs or depresses both humoral and cellular immunity by reducing generation of primary and secondary immunoglobulin, and production of inflammatory and anti-inflammatory cytokines.

The elevated level of cortisol depresses the production of L-selectin expression on the surface of neutrophil that controls recruitment of neutrophil and restricts movement of neutrophils with reduced phagocytic activity. The induction of HSPs during heat stress enhances and facilitates the innate immune cells to act against the invading pathogen. The movement of circulating white blood or immune cells is reduced into the mammary glands during heat stress and results more incidents of mastitis in dairy animals. Moreover, the production of IgG and IgM decreases during summer. Heat stress also impairs the digestive process, thereby leading to reduced absorption of IgG in calves. HSPs play an important role in antigen processing and presentation in addition to their chaperonic activities, thereby activating adaptive and innate immune response during heat stress. During heat stress, the expression of TLR genes was also observed to be altered which could be indicative of the adaptive strategy adopted by animals to combat the deleterious impact of heat stress on their immune system. Further, the animal maintains the balance between Th1, which is responsible the CMI, and Th2 which favours the HMI under normal conditions. The increased levels of glucocorticoids during heat stress disrupt the Th1–Th2 balance, shifting from Th1 to Th2 which enhances the HMI. Thus, prolonged heat stress negatively affects the immune system by modifying balance between innate and adaptive immune system and also making the animals more prone to infectious diseases.

28.8 Heat Balance

Homeothermic animals maintain a stable core body temperature by balancing the heat generated by metabolism and lost to the environment. The heat balance is accomplished by physiological, morphological and behavioural thermoregulatory mechanisms. The animal losses heat continuously as sensible heat from the body surface through conduction, convection and radiation during stressful environmental conditions. Further, the animal losses heat incessantly as evaporative (insensible) heat through the respiratory tract and skin surface. The exact quantity of heat loss depends upon the thermal demand of the surrounding environment and the resistance to heat flow in the tissue, skin and its plumage. The environmental heat demand is a function of meteorological factors, reflecting the cooling power of the surroundings which equals the rate of heat flow from an animal to a specific environment. The thermal balance of any animal is determined by the net heat exchange by heat transfer mechanisms, together with metabolic heat production (M). The general equation describing heat balance is (Willmer et al. 2005):

$$M = h_{\mathrm{cond}}(T_{\mathrm{b}} - T_{\mathrm{a}}) + h_{\mathrm{conv}}(T_{\mathrm{s}} - T_{\mathrm{a}}) + h_{\mathrm{ad}}(T_{\mathrm{s}} - T_{\mathrm{sur}}) + E + S$$

where h values are the various heat transfer coefficients (conduction, convection and radiation), T_b is body temperature, T_a is ambient temperature, T_s is surface temperature of the body, T_{sur} is surrounding surface temperature, E is the evaporative heat loss and S is heat storage. This equation may be shortened for a constant condition where there is no change in animal body temperature, no heat storage and no difference between T_s and T_{sur}, as:

$$M = h(T_{\mathrm{b}} - T_{\mathrm{a}}) + E$$

where h is the total heat transfer coefficient. In a thermally stable environment, in the ectotherm, E and M are negligible and the equation is simplified as $T_b = T_a$.

The thermal balance of an animal is determined by metabolic heat production and its dissipation via evaporation, radiation, convection and conduction. The autonomic thermoeffectors play the vital physiological mechanisms to maintain normothermia in the absence of behavioural thermoregulations. Autonomic mechanisms activate the thermoregulatory pathways in the anterior hypothalamus to retain equity between heat generation and loss. The hypothalamus activates the body heat production or dissipation in response to environmental conditions. The adaptive thermoregulation maintains a balance between thermogenesis and thermolysis processes which are increased during cold or heat responses, respectively. Thermogenesis is the process by which body heat is produced by the alterations in metabolism, muscle activity and hormones levels. However, thermolysis refers to the overall process to dissipate excessive body heat by vasodilatation, sweating and panting.

28.9 Thermal/Heat Exchange Mechanisms

Approximately, 75% of heat losses are continuous by sensible means of convection, conduction and radiation, and 25% as latent or hidden/insensible ways of evaporative heat loss through the respiratory system and skin. During variation in environmental temperature, initially animals manage to maintain homeothermy by regulating its surface temperature through the process of vasodilatation or vasoconstriction. In addition, the animal modifies its exposed surface area through behavioural responses by changing its posture. Heat dissipation depends on the surrounding conditions of an animal and is primarily determined by peripheral temperature with the underlying changes in temperature gradient. The body surface area of the endotherms is poorly covered with coats or even lacks insulation with a higher range of cutaneous blood flow, and these areas are referred to as 'thermal windows', such as ears, feet and nose of mammals, and bills, feet, comb and wattles of birds. Further, heat transfer or dissipation mechanisms greatly vary among animals, depending on several internal and external factors like estrus, pregnancy, parturition, lactation and environmental conditions. The T_a, RH, wind speed, solar radiation and shade also highly influence the body temperature. As soon as the temperature balance between the surface of the animal and the environment is decreased during hot environmental conditions, the animal enhances its evaporative heat loss to make up the reduced sensible heat loss. The feed intake and metabolism by the animal increase with reduced environmental temperature. In addition to this, the enhanced shivering or non-shivering thermogenesis are vital adaptive responses adopted by animals. Therefore, to maintain a constant core body temperature, heat produced and gained must be equal to the heat lost from the body. There are four major pathways (convection, conduction, radiation and evaporative heat loss) that aids in maintaining homeothermy in animals.

28.9.1 Radiation

Radiation is the movement of heat through release of electromagnetic energy from warm body to cold environment without physical contact. These electromagnetic waves are divided into short-waves or waves from the sun and long waves which radiate from the environment. Radiant energy heats the air indirectly by heating solid surfaces, for example soil, clouds, water, trees and animals. The loss or gain of heat ensued from the body by infrared waves depends on temperature and received infrared thermal radiation from its surroundings. However, heat transfer by the radiation is influenced by surface area of the animal, skin temperature, surrounding air and the emissivity of the animal's skin. The degree of radiant heat transfer is very complex where exposure is a function of direct sunlight to the surface of the animal and the level of reflection from the surroundings. Further, the intensity of radiation depends on the colour and presence or absence of vegetation on the surface. In animals, solar radiation is a function of surface area exposed to the radiation and the colour and structure of their coat. The black coat colour absorbs more heat (absorbance ~1.00) than red (0.65) and white fur (0.37).

28.9.2 Conduction

The dissipation of heat from warmer to cooler objects through direct physical contact can be termed as conduction. This depends on the temperature gradient between the contacting surfaces and also on thermal conductivity. The process of conduction transfers the core heat to the skin surface and from the periphery to surroundings. If an animal is lying on a cool surface, the conductive heat transfer is significantly higher than standing depending upon the thermal conductance, temperature gradient and area of contact. Conduction plays a minimal role in heat dissipation mechanism during thermal stress in livestock due to the reduced thermal gradient.

28.9.3 Convection

It is transfer of heat from one molecule to another or could be considered as a specialized conduction where the heat from warm body is moved away from the area by a current of air or water. The efficiency of convective heat loss is influenced by velocity of air, air temperature, body surface area and surface temperature. The factors that resist flow of air will reduce the rate of convective heat transfer because animal lose heat to the environment through air. The animal's fur entraps a layer of air close to the skin which prevents the passive convection. However, convection facilitates in respiratory heat loss by increased air flow via nasal routes where the upper respiratory tract takes away extra quantity of body heat. In addition, convective heat transfer is one of the major approaches adopted by animals to transfer core body heat to the periphery.

28.9.4 Evaporation

Evaporation is the process of removal of adequate heat into a liquid which turns it into a gas. The level of heat transferred during hot environmental conditions through conduction, convection and radiation is inadequate. The evaporation is the principal way of heat loss mechanism in animals when environmental temperature and radiant heat are high or equivalent to skin temperature.

28.10 Limits of Environmental Temperatures

The approach for thermal preference is highly related with metabolic activity, and its efficiency is optimal within a specific range of core body temperature (T_c). The cell integrity or function is affected when core temperature increases or decreases above the acceptable physiological limits. The temperature preference seeking behaviour in animals reduces the temperature difference between the animal and its environment, thereby decreasing the temperature gradient. This approach persuades minimum metabolic energy expenditure in animals to maintain normothermia. However, there is an overlap between the range of preferred environmental temperature and the comfortable temperature zone. Further, the thermal preference varies among species depending upon the time of day due to daily variations in their requirement for body heat production. The day and night variations in preferred T_a are important in heterotherm animals where there is a great variation in the T_c. There is a necessity for the animal to choose appropriate temperature especially during the torpor phase wherein the animal which tries to save energy. The T_a is an important factor that influences thermoregulatory mechanism of animals at different levels.

28.10.1 Thermoneutral Zone

Thermoneutral Zone (TNZ) is defined as 'the range of environmental temperature within which body temperature is maintained at a constant with minimal effort from thermoregulatory mechanism'. The TNZ can be differentiated by nature of blood vessels of the skin over the whole body which are neither all vasodilated nor vasoconstricted; evaporative heat loss is minimal, piloerection and behavioural responses to cold or heat are absent. TNZ varies with the age, species, breed, insulation, level of nutrition, earlier experience of temperature acclimation or acclimatization, production level, housing conditions, behavioural responses and time of the day. Generally, TNZ is very narrow for the young animals in comparison to adult animals which varies widely within the same species. Further, it is the temperature zone at which the animal may perform at its maximum. TNZ and heat tolerance threshold level vary in sheep from 5 to 25 °C depending upon the breeds and climatic regions. When the environmental temperature reaches the limits of the TNZ in opposite directions, it approaches the lower and upper critical temperatures (UCT). The ambient temperature (T_a) below which the rate of heat production of a resting homeotherm increases to maintain thermal balance is known as the lower critical temperature (LCT). Hence, the normal metabolic rate is insufficient to re-establish homeostasis and the body has to generate additional heat while environmental temperature declines. Where, the metabolic rate of animal raise from the basal level to meet environmental demands for heat either by shivering or non-shivering thermogenesis. The UCT is the T_a above which the rate of evaporative heat loss of a resting animal is increased to sustain thermal balance. Endotherms have relatively low variation in UCT compared to the variation in the LCT. The UCT can be demarcated by increased metabolic rate and evaporative heat loss along with minimal tissue thermal insulation. The heat production is increased or decreased when temperature is below or above the critical limits.

28.10.2 Lethal Body Temperature

Lethal body temperature, generally considered as the extreme T_c at which 50% of the experimental animals die. The lower lethal body temperature varies from 15 to 20 °C in majority of the species that is an average of 20 °C below the normal core temperature. The young animals have lower lethal temperatures than the adult animals of the same species. The upper lethal body temperature is life-threatening than the lower lethal temperature. Majority of the mammals die when their T_c reaches by 42–45 °C which is 3–6 °C above the normal body temperature.

28.10.3 Hypothermia

Hypothermia is an abnormally low T_c which develops as a consequence of continued exposure to cold along with an inability to conserve and enhance heat producing mechanisms. Hypothermia can be primary or secondary where primary hypothermia commonly occurs due to the exposure of animals possessing normal heat production to cold environment. Secondary hypothermia results as a consequence of alterations in heat production which may be due to illness, injury or drugs. Naturally, secondary hypothermia results in morbidity and mortality in particularly ill animals. The ability to withstand lowered body temperatures varies among species and is life threatening if not tackled effectively.

28.10.4 Hypothermic Spiral

The thermoregulatory mechanism is impaired when T_c reduces lower than 94 °F (34.4 °C) and animals typically stop to shiver or seek heat. During such conditions, the peripheral vasodilation is predominated rather than vasoconstriction and results in continuous loss of core temperature with reduced metabolic rate. Simultaneously, severe hypothermia depresses the central nervous system making the hypothalamus less responsive to hypothermia, and cessation of thermoregulation ceases as the T_c falls below 88 °F (31.1 ° C).

28.10.5 Fever and Hyperthermia

Fever is an increase in T_c above the normal range which is caused by microorganisms and it is a beneficial effect which stimulates the body defence or immunological mechanisms. The temperature set point of the hypothalamus is enhanced to facilitate the body to deploy the heat conserving and producing mechanisms that may elevate temperatures of 41 °C (106 °F) in mammals. The point at which the body temperature reaches maximum and re-establishes the balance between heat loss and heat production, and temperature is precisely regulated at the new high level. The various heat loss mechanisms of the body are stimulated, and body temperature returns to its normal level at the end of fever. Therefore, shivering and cold are characteristics of initial phase of fever and it is self-limiting. Hyperthermia is an increase in body temperature above normal range while body generates or absorbs more heat than it dissipates. The heat stroke, an indicator of hyperthermia, occurs when the heat production is higher than the evaporative capacity or evaporative mechanisms are impaired due to loss of body fluid and decreased blood volume.

28.10.6 Hibernation and Estivation

Hibernation and estivation are the important physiological and behavioural responses of homeothermic animals to avoid or escape from the extremes of environmental conditions. Hibernation refers to a cessation of coordinated locomotor activity and a reduction in body temperature, total metabolism, heart beat and respiration during winter. Further, the animals have the ability to spontaneously reoccur in normal homeothermic condition without any external heating. These hibernating animals awake from their dormant state periodically for urination as well as when body temperature reduces to level near freezing point. The brown fat is a connective tissue, and colour is due to presence of cytochrome pigments with higher level of mitochondria and usually found in hibernating animals. The new born animals of some species also pose brown fat which disappears within the few months of their life. Generally, brown fat accumulates in the subcutaneous region between the scapulae, kidneys and myocardium. Brown fat depots are the source of non-shivering thermogenic activities to generate more heat and being distinguished from white fat by its colour and metabolic effects. Further, activation of brown fat utilizes more amount of oxygen to generate high level of energy or heat. Estivation refers to a torpid sleeping state during the summer in harmony to hibernation during winter. The major advantage of estivation is to reduce the metabolic rate and T_c to prolong the period of survival of the animal with its energy reserves and preserves a significant quantity of water. However, the true torpor is not a state in which endothermy and thermoregulation are shunned but one in which they are regulates at a new level with a new critical minimum temperature. When environmental temperature drops below new critical temperature, the metabolic rate will be enhanced to maintain the critical body temperature. Animals such as the desert tortoise choose to sleep in their burrows when temperature is at its maximum and it will be active when the temperature drops down. Similarly, small endothermic animals of temperate climates also use torpor as an escape from transient seasonal or night low temperatures.

28.11 Thermoregulation

The mammals and birds maintain relatively a constant T_c to prevent the alternations in the physiological and chemical responses. In general, T_c is maintained within a narrow range of 36–39 °C depending upon the specific mammalian species such as cats and dogs 37–38 °C; sheep and goats 39 ° C and cattle 38.4 °C. The constant core temperature is maintained by many homeostatic mechanisms when the TNZ of specific species is altered. The reduction in core and skin temperatures activates heat production by shivering or activation of brown adipose tissue, vasoconstriction to conserve heat and warmth-seeking behaviour for thermal comfort. In contrast, as soon as the core or skin temperature increases, heat dissipation mechanism is activated through evaporative cooling, sweating, panting and skin vasodilatation. Further, the reliability of T_c is not absolute where temperature homeostasis may be compromised in fever and hibernation to support survival by eliminating pathogens and preserving nutrients, respectively.

The thermal receptors for cold and warm are present throughout the body, and the signals are traversed via A-delta and C fibres for cold and warm receptors, respectively. The thermoregulatory process occurs by three pathways, afferent thermal sensing from the periphery, central regulation in the hypothalamus and efferent responses. The peripheral body temperatures are continuously fluctuating, whereas the posterior hypothalamic thermoregulatory centre sustains a quite constant T_c. The cellular metabolism enhances the heat increment in the body which is transferred to conductive tissues then to the skin and is subsequently lost to the environment. In general, heat is transferred from the animal's body core to the skin via a network of blood vessels that are controlled by the autonomic nervous system. The rate of blood flow through these arteriovenous anastomoses fluctuates during climatic extremes wherein the vasodilation (during warm environment) facilitates the heat loss by increased blood flow and vasoconstriction (during cold environment) restores the T_c with decreased blood flow.

The body functions are primarily regulated by T_c where the increase or decrease in temperature alters the biochemical processes that occur in the body. Thus, it is essential to maintain a relatively constant T_c for the optimal function of all the tissues in the body. The animals are classified as homeotherms (or) warm blooded animals and poikilotherms (or) cold blooded animals based on the constancy of the T_c. Homeotherms are the animals having the capability of regulating their body temperature within a limited range nearly around 37–40 °C even when the external temperature varies (mammals and birds). The level of heat generation is 8–10 times greater in homeotherms than poikilotherms of the similar body size and temperature. Poikilotherms are the animals that do not have a control over their body temperature and varies with the environmental temperature. They are also called as temperature conformers (reptiles, amphibia, fishes and invertebrates).

Animals are also classified based on the source of body heat production as endotherms, ectotherms and heterotherms. Endotherms are the animals which could generate their own heat through metabolic heat production (birds, mammals) and maintain their body temperature above T_a with well isolated fur, feathers or fat which make them to conserve heat against high temperature (lower vertebrates and insects). They are having higher resting metabolic rate of five times greater in comparison to ectotherms of identical body size and temperature. Whereas, ectotherms are the animals that absolutely depend on the environment for their heat production. These animals are low metabolic heat producers with features of high thermal conductance and regulate their body temperature by absorbing heat from environment (crab). The heterotherms are animals capable of varying their degree of endothermic heat production. They do not regulate body temperature in the narrow limits (monotremes-egg laying mammals and few large fishes).

28.11.1 Selective Brain Cooling

Most of the domestic mammals (except horse) are highly competent to decrease the brain temperature lower than T_c when threshold temperature of brain exceeds which is called as selective brain cooling (SBC). The SBC is achieved through a specialized anatomical feature wherein the carotid rete, passing through the cavernous sinus, receives cooler venous blood from the nose (provided the hemodynamic conditions in the effluent veins are appropriate). This allows heat to be exchanged from arterial blood (carotid rete) to the cooler venous blood (cavernous sinus), thus cooling the incoming carotid arterial blood before entering the circle of Willis to supply the brain. Therefore, SBC is a protective mechanism which maintains brain temperature against the elevated T_c during hot environment and fever.

28.11.2 Role of Animal Skin in Thermoregulation

The integumentary system is the set of organs which forms the external covering of the body and protects the animal against various hazards such as invasion by microorganisms, desiccation, abrasion, chemicals and environmental factors. The integumentary system/skin is important in the regulation of body homeostasis through controlling thermoregulation, sensory reception, biochemical synthesis, absorption and protection. The cold and heat sensory receptors are present in the skin to perceive thermal variations in the environment. As the body temperature increases, the hypothalamus directs neuronal signal to the sweat glands in the skin to secrete sweat for cooling the body. In addition, hypothalamus facilitates the dilation of the peripheral blood vessels of skin to accommodate high blood flow which favours heat convection from the skin surface. In contrary, when body temperature decreases, the sweat glands constrict to reduce the production of sweat. The skin also functions as a mini-excretory system for urea, salts and water in addition to synthesizes of vitamin D.

The physiological properties and functions of the skin are varying among the different breeds of animals. The physical, environmental, physiological factors and hair coat's optical properties are actively involved in skin temperature dynamics and are affected by evaporative cooling. The sweating rates are varied significantly within and between breeds which is attributed to genetic variation in thermoregulation among the individuals. The evaporative cooling is highly influenced by wind velocity, RH and solar radiation. In addition, the physical and optical characteristics of hair coat such as hair coat density and thickness, hair length and colour also influence evaporative cooling. Black/dark coloured skin and hair enhance solar absorption and elevate heat load on the skin surface. Further, hair coat density hinders evaporation of water from the skin surface by covering a thin film of water at the skin-hair coat interface. The sweating rates also differ between high-producing dairy and feedlot animals in their natural habitats.

28.11.3 Thermoregulation in Birds

Birds are endothermic and depend on high heat production to maintain the higher core body temperature to which birds are forced to distribute significantly more energy into thermoregulatory process. The insulating features in birds play vital role in conservation of internally produced heat and facilitate the thermal conductance which is an energy demanding process. The birds maintain a significantly constant body temperature of 41–42 °C at rest and inactive phase in a wide range of environmental conditions (tropical, temperate and polar) through physiological, morphological and behavioural responses. The endothermic characteristics facilitate the survival of birds in different environmental situations such as aerial, aquatic and terrestrial. The T_c of birds is high with 3–4 °C above mammalian body temperature. The lack of sweat glands and the high efficiency of the thermal insulation via feathers protect the birds against cold but they are more vulnerable to heat stress. In birds, the air sacs are extensions of the lungs that extend into the body cavities. The air of pulmonary ventilation cools the body of birds more than that of mammals due to the larger gradient and closeness of the air sacs to the vital organs.

The effect of T_a on avian body temperature regulation is significant and reported across seasons and different life stages. During winter, birds have to generate more heat so as to maintain normal body temperature. In the summer and spring, they have to dissipate excessive heat, especially during the breeding season. The thermoregulatory system maintains a constant body temperature with fluctuating environmental temperatures in birds. This thermoregulatory mechanism is comprised of a sensory part which senses the variations in environment and an integrating part, the thermoregulatory centre in preoptic anterior hypothalamus which perceives temperature fluctuations through peripheral thermoreceptors. Therefore, the T_c is maintained by heat generation or heat loss mechanisms in relation to thermal status of bird. In addition, the assertive part consisting of neuroendocrine signals, controls the downstream mechanisms to sustain the body temperature by shivering and non-shivering thermogenesis, evaporative heat loss, peripheral vasoconstriction or vasodilation and behavioural changes. Shivering thermogenesis is the major activity in contribution of thermogenesis in the birds. The heat production below LCT, increased through shivering thermogenesis, is characterized by asynchronous muscle contractions via production of chemical energy through the hydrolysis of ATP as heat instead of kinetic energy. In addition, contraction or tremor amplitude during shivering is low which prevents the convective heat loss. The non-shivering thermogenesis also contributes towards heat production in birds. However, when environmental temperature is above UCT, heat needs to be dissipated to restore normal body temperature which is achieved primarily through evaporation. The birds are well adapted to hot and arid environments and dissipate excessive heat through evaporation of water via cutaneous surfaces or gular fluttering. Majority of passerines dissipate heat by respiratory evaporative heat loss which is energy demanding process in contrast to cutaneous evaporation. Hyperthermia augments the risk of physiological damage, and hypothermia could enhance predation risk in certain conditions.

The thermal gradient between the body and environment during hot conditions hinders the heat dissipation from the body. Additionally, the availability of feed and water is highly essential during hot environment for birds due to the high energy demanding activity of panting and to facilitate or potentiate the major heat dissipation mechanism of evaporative cooling. However, during the dry season in arid regions, water is in short supply, thus making the birds prone to hyperthermia. Therefore, the birds focus to save water which is size dependent and its efficiency reduces with body size, particularly for long bouts of heat stress. Water loss in large sized birds exceeds 1 kg during hyperthermic conditions.

The maintenance of constant body temperature is an energy demanding process and to reduce the energy cost of thermoregulation, birds use a variety of morphological and behavioural characters to modify the rates of heat loss and heat gain. The unfeathered body surface areas are the major important site for heat exchange with the environment. During cold to minimize heat loss, the arteries and veins in the unfeathered areas play a crucial role of counter-current heat exchange system to retain heat. Standing on one leg while tucking the other among its breast feathers, reduces exposure of the limb, thereby reducing heat loss from the body. Further, birds' fluffs out their feathers that increases the thickness of their coat, thereby effectively enhancing their insulation. Additionally, they also alter their posture or orient towards the sun, thereby reducing heat loss from the body. The birds are capable of sustain their body temperature at lower level when they are inactive, and this regulated hypothermia facilitating a significant energy savings. Further, hummingbirds, swifts and poorwills enter a state of torpor wherein their body temperature may drop as much as 50 °F (10 °C) for several hours during the night or days in extreme weather conditions.

28.12 Measurement of Heat Stress in Animals

28.12.1 Temperature Humidity Index

It is essential to evaluate the impact of environmental factors on animal production. Livestock weather safety index (LWSI) is one of the methods for estimation of level of heat stress in livestock species. The LWSI is based on the temperature humidity index (THI) which is uncomplicated, reliable and easy approach to assess the heat load in stressed animals. THI is a combination of T_a and RH into a single value to reveal the degree of stress and to establish its influence on animal production. The THI formulas differ according to region and parameters obtained to measure it. Thom (1959) established THI based on T_a and RH as follows: THI = 9/5 × [(T_a × 17.778) − (0.55 − (0.55 × RH/100)) × (T − 14.444)], value of THI <72 indicates thermoneutral conditions and THI 76–78.5 represents mild to moderate heat stress. Another THI was developed by inclusion of wet and dry bulb air temperatures for a specific day, THI = 0.72 (W °C + D °C) + 40.6, where W °C = wet bulb and D °C = dry bulb. The THI values of 70 or less = comfortable, 75–78 = stressful, and values above 78 = extreme distress (McDowell et al. 1976). The other equation to estimate THI was equated as THI = (Dry-Bulb Temperature °C) + (0.36 Dew Point Temperature °C) + 41.2. As per this formula, THI values exceeding 72 indicates mild stress, 80 reflects medium stress and values above 90 signifies severe heat stress in cattle. When the T_a is measured in Fahrenheit (°F), the equation to determine THI is as follows (LPHSI 1990); THI = db °F − {(0.55 − 0.55 RH) (db °F − 58)}, where db °F is the dry bulb temperature in °F and RH is the relative humidity (%)/100. The THI values specify as follows: values <82 = absence of heat stress; 82 to <84 = moderate heat stress; 84 to <86 = severe heat stress and over 86 = extreme severe heat stress.

28.12.2 Black Globe Temperature and Humidity Index

The severity of heat stress can also be evaluated with few other indices based on the availability meteorological data such as black globe temperature and humidity index (BGHI), thermal comfort index (TCI) and global comprehension index (GCI). Where, BGHI = BGT + (0.36 × DPT) + 41, with BGT = black globe temperature (°C) and DPT = dew point temperature (°C). The values of BGHI up to 74 is a comfortable condition, 74–78 as alert, 79–84 as danger and above 84 as emergency zone. Further, many numbers of computation methods were developed over the years to establish an accurate method encompassing the major environmental factors apart from the T_a and RH. Barbosa and Silva (1995) estimated the thermal comfort with the following equation: TCI = (0.6678 × AT) + (0.4969 × PVP) + (0.5444 × BGT) + (0.1038 × WS), where AT = air temperature (°C), PVP = partial vapor pressure (kPa), BGT = black globe temperature (°C) and WS = wind speed (m/s).

28.12.3 Heat Load Index

The heat load index (HLI) was developed to measure heat load in animals based on relative humidity (RH%), windspeed (WS, m/s) and black globe temperature (BGT, °C). Where, BGT cannot be measured directly and is predicted using air temperature (Tdb, °C) and solar radiation (SR, W/m^2) as BGT = 1.33 × Tdb − 2.65 × Tdb 0.5 + 3.21× log10(SR + 1) + 3.5. Further, HLI has two portions based on a black globe temperature threshold of 25°C; HLIBGT >25 = 8.62 + 0.38 RH + 1.55 BGT − 0.5 WS + e(2.4 − WS) where e is the base of the natural logarithm, and HLIBGT <25 = 10.66 + 0.28 RH + 1.3 BGT − WS. As per this equation, HLI = 86 is the threshold for Angus steers under open area and 96 for *Bos indicus,* whereas it is around 89 for white coat animals. However, the threshold of HLI is a minimum of 81 for sick animals (Gaughan et al. 2008).

28.12.4 Wind-Chill Index

Wind-chill index (WCI) was developed relating ambient temperature (T_a) and wind speed to the time for freezing water for human (Siple and Passel 1945). Further, it has been extended to the domestic animals, WCI (kcal/m^2/h) = (10$\sqrt{v}$ − v + 10.5) × (33 − T_a) where v is the wind velocity (m/s) and T_a is air temperature (°C). The WCI (kcal/m^2/h) was classified in to five levels, no chill (<300.0), low (<300.1–350.0), moderate (350.1–400.0), high (400.1–450.0) and extreme (>450.1).

28.12.5 Comprehensive Climate Index

The comprehensive climate index (CCI) has developed for a wide range of environmental conditions with an adjustment to ambient temperature (T_a), relative humidity, wind speed and radiation. CCI = AT + FRH + FWS + FSR (Mader et al. 2010), where FRH corresponds to the correction factor for AT due to relative humidity, FWS is correction factor for AT due to wind speed and FSR is the correction factor for AT due to solar radiation. The level of thermal stress is categorized as no stress (CCI ≤25), mild (>25 and ≤30), moderate (>30 and ≤35), severe (>35 and ≤40), extreme (>40 and ≤45) and extreme danger (CCI >45).

28.12.6 Tunica Dartos Index

The tunica dartos index (TDI) was developed to judge the ability of ram's heat tolerance during higher ambient temperature. The TDI was predicted by three ways as TDIa, TDIb and TDIc. TDIa referred as percentage of change in scrotal length in association with maximum scrotal length minus the testis length. TDIb indicated the variation between rectal temperature (RT) and scrotal skin temperature in percentage of RT, and TDIc is the combination of TDIb and TDIa. The three formulae are as follows: TDIa = [(Max. SCL − Min. SCL)/(Max. SCL − TL)] × 100; TDIb = (RT − SST)/RT × 100 and TDIc = [(RT − SST)/RT × (Max. SCL − Min. SCL)/(Max. SCL − TL)] × 100. Where RT—average rectal temperature in summer, SST—average scrotal skin temperature in summer, Max. SCL—average maximum scrotal length in summer, Min. SCL—average minimum scrotal length in winter and TL—average testis length during the winter and summer seasons.

28.12.7 Infrared Thermography

Infrared thermography (IRT) is a non-invasive and non-contact method of measurement of temperature which is rapid without restraining stress and also provides an opportunity for automation. Infrared thermographic cameras measure temperature of animals that radiates heat from the external body surfaces. Heat emissions are displayed as a thermogram of pixels in varying colours or shades that indicate different infrared temperature. In IRT, emissivity of objects varies from 0 to 1 and the emissivity of cattle external body surfaces varies between 0.93 and 0.98 depending on skin and hair colour, and density of hair. The IRT of the eyes specifically the skin around the inner corner of the eye socket indicates core body temperature of animal. In fact, the eyes are positioned close to the hypothalamic thermo-sensitive area which facilitates quick response. The blood vessels of eyes are identical to the brain and choroid vessels which are similar to the small intestine and kidney. The IRT of eyes moderately depict the rectal and vaginal temperature. Therefore, changes in the eye and rectal temperature IRT may indicate similar physiological responses to stress.

The IRT of external body surfaces invariably specifies surface temperature which is highly associated with peripheral temperature. IRT data are more dependable than routine measurement of rectal temperature. The IRT of limbs indicates the peripheral temperature which is highly fluctuating due to the formation of buffer between the core body and the environmental temperature through convection. IRT of external body surface area differs based on distinct tissue metabolism, blood flow, conduction and the ability of an object to absorb and emit radiation. Limbs are major route of heat dissipation in animals. The fluctuations in peripheral IRT of limbs are highly noticeable at the coronary band due to muscular activity involved in the movement, weight bearing and more blood circulation to provide nutrition. Therefore, IRT of external body surfaces is highly important to assess the environment induced changes in the animals' body temperature.

28.13 Heat Tolerance

Heat stress is a considerable issue in livestock production system which negatively affects the feed intake and well-being of the animal, apart from reduced performance. Heat tolerance is the ability of an animal to maintain its genetic potential of optimum production under hot environmental conditions. Heat balance is a complex phenomenon affected by many factors such as climate (environmental temperature, relative humidity, wind speed, radiant heat and altitude), animal (age, genotype, type of hair coat, level of acclimatization, health status, production level) and management (housing, provision of shade and fans). The level of heat tolerance varies among species, breeds and also between animals; for example, Holsteins are less tolerant than Jersey cows, beef cattle with black hair are more prone to heat stress by direct solar radiation than those with lighter hair. The lactating animals are more sensitive to heat stress than dry cows due to high metabolic heat production. Further, heat tolerance may be described as the rate of decline in milk, fat and protein yields per unit increase of THI based on the production ability. The decline in production in heat-tolerant cows is slower with regard to increasing heat stress when compared to heat stress susceptible cows. Goats are emerging as being more tolerant to heat stress than sheep, while both small ruminant species are superior to cattle. Sheep and goat have an advantage due to their morphological and physiological traits to withstand heat stress and dissipate body heat when compared to cattle. There is genetic variation in the performance of animals under heat stress conditions which can be identified for selection of heat-tolerant animals by genomic breeding. At present, genomic best linear equitable assessment is referred to determine genetic breeding value (GEBV) for heat tolerance in association with milk production and its composition. However, GEBV represents lower heat resilient for milk production when the THI is beyond 60.

28.14 Improving Resilience to Heat Stress

The impact of heat stress on animal production is evaluated by assessing the animal comfort. This can be measured using indices like Iberian heat tolerance test (HTC). HTC is used to evaluate heat tolerance capacity of cattle by measuring rectal temperature that rise above 101.0 °F (38.33 °C). Higher HTC value indicates higher heat-tolerant capacity of the animal. Further, this index paid pathway for the development of other models such as coefficient of adaptability, biochemical index of heat tolerance and discomfort index or milk production decline index. The genetic resilience can be achieved through repeated selection for better performance and heat tolerance in animals. The heat tolerance capacity significantly varies between species/breeds in addition to individuals. The natural selection procedures help in the development of breeds with a better ability to withstand heat stress in tropical regions. The heat-tolerant tropical local or indigenous breeds are distinguished by their small size, low productivity and distinct morphological characters such as skin or hair type, sweating capacity, tissue insulation and special appendages in comparison with normal breeds. The sweating rate is also differing between animals of same breed due to genetic variation in thermoregulation. The heat tolerance capacity of local breeds may be propagated by cross breeding with an exotic breed to enhance the performance level. This cross breeding is successfully adopted in beef cattle and meat chicken to improve heat tolerance.

28.15 Responses of Animals to Heat

Animals exhibit varied behavioural, physiological, biochemical, cellular and molecular responses against heat stress. The shade seeking behaviour and reduced feed intake are the primary responses exhibited by an animal during heat stress condition. Additionally, animals, especially buffaloes wallow often and/or standing near the water bodies to enhance the conductive heat loss to water in order to reduce heat stress. The increase in T_c increases the skin temperature by cutaneous vasodilatation and promotes a thermal exchange gradient to expedite heat loss by inhibiting of sympathetic vasoconstrictor tone. The temperature may decrease vasoconstrictor tone either via a rise in temperature at central nervous system (CNS) or through impulses mediated by thermoreceptors in the skin and other parts of the body. Above an environmental temperature of about 31 °C, skin vasodilatation no longer increases heat dissipation, and a rise in body temperature would be observed unless heat loss can be augmented by other assisted means. The evaporative heat loss is an effective means of heat dissipation mechanism in animals. One calorie is required to increase the temperature of 1 g of water by 1 °C whereas to evaporate the same quantity of water from the body needs around 600 cal. The heat generated in resting animals is dissipated by evaporation of water from the skin surface and respiratory tract which accounts for around 25% of heat dissipation at normal T_a and RH. The evaporative heat loss from cutaneous and respiratory system is constant at thermoneutral conditions. The increase in T_a enhances blood flow to the skin that leads to higher evaporative heat loss by sweating.

28.15.1 Sweating

There are two types of sweat glands, eccrine and apocrine glands. The eccrine sweat glands open on the surface of the skin via duct and are supplied by cholinergic fibres present in sympathetic nerves. The apocrine glands are associated with hair follicles and ducts open into the hair follicle which are not supplied by secretory nerves and are sensitive to epinephrine carried in the bloodstream. In many domestic animals, apocrine sweat glands are important for evaporative heat loss. The apocrine glands are controlled by the alpha-adrenergic system, in comparison to horses which are beta-adrenergic. The blood flow to capillary beds affects the rate of sweat production during heat stress in animals due to their close association. The sweat gland numbers vary regionally in the order of the neck, flank, back, thigh, forehead and abdominal regions. However, the evaporation of moisture from the skin is highly associated with the functionality of the glands rather than their distribution over the skin. Thermoregulatory sweating is achieved in two approaches, by the increased CNS temperature and spontaneous stimulation of warmth receptors in the skin and other parts of the body. The rate of sweating depends on the rate of increase in core body temperature and skin temperature. The degree of relevance of sweating as a heat loss mechanism varies among species where sweating is minimal and panting is more in dogs. The evaporative heat loss from the skin surface of cows is around 150 $g/m^2/h$ at an environmental temperature of 40 °C. Sweating rate is very low in sheep and maximum sweat secretion in shorn sheep is 32 $g/m^2/h$ during heat stress where evaporative heat loss is more important.

28.15.2 Respiratory Frequency

The major function of respiratory system is to eliminate carbon dioxide (CO_2) from the tissues and supply of oxygen. Respiration rate is an indicator of heat stress or heat load on the animal during hot environmental conditions. The respiratory tract acts as a major heat loss route in heat stressed animals when the other heat dissipation mechanisms become inadequate. The normal respiration rate among domesticated animals ranges from 20 to 30 breathes/min at TNZ. The increase in the respiration rate by 80–120 breathes/min in cattle indicates that they are under moderate to high thermal stress and above 120 breathes/min is designated as an excessive heat load. During heat stress, respiration is associated with the level of heat load in animals that increases ventilation of the dead space by increasing the frequency and reducing the tidal volume. When animals pant with closed mouth, body heat is exchanged via the upper respiratory tract which is transported to the nasal mucosa by blood. On dissipating heat via respiration, the cool blood drains into the venous sinuses at the base of the skull. The respiratory frequency reduces at the elevated body temperatures with increased tidal and minute volumes. The increase in minute volume establishes the continuous and enhanced respiratory evaporation even the increase occurs at the expense of an overventilation of the alveoli. The enhanced ventilation reduces CO_2 level and results in increased blood pH. The transformation of respiration rate from rapid and shallow to slower and deeper breathing indicates that the physiological mechanism allows maximum evaporative cooling with least interruption in the blood gases. The last phase of respiratory frequency is open mouth panting with protrusion of tongue which is synchronized with highest respiratory frequency that is regulated by the airway resistance.

28.16 Responses of Animals to Cold

The decrease in environmental temperature activates thermogenic activities, which prevents the decline in T_c in homeotherms. The regulatory responses are elicited against the drop in body temperature by reduction of heat loss via insulation. When this regulation is inefficient in restoring T_c, metabolic increment as chemical regulation is intensified to protect the body temperature. Animals reduce the heat loss by changing the posture, intensifying insulation properties of fur via piloerection and increasing fur growth and subcutaneous fat deposition. Vasoconstriction occurs at the extremities of the animals once the environmental temperature drops. The vasoconstriction reduces blood flow to skin and periphery resulting in reduction in temperature which neutralizes or reduces the temperature gradient between the skin and the environment. Further, the functional insulation of the skin is increased due to reduction in the convectional heat loss by decreased blood supply to skin. In addition, the sensory receptors transmit signals of lower temperature to the CNS which primarily induces the peripheral vasoconstriction. Moreover, the close arrangement of arteries and veins assists in the heat conservation where cold venous blood is transported centrally alongside to warm arterial blood from periphery. The continuous heat exchange between the returning venous blood is warmed, and the arterial blood is cooled which effectively reduces heat losses by countercurrent heat exchange mechanism in cold environment.

28.17 Thermoregulation During Cold and Heat

The maintenance of T_c at normal range during cold when the environmental temperature goes below the LCT depends upon the ability of animal to increase the metabolic rate. The increment in metabolic rate of small mammals is higher which is proportionate to the three-fourth power of body weight (six times) as that of basal metabolic rate. New born animals have maximum metabolic rate by a factor five and the metabolic rate reduces as they grow. However large animals maintain homeothermy during cooler temperatures, primarily by taking advantage of their insulation capacity than increasing metabolic rate. Sheep are more tolerant to cold than the high-producing cows. The hypothermia causes cold injuries in the extremities such as the ears but the arteriovenous anastomoses give some protection against frostbite. However, maintenance of homeothermy is more critical during hot environmental conditions than a cold. The heat tolerance capacity of the animals depends on the evaporative cooling mechanisms where the sweating species tolerate higher environmental temperatures than panting species. Animals establish a new equilibrium of body temperature and continue as the heat load is moderate. When the heat load turn into severe, animal losses its capacity to control T_c, which gradually increases and results in hyperthermia. Hyperthermia occurs at relatively low T_a when RH and solar radiation are high with increased metabolic heat production. The intensification of hyperthermia eventually causes failure in sweating and respiratory mechanisms and finally results in a breakdown of thermoregulation.

28.17.1 Impact of Heat/Cold on Metabolism

The metabolic heat is the principal source of heat aggregation in animals which is generated within the body for each and every biochemical reaction associated with body function such as growth, lactation and pregnancy. The metabolic heat increment is necessary during cold to protect body temperature while metabolic heat has to be eliminated away from the body during warm periods. Whenever the animals are not able to transfer sufficient heat to their surroundings and accumulate it within the body, it results in hyperthermia. Animals that are well adapted to hot conditions consistently decrease heat production or enhance heat loss mechanisms to sustain the homeothermy. The environmental temperatures influence the metabolic and endocrine profile of animals. Exposure of animals to cold or heat stress reduces the productivity and feed efficiency. The susceptibility of animals to cold stress varies with the stage of life, production phase and breed. The energy requirement increases with decreasing temperatures in winter to elevate resting heat production to maintain the T_c by shivering or non-shivering thermogenesis.

Shivering is an involuntary function of the body and consists of muscle contractions usually preceded by an increased muscular tone. The circuit consisting of gamma motoneurons, muscle spindles and muscle afferent fibres is apparently of great importance in the control of shivering. The shivering thermogenesis is initiated and regulated by peripheral and central temperatures where peripheral cooling activates shivering without any change in brain temperature. The local cooling of the anterior hypothalamus or the spinal cord also induces shivering thermogenesis with a constant environmental temperature. The non-shivering thermogenesis occurs through calorigenic effect of epinephrine and nor-epinephrine that is secreted during cold stress. Further, the higher level of thyroxine during cold potentiates the calorigenic action of epinephrine. Therefore, thermal stress caused by the variations in T_a above and below TNZ results in decreased performance of animals. Heat stress induces a decrease in feed intake which could be in an effort of animal to reduce the metabolic heat production. Reduced feed intake accounts for around 30–50% of the total decline in milk production. Therefore, the impacts of heat stress on livestock productivity are probably arbitrated by alterations in the metabolism, which could not be assumed on the level of nutrition nevertheless heat stress modifies metabolism independent of nutrient intake. Heat stress influences the post-absorptive metabolisms that are independent of decreased feed intake and energy balance in livestock. The changes may be an adaptive approach when the animal attempts to sustain T_c during extreme environmental conditions. In general, milk yield of dairy cow is reduced during cold stress and the initial phase of lactation is highly susceptible. Further, the milk production decreases when the T_a falls below −5 °C depending upon the breed, level of feed intake and acclimatization.

28.17.2 Adjustment of Energy Requirements During Heat/Cold Stress

The homeotherms are able to adjustments their energy requirement for maintenance during different environmental conditions. The maintenance energy requirement of animal depends on level of heat stress or the severity of heat stress that could vary among animals based on the acclimatization, availability of feed, level of productivity and diurnal fluctuations in radiant heat load. The severe heat stress condition increases the maintenance requirements by the enhanced panting and alterations in tissue metabolism due to elevated tissue temperature with reduced metabolic rate. However, severe heat stress reduces appetite of animal that results in reduced productivity and metabolic heat production. The adjustment of energy requirement of animal's maintenance during cold stress depends on the level of acclimatization ability to cold conditions. The animals acclimatized to cold stress have an increased metabolic rate with an enhanced capacity to increase their rate of metabolic heat production to prevent hypothermia during severe cold stress. The enhanced heat generation during cold stress requires an effective usage of energy substrates either from diet or tissue reserves.

28.18 Neuron-Endocrine Response to Stress in Animals

The physiological responses of animals to stress activate endocrine, autonomic and CNS responses along with redistribution of blood flow. Different systems act in a synergistic way in accordance to the level of stress, to sustain the homeostasis by stimulating physiological mechanisms to reduce the adverse impacts. The stress regulating systems differ between individuals depending upon their earlier experience, physiological status, genetic predisposition, extent and severity of stress. Hence, stress brings about certain changes in the neuroendocrine reactions that stimulate different hormonal axis and secretion of hormones which promote the adaptive and behavioural responses of animals. These stress hormones regulate the energy supply for muscular and neural activities, increase the awareness of the environment, improve the glucose concentration to brain, modifications in cardiovascular and respiratory functions, modulation in immune responses and finally lead to reduction in productive and reproductive performance. The environmental stressors mainly activate two primary neuroendocrine adaptive mechanisms which include sympathetic-adrenal-medullary (SAM) and the hypothalamic–pituitary–adrenal (HPA) axis. These axes act synergistically to elicit different stress responses in combination with interplay of adaptive responses of different organs and receptors to conquer extreme stress conditions. Figure 28.1 describes the different target endocrine glands and other components involved in the different adaptive mechanisms.

Neuroendocrine response is one of the principal defence mechanisms instituted by the animal to counter stress that stimulate HPA axis through sensory organs with the integration of brain centre. HPA controls the thermoregulation of animals by secreting different neurotransmitters and hormones. Further, HPA axis is activated directly or indirectly by heat stress, drought and nutritional stress and as well as disease conditions which enhances glucocorticoid secretion. The function and regulation of HPA axis are the foremost important prerequisite in the adaptation process of animals. The activation of SAM axis establishes immediate activation of the autonomic nervous system for the release of catecholamines, adrenaline and nor-adrenaline. The adrenal and thyroid hormones are essential in thermoregulation and metabolic response of animals during

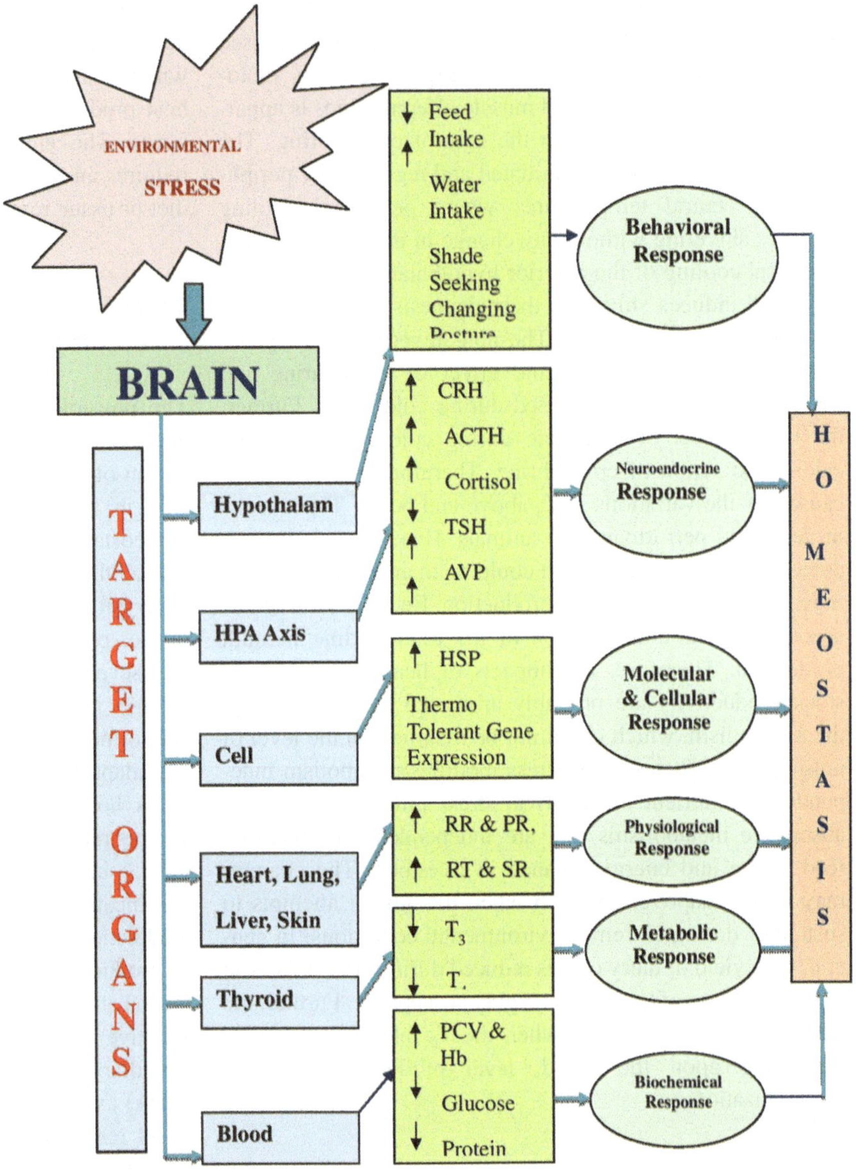

Fig. 28.1 Overview of neuroendocrine response during heat stress in livestock. High environmental temperature stimulates brain, pituitary, adrenal glands to release various signalling molecules and hormones into circulating system. Physiological, haemato-biochemical changes occur in animals in response to heat stress to establish thermal equilibrium

stress. The corticotrophin-releasing hormone (CRH), adrenocorticotropic hormone (ACTH) and glucocorticoids are the principal stress hormones released from HPA axis. The HPA axis enhances ACTH concentration on activation which is consequently improved the level of glucocorticoids particularly cortisol, the stress-relieving hormone. ACTH is one of the major pituitary hormones that supports growth and development of adrenal cortex and enhances the synthesis and secretion of glucocorticoids. The glucocorticoids production mainly depends on the integrity of the HPA axis and the secretory level of ACTH from the anterior pituitary. The adrenal cortex synthesizes cortisol which regulates the behavioural and neuroendocrine activities during stress. Cortisol is believed to be an important biomarker to quantify the level of stress among all the species of animals. Increased cortisol levels induce the hepatic gluconeogenesis which enhances the production of glucose from non-carbohydrate sources to establish energy homeostasis and to restore the life sustaining activities. The cortisol is the primary stress-relieving hormone that regulates the stress response in ruminants. Glucocorticoids are the downstream effectors of the HPA axis and regulate the physiological process by stimulating the intracellular receptors. Glucocorticoids are released in the circulatory system with the carrier proteins, and the carrier protein maintains the bioavailability of glucocorticoids for the immediate response to stress. Higher level of glucocorticoids supports the animal's survival in the extreme stress conditions. Glucocorticoids regulate the mobilization of energy expenditure in the body to sustain the energy homeostasis. However, glucocorticoids secretion is highly variable in many species during stress and also produced in a diurnal pattern that is influenced by many factors such as genetics.

The association between the CNS and pituitary regulates the SAM axis stimulation which releases β-endorphin that facilitates glucocorticoids and catecholamines to interact with a wide range of cells to modify the metabolic and immune functions. Catecholamines secreted in response to stress, regulate the adaptive mechanisms such as 'flight and fright' responses. In addition, catecholamines coordinate the cardio-pulmonary system by enhancing the cardiac output and respiration rate, sweating rate and redistribution of blood flow to the respiratory system and other vital organs to escalate different stress responses. Catecholamines act on adrenergic receptors of the visceral organs and smooth muscles to induce signalling pathways to modify different endocrine functions. Further, it alters the effects on afferent sensory nerves that impact CNS. These accelerated responses of catecholamines are highly essential for survival. However, the continued high level of catecholamines may result in pathological conditions such as cardiac hypertrophy, hypertension and post-traumatic stress disorder. The preganglionic neurons of the sympathoadrenal system secrete the neurotransmitter acetylcholine that activates the postganglionic neurons to produce nor-epinephrine. The preganglionic neurons of spinal cord are entered into the adrenal medullary ganglia of the sympathoadrenal system, and the terminals connect the endocrine cells known as chromaffin cells. The activation of chromaffin cells by acetylcholine increases the release of catecholamines in to the peripheral blood circulatory system.

The acute thermal stress activates SAM which is associated with shift in the water and electrolyte balance that is essential to support the evaporative water loss. The enhanced level of vasopressin (antidiuretic hormone) facilitates the conservation of water and increases water intake to compensate the water losses through respiratory tract and skin. The baroreceptors in the atrium and greater blood vessels, and hypothalamic osmoreceptors are activated by consistent variation in body fluids during heat stress which further increases vasopressin secretion to limit the dehydration. Whereas during cold stress polyurea, characterized by inhibition of vasopressin, facilitates more water loss through urination and prevent heat loss from tissues into water. In addition, vasopressin or antidiuretic hormone secreted from hypothalamus which enhances the effects of corticotropin-releasing factor on ACTH release from the anterior pituitary. Furthermore, vasopressin plays a vital function in the maintenance of concentration of ACTH during prolonged heat stress. The renin–angiotensin–aldosterone system also aids in maintenance of electrolyte homeostasis and hypovolemia as a result of dehydration during heat stress. Hypovolemia reduces the blood flow to the kidney and activates juxtaglomerular apparatus which increases the secretion of renin. The increased level of renin stimulates the production of angiotensin which consecutively enhances aldosterone from the adrenal cortex. The elevated aldosterone favours re-absorption of water and electrolytes especially sodium in the kidney to prevent the excretion of more water.

The hypothalamic–pituitary–thyroid (HPT) axis is most important in coordination of energy utilization by adjusting the basal metabolic rate through the actions of thyroid hormones. The HPT axis is situated in the medial region of the paraventricular nucleus of the hypothalamus that secretes and releases thyrotropin-releasing hormone (TRH) into the pituitary. The TRH triggers the secretion of TSH from the anterior pituitary, and consequently the production and release of thyroid hormones. Thyroid hormones, triiodothyronine (T3) and thyroxine (T4) are the principal metabolic hormones in animals, and the acute and chronic stress influences the HPT axis. The activation of HPT enhances TSH due to the direct stimulatory effect of glucocorticoids on the pituitary thyrotrope. However, prolonged stress habitually lowers the HPT activity in animals to achieve lower metabolic heat production during heat stress. The decreased HPT activity is transmitted to the hypothalamus by glucocorticoids to lower the TRH production. Further, elevated level of somatostatin as an effect of increased intrahypothalamic CRH release also regulates the lower TSH secretion during heat stress. The decreased TSH production impairs conversion ratio of T4 to T3 in the stressed animal. Therefore, thyroid activity is decreased during heat stress and results in lower level of T3 and T4 and it requires many days to reach normal level. However, the reduced level of thyroid hormones is not an instant response to acute heat stress but alternatively associated in the acclimatization of animals to a continued thermal load. Similarly, the lower level of thyroid hormones is highly associated with a decrease in metabolic rate and reduction in cellular heat production.

Growth hormone, produced in the anterior pituitary gland, is involved in energy partitioning along with an initiation and maintenance of lactation in animals. The concentration of growth hormone decreases during acute and chronic heat stress. However, prolactin levels, secreted by the anterior pituitary, increase during heat stress which may be involved in potassium and sodium turnover, and water metabolism. Heat stress impairs the reproductive performance of animals through the hypothalamic–pituitary–gonadal axis by inhibiting gonadotropin-releasing hormone in the hypothalamus. The secretion of gonadotropins, follicle stimulating hormone and luteinizing hormone (LH) are reduced in the anterior hypophysis which in turn affects the production of sex steroids. The decreased concentration of LH impairs the development of dominant follicle and results in decreased production of estradiol which leads to poor expression of estrus and low fertility in females.

28.19 Endocrine Response of Birds to Stress

The preoptic anterior hypothalamus induced by thermoreceptors stimulates the hypothalamic paraventricular nucleus which resulted in higher secretion of TRH. The TRH activates the thyrotrophs in the anterior pituitary to secrete thyroid stimulating hormone (TSH) which consequently activates adenylate cyclase and cAMP production enhanced thyroid hormones. The pituitary secretion of TSH is controlled through a negative feedback mechanism of T3. Further, transformation of T4 to T3 is increased during cold conditions by the enzyme deiodinase in tissues and liver and results in higher concentration of circulating T3. The binding ability of T3 with nuclear and mitochondrial receptors in tissues enhances the expression pattern of genes in relation to metabolic rate and respiration. Thyroid hormones are essential for the regulation of thermogenesis below TNZ. The potential thermoregulatory mechanisms to enhance the capacity for thermogenesis in birds include increasing shivering and non-shivering thermogenesis. The enhancement of mass-specific aerobic enzyme capacity of muscle tissue and increasing mitochondrial density helps in aerobic metabolism. Further, enhancing the capacity to uncouple mitochondria from energy production, causing energy that would have been used to phosphorylate ADP to ATP to be released as heat. Thyroid hormones involve in the regulation of non-shivering thermogenesis whereas shivering thermogenesis is principally controlled directly by neuronal mechanisms in birds.

Glucocorticoids are essential in the maintenance of physiological and energy homeostasis in the birds. The primary stressor hormone of bird is corticosterone which is secreted from the adrenal glands in response to stress on activation by ACTH from the hypothalamus. The corticosterone concentration increases across captive and wild species of birds in response to quick changes in T_a due to cold and heat. Thus, corticosterone is prerequisite for the production of glucose from liver glycogen or fat reserves which catalyse internal energy reserves that may stimulate food seeking behaviour to provide the higher metabolic rate in cold conditions. The mitochondrial glucocorticoid receptors are present in avian muscle cells that enhance the regulation of mitochondrial function for the energy production. During heat stress, both glycogenolysis and gluconeogenesis pathways are activated in the liver thereby providing endogenous energy sources under stress. Further, the HPA axis augments the avian thermoregulation in correlation with the HPT axis especially secreting CRH. The increased level of CRH induces thermogenesis and increases the T_c in chicks along with higher concentration of TSH and circulating T4 and T3 which indicates an interaction between CRH and thyroid hormones on thermoregulation. The hypothalamic orexigenic neuropeptide Y responds to cold exposure in poultry. Further, the higher level of melatonin secreted form pineal gland increases the T_c which induce enhanced cold resistance, thermal insulation and maximal heat production in chicken and quails.

28.20 Biological Rhythms

The body as a whole or its components is in a dynamic state where the physiological process repeats itself with more or less constant time intervals. The phenomenon is defined as a 'biological rhythm' which is influenced by a 'biological clock'. The variations in rhythms may be as a consequence of interactions with external physical cycles, such as daylength, temperature, humidity and atmospheric pressure. Hence, the biological rhythms may be internal (endogenous) which is regulated by the internal biological clock such as body temperature cycle and external (exogenous) factors which is governed by the coordination of internal cycles with external stimuli such as sleep/wakefulness and day/night. These stimuli are known as zeitgebers which include environmental time signals such as sunlight, food, noise and social interaction. The zeitgebers are supportive in resetting of the biological clock to a 24-h day.

The animals accomplish behavioural patterns in a day at a constant time interval which are triggered by habit and daily fluctuations in environmental conditions. However, some physiological events are having their own individual biological clocks such as heart rate, body temperature and metabolism. These clocks get out of phase with each other besides the time and adjusted to 24-h day and night cycles and are termed as 'circadian rhythms' (Latin words *circa* 'about' and *diem* 'day'). Circadian rhythms are elucidated as innate self-sustaining oscillations which are brought into a specific rhythm by the environmental factors especially light and temperature. A cycle or rhythm is established during an event occurs regularly through a peak and a trough over a particular period of time. The biological rhythms are classified into three types as per the variations in the duration of time period that is the time interval separating one peak or trough from the next in a repeating cycle. A circadian rhythm that is synchronized with the day or night cycle is called as 'diurnal rhythms'. Further, biological rhythms such as feeding cycles which have shorter rhythm (shorter than a circadian rhythm) are termed as 'ultradian rhythms'. The biological rhythms with a cycle of more than 24 h are referred as 'infradian rhythms' such as estrus cycle in animals.

28.21 Adaptation

Animal adaptation is the alterations in the genetic and physiology which develop in an animal in response to internal and external stimuli. Adaptation is the evolutionary process whereby an organism becomes better able to live in its habitat(s) when they are continuously exposed to drastic environmental changes. These animals establish functional and structural changes that enhance their potential to survive without stress in a unique environment. Therefore, animals with higher adaptive traits will be able to survive in harsh environmental conditions. Generally, adaptation is a long and slow process that takes generations to accomplish, which is rarely reversible.

28.21.1 Types of Adaptation

There are different types of adaptation as follows:

Genetic adaptation: It refers to the heritable animal characteristics that are transformed from one generation to the other, which favour survival of a population in a particular environment. This may involve evolutionary changes over many generations (selection by nature) or acquiring specific genetic properties (selection by man).

Physiological adaptation: It is the capacity and process of adjustment of the animal by itself, to other living things and to its external physical environment. Physiological adaptation signifies the changes that occur within an individual, over shorter or longer periods of time. This includes all the physiological changes that altered respiration rate, heartbeat, temperature, etc. exhibited by the animal which aids in maintaining homeostasis.

Biological adaptation: It refers to the changes with respect to morphological, anatomical, physiological, biochemical and behavioural characteristics of the animal which promote welfare and survival of the animal in a given environment.

Phenotypic adaptation: Phenotypic adaptations are the modifications that develop during the course of life of an individual due to changes in genetic or environment which are non-heritable. Phenotypic adaptation is the changes in physical (development of callous) or behaviour of animals (domestication of wild animal).

Nutritional adaptation: The availability nutrients for animals depend upon the climatic and ecological changes that have occurred in a particular place. The feed and fodder production are predominately reduced in climate change perspectives. The soil and water in the coastal areas of tropical regions contain low calcium due to leaching which is reflected in the pastures with low levels of this mineral. Therefore, it is very vital for the animals to adjust to the varying feed and fodder availability especially in the climate change scenario.

28.21.2 Adaptation Characteristics

Adaptation is an adjustment that decreases the physiological strain created by a constituent of stressful environmental factors. The degree of environmental stress may be quantified indirectly by the responses of the animal as strain or product of adaptation. Adaptation includes physiological alterations in the animal within its lifetime to the environment or genetic adaptation that involves forces of selection across generations. These adaptations may be restricted to a particular area of tissues or impacts the whole animal. Adaptation to hot environments can involve physical, behavioural, physiological and morphological changes. The physical adjustments are carried out by genetic selection or heritable characteristics which could be transferred from generation to generation. The subtropical cattle have dewlaps and longer limbs than temperate cattle to facilitate the heat dissipation. Variation in skin colour is also an adaptation feature to hot environments to modify the absorption of solar radiation. The light-coloured coat of animal enhances the reflection of solar radiation away from the body during heat stress, and dark coat colour facilitates heat absorption in winter. The development of larger layers of subcutaneous fat and thermal insulation may result in a poorer heat tolerance but the same may be better adapted to cold. In addition, the hair cover and shedding are also influenced by seasonal changes; most animals shed their winter hair as the summer approaches. The body size has an impact on adaptation to hot or cold environments. The superfluous large skin folds form larger surface area to favour heat loss per unit of weight in tropical animals and compact body conserves heat loss. The ears, dewlap, navel flap and vulva are larger, grooved and loosely attached to the body in heat-tolerant animals whereas temperate or cold-tolerant animals are more compact with no dewlaps and small hairy ears. The metabolic rate of heat adapted animals is lower than temperate species along with enhanced blood flow to the peripheries to enable heat loss via conduction and convection.

Acclimatization: Acclimatization is a long-term adaptive physiological adjustment which results in an increased tolerance to continuous or repeated exposure to complex climatic stressors (normally produced under field conditions). When an animal voluntarily migrates from a mountain valley to a high altitude, its lung ventilation rate typically will increase initially to acquire adequate oxygen. After few days or weeks, lung ventilation begins to drop back towards the sea level rates as other physiological mechanism that facilitates gas exchange at high altitude. After several days, the individual is said to be acclimatized to a new high-altitude condition. The acclimatization is a rapid phenomenon wherein a series of physiological and/or biochemical adjustment takes place within the animal as a result of exposure to new environmental conditions. The typical characteristics of well-adapted animals are minimum weight loss when exposed to stressors, normally maintained reproductive rate, high resistance to diseases and high longevity with low mortality rate.

Acclimation: It refers to the adaptive changes that take place in response to a single climatic variable (normally produced in a laboratory or climatic chamber). For example, if an animal is placed in a hypobaric chamber by simulating high-altitude conditions, the animal becomes acclimated to the experimental conditions within a few days. The acclimatization and acclimation may be reversible.

Habituation (general): It is a gradual quantitative change of response which may lead to a loss of response, as a result of repeated stimulation.

Habituation (specific): It is the gradual reduction in sensation associated with a given repeated stimulus specific to the part of the body which has been repeatedly stimulated.

Learning: It is the acquisition of a new response, or a qualitative change of an existing response, or an inhibition or facilitation of an existing response by a new stimulus.

Conditioning: It is the transfer of an existing response to a new stimulus.

28.22 Heat Stress Amelioration Strategies

The combination of high environmental temperature, relative humidity, solar radiation and air movement exceeds thermoneutral zone which causes heat stress. The high-producing dairy animals are more susceptible to heat stress due to their high metabolic heat production and more feed consumption. It is highly essential to maintain the optimum temperature to maximize productive performance. Hence, the productivity of animals during heat stress may sustained through the physical modifications of environment, nutritional management and genetic development of heat-tolerant breeds.

28.22.1 Physical Modification of Environment

The optimum maintenance of livestock environment is highly essential due to the changing climatic point of view. This management aspect is a challenge to reduce the detrimental effects of environmental conditions on animal production. Further, it is also important to reduce the cost of environmental protection methods for domestic animals. The physical modification of environment including provision of shade, shelter with cooling mechanisms is important in tropical and subtropical areas to maintain productivity and reproduction during heat stress. The simple and basic attempt to reduce heat load from direct solar radiation in cattle is the use of shades which can be natural or artificial. Trees are most effective in providing shade since they protect from the sun and capture radiation by evaporative humidity in the leaves. The painting of upper part of the shade unit or roof with white colour and installation of 2.5 cm thick isolating material reduce solar radiation. Further, the height of shades at the corral must be 3.6–4.2 m^2 in order to ensure reduction in solar radiation. The shading units must be high enough from the ground to facilitate air circulation and tractor access for corral cleaning. The shades have to be placed in the centre of the corral that could prevent the accumulation of humid beneath the structure. The cooling systems are more effective in reducing heat load from dairy cows through evaporation. However, the ameliorative responses vary with techniques where cooling has consistently improved feed intake and milk production during heat stress in dairy cows. The sprinkling and ventilation of dairy cows also enhanced the feed intake with less quantity of water and sustained the milk, fat and protein production during heat stress.

28.22.2 Nutritional and Feeding Managements

Nutritional intervention is one of the important heat stress ameliorative procedures to combat heat stress and to maintain the production performance during hot environmental conditions. The modification of ration balancing is essential in reducing the adverse effects of heat stress in dairy cows. Animals should be fed at cool hours of the day with proper time intervals, and it is important to modify the composition of macro and micro-nutrients of feed in addition to supplementation of vitamins, minerals and feed additives. The feeding of animals during the cooler periods of the day improves the feed intake that helps in minimizing of metabolic and climatic heat load. The frequent feeding evades the diurnal variations in ruminal metabolites and improves the effective feed utilization in the rumen. The encouragement of grazing at cooler periods of the day, like early mornings and late evening, would reduce negative effects of heat stress. Further, nutrient requirements change during heat stress which needs to be modified to ensure normal feed intake and performance.

The digestive and metabolic processes are the additional sources of heat production from the animals during heat stress where high fibre content of feed increases the heat production. The rectification in the dietary fibre content is important to cut down heat augmentation in rumen fermentation. Therefore, low fibre content of feed is advocated to prevent high heat generation in rumen during heat stress. The feed intake is improved when the neutral detergent fibre (NDF) level of roughage ranging between 27% and 35% in the ratio which was also reflected as reduced the respiration rate and rectal temperature in animals during heat stress. Further, reduction of dietary roughage NDF from 18% to 12% on dry matter basis significantly reduced the rectal temperature. However, supply of sufficient dietary high-quality fibre forage is essential to establish normal rumen activity. Further, highly fermentable carbohydrates may help in sustaining the feed intake in heat stress, and care must be taken to counterbalance the high-grain diets which results in rumen acidosis. However, it is necessary to maintain the ideal rumen function with an adequate level of 18% acid detergent fibre and 28% NDF on dry matter basis of the feed.

Energy is the foremost important nutrient, and it has to be enhanced in the diet in addition to concentrate to decrease forage level. The energy level of the feed should be higher to meet the maintenance requirement as well as to support the extra demand for thermoregulation during heat stress. The reduction of forage to concentrate ratio enhances the efficiency of nutrients utilization in heat stressed animals. The addition of extra fat in the feed enhances the net energy intake during heat stress due to higher energy density with lower metabolic heat generation than fibre and starch. The inclusion of fat in the diet must be limited up to the level of 5% without any adverse effects on ruminal microbes. Heat stressed animals are in negative nitrogen balance due to decreased feed intake. Hence, good quality of protein source with crude protein (CP) content of 16% with low degradability is ideal because highly degradable CP increases the endogenous heat generation in ruminants. The high content of undegradable protein, calcium soaps of fatty acids and monopropylene glycol in feed enhance the performance of animals with reduced plasma urea. The supplementation of dietary essential amino acids is also necessary for the restoration of protein producing machineries or processes such as transcription and translation to maintain the production performance in heat stressed animals. The inclusion of lipoic acid enhances the thermotolerance by promoting insulin and antioxidant status during heat stress. The feeding of rumen protected glutamine, arginine, tryptophan and citrulline intensifies the immune status particularly cell-mediated immune response in heat stressed ruminants.

28.22.2.1 Water

Water is an essential element in the body of animals and is essential for the maintenance of physiological functions including tonicity of tissue, lubrication, thermoregulation, nutrient transport and excretion. Water metabolism is highly associated with the thermoregulatory mechanisms of the ruminants in the regulation of homoeothermic status. Water is the primary heat carrier medium in the removal of excessive heat load from the core body through evaporative heat loss. Water requirements of animals are regulated by dry matter intake, environmental temperature and loss of water through evaporation, urine, faces and milk. Heat stress enhances the water requirement to facilitate higher heat dissipation. Ruminants are experiencing moderate to severe water restriction during different environmental conditions and the demand for water increase due to high ambient temperature and solar radiation. Sheep and goats are better adapted for the drought conditions particularly goats are having high potential to conserve water. However, goats drink doubled their water requirement to enhance heat loss by sweating and panting during heat stress conditions. Therefore, the best way to reduce heat stress is to provide clean fresh cool drinking water ad libitum to ensure optimum performance.

28.22.2.2 Vitamins and Minerals

The decrease in feed intake during hot environmental conditions influences the requirement of vitamins and minerals which are associated with health status and immunity. Hence, it is advocated to include vitamins and minerals in the diet to minimize the effects heat stress in animals. The incorporation of selenium, copper and zinc, in addition to Vitamin A and E more than NRC recommendation, may improve the immunity and health status of animals during heat stress. The cations requirements are increased by the kidney especially Na^+ and K^+ during heat stress in animals due to their higher rate of excretion up to 80% and 18%, respectively. $NaHCO_3$, K_2CO_3 and $KHCO_3$ are the source of Na^+ and K^+ and its inclusion in the diet improved the feed intake in heat stressed animals. The feed additives, fungal cultures and plant extracts enhance the feed intake and favour the rumen metabolism and thermoregulation during heat stress. The inclusion of yeast increases the nutrient digestibility and feed efficiency by maintaining rumen pH. The administration of plant extract daidzein alleviates the heat stress in ruminants and enhances the antioxidant potential with higher level of glutathione peroxidase.

28.22.3 Genetic Selection of Heat-Tolerant Breeds

The current research findings and approaches in together positively improved the physical modification of environment and nutritional management procedure in ameliorating the impacts of heat stress on animal performance. However, long-term approaches are essential to establish heat resilient breeds or animals in the view of variances in thermal tolerance among livestock species. The selective breeding of dairy animals for high milk production has increased the susceptibility to heat stress with compromised summer production and reproduction. The selection for high milk production resulted in decreased ability of thermotolerance and depression in fertility. Therefore, the recognition of heat-tolerant animals among high-producing breeds will be more effective to sustain the productivity and survivability during heat stress. The cattle having shorter hair with higher diameter and lighter coat colour are more adapted to hot environmental conditions in comparison to those with longer hair coats and darker colours. This phenotypic character has been identified in tropical *B. taurus,* and this dominant gene facilitates higher sweating rate, lower rectal temperature and respiratory rate in homozygous cattle in the tropical regions. The genes of heat shock proteins which are highly in association with thermotolerance could be distinguished biomarkers in the marker-assisted selection programmes. The incidence of polymorphisms in relationship with heat-tolerant genes is expressed in different breeds such as HSP90AB1 in Thai native cattle, Sahiwal and Frieswal; HSF1 gene, HSP70A1A gene, HSP70A1A gene and HSBP1 in Chinese Holstein cattle. Apart from HSPs, few more thermo-tolerant genes have been identified in livestock species that are changing in their expression pattern during heat stress like insulin-like growth factor-1 (IGF), toll-like receptors (TLR), etc. In addition, the genes that are economically important were perceived in heat stressed animals such as ATP1B2, thyroid hormone receptor, interleukins, fibroblast growth factor, protein kinase C, NADH dehydrogenase, phosphofructokinase and glycosyl transferase. Further comprehensive research findings are essential to elucidate the expression pattern of these genes in diversified animal species in prior to be designated as biological markers that may be used in marker assisted selection programmes to develop thermo-tolerant breeds.

Learning Outcomes

- The chapter highlighted the various adaptive responses of farm animals and birds to cope with environmental challenges.
- Special emphasis was given to understand all important definitions and terms pertaining to environmental physiology of farm animals which could be very useful for the readers to prepare themselves for any competitive exams.
- Further the chapter also highlights the various indices to quantify heat stress response and identified different associated biomarkers.
- The chapter also signified the importance of channelizing the energy resources towards adaptation pathway to activate life sustaining process in heat stressed animals and thereby leading to compromised production performance.
- Lastly the chapter describes in brief the various strategies which could be implemented to ensure animal welfare during heat stress exposure.

Exercises

Objective Questions

Q1. What is the study of animals in their natural habitat?
Q2. Climate comprised of the atmospheric variables over a period of how many years?
Q3. What are the components of environment?
Q4. What is higher threshold temperature for animal health and welfare?
Q5. What are cardinal signs of heat stress?
Q6. What are the methods of sensible heat loss?
Q7. How does the animal dissipate heat as sensible heat loss?
Q8. At what temperature an animal can perform its maximum?
Q9. In which type of animals do you find brown fat?
Q10. What are homeotherms?
Q11. Define Poikilotherms?
Q12. How animals are classified based on the source of body heat production?
Q13. What is the method to estimate the level of heat stress in animals?
Q14. Enlist the types of sweat glands?
Q15. What is the evaporate rate in the skin of heat stressed cattle at 40 °C?
Q16. Enlist the major ways of heat production during cold stress?
Q17. What is the primary endocrine indicator of heat stress?

Subjective Questions

Q1. Impact of heat stress on animal production performance?
Q2. What are the environmental factors which affects animal performance?
Q3. Impact of high altitude on animal performance?
Q4. What are the thermal exchange methods?
Q5. Define thermoneutral zone?
Q6. What is the role of skin in animal thermoregulation?
Q7. How is heat stress measured?
Q8. Describe in brief the response of animals to stress?
Q9. What are the neuron-endocrine response to stress in animals?
Q10. What is the impact of heat stress in female animals?
Q11. Explain the cellular response to heat stress?
Q12. What are the heat stress ameliorative measures?

Answer to Objective Questions

A1. Whole-animal physiology/physiological ecology/environmental physiology.
A2. Over a period of 30 years.
A3. Ambient temperature, relative humidity, radiant heat, precipitation, atmospheric pressure and wind velocity.

A4. 30 °C with relative humidity below 80% and 27 °C with RH above 80%.
A5. Increased rectal temperature, respiration and heart rate.
A6. Convection, conduction and radiation.
A7. Respiratory tract and skin.
A8. Thermoneutral zone.
A9. Hibernating animals.
A10. Animals having the ability to control their body temperature.
A11. Animals do not have a control over their body temperature and also called as temperature conformers.
A12. Endotherms, ectotherms and heterotherms.
A13. The temperature humidity index (THI) is a simple, reliable and easy method to assess the potential level of heat stress in animals.
A14. There are two types of sweat glands, eccrine and apocrine glands.
A15. The evaporative heat loss from the skin surface of cows is around 150 $g/m^2/h$.
A16. Shivering and non-shivering thermogenesis.
A17. Cortisol.

Answer to Subjective Questions

A1. Heat stress affects feed intake, diverts more energy towards maintenance of homeostasis, decrease in body weight, decrease in milk production.
A2. Environmental temperature, relative humidity, solar radiation, wind velocity, precipitation.
A3. High-altitude environment, extreme cold, mountainous terrain, reduced oxygen, high solar radiation, short vegetations.
A4. Radiation, conduction, convection, evaporation.
A5. Comfort zone, upper critical temperature, lower critical temperature, lethal body temperature.
A6. Integumentary system, hair coat types, blood vessels, sweat glands.
A7. THI, BGHI, TCI, GCI.
A8. Respiration, sweating, cardiovascular system.
A9. HPA axis, HPT axis, autonomic and CNS.
A10. Heat stress, GnRH-LH and FSH-estradiol, progesterone, low fertility.
A11. Heat stress, activation of HSFs, synthesis of HSPs.
A12. Physical modification of environment, nutritional management, genetic selection.

Further Reading

Afsal A, Sejian V, Bagath M, Krishnan G, Devaraj C, Bhatta R (2018) Heat stress and livestock adaptation: neuro-endocrine regulation. Int J Vet Anim Med 1(2):108

Barbosa OR, Silva RG (1995) Thermal comfort indices for sheep. Boletim de Indústria Animal 52:29–35

Collier RJ, Collier JL (2012) Environmental physiology of livestock. Wiley-Blackwell, Ames, IA

Collier RJ, Gebremedhin KG (2015) Thermal biology of domestic animals. Annu Rev Anim Biosci 3:10.1–10.20

Conte G, Ciampolini R, Cassandro M, Lasagna E, Calamari L, Bernabucci U, Abeni F (2018) Feeding and nutrition management of heat-stressed dairy ruminants. Ital J Anim Sci 17(3):604–620

Daghir NJ (2008) Poultry production in hot climates. Wallingford, Cambridge

Daghir NJ (2009) Poultry production in hot climates. Cromwell Press, Trowbridge

Ewing S, Lay D, Von Borell E (1999) Farm animal well-being: stress physiology. Prentice Hall, Upper Saddle River, NJ

Gaughan JB, Mader TL, Holt SM, Lisle A (2008) A new heat load index for feedlot cattle. J Anim Sci 86(1):226–234

Habeeb AA, Gad AE, Atta MA (2018) Temperature-humidity indices as indicators to heat stress of climatic conditions with relation to production and reproduction of farm animals. Int J Biotechnol Recent Adv 1(1):35–50

Hafez ESE (1968) Adaptation of domestic animals. Lea & Febiger, Philadelphia

Krishnan G, Bagath M, Pragna P, Vidya MK, Aleena J, Archana PR, Sejian V, Bhatta R (2017) Mitigation of heat stress impact in livestock reproduction. In: Theriogenology. Intech Open Science, London, pp 63–86

Livestock and Poultry Heat Stress Indices (LPHSI) (1990) Agriculture engineering guide. Clemson University, Clemson, SC

Mader TL, Johnson LJ, Gaughan JB (2010) A comprehensive index for assessing environmental stress in animals. J Anim Sci 88(6): 2153–2165

McDowell RE, Hooven NW, Camoens JK (1976) Effect of climate on performance of Holsteins in first lactation. J Dairy Sci 59:965–971

Moberg GP, Mench JA (2000) The biology of farm animal stress: basic principles and implications for animal welfare. CABI Publishing, Wallingford

Nilsson JA, Molokwu MN, Olsson O (2016) Body temperature regulation in hot environments. PLoS One 11:20160822

Philip C, Piggins D (1992) Farm animals and the environment. CAB International, Wallingford

Romanovsky AA (2014) Skin temperature: its role in thermoregulation. Acta Physiol 210:498–507

Schmidt-Nielsen K (1997) Animal physiology: adaptation and environment, 5th edn. Cambridge University Press, Cambridge

Seixas L, de Melo CB, Tanure CB, Peripolli V, McManus C (2017) Heat tolerance in Brazilian hair sheep. Asian Australas J Anim Sci 30(4):593–601

Sejian V, Bhatta R, Gaughan J, Malik PK, Naqvi SMK, Lal R (2017) Climate change impact on livestock: adaptation and mitigation. Springer, New Delhi

Sejian V, Bhatta R, Gaughan JB, Dunshea FR, Lacetera N (2018) Review: Adaptation of animals to heat stress. Animal 12(S2):s431–s444

Sejian V, Gaughan JB, Baumgard L, Prasad CS (2015) Climate change impact on livestock: adaptation and mitigation. Springer, New Delhi

Sejian V, Naqvi SMK, Ezeji T, Lakritz J, Lal R (2012) Environmental stress and amelioration in livestock production. Springer-Verlag, Berlin

Siple PA, Passel CF (1945) Measurements of dry atmospheric cooling in subfreezing temperatures. Proc Am Philos Soc 89(1):177–199

Thom EC (1959) The discomfort index. Weatherwise 12:57–61

Willmer P, Stone G, Johnston I (2005) Environmental physiology of animals. Blackwell Science, Oxford

Thermoregulation in Birds

29

G. Krishnan, C. Devaraj, M. V. Silpa, and V. Sejian

Abstract

The elevated or low temperature reduces growth, egg and meat production in birds as a result of reduced feed intake and impaired intestinal development. Further, heat/cold stress affects the digestion, absorption of nutrients and impaired immune functions. The increased respiration eliminates more CO_2 that lowers the formation of HCO_3 and results in thin eggshell production and higher incidence of egg breakage during heat stress. Heat stress modifies the protein/fat ratio, increases lipid peroxidation, mobilization of minerals and vitamins from tissues, results in decreased meat quality of broilers and reduced nutritive value of eggs. Heat/cold stress activates both the sympathetic nervous system to secrete catecholamines, which increase respiration and peripheral vasodilation and the hypothalamic–pituitary–adrenal (HPA) axis which releases corticotrophin-releasing hormone (CRH) from hypothalamus. The higher level of CRH stimulates the secretion of adrenocorticotrophic hormone (ACTH) from the pituitary which activates the adrenal glands to enhance release of corticosteroid which stimulates gluconeogenesis to generate energy required during heat stress. Heat stress stimulates the production of highly conserved heat shock proteins that help in protein folding and unfolding, assembly of multi protein complexes, transport and sorting of proteins into subcellular compartments, minimization of protein aggregation and protecting the cells from apoptosis.

G. Krishnan · C. Devaraj · V. Sejian (✉)
Centre for Climate Resilient Animal Adaptation Studies, ICAR-National Institute of Animal Nutrition and Physiology, Bangalore, Karnataka, India

M. V. Silpa
Institute of Animal Breeding and Genetics, Justus-Liebig-Universität Gießen, Gießen, Germany

© The Author(s), under exclusive license to Springer Nature Singapore Pte Ltd. 2023

P. K. Das et al. (eds.), *Textbook of Veterinary Physiology*, https://doi.org/10.1007/978-981-19-9410-4_29

Graphical Abstract

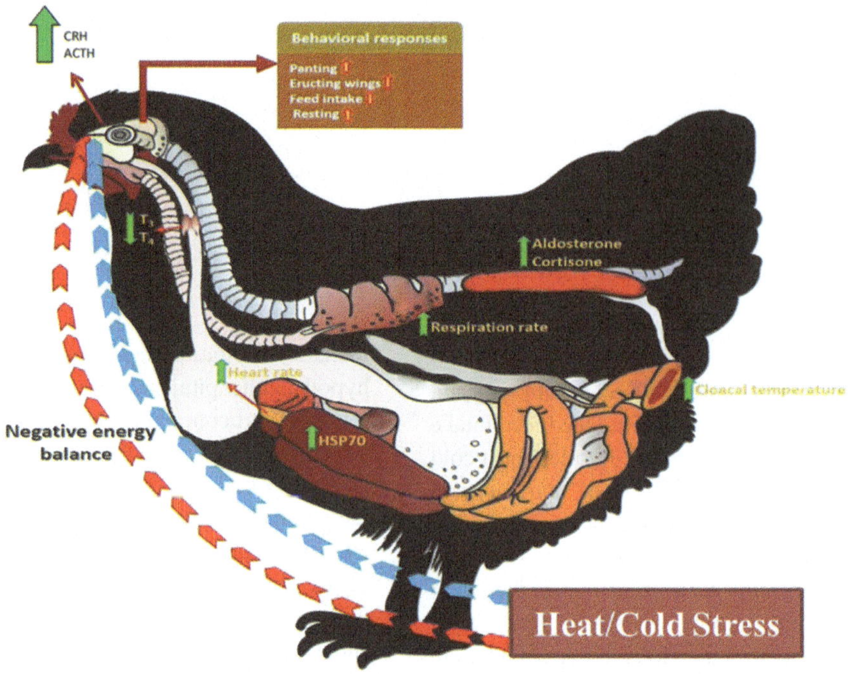

Description of the graphic: High or low environmental temperature initiates various physiological, behavioural and endocrine responses to maintain the thermal homeostasis in birds. The activated sympathetic nervous system enhances the secretion of catecholamines which increases respiratory frequency and vasodilation of peripheral blood vessels. Heat/cold stress activates the HPA axis which releases CRH from the hypothalamus, and higher level of CRH stimulates the secretion of ACTH from the pituitary that enhances release of corticosteroid from adrenal glands. The increased corticosteroid favours in the gluconeogenesis and production of glucose to meet the high energy demand during heat/cold stress. Heat/cold stress increases the generation of superoxide and other ROS. The increased lipid peroxidation compromises biological membrane integrity and functions

Keywords

Birds · Egg · Heat stress · Stress biomarker · Thermoregulation

Learning Objectives

- Learning the heat/cold stress responses of birds.
- Understanding the thermoregulatory mechanism of birds.
- Studying the impact of heat/cold stress on growth, production and reproduction in birds.

29.1 Introduction

Climate change is one of the biggest daunting challenges faced globally, with severe and widespread deleterious effects, warming the planet, threatening ecological systems and disrupting the life cycle of several species. Global climate change is mostly caused by greenhouse gas (GHG) emissions that result in warming of the atmosphere. The changing environment jeopardizes the achievement of human development and poses as a major threat to the poor who rely heavily on the natural resource base for their livelihoods. The world average temperature was projected to rise in the range of 0.3–4.8 °C by 2100. Even 1 °C raise in temperature can hamper the delicate balance of ecosystems and affect plants and animals that inhabit them. Therefore, this urges the need to warrant implementation of immediate actions to build the resilience of agricultural production systems.

Poultry industry is in leading position among the agricultural and allied sectors. Poultry meat and eggs are the good source protein which is beneficial for millions of people in below poverty. The consumption of poultry meat and eggs has advantageous over other livestock products since there is no religious taboos and affordable to everyone including economically poor peoples. Poultry meat contains less saturated fats and more micronutrients. Heat or cold stress has arisen as one of the major constraints for poultry production system particularly in the hot and humid tropical and temperate regions. Absence of sweat glands makes thermoregulation becomes challenging for the birds during hot weather, thus compromises their production potential. Moreover, chronic heat stress causes bird mortality. Therefore, this chapter elaborate in detail about the negative impacts of heat or cold stress on poultry production and thermoregulatory responses in birds.

29.2 Thermoregulatory Mechanism in Birds

In the current trend of changing climate, birds are posed with challenges to maintain thermoregulation such as increased occurrence of extreme weather events, lower ability to predict climate and much increased mean temperature. Across wide ranges of changes in environment, birds usually maintain a stable body temperature of about 41.02 ± 1.29 °C during active phase at rest, through various physiological, behavioural and morphological adaptations. Thermoregulatory system of birds is comprised of three parts, namely a sensory part comprising of thermoreceptors, osmoreceptors and baroreceptors, an integrating part consisting of thermoregulatory centre occupying preoptic anterior hypothalamus and the command part involving temperature control mechanisms such as non-shivering thermogenesis, shivering thermogenesis, evaporative heat loss mechanism (cutaneous and respiratory heat loss), peripheral vasodilatation or constriction and behavioural mechanism of adaptation. By evolution, birds are benefitted with its morphological features such as feathers and plumage, changes in peripheral blood circulation and bodily fat distribution for thermoregulation.

Majorly, behavioural method of thermoregulation is expressed by the unfeathered portions of foot, leg, head comb, wattle and distal segments of wings. For heat loss, the behavioural modifications exhibited by the birds include separating and holding the wings from the body for enhanced air circulation, lying down with forward stretched head and neck, increased preference for sitting on the soil and increased frequency of squashing of breast feathers. Thermal panting behaviour is commonly observed in the birds experiencing heat stress. Ruffling of head and back feathers, avoiding sun's direction and facing away from the sun and shade-seeking are the kind of behaviours observed in many birds. Excessive drinking of water, tucking of head under the wings and splashing excess quantities of water over and above their feathers are also expressed by birds as a method of behavioural adaptation to heat stress. In contrary, there are also means of heat conservation behaviour elicited by birds such as hunching up with feathers stretched out, completely fluffing out and sleeping most of the time with tucked head under the wing and sunbathing. A unique phenomenon of torpidity or hibernation is shown by few bird species. Huddling and seeking warm environment nature of behaviour are observed in case of small birds. At extreme conditions, birds also utilize migration as a method of adaptation to tackle changing weather.

Basically, below the low critical temperature, heat generation is attained by non-shivering and shivering thermogenesis and in conditions above critical temperature, heat loss happens by evaporative cooling mechanism majorly through panting, cutaneous evaporation, gular flutter and lingual flutter. A novel thermoregulatory mechanism that enhances thermal tolerance was identified in desert inhabited passerines, called as vocal panting. It is a very high pitched, quick rhythm vocals generated during panting. These heat-calls are observed to be produced as a side effect of vibration of a segment of respiratory tract. This behaviour increases the area available for evaporative water loss and believed to be associated with thermoregulatory mechanism for improving heat tolerance.

Thermogenesis is governed by thyroid hormones. Particularly, thyroid hormones regulate thermoregulation below thermoneutral zone (facultative thermogenesis) and are involved mainly in non-shivering thermogenesis. Association of corticosterone with body temperature and environmental temperature extended over summer and winter. Other hormones such as neuropeptide Y, melatonin, leptin, ghrelin and arginine vasotocin are also reported to have a role in avian thermoregulation.

29.3 Impact of Heat Stress and Cold Stress on Production Performance in Birds

The exposure of birds to hot or cold environmental conditions in tropical and subtropical regions or temperate regions adversely affects their productive performance and results huge economic losses in the poultry industries. The thermoneutral zone of birds varies between 18 and 22 °C for their optimum growth performance and egg production (Fig. 29.1). The optimum temperature of 30 °C is the critical point for birds wherein they maintain their production by compensating reduction in dry matter intake through their efficient feed conversion ability. However, beyond 30 °C dry matter intake decreases where birds are incompetent in maintaining body temperature and energy homeostasis, and thus leads in production loss with enhanced mortality. The heat or cold stress activates the hypothalamo–hypophyseal–adrenocortical (HPA) axis in birds and results in enhanced release of corticosteroids. The higher concentration of corticosteroids in birds enhances catabolic effects with increased oxidative stress and results in muscle wasting and reduced growth. Further, heat stress with high relative humidity disrupts the homeostasis of birds. Birds try to decrease its

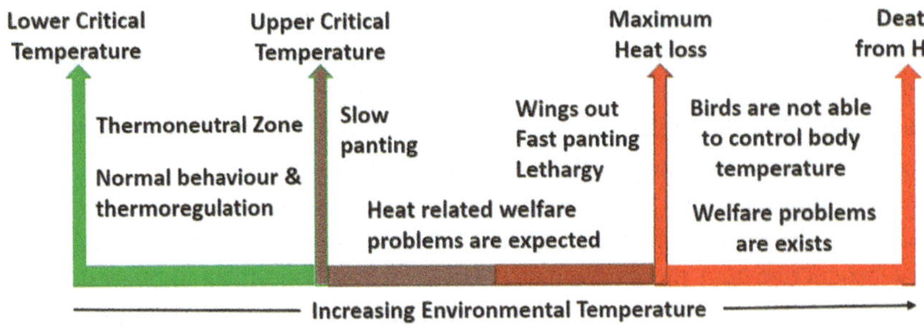

Fig. 29.1 Thermoregulation of in birds during heat stress. When the environmental temperature exceeds lower critical temperature, the thermogenic mechanisms are initiated to increase heat production. Similarly, ambient temperature beyond upper critical temperature initiates thermolytic mechanisms to decrease heat production and enhances the heat dissipation away from the body. Inadequate heat stress amelioration measures adopted at this juncture would lead to drastic reduction in production which may also culminate with mortality in birds

metabolic heat generation by decreasing feed intake with consequences of poor performance during hot condition. Birds dry matter intake increases during cold stress condition in order to maintain the body temperature. The impact of heat or cold stress is more severe in birds with larger body size and growth rate. The elevated or low environmental temperature affects the growth rate and meat production in broiler birds as an outcome of impaired dry matter intake and intestinal development. Heat or cold stress affects the functions of digestive enzymes and absorption of nutrients with impaired immune functions in the intestine. The stocking density is major factor to be considered in view productivity and welfare in broiler production during high ambient temperature. Further, heat or cold stress modifies the nutritional composition of broiler meat by lowering breast muscle yield with higher abdominal and intermuscular fat deposition.

The decrease in the dry matter intake during hot condition depends on the severity of heat stress. The reduction in dry matter intake increases with age in meat type birds and reaches 50%, whereas reduction is around 30–50% in layers. Birds reared in cage systems are more prone to heat stress compared to deep litter systems. Further, the elevated environmental temperatures reduce the secretory pattern and function of digestive enzymes trypsin, chymotrypsin and amylase which results in impairment in the digestive processes and absorption of nutrients. The insufficient nutrients and energy demand directly affect egg and meat production. The decrease in quality and quantity of egg production in layers is an effect of decreased dry matter intake and feed conversion ratio. It is also well established that heat stress enhances yolk and serum cholesterol levels in layers. The majority of energy is diverted towards restoration of thermoregulatory mechanisms in layers exposed to heat stress which depresses body weight, number of eggs, egg weight and shell quality (Fig. 29.2). Further, panting in birds during heat stress causes increased production loss of carbon dioxide which is necessary for the development of calcium carbonate that is responsible in formation of eggshell. The low levels of carbon dioxide causes reduction in eggshell thickness with higher incidents of broken eggs. The hen-day egg production also decreases significantly in layers when they are experiencing a continuous heat stress during their peak production.

The reduced dry matter intake and increased mortality in layers during hot conditions may also result in decreased egg production. Further, increased mortality in layers and broilers during hot environment particularly in summer have also been reported. The high environmental temperature reduces protein synthesis with enhanced protein breakdown in broilers. The alterations in protein/fat ratio, increased lipid peroxidation, decreased levels of minerals and vitamins in tissues result in compromised meat quality of broilers and reduced nutritive value of eggs in layers. The meat quality is affected by high drip loss and high occurrence of pale, soft and exudative meat which does not appeal the consumer preference and causes huge economic loss. Therefore, poor production performances of birds during heat stress are related with reduced appetite and lower dry matter intake in birds so as to prevent further heat increment. Further, heat stress damages the intestinal membrane, thereby hampering the absorption of vital nutrients leading to impaired digestibility and metabolism. Additionally, heat stress depresses the reproductive performance male birds by reducing seminiferous epithelial cell differentiation and intracellular iron exchange that result in decreased semen quality and quantity. In addition, semen volume, spermatozoa concentration and consistency are depressed during hot environmental conditions. The occurrence of infertility and low hatchability are more common in heat stressed female birds. Further, heating of fertile eggs beyond the optimum temperature during incubation resulted in impaired tissue growth at various phases of incubation with asymmetrical skeletal development. The heat stressed embryos were developed with shorter face length, reduced lung weight and unsteady gait which resulted in weaker chicks with higher mortality.

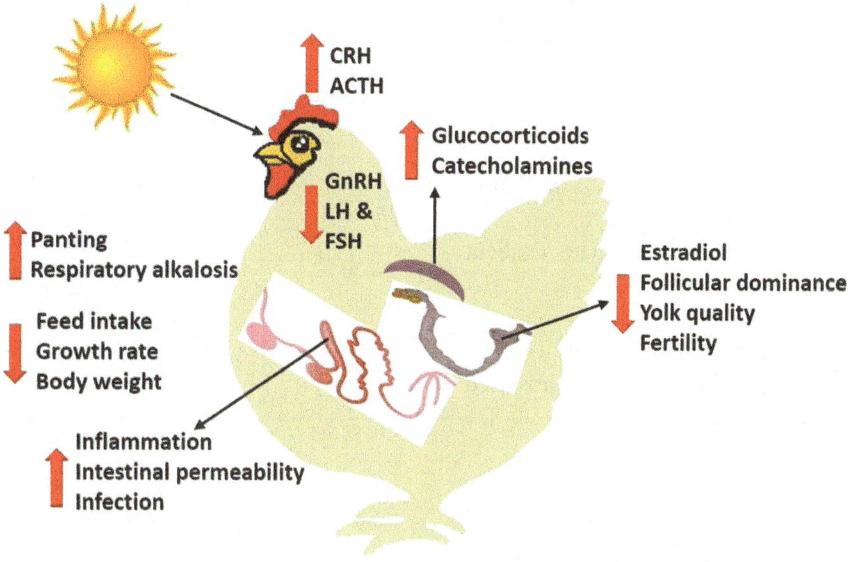

Fig. 29.2 Impact heat stress on productive and reproductive performance of birds during heat stress. Heat stress decreases the feed intake, modifies secretory pattern and activity of digestive enzymes resulting in impaired digestive processes and absorption of nutrients. The insufficient nutrients and energy demand directly affect egg production in layers and the growth rate in broilers. Further, heat stress activates the HPA axis that leads to increase in CRH and ACTH release, consequently increasing glucocorticoids particularly corticosterone in birds. In addition, the decrease in GnRH and simultaneously reduction in LH and FSH impairs the reproductive performance in birds. The majority of energy is diverted towards restoration of thermoregulatory mechanisms in layers during heat stress which depresses body weight, number of eggs, egg weight and shell quality

29.4 Energy Expenditure During Thermoregulation in Birds

The thermoregulatory mechanism of birds varies from mammals with regard to their higher metabolic rate and reduced ability for heat dissipation due to plumage and absence of sweat glands. Metabolic heat production occurs especially during digestive processes starting form feed intake, digestive break down of feed resources, its absorption and utilization. During pleasant and comfortable environmental conditions, the metabolic heat generated ensures normal productivity in birds, indicating their comfort zone. However, several factors influence the impact of metabolic heat produced. For example, the maintenance metabolism is higher in growing birds which therefore contributes severely to the heat increment. The composition and texture of feed also influence the energy expenses in association with feeding activities. In addition, the inability to convert feed into protein and lipid above maintenance requirement contributes 20–25% of heat production. Heat production in broilers mainly dependent on genetics that the breeds preferred for fast growth with a low feed conversion ratio (FCR) produces lower heat than either of slow growth with a low FCR or slow growth with a high FCR. Further, heat production is high in broilers due to their higher growth rate that is associated with more feed intake.

Heat stress manifests a wide range of physiological, behavioural, neuroendocrine and molecular responses in birds to restore their core body temperature within the normal limits. The metabolic activities regulate the physiological functions of maintenance, growth and egg production which increases the heat production and it depends upon the species, breed, body weight, production, feed intake and feed quality in birds. The regulation of body temperature in birds is established by radiation, conduction, convection, evaporation and excretion. The heat loss through radiation is an electromagnetic process where body temperature is transmitted to cooler surroundings without a medium by air. The evaporative heat loss mechanism is a significant way of heat dissipation in birds through respiratory tract which majorly rely on panting and nasal cavity acts as heat exchanger. In addition, a small quantity of heat also lost through skin surface and comb of birds. The radiative heat loss occurs from featherless regions of birds by raising their wings during hot environmental conditions. The high wind velocity favours the removal of heat load from bird's body by convection where there is optimum ventilation. Further, birds drink double the quantity of water during heat stress to eliminate the heat load through urine and wet faeces. Further, few responses are also evolved to establish for a long period of time to acclimatize the bird to that particular environmental condition which falls within the upper limit of the thermoregulatory zone. However, climatic extremes beyond the threshold limit in birds may lead to deleterious situation in birds wherein none of the adaptive mechanisms can sustain the bird's survivability, finally culminating to death.

Know More..............
- Birds' feathers provide remarkable insulation against the cold.
- Emperor penguins are the most climate resilient birds.
- Birds lack sweat glands.
- Average body temperature of birds: 40 °C.
- Gular fluttering is common mechanism in birds used to facilitate heat loss.

29.5 Thermoregulatory Mechanism in Birds

29.5.1 Behavioural Responses

Birds are homeotherms which can regulate its core body temperature relatively constant within a narrow range of 41–42 °C. The ideal temperature varies between 19 and 22 °C for optimum production performance in layers. When the ambient temperature and relative humidity exceed beyond the comfort zone, birds lose their capacity to effectively eliminate heat load away from the body that brings about changes in behavioural, physiological and immunological responses which severely affects their productive performance (Table 29.1).

The increase in the core body temperature of birds increases the respiratory rate. The enhanced respiratory frequency leads to severe panting and loss of more water from the respiratory tract as evaporation and results in dehydration and respiratory alkalosis. In addition, air sacs in birds facilitate air circulation on surface to enhance the heat dissipation through evaporation along with more gas exchange. The birds spend more time on resting with their wings elevated, covering the body in the litter materials, increasing drinking and panting frequency with restricted or reduced feeding, moving and walking. The heat dissipation processes are difficult in birds due to their feather coverage trapping heat to prevent hypothermia. The capillary blood circulation is redistributed throughout the body to increase the sensible heat loss during hot condition.

Birds elicit different behavioural responses during heat stress to prevent further heat increment by panting and wing droop which facilitate heat elimination from the core body to the periphery and the surrounding environment. The impairment in the release of hormones during heat stress activates various behavioural changes such as panting, sand bathing and standing with wings drooped and lifted slightly from the body to enhance heat dissipation. Heat stress evokes various behavioural responses to heat stress such as preening, feather ruffle and pecking among laying hens and broilers. Though, birds respond to heat stress similarly, however depending upon the intensity and duration, the responses exhibited by each bird in a flock may vary. Such variations among individuals and/or breeds indicate supremacy of the individual and/or breed to combat heat stress. For instance, enhanced alert behaviour, feather ruffle and crowing are frequently found in White Leghorn, whereas a relaxed behaviour and preening can be observed in Red Jungle fowl during heat stress. Further, the conditions that create stress to birds such as transportation, stocking density and social interactions are also influence their normal behaviour.

Table 29.1 The physiological response of birds at different ambient temperature

Ambient temperature (°C)	Physiological response
13–18 °C	Birds do not require any extra effort to maintain normal core body temperature and physiological functions within this temperature range.
18–24 °C	Optimum temperature range where the birds can perform maximum.
24–30 °C	A small drop in feed intake however, if sufficient nutrients are provided and intake is adequate then production is maintained. However, egg size may decrease and impaired shell quality may be observed when the temperature reaches the upper limit.
30–32 °C	The feed intake decreases further with lower body weight gain in broilers. The egg size and production drop with thin shell.
32–35 °C	The feed intake reduces drastically with typical symptoms of heat stress in birds.
35–38 °C	Pronounced or severe heat stress symptom. Feed intake and production are severely affected with enhanced water consumption.
Above 38 °C	Immediate measures needed to cool the birds and ensure survival of birds and it is very essential to restore core body temperature.

29.5.2 Physiological Adaptability in Birds

High environmental temperature impacts the physiological functions of poultry species which eventually affect the production performance of birds. Poultry species, like other mammalian animals, can undergo a number of complex and sequential physiological events which enable them to minimize the adverse impacts of heat stress and preserve heat homeostasis (Fig. 29.3). Birds exposed to higher ambient temperature exhibit various behavioural and physiological changes which allow them to adapt to the harsh climatic

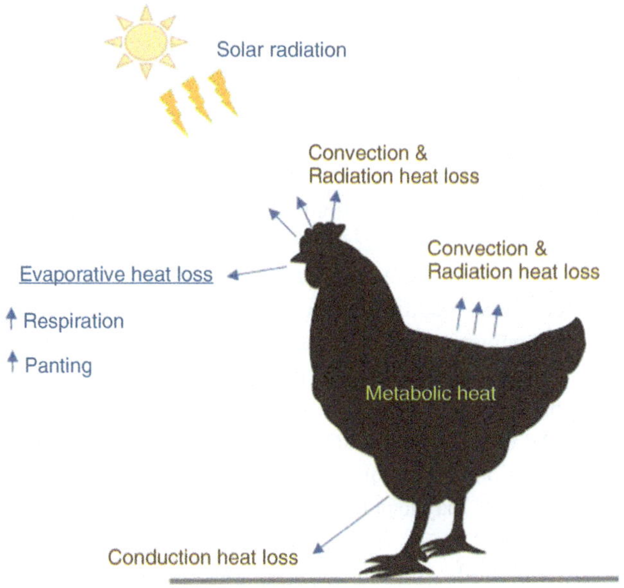

Fig. 29.3 Physiological responses to heat stress in poultry. When birds exposed to heat exposure, HPA axis get activated and birds elicit physiological responses to loss heat load through conduction, convection, evaporation and radiation

condition. Under hot and humid environmental conditions, birds maintain their body temperature through heat dissipation via conduction, radiation, convection and evaporation to the surrounding environment. In a thermoneutral environment, birds dissipate heat through non-evaporative cooling, i.e. radiation, conduction and convection. It is also known as sensible heat loss and is energetically efficient mechanism. In brief, birds increase the blood flow to the body surface area such as wattles, comb, shanks and unfeathered areas under wings thereby heat is lost from the body to the surrounding environment through temperature gradient. The amount of heat dissipation through conduction, convection and radiation depends on the temperature gradient between the birds and its surrounding environment. At the same time, nutrient digestibility is reduced since the blood flow and energy are diverted away from vital organs to body surface area, thereby lowering the metabolism and heat production. Increased respiratory frequency during heat stress leads to loss of more carbon dioxide and disturbances in acid base balance.

Rectal temperature reflects the core body temperature in birds and it is one of the biomarkers of heat stress since it indicates the thermal balance in birds. Therefore, changes in rectal temperature in response to heat stress are used to evaluate the degree of adaptability of birds to heat stress conditions. Heat stress conditions significantly increase the body temperature in birds which are exposed to high ambient temperature. It is reported that the rectal temperature and body temperature were significantly increased in various species of birds exposed to heat stress than the birds reared in normal conditions.

When ambient temperature exceeds the thermoneutral zone, birds switch their cooling mechanism to the evaporative heat loss. Generally, all animals when exposed to heat stress starts increasing their respiration rate and effectively reduce the heat load by evaporation of moisture from the respiratory tract. This is particularly important for birds since they don't have sweat glands and must solely depend on heat loss through respiratory evaporative cooling. Birds tend to increase the heat loss across the respiratory tract surfaces by increasing their respiration rate and decreasing the tidal volume (panting) with rapid vibration of gular membranes (known as gular flutter) to the maximum possible level to maintain their body temperature. However, panting requires increased energy for the increased muscular activity. Therefore, reduced energy efficiency also associated with heat stress. Birds have a special anatomical structure called as air sacs which are very useful during panting, as they enhance air circulation on surfaces of respiratory tract contributing to increased gaseous exchange with the surrounding air and consequently, enhancing evaporative loss of heat. Birds have feathers that help them regulate their body temperature. Feathers provide great insulation during cold weather but inhibit heat loss during summer condition. During heat stress condition, birds' water intake increases in order to equilibrate the body temperature through conductive heat loss when the water warms the body.

29.5.3 Neuroendocrine Response

Neuroendocrine system performs a major role in sustaining homeostasis and normal physiological functions of birds during heat stress. Heat stress activates the sympathoadrenal medullary (SAM) axis to restore and control homeostasis. The sympathetic nerves sense the increase in environmental temperature and communicate the information to the adrenal medulla which enhances the secretion of catecholamines. The enhanced level of catecholamines increases respiratory frequency and vasodilation of peripheral blood vessels to ameliorate heat stress. Further, the prolonged high temperature activates the hypothalamic–pituitary–adrenal (HPA) axis which releases corticotrophin-releasing hormone (CRH) from hypothalamus. The higher level of CRH stimulates the secretion of adrenocorticotrophic hormone (ACTH) from the pituitary which activates the adrenal glands to enhance release of corticosteroid. The increased corticosteroid favours in the gluconeogenesis to meet the high energy demand during heat stress.

Thyroid hormones play a vital role in the thermoregulation of birds. The stimulation of thermoreceptors activates the hypothalamus–pituitary–thyroid axis leading to increase in thyrotropin releasing hormone (TRH). TRH activates the thyrotropes in the anterior pituitary to secrete thyroid stimulating hormone which acts on the thyroid gland and increases the synthesis of thyroxine (T4). T4 is the source of production of biologically active triiodothyronine (T3) by deiodinase enzymes. However, heat stress inhibits the conversion of T4 to T3 and results in reduced T3 concentration that decreases the metabolic heat increment in birds. Further, decrease in TRH results in reduction of T3 and T4 from the thyroid gland which regulates the basal metabolic rate. Therefore, T3 concentration decreases in heat stressed birds due to a decrease in peripheral deiodination of T4 to T3, whereas T4 level is inconsistent. Heat stress also affects secretory pattern of gonadotrophin-releasing hormone in birds which impairs secretion of steroid hormones, progesterone, testosterone and oestradiol. Additionally, the biological activity of gonadotropin is depressed in layers during heat stress. Therefore, the neuroendocrine changes are responsible for decrease in growth rate and reproductive performances in heat stressed birds.

29.5.4 Blood Biochemical Response

Heat stress severely affects the metabolic status and physiological equilibrium in birds that leads to health problems and high mortality. The levels of blood total lipids and cholesterol are reducing with increasing environmental temperature. The haematological values such as haemoglobin concentration (g/dL) and haematocrit percent (%) are decreasing during heat stress depending upon the age of birds. The increase in water consumption may result in reduction of red blood cells which consequently decrease haemoglobin concentration and packed cell volume in birds during heat stress. The increase in corticosterone level induces lymphoid organ involution and modifies the features of heterophil and lymphocyte in birds. Heat stress significantly decreases plasma total protein, albumin and globulins in heat stressed birds. Plasma total protein also varies between sexes and different seasons with highest during thermoneutral temperature or winter and lowest in summer. The prolonged exposure of broilers birds to heat stress elevates the levels of alanine aminotransferase and aspartate aminotransferase enzyme activities. Heat stress also enhances levels of lactate dehydrogenase, glutamic oxaloacetic transaminase and creatine kinase which indicates tissue damage in liver and muscle. The plasma glucose level increases in heat stressed birds that could be a result of enhanced gluconeogenesis. Malondialdehyde (MDA) level also increases in heat stressed birds due to enhanced lipid peroxidation. These changes in the blood parameters are part of the thermoregulatory responses accomplished by birds to facilitate them to withstand heat stress.

The increase in respiration rate during heat stress results in respiratory alkalosis which disrupts the acid base balance and leads to increase blood pH in association with reduced pCO_2. Respiratory alkalosis is associated with decreased growth rate in broilers. Whereas, the metabolic alkalosis occurs when there is a disturbance in the fixed acids and bases in the extracellular fluid. The blood HCO_3 level decreases in panting birds under acute heat stress and more expulsion of CO_2 due to elevated respiration rate helps to bring down the body temperature. The higher level of dehydration that occurs due to more water excretion in faces and urine causes further disturbances in acids base balance.

Heat stress modifies the electrolyte balance which is vital for the maintenance of acid base balance, cellular homeostasis, synthesis of tissue protein, electrical potential of cell membranes, enzymatic reactions and osmotic pressure. However, the enhanced excretion of fluid in the urine with more concentration of electrolytes and the disproportion of dietary Na, K and Ca may lead to metabolic alkalosis with increased blood pH and HCO_3. Heat stress impairs function of immune system of birds and enhances the disease vulnerability of birds and leads to high mortality and morbidity.

29.5.5 Metabolic Response

As an adaptive response, birds significantly reduce their voluntary feed intake, digestibility and nutrient utilization to maintain constant body temperature during both acute and chronic heat stresses. This response is considered as the main adaptive mechanism to reduce metabolic heat production (Fig. 29.4). Liver, the important metabolic organ plays a significant role during heat stress condition to maintain the homeostasis by regulating the metabolism of carbohydrate, protein and lipid, thereby controlling the plasma levels of metabolites. However, heat stress impairs the activity of liver, thereby altering the plasma metabolite balance. Proteins, especially amino acids are very essential for growth and concentration of amino acids in the body also regulate feed intake. Heat stress was reported to decrease the amino acids in plasma and tissues of chicken during prolonged heat exposure. The altered free amino acids in various tissues like brain and skeletal muscle vary from that in plasma. Similarly, all the altered free amino acids in the various parts of the brain, except for proline and cystathionine are different which indicate that alterations in free amino acid contents due to heat stress may be tissue-specific, which is in agreement with the fact that enzymatic activity related to protein metabolism and protein synthesis is tissue-specific.

Glucose is very much essential nutrient since it is considered as a major source of energy for all the tissue, particularly for the brain. Glucose can be converted into a number of intermediate metabolites and is used in synthesis reactions, or utilized for the production of ATP. During stressful condition, liver can generate glucose either by glycogenolysis (from stored glycogen) or gluconeogenesis. Constant supply

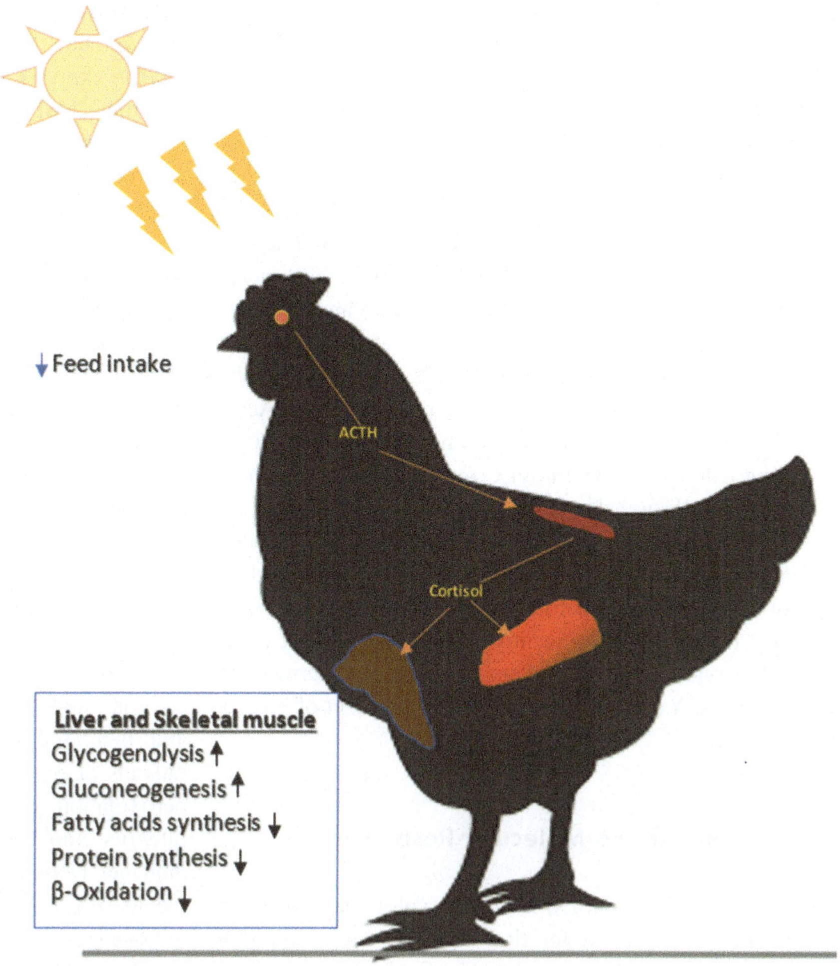

Fig. 29.4 Metabolic responses to heat stress in poultry. Heat stress causes decline of feed intake in birds to reduce metabolic heat production. Heat stress also reduces protein and fatty acid synthesis; enhances glycogenolysis and gluconeogenesis in order to provide energy to birds

of energy is crucial to tackle heat stress. Elevated environmental temperature increases the rate of glycogenolysis and gluconeogenesis in liver of heat stressed birds. Further, it is essential to sustain the absorption of glucose from intestinal lumen. Glucose transportation in epithelial and non-epithelial cells is carried out by sodium dependant glucose transporter (SGLT) and glucose transporter (GLUT) systems, respectively. Heat stress decreases the uptake of glucose from digestive tract. The reduced levels of glucose in serum of heat stressed birds indicate that the birds were in negative energy balance. Research data suggested that many of the metabolites involved in glucose metabolism have been increased in response to heat stress such as glucose 6 phosphate (G6P) and glucose. The activities of metabolic enzymes involved in glucose metabolic pathways are also increased during heat stress exposure. Phosphoglucomutase, an enzyme that converts glucose 1 phosphate to G6P was significantly increased during exposure of birds to elevated temperature. It was reported that heat stress enhances the gluconeogenesis in liver which was indicated by the higher levels of fructose-6-phosphate (F6P) and fructose-bisphosphatase 2, the enzyme that transform F6P to G6P.

Amino acids are considered as a main source of energy for the liver. In birds, body protein deposition is reduced under elevated temperature condition. Heat stress intensity and duration affect the protein metabolism differently. Acute heat stress depresses the protein synthesis, lowers the protein deposition and increases the protein catabolism in heat stressed animals. Acute heat stress alters the concentration metabolites in plasma, many of which are involved in protein metabolism specifically glycine, serine, threonine, arginine, proline, phenylalanine, cysteine and methionine metabolism in broiler chicken. However, chronic heat stress decreases protein synthesis, catabolism and levels of certain amino acids in plasma. Most of the amino acids are reduced during heat stress condition in chicken except cysteine. The increased serum uric acid levels in heat stressed broiler chicken suggest that the protein might be degraded for energy supply which is supported by the increased levels of some amino acids such as proline, L-cysteine, methionine and threonine in heat stressed birds even though feed intake had been reduced significantly. However, the concentration of glycine decreased in heat stressed birds possibly due to increased uric acid metabolism since glycine is required for uric acid synthesis. Non-protein amino acids such as ornithine and citrulline levels also increased in heat stressed birds that indicate its role in stress relief and thermoregulation.

Lipids are stored in adipose tissue as triglycerides. Triglycerides when broken down metabolically and release free fatty acids can be utilized for various purposes. Fatty acids like as stearic acid, arachidonic acid, palmitic acid, linoleic acid and oleic acid concentration are declined in birds exposed to heat stress. However, certain fatty acids (like myristate, myristoleate, and palmitoleate) and enzymes involved in fatty acid synthesis were elevated during heat stress exposure in chicken. Acetyl-CoA carboxylase alpha enzyme that converts acetyl-CoA to malonyl-CoA; Acyl-CoA synthetase, which converts myristate to myristoyl-CoA and palmitate to palmitoyl-CoA were increased in heat stress birds (Jastrebski et al. 2017). Concentration of non-esterified fatty acids (NEFA) in plasma also decreased in response to high environmental temperature due to reduction of lipolytic enzyme activity and lipolysis in heat stressed animals. Heat stress significantly influences the TCA cycle since the concentration of α-ketoglutaric acid (intermediate product of TCA cycle) and pyruvate (end product of glycolysis) increased in heat stressed birds.

29.5.6 Cellular and Molecular Response

Heat Shock Proteins (HSPs) are highly conserved molecular chaperones well known for their roles as stress response proteins that helps in protein folding and unfolding, assembly of multi-protein complexes, transport and sorting of proteins into subcellular compartments, minimization of protein aggregation and protection of cells against apoptosis. The HSPs are controlled at the transcription level in all organisms however stress factors activate a specific heat shock transcription factor (HSF) mainly HSF-1. Further, accumulation of denaturated proteins in the cytosol stimulates HSF-1 as a response to stress. The phosphorylation favours the trimer formation of phosphorylated HSF-1 which enters into the nucleus and binds to promoter region (heat shock element) and enhances *HSP* gene expression. The increased production of HSPs selectively binds to the degenerated proteins or aggregated proteins and newly synthesized polypeptides. However, formation of HSP-HSF complex decreases HSF-1 production in negative feedback regulatory mechanism to maintain the internal homeostasis. There are many heat shock protein families which are organized based on their molecular weights as HSP110, HSP90, HSP70, HSP60, small molecule HSPs and ubiquitin. Heat stress activates many genes of HSPs families where the higher expression of Hsp70 promotes lipogenesis and heat tolerance HSPs are essential in protecting and repairing cells and tissues during stress. The production of HSPs is highly essential to restore the survival of stressed cells and maintenance of the internal environment. The exposure of birds to heat stress enhances HSP70 expression which has a wide range of chaperonic activities. HSP70 plays a vital role in cellular protection during hot condition stress and develops the thermotolerance in birds. These chaperones help in maintaining and sustaining the native structure of proteins by protection and stabilization of stress labile proteins, renaturation of denaturated proteins and preventing protein aggregation. They are responsible for refolding of misfolded and aggregated protein.

Genome-wide transcriptomic studies have revealed many pathways that are modified during heat stress in broilers. The pathway analyses determined a wide range of affected cellular responses such as apoptosis, cell cycle, DNA repair, membrane trafficking and immune function. Heat stress activates the genes that are involved in lipid metabolism and increased numerous genes encode enzymes involved in different phases of fatty acid synthesis in birds. Heat stress stimulates the expression of NADPH oxidases and enhances generation of superoxide and other reactive oxygen species (ROS) by transferring electrons across cell membranes. ROS are continually produced in vivo due to metabolic activities and increased significantly during heat stress. The free radicals affect the cellular metabolism by altering the lipid peroxidation reactions compromising biological membrane integrity and functions. The higher level of free radicals damages DNA and proteins and makes them more susceptible to oxidative damage. Further, gene ontology analyses unveiled many pathways are affected during heat stress which were associated with ion channel activities, redox balance and lipid metabolism. The network analyses also determined the interactions of many of these genes with heat shock proteins (HSPs) that regulate heat stress responses.

29.5.7 Biomarkers for Heat Stress in Poultry Birds

The biomarkers are the indices used to assess animals/birds or may be also considered as biological measures of a biological state. Heat stress affects the health, welfare and productive performances of birds (Fig. 29.5). The reduction in glucose and reduction in proteins, calcium, phosphorus and alkaline phosphatase concentrations are some of the biochemical biomarkers in birds during heat stress. Cholesterol content is elevated in liver, breast and thigh muscles of heat stressed birds. Further, creatine kinase and lactate dehydrogenase are more consistent isozyme in a tissue or organ activity which was increased in heat stress and indicating muscle damage or myopathy. The heterophil to lymphocyte ratio is one of a reliable stress estimator in birds which increases due to environmental or heat stress.

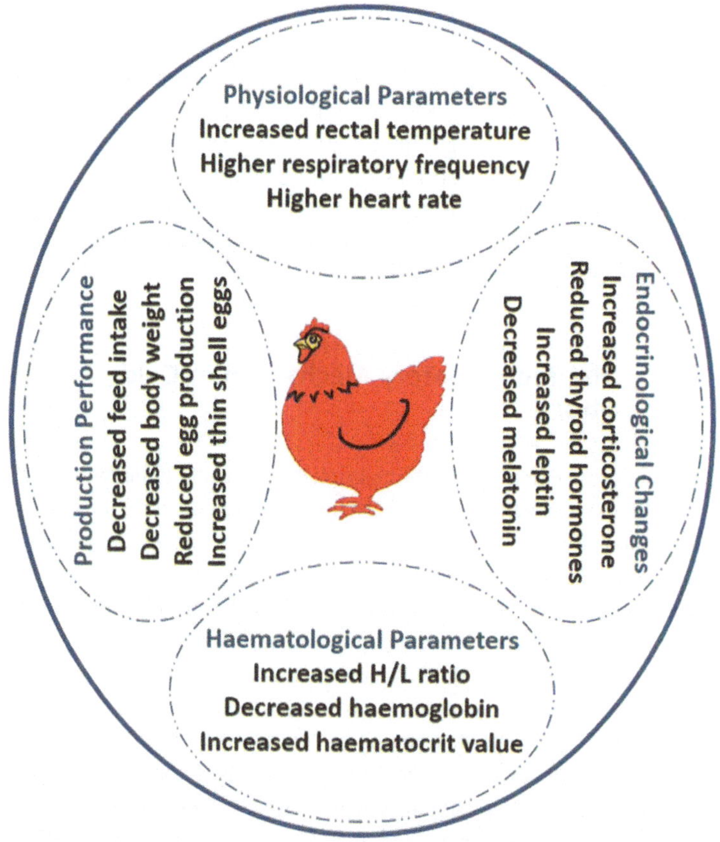

Fig. 29.5 Biomarkers used to determine the heat stress and health status in birds. Biomarkers are biological measures of a biological state. Heat stress affects the health, welfare and productive performances of birds. The changes in the physiological, metabolic process and cellular activities are expressed as intermediate and end metabolites

Further, heat stress stimulates the HPA axis and increases the release of glucocorticoids particularly corticosterone which is an indicator of stress in birds. The higher level of glucocorticoids during heat stress causes a quick influx of heterophils into the blood from bone marrow that increased the concentration of circulating heterophils. Thyroid activities are reduced during heat stress to prevent increment and results in reduced concentration of T3 in birds. The HSPs are important for protecting cells and protein folding and formation during heat stress. The abundance of relative expression mRNA of various HSPs particularly *HSP70* has been closely associated with heat stress in birds. Further, heat stress enhances the production of mitochondrial ROS and decreases avian uncoupling protein. The imbalance between ROS and antioxidants in heat stressed broilers results in oxidative stress. Malondialdehyde is a biomarker used to estimate oxidative stress in poultry by indirectly quantifying the level of peroxidation due to ROS. Additionally, the primary redox pathways; superoxide dismutase, catalase and glutathione peroxidase, significantly decrease in heat stressed birds due to higher production of ROS. The lipid peroxidation is a major consequence of heat stress where protein carbonyl, MDA, 8-hydroxy-2′-deoxyguanosine and advanced glycation end products are considered as biomarkers of protein, lipid, DNA and carbohydrate oxidation in birds.

29.6 Thermoregulation During Flight

Flight is the most energy consuming process, in the progress of which the basal metabolic rate increases in a multitude fashion in comparison with resting state and hence results in ample amount of heat generation. All possibilities are there for development of hyperthermia in birds during flight. This sort of flight induced heat load, triggers responses in birds to have increased the evaporative water loss and converted heat loss from unfeathered areas of the bird to the environment. Flight instigated hyperthermia is beneficial for improving the thermal gradient between the birds and the environment and hence for heat dissipation to the environment. Another important function of hyperthermia during flight is that it serves as mechanism of heat storage. Bird can get rid of this heat storage during flight in later periods. It is established that in flying birds, core temperature is independent of the ambient temperature. The degree of elevation of core temperature is independent of ambient temperature but is positively correlated with the rate of flight. Flight induced hyperthermia is not the result of inadequacy in thermoregulatory mechanism but it is an adaptive response to flight, a strenuous exercise. The thermal resistance caused by skin and the feathers could possibly give an explanation for paradoxical increase in skin temperature of flying birds.

This is also an adaptive response against an inevitable rise in heat flow.

Leg trailing is the behavioural method of thermoregulation exhibited by most of the birds during flight. The unfeathered portions of bird's legs play an important role in non-evaporative cooling (radiation and convection) during flight. Hovering mode of flight is also exhibited by some species of birds, which will increase the air flow and helps in heat dissipation during flight.

29.7 Thermoregulatory Mechanism in Cold Stress

It is widely accepted that the animals inhabiting higher altitudes or cold environments obey Bergmann and Allen's rule of adaptation. Bergmann's rule states that the body surface area of an animal determines its heat dissipation capacity, the volume determines its ability to generate heat and the surface area to volume ratio fixes its thermoregulation capacity. In simple terms, it can be understood that endothermic animals will have lower surface area to volume ratio in high altitude regions. Allen's rule proclaims that among the endotherms, length of extremities reduces as the altitude increases as a means of thermoregulatory adaptation mechanism. It is well established that both the rules hold well in mammals and birds. According to both of these thermoregulation hypothesis, larger body size along with smaller extremities helps the animals to conserve more heat in cold environmental conditions and in contrary, smaller body size with larger extremities helps in regulation of heat dissipation in warmer environments.

Morphological traits such as plumage are playing role in cold adaptation of birds, however, unfeathered portions of leg and feet are critical during cold exposure. Plumage of birds acts as a critical barrier between the bird's body surface and the external environment and helps in meeting out 90% of the animal's requirement of insulation. Feathers are vital importance for birds' insulation to support thermoregulation of birds. Heat transfer through plumage will occur by conduction and convection via air, conduction through solid portions of feathers and through radiation. The thermoregulatory capacity of feathers is associated with various traits like plumage morphology, density and depth. During cold stress, birds significantly reduce the amount of heat dissipation from legs and feet, which can be clearly explained by the vascular counter current mechanism of heat transfer and conservation. Bill size also has the potential to influence the heat conservation, in the context of employing thermoregulatory behaviour of back rest to keep birds warmer under cold conditions. As per Allen's rule of adaptation, higher latitudes with lower temperature tend to bear smaller bills than the birds from lower latitudes. The uninsulated and vascularised bills of birds can serve as thermal window for heat exchange.

It is well accepted fact that the thermal needs of birds in response to cold stress are met out largely by shivering thermogenesis. Under immediate exposure to cold environment, sudden increase in metabolic rate associated rigorous activity of muscles leads to shivering thermogenesis and is very much essential for thermoregulation for cold adaptation. But in progress, chronic cold exposure gradually replaces shivering thermogenesis with non-shivering thermogenesis and in fact improves the cold adaptation capacity. Non-shivering thermogenesis generates heat by processes involving chemical energy without muscular contraction. Under very severe cold environments, both shivering and non-shivering thermogenesis mechanisms help the bird to cope with heat generation.

A foremost physiological adaptation to cold stress in birds is accumulation of body fat. This fat accumulation acts as an extra fuel for thermoregulation of birds under cold stress, serves as a reservoir to tackle long winter nights and most importantly helps in to act as a response against sudden drop in temperature. Above all, accumulation of extra layers of fat acts as an added insulation against cold. Another important cold adaptation mechanism observed in birds is development of winter plumage with increase in quantity of feathers for better insulation and was covered briefly in preceding sections. As an extreme response to cold stress, bird chooses migration as a method of adaptation to avoid deleterious effects of winter and for better survival. Migratory behaviour in birds is exhibited as three ways of shifting the positions, such as latitudinal shift, altitudinal shift and habitat shift and a bird chooses the combinations of any of these shifts as an adaptation mechanism in order to provide themselves with maximum chance of survival.

29.8 Differences in Thermotolerance Between Temperate and Tropical Birds

Characterisations of thermoregulation both in tropical and temperate birds are very much important to predict their vulnerability to climate change. Upper critical temperature (UCT) is used to assess the latitudinal variation in heat tolerance and susceptibility to heat stress. Temperate birds have narrower thermal safety margins (i.e. they are experiencing maximum air temperatures closer to their UCTs) than their temperate counterparts and were therefore projected to be at greater risk from climate warming (Pollock et al. 2021).

Heat tolerance limits (HTL): the air temperature at which an endotherm loses the ability to regulate its body temperature. Tropical birds have higher HTL (ΔHTL = 2.2 °C; 45.2 vs. 43.0 °C) and upper critical temperature (ΔUCT = 1.1 °C; 38.7 vs. 37.6 °C) which indicate that tropical birds have better thermotolerance ability than the temperate birds. However, these differences between tropical and temperate birds do not seem to impact susceptibility to thermal stress neither thermal safety margins nor thermal tolerances (the difference between HTL and Tmax) vary between temperate and tropical species. Temperate birds show consistent seasonal changes in thermoregulatory traits especially during winter seasonal changes showed by temperate birds seem to be primarily aimed to conserve heat making them acclimatized to cold temperature (Pollock et al. 2019). However, tropical birds do have capability of seasonal adjustment and are rather more phenotypic flexible than the temperate birds. Temperate birds have wider TNZ, whereas tropical birds showed idiosyncratic patterns of seasonal variation in LCT, UCT and TNZ. Total feather mass and density of downy feathers were significantly high in temperate birds which provide insulation to tolerate cold temperature and making the survivability of these birds in warmer tropical regions very difficult. However tropical birds have better adaptive potential to combat heat stress and produce optimally, thereby making them more resilient to climate change.

Learning Outcomes

- High or low ambient temperature initiates various physiological, behavioural and endocrine responses to maintain the thermal balances in birds.
- Heat/cold stress affects the growth performance, egg and meat production in birds as a result of reduced feed intake and impaired digestion and absorption of nutrients.
- Heat/cold stress stimulates the production of highly conserved molecular heat shock proteins that helps in protein folding and unfolding and maintenance of cellular homeostasis.

Exercises

Objective Questions

Q1. What is the temperature range of thermoneutral zone for poultry?
Q2. What is the crucial temperature point in poultry for thermoregulation?
Q3. How heat and cold stress activates the stress response in poultry?
Q4. Impacts of heat stress on productive performance of poultry?
Q5. Why heat stress is very critical in birds?
Q6. The first sign of heat stress in poultry?
Q7. What does enhanced panting lead to?
Q8. What is the ideal temperature for optimum production in birds?
Q9. Which is the stress hormone in birds?
Q10. Mention is the thermogenic hormone in birds?
Q11. What is the biomarker for oxidative stress?
Q12. Haematological indicator of heat stress in birds?
Q13. Which is the cellular and molecular biomarker of heat stress in birds?

Subjective Questions

Q1. What is the impact of heat stress on productive performance of poultry?
Q2. How heat stress impairs the productive performance in birds?
Q3. How the egg shell quality is affected during heat stress?
Q4. What are the heat dissipation of mechanism in birds?
Q5. How the acid base balance is disturbed in birds during heat stress?
Q6. How does metabolic acidosis occur in birds during heat stress?
Q7. What is the neuroendocrine response in birds?
Q8. What are changes occurring in the blood biochemicals in heat stressed birds?
Q9. Explain the molecular response to heat stress in birds?
Q10. What is major impact of lipid peroxidation in heat stressed birds?

Answer to Objective Questions

A1. 18–22
A2. 30
A3. Activates the hypothalamo–hypophyseal–adrenocortical (HPA) axis
A4. Reduced feed intake, decrease in body weight, reduction in egg production and shell quality
A5. Lack of sweat glands and higher metabolic rate
A6. Increased respiration rate
A7. Respiratory alkalosis
A8. 19–22
A9. Corticosterone
A10. Thyroid hormones
A11. Malondialdehyde
A12. Decrease in haemoglobin (g/dL) and haematocrit percent
A13. Heat shock proteins in particularly HSP70

Answer to Subjective Questions

A1. Decrease in feed intake, reduced body weight gain, decrease in number eggs and poor egg shell quality.
A2. Reduced feed intake, decreased digestive enzymes and biological activities, inflammation in the digestive tissues.
A3. The increased CO_2 elimination results in decreased HCO_3 which is essential for the formation of egg shell.

A4. Evaporative heat loss, convection, conduction and radiation.

A5. The increase in respiration rate during heat stress results in respiratory alkalosis which disrupts the acid base balance and leads to increase blood pH in association with reduced pCO_2.

A6. The enhanced excretion of fluid in the urine with more concentration of electrolytes and the disproportion of dietary Na, K and Ca may lead to metabolic alkalosis with increased blood pH, HCO_3 and base excess.

A7. The activation of hypothalamic–pituitary–adrenal axis enhanced the glucocorticoids particularly corticosterone and decrease in triiodothyronine.

A8. The levels of blood total lipids and cholesterol are reducing with increasing environmental temperature. The haematological values such as haemoglobin concentration (g/dL) and haematocrit percent (Ht%) are decreasing during heat stress depending upon the age of birds.

A9. Heat stress enhances the production of HSPs which selectively bind to the degenerated proteins or aggregated proteins and newly synthesized polypeptides.

A10. The lipid peroxidation is a major consequence of heat stress where protein carbonyl, MDA, 8-hydroxy-2′--deoxyguanosine and advanced glycation end product are considered as biomarkers of protein, lipid, DNA and carbohydrate oxidation in birds.

Further Reading

Ajakaiye JJ, Perez-Bello A, Mollineda-Trujillo A (2011) Impact of heat stress on egg quality in layer hens supplemented with l-ascorbic acid and dl-tocopherol acetate. Vet Arhiv 81:119–132

Chowdhury VS (2018) Heat stress biomarker amino acids and neuropeptide afford thermotolerance in chicks. J Poult Sci 56:0180024

Chowdhury V, Tomonaga S, Nishimura S, Tabata S, Furuse M (2012) Physiological and behavioral responses of young chicks to high ambient temperature. J Poult Sci 49:212–218

Deeb N, Cahaner A (1999) The effects of naked neck genotypes, ambient temperature, and feeding status and their interactions on body temperature and performance of broilers. Poult Sci 78(10):1341–1346

Etches RJ, John TM, Verrinder Gibbins AM (1995) In: Daghir NJ (ed) Behavioural, physiological, neuroendocrine and molecular responses to heat stress. CAB International, pp 31–65

Goel A (2021) Heat stress management in poultry. J Anim Physiol Anim Nutr 105:1–10. https://doi.org/10.1111/jpn.13496

Hafez ESE (1964) Behavioral thermoregulation in mammals and birds. Int J Biometeorol 7(3):231–240

Imik H, Atasever MA, Urgar S, Ozlu H, Gumus R, Atasever M (2012) Meat quality of heat stress exposed broilers and effect of protein and vitamin E. Br Poult Sci 53:689–698

Jastrebski SF, Lamont SJ, Schmidt CJ (2017) Chicken hepatic response to chronic heat stress using integrated transcriptome and metabolome analysis. PLoS One 12(7):e0181900

Liu L, Ren M, Ren K, Jin Y, Yan M (2020) Heat stress impacts on broiler performance: a systematic review and meta-analysis. Poult Sci 99:6205–6211

Lu Z, He X, Ma B, Zhang L, Li J, Jiang Y, Zhou G, Gao F (2018) Serum metabolomics study of nutrient metabolic variations in chronic heat-stressed broilers. Br J Nutr 119(7):771–781

Mack LA, Felver-Gant JN, Dennis RL, Cheng HW (2013) Genetic variation alter production and behavioral responses following heat stress in 2 strains of laying hens. Poult Sci 92:285–294

Marder J, Arad Z (1989) Panting and acid-base regulation in heat stressed birds. Comp Biochem Physiol A Physiol 94(3):395–400

Pollock HS, Brawn JD, Agin TJ, Cheviron ZA (2019) Differences between temperate and tropical birds in seasonal acclimatization of thermoregulatory traits. J Avian Biol 50(4). https://doi.org/10.1111/jav.02067

Pollock HS, Brawn JD, Cheviron ZA (2021) Heat tolerances of temperate and tropical birds and their implications for susceptibility to climate warming. Funct Ecol 35(1):93–104

Ranjan A, Sinha R, Devi I, Rahim A, Tiwari S (2019) Effect of heat stress on poultry production and their managemental approaches. Int J Curr Microbiol App Sci 8(2):1548–1555

Ruuskanen S, Hsu BY, Nord A (2021) Endocrinology of thermoregulation in birds in a changing climate. Mol Cell Endocrinol 519:111088. https://doi.org/10.1016/j.mce.2020.111088

Saeeda M, Abbas G, Alagawany M, Kamboh AL, El-Hack MEA, Khafaga AF, Chaoa S (2019) Heat stress management in poultry farms: a comprehensive overview. J Thermal Biol 84:414–425

Smit B, Zietsman G, Martin RO, Cunningham SJ, McKechnie AE, Hockey PAR (2016) Behavioural responses to heat in desert birds: implications for predicting vulnerability to climate warming. Clim Change Responses 3:9. https://doi.org/10.1186/s40665-016-0023-2

Vandana GD, Sejian V, Lees AM, Pragna P, Silpa MV, Maloney SK (2021) Heat stress and poultry production: impact and amelioration. Int J Biometeorol 65:163–179

Wang S, Edens FW (1993) Stress-induced heat-shock protein synthesis in peripheral leukocytes of turkeys (Meleagris gallopavo). Comp Biochem Physiol 106(3):621–628

Appendix

See Tables A.1 and A.2.

Table A.1 Physiological parameters in different domestic and wild animals

Species	Life span (years)	Body temperature (°C)	Pulse rate (per min)	Respiration rate (per min)
Cattle	15–25	38–39.3	38.0–39.3	30–32
Buffalo	15–25	38–39	50–70	30–40
Horse	25–30	37.2–38.2	28–40	9–10
Goat	15–18	38.5–39.7	70–80	12–15
Sheep	10–12	38.3–39.9	70–80	12–15
Pig	15–20	38.7–39.8	60–90	8–18
Dog	10–13	37.5–39	60–80	14–16
Cat	12–18	38.1–39.2	110–130	20–30
Rabbit	7–10	38.6–40.1	200–210	50–60
Rat	2.5–3.5	38–39	330–480	85–90
Mice (*Mus musculus*)	9–12	37–38	310–840	80–230
Tiger (*Panthera tigris sumatrae*)	8–10	38–39.5	56–97	7–40
Lion (*Panthera leo*)	8–10	38–39	42–76	14–32
Leopard (*Panthera pardus*)	12–17	38.5–39.5	65–70	13–15
Asian elephant (*Elephas maximus maximus*)	48–50	36–37	25–30	72–98
African elephant (*Loxodonta africana africana*)	60–70	36–37	25–30	72–98
Giraffe (*Giraffa camelopardalis*)	25–28	37–38	58–59	12–13
Fowl	10–20	40.6–43	120–160	15–30

© The Author(s), under exclusive license to Springer Nature Singapore Pte Ltd. 2023

P. K. Das et al. (eds.), *Textbook of Veterinary Physiology*, https://doi.org/10.1007/978-981-19-9410-4

Table A.2 Some hematological profiles in different domestic and wild animals

Species	Blood volume (mL/kg bwt)	Hemoglobin (g %)	PCV (%)	TEC ($\times10^6/\mu$ L)	TLC ($\times10^3/\mu$ L)
Cattle	55 (52–77)	8–15	40	6–8	7–10
Buffalo	82.0	9.57–17.05	30.25–50.08	7.81–8.36	7.35–16.94
Horse	76	11.5–16	33.4	7–10	8–11
Goat	70	8–12	34	13–14	8–12
Sheep	60	9–15	32	10–13	7–10
Pig	65	10–16	41.5	6–8	15–22
Dog	86 (79–90)	12–18	45.5	6–8	9–13
Cat	55 (47–66)	10–15	40	6–8	10–15
Rabbit	56 (44–70)	8.9–15.38	41.4	4.37–7.43	2.71–12.23
Rat	55–70	8.6–16.5	10–48	2.9–6.8	3.6–14.8
Mice (*Mus musculus*)	79 (78–80)	12.92 ± 0.34	39.81 ± 1.64	7.42 ± 0.85	4.98 ± 1.26
Tiger (*Panthera tigris sumatrae*)	–	7.8–13.8	36–45	4.66–9.15	6.2–11.05
Lion (*Panthera leo*)	–	8.9–14.6	26.8–44.1	5.1–8.3	7.2–25.6
Leopard (*Panthera pardus*)	–	12.98 ± 0.85	38.24 ± 1.94	7.06 ± 0.56	13.1 ± 0. 9
Asian elephant (*Elephas maximus maximus*)	–	9.8–15.2	29.4–40.7	1.9–3.2	7.9–21.9
African elephant (*Loxodonta africana africana*)	–				
Giraffe (*Giraffa camelopardalis*)	–	9.4–19.7	27–56	5.8–19.1	4.0–24.5
Fowl	60	10.1	29	2.5–3.2	20–30

Glossary

A-band Is the dark band in muscle fiber which is the complex structural assembly of the protein myosin and associated non-myosin components.

Absolute refractory period The period after the occurrence of an action potential during which the membrane is non-responsive to a given stimulus.

Acclimation It refers to the adaptive changes that take place in response to a single climatic variable (normally produced in a laboratory or climatic chamber).

Acclimatization Acclimatization is a long-term adaptive physiological adjustment which results in an increased tolerance to continuous or repeated exposure to complex climatic stressors (normally produced under field conditions).

Achlorhydria It is a clinical condition in which the stomach is unable to produce hydrochloric acid. It is caused due to a variety of reasons such as pernicious anemia, *Helicobacter pylori* infection, gastric bypass, hypothyroidism, radiation exposure of gastric mucosa.

Acidosis An abnormally high level of acid in the body fluid.

Acute phase proteins (APPs) They are inflammatory mediators or inhibitors mainly produced from liver. Their concentrations increased (or decreased) at the rate of 25% or more at the time of inflammation. They therefore act as a suitable biomarker of inflammation.

Adducin It is a calcium/calmodulin binding protein existing as αβ dimer that enables capping of actin for spectrin–actin interactions at the junctional complex.

Adenohypophysis The anterior, predominantly glandular part of the pituitary gland secretes various hormones like gonadotropin, somatotropin, ACTH, TSH, prolactin, etc.

Adipokines The cell signalling proteins released from adipose tissues. These are the cytokines.

Adipophilin It is called as lipid droplet proteins involved in the production of cytoplasmic lipid droplets during secretary differentiation of mammary alveolar epithelium and plasy a pivotal role in both formation and secretion of milk lipids.

Adjuvants Adjuvants are the substances that enhance the immunogenicity of an antigen after causing sustained release of the antigen, stimulating cytokines and chemokines release, recruitment of immune effector cells, helping in antigen processing and presentation.

After hyperpolarization A slight, transient hyperpolarization that sometimes occurs at the end of an action potential.

Agonistic Fighting or aggressive or defensive social behavior among the same species, like a threat, attack, retreat, etc.

Agouti-related protein Agouti-related protein, also called agouti-related peptide, is a neuropeptide produced in the brain by the AgRP/NPY neuron. It is synthesized in neuropeptide Y-containing cell bodies located in the ventromedial part of the arcuate nucleus in the hypothalamus.

Alkali reserve It is the capacity of blood to combine with CO_2 and expressed as the volume of CO_2 (mL/100 mL plasma).

Alkaline tide It is a state of metabolic alkalosis developed after heavy meal due to the diffusion of HCO_3^- into the venous blood. During every hydrogen ion secretion, one bicarbonate ion enters into the blood in exchange for chloride by Cl^-/HCO_3^- exchange (CHE) transporter.

Allometric growth Any organ or body part that grows at a different rate than the general body growth.

Allomone A chemical substance released from one species (animal) can influence another animal or species, like, a pheromone.

Anaphylactic reactions It is a localized or systemic fatal host response occurred in response to allergens and mediated by pharmacologically active substances released from mast cells which cause vasodilation, smooth muscle contraction, mucus production, and sneezing.

Angioedema Area of swelling (oedema) of the lower layer of skin and tissue just under the skin or mucous membranes.

Antimicrobial peptides (AMPs) They are used by many organisms as first line of defense against pathogens. They are multifunctional peptides with bacteriostatic, bactericidal, and cytolytic properties. They are promptly synthesized after infection and kill a wide range of pathogens.

Apnea The cessation of breathing.

© The Author(s), under exclusive license to Springer Nature Singapore Pte Ltd. 2023

P. K. Das et al. (eds.), *Textbook of Veterinary Physiology*, https://doi.org/10.1007/978-981-19-9410-4

Apoptosis A form of programmed cell death by fragmentation of DNA caused by activation or absence of any stimulus or by a genetically regulated physiological process. It may be termed cell suicide or programmed cell death.

Apotransferrin It is a beta globulin that binds with two molecules of ferric iron per molecule. It is responsible for carrying iron in circulation.

Aquaporins (AQP) These are the family of 28 kDa membrane proteins which help in the transport of water and small solutes across the epithelial cells. AQP are activated by arginine-vasopressin via V2-R receptors and translocated to the cellular membrane for water transport.

Astrocytes Are star shaped neuroglial cells with multiple radiating cytoplasmic processes remaining in the CNS.

Atrial natriuretic peptide (ANP) A peptide hormone released from the cardiac atria that promotes urinary loss of Na^+ in mammals.

Aunting behavior (primates) It is alloparenting performed by any group member other than the mother.

Azoospermia Presence of no spermatozoa in the semen.

B cells Lymphocytes that matured in the bone marrow (or liver during fetal life) or but in birds, in bursa of Fabricius (birds).

Biological rhythm The body as a whole or its components is in a dynamic state where the physiological process repeats itself with more or less constant time intervals. The phenomenon is defined as a "biological rhythm" which is influenced by a "biological clock."

Blepharoplast Basal body of flagella.

Blood–testes barrier A specialized barrier that is found between blood capillaries and seminiferous epithelium.

Blood–brain barrier (BBB) Is a special kind of barrier in the CNS which isolates neuronal tissue of CNS from the general circulation.

Broca's area Is a small area on the left side of the brain (sometimes on the right in left-handed individuals) is important in language processing.

Calsequestrin Is the main Ca^{2+} binding protein in the sarcoplasmic reticulum and serves as the main Ca^{2+} storage and buffering protein.

Capacitation Process of changes of spermatozoa in the female genital tract to get competence for fertilization.

Castration Defunct or sterilize the male animal's testis (or testes) by surgical (orchidectomy), chemical or mechanical methods. It may be unilateral (one testis) or bilateral (both testes). Related words—**Spaying**: Loss of ovary function (ies) in female animals; **Emasculation**: Removal of both testes and the penis.

Cerebrospinal fluid Is a colorless fluid found in the ventricles of the brain, central canal of the spinal cord and in the subarachnoid space surrounding the outer surface of the brain and spinal cord.

Ceruloplasmin It is a glycoprotein synthesized in the liver and involved in transport of copper.

Chaperone A group of proteins that aid macromolecules, like proteins and nucleic acids, assemble and fold into a complex structure.

Chemokines These are chemotactic cytokines and attract the leukocytes to the site of infection. Structurally chemokines are subdivided into four families based on the N terminal cysteine residue.

Cholangiocytes These are the epithelial cells that form a three-dimensional network of bile ducts. The hepatic bile is modified at the biliary tract through the secretion and reabsorption by the cholangiocytes.

Christiesomes These are the cell fragments containing endoplasmic reticulum, mitochondria, and lipid droplets identified in goat milk which are involved in triglyceride synthesis.

Chromatin A complex of nucleic acids (DNA) and proteins, primarily histones, in the eukaryotic cell nucleus that stains with basic dyes and condenses to form chromosomes during cell division.

Chyme It is a thick semisolid mass of partially digested food together with the digestive secretions formed in the stomach and intestine during digestion.

Circadian rhythm The biological process of an organism that is controlled internally (endogenous) and adjusted with the environment (entrainment) displays physiologically and specific behavior patterns in an oscillation of about 24 h sleep-wake cycle. Related words—**Diurnal rhythms**: When repetition (cycle) once in 24 h but displayed only in the day-time due to the influence of stimuli like daylight; **Nocturnal**: A kind of circadian rhythm that occurs only at night; **Ultradian rhythms**: Repeated oscillation (cycle) in 24 h have a shorter period and higher frequency than circadian rhythm like, include heartbeats, pulse, breathing, temperature regulation, eye blinking, etc.; **Infradian rhythms**: Cycle exhibits longer than 24 h like duration of a sexual cycle; **Crepuscular rhythm**: That exhibit only at twilight; **Biological rhythm**: The biological event that displays in a specific time interval, viz. circadian rhythm, diurnal rhythms, etc.

Circulating cell-free fetal DNA (ccffDNA) It is fetal DNA that circulates freely in the maternal blood. Analysis of ccffDNA may provide earlier diagnosis of fetal conditions.

Cloning Process of producing individual organisms with either identical or virtually identical DNA, either by natural or artificial means.

Clutch (of bird) Eggs lay continuously or at a stretch in the nest by a bird. Clutch size means the number of eggs laid in a group in the nest by a bird.

Cocaine and amphetamine-related transcript (CART) It is a neuropeptide protein. CART appears to have roles in reward, feeding, and stress, and it has the functional properties of an endogenous psychostimulant.

Colony stimulating factor (CSF) These are responsible for differentiation of leukocytes in the bone marrow. There are four families of CSF, namely granulocyte colony-stimulating factor (G-CSF), macrophage colony-stimulating factor (M-CSF), granulocyte-macrophage colony-stimulating factor (GM-CSF), and multiple-colony-stimulating factor (also called IL 3).

Crop It is a pouch like structure originating as a dilatation of the cervical esophagus with food storage function. The crops are well developed in omnivores and herbivores/ granivorous birds compared to carnivorous birds. The crop is absent in ostriches, gulls, owls, geese, and penguins.

Cryoprotectants A cryoprotectant is a chemical that protects biological tissue from harm caused by freezing.

Cystoscope An instrument equipped with light (endoscope) used to examine and medication into the interior of the urinary bladder, ureter, and urethra.

Damage associated molecular patterns (DAMPs) DAMPs are endogenous molecules released in response to stress or tissue injury and are potent stimulators for non-infectious inflammation. They are also recognized by PRRs. Fibronectin, fibrinogen, defensins, etc. are the examples of DAMPs.

Delta cells (δ-cells or D cells) These are somatostatin-producing cells of the stomach, intestine, and pancreatic islets. D cells have close connection with gastrin-producing G cells, and somatostatin inhibits gastrin release.

Dendritic cells (DCs) These are the antigen presenting cells reside in the skin and mucosal surfaces. They take the antigen by means of endocytosis, phagocytosis, pinocytosis, and macropinocytosis and carry the antigen from peripheral lymph nodes and present it to primary lymph nodes.

Dendritic spines It is a small membranous protrusion from a neuron's dendrite that typically receives input from a single axon at the synapse.

Dog erythrocyte antigen (DEA) The major determinant of canine blood group is dog erythrocyte antigen (DEA), and there are eight major blood groups in the dog, labeled as DEA 1 to 8. The major antigens are DEA 1.1 and DEA 1.2. Dogs can be positive for either DEA 1.1 or 1.2 or are negative for both.

Eclampsia The condition of acute convulsion with high blood pressure in late pregnancy or during labor, or within 24 h after parturition.

Ectopic Occurs unusually or is present in an abnormal position.

Ectotherms Ectotherms are the animals that entirely depend on the environment for their heat production.

Embryo transfer technology (ETT) Embryo transfer is a process in which embryos are harvested from donor females and transferred into the uterus of recipients, who act as a foster mother for the embryonic development for the rest of the pregnancy.

Emphysema Abnormal permanent enlargement and destruction of the alveoli, which are small air sacs deep in the lungs.

End-diastolic volume (EDV) The volume of blood in the ventricle at the end of diastole when filling is complete.

Endotherms Endotherms are the animals which could generate their own heat through metabolic heat production (birds, mammals).

End-systolic volume (ESV) The volume of blood in the ventricle at the end of systole when emptying is complete.

Enterochromaffin (Enteroendocrine cells) These are small polygonal cells found predominantly in the small intestine and appendix, but also scattered in the colon, rectum, and stomach. In gastric mucosa, these cells are found among the parietal and chief cells.

Ependymal cells Are the lining cells of the ventricles of the brain and the central canal of spinal cord are. These cells are ciliated columnar type and are situated between brain extracellular fluid and cerebrospinal fluid (CSF).

Epitopes Epitopes are the antigenic determinants, a small site of an antigen that can activate immune response by activating T or B cells.

Erythropoietin It is a glycoprotein synthesized primarily from kidney. It stimulates the production of pro-erythroblasts from PHSC and increases the number of erythrocytes in the circulation. Tissue oxygenation is the basic regulator for erythropoietin synthesis.

Estivation Estivation refers to a torpid sleeping state during the summer in harmony to hibernation during winter. The major advantage of estivation is to reduce the metabolic rate and Tc to prolong the period of survival of the animal with its energy reserves and preserves a significant quantity of water.

Eustachian tube (auditory tube) It is an auditory tube extending from the middle ear to the anterior portion of throat behind the nose. It links the nasopharynx with middle ear. It helps to maintain the air pressure. The secretions from this area protect the middle air from infections.

Eutherian The species of mammals having placenta in the females.

Eutocia Normal parturition, characterized by uterine contractions followed by progressive cervical dilatation and fetal expulsion.

Extravasation Force out or leakage from the vessel to the surroundings.

Feedback inhibitor of lactation (FIL) It is a glycoprotein that acts as local inhibitory factor for milk secretion. FIL prevents the differentiation of mammary secretary epithelial cells, suppresses the synthesis of milk protein and lactose, and induces apoptosis in the mammary gland.

Ferritin It is a globular protein that stores iron inside the cells in a soluble non-toxic form.

Fetal hemoglobin It is composed of two alpha and two gamma chains. It has more oxygen affinity than adult Hb. At birth 41–100% of total Hb is HbF and diminished after 2–3 months and remains very small amount in circulation (<2%).

Fight or flight response Physiological reaction that occurs in response to a perceived harmful event, attack, or threat to survival.

Fight or flight responses It is an automatic physiological reaction to an event that is perceived as stressful or frightening. The perception of threat activates the sympathetic nervous system and triggers an acute stress response that prepares the body to fight or flee.

Flehmen response It is a behavior in which an animal curls back its upper lip exposing its front teeth, inhales with the nostrils usually closed, and then often holds this position for several seconds

Follistatin An autocrine glycoprotein hormone is synthesized by folliculostellate (FS) cells of the anterior pituitary and acts primarily on TGF-β superfamily activin to reduce its activity resulting from inhibiting follicle-stimulating hormone (FSH). It is also known as activin-binding protein or FSH-suppressing protein (FSP).

Frank-Starling law of the heart Intrinsic control of the heart, such that increased venous return resulting in increased end-diastolic volume leads to an increased strength of contraction and increased stroke volume; that is, the heart normally pumps out all the blood returned to it.

Freemartin The sterile male twin or infertile female twin with an opposite sexed fetus due to hormonal influence during fetal growth. Found in cattle and rarely in sheep, goats, and pigs

Functional syncytium A group of cells that are interconnected by gap junctions and function electrically and mechanically as a single unit.

Furstenberg's rosette These are the series of 6–10 longitudinal folds situated just above the streak canal. It acts as local defence factor against pathogen after recruiting leukocytes specially lymphocytes and plasma cells. It also secretes keratin which occludes the lumen of teat canal between milking and prevents bacterial entry.

Galactopoiesis It is defined as the maintenance of established lactation. The term may often use to describe the enhancement of established lactation.

Galactosemia It is occurred due to galactose-1-phosphate uridylyltransferase which helps in utilization of galactose. The symptoms of galactosemia are cataract, hepatomegaly, splenomegaly, ascites, and feeble mindedness.

Gastric mucosal barrier Gastric mucosal barrier protects the stomach mucosa against gastric acid and other noxious agents. There are three levels of protective mechanism, pre-epithelial, epithelial, and sub-epithelial.

Gastrin cells (G cells) These are flask-shaped cells with microvilli at the apical surface found in the pyloric mucosa. G cells secrete gastrin in response to peptides and amino acids. The neurotransmitters helped in gastrin release are gastrin-releasing peptide (GRP) and bombesin.

Gastrin release peptide (GRP) It is a polypeptide synthesized from postganglionic non-cholinergic neurons of the enteric nervous system and stimulates gastrin release.

Genistein An isoflavone compound acts as an angiogenesis inhibitor, and phytoestrogen has a concentration depending on several activities in mammals.

Geophagia It is the intentional practice of eating earth or soil-like substances such as clay, chalk, or termite mounds.

Ghrelin A peptide hormone is synthesized by ghrelinergic cells in the gastrointestinal tract from an empty stomach. Ghrelin stimulates feed intake by promoting gastric motility. It also stimulates the secretion of gastric HCl hence known as the satiety hormone, hunger hormone, or lenomorelin (INN).

Glucagon-like peptide-1 (GLP-1) It is a peptide released from intestinal L cells. It helps in the inhibition of gastric emptying and induction of satiety.

Glucose-dependent insulinotropic polypeptide or Gastric inhibitory polypeptide (GIP) It is a peptide secreted from K cells in the duodenum and jejunum. The major functions of GIP include stimulation of insulin secretion, induction of satiety, and stimulation of lipoprotein lipase.

Glycophorins It is a sialoglycoprotein of erythrocyte membrane with two domains. N-terminal domains are situated toward outer surface and they act as blood group antigens (ABO and MN), whereas C-terminal domain faces the cytoplasm and interacts with the cytoskeleton. Due to high sialic acid content, glycophorins are the major contributor to the surface negativity of RBC membrane.

Glycoprotein The conjugated/compound protein with carbohydrate(s) components (as non-protein or prosthetic groups) attached to the polypeptide chain.

Gonadostat hypothesis During the prepubertal stage of development, the gonads release low but constant gonadal steroids that exert negative feedback on the hypothalamic neurons for GnRH release. Puberty occurs when the hypothalamus is desensitized to gonadal steroids. The hypothesis can explain the sexual quiescent before the onset of puberty.

Gular flutter Rapid fluttering of thin skin of floor of mouth and upper throat of bird to facilitate heat loss.

Gyri and sulci The surfaces of hemisphere are highly convoluted having numerous elevations and depressions known as gyri and sulci, respectively.

Habituation A gradual quantitative change of response which may lead to a loss of response, as a result of repeated stimulation.

Handmade cloning Handmade cloning is microinjection-free modified method of SCNT in which enucleation of oocyte is done by cutting of oocytes by fine microblade.

Hapten They are low molecular weight (below 10,000 Da) substances which can react with its corresponding antibody but unable to induce immune response by their own. Their immunogenic property can augmented by carrier molecules (albumin or globulin). Example: pneumococcal capsular polysaccharide, polysaccharide C of Streptococci and cardiolipin antigens, etc.

Haptoglobin It is an acute phase protein that synthesized in liver and involved in transportation of hemoglobin.

Harderian glands These are pigmented lacrimal glands situated at the posterior side of the ocular globes. It is seen in many birds and mammals except carnivores. The secretions of these glands are rich in lipid and porphyrin that lubricate the eyes and eyelids.

Hemopexin It is a plasma protein of globulin class that is involved in transportation of heme.

Homeorhesis The orchestrated or coordinated control in metabolism of body tissues necessary to support a physiological state.

Hepcidin it is a cysteine rich peptide synthesized by the hepatocytes and involved in compensatory regulation of iron absorption. It inhibits iron transport by blocking ferroportin in enterocytes, hepatocytes, and macrophages during iron deficiency.

Hermaphrodite An organism having both male and female genitalia.

Heterotherms Heterotherms are animals capable of varying their degree of endothermic heat production.

Hibernation Hibernation refers to a cessation of coordinated locomotor activity and a reduction in body temperature, total metabolism, heartbeat, and respiration during winter.

Homeorhesis An orchestrated or controlled change in the homeostasis to support a particular physiological state.

Homeostasis The maintenance and existence of a nearly constant internal environment.

Homeotherms Homeotherms are the animals having the ability to control their body temperature within a narrow range nearly around 37–40 °C even when the external temperature varies (e.g., mammals and birds).

H-Y antigen A cell surface antigen (protein) positioned on the Y chromosome is regarded as the major male sex determiner.

Hyperaemia Increase in the inflow of blood in an area of the body compared to other sites.

Hyperchlorhydria (sour stomach/acid stomach) It is a clinical condition in which the gastric HCl production is more than normal. It usually occurs due to higher gastrin production.

Hyperthermia Hyperthermia can be defined as a rise in body temperature above the hypothalamic set point when heat-dissipating mechanisms are impaired or overwhelmed by external or internal heat.

Hypochlorhydria (HCH) It is characterized by reduced secretion of gastric acid. The predominant causes of HCH are chronic atrophic gastritis, Helicobacter pylori infection, or autoimmune disorders.

Hypothermic spiral The thermoregulatory mechanism is impaired when Tc declines lower than 94 °F (34.4 °C) and animals typically stop to shiver or seek heat.

Imprinting behavior It is a form of learning in which a very young animal fixes its attention on the first object with which it has visual, auditory, or tactile experience and thereafter follows that object.

In vitro embryo production (IVEP) (IVEP) In vitro embryo production (IVEP) refers to the processes of in vitro oocyte maturation (IVM), in vitro fertilization (IVF), and the early days of in vitro embryo culture (IVC).

Innate lymphoid cells (ILC) These cells are involved in inflammation. They don't have antigen specificity due to lack of T cell receptor or any other cell surface markers. Their primary role is to produce cytokines.

Insulin-like factor-3 The protein (an insulin-like trophic hormone) encoded by the INSL3 gene has a role in testicular descent in fetal life.

Insulin-like growth factor 1 (IGF-1) Also known as somatomedin C, secreted from the liver by the influence of growth hormone maximum before puberty having insulin-like molecular structure binds with binding protein (IGFBP-3) under the regulation of insulin and regulates the cell growth and differentiation.

Interferons (IFN) They were emerged as antiviral proteins but in later their roles in immunomodulation and cancer immunology have been identified. IFNs are classified into three types, type I (IFN-α and IFN-β), type II (IFN-γ), and type III (IFN-λ1, 2 and 3).

Interleukins They are so named with a thought that it was synthesized by leukocytes but later it was found that interleukins can be produced from a variety of cells. They play pivotal role in hematopoiesis, activation, and differentiation of immune cells.

Intermediary metabolism The sum of all intracellular processes that transform the nutrients into cellular components. It includes the catabolism (breakdown of macromolecules) and anabolism (synthesis of macromolecules).

Interstitial cells of Cajal (ICC) These are bipolar cells or spindle-shaped cells with elongated processes associated with the electrical activity of GI smooth muscles. ICC resembles Purkinje cells of heart with rhythmic oscillating properties hence called "pacemakers of the guts."

Kai blood types Two new blood groups, namely Kai 1 and Kai 2 were reported in the dogs mostly found in North America. It was so named as Kai meaning "dogs" in Korean. These antigens were biochemically characterized through ELISA, and it was reported that both of these antigens didn't co-exist but both could be absent.

Kallmann syndrome A genetic disorder (in humans) causing hypogonadism resulting in delayed puberty and total lack of sense of smell (anosmia) or a reduced sense of smell.

Karyogamy Fusion of cell nuclei during fertilization.

Kisspeptin The protein has a role in commencing the secretion of gonadotropin-releasing hormone (GnRH) through G-protein coupled receptor (GPR54), particularly during puberty, and also has a role in tumor suppression and kidney function.

Kupffer cells (Littoral cells) They are specialized phagocytic cells situated adjacent to the sinusoidal endothelial cells of liver. These are the largest proportion of tissue resident macrophages. The major function of these cells is the phagocytosis of pathogens. They also synthesize inflammatory cytokines like TNF-alpha, oxygen radicals, and proteases.

Lactogenesis The process of initiation of milk secretion from mammary gland.

Lactation curve It is the graphical representation of milk production in respect to time. A typical lactation curve of cow is "bell shaped" which follows a rapid accelerating phase and reached peak and then decline till the end of lactation.

Lactoferrin (LF) It is an iron-binding glycoprotein composed of a single polypeptide chain comprising of 689 amino acid residues. It has molecular weight between 76 and 80 kDa. Bovine LF can act as antibacterial, antiviral, immune modulator antioxidant, anticancer, and anti-allergic agent.

Lactogenesis Lactogenesis is the biological process of onset of milk secretion which includes the enzymatic and cytological differentiation of mammary alveolar cells in early pregnancy to full lactation after parturition.

Lactose Intolerance It is occurred due to the deficiency of lactase enzyme which hydrolyzes lactose into glucose and galactose. Lactase activity developed in infants which disappeared after weaning. It leads to lactose malabsorption including diarrhea, bloating and flatulence, abdominal pain, and gaseous accumulation in the intestine.

Lactose synthetase It is the principle enzyme of lactose synthesis in the mammary epithelium. It is a complex of two proteins combines reversibly, in 1:1 stoichiometry.

Lactotroph The pituitary cell that produces prolactin.

Leptin A peptide hormone chiefly produced by fat cells and enterocytes in the small intestine acts upon the arcuate nucleus of the hypothalamus to regulate appetite and energy metabolism by reducing burn stored fat in adipose tissue as reduced estrogen and controls the onset of puberty, acting with kisspeptin. It also involves in regulation of the inflammatory response.

Luebering–Rapoport pathway In this pathway, mature RBC produces 2,3-diphosphoglycerate (2,3-DPG) which helps to release oxygen from the hemoglobin to make it available for tissue utilization.

Major histocompatibility complex (MHC) These are the cell surface proteins encoded by a group of genes present in chromosome 6 in human. The main function of MHC is to discriminate between self and non-self.

Mammogenesis It is defined as the growth and development of the mammary gland. It occurs through a series of structural and functional development, differentiation, and involution associated with growth and reproductive stages of animals and regulated by hormones and growth factors.

Margination The movement of leukocytes from central to periphery of the blood vessels during inflammatory conditions.

Marsupials The non-placental mammals, having no true placenta but usually have a pouch on the female's abdomen that covers the teats and provides to carry the young, under order Marsupialia (like kangaroos, wombats, bandicoots, opossums).

Mast cells Mast cells were identified by Ehrlich in 1878 and initially named as "Mastzellen" meaning well fed cells due to their highly granular cytoplasm. They are mesenchymal cells phenotypically similar with the basophils and derived from myeloid stem cells and residing in the skin and mucosal tissues. Mast cells play pivotal role in inflammation and allergic reactions (type I hypersensitivity).

Mesoclimate The climate of small areas of the earth's surface which may not be representative of the general climate of the district.

Methemoglobin It is the true oxide of hemoglobin as ferrous iron is converted to ferric iron during the formation of methemoglobin. The oxidizing agents such as ferricyanide and nitrites react with hemoglobin to form methemoglobin. Methemoglobin is unable to carry oxygen. Under normal condition, a small amount of methemoglobin is formed in the circulation but the reducing agents such as glutathione, ascorbic acid decrease its accumulation.

Microarray The laboratory tool used to detect the expression of several fragments of genes at the same time. DNA microarrays are extensively used along with RNA, protein, peptide, and carbohydrate microarrays.

Microfold (M) cells These are the specialized epithelial cells found in the intestinal lymphoid tissue (Peyer's patches). M cells help in antigen transport across the epithelial cells. M cells engulf luminal pathogens and their antigens by phagocytosis and presented to dendritic cells at lamina propria.

Microglia Are small cells with long thin tortuous processes which look like spines and are phagocytic in function and are motile.

Microtubule The globular proteins, alpha- and beta-tubulin made microscopic hollow cylindrical tube, the largest structures in the cytoskeleton of some eukaryotic cells, stretched throughout the cell to provide the cellular shape and keep its organelles in place also transporting materials within cells as well as involve in cellular movement and cell division (spindle).

Micturition The process of voiding urine.

Migrating motor complex (MMC) It is a cyclic motor pattern of the GI tract exhibited during the interdigestive state. It is appeared as clusters of contractions divided into four phases that propagate over a longer intestinal segment.

Milk somatic cells (SCC) These are body derived cells composed of mammary epithelial cells (70–75%) and leukocytes (25–30%) secreted in milk. In normal healthy quarter, the total somatic cell populations are below 100×10^3 cells/mL of milk which may increase up to several folds during intramammary infection. They are the cellular defence factors protecting the mammary gland against pathogens.

Monoclonal antibodies Antibodies that bind to a single epitope are called monoclonal antibodies. Monoclonal antibodies are produced from same clone of B cells. Monoclonal antibodies have high specificity and reproducibility.

Monocular vision or periscopic vision It is a type of vision characterized by seeing with only one eye at a time. Animals with laterally placed eyes view the objects with one eye at a time independently of the other eye. This is due to wider visual angle between optic axis and median line of the eye, e.g., amphibians and reptiles.

Mononuclear Phagocytes System (MPS) MPS consists of circulating monocytes and tissue macrophages. After the maturation of the monocytes, they migrate in the tissue and differentiated into tissue macrophages and causing the destruction of the pathogens.

Monotocous Producing single offspring or young at a birth.

Monotreme An egg-laying mammal under the order Monotremata.

Motilin It is a gut hormone released from enteroendocrine cells (Mo cells) in the upper small intestine and increases gastrointestinal motility.

Mucosal-associated lymphoid tissue (MALT) They are situated in the mucous membranes of the gastrointestinal, respiratory, and urogenital systems. They resemble the structure of lymph nodes and populated with plasma cells.

Mullerian duct The primordial structure that gives rise to the female reproductive tract.

Multiple Ovulation Embryo Transfer Technology (MOET) It is a process in which a multiple eggs of animal are fertilized and the embryos are collected by flushing.

Myelodysplasia It is characterized by defective formation of blood cell precursors in the bone marrow. Leukemia and thrombocytopenia are also occurred in addition to anemia in myelodysplasia.

Myoepithelial cells These are modified smooth muscle with long cytoplasmic projections that surround each alveolus and small ducts. Due to their long cytoplasmic projections, they are also termed as basket cells.

Myogram Myogram is the graphical representation of the phenomena of muscular contractions.

Myometrium The middle layer in the uterus that is mainly composed of smooth muscle tissue.

Myosin Contractile proteins form the thick filament and attach with actin during muscle contraction.

Natural killer (NK) cells NK cells are responsible for cell mediated immune response due to their cytotoxic activity. Upon activation, NK cells release perforins and granzymes from their granules and cell lysis occurs. Activated NK cells also secrete TNF-α, IFN-γ, interleukins (IL-5, IL-10, IL-13), and chemokines.

Neuropeptide Y (NPY) A neurotransmitter produced mainly in the neurons of the sympathetic nervous system along with different locations including the hypothalamus and performs various physiological and homeostatic processes, like increasing food intake and energy storage, reducing blood pressure, anxiety, stress, reducing pain perception, and affects the circadian rhythm and HPA axis by modulating stress.

Neutrophil derived microparticles (NDMPs) These are spherical microvesicles of 50–1000 nm diameter containing mRNA, microRNA, cell adhesion molecules (CD11a and L-selectin), and inflammatory proteins surrounded by lipid bilayer.

Neutrophil extracellular traps (NETs) These are extracellular meshes composed of chromatin and neutrophil granular proteins that entrap the pathogens and immobilize them. The process is called NETosis.

Niche Niche is a portion of a tissue that creates a unique microenvironment.

Nictitating membrane (third eyelid) A transparent third eyelid present in the medial canthus. It is present in fish, amphibians, reptiles, birds, and mammals (cats, camels, polar bears, seals, and aardvarks) but rare in primates. It protects the eye, and its glands also produce tears.

Nymphomania Abnormal estrous behavior patterns are the most noticeable sign of cystic ovarian disease.

Oligodendrocytes Are neuroglial cells form the myelin sheath in the nerve fibers in CNS.

Orexin It is a neuropeptide released from hypothalamus and gastric mucosa. It stimulates feed intake and gastric HCl secretion.

Organ of Corti It is the receptor organ of hearing situated within the cochlea. It is composed of three rows of outer hair cells and one row of inner hair cells. It helps in transduction of auditory signals and converts mechanical energy into electrical energy to facilitate hearing.

Osmoconformers Organisms whose internal bodily fluids are isotonic to their surroundings.

Ovum pick up Ovum pick up is a non-invasive procedure for recovering oocytes from antral follicles in live animals

Oxidative injury The molecular oxygen can be reduced to form highly reactive oxygen species (ROS) such as hydrogen peroxide (H_2O_2), hydroxyl radical, superoxide anion, hypochlorous acid (HOCl), and nitric oxide (NO). Low concentrations of these free radicals are required for phagocytosis, signal transductions, and the biosynthesis of prostaglandins but higher concentrations of free radicals can be detrimental for the cells in terms of DNA damage, inactivation of enzymes, oxidation of hormones, lipids peroxidation, and membrane disturbance.

Oxyntomodulin It is peptide hormone released from the L cells of duodenum and stimulates gastric acid secretion.

Palade granule Ribosome is also known as Palade granule.

Paneth cells They are situated at the base of the intestinal crypt responsible for the secretion of antimicrobial peptides (AMPs) like alpha defensins, secretory phospholipase A2 (sPLA2), lysozyme, cathelicidins, C-type lectin regenerating islet-derived protein IIIγ (RegIIIγ), and angiogenin 4. These AMPs help to protect the epithelial barrier.

Panhypopituitarism The condition of inadequate or absent production of all anterior pituitary.

Passerine birds The birds belonging to the order Passeriformes are called Passerine birds. It is also called **perching birds**; most of the birds are these types, generally small and living near the ground with feet having four toes arranged to allow for gripping the perch.

Passerine Birds have the arrangement of their toes (three pointing forward and one back) to facilitate perching. They are also termed **perching birds** or **songbirds** and belong to the order Passeriformes, half of the almost all bird species.

Pathogen associated molecular patterns (PAMP) These are the distinct structures of the pathogens which can be recognized by the immune cells through their pattern recognition receptors (PRRs). Examples are zymosan, peptidoglycans, surface glycoproteins, etc.

Peptide yy (pyy) It is a short (36-amino acid) peptide released from cells in the ileum and colon in response to feeding.

Peroxiredoxin 2 (Prx2) It is one of the most abundant proteins in erythrocytes reducing H_2O_2 and alkyl hydroperoxides to water and alcohol.

Pheromone The chemical substance secreted into the environment by the animal, bird, and insect influencing the behavior or physiology of others of its species. Animals have generally used it in reproduction, communication, and feeding practices.

Photoperiod The interval in 24 h (a day) during which an organism is exposed to light or receives illumination, i.e., day length.

Photopigments These are the transmembrane proteins of photoreceptors coupled with G-protein. The function of photopigments is to alter the membrane potential of photoreceptors triggered by the light, e.g., opsins and retinene.

Pinocytosis The ingestion of liquid into a cell by the budding of small vesicles from the cell membrane.

Placentophagia It refers to consumption of the afterbirth by either mothers or—in species with male allomaternal care—males as well. It is almost universal in both carnivorous and herbivorous mammals.

Plasmodesmata Plasmodesmata are cytoplasmic bridges in the shape of minute pores that connect adjoining plant cells to form a symplast or continuous protoplasm.

Platelet Small subcellular fragments that circulate in blood, where they promote homeostasis.

Pleomorphic Cells or nuclei of the cell having variable shape and size.

Plumage The color and arrangement pattern of feathers in birds.

Pluripotential hematopoietic stem cells (PHSCs) They are the cells from which all the cells of the hematopoietic systems are generated. PHSCs have the capability to undergo mitotic cell divisions, and a portion of these cells remains undifferentiated and maintains a constant pool of PHSC throughput life, whereas other portion of PHSCs are programmed for differentiation to produce specialized.

Poikilotherms Poikilotherms are the animals that do not have a control over their body temperature and vary with the environmental temperature. They are also called as temperature conformers (reptiles, amphibia, fishes, and invertebrates).

Point of inflection A point at which the accelerating and decelerating phase meet, the growth rate starts to decrease.

Polyclonal antibodies Antibodies that are secreted by different B cell lineages within the body

Polyclonal antibodies Polyclonal antibodies are the heterogeneous immunoglobulin mixture against a single antigen and can bind with different epitopes of a single antigen. Polyclonal antibodies are produced from different B cell clones.

Polycythemia Vera It is characterized by the proliferation of erythroid precursors occur autonomously independent of erythropoietin. Splenomegaly, leukocytosis, and thrombocytosis are the common clinical manifestations of polycythemia vera. It is generally occurred in dogs and cats of middle age groups (6–7 years).

Polygynous It refers to the mating system in which females have a single mate during some specified period of time or their entire life and males mate with multiple females.

Polyphagia It is an abnormally strong sensation of hunger or desire to eat often leading to or accompanied by overeating.

Prebiotics Non-digestible food ingredient/supplement that selectively stimulates the growth of some or all of the non-pathogenic favorable organisms (bacteria) in the gut of the host.

Preening Preening is a maintenance behavior, found in birds involved the use of the beak to position feathers, and interlock feather barbules to separate.

Primary/azurophilic granules They are the largest granules of neutrophils and are appeared first during promyelocyte stage and they have affinity for basic dye azure A. They contain myeloperoxidase (MPO), proteases lysozyme, elastase, cathepsins, and acid phosphatase. The main function of primary granular contents is antimicrobial destruction.

Probiotics Combination of live microorganisms (bacteria and yeast) that enhance/boost the gut health as well as overall performance when consumed.

Proctodeum The caudal most portion of the cloaca where cloacal bursa opens. The proctodeum opens outside through vent.

Progenitor cells A progenitor cell is a unipotent cell that can be differentiated into its target cells (e.g., erythroid progenitors can give rise to erythrocytes). Unlike stem cells they have limited self-renewal capacity.

Protamines The major nuclear proteins of spermatozoa, having low molecular weight arginine along with lysine amino acids which replace later histones in the haploid phase of spermatogenesis, essential for sperm head condensation, maintaining and stabilization of DNA, are protamines. Mutations in the protamine genes cause infertility and are considered a biomarker of sperm quality.

Proteomics The identification and study of the proteins of a cell, tissue, or organism to determine their structure, interactive networks, and function.

Ptyalism Hypersecretion of salivary glands is called ptyalism. It is characterized by drooling of saliva. Ptyalism is secondary to swallowing disorders in animals. The causes of ptyalism are toxins, drugs, and poison, such as organophosphorus compounds, glossitis, stomatitis, convulsive disorders, nervousness, motion sickness, linear foreign body ingestion, and oral tumor.

Pyknosis Degenerative changes and shrinking of the nucleus due to clumping of the chromosomes, associated with hyperchromatic.

Reactive oxygen species (ROS) Oxygen-containing unstable molecule or free radical, which readily react with other molecules, and affect proteins, DNA, RNA even cell death in excess, like, singlet oxygen, superoxide, peroxides, hydroxyl radical and alpha-oxygen.

Reactive oxygen species Used to denote reactive molecules derived from oxygen.

Recombinant gene transfer technology Transfer of desirable gene sequences from other than sources via appropriate vector to obtain improved and desired characteristics in living organism.

Renin A hormone released from mammalian kidneys in response to a decrease in NaCl/ECF volume/arterial blood pressure; activates angiotensinogen.

Restitution It is the process of migration of new cells from the gastric pits to replace the damaged cells within a very short period of time.

Reticular formation: Is the phylogenetically primitive network of small neurons occupying the midventral portion of the medulla and midbrain, extending throughout the brainstem and into the spinal cord.

Rheopectic When shear is applied to some colloids, they solidify more quickly.

Rhodopsin It is a biological pigment and G-protein coupled receptor found in the rod cells. Enzyme rhodopsin kinase joins 11-cis retinal and scotopsin to reform rhodopsin.

Rooster An adult male domestic fowl. A hen with a right functional ovary may also be called a rooster.

Rubriblast See proerythroblast.

Satellite cells Are neuroglial cells in the nervous system help in regulation of the external chemical environment around the neurones of the PNS.

Secondary/specific granules These granules are unique to neutrophils hence called specific granules. They are appeared during metamyelocyte stage. The secondary granules contain matrix metalloproteinases (MMPs) (MMP 3, 8, 9), gelatinase, collagenase, lactoferrin, alkaline phosphatase, histaminase, plasminogen activator, and β2 microglobulin. The secondary granules help in microbial destruction and neutrophil migration.

Serine proteases A class of proteolytic enzymes that possess serine at the active site.

Silent heat It is defined as the maintenance of ovarian functions without the presence of vulvar edema, serosanguinous vaginal discharge, and charm for male dogs.

Somatic cell nuclear transfer (SCNT) Somatic cell nuclear transfer or cloning is a technique in which the nucleus of a somatic cell is transferred to the cytoplasm of an enucleated egg (an egg that has had its own nucleus removed).

Spermatheca The small cavity or sac in the lower group of female animals to hold the spermatozoa before fertilization.

Spermatogonial stem cells Spermatogonial stem cells (SSCs) are the most primitive spermatogonia in the testicular tissue. SSCs maintain highly productive spermatogenesis by self-renewal and ongoing creation of daughter spermatogonia that develop into spermatozoa.

Spermiogenesis The development of mature and motile spermatozoa from spermatid as occurred in the last stage of spermatogenesis.

Stem cell An unspecialized or undifferentiated cell that can generate one or more specialized or differentiated cells like blood cells. It is used in therapeutic due to its regenerating and repairing properties of the damaged tissue.

Stereo vision In this type of vision, 3D images are created from multiple 2D views. Primates, cat, and other felines have three-dimensional view because of small angle between optic axis and median line of the eye with each eye viewing the same object at a different angle.

Streak canal (Ductus papillaris) The distal opening of the teat cistern is called streak canal or teat canal through which milk is removed.

Superantigens These antigens are able to activate a large proportion of T cells (up to 25%) in comparison to conventional antigens that are able to induce only 1–2% T cells. The super antigens cause hyper activation of immune system. Pyrogenic exotoxins (leads to shock) and enterotoxins (leads to food poisoning) of Staphylococci are the examples of super antigens.

Superovulation Superovulation, also known as superstimulation, is a treatment that aims to boost the donor animal's ovulation rate and thus the number of available oocytes without interfering with the physiological and endocrinological processes involved in oocyte maturation, ovulation, and fertilization, as well as embryonic and fetal development.

Superoxide dismutase (SOD) It is a copper and zinc containing metalloenzyme that catalyzes the transfer of an electron from one superoxide anion (donor) to another (recipient) thus the donor molecule becomes dioxygenated and the recipient rapidly reacts with two hydrogen ions to form hydrogen peroxide. It protects the cell from oxidative injury.

Symplast Plasmodesmata between adjacent plasma cells form continuous protoplasm known as symplast.

Syncytium A cytoplasmic mass or a single cell containing multiple nuclei formed either by fusion of cells or by division of nucleus.

Syntaxin Is a family of membrane integrated Q-SNARE proteins participating in exocytosis.

Tapetum lucidum It is a biologic reflector system situated behind the retina in many vertebrate species. It allows the retina to make optimal use of available light by photon-photoreceptor stimulation. It helps in night vision in some nocturnal animals.

Telodendria Are the short branches known as at the termination the axons as well as the axon collaterals.

Telomeres They are non-coding specialized repetitive DNA sequences located at the ends of chromosomes.

Temperate climate The climate of the "middle" latitudes; the variable, climate between the extremes of tropical climate and polar climate.

Thrombocytopathies It is the impairment of platelet functions. It is either congenital (Von Willebrand disease, Glanzmann thrombasthenia, Canine thrombopathia) or acquired (multiple myeloma, chronic kidney diseases, and drug induced).

Thrombocytopenia It is the pathological condition which leads to decrease platelet counts. It is either congenital or acquired (drug induced, Rickettsial diseases, or immune mediated).

Tonoplast Selectively permeable membrane surrounding the sap vacuoles is known as tonoplast.

Totipotent Ability of cell(s) to differentiate into the specialized cell(s), which may be a new individual, organ, or tissue.

Transcriptomics Study of the transcriptome—the complete set of RNA transcripts that are produced by the genome, under specific circumstances.

Transferrin It is a plasma glycoprotein synthesized from liver that binds reversibly with iron in ferric form. Erythroid precursor cells contain transferrin receptor, and the transferrin bound iron is internalized inside these cells through receptor mediated endocytosis.

Transforming Growth Factor β (TGF-β) It is produced from a variety of cells including T cells and monocytes. The main function of TGF-β is the inhibition of cellular growth and production of extracellular matrix. TGF-β also acts as negative regulator of T cell and macrophage activation.

Transgenesis The process of introducing an exogenous or modified gene (transgene) into a recipient organism of the same or different species from which the gene is derived.

Translocation Changing the location of any substance.

Transporter protein The protein remaining undyingly across the biological membrane transports ions, small molecules, or macromolecules with a channel or through a carrying mechanism by facilitated diffusion or active transport process from one side to the other.

Treppe Stepwise increase in contraction tension.

Urine marking It is a natural, instinctive behavior in dogs, but it is not appropriate inside the house. Dogs, especially sexually intact male dogs, urinate on objects to leave a message for other dogs (e.g., claiming their territory). Urine marking behavior usually begins when the dog reaches sexual maturity.

Urodeum It is the cloacal compartment lies between coprodeum and proctodeum that stores urine. The ductus deferens opens into the urodeum.

Vasoactive intestinal peptide (VIP) It is neurohormone secreted from postganglionic non-cholinergic neurons of the enteric nervous system. The major functions of VIP are relaxation of GI smooth muscles, stimulation of pancreatic and biliary secretions, and inhibition of gastric acid secretion.

Ventilation-perfusion ratio (VA/Q.) It is the quantitative measurement or ratio of all the air entering the alveoli to the total blood flowing to both the lungs per minute.

Viviparous The animals that can produce living offspring that develops inside the mother's body. It happens in mammals and some reptiles and fishes. The animals able to produce eggs are called oviparous.

Von Willebrand disease It is the deficiency of von Willebrand's factor. It is one of the most common inherited bleeding disorders in canines with almost all the breeds.

Xenotransfusion A form of xenotransplantation was initially defined as the transfer of blood from one species into the veins of another.

Xerostomia (dry mouth) Xerostomia or dry mouth is a clinical condition developing due to hyposalivation. Xerostomia causes discomfort during eating and oral infections. It is uncommon in dogs and cats but can be occurred in animals under frequent radiation exposure. Administration of drugs like atropine, severe dehydration, fever, and anesthesia may also cause hyposalivation. Immune mediated keratoconjunctivitis sicca in canines can also lead to xerostomia.

Yolk Part of the egg nourishes the embryo during its formation.

Z-line Neighboring, parallel lines that define a sarcomere.

Zygote Fertilized egg cell that results from the union of a female gamete (egg or ovum) with a male gamete (sperm).

Zymogen An inactive substance that can be converted into an enzyme when activated by another enzyme.

α2-Macroglobulin They are globulin fractions of plasma protein synthesized mainly in liver but also locally by macrophages, fibroblasts, and adrenocortical cells. They are the proteases inhibitor and involved in inflammation.

α-Granules They are the largest and most abundant granules of the platelets (50–80/platelet). α-granules are filled with proteins like β-thromboglobulin and thrombospondin to support platelet adhesion and aggregation. Besides these, the α-granules contain some mitogenic proteins such as transforming growth factor-β (TGF-β) and platelet-derived growth factors (PDGF).

Index

A

© The Author(s), under exclusive license to Springer Nature Singapore Pte Ltd. 2023

P. K. Das et al. (eds.), *Textbook of Veterinary Physiology*, https://doi.org/10.1007/978-981-19-9410-4

B

C

D

F

M

U

V

The manufacturer's authorised representative in the EU is Springer Nature Customer Service Centre GmbH, Europaplatz 3, 69115 Heidelberg, Germany. If you have any concerns regarding our products, please contact ProductSafety@springernature.com

Printed and bound by CPI Group (UK) Ltd, Croydon, CR0 4YY

01/10/2026

02225884-0002